GLOSSARY FOR HORTICULTURAL CROPS

GLOSSARY FOR HORTICULTURAL CROPS

JAMES SOULE
Professor of Horticultural Science
Department of Fruit Crops
College of Agriculture
Institute of Food and Agricultural Sciences
University of Florida, Gainesville

Sponsored by the
AMERICAN SOCIETY
FOR HORTICULTURAL SCIENCE

JOHN WILEY & SONS
New York • Chichester • Brisbane • Toronto • Singapore

Copyright © 1985 by John Wiley & Sons, Inc.

All rights reserved. Published simultaneously in Canada.

Reproduction or translation of any part of this work beyond that permitted by Section 107 or 108 of the 1976 United States Copyright Act without the permission of the copyright owner is unlawful. Requests for permission or further information should be addressed to the Permissions Department, John Wiley & Sons, Inc.

Library of Congress Cataloging in Publication Data:

Soule, James.
   Glossary for horticultural crops.

  Bibliography: p.
  Includes indexes.
   1. Horticulture—Terminology. I. Title.
SB317.58.S68 1984   635'.014   84-20927
ISBN 0-471-88499-5

Printed in the United States of America

10 9 8 7 6 5 4 3 2 1

PREFACE

This is an era of increasing specialization in virtually all sciences. Old areas of traditional horticulture are expanding and new ones like genetic engineering and molecular biology are developing rapidly. An ever increasing flood of knowledge is already too vast for assimilation in all but the most restricted fields. No one can contend today that a person can be fully acquainted with every aspect of the plant and plant-related sciences. Neverless, broad-based training to accompany specialization is necessary indeed. The widest possible perspective must also be maintained if all of us engaged in horticulture are to find solutions for the many problems and the challenges that are inherent in the complex interrelationships of whole plants grown in given environments. It is vital in this situation to maintain channels of communication among the horticultural disciplines and other interests so that we may create mutual understanding throughout this vast field.

This glossary was compiled as a means of bridging this communications gap by providing readers with ready access to technical terms over a broad spectrum of the plant and plant-related sciences. It is based in part on glossaries assembled in 1975, 1977, and 1978 in collaboration with Dr. W. B. Sherman for our courses in fruit crops, in part on contributions from numerous friends, and on extensive reading, plus a lifetime of experience in the teaching, research, and business aspects of tropical and subtropical plants. There are several thousand terms in this volume, yet each, with its cogeners, is readily accessible in a combination of related terms grouped alphabetically in sections, cross-referenced within and between sections, and indexed by terms and crops. The terms, as well as the numerous illustrations that accompany them, are numbered to facilitate future expansion of any section. All of these features were included to provide the glossary with flexibility as a working reference useful to students, teachers, research scientists, extension specialists, administrators, growers, other professionals, and the sizable segment of the population that enjoys plant culture as an avocation. It is hoped this volume will be a valuable tool for students who are studying terms on their own, as assigned to them by their professors, as a ready source of information for research scientists, extension specialists, administrators, or growers, as a dictionary for anyone who is interested in acquiring the terminology of a new or unfamiliar research area, as a guide for dooryard plant enthusiasts who wish to master the meanings of technical terms useful to them, and finally as a delightful (and educational) way for "browsers" to learn a great deal about the subjects presented.

Every effort has been made to ensure accuracy in the definitions as they are understood at present. My good friend, the late Dr. G. H. M. Lawrence, in conversations at the Bailey Hortorium and subsequently in his classic <u>Taxonomy of Vascular Plants</u> (1951), emphasized the propensity of botanists to use several terms in the same sense or the same term in several senses in successive publications. This is certainly true in other areas of science. Terminology in nearly every field is changing continually as knowledge expands. It is generally acknowledged, however, that the more familiar is a field, the greater is the likelihood that definitions of its terms will be relatively short. What may be crystal clear to one

PREFACE

reader may be pure Greek to another, meaning that it is essential in a book of this nature to provide a balance among accuracy, clarity, and brevity in the definition of any given term. A glossary should therefore undergo continuous scrutiny and be revised periodically so that mistakes can be corrected, changes made in wording, and new or previously overlooked terms added.

My concept of horticulture as the subject of this book is that of a huge, complex, amorphous entity in which the common bond is the culture of plants as individuals. An enormous variety of fruits and vegetables is cultivated around the world for their value as food, beverages, oils, nuts, spices, herbs and condiments, rubber, and drugs and of ornamentals grown for their esthetic qualities. Mankind has gathered and grown for countless millenia and still use fruits and vegetables to add variety, color, flavor, stimulation, and the necessary vitamins and minerals to supplement a basic regimen and flowering and foliage plants to brighten their surroundings. Nearly every herbaceous or woody cultigen known and many wild forms are included among the three major commodity areas and their numerous specializations. Horticulture in its broad sense is a focal point for the plant and other sciences. The collective knowledge of morphology, taxonomy, plant physiology, biochemistry, biophysics, genetics, plant breeding, entomology, nematology, plant pathology, soil science, engineering, meteorology, computer science, and business, among others, can all be applied to a greater or lesser degree to the arts and skills of propagation and other details of crop culture, handling and disposal. Many persons who study, carry out breeding trials, perform laboratory, greenhouse, and field experiments, grow, cultivate, harvest, and sell plants or their produce or simply enjoy them for their pleasure are involved. Thus horticulture, whether pursued as a vocation or avocation, plays a profound role in everyone's life.

A great many individuals have most generously given me assistance and encouragement in every possible manner during the many months required to assemble this volume. These valuable contributions receive the accolade due each of them in the acknowledgments section that follows this preface.

Finally, let me express my heartfelt gratitude for their patience and devotion to all of those who assisted me in the arduous task of several years' work in preparing this manuscript in final camera-ready form: Miss Mercedes Cantu typed the original version, then acted as general coordinator and supervisor of several young people who assisted her in retyping, proofreading and collating of the text and indexes. Mrs. Roxann Mayros typed the front matter, selected references, indexes, figure legends and made corrections in the manuscript prior to initial camera-ready submission. Mrs. Cavelle B. Grimes began the final typing, completed the first three chapters, and was kind enough to train Mrs. Eleanor J. Rives when other duties prevented her from continuing the work. Mrs. Rives then took over and has cheerfully retyped and corrected the final three chapters, selected references, front matter and indexes using a combination of a word processor, separate software programs for the text, etc., and indexes, and typewriter to complete the manuscript for final submission.

James Soule

Gainesville, Florida
June 1985

ACKNOWLEDGMENTS

I acknowledge most gratefully the generous assistance of numerous individuals, institutions, agencies, and publishers in the preparation of this book. Particular thanks for their contributions are extended to the following:

Dr. James A. Coffelt, Insect Attractants Laboratory, U.S. Department of Agriculture, Gainesville, Florida, for the loan of reference materials on insects and certain other animal pests, for definitions of pheromones and grubs, and for many helpful suggestions provided during his painstaking review of the sections on animal pests.

Dr. Dermot P. Coyne, Department of Horticulture, University of Nebraska, Lincoln, Nebraska, for his enthusiastic interest and support throughout the preparation of this book, especially for the thorough review of the latter in which he and Drs. R. D. Uhlinger, Richard Sutton, Sotoro Salac, Jay Fitzgerald, Ellen Paparozzi, E. J. Kinbacher, and Peter Fuller of his department and David Wysong, Department of Plant Pathology, made numerous valuable notations for suggested changes in definitions in the manuscript, and for reference material used in the definitions of cultivar, provisions of the Federal Seed Act, Guidelines for classifying cultivated plant populations, and the Plant Variety Protection Act.

Dr. Malcolm A. Dana, Department of Horticulture, University of Wisconsin, Madison, Wisconsin, for his review of the chapter on morphology and anatomy and particularly the definition of adventitious roots.

Dr. Thomas A. Fretz, Department of Horticulture, Virginia Polytechnic Institute, Blacksburg, Virginia, who, as chairman of the American Society for Horticultural Science Publications Committee (1980-1981), sent copies of the 1978 <u>Glossary for Horticultural Crops</u> (by Soule and Sherman) to the members of that committee in the fall and winter of 1980-1981; his and the members' suggestions and subsequent whole-hearted support were instrumental in the decision to compile this glossary in its present form.

Dr. Ricardo E Gomez, Federal Extension Service, U.S. Department of Agriculture, Washington, D.C., for assistance in obtaining permission for use of material from U.S. Department of Agriculture publications.

Dr. William Grierson, formerly Head Harvesting and Handling Section, Agricultural (Citrus) Research and Education Center Lake Alfred, Florida, and long-time close friend, for material, counsel, and assistance over the last 30 years and his contributions to our manual on Citrus Maturity and Packinghouse Procedures (Soule and Grierson, 1978) and other publications used as background information in compiling the glossary.

Dr. A. A. De Hertogh, Department of Horticulture, North Carolina State University, Raleigh, North Carolina for permission to use definitions for terms abstracted from A. A. Hertogh et al., 1971, HortScience 6(2), Acta Hort., 47, 1975, and Holland bulb forcer's guide, 1981, for reviewing the sections on bulbous crops, and steadfast support throughout the compilation of the glossary.

Dr. James R. Ice, Managing Editor, AVI Publishing Company, Westport, Connecticut, for his reviews of the 1978 and present glossary and encouragement in many ways.

Dr. Jules Janick, Department of Horticulture, Purdue University, West Lafayette, Indiana, former Science Editor for the American Society for Horticultural Science, for his long-time support, sagacious counsel,

ACKNOWLEDGMENTS

and assistance and for furnishing names of reviewers, guiding me to a publisher, and his enthusiastic advocacy of the glossary's approval by the Society.

Dr. J. N. Joiner, Department of Ornamental Horticulture, University of Florida, Gainesville, Florida, for loan of his book on foliage plants used in compiling definitions for greenhouses and similar structures and foliage plant mixes.

Mr. Robert F. Kasmire, Department of Vegetable Crops, University of California, Davis, California, for use of personal material on MUM, Plan III, and precooling formulas and his valuable suggestions for the organization of the chapter on postharvest handling and marketing.

Dr. J. F. Kelly, Department of Horticulture, Michigan State University, East Lansing, Michigan, for his review of the glossary and for kindly furnishing definitions for vegetable and vegetable crops and pointing out several errors; also Dr. Frank Dennis, for his suggestions and review of the glossary.

Dr. Jose E. O. de Lima, Limeira, Sao Paulo, Brazil, for generously furnishing drawings of a navel orange flower and fruit.

Dr. D. N. Maynard, Department of Vegetable Crops, University of Florida, Bradenton, Florida, chairman of the American Society for Horticultural Science Publications Committee (1981-1983), and members of that Committee, for continuous encouragement throughout the preparation of the glossary and their advocacy of its approval by the Society.

Dr. P. L. Martin, University of Florida Presses, Gainesville, Florida, for providing copies of Watkins and Sheehan 1975 _Florida Landscape Plants_ from which selected illustrations were excised for use in the glossary.

Mr. Lewis S. Maxwell, Tampa, Florida, a long-time friend who dates back to Soule's Nursery days in the late 1930s, for his gift of photographs of avocado flowers reproduced as line drawings in the glossary.

Dr. Terrell Nell, Department of Ornamental Horticulture, University of Florida, Gainesville, Florida, for the loan of books and for valuable suggestions on floricultural crops.

Dr. D. H. Roberts, Department of Plant Pathology, University of Florida, Gainesville, Florida, for the generous loan of class notes, the glossary from his and Dr. Boothroyd's textbook on fundamentals of plant pathology, and his copy of the Commonwealth Mycological Institute's plant pathologists pocketbook, all of which were used to the fullest extent in the sections on plant pests and nematodes, and for his review of the plant pests section.

Dr. R. W. Robinson, New York (Cornell) Agricultural Experiment Station, Geneva, New York, for the correction of certain disease names and useful comments on suggested changes in the glossary.

Dr. Thomas J. and Mrs. Marion R. Sheehan, Department of Ornamental Horticulture, University of Florida, Gainesville, Florida, for their loan of original drawings from which photographs were made for use in the glossary and for useful information on orchid terms.

Dr. Robert E. Stall, Department of Plant Pathology, University of Florida, Gainesville, Florida, for his loan of the 1980 article by Dye et al. on plant pathogenic bacteria and pathovar names.

Dr. W. F. Wardowski, Citrus Packinghouse Extension Specialist, Department of Fruit Crops, Citrus Research & Education Center Lake Alfred, Florida, for his generous assistance as an integral member of the Grierson-Soule-Wardowski research-teaching-extension team in citrus harvesting and handling.

Dr. Barbara D. Webster, Department of Agronomy and Environmental Sciences, University of California, Davis, California, for her review of

ACKNOWLEDGMENTS

the chapter on horticultural taxonomy and plant breeding.

My colleagues in the Department of Fruit Crops, University of Florida, Gainesville, Florida, who individually and collectively have given enthusiastic, solid support throughout this endeavor: Dr. Wayne B. Sherman, my erstwhile coauthor, for use of his personal notes, numerous definitions, and two illustrations in the sections on biosystematics, plant breeding, and fruit setting; Dr. Karen E. Koch, for furnishing me with definitions for terms on metabolic sites and processes, her review and perceptive criticisms of the chapter on horticultural physiology and crop ecology, loan of the photograph of individual grapefruit juice vesicles reproduced as a line drawing, and answers to innumerable questions as the work progressed; Dr. Frederick S. Davis, for loan of books on general botany and water relations and his review of the pages on stress physiology; Dr. Paul M. Lyrene, for furnishing terms and definitions for the section on plant breeding; Dr. David W. Buchanan, for loan of books on many occasions; and Dr. Michael J. Burke, for loan of his books on stress physiology.

Dr. Mary M. Conway, Life Science Editor, John Wiley & Sons, New York, whose immense talents and broad background have been of inestimable value in answering countless questions and providing thoughtful counsel in the difficult and laborious process of preparing this glossary as camera-ready material for publication.

L. H. Bailey Hortorium, Cornell University, Ithaca, New York, (D. M. Bates, Director), for permission to use illustrations (redrawn and adapted) and definitions abstracted from articles by L. H. Bailey in Gentes Herbarum, volumes III to VIII, 1935 to 1949.

Euphytica (Netherlands Journal of Plant Breeding) (A. C. Zeven, Managing Editor), Wageningen, The Netherlands, for permission to abstract definitions for carpel development from G. P. Chapman, 1964, Euphytica 13.

Florida Department of Natural Resources, Bureau of Aquatic Resources and Control (Dennis M. Riley, Chief) for permission to use definitions on habitats and life forms abstracted from Tarver et al., 1978.

Florida State Horticultural Society (L. W. Timmer, Editor, Florida Agricultural Research & Education Center, Lake Alfred) for permission to use definitions abstracted from articles by H. H. Bryan et al., D. Carnell and K. Thompson, J. W. Prevatt et al., and L. K. Shaw et al., in Proc. Florida State Horticultural Society, 93, 1980.

Dr. David T. Goldman, Editor, U.S. Department of Commerce, National Bureau of Standards, NBS Special Publication 330, 4th ed., The International System of Limits (SI), for permission to reproduce tables from this publication.

Drs. M. G. Holmes, W. H. Klein and J. Sager, Smithsonian Environmental Research Laboratory, Research Center, Rockville, Maryland, for the use of definitions from their paper "Photons, flux, and some light on philology" published in the February 1985 issue of HortScience.

Inter-American Institute of Agricultural Sciences, Turrilba, Costa Rica (Julio Escoto B., Head, Technical Communication Unit) for permission to use the definition for high humidity beds abstracted from Hardy, 1960. Cacao manual.

Drs. Donald T. Krizek, Plant Stress Laboratory, U.S. Department of Agriculture, ARS, Beltsville, Maryland, and J. C. McFarlane, Corvallis Environmental Research Laboratory, Environmental Protection Agency, Corvallis, Oregon, for permission to use definitions from their paper "Controlled-environment guidelines", published in the October 1983 issue of HortScience.

Mr. J. Marshall, J. M. Dent & Sons, Ltd., London, for permission to redraw two examples of topiary from M. L. Gothein's A history of garden

ACKNOWLEDGMENTS

art.

Elspeth Napier, Editor, Royal Horticultural Society, London, for permission to redraw an example of a spindle tree from one of the Society's publications.

New Jersey Agricultural Experiment Station, Rutgers University, New Brunswick, New Jersey (G. M. Markle) for permission to use material from Magness, Markle and Compton, 1971. Food and feed crops of the United States.

Dr. M. J. Savage, Department of Soil Science and Agrometeorology, University of Natal, Pietermaritzburg, South Africa, for permission to use definitions from his papers "Modification for guidelines for reporting studies in controlled environment chambers and "Use of the international system of units in the plant sciences", published in the June 1978 and August 1979 issues, respectively, of HortScience.

Tea Research Foundation of Kenya (W. Urieno, Director) for permission to redraw illustrations from Tea Estate Practice 1966.

Drs. R. W. Thimijan, Beltsville Agricultural Research Center, ARS, U.S. Department of Agriculture, Beltsville, Maryland, and R. D. Heins, Department of Horticulture, Michigan State University, East Lansing, Michigan, for permission to use definitions from their paper "Photometric, radiometric and quantum light units of measure: A review of procedures for interconversion" published in the December 1983 issue of HortScience.

University of California, Division of Agricultural Sciences, Berkeley, California (William Wade, Manager, Agricultural Science Publications) for permission to use material from publications by Baker, 1957, Chapman and Pratt 1961, Guillou 1960, and Hutchins, 1967.

University of Florida, Institute of Food and Agricultural Sciences, Florida Agricultural Experiment Station, Florida Cooperative Extension Service, Gainesville, Florida (Mrs. JoAnn Pierce, Acting Chairman, Editorial Department), for permission to use material from publications by Carlisle and NeSmith, 1973, Dunn, 1979, Harrison, 1977, Harrison et al., 1967, Ruehle, 1958, 1963, Smajstrla and Harrison, Tucker, 1978, Department of Entomology and Nematology, 1978, West, 1957, Wolfe and Lynch, 1940, and Teem, 1977.

University of Hawaii, College of Tropical Agriculture and Human Resources, Hawaii Agricultural Experiment Station, Honolulu, HI (Charles DeLuca, Director, Agricultural Publications) for permission to use material from Goto and Fukunaga, 1956.

United States Department of Agriculture Office of Information, Publishing Division, Washington, D.C. (H. Nelson Fitton, Chief) for permission to use material from Ashby 1970, Harvey, Smith, and Kaufman, 1972, Little and Wadsworth, 1964, Lutz and Hardenburg, 1968, Markeson, 1959, McGregor, 1976, McCulloch, Cook, and Wright, 1968, Pierson, Ceponis, and McCulloch, 1971, Redit, 1969, Ramsey and Smith, 1961, Ramsey Friedman, and Smith, 1959, Smoot, Houck, and Johnson, 1971, Smith, McCulloch, and Friedman, 1966, Yearbook of Agriculture, 1961, Division of Pomology, 1896, Agricultural Marketing Service, 1980, Marketing Information Division, 1960, Solntseva, 1976, Agricultural Marketing Service, 1976, U.S. Department of Agriculture, 1980, U.S. Congress, 1970, U.S. Department of Agriculture, 1982, and Khokhlov, 1976.

Drs. A. E. Watada, U.S. Department of Agriculture Horticultural Crops Quality Laboratory, Beltsville, Maryland, and his coauthors, members of the American Society of Horticultural Science for kindly furnishing two figures and permission to use definitions from their paper "Terms for the description of developmental stages of horticultural crops" published in the February 1984 issue of HortScience.

I also express my gratitude to the following publishers and individ-

ACKNOWLEDGMENTS

ual authors who have generously granted permission to use illustrations and definitions of terms from their publications, with express acknowledgment of the rights and limitations on publication stipulated in these permissions:

Academic Press, Inc., New York, and the respective authors for definitions (abstracted) from Advances in Virus Research, vol. 20,©1976: R. R. Granados. Infection and replication of insect pathogenic viruses in tissue culture on p. 532; H. M. Mazzone and G. H. Tignor, Insect viruses: Serological relationships on p. 530 and 532; and R. J. Shepard, DNA viruses of higher plants on p. 513; from P. de T. Alvim and T. T. Kozlowski (Eds.), Ecophysiology of tropical crops,©1977: P. de T. Alvim, Cacao, on p. 370 and 376; D. P. Bartholomew and S. D. Kadzimin, Pineapple, on p. 376; and J. L. Montieth, Climate, on p. 338, 370, and 373; from P. J. Kramer, Water relations of plants,©1983, on p. 399 and 401-405; from R. A. Larson, Ed., Introduction to floriculture,©1980, lists of plants on p.5, 6, and definitions on p. 420, 559- 565, 567, 580, 587, 592, 619, and 623; and from J. W. Wright, Introduction to forest genetics, ©1976, on p. 315, 316, 370, and 372.

Academic Press, Inc. (London) Ltd., and the author for definitions (abstracted) from A. R. Rees, The growth of bulbs,©1972, for definitions on p. 565-568.

American Society of Agronomy, Madison, Wisconsin, for definitions (abstracted) from J. R. Harlan, Crops and man,©1975, on p. 3, 310, and 317.

American Society of Heating, Refrigerating and Air Conditioning Engineers, Atlanta, Georgia, for definitions from ASHRAE, 1981, Fundamentals volume, chap. 35, Terminology,©1981, on p. 588, 589, 591, and 593.

Bibliographisches Institut AG, Mannheim and Longman Group Ltd., Harlow, England, for an illustration (redrawn) on p. 251 from W. Manshard, Tropical Agriculture,©1974.

Cambridge University Press, New York, for illustrations (redrawn) and definitions (abstracted) from E. J. H. Corner, The seeds of dicotyledons (2 vol.),©1976, on p. 145, 182, and 184-193 and p. 125 and 163-167, respectively; from P. W. Richards, The tropical rain forest,©1952, for an illustration on p. 255 and definition (abstracted) on p. 242; and from A. Young, Tropical Soils and Soil Survey,©1976, for definitions on p. 472-476, 479, 481, and 482.

Commonwealth Agricultural Bureaux, Farnham Royal, United Kingdom, for definitions (abstracted) from E. H. Hewitt, Soil and water culture methods used in the study of plant nutrition, Commonwealth Bureau of Horticultural and Plantation Crop Technical Communication 22 (revised),©1966, on p. 490, 491, and 493-496; and B. Mosse, Graft-incompatibility in fruit trees, C.B.H.P.C. Tech. Communication 28.©1962, on p. 449.

Commonwealth Mycological Institute, Richmond-upon Thames, Surrey, United Kingdom, for definitions from C.M.I. Plant pathologist's pocketbook,©1968, on p.504-506, 510-514, 516, 521-530, 533-538, 630, 638, 639, 642, 643, 645, 646 and 650; and D. W. Dye, J. F. Bradbury, M. Goto, A. C. Hayward, R. A. Elliott, and M. N. Schroth, International standards for naming pathogenic bacteria and a list of pathovar names and pathotype strains, Review of plant pathology 59(4): 153-168,©1980, on p. 511, 512, 517, 518, and 629-633.

Cornell University Press, Ithaca, N.Y., for illustrations (redrawn) and definitions (abstracted) from H. B. Tukey, Dwarfed fruit trees,©1964, on p. 546, 547, 555, and 556 and p. 416, 540, 541, 543, and 544, respectively.

E. J. H. Corner, The natural history of palms,©1966, for illustrations (redrawn) on p. 14, 36, 38, 45, 74, 79, 83, 227, 253, 257, and 296

ACKNOWLEDGMENTS   xii

and 22, 23, 61, 162, 241-243, 267 and 372, respectively.

David & Charles (Holdings) Ltd. Newton Abbot, Devonshire, England, for illustrations (redrawn) and definitions (adapted) from W. T. Stearn. Botanical latin,©1966, illustrations on p. 34, 293, 295 and 298 and definitions on p. 20-26, 59-65, 100, 102, 118, 121, 122, 125-127, 129, 161, 162, 165, 167, and 263-279.

Dover Publications, Inc., New York, for illustrations (redrawn) from T. Johnson, The herbal: or general history of plants by John Gerard (unabridged and enlarged republication of 1633 edition with pages from 1626 edition, revised by T. Johnson),©1975, on p. 12, 55, 67, 113, 250, 252, 254, and 281; and from C. F. Millspaugh, American medicinal plants (2 vol.) (unabridged replication of 1892 edition in one volume with new tables),©1974, on p. 56 and 300.

W. H. Freeman and Company, New York, for definitions (abstracted) and illustrations (redrawn) from P. W. Atkins, Physical chemistry,©1978, definitions on p. 328, 329, and 333; P. Echlin, Pollen, Scientific American 218 (April): 80-90,©1968, illustrations on p. 140, 141, and 142; A. S. Foster and E. M. Gifford, Jr., Comparative morphology of vascular plants, 2nd ed.,©1974, illustrations on p. 140 and 217 and p. 119, respectively; Jules Janick, Horticultural science, 2nd ed.©1972, illustrations on p. 547 and 548 and definitions (abstracted) on p. 539-542; H. J. Oosting, The study of plant communities,©1948, definition on p. 308; D. A. Roberts and C. W. Boothroyd, Fundamentals of plant pathology,©1972, 1975, definitions on p. 245, 504, 509-518, 520-531, 533, 537, and 626; L. Stryer, Biochemistry,©1975, definitions on p. 354 and 363; and M. N. Westwood, Temperate-zone pomology,©1978, illustrations on p. 29, 30, 549, and 554, and definitions on p. 19, 27, 274, 358, 359, 370, 372, 393-395, 539-542, 569, 570, 576, 579.

Gustav Fisher Verlag, Stuttgart, and Longman Group Ltd., Harlow England, for illustrations (redrawn) on p. 33, 69, 141, 207, 212, 223, 249, 254, 260, 261, 288, 289, and 301 and definitions (abstracted) on p. 10, 19, 20, 59, 60, 64, 100-103, 120, 122, 161, 162, 195, 199, 238-246, 376, 377, and 407, from Strasburger's textbook of botany, 30th ed. (New English edition, edited by P. R. Bell and D. E. Coombe.),©1976.

Harper and Row Publishers, Inc., New York, for illustrations (redrawn) and definitions from A. E. Radford, W. C. Dickison, J. R. Massey, and C. R. Bell, Vascular plant systematics,©1974, illustrations on p. 136, 137, 146, 147, and 293-298 and definitions on p. 120 and 121.

Mrs. H. E. Hayward, Riverside, California, for illustrations (redrawn) from H. E. Hayward, The structure of economic plants,©1938, on p. 98, 99, 157, and 176.

Hilltop Orchards and Nurseries, Inc., Hartford, Michigan, for an illustration (redrawn) from Hilltop Orchards and Nurseries, Inc. 1982 Catalog,©1982, on p. 50 and definition (abstracted) on p. 26.

Horticultural Publications, Rutgers University, New Brunswick, New Jersey, for a definition (abstracted) on p. 407 from N. F. Childers, Modern fruit science, 6th ed.,©1975.

Horticulture Education Association (Hon. Editor Dr. W. M. Dullforce), Kegworth, England, for an illustration (redrawn) on p. 546 and definitions (abstracted) on p. 540 and 542, from Scientific Horticulture 11:67-74, 1955.

Miss Joan Hutchinson, 12 Kenmore Close, Kent Row, Kew, England, for illustrations (redrawn) and definitions (abstracted) from J. Hutchinson, Evolution and phylogeny of flowering plants,©1969, on p. 42, 56, 78, 80, 81, 85, 89, 132, 134, 137, 143, 146, 151, 170, 175, 176, 178, 183, 286, 294- 296, and 298-300 and on p. 121, 123, 239, 270, and 276, respectively.

International Bureau for Plant Taxonomny and Nomenclature, Utrecht,

ACKNOWLEDGMENTS xiii

The Netherlands, for definitions from J. S. L. Gilmour, F. R. Horne, E. L. Little, Jr., F. A. Stafleu, and R. H. Richens, Eds., International code of nomenclature of cultivated plants,©1969, on p. 304, 305, and 311; and F. A. Stafleu, Ed., International code of botanical nomenclature,©1972, on p. 305 and 306.

Dr. P. J. Kramer, Department of Botany, Duke University, Durham, North Carolina, for kindly giving permission to use definitions (abstracted) from his Water Relations of Plants,©1983, Academic Press, Inc., New York, on p. 399 and 401-405.

Dr. J. Levitt, Department of Plant Biology, Carnegie Institute of Washington, Stanford, California, for kindly giving permission to use his Responses of Plants to Environmental Stresses,©1972, Academic Press, New York, as general background for definitions on stress physiology in IV.5., p. 396-416.

Linnean Society of London (Academic Press, New York) for definitions abstracted from J. Burley and B. T. Styles, Eds., Tropical trees, Variation, breeding, and conservation, Linnean Society (London) Symposium Series No. 2,©1976: J. Burley. Genetic systems and genetic conservation of tropical pines, on p. 19 and 23; D. H. Janzen, Why do bamboos wait so long to flower? on p. 10 and 268; and K. Jong. Cytology of the Dipterocarpaceae, on p. 308.

Longman Group Ltd., Harlow, England, for illustrations (redrawn) and definitions (abstracted) from C. W. S. Hartley, The oil palm 22nd ed., ©1977, illustrations on p. 15 and 258 and definitions on p. 243 and 434; E. M. Lind and M. E. Morrison, East African vegetation,©1974, illustrations on p. 182, 250, 252, 253, 262 and definition on p. 238; K. A. Longman and J. Jenik, Tropical forest and its environment,©1974, illustrations on p. 258 and 259 and definitions on p. 10, 19, 23, 100, 239-245, and 267; J. W. Purseglove, Tropical crops, Dicotyledons,©1968, illustration on p. 156; and Monocotyledons, 2 vol.,©1972, illustration on p. 135; E. W. Russell, Soil conditions and plant growth, 10th ed.,©1973, definitions on p. 354, 472, 474, 476, 477, 482, 484, and 487; N. W. Simmonds, Principles of crop improvement,©1979, definition on p. 316.

The Macmillan Publishing Company, New York, for illustrations (redrawn) and definitions (abstracted) from L. H. Bailey, Standard cyclopedia of horticulture (6 vol. as 3),©1928, illustrations on p. 17, 42, 57, 72, 109-112, 114, 116, 117, 132, 149, 156, 168, 253, 282, and 288 and definitions on p. 103 and 423; H. H. Hume, The cultivation of citrus fruits, ©1926, illustrations on p. 133, 180, and 181; G. H. M. Lawrence. Taxonomy of vascular plants,©1951, illustrations on p. 293-300 and definitions on p. 22, 23, 26, 62-65, 100, 102, 122, 123, 129, 160, 161, 163, 241-243, 267-269, 271, 272, 274-276, 278, 279, 302-307 and 407; G. H. M. Lawrence, An introduction to plant taxonomy,©1955, illustrations on p. 105, 106, 149, 150, 152, 283-291, 292, 294, 295, and 468 and definitions on p. 10, 19, 22, 23, 25-27, 59, 60, 63, 100-103, 118-129, 158-161, 163, 267-269, 271, 272, 276-279, 309, and 385-387; T. L. Lyon and H. O. Buckman, The nature and properties of soils, 4th ed.,©1946, definitions on p. 472, 478, and 482; E. D. Merrill, Plant life of the Pacific world,©1945, illustration on p. 256; W. C. Muenscher, Weeds, 2nd ed.,©1955, illustrations on p. 48, 72, 115, 253, and 290 and a definition on p. 272; J. J. Ochse, M. J. Soule, Jr., M. J. Dijkman, and C. Wehlburg, Tropical and subtropical agriculture (2 vol.)©1961, illustrations on p. 46, 54, 58, 180, 181, 467, 551, and 553 and definitions on p. 303, 475, 476, 478, 479, 508, 540, 541, 576, 577, and 674-677; S. Schuler, America's great private gardens,©1967, illustration on p. 556 and definition on p. 544; and M. W. Strickberger, Genetics, 2nd ed.,©1976, definition on p. 323; and E. P. Christopher, The pruning manual,©1957, illustrations (redrawn) on p. 58 and 545.

ACKNOWLEDGMENTS

xiv

G.-P. Maisonneuve et Larose, Paris, for illustrations (redrawn with legends in English) from J. Braudeau, Le cacaoyer,©1969, on p. 28, 37, and 49; J. Champion, Le Bananier,©1963, on p. 456; R. Coste, Le cafeier,©1968, on p. 49; Y. Fremond, R. Ziller, and M. de Nuce Lamothe, Le cocotier, ©1963, on p. 172 and 179; J. C. Praloran, Les agrumes,©1971, on p. 44, 97, 250, and 282; C. Py and M.-A. Tisseau, L'ananas, ©1965, on p. 70, 84, 153, 174, and 456; and C. Surre and R. Ziller, Le palmier a huile,©1963, on p. 15, 97, and 179.

McGraw-Hill Book Company, New York, for illustrations (redrawn) and definitions (abstracted) from A. J. Eames, Morphology of the angiosperms, ©1961, illustration on p. 215 and definitions on p. 200, 388, and 390; A. J. Eames and L. H. MacDaniels, An introduction to plant anatomy, 2nd ed., ©1947, illustrations on p. 66, 209-212, 217-220, 223, 224, 226, 228-232, 237, 295, and 296 and definitions on p. 59, 162, 194, 195, 197-200, 202, 203, and 205; F. D. Heald, Manual of plant diseases, 3rd ed.,©1933, illustrations on p. 253, and 289 and names of diseases on p. 240, 244, 510, and 518-520; J. B. Hill, L. O. Overholts, H. W. Popp, and A. R. Grove, Jr., Botany, 3rd ed.,©1960, illustrations on p. 42, 151, 171, 182, 184, 208, 213, 215, 233, and 234 and definitions on p. 23, 59, 94, and 203; P. Maheshwari, An introduction to the embryology of the angiosperms,©1950, illustration on p. 141 and definitions on p. 118 and 124; W. H. McAdams, Heat transmission, 2nd ed.,©1942, a definition on p. 592; and H. C. Thompson, Vegetable crops, 4th ed.,©1949, names of vegetable crop groups and vegetables on p. 7 and 8.

Meister Publishing Company, Willoughby, Ohio, for definitions (abstracted) from Farm chemicals handbook, 1983,©1983, on p. 490-493, 495-497, 534, and 537.

Martinus Nijhoff Publishers BV, The Hague, Netherlands, for illustrations (redrawn) and definitions (abstracted from C. G. G. J. van Steenis, Ed., Flora Malesiana, Series 1, Spermatophyta, vol. 4,©1948-1954, vol. 5©1955-1958, illustrations from vol. 4^1 on p. 252 and 262, and definitions from vol. 4^1 on p. 239 and 241-243, from vol. 4^2 on p. 309 and 310, and from vol. 5 on p. 246 and 307-311.

Northern Nut Growers Association, Hamden, Connecticut, (R. A. Jaynes, Ed.), for illustration (redrawn) from R. A. Jaynes, Nut tree culture in North America,©1979, on p. 171.

Oxford University Press, New York, for illustrations (redrawn) and definitions (abstracted) from John Hutchinson, The families of flowering plants, 3rd ed.,©1973, on p. 134-136, 138, 143, and 144, and p. 119, 120, 302, 303, 305, and 310, respectively.

The Packer, Shawnee Mission, Kansas, for definitions from C. Risch, Ed., Produce glossary, p. A-5, A-7, A-9, and A-11-12, In The Packer's 1981 produce availability and merchandising guide,©1981, on p. 575, 578, 581-588, 591, 595-597, 609, 610, 616-626, and 670.

Pergamon Press Ltd., Oxford, United Kingdom, and the authors for an illustration (redrawn) and definitions (abstracted) from K. Faegri and L. van der Pijl, The principles of pollination ecology, 3rd ed.,©1979, on p. 144, and p. 385-393, and 395, respectively.

Lawrence Pollinger Ltd., London and the estate of Raymond Bush, for an illustration (redrawn) and definition (adapted from illustration) from R. Bush, Tree Fruit growing.©1943, on p. 556 and p. 540, respectively.

Prentice-Hall, Inc., Englewood Cliffs, New Jersey, for illustrations (redrawn) and definitions (abstracted) from H. T. Hartmann and D. E. Kester, Plant propagation, 3rd ed.©1975, illustrations on p. 460, 465, and 466, and definitions on p. 27, 164, 242, 373, 417-419, 422, 425, 427-439, 441-449, 451, and 452; and J. N. Joiner, Ed., Foliage plant production, ©1981, definitions on p. 419, 429, 431, 561, 564, 565, and 623.

ACKNOWLEDGMENTS

Science Kit and Boreal Laboratories, Tonawanda, New York, for illustrations (redrawn) from S.K.B.L. Turtox key card for pollen grains, ©1980, on p.142.

Smithsonian Institution Press, Washington, D.C., for illustrations (redrawn) from P. B. Tomlinson and A. M. Gill, Growth habits of tropical trees: Some guiding principles, p. 129-143, In B. J. Meggers et al. Eds., Tropical forest ecosystems in Africa and South America: A comparative review,©1973. on p. 39, 53, 70, and 83; and J. Cuatrecasas, Cacao and its allies, A. taxonomic revision of the genus Theobroma. Contributions from the United States National Herbarium, vol. 35, part 6, p. 369-614, 1964, on p. 23, 24, 41, 83, 172, and 300.

Springer-Verlag, New York, and Heidelberg, and the respective authors for illustrations (redrawn) and definitions (abstracted) from F. Hallé, R. A. A. Oldeman, and P. B. Tomlinson, Tropical trees and forests, An architectural analysis,©1978, illustrations on p. 17, 38, 40, 44, 50-52, 54, 57, 58, 84, and 249, and definitions on p. 20, 22-27, 241, 244, 268, and 372; W. Larcher, Physiological plant ecology, 2nd ed.,©1980, definitions on p. 370, 371, 373, 401, and 403-406; and L. van der Pijl, Principles of dispersal in higher plants, 2nd ed.,©1972, definitions on p. 164, 166, 392, and 393.

Termacarphi Pty., Ltd., Wantirna, Victoria, Australia, for a definition (abstracted) from T. P. O'Brien and M. E. McCully, The study of plant structure, Principles and selected methods,©1981, on p. 332.

University of Florida Presses, Gainesville, Florida, for illustrations from J. V. Watkins and T. J. Sheehan, rev. ed.,©1975, by the Board of Regents of the State of Florida, on p. 13, 14, 16-18, 38, 39, 41-43, 45, 46, 50-52, 54, 55, 57, 72, 75, 78, 81, 86, 96, 108, 254, 255, 258, 259, 295, and 301.

University of Miami Press, Baltimore, Maryland, for illustrations (redrawn) from M. J. Dijkman, Hevea, Thirty years of research in the Far East,©1951, on p. 18, 45, 135, 169, 222, and 282.

Van Nostrand Reinhold Company, Inc., New York, for illustrations (redrawn) and definitions (abstracted) from T. J. Sheehan and M. R. Sheehan, Orchid genera illustrated,©1979, on p. 96, 149, 153, and 154, and p. 118, 122, and 126, respectively.

Wadsworth Publishing Company, Belmont, California, for definitions (abstracted) from F. B. Salisbury and C. W. Ross, Plant physiology, 2nd ed.,©1978, on p. 240, 244, 330, 331, 334-336, 340, 343, 349-360, 362, 363, 365, 366, 368-376, 398, 399, 405, 407, and 593.

John Wiley & Sons, Inc., New York, and Dr. N. C. Turner, for illustrations (redrawn) and definitions (abstracted) from K. Esau, Plant anatomy, 2nd ed.,©1965, illustrations on p. 210, 213, 214, 216, 217, 221, 225-227, 230- 232, 235, 236, 256, and 294, and definitions on p. 60, 62, 63, 94, 119- 121, 124, 125, 161-163, 165, 194, 195-206, 389, and 390; M. Jakob and G. A. Hawkins, Elements of heat transfer, 3rd ed.,©1958, definitions on p. 589; C. M. Peairs and R. H. Davidson, Insect pests of farm, garden and orchard, 5th ed.,©1956, definitions on p. 498, 501-503, 504, 505, 507, 508, and 660-662; D. H. Urquhart, Cocoa,©1961, definition on p. 425; I. J. Condit, The fig,©1947, illustrations on p. 116 and 174 and definitions on p. 103, 161 and 388-390; F. E. Lloyd, The carnivorous plants,©1942, definition on p. 241; E. V. Wulff, An introduction to historical plant geography,©1943, definitions on p. 307, 309, and 311; N. C. Turner, Drought resistance and adaptation to water deficits in crop plants. p. 343-372. In H. Mussell and R. C. Staples, Eds., Stress physiology in crop plants, ©1979, definitions on p. 369-398, 400, 406, and 407; and from N. C. Turner and P. J. Kramer, (Eds.), Adaptations of plants to water and high temperature stress,©1980: P. J. Kramer, Drought stress and the origin of

ACKNOWLEDGMENTS

adaptation, definitions on p. 397-401 and N. C. Turner and M. M. Jones, Turgor maintenance by osmotic adjustment: A review, on p. 402 and p. 406.

Worth Publishers, Inc., New York, for definitions (abstracted) from N. L. Allinger, M. P. Cava, D. C. Jongh, C. R. Johnson, N. A. Lebel, and C. L. Stevens, Organic chemistry,©1971, on p. 330, 341-343, 348, and 349; A. L. Lehninger, Biochemistry,©1970, on p. 332, 335, 344, 352, 353, 355, 360-363, and 366-369; and A. L. Lehninger, Biochemistry, 2nd ed.,©1975, on p. 331.

J. S.

ABBREVIATIONS FOR AUTHORS CITED FOR TERMS

AL	Allinger et al. (Worth Publishers)
AS	Association of Official Seed Analysts
ASHRAE	American Society of Heating, Refrigeration & Air Conditioning Engineers
AT	Atkins (W. H. Freeman & Co.)
AV	Alvim [Academic Press (NY)]
AY	Ashby (U.S. Dept. Agr., Agr. Handbook)
B & B	Britton and Brown (New York Botanical Gardens)
B & C	Bell & Coombe (Gustav Fischer Verlag)
B & K	Bartholomew & Kadzimin [Academic Press (NY)]
BA	Baker (Calif. Agr. Exp. Sta. Man.)
BE	Berry (Southern Freight Tariff Bureau, Atlanta)
B-GH	Bailey-Gentes Herbarum (Bailey Hortorium)
BI	Burkill (Ministry of Agr. & Cooperatives, Kuala Lumpur)
BL 68	Block (1968) (Southern Freight Tariff Bureau, Atlanta)
BL 69	Block (1969) (Southern Freight Tariff Bureau, Atlanta)
B-M	Bailey (Cyclopedia of Horticulture) (Macmillan & Co.)
BN	Bryan et al. (Fla. State Hort. Soc.)
BR	Braudeau (Maisonneuve & Larose)
BS	Bush (Lawrence Pollinger Ltd., London)
BU	Burley (Linnaean Soc. London)
C & N	Carlisle & NeSmith (University Florida Soil Science Dept. and U.S. Soil Conservation Serv.)
CH	Champion (Maisonneuve & Larose)
CI	Christopher (Macmillan Co.)
CMI	Commonwealth Mycological Institute
CN	Condit (John Wiley & Sons)
CO	Cohen (AVI)
CP-G	Chapman, G. P. (Euphytica)
CP-H	Chapman & Pratt (Univ. Calif., Agr. Science Div., Berkley)
CR 66	Corner (Cambridge Univ.)
CR	Corner (Cambridge University Press)
CS	Childers (Horticultural Publications)
CT	Coste (Maisonneuve & Larose)
CU	Cuatrecasas (Smithsonian Institution Press)
DB	Du Breuil (Paris)
DK	Dijkman (Univ. Miami Press)
DR	Darwin (D. Appleton & Co., NY)
DU	Dunn (Univ. Florida, Dept. of Entomology and Nematology)
DV	Davis (Florida Dept. of Conservation)
DY	Dye et al. (Rev. Plant Path.)
E & M	Eames and MacDaniels (McGraw-Hill)
EA	Eames (McGraw-Hill)
EC	Echlin (Scientific American, W. H. Freeman & Co.)
ES	Esau (John Wiley & Sons)
F & G	Foster and Gillford (W. H. Freeman & Co.)

*Names listed here in parentheses are cited in the Acknowledgments and/or Selected References.

ABBREVATIONS FOR AUTHORS CITED FOR TERMS

F & P	Faegri and van der Pijl (Pergamon)
F & R	Friedman and Radspinner (U.S. Dept. Agr.)
FC	Farm Chemicals (Meister)
FDOC	Florida Dept. of Citrus
FR	Freeman (Univ. of Florida, Dept. of Fruit Crops)
FZL	Fremond, Ziller & Lamothe (Maisonneuve & Larose)
G & B	Goldman & Bell (SI Units 1981)
G & F	Goto & Fukunaga (Univ. of Hawaii Agr. Extension Serv. Cir.)
G 62	Gray (1862) (Ivison, Phinney & Co., NY)
G 68	Gray (1868) (Ivison, Blakeman, Taylor & Co., NY)
G-D	Gray, D. (AVI)
GCWG	GCWG (Growth Chamber Working Group ASHS)
GE	Geraldson (Proc. Fla. State Hort. Soc.)
GL	Gilmour (Internat. Bureau for Plant Taxonomy & Nomenclature, Utrecht)
GO	Gothein (J. M. Dent & Sons, London)
GR	Granados [Academic Press (NY)]
GSK	Grierson, Soule & Kawada (AVI)
GU	Guillou (Univ. Calif. AES)
H & K	Hartmann & Kester (Prentice-Hall)
HA	Hardy (Inter-American Institute of Agr. Sciences, Turrialba)
HAR	Harrison (Fla. Coop. Extens. Serv.)
HC	Hutchins (Univ. Calif. Agr. Exp. Sta. Cir.)
HD	Hedden et al. (Proc. Internat. Soc. Citriculture 1977)
HD 17	Hedrick (New York Dept. of Agriculture)
HD 21	Hedrick (New York Dept. of Agriculture)
HD 25	Hedrick (New York Dept. of Agriculture)
HE	Heald (McGraw-Hill)
HI	Hill et al. (McGraw-Hill)
HKS	Holmes, Klein & Sager, HortScience
HM	Hume (1907 Orange Judd; 1926 Macmillan)
HN	Harlan (American Soc. Agronomy)
HOBJ	Harrison et al. (Fla. Agr. Extension Serv. Cir.)
HON	Hilltop Orchards & Nurseries
HOT	Hallé, Oldeman & Tomlinson (Springer Verlag)
HR	de Hertogh (HortScience; Acta Hort.; Netherlands Flower-Bulb Institute, New York)
HS 69	Hutchinson (1969) (Miss Joan Hutchinson, Kew)
HS 73	Hutchinson (1973) (Oxford University Press)
HSK	Harvey et al. (U.S. Dept. Agr., Agr. Handbook)
HT	Hartley (Longman Group Ltd. (London)
HW	Hewitt (Commonwealth Agr. Bureaux)
HY	Hayward (Mrs. H. E. Hayward)
I	Isenberg (AVI)
J	Janick (W. H. Freeman & Co.)
JG	Jong [Linnean Society (London)]
JH	Johnson (Dover Publications)
JK	Jakob & Hawkins (John Wiley & Sons)
JM	Jamison (National Perishable Freight Committee, Chicago)
JN	Jensen (Bioscience)
JR	Joiner (Prentice-Hall)
JY	Jaynes (Northern Nut Growers Association)
JZ	Janzen [Linnean Society (London)]
K & M	Krizek & McFarlane (HortScience)
K	Kasmire (personal)
KH	Khokhlov (U.S. Dept. Agr., Agr. Res. Service)

ABBREVATIONS FOR AUTHORS CITED FOR TERMS

KK	Koch (personal)
KR	Kramer [Academic Press (NY)]
L & B	Lyon & Buckman (Macmillan Publishing Co.)
L & H	Lutz & Hardenburg (U.S. Dept. Agr., Agr. Handbook)
L & J	Longman & Jenik [Longman Group Ltd. (London)]
L & M	Lind & Morrison [Longman Group Ltd. (London)]
L & W	Little & Wadsworth (U.S. Dept. Agr., Agr. Handbook)
L 51	Lawrence (Macmillan Co.)
L 55	Lawrence (Macmillan Co.)
LA	Larcher (Springer Verlag)
LB	Lubbock [Macmillan & Co. (London)]
LE 70	Lehninger (1970) (Worth Publishers)
LE 75	Lehninger (1975) (Worth Publishers)
LL	Lloyd (John Wiley & Sons)
LR	Larson [Academic Press (NY)]
LS	Lucas (Die Lehre von Baumschnitt)
M & B	Mohr and van Baren (Uitgeverij/van Hoeve, The Hague)
M & L	Maynard & Lorenz (AVI)
MA	McAdams (McGraw-Hill)
MCW	McCulloch, Cook and Wright (U.S. Dept. Agr., Agr. Handbook)
ME	Mosse (Commonwealth Agr. Bureaux)
MG	McGregor (U.S. Dept. Agr., Agr. Handbk.)
MH	Maheshwari (McGraw-Hill)
MK	Markeson (U.S. Dept. Agr., Farmer Coop. Serv.)
ML	Millspaugh (Dover Publications)
MMC	Magness, Markle & Compton (New Jersey Agr. Exp. Sta. Bul.)
MNS	Market News Service (U.S. Dept. Agr., AMS)
MO	Monteith [Academic Press (NY)]
MP	Meister Publishing Co.
MR	Merrill (Macmillan Publishing Co.)
MRW	McCain, Raabe & Wilhelm (Calif. Coop. Extens. Serv. Leaflet)
MS	Manshard (Bibliographisches Institut AG, Mannheim)
MU	Muenscher (Macmillan Publishing Co.)
MZ	Mazzone & Tignor [Academic Press (NY)]
N	Nybom (Natl. Acad. Science)
O & M	O'Brien & McCully (Termarcarphi Pty, Ltd.)
O 31	Ochse (1931) (G. Kolff & Co.)
OS	Oosting (W. H. Freeman & Co.)
OSDW	Ochse et al. (1961) (Macmillan Co.)
P & D	Peairs & Davidson (John Wiley & Sons)
P & T	Py & Tisseau (Maisonneuve & Larose)
P & V	Page & Vigoreux (SI units 1974)
PA	Pantastico et al. (AVI)
PACA	(U.S. Dept. Agr. 1982)
PCM	Pierson et al. (U.S. Dept. Agr., Agr. Handbook)
PG 68	Purseglove (1968) [Longman Group Ltd. (London)]
PG 72	Purseglove (1972) [Longman Group Ltd. (London)]
PJ	van der Pijl (Springer Verlag)
PR	Praloran (Maisonneuve & Larose)
PV	Prevatt (Proc. Fla. State Hort. Soc.)
R & L	Ruehle & Ledin (Fla. Agr. Extens. Bul.)
R & S	Ramsey & Smith (U.S. Dept. Agr., Agr. Handbook)
R	Roberts (personal notes)
RB	Roberts & Boothroyd (W. H. Freeman & Co.)
RB 24	Robbins (P. Blakiston's Son & Co.)
RD	Radford et al. (Harper & Row)

ABBREVATIONS FOR AUTHORS CITED FOR TERMS

RE	Rees [Academic Press Inc. (London) Ltd.]
RFS	Ramsey et al. (U.S. Dept. Agr., Agr. Handbook)
RH	Ruehle (1958, 1963) (Fla. Agr. Extens. Bul., Agr. Exp. Sta. Bul.)
RHS	Royal Horticultural Society
RI	Richards (Cambridge Univ. Press)
RS	Risch (The Packer)
RT	Redit (U.S. Dept. Agr., Agr. Handbook)
RU	Russell [Longman Group Ltd., (London)]
S & G	Soule & Grierson (Univ. Florida Dept. Fruit Crops)
S & H	Smajstrla & Harrison (Fla. Coop. Extens. Serv. Cir.)
S & R	Salisbury & Ross (Wadsworth Publishing Co.)
S & Z	Surre & Ziller (Maisonneuve & Larose)
SA	Sargent (Houghton Mifflin Co., Boston)
SB	Strickberger (Macmillan Publishing Co.)
SC	Schuler (Macmillan Publishing Co.)
SE 78	Savage (HortScience 1978)
SE 79	Savage (HortScience 1979)
SF	Stafleu (A. Oosthoek's Uitgeversmaatschappii N.V., Utrecht)
SHJ	Smoot et al. (U.S. Dept. Agr., Agr. Handbook)
SI	Scientific Horticulture (Hort. Ed. Assn.)
SK	Smock (AVI)
SKB	(Science Kit & Boreal Labs Inc.)
SM 79	Simmonds (1979) [Longman Group Ltd. (London)]
SMF	Smith et al. (U.S. Dept. Agr., Agr. Handbook)
SN	Sanford (American Potash Institute, Washington)
SP	Shepherd [Academic Press (NY)]
SR	Schneider (L. Upcott Gill, London)
SS	Sheehan & Sheehan (Van Nostrand Reinhold)
ST	Stearn (David & Charles Publishers, Newton Abbot, Devon, UK)
v.ST	van Steenis (P. Noordhoff Ltd., Groningen)
SV	Solntseva (U.S. Dept. Agr., ARS)
SW	Shaw (Proc. Fla. State Hort. Soc.)
SY	Stryer (W. H. Freeman & Co.)
SYB	Soule, Yost & Bennett (1966, 1969, Trans. ASAE, U.S. Dept. Agr.), (Trans. Am. Soc. Agr. Engineers)
T & G	Tomlinson & Gill (Smithsonian Institution Press)
T & H	Thimijan & Heins (HortScience 1983)
T & K - KR	Turner & Kramer, Kramer (John Wiley & Sons 1980)
T & K - T & J	Turner & Kramer - Turner & Jones (John Wiley & Sons 1980)
T	Tucker (Florida Coop. Extens. Serv. Cir.)
TA	Tarver et al. (Fla. Dept. Natural Resources)
TI	Tai (Jamaica Banana Board)
TK	Tukey (Cornell Univ. Press)
TM	Teem (Univ. of Florida, Dept. of Agronomy)
TP	Thompson (McGraw-Hill)
TRI	Tea Research Institute of East Africa
TU	Turner (John Wiley & Sons 1979)
UF	(Fla. Insect Control Guide 1978)
UQ	Urquhart (John Wiley & Sons)
US - GN	U.S. Dept. Agr. (Grade names)
US - P	U.S. Dept. Agr., Div. of Pomology
US - S	U.S. Dept. Agr. Yearbook: Seeds
US - SG	U.S. Dept. Agr. (Standards for Grades)
V & R	Volin & Ramos (Fla. State Hort. Soc. 1980)
W & L	Wolfe & Lynch (Florida Agr. Exp. Sta. Bul.)

ABBREVATIONS FOR AUTHORS CITED FOR TERMS

W	Watkins & Sheehan (1975) (Univ. of Florida Presses)
WALK	Wehlburg, Alfieri, Langdon & Kimbrough (Fla. Dept. Agr. & Consumer Services, Div. Plant Industry Bul.)
WE	West (Florida Agr. Exp. Sta. Cir.)
WF	Wulff (John Wiley & Sons)
WH	Whitney & Sumner (Proc. International Soc. Citriculture 1977)
WR	Wright [Academic Press (NY)]
WS	Westwood (W. H. Freeman & Co.)
WT	(Watada et al., HortScience 1984)
Y & S	Young & Soule (Plant Prop. Lab Manual 1978)
Y	Young, A. (Cambridge Univ. Press)

CONTENTS

Abbreviations for Authors   xvii

I. HORTICULTURAL CROPS

0. General   3
1. Fruit Crops   3
2. Ornamental Horticulture and Commercial Floriculture   4
3. Vegetable Crops   6

II.0. MORPHOLOGY and ANATOMY

0. General   10
        Fig. II.0.01-07   12

1. Vegetative Structures   19
    1.1. Buds and stems   19
       1.1.1. Buds   19
       1.1.2. Vernation (prefoliation)   20
       1.1.3. Stems   22
       Fig. II.1.1.1.01-.06; 1.1.2.01-.03; 1.1.3.01-.22   28

    1.2. Leaves   59
       Fig. 1.2.01-.28   66
    1.3. Roots   94
       Fig. 1.3.01-.04   96

2. Flowering and Fruiting Structures   100
    2.1. Inflorescences   100
       Fig. II.2.1.01-.04   104
    2.2. Flowers   118
       Fig. II.2.2.01-.28   130
    2.3. Fruits   158
       2.3.1. Fruit types   158
       2.3.2. Fruit parts (exclusive of seeds)   161
       2.3.3. Seeds   163
       Fig. II.2.3.1.01-.07; 2.3.2.01-.07; 2.3.3.01-.12   168

3. Anatomy   194
        Fig II.3.01-.31   207

CONTENTS

4. Modifications 238

    4.1. Life forms 238
    4.2. Habitats 238
    4.3. Genotypic modifications 241
    4.4. Phenotypic modifications 246
            Fig. II.4.01-.14 249

5. General Morphological Terms 263

    5.1. Color 263
    5.2. Direction and size 266
    5.3. Duration 267
    5.4. Apex, base, margins 268
        5.4.1. Apex 268
        5.4.2. Base 269
        5.4.3. Margins 269
    5.5. Shapes 271
    5.6. Surface features 273
    5.7. Texture 278

6. Bryophytes, Pteridophytes, and Gymnosperms 279

            Fig. II.5.01-.20, .6.01 281

III. HORTICULTURAL TAXONOMY and PLANT BREEDING

1. Taxonomy 302

    1.0. Horticultural Taxonomy 302
    1.1. Botanical categories, classification, nomenclature 302
    1.2. Biosystematics 307

2. Plant Breeding 311

    2.0. General 311
    2.1. Breeding and selection systems, gene pools 316
    2.2. Terms 317

            Fig. III.2.2.01 327

IV. HORTICULTURAL PHYSIOLOGY and CROP ECOLOGY

0. General 328

1. Metabolic Sites and Processes 350

    1.1. Cell structure, organelles 350
    1.2. Major metabolic processes 352

2. Hormones and Growth Regulators, Other Metabolic Products 357

    2.1. Hormones and growth regulators 357

CONTENTS

2.2. Other metabolic products	359
3. Growth and Development	370
3.1. Growth cycles, juvenility	370
3.2. Photoperiodism, thermoperiodism	373
3.3. Tropism, nastic movements, nutation	376
4. Fruit Setting	378
4.1. Amphimixis (syngamy), apomixis	378
Fig. IV.4.1.01-.03	382
4.2. Carpel development	385
4.3. Dichogamy, dicliny, hercogamy, heterostyly	385
4.4. Pollination	387
4.5. Pollinators, pollinator adaptations	391
4.6. Polyembryony	392
4.7. Dispersal	392
4.8. Other terms	393
5. Water Relations, Cold, and Other Forms of Stress	396
5.1. General terms	396
5.2. Water relations (including stress aspects)	398
5.3. Cold stress	407
5.4. Other forms of stress	412
5.4.1. High temperature (heat)	412
5.4.2. Light stress	413
5.4.3. Salt (ionic), pressure, and flooding (waterlogging) stresses	414
5.4.4. Atmospheric pollution and mechanical stresses	415

V. PROPAGATION, NURSERY HANDLING, SOILS, CROP PRODUCTION

1. Propagation and Nursery Handling	417
1.0. General	417
1.1. Nursery facilities, structures, equipment	418
1.2. Materials and supplies	427
1.3. Generative (sexual) propagation	432
Fig. V.1.3.01	440
1.4. Vegetative (asexual) propagation	441
Fig. V.1.4.01-.17	453
1.5. Micropropagation	470
Fig. V.1.5.01	471
2. Soils	472
Fig. V.2.01	483
3. Crop Production (Fruits, Vegetables, and Woody Ornamentals)	484
3.0. General	484
3.1. Site preparation	484
Fig. V.3.1.01	488

CONTENTS

```
        3.2. Mineral nutrition                                      490
        3.3. Pests, pest management                                 497
             3.3.0. General                                         497
             3.3.1. Pests - Animals                                 498
             3.3.2. Pests - Plants                                  509
             3.3.3. Pest management                                 531
        3.4. Plant and crop-size control                            539
                  Fig. V.3.4.01-.12                                 545
        3.5. Water regulation                                       557

   4. Crop Production (Commercial Floriculture)                     559

        4.1. Culture and shipment to market                         559
        4.2. Pests                                                  565
             4.2.1. Parasitic diseases, including animals           565
             4.2.2. Physiological disorders (nonparasitic diseases) 566
```

VI. POSTHARVEST HANDLING and MARKETING

```
1. Maturity                                                         569

              Fig. VI.1.01, .02                                     573

2. Harvesting                                                       575

3. Packinghouse                                                     577

4. Containers                                                       585

5. Refrigeration, Storage                                           588

6. Carriers (Loading, Routing, Services)                            594

7. Laws, Rules, and Regulations                                     597

        7.1. Perishable Agricultural Commodities Act (PACA)         597
        7.2. Other laws, rules, and regulations                     609

8. Destination Markets, Business Terminology                        616

9. Market Diseases of Fruits and Vegetables                         626

        9.1. Pathological (parasitic) disorders                     626
        9.2. Animal pests that cause market disorders               659
        9.3. Physiological market disorders                         663

10. Processing                                                      674


Selected References                                                 679

Index of Terms                                                      693

Index of Crops                                                      833
```

GLOSSARY FOR HORTICULTURAL CROPS

TERMS

I. HORTICULTURAL CROPS   3

0. General   3
1. Fruit Crops   3
2. Ornamental Horticulture and Commercial Floriculture   4
3. Vegetable Crops   6

I.0. General

.001 Horticultural crops: the culture of nearly all cultivated or potentially cultivatable (i.e., products now gathered from the wild, as Brazil nut) plants, including primarily angiosperms but also gymnosperms, pteridophytes, and bryophytes (the last three principally as ornamentals); an immense and heterogeneous group often separated by use categories (e.g., by Heyne, Burkill, and others in their compilations of useful products for various parts of the world), such as fruits, vegetables, oil and nut crops, beverage crops, spice, condiment, and essential oil crops, rubber, and ornamentals; nonherbaceous oil and nut crops, beverage crops, rubber, and perennial spice, condiment and essential oil crops are customarily included among fruit crops, and annual or biennial (and usually) herbaceous spice, condiment, and essential oil crops, among vegetable crops, along with muskmelons and watermelons (as well as strawberries in some states); ornamentals include nursery plants of all types and turf crops and commercial floriculture, bulbous plants, cut flowers, pot plants, bedding plants, hanging baskets, and foliage plants.

.002 Horticulture: the growing of flowers, fruits, and vegetables and of plants for ornament and fancy (Bailey 1928); the culture of plants given individual care and attention (in contrast to agronomy, the culture of field and pasture crops, or forestry, the culture of trees for timber, pulpwood, etc.); includes three major subdivisions: fruit crops, ornamental horticulture and commercial floriculture, and vegetable crops (see Horticultural crops, I.0.001).

.003 Weed: a plant adapted to a habitat disturbed by man, genetically labile and phenotypically plastic, useful in some respects (e.g., as occupying and covering waste areas and thereby reducing erosion) and harmful in others (e.g., as competitors of crop plants for nutrients). (HN)

I.1. Fruit Crops

.000 Fruit Crops: perennial woody (sometimes suffrutescent), deciduous, or evergreen trees, shrubs, vines (and lianes), herbs, palms, and angiosperms, except for a few gymnosperms; usually distinguished

I.1. HORTICULTURAL CROPS

according to their general area of culture as temperate zone (deciduous), subtropical, and tropical crops.

.001 Beverage crops: trees, shrubs, or vines grown for the stimulating alkaloids contained in their seeds (e.g., coffee, cocao, kola, and guarana) or leaves (tea and maté).

.001a. Tea: properly the brewed infusion of black, oolong or green tea made from the partially grown leaves and terminal buds of Camellia sinensis; also a generic term for the brewed infusion of leaves, bark, and roots of many other plants used for beverages or medicinal purposes.

.002 Brambles: (see Small fruits), I.1.012).

.003 Citrus: Citrus species and relatives in the Rutaceae.

.004 Deciduous crops: (see Temperate zone crops, I.1.016).

.005 Essential oil crops: (see Spice crops, I.1.013).

.006 Fruit: botanically, a ripened ovary and adnate parts thereof; the edible, more or less succulent product of a perennial or woody plant grown or used for food in fresh, cooked, or processed (i.e., as juice and preserves) form.

.007 Nut crops: true nuts (hard-shelled indehiscent fruit derived from a compound ovary) or nutlike fruits or seeds, such as walnut, pecan, chestnut, pistachio, filbert, almond, cashew, Brazil nut, macadamia, coconut, and piñon nut, grown for eating as such or used in confectionery and the like. (MMC)

.008 Oil crops: Trees or shrubs from whose fruit (mesocarp) or seeds oil may be extracted for use in cooking or other purposes, such as coconut, oil palm, olive and tung (herbaceous annuals, such as soybean, cottonseed, peanut, linseed, sesame, safflower, sunflower, and castor bean, are excluded, being grown as agronomic crops).

.009 Pome fruits: crop or tree belonging to the Pomoideae subfamily of the Rosaceae, such as apple, pear, quince, medlar, loquat, and hawthorn. (MMC)

.010 Pomology: (from pome fruits) fruit growing (Bailey); now used mainly for temperate zone fruit and nut crops.

.011 Rubber: natural rubber is a substance obtained from the milky juice (latex) of perhaps 400 plants in a dozen or more families and usually characterized by its elasticity, although its properties vary widely, depending on the source and preparation (i.e., in a crude state, mixed, or vulcanized); also a term applied to Hevea brasiliensis which furnishes about 90% of the world's natural rubber.

.012 Small fruits: brambles (blackberries and raspberries, Rubus spp.) and strawberries (Fragaria spp.) in the Rosaceae; blueberries, cranberry (Vaccinium); currants and gooseberries (Ribes). (MMC)

.013 Spice (essential oil) crops: perennial trees, shrubs, or vines whose aromatic unopened flower buds (cloves), fruit (black pepper, allspice, vanilla), pulp (tamarind), seeds (nutmeg), aril (mace), bark (cinnamon, cassia), or rhizomes (ginger) are used in cooking to season food and flavor sauces; many of the spices, including others not listed, plus eucalyptus and cajeput, furnish essential oils, as do many of the herbaceous spice, condiment, and essential oil crops referred to under Vegetable Crops (I.3) for various pharmaceutical and cosmetic purposes. (MMC)

.014 Stone fruits: crops or trees belonging to the genus Prunus in the Rosaceae, such as peach, apricot, cherry, plums, and almond.

.015 Subtropical crops: fruits, etc., grown in areas in which occasional light freezes occur; a period of dormancy induced by cool weather

I.1. HORTICULTURAL CROPS

(i.e., temperatures below 10 to 13°C (50 to 55°F) for a month or two) is generally essential for proper flowering or best fruit quality in most instances; examples are orange, lemon, lychee, longan, and kiwi (= Chinese gooseberry). (MMC)

.016 Temperate zone (deciduous) crops: fruit or nut crops that require a more or less extended chilling period for proper vegetative growth and flowering; usually grown in areas with prolonged cold weather, often well below freezing; may be grown at high altitudes (i.e., 2000-3500 m) in tropical areas or, in the case of certain crops, at lower altitudes by substitution of defoliation and dry weather to satisfy the chilling requirement; examples are apple, pear, peach, cherry, almond, walnut, grape, brambles, strawberries, and pecan, plus the warm-temperate fig, olive, pistachio, and pomegranate. (MMC)

.017 Tropical crops: fruits, etc., grown in areas free from frost, with growth ceasing in most crops at temperatures below about 20°C (68°F); dormant periods, dictated by dry weather, are essential for many crops, e.g., grapefruit, lime, mango, avocado, coffee, kola, and cashew, whereas pineapple, banana, papaya, cacao, tea, coconut, oil palm, and black pepper, prefer rainfall well distributed throughout the year. (MMC)

I.2. Ornamental Horticulture and Commercial Floriculture

.000 Ornamental horticulture (broad sense): the cultivation of plants of all kinds (including many fruit and vegetable crops) for show and to satisfy the eye (i.e., for esthetic purposes) rather than for commercial production of food; divided into three major subdivisions with regard to their commercial and other uses and market outlets (numerous specialists in individual crops): (a) commercial floriculture, (I.2.002); (b) nursery plants, (I.2.011); and (c) turf crops, (I.2.013).

.001 Commercial floriculture: the culture of greenhouse and field crops grown as

.001a. Bedding plants: any herbaceous plant used in home landscaping (including flowers, herbs, ground covers, perennials, gesneriads), some Palmae, and small fruits, and/or in garden beds, planters for porches and patios or window boxes. (LR)

.001b. Bulbous plants: members of a group that includes bulbs, corms, tubers, tuberous roots, and rhizomes, of which narcissus (daffodils, jonquils), Dutch iris, hyacinths, and tulips are bulbs, and crocus, corms; forced commercially as cut flowers and flowering pot plants or sold as pot plants for consumer forcing. (LR)

.001c. Cut flowers: mainly chrysanthemums, carnations, roses, snapdragons, orchids, gladiolus, plus acacia, anemone, anthuriums, asters, camellias, cornflowers, shasta and other daisies, larkspur, ginger (Costus), dahlia, delphinium, freesias, gardenias, baby's breath, wax flower, statice, stocks, peonies, ranunculus, stephanotis, bird of paradise, violets, calla lilies, and zinnias. (LR)

.001d. Foliage plants: include more than 1000 species of monocotyledons, dicotyledons, gymnosperms, and pteridophytes, perennial herbs, vines (or lianes), palms, shrubs, and trees, grown in

I.2. HORTICULTURAL CROPS

containers for their attractive foliage, and sometimes flowers and fruit, notably species of Araceae (aroids), Marantaceae, Moraceae (especially Ficus), Agavaceae, Araliaceae, Acanthaceae, Bromeliaceae (bromeliads), Gesneriaceae (gesneriads), Palmae, conifers (e.g., Araucaria), and ferns for indoor or patio use; sold in supermarkets, florist shops, and nurseries; a large and expanding segment of commercial floriculture. (LR)

.001e. Hanging baskets: flowering plants, foliage plants, herbs, vegetables, and fruits. (LR)

.001f. Pot plants: mainly azaleas, pot chrysanthemums, gloxinias, African violets and other gesneriads, poinsettias, Easter lilies, hydrangeas, cyclamen, begonias, kalanchoe, plus calceolaria, Star-of-Bethlehem, peppers, Christmas cherry, bleeding heart, Persian violet, yellow pachystachys, geraniums, primroses, pot roses, Christmas cactus, cinerarias, and Cape primrose. (LR)

.002 Gardener: a person who cares for a landscape planting or garden.

.003 Ground covers: low, usually nongrass plants, such as ivies, liriope, Ophiopogon, Pachysandra, and Vinca spp.; used where essentially complete coverage of ground areas is desired and where traffic is light; may require periodic pruning or thinning out but are not mown as most lawn grasses and are usually less subject to pest problems, thus being low maintenance; often used in conjunction with turf species in lawns or parks (i.e., in different areas according to the traffic or accessibility for maintenance); (see Turf crops, 1.2.011).

.004 Landscape architect: the profession of arranging and modifying the effects of natural scenery in a given tract to achieve the best esthetic effect, considering the use (i.e., parks, estates, grounds of government buildings or other large structures, and roadside plantings) to which the tract is to be put; a person with formal training in architecture and landscape design (but not necessarily with detailed knowledge of plant materials and their culture).

.005 Landscape designer: an ornamental horticulturist with training or experience in landscape design less intensive or rigorous than that of a landscape architect.

.006 Landscape plants: (see Nursery plants, 1.2.009).

.007 Lawn grasses: (see Turf, 1.2.011).

.008 Nurseryman: a person who propagates, grows, sells, and sets out plants (the last as an added service), along with their care and maintenance.

.009 Nursery plants (woody ornamentals, landscape plants): palms, vines, deciduous and narrow-leaved (coniferous), and broad-leaved evergreen shrubs and trees, fruit crops, nut crops, perennial herbs (e.g., bulbous and rhizomatous plants, aquatics, ground covers, and turf), garden annuals and biennials (including vegetables), grown in greenhouses, shade (lath, screen) houses, containers, or the field for landscape purposes and home gardens.

.010 Permaculture: edible plants in the landscaping of homes.

.011 Turf crops: low, usually perennial, plants, such as grasses, and creeping legumes, grown for lawns, parks, parkways, roadsides, and golf courses, where an essentially complete cover of the ground area is desired; may be seeded (e.g., rye grasses, bluegrass, bahiagrass, bermudagrass, dichondra, and bent) or sodded (e.g., St. Augustine grass, special strains of bermudagrass,

I.2. HORTICULTURAL CROPS

bent, centipedegrass, and zoysiagrasses), the latter used when quick (plugs or strips) or complete (solid sod) coverage is desired or the particular turf species does not produce seeds, is a hybrid form that does not come true from seed, or is difficult to germinate without intensive care; also see Ground covers (I.2.003).
.012 Woody ornamentals: (see Nursery plants, I.2.009).

I.3. Vegetable Crops

.000 Vegetable crops: the culture of vegetables, which may be on the scale of a few plants in a home garden to thousands of hectares in commercial operations; an immense group in the aggregate, with thousands of species, varieties, and hybrids and countless cultivars; classified into about a dozen categories according to their principal uses.
.001 Vegetable: an edible plant or plant part eaten cooked or raw as a main part of a meal, side dish, or appetizer; includes herbaceous garden plants and herbaceous parts of some woody perennials (e.g., tender shoots or young leaves, inflorescences, or fruits).
.002 Asparagus: (see Miscellaneous crops, I.3.011).
.003 Beans: (see Peas and beans, I.3.014).
.004 Bulb crops: species of Allium, including onion, garlic, leek, shallot, and chive, whose bulbs are eaten raw or cooked or they and their leaves are used as flavoring for other vegetables, meat, fish, and sauces. (TP, MMC)
.005 Cole crops: plants belonging to the genus Brassica (Cruciferae), including cabbage, cauliflower, broccoli, brussels sprouts, kohlrabi, and Chinese cabbage, whose leaves, unopened flower buds, inflorescences, or swollen stems are used as cooked or raw vegetables. (TP, MMC)
.006 Condiments, spices, and essential oils: annuals or biennials whose plants or plant parts are used to impart special flavors to foods; essential oils are also used in confections, pharmaceuticals, and cosmetics; they include anise, basil, bay, capsicum peppers (paprika, cayenne pepper, and chili pepper), caraway seed, cardamon, celery seed, chervil, coriander seed, cumin seed, dill, fennel seed, fenugreek, horseradish, marjoram, oregano, mints (peppermint, spearmint, etc.), mustard seed, poppy seed, rosemary, saffron, sage, savory, tarragon, thyme, turmeric, camomile, angelica, wormwood, borage, caper, costmary, wintergreen, hops, licorice, lavender, horehound, pennyroyal, and rue. (MMC)
.007 Cucurbits: the herbaceous or woody vines (lianes) of the Cucurbitaceae include cucumbers, pumpkins (malangas, calabazas), squashes, chayote, muskmelons (usually miscalled cantaloupes, a quite different fruit rarely grown in the United States), watermelons, and various gourds (mostly used as utensils in some countries or for fall decorations in the United States but some for food, especially when young); a large and diverse group used as vegetables or pickles (e.g., cucumbers) and as dessert (or breakfast) fruits (e.g., muskmelons and watermelons). (TP, MMC)
.008 Essential oils: (see Condiments, spices, and essential oils, I.3.006).
.009 Globe artichoke: (see Miscellaneous crops, I.3.011).

I.3. HORTICULTURAL CROPS

- .010 Greens: (see Pot herbs, I.3.016).
- .011 Miscellaneous crops, including perennials: a "catchall" for vegetable crops not assignable to one of the other groups, such as sweet corn, popcorn, okra (annuals), and asparagus, rhubarb, globe artichoke (perennials).
- .012 Okra: (see Miscellaneous crops, I.3.011). (TP, MMC)
- .013 Olericulture: vegetable gardening (see Vegetable crops, I.3.000).
- .014 Peas and beans (pulses): members of the Leguminosae whose seeds plus pods (strictly, legumes) are cooked alone or in combination with other vegetables and as ingredients for soup; examples are garden peas, beans, garden beans, lima beans, tepary beans, cowpeas, and chickpeas. (TP, MMC)
- .015 Popcorn: (see Miscellaneous crops, I.3.011).
- .016 Pot herbs or greens: vegetables prepared for consumption by boiling, as spinach, New Zealand spinach, kale, chard, mustard, collards, and dandelion; sometimes used fresh in salads. (TP, MMC)
- .017 Pulses: (see Peas and beans, I.3.014).
- .018 Rhubarb: (see Miscellaneous crops, I.2.011).
- .019 Root crops: plants whose swollen roots (or, in some cases, the lower stem or hypocotyl plus tap root), such as carrot, beet, parsnip, turnip, rutabaga, radish, and celery, are cooked or eaten raw (e.g., carrot, radish). (TP, MMC)
- .020 Salad crops: green leafy vegetables, including many pot herbs, usually served cold with oil, vinegar, and various other condiments, as well as other vegetables or fruit cut up and mixed with or laid on top of the salad; as lettuce (several distinct forms), endive, celery, chicory, water or garden cress, and parsley. (TP, MMC)
- .021 Solanaceous fruits: plants belonging to the genus Solanum and related genera of the Solanaceae, including tomato, eggplant, and pepper (Capsicum), whose fruit are cooked, eaten raw, or used as flavorings in other culinary dishes. (TP, MMC)
- .022 Spices: (see Condiments, spices and essential oils, I.3.006).
- .023 Starchy root and tuber crops: a diverse group whose enlarged stolons or tubers (e.g., potato, Jerusalem artichoke) or roots (e.g., sweet potatoes, yams (Dioscorea) of several species, aroids of several genera, and cassava) provide a source of carbohydrates. (MMC)
- .024 Sweet corn: (see Miscellaneous crops, I.3.011).

II.0. MORPHOLOGY and ANATOMY   9

0. General   10
        Fig. II.0.01-07   12
1. Vegetative Structures   19

        1.1. Buds and stems   19
        1.1.1. Buds   19
        1.1.2. Vernation (prefoliation)   20
        1.1.3. Stems   22
               Fig.II.1.1.1.01-.06, 1.1.2.01-.03, 1.1.3.01-.22   28
    1.2. Leaves   59
         Fig. 1.2.01-.28   66
    1.3. Roots   94
         Fig. 1.3.01-.04   96

2. Flowering and Fruiting Structures   100

    2.1. Inflorescences   100
         Fig. II.2.1.01-.04   104
    2.2. Flowers   118
         Fig. II.2.2.01-.28   130
    2.3. Fruits   158
         2.3.1. Fruit types   158
         2.3.2. Fruit parts (exclusive of seeds)   161
         2.3.3. Seeds   163
                Fig. II.2.3.1.01-.07, 2.3.2.01-.07, 2.3.3.01-.12 168

3. Anatomy   194
        Fig. II.3.01-31   207

4. Modifications   238

    4.1. Life forms   238
    4.2. Habitats   238
    4.3. Genotypic modifications   241
    4.4. Phenotypic modifications   246
         Fig. II.4.01-.14   249

5. General Morphological Terms   263

    5.1. Color   263
    5.2. Direction and size   266
    5.3. Duration   267
    5.4. Apex, base, margins   268
         5.4.1. Apex   268
         5.4.2. Base   269
         5.4.3. Margins   269
    5.5. Shapes   271
    5.6. Surface features   273
    5.7. Texture   278

6. Bryophytes, Pteridophytes, and Gymnosperms   279
        Fig. II.5.01-.20,.6.01   281

II.0. MORPHOLOGY and ANATOMY

II.0. General

.000 **Horticultural morphology:** the study of plant forms and structure, the relationships of the parts to one another, and their respective roles and functions, with special reference to horticultural crops; anatomy, cytology (study of cells), and embryology (structure and development of the embryo) represent component areas generally designated as separate sciences.

.001 **Dimorphic:** occurrence in two forms, e.g., butternut squash with regular (dumbbell) and cylindrical-shaped fruits on the same plant or southern red cedar leaves. (Fig. II.0.01a)

.002 **Disc, disk:** several distinct usages, such as the flattened extremity of a tendril (as Virginia creeper); a type of flower as in many composites; the flattened extremity of a receptacle as in composite inflorescences; a more or less fleshy or elevated development of the receptacle or of coalesced nectaries or staminodes about the pistil; variable in shape and form. (Fig. II.0.01b)

.003 **Forb:** any herbaceous plant other than a grass. (L & J)

.004 **Habit:** the general shape and growth characteristics of a plant, whether upright, trailing, spreading, or pyramidal.

.005 **Hardwooded:** woody, as opposed to succulent or herbaceous.

.006 **Herb:** seed plant that does not develop wood or lignified tissue, as that of a shrub or tree, but is more or less soft or succulent. (Fig. II.0.01c,d, .02a,b)

.007 **Herbaceous:** not woody; dying down each year of temperate-zone and some tropical plants; said also of soft branches before they become woody. (Fig. II.0.01c,d) (L 55)

.008 **Hiatus:** an opening, gap, or break.

.009 **Iteroparous:** a plant whose numerous shoots form a dense clump, like bamboo. (Fig. II.0.02a) (JZ)

.010 **Legume:** a member of the pea family (<u>Fabaceae</u>); also a type of fruit.

.011 **Liana, liane:** (see Vine, II.0.026). (Fig. II.0.02c,d)

.012 **Macrobiotic:** that which is long-lived.

.013 **Masticatory:** substance to be chewed to increase the flow of saliva, like betel nut.

.014 **Mesobiotic:** average or intermediate-lived.

.015 **Microbiotic:** short-lived.

.016 **Ombrophilous:** a plant capable of withstanding much rain, like those in the humid tropics. (L & J)

.017 **Palm:** any plant of the palm family (<u>Palmae</u> = <u>Arecaceae</u>). (Fig. II.0.03)

.018 **Plastochron:** the interval occuring between two successive events in a series of periodic repetitions; can be used with reference to the development of successsive leaves, internodes, and axillary buds, stages of vascularization in a shoot, plant parts, or the plant as a whole (i.e., the age of the plant). (Fig.II.0.04) (B & C)

.019 **Pleimorph, pleiomorphic:** (see Pleiomorphism, V.3.3.2.325).

.020 **Polymorphic:** occurring in more than two forms; variable as to habit or some morphological feature. (Fig. II.0.05)

.021 **Prefixes:**
    Bi-:   two, bis = twice.
    Di-:   two.
    E-, Ex-:   without, off, from, beyond.

II.0. MORPHOLOGY and ANATOMY

         Epi-:   on, upon, above, over.
         Infra-:   below, beneath, on the underside.
         Inter-:   between.
         Intra-:   within, on the inside.
         Neo-:   new, recent.
         Ob-:   over, against, inverted.
         Paleo-, palaeo-:   ancient, old.
         Peri-:   about, around, surrounding.
         Poly-:   many.
         Tri-:   three.
         Uni-:   one.

.022 **Rattan:** any of several genera of climbing palms, such as Calamus and Daemonorops, belonging to the subfamily Lepidocaryoideae; also a portion of stem on one of these plants used for walking sticks or wicker work. (Fig. II.0.03a)

.023 **Shrub:** woody plant that remains low and produces stems from the base, not treelike nor with a single bole; a descriptive term not subject to strict circumspection. (Fig. II.0.06a,b)

.024 **Teratological:** referring to a monstrosity or serious malformation.

.025 **Tree:** perennial woody plant with essentially a single stem, usually branching at some distance from the ground; most often applied to dicotyledons. (Fig. II.0.06c,d, .07)

.026 **Vine:** any plant with a stem that requires support and that climbs by tendrils or other means or trails or creeps along the ground; a climber; woody vines are called lianas or lianes. (Fig. II.0.0.02c,d)

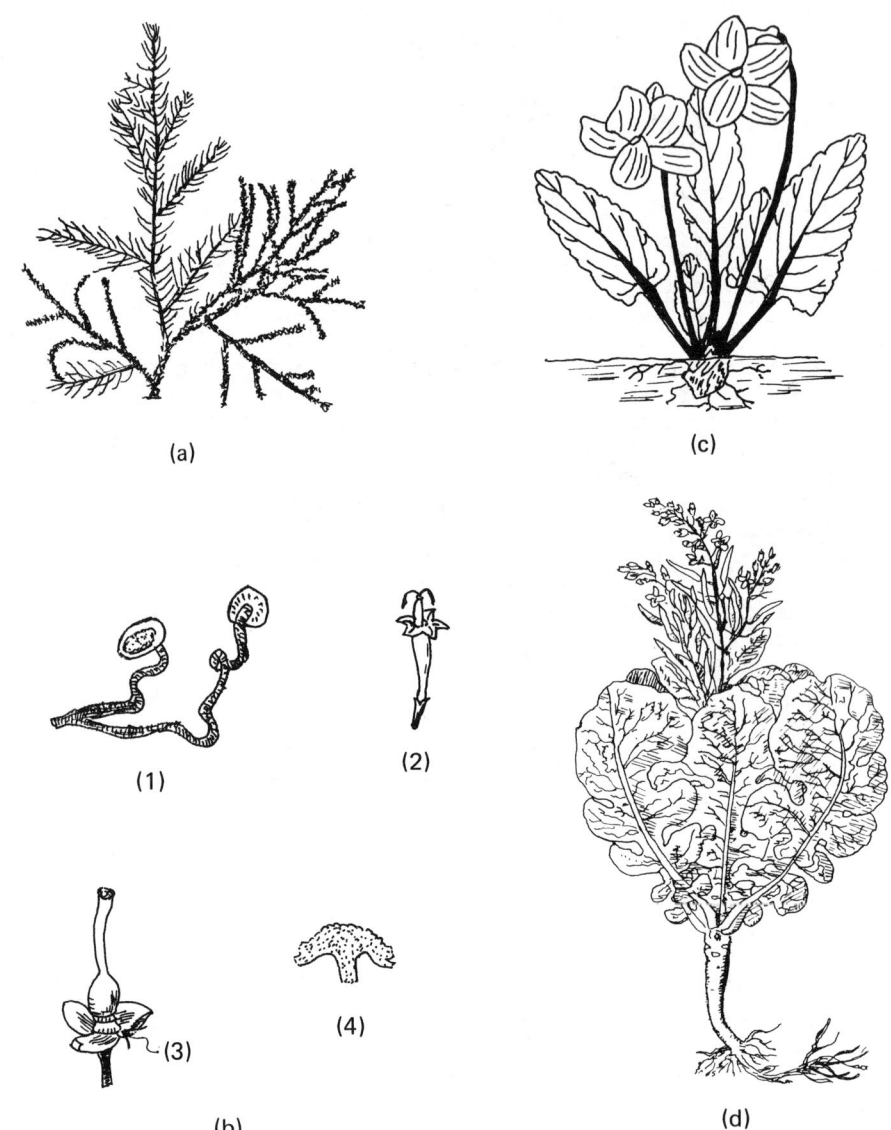

Fig. II.0.01. (a) Dimorphic leaves of southern red cedar (Juniperus silicicola); (b) discs as on (1), the flattened extremities of Virginia creeper tendrils, (2) a type of flower in many composites, (3) a fleshy extension of the receptacle in Citrus and many other flowers, and (4) the flattened extremity of a receptacle; (c), (d), herbs, herbaceous, typified by (c) violet (Viola sp.) and (d) collard (Brassica oleracea). (b), (c) Redrawn from Gray, 1862; (d) redrawn from Johnson, 1633; see Acknowledgments and Selected References.

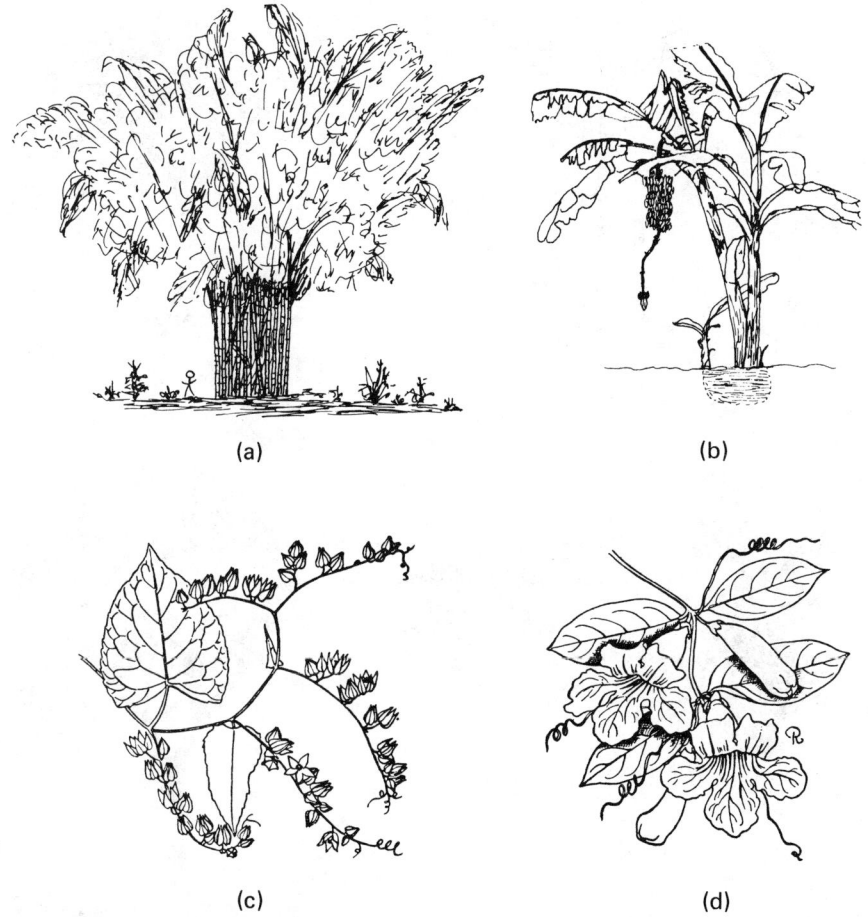

Fig. II.0.02. Herbs, herbaceous: (a) bamboo, a giant herb that grows in a dense clump, i.e., iteroperous; (b) banana (Musa sp.); vines, lianes: (c) coral vine (Antigonon leptopus), a herbaceous climber; (d) painted trumpet (Clytostoma callistegioides), a typical bignoniaceous liane. (b) Adapted from Tai, 1958; (c),(d), from Watkins and Sheehan, 1975; see Acknowledgments and Selected References.

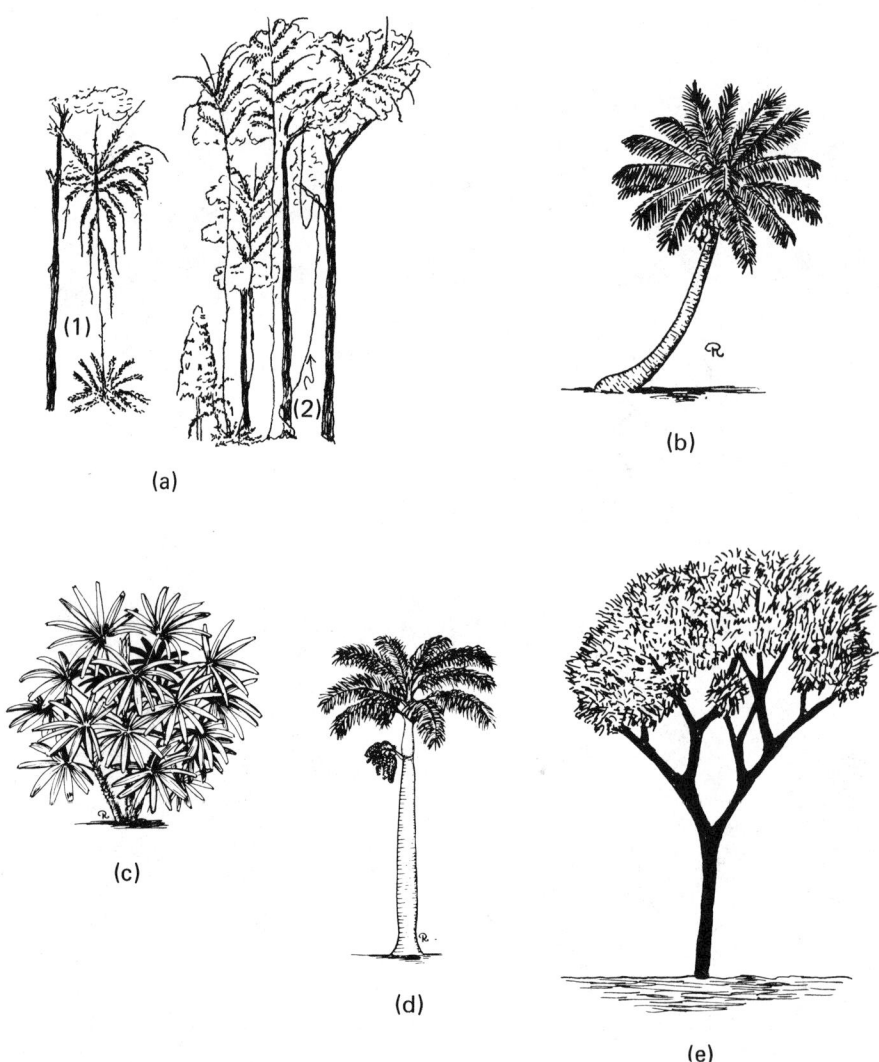

Fig. II.0.03. Palms: (a) Young (1) and old (2) rattans in a Malaysian forest; (b) coconut, "one of God's greatest gifts to mankind" (Burkill 1966); (c) lady palm (Rhapis flabelliformis), a soboliferous (cluster) shrubby type; (d) the stately royal palm (Roystonea regia); and (e) doum palm (Hyphaene thebaica), one of the few naturally branching forms in the Palmae. (a), (e) Redrawn from Corner 1966; (b), (c), (d), from Watkins and Sheehan, 1975; see Acknowledgments and Selected References.

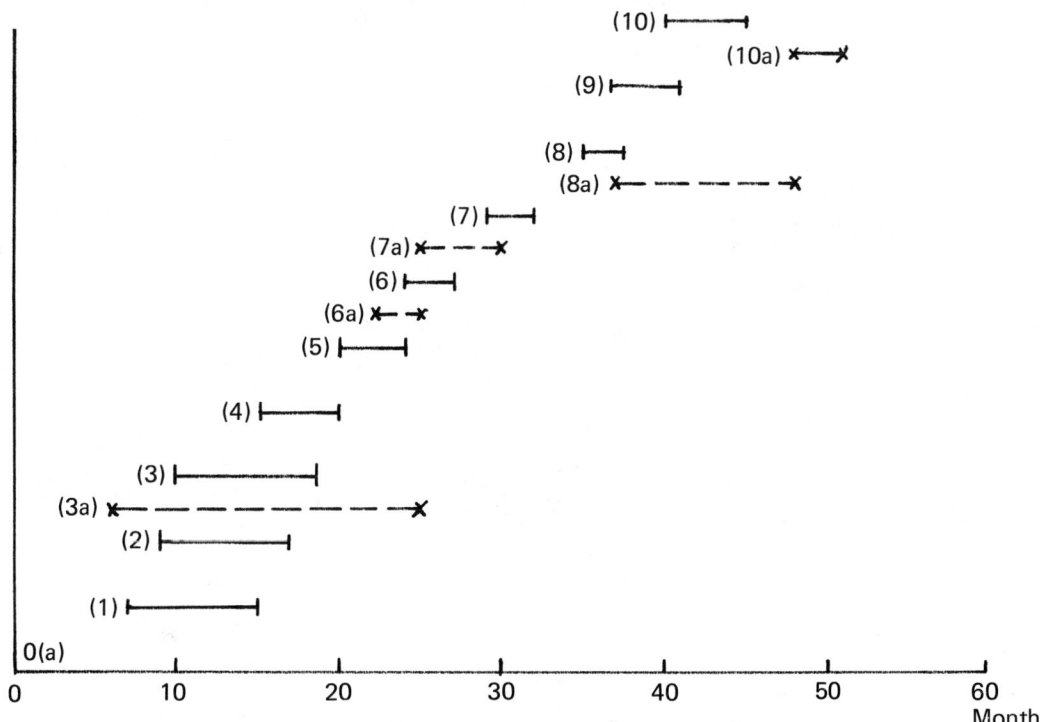

Fig. II.0.04. Plastochron for oil palm inflorescences (————) and leaves (x———— ————x). Developmental stages for inflorescences: (0) initiation [same as 0(a)]; (1) outer spathe initiated; (2) inner spathe initiated; (3) sex determination and initiation of first bracts; (4) fourth bract initiated; (5) spikelets initiated; (6) spikelet differentiation distinct; (7) abortion (if it occurs); (8) anthesis; (9) fruit set; (10) ripening and harvest. Leaf developmental stages: [0(a)] initiation; [3(a)] juvenile phase, [6(a)] spear leaf (first (unexpanded) leaf visible); [7(a)] rapid elongation phase; [8(a)] adult phase; [10(a)] senescence. Adapted from Hartley, 1977, and Surre and Ziller, 1963; see Acknowledgments and Selected References.

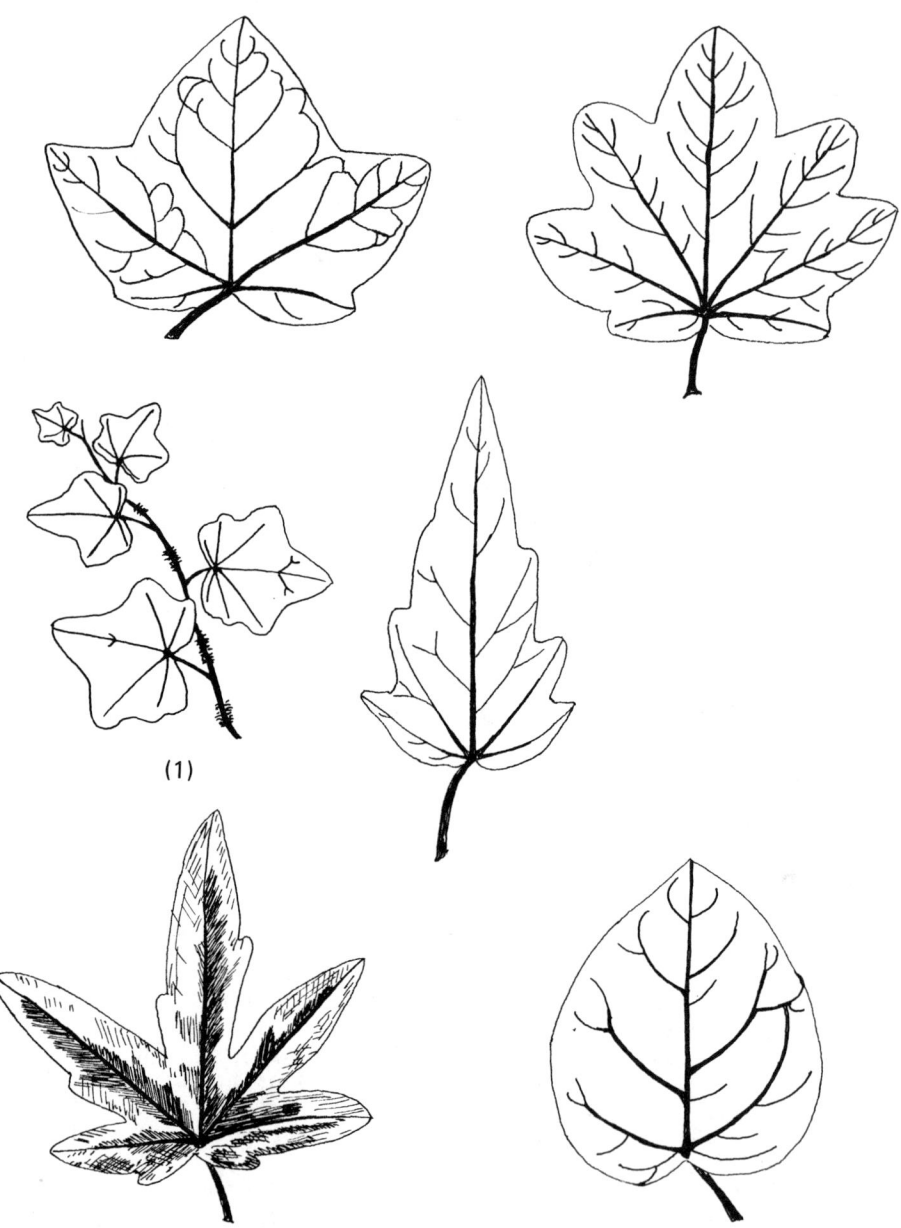

Fig. II.0.05. Polymorphic leaves of English ivy (<u>Hedera helix</u>), of which the "usual" form is at (1). From Watkins and Sheehan, 1975; see Acknowledgments and Selected References.

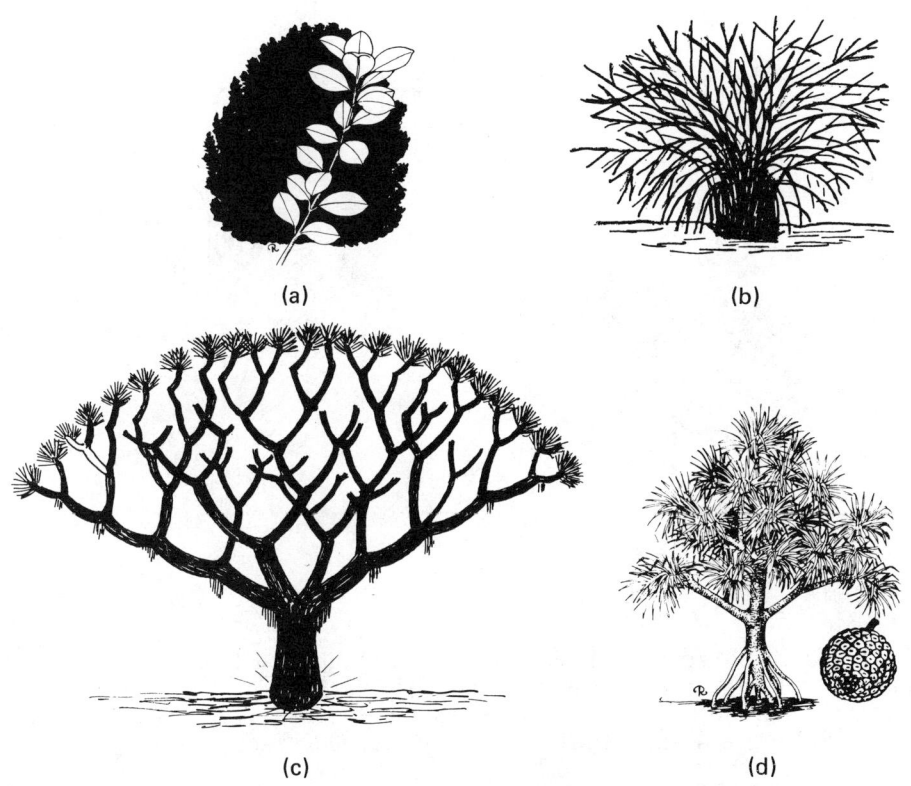

Fig. II.0.06.   Shrubs: (a) boxwood (<u>Buxus microphylla</u>); (b) peony (<u>Paeonia</u> sp.).   Trees: (c) Dragon's blood (<u>Dracaena draco</u>), one of the comparatively few monocots with secondary growth; (d) pandanus (<u>Pandanus utilis</u>), a distinctive monocot tree with stilt roots and dense terminal spiral cluster of leaves on its branches; several other pandans are utilized as ornamentals. (a), (d) From Watkins and Sheehan, 1975; (b) redrawn from Bailey, 1928; (c) redrawn from Hallé et al., 1978; see Acknowledgments and Selected References.

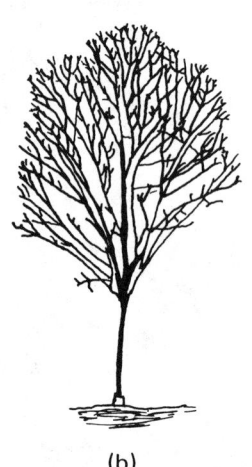

Fig. II.0.07. Trees: (a) China-fir (Cunninghamia lanceolata); (b) 'Tjir 1' rubber (Hevea brasiliensis), one of the best of the pre-World War II Far Eastern clones. (a) From Watkins and Sheehan, 1975; (b) redrawn from Dijkman, 1951; see Acknowledgments and Selected References.

II. MORPHOLOGY and ANATOMY

II.1. Vegetative Structures

II.1.1. Buds and stems

II.1.1.1. Buds

.000 Buds: undeveloped shoots or other appendages or organs, often protected by leaflike bracts or scales and hairs. (L 55)
.001 Accessory: more than one bud in an axil (see also II.2.3.1.001). (Fig. II.1.1.1.01a)
.002 Alternate: placed singly and alternately one above the other on some common body, like buds on a stem. (Fig. II.1.1.1.01b) (ST)
.003 Apical: (see Terminal, II.1.1.1.031).
.004 Axillary: arising out of the axil, the angle between a stem and a leaf. [Fig. II.1.1.1.01c, d(2), e] (L 55)
.005 Cauliflorous: floral buds borne on the trunk and larger branches, as in cacao or jaboticaba, amoung others (see II.2.1.015). (Fig. II.1.1.1.01e). (L & J)
.006 Decussate: arranged in pairs that alternately cross one another, like the buds and subtending leaves of many plants. (Fig. II.1.1.1.02f) (L 55)
.007 Epicormic: vegetative growth arising from the trunk of a tree after the branches have been lopped off (buds may be latent or adventitious); compare with cauliflorous (II.1.1.1.005;.02g) (JZ, HOT)
.008 Fibonacci series (Schimper and Braun series): (see Parastichy, II.1.1.1.016)
.009 Floral: containing rudimentary floral parts in successive layers overarching the apex of the growing point and generally larger, plumper, and more rounded at the tip than vegetative or mixed buds. (Fig. II.1.1.1.02h)
.010 Grape eye: a compound bud with a primary and several secondary axes, of which the secondary do not usually develop unless the primary fails to do so. (Fig. II.1.1.1.02i) (WS)
.011 Infrapetiolar: more or less completely enveloped by the petiole base, as in sycamore (Platanus). (Fig. II.1.1.1.03j) (G 68)
.012 Lateral: fixed near (or on) the side of anything, like a lateral bud, an anther to a filament, or in an axillary position (II.1.1.1.004; II.2.3.2.004). [Fig. II.1.1.1.01d(3)] (L 55)
.013 Mixed: containing both vegetative and floral structures, as in apple (Malus) and pear (Pyrus). (Fig. II.1.1.1.03k) (L 55)
.014 Opposite: placed on opposite sides (i.e. $180°$ apart) of some other body or thing on the same plane; subopposite is used when buds are not exactly aligned. (Fig. II.1.1.1.03-0) (L 55)
.015 Orthostichy: arrangement of organs (e.g., buds) in vertical rows. [Fig. II.1.1.1.05r(3)] (B & C)
.016 Parastichy: growth spirals of organs (e.g., buds, scales, and flowers in heads); arranged in a Fibonacci series (Schimper and Braun series) which has the form 1/2 (Gramineae), 1/3 (Cyperaceae), 2/5 (Rosa), 3/8 (Aster, Brassica, Plantago), 5/13 (Sempervivum), 8/21 (Pinus sylvestris), 13/34, 21/55, 34/89 (Sunflower); (see also Phyllotaxis, II.1.1.1.017). (Fig. II.1.1.1.04, .05) (B & C)
.017 Phyllotaxis: arrangement of organs on an axis; (see Parastichy, II.1.1.016; Spirodistichy, II.1.1.025). (Fig. II.1.1.1.04, .05) (B & C)
.018 Primordium: the first recognizable but undifferentiated stage of a developing organ within a bud (or a meristematic region). (B & C)

II.1.1.1. MORPHOLOGY and ANATOMY

.019   Pseudoterminal: an axillary bud taking the place of the terminal bud proper which has abscised or failed to develop, as in chestnut (Castanea). (Fig. II.1.1.1.03m) (L 55)
.020   Rank: (see Rows, II.1.1.1.022).
.021   Rosulate: (see II.1.2.094).
.022   Rows, series: arranged in rows that are not necessarily opposite one another; rank denotes a vertical row. (Fig. II.1.1.1.05) (B & C)
.023   Schimper and Braun series: (see Fibonacci series = Parastichy, II.1.1.1.016).
.024   Series: (see Rows, II.1.1.1.022).
.025   Spirodistichy: the gradual development of a spiral arrangement from a distichous one, as in Pandanus, Dracaena, Gasteria, and the like. (Fig. II.1.1.1.06) (B & C)
.026   Stipitate: stalked, as in witch hazel (Hamamaelis) or grape. (Fig. 1.1.1.03n) (L 55)
.027   Subopposite: (see Opposite, II.1.1.1.014).
.028   Subterminal: adjacent to the terminal, as in avocado (Persea americana). [Fig. II.1.1.1.0d(2)] (G 62)
.029   Supernumerary: exceeding the usual number, as in supernumerary (and accessory) buds. (Fig. II.1.1.1.01a) (HOT)
.030   Superposed buds: a vertical arrangement of two or more buds in an axil; two configurations with respect to the sequence of development and opening; acropetal, as in Carya, Juglans, Bignoniaceae, Leguminosae, and Rubiaceae, among others, or basipetal, as in the Simaroubaceae. [Fig. II.1.1.1.01a(2),(3)]
.031   Terminal, apical: borne at the apex (distal end) of a stem. [Fig. II.1.1.1.01d(1)] (HOT, G 62)
.032   Transverse buds: a horizontal arrangement of two or more buds in an axil, as in Prunus spp. and Araceae, among others. [Fig. II.1.1.1.01a(1)] (HOT)
.033   Vegetative: a bud that contains rudimentary leaf and stem parts only. (Fig. II.1.1.1.03o, p) (L 55)

II.1.1.2. Vernation (prefoliation)

.000   Vernation (prefoliation): disposition or arrangement of leaves in the bud.
.001   Alternative: when the inner of the pieces in two rows is covered by the outer in such a way that each of the exterior rows overlaps half of two of the interior. (Fig. II.1.1.2.01a) (ST)
.002   Amplex: (see Equitant, II.1.1.2.012).
.003   Circinate: when the parts are rolled spirally downward, as in a crosier. (Fig. II.1.1.2.01b) (ST)
.004   Cochlear: when one piece is larger than the others, hollowed like a helmet or bowl, and covers all of the others. (Fig. II.1.1.2.01c) (ST)
.005   Conduplicate: when the sides are applied or laid parallel to one another, folded. (Fig. II.1.1.2.01d) (ST)
.006   Contorted: each piece is oblique in figure and overlaps its neighbor by one margin; the other margin is overlapped in like manner by that which stands next to it. (Fig. II.1.1.2.01e) (ST)
.007   Contorto-duplicate: twisted and doubled or folded. (ST)
.008   Convolute: when one piece (or margin) is wholly rolled up in another. (Fig. II.1.1.2.01f) (ST)

II.1.1.2. MORPHOLOGY and ANATOMY 21

.009   Corrugate: (see Crumpled, II.1.1.2.010).
.010   Crumpled, corrugate, wrinkled: when the parts are folded together irregularly in every direction. (Fig. II.1.1.2.01g,h) (ST)
.011   Curvative: when the margins are slightly curved backward or forward without any sensible twisting. (Fig. II.1.1.2.01i) (ST)
.012   Equitant, amplex: when the parts overlap one another in a parallel manner and entirely without any involution; (compare obvolute = half-equitant). (Fig. II.1.1.2.02j). (ST)
.013   Half-equitant: (see Obvolute, II.1.1.2.017).
.014   Imbricate: when parts lay over one another in regular order, like tiles on a roof, i.e., overlap one another without involution. (Fig. II.1.1.2.02-1) (ST)
.015   Induplicate: with margins abruptly bent inward and the external face of these edges applied to one another without twisting. (Fig. II.1.1.2.02k, .03s) (ST)
.016   Involute: when the edges are rolled inward spirally on each side (i.e., rolled over the upper or outer surface). (Fig. II.1.1.2.02m) (ST)
.017   Obvolute, half-equitant, semiamplex: when the margins of one part (leaf or petal) alternately overlap those of the part that is opposite. (Fig. II.1.1.2.02n) (ST)
.018   Plain: not covered, bent, or folded, as in the manner of unfolding. (Fig. II.1.1.2.02o) (ST)
.019   Plaited, pleated, plicate: folded together like the plaits of a closed fan. (Fig. II.1.1.2.02p) (ST)
.020   Prefloration: (see Estivation, II.2.2.032).
.021   Prefoliation: (= Vernation, II.1.1.2.000).
.022   Ptyxis: manner of unrolling or unfolding a single leaf or other organ (a source of confusion because the same terms for vernation and estivation are used).
.023   Reclinate: when the parts are bent down on the stalk. (Fig. II.1.1.2.02q) (ST)
.024   Reduplicate: when the margins are bent abruptly outwards and their faces touch without overlapping. (Fig. II.1.1.2.02r, .03t) (G 62)
.025   Replicate: when the upper part is curved back and applied to the lower. (Fig. II.1.1.2.03u) (ST)
.026   Revolute: when the edges are rolled backward spirally on each side (i.e., rolled over the lower or inner surface). (Fig. II.1.1.2.03v) (ST)
.027   Supervolute: when one edge is rolled inward and enveloped by the opposite edge rolled in an opposite direction. (Fig. II.1.1.2.03w, x) (ST)
.028   Twisted: (see Contorted, II.1.1.2.006).
.029   Valvate: parts applied to one another by the margins only. (Fig. II.1.1.2.03y) (ST)
.030   Vexillary: when one piece is much larger than the others and is folded over them; the latter is arranged face to face (the type of imbricate estivation in truly papilionaceous flowers. Gray 1862). (Fig. II.1.1.2.03z) (ST)
.031   Wrinkled: (see Crumpled, II.1.1.2.010).

II.1.1.3. Stems

.000 Stem: typically, the aerial axis of a plant, which serves as mechanical support for leaves, flowers, and fruits and is divided into nodes and internodes; also modified as underground or storage structures, such as bulbs, corms, rhizomes, and tubers.
.001 Acaulescent: stem inconspicuous or absent. (Fig. II.1.1.3.01a) (L 55)
.002 Acropetal: arising or developing in a longitudinal plane from a lower toward a more apical position, as in the sequence of branching or development of other organs. (Fig. II.1.1.3.01b) (L 55)
.003 Acrotony (branching): branching at or toward the tip of a stem. (Fig. II.1.1.3.02a) (HOT)
.004 Alate: winged with corky excrescences; e.g., winged elm (Ulmus alata) or Euonymus. (G 62)
.005 Anisotony: unequal growth in dichotomous branching. (Fig. II.1.1.3.02b) (CR 66)
.006 Arborescent: tree or treelike habit.
.007 Armed: provided with any strong or sharp defense, such as spines, thorns, or barbs; (see Genotypic modifications, II.4.3.) (Fig. II.1.1.3.02c) (L 51, G 62)
.008 Arundinaceous: resembling a reed or cane.
.009 Ascendant, ascending: having a direction upward with an oblique base. (Fig. II.1.1.3.03a) (ST)
.010 Axil: upper angle that a petiole or peduncle makes with the stem that bears it. (L 55)
.011 Axis: main stem. (L 55)
.012 Basal: fixed at the base of anything. (ST)
.013 Basipetal: developing in a longitudinal plane from an apical or distal point toward the base, as in the sequence of branching or development of other organs. (Fig. II.1.1.3.03b) (L 55)
.014 Basitony (branching): branching at or toward the base of the stem. (Fig. II.1.1.3.03c) (HOT)
.015 Bole: unbranched stem of a forest tree. (L 55)
.016 Branched: (see Ramose, II.1.1.3.0/6).
.017 Branching by apposition (sympodial): a type of plagiotropic branching in which the terminal meristem (apical bud) remains alive and continues vegetative growth but branch extension takes place from an axillary meristem, as in Terminalia catappa, Bucida spp., (the so called "pagoda branching," which gives such trees a distinctive appearance). (Fig. II.1.1.3.03d) (HOT)
.018 Branching by substitution (sympodial): a type of determinate plagiotropic branching in which the termianl bud develops into an inflorescence or aborts; further extension of the axis is from growth of a lateral bud, as in Mangifera indica, Ailanthus spp., Catalpa spp., and Diospyros spp., among others. (Fig. II.1.1.3. 03e) (HOT)
.019 Branch system: the way in which a plant is constructed with respect to the morphological relationship of main and lateral or secondary axes; in the case of trees and other treelike plants the trunk or main axis is, in general, strictly orthotropic, whereas the lateral (branch) axes may be orthotropic, reversibly plagiotropic, irreversibly plagiotropic, or phyllomorphic. (Fig. II.1.1.3.04, .13a, .15c, .18e) (HOT)
.020 Burr knot: swelling or protrusion on a stem that contains latent (adventitious) buds; common in apple and quince. (Fig. II.1.1.3. 05a)

II.1.1.3. MORPHOLOGY and ANATOMY

.021 **Caudex**: stem of woody monocotyledons, like palms and some aroids. (The stem of grasses is a culm.) (Fig. II.1.1.3.05b) (L 55)
.022 **Cauline**: arising from the stem. (L 55)
.023 **Centric**: (see Cylindrical, II.1.1.3.033).
.024 **Chupon**: upright stem (sucker) of a cacao (Theobroma cacao) seedling characterized by phyllotaxy of 3/8. [Fig. II.1.1.3.05d(1); .13b (1)] (CU)
.025 **Cladophyll**: modified stem that resembles a leaf (in form and appearance but arises in the axil of a minute, bractlike, often caducous, true leaf); found typically in plants adapted to a xeric habitat (see Genotypic modifications II.4.3.). (Fig. II.1.1.3. 06a) (L 51)
.026 **Clambering**: vine growing over an obstruction without means of self-support. (Fig. II.1.1.3.06c) (L 55)
.027 **Climbing**: (see Scandent, II.1.1.3.085).
.028 **Columnar**: aerial axis is more or less cylindrical and unbranched, as in many monocotyledons. (Fig. II.1.1.3.05c, .09b) (G 62)
.029 **Continuous growth**: said of a plant that does not show morphological segmentation of its stem or branches, as in palms and red mangrove. (Fig. II.1.1.3.07a,b) (HOT)
.030 **Coppice**: a small wood or thicket grown for the purpose of periodic cutting or the periodic removal of branches in like manner from forest trees or shrubs, as in the collection of bark from wild Cinnamomum spp. or a part of a landscape planting. (JZ, L & J)
.031 **Crown**: that part of the stem or a group of stems at which the junction with the roots occurs (also known as root crown); the aerial parts of trees or palms; the usually green extension of stem proper in palms of the subfamily Arecoideae formed by the long, clasping petioles (properly a crownshaft). (Fig. II.1.1.3.07c, .08a) (L 55)
.032 **Culm**: stem of grasses. (Fig. II.1.1.3.08b) (L 51)
.033 **Cylindrical, centric**: having a nearly true cylindrical figure, like many stems. [Fig. II.1.1.3.08b(2)] (ST)
.034 **Decumbent**: reclining on the earth and rising again from it at the apex, like some bramble (Rubus) canes and cranberry stems. (Fig. II.1.1.3.09a) (L 55)
.035 **Deliquesent**: secondary branches arise from a vertical axis some distance from the ground and subdivide successively several times, as in trees or treelike ferns. (Fig. II.1.1.3.09c) (HI)
.036 **Dichotomous**: a pair of secondary branches arises from a vertical axis some distance from the ground and subdivides successively into pairs several times, as in Hyphaene, said pairs being mirror images; can also occur in plagiotropic stems (e.g., Nypa). (Fig. II.1.1.3.09d) (CR 66)
.037 **Distichous**: when things are arranged in two rows, one opposite the other, so that they lie in a single plane, as in the arrangement of leaves on a plagiotropic branch; note, however, that leaf attachment may be alternate, opposite, or decussate. (Fig. II.1. 1.3.10a; 1.2.08c). (HOT)
.038 **Erect**: pointing toward the zenith like many stems. (Fig. II.1.1.3. 10b) (ST)
.039 **Excurrent**: numerous horizontal branches arise from a single vertical monopodial stem in a conical shape, as in conifers and some dicotyledonous trees. (Fig. II.1.1.3.10b) (HI)
.040 **Fasciated**: exhibiting fasciation. (Fig. II.5.09h) (ST)
.041 **Fasciation**: a malformation in stems that results in their enlargement and flattens as if several stems were fused. (ST)

II.1.1.3. MORPHOLOGY and ANATOMY

.042 **Fistulous, fistular:** hollow cylindrical or terete body closed at both ends like a stem or petiole. (Fig. II.1.1.3.10c) (ST)
.043 **Floricane:** flowering and fruiting stem; especially of a bramble (Rubus). [Fig. II.1.1.3.11a(2)] (B-GH)
.044 **Haulm:** the residue after a crop has been harvested, as after pulses and the like or the litter or straw from grains. (DR)
.045 **Horizontal:** when the plane points to the zenith, the apex to the horizon, as in some stems. (Fig. II.1.1.3.12b) (ST)
.046 **Implicative:** intertwined; intertwisted; entwined. (Fig. II.1.1.3.12c) (ST)
.047 **Irreversibly plagiotropic:** the situation in which a plagiotropic branch will not revert to an orthotropic condition, as in Coffea, Piper nigrum, and many gymnosperms (e.g., Norfolk Island pine), among others; often characterized by sympodial branching (II.1.1.3.099). (Fig. II.1.1.3.13a) (HOT)
.048 **Jorquette:** collective term for the fan (lateral) branches, usually three to five, which arise near the apex of the main axis of a stem or chupon of a cacao seedline, characterized by phyllotaxy 1/2. [Fig. II.1.1.3.05d, .13b(2)] (CU)
.049 **Lammas shoot:** abnormal midsummer growth from a (normally) dormant bud caused by excess moisture or other environmental stress. (Lammas Day is August 1, lammastide, that time of year, hence the name). (HOT)
.050 **Limb:** a secondary or lateral branch (stem) of a woody plant; also the expanded portion of a petal or leaf (usually termed "blade" in the latter).
.051 **"Long shoot":** ordinary, usually long-lived shoot characterized by nodes often widely spaced; the common type of stem found in many woody and some herbaceous dicots. [Fig. II.1.1.3.04a(1)]. (HOT)
.052 **Macrophyll (megaphyll):** a branch that is limited in growth and has become leaflike in form (obsolete).
.053 **Microphyll:** the lateral outgrowth of a stem, as in club mosses (obsolete).
.054 **Modular construction:** new meristems repeat the construction of the parent (initial) meristem precisely in a morphologically qualitative sense, as in castor bean (Ricinus) or cassava (Manihot). (Fig. II.1.1.3.14b) (HOT)
.055 **Module:** axis in which the entire sequence from initiation of the meristem to the onset of flowering and fruiting (aerial differentiation) is carried out. (Fig. II.1.1.3.14b) (HOT)
.056 **Monoaxial:** the seedling meristem is the only aerial meristem produced and remains active throughout the life of the tree (e.g., oil palm and coconut). (Fig. II.1.1.3.14c) (HOT)
.057 **Monocaulous:** a plant with a single stem, a tree. (Fig. II.1.1.3.14c, .15a) (HOT)
.058 **Monopodial:** indeterminate growth of an axis, with branches (i.e., polyaxial) or without branches (monoaxial). (Fig. II.1.1.3.15a) (HOT)
.059 **Neoformation (leaf and stem development):** initiation of leaves and buds after growth has begun, the continued extension of primordia that takes place immediately after initiation without an intervening rest period; see Preformation (II.1.1.3.069). [Fig. II.1.1.3.15b(1)] (HOT)
.060 **-nodal:** the point of insertion of a leaf: the joint of a stem; prefixes can be uni-, bi-, tri-, quadri-, poly-, etc., for one, two, three, four, or many leaves at a node.

II.1.1.3. MORPHOLOGY and ANATOMY

.061 Novirame: flowering or fruiting shoot arising from a primocane, sometimes encountered in brambles (Rubus). (Fig. II.1.1.3.12a) (B-GH)
.062 Orthotropic (stem): a geotropic (vertical) axis, typically with spiral or decussate phyllotaxis (i.e., leaf and branch arrangement alternate or opposite), often completely vegetative and indeterminate, and three-dimensional branching; the characteristic configuration for most tree trunks. [Fig. II.1.1.3.04a, c(1)] (HOT)
.063 Parcifrond: leafy shoot from a Rubus floricane (II.1.1.3.043). (B-GH)
.064 Phylloclad: a branch, more or less flattened, functioning as a leaf, as in cacti (sometimes termed "cladophyll"); (see Genotypic modifications, II.4.3). (Fig. II.1.1.3.06b) (L 55)
.065 Phyllomorphic: branches that resemble (mimic) compound leaves to the extent that entire branches instead of individual leaves will abscise on senescence, as in Castilla and Phyllanthus spp. (Fig. II.1.1.3.15c) (HOT)
.066 Plagiotropic (stem): a diageotropic (more or less horizontal) axis, with two-dimensional branching, distichous phyllotaxis or secondarily arranged so (by twisting of internodes or petioles or both in plants with decussate leaf arrangement), dorsiventral symmetry, and bearing lateral or terminal inflorescences (branches indeterminate or determinate). [Fig. II.1.1.3.04b, c(2)] (HOT)
.067 Polyaxial stem: several to many aerial meristems, in addition to the original seedling meristem are present; two different configurations, the differentiated polyaxial in which separate meristems are involved in trunk and branch formation, as in oak, maple, ash, beech, mango, avocado, rubber, cacao, coffee and many legumes, and the mixed polyaxial in which the same meristem contributes trunk and branch(es) to the construction of the tree, as in elm. (Fig. II.1.1.3.16a, b) (HOT)
.068 Polycaulous: a plant with several stems, a shrub. (HOT)
.069 Preformation (leaf and stem development): initiation of leaves and stems within a bud, a large part of the shoot being present as primordia in the bud and with a rest period prior to expansion. [Fig. II.1.1.3.15b(2)] (HOT)
.070 Prickle: small and weak spinelike body borne irregularly on the bark or epidermis, (see Genotypic modifications, II.4.3.027). (Fig. II.1.1.3.16d) (L 55)
.071 Primocane: the first season's shoot or cane of a biennial woody stem, as in brambles (Rubus). (Fig. II.1.1.3.11a) (B-GH)
.072 Procumbent: trailing over the surface of the ground. (Fig. II.1.1.3.16c) (L 55)
.073 Prolepsis, proleptic: rhymthic or discontinuous growth; characterized by bud scales or scale leaves and delayed leaf formation. [Fig. II.1.1.3.17a(1)] (HOT)
.074 Prostrate, prone: lying flat on the ground or any other thing. (L 55)
.075 Rameous: of or belonging to the branches. (ST)
.076 Ramose, branched: divided into many branches, like many shrubs. (Fig. II.1.1.3.17b) (G 62)
.077 Reaction wood: (see II.3.155).
.078 Reclining, reclinate: falling gradually back from the perpendicular. (ST)
.079 Reiteration (reiteration branching): the process of architectural adjustment by which a damaged tree accommodates itself to its environment (i.e., the regrowth of a tree tends to follow the

II.1.1.3. MORPHOLOGY and ANATOMY

original branch-system pattern, as modified by the proximity of other trees around the space it has available). (Fig. II.1.1.3.18b) (HOT)

.080 Repent: creeping, prostrate, and rooting. (Fig. II.1.1.3.18c) (L 55)

.081 Reversibly plagiotropic: the situation in which a plagiotropic branch will revert to an orthotropic condition and continue that axis if the main trunk is destroyed (e.g., in Ceiba). (Fig. II.1.1.3.18e) (HOT)

.082 Rhizomatous branching: undergound, plagiotropic branching, which may be dichotomous in some instances, as in bamboos, some palms (e.g., Nypa), some gingers, and stoloniferous grasses. (Fig. II.1.1.3.18d) (HOT)

.083 Rhizome: (see V.1.4.060).

.084 Rhythmic growth: periodic growth of a plant axis (see Unit of Morphogenesis, II.1.1.3.103, and Unit of extension, II.1.1.3.102). (Fig. II.1.1.3.14b) (HOT)

.085 Scandent: climbing (see Genotypic modifications, II.4.3.002). (Fig. II.1.1.3.19a) (L 55)

.086 Seedstalk: a flowering and fruiting stem, particularly of biennials whose commercial product is allowed to remain for the purpose of obtaining a seed crop the second year. (Fig. II.1.1.3.19b) (DR, G 62)

.087 "Short shoots": short, often short-lived branch characterized by closely spaced nodes; spurs or spurlike (II.1.1.3.093), as in some temperate zone fruit trees. [Fig. II.1.1.3.14a(2)] (HOT, HON)

.088 Soboliferous: plant forming clumps from underground rhizomes. (Fig. II.1.1.3.19c) (L 55)

.089 Spine: strong, sharp-pointed woody body, mostly arising from the woody portion of a stem; synonymous with thorn (see Genotypic modifications, II.4.3.042). (Fig. II.1.1.3.20a) (L 55)

.090 Spinescent: refers to a plant that develops spines or has such a tendency; a branch tip modified to a needlelike point (see Genotypic modifications, II.4.3.042). (Fig. II.1.1.3.20b) (G 62)

.091 Spinulose: covered with small spines; sometimes glandular and stinging or burning, hence the obsolete term, "urent," which is rarely used (see Surface features, II.5.6.146). (Fig. II.1.1.3.20c) (L 51)

.092 Spiny: bearing several to many spines; thorny. [Fig. II.1.1.3. 20a(1)] (G 62)

.093 Spur: two separate meanings: (1) tubular or saclike projection from a blossom, as of a petal or sepal, and usually containing a nectar-secreting gland; (2) a short, leafy flowering and fruiting branch, often with leaves and flowers in fascicles (clusters) at closely spaced nodes, characteristic of certain Rosaceae and some other plants. [Fig. II.1.1.3.14a(3)] (L 51)

.094 Stolon: (see II.4.3.061).

.095 Straight: not wavy or curved or deviating from a given direction in any way, as in many stems. (ST)

.096 Strict: very straight; straight in excess, as in certain stems. (Fig. II.1.1.3.21a) (ST)

.097 Suffrutescent: stem woody at base, herbaceous at the tip. (Fig. II.1.1.3.21b) (L 55)

.098 Syllepsis, sylleptic: continuous development of stems without any evident intervening rest period; characterized by no bud scales

II.1.1.3. MORPHOLOGY and ANATOMY

.099
.100
.101
.102
.103
.104
.105

    or scale leaves and immediate leaf formation. [Fig. II.1.1.3. 17a (2)] (HOT)

.099 Sympodial: an apparent main axis not developed from a terminal bud but made up of successive secondary axes that result from development of an axillary (lateral) bud, as in grape (<u>Vitis</u>); represented in certain monocotyledons by successive growth of individual secondary axes from a rhizome, as in <u>Musa</u>; also represented by successive alternation of jorquette and chupon growth in cacao seedlings. (Fig. II.1.1.3.21c) (HOT)

.100 Tiller: branch arising from the base of a monocot, especially a grass. (Fig. II.1.1.3.22a) (HOT)

.101 Tubular: approaching a cylindrical figure and hollow, like a grass stem. (Fig. II.1.1.3.22c) (HOT)

.102 Unit of extension (UE): the cycle of extension of a stem that begins with expansion of long internodes associated with scale leaves and finishes with short internodes associated with foliage leaves; essentially what is denoted as a "flush" with reference to periodic growth of many plants. (Fig. II.1.1.3.14b) (HOT)

.103 Unit of morphogenesis (UM): The cycle of activity, or initiation of an apical meristem in rhythmic growth, the cycle beginning with foliage leaves and ending with scale leaves (which may subtend floral structures); this cycle overlaps that of the cycle of extension (and indeed must precede the latter). (Fig. II.1.1.3.14b) (HOT)

.104 Watersprout: a vigorous shoot that arises from the lower trunk or large branches of a tree and rapidly grows through the tree to the top of the crown, thus tending to crowd, suppress, or overgrow the regular branches; it is usually strongly juvenile and slow to flower and bear fruit until branches have formed. (Fig. II.1.1.3.22b) (WS, H & K)

.105 Woody: stems which live for a successive number of years and whose secondary wall thickening is lignin (note, however, that stems of numerous tropical plants live for many years but are not woody). (L 55)

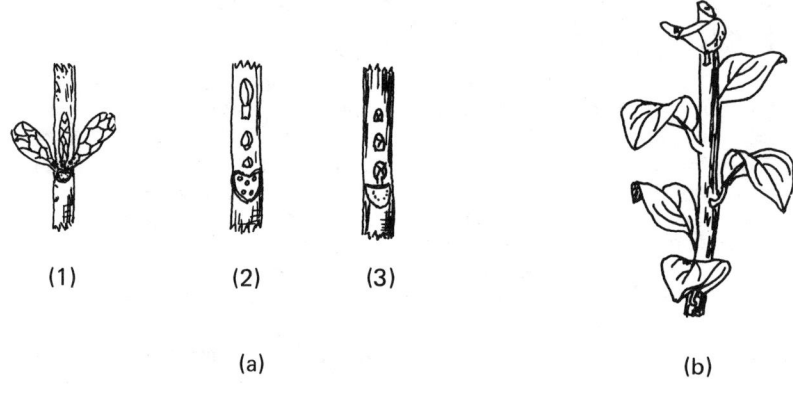
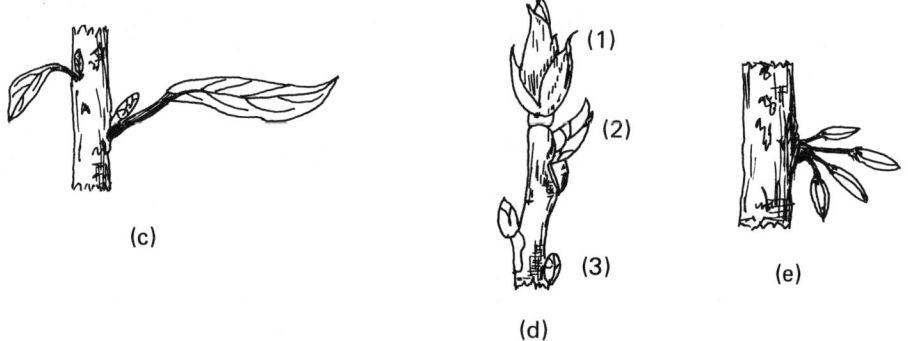

Fig. II.1.1.1.01. Buds: (a) accessory in (1) transverse, (2) acropetal, and (3) basipetal superposed arrangements; (b) alternate, subtended by leaves; (c) axillary; (d) (1) terminal, (2) subterminal, and (3) lateral; (e) cauliflorous. (a), (c) Redrawn from Gray, 1862; (d) redrawn from Gray, 1868; (e) adapted from Braudeau, 1969; see Acknowledgments and Selected References.

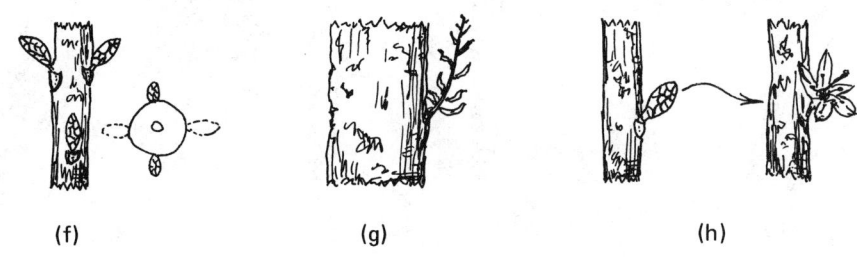

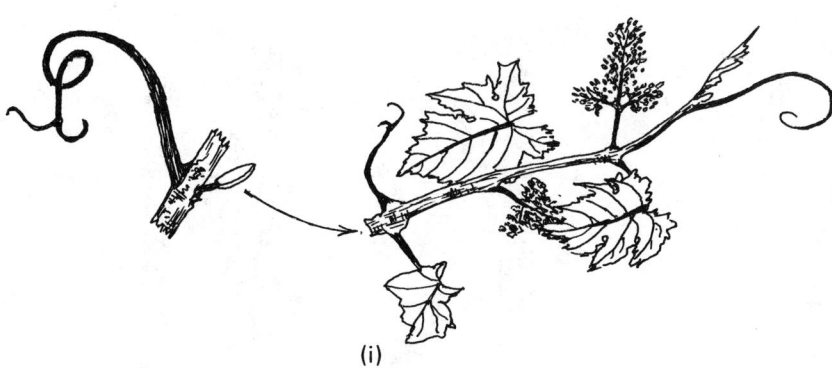

Fig. II.1.1.1.02. Buds: (f) decussate; (g) epicormic; (h) floral; (i) grape eye. (i) Redrawn from Westwood, 1978; see Acknowledgments and Selected References.

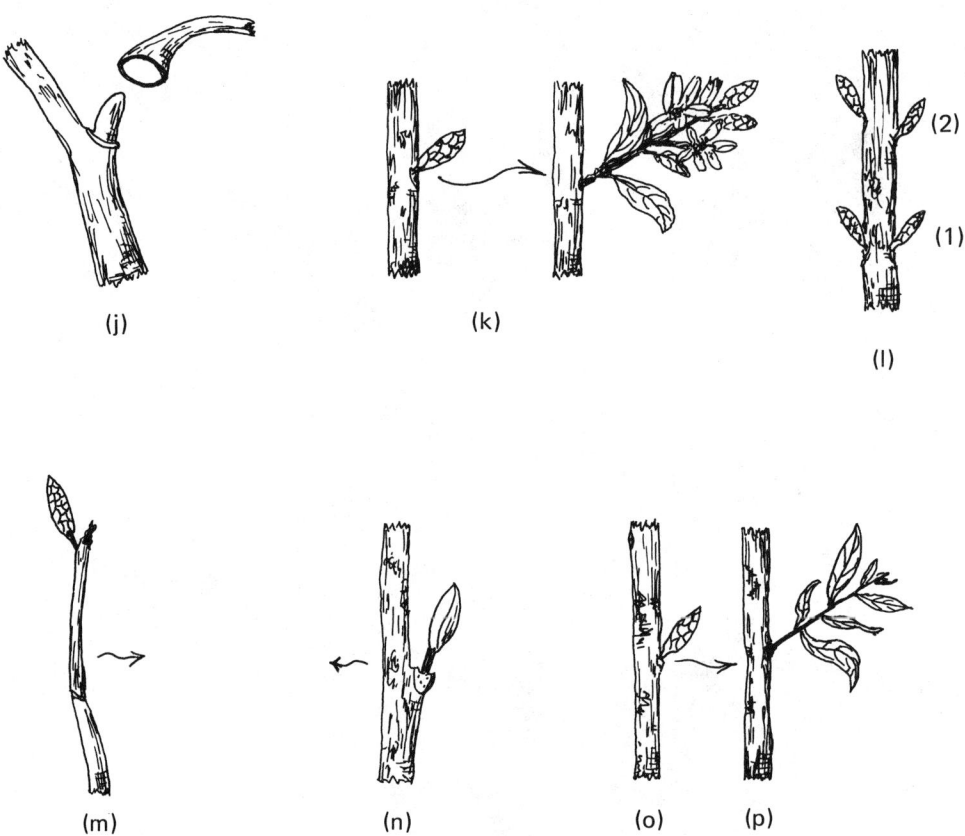

Fig. II.1.1.1.03. Buds: (j) infrapetiolar; (k) mixed; (l) opposite (1), subopposite (2); (m) pseudoterminal; (n) stipitate; (o), (p) vegetative as bud (o) and resulting branch (p). (j), (l), (m) Redrawn from Gray, 1862; (n) adapted from Westwood, 1978; see Acknowledgments and Selected References.

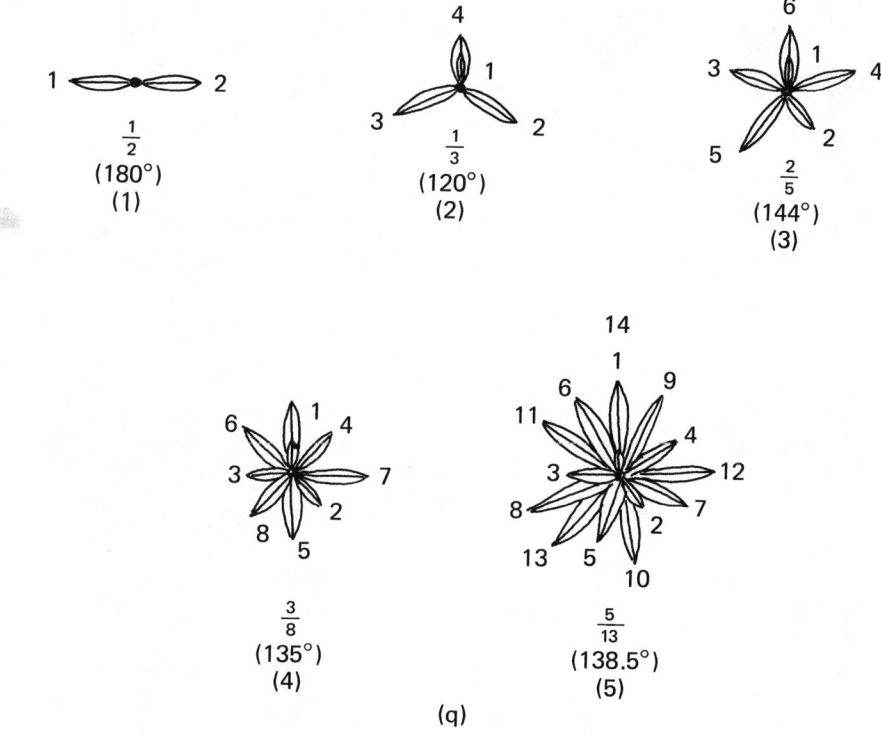

Fig. II.1.1.1.04. Buds: (q) parastichy, phyllotaxis of grasses, and many other monocots, distichous leaves of many dicots (1), sedges (2), and many dicots, and gymnosperms (3), (4), (5).

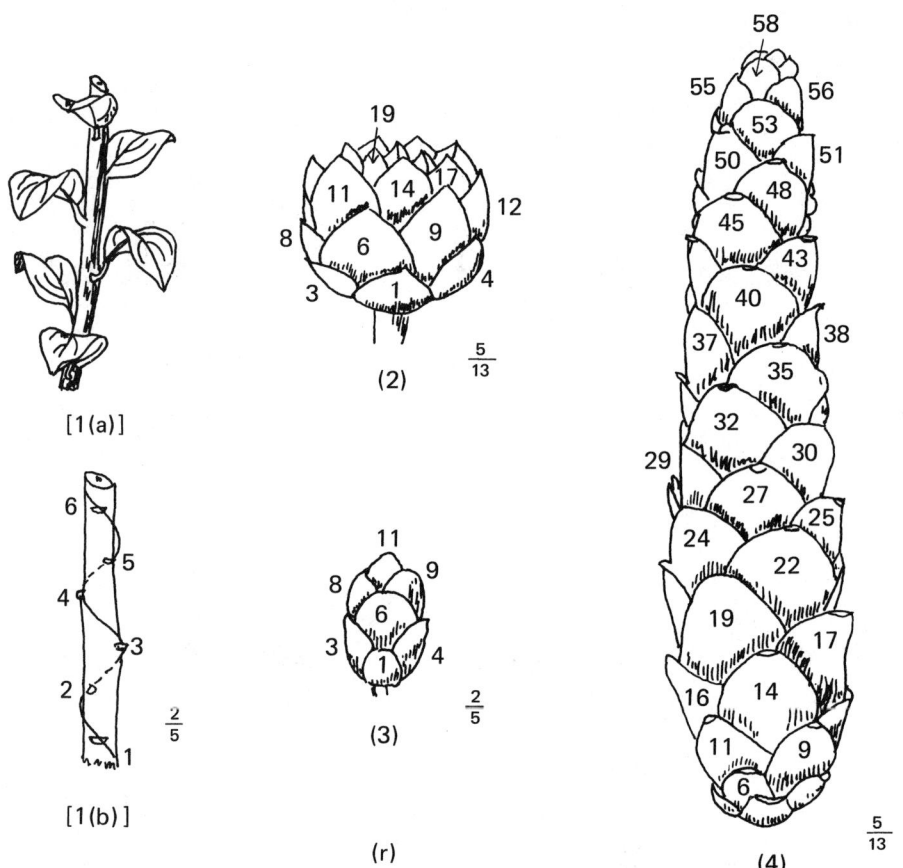

Fig. II.1.1.1.05. Buds: (r) parastichy, orthostichy, phyllotaxis, rows, series, with 2/5 phyllotaxis in apple (1) and larch cone (3) and 5/13 in houseleek (2) and white pine (4); orthostichies are shown in leaves 1 and 14 of houseleek (2), 1, 6, and 11 of the larch cone (3), and 14, 27, 40, and 53 in white pine (4); leaves or scales are in rows or series in (2), (3), and (4), with a rank in (3). (r) Redrawn from Gray, 1862; see Acknowledgments and Selected References.

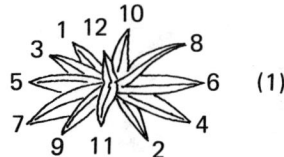

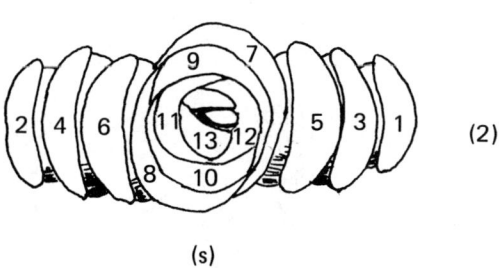

Fig. II.1.1.1.06. Buds: (s) spirodistichy with leaf sequences numbered to show the transition from a distichous to a spiral arrangement in *Crinum powellii* (1) and *Aloe serra* (2), the latter in transverse section. Redrawn from Bell and Coombe, 1976; see Acknowledgments and Selected References.

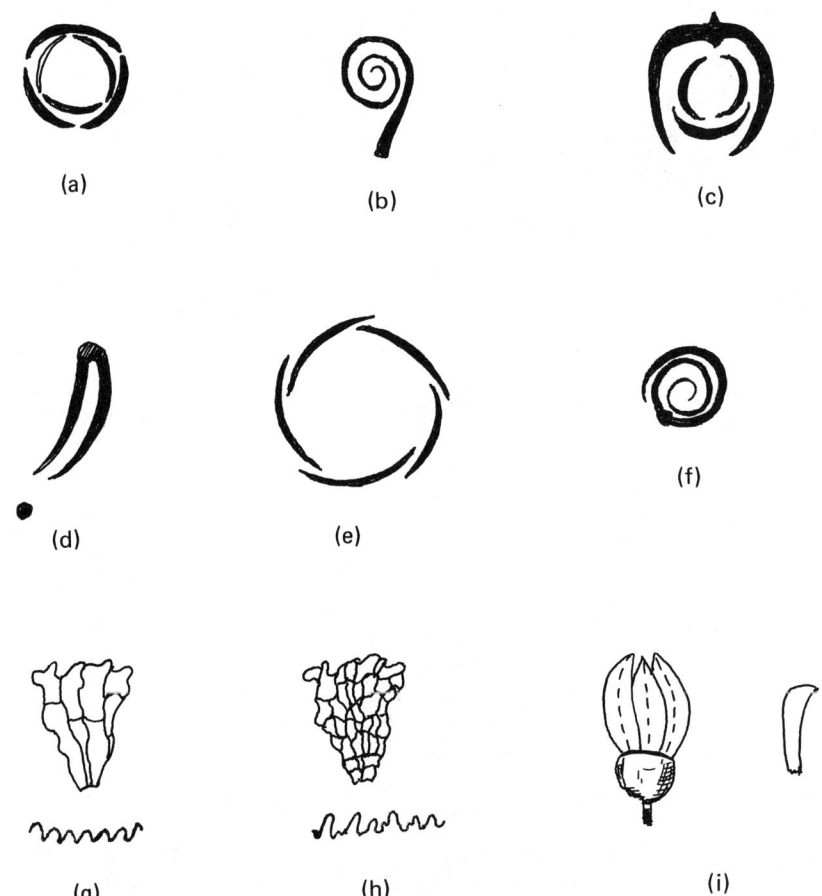

Fig. II.1.1.2.01. Vernation (estivation): (a) alternative; (b) circinate, as in ferns and sundew (<u>Drosera</u>); (c) cochlear; (d) conduplicate, as in oaks and magnolias; (e) contorted; (f) convolute, as in apricot, cherries, banana, ensete (<u>Ensete</u> spp.). and phlox; (g), (h) corrugated, crumpled, as in pomegranate; and (l) curvative. (a), (c) Redrawn from Stearn, 1966; (b), (d)-(i) redrawn from Gray, 1862; see Acknowledgments and Selected References.

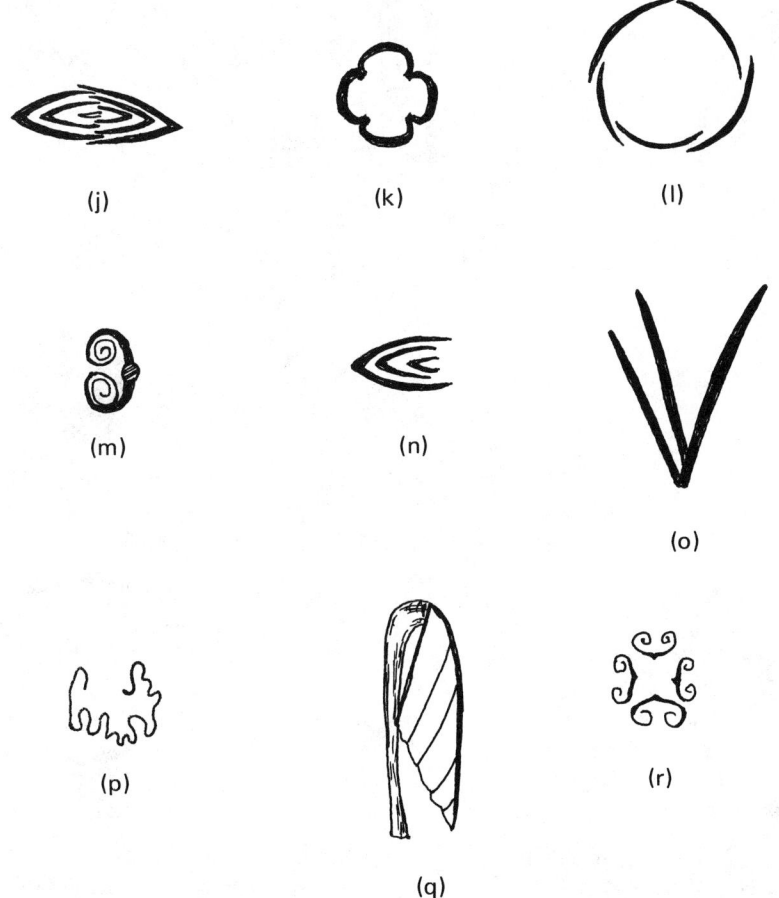

Fig. II.1.1.2.02. Vernation (estivation): (j) equitant, as in Iris leaves; (k) induplicate; (l) imbricate, a common type of estivation; (m) involute, as in violets and water lilies; (n) obvolute (half-equitant); (o) plain; (p) plaited (plicate, pleated), as in maples, currants, grapes, palms, and tobacco; (q) reclinate, as in tulip poplar (Liriodendron) and stamens in some species of Moraceae and Myrtaceae; (r) reduplicate. (j)-(r) Redrawn from Gray, 1862; see Acknowledgments and Selected References.

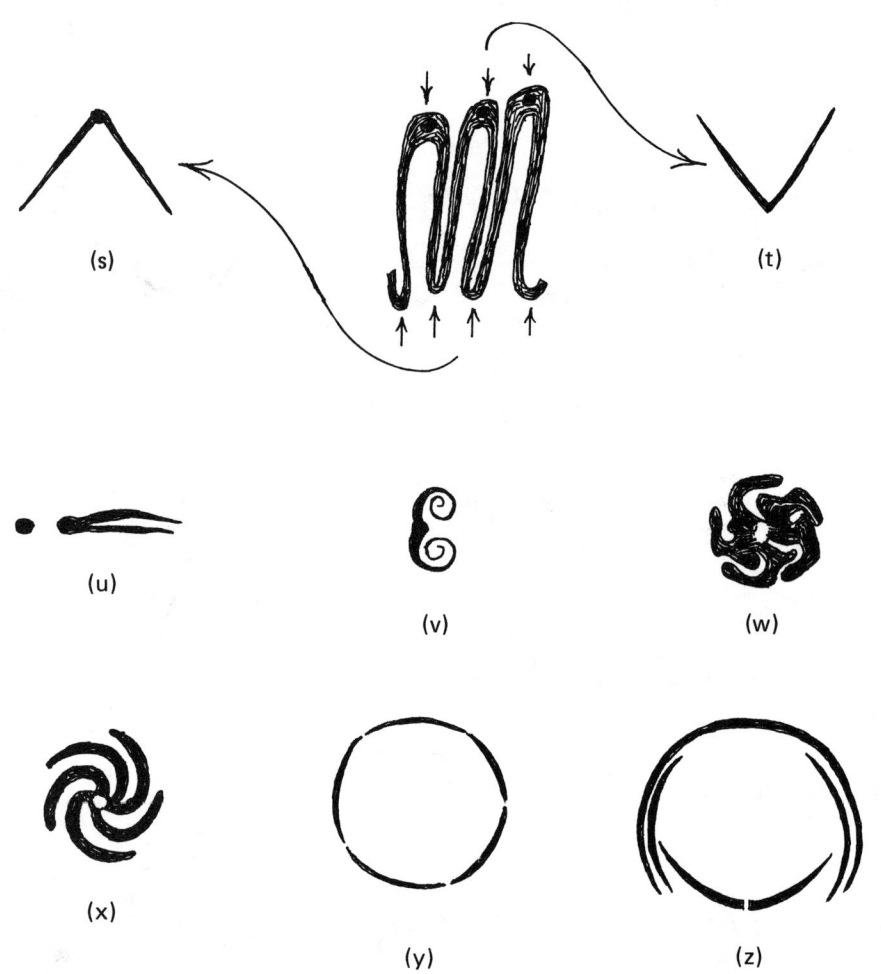

Fig. II.1.1.2.03. Vernation (estivation): (s) induplicate and (t) reduplicate types of vernation in palm leaves, in which the former involves splitting the folds opposite the midrib, as in arecoid and cocoid palms and rattans, whereas the latter splits on the midrib side and does not contain a midrib as in date and all fan-leafed palms; (u) replicative; (v) revolute, as in Rhododendron spp. and rosemary (Rosemarinus spp.); (w) supervolute plaited; (x) supervolute; (y) valvate, a common type of estivation; (z) vexillary, the estivation pattern in the Fabaceae. (s),(t) Adapted from Corner, 1966; (u)-(z) redrawn from Gray, 1862; see Acknowledgments and Selected References.

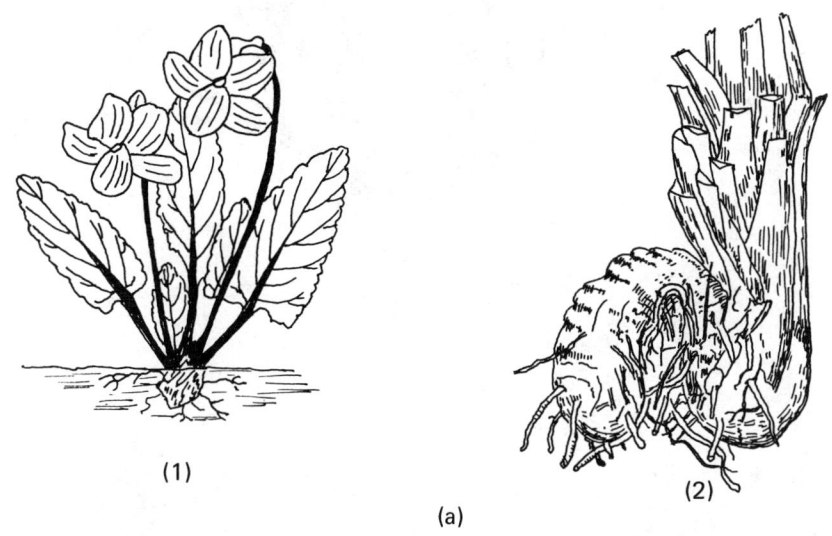

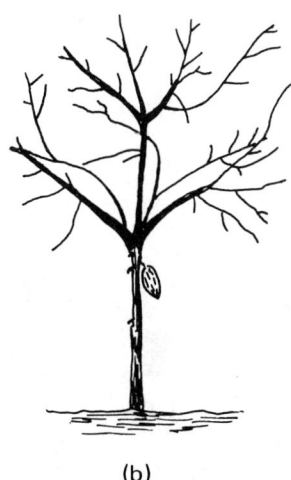

Fig. II.1.1.3.01. Stems: (a) acaulescent, found in many herbaceous plants, e.g., violet (1) and some woody ones, such as Sabal minor (2); (b) acropetal growth in a young cacao (Theobroma cacao) tree. (a)(1) Redrawn from Gray, 1862; (a)(2) redrawn from Bailey, 1944b; (b) redrawn from Braudeau, 1969; see Acknowledgments and Selected References.

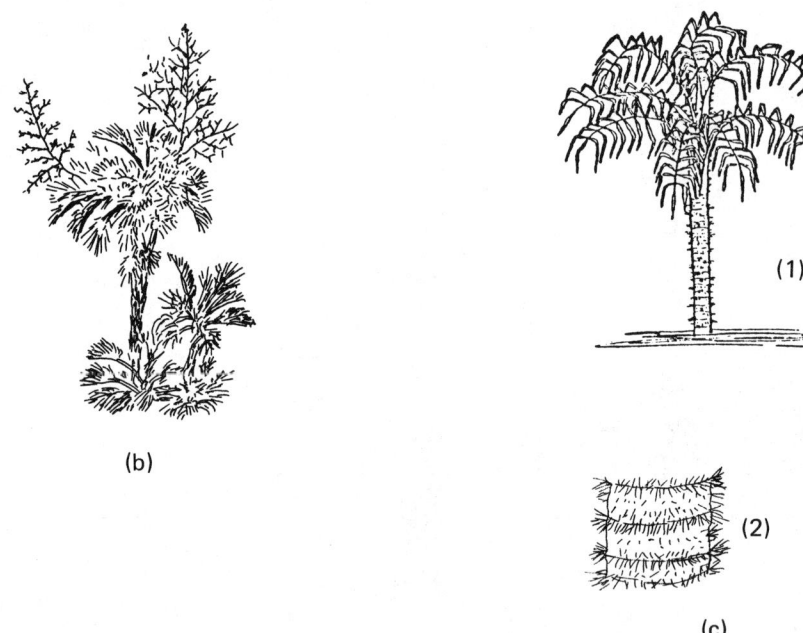

Fig. II.1.1.3.02. Stems: (a) acrotony in Dracaena draco, an arborescent monocot; (b) anisotony, Nannorrhops ritchiana of India; (c) armed palm (1), gru-gru (Acrocomia sp.) with vicious spines on the trunk (2) and petioles. (a) Redrawn from Hallé et al., 1978; (b) redrawn from Corner, 1966; (c)(1) from Watkins and Sheehan, 1975; see Acknowledgments and Selected References.

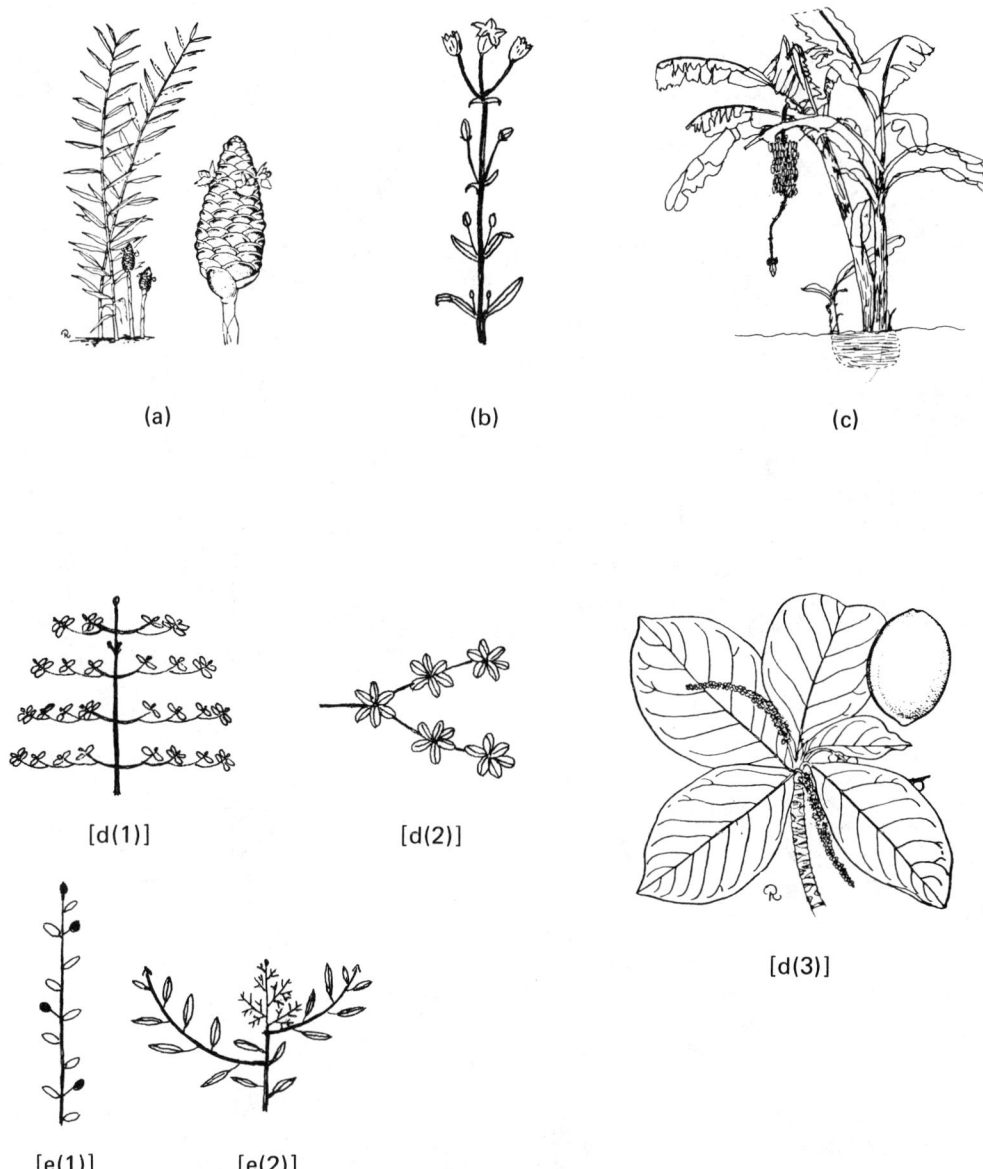

Fig. II.1.1.3.03. Stems: (a) ascending, ascendant, as pine-cone ginger (Zingiber zerumbet); (b) basipetal growth; (c) basitony, banana (Musa sp.); (d) branching by apposition, tropical almond (Terminalia catappa), diagram of tree (1), top view of branch (2), and branch tip (3); (e) branching by substitution as leaf-opposed (1) or terminal inflorescences (2), as in mango (Mangifera indica). (a), (d)(3) From Watkins and Sheehan, 1975; (b) redrawn from Gray, 1862; (c) adapted from Tai, 1958; (d)(1), (2), (e) redrawn from Tomlinson and Gill, 1973; see Acknowledgments and Selected References.

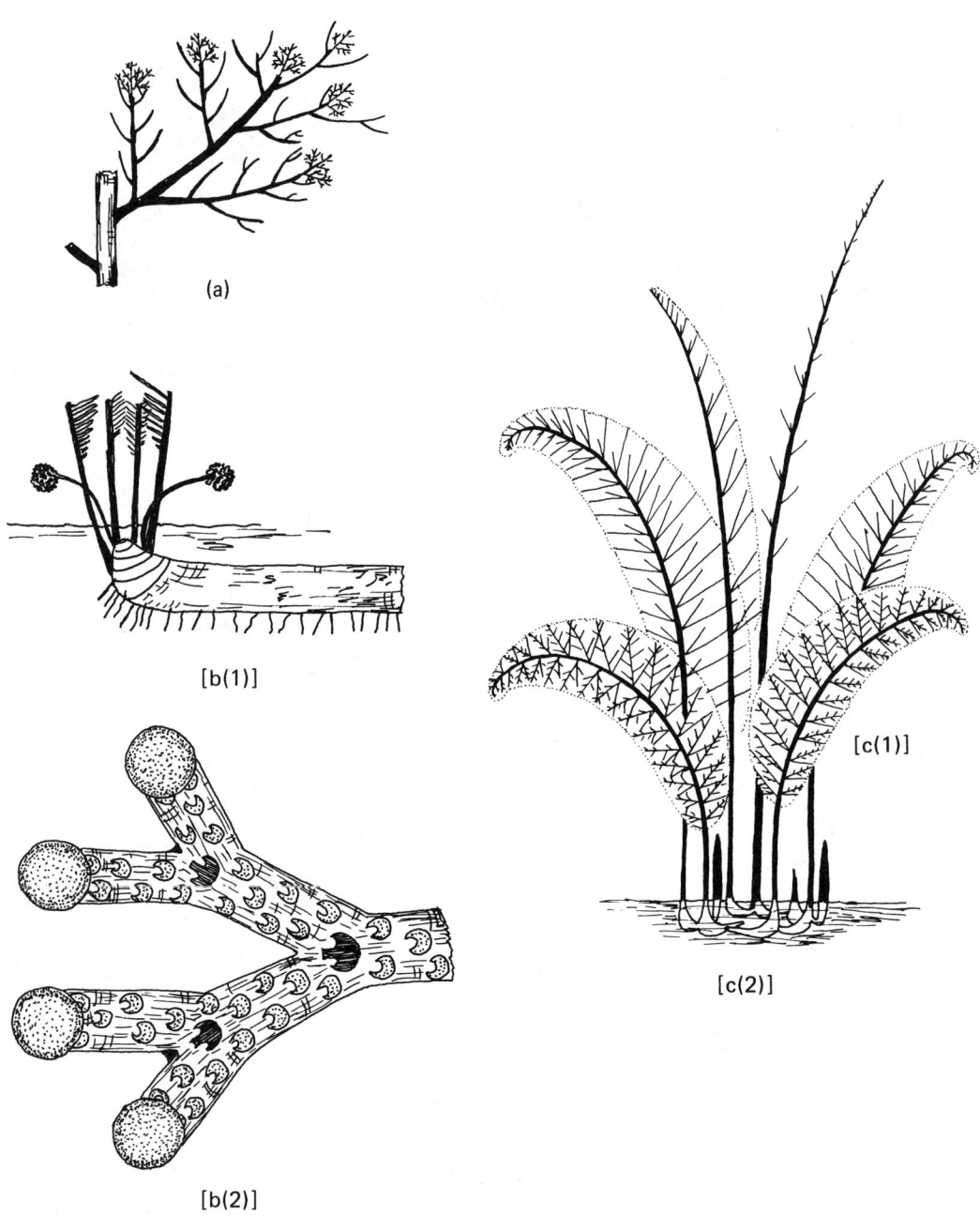

Fig. II.1.1.3.04. Stems, branch systems: (a) orthotropic only, as in mango (Mangifera indica); (b) plagiotropic only, as in nipa palm (Nypa = Nipa fruticans) in side (1) and top (2) views; (c) orthotropic (upright stems or culms) (1) and plagiotropic (horizontal stems) (2), as in bamboo. (a) Adapted from, (b), (c), redrawn from Hallé et al., 1978; see Acknowledgments and Selected References.

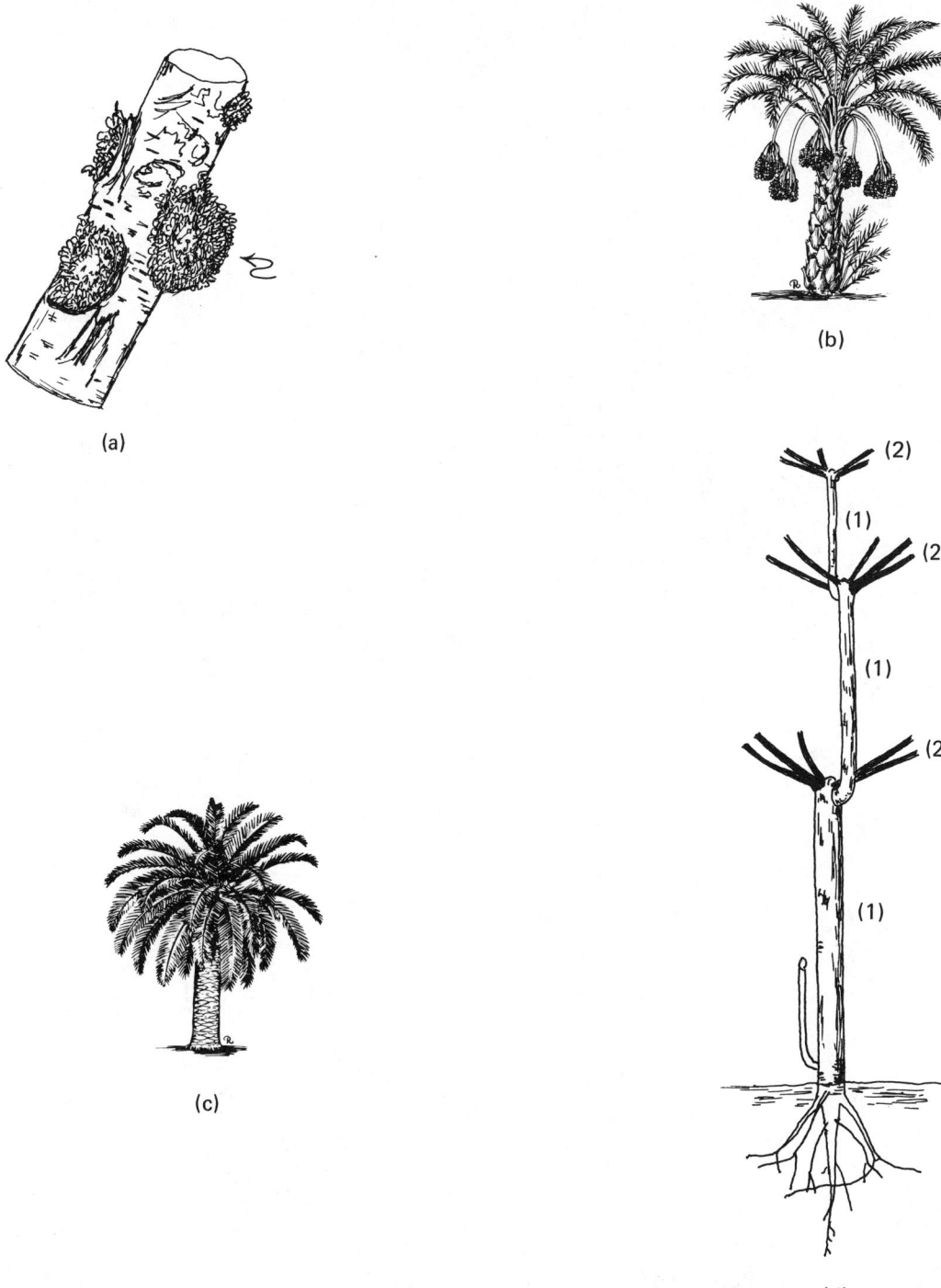

Fig. II.1.1.3.05. Stems: (a) burr knots on two-year-old MM106 apple rootstock in Florida; (b) caudex, date palm (Phoenix dactylifera); (c) columnar, Canary Island date palm (P. canariensis); (d) seedling cacao (Theobroma cacao) tree showing upright (orthotropic) stem or chupon (1) and the lateral (plagiotropic) fan branches (2), collectively a jorquette, which arise just beneath the apical bud of each chupon. (b), (c) From Watkins and Sheehan, 1975; (d) from Cuatrecasas, 1974; see Acknowledgments and Selected References.

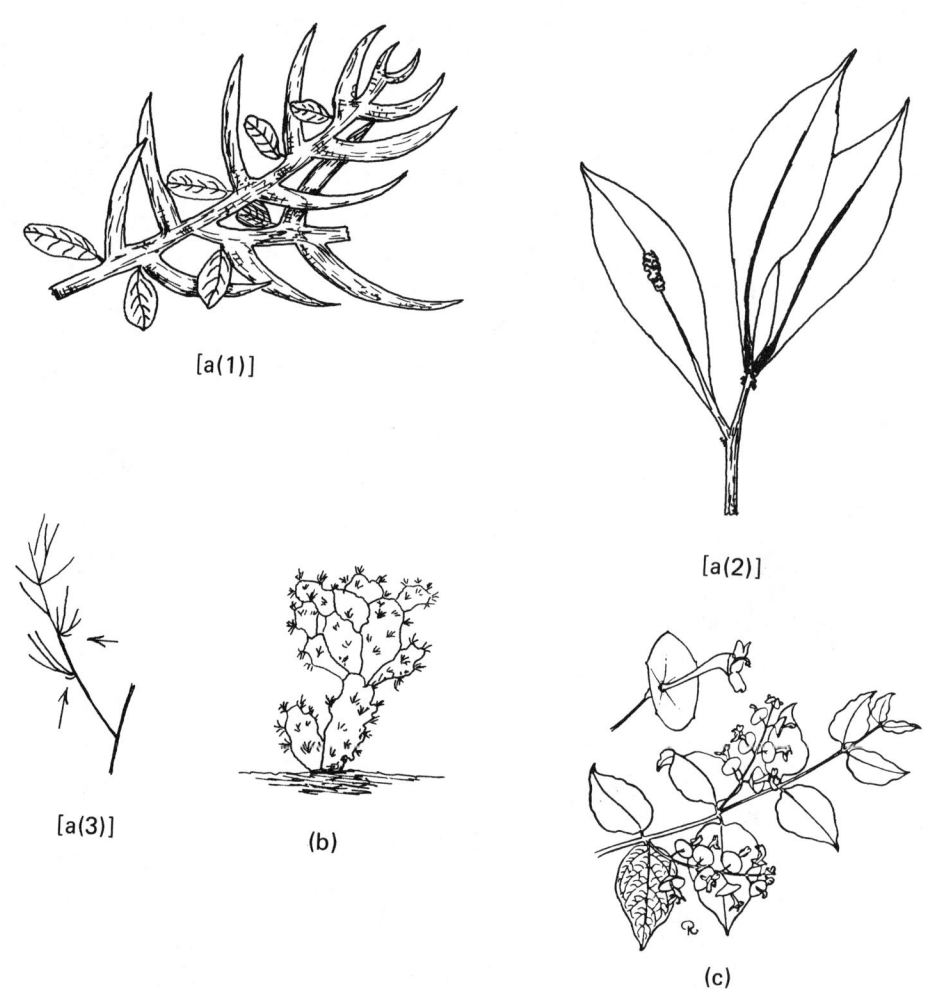

Fig. II.1.1.3.06. Stems (a), (b) modified stems: (a) Cladophylls (cladodes), (1) (Cocculus balfouri), (2) Ruscus sp., (3) Asparagus officinalis (arrows pointing to true leaves as scales); (b) phylloclads, as in Opuntia sp.; (c) clambering shrub (climbs but requires support), Chinese hat-plant (Holmskioldia sanguinea). (a)(1) Redrawn from Hutchinson, 1969; (2) redrawn from Hill et al., 1960; (3) redrawn from Robbins, 1924; (b) redrawn from Bailey, 1928; (c) from Watkins and Sheehan, 1975; see Acknowledgments and Selected References.

Fig. II.1.1.3.07. Stems (a), (b) continuous growth: (a) Chinese fan palm (<u>Livistona chinensis</u>); (b) trunk of Beatrice or step palm (<u>Archontophoenix alexandrae</u> var. <u>beatricae</u>) showing unusually prominent leaf scars (white rings); (c) crown, like the crownshaft on Alexandra palm (<u>Archontophoenix alexandrae</u>) (1), with enlarged view showing two inflorescence bases below the crownshaft of <u>A. cunninghamiana</u> (2). (a), (c)(1) From Watkins and Sheehan, 1975; (b), (c)(2) redrawn from Bailey, 1935; see Acknowledgments and Selected References.

43

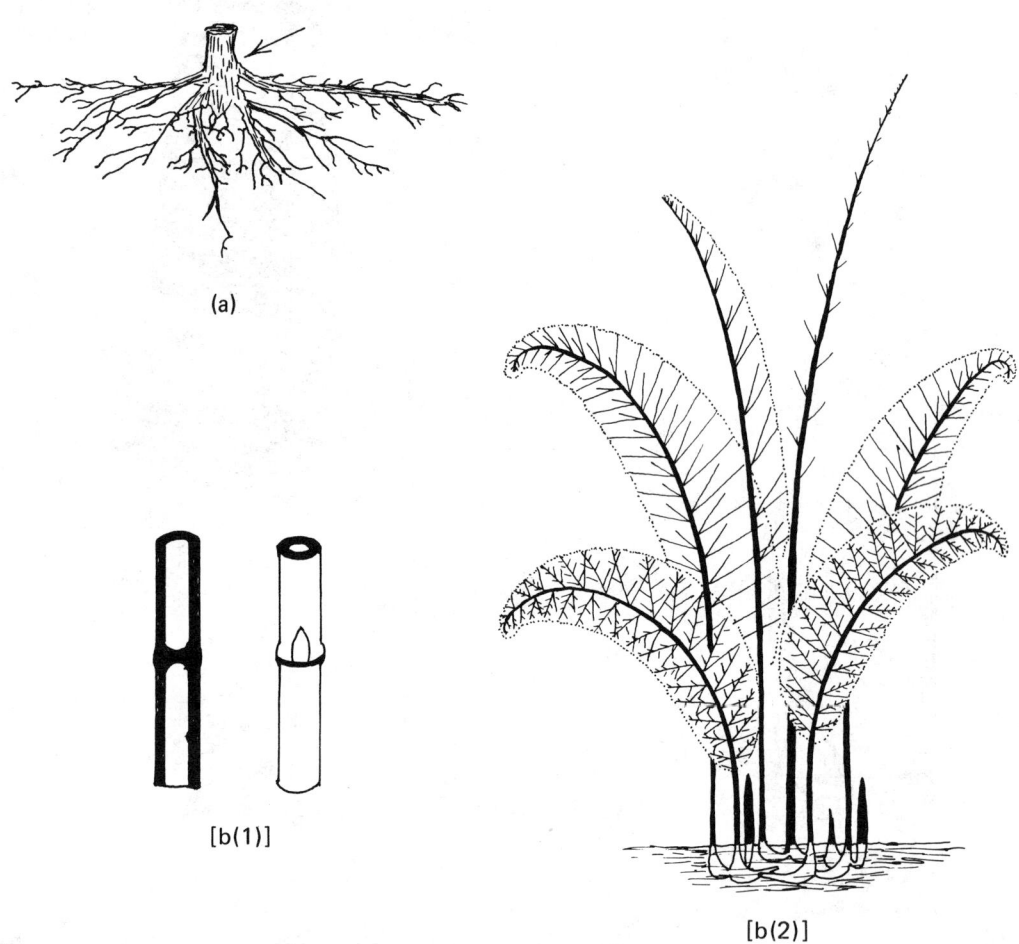

Fig. II.1.1.3.08. Stems, crowns: (a) rootcrown of citrus trees; (b) culms of bamboo clump (1) and details of structure of a node (2). (a) Redrawn from Praloran, 1971; (b)(1) redrawn from Hallé et al., 1978; see Acknowledgments and Selected References.

Fig. II.1.1.3.09. Stems: (a) decumbent shrub, shore juniper (<u>Juniperus conferta</u>); (b) columnar trunk (bole), cabbage palm (<u>Sabal palmetto</u>); (c) deliquescent tree, 'Tjir 1' rubber (<u>Hevea brasiliensis</u>); (d) dichotomous, doum palm (<u>Hyphaene thebaica</u>). (a), (b) From Watkins and Sheehan, 1975; (c) redrawn from Dijkman, 1951; (d) redrawn from Corner, 1966; see Acknowledgments and Selected References.

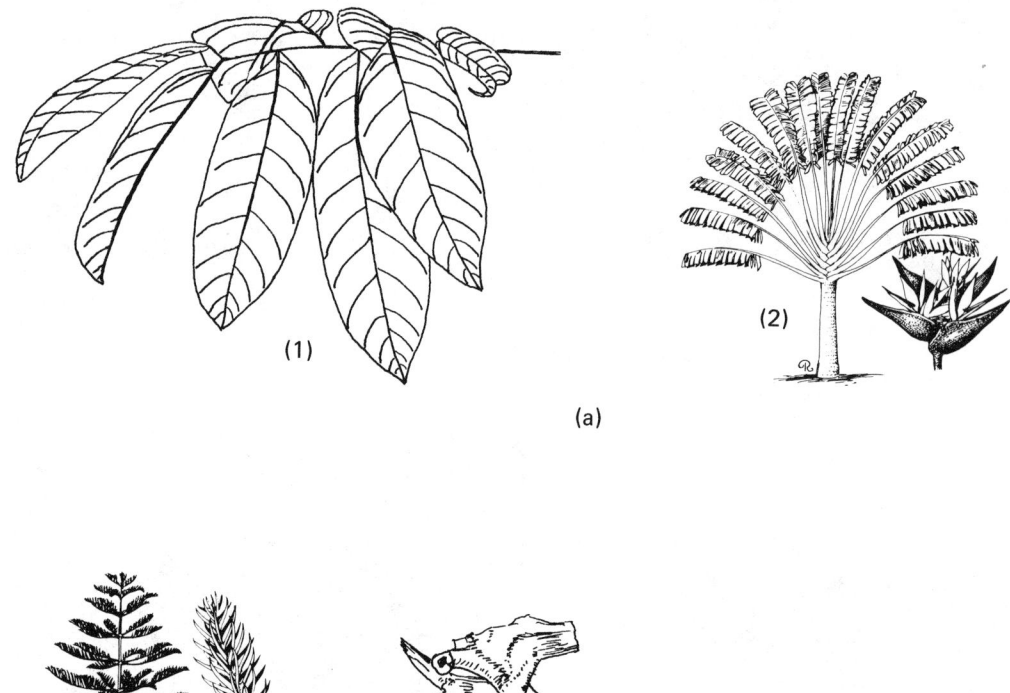

Fig. II.1.1.3.10. Stems: (a) distichous arrangement, (1) soursop (<u>Annona muricata</u>) and (2) traveler's "palm" (<u>Ravenala madagascariensis</u>, actually a relative of banana); (b) erect, excurrent, Norfolk Island pine (<u>Araucaria heterophylla</u>); (c) fistular, as in papaya (<u>Carica papaya</u>) petiole. (a)(1) Redrawn from Ochse, 1931; (a)(2), (b) from Watkins and Sheehan, 1975; (c) adapted from Ochse et al., 1961; see Acknowledgments and Selected References.

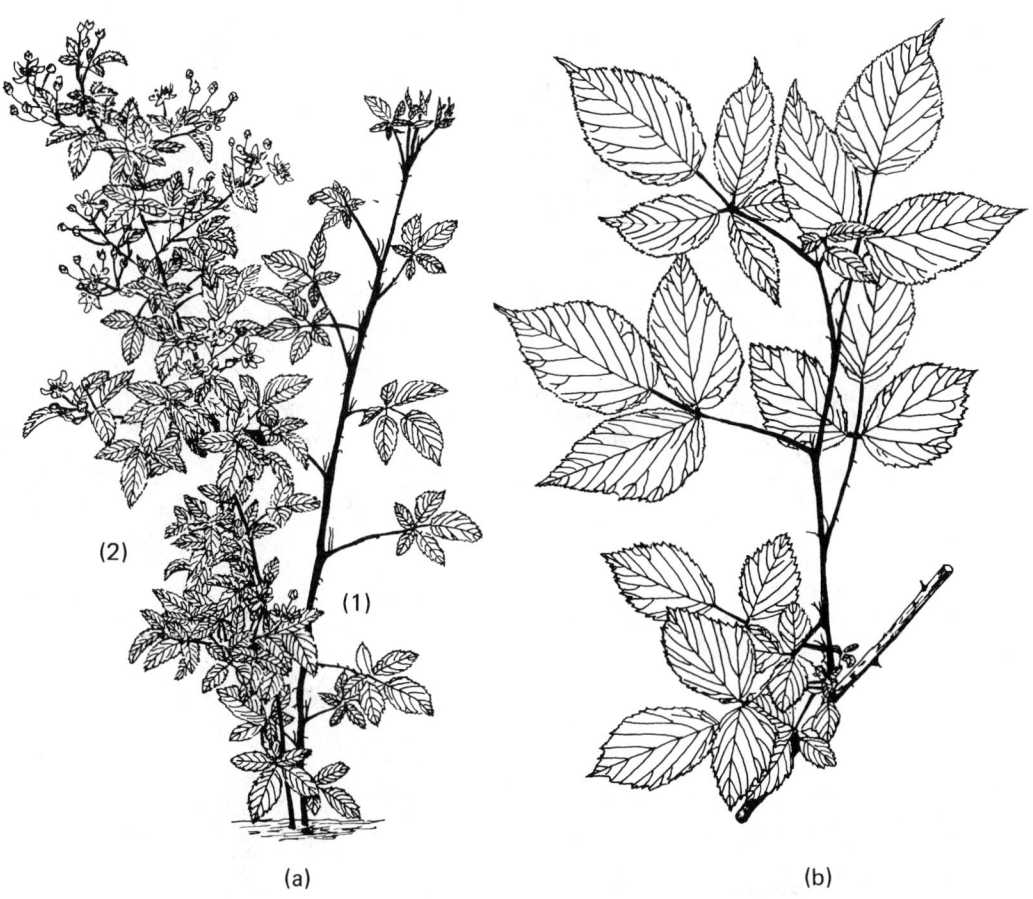

Fig. II.1.1.3.11. Stems: details of biennial succession in Rubus: (a) (1) primocane, the first year's growth, (2) floricane, the flowering shoot of the second year's growth; (b) parcifrond, leafy shoot from a floricane. Redrawn from Bailey, 1941; see Acknowledgments and Selected References.

Fig. II.1.1.3.12. Stems: (a) novirame, blooming terminal or lateral of a primocane, infrequent in occurrence and usually with an abnormal type of inflorescence; (b) horizontal, e.g., Juniperus horizontalis; (c) implicative, dodder (Cuscuta sp.) on alfalfa stem. (a) Redrawn from Bailey, 1941; (c) redrawn from Muenscher, 1955; see Acknowledgments and Selected References.

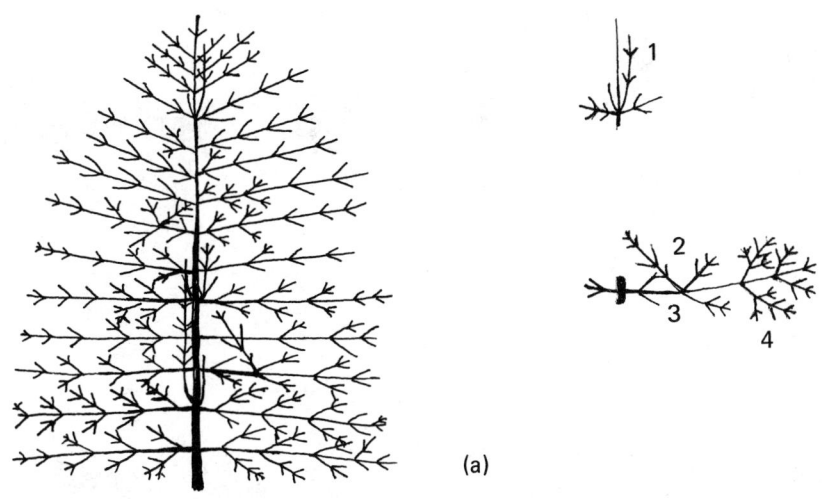

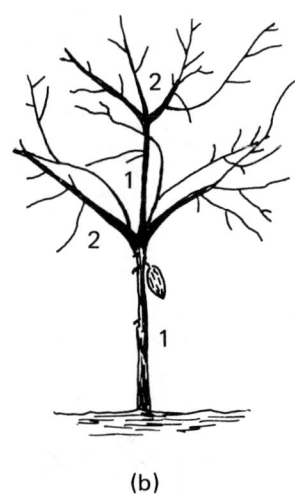

Fig. II.1.1.3.13. Stems: (a) irreversibly plagiotropic, Arabian coffee (Coffea arabica), with (1) orthotropic (upright) branch (must be used if an upright plant is to be obtained by vegetative propagation), (2) adventitious shoot on (3) primary plagiotropic (lateral) and (4) secondary and tertiary branches; (b) young cacao tree with (1) chupon and fan branches = jorquette (2). (a) Redrawn from Coste, 1968; (b) redrawn from Braudeau, 1969; see Acknowledgments and Selected References.

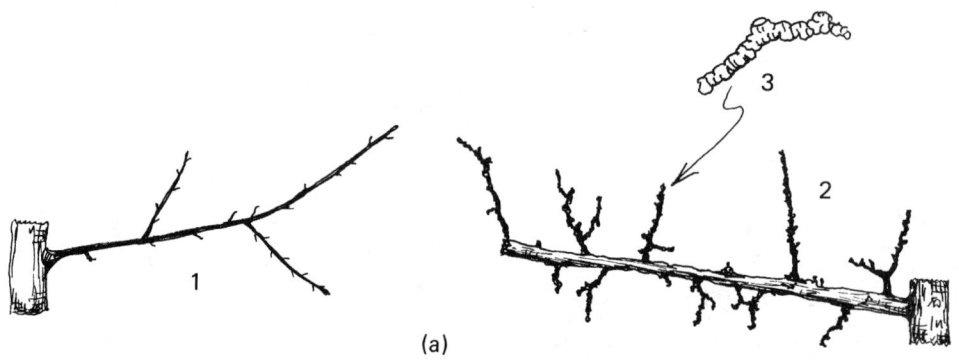

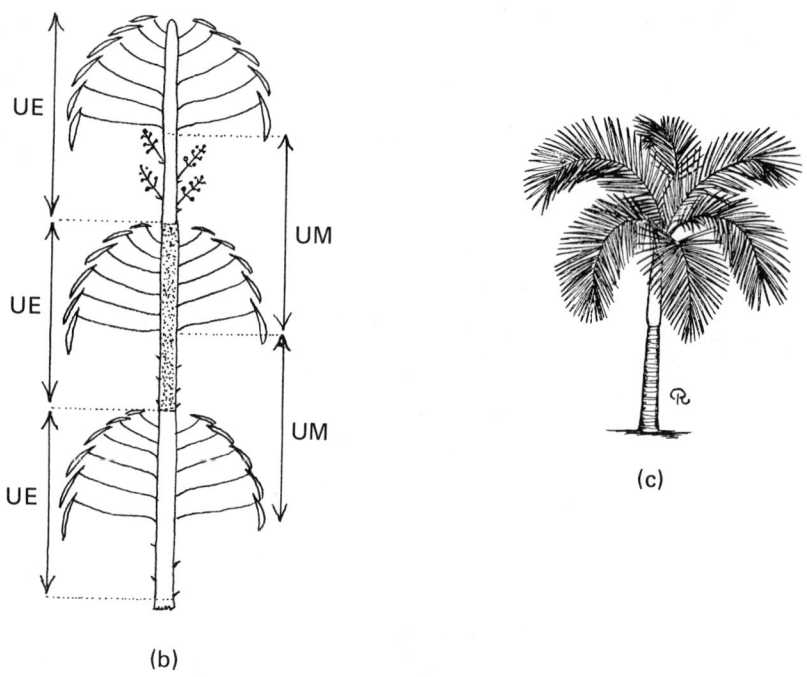

Fig. II.1.1.3.14. Stems: (a) long shoot (1) and short shoots (2) with detail of spur (3) on an apple tree; (b) modules, modular construction, and rhythmic growth in young rubber (Hevea brasiliensis) tree; UE are units of extension (i.e., morphologically distinct units) and UM, units of morphogenesis (i.e., single activity periods of the apical meristem); (c) monoaxial, Seaforth palm (Ptychosperma elegans). (a) Adapted from Hilltop Orchards & Nurseries catalog, 1982; (b) redrawn from Hallé et al., 1978; (c) from Watkins and Sheehan, 1975; see Acknowledgments and Selected References.

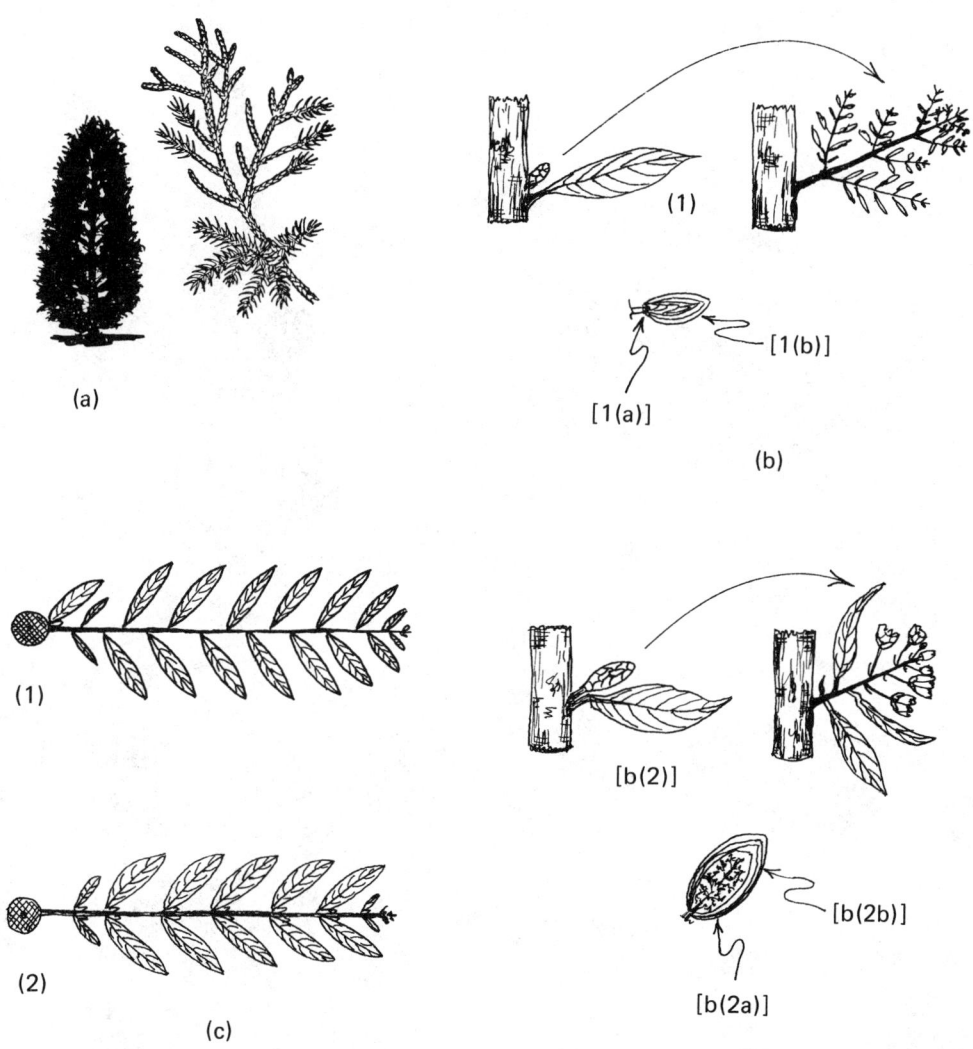

Fig. II.1.1.3.15. Stems: (a) monocaulous, monopodial tree (i.e., with single trunk and indeterminate growth), Japanese juniper (<u>Juniperus chinensis</u>); (b) neoformation (1), in which primordia are developed from apical meristem [1(a)] and a bud [1(b)] and initiate growth without a rest period; preformation (2) in which primordia [2(a)] are developed and undergo a rest period within the bud [2(b)] before growth begins; (c) phyllomorphic branch, where a branch with simple leaves (2) mimics a compound leaf (1) even to the extent of dropping as a unit. (a) From Watkins and Sheehan, 1975; (b), (c) adapted from Hallé et al., 1978; see Acknowledgments and Selected References.

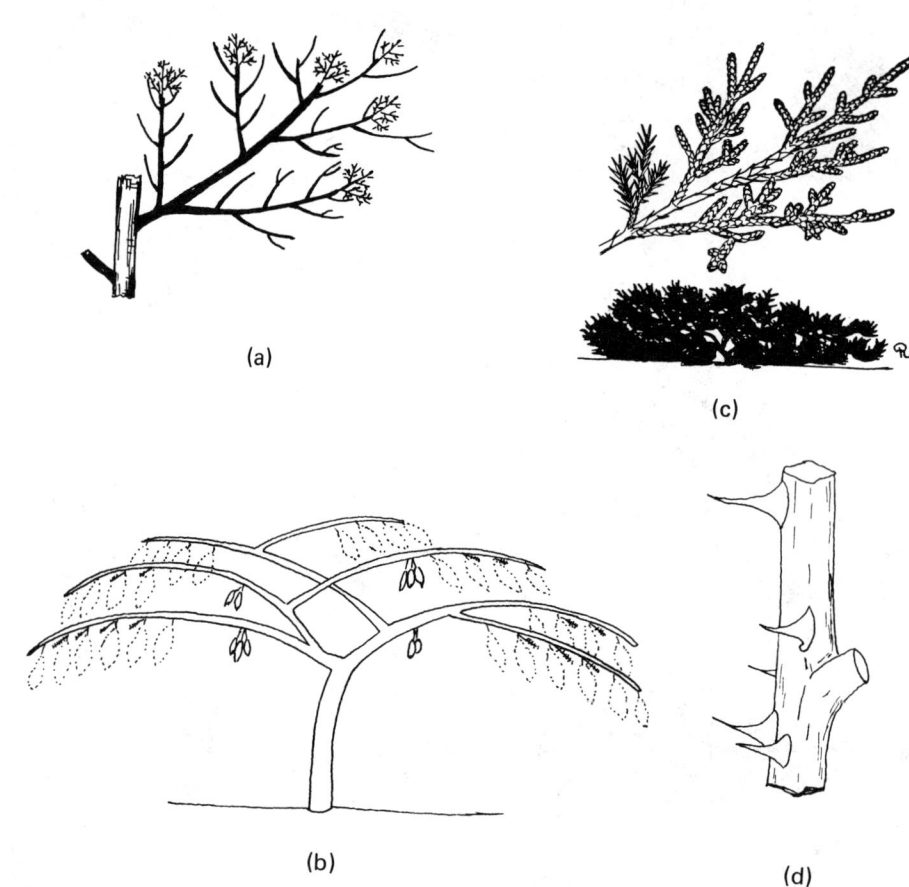

Fig. II.1.1.3.16. Stems; (a), (b) polyaxial: (a) differential polyaxial, mango (<u>Mangifera indica</u>); (b) mixed polyaxial, carambola (<u>Averrhoa carambola</u>); (c) procumbent, short juniper (<u>Juniperus conferta</u>); (d) prickles, on <u>Rubus</u> sp. (a), (b) Adapted from Hallé et al., 1978; (c) from Watkins and Sheehan, 1975; (d) redrawn from Bailey, 1941; see Acknowledgments and Selected References.

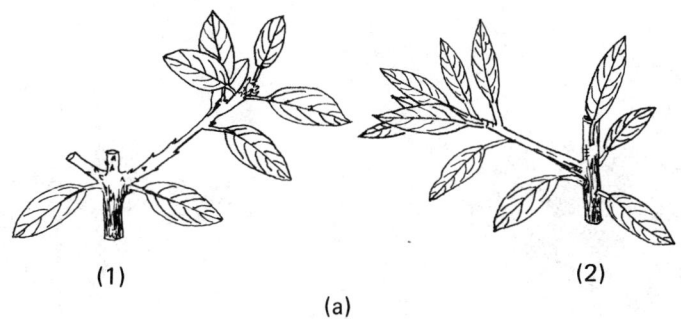

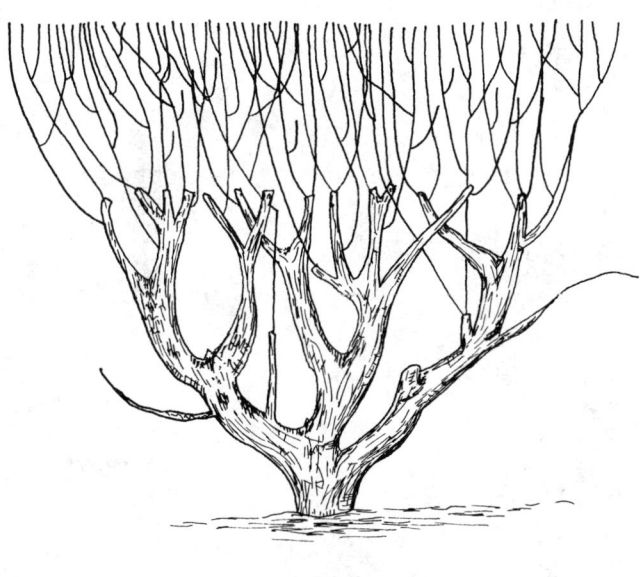

Fig. II.1.1.3.17. Stems: (a) prolepsis (1) and syllepsis (2) in avocado (Persea americana); (b) ramose, typified by dense branch structure of tea (Camellia sinensis) given successive prunings to develop a flattopped hedge. (a) Redrawn from Tomlinson and Gill, 1973; (b) redrawn from TRI East Africa, 1967; see Acknowledgments and Selected References.

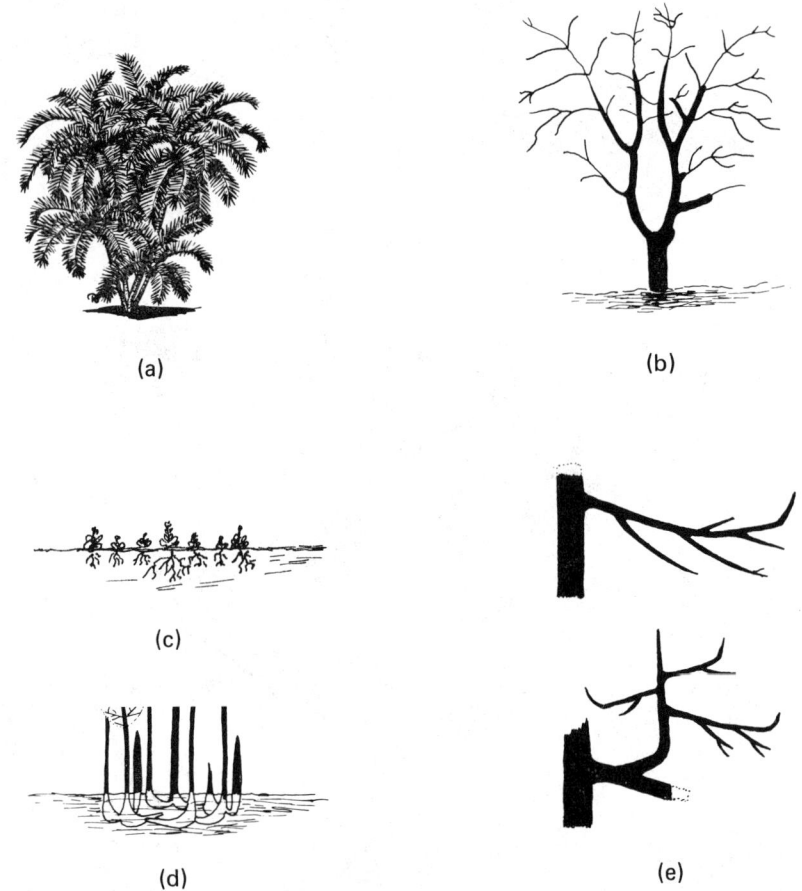

Fig. II.1.1.3.18. Stems: (a) reclining, reclinate, Senegal date palm (Phoenix reclinata); (b) reiteration, shown in Arabian coffee pruned to candelabra (Costa Rica) system; (c) repent, bugle weed (Ajuga reptans); (d) rhizomatous branching, bamboo; (e) reversibly plagiotropic, where a plagiotropic branch turns up and forms an upright (orthotropic) stem to replace the original axis, as in kapok (Ceiba). (a) From Watkins and Sheehan, 1975; (b) redrawn from Ochse et al., 1961; (d) adapted from Hallé et al., 1978; see Acknowledgments and Selected References.

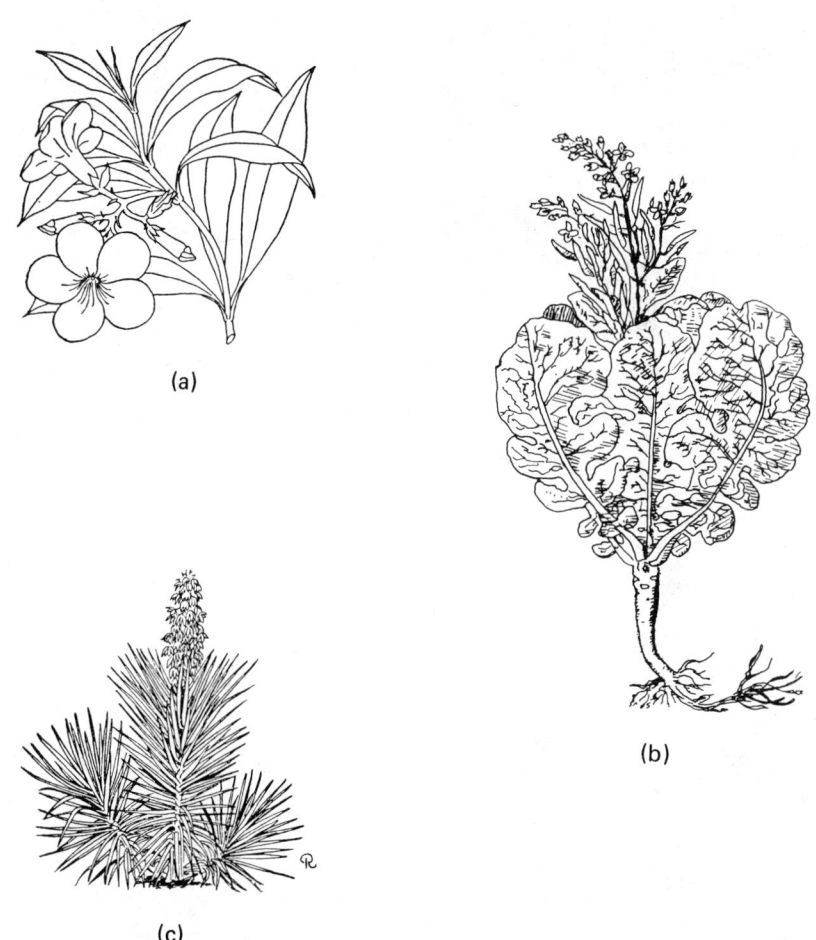

Fig. II.1.1.3.19. Stems: (a) scandent, allamanda (Allamanda cathartica) which can be trained as a vine (i.e., on a trellis) or as a shrub; (b) seedstalk, flowering portion (arrow) of collard plant which is allowed to bloom (remain the second year) only if seeds are desired; (c) soboliferous, Spanish bayonet (Yucca aloifolia). (a), (c) From Watkins and Sheehan, 1975; (b) redrawn from Johnson, 1633; see Acknowledgments and Selected References.

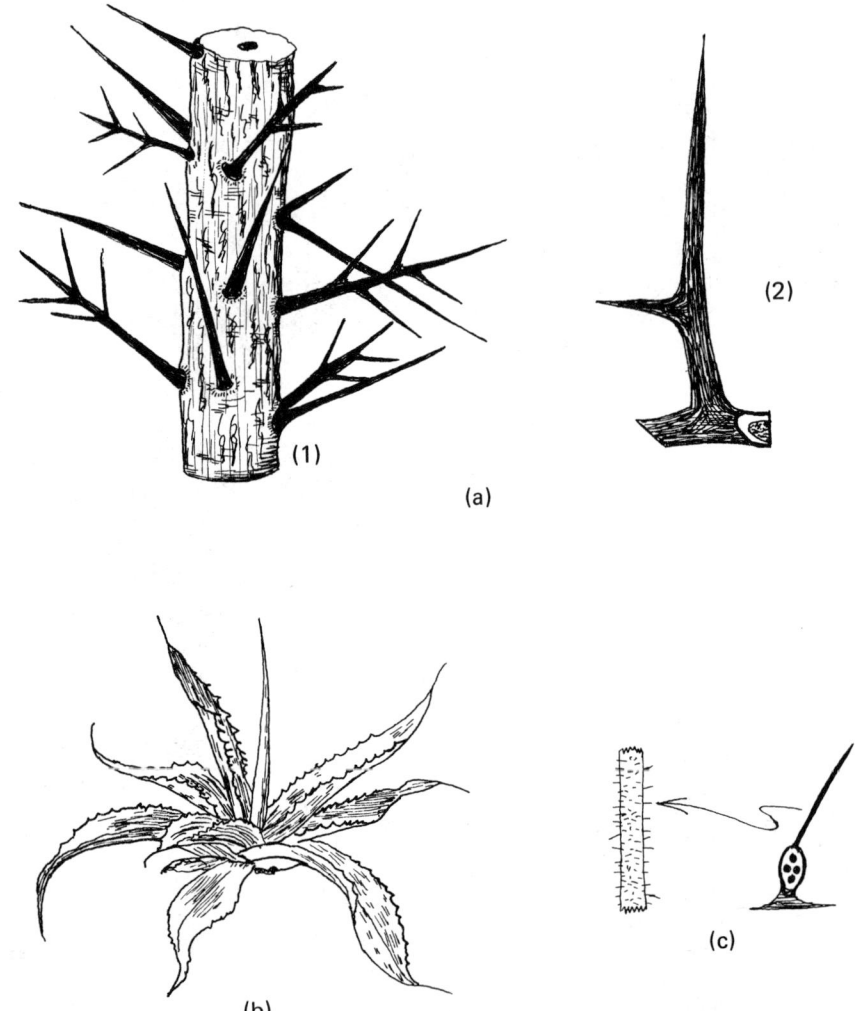

Fig. II.1.1.3.20. Stems: (a) spine, spiny, (1) paniala (<u>Flacourtia cataphracta</u> = <u>jangomas</u>) and (2) honey locust (<u>Gleditsia triacanthos</u>); (b) spinescent, agave leaves; (c) spinulose, stinging nettle (<u>Urtica dioica</u>) with enlarged hair, of which the bulbous base contains formic acid and other toxic substances at right. (a)(1) Redrawn from Hutchinson, 1969; (a)(2) redrawn from Gray, 1862; (b) redrawn from Gray, 1868; (c) adapted from Millspaugh, 1892; see Acknowledgments and Selected References.

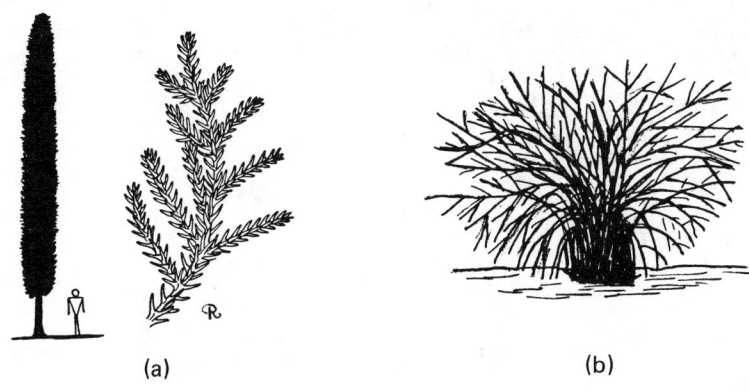

Fig. II.1.1.3.21. Stems: (a) strict, Italian cypress (Cupressus sempervirens); (b) suffrutescent, peony (Paeonia sp.); (c) sympodial, dragon tree (Dracaena draco). (a) From Watkins and Sheehan, 1975; (b) adapted from Bailey, 1928; (c) redrawn from Hallé et al., 1978; see Acknowledgments and Selected References.

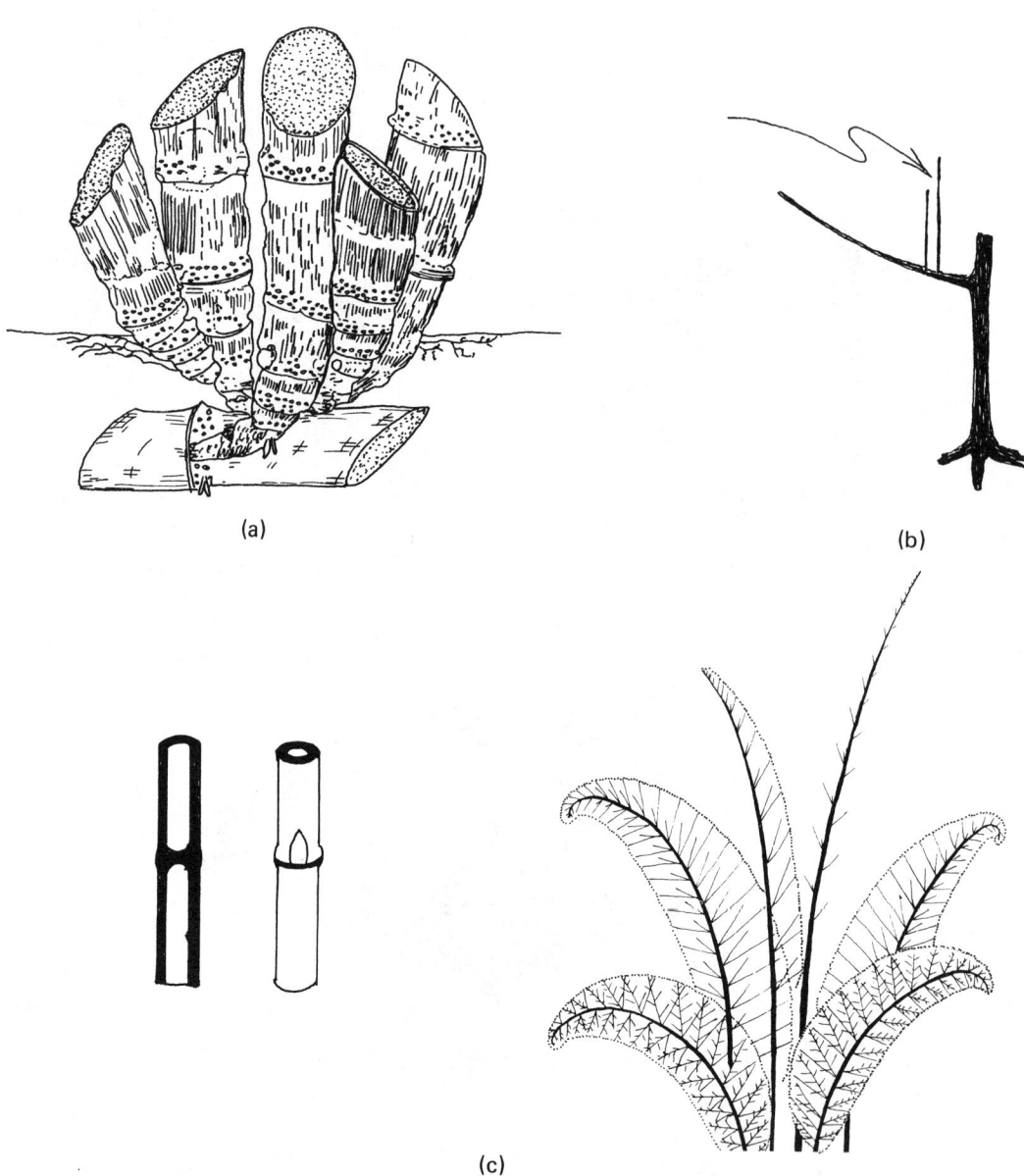

Fig. II.1.1.3.22. Stems: (a) tiller, sugar cane (<u>Saccharum</u> sp.); (b) watersprouts (arrow) on fruit tree; (c) tubular, as in long arching culms of bamboo, which may range in diameter from pencil size to 30 cm or more and in height from one to nearly 30 m; an external view of node at left. (a) Redrawn from Ochse et al., 1961; (b) adapted from Christopher, 1957; (c) adapted from Hallé et al., 1978; see Acknowledgments and Selected References.

II.1.2. MORPHOLOGY and ANATOMY

II.1.2. Leaves

.000 Leaves: one of the major organs of plants; several distinct types, such as foliage leaves which include cataphylls (undeveloped leaves), scales and bracts (protective leaves), and perianth and sporophylls (floral leaves); true foliage leaves, like those found in ferns and seed plants, are expanded organs, often with special modifications like tendrils and traps (carnivorous), that form most of the external covering and serve as the principal sites of sugar manufacture via photosynthesis, and perform the vital functions of moisture and gas exchange with the atmosphere. Foliage leaves typically have three principal parts: blade (flat expanded part), petiole (leaf stalk, which may be absent; the leaf is then sessile), and a pair of stipules (not present in some families); the blade may be entire or more or less dissected, thus simple, or have few to many segments, thus compound; these leaves are lateral outgrowths of a stem, arising in definite acropetal or basipetal succession from a growing point (apical meristem), thus marking the nodes of a stem, and have one or more buds in the axil, some of which may be floral, and one, a vegetative (branch) bud; most have dorsiventral symmetry, although modified in some plants as centric (cylindrical or terete), isobilateral, or inverted dorsiventral. Protective and floral leaves are modified to perform specific functions but arise in the same manner as foliage leaves, and like the latter, differ from stems, in that, growth is limited except in those instances in which an intercalary meristem is present (e.g., many monocots). (G 62, G 68, HI, B & C).

.001 Abscission: defoliation or loss of an organ, the actual break occurring in a separate layer; a protective layer of scar tissue and periderm develop after the leaf or branch has fallen. (Fig. II.1.2. 01a,b) (E & M)

.002 Acerose: narrow, stiff, pointed like a needle. (Fig. II.1.2.01c) (G 62)

.003 Acyclic: (see Spiral, II.1.2.104).

.004 Alate, winged: having a thin broad margin or wing, like a Citrus petiole; also stems such as winged elm or Euonymus. (Fig. II.1.2. 01d) (G 62)

.005 Alternate: a leaf node or leaflets not opposed along a rachis; i.e., a spiral arrangement along with the axis. (Fig. II.1.2.01e) (G 62)

.006 Amplexicaul, stem-clasping: the same as embracing, except applied only to stems; semiamplexicaul is the same but to a lesser degree. (Fig. II.1.2.02a) (G 62)

.007 Angular: having projecting longitudinal angles, like some petioles. (Fig. II.1.2.02b) (ST)

.008 Anisophyllous: diversity of leaf structure (compare with Anisophylly, II.1.2.009). (Fig. II.1.2.02c) (B & C)

.009 Anisophylly: leaves of different sizes and petiole lengths are "facultative" if induced by gravity or "obligate" if the condition is genetic (i.e., an inherited trait). (Fig. II.1.2.02d) (B & C)

.010 Apex terms: (see II.5.4.1).

.011 Appendage: attached secondary or subsidiary part, sometimes projecting or hanging. (Fig. II.1.2.02e) (L 55)

.012 Arced, curved: bent to represent the arc of a circle. (ST)

II.1.2. MORPHOLOGY and ANATOMY

.013 Articulate: when one body is united with another, as a petiolule to a rachis or the blade of a unifoliolate leaf to its petiole. (Fig. II.1.2.03a) (G 62)

.014 Articulated, jointed: falling in pieces at the joints; separating readily at the joints or having the appearance of being jointed. (Fig. II.1.2.03a) (ST)

.015 Base terms: (see II.5.4.2).

.016 Bifacial leaf: a leaf with dorsiventral differentiation; two types: the regular type (most leaves) with palisade cells as layers under the upper epidermis, vascular traces and fiber groups near the center, and spongy mesophyll overlying the lower epidermis with stomata in it, and the inverted type (e.g., Allium ursinum) in which the sequence is reversed. (Fig. II.1.2.04a,b) (B & C)

.017 Bifoliolate, binate: a compound leaf with two leaflets (see Paired, II.1.2.069). (Fig. II.1.2.03b) (G 62)

.018 Bijugate: (see Paired, II.1.2.069).

.019 Binate: (see Bifoliolate, II.1.2.017).

.020 Blade: expanded portion of a leaf. (Fig. II.1.2.03c) (G 62)

.021 Calyptra: hood or lid, particularly the lid in eucalyptus fruits or of stipules that form the cover over buds in Moraceae. (Fig. II.1.2.03d) (L 55)

.022 Canaliculate, channeled: long and concave, resembling a gutter or channel. (Fig. II.1.2.05a) (ST)

.023 Cataphyll: a scale, as on buds and underground stems, for protection, storage, or both; an undeveloped leaf at the beginning of growth. (Fig. II.1.2.05b) (ES)

.024 Channeled: (see Canaliculate, II.1.2.022).

.025 Cladophyll: (see Genotypic modification, II.4.3.013; see also II.1.1.3.025).

.026 Compound: having various divisions, like a leaf with two or more leaflets or a fruit produced from two or more pistils. (Fig. II.1.2.05c) (G 62)

.027 Connate-perfoliate: opposite sessile leaves connate by their bases, the axis seemingly passing through them. (Fig. II.1.2.05d) (G 62)

.028 Continuous: the reverse of interrupted. (G 62)

.029 Costapalmate: palmate leaf whose petiole continues through the blade as a distinct midrib, thus producing a condition midway between palmate and pinnate; found in certain palms, e.g., Sabal. (Fig. II.1.2.06a) (B-GH)

.030 Cotyledon(s): (see II.2.3.3.014c).

.031 Crowded: when the parts are pressed closely around each other. (Fig. II.1.2.07a) (ST)

.032 Curved: (see Arced, II.1.2.012).

.033 Cushioned: (see Pulvinate, II.1.2.087).

.034 Cylindrical unifacial leaf: (see Unifacial leaf, II.1.2.119). (Fig. II.1.2.04e)

.035 Decompound: twice pinnate compound leaf, as in many leguminous plants and Moringa; with the petiole or rachis divided into secondary units (=petiolules, rachillae). (Fig. II.1.2.07b; .21b) (G 62)

.036 Decurrent: extending downward; said of a leaf whose base extends down to form a wing or ridge along the stem. (Fig. II.1.2.07c) (G 62)

.037 Decussate: an arrangement of opposite leaves in which pairs are borne alternately at right angles. (Fig. II.1.2.07d) (G 62)

.038 Digitate: compound leaf with leaflets arising from a common point on the petiole (see Palmate, II.1.2.071). (Fig. II.1.2.08a) (G 62)

II.1.2. MORPHOLOGY and ANATOMY 61

.039 **Disciform:** flat and circular like a disk. (Fig. II.1.2.08b) (G 62)
.040 **Embracing:** with the base clasping, as in some leaves (see also Sheathing, II.1.2.101). (Fig. II.1.2.08d) (G 62)
.041 **Even-pinnate:** (see Paripinnate, II.1.2.072).
.042 **Extension growth:** successive elongation (and lateral expansion) of a monocot leaf from the apex of the blade to the basal portion of the petiole and sheath by continued activity of the intercalary meristem present. (Fig. II.1.2.09) (CR 66)
.043 **Falcate, sickle- or scythe-shaped:** plane and curved with parallel edges like a scythe or sickle blade, such as the mature leaves on many eucalypts. (Fig. II.1.2.10c) (G 62)
.044 **Fan-shaped:** (see Flabelliform, II.1.2.047).
.045 **Fascicled:** when several things appear to proceed from a common point from great suppression of the internodes in an alternate arrangement; = fasciculate. (Fig. II.1.2.10a) (G 62)
.046 **Fasciculate:** two to several leaves crowded together that appear to arise from the same point but actually alternate in arrangement with short internodes, as in Punica and pine. (Fig. II.1.2.10b) (G 62)
.047 **Flabelliform, fan-shaped:** plaited like the rays of a fan. (Fig. II.1.2.10d) (ST)
.048 **Flattened unifacial leaf:** (see Unifacial leaf, II.1.2.119). (Fig. II.1.2.04g)
.049 **Furcate, forked:** having long terminal lobes like the prongs of a fork. (Fig. II.1.2.10e) (ST)
.050 **Granular:** divided into little knobs or knots. (ST)
.051 **Half-terete:** (see Semiterete, II.1.2.098).
.052 **Heterophylly:** the existence of two or more forms of leaf on the same plant or on different plants of the same species, often as a result of juvenile versus mature growth; can be facultative (induced by environmental conditions) or obligate (hereditary) (see also Anisophylly, II.1.2.009). (Fig. II.1.2. 11a) (G 62)
.053 **Imparipinnate:** when the petiole (rachis) is terminated by a single leaflet or tendril; odd-pinnate. (Fig. II.1.2.11b,c, .12a) (G 62)
.054 **Infrafoliar:** below the (present) leaves, like the inflorescence in Arecoid palms. (Fig. II.1.2.13c) (B-GH)
.055 **Interpetiolar:** between or among the parts of the petiole, as in certain stipules (II.1.2.109). (Fig. II.1.2.13a) (G 62)
.056 **Interrupted:** when any symmetrical arrangement is disturbed or destroyed by local causes; e.g., a pinnate leaf with some pinnae much smaller or wanting. (Fig. II.1.2.13b) (ST)
.057 **Intrafoliar:** immediately above or in front of the leaves, such as inflorescences and fruit borne in axils of leaves. (Fig. II.1.2. 13d) (G 62)
.058 **Inverted bifacial leaf:** (see Bifacial leaf, II.1.2.016). (Fig. II.1.2.04b)
.059 **Involute:** rolled inward; the reverse of revolute. (Fig. II.1.2.14a) (G 62)
.060 **Jointed:** (see Articulate, II.1.2.013).
.061 **Ligule, ligulate:** (see II.5.5.032).
.062 **Lobe:** any part or segment of an organ; specifically a part of a leaf, petal, or sepal that represents a division about midway between the margin and the center. (Fig. II.1.2.14e) (G 62)
.063 **Margin terms:** (see II.5.4.3).
.064 **Multijugate:** (see Paired, II.1.2.069).
.065 **Needlelike unifacial leaf:** (see Unifacial leaf, II.1.2.119).

II.1.2. MORPHOLOGY and ANATOMY 62

.066 Ocrea (ochrea): tubular sheath around the base of the petiole that consists of a single stipule or a pair of coherent stipules, as in Polygonaceae and some palms; may be expanded in the latter. (Fig. II.1.2.14c,d) (G 62)
.067 Odd-pinnate: (see Imparipinnate, II.1.2.053).
.068 Opposite: two leaves at a node or two leaflets on a rachis on opposite sides of the axis. (Fig. II.1.2.15a) (L 51)
.069 Paired: a bifoliolate leaf; e.g., bijugate, trijugate and multi-jugate refers to compound leaves with two, three, or many pairs of leaflets, respectively. (Fig. II.1.2.15b) (G 62)
.070 Palaceous: when the foot-stalk adheres to the margin; used in contrast to peltate. (Fig. II.1.2.15c). (ST)
.071 Palmate: leaf lobed or divided or ribbed in palmlike or handlike fashion = digitate, although the latter term is properly applied to a compound leaf. (Fig. II.1.2.16b) (G 62)
.072 Paripinnate: even-pinnate compound leaf without a terminal leaflet; i.e., when the petiole (rachis) is terminated by neither leaflet nor tendril, as in many Sapindaceae; some leaves, however, may be spuriously paripinnate, as in some Anacardiaceae. (Fig. II.1.2.11c, .12b) (G 62)
.073 Pedate: (see II.5.4.3.029). (Fig. II.1.2.16a)
.074 Peltate: fixed to the stalk by the center or at some point distinctly within the margin, like the leaves of garden nasturtium and water lilies. (Fig. II.1.2.15d) (ST)
.075 Perfoliate: when the two basal lobes of an amplexicaul leaf are united so that the stem appears to pass through the leaf. (Fig. II.1.2.05e) (G 62)
.076 Perforated: when irregular spaces are left open in the surface of anything so that it is pierced with holes, as in leaves of some aroids. (Fig. II.1.2.16c) (ST)
.077 Petiolar: inserted on the petiole, like glands. (Fig. II.1.2.16d) (G 62)
.078 Petiole: leaf-stalk; subtended by one or more buds on the axis in the upper angle at the point of attachment. [Fig. II.1.2.17a(1)] (G 62)
.079 Petiolule: petiole of a leaflet. [Fig. II.1.2.17a(1)] (G 62)
.080 Phylloclad: (see Genotypic modifications, II.4.3.036). (Fig. II.1.1.3.06b)
.081 Phyllode: (see Genotypic modifications, II.4.2.037). (Fig. II.1.2.10c)
.082 Phyllome: general term for leaves, scales, bracts, and floral appendages. (ES)
.083 Pinnate: featherlike; leaf with leaflets borne on a rachis, the continuation of the petiole. (Fig. II.1.2.17b) (G 62)
.084 Plane shape and outline terms: (see II.5.5.).
.085 Prophyll: the first cataphylls on a lateral branch; scale leaves. (Fig. II.1.2.18a) (ES)
.086 Pulvinate, cushioned: convex or rather flattened; having a pulvinus. (Fig. II.1.2.18b,c) (G 62)
.087 Pulvinus: cushionlike enlargement (a group of motor cells) of a petiole at the point of insertion on the stem or of a petiolule at its insertion on the rachis; often forms a definite articulation; occasionally found on both ends of a petiole; actually common in many plants, e.g., palms and many legumes; acts to orient the leaf or leaflets in response to a stimulus, usually light. (Fig. II.1.2.18b,c) (ES)

II.1.2. MORPHOLOGY and ANATOMY

.088 Quinate, quinquefoliolate: compound leaf with five leaflets arising from the same point. (Fig.II.1.2.18d) (G 62)
.089 Rachilla (pl. rachillae): secondary unit of a rachis; applied particularly in the Gramineae (Poaceae), although used for secondary and higher orders of branching in compound leaves or complex inflorescences. [Fig. II.1.2.06b(3)] (ES)
.090 Rachis (pl. rachises, rachides): axis bearing leaflets or flowers; the extension of the petiole or peduncle, to which the petiolules or pedicels, respectively, are attached if present. [Fig. II.1.2.06b(2),17b] (G 62)
.091 Resupinate: a leaf inverted in position by twisting the petiole, as in Alstroemeriaceae (see also II.2.2.068). (L 55)
.092 Revolute: rolled backward from the direction ordinarily assumed by other similar bodies, as in certain tendrils, some leaf tips, or occasionally a leaf margin. (Fig. II.1.2.14b) (G 62, ST)
.093 Rosette: leaves radiating from a crown or center and usually at or close to the ground, as in pineapple and many herbaceous perennials. (Fig. II.1.2.19a) (G 62, L 51)
.094 Rosulate: leaves borne in a basal rosette. (Fig. II.1.2.19a) (L 51)
.095 Scattered: (see Sparse, II.1.2.103).
.096 Scythe-shaped: (see Falcate, II.1.2.043).
.097 Secund, one-sided: having all of the parts turned one way because of twists in their stalks, like certain leaves and flowers. (Fig. II.1.2.19b) (G 62, ST)
.098 Semiterete, half-terete: flat on one side, terete on the other, like some petioles. (Fig. II.1.2.19c) (G 62)
.099 Sessile: sitting close on the body that supports it and without a sensible stalk, as in some leaves and flowers. (Fig. II.1.2.19d) (G 62, ST)
.100 Sheathing, sheath: surrounding a stem or other body by the convolute base; any long or more or less tubular structure surrounding an organ or part; the base of a leaf when sheathing (close fitting) or investing a stem or branch, as in many monocots; any sheathlike spathe; an ocrea or sheathing stipule. (Fig. II.1.2.19e, .14d) (G 62)
.101 Sickle-shaped: (see Falcate, II.1.2.043)
.102 Simple: leaf neither compound nor costapalmate. (Fig. II.1.2.03c) (L 51)
.103 Sparse, scattered: used in the opposite (contrary) sense to whorled, opposite, or ternate; e.g., an alternate arrangement with long in-ternodes. (Fig. II.1.2.20a) (G 62, ST)
.104 Spiral, acyclic: arranged in a spiral manner around some common axis, like alternate leaves on a stem. (Fig. II.1.2.20b) (G 62)
.105 Spuriously compound: simple leaves with axillary buds arranged on branches to appear compound; the illusion is extended by the branches which drop as units, as in Phyllanthus; these branches are known as "phyllomorphic" (II.1.1.3.065)
.106 Stalk: the supporting unit of any organ, like a petiole, peduncle, pedicel, filament, or stipe. (L 51)
.107 Stipel: stipule of a leaflet. (Fig. II.1.2.20c) (G 62)
.108 Stipulate: with or bearing stipules. (Fig. II.1.2.20d) (G 62)
.109 Stipule: basal appendages of a leaf; normally two, various in form, often bractlike; may be present or absent in a family. (Fig. II.1.2.20d) (G 62)
.110 Subopposite: opposite leaf or leaflet arrangement in which one is slightly above or below the other of a pair. (Fig. II.1.2.20e) (L 55)

II.1.2. MORPHOLOGY and ANATOMY

.111 Supradecompound: having various decompound divisions or ramifications, like a thrice compound leaf. (II.1.2.06b) (ST)
.112 Terete: cylindrical, centric; the opposite of angular, like a petiole. (Fig. II.1.2.20f) (L 51)
.113 Ternate: when three leaves, sepals, or petals are in a ring around a common axis; (see Trifoliolate, II.1.2.115). (Fig. II.1.2.21a) (L51)
.114 Trifoliate: three-leaved. (L 51)
.115 Trifoliolate: ternately compound leaf; three leaflets that may be arranged palmately or pinnately. (Fig.II.1.2.21a) (G 62)
.116 Trijugate: (see Paired, II.1.2.069).
.117 Triternate: when the common petiole divides into three secondary petioles, each subdivided into three tertiary petioles and each of which bears three leaflets. (Fig. II.1.2.21b) (G 62)
.118 Umbrella-shaped, umbraculiform: resembling an expanded umbrella; hemispherical and convex with rays or plaits proceeding from a common center. (Fig. II.1.2.21c) (ST)
.119 Unifacial leaf: a leaf with dorsiventral symmetry denoted unifacial (one-faced) or equifacial (equal-faced) in contrast to the usual bifacial configuration; several types: unifacial cylindrical, as in _Allium sativum_ and _Juncus effusus_; unifacial flattened, as in _Iris_ (which differs in having two instead of the usual single row or circle of vascular bundles); equifacial flattened (_Lactuca serriola_); equifacial needlelike, as in _Pinus_; and equifacial cylindrical, as in _Sedum album_ (the last two differing in that the needlelike type is flat on one side instead of cylindrical). (Fig. II.1.2.04c-h) (B & C)
.120 Unifoliolate: compound leaf reduced to a single leaf, distinguishable from a simple leaf by the presence of a point of articulation at the junction between the petiole and blade. (Fig. II.1.2.12c)
.121 Venation: arrangement of veins in a leaf or vascular traces in a petal, sepal, or bract; several types with a great many variations and intermediate forms. (Figs. II.1.2.22 - II.1.2.27)
.121a Anastomose: term applied to veins, whose ramifications are united at the points at which they come in contact. (Fig. II.1.2.22a) (G 62)
.121b. Costate, uniribbed: when there is only one rib (midrib) as in most leaves. (Fig. II.1.2.22b) (ST)
.121c. Curve-ribbed: when the ribs describe a curve and meet at the apex. (Fig. II.1.2.22c) (ST)
.121d. Palmate: three or more main veins diverge from the point of petiole attachment. (Fig. II.1.2.22d) (G 62)
.121e. Palmate-pinnate: one main vein or midrib with several to many secondary veins arranged along the midvein. (Fig. II.1.2.23e) (ST)
.121f. Palmiform: when the ribs are arranged as in a palmate leaf. (Fig. II.1.2.23f) (ST)
.121g. Palmiribbed: when the ribs are palmate. (Fig. II.1.2.23g) (G 62)
.121h. Parallel: veins are more or less parallel to the margins. (Fig. II.1.2.24h) (ST)
.121i. Parallel-ribbed, straight-ribbed: when the lateral ribs are straight; same as penniparallel. (Fig. II.1.2.24i) (ST)
.121j. Pedatiribbed: when the ribs are pedate (see II.5.4.3.029). (Fig. II.1.2.24j) (ST)
.121k. Peltiribbed: when the ribs are peltate (see II.1.2.079). (Fig. II.1.2.24k) (ST)

II.1.2. MORPHOLOGY and ANATOMY 65

.121l. Penniform: when the ribs are disposed as in a pinnate leaf but are confluent at the point. (Fig. II.1.2.24l) (ST)
.121m. Penniparallel (penninerved): parallel veins arise at an angle (often perpendicular or nearly so) from a midvein (midrib), as in Musa. (Fig. II.1.2.25m) (L 51)
.121n. Penniribbed: when the ribs are pinnate. (Fig. II.1.2.25n) (ST)
.121o. Pinnate: (see II.1.2.083). (Fig. II.1.2.25o)
.121p. Reticulated: when the veins are like lace in a network. (Fig. II.1.2.26p) (ST)
.121q. Ribbed: having several ribs. (Fig. II.1.2.26q) (ST)
.121r. Straight-ribbed: (see Parallel-ribbed, II.1.2.121i).
.121s. Triple-ribbed: when two lateral ribs emerge from a middle one a little above its base. (Fig. II.1.2.26s) (ST)
.121t. Triribbed: when there are three ribs, all proceeding from the base. (Fig. II.1.2.26t) (ST)
.121u. Uniribbed: (see Costate, II.1.2.121b).
.121v. Venose: when the lateral veins are variously divided; when the veins are prominent on the leaf surface. (Fig. II.1.2.27v) (ST)
.122 Vertebrate: when a leaf is contracted at intervals with an articulation at each contraction. (Fig. II.1.2.28a) (ST)
.123 Verticillate: leaves arranged in whorls, or seemingly but falsely so, as in Labiatae and many other families. (Fig. II.1.2.28b) (G 62)
.124 Whorl: arrangement of three or more leaves at a node = verticillate. (Fig. II.1.2.28c) (G 62)
.125 Winged: (see Alate, II.1.2.004).

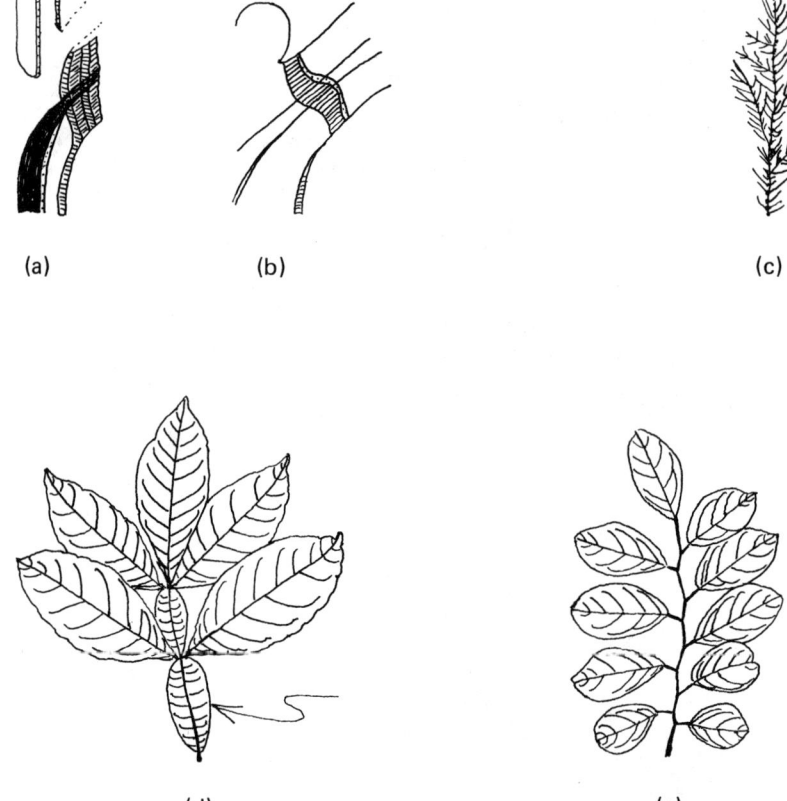

Fig. II.1.2.01. Leaves: (a) abscission (Castanea second-year stage); (b) abscission (Catalpa); (c) acerose (Juniperus silicicola); (d) alate (Citropsis gilletiana); (e) alternate (leaflets of Murraya paniculata). (a), (b) From Eames and MacDaniels, 1947; see Acknowledgments and Selected References.

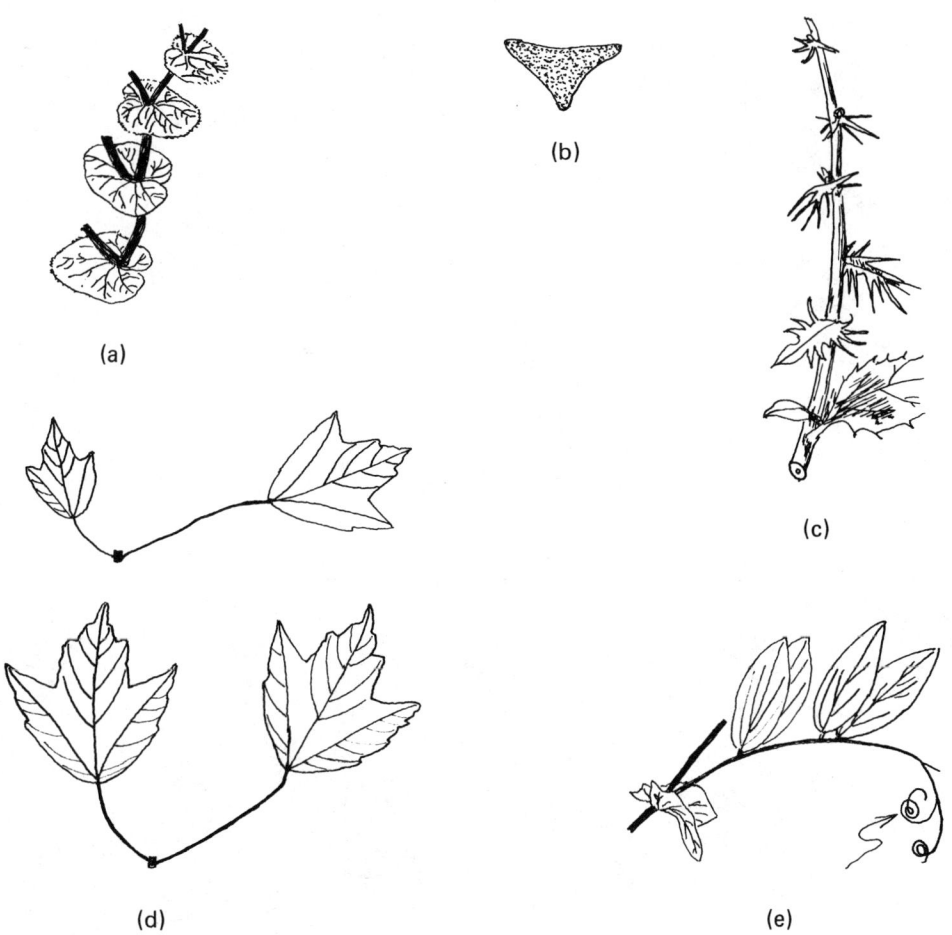

Fig. II.1.2.02. Leaves: (a) amplexicaul (candy alexander); (b) angular (Cyperus stem t.s.); (c) anisophyllous (barberry); (d) anisophylly (Acer rubrum); (e) appendage (tendril of Lathyrus leaf). (a) From Johnson, 1633; (c), (e) from Gray, 1862; see Acknowledgments and Selected References.

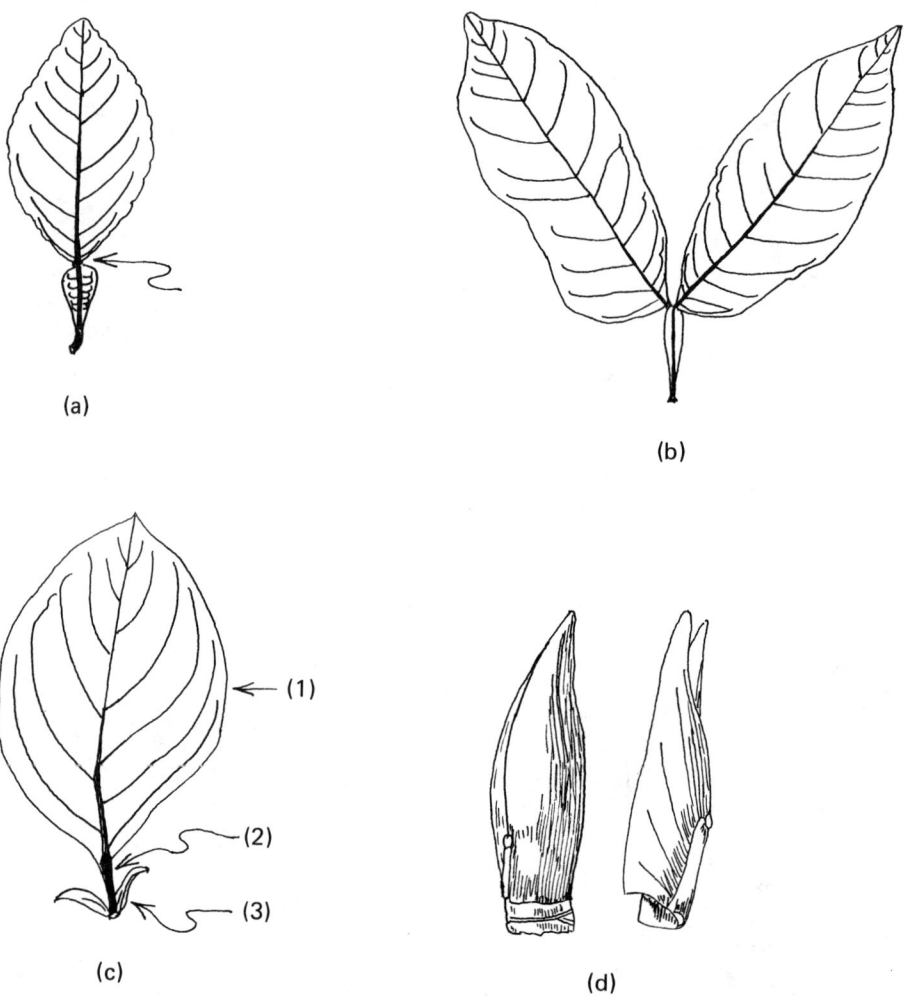

Fig. II.1.2.03. Leaves: (a) articulate, articulated (Citrus sinensis leaf) with joint (arrow) between blade and petiole; (b) bifoliolate (from Melicocca bijuga leaf); (c) simple leaf (quince) with (1) blade, (2) petiole, and (3) stipules; (d) calyptra (Magnolia sp.). (c), (d) From Gray, 1862; see Acknowledgments and Selected References.

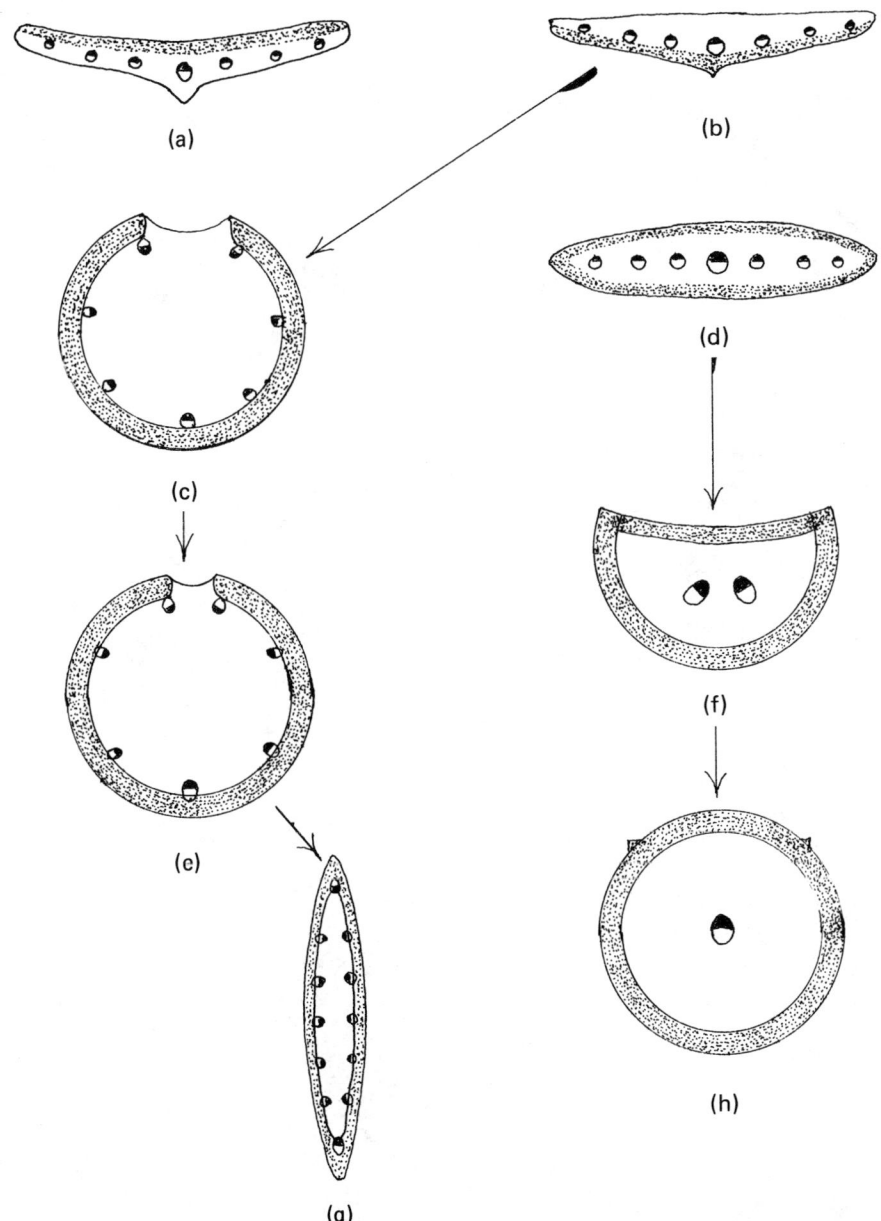

Fig. II.1.2.04. Leaves (transverse sections): (a) bifacial (normal form); (b) inverted bifacial (Allium ursinum); (c) unifacial cylindrical (Allium sativum); (d) equifacial flattened (Lactuca serriola); (e) unifacial cylindrical (Juncus effusus); (f) equifacial needlelike (Pinus); (g) unifacial flattened (Iris); (h) equifacial cylindrical (palisade parenchyma stippled, xylem black). Redrawn from Bell and Coombe, 1976; see Acknowledgments and Selected References.

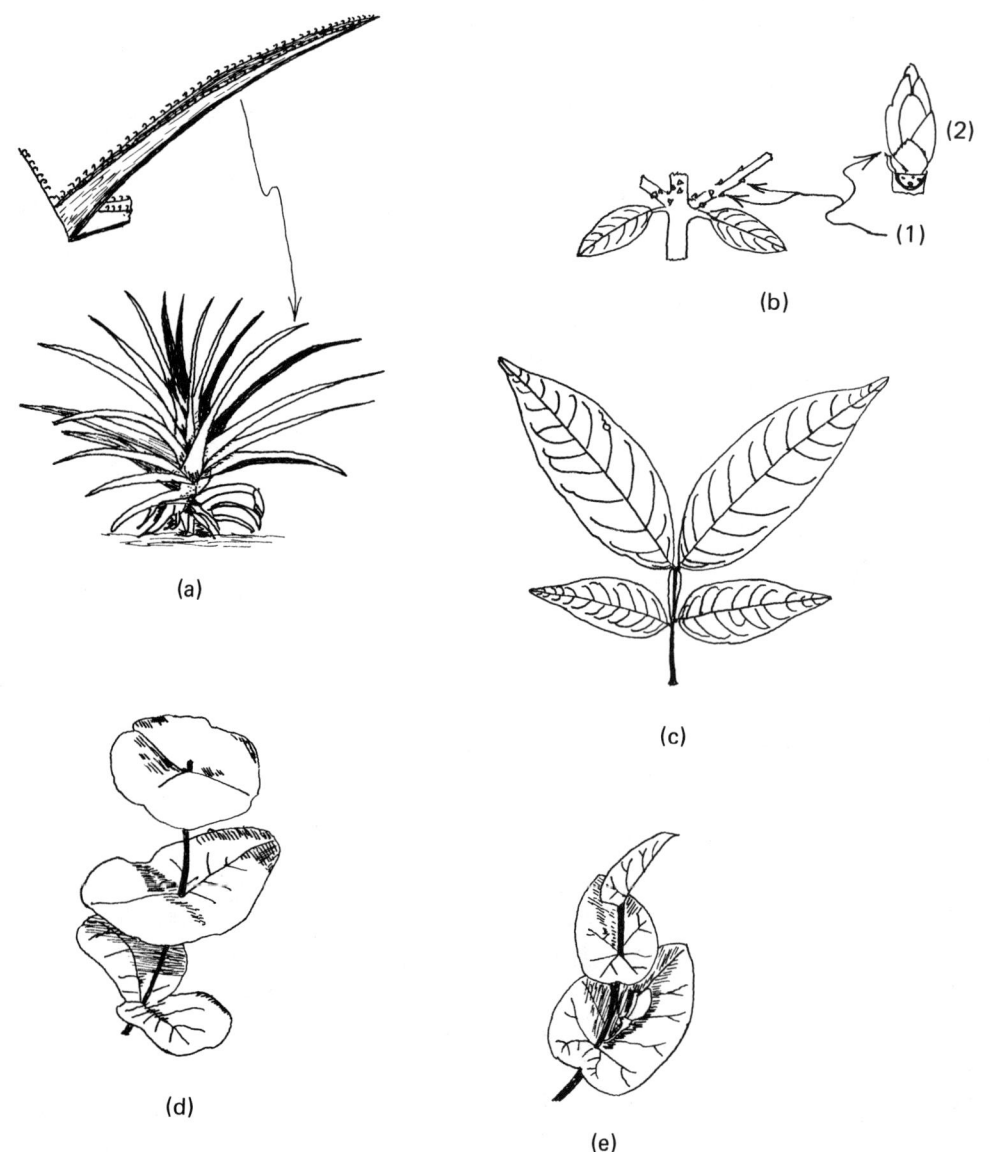

Fig. II.1.2.05. Leaves: (a) canaliculate (pineapple leaf); (b) cataphyll as (1) scale leaves and (2) bud scales; (c) compound (Spanish lime); (d) connate-perfoliate (honeysuckle); (e) perfoliate (Baptisia perfoliata). (a) Pineapple plant redrawn from Py and Tisseau, 1965; (b)(1) redrawn from Tomlinson and Gill, 1973; (b)(2), (d), (e) redrawn from Gray, 1862; see Acknowledgments and Selected References.

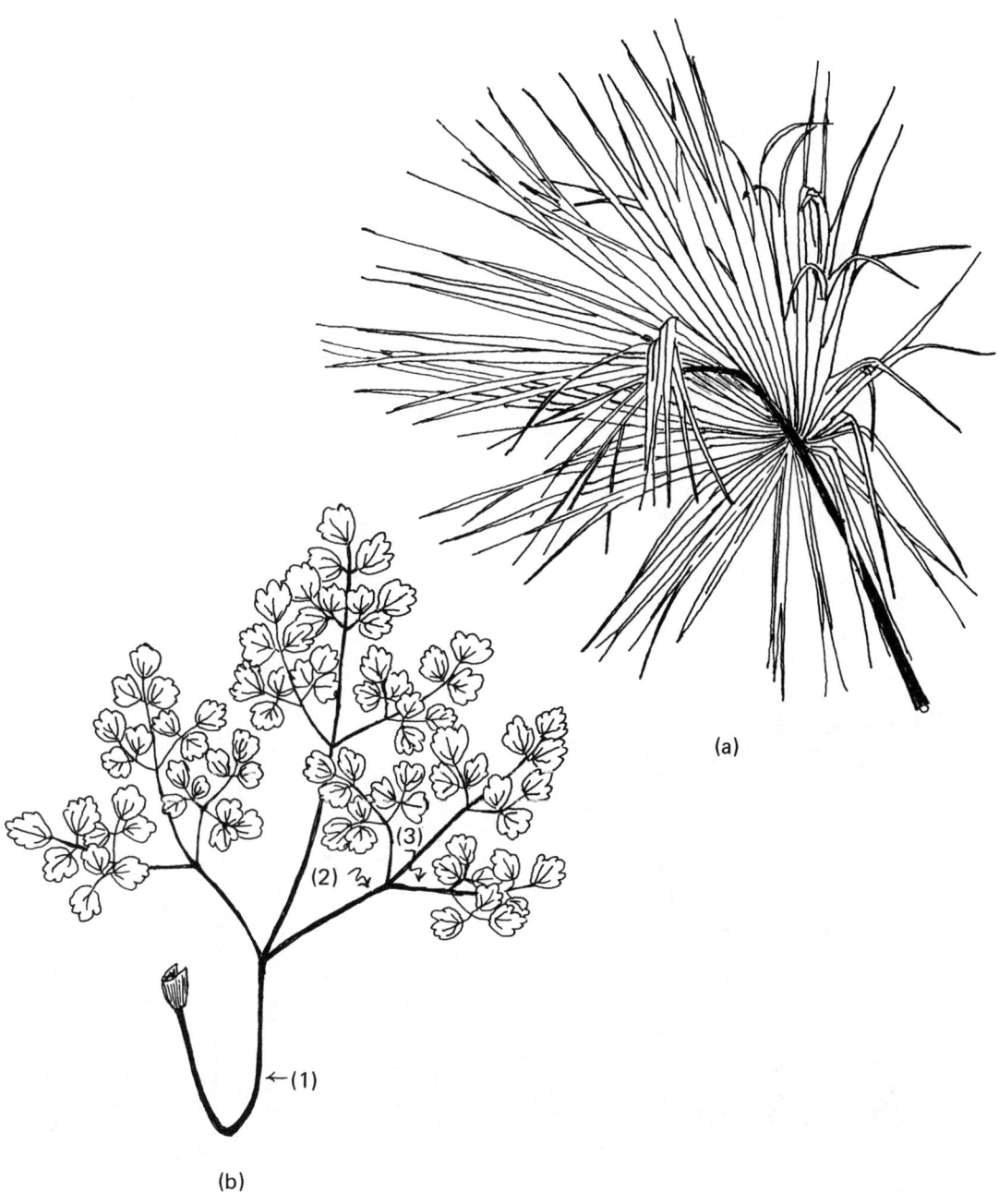

Fig. II.1.2.06. Leaves: (a) costapalmate leaf of Sabal palmetto; (b) supradecompound leaf of Thalictrum cornuti with (1) petiole, (2) rachis, and (3) rachilla as successive subdivisions of the axis. (a) Redrawn from Bailey, 1944b; (b) redrawn from Gray, 1868; see Acknowledgments and Selected References.

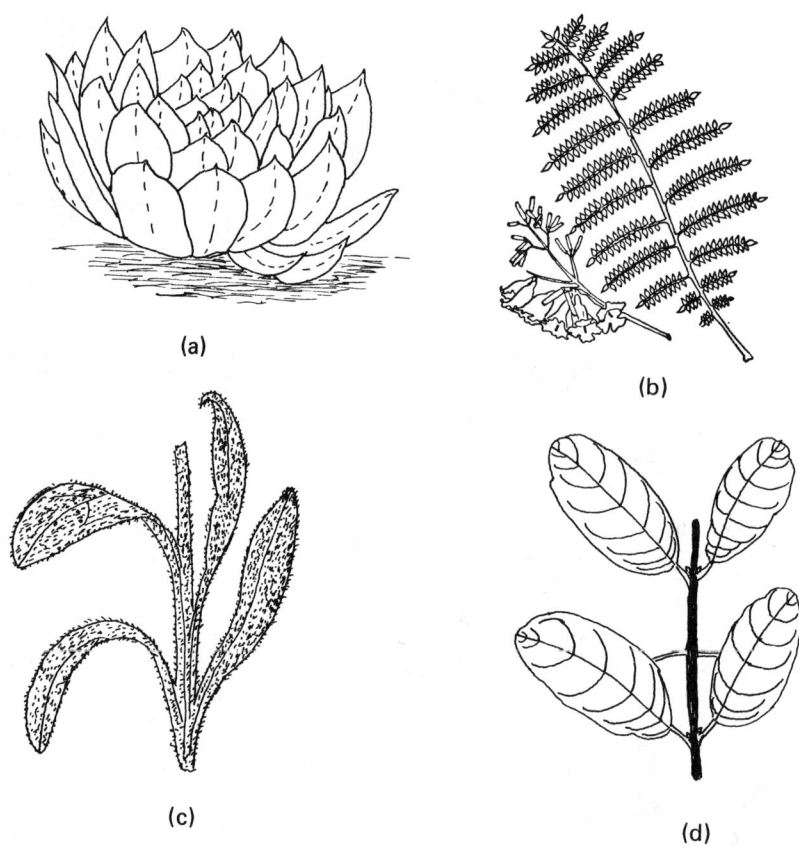

Fig: II.1.2.07. Leaves: (a) crowded leaves of Sempervivum; (b) decompound leaves of Jacaranda acutifolia; (c) decurrent leaves of mullein; (d) decussate leaves of feijoa. (a) Redrawn from Bailey, 1928; (b) from Watkins and Sheehan, 1975; (c) adapted from Muenscher, 1955; see Acknowledgments and Selected References.

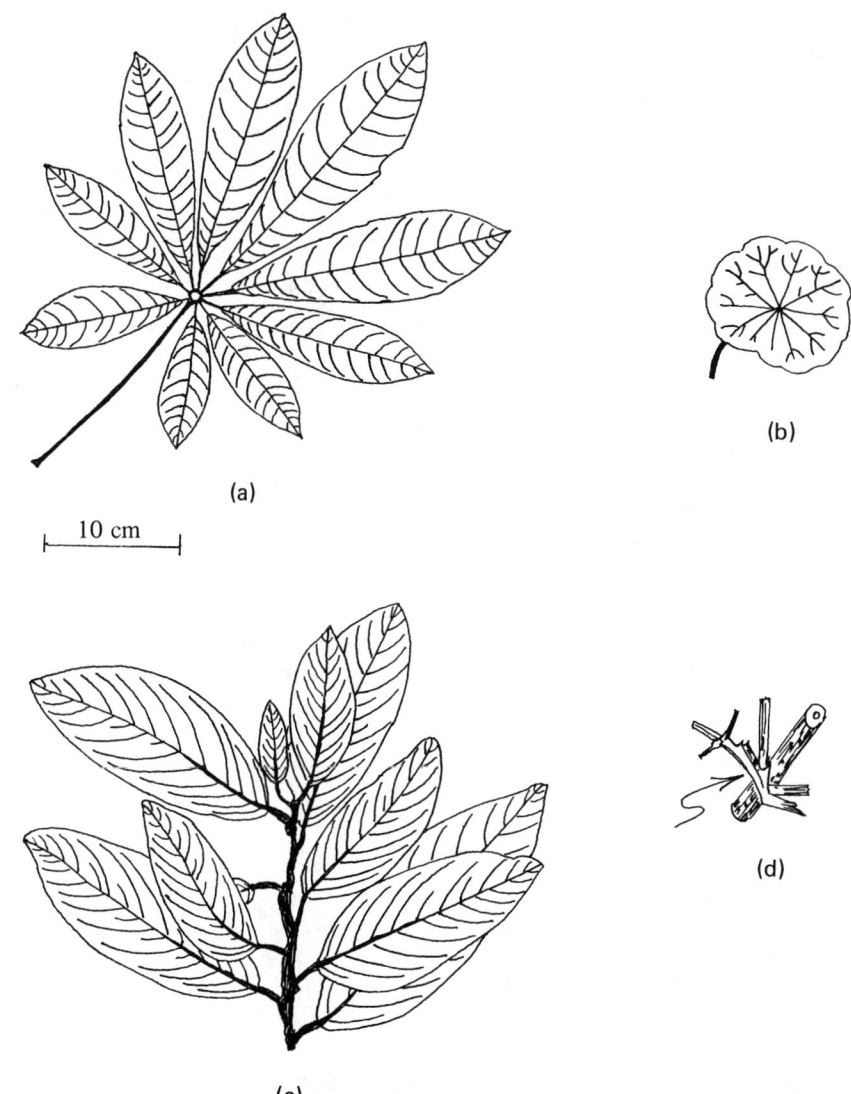

Fig. II.1.2.08. Leaves: (a) digitate leaf of Pachira aquatica; (b) disciform leaf of nasturtium; (c) distichous leaves of sugar-apple (Annona squamosa); (d) embracing leaf bases of poison hemlock (Conium maculatum). (b), (d) From Gray 1862; (c) from Little and Wadsworth, 1964; see Acknowledgments and Selected References.

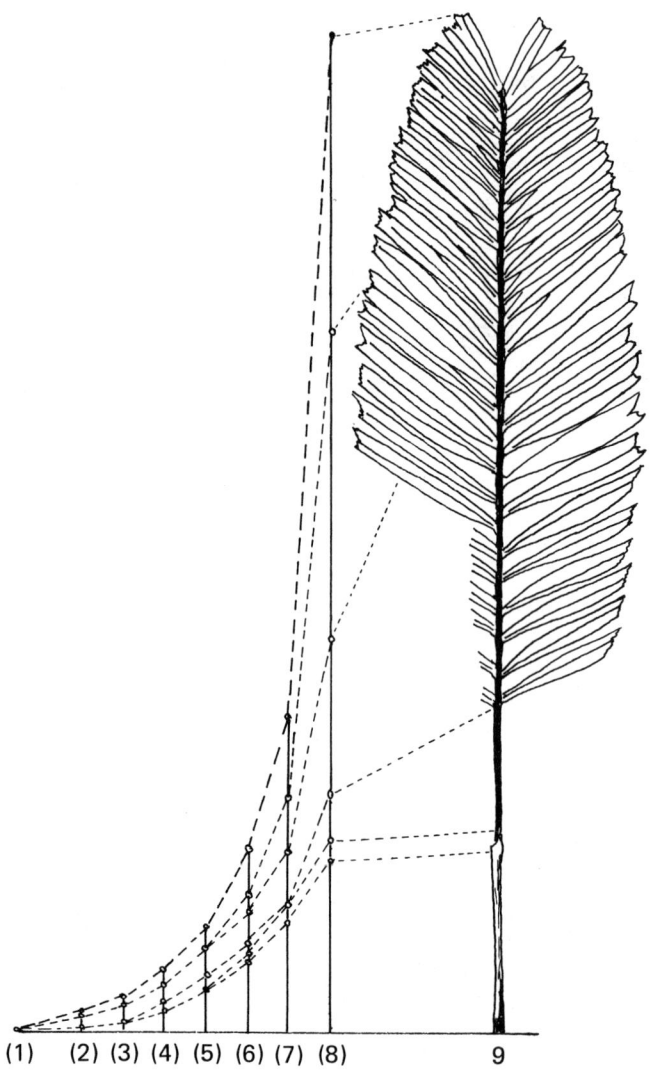

Fig. II.1.2.09. Leaves: extension growth of a Ptchychosperma palm leaf where (1) represents the first leaf bud, (2) to (8) successively older leaves within the crown, (8) the spear leaf (unexpanded and the first one externally visible), and (9) the newest mature, fully expanded leaf; the seven additional ones that make up the full complement of 16 leaves are not shown. Redrawn from Corner, 1966; see Acknowledgments and Selected References.

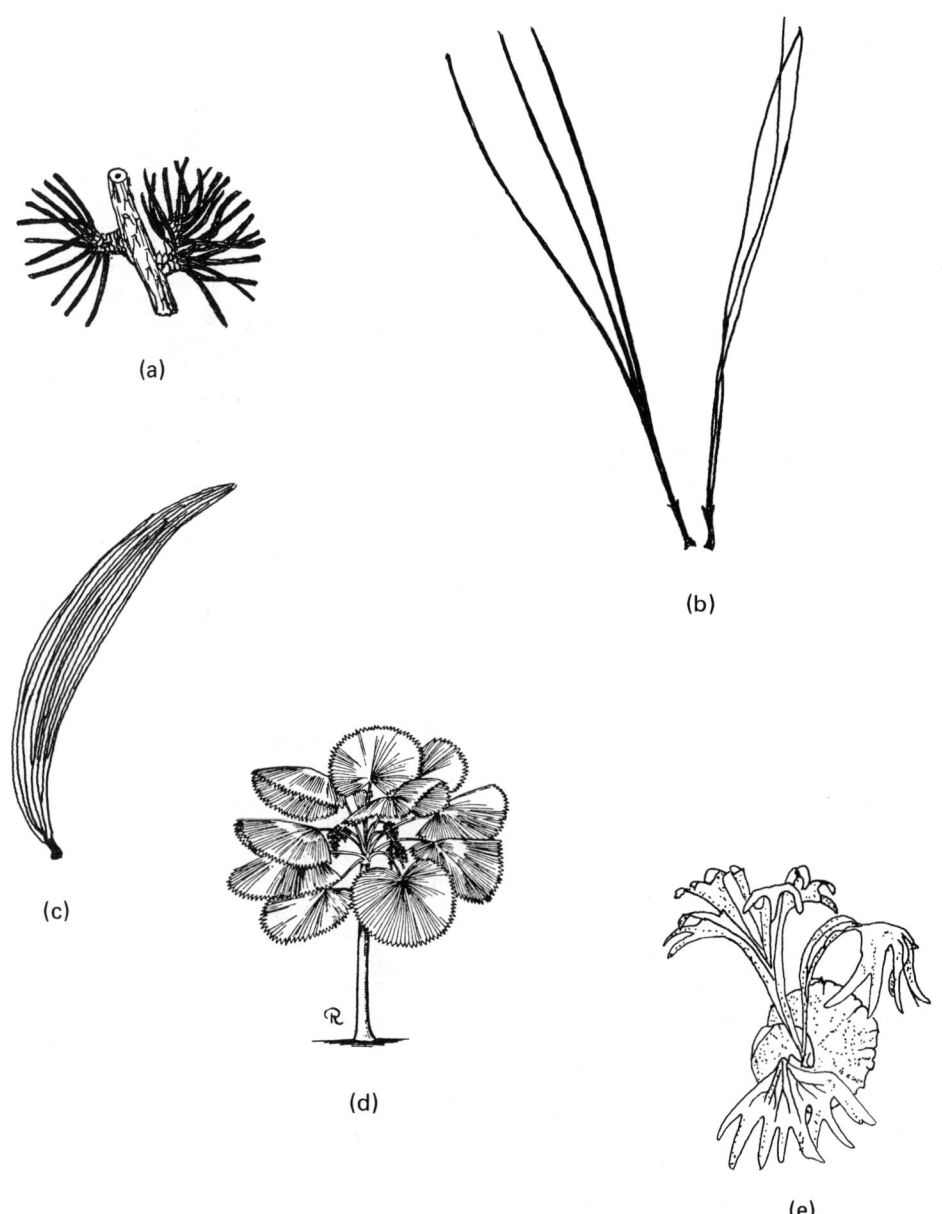

Fig. II.1.2.10. Leaves: (a) fascicled larch leaves; (b) fasciculate loblolly pine (Pinus taeda) leaves; (c) falcate Acacia phyllode; (d) flabelliform leaves of Licuala grandis; (e) furcate leaves of staghorn fern. (a) Redrawn from Gray, 1862; (c) redrawn from Bailey, 1928; (d) from Watkins and Sheehan, 1975; (e) adapted from Watkins and Sheehan, 1975; see Acknowledgments and Selected References.

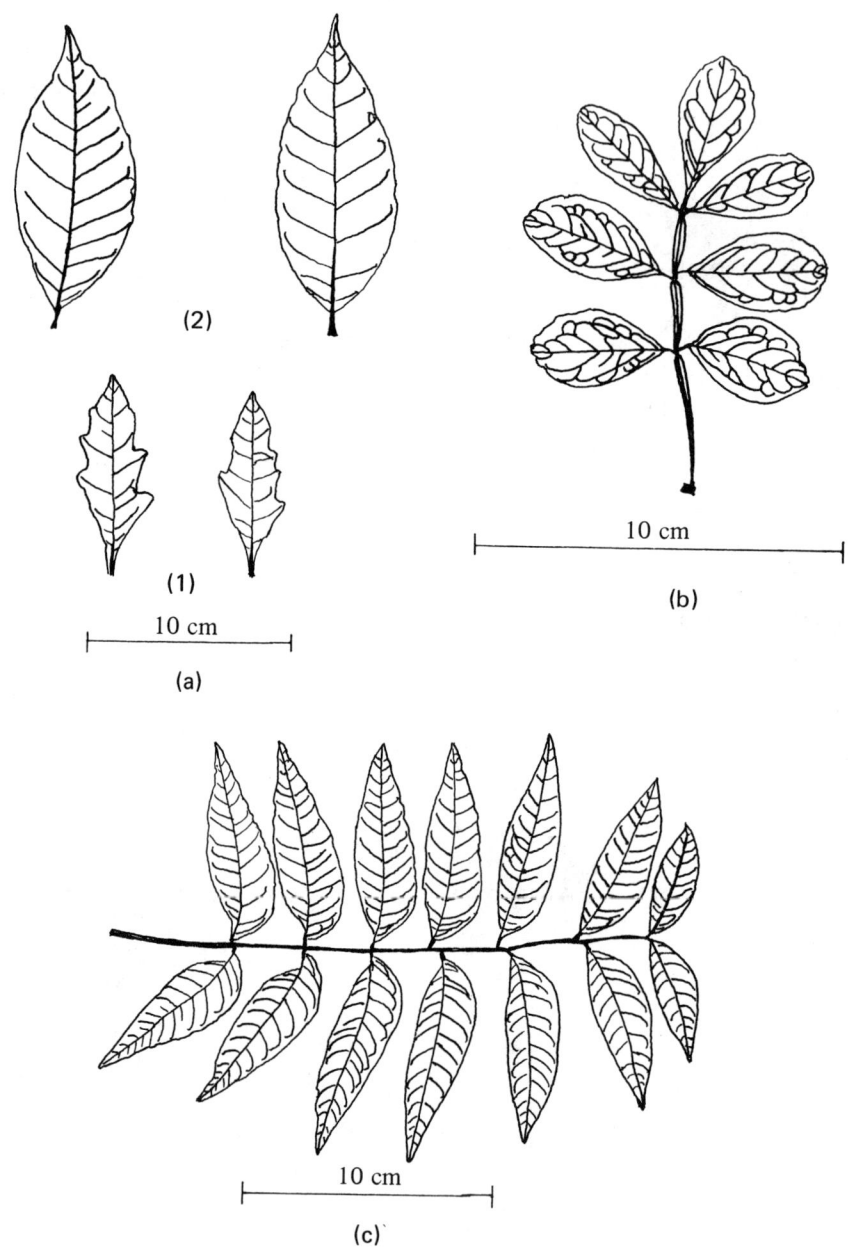

Fig. II.1.2.11. Leaves: (a) heterophylly exhibted by (1) juvenile and (2) mature Clymenia polyandra leaves; (b) imparipinnate Indian wood-apple (Feronia limonia) leaf with opposite leaflets; (c) spuriously paripinnate leaf of Chinese pistachio (Pistacia chinensis), a common occurrence, although the leaf is supposed to be imparipinnate.

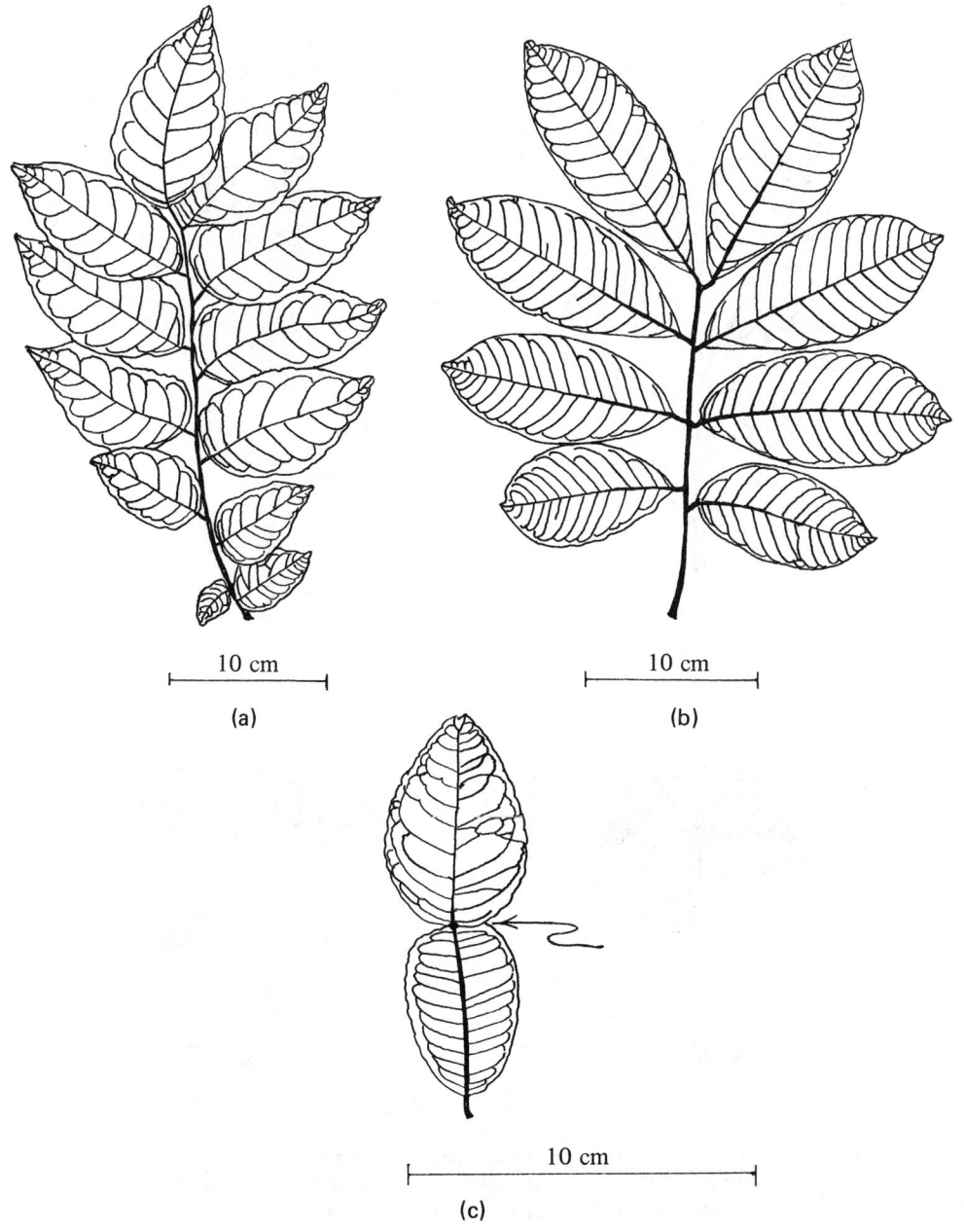

Fig. II.1.2.12. Leaves: (a) imparipinnate wampee (Clausena lansium) leaf with alternate leaflets; (b) paripinnate akee (Blighia sapida) leaf; (c) unifoliolate Mauritius papeda (Citrus hystrix) leaf, with articulation at the junction between the petiole and blade, a compound leaf reduced to one leaflet.

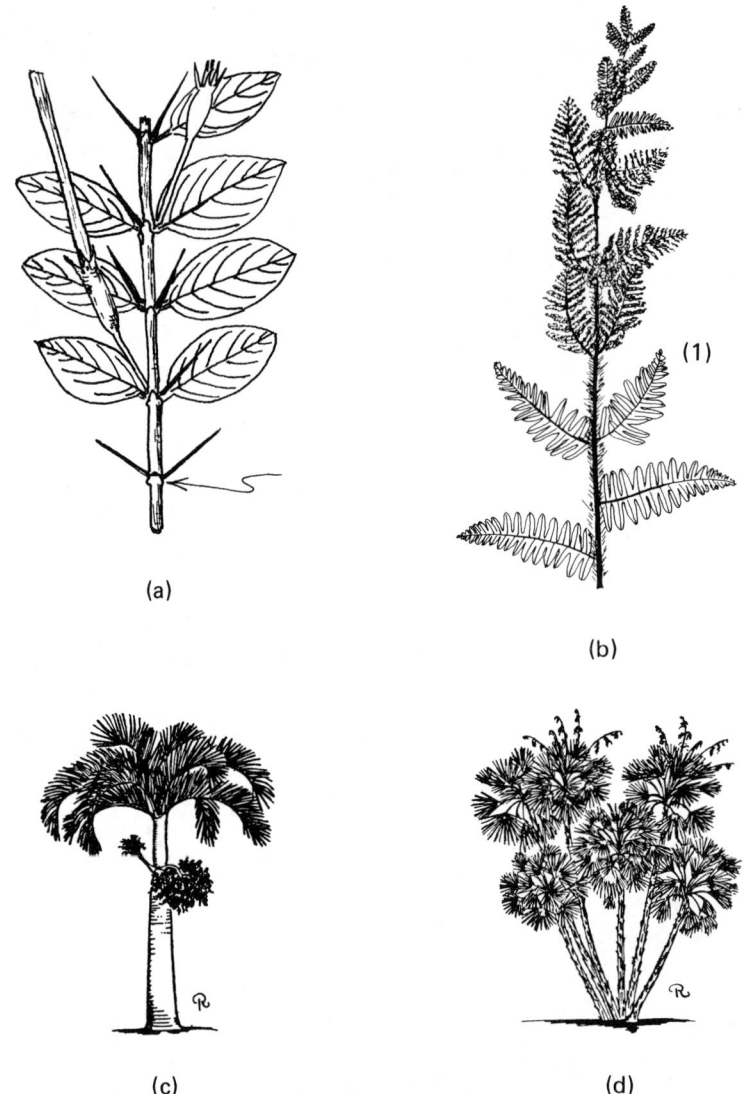

Fig. II.1.2.13. Leaves: (a) interpetiolar stipules on Catesbaea spinosa; (b) interrupted leaflets, the center portion (with an enlarged view of a single upper leaflet at left) bearing spores, with normal leaflets above and below on Osmunda claytoniana; (c) infrafoliar inflorescences borne in the axils of fallen leaves on Manila palm (Veitchia merrillii); (d) intrafoliar inflorescences borne in the axils of present leaves on paurotis palm (Paurotis wrightii). (a) Redrawn from Hutchinson, 1969; (b) redrawn from Schneider, 1894; (c), (d) from Watkins and Sheehan, 1975; see Acknowledgments and Selected References.

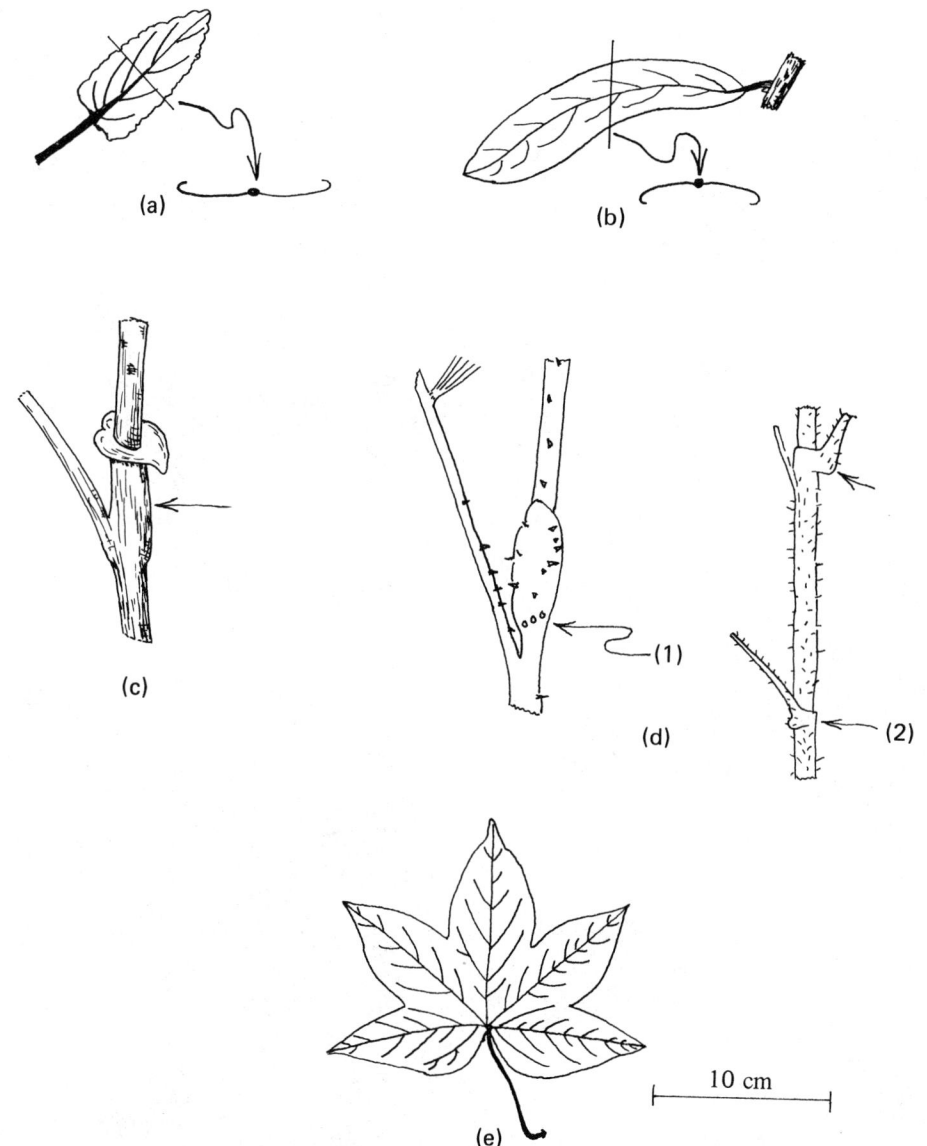

Fig. II.1.2.14. Leaves: (a) involute leaf of violet; (b) revolute leaf of live oak (Quercus virginiana); (c), (d) ocrea (ochrea); (c) is the usual form in some Polygonaceae and (d) in palms, which may harbor ants (1) or be geniculate (2); (e) lobed leaf of sweet gum (Liquidambar styraciflua). (a), (c) Redrawn from Gray, 1862; (d) redrawn from Corner, 1966; see Acknowledgments and Selected References.

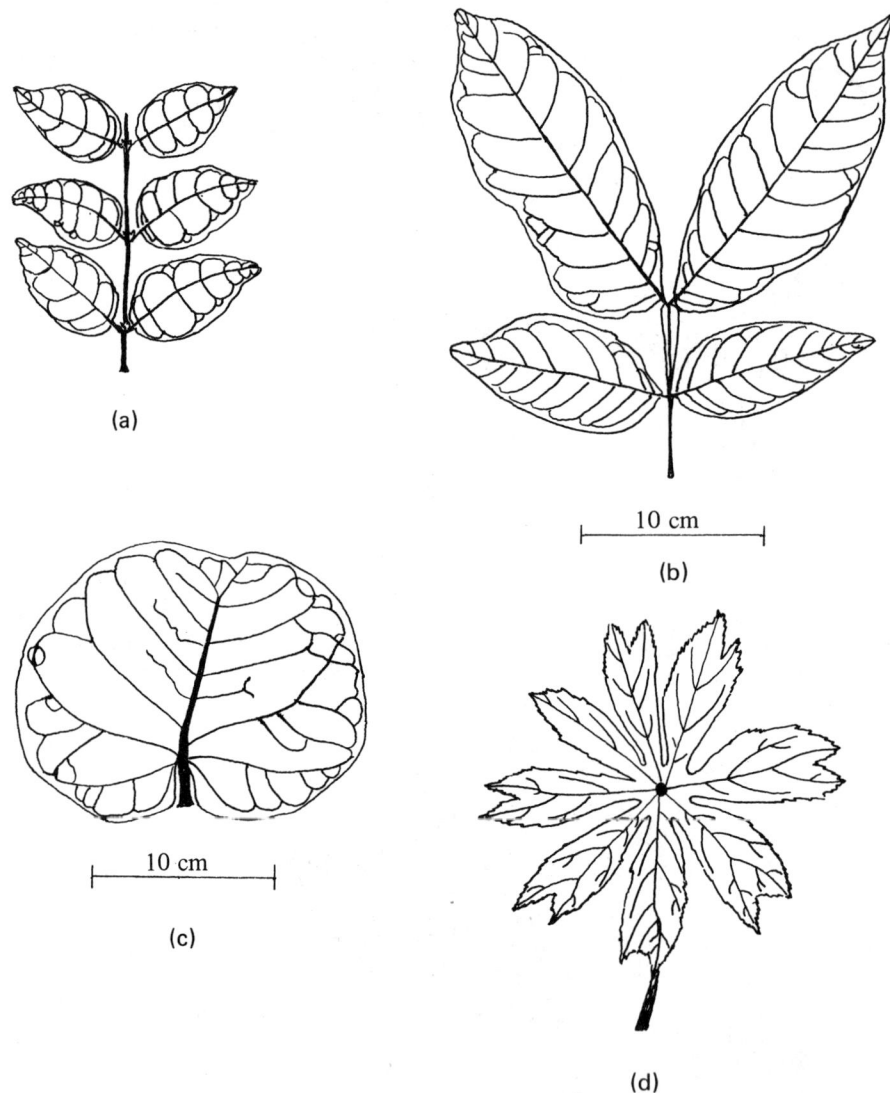

Fig. II.1.2.15. Leaves: (a) opposite leaves of Surinam cherry (Eugenia uniflora); (b) paired leaflets (bijugate leaf) of mamoncillo (Melicocca bijuga); (c) palaceous leaf of seagrape (Coccoloba uvifera); (d) peltate leaf of Podophyllum peltatum. (d) Redrawn from Hutchinson, 1969; see Acknowledgments and Selected References.

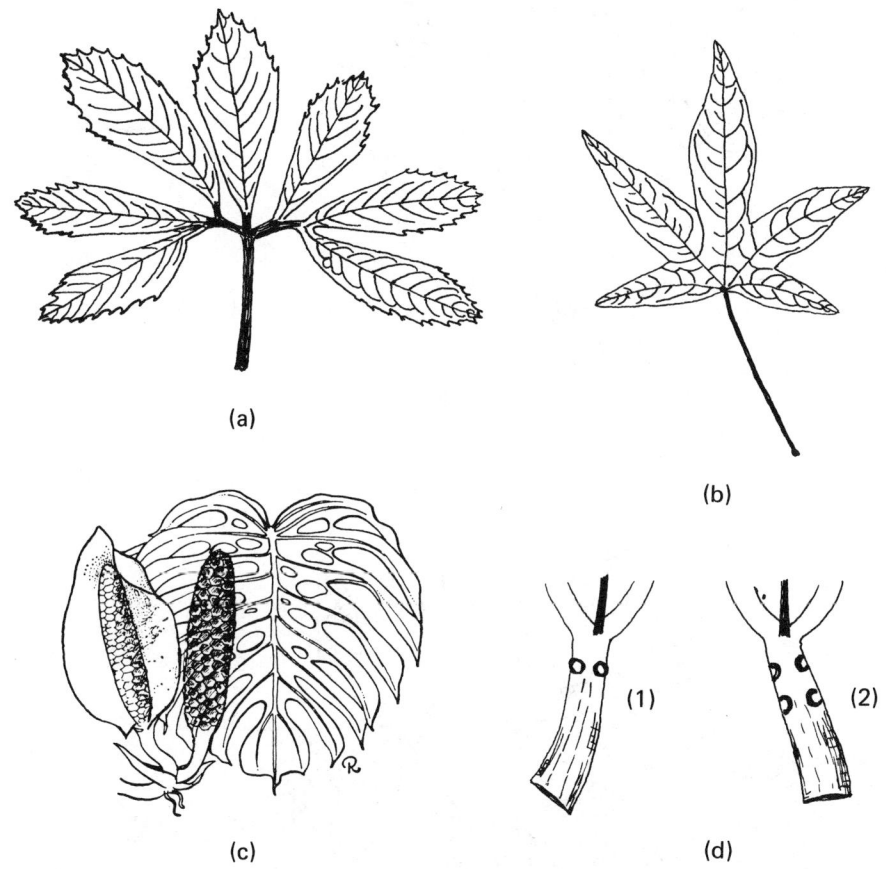

Fig. II.1.2.16. Leaves: (a) pedate leaf of Helleborous niger; (b) palmate leaf of sweet gum (Liquidambar styraciflua); (c) perforated leaf of Monstera deliciosa: (d) petiolar globose (1) and reniform (2) glands on peach (Prunus persica) leaves. (a) Redrawn from Hutchinson, 1969; (c) from Watkins and Sheehan, 1975; see Acknowledgments and Selected References.

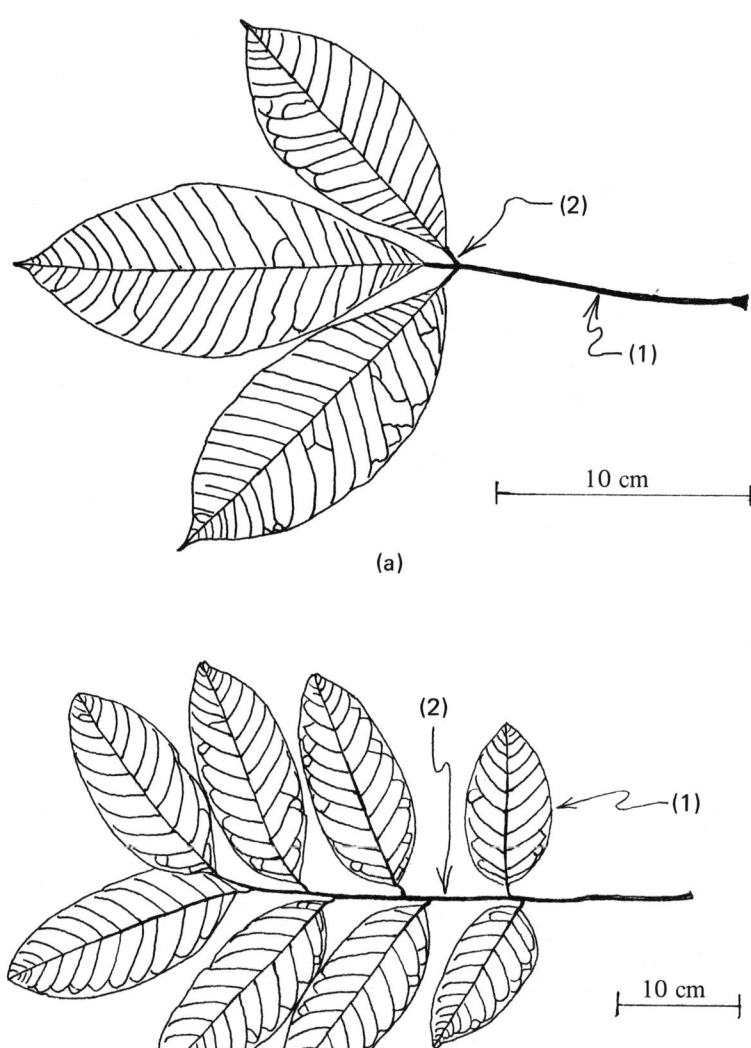

Fig. II.1.2.17. Leaves: (a) petiole (1) and petiolule (2) of a digitate (trifoliolate) rubber (<u>Hevea brasiliensis</u>) leaf; (b) a pinnate leaf, rambutan (<u>Nephelium lappaceum</u>), with leaflets (1) arranged along a rachis (2).

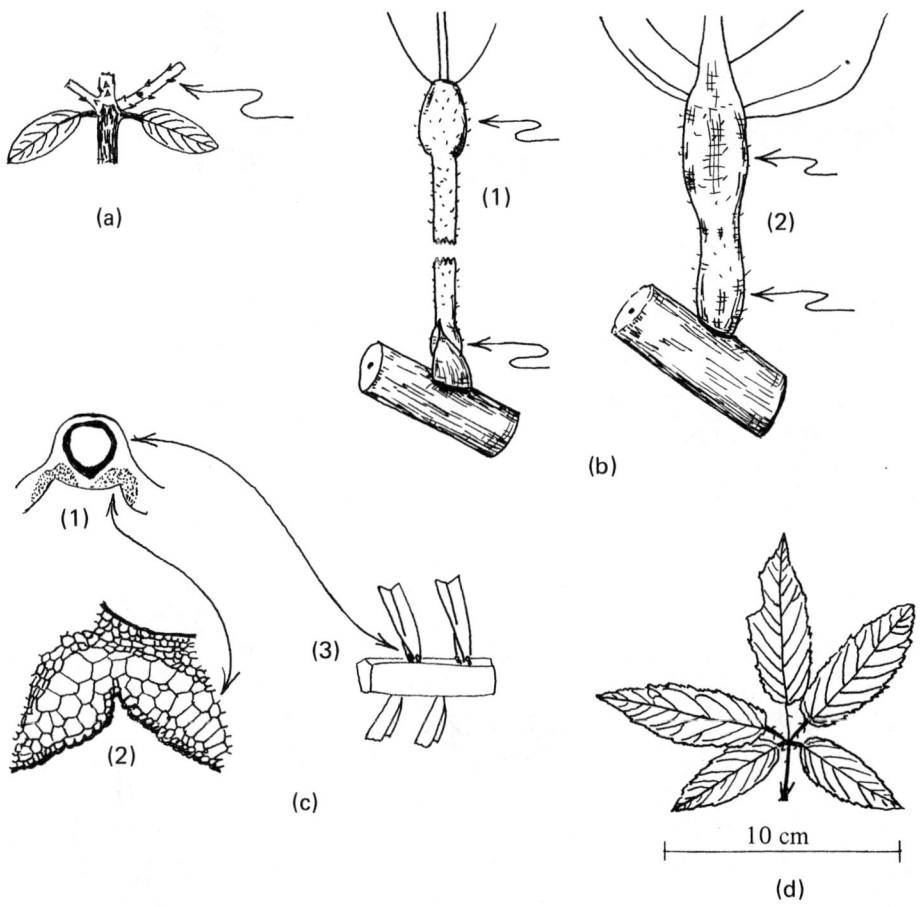

Fig. II.1.2.18. Leaves: (a) prophylls along the base of an avocado (Persea americana) branch; (b) pulvini at ends of the long petiole (i.e., bipulvinate) typical of a leaf borne on an orthotropic seedling stem (1) and those on the short petiole typical of a fan leaf borne on a plagiotropic branch (2) of cacao (Theobroma cacao); (c) pulvini on a coconut (Cocos nucifera) leaf midrib shown in cross section (1) with the pulvini (stippled areas) enlarged (2) and their position at the base of each leaflet (3); the purpose is to set the angle at which the latter is exposed to light; (d) a quinate leaf typified by dewberry (Rubus trivialis). (a) Redrawn from Tomlinson and Gill, 1973; (b) adapted from Cuatrecasas, 1964; (c) redrawn from Corner, 1966; see Acknowledgments and Selected References.

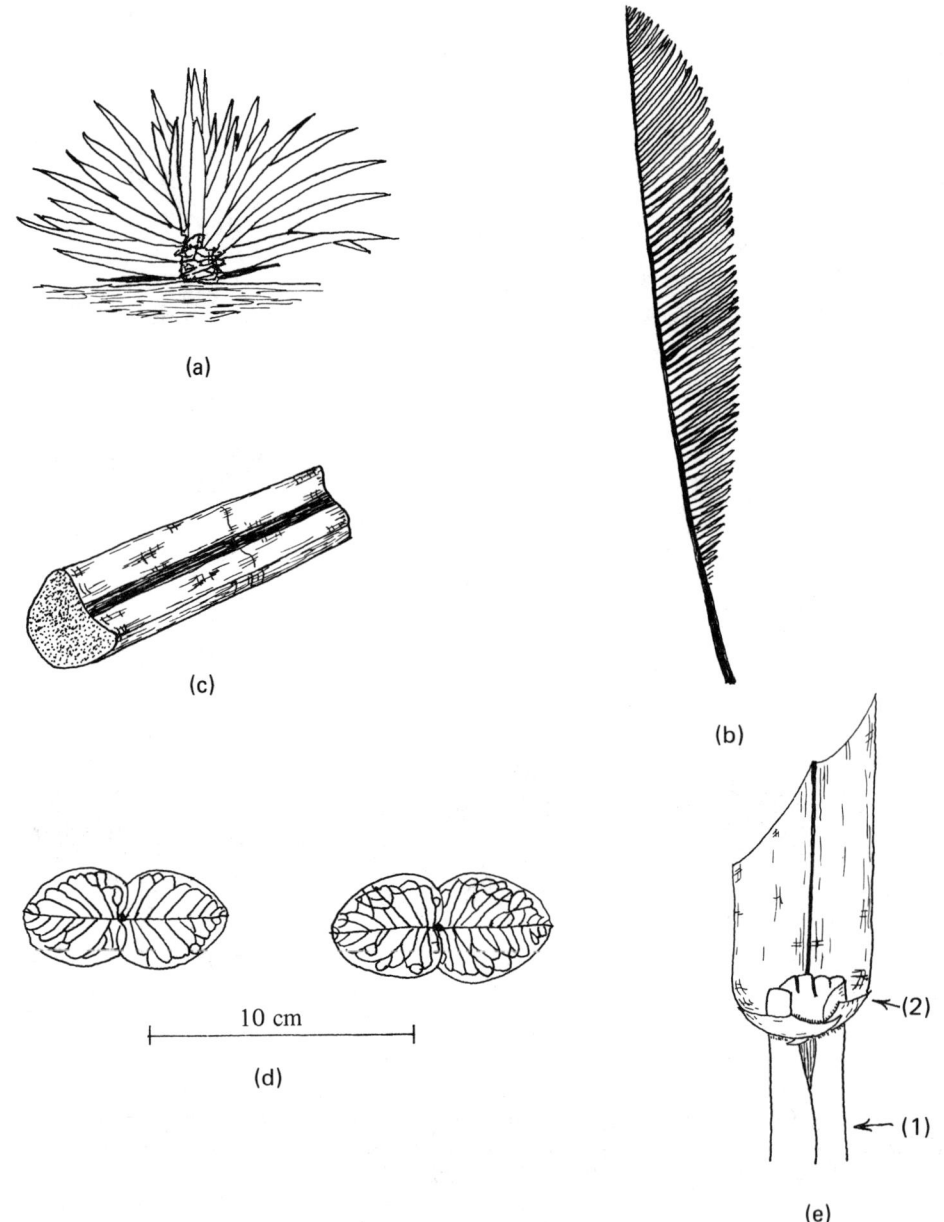

Fig. II.1.2.19. Leaves: (a) pineapple (Ananas comosus) leaves arranged in a rosette (i.e., rosulate); (b) secund arrangement as in a young nipa (Nypa = Nipa fruticans) leaf; (c) semiterete petiole typical of many palms; (d) sessile (and opposite) leaves of Eucalyptus cinerea; (e) sheathing petiole (1), with ligule (2) at the top, found in many grasses. (a) Redrawn from Py and Tisseau, 1965; (b) redrawn form Hallé et al., 1978; (e) redrawn from Robbins, 1924; see Acknowledgments and Selected References.

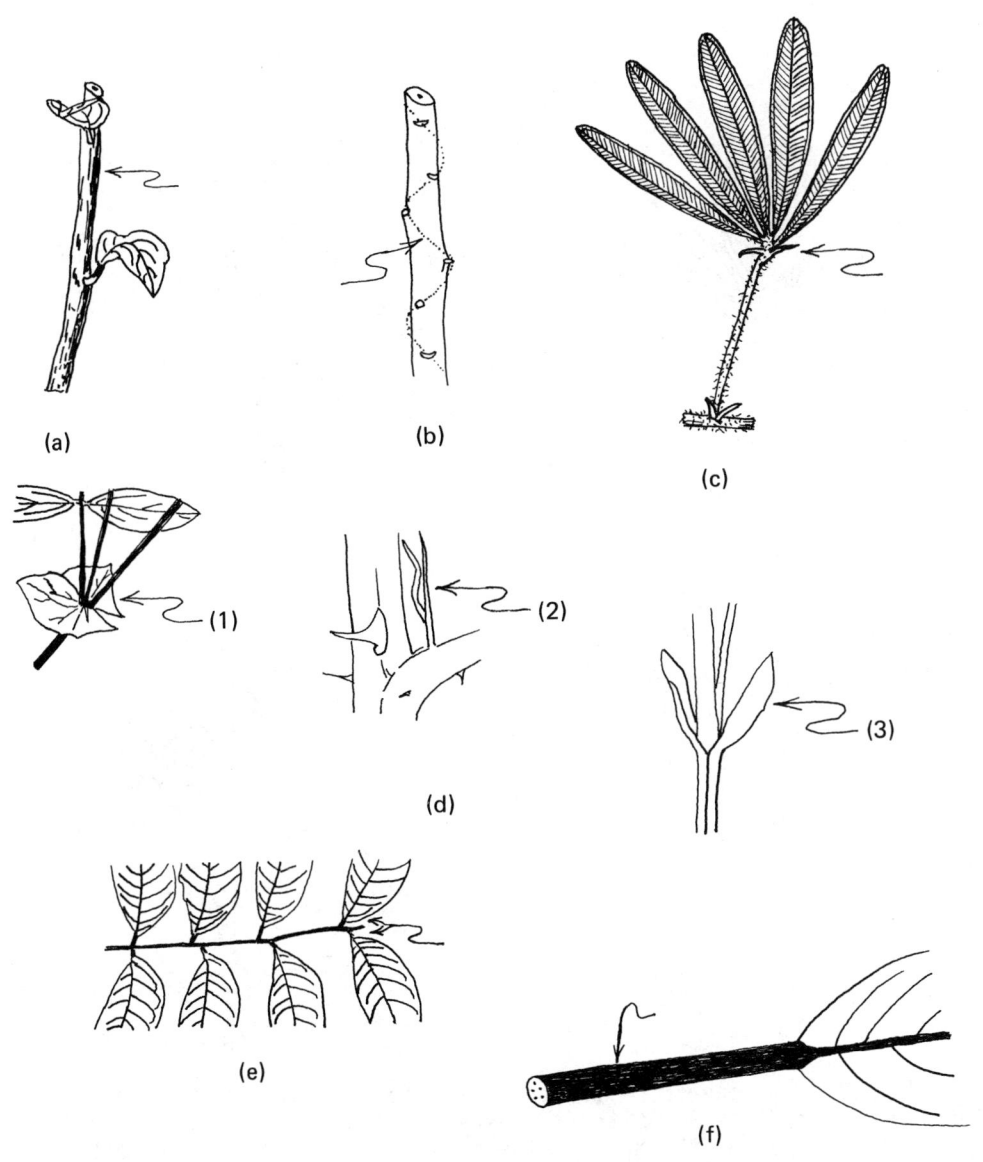

Fig. II.1.2.20. Leaves: (a) sparse or scattered leaves on a stem; (b) spiral position of leaves around a stem (may be right-handed, as shown, or left-handed); (c) stipels (stipules subtending leaflets) here in Lupiniphyllum lupinifolium subtending only the outer pair of leaflets of a digitate leaf; (d) stipules at the base of the petiole in Lathyrus (1) or on the petiole (2) or axis (3) in Rubus ; (e) subopposite leaflets in Chinese pistachio (note that the lowest pair is opposite); (f) a terete petiole (or petiolule). (a), (b), (d)(1) Redrawn from Gray, 1862; (c) adapted from Hutchinson, 1969; (d)(2), (d)(3) redrawn from Bailey, 1941; see Acknowledgments and Selected References.

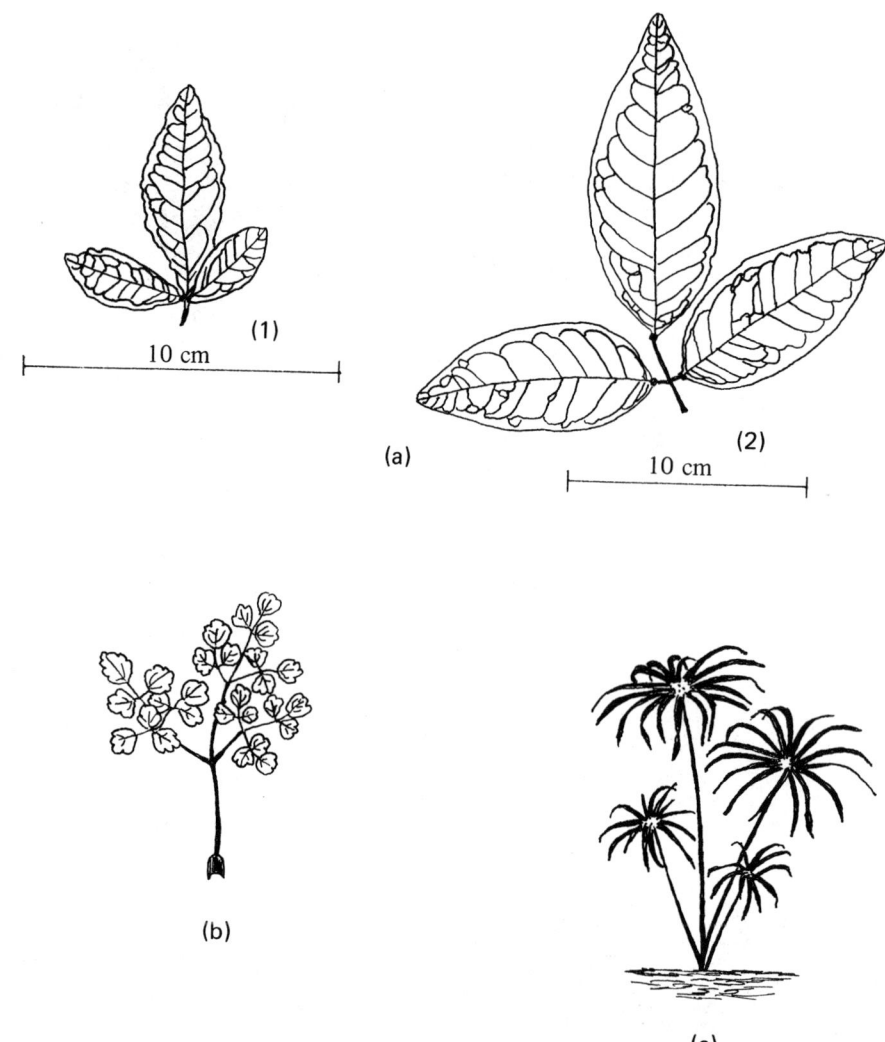

Fig. II.1.2.21. Leaves: (a) trifoliolate leaves, (1) 'Morton' citrange (Poncirus trifoliate X Citrus sinensis) is digitate where the points of articulation are at the apex of the petiole and (2) Glycosmis citrifolia, pinnate, where the terminal leaflet is attached to a rachis with the point of articulation at the base of the blade rather than at the base of the petiole; (b) a triternate (decompound) leaf whose petiole is divided into three rachises, each of which has three rachillae subdivided into three subrachillae and each of the latter has three leaflets; (c) umbrella-shaped leaves, typified by Cyperus alternifolius. (b) Adapted from Gray, 1862; (c) adapted from Watkins and Sheehan, 1975; see Acknowledgments and Selected References.

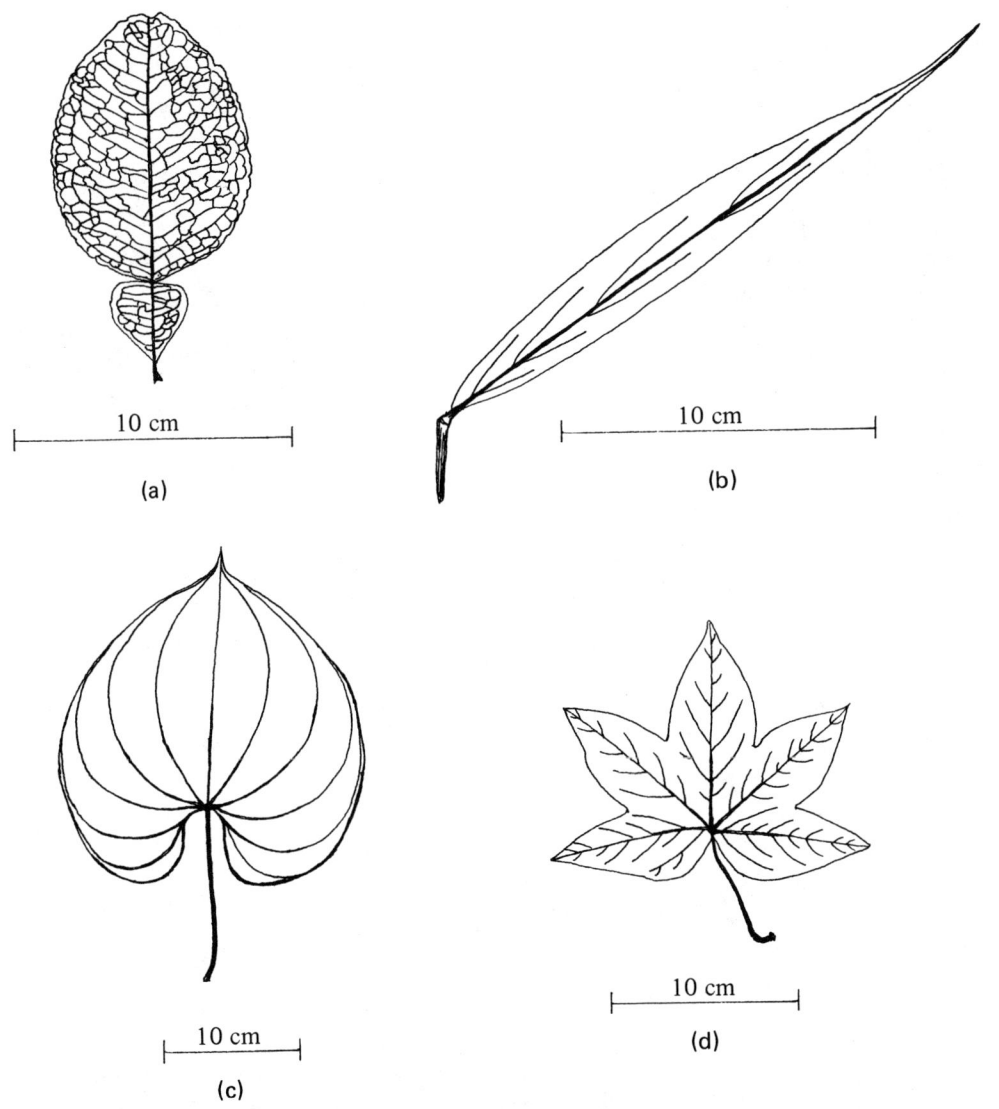

Fig. II.1.2.22. Venation: (a) anastomose (Citrus grandis); (b) costate (Zingiber officinale); (c) curve-ribbed (Dioscorea bulbifera); (d) palmate (Liquidambar styraciflua).

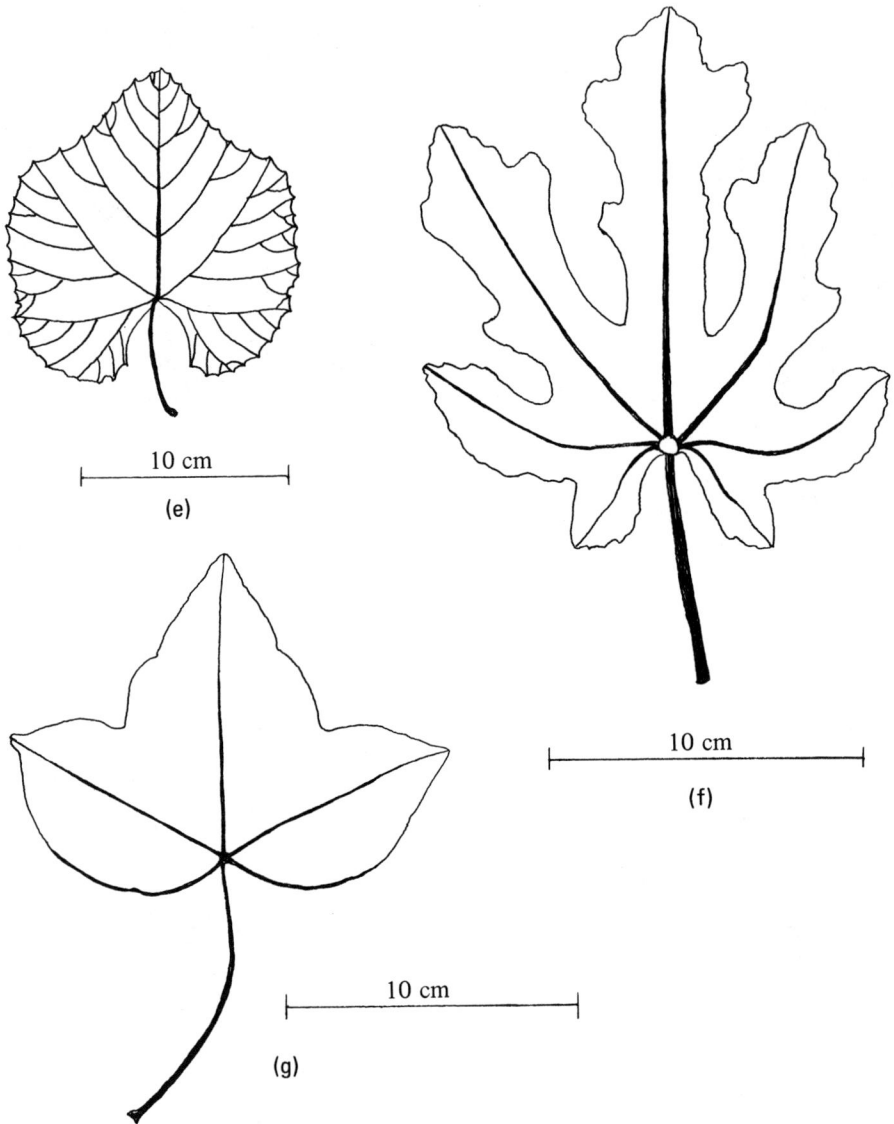

Fig. II.1.2.23 Venation: (e) palmate-pinnate (<u>Vitis</u> <u>simpsonii</u>); (f) palmiform (<u>Ficus</u> <u>carica</u>; (g) palmribbed (<u>Hedera</u> <u>canarienis</u>)

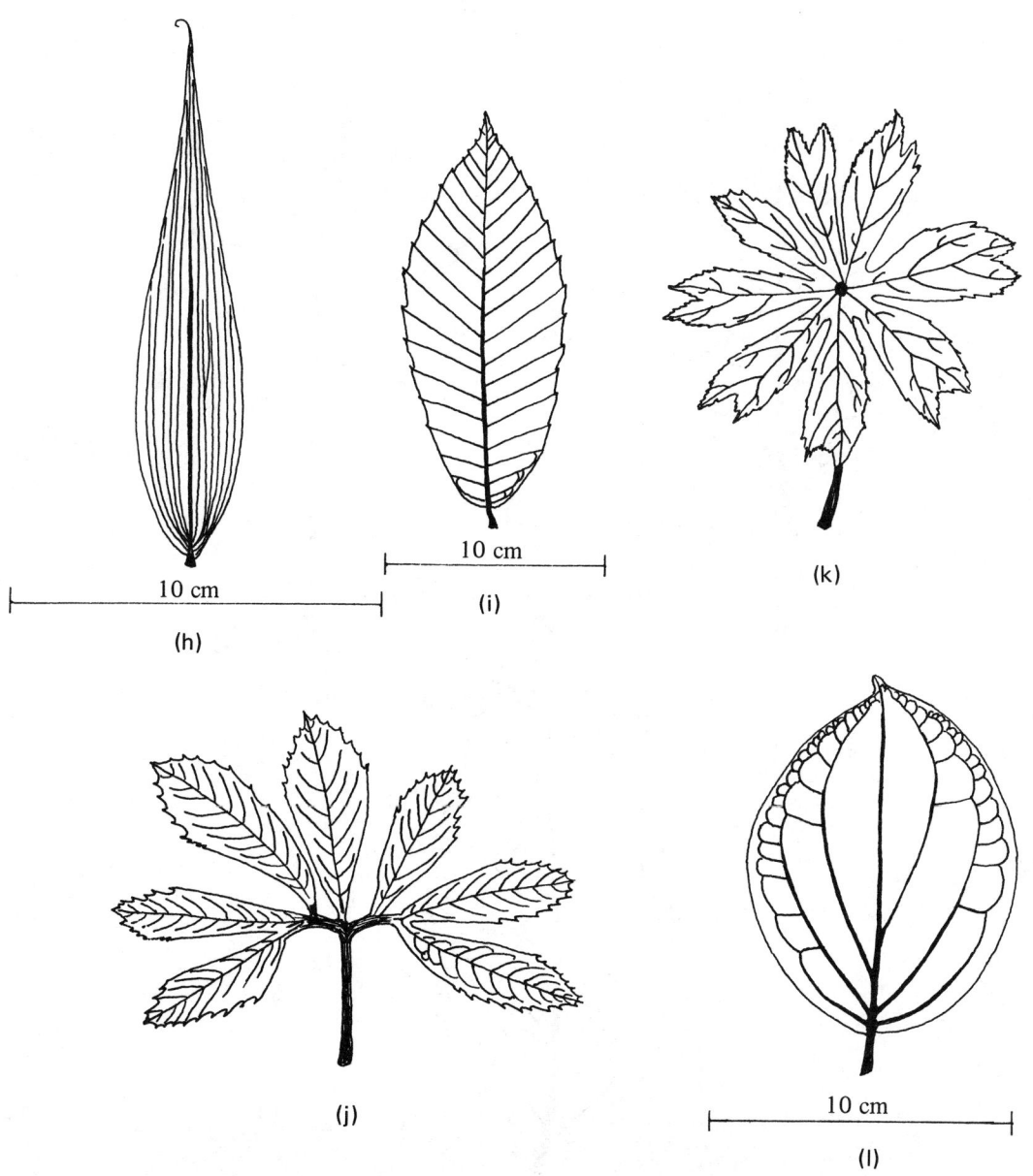

Fig. II.1.2.24. Venation: (h) parallel (Gloriosa greeneae hyb.); (i) parallel-ribbed (Castanea mollissima); (j) pedatiribbed (Helleborus niger); (k) peltiribbed (Podophyllum peltatum); (l) penniform (Piper nigrum). (j), (k) Redrawn from Hutchinson, 1969; see Acknowledgments and Selected References.

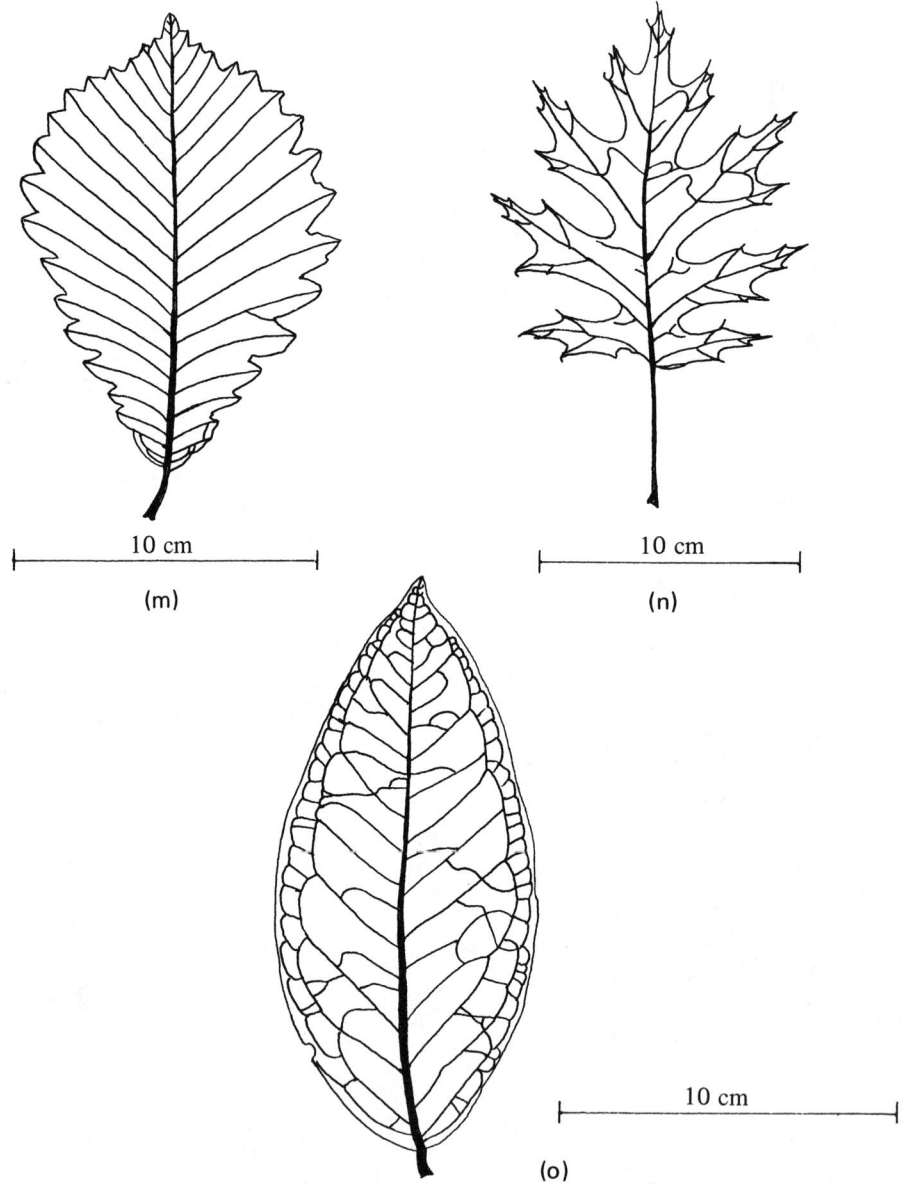

Fig. II.1.2.25. Venation: (m) penniparallel (<u>Quercus prinus</u>); (n) penniribbed (<u>Quercus shumardii</u>); (o) pinnate (<u>Syzygium malaccensis</u>).

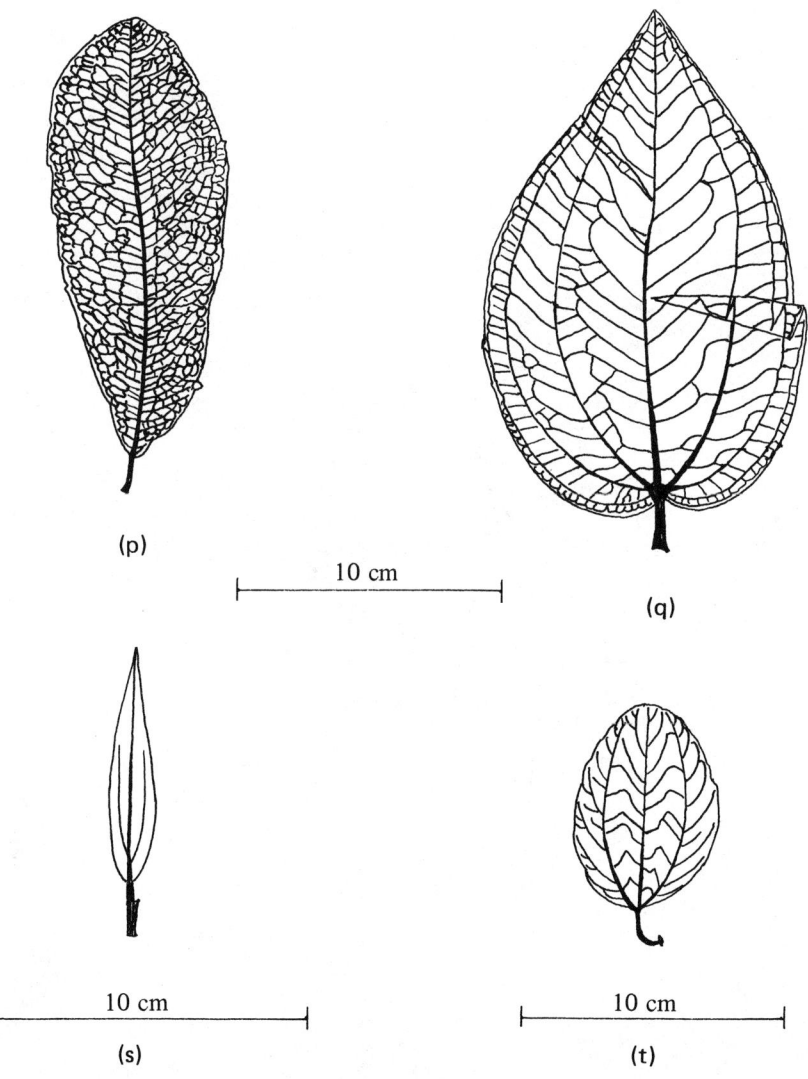

Fig. II.1.2.26. Venation: (p) reticulated (Macadamia integrifolia); (q) ribbed (Tibouchina semidecandra); (s) triple-ribbed (juvenile Zingiber officinale); (t) triribbed (Zizyphus mauritiana).

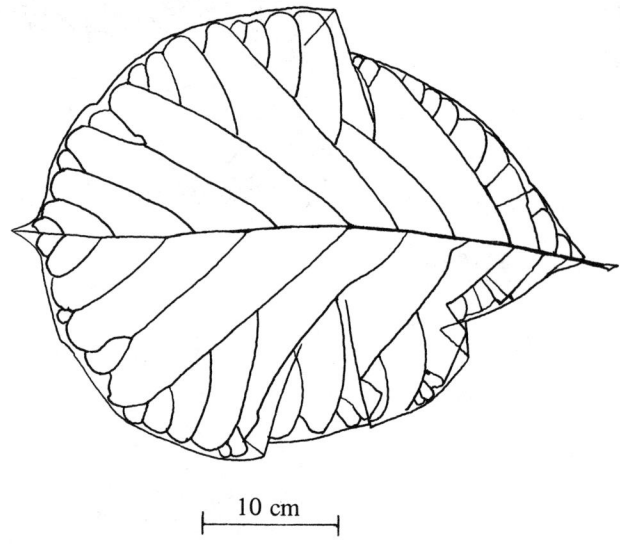

Fig. II.1.2.27. Venation: (v) venose (<u>Tectona grandis</u>).

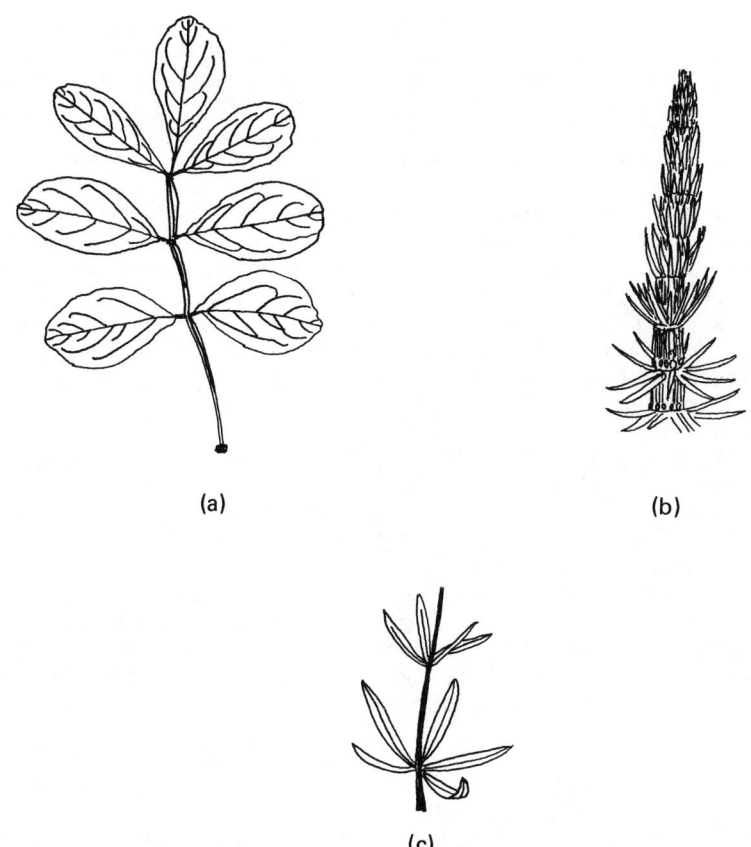

Fig. II.1.2.28. Leaves: (a) vertebrate leaf of Indian wood-apple (<u>Feronia limonia</u>) comprised of a winged petiole and jointed rachises which also have articulations at each point where leaflets are attached; (b) verticillate arrangements typified by <u>Hippurus vulgaris</u>, where each whorl has numerous leaves in successive tiers; (c) whorls with four or five leaves at each node, as in <u>Galium</u> (the number varies from three or four to about eight in most verticillate = whorled arrangements). (b), (c) Redrawn from Gray, 1862; see Acknowledgments and Selected References.

II.1.3. Roots

.000 Roots: typically the subterranean axis of a plant with the primary functions of anchorage and absorption of water and other nutrients.
.001 **Adventitious**: all roots, other than the primary root of a germinated seed, arise adventitiously (i.e., sporadically, spontaneously and unpredictably) and may do so from stem or root tissues; this concept implies that roots in general do not arise from essentially fixed or predictable locations as do lateral stem buds (see Cladogenous, II.1.3.004). (Fig. II.1.3.01)
.002 **Aerial**: roots produced above ground are in three general types: short, holdfast roots, some of which are provided with special devices, such as a disc, for adhesion (e.g., Hedera spp.); branch prop roots that descend from lateral branches of certain Ficus, become anchored, and enlarge with age; and trunk prop roots that arise at the base of the stem of certain palms and mangroves and grow so that the plants appear to be on stilts. Trunk prop roots are developed on plants that grow in wet or periodically inundated habitats, whereas branch prop roots usually appear on plants in normal mesophytic situations. (Fig. II.1.3.a, c-e) (G 62)
.003 **Allorhizic**: taproot plus lateral roots, as contrasted with homorhizic (II.1.3.009). (ES)
.004 **Cladogenous**: term used by some authors for adventitious roots that arise from young stem tissues in dicotyledons, monocotyledons, and lower vascular plants. (ES)
.005 **Feeder**: fine rootlets that form the ultimate branches of a root system and serve as the zone in which nutrients are absorbed; distinguishable from older roots by their generally light color, small size, lack of fully developed cork or suberin layers, and, frequently, the presence of root hairs. [Fig. I.1.3.02b(2), d(3)] (ES)
.006 **Fibrous**: long slender roots, all about equal size; usually of adventitious origin; typical of monocotyledons and some dicotyledons. (Fig. II.1.3.02d, .03) (HI)
.007 **Fleshy**: primary or lateral roots enlarged by storage of starch, inulin, and the like = tuberous root (II.1.3.017). (Fig. II.1.3.04b) (G 62)
.008 **Graminoid roots**: finely branched roots less than 0.1 mm in diameter and covered with dense root hairs.
.009 **Homorhizic**: stem-borne adventitious roots, all of which are equivalent (see Allorhizic, II.1.3.003). (ES)
.010 **Magnoloid roots**: coarsely branched roots more than 0.5 mm in diameter, with few or no root hairs (see Graminoid roots, II.1.3.008).
.011 **Mycorrhizae**: (see V.3.3.2.264).
.012 **Primary**: the root of the young seedling; sometimes fleshy or swollen; short-lived in monocotyledons, often persisting in dicotyledons and becoming the taproot. [Fig. II.1.3.02a, c(4), .04a(2)] (ES)
.013 **Radical**: arising from the root. (L 51)
.014 **Root hairs**: (see II.3.162). [Fig. II.1.3.04a(1)]
.015 **Secondary**: branch roots from a primary root; also the first division of a fibrous (adventitious) root. [Fig. II.1.3.02a, b(2), c(5,6), d] (ES)
.016 **Tap root**: (see II.3.199). [Fig. II.1.3.02a, b(1); .04a(2), b]

II.1.3. MORPHOLOGY and ANATOMY

.017 Tuberous: fleshy storage root, with buds, if any, at the base of the stem proper or near the transition from stem to root morphology (see Genotypic modifications II.4.3.063). (Fig. II.1.3.04c) (G 62)

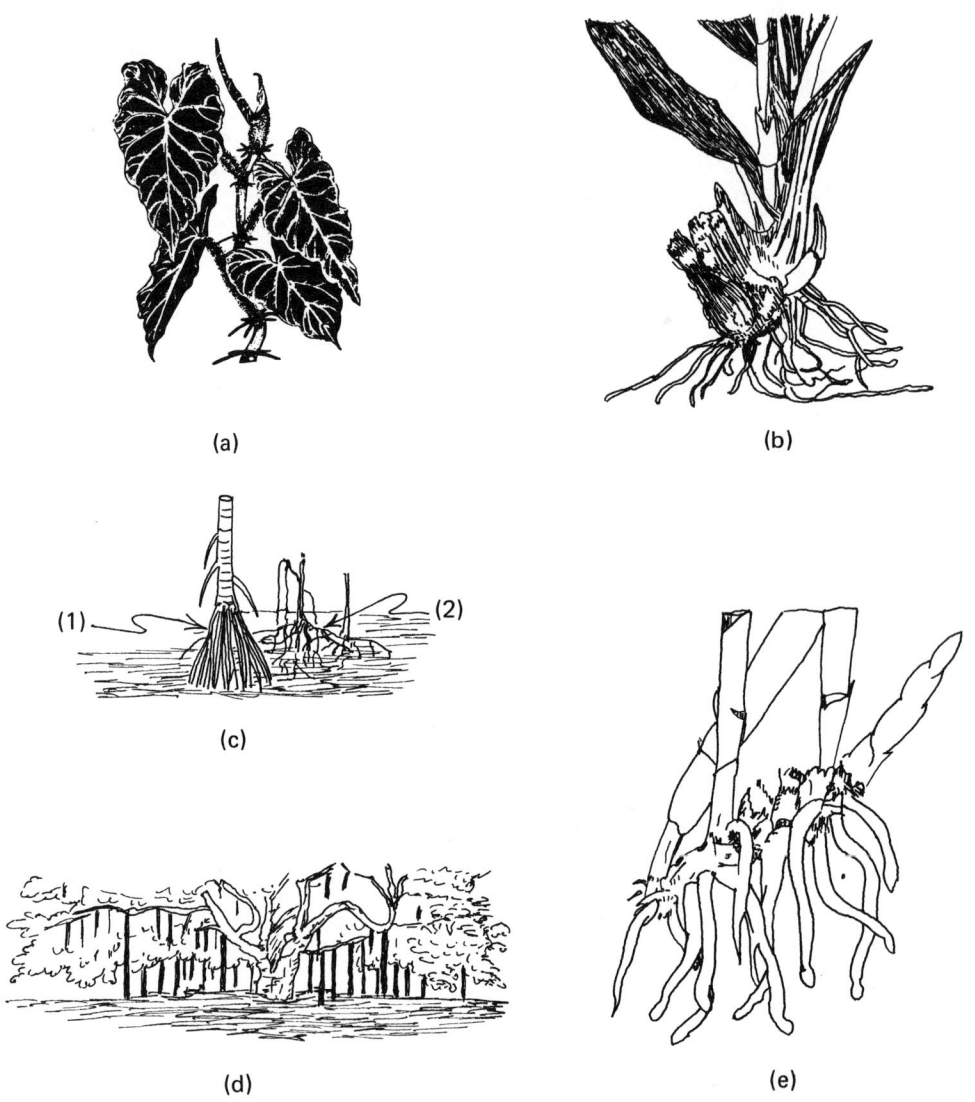

Fig. II.1.3.01. Adventitious roots: (a) aerial roots on vining-type Philodendron sp; (b) terrestrial roots on "nun's orchid" (Phaius tankervilliae); (c) stilt roots on (1) screw pine (Pandanus sp.) and (2) red mangrove (Rhizophora mangle); (d) prop roots on an old banyan (Ficus indica) tree; (e) aerial roots on Cattleya orchid. (a) From Watkins and Sheehan, 1975; (b), (e) adapted from Sheehan and Sheehan, 1979; (c), (d) redrawn from Gray, 1862; see Acknowledgments and Selected References.

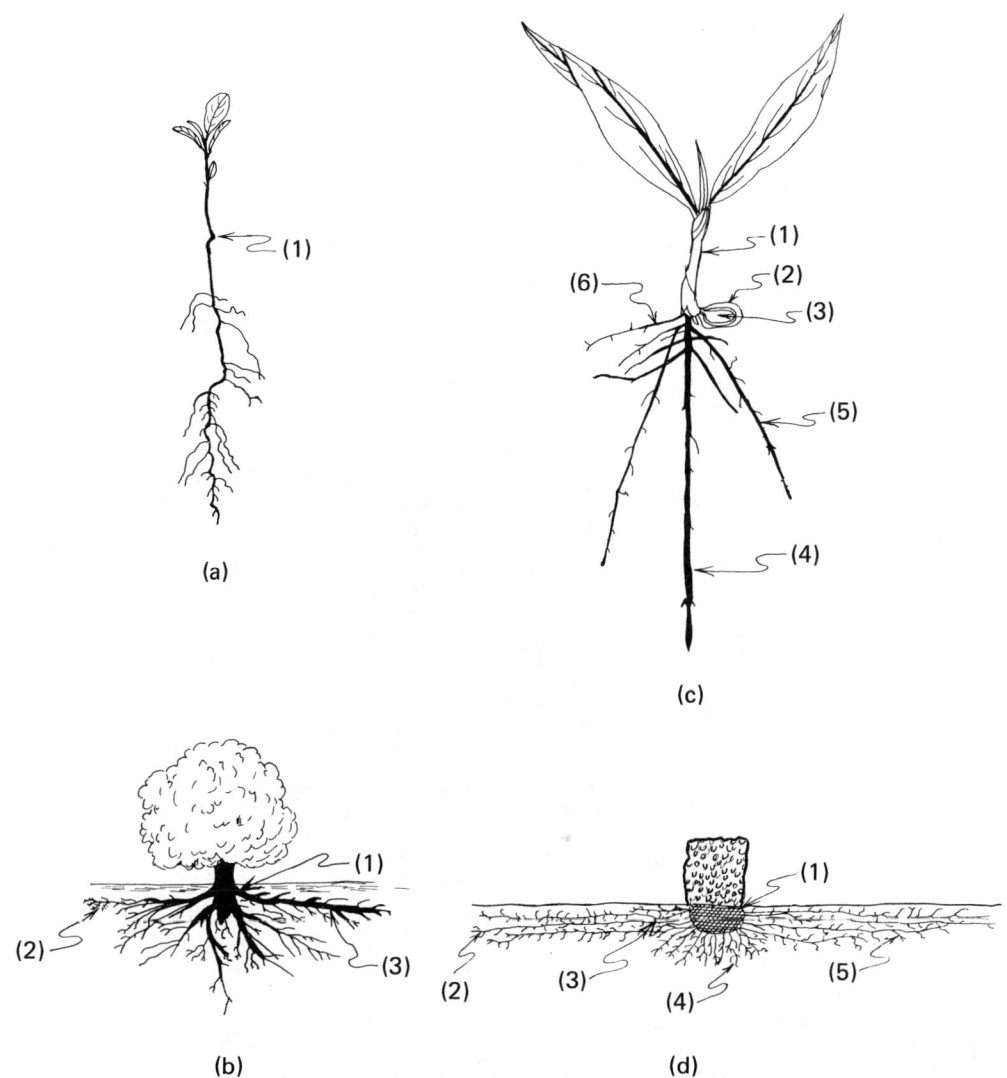

Fig. II.1.3.02. Roots: (a) two-month-old sour orange (Citrus aurantium) seedling, with stem-root junction at (1), the primary (future tap) root with numerous secondary roots branching from it; (b) old sweet orange (Citrus sinensis) tree with (1) short taproot, (2) feeder roots as ultimate branches of (3) lateral secondary roots; (c) two-month-old oil palm (Elaeis guineensis) seedling with (1) plumule, (2) seed shell with (3) haustorium (cotyledon) drawing nourishment from the endosperm, (4) radicle (true primary root), (5) "primary root" (actually a secondary root), and (6) an adventitious root, more of which will develop later and form the entire root system; (d) old oil palm plant with (1) underground part of stem, (2) long "primary" roots, (3) mass of feeder (quarternary) roots only a few mm long, (4) short "primary" roots, and (5) secondary roots (compare this fibrous root system with that of corn in Fig. II.1.3.03). (b) Redrawn from Praloran, 1971; (c), (d) redrawn from Surre and Ziller, 1963; see Acknowlegments and Selected References.

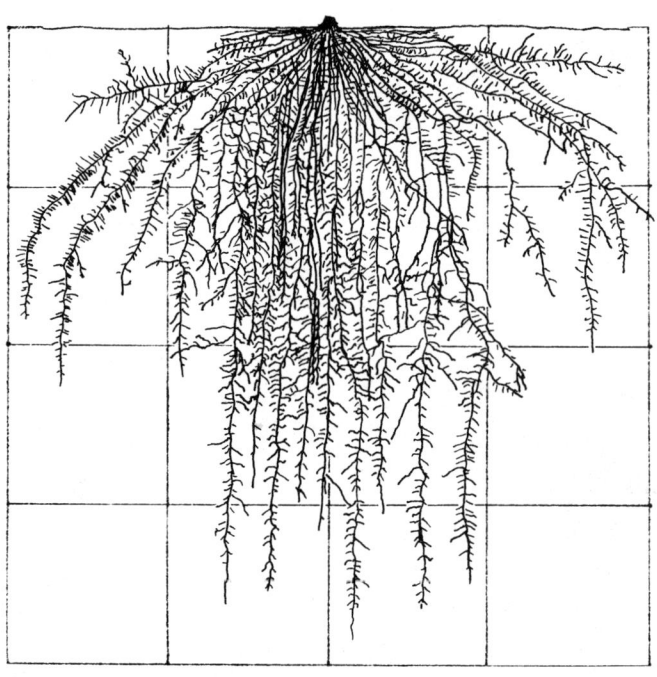

Fig. II.1.3.03. Roots: mature root system of corn (Zea mays) plant (squares are 50 x 50 cm). Redrawn from Hayward, 1938; see Acknowledgments and Selected References.

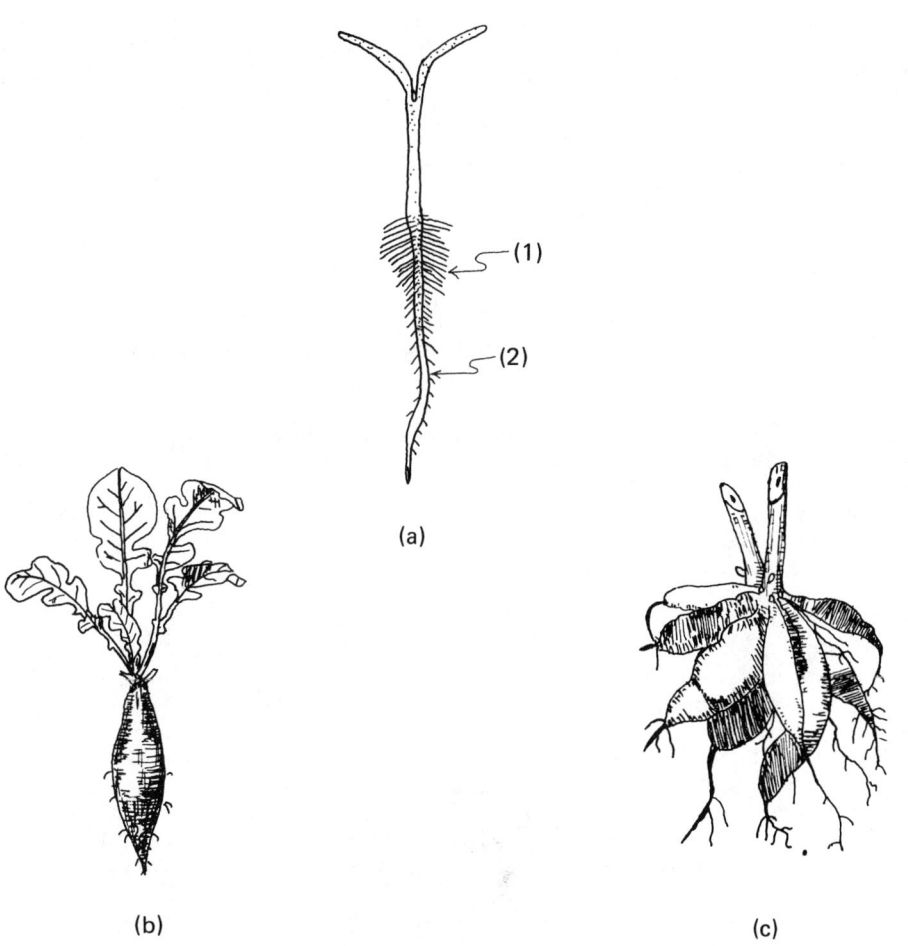

Fig. II.1.3.04. Roots: (a) young tomato (Lycopersicon esculentum) seedling, with (1) root hairs and (2) primary root; (b) fusiform fleshy root of radish (Raphanus sativa); (c) tuberous roots of dahlia. (a) Redrawn from Hayward, 1938; (b), (c) redrawn from Gray, 1862; see Acknowledgments and Selected References).

II.2.1. MORPHOLOGY and ANATOMY

II.2. Flowering and Fruiting Structures

II.2.1. Inflorescences

.000 Inflorescences: arrangement of flowers on the floral axis. (Figs. II.2.1.01 to II.2.1.14)
.001 Aggregated, conglomerated, clustered: collected in parcels, each of which has a rounded figure. (Fig. II.2.1.05a) (ST)
.002 Ament: (see Catkin, II.2.1.013).
.003 Axillary: (see Solitary, II.2.1.065).
.004 Basiflory: (see Cauliflory, II.2.1.015).
.005 Boat-shaped: (see Cymbiform, II.2.1.023).
.006 Bostryx: helicoid (monochasial) cyme (e.g., Hypericum, Hemerocallis). (Fig. II.2.1.05b) (B & C)
.007 Bract: foliage leaf much reduced in size or of different form or character; bracteate = with bracts. [Fig. II.2.1.01(11)] (L 51)
.008 Bracteate: (see Bract, II.2.1.007).
.009 Bracteole: minute or miniature bract; also a bract subtending an individual flower of a cluster as in composites. [Fig. II.2.1.01(10)] (L 51)
.010 Broom-shaped: (see Muscariform, II.2.1.048).
.011 Brush-shaped: (see Muscariform, II.2.1.048).
.012 Capitulum: (see Head, II.2.1.037).
.013 Catkin = ament: flexous spikelike assemblage of bracteate dichasia or cymules; actually a rather complex inflorescence that is decidedly not primitive. (Fig. II.2.1.05c) (L 55)
.014 Cauliflorous: (see Cauliflory, II.2.1.015).
.015 Cauliflory (cauliflorous): flowers borne on the trunk or larger branches (i.e., on older wood rather than at or near the tips of stems); several distinct subdivisions: basiflory, flowers borne at the base of the trunk; flagelliflory, flowers borne on pendulous twigs that arise from the lower part of the trunk; ramiflory, flowers borne on the larger branches and leafless twigs but not on the trunk; rhizoflory, flowers borne on roots (rare); and trunciflory, flowers borne directly on the trunk but not on branches or twigs (equivalent to cauliflory in a narrow sense). (Fig. II.2.1.05d) (L & J)
.016 Clustered: (see Aggregated, II.2.1.001).
.017 Coflorescence: lateral inflorescence in a synflorescence (II.2.1.070). [Fig. II.2.1.06a(2)] (B & C)
.018 Compound dichasium: (see Dichasium, II.2.1.030).
.019 Conglomerated: (see Aggregated, II.2.1.001).
.020 Corymb: indeterminate, flat- or convex-topped inflorescence, a modified panicle. [Fig. II.2.1.02(4),(5), .05e] (L 55)
.021 Corymbose: flowers arranged to resemble a corymb. (Fig. II.2.1.06b, .12b) (G 62)
.022 Cymba: woody durable bootlike spathe or spathe valve that encloses the inflorescence and opens and persists, as in many palms. [Fig. II.2.1.07a(1)] (L 51)
.023 Cymbiform, boat-shaped: concave, tapering toward both ends, with an external keel. (Fig. II.2.1.07b) (L 51)
.024 Cyme: determinate inflorescence reduced from a compound dichasium to produce a more or less flat-topped or convex cluster. [Fig. II.2.1.03(3), .06d, .09f] (L 55)

II.2.1. MORPHOLOGY and ANATOMY

.024a. Cincinnus (glomerule): much modified helicoid cyme in which the peduncle and pedicels are shortened or absent. [Fig. II.2.1.03(8)]
.024b. Fascicled cyme: flowers are much congested.
.024c. Helicoid cyme: flowers are lost from the same side of each dichotomy. [Fig. II.2.1.03(7), .07e]
.024d. Scorpioid cyme: nearly straight inflorescence resembling a raceme from loss of flowers from the opposite side of each dichotomy. [Fig. II.2.1.03(4), .07d]
.024e. Verticillaster (verticillate inflorescence): mixed inflorescence that consists of a pair of much condensed, nearly sessile cymes disposed around the axis like a true verticil (II.2.1.080). [Fig.II.2.1.07c(1)]
.025 Cymose: flowers in a cymelike inflorescence that may or may not be a true cyme. (Fig. II.2.1.06c) (G 62)
.026 Cymose-paniculate: compound inflorescence in which the primary axis is paniculate and secondary axis cymose. (Fig. II.2.1.04a)
.027 Cymule: diminutive, usually few-flowered cyme. [Fig. II.2.1.07c(2)] (L 55)
.028 Determinate: terminal or central flower develops first, thereby arresting growth by elongation of the primary axis; the opposite of indeterminate. (Fig. II.2.1.03a) (L 55)
.029 Determinate panicle: a branched inflorescence in which the terminal flowers of the main and successively lower axes bloom first; that is, the order of flower opening is basipetal. (B & C)
.030 Dichasium: a determinate cluster of three flowers that arises from a common peduncle by dichotomous branching immediately below a terminal flower; a compound dichasium results when successive dichotomous branching of lateral pedicels or peduncles occurs. [Fig. II.2.1.03(1),(2), .08b,c] (L 55)
.031 Discoid: disklike; a head without ray florets, as in some composites. (Fig. II.2.1.09d) (G 62)
.032 Drepanium: a one-sided (monochasial) cyme with successive branches away from the main axis (e.g., _Juncus_ spp.). (Fig. II.2.1.09b) (B & C)
.033 Epicalyx: (see Involucre, II.2.1.042).
.034 Flagelliflory: (see Cauliflory, II.2.1.015).
.035 Floret: a small flower, specifically one of the small flowers that collectively form the head (II.2.1.037) in _Compositae_. [Fig. II.2.1.09a(2)] (L 55)
.036 Glomerule: headlike, compacted, or sessile cyme (see Cyme-Cincinnus, II.2.1.024a). (L 55)
.037 Head = capitulum: sessile or nearly sessile flowers on a short or flattened axis; may be derived from an umbel by loss of the lower pedicels or a spike by compression of the axis; may be naked or subtended by a cluster of bracts or bracteoles. [Fig. II.2.1.02(7), .03(6), .09e] (L 55)
.038 Heterogamous: inflorescence (or plant) with two or more kinds or forms of flower. (Fig. II.2.1.09c) (G 62)
.039 Homogamous: head or cluster with flowers all of one kind, with stigmas becoming receptive at the time pollen dehiscence occurs in the same flower. (Fig. II.2.1.09d) (G 62)
.040 Indeterminate: terminal or central flower opening last and primary axis often elongating as flowers develop from the lower to the upper portions. (Fig. II.2.1.02) (L 55)

II.2.1. MORPHOLOGY and ANATOMY

.041 Indeterminate panicle: branched inflorescence in which the lowest flower on each successive higher lateral and the main axis blooms first; that is, the order of flowering is acropetal. [Fig. II.2.1.02(3), .08h, .11c] (L 55)
.042 Involucre: one or more whorls or small leaves or bracts (phyllary) standing close beneath a flower or flower cluster; denoted "epicalyx" in Malvaceae and some Rosaceae (e.g., Fragaria), where it appears on or just below the calyx; accrescent in some plants and forms the outer covering of a fruit, as in Corylus. (Fig. II.2.1.01(6),.10) (L 55)
.043 Lax, loose: when the parts are distant from one another in a light, open kind of arrangement. (ST)
.044 Leaf-opposed: inflorescence borne opposite a leaf, as in Vitis, which has an alternate leaf arrangement; an unusual situation in which intercalary growth of the internode carries the bud from its normal axillary position up to opposite the next higher leaf. (Fig. II.2.1.11a) (L 51)
.045 Loose: (see Lax, II.2.1.043).
.046 Monochasium: a simple determinate inflorescence that consists of a terminal and one lateral flower, instead of two lateral flowers, as in a dichasium. (Fig. II.2.1.07e, .09b, .13a) (L 55)
.047 Monotelic: (see Synflorescence, II.2.1.070)
.048 Muscariform, brush- (broom)- shaped: shaped like a brush or broom. (Fig. II.2.1.11b) (G 62)
.049 Panicle: indeterminate branch system in which the primary axis bears branched secondary axes and pedicellate flowers. [Fig. II.2.1.02(3), .08h, .11c] (L 55)
.050 Paniculate: paniclelike, whether a true panicle or not. (Fig. II.2.1.06a, .11d) (G 62)
.051 Paniculiform: in the form of a panicle, like the inflorescences of certain Rubus spp. [Fig. II.2.1.06a(1), .11d] (B - GH)
.052 Paraclade: a lateral inflorescence (= coflorescence) which repeats the symmetry of the main axis in complex inflorescences (see Synflorescence, II.2.1.070). [Fig. II.2.1.06a(2)] (B & C)
.053 Peduncle: a stalk of a flower cluster (inflorescence) or of a solitary flower when it is the remaining member of an inflorescence; the term rachis, also used for compound leaves, is often applied to the extension of the peduncle as the primary axis of a panicle or other multibranched inflorescence. [Fig. II.2.1.01(9)] (L 55)
.054 Phyllary: a bract or bracteole that forms part of an involucre (II.2.1.042). (Fig. II.2.1.10b, .12a) (L 55)
.055 Pirostele: compound stem from which flowers of Pyrus arise. [Fig. II.2.1.12b(1)] (B-GH)
.056 Polytelic: (see Synflorescence, II.2.1.070)
.057 Prophyll: (= bracteole, bractlet, II.2.1.009)
.058 Raceme: indeterminate, single axis bearing pedicellate flowers. [Fig. II.2.1.02(2), .08f] (L 55)
.059 Racemose, racemiform: flowers in a racemelike inflorescence that may or may not by a true raceme. (Fig. II.2.1.08d, .12d) (L 55)
.060 Radiate, radiant: diverging from a common center like rays, like the ligulate florets of a composite flower. [Fig. II.2.1.09a(2)] (G 62)
.061 Ramiflory: (see Cauliflory, II.2.1.015).
.062 Rhipidium: a one-sided (monochasial) cyme with successive branches alternately away and toward the main axis (e.g., Iris). (Fig. II.2.1.13a) (B & C)
.063 Rhizoflory: (see Cauliflory, II.2.1.015).

II.2.1. MORPHOLOGY and ANATOMY

.064 Scape, scapose: acaulescent (usually herbaceous) plant that bears a flower or flowers on a long leafless stalk (or scape). [Fig. II.2.1.12e(1)] (L 55)

.065 Solitary = axillary: flowers that appear singly in the axils of ordinary foliage leaves, which may, however, not be present. (Fig. II.2.1.08a) (L 55)

.066 Spadix: a term applied to two different types of inflorescence; that of aroids is thick or fleshy and spikelike; that of palms has sessile or partially imbedded flowers borne on a single rachis, on a series of spikes emanating from the peduncle, or on spikelets with paniculate branching from the peduncle, but rarely thick or fleshy. [Fig. II.2.1.13b(1), c(1)] (G 62)

.067 Spathe: bract or leaf subtending a spadix, often compound or multiple in palms. [Fig. II.2.1.07a(1), .13b(2), c(2)] (G 62)

.068 Spike: indeterminate single axis bearing sessile flowers, as in Piper nigrum and many herbaceous ornamentals. (Fig. II.2.1.14d) (L 55)

.069 Syconium: compound inflorescence with staminate and pistillate flowers borne around the inside of a hollow receptacle; specifically, the inflorescence of Ficus. (Fig. II.2.1.13d) (CN)

.070 Synflorescence: compound inflorescence consisting of a terminal inflorescence and lateral inflorescences; two main configurations, the monotelic, in which the main inflorescence is determinate with a single terminal flower and the same morphological organization is repeated in the lateral inflorescences, and the polytelic in which the main inflorescence axis ends in a many-flowered inflorescence and this morphological organization is repeated in the lateral inflorescences. (Fig. II.2.1.16a) (B & C)

.071 Tassel: inflorescence at the top of a corn plant comprised of staminate (pollen-bearing) flowers. (Fig. II.2.1.14a) (B - M)

.072 Thyrse, thyrsus: compound inflorescence composed of a paniculate or racemose primary axis (indeterminate) with secondary axes of simple or compound dichasia (determinate), as in grape. [Fig. II.2.1.02(8), .14c] (L 55)

.073 Trunciflory: (see Cauliflory, II.2.1.015).

.074 Umbel: flowers pedicellate, the latter arising from a common point at the apex of the stem or peduncle and often subtended by a number of bracts; may be determinate or indeterminate and simple or branched (compound). [Fig. II.2.1.02(6), .03a(5)b, .08g] (G 62)

.075 Umellate, umbelliform: like an umbel or resembling one. (Fig. II.2.1.12c) (G 62)

.076 Umbellate-paniculate: compound inflorescence in which the primary axis is paniculate and the secondary axes, umbellate, as in some Araliaceae. (Fig. II.2.1.04d)

.077 Umbellate-racemiform: compound inflorescence in which the primary axis is racemiform and the secondary axes, umbellate, as in some Araliaceae. (Fig. II.2.1.04c)

.078 Umbellate-spicate: compound inflorescence of spikes arranged in umbellate form, as in some Piperaceae. (Fig. II.2.1.04b)

.079 Umbelliform: (see Umbellate, II.2.1.075).

.080 Verticil, verticillate, whorled: said of several flowers, flower parts, or inflorescences arranged about the same point on an axis. [Fig. II.2.1.02b(2)] (G 62)

.081 Verticillaster: (see Cyme, II.2.1.024e).

.082 Whorled: (see Verticil, II.2.1.080).

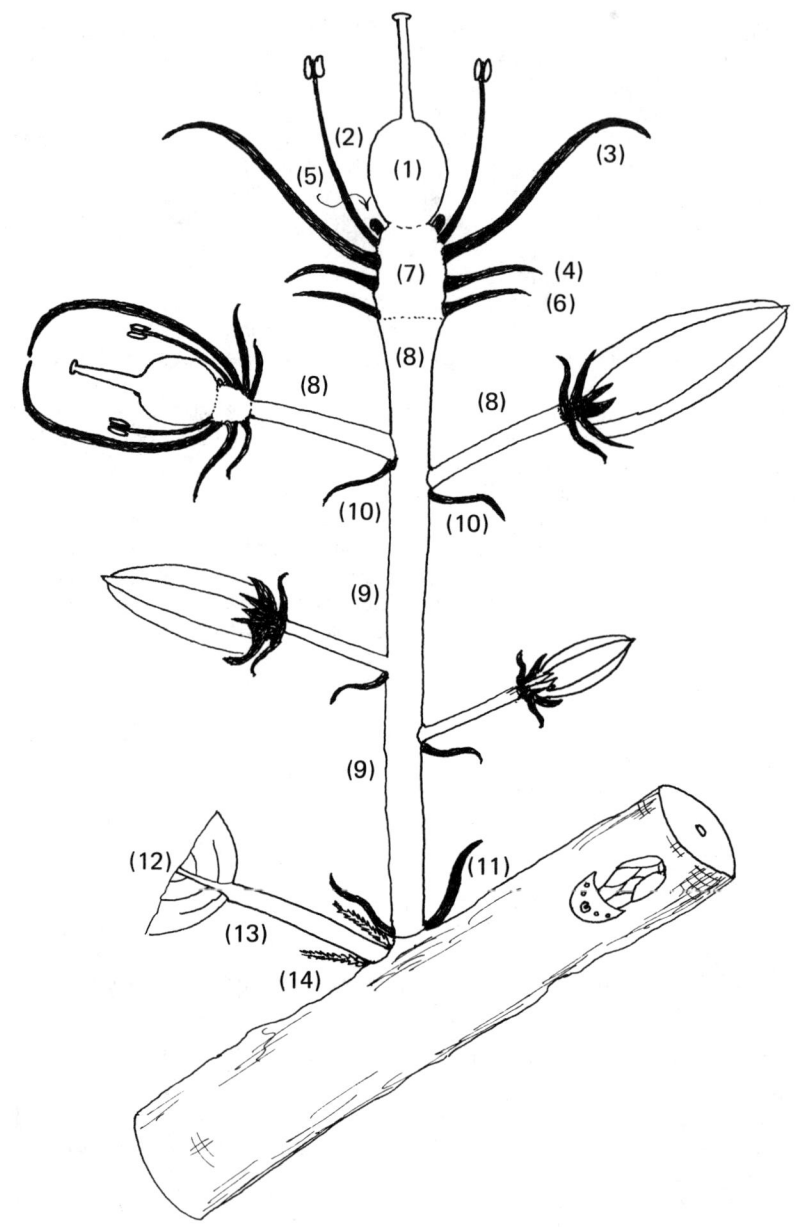

Fig. II.2.1.01. Inflorescences: diagram with (1) gynoecium, (2) androecium, (3) corolla, (4) calyx, (5) disk, (6) involucre, (7) receptacle, (8) pedicel, (9) peduncle (rachis), (10) bracteole, (11) bract, and leaf with (12) blade, (13) petiole, (14) stipules.

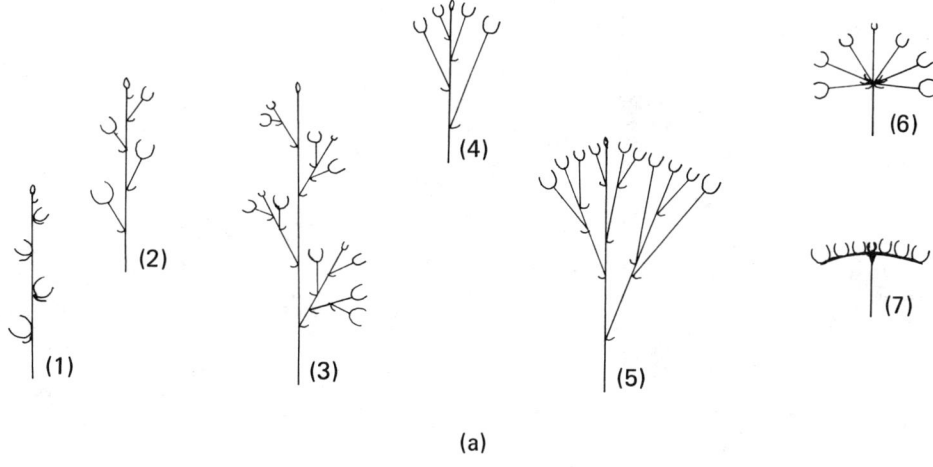

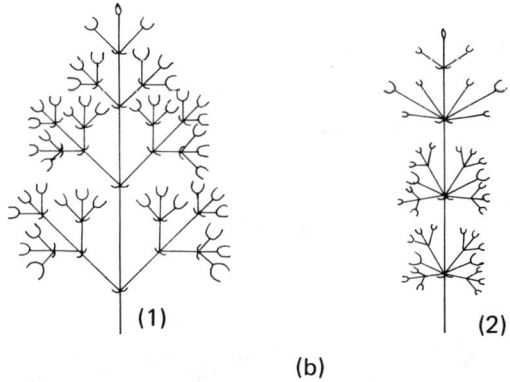

Fig. II.2.1.02. Inflorescence types (diagrammatic): (a) indeterminate, (1) spike, (2) raceme, (3) panicle, (4) corymb, (5) compound corymb, (6) umbel, (7) head; (b) compound: (1) thyrse, thyrsus (primary axis paniculate, secondary axis dichasial, simple or compound), (2) verticillate (primary axis spicate or racemose, secondary axis cymose or dichasial). Adapted from Lawrence, 1955; see Acknowledgments and Selected References.

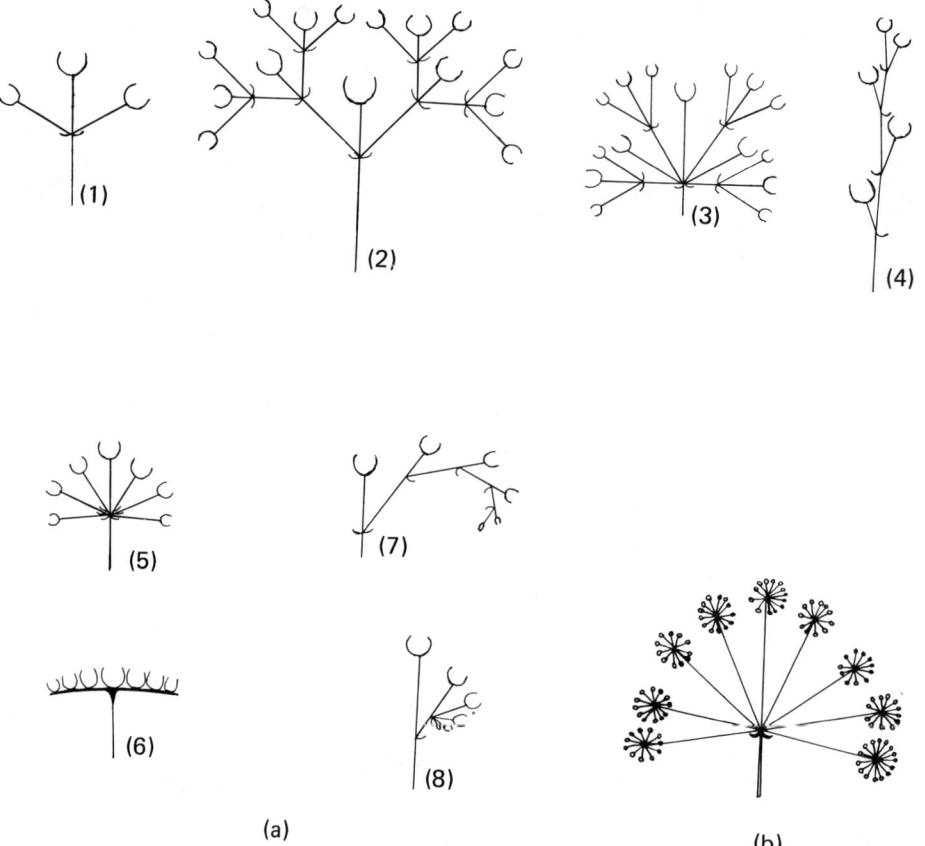

Fig. II.2.1.03. Inflorescence types: (a) determinate: (1) simple dichasium, (2) compound dichasium, (3) cyme, (4) scorpioid cyme, (5) umbel, (6) head, (7) helicoid cyme, (8) cincinnus (glomerule); (b) compound umbel (i.e., an umbel of umbels). (a) Adapted from Lawrence, 1955; see Acknowledgments and Selected References.

Fig. II.2.1.04. Inflorescence types, compound inflorescences: (a) cymose-paniculate (primary axis, paniculate, secondary axes, cymose); (b) umbellate-spicate (primary axis, an umbel of spikes, secondary axes umbellate); (c) umbellate-racemiform (primary axis, an umbel of racemes, secondary axes, umbellate); (d) umbellate-paniculate (primary axis, an umbel of panicles, secondary axes, umbellate). (Found in Araliaceae.)

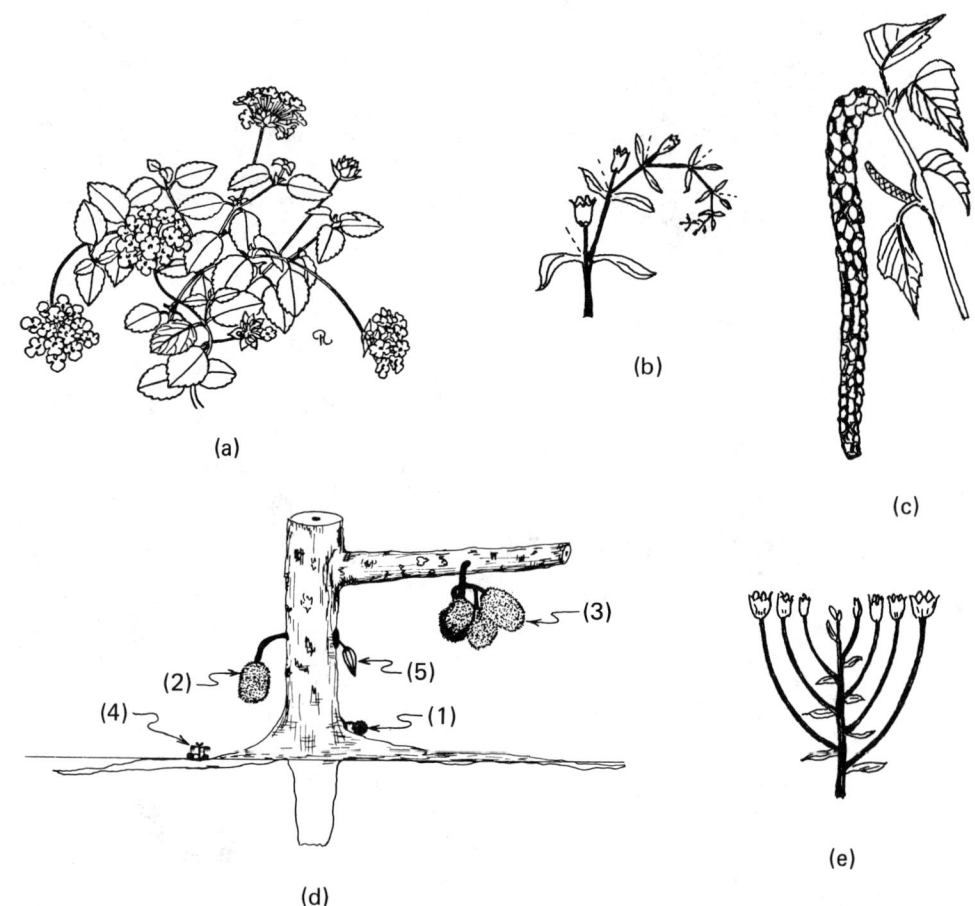

Fig. II.2.1.05. Inflorescences: (a) aggregated as heads of weeping lantana (<u>Lantana montevidensis</u>); (b) bostryx (helicoid cyme) typical of some <u>Hypericum</u> spp.; (c) catkin of white birch (<u>Betula</u> sp.); (d) diagram of cauliflory, subdivided as (1) basiflory, (2) flagelliflory, (3) ramiflory (as in <u>Artocarpus heterophyllus</u>), (4) rhizoflory (rare but does occur in some figs and palms), (5) trunciflory (differs from (2) and (3) in that the pedicel arises directly from the trunk, e.g., cacao (<u>Theobroma cacao</u>), rather than from a short branch); (e) corymb. (a) From Watkins and Sheehan, 1975; (b), (c), (e) redrawn from Gray, 1862; see Acknowledgments and Selected References.

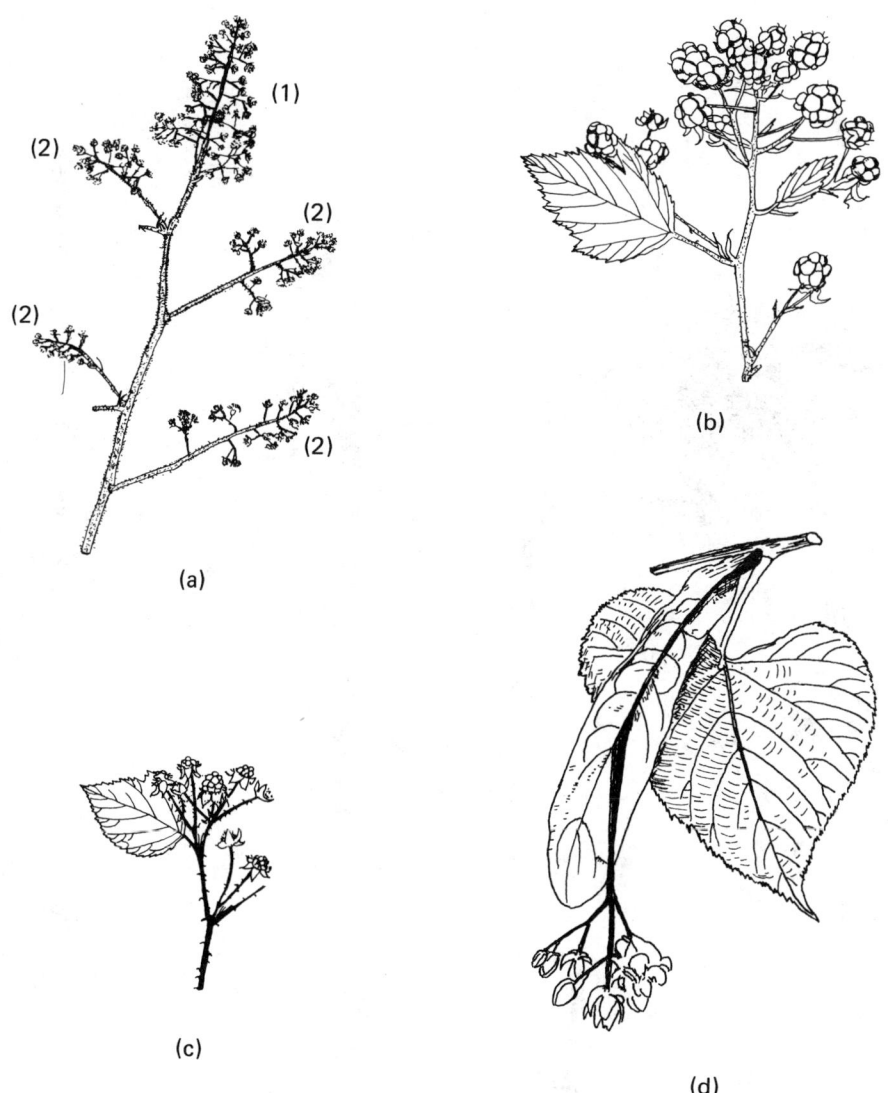

Fig. II.2.1.06. Inflorescences: (a) synflorescence (complex inflorescence) shown here in Rubus trichomallus with a terminal paniculiform terminal inflorescence (1) and four lateral inflorescences-coflorescences (2) which are smaller versions, or paraclades, of the terminal kind; (b) corymbiform inflorescence of Rubus hanesii; (c) cymiform inflorescence (Rubus sp.); (d) cyme of basswood (Tilia americana), unusual in the peduncle being adnate below its midpoint to the subtending bract. (a), (b), (c) Redrawn from Bailey, 1944c, 1945, 1941, respectively; (d) redrawn from Bailey, 1928; see Acknowledgments and Selected References.

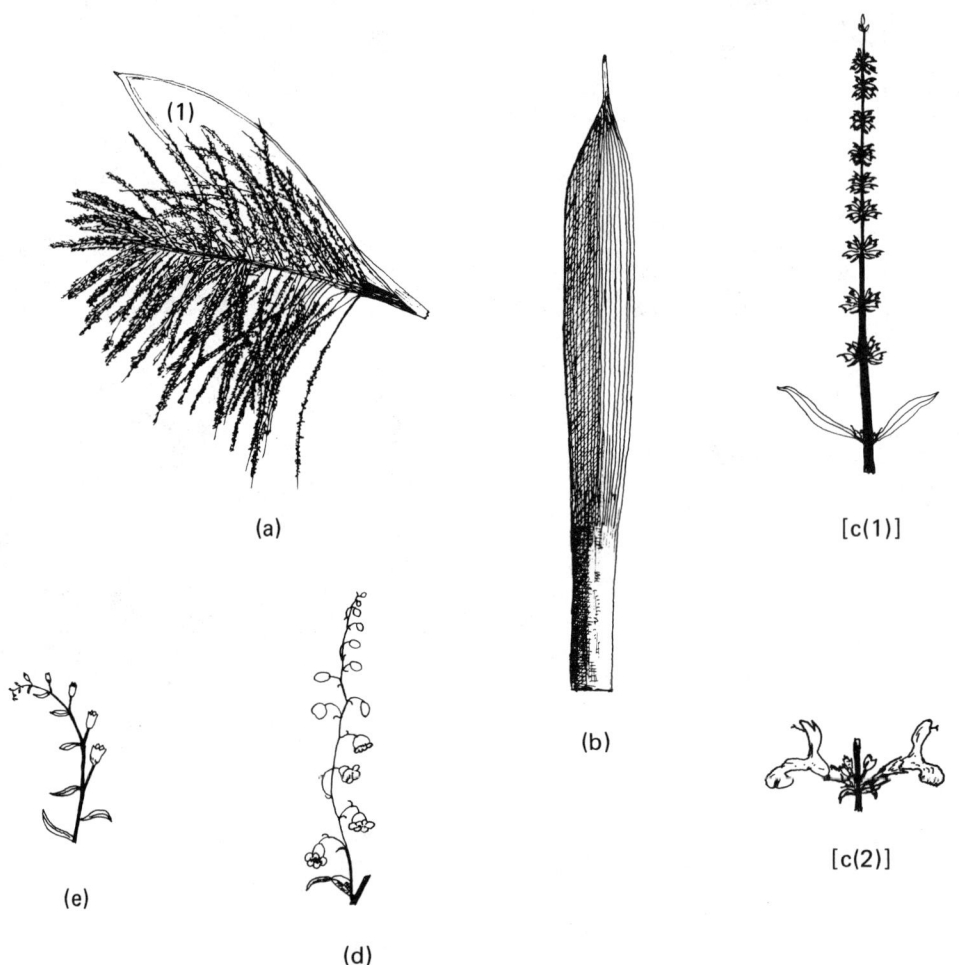

Fig. II.2.1.07. Inflorescences: (a) cymba (spathe) (1) encloses inflorescence of pindo palm (Butia capitata); (b) outer (adaxial) side of the large boat-shaped spathe of queen palm (Arecastrum romanzoffianum); (c) verticillaster (1) with enlarged cymules (2) that form the individual clusters of flowers along the rachis; (d) scorpioid cyme or monochasium, typified by lily of the valley; (e) helicoid cyme. (a), (b) Redrawn from Bailey, 1936; (c)(2) redrawn from Lawrence, 1955; (d), (e) redrawn from Gray, 1862; see Acknowledgments and Selected References.

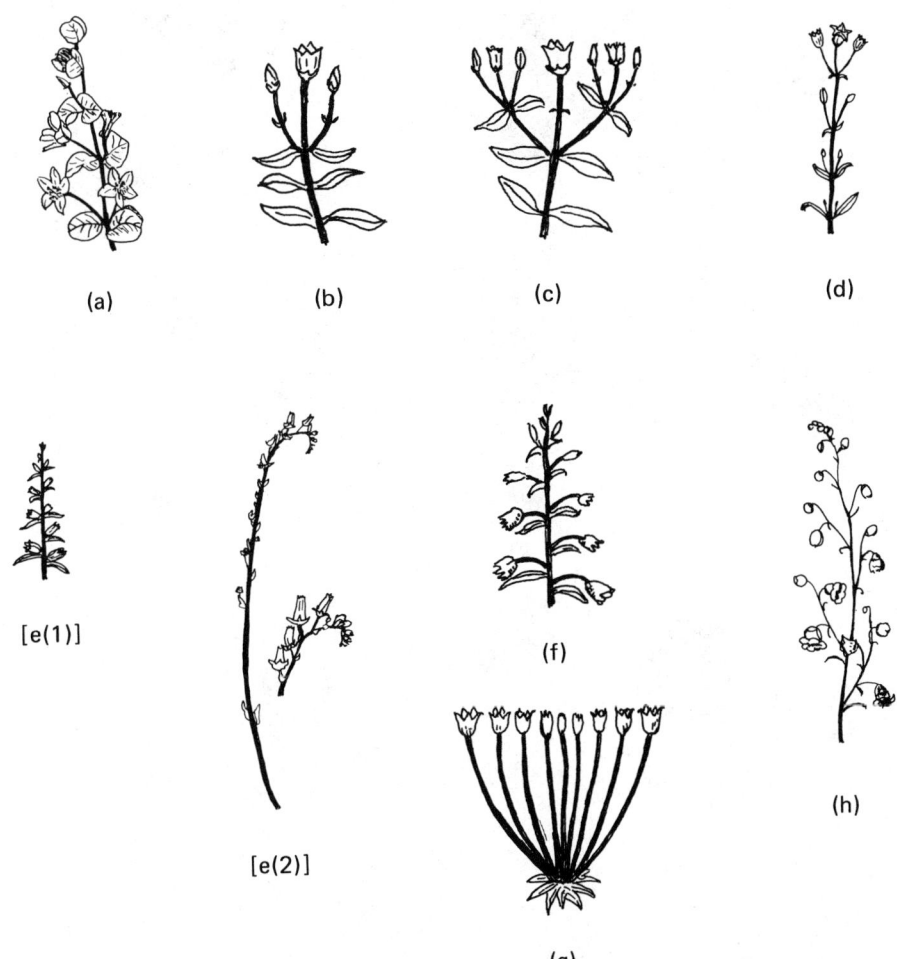

Fig. II.2.1.08. Inflorescences: (a) solitary flowers (moneywort); (b) simple dichasium; (c) compound dichasium; (d) racemiform (like a raceme but determinate); (e) spike, regular (1) and secund [one-sided, as in Cotyledon secunda (2)]; (f) raceme; (g) umbel (indeterminate); (h) panicle. (a)-(e)(1), (f)-(g) Redrawn from Gray, 1862; (e)(2) redrawn from Bailey, 1928; see Acknowledgments and Selected References.

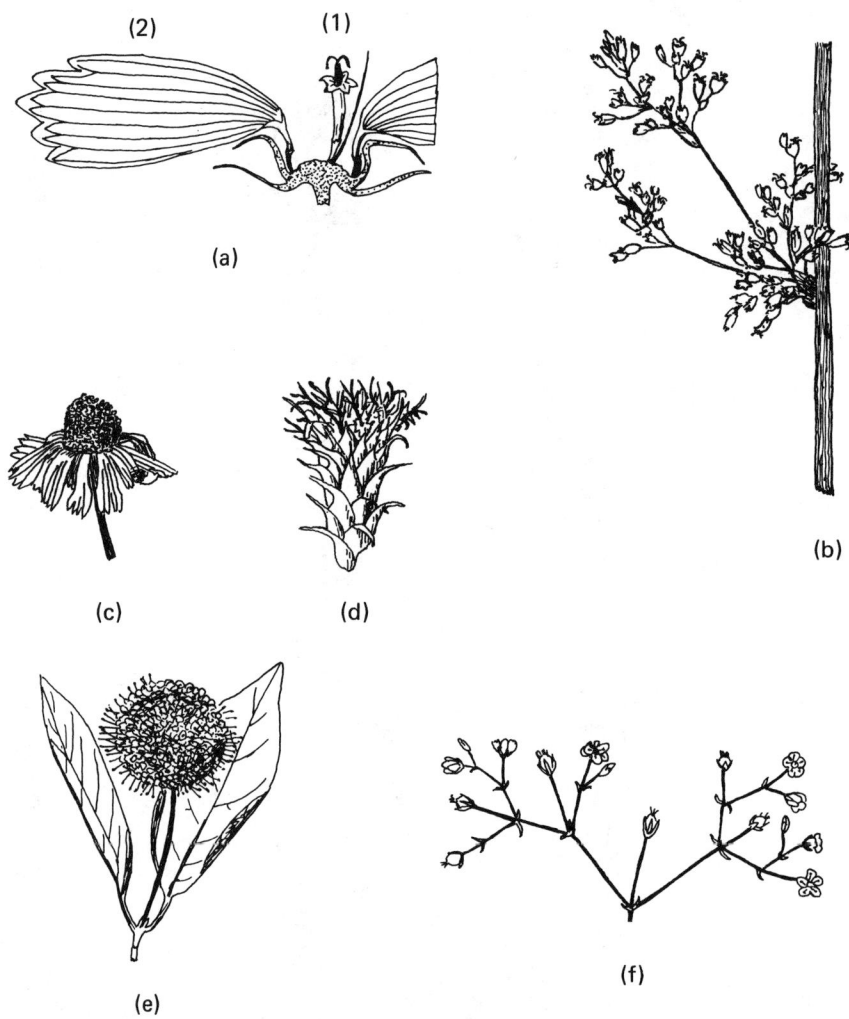

Fig. II.2.1.09. Inflorescences: (a) transverse section of coreopsis flower with single tubular perfect head flower (1) and neutral ray flower (2); (b) drepanium (Juncus effusus); (c) heterogamous inflorescence of Helenium autumnale with perfect head flowers and neutral ligulate ray flowers; (d) homogamous discoid inflorescence of Liatris squarrosa, all flowers tubular and perfect; (e) head of buttonbush (Cephalanthus occidentalis); (f) lax cyme of Alsine michauxii. (a), (c)-(f) Redrawn from Gray, 1862; (b) redrawn from Bailey, 1928; see Acknowledgments and Selected References.

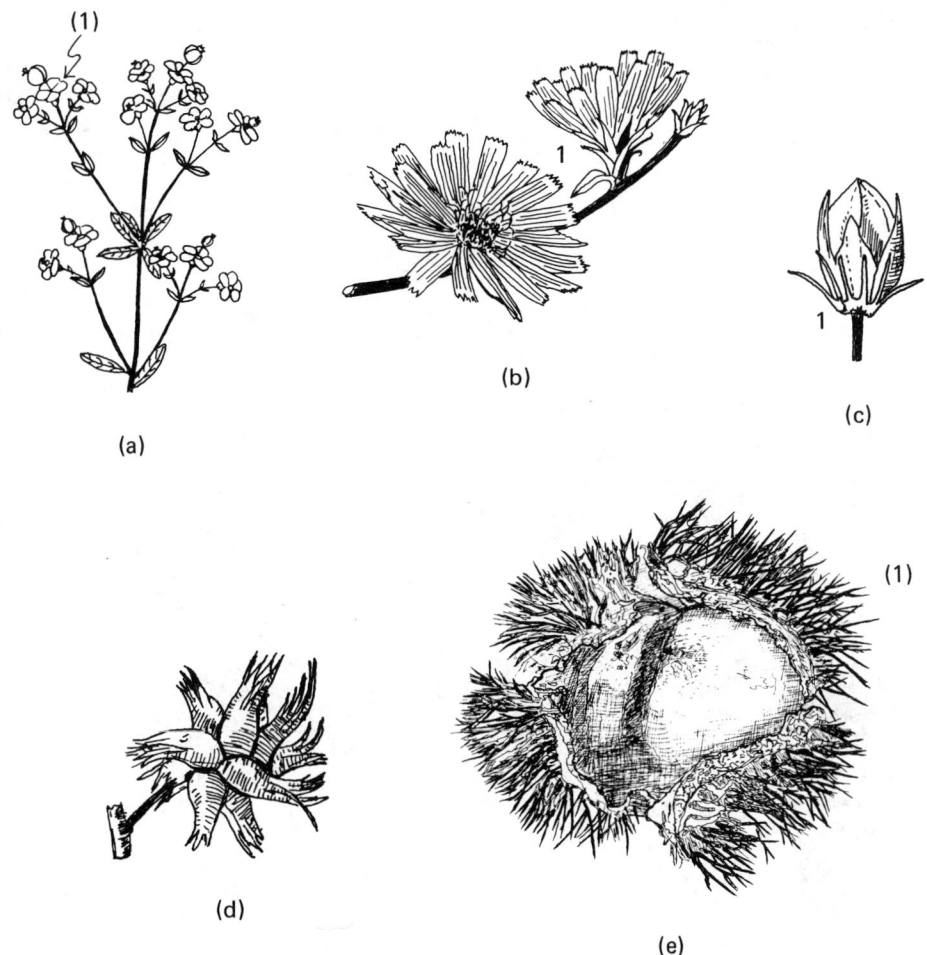

Fig. II.2.1.10. Involucres: (a) petaloid (1) in Euphorbia corollata; (b) phyllaries (1) as in cichory inflorescence; (c) epicalyx (1) of Hibiscus moscheutos; (d) accrescent, leaflike leathery involucre, filbert (Corylus avellana); (e) accrescent, spiny involucre forming the outer covering of a Chinese chestnut (Castanea mollissima) burr [note that spines (1) are branched and stiletto-pointed]. (a)-(c) Redrawn from Gray, 1862; (d) redrawn from Johnson, 1633; see Acknowledgments and Selected References.

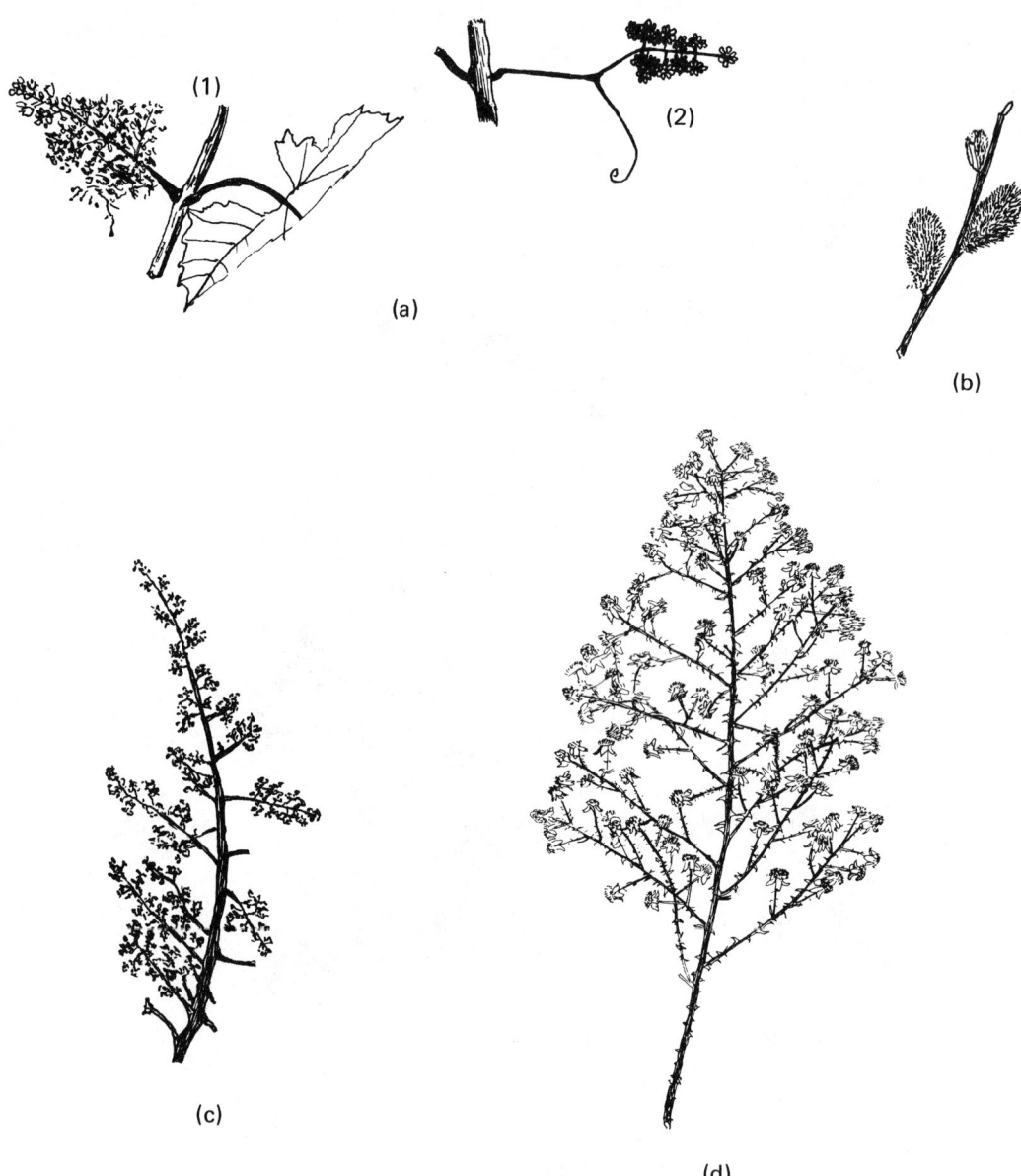

Fig. II.2.1.11. Inflorescences: (a) leaf-opposed, frost (Vitis vulpina) (1) and muscadine (V. rotundifolia) (2) grapes; (b) muscariform, catkin, Salix discolor (pussy willow); (c) small panicle, mango (Mangifera indica); (d) gigantic paniculiform inflorescence, Rubus effertus. (a)(1) Redrawn from Bailey, 1928; (b) redrawn from Sargent, 1905; (c) redrawn from Ruehle and Ledin, 1960; (d) redrawn from Bailey, 1944b; see Acknowledgments and Selected References.

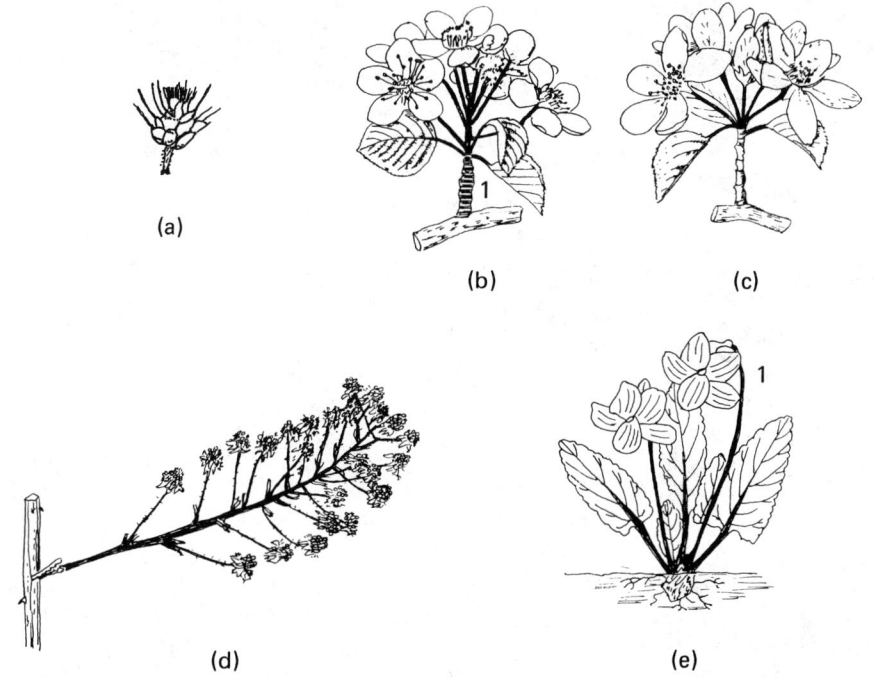

Fig. II.2.1.12. Inflorescences: (a) phyllaries in blessed thistle (Cnicus benedictus); (b) corymbiform inflorescence of pear (Pyrus spp.) with central column or pirostele (1); (c) the very different umbelloid one of apple (Malus spp.); (d) racemiform inflorescence, Rubus bractealis; (e) scape (1), violet (Viola sagitatta). (a) Redrawn from Muenscher, 1955; (b), (c) redrawn from Bailey, 1949; (d) redrawn from Bailey, 1947; (e) redrawn from Gray, 1862; see Acknowledgments and Selected References.

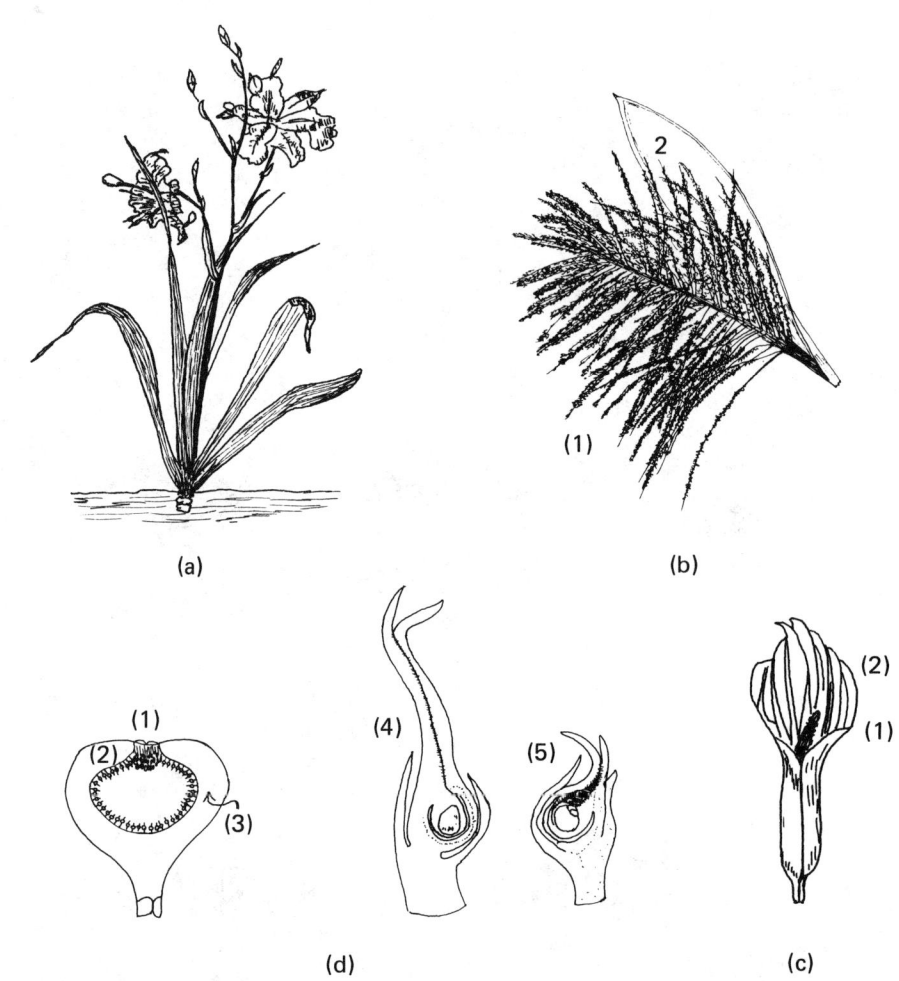

Fig. II.2.1.13. Inflorescences: (a) rhipidium, iris (Iris japonica); (b) multiple spikelets on a slender main rachis form the spadix (1) and heavy woody cymba, the spathe (2), of pindo palm (Butia capitata); (c) the spadix (1) in Indian turnip (Arisaema triphylla) is thick and fleshy and the spathe (2) thin and nonwoody; (d) syconium (1)-(3), monoecious inflorescence of fig (Ficus spp.), with staminate flowers (1) near the ostiole (2) and pistillate flowers (3) lining the rest of the cavity; the latter are long-styled (4) in edible fig (Ficus carica) and short-styled (5) in the caprifig (figures are semidiagrammatic). (a) Redrawn from Bailey, 1928; (b) redrawn from Bailey, 1936; (c) redrawn from Gray, 1862; (d) redrawn from Condit, 1947; see Acknowledgments and Selected References.

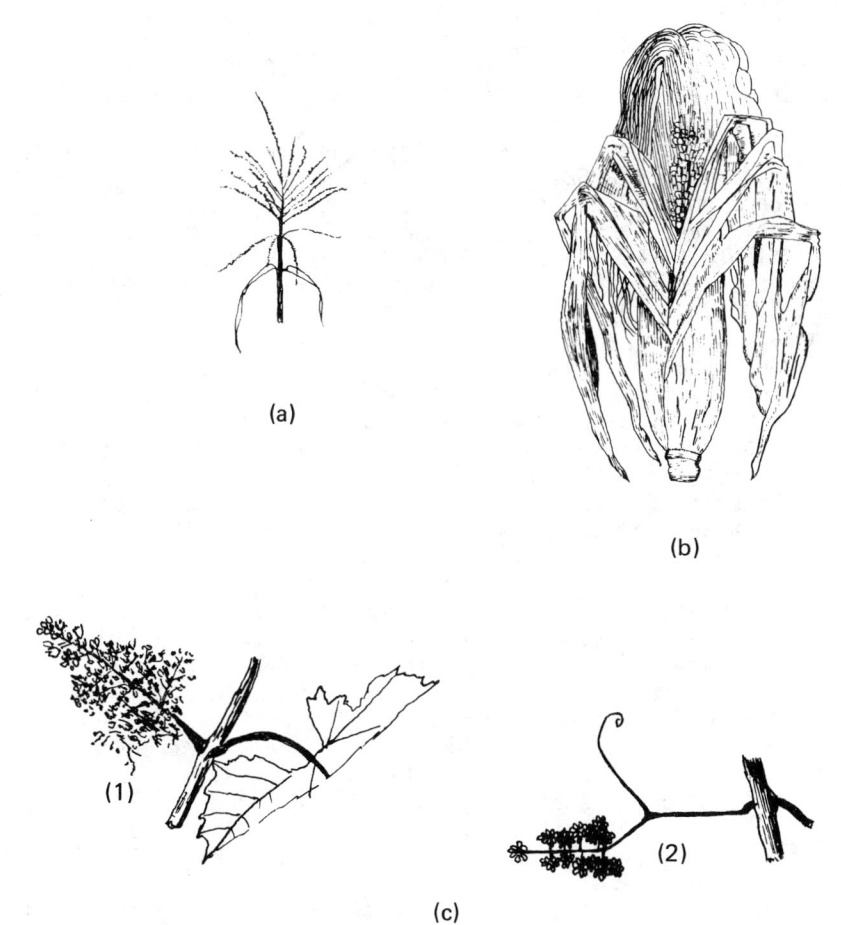

Fig. II.2.1.14. Inflorescences: (a) tassel (staminate) and (b) ear (spike, pistillate) inflorescences of corn (Zea mays); (c) thyrse (thyrsus) of frost (1) and muscadine (2) grapes. (a), (b), (c)(1) Redrawn from Bailey, 1928; see Acknowledgments and Selected References.

II.2.2. Flowers

.000 Flowers: an axis bearing one or more pistils, one or more stamens, or other structures associated with a reproductive organ. (Fig. II.2.2.01, .02, .24)
.001 Abaxial: away from the axis or central line, turned toward the base; dorsal. (L 55)
.002 Accrete: fastened to another body and growing with it, like epifoliar flowers or an accrescent calyx on a fruit. (Fig. II.2.2.15i) (ST)
.003 Adaxial: toward the axis or center, turned toward the apex; ventral. (L 55)
.004 Adherent, clinging: united laterally by the whole surface with another organ, like sepals to petals in some flowers. (Fig. II.2.2.22c) (ST)
.005 Adnate: adhering to the face of a thing, like stamens to petals; the fusion of unlike parts. (Fig. II.2.2.19b) (L 55)
.006 Addressed, appressed: closely and flatly pressed against. [Fig. II.2.2.23a(3)] (L 55)
.007 Androecium: stamens collectively; the male reproductive component of a flower. (G 62)
.007a. Anther: pollen-bearing element. [Fig. II.2.2.01a(6)] (G 62)
.007b. Anther cap: a cap over the anther in a monandrous orchid flower. [Fig. II.2.2.25b(1), d(1)] (SS)
.007c. Archesporium: cell from which pollen grains are formed in the microsporangium. (MH)
.007d. Arrangement of the androecium in a flower:
.007d-1. Antipetalous (stamens): (see Diplostemonous, II.2.2.007d-6).
.007d-2. Antisepalous (stamens): (see Diplostemonous, II.2.2.007d-6).
.007d-3. Bundles: (see Polyadelphous, II.2.2.007d-15).
.007d-4. Diadelphous: stamens connate by their filaments in two groups. (Fig. II.2.2.04c) (G 62)
.007d-5. Didynamous: four separate stamens in two pairs, one shorter than the other. (Fig. II.2.2.05c) (G 62)
.007d-6. Diplostemonous: having the androecium in two whorls, the stamens thus being twice as many as the petals; the normal arrangement is for stamens of the inner whorl to be antipetalous (opposite or in front of the petals) and the outer whorl, antisepalous; the reverse arrangement is denoted obdiplostemonous. (Fig. II.2.2.06c) (G 62)
.007d-7. Distinct (separate): stamens more or less the same length and not connate or coherent by their anthers or filaments. (Fig. II.2.2.05c, .06c) (G 62)
.007d-8. Epipetalous: stamens borne on or arising from the corolla, as in many families with sympetalous flowers, such as Apocynaceae, Rubiaceae, and Solanaceae. (Fig. II.2.2.05e, .09a) (G 62)
.007d-9. Episepalous: stamens borne on or apparently arising from the calyx as in Onagraceae. (Fig. II.2.2.22c) (G 62)
.007d-10. Exserted: produced or projected beyond some enclosing organ or part, as in stamens or styles in certain flowers. (G 62)
.007d-11. Fascicles: (see Polyadelphous, II.2.2.007d-15).
.007d-12. Monadelphous: stamens connate by their filaments in a single group, as in Malvaceae and Meliaceae. (Fig. II.2.2.04a,b,d, .22d) (G 62)
.007d-13. Obdiplostemonous: (see Diplostemonous, II.2.2.007d-6).

II.2.2. MORPHOLOGY and ANATOMY

.007d-14. Phalanges: (see Polyadelphous, II.2.2.007d-15).
.007d-15. Polyadelphous: stamens connate by their filaments in more than two groups, the latter often denoted as bundles, fascicles, or phalanges. (Fig. II.2.2.04f) (G 62)
.007d-16. Separate (stamens): (see Distinct, II.2.2.007d-7).
.007d-17. Synandrium: stamens coherent by their anthers, as in some aroids (Araceae). (Fig. II.2.2.06a) (L 55)
.007d-18. Synantherous, syngenesious: with anthers cohering into a tube or ring, as in Sinningia and Compositae. (Fig. II.2.2.04b, e, .06b,e) (G 62)
.007d-19. Syngenesious: (see Synantherous, II.2.2.007d-18).
.007d-20. Tetradynamous: six stamens in three pairs, with one pair shorter than the other two, as in the Cruciferae. (Fig. II.2.2.05f) (G 62)
.007e. Caudicle: little stem, stemlet; stem of a pollinium in orchids. [Fig. II.2.2.24d(3), .25d(2),e] (DR)
.007f. Connective: extension of the filament as in the tissue joining the cells of an anther. (Fig. II.2.2.08f) (G 62)
.007g. Contabescence: stamen abortion. (Fig. II.2.2.05c)
.007h. Dehiscence: release of pollen grains. (G 62)
.007h-1. Apical tubes: opening by anthers produced into tubes, as in Vaccinium. (Fig. II.2.2.10d) (G 62)
.007h-2. Extrose: opening of anther slits or flaps along the surface facing outward (see also II.2.2.033). (Fig. II.2.2.10g) (G 62)
.007h-3. Flaps (valves): opening by individual flaps for each locule, as in Lauraceae. (Fig. II.2.2.10f,g) (G 62)
.007h-4. Introrse, anterior: opening of anther slits or flaps along the surface facing inward. (Fig. II.2.2.10g) (G 62)
.007h-5. Longitudinal: lengthwise splitting of each anther sac (most common). (Fig. II.2.2.10a) (G 62)
.007h-6. Poricidal: opening by apical pores. (Fig. II.2.2.10d,e) (G 62)
.007h-7. Transverse: crosswise splitting of anther sac. (L 55)
.007i. Didymous, twins: growing in pairs or divided into two equal parts, like stamens. (Fig. II.2.2.05a) (G 62)
.007j. Dorsal, dorsifixed: fixed on the back of anything, as a filament on an anther. (Fig. II.2.2.07c) (G 62)
.007k. Endothecium: the outermost layer of the anther wall, located below the epidermis. [Fig. II.2.2.11b(4)] (F & G, ES)
.007l. Farina: pollen (obsolete).
.007m. Filament: stalk of the stamen. [Fig. II.2.2.01a(7).11b(1)] (G 62)
.007n. Geniculate, knee-jointed: bent abruptly like a knee. (Fig. II.2.2.08h,.17e) (L 55)
.007o. Horseshoe-shaped (hippocrepiform): bent almost into a circle but with an opening, like certain embryos and stamens. (Fig. II.2.2.08e) (H 73)
.007p. Innate: adhering to the apex of a thing, like an anther to a filament. (Fig. II.2.2.07a) (G 62)
.007q. Knee-jointed: (see Geniculate, II.2.2.007n).
.007r. Lobes (sacs or locules): normally two in angiosperms, one in some families (e.g. Malvaceae), transversely locellate in Loranthaceae, each containing two sporangia. (Fig. II.2.2.07,.10) (G 62, H 73)
.007s. Microsporangium: pollen sac or pollen mother cell that contains archesporia; the process of forming microspores is called gametogenesis. (Fig. II.2.2.11a) (ES)

II.2.2. MORPHOLOGY and ANATOMY

.007t.   Microspores (pollen grains): dinucleate and trinucleate cells covered with a variously scored, furrowed, pitted, or ridged outer wall (the extine) and a thin inner wall (the intine). (Fig. II.2.2.12, .13) (L 55)
.007t-1.   Aperturate: (see Sporoderm, II.2.2.007t-10).
.007t-2.   Colpate, colpi: (see Sporoderm, II.2.2.007t-10).
.007t-3.   Extine: (see Sporoderm, II.2.2.007t-10).
.007t-4.   Granular: individual grains distinct or separate. (Fig. II.2.2.13d)
.007t-5.   Intine: (see Sporoderm, II.2.2.007t-10).
.007t-6.   Massula: a coherent group of pollen grains produced from a single mother cell; e.g., the pollinium of an orchid. (MH)
.007t-7.   Palynology: the study of the structure of pollen grains and spore walls. (ES)
.007t-8.   Pollen grains: (see Microspores, II.2.2.007t).
.007t-9.   Pollinium: grains of a pollen sac agglutinated into a single, often hard and horny mass, as in orchids (Orchidaceae) and milkweed (Asclepiadaceae); furnished in the former with a pedicel (caudicle) and viscid disc (viscidium). [Fig. II.2.2.24d(2),(3), .25d(2),e, .26d-f] (L 55, DR)
.007t-10.   Sporoderm: the wall of an individual microspore (individual pollen grain), with highly complex sculpturing; the extine is the outer wall, the intine, the inner wall; the wall may be provided with pores, then aperturate or with furrows, colpate. (Fig. II.2.2.12d; .13a) (ES)
.007t-11.   Tetrad: clusters of four coherent or connate pollen grains. [Fig. II.2.2.11b(7),.12b(2),c,d] (L 55)
.007t-12.   Vesiculate pollen: pollen grains in which the extine (outer wall) is lifted up to form an air bladder and thereby provides greater buoyancy for movement by wind. [Fig. II.2.2.12d(5)] (B & C)
.007u.   Microsporocyte: Microspore mother cell which undergoes two meiotic divisions to produce four microspores. [Fig. II.2.2.11a(3), .12b] (ES)
.007v.   Petaloid: having the color and texture of a petal, as in stigmas or modified stamens in certain flowers. (Fig. II.2.2.09b) (G 62)
.007w.   Positions: manner of attachment to filament.
.007w-1.   Apical: similar to basifixed but has the anthers tapering to a point with a prominent connective between the lobes, as in the Oleaceae. (Fig. II.2.2.07d, .08a) (H 73)
.007w-2.   Apicifixed: attached at the apex of the filament. (Fig. II.2.2.08b,d) (G 62)
.007w-3.   Basifixed: terminal on filament or seemingly so. (Fig. II.2.2.07a,h) (L 55)
.007w-4.   Divergent: anther lobes spread at an angle from the apex. (Fig. II.2.2.07d) (RD)
.007w-5.   Dorsifixed: anther lobes affixed on the back to the filament. (Fig. II.2.2.08c) (L 55)
.007w-6.   Forked: branched anthers typical of stamens in Vaccinium. (Fig. II.2.2.10e) (G 62)
.007w-7.   Horseshoe-shaped: inverted "U"-shaped, as in anthers of Bixa orellana. (Fig. II.2.2.08e) (H 73)
.007w-8.   Lateral: (see II.1.1.1.012).
.007w-9.   Medifixed: attached by the middle. (Fig. II.2.2.07b) (L 55)
.007w-10.   Oblique: one anther situated higher than the other. (Fig. II.2.2.07e) (RD).

II.2.2. MORPHOLOGY and ANATOMY

.007w-11. Parallel: anthers situated side by side. (Fig. II.2.2.07f) (RD)
.007w-12. Transverse: anthers situated at right angles to the axis. (Fig. II.2.2.07g) (RD)
.007w-13. Twisted: sinuous; curved or bent in the form of a spiral, as in anthers of certain plants (e.g., Cucurbitaceae and some Palmae). (Fig. II.2.2.05f) (G 62)
.007w-14. Versatile: (see II.2.2.007c,ad).
.007x. Sporangium: comprised of the microsporangium plus accessory tissues; two sporangia in an anther lobe usually become uniloculate prior to anthesis from disintegration of tissues separating the sacs. (Fig. II.2.2.11b) (L 55)
.007y. Staminode: sterile, nonfunctional, often antherless stamen, sometimes variously modified in shape or form. [Fig. II.2.2.01i(2), .09d(1), e(2), f(1)(ii)] (L 55)
.007z. Stomium: the opening in the anther wall where dehiscence occurs; may be between two locules, oriented transversely near the apex of the anther lobe, or lateral. (Fig. II.2.2.11c) (ES)
.007aa. Tapetum: the innermost of the parietal layers in the anther wall; apparently secretory tissue concerned with the nutrition of the pollen mother cells and young microspores. [Fig. II.2.2.11a(4), b(6)] (ES)
.007ab. Twins: (see Didymous, II.2.2.007i).
.007ac. Twins: (see Geminate, II.2.2.039).
.007ad. Versatile: adhering slightly by the middle so that the two halves are nearly balanced and swing backward and forward, as in anthers. (L 55)
.008 Anisogynous: having fewer carpels than sepals. (Fig. II.2.2.03a) (G 62)
.009 Anisomerous: floral organs in each whorl unequal in number and size. (Fig. II.2.2.03b, .24c) (H 69)
.010 Annulus: a ring; a ringlike part, structure, marking, or the like, as an annular disk. [Fig. II.2.2.03c, .27c(3)] (G 62)
.011 Anterior: (see Introrse, II.2.2.046).
.012 Anticous: (see Introrse, II.2.2.046).
.013 Bisexual, hermaphrodite: contains both androecium and gynoecium. (L 55)
.014 Centrifugal: developing from the center outward. (L 55)
.015 Centripetal: developing from the outside toward the center. (L 55)
.016 Coherent, confluent: in general the fastening (but not fusion) of homogeneous (like) parts; also there is no difference in practice among coherent, adherent, and accrete, as in many flower parts. (Fig. II.2.2.05g) (ST)
.017 Column: body formed of the union of stamens and pistil in orchids or of the stamens only, as in mallows (Malvaceae). (Fig. II.2.2.06f, .25b-d) (L 55)
.018 Complete: contains all four whorls, calyx, corolla, androecium (stamens), and gynoecium (pistils). (Fig. II.2.2.01a,b) (L 55)
.019 Compressed: flattened lengthwise like a peapod. (Fig. II.2.2.28a) (ST)
.020 Connate: the fusion of like parts (see Adnate, II.2.2.005). (L 55)
.021 Connivent, convergent: having a gradually inward direction, as in many petals. (Fig. II.2.2.09c) (ST)
.022 Corona = crown, coronet: any appendage or extrusion that stands between the corolla and stamens, as in flowers of Passiflora, is on the corolla (as in some amaryllids), or is an outgrowth of the

II.2.2. MORPHOLOGY and ANATOMY

.023
.024
.025
.026
.027
.028
.029
.030
.031
.032
.033
.034
.035
.036
.037
.038
.039
.040
.041
.041a.
.041b.
.041c.
.041d.
.041e.
.041f.
.041g.
.041h.

staminal part or circle (as in milkweeds); (see also Crown, II.1.1.3.031). [Fig. II.2.2.26a(3), .27a(2)] (L 55, G 62)
Crateriform, goblet-shaped: concave, hemispherical, a little contracted at the base. [Fig. II.2.2.22b(1)] (ST)
Crown: (see Corona, II.2.2.022).
Crowned: situated on the top of anything. (G 62)
Cyclic: having the parts arranged in cycles or whorls; commonly applied to flowers. (L 55)
Diandrous: with two stamens, as in certain orchid flowers. (Fig. II.2.2.24e,f) (SS)
Diclinous, dicliny: unisexual. (Fig. II.2.2.01c,d) (L 55)
Dimidiate, halved: only half or partially formed. (Fig. II.2.2.10c) (G 62)
Distant, remote: used in contradistinction to imbricated, dense, or approximated. (ST)
Epiphyllous (epifoliar): inserted on the leaf, as an epiphyllous flower or inflorescence. (Fig. II.2.2.03e,f) (G 62)
Estivation (aestivation), prefloration: arrangement of floral parts in the bud (see Vernation, II.1.1.2 for terms). (G 62)
Extrafloral: parts outside the flower proper, as in accessory parts (e.g., bracts and receptacle). (G 62)
Extrorse, posticous: turned away from the axis to which it appertains, or opening to the outside as some anthers in dehiscence. (ST)
Filiform, thread-shaped: slender like a thread; e.g., a filament. [Fig. II.2.2.10e(2)] (G 62)
Floral: of or belonging to the flower. (G 62)
Foliaceous, leaflike: having the form or texture of a leaf. (Fig. II.2.2.22d) (ST)
Free, distinct, separate: when there is no cohesion between parts, as in a freestone peach or nectarine, where the flesh separates readily from the stone; also in many flowers (e.g., polysepalous or polypetalous organs). (Fig. II.2.2.06c) (L 55)
Geminate, twins: growing in pairs. (L 55)
Glume: small, chafflike bract, as in grasses. [Fig. II.2.2.24b(1)(i)] (L 55)
Gynoecium: the pistil (or pistils) collectively; the female reproductive unit of the flower. [Fig. II.2.2.01a(1-4), .14] (L 55)
Androgynophore (gynandrophore): structures formed by stamens that arise hypogynously from a gynophore apex and usually bear the pistil, as in Passiflora. [Fig. II.2.2.27a(1)] (B & C)
Apocarpous: pistils (carpels), separate; the term is frequently applied to a gynoecium of separate pistils, as in Magnolia, Annona, and Ranunculus. [Fig. II.2.2.14a,b, .22b(2)] (G 62)
Beaked: (see Proboscoid, II.2.2.041w).
Beaked: (see Rostrate, II.2.2.041y).
Carpel: foliar, usually an ovule-bearing unit of a simple ovary or one of the foliar units of a compound ovary or pistil, in which two or more carpels are combined by connation in its orgin and development. (Fig. II.2.2.14a) (L 55)
Cochleate: twisted in a short spire like a snail shell. (Fig. II.2.2.15f) (G 62)
Dissepiment: a partition or septum that divides the locules of a compound ovary or a fruit derived from a compound ovary. [Fig. II.2.2.14d(2)] (G 62)
Gamete: a mature male or female reproductive cell. (Fig. II.2.2.13b(3), .16b(9)] (L 51)

II.2.2. MORPHOLOGY and ANATOMY 123

.041i. Gametogenesis: the process of the formation of male and female gametes. (Fig. II.2.2.16a) (L 51)
.041j. Gametophyte: the phase in the life cycle of a plant that bears the gametes. (L 51)
.041k. Globose: forming a nearly true sphere. (G 62)
.041l. Gynandrophore: (see Androgynophore, II.2.2.041a).
.041m. Gynophore: stipe (II.2.2.041ab) of an ovary prolonged within the perianth. (L 55)
.041n. Gynostemium: fusion of the upper parts of the gynoecium and androecium to form a single central structure, as in the Orchidaceae and Asclepiadaceae. (Fig. II.2.2.24f, .25b) (SS)
.041o. Lamellate: provided with many finlike blades or cross partitions, as in the ovary of Nymphaeaceae and certain stamens. (Fig. II.2.2.15g) (L 55)
.041p. Locule: chamber or space within a simple ovary (often erroneously termed cell). [Fig. II.2.2.02(6),(16)] (L 55)
.041q. Megasporangium: (see Ovule, II.2.2.041t-4).
.041r. Megaspore mother cell, megasporocyte:(see Ovule, II.2.2.041t-4).
.041s. Megasporophyll (of an angiosperm flower): (see Carpel, II.2.2.041e).
.041t. Ovary: ovule-bearing unit of the pistil; simple if formed from a single carpel, compound if formed from more than one. [Fig. II.2.2.01a(3), .25b(4), c(2)] (L 51)
.041t-1. Placenta: zone or area occupied by ovules; the intrusion of the ovary wall into the locule. (Fig. II.2.2.14g) (L 55)
.041t-2. Placentation: mode of ovule arrangement within an ovary. (Fig. II.2.2.14c-g, .15) (L 55)
.041t-2-a. Axile: ovules attached to a septum (partition that results from the union and fusion of two or more single, unicarpellate, unilocular pistils) or septa in a compound ovary; ovules are usually in vertical rows. [Fig. II.2.2.02(6),(16), .14c,d] (L 55)
.041t-2-b. Axile-parietal: anomalous arrangement found in pomegranate (Punica) in which placentation is axile in the lower half and parietal in the upper half of the ovary. (Fig. II.2.2.14d) (H 69)
.041t-2-c. Basal: ovules reduced to one and located at the bottom of the locule of a simple ovary or several ovules in a unilocular compound ovary. (Fig. II.2.2.15c) (L 55)
.041t-2-d. Erect: one or more ovules standing straight in the bottom of a locule. (Fig. II.2.2.15c) (L 55)
.041t-2-e. Free-central: ovules attached to a free-standing central column in a compound ovary; free-basal is similar except that the column is shorter and more or less globose. (Fig. II.2.2.15a,b) (L 55)
.041t-2-f. Lamellate: ovules scattered over the surfaces of the partitions in a compound ovary. (Fig. II.2.2.15g) (L 55)
.041t-2-g. Marginal (ventral): ovules in a double row along the inner (adaxial) wall of a simple ovary. (Fig. II.2.2.15e) (L 55)
.041t-2-h. Parietal: ovules attached to the inner side of the ovary wall in a compound ovary. (Fig. II.2.2.14e-g) (L 55)
.041t-2-i. Pendulous: ovules reduced to one and suspended from the top of a simple ovary. (Fig. II.2.2.15d) (L 55)
.041t-3. Positions of ovary: (relative to other floral parts). (Fig. II.2.2.19) (L 55)

II.2.2. MORPHOLOGY and ANATOMY

.041t-3-a.   Half-inferior (subinferior): ovary is situated in a more or less intermediate position between superior and inferior. (Fig. II.2.2.19c)

.041t-3-b.   Inferior: ovary is below the apparent area of perianth and androecial attachment. (Fig. II.2.2.19d)

.041t-3-c.   Subinferior: (see Half-inferior, II.2.2.041t-3-a).

.041t-3-d.   Superior: ovary situated above the zone of attachment of the perianth and androecium. (Fig. II.2.2.19a,b)

.041t-4.   Ovule: organ bearing the megasporangium (tissue that produces the megasporocyte) in the ovary, developing into a seed (II.2.3.3) after fertilization (when this is required); consisting of the integuments and nucellus, in which the embryo sac is developed. [Fig. II.2.2. 01a(4), .16b,c] (ES)

.041t-4-a.   Chalaza: basal part of the ovule where it is attached to the funiculus and the point at which vascular tissues enter and spread into the ovule (see also Seeds, II.2.3.3). [Fig. II.2.2.16b(3)] (ES)

.041t-4-b.   Embryo sac: the female gametophyte that arises from megaspore by gametogenesis; typically contains eight nuclei. [Fig. II.2.2.16b(8)] (MH)

.041t-4-b-1.   Antipodals: three nuclei normally at the chalazal end of the embryo sac. [Fig. II.2.2.16b(12)]

.041t-4-b-2.   Egg (female gamete): normally located at the micropylar end of the embryo sac; unites with one of the sperm nuclei in fertilization. [Fig. II.2.2.16b(9)]

.041t-4-b-3.   Polar nuclei: two nuclei near the center of the embryo sac which fuse with a sperm nucleus to form the endosperm (3n). [Fig. II.2.2.16b(11)]

.041t-4-b-4.   Synergids: two subsidiary nuclei, one on either side of the egg. [Fig. II.2.2.16b(10)]

.041t-4-c.   Foramen: (see Micropyle, II.2.2.041t-4-i).

.041t-4-d.   Funiculus (funicle): stalk on which the ovule is borne. [Fig. II.2.2.16b(2)] (ES)

.041t-4-e.   Hilum: the mark at the point of attachment of the funiculus to the ovule. (L 55)

.041t-4-f.   Integuments: layers of tissue that envelop the ovule and later develop into the seed coat(s) (see also Seed, II.2.3.3). [Fig. II.2.2.16b(5),(6)] (ES)

.041t-4-g.   Macrospore: (see Megaspore, II.2.2.041t-4-h).

.041t-4-h.   Megaspore (macrospore): spore with the property of giving rise to the female gametophyte (embryo sac); one of the four cells produced by two meiotic divisions of the megaspore mother cell (megasporocyte) of the megasporangium.

.041t-4-i.   Micropyle (foramen): opening or "mouth" of the ovule. (Fig. II.2.2.16b(4)] (L 55)

.041t-4-j.   Nucellus: body tissue of the ovule. [Fig. II.2.2.16b(7)] (ES)

.041t-4-k.   Positions of ovules inside an ovary. (L 55)

.041t-4-k-1.   Amphitropous (hemianatropous or half-inverted): the developing ovule is turned 90° on the funiculus so that the latter is fused to the nucellus for a part of the turn. [Fig. II.2.2.16c(2)] (L 55)

.041t-4-k-2.   Anatropous (inverted): the ovule is turned 180° during development so that the funiculus is fused to the nucellus along one side and the micropyle is close to the

II.2.2. MORPHOLOGY and ANATOMY 125

.041t-4-k-3. placental surface (see also Seed, II.2.3.3). [Fig. II.2.2.16c(3)] (L 55)
Campylotropous (incurved): the ovule is curved so that the micropyle almost meets the funiculus (see also Seed, II.2.3.3). [Fig. II.2.2.16c(4)] (L 55)
.041t-4-k-4. Obcamplyotropous: the ovule is curved so that the micropyle almost meets the funiculus which is fused around the nucellus. [Fig. II.2.2.16c(5)] (CR 76)
.041t-4-k-5. Orthotropous (upright): the ovule stands erect and the micropyle is at the end opposite the funiculus. [Fig. II.2.2.16c(1)] (L 55)
.041u. Pistil: collective term for ovary, style, and stigma (see Gynoecium, II.2.2.041). (L 55)
.041v. Polycarpic: having a gynoecium that forms two or more distinct ovaries = an aggregate, collective fruit (see Syncarpelly, II.2.3.2.039). (Fig. II.2.2.28b) (ES)
.041w. Proboscoid, beaked: having a hard terminal horn, like some pistils and fruits. (Fig. II.2.2.28d) (ST)
.041x. Rostellum: a small beak; such as that in orchid flowers as part of the column. [Fig. II.2.2.25b(2)] (L 55)
.041y. Rostrate, rostellate, beaked: terminating gradually in a hard, long, straight point (applied particularly to prolongations of pistils and fruit). (Fig. II.2.2.28d) (L 55)
.041z. Spiral: resembling in direction the spires of a corkscrew or other twisted object, like some styles or anthers. (Fig. II.2.2.05e, .28c) (ST)
.041aa. Stigma: the part of the pistil that receives the pollen; variable in shape and form according to the mode of pollination, with as many stigmatic surfaces provided as carpels unless fusion of the surfaces has taken place. [Fig. II.2.2.01a(1), .18, .25b(3)] (G 62)
.041ab. Stipe: the stalk of a pistil or other small organ. [Fig. II.2.2.14c(2), .15h(1)] (L 55)
.041ac. Stipitate: elevated on a stalk (stipe) which is neither a petiole nor a peduncle, as in Passiflora and some other flowers. [Fig. II.2.2.14c, .15h(2)] (ST)
.041ad. Style: narrow constricted "neck" between the stigma and ovary, sometimes absent when the stigma is sessile; often with internal cavities (stylar canals) to facilitate penetration of pollen tubes. [Fig. II.2.2.01a(2), .17] (L 55)
.041ae. Syncarpous: two or more carpels fused in a compound pistil. (Fig. II.2.2.14c-f) (L 55)
.042 Halved: (see Dimidiate, II.2.2.029).
.043 Hermaphrodite: (see Bisexual, II.2.2.013).
.044 Hypanthium (floral cup): structure formed by adnation of sepals, petals, and stamens, plus receptacle (in part), as in Rosaceae. [Fig. II.2.2.19b, .27a(4)] (L 55)
.045 Hypsophyll: floral bract. [Fig. II.2.2.27a(5), b(2)] (ES)
.046 Introrse, anterior, anticous: turned toward the axis to which it appertains. (Fig. II.2.2.10g) (G 62)
.047 Leaflike: (see Foliaceous, II.2.2.037).
.048 Lemma: a small greenish bract that is part of a grass floret. [Fig. II.2.2.24b(2)(ii)] (L 55)
.049 Locular, septate, partitioned: divided by internal partitions into chambers or compartments, as in anthers and ovaries. (Fig. II.2.2.08g) (L 55)

II.2.2. MORPHOLOGY and ANATOMY

.050 Monandrous: with one stamen, like most orchid flowers. (Fig.II.2.2. 25a,b) (SS)
.051 Monoclinous, monocliny: having both androecium and gynoecium on the same flower (see Diclinous, II.2.2.028) (G 62)
.052 Nectary: (see II.3.116). [Fig. II.2.2.24d(1)]
.053 Neutral: contains neither androecium nor gynoecium. (Fig. II.2.2. 01i) (G 62)
.054 Nodding: inclining very much from the perpendicular so that the apex is directed downward, as in certain flowers. (Fig. II.2. 2.21i) (ST)
.055 Nude: androecium or gynoecium or both are present but calyx and corolla are absent. (Fig. II.2.2.01h) (L 55)
.056 Palate: (see Personate, II.2.2.061c-14-r).
.057 Palea: one of the greenish bracts that enclose a grass floret. [Fig. II.2.2.24b(2)(i)] (L 55)
.058 Pappus: (see II.5.6.088).
.059 Partitioned: (see Locular, II.2.2.049).
.060 Pedicel: stalk supporting a single flower. [Fig. II.2.2.01a(12)] (G 62)
.061 Perianth: collective term for the calyx and corolla. [Fig. II.2.2. 01a(8),(9), .20] (G 62)
.061a. Calyx: outer envelope of usually green leaflike or bractlike structures of the perianth, individually termed sepals. (Fig. II.2.2.01a(9)] (G 62)
.061a-1. Chorisepalous: (see Polysepalous, II.2.2.061a-3).
.061a-2. Gamosepalous (synsepalous): sepals fused or connate anywhere along their length; the tubular part is denoted the calyx tube. (Fig. II.2.2.20e,f, .22c) (G 62)
.061a-3. Polysepalous (chorisepalous): sepals distinct. [Fig. II.2.2. 14b(2)] (G 62)
.061a-4. Sepal: (see Calyx, II.2.2.061a).
.061a-5. Synsepalous: (see Gamosepalous, II.2.2.061a-2).
.061b. Coalescent: incomplete fusion or sticking together of like parts, such as sepals and petals. (L 55)
.061c. Corolla: second or inner envelope or the perianth composed of petals. [Fig. II.2.2.01a(8)] (G 62)
.061c-1. Actinomorphic: radially symmetrical in form; several distinct corolla patterns, certain of which may be considered zygomorphic in some plants (or vice-versa). (L 55)
.061c-1-a. Bell-shaped: (see Campanulate, II.2.2.061c-1-b).
.061c-1-b. Campanulate: bell-shaped; broad tube with flaring limb or lobes. (Fig. II.2.2.21e) (G 62)
.061c-1-c. Cotyloform: like rotate (wheel-shaped) but with an erect limb. (Fig.II.2.2.22a) (ST)
.061c-1-d. Cruciform: cross-shaped, as in Cruciferae. (Fig. II.2.2. 20d) (G 62)
.061c-1-e. Cupola-shaped: slightly concave with a nearly entire margin. [Fig. II.2.2.27a(3)] (ST)
.061c-1-f. Cup-shaped: (see Cyathiform, II.2.2.061c-1-g).
.061c-1-g. Cyathiform, cup-shaped: same as urceolate but not contracted at the margin. [Fig. II.2.2.27(1)] (G 62)
.061c-1-h. Cylindraceous: more or less like a cylinder; a cylindrical tube with lobes small and flaring (tubular). (Fig. II. 2.2.23e) (G 62)
.061c-1-i. Funnelform (infundibular): funnel-shaped; with the tube expanding gradually from the bottom upward. (Fig. II. 2.2.21j) (G 62)

II.2.2. MORPHOLOGY and ANATOMY

.061c-1-j.   Goblet-shaped: (see Crateriform, II.2.2.023).
.061c-1-k.   Hypocrateriform, salver-shaped: figure (calyx, corolla, etc.) with a long, slender tube and flat limb. (Fig. II.2.2.20f,g) (G 62)
.061c-1-l.   Infundibular: (see Funnelform, II.2.2.061c-1-i).
.061c-1-m.   Olliform: (see Vase-shaped, II.2.2.061c-1-w).
.061c-1-n.   Pitcher-shaped: (see Urceolate, II.2.2.061c-1-v).
.061c-1-o.   Rotate, wheel-shaped: a calyx, corolla, or other organ with a short tube and spreading segments. (Fig. II.2.2.21a,b, .22d) (G 62)
.061c-1-p.   Salverform: long tube and shorter limb at a right angle. (Fig. II.2.2.21h) (G 62)
.061c-1-q.   Salver-shaped: (se Hypocrateriform, II.2.2.061c-1-k).
.061c-1-r.   Subglobose: similar to urceolate but rounded in outline. (Fig. II.2.2.21-1) (ST)
.061c-1-s.   Trumpet-shaped: (see Tubiform, II.2.2.061c-1-t).
.061c-1-t.   Tubiform, trumpet-shaped: hollow and dilated at one end like a trumpet. (Fig. II.2.2.06e) (ST)
.061c-1-u.   Tubular: more or less cylindrical tube, limb small. [Fig. II.2.2.22c, .23e, i(2)] (L 55)
.061c-1-v.   Urceolate, pitcher-shaped: large urn-shaped tube, usually somewhat constricted above the middle, and with small, more or less flaring lobes. (Fig. II.2.2.21m) (L 55)
.061c-1-w.   Vase-shaped, olliform: formed like a flower pot; an inverted truncate cone. (Fig. II.2.2.20c) (ST)
.061c-2.   Alae: the upper two petals of a flag-type flower in the Leguminosae. [Fig. II.2.2.20d, .23h(2)] (G 62)
.061c-3.   Carina (keel): boatlike structure formed by coalescence of the two lower lateral petals along their lower edges in a legume flower. [Fig. II.2.2.20d, .23h(3)] (G 62)
.061c-4.   Choripetalous: (see Polypetalous, II.2.2.061c-11).
.061c-5.   Epicoralline (= epipetalous): inserted on the corolla, like stamens in certain families (see II.2.2.007d-8). (G 62)
.061c-6.   Galea: upper (proximal) lip of a labiate corolla; helmet-shaped. [Fig. II.2.2.20f, .23b(1)] (G 62)
.061c-7.   Gamopetalous (sympetalous): petals connate at their base or anywhere along their length. (Fig. II.2.2.20e) (G 62)
.061c-8.   Irregular: a flower with neither bilateral nor radial symmetry. (Fig. II.2.2.20i) (L 55)
.061c-9.   Labellum: a lip, particularly the lip in an orchid flower. [Fig. II.2.2.24c(1), e(1),f(1)] (L 55)
.061c-10.   Patent, spreading: having a gradually outward direction, like petals from the ovary. (Fig. II.2.2.21b,d,e) (G 62)
.061c-11.   Polypetalous (choripetalous): petals distinct. [Fig. II.2.2.21i, .22a, b,e] (G 62)
.061c-12.   Synpetalous: (see Gamopetalous, II.2.2.061c-7).
.061c-13.   Vexillum: the median petal of a flag-type flower in the Leguminosae. [Fig. II.2.2.23e(1)] (G 62)
.061c-14.   Zygomorphic: bilaterally symmetrical in form, as in the Labiatae, Caesalpiniaceae, and Compositae. (Fig. II.2.2.20e-h, .23,h) (L 55)
.061c-14-a.   Bilabiate: having two lips (each with two or three lobes), as in mints (Labiatae) and figwort (Scrophularia). (Fig. II.2.2.20e, .23g) (G 62)
.061c-14-b.   Calcarate: spurred (see Spur, II.1.1.3.093). (Fig. II.2.2.23c,d) (G 62)

II.2.2. MORPHOLOGY and ANATOMY

.061c-14-c. Calceolate: slipperlike; having the form of a round-toed shoe. (Fig. II.2.2.24e)
.061c-14-d. Carinate: keeled; provided with a projecting central, longitudinal line or ridge on the lower or under surface. (Fig. II.2.2.20d, .23h(3)] (G 62)
.061c-14-e. Caryophyllaceous: (see Unguiculate, II.2.2.061c-14-w).
.061c-14-f. Clawed: (see Unguiculate, II.2.2.061c-14-w).
.061c-14-g. Corniculate (horned): bearing or terminating in a small hornlike protuberance or process, as in flowers of Asclepiadaceae. [Fig. II.2.2.26a(4)] (G 62)
.061c-14-h. Coronate: with a corona or crown. [Fig. II.2.2.26a(3), .27 a(2), b(1)] (G 62)
.061c-14-i. Cucullate: hooded, or hood-shaped, as in flowers of Asclepiadaceae. (Fig. II.2.2.26) (G 62)
.061c-14-j. Galeate: with a hood or helmet-shaped part or structure, as found in the upper lip of some corollas. (Fig. II. 2.2. 20f; .23b) (G 62)
.061c-14-k. Gibbous: with a swelling or bulging on one side or near the base, like a corolla. (Fig. II.2.2.23c) (G 62)
.061c-14-l. Hooded: (see Cucullate, II.2.2.061c-14-i).
.061c-14-m. Keeled: (see Carinate, II.2.2.061c-14-d).
.061c-14-n. Labiate, bilabiate: term applied to a synsepalous calyx or sympetalous corolla separated into two unequal divisions, one anterior, the other posterior, with respect to the axis; bilabiate is the more common term. (Fig. II.2.2.20, .23b) (G 62)
.061c-14-o. Ligulate: (see II.5.5.032).
.061c-14-p. Papilionaceous: pealike, shaped like a butterfly with a standard; wings and keel, as in the Fabaceae. (Fig. II. 2.2.20d, .23h) (G 62)
.061c-14-q. Peloric, pelory: a flower with abnormal structural regularity appearing on a plant that typically has irregular (zygomorphic) flowers. (Fig. II.2.2.20j) (G 62)
.061c-14-r. Personate, ringent: term applied to a sympetalous corolla in which the limb is unequally divided, the upper lip being arched and the lower lip (palate) prominent and pressed against it so that the opening resembles the mouth of a gaping animal when the lips are compressed (e.g., the appearance of a snapdragon flower when pressed laterally). (Fig. II.2.2.23a,c) (G 62)
.061c-14-s. Ringent: (see Personate, II.2.2.061c-14-r).
.061c-14-t. Saccate: (see Calceolate, II.2.2.061c-14-c).
.061c-14-u. Tumid: swollen, somewhat inflated. (G 62)
.061c-14-v. Turgid: slightly swollen. (L 55)
.061c-14-w. Unguiculate: tapering below into a claw or stalklike base; said of certain petals, like those of the pink and wallflower. (Fig. II.2.2.23f) (G 62)
.061d. Positions of perianth relative to ovary. (G 62)
.061d-1. Epigynous: perianth and stamens borne or arising from the ovary; here the ovary is inferior and the flower is not perigynous. (Fig. II.2.2.19d)
.061d-2. Hypogynous: perianth and stamens are borne on the receptacle (torus) or under the ovary; = hypogamous. (Fig. II.2.2.19a)
.061d-3. Perigynous: perianth is borne on or arises around the ovary but neither beneath nor above it; e.g., when the calyx, corolla, and stamens all arise from the edge of a cup-shaped hypanthium. (Fig. II.2.2.19b,c)

II.2.2. MORPHOLOGY and ANATOMY

.061e. Sepalody: resembling, or of the nature of, a sepal. (L 55)
.061f. Tepal: components of the perianth are similar in size, form, and coloration so that sepals cannot be distinguished from the petals, as in Musaceae and Orchidaceae, among monocotyledons. (Fig. II.2.2.20c) (L 55)
.062 Pistillate: flower with gynoecium only; may, however, have a calyx and/or a corolla but lacks an androecium. (Fig. II.2.2.01c,g, .27e) (L 55)
.063 Pistillode: sterile, functionless pistil. [Fig. II.2.2.01i(1)] (L 55)
.064 Posterior: at or toward the back; opposite the front; toward the axis. (L 51)
.065 Posticous: (see Extrorse, II.2.2.034).
.066 Receptacle (torus): the more or less enlarged or elongated end of the stem or floral axis on which some or all of the flower parts are borne. (L 51)
.067 Remote: (see Distant, II.2.2.030).
.068 Resupinate: inverted in position by a twisting of the stalk, as in orchid flowers, or of the petiole, as in Alstroemeriaceae. (Fig. II.2.2.20g; compare with Fig. II.2.2.20b,c) (L 55)
.069 Rosaceous: having the same arrangement as the petals of a single rose. (Fig. II.2.2.21k) (G 62)
.070 Secund, unilateral: one-sided; arranged on, or turned toward, one side only; excentric. (Fig. II.2.2.03d) (L 51)
.071 Sepaloidy: metamorphosis of other floral organs into sepals. (L 55)
.072 Septate: (see Locular, II.2.2.049).
.073 Spreading: (see Patent, II.2.2.061c-10).
.074 Staminate: flower with androecium only. (Fig. II.2.2.01d,f, .27d,f) (L 55)
.075 Stellate: the same as verticillate except the parts are narrow and acute. (Fig. II.2.2.03g) (ST)
.076 Sterile: lacking functional sexual organs. (Fig. II.2.2.01i) (L 55)
.077 Strombus-shaped: twisted in a long spire. (Fig. II.2.2.03h) (ST)
.078 Subconfluent: almost confluent; in other words, coalescent. (L 51)
.079 Thread-shaped: (see Filiform, II.2.2.035).
.080 Torus: (see Receptacle, II.2.2.066).
.081 Unilateral: (see Secund, II.2.2.070).
.082 Unilocular: anther or ovary with one locule. (Fig. II.2.2.10c) (L 55)
.083 Ventricose: swollen unequally on one side. [Fig. II.2.2.23g(1)] (L 55)
.084 Verticillate, whorled: (see II.2.1.080).
.085 Wheel-shaped: (see Rotate, II.2.2.061c-1-o).
.086 Whorled: (see Verticillate, II.2.1.080).

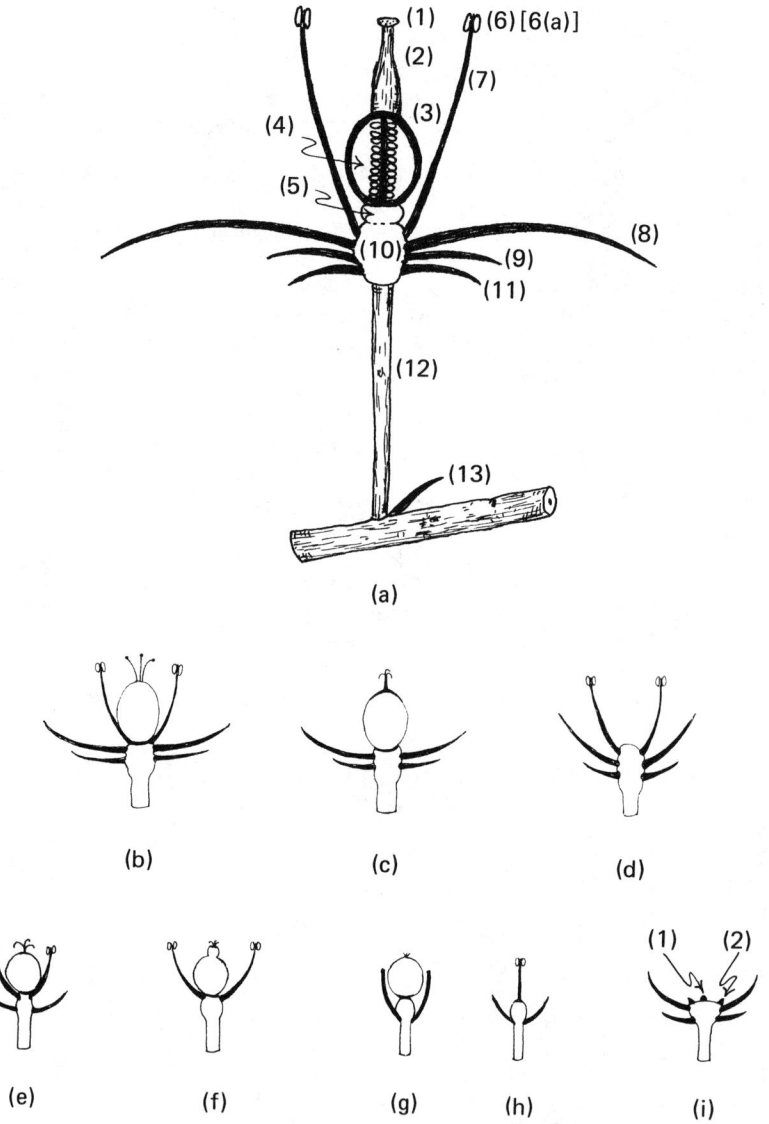

Fig. II.2.2.01. Flowers (diagrammatic): (a) flower parts': (1)-(4) gynoecium; (2) stigma, (2) style, (3) ovary with (4) ovules, (5) disk; (6)-(7) androecium, (6) anther with connective [6(a)] between the lobes; (7) filament; (8)-(9) the perianth: (8) petal (collectively the corolla), (9) sepal (collectively the calyx), (10) receptacle; (11) involucre, (12) pedicel, (13) bract subtending (12); (b) complete (perfect, monoclinous; (c) pistillate (diclinous), (d) staminate (diclinous); (e) apetalous, perfect; (f) nude (no perianth); perfect; (g) nude, pistillate and bracteate (i.e., with an involucre); (h) nude, staminate, bracteate; (i) neuter, with pistillode (1) and staminodes (2) (either or both may be absent, however).

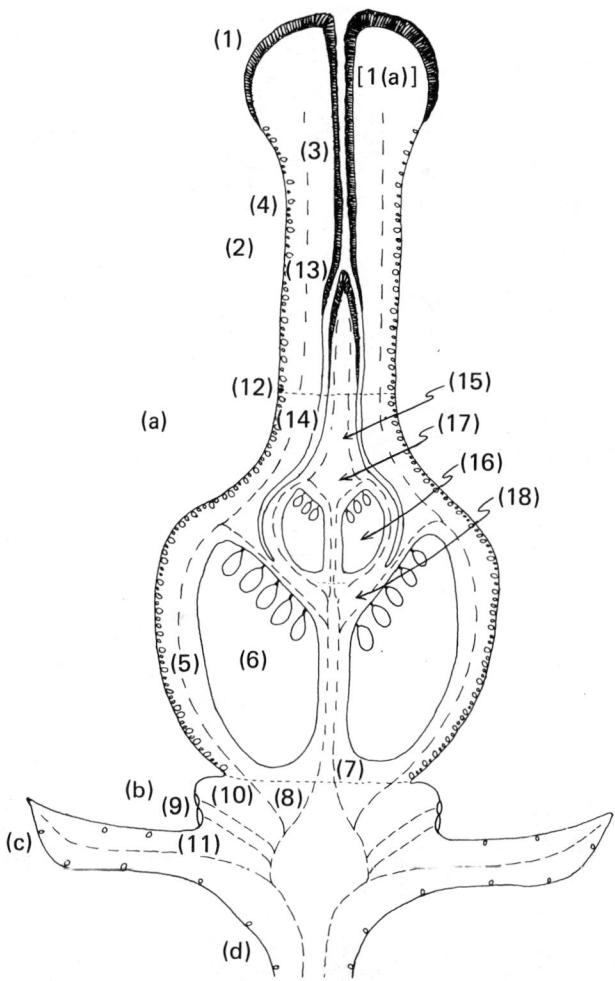

Fig. II.2.2.02. Flowers, navel orange: (a) [gynoecium; (b) disk, (c) calyx; and (d) upper end of pedicel just after petal fall (l.s.; ovary diameter approx. 7 mm). Primary fruit (1-12): (1) stigma with papillae over the outer surface and lining stylar canal [1(a)]; (2) style, with vascular bundles (3), (4) oil glands, also just under the epidermal layers of secondary fruit (not shown), (5) lateral vascular bundles, (6) locule of ovary with ovules axile, (7) axial vascular bundles in central axis, (8) abscission zone of primary fruit, (9) remnants of stamens and petals, (10) stamen and petal vascular traces, (11) sepal vascular trace, (12) abscission zone of style. Secondary fruit (13-18): (13) stigma with papillae over outer surface, (14) style with (15) vascular bundles, (16) locule of ovary with ovules axile, (17) axial vascular bundles, (18) abscission zone (note that the entire fruitlet with abscise if this zone develops during any stage of development, whereas abscission of the style does occur normally). Redrawn with the kind permission of Jose Lima; see Acknowledgments.

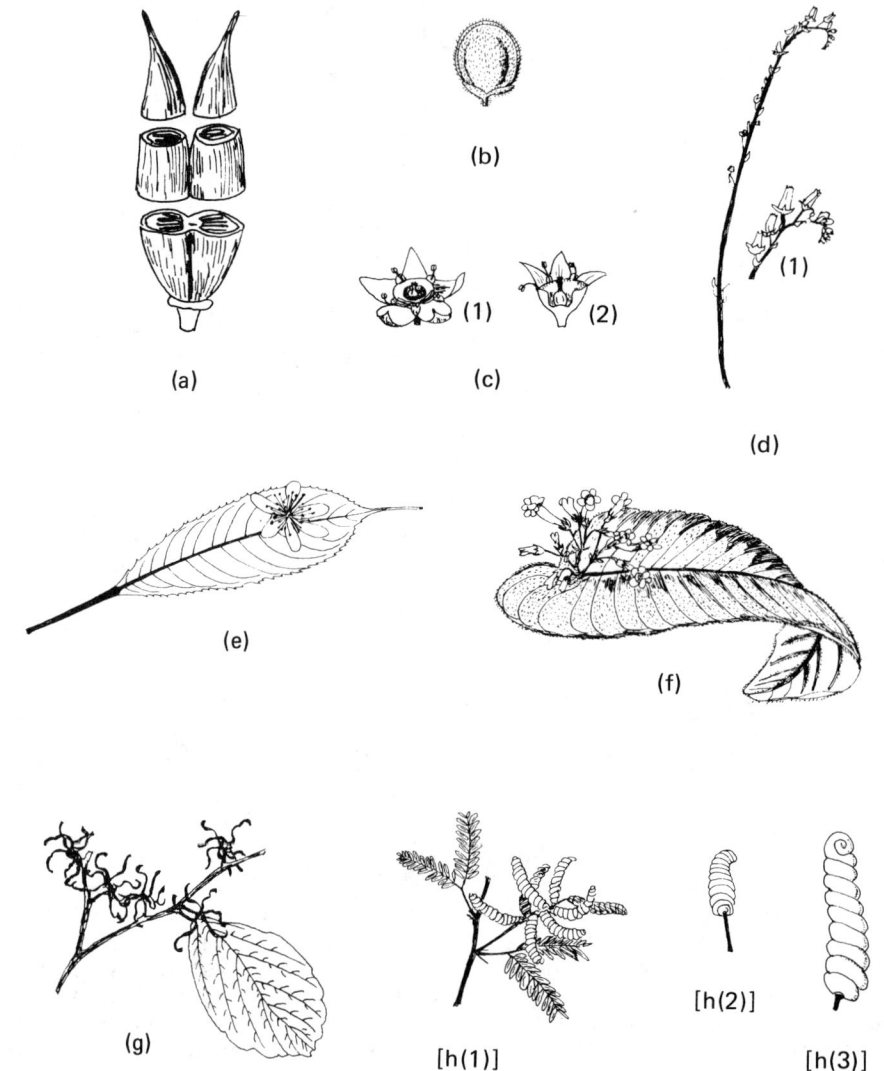

Fig. II.2.2.03. Flowers: (a) anisogynous (saxifrage); (b) anisomerous (calyx of Davilla grandiflora, where two large sepals protect the rest of the flower); (c) annular disk (1) and longitudinal section (2), buckthorn; (d) secund (Cotyledon secunda scape and detail (1) at right); (e) epifoliar (Phylloclinum paradoxum); (f) epiphyllous (Acanthonema strigosum); (g) stellate petals (witch hazel); (h) strombus-shaped, (1) Prosopis pubescens, (2) Acacia obliqua, (3) Prosopis strombulifera. (a), (c) Redrawn from Gray, 1862; (b), (e), (f), (h)(2), (3) redrawn from Hutchinson, 1969; (d) redrawn from Bailey, 1928; (g), (h)(1) redrawn from Sargent, 1905; see Acknowledgments and Selected References.

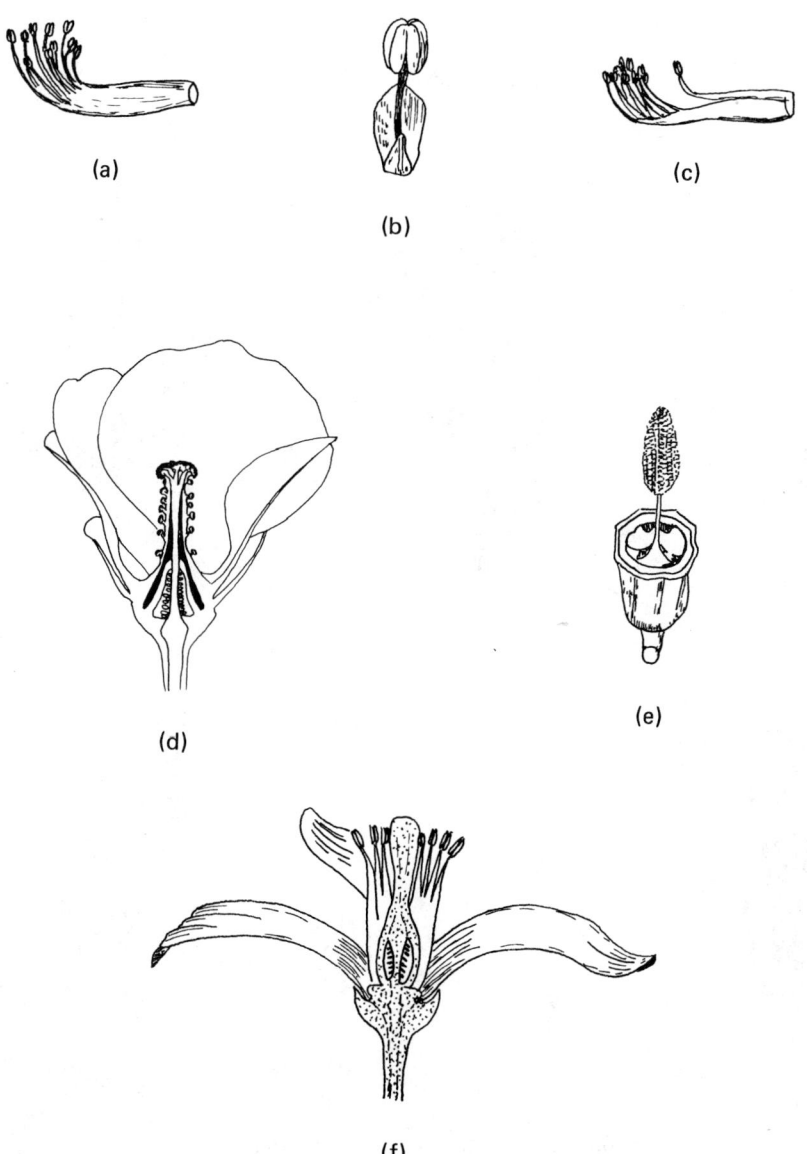

Fig. II.2.2.04. Flowers, staminal arrangement: (a) monadelphous (lupine); (b) monadelphous and syngenesious (Salix purpurea); (c) diadelphous (pea); (d) monadelphous (okra, l.s.); (e) triadelphous and syngenesious (gourd); (f) polyadelphous in bundles or fascicles (sweet orange). (a)-(c), (e) Redrawn from Gray, 1862; (d) redrawn from McGregor, 1976; (f) redrawn from Hume, 1907; see Acknowledgments and Selected References.

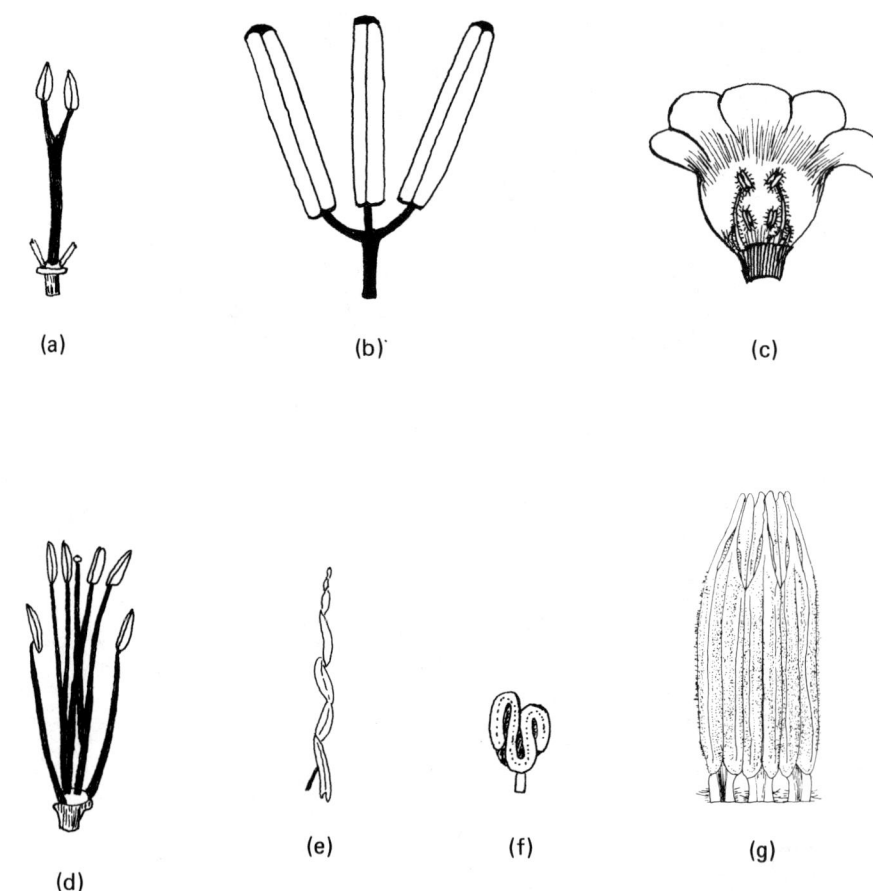

Fig. II.2.2.05. Flowers, staminal arrangement: (a) didymous (twins) (Streptanthus hyacinthoides); (b) tridymous (Typha angustifolia); (c) didynamous (visible anthers), distinct, contabescence (missing anther marked by +, Gerardia purpurea); (d) tetradynamous (mustard); (e) epipetalous, spiral (Tarenna spiranthera); (f) sinuous (twisted, melon); (g) coherent anthers (tomato). (a), (c), (d), (f) Redrawn from Gray, 1862; (b) redrawn from Hutchinson, 1973; (e) redrawn from Hutchinson, 1969; (g) redrawn from McGregor, 1976; see Acknowledgments and Selected References.

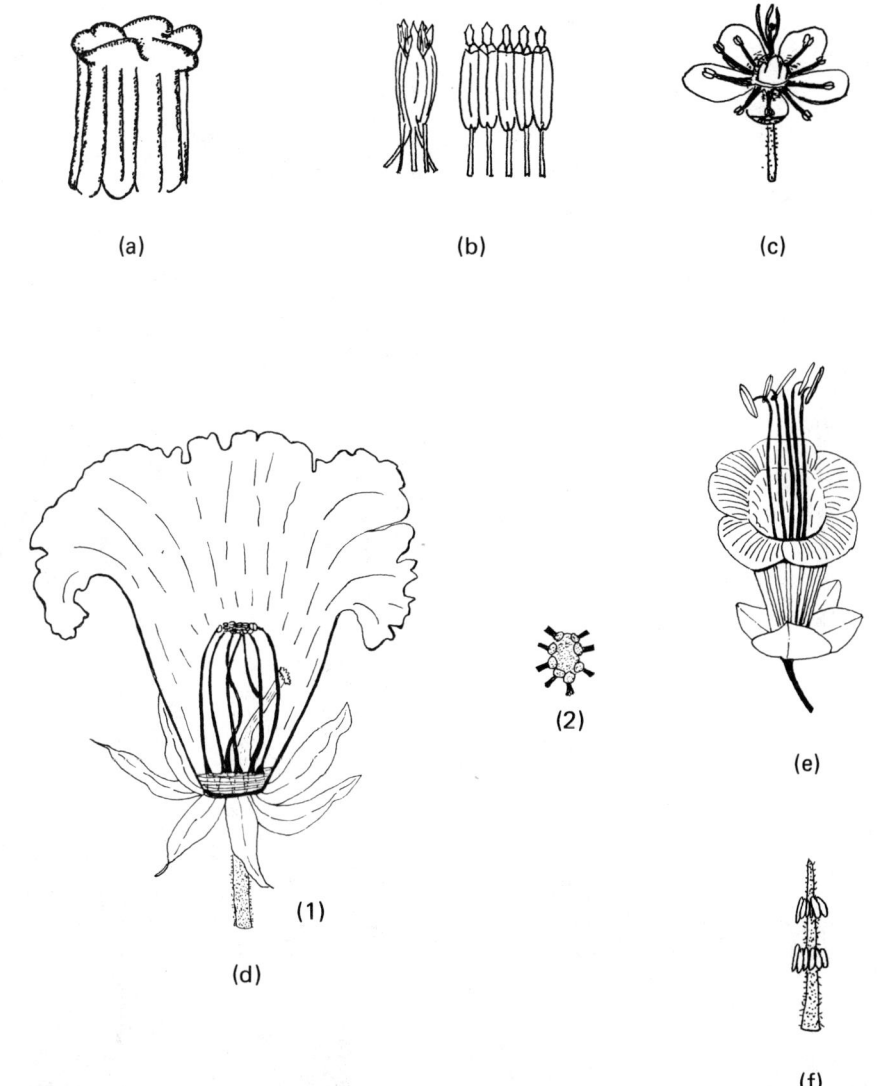

Fig. II.2.2.06. Flowers, staminal arrangement: (a) synandrium (Xanthosoma sagittifolium); (b) syngenesious (composite, as tube and spread out); (c) diplostemonous, distinct (Moehringia laterifolia); (d) synantherous (gloxinia) in l.s. (1) and detail of anther ring (2); (e) exserted (tubiform corolla, Cobaea scandens); (f) staminal column (Hevea brasiliensis). (a) Redrawn from Purseglove, 1972; (b), (c) redrawn from Gray, 1862; (e) redrawn from Hutchinson, 1973; (f) redrawn from Dijkman, 1951; see Acknowledgments and Selected References.

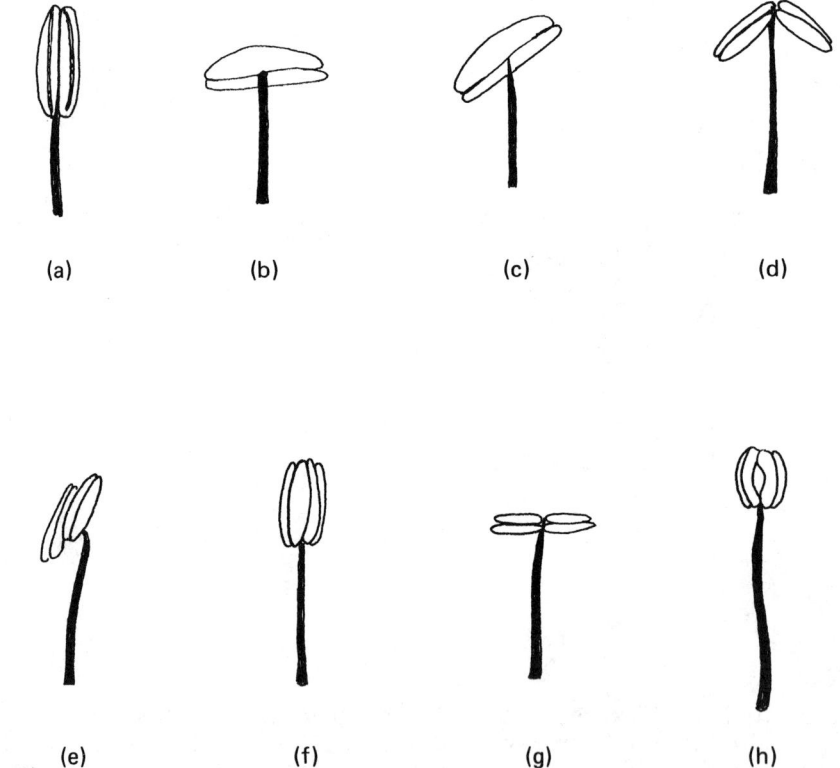

Fig. II.2.2.07. Flowers, anther positions: (a) basifixed innate; (b) medifixed (not versatile); (c) dorsifixed (versatile); (d) divergent; (e) oblique; (f) parallel; (g) transverse; (h) basifixed. (a), (b), (c) Redrawn from Hutchinson, 1973; (d)-(g) redrawn from Radford et al., 1974; see Acknowledgments and Selected References.

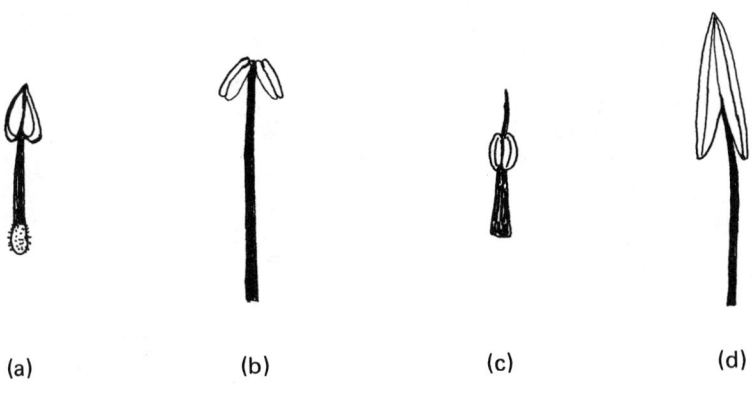
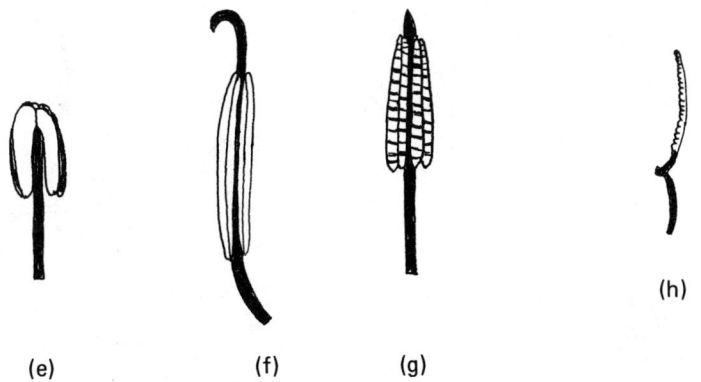

Fig. II.2.2.08. Flowers, anther positions: (a) basifixed apiculate; (b) apicifixed; (c) adnate; (d) apicifixed (Ipomoea mauritiana); (e) horseshoe-shaped (Bixa orellana); (f) uncinate (produced connective, Nelumbo); (g) locellate (Aegiceras corniculatum); (h) geniculate filament (Melastoma malabathricum). (b), (c) Redrawn from Radford et al., 1974; (a)-(d), (h) redrawn from Hutchinson, 1969; see Acknowledgments and Selected References.

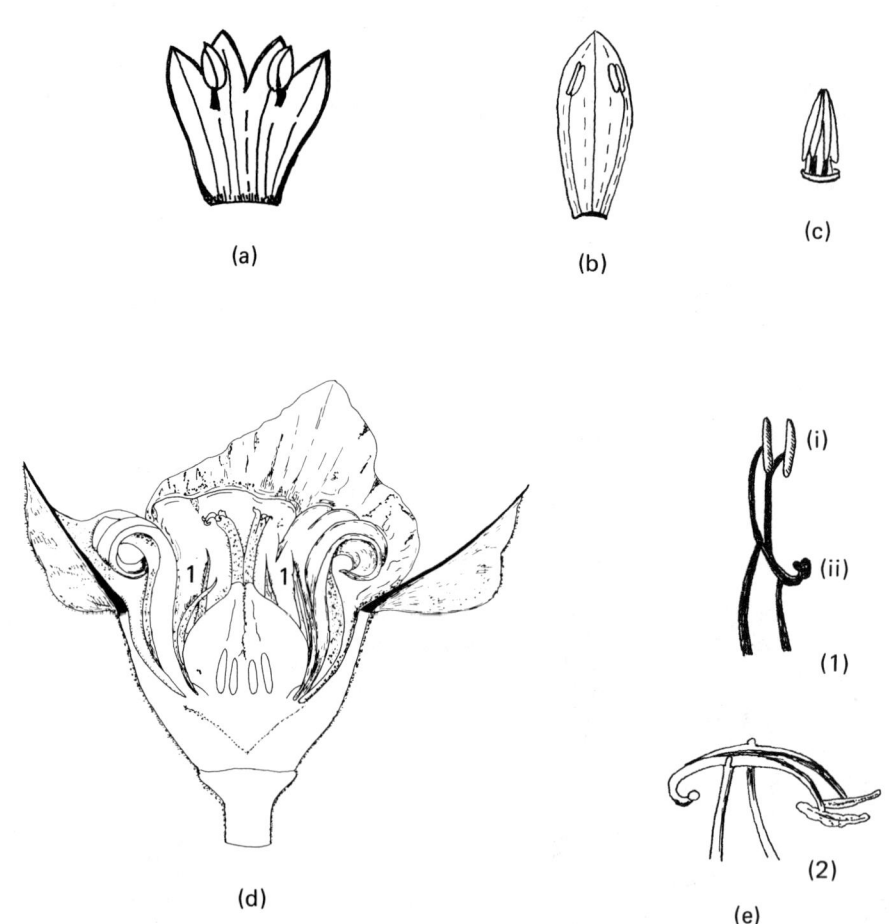

Fig. II.2.2.09. Flowers, stamens: (a) epipetalous and apiculate (<u>Ligustrum massalongianum</u>); (b) petaloid; (c) connivent anthers; (d) staminodes (1) (<u>Diospyros kaki</u>); (e) a pair of versatile anthers in natural position (1) with an expanded connective and fertile (i) and sterile (ii) anthers and when moved by a bee (2) (<u>Salvia officinalis</u>). (a) Redrawn from Hutchinson, 1973; (c) redrawn from Gray, 1862; (d) redrawn from McGregor, 1976; (e) redrawn from Lubbock, 1894; see Acknowledgments and Selected References.

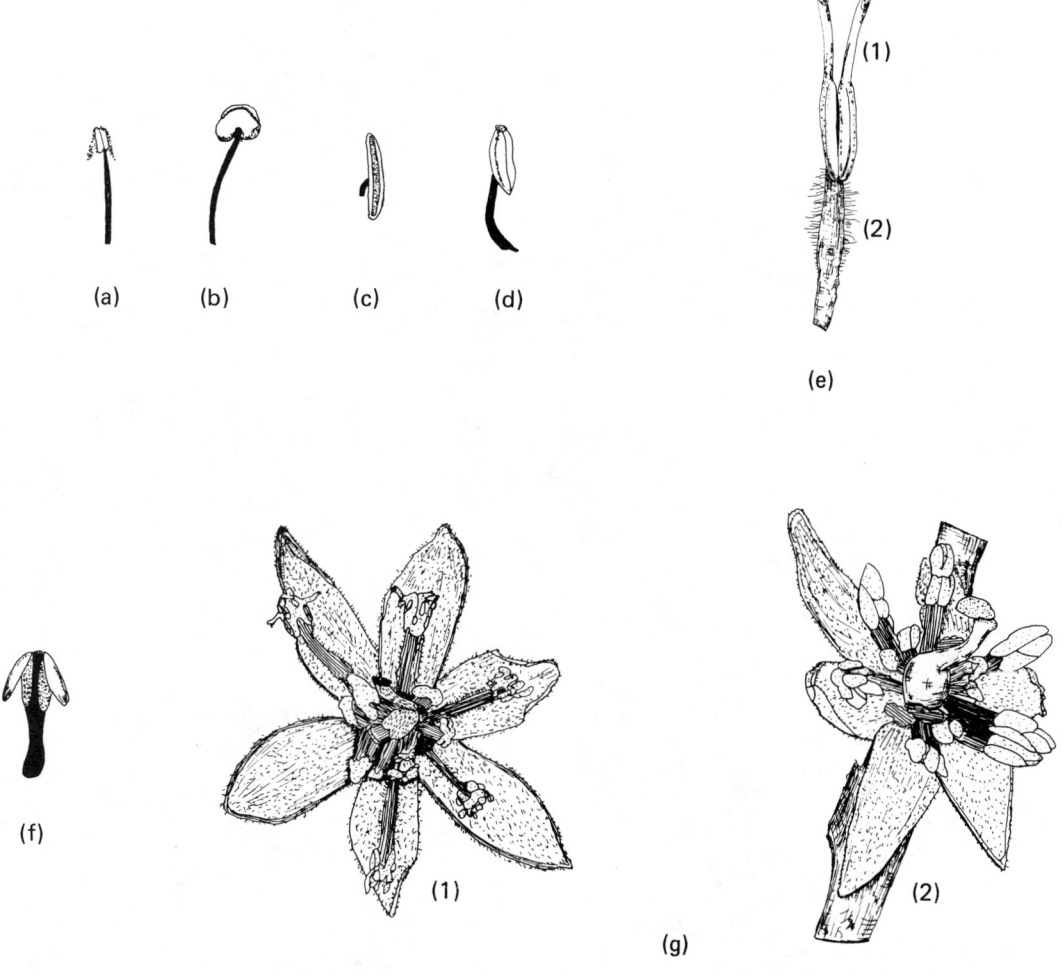

Fig. II.2.2.10. Flowers, anther dehiscence: (a) longitudinal (most common mode); (b) transverse and one-celled (mallows); (c) unilocular and dimidiate (globe amaranth); (d) terminal pore (Pyrola); (e) apical pores (1) with filiform hairs (2) on forked filament ('Tifblue' rabbiteye blueberry); (f) flaps (barberry); (g) avocado, introrse in outer whorls, extrorse in innermost whorl, dehiscence by flaps in staminate stage (1) preceded by pistillate stage (2), a situation in which the same flower opens twice. (a)-(d), (f) Redrawn from Gray, 1862; (e) redrawn from McGregor, 1976; (g) redrawn from Maxwell photo given to Soule; see Acknowledgments and Selected References.

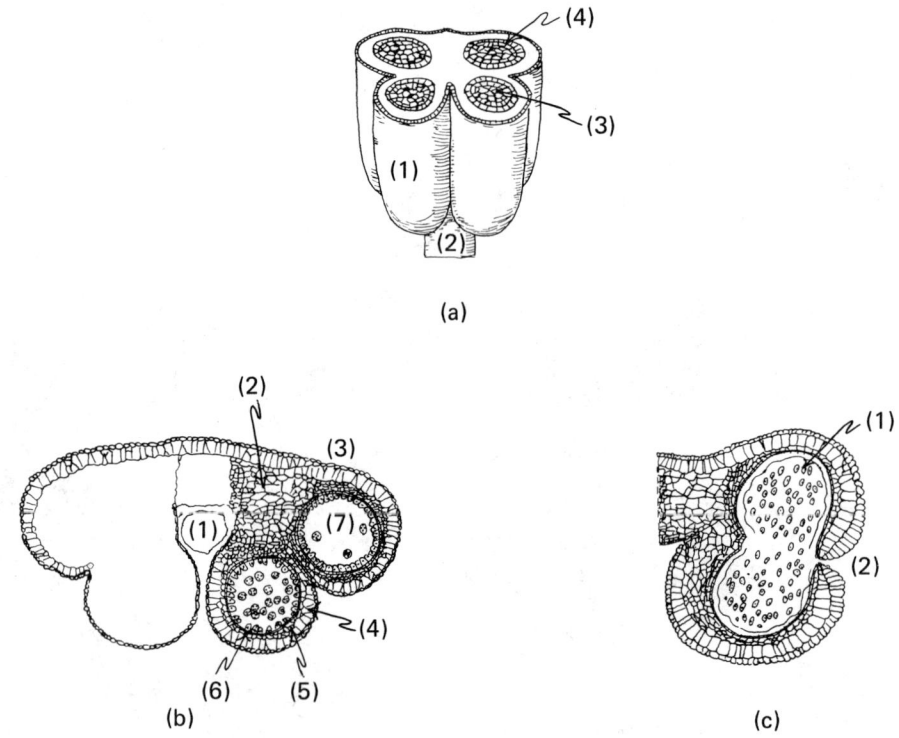

Fig. II.2.2.11. Flowers: (a) isometric view of lower portion of (1) young anther or microsporangium with (2) filament below and (3) microspore mother cells surrounded by (4) tapetum; (b) lily anther with four locules, (1) filament, (2) connective joining the two pairs of locules (lefthand pair outlined only), (3) epidermis, microsporangium wall consisting of (4) endothecium, (5) middle layers, (6) tapetum, and (7) microspore tetrads of each locule; (c) a later stage showing (1) pollen grains that will escape through the stomium (2). (a) Redrawn from Echlin, 1968; (b), (c) redrawn from Foster and Gifford, 1974; see Acknowledgments and Selected References.

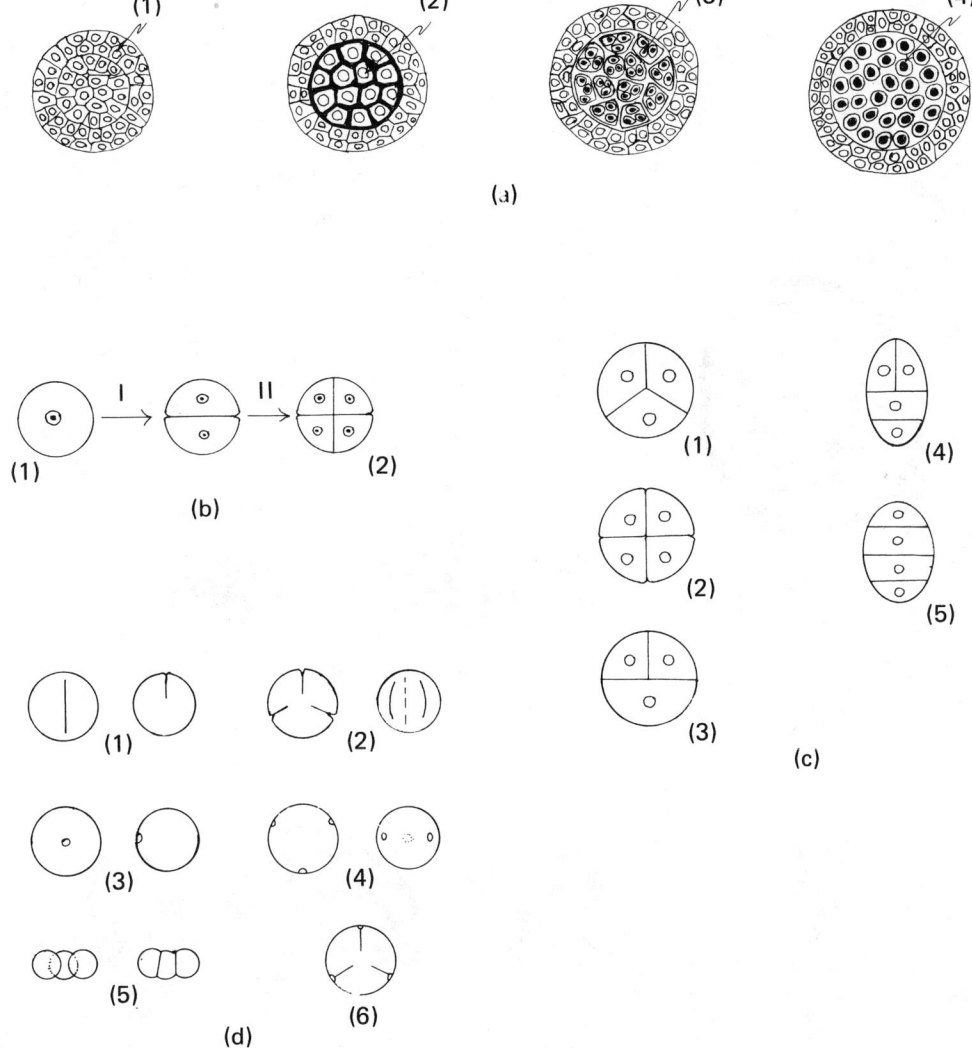

Fig. II.2.2.12. Flowers: (a) diagram of microspore development as (1) sporogenous tissue, (2) mother cells, (3) tetrads, and (4) microspores; (b) microsporogenesis from the microsporocyte, (1) microspore mother cell with two divisions (one meitotic, one mitotic) to form the tetrad of microspores; (c) configuration of tetrads, (1) tetrahedral, (2) isobilateral, (3) decussate, (4) T-shaped, (5) linear; (d) polar (left) and equatorial (right) views of (1) monocolpate, (2) tricolpate, (3) monoporate, (4) triporate, (5) vesiculate, and (6) tricolporate pollen grains. (a) Redrawn from Echlin, 1968; (b) redrawn from Sherman (personal notes); (c) redrawn from Maheswari, 1950; (d) redrawn from Bell and Coombe, 1976; see Acknowledgments and Selected References.

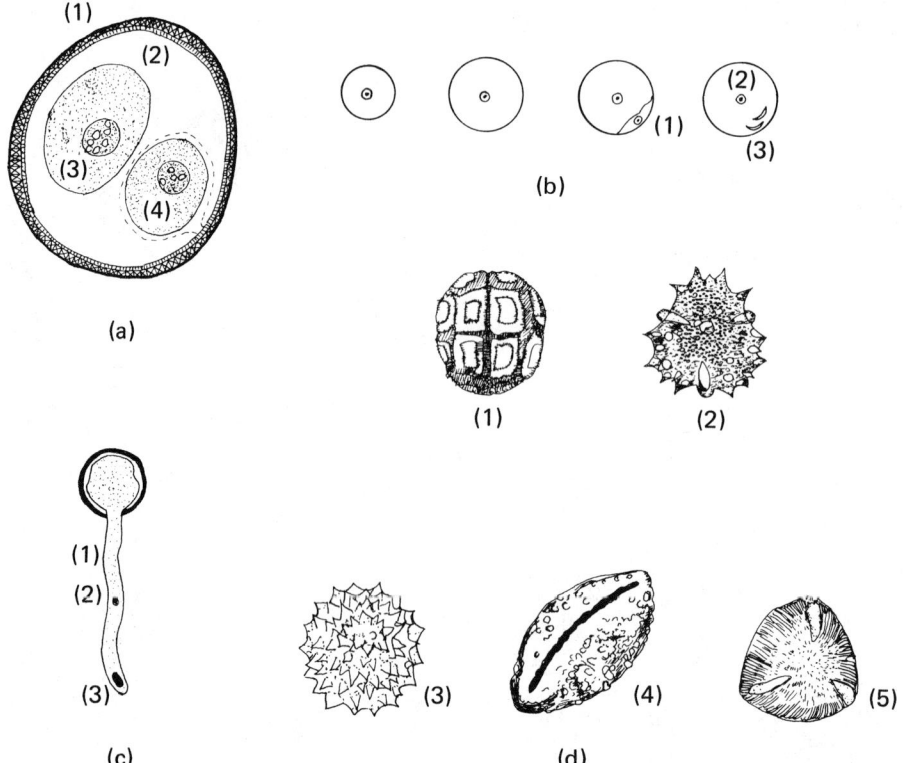

Fig. II.2.2.13. Flowers: (a) pollen grain (enlarged t.s.) showing the grain wall (sporoderm) as the (1) extine and (2) intine, (3) tube-cell nucleus, and (4) generative-cell nucleus; (b) diagram of development of a microspore with separation of the nucleus to form (1) a generative cell in the third stage, (2) the pollen tube nucleus, and (3) two sperm in the fourth stage; (c) a germinating pollen grain with (1) tube, (2) generative-cell nucleus, and (3) tube-cell nucleus; (d) pollen grains, (1) Acacia longifolia, (2) Tanacetum camphoratum, (3) Ambrosia trifida, (4) Liriodendron tulipifera, and (5) Rosa rugosa. (a), (c) Redrawn from Echlin, 1968; (b) redrawn from Sherman (personal notes); (d) redrawn from Science Kit & Boreal Labs Inc., 1980; see Acknowledgments and Selected References.

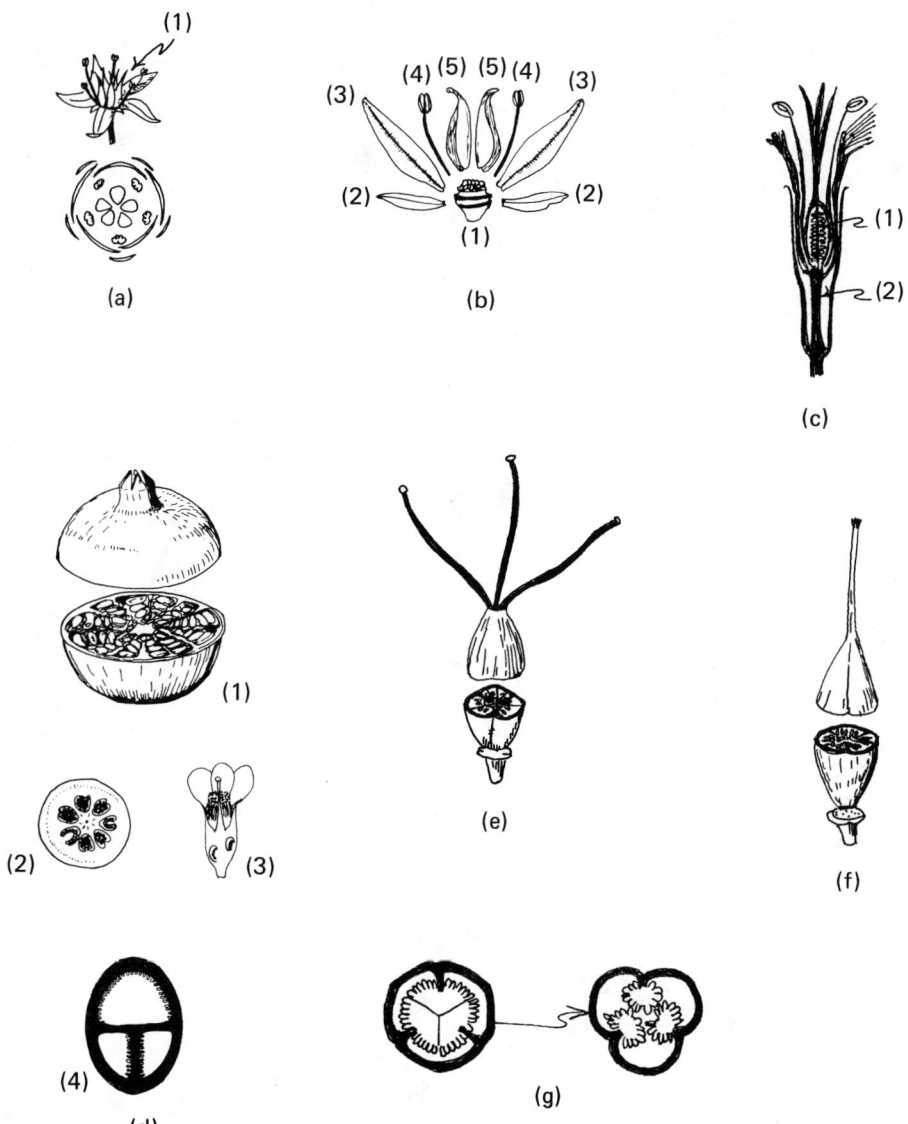

Fig. II.2.2.14. Flowers: (a) apocarpous gynoecium, Crassula (1), with floral diagram below; (b) (1) receptacle and flower parts, (2) sepals, (3) petals, (4) stamens, and (5) pistils of stonecrop (Sedum ternatum, expanded view); (c) syncarpous (compound) gynoecium of Silene pennsylvanica with axile placentation and stipe (2); (d) pomegranate with fruit (1) and ovary (2) in cross section, flower (3), and ovary (4) in longitudinal section; the last two show parietal placentation in the upper and axile in the lower half; (e), (f), (g) three species of Hypericum with a syncarpous gynoecium of three locules and parietal placentation, (e) with separate styles, (f) with a compound style and three small stigmas, and (g) t.s. of ovary (left), where the placentation appears axile but is clearly parietal in the fruit (right). (a)-(c), (e)-(g) Redrawn from Gray, 1862; (d)(1), (2) redrawn from Hutchinson, 1969, (3) from Hutchinson, 1973; see Acknowledgments and Selected References.

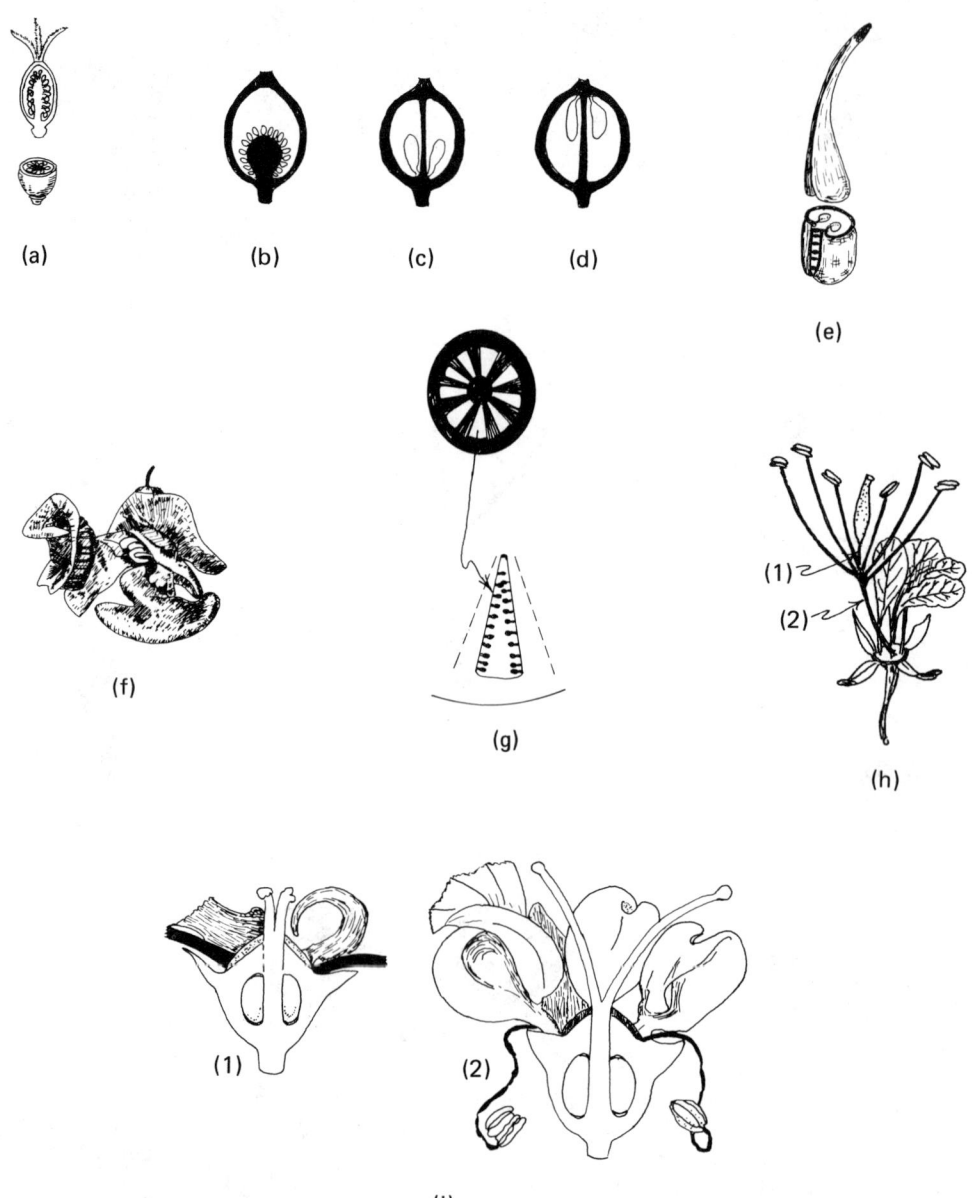

Fig. II.2.2.15. Flowers, (a)-(f), placentation: (a) free central (l.s. with t.s. below); (b) free-basal; (c) basal-erect; (d) apical-pendulous; (e) ventral; (f) cochleate (Phaseolus caracalla); (g) lamellate (enlarged section below); (h) flower with (1) stipe and (2) stipitate receptacle (Gynandropsis); (i) accrete style (compare (1) staminate stage with (2) pistillate stage) in coriander. (a), (e), (f), (h) Redrawn from Gray, 1862; (b)-(d) redrawn from Hutchinson, 1973; (g) redrawn from Faegri and van der Pijl, 1979; (i) redrawn from McGregor, 1976; see Acknowledgments and Selected References.

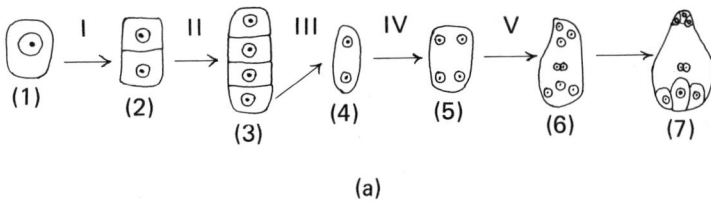

(a)

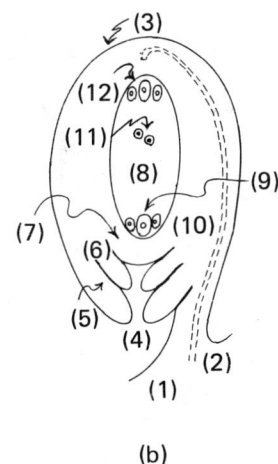

(b)

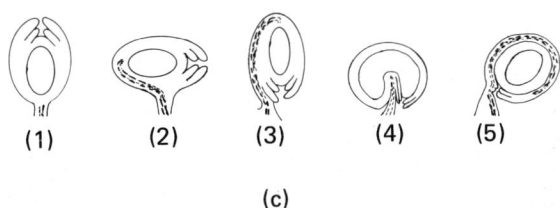

(c)

Fig. II.2.2.16. Flowers: (a) gametogenesis, (1) megaspore mother cell (megasporocyte) which undergoes two divisions (one meiotic and one mitotic) to form four megaspores (3) one of which undergoes three mitotic divisions [(4) to (6)] to form the embryo sac (7); (b) anatropous ovule, (1) placenta, (2) funiculus with vascular bundle leading to the (3) chalaza, (4) micropyle, (5) outer integument, (6) inner integument, (7) nucellus (body tissue of ovule), (8) embryo sac, (9) egg (female gamete), (10) two synergids, (11) two polar nuclei, (12) three antipodals; (c) ovule positions, (1) orthotropous, (2) amphitropous, (3) anatropous, (4) campylotropous, (5) obcampylotropous. (a), (b), (c)(1)-(4) Redrawn from Sherman (personal notes); (c)(5) adapted from Corner, 1976; see Acknowledgments and Selected References.

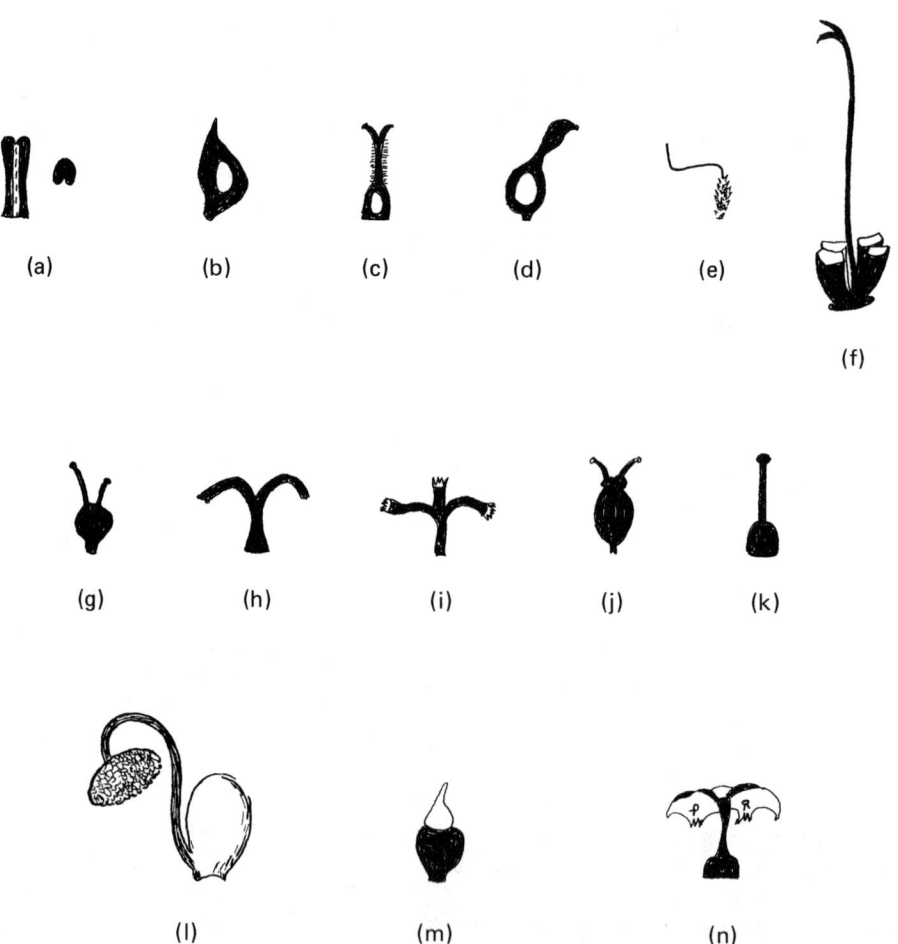

Fig. II.2.2.17. Flowers, styles: (a) conduplicate (t.s. at right); (b) excentric; (c) fimbriate; (d) flabellate: (e) geniculate; (f) gynobasic (Lamium album); (g) heterostylous; (h) homostylous; (i) petaloid; (j) stylopodic; (k) terete; (l) basal (Stylobasium lineare); (m) tuberculate; (n) umbraculate. (a)-(e), (g), (k), (m), (n) Redrawn from Radford et al., 1974; (f), (n) redrawn from Hutchinson, 1969; see Acknowledgments and Selected References.

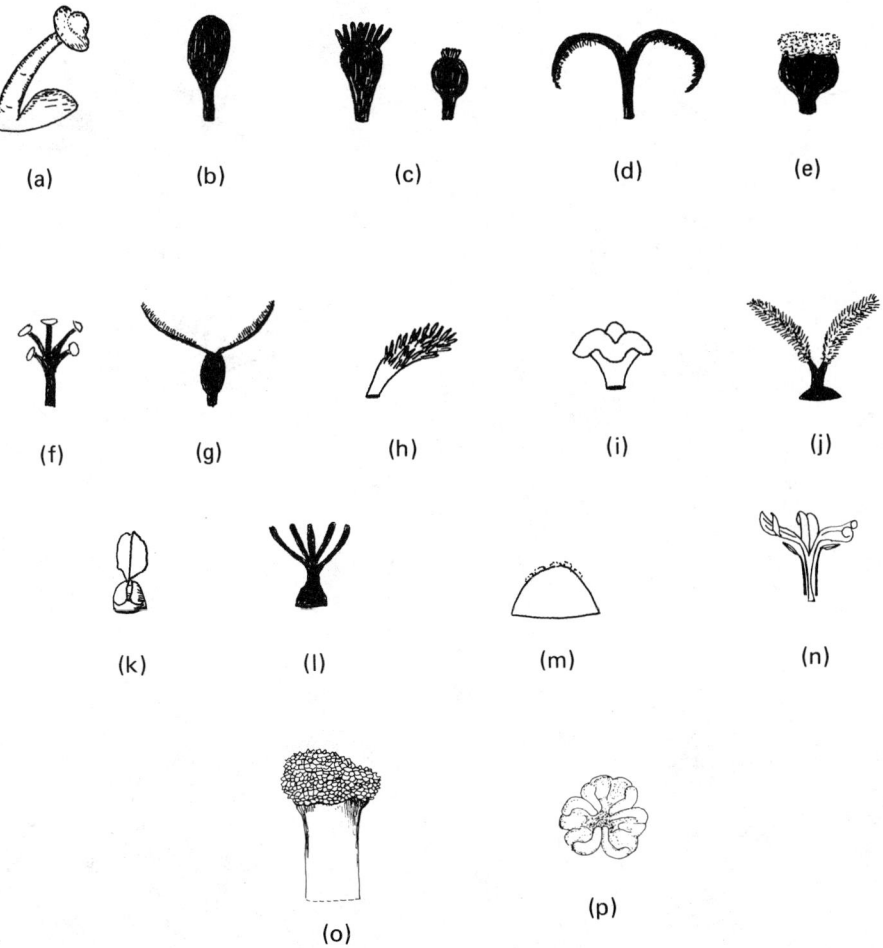

Fig. II.2.2.18. Flowers, stigmas: (a) capitate; (b) clavate; (c) crested; (d) decurrent; (e) diffuse; (f) discoid; (g) fimbriate; (h) lineate; (i) lobed; (j) plumose; (k) connate (Apocynum androsaemifolium): (l) terete; (m) sessile; (n) petaloid (Iris cristata); (o) side view of tomato; (p) top view of banana yucca (Yucca baccata). (a) Redrawn from Lawrence, 1955; (b)-(j), (l), (m) redrawn from Radford et al., 1974; (k), (n), redrawn from Gray, 1862; (o), (p) redrawn from McGregor, 1976; see Acknowledgments and Selected References.

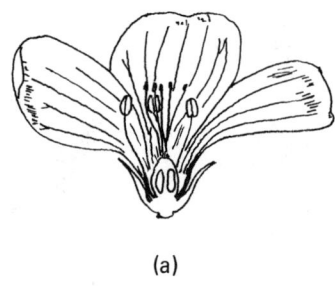

(a)

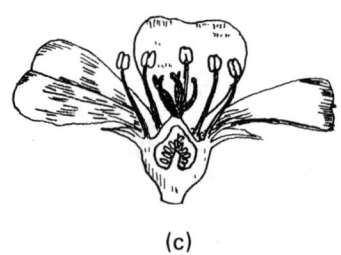

(c)

(b)

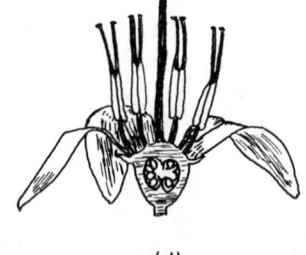

(d)

Fig. II.2.2.19. Flowers, positions of ovary relative to other floral parts: (a) superior (flax); (b) superior (cherry); (c) half-inferior (purslane); (d) inferior (cranberry); positions of perianth (and stamens) relative to ovary: (a) hypogynous; (b) perigynous (with hypanthium = floral cup); (c) perigynous; (d) epigynous. Redrawn from Gray, 1862; see Acknowledgments and Selected References.

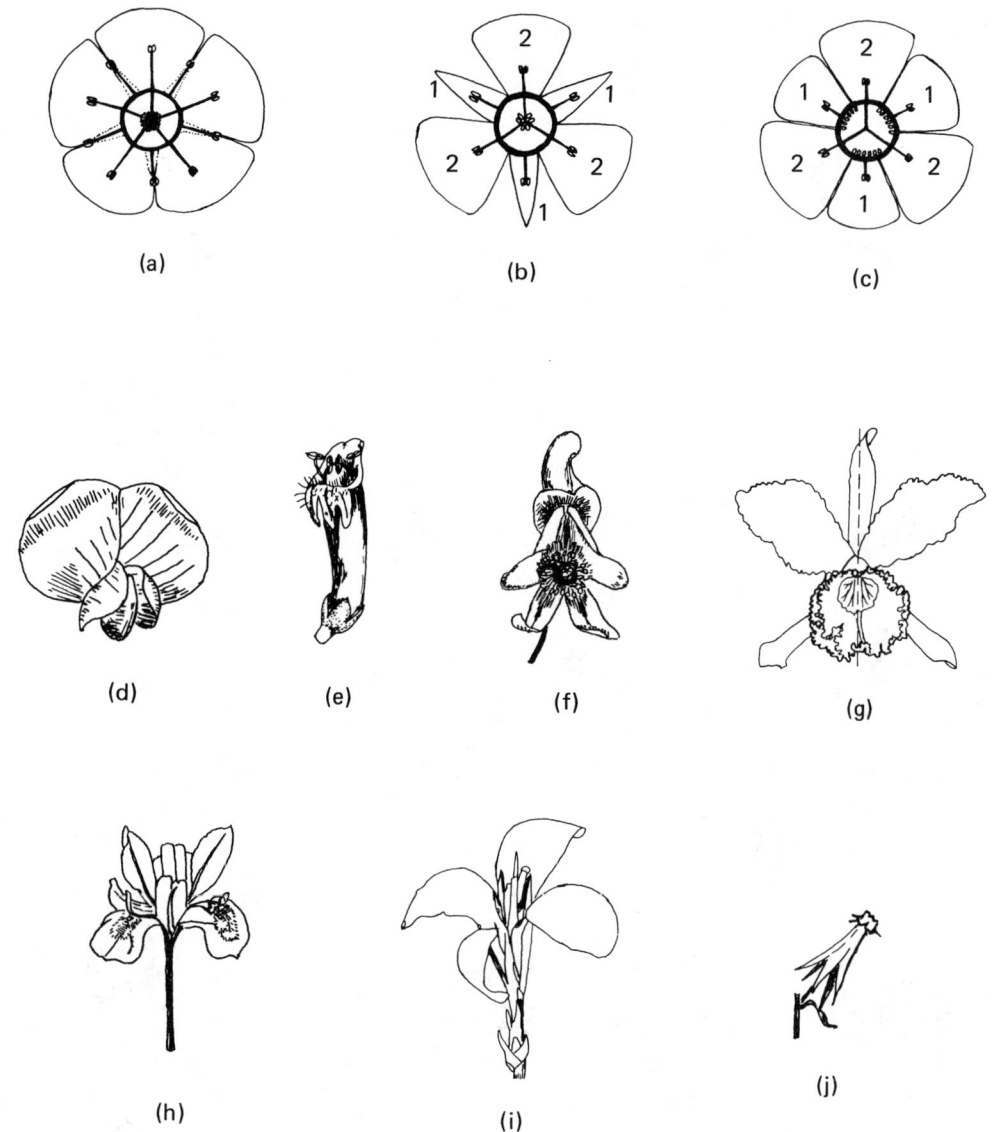

Fig. II.2.2.20. Flowers, perianth: (a) actinomorphic, polypetalous (dicot flower with diplostemonous or obdiplostemonous stamens); (b) actinomorphic, polypetalous, (1) sepals and (2) petals distinct (monocot flower); (c) actinomorphic, tepals = (1) sepals and (2) petals similar; (d) zygomorphic, papilionaceous; (e) zygomorphic, tubular, bilabiate, synsepalous, sympetalous; (f) zygomorphic with galea (monkshood); (g) zygomorphic (bilateral symmetry), resupinate [compare with (c), where perianth is in normal position, Cattleya orchid]; (h) zygomorphic (Iris); (i) irregular (Canna); (j) pelory (toadflax, compare with normal form in Fig. II.2.2.23c). (d)-(f), (h) Redrawn from Gray, 1862; (e) redrawn from Lawrence, 1955; (g) redrawn from Sheehan and Sheehan, 1979; (i), (j) redrawn from Bailey, 1928; see Acknowledgments and Selected References.

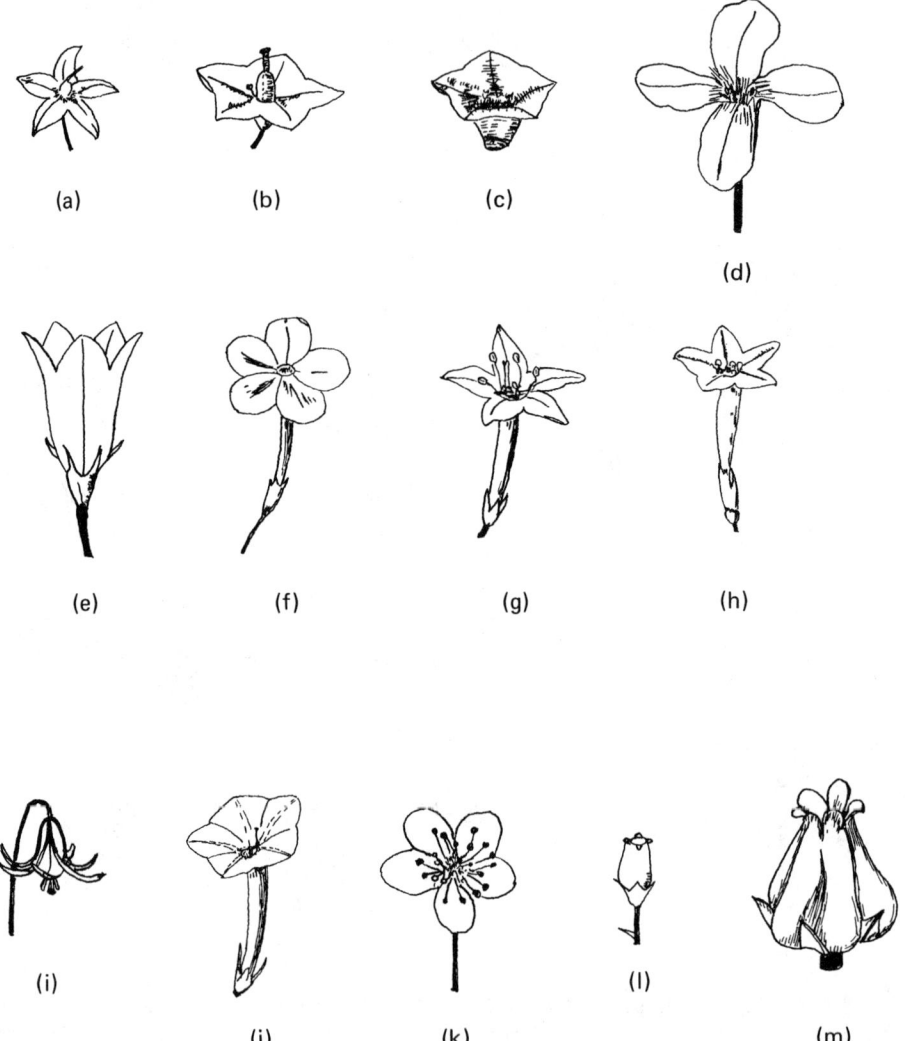

Fig. II.2.2.21. Flowers, actinomorphic corolla types: (a) rotate (bittersweet); (b) rotate (wheel-shaped, potato); (c) vase-shaped (ground cherry); (d) cruciform (mustard); (e) campanulate (harebell); (f) hypocrateriform (phlox); (g) salver-shaped (gilia); (h) salverform (cypress-vine); (i) liliaceous, nodding (dog-tooth violet); (j) infundibular (funnel-shaped, scarlet morning-glory); (k) rosaceous (pear); (l) subglobose (blueberry); (m) urceolate. (a)-(j), (l) Redrawn from Gray, 1862; (k) redrawn from Bailey, 1949; (m) redrawn from Lawrence, 1955; see Acknowledgments and Selected References.

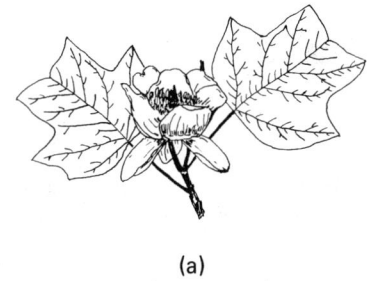
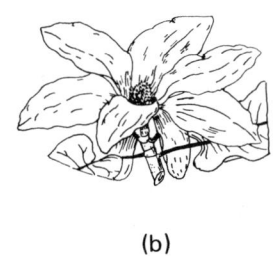

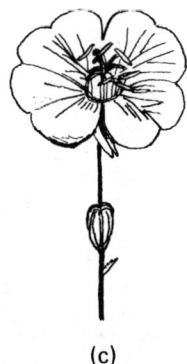

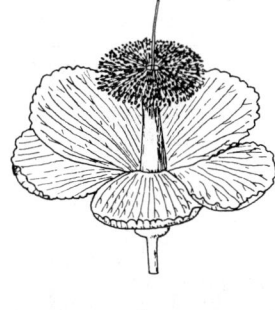

Fig. II.2.2.22. Flowers, actinomorphic corolla types: (a) cotyloform (Liriodendron tulipifera); (b) crateriform (Magnolia macrophylla); (c) tubular with gamosepalous tubular calyx adherent to the ovary and (apparently) episepalous stamens (evening primrose); (d) foliaceous corolla (squash), (e) rotate (baobab) with staminal column. (a), (b) Redrawn from Sargent, 1905; (c) redrawn from Gray, 1862; (d) redrawn from from Hill et al., 1960; (e) redrawn from Hutchinson, 1969; see Acknowledgments and Selected References.

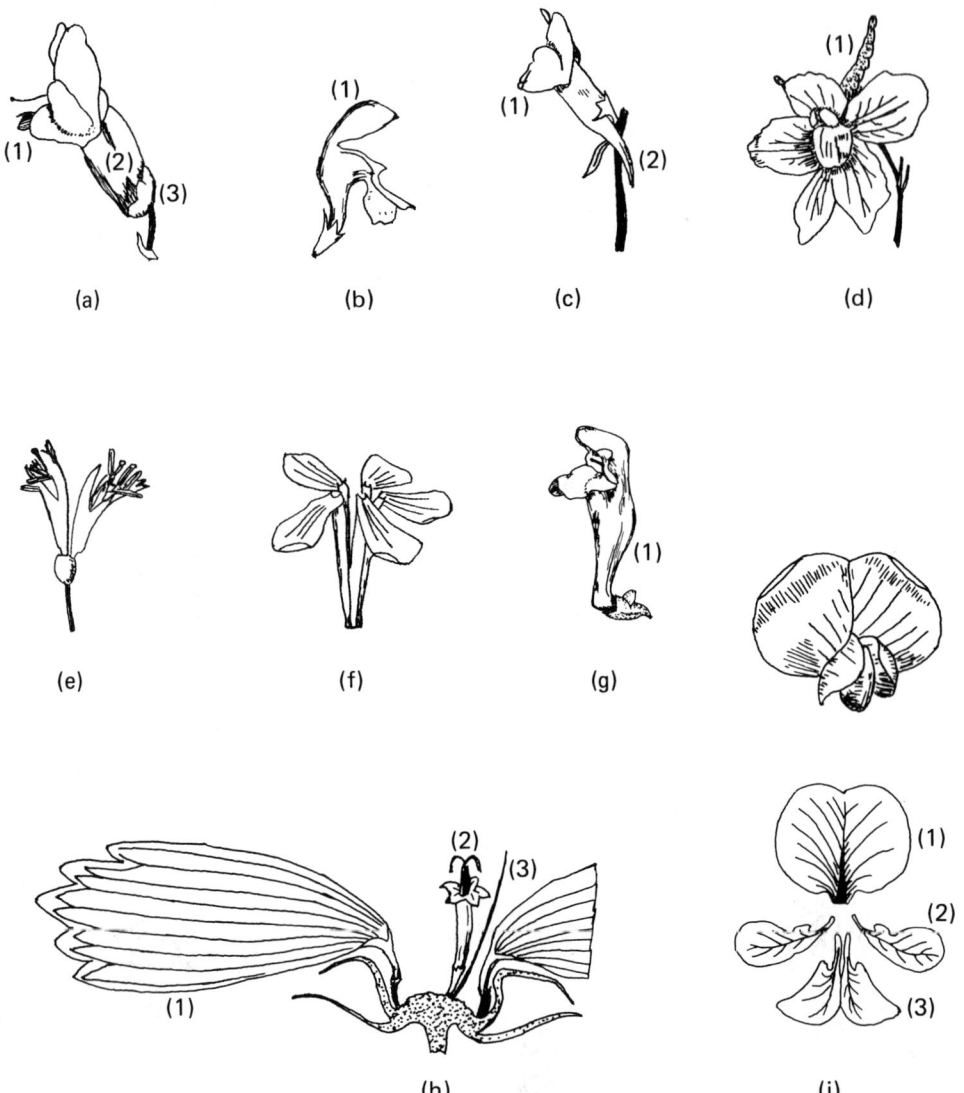

Fig. II.2.2.23. Flowers, zygomorphic corolla types: (a) personate (snapdragon) with (1) palate, (2) tumid tube, and (3) adpressed calyx; (b) labiate (ringent, Lamium) with (1) galea; (c) personate (Linaria) with (1) palate and (2) spurred = calcarate and gibbous corolla; (d) calcarate = spurred calyx (1) of larkspur; (e) tubular (Lonicera oblongifolia); (f) unguiculate (caryophyllaceous, soapwort); (g) bilabiate (Scutellaria) and (1) ventricose; (h) papilionaceous (Robinia pseudacacia) flower (above) and its parts (below), as (1) vexillum, (2) wings = alae, and (3) carina = keel; (i) composite with (1) ligulate = ray and (2) tubular (actinomorphic) disk flowers, the latter with (3) chaff = bract. (a)-(f), (h), (i) Redrawn from Gray, 1862; (g) redrawn from Lawrence, 1955; see Acknowledgments and Selected References.

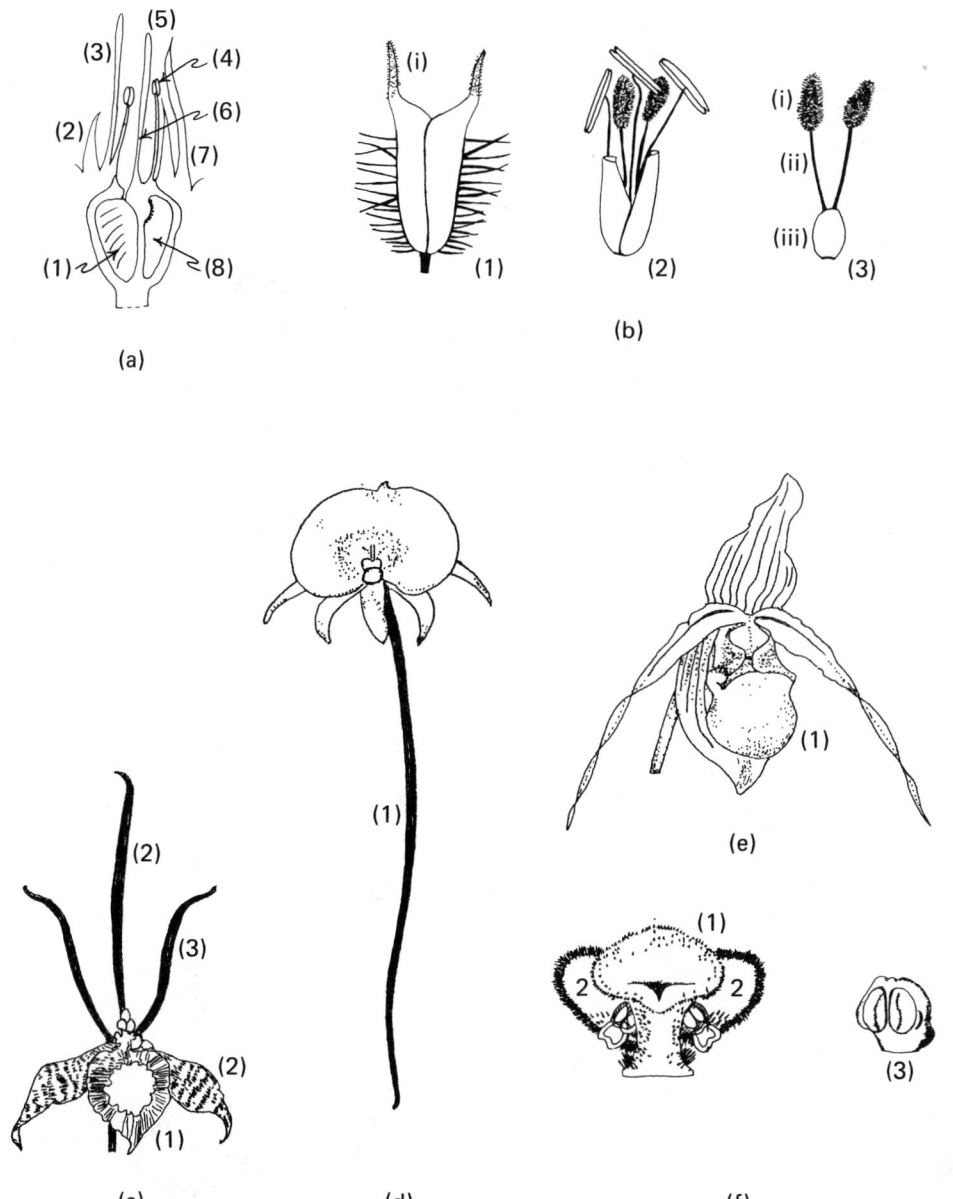

Fig. II.2.2.24. Flowers: pineapple (l.s.) with (1) gland, (2) sepal, (3) petal, (4) stamen, (5) stigma, (6) style, (7) bract, (8) ovary (with ovules on placenta); (b) timothy, with (1) spikelet with (i) glume, (2) floret with (i) palea and (ii) lemma, (3) ovary with (i) plumose stigma, (ii) style, and (iii) ovary; (c)-(f) orchids: (c) anisomerous Oncidium (monandrous) with distinct (1) labellum, (2) sepals, and (3) petals; (d) Anagraecum (monandrous) with (1) long spur = nectary; (e) Phragmipedium, with (1) calceolate labellum; (f) gynandrium = gynostemium of diandrous Phragmipedium with (1) anther cap, (2) pollinia, and (3) enlargement of pollinia. (a) Redrawn from Py and Tisseau, 1965; (b) redrawn from Robbins, 1924; (c)-(f) redrawn from Sheehan and Sheehan, 1979; see Acknowledgments and Selected References.

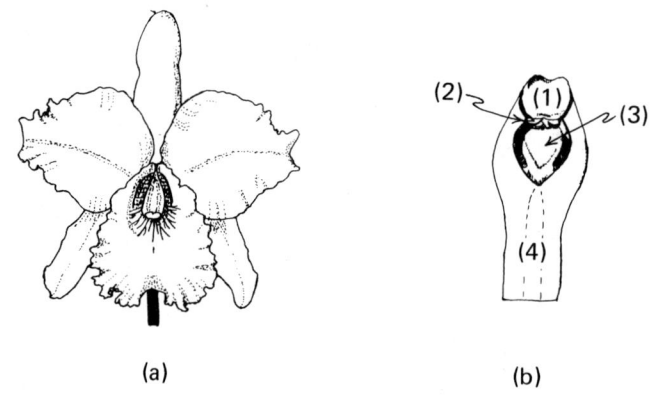
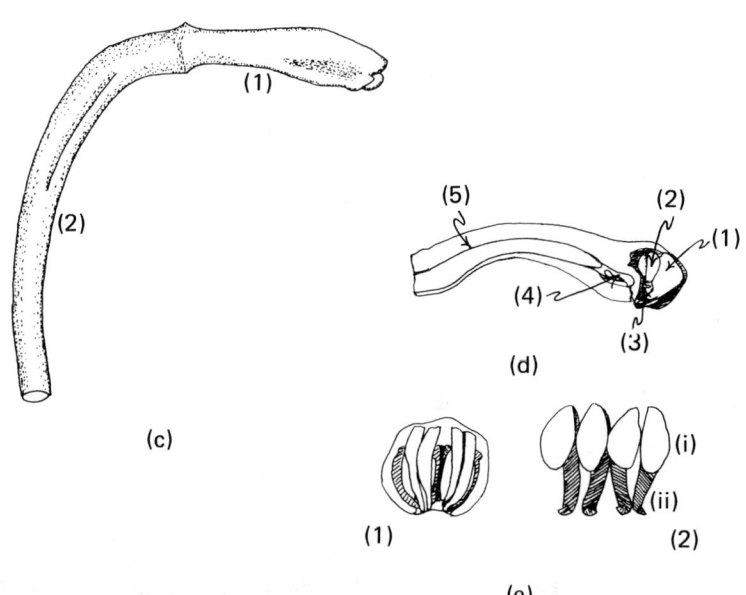

Fig. II.2.2.25. Flowers, orchids: (a) Cattleya (monandrous) flower; (b) ventral view of gynandrium = column with (1) anther cap, (2) rostellum, (3) stigmatic surface, and (4) ovary; (c) side view of (1) column and (2) ovary; (d) longitudinal section of column with (1) anther cap, (2) pollinia, (3) rostellum, (4) stigmatic surface, and (5) stylar canal; (e) pollinia: anther cap with pollinia in place and caudicles folded over them (1) and a separate view (2) of four pollinia (i) with each caudicle (ii) stretched out. Redrawn from Sheehan and Sheehan, 1979; see Acknowledgments and Selected References.

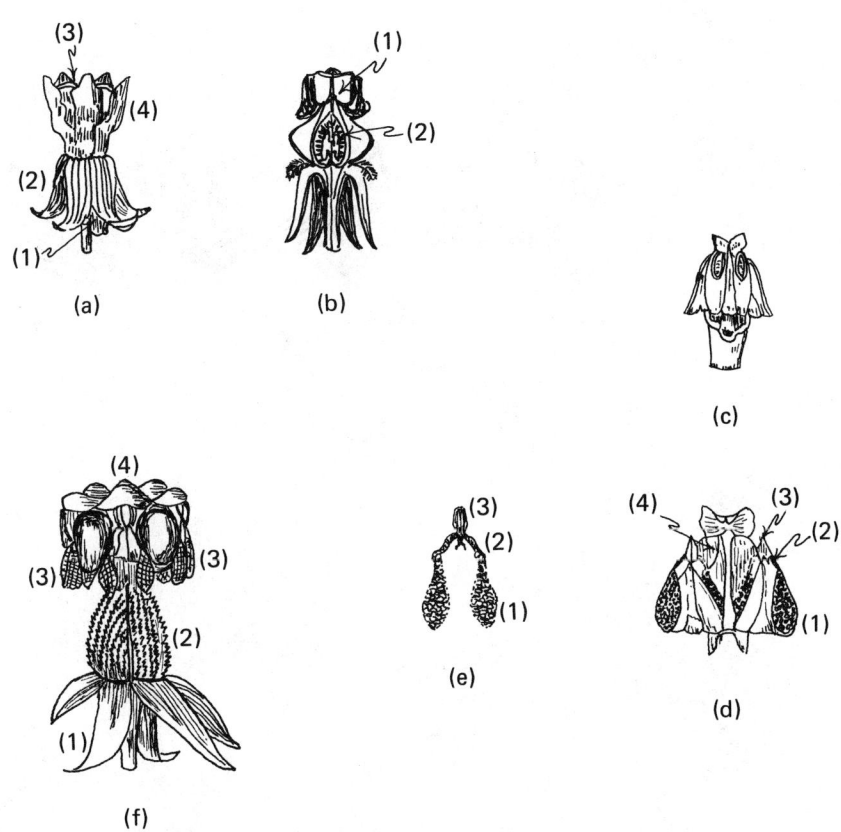

Fig. II.2.2.26. Flowers, milkweed (<u>Asclepias cornuti</u>): (a) expanded flower with (1) calyx, (2) corolla, (3) reflexed staminal crown, and (4) corniculate appendages; (b) vertical section of flower (hooded appendages removed) with (1) thick stigmas and (2) ovaries; (c) back view of stamen (appendage cut away); (d) stamen more magnified with pollen-masses [(1) pollinia] cohering by their (2) caudicles each to a (3) gland from the summit of the (4) stigmatic body; (e) pair of detached (1) pollinia suspended by their (2) caudicles from the (3) gland; (f) flower with (1) calyx, (2) fertilized enlarging ovaries, with (3) pollinia hanging by their stalks, and (4) large stigma common to the two ovaries. Redrawn from Gray, 1862; see Acknowledgments and Selected References.

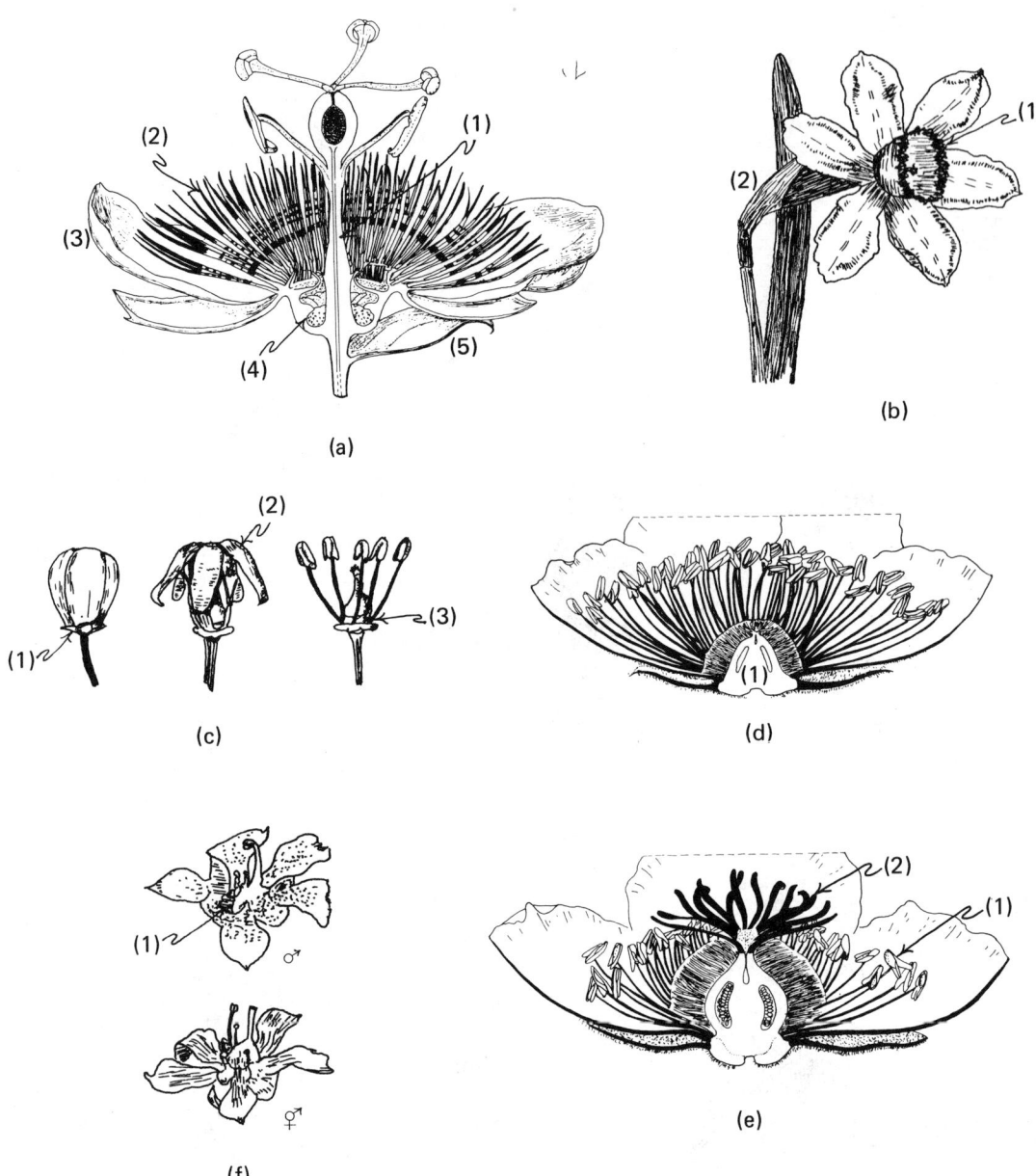

Fig. II.2.2.27. Flowers: (a) purple passion fruit (Passiflora edulis) flower with (1) androgynophore, (2) corona, (3) cupola-shaped corolla, (4) hypanthium, and (5) hypsophyll (bract); (b) Narcissus incomparabilis flower with (1) cyathiform corona and (2) hypsophyll; (c) grape flower with (1) sepals, (2) calyptrate, caducous corolla, and (3) annulate, nectariferous disk; (d), (e) kiwi (Actinidia chinensis) flowers: (d) staminate with (1) abortive ovary; (e) pistillate with (1) abortive stamens, and (2) linear stigmas; (f) mango flowers, staminate (above) with one fertile stamen and (1) staminodes and perfect (below). (a), (d), (e) Redrawn from McGregor, 1976; (b), (c) redrawn from Bailey, 1928; (f) redrawn from Purseglove, 1968; see Acknowledgments and Selected References.

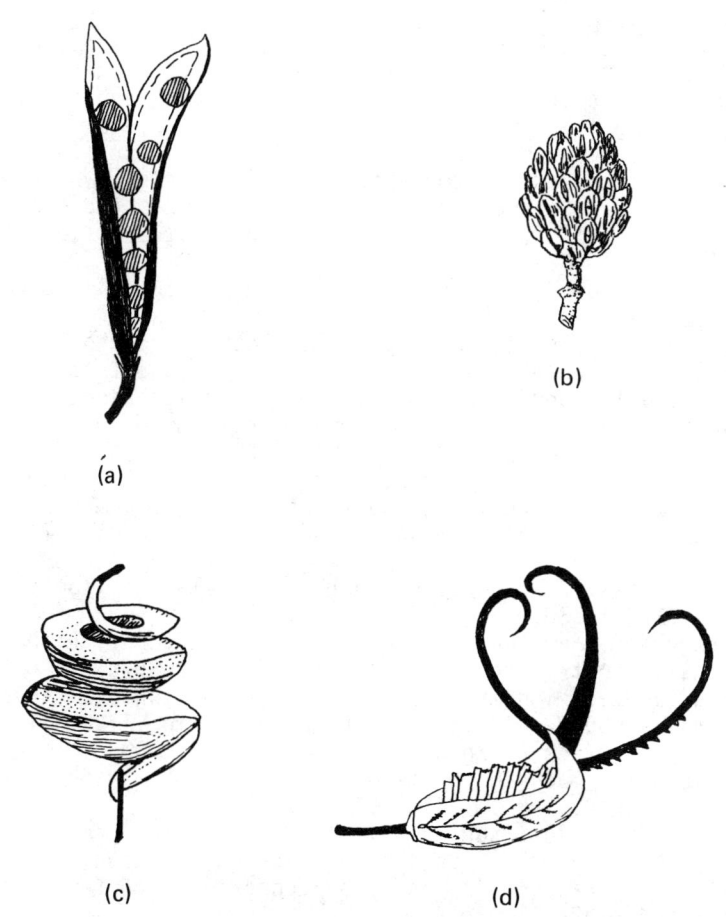

Fig. II.2.2.28. Flowers: (a) compressed, as the pod (legume) of a sweet pea; (b) polycarpic, as in Magnolia macrophylla (shown here in fruit); (c) spiral, as in alfalfa (Medicago sativa) ovary and fruit; (d) proboscoid, rostrate, as flower (and fruit) of unicorn plant (Proboscidea altheifolia). (a), (b) Redrawn from Gray, 1862; (c) redrawn from Hayward, 1938; (d) redrawn from Hutchinson, 1969; see Acknowledgments and Selected References.

II.2.3. Fruits

II.2.3.1. Fruit types*

.000 Fruit: a ripened or maturing ovary (pistil) with parts adnate; the seed-bearing organ; may be simple or compound or accompanied by other parts of the flower or floral axis. (Some fruits, e.g., cucumbers, okra, sweet corn, peas, and beans, are harvested when immature.) (G 62)

.001 Accessory: a type of aggregate fruit in which the conspicuous and often fleshy parts are composed of tissues external to the pistil, as in Annona and Fragaria. [Fig. II.2.3.1.01d, .04i(2); .05e; .06a-c; .07] (G 62)

.002 Achene: dry, indehiscent, usually one or two-seeded, wingless, with pericarp thin and adnate to the seed or thick and hard, sometimes bony; fruit small and from a unicarpellate (and simple) ovary, as in Fragaria, Ranunculus, Anemone, Clematis, Compositae, Urticaceae, and Moraceae. (Fig. II.2.3.1.01a-g) (G 62)

.003 Aggregate: product of connation or coherence of two or more pistils that were distinct in a flower, as in Rubus and Annona (see also Accessory, II.2.3.1.001, and Multiple, II.2.3.1.023.) (Fig. II.2.3.1.06) (G 62)

.004 Baccate: fleshy or berrylike, whether botanically a berry or not. (G 62)

.005 Berry: product of a single pistil, fleshy throughout, usually indehiscent, and homogeneous in texture. (Note that many fruits are erroneously termed a berry, which designation is often applied indiscriminately to any fleshy fruit regardless of its actual structure.) (Fig. II.2.3.1.05a) (G 62)

.006 Capsule: dry, usually dehiscent (occasionally explosively so), one- to many-seeded product of a single compound ovary; often classed according to manner of dehiscence. (Fig. II.2.3.1.02g-n, .03a-c) (G 62)

.006a. Circumscissile: pod opening by a circular horizontal line. (Fig. II.2.3.1.03d,e)

.006b. Irregular: opening without regular pattern, either longitudinally or transversely. (Fig. II.2.3.1.03b,c)

.006c. Loculicidal: longitudinal lines of opening are directly through the capsule wall into each locule. (Fig. II.2.3.1.02i,k)

.006d. Operculum: (see Poricidal capsule, II.2.3.1.006e).

.006e. Poricidal: opening by means of pores that may be provided with a lid (operculum). (Fig. II.2.3.1.03a)

.006f. Septicidal: longitudinal lines of opening coincide with the points on the capsule wall where the septa join it; the line of separation is in the approximate center of each septum. (Fig. II.2.3.1.02g,h,j)

.006g. Septifragal: a modification of loculicidal or septicidal dehiscence in which the valves fall away to leave the dissepiments or septa attached to the axis. (Fig. II.2.3.1. 02,l,m)

.007 Carpophore: a slender prolongation of the floral axis that supports a carpel or carpels. [Fig. II.2.3.1.03k(1)] (L 55)

*Note that many fruits do not fit into any of the categories listed here; names are applied in a purely descriptive sense.

II.2.3.1. MORPHOLOGY and ANATOMY

.008 Caryopsis: small, unilocular indehiscent fruit with thin, membranous, closely adherent pericarp so that fruit and seed are incorporated in one body, as in wheat, barley, maize, and other grasses. (Fig. II.2.3.1.01k-o) (G 62)
.009 Coccus: one of the separable valves of a schizocarp, as that of Malvaceae. [Fig. II.2.3.1.02g(1)] (G 62)
.010 Commisure: the face by which mericarps cohere (see Mericarp, II.2.3.1.022). [Fig. II.2.3.1.03k(2)(b)] (L 55)
.011 Compound fruit: produced from two or more pistils; the fruit are denoted aggregate if the pistils are in the same flower or multiple if the pistils are from separate flowers; either type may also include nongynoecial tissues, in which case it is also accessory. (Fig. II.2.3.1.06, .07) (L 55)
.012 Cremocarp: (see Schizocarp, II.2.3.1.032).
.013 Cupule: cupshaped involucre in which the bracts are indurated and coherent, especially in oak (Quercus). (Fig. II.2.3.1.04d) (G 62)
.014 Drupaceous: drupelike, whether actually a drupe or not. (L 55)
.015 Drupe: simple fruit with soft exterior, fleshy, usually indehiscent, with heterogeneous texture and the center with a hard, bony, or cartilaginous endocarp enclosing the seed proper; examples include stone fruits, mango, coconut, oil palm, coffee, and acerola, some of which have fleshy mesocarp only in the early stages of development. (Fig. II.2.3.1.05d,f-h) (G 62)
.016 Drupelet: miniature drupe, such as that in Rubus or fig (the "seeds" in the latter). (Fig. II.2.3.1.06a,c,d, .07d) (G 62)
.017 Follicle: dry, usually dehiscent, one- to many-seeded; product of a unicarpellate ovary; dehiscing by the ventral suture only; examples are macadamia and Cola, the latter being multiple; also Asclepiadaceae, Paeoniaceae, and Helleboraceae. (Fig. II.2.3.1.02a,b) (G 62)
.018 Hesperidium: exterior a firm, hard, or leathery rind, fleshy, usually indehiscent, heterogeneous in texture, divided into several segments by septa; usually considered a special form of berry (II.2.3.1.005); note also that the flesh (juice sacs) is produced from the endocarp; fruit type of citrus. (Fig. II.2.3.1.05b) (L 55)
.019 Indehiscent: not regularly opening; most often applied to certain fruit types; e.g., berries, drupes, and pomes. (Fig. II.2.3.1.05) (G 62)
.020 Legume: similar to a follicle, except that dehiscence is by two longitudinal sutures, ventral and dorsal. (Fig. II.2.3.1.02c-e) (G 62)
.021 Loment: similar to legume except that dehiscence is by transverse sutures; found in Leguminosae, especially Fabaceae and Mimosaceae. (Fig. II.2.3.1.02f) (G 62)
.022 Mericarp: peculiar seedlike fruits of Umbelliferae, a type of schizocarp with two carpels (i.e., two mericarps cohering to a commissure and suspended or supported from a stylopodium by a wiry carpophore). [Fig. II.2.3.1.03k(2)] (L 55)
.023 Multiple fruit: fruit produced from two or more pistils in different flowers; a fruit produced from pistils in an inflorescence rather than those in a single flower (see Compound fruit, II.2.3.1.011); examples are mulberry, breadfruit, fig, and pineapple. (Fig. II.2.3.1.07) (G 62)
.024 Nut: dry, indehiscent, usually one- to two-seeded; fruit usually large and from a two- to many-loculed ovary; often used indiscriminately to denote any fruit or seed with a hard shell; e.g.,

II.2.3.1. MORPHOLOGY and ANATOMY

peanut, Brazil nut, coconut, or macadamia nut, these being a legume, seed, drupe, and follicle respectively. (Fig. II.2.3.1.04 d-h) (G 62)

.025 Operculum: lid, or covering flap, of a pyxidium (pyxis) in seed plants; the calyx limb in Eucalyptus. (Fig. II.2.3.1.03d,e) (L 55)

.026 Pepo: similar to a hesperidium but lacking septa and with placentation parietal rather than axile; the fruit of Cucurbitaceae. (Fig.II.2.3.1.05c) (G 62)

.027 Pod: dehiscent dry pericarp, such as that on a bean; a general descriptive term sometimes used for fleshy fruits resembling a pod (e.g., cacao). (Fig. II.2.3.1.02c-f, .04h) (G 62)

.028 Pome: a fruit in which the papery or cartilaginous endocarp is embedded in the mesocarp, fused with and completely enveloped by the enlarged fleshy receptacle or the fused bases of the sepals; the ripened ovary here is only a small part of the total structure; an accessory fruit, as is plainly evident in apple, pear, and quince. (Fig. II.2.3.1.05d) (G 62)

.029 Pyxis (pyxidium): dry, usually dehiscent, one- to many-seeded; product of a two- to many-carpellate ovary; splitting and releasing seeds by circumscissile dehiscence, as in Lecythis, some Solanaceae, Amaranthaceae, Primulaceae, Plumbaginaceae, and Myrtaceae. (Fig. II.2.3.1.03d,e) (G 62)

.030 Replum: the placental framework, or thin, false dissepiment, separating the two valves of some unicarpellary fruits, like siliques (Cruciferae), and some legumes, from which the valves fall away at maturity. (Fig. II.2.3.1.03g) (L 51)

.031 Samara: dry, indehiscent, usually one-seeded and winged; can be single as in Ulmus or double as in Acer; a type of schizocarp. (Fig. II.2.3.1.04a-c) (G 62)

.032 Schizocarp (= cremocarp): dry, usually dehiscent, one- to many-seeded; product of two- to many-carpellate ovary; splitting into halves on dehiscence; the fruit type of the Umbelliferae. (Fig. II.2.3.1.03i-k, .04a-c) (L 55)

.033 Silicle: short capsule, usually not more than twice as long as broad, with longitudinal dehiscence, as in Cruciferae (e.g., Capsella and Iberis). (Fig. II.2.3.1.03h) (G 62)

.034 Silique: long and slender capsule with longitudinal dehiscence, as in Cruciferae. (Fig. II.2.3.1.03f,g) (G 62)

.035 Stylopodium: a disclike enlargement at the base of the style, as in some umbellifers. [Fig. II.2.3.1.03j(1)] (L 55)

.036 Syconium: the "fruit" of Ficus in which the receptacle by involution has developed into a hollow, more or less fleshy axis, with a small opening (ostiole; see II.2.3.2.025) through which diminutive wasps pass to effect pollination or deposit their eggs; the true fruits are numerous drupelets on the inner surface of the axis. Common-type fig cultivars ordinarily develop fruit parthenocarpically. (Fig. II.2.3.1.07d) (L 55)

.037 Syncarp (syncarpium): multiple fruit produced from connation or coherence of pistils, plus nonovarian tissues from several to many flowers; a fruit produced from a more or less entire inflorescence, as in Artocarpus, Ananas, and mulberry (Morus). (Fig. II.2.3.1.07a-c) (SA)

.038 Utricle: dry, indehiscent, usually one- to two-seeded, wingless, with the pericarp thin, loose, and free from the seed, as in Chenopodiaceae, Amaranthaceae, and some Gramineae. (Fig. II.2.3.1.01h-j) (G 62)

II.2.3.2. MORPHOLOGY and ANATOMY

II.2.3.2. Fruit parts (exclusive of seeds)

.001 Alate: winged. (Fig. II.2.3.2.01c,d)
.002 Anisopterous: unequally winged fruit. (Fig. II.2.3.2.01a,b) (HS 69)
.003 Apocarpous: a fruit developed from a single carpel in Levina's classification. (Fig. II.2.3.2.02a) (ES)
.004 Axillary (lateral): borne in the axils of present or fallen leaves. (Fig. II.2.3.2.03d) (L 55)
.005 Bladdery: (see Inflated, II.2.3.2.017).
.006 Burr, bur: applied to an involucre, persistent calyx or pericarp, which provides a rough (e.g., filbert, Corylus) or spiny (e.g., Castanea or Cenchrus) protective envelope around the fruit. (Fig. II.2.3.2.02b) (G 62)
.007 Choricarpelly: in Winkler's classification, an aggregate fruit with free carpels. (Fig. II.2.3.2.02e; .03a, .2.3.1.06d) (ES)
.008 Conical, pyramidal: having the figure of a true cone; pyramidal is the same except for its angular sides. (Fig. II.2.3.2.02c) (ST)
.009 Conoidal: resembling a conical figure. (Fig. II.2.3.2.02e,g) (G 62)
.010 Depressed: flattened vertically. (Fig. II.2.3.2.02h) (G 62)
.011 Ellipsoid: elliptical solid figure. (Fig. II.2.3.2.02i) (G 62)
.012 Epichlamydy: in Winkler's classification, a free fruit developed from a hypogynous flower. (Fig. II.2.3.2.03b) (ES)
.013 Geocarpic, geocarpus: an uncommon protective adaptation in which fruits are produced underground; the most familiar example is peanut (Arachis) but also found in a number of other families (cf. Hutchinson 1969, 1973). (Fig. II.2.3.02h)
.014 Globoid: (see Spheroid, II.2.3.037).
.015 Husk: outer covering of certain fruits (like Juglans), usually derived from the perianth or involucre. (Fig. II.2.3.2.04a) (L 51)
.016 Hypochlamydy: in Winkler's classification, a cup fruit developed from a perigynous or epigynous flower. (Fig. II.2.3.2.04b) (ES)
.017 Inflated, bladdery: thin, membranous, translucent, swollen equally like a balloon. (Fig. II.2.3.2.02c) (G 62)
.018 Infructescence: a ripened inflorescence; a fruit derived from an inflorescence (i.e., a multiple fruit, II.2.3.1.023). (Fig. II.2.3.2.04d) (B & C)
.019 Lachrymiform, tear-shaped: same as pyriform but the sides are not contracted. (Fig. II.2.3.2.04e) (ST)
.020 Lysicarpous: in Levina's classification, a fruit with free central placentation. (Fig. II.2.3.2.04c) (ES)
.021 Lysigeny: (see II.3.1.04).
.022 Napiform, turnip-shaped: having the figure of a depressed sphere. (Fig. II.2.3.2.03e) (G 62)
.023 Oblate: spheroid flattened or depressed at the poles, like a grapefruit or mandarin. (Fig. II.2.3.2.06d) (L 55)
.024 Obovoid: solid obovate (reverse egg-shaped) in outline. (Fig. II.2.3.2.07a) (L 55)
.025 Ostiole: an eye, an opening into a cavity, like that in a fig syconium. (See Fig. II.2.3.1.07d) (CN)
.026 Ovoid: ovate (egg-shaped) solid figure. (Fig. II.2.3.2.04d, .06a) (L 51)
.027 Paracarpous: in Levina's classification, a fruit with parietal placentation. (Fig. II.2.3.2.03c) (ES)
.028 Pear-shaped: (see Pyriform, II.2.3.2.035).
.029 Pericarp: the wall of a ripened ovary = fruit; sometimes used loosely to designate a fruit. (Fig. II.2.3.2.05a) (L 51)

II.2.3.2. MORPHOLOGY and ANATOMY

.029a   Albedo: spongy membranous mesocarp in a hesperidium (citrus fruit). (Fig. II.2.3.2.05d) (ES)

.029b   Cling, clingstone: the condition in certain fruits, such as peach or nectarine in which the mesocarp adheres closely to the bony endocarp (stone).

.029c   Endocarp: the inner layer, variously developed as a simple membrane around the cavity enclosing a single seed, as in avocado (Persea americana); a bony or cartilaginous structure or "stone" enveloping the seed in a drupe, as in stone fruits, coconut, oil palm, mango, and acerola; multicellular saclike protrusions, as in citrus, or fleshy protrusions into the locules, as in banana; the last two forms are denoted "pulpa" (and often intermingled with outgrowths from the seed coats so that the true origin is not always apparent). [Fig. II.2.3.2.05a(3), b(2)(a), c(1)(a), f(2)] (E & M, G 62, B & C)

.029d   Exocarp: the outer layer. [Fig. II.2.3.2.05a(1)] (G 62)

.029e   Free, Freestone: the condition in which the mesocarp separates readily from the bony endocarp; intermediate states between free and cling are denoted as semifree or semicling.

.029f   Germ-pore: one or more soft regions in the stone or hard endocarp wall of a fruit opposite the embryo, which may be proximal, as the "eyes" in a coconut shell, or oblique, as in oil palm; the purpose is to permit germination without complete disintegration of the endocarp in these indehiscent fruits. (Fig. II.2.3.2.05b,c) (CR 66)

.029g   Mesocarp: the middle layer or layers; sometimes fleshy, as in stone fruits, mango, acerola, and oil palm; sometimes fibrous, as in coconut; rarely spongy-membranous, as in the albedo and segment membranes of citrus. [Fig. II.2.3.2.05a(2)] (G 62)

.029h   Sarcocarp: fleshy mesocarp. [Fig. II.2.3.2.05a(2)] (G 62)

.029i   Stone: a bony endocarp, as in stone fruits (Prunus). [Fig. II.2.3.2.05a(3)] (G 62)

.030   Prolate: an ellipsoid whose polar diameter is longer that the equatorial; the reverse of oblate, as in lemon or lime fruits. (Fig. II.2.3.2.07b)

.031   Pulpa: (see Endocarp, II.2.3.2.029c)

.032   Pungent: acrid to the taste.

.033   Pyramidal: (see Conical, I.2.3.2.008).

.034   Pyrene: the nutlet, the seed and its enclosing bony endocarp, in a drupe or drupelet. (Fig. II.2.3.2.02f) (G 62)

.035   Pyriform, pear-shaped: differs from turbinate in being more elongated. (Fig. II.2.3.2.06b,c) (ST)

.036   Schizolysigenous: (see II.3.169).

.037   Spheroid, globoid: spherical solid figure, a little depressed at both ends (thus similar to oblate but to a lesser degree). (Fig. II.2.3.2.04d, .07c,d) (ST)

.038   Superposed: parts one above the other, like seeds in a fruit. (Fig. II.2.3.2.07e) (G 62)

.039   Syncarpelly: in Winkler's classification, a unit fruit with united carpels. (Fig. II.2.3.2.03d) (ES)

.040   Syncarpous (in narrow sense): in Levina's classification, a fruit with axile placentation. (Fig. II.2.3.2.03e) (ES)

.041   Tear-shaped: (see Lachrymiform, II.2.3.2.019).

.042   Top-shaped: (see Turbinate, II.2.3.2.043).

.043   Turbinate, top-shaped: inversely conical with a contraction toward the point. (Fig. II.2.3.2.07f) (ST)

.044   Turnip-shaped: (see Napiform, II.2.3.2.022).

II.2.3.2. MORPHOLOGY and ANATOMY

.045 Valve: the separable parts of a pod, capsule, or other dehiscent fruit. (Fig. II.2.3.2.07g) (L 51)
.046 Vesicle: small bladdery sac or cavity filled with air or fluid, like juice vesicles in citrus fruit. (Fig. II.2.3.2.05d) (L 55)

II.2.3.3. Seeds

.000 Seed: the mature ovule; sometimes erroneously applied to seedlike fruits, like an achene or caryopsis in which the pericarp adheres closely to the testa. (G 62)
.001 Accumbent: lying against an object; e.g., when cotyledons have their edges placed against the radicle. (Fig. II.2.3.3.02a) (G 62)
.002 Albumen: (see Endosperm, II.2.3.3.015).
.003 Annular: embryo coiled around the periphery of the seed. (Fig. II.2.3.3.02b) (G 62)
.004 Antiraphe: the side of the seed opposite the raphe (II.2.3.3.034), often developed or exaggerated in certain seed forms, like the campylotropous. [Fig. II.2.3.3.08a(3)] (CR 76)
.005 Ategmic: seeds lacking a seed coat, as in Santalales. (CR 76)
.006 Bitegmic: literally, with two seed coats, inner and outer (see Testa, II.2.3.3.041). (CR 76)
.007 Chalaza: the basal part of the ovule where the latter is attached in the funiculus and the point of entry for the vascular tissues; normally punctiform but sometimes modified. [Fig. II.2.3.3. 10b(5)] (CR 76)
.007a Pachychalazal: see Seed forms, II.2.3.3.036a-5).
.007b Perichalazal: see Seed forms, II.2.3.3.036a-6).
.008 Coiled: embryo wound in a spiral. (Fig. II.2.3.3.02d) (G 62)
.009 Coma: tuft of hairs attached to a seed. (Fig. II.2.3.1.01f,g)(G 62)
.010 Crested: having an elevated, irregular, or notched ridge resembling the crest of a helmet; chiefly applied to seeds. (ST)
.011 Curved: embryo in a partial circle. (Fig. II.2.3.3.02e,f) (G 62)
.012 Dorsal seed form: a type of anatropous seed form (II.2.3.3.036a) in which the vascular bundle extends around the dorsal side of the seed. (Fig. II.2.3.3.05c,d) (CR 76)
.013 Elaiosome: external oil-secreting body on seeds of certain plants, particularly those inhabited or visited by ants (i.e., Myrmecophytes, II.4.2.024). (CR 76)
.014 Embryo: rudimentary plant within a seed, developed from a zygote (= sexual embryo) or from other nuclei in the embryo sac or cells of the nucellus or integuments (=apomictic embryo) (see also IV.4.1.006a-1). [Fig. II.2.3.3.02, .03f-i, .07c(5), d(4), .10c(4), .11b(1), d(3) .12b(5)] (L 55)
.014a. Coleoptile: sheath over the plumule of grain. [Fig. II.2.3.3.01b(5)] (ES)
.014b. Coleorhiza: sheath over the radicle of grain. [Fig. II.2.3.3.01b(9)] (G 62)
.014c. Cotyledon: seed leaf; the primary leaf (in monocotyledons) or leaves (in dicotyledons); may have several in gymnosperms and some dicotyledons (e.g., Cola spp.). [Fig. II.2.3.3.01a(1), b(2), .03f-i(8), .07e(2)]. (G 62)
.014d. Hypocotyl: axis of the embryo below the cotyledons which has the radicle at its basal end. [Fig. II.2.3.3.01a,b, .03f-i(9)] (L 55)
.014e. Monoembryonic, monoembryony: (see IV.4.6.004).

II.2.3.3. MORPHOLOGY and ANATOMY

.014f. Nucellar embryony: (see adventitious embryony, IV.4.1.006a-1).
.014g. Nucellus: body tissue of ovule. [Fig. II.2.3.3.03f-i(2),.7f(9), .10b(4),.12d(5)]
.014h. Plumule: the rudimentary stem. [Fig. II.2.3.3.01a(2),b(6), .03f-i(11)] (G 62)
.014i. Polyembryony: (see IV.4.6).
.014j. Radicle: rudimentary root that develops later into the primary root. [Fig. II.2.3.3.01a(7),b(8)] (G 62)
.014k. Scutellum: cotyledon in grain (Gramineae). [Fig. II.2.3.3. 01b(2)] (G 62)
.015 Endosperm (albumen): storage tissue that contains starch, inulin, and oil in many seed as food supply for the embryo; derived from fusion of the polar nuclei and a sperm nucleus during fertilization (see Fruit setting, Syngamy, IV.4.1.026); may be smooth or ruminate. [Fig. II.2.3.3.01b(3),d(5), .03f-i(4), .07c(4), .09a(1), .10c(5), .11d(4), .12b(4)] (G 62)
.016 Endostome: foramen (micropyle) of the inner coat (integument) of the ovule (see Exostome, II.2.3.3.018). [Fig. II.2.3.3.10b(3), .12b(3)] (G 62)
.017 Excentric: embryo at one side or corner of the ovule, as in many grains. (Fig. II.2.3.3.02g) (G 62)
.018 Exostome: foramen (micropyle) of the outer seed coat (integument) of the ovule; a term used mainly to indicate the location of an aril (e.g., exostomal = micropylar aril, II.2.3.3.042b). (G 62, CR 76)
.019 Funiculus (funicle): stalk on which the ovule is borne; sometimes elongating and becoming threadlike (as in Magnolia); often fused in various configurations to the ovule (II.2.2.041t-4; see also Seed forms, II.2.3.3.036). (Fig. II.2.3.3.01c) (G 62)
.020 Germination: resumption of growth by the embryo after a longer or shorter period of dormancy in the seed; favorable conditions include supply of water, oxygen, heat, and light (or its absence), plus after-ripening, in some instances, and removal of inhibitors in the fruit, seed coats, or embryo. (H & K)
.020a False vivpary: immediate or slightly delayed growth of certain vegetative structures, as bulbils, bulblets, and subterranean stolons. (PJ)
.020b. Ovipary: the "normal" process that takes place after separation of the seed from the mature fruit and removal of inhibitors to growth, including after-ripening (see Dormancy, IV.3.1.010).
.020c. Ovovivipary: immediate germination after separation of the seed from the mature fruit; differs from ovipary in that after-ripening is not required.
.020d. Vivipary: germination of seeds within the fruit while it is still attached to the parent plant; e.g., late-season grapefruit or red mangrove. (Fig. II.2.3.3.01e,f) (G 62, PJ)
.021 Hilum (umbilicus): scar on a seed marking the place where the funiculus was attached to the ovule; sometimes exaggerated, as in the hilar seed form where the hilum extends around most of the seed (as shown by the tracheid bar in Mucuna). [Fig. II.2.3.3. 01a(6), .08b(3), c(3), .10a(1)] (G 62)
.022 Hypostase, hypostasis: (see Suspensor, II.2.3.3.038).
.023 Incumbent: folded inward and lying on, e.g., when the radicle rests on one side of a cotyledon instead of along the edge. (Fig. II.2. 3.3.02h) (G 62)

II.2.3.3. MORPHOLOGY and ANATOMY

.024 Kneecap-shaped: (see Patelliform, II.2.3.3.029). (G 62)
.025 Lachrymiform, tear-shaped: (see II.2.3.2.019).
.026 Lenticular, lens-shaped: resembling a double convex lens, like many seeds. (G 62)
.027 Micropyle (foramen): minute scar or pore that indicates the opening into the ovule (see Exostome, II.2.3.3.018 and Endostome, II.2.3.3.016). (Fig. II.2.3.3.01, .03-.12) (G 62)
.028 Opposite: pointing directly to a particular place, like the radicle to the hilum. (Fig. II.2.3.3.01a) (G 62)
.029 Patelliform, kneecap-shaped: broad, round, thick; convex on the lower and concave on the upper surface, like some seeds. (Fig. II.2.3.3.02) (ST)
.030 Peripheric: situated around the perimeter like the embryo in certain seed. (Fig. II.2.3.3.02i) (G 62)
.031 Perisperm: nutritive tissue derived from the nucellus, thus outside the embryo sac proper, as in the case of endosperm (with which it is often confused). [Fig. II.2.3.3.12e(3)] (ES)
.032 Preraphe: the part of the seed between the micropyle and hilum; normally short but lengthened in some ovules and during seed development (see Hemianatropous seed form, II.2.3.3.036a-3). (Fig. II.2.3.3.04b, .05a,b) (CR 76)
.033 Pseudoarillate: seed with a pseudoaril (see Aril, II.2.3.3.042b). (Fig. II.2.3.3.12) (CR 76)
.034 Raphe: commonly defined as the ridge formed on a seed by fusion of the funiculus with the nucellus of an anatropous ovule; sometimes enlarged or exaggerated, as in the obcampylotropous seed form, where the raphe extends nearly around the nucellus. [Fig. II.2.3.3.04e, .08a(3)] (CR 76)
.035 Ruminate: when a hard body (e.g., endosperm) is pierced in various directions by narrow cavities filled with dry cellular matter. [Fig. II.2.3.3.09a(1), .10c(1),(5)] (G 62)
.036 Seed forms: development of ovules into several distinct seed forms, as recognized by Corner (1976). (Fig. II.2.3.3.04, .05) (CR 76)
.036a. Anatropous: several distinct seed forms are developed from this type of ovule. (Fig. II.2.3.3.04c) (CR 76)
.036a-1. Campylotropous: exaggerated development of the antiraphe which results in a pronounced curvature of the seed, as in Psidium. (Fig. II.2.3.3.02c, .04d, .06a,b) (CR 76)
.036a-2. Hemianatropous: characterized by a preraphe with a longitudinal vascular bundle between the micropyle and hilum, the latter being far removed from the former, as in Connaraceae. (Fig. II.2.3.3.04b) (CR 76)
.036a-3. Hilar: the greater part of the circumference of the seed, which is usually flattened, is made up of the extended hilum, as in Mucuna; can also be derived from a campylotropous ovule. (Fig. II.2.3.3.04f; .11a-c) (CR 76)
.036a-4. Obcamplyotropous: the raphe of the anatropous ovule enlarges more than the antiraphe, as in seeds of Bauhinia and Barkleya. (Fig. II.2.3.3.04e, .06e-h) (CR 76)
.036a-5. Pachychalazal: extension of the chalaza in all directions with intercalary growth; the multiplication of cells in which the two integuments join the nucellus and chalaza to form a single wall, the integuments persisting only at the micropylar end in more or less vestigial form, as in Flacourtiaceae, Lauraceae, Meliaceae, Rosaceae, Sapindaceae, Ebenaceae, Euphorbiaceae, and also Macadamia. (Fig. II.2.3.3.04g, .08b) (CR 76)

II.2.3.3. MORPHOLOGY and ANATOMY

.036a-6. Perichalazal: the inner integument is attached to the outer integument along the entire course of the vascular bundle, which extends around the periphery from funicle to micropyle; thus a perichalaza surrounds the nucellus as a hoop or band, as in Annonaceae, Swietenia, Lansium, and some Vitaceae. (Fig. II.2.3.3.04h, .09a-d, .12d) (CR 76)

.036b. Orthotropous: seeds developed from orthotropous ovules, as in Urticaceae, Proteaceae, Flacourtiaceae, Piperaceae, and Polygonaceae. (Fig. II.2.3.3.04a) (CR 76)

.037 Strophiole: small swelling on the raphe (see Aril, II.2.3.3.042b). (PJ)

.038 Suspensor (= hypostase, hypostasis): a cell or group of cells developed from a zygote which attaches the embryo to the embryo sac. [Fig. II.2.3.3.03f-i(5), .10b(6)] (G 62)

.039 Tegmen: seed coat developed from the inner integument of the ovule (see Testa, II.2.3.3.041); a diverse and often highly complex structure with several more or less distinct layers. (Fig. II.2.3.3.04, .07, .08, .12)(CR 76)

.040 Tegmic seeds: those in which the mechanical (protective or strengthening) layer is found in tissues of the tegmen. (Fig. II.2.3.3.03, .12e) (CR 76)

.040a. Endotegmic: mechanical layer in the inner epidermis, consisting of short palisade or cuboid sclerotic cells (e.g., Piperaceae, Saururaceae, and Nandinaceae). [Fig. II.2.3.3.03e, .12d(5)]

.040b. Exotegmic: mechanical layer present in the outer epidermis as a palisade (e.g., Malvales), layer or fibers (e.g., Celastraceae), or with stellately lobed cells (e.g., Guttiferae, Hypericaceae, and Geraniaceae). [Fig. II.2.3.3. 03d, .08b(4), c(4)]

.040c. Mesotegmic: mechanical layer in the mesophyll (rare).

.041 Testa: commonly defined as the outer covering of the seed (e.g., seed coats) which is developed from the outer and inner integuments of the ovule; in a strict sense, the seed coat develops from the outer integument only; a diverse and often highly complex structure in which several more or less distinct layers may be recognized and from which a variety of outgrowths may arise. [Fig. II.2.3.3.03, i(1), .04b(1), .07f(7), .08a, .12d(3,4)] (CR 76, G 62)

.042 Testal outgrowths: structures arising from one or more layers of the testa, usually the exotesta, but not forming a part of the testa proper. (CR 76)

.042a. Alate: an outgrowth from an aril or raphe, to form a wing. (Fig. II.2.3.3.11d,e) (CR 76)

.042b. Aril: an outgrowth, often fleshy, arising near the top of the funiculus, thus around and next to the seed proper; sometimes intermingled with tissue arising as an arillode; now considered a general term, qualified as to funicular, hilar, exostomal or micropylar, chalazal, and raphe origin, to which separate names are often applied. [Fig. II.2.3.3.05a(2), .08a-c, .09b,d(2), .11b,d] (CR 76)

.042c. Arillate: strictly, with an aril; often used in reference to any outgrowth of the testa. [Fig. II.2.3.3.10b(5)] (CR 76)

.042d. Arillode: encircling structure or outgrowth around the exostome. (CR 76)

.042e. Arillostome: outgrowth of the testa at the hilar end of the seed (similar to an embryotega but differing in persisting on the mature seed). [Fig. II.2.3.3.10c(3), d(3)] (CR 76)

II.2.3.3. MORPHOLOGY and ANATOMY

.042f. Caruncle: small swelling near the micropyle found in some Euphorbiaceae; e.g., seeds of castor bean (Ricinus). (G 62)
.042g. Embryotega: caplike callosity near the hilum of certain seeds and detached by the radicle in germination. (CR 76)
.042h. Hairs: (see also Surface features, II.5.6.). (Fig. II.2.3.3.07e,f)
.042i. Sarcotesta: descriptive term for the fleshy and edible testa or outer part of it, as in papaya; variable in construction and in the layer of the testa involved; sometimes intermingled or confused with the aril or occasionally, as in Connaraceae, Meliaceae, and Sapindaceae, continuous with the aril which is also present; sometimes intermingled with pulpa from the endocarp (inner fruit wall). (Fig. II.2.3.3.01d, .07a-d, .09e) (CR 76)
.043 Testal seeds: those in which the mechanical layer is found in tissues of the testa. (Fig. II.2.3.3.12c,d) (CR 76)
.043a. Endotestal: mechanical layer in the inner epidermal layer, like a palisade (e.g., Myristicaceae, Dilleniaceae, or Cruciferae); tracheidal, the tangentially elongate cells with spiral or annular thickening (e.g., Magnoliaceae or Vitaceae); or a single layer of cuboid sclerotic cells (e.g., Grossulariaceae). (Fig. II.2.3.3.03c)
.043b. Exotestal: mechanical layer in the outer epidermal layer, like a palisade of radially elongate cells (e.g., Ranunculaceae, Rhamnaceae, Leguminosae, and Winteraceae) or a layer of fibers (various genera of different families). (Fig. II.2.3.3.03a)
.043c. Mesotestal: mechanical layer in the mesophyll, like an outer hypodermal palisade (e.g., Paeoniaceae), cuboid sclerotic cells in the mesophyll generally (e.g., Myrtaceae, Rosaceae, or Theaceae), or fibers in the mesophyll generally (Annonaceae). [Fig. II.2.3.3.01d, .03b, .06a(2), .07c(2,3)]
.044 Umbilicus: (see Hilum, IIi.2.3.3.021).
.045 Unitegmic: literally, with only a single seed coat; used without distinction as to its derivation, except that it means uniintegumented and not provided only with a tegmen; neither does it refer to testal seeds of bitegmic ovules which lose the tegmen in the course of development. (Fig. II.2.3.3.12c-e) (CR 76)
.046 Vermicular: (see Worm-shaped, II.2.3.3.047).
.047 Worm-shaped, vermicular: thick and almost cylindrical but bent in different places. (ST)
.048 Zygote: (see Embryo, II.2.3.3.014).

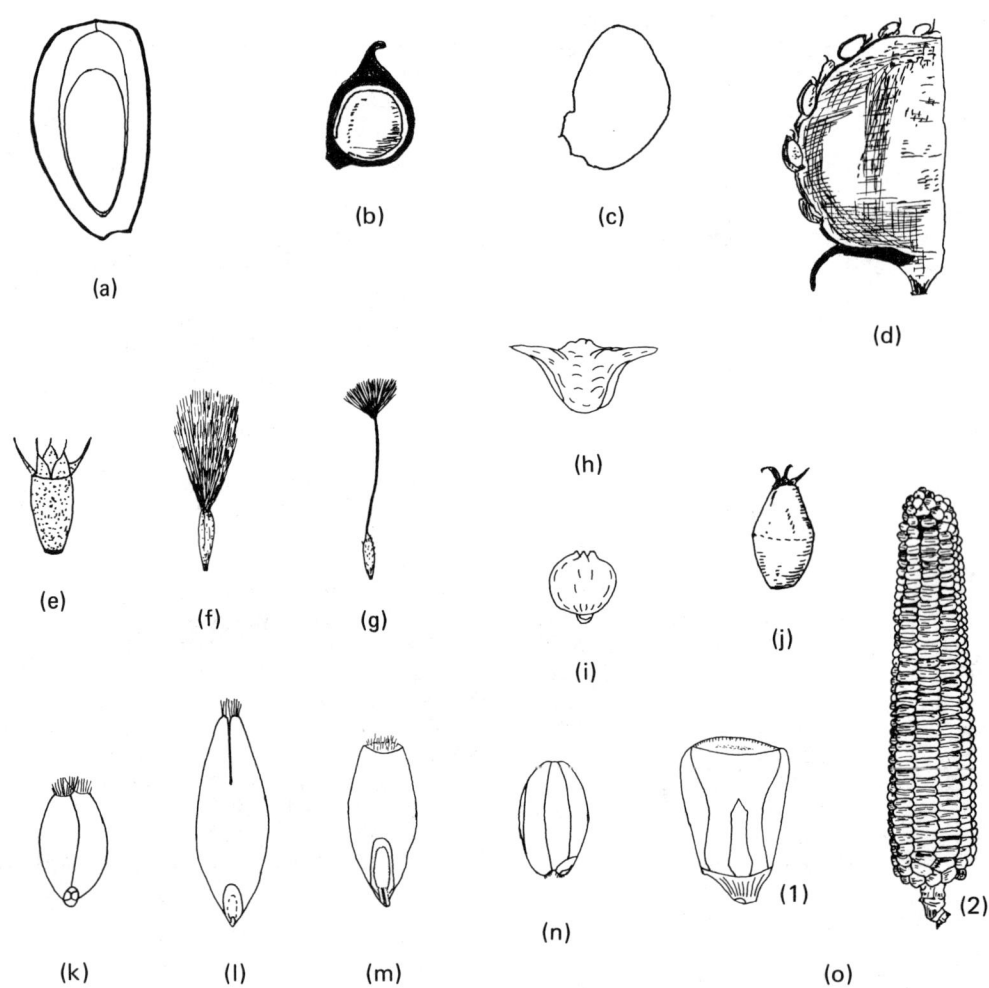

Fig. II.2.3.1.01. Fruit types: (a)-(c), achene, (a) sunflower; (b) common buttercup (l.s.); (c) strawberry; (d) accessary fruit, strawberry with achenes on surface; (e)-(g) achene (cont.), (e) sneezewood (<u>Helenium</u>); (f) sow thistle with pappus (coma); (g) dandelion; (h)-(j) utricle, (h) prickly-seeded spinach; (i) smooth-seeded spinach; (j) amaranth (forms a small pyxis on dehiscence); (k)-(o) caryopsis, (k) wheat; (l) barley; (m) rye; (n) rice, and (o) corn, as (1) individual fruit and (2) multiple fruit (ear). (a), (c) Redrawn from McGregor, 1976; (b), (d)-(g), (j) redrawn from Gray, 1862; (h)-(o) (1) redrawn from Robbins, 1924; (o)(2) redrawn from Bailey, 1928; see Acknowledgments and Selected References.

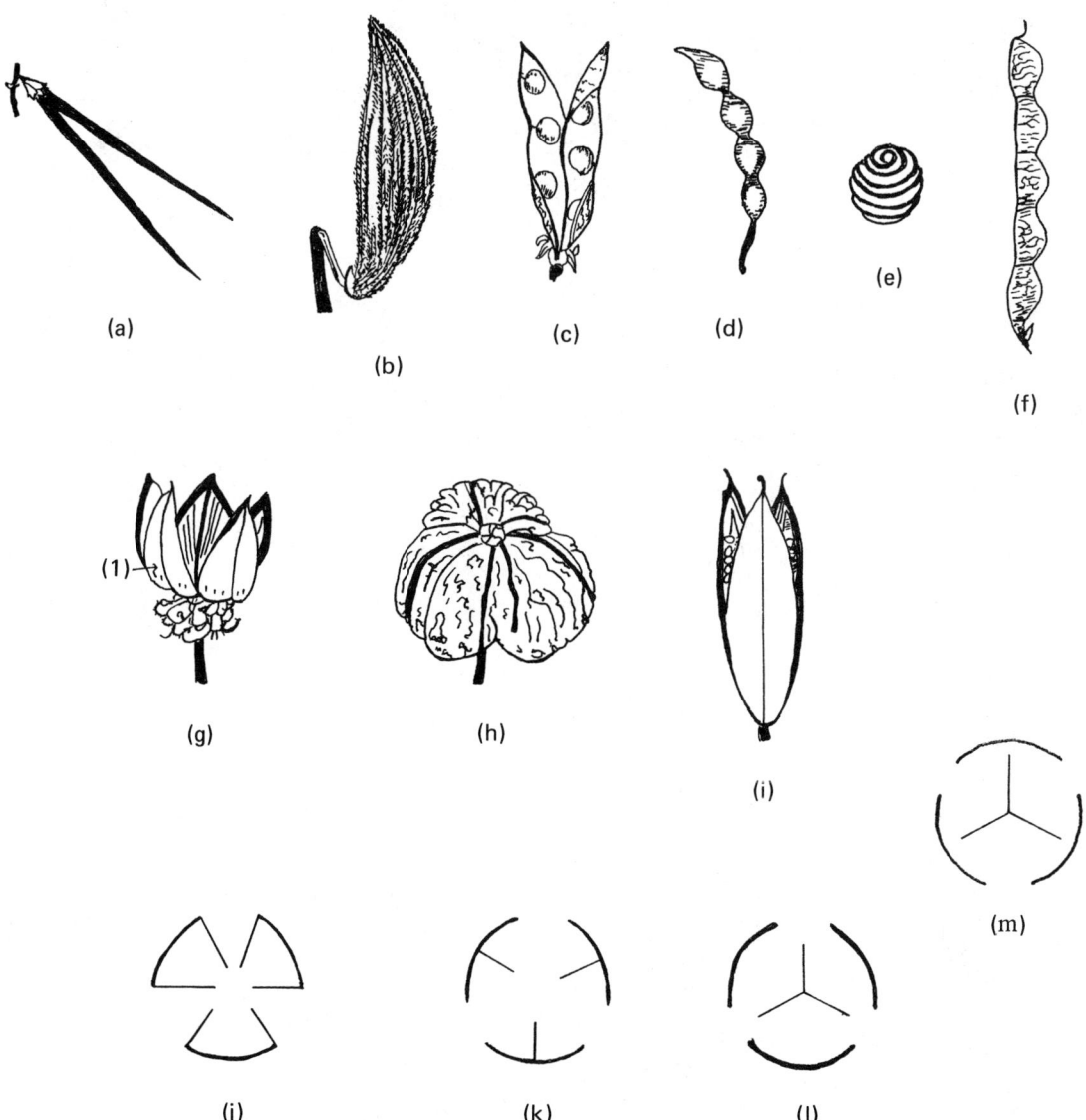

Fig. II.2.3.1.02. Fruit types: (a), (b), follicle, (a) dogbane (*Apocynum androsaemifolium*); (b) common milkweed (*Asclepias cornuti*); (c)-(e) legume, (c) pea; (d) *Sophora*; (e) *Medicago scutella*; (f) loment: *Desmodium*; (g)-(n) capsule, (g) mallow, with coccus (1), and (h) rubber (*Hevea brasiliensis*) with (j) septicidal dehiscence; (i) *Iris* with (k) loculicidal dehiscence; (1) septifragal (septicidal form) dehiscence; (m) septifragal (loculicidal form) dehiscence. (a)-(g), (i)-(m) Redrawn from Gray, 1862; (h) redrawn from Dijkman, 1951; see Acknowledgments and Selected References.

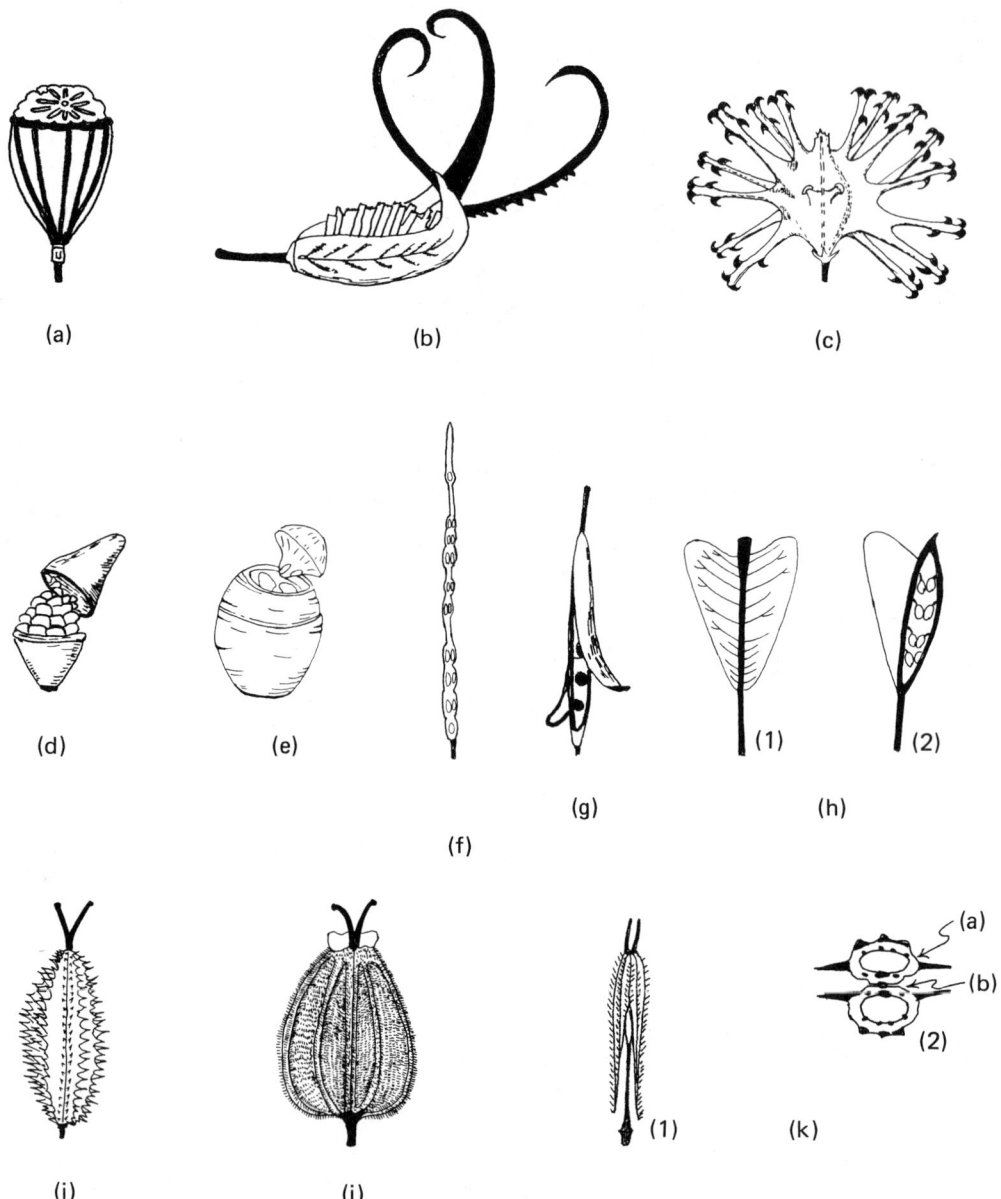

Fig. II.2.3.1.03. Fruit types: (a)-(c) capsule, (a) poricidal (poppy); (b), (c) tardily dehiscent: (b) unicorn plant (Proboscidea altheifolia), (c) grapple plant (Harpagophytum procumbens); (d)-(e) pyxis, (d) purslane; (e) paradise nut (Lecythis ollaria); (f)-(g) silique: (f) radish; (g) cardamine (dehiscing); (h) silicle, shepard's purse (Capsella) as (1) mature fruit, (2) same in median l.s.; (i)-(k) schizocarp (= cremocarp), (i) carrot (Daucus); (j) anise (Pimpinella) with (1) stylopodium; (k) Angelica with (1) carpophore and (2) two mericarps with (a) oil canals (= vittae) and (b) commisure. (a)-(c), (e)-(f), (i)-(j) Redrawn from Hutchinson, 1969; (d),(g),(h),(k) redrawn from Gray, 1862; see Acknowledgments and Selected References.

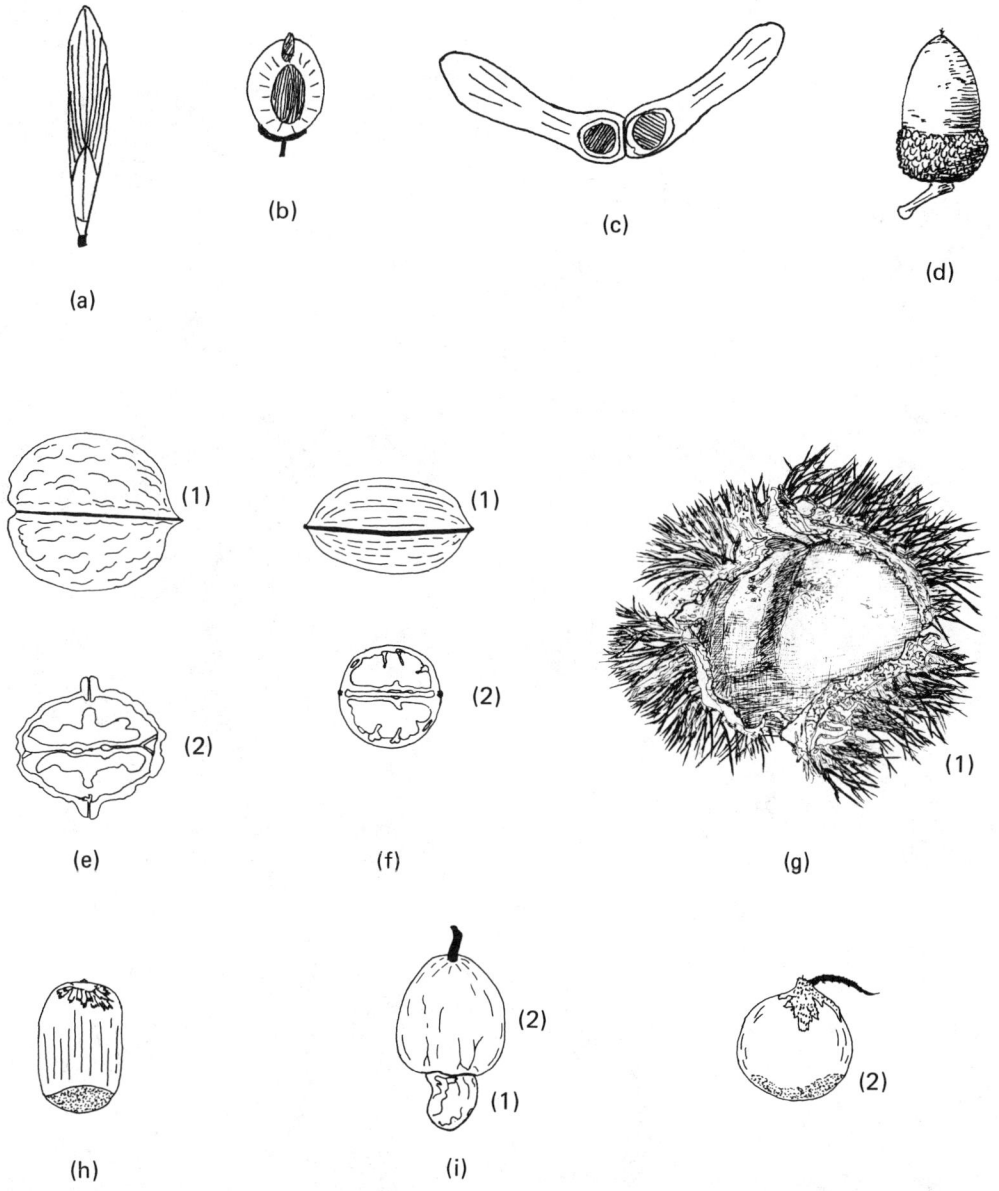

Fig. II.2.3.1.04 Fruit types: (a)-(c) samara (a type of schizocarp), (a) white ash; (b) american elm; (c) maple (double); (d)-(h) nut, (d) acorn (white oak); (e) 'Mission' walnut (Juglans regia) (1) with (2) t.s. below; (f) 'Idlewild' pecan (Carya illinoensis) (1) with (2) t.s. below; (g) chestnut (Castanea) with (1) burr (C. mollissima) holding several nuts and (2) 'Murrell' nut (C. dentata) below; (h) 'Istrian' filbert (Corylus avellana); (i) cashew (Anacardium occidentale) with (1) nut and (2) "apple." (a), (b), (d) Redrawn from Gray, 1862; (c) from Hill et al., 1960; (e), (f), (g)(2), (h) redrawn from U.S. Dept. Agriculture, 1896; (i) redrawn from Jaynes, 1979; see Acknowledgments and Selected References.

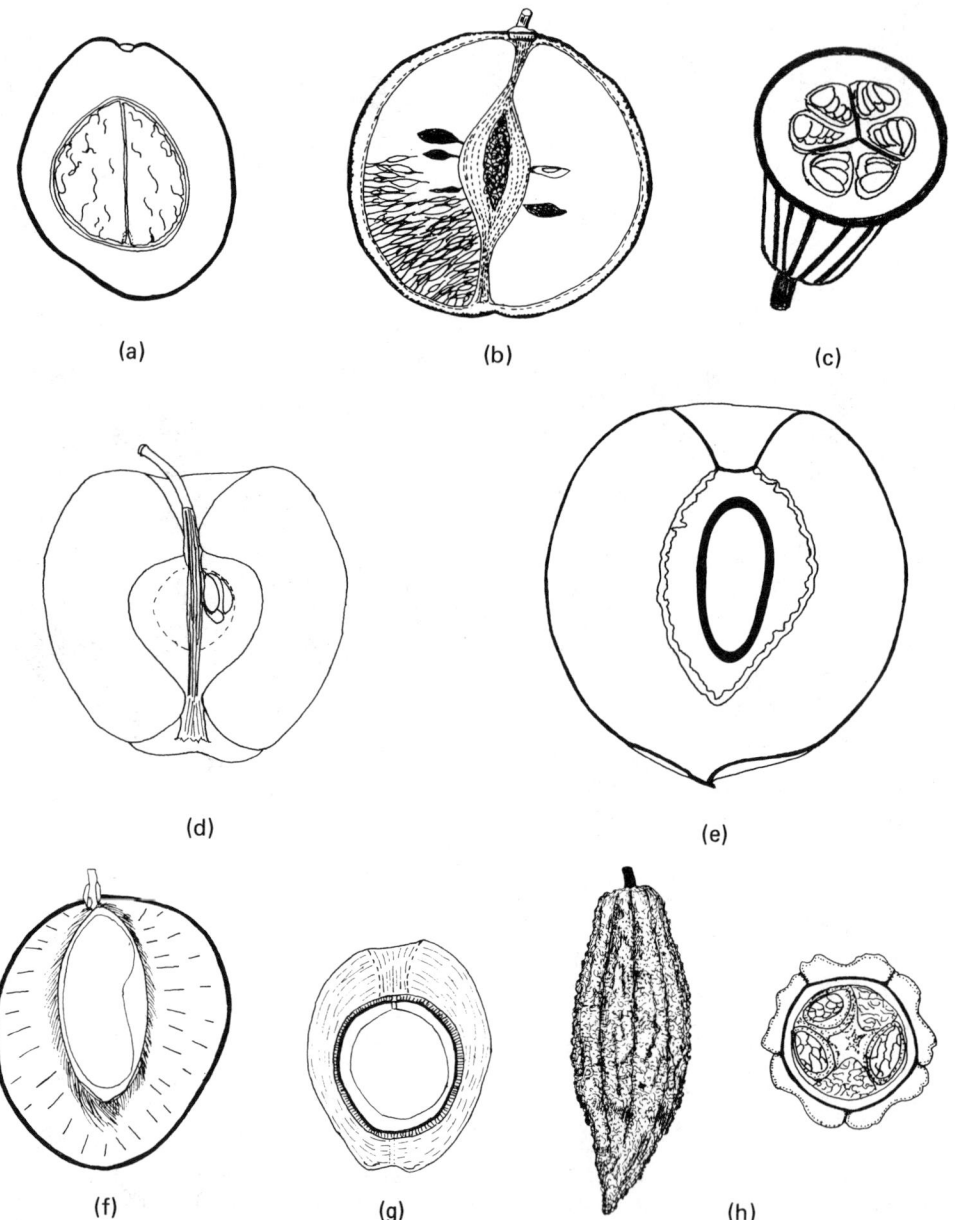

Fig. II.2.3.1.05. Fruit types: (a) berry ('Waldin' avocado, l.s.) (b) hesperidium (sweet orange, l.s.); (c) pepo (gourd); (d) pome, apple; (e)-(h), drupe, (e) peach; (f) mango; (g) coconut; (h) cacao (t.s. at right), also called a "pod". (a) Redrawn from Ruehle, 1963; (c) redrawn from Gray, 1862; (d) redrawn from Hedrick, 1917; (e) redrawn from Robbins, 1924; (g) redrawn from Fremond, Ziller, and Lamothe, 1963; (h) redrawn from Cuatrecasas, 1964; see Acknowledgments and Selected References.

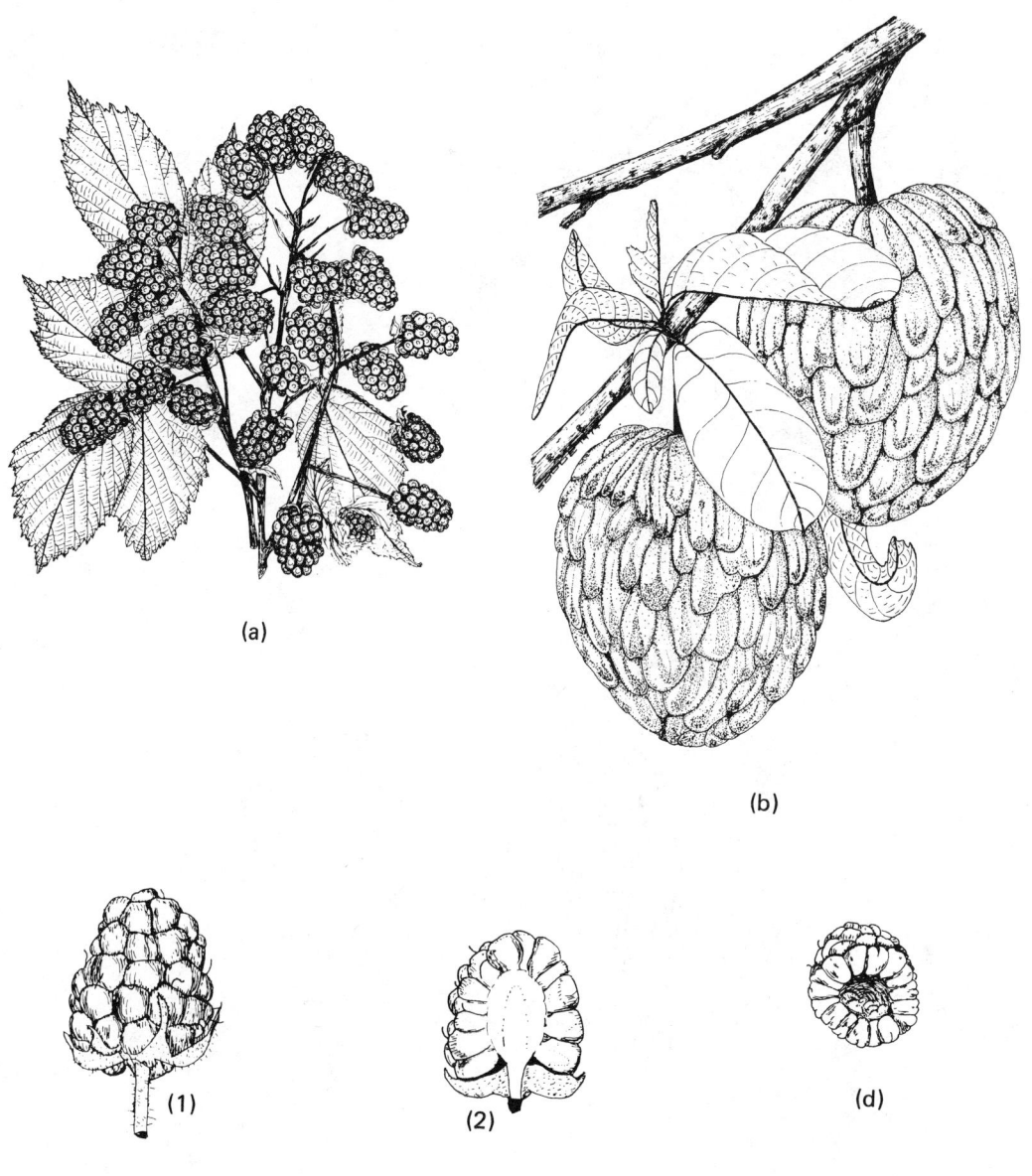

Fig. II.2.3.1.06. Fruit types, aggregate fruits: (a) 'Kittatinny' blackberry (Rubus allegheniensis); (b) ilama (Annona diversifolia); (c) blackberry with l.s. at right; (d) raspberry. [Note that (a)-(c) are also accessory fruits, whereas (d) is not.] (a) Redrawn from Hedrick, 1925; (c), (d) redrawn from Bailey, 1941; see Acknowledgments and Selected References.

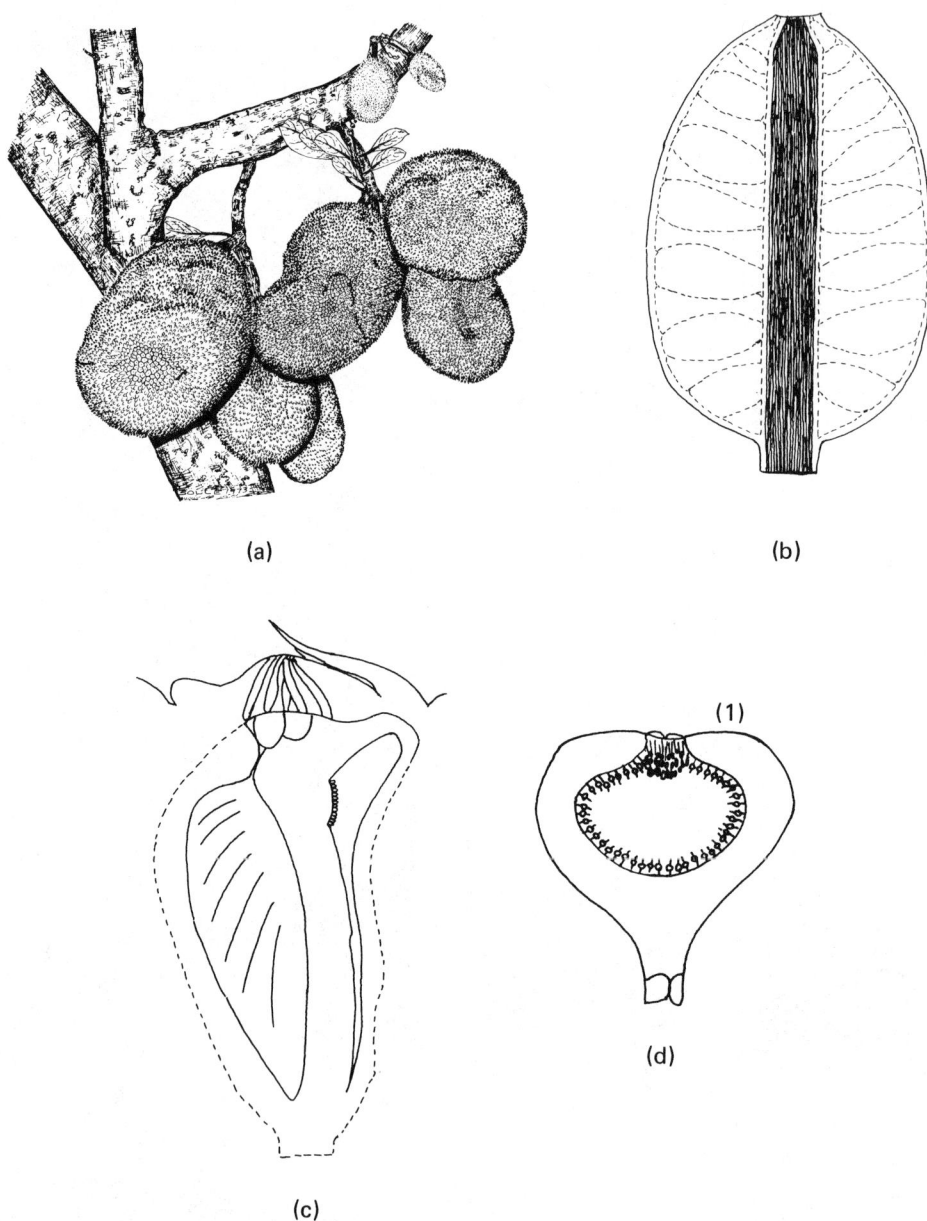

Fig. II.2.3.1.07. Fruit types, multiple (and accessory) fruits: syncarp: (a) jackfruit (Artocarpus heterophyllus); (b), (c) pineapple (Ananas comosus); (b) whole fruit (crown cut off, l.s.); (c) individual fruitlet (l.s.); (d) fig (Ficus carica) syconium with (1) ostiole. (b),(c) Redrawn from Py and Tisseau, 1963; (d) redrawn from Condit, 1947; see Acknowledgments and Selected References.

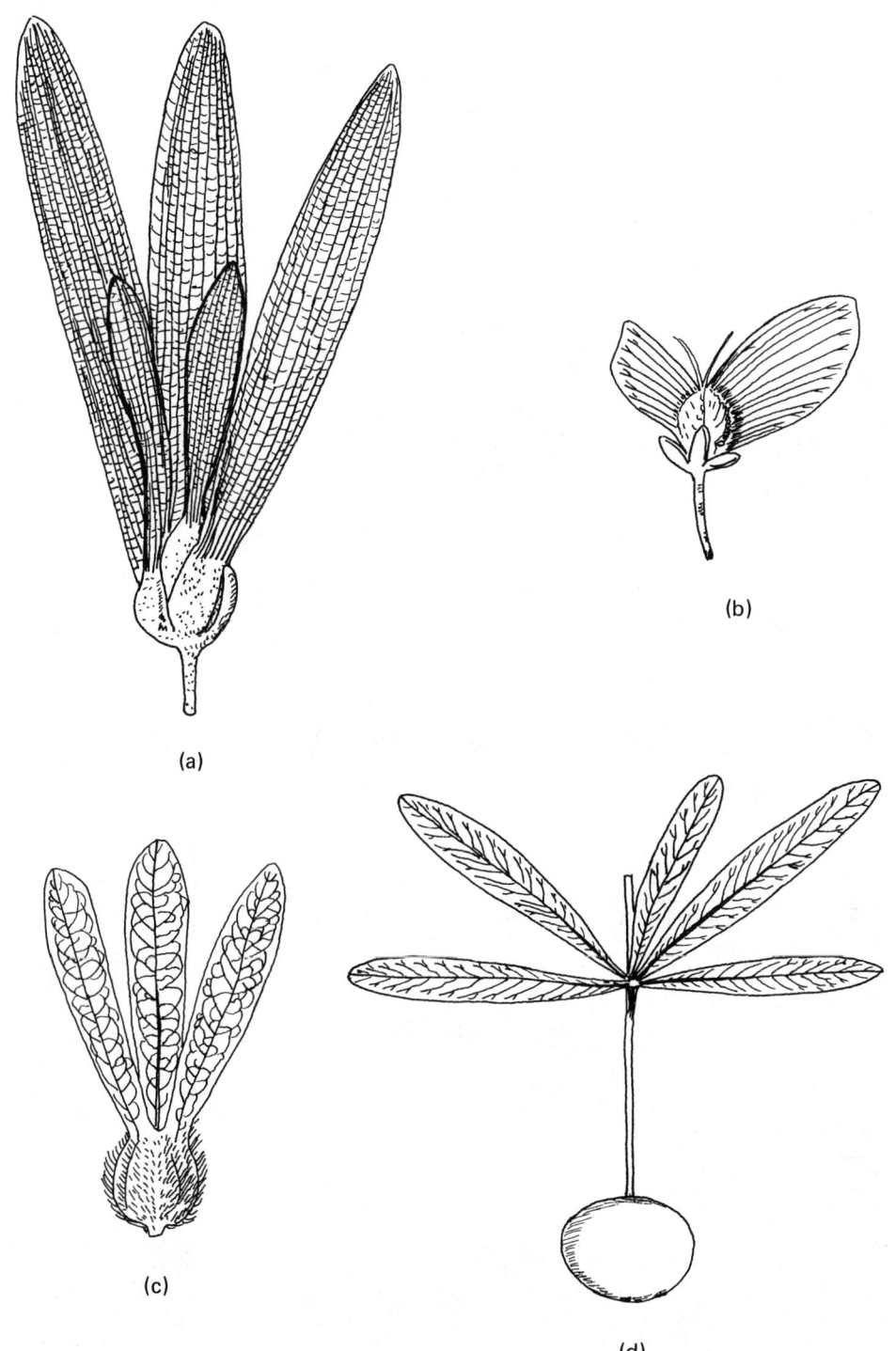

Fig. II.2.3.2.01. Fruit parts: (a), (b) anisopterous, (a) <u>Shorea rigida</u>; (b) <u>Acridocarpus orientalis</u>; (c), (d) alate, (c) <u>Triplaris surinamensis</u>; (d) <u>Melanorrhoea glabra</u>. [Note that wings in (a) and (c) are persistent sepals, outgrowths of the pericarp in (b), and persistent petals in (d).] Redrawn from Hutchinson, 1969; see Acknowledgments and Selected References.

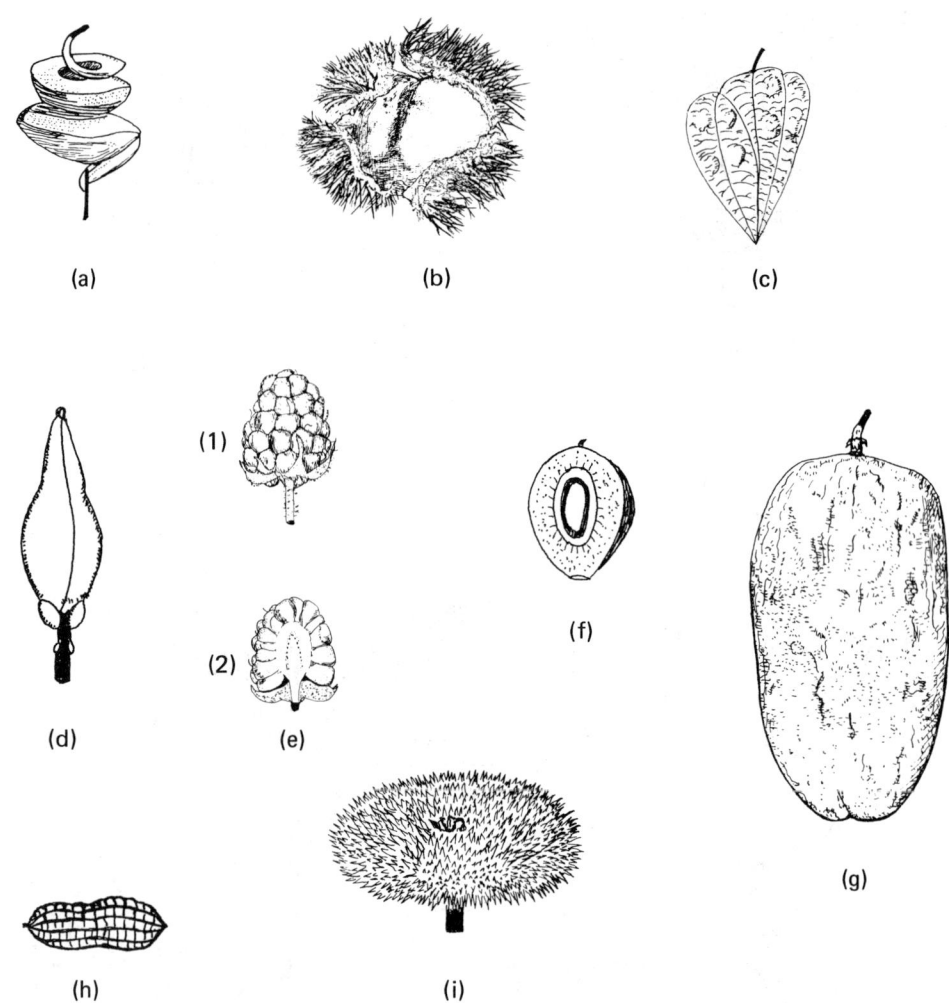

Fig. II.2.3.2.02. Fruit parts: (a) apocarpous, alfalfa (Medicago sativa); (b) bur (burr), Chinese chestnut (Castanea mollissima); (c) inflated, bladdery, pyramidal, golden rain tree (Koelreuteria paniculata); (d) conical, Physorrhynchus brabhuicus, (e) conoidal, blackberry (ts below); (f) pyrene, blackberry; (g) ellipsoid: giant granadilla (Passiflora quadrangularis); (h) geocarpic, geocarpous, peanut; (i) depressed, Apeiba echinata. (a) Redrawn from Hayward, 1938; (c), (d), (h), (i) redrawn from Hutchinson, 1969; (e) redrawn from Bailey, 1941; (i) redrawn from Gray, 1862; see Acknowledgments and Selected References.

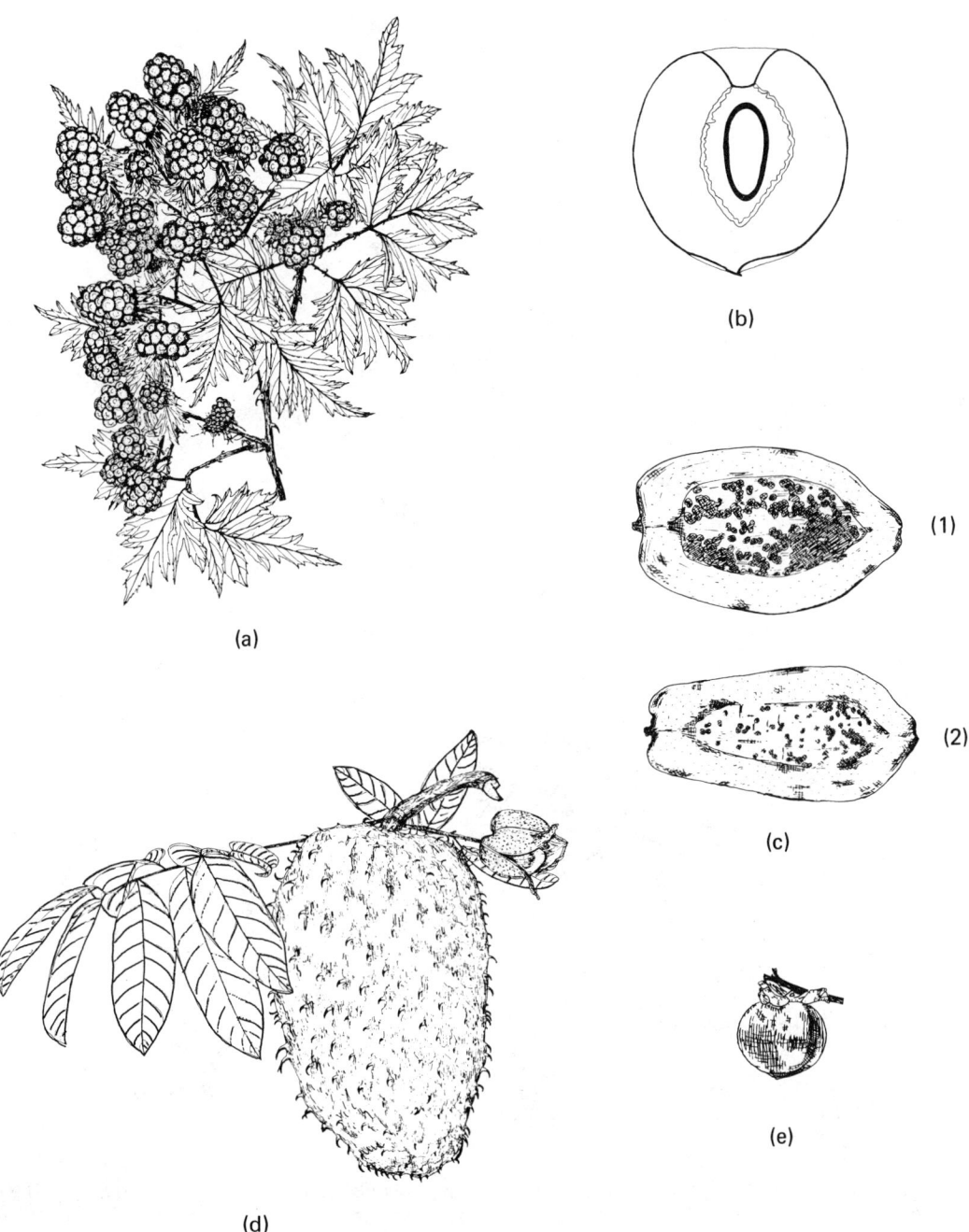

Fig. II.2.3.2.03. Fruit parts: (a) choricarpelly, 'Oregon Evergreen' blackberry (Rubus laciniatus); (b) epichlamydy, 'Elberta' peach (Prunus persica); (c) paracarpous, papaya (Carica papaya) with (1) pistillate, (2) bisexual; (d) axillary, syncarpelly, soursop (Annona muricata); (e) napiform, syncarpous, Japanese persimmon (Diospyros kaki). (a) Redrawn from Hedrick, 1925; (b) redrawn from Hedrick, 1917; (c) redrawn from Wolfe and Lynch, 1940; (d) redrawn from Ochse, 1931; (e) redrawn from Ochse et al., 1961; see Acknowledgments and Selected References.

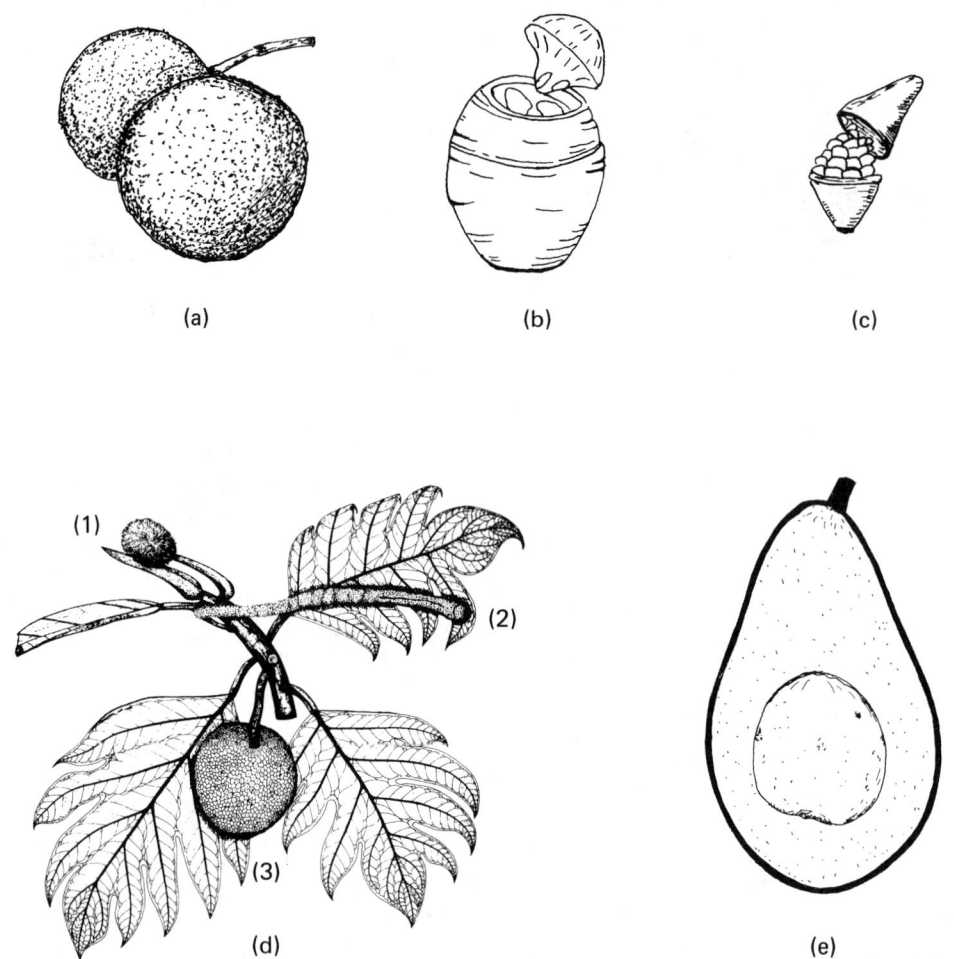

Fig. II.2.3.2.04. Fruit parts: (a) husk, black walnut (Juglans nigra); (b) hypochlamydy, paradise nut (Lecythis ollaria); (c) lysicarpous, purslane (Portulaca oleracea); (d) infructescence, breadfruit (Artocarpus altilis) with (1) pistillate inflorescence, (2) staminate inflorescence, and (3) globose to ovoid fruit; (e) lachrymiform, 'Hall' avocado (Persea americana). (a) Redrawn from Sargent, 1905; (b) redrawn from Hutchinson, 1969; (c) redrawn from Gray, 1862; (e) redrawn from Ruehle, 1963; see Acknowledgments and Selected References.

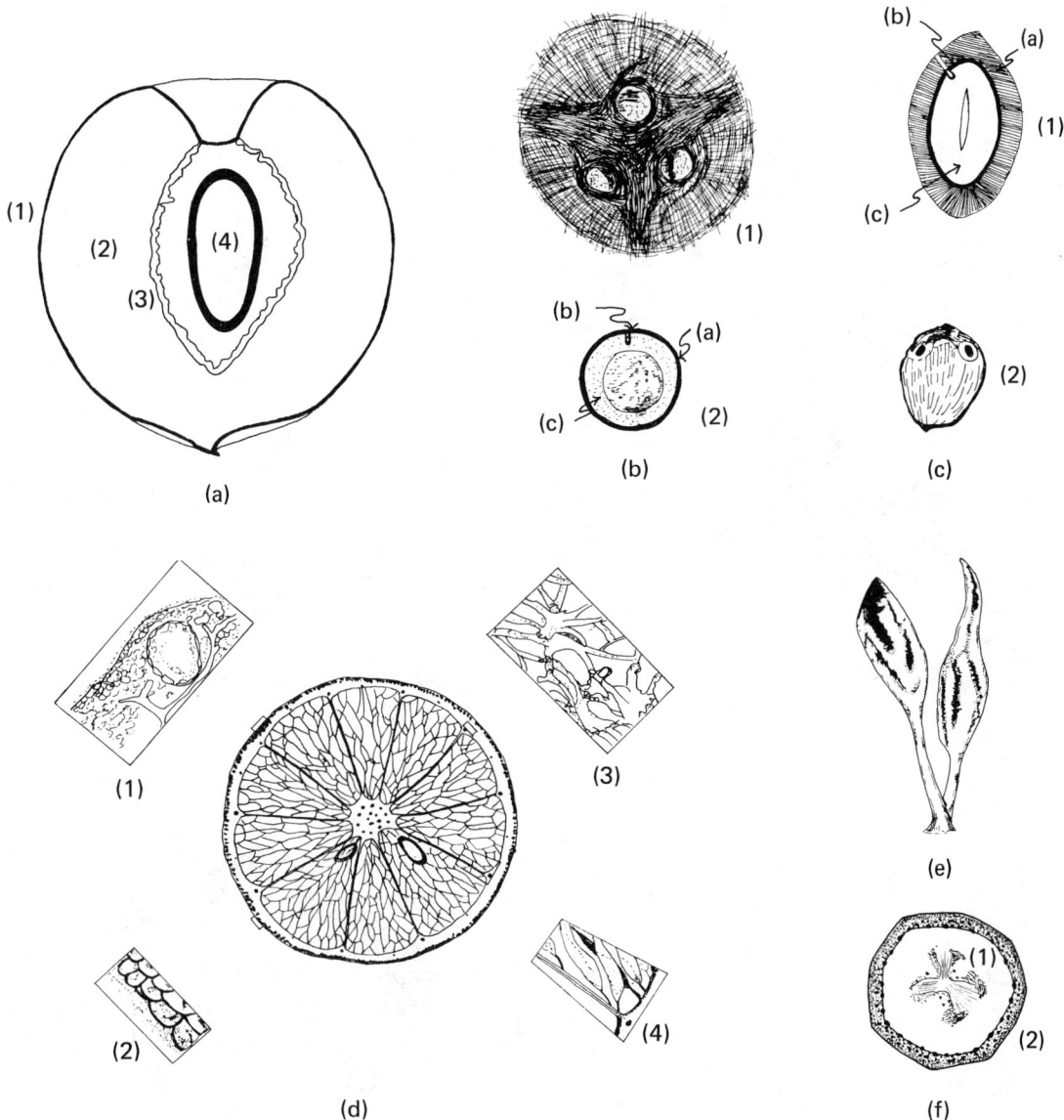

Fig. II.2.3.2.05. Fruit parts: (a) pericarp, peach, with (1) exocarp, (2) mesocarp, sarcocarp, (3) endocarp, (4) seed; (b), (c) germ pores in palm fruits, (b) coconut with (1) view of stem end of nut, (2) with three germ pores (ts), with (a) endocarp (shell), (b) germ pore, and (c) endosperm (meat); (c) oil palm with (1) ls with (a) shell, (b) oblique germ pore, and (c) thick endosperm, and (2) nut with two of three eyes visible; (d) hesperidium with enlarged views of (1) flavedo (with underlying oil gland), (2) cuticle, (3) albedo, and (4) juice vesicles; (e), (f) pulpa, (e) juice vesicles of citrus; (f) banana [ts with (1) pulpa and (2) vascular bundles in rind (meso- and exocarp)]. (a) Redrawn from Hedrick, 1917; (b) redrawn from Fremond, Ziller, and Lamothe, 1963; (c) redrawn from Surre and Ziller, 1963; (d) redrawn from Soule and Grierson, 1981; (e) courtesy of K. E. Koch; see Acknowledgments and Selected References.

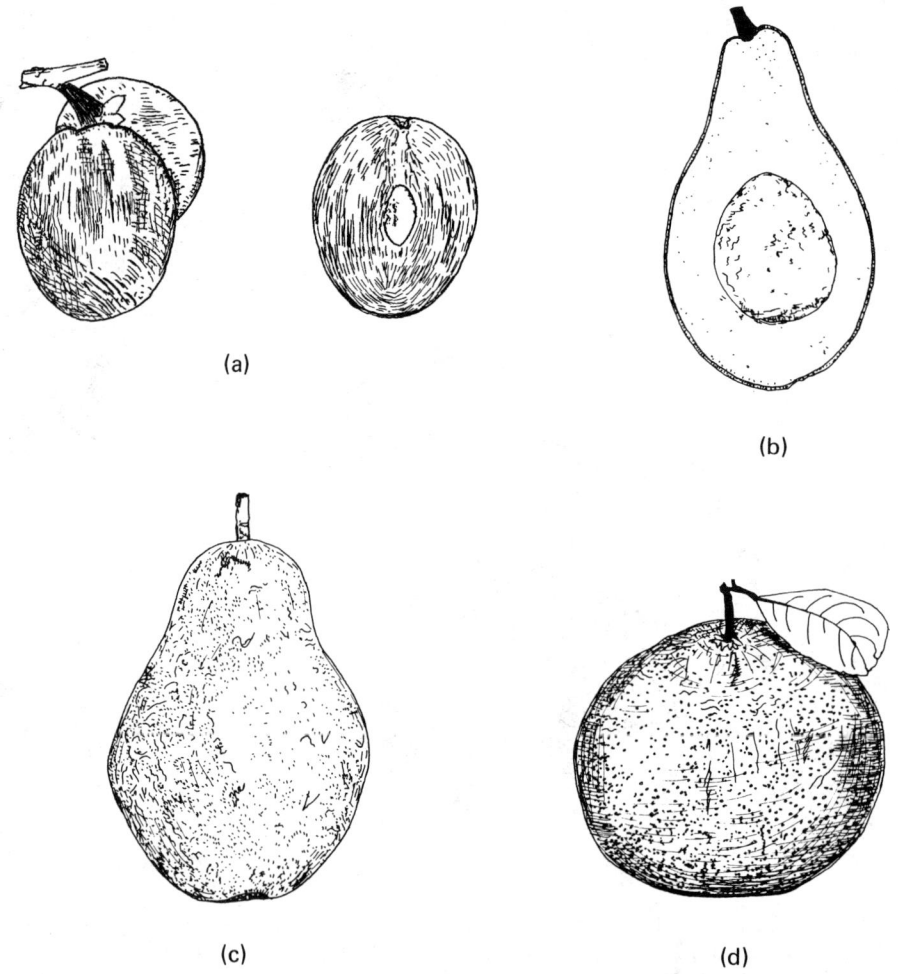

Fig. II.2.3.2.06. Fruit parts: (a) ovoid, sapodilla (Manilkara sapota), ts at right; (b), (c) pyriform, (b) 'Lula' avocado (ls); (c) 'Bartlett' pear (Pyrus communis); (d) oblate, grapefruit (Citrus paradisi). (a) Redrawn from Ochse et al., 1961; (b) redrawn from Ruehle, 1963; (c) redrawn from Hedrick, 1921; (d) redrawn from Hume, 1926; see Acknowledgments and Selected References.

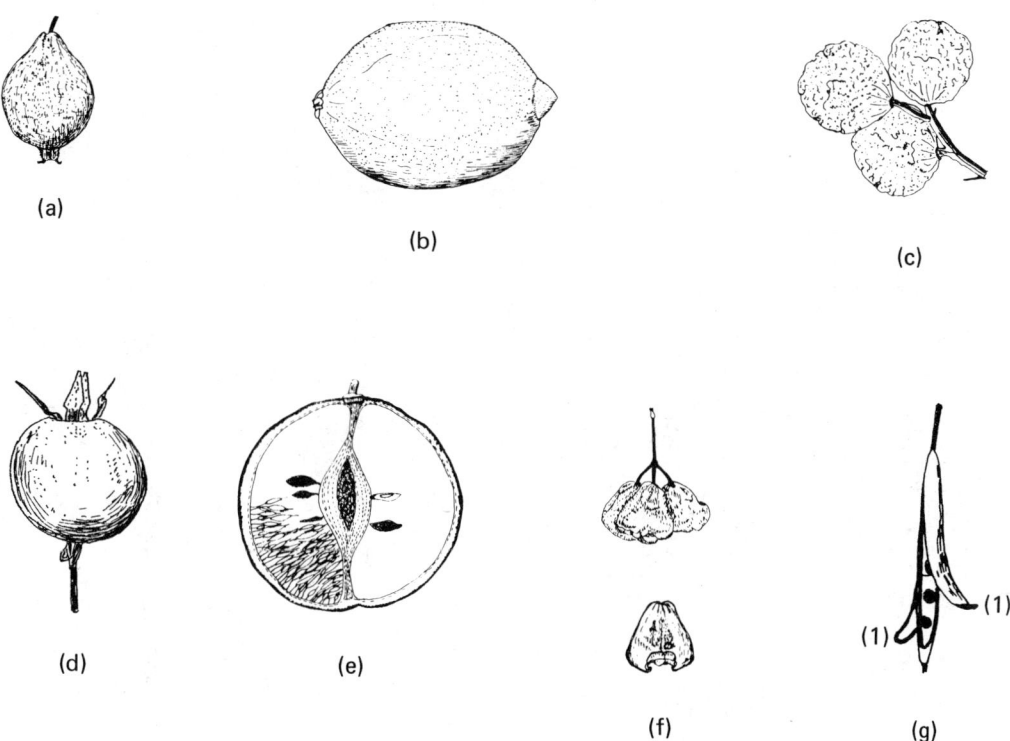

Fig. II.2.3.2.07. Fruit parts: (a) obovoid, guava (Psidium guajava); (b) prolate, lemon (Citrus limon); (c), (d) spheroid, globoid, (c) trifoliate orange (Poncirus trifoliata); (d) pitomba (Eugenia lushnathiana); (e) superposed (seeds), sweet orange; (f) turbinate, Semarang rose-apple (Syzygium javanicum); (g) valve (1), cardamine. (a), (d), (f) Redrawn from Ochse et al., 1961; (b), (c) redrawn from Hume, 1926; (g) redrawn from Gray, 1862; see Acknowledgments and Selected References.

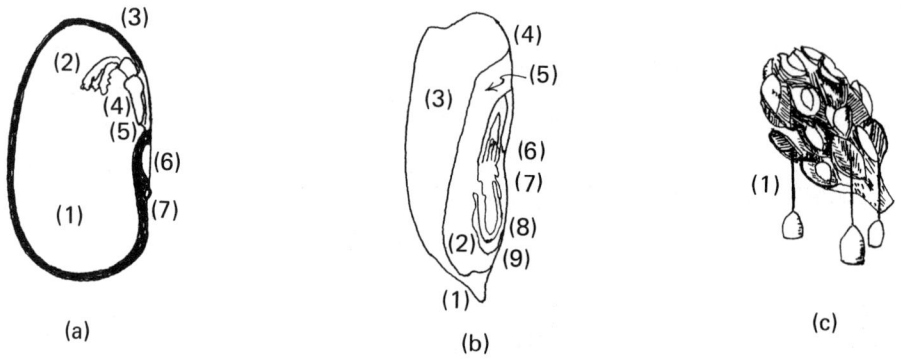
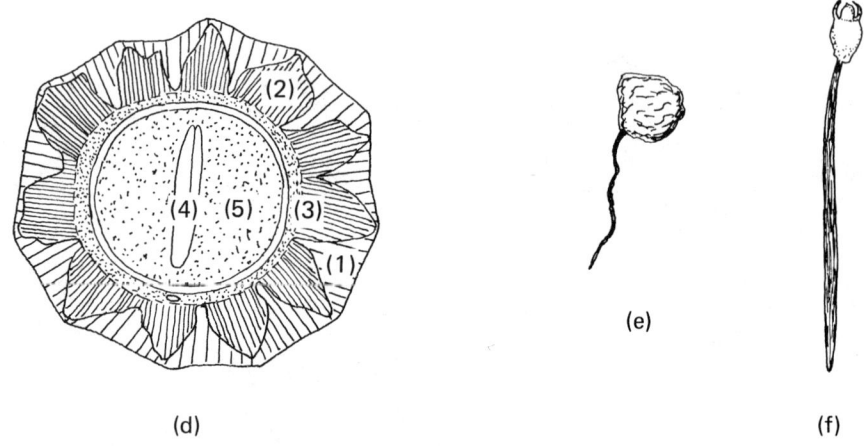

Fig. II.2.3.3.01. (a)-(d) Seeds, seedlike fruits: (a) bean (ls), with (1) cotyledons, (2) plumule, (3) seed coats, (4) hypocotyl, (5) radicle, (6) hilum, (7) micropyle; (b) corn (caryopsis, ls), with (1) peduncle, (2) scutellum (cotyledon), (3) endosperm, (4) pericarp, (5) coleoptile covering (6) plumule, (7) hypocotyl, (8) radicle with (9) coleorhiza covering it; (c) magnolia fruit, with (1) funiculus (funicle); (d) papaya (ts, X 10), with (1) sarcotesta, (2) mesotesta, (3) inner mesotesta, (4) embryo, (5) endosperm; (e), (f) vivipary, (e) grapefruit (late season ca. X 3/2); (f) mangrove (ca. X 1/3). (a), (b) Redrawn from Hill et al., 1960; (c) redrawn from Gray, 1862; (d) redrawn from Corner, 1976; (f) redrawn from Lind and Morrison, 1974; see Acknowledgments and Selected References.

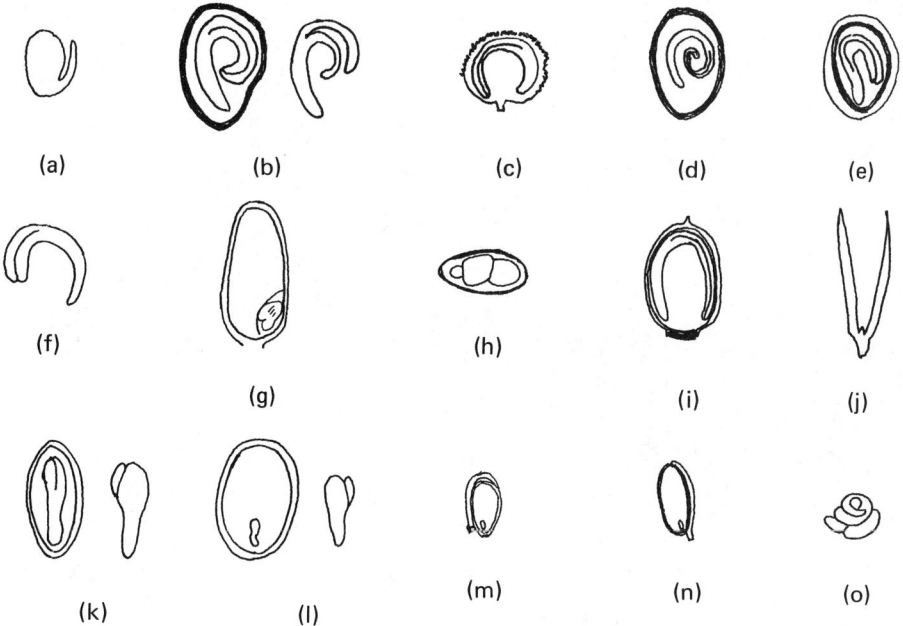

Fig. II.2.3.3.02. Cotyledon and embryo sizes and arrangements of seeds and seedlike fruits: (a) accumbent (Arabis); (b) annular (large embryo, potato); (c) campylotropous (chickweed); (d) coiled (Nephrophyllum); (e) curved (American linden); (f) curved (Menispermum); (g) excentric (rice); (h) incumbent (Capsella); (i) peripheric (Mirabilis); (j) large embryo (pumpkin); (k) medium embryo (linear, barberry); (l)-(n) small embryos: (l) (peony), (m) (Aquilegia), (n) (Epiphegus); (o) spirally twisted (Cuscuta). (a)-(c), (e), (g)-(o) Redrawn from Gray, 1862; (d) redrawn from Hutchinson, 1969; see Acknowledgments and Selected References.

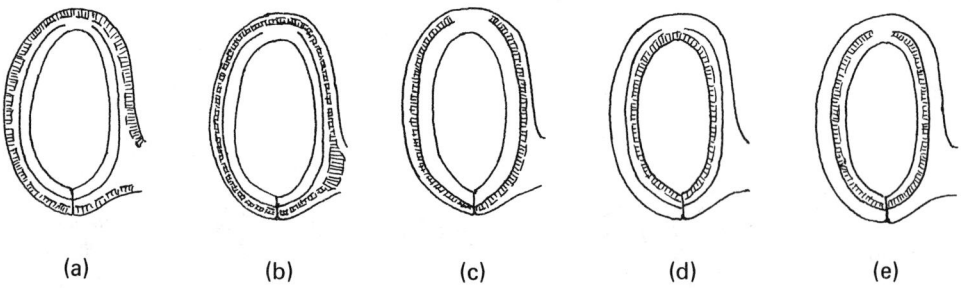

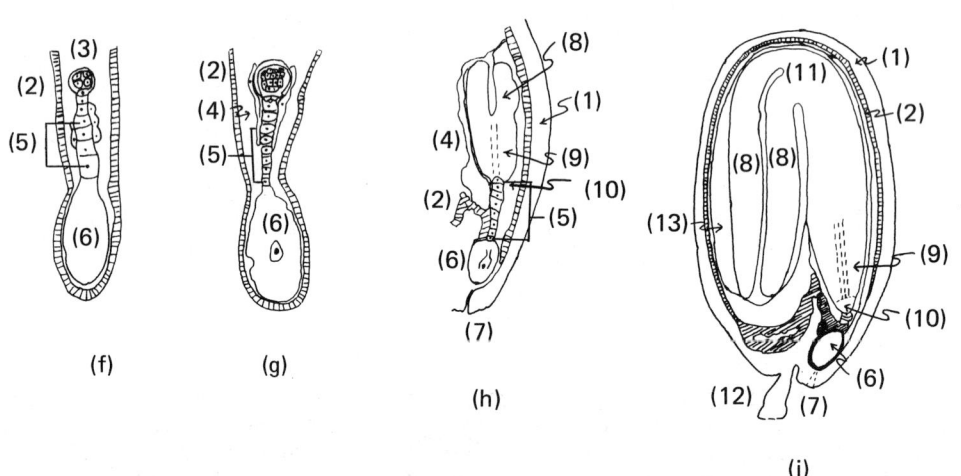

Fig. II.2.3.3.03. (a)-(e) Seed structure: location of mechanical layers (hatched) in the testa (outer) and tegmen (inner seed coat) of seeds derived from an anatropous ovule (ls, not to scale), (a) exotestal; (b) mesotestal; (c) endotestal; (d) exotegmic; (e) endotegmic (mesotegmic rare). (f)-(i) Development of Capsella zygote (fertilized egg) from ovule to seed: (f)-(h) early stages; (i) immature seed (all ls highly magnified); (1) = outer integument (testa), (2) = nucellus (hatched), (3) = embryo, (4) = endosperm, (5) = suspensor (= hypostase, several cells) with (6) = large basal cell, (7) = micropyle, (8) = cotyledon, (9) = hypocotyl, (10) = radicle, (11) = plumule, (12) = funiculus (funicle), (13) = embryo sac cavity. (a)-(e) Redrawn from Corner, 1976; (f)-(i) redrawn from Hill et al., 1960; see Acknowledgments and Selected References.

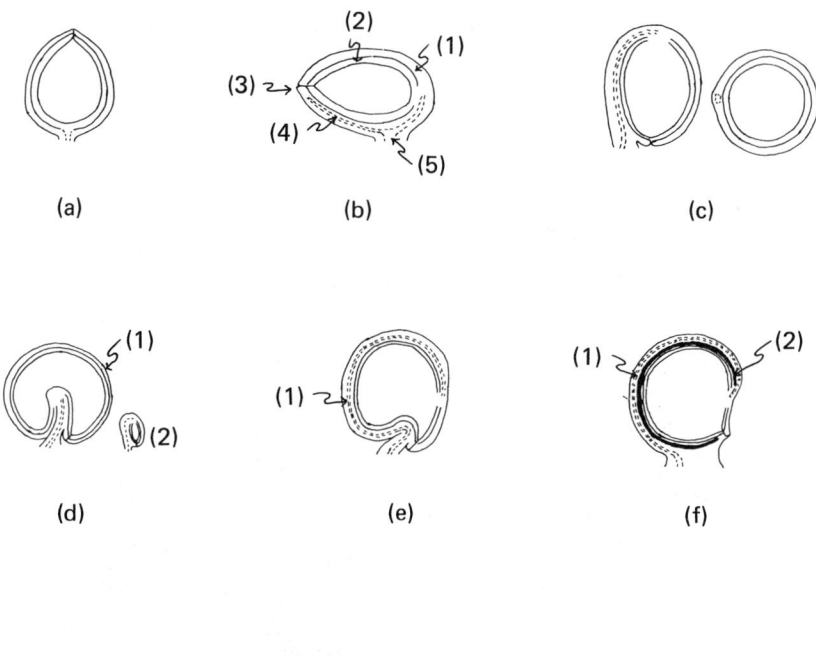

Fig. II.2.3.3.04. Seed forms (ls, diagrammatic) showing testa, tegmen, and vascular bundles (dotted lines): (a) orthotropous; (b) amphitropous (hemianatropous) or preraphe seed, with (1) testa, (2) tegmen, (3) micropyle, (4) vascular bundles, and (5) funiculus; (c) anatropous (ts at right); (d) campylotropous, with (1) exaggerated antiraphe, derived from an anatropous (2) or a campylotropous ovule; (e) obcampylotropous with exaggerated raphe (1); (f) hilar seed with (1) exaggerated hilum and (2) tracheid bar around most of the seed; (g) pachychalazal (chalaza thickened by intercalary growth to form the greater part of the seed coat, ts); (h) perichalazal (chalaza surrounding the nucellus as a hoop or band, ts), with similar ls between (g) and (h). (Note that all forms except (a) and (b) may be derived from an anatropous ovule.) Redrawn from Corner 1976; see Acknowledgments and Selected References.

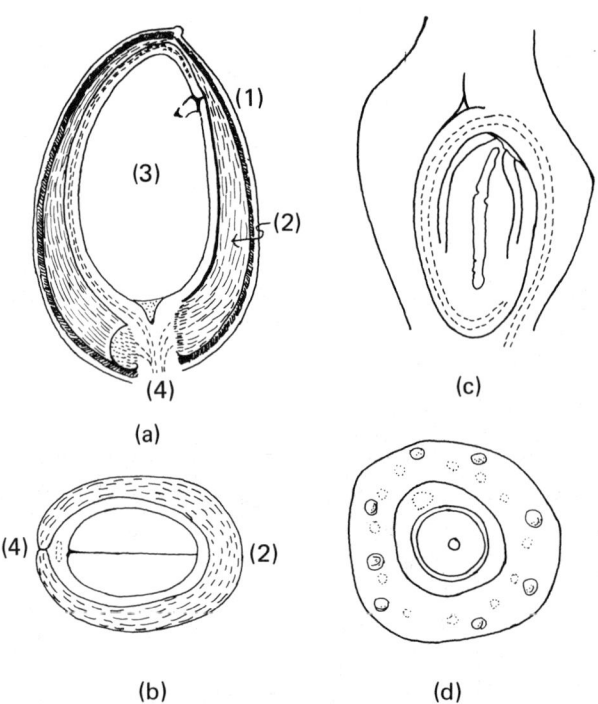

Fig. II.2.3.3.05. Preraphe and dorsal seed forms: (a), (b) Preraphe; Rourea minor, (a) fruit, with (1) endocarp, (2) aril, (3) embryo opposite micropyle, and (4) vascular bundles (ls); (b) seed, with (2) aril, and (4) vascular bundle (dotted ellipse) in preraphe (ts X2). (c), (d) Dorsal, ovary of open Cinnamomum iners flower, (c) (ts); (d) (ls X30). Redrawn from Corner, 1976; see Acknowledgments and Selected References.

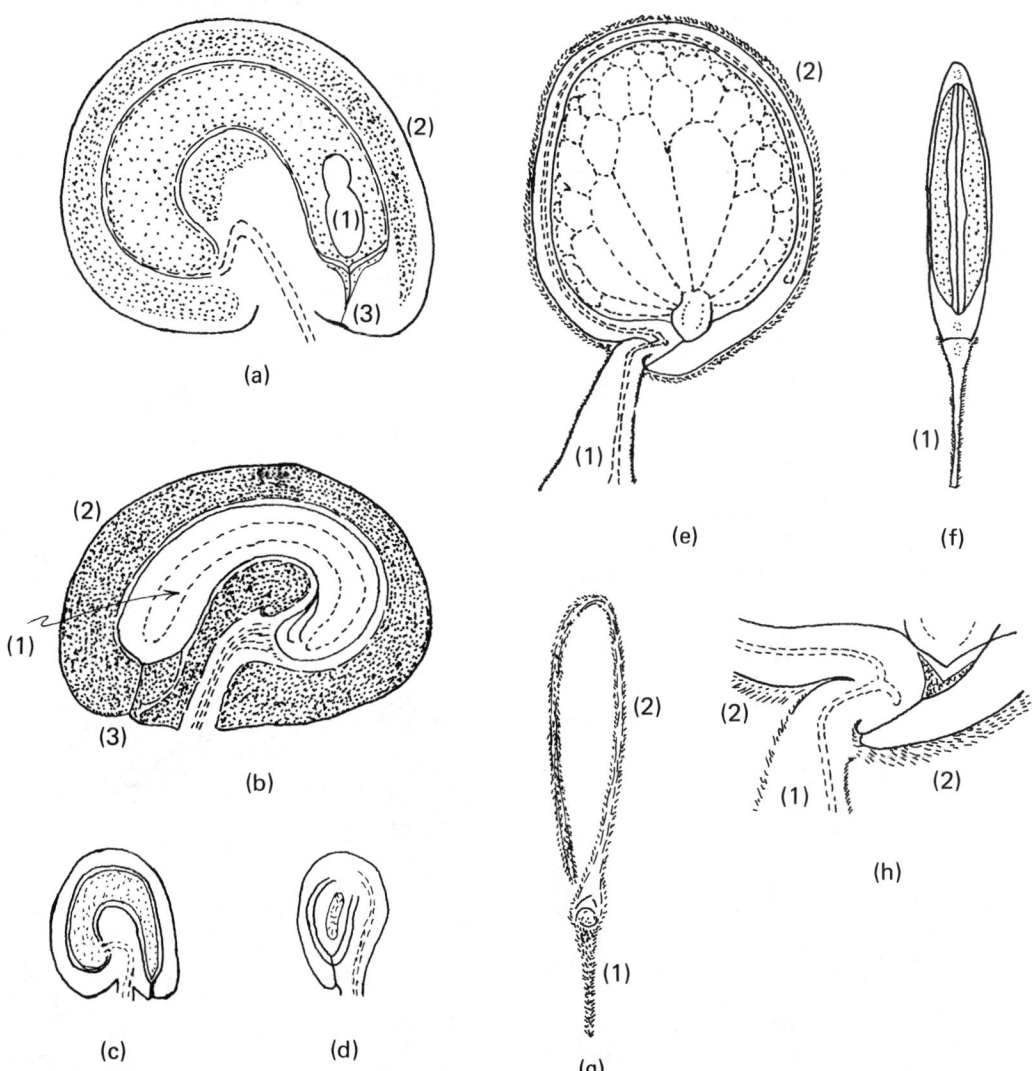

Fig. II.2.3.3.06. Campylotropous and obcampylotropous seeds: (a)-(d) Campylotropous, (a) guava (Psidium guajava, ls, X15) developed from (c), a campylotropous ovule, with (1) embryo, (2) mesotesta, and (3) micropyle; (b) Cattley guava (P. cattleianum, ls X50) derived from (d), an anatropous ovule. (e)-(h) Obcampylotropous (Bauhinia integrifolia): (e) seed (median ls, X2), with (1) funicle and (2) endocarp hairs; (f) seed (transmedian ls, X2); (g) hoop of detached hairs (X2); (h) enlarged view (X5) of micropylar end of (e). Redrawn from Corner, 1976; see Acknowledgments and Selected References.

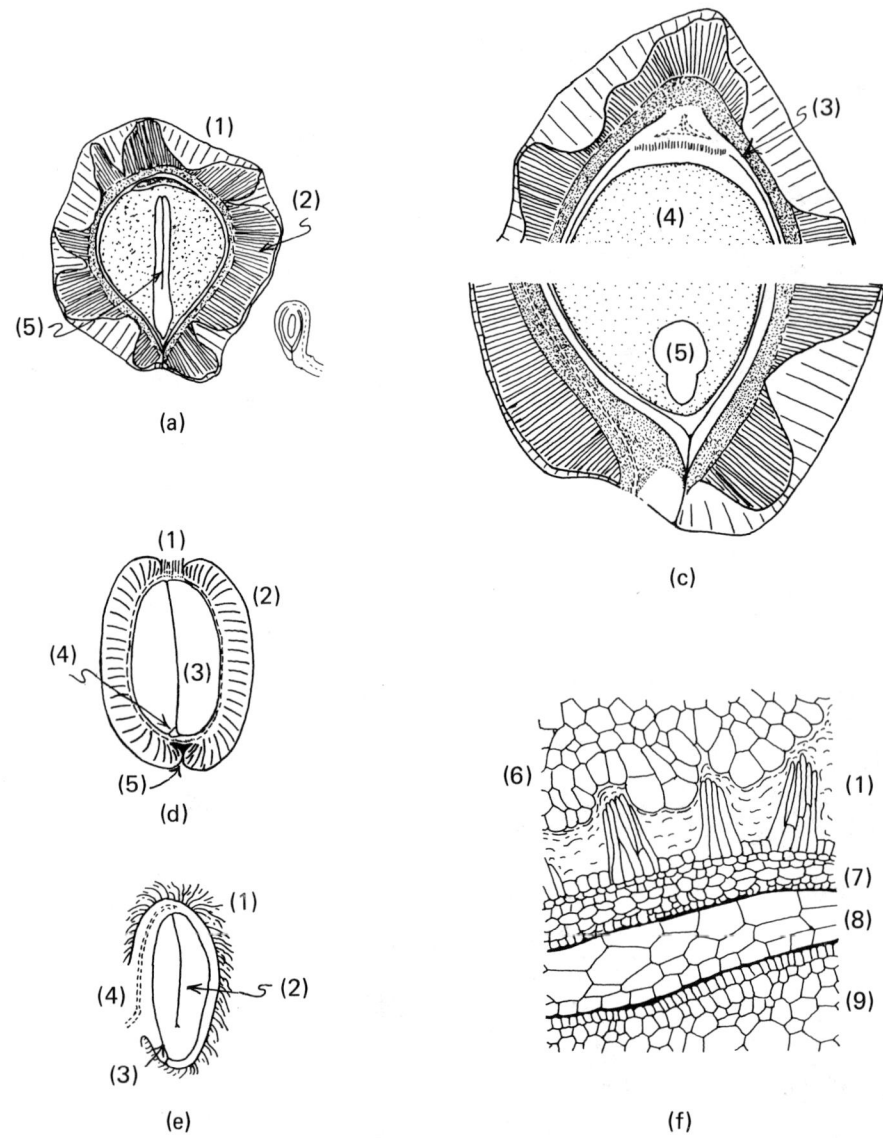

Fig. II.2.3.3.07. (a)-(d) Sarcotestal seeds: (a)-(c) papaya, (a) ls with ovule at right (X10); (b) chalazal end; and (c) micropylar end of fully grown but immature seed, (ls, X25), with (1) sarcotesta, (2) mesotesta, (3) inner mesotesta, (4) endosperm, and (5) embryo; (d) rambutan (Nephelium lappaceum), (ls, X2), with (1) funicle, (2) sarcotesta, (3) endosperm, (4) small embryo, and (5) micropyle. (e), (f) Seed hairs, Indian wood-apple (Feronia limonia): (e) mature seed (ls, X8), with (1) hairs, (2) cotyledons, (3) micropyle, and (4) funicle; (f) young seeds, (ts, X225), with (1) incipient hairs, (6) endocarp, (7) testa from which hairs arise, (8) tegmen, and (9) nucellus. Redrawn from Corner, 1976; see Acknowledgments and Selected References.

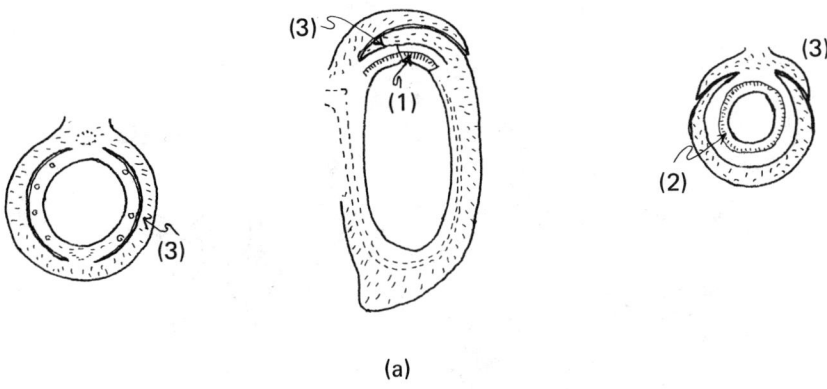

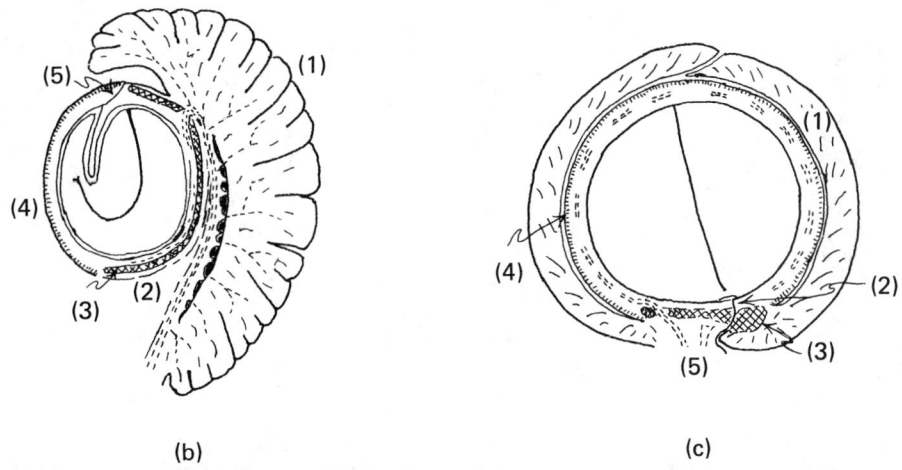

Fig. II.2.3.3.08. Pachychalazal seed forms: (a) Aglaia, with (1) vestigial tegmen (hatched) lateral development of testa around (2) pachychalaza, and (3) perichalazal aril from raphe-antiraphe (ls), ts views of (a) at points indicated by double arrows; (b) Alectryon sp., with (1) raphe-aril, (2) pachychalaza (dashed line), (3) sclerotic hilum, (4) exotesta, and (5) micropyle; (c) Euphoria malaiensis, with (1) exostomal aril, (2) micropyle, (3) sclerotic hilum, (4) exostesta, and (5) funicle, with vascular bundles. Redrawn from Corner, 1976; see Acknowledgment and Selected References.

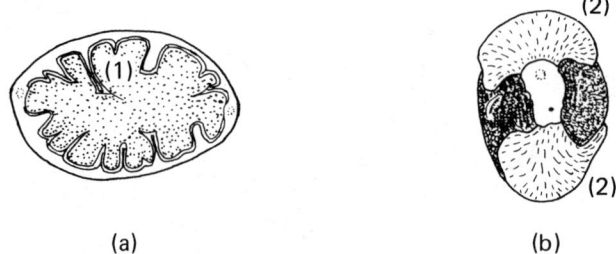
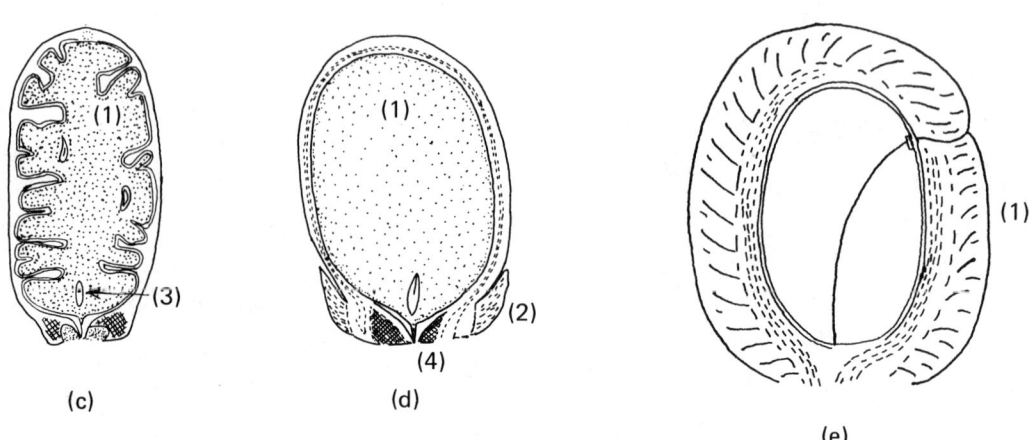

Fig. II.2.3.3.09. (a)-(d) Perichalaza seed form: <u>Xylopia</u>, (a) seed, (ts, X10), with (1) ruminate endosperm; (b) micropylar view (X10), with (2) aril; (c) seed (median ls, X10), with (3) small embryo; (d) (transmedian ls, X10) with (2) aril and (4) micropyle; (e) sarcotesta (1) around rambutan (<u>Nephelium lappaceum</u>) seed, included here to show how this structure differs from that of an aril. Redrawn from Corner, 1976; see Acknowledgments and Selected References.

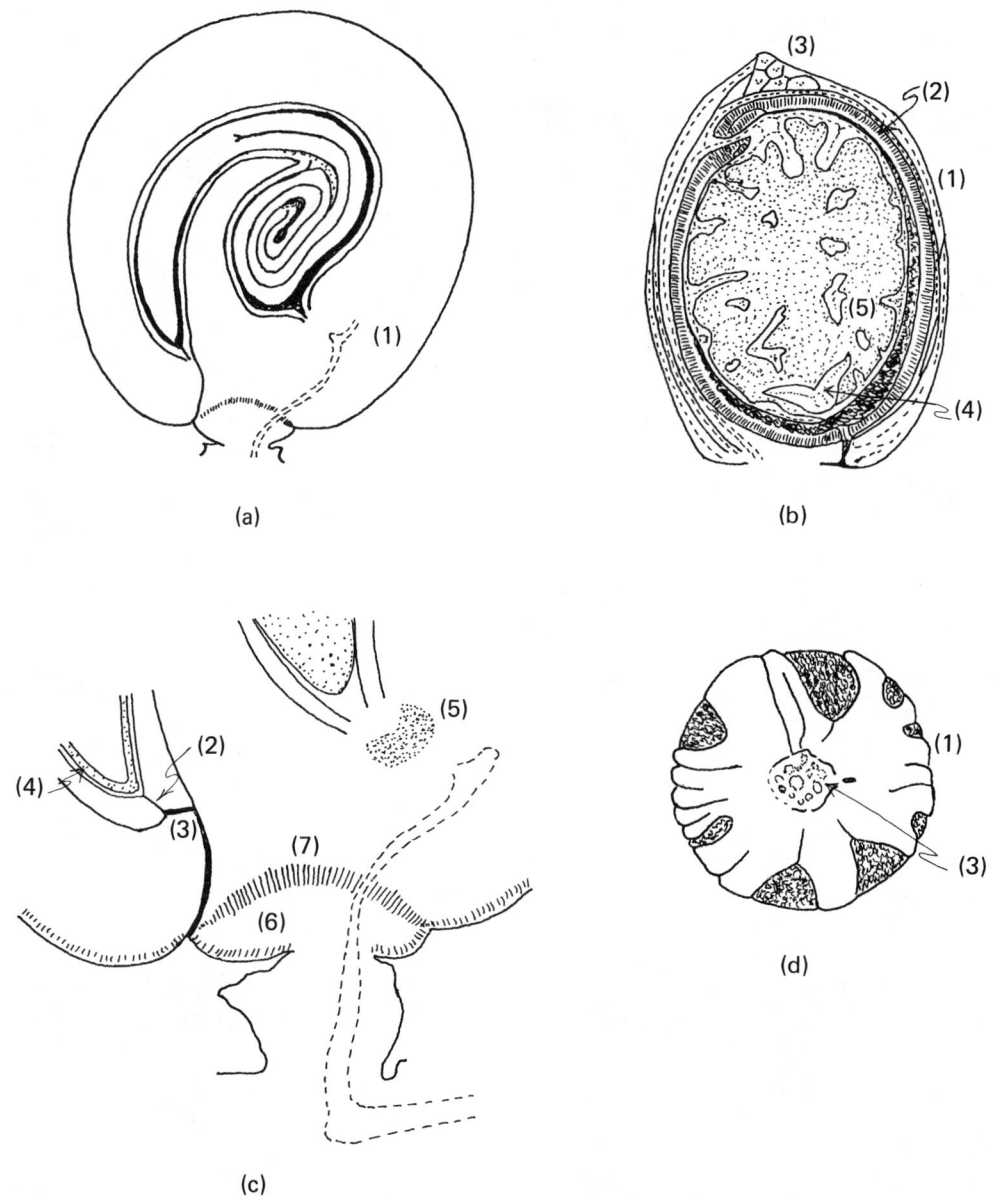

Fig. II.2.3.3.10. Arillate seeds: (a), (b) Dodonaea viscosa, (a) mature seed, with (1) funicular aril in hilum (ls, X18); (b) hilar region of fully grown immature seeds, with (2) micropyle, (3) endostome, (4) trace of nucellus, (5) chalaza, (6) hypostase, and (7) hilar palisade (ls, X50); (c), (d) nutmeg (Myristica fragrans), (c) mature seed (ls, x2), with (1) well developed aril, (2) endotesta, (3) arillostome, (4) winglike embryo, and (5) ruminated endosperm; (d) surface view of hilar end with (1) aril and (3) arillostome (X2). Redrawn from Corner, 1976; see Acknowledgments and Selected References.

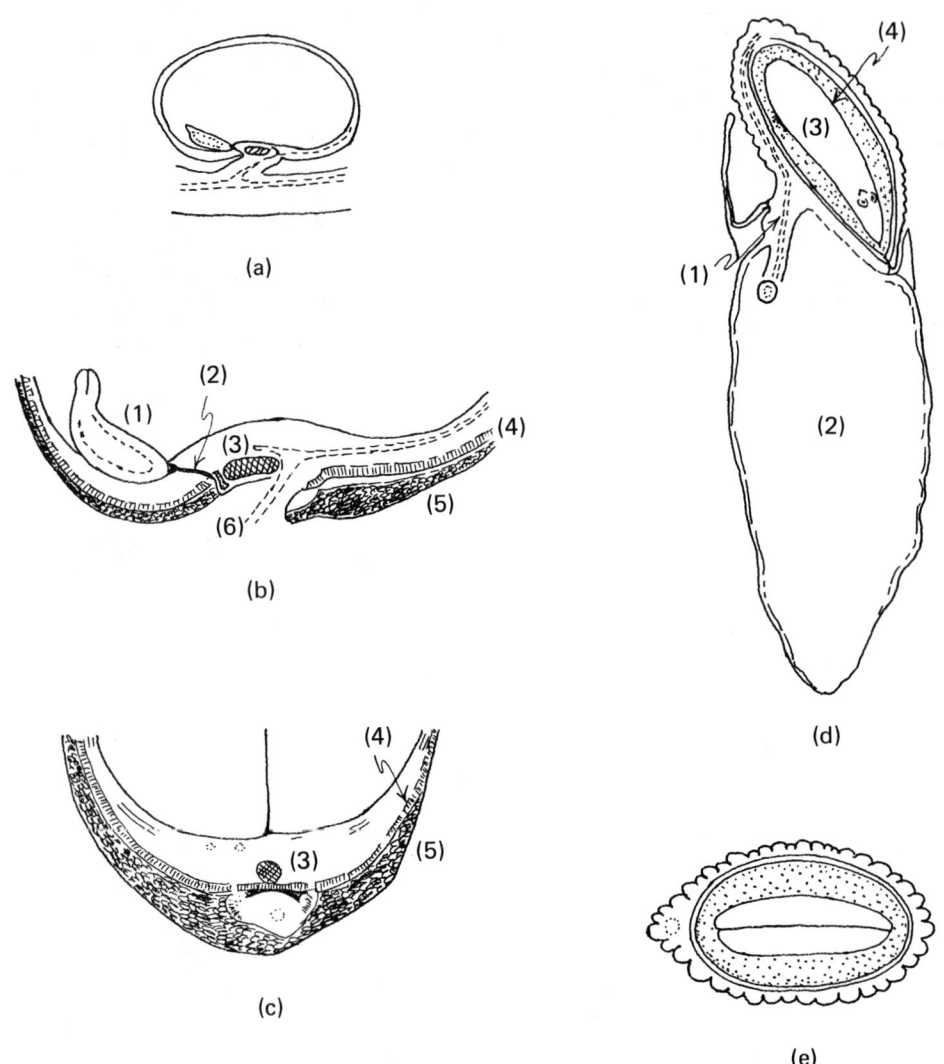

Fig. II.2.3.3.11. Hilar (a)-(c) and alate (d), (e) seeds: Hilar seed form, (a) Erythrina subumbrans (ls, X12); (b) basal part enlarged (ts, X6), with (1) embryo, (2) micropyle, (3) tracheid bar (cross-hatched), (4) palisaded exotesta (hatched), with (5) rim-aril (outer dark area) adherent to it and the top of the (6) funicle; (c) the ls counterpart of (b). (d). (e) Alate: (d) Catha edulis (ls), with (1) funicle, (2) winglike aril, (3) embryo, and (4) endosperm (stippled); (e) same, ts (X25). Redrawn from Corner, 1976; see Acknowledgments and Selected References.

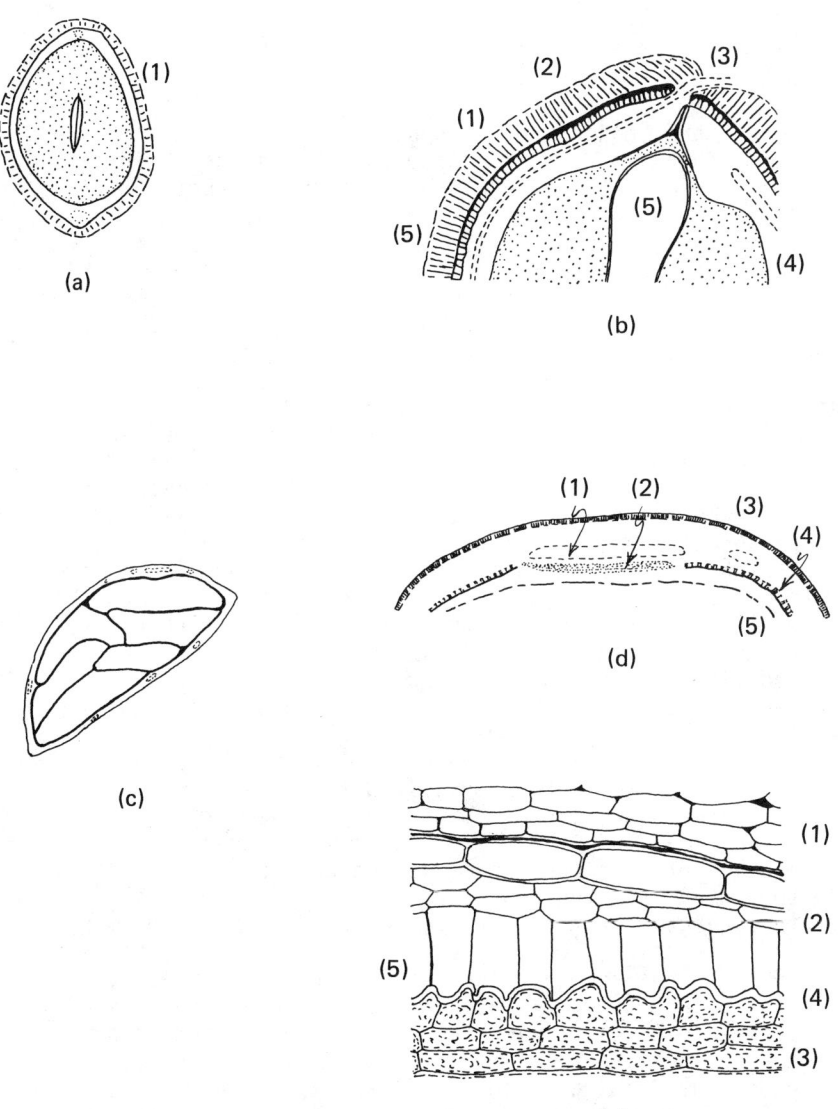

Fig. II.2.3.3.12. Pseudoarillate and unitegmic seeds: Pseudoarillate, (a) mature Diospyros sp. seed invested by (1) pseudoaril (ts, X5); (b) micropylar end showing (1) pseudoaril, (2) exotestal palisade (hatched), (3) trace of tegmen at the endostome, (4) endosperm (stippled), and (5) embryo (median ls, X10); (c)-(e) Unitegmic seeds: (c), (d) testal (tegmen absent); (c) Canarium indicum seed (ts, X5); (d) perichalaza (1) with (2) hypostase (stippled), two epidermal layers of testa (3), (4), and nucellar remains (5), (ts, X25); (e) tegmic (testa absent), Piper nigrum (ls, X500), micropylar end of immature seed coat with adjacent (1) pericarp, (2) tegmen, and (3) immature perisperm separated by (4) crenulate cuticle from (5) endotegmen. Redrawn from Corner, 1976; see Acknowledgments and Selected References.

II.3. Anatomy

.000 Anatomy: the branch of morphology that treats of the character, size, form, and position of plant organs and their parts, particularly the details of their internal structure. (Fig. II.3.01, .02)
.001 Actinocytic: (see Stoma, II.3.190).
.002 Aerenchyma: (see Parenchyma, II.3.125).
.003 Aliform: (see Axial parenchyma, II.3.014).
.004 Amphicribal: (see Vascular bundle arrangement, II.3.210).
.005 Amphivasal: (see Vascular bundle arrangement, II.3.210).
.006 Anisocytic: (see Stoma, II.3.190).
.007 Annular (thickening): (see Tracheid, II.3.202).
.008 Anomalous secondary growth: examples are lobed xylem, the lobes alternating with phloem, as in Bignoniaceae; includes phloem (intraxylary), as in Strychnos and Thunbergia; successive supernumerary cambia and alternate layers of xylem and phloem, as in Chenopodiaceae, Amaranthaceae, and Nyctaginaceae; intensified growth of parenchyma away from the cambium, as in vascular tissue from pith or xylem parenchyma in Bauhinia and many Bignoniaceae; anomalous growth contributes to the flexibility and great strength of lianes. (Fig. II.3.03) (E & M)
.009 Anomocytic: (see Stoma, II.3.190.)
.010 Apical meristem: (see Meristems, II.3.108).
.011 Apoplasm: intercellular space.
.012 Apotracheal: (see Axial parenchyma, II.3.014).
.013 Astrosclereid: (see Sclereids, II.3.170).
.014 Axial parenchyma (wood): two basic types of distribution of axial parenchyma in wood of angiosperms; apotracheal, in which the position of the parenchyma is independent of the vessels, although they may be touching, and which may be diffuse, dispersed throughout the growth ring, banded, in bands, or marginal, at the end, terminal, or the beginning, initial, of the seasonal increment; paratracheal, two kinds of elements associated with each other, as scanty, vasicentric, surrounding vessels, aliform, vasicentric with winglike tangential extensions, or confluent, coalesced aliform. (Fig. II.3.04) (ES)
.015 Banded: (see Axial parenchyma, II.3.014).
.016 Bark: the softer, outer envelope of a stem or root, including the layers external to the cambium or wood and from which they may be more or less readily peeled; more strictly, the corky layers formed on the outer surface of woody plants, i.e., the rhytidome (II.3.160), or may refer to the periderm (II.3.128) as such.
.017 Bast: (see Extraxylary fibers, II.3.063).
.018 Bicollateral: (see Vascular bundle arrangement, II.3.210).
.019 Brachysclereid: (see Sclereids, II.3.170).
.020 Branch gap: the parenchymatous region adaxial from a diverging branch trace. (Fig. II.3.05) (E & M)
.021 Branch trace: vascular strand or two strands that lead from the stem to a branch. (Fig. II.3.05) (E & M)
.022 Bundle sheath: a layer of compactly arranged parenchyma around a vascular bundle; an endodermis in angiosperm leaves; extensions of the sheath may have collenchyma thickening in dicots; and monocots have fibers associated with their vascular bundles. [Fig. II.3.21b(1)] (E & M, ES)
.023 Callose: carbohydrate (glucose residues) deposits on the surface of

II.3. MORPHOLOGY and ANATOMY

the sieve area of sieve tubes in phloem. (ES)
.024 Callus: (see V.1.4.008; also, II.5.6.017).
.025 Calyptrogen: initials of rootcap in monocots; i.e., a type of meristem. (Fig. II.3.19c,d) (ES)
.026 Cambium: a secondary meristem, which may be fascicular, the ordinary secondary meristem in many herbs and lianes (woody vines), which is separated by broad parenchymatous rays, or interfascicular, in which arcs of cells in the parenchymatous rays are meristematic; the latter may join the former to make a complete cylinder or the interfascicular cambium may continue to produce rays, which are characteristic of lianes; there is also an intermediate form, as in many herbaceous annual and biennials, from which secondary xylem and phloem tissues are proliferated, or in sunflower, secondary xylem only; there are two types of cells involved: spindle-shaped fusiform initials and nearly isodiametric, relatively small ray initials. (Fig. II.3.06a-c, .07a,b) (E & M, ES)
.027 Casparian strip: (see Endodermis, II.3.052).
.028 Chlorenchyma: (see Parenchyma, II.3.125).
.029 Collateral: (see Vascular bundle arrangement, II.3.210).
.030 Collenchyma: a specialized form of parenchyma, the supporting, live tissue of young organs, is located in strands or continuous cylinders near the surface of the cortex in stems and petioles and along the veins of foliage leaves; short prismatic to much elongated cells with unevenly thickened primary walls. (Fig. II.3.04g-i) (ES)
.031 Colleter: cells that produce a sticky excretion, as on bud scales. (ES)
.032 Companion cells: specialized parenchyma cells, associated with sieve tubes and retaining their nuclei at maturity. [Fig. II.3.26b(4), c(2)] (ES)
.033 Concentric (phloem): (see Phloem positions, II.3.136).
.034 Confluent: (see Axial parenchyma, II.3.014).
.035 Contractile root(s): about 450 species of 135 genera in 82 families have taproots, lateral roots, and/or adventitious roots that undergo 10 to 70% shortening by passive contraction of the cortical cells, which serves to pull the growing points into the soil or nearer to the surface; a type of secondary thickening in the roots of many seedlings, bulbs, and rhizomes. (Fig. II.3. 17b,c) (ES, B & C)
.036 Cortical (fiber): (see Fibers, II.3.063).
.037 Crystals: mainly calcium oxalate and anhydrides of silica, the former present in cells as solitary rhombohedral or octahedral (prismatic) crystals; several distinct types: (a) druse, compound calcium oxalate crystals; (b) sphaerite, hydrous aluminum phosphate; (c) raphide, elongated crystals aggregated into bundles; (d) styloid, slender-pointed processes. (Fig. II.3..08a-d) (ES)
.038 Cuticle: a multilayered "protective skin" of complex orgin, structure, and development overlying the epidermis, usually separated from the latter by a pectin layer; consists of an inner layer of cutin (a heterogeneous polymer of fatty acids) and cellulose and an outer layer entirely or mainly of cutin, with waxy materials embedded in and over the cuticular surface; epicuticular waxes are a complex of secondary and primary alcohols, paraffins, aldehydes, fatty acids and diketones, the composition of which changes with the development of the organ (e.g., leaf or fruit).

II.3. MORPHOLOGY and ANATOMY

[Fig. II.3.12a(2), .15a(5)] (ES, FR)
.039 Cystolith: a calcium carbonate outgrowth from a cell wall impregnated with the mineral; occurs in ground parenchyma and epidermal cells in hairs or lithocysts. [Fig. II.3.08e(2)] (ES)
.040 Dedifferentiation: (see Meristems, II.3.108).
.041 Definitive callose: fusion of adjacent callose masses which indicates the cessation of sieve element activity (see Callose, II.3.023, and Phloem, II.3.135). (ES)
.042 Dermal tissues: the outer layers of a plant (see Epidermis, II.3.054, and Periderm, II.3.128).
.043 Dermatocalyptrogen: initials in which rootcap and epidermis have a common origin, as in dicots. (Fig. II.3.19e) (ES)
.044 Dermatogen: (see Histogen theory, II.3.080).
.045 Diacytic: (see Stoma, II.3.190).
.046 Diffuse: (see Axial parenchyma, II.3.014).
.047 Diffuse porous woods: a type of wood in angiosperms in which the vessels are of equal diameter and uniformly distributed throughout the wood (see Ring porous wood, II.3.161). (Fig. II.3.10c) (ES)
.048 Dilatation: the tangential growth of permanent tissues outside the cambium by anticlinal (radial) division of cells in the primary cortex, primary phloem, and epidermis so that the surface of a stem or root remains essentially smooth (i.e., without fissures), as in many trees, shrubs, and lianes. (Fig. II.3.09a) (B & C, ES)
.049 Disjunctive (parenchyma, tracheids): separation of cells next to an expanding vessel during differentiation. (ES)
.050 Druse: (see Crystals, II.3.037)
.051 Endarch (xylem): (see Xylem differentiation, II.3.222).
.052 Endodermis: boundary or outermost layer of the vascular cylinder, with the pericycle (II.3.127) inside it; best known in roots but does occur in some herbaceous angiosperms and in rhizomes but not in the aerial axes of woody dicots or gymnosperms; distinguished by a <u>casparian strip</u>, a band of lignin and suberin in the radial and transverse walls; a <u>starch sheath</u> (so known because the cells contain starch granules), a type of endodermis, is found in young stems of many angiosperms (including woody dicots). [Fig. II.3.11a, .16c(3)] (ES)
.053 Ephedroid perforation: (see Vessel member, II.3.217).
.054 Epidermis: a dermal tissue; a continuous layer on the outer part of the plant body in the primary state; cells variable in shape, often tabular; others are guard cells of stomata and various trichomes, including root hairs; may contain secretory and sclerenchyma cells; epidermal cells on aerial plant parts have cuticle on their outer walls and cutinization of some or all of the other walls. [Fig. II.3.12a(1), .13a(1), .21a(13), .31] (ES)
.055 Epithelium: tissue lining a tube or cavity, the cells forming almost an unbroken sheet or membrane; composed of parenchyma cells. (ES)
.056 Epitrophy: increased increments of growth on the upper side of horizontal or ascending branches or roots. (ES)
.057 Exarch (xylem): (see Xylem differentiation, II.3.222).
.058 Excretion: the separation of products eliminated from metabolic processes of a plant (see Secretion, II.3.177). (ES)
.059 Exodermis: specialized layer beneath the epidermis of a root; resembles endodermis and contains suberin; the term for the hypodermis of a root. [Fig. II.3.13a(2)] (ES)
.060 Extraxylary fiber: (see Fibers, II.3.063).

II.3. MORPHOLOGY and ANATOMY

.061 Fascicular: (see Cambium, II.3.026; Secondary growth of vascular system, II.3.173).
.062 Feeder root: (see Tap root, II.3.199).
.063 Fibers: found as separate strands of bundles in cortex and phloem, sheaths, or bundle caps associated with vascular bundles (e.g., abaca, Musa textilis), and in groups or scattered in xylem and phloem; xylary fibers, an integral part of xylem such as fiber-tracheids, libriform types, the latter most specialized, may be septate; extraxylary fibers of three types: phloic or phloem fibers, originating in the primary or secondary phloem, cortical fibers, originating in the cortex, and perivascular fibers on the periphery of the vascular cylinder inside the innermost cortex but apparently not originating in the phloem (extraxylary fibers are found in ramie, kenaf, and jute, among others, as "bast" fibers important in commerce). [Fig. II.3.16c(2), .23a,b,e, .24e-g] (ES)
.064 Fiber-tracheid: (see Fibers, II.3.063).
.065 Filiform sclereid: (see Sclereids, II.3.170).
.066 Foraminate (phloem): (see Phloem positions, II.3.136).
.067 Fundamental tissues: living cells (parenchyma and collenchyma), capable of growth and division, in the cortex, leaf mesophyll, and vascular tissues; (see Parenchyma, II.3.125, Collenchyma, II.3.030). (ES)
.068 Fusiform initial: (see Cambium II.3.026).
.069 Gland: secretory or excretory cells on or in a plant. (Fig. II.3.31f,g) (ES)
.070 Glandular hair: a hair with a gland. (Fig. II.3.042f,g)
.071 Ground meristem: (see Meristems, II.3.108).
.072 Ground tissues: (see Fundamental tissues, II.3.067).
.073 Growth in axis diameter: a combination of cell division (anticlinal and periclinal) and enlargement; diffuse or more or less restricted to cortex and pith in dicots and gymnosperms; cell division in monocots localized in peripheral mantlelike growth, cells in radial series; obconical form of axis if there is an acceleration of growth in width, combined with an increase in the size of the apical meristem (seen most clearly in the early development of palms). (ES)
.074 Growth layer (growth ring): distinct zonations of wood as a result of periodic cambial activity; "spring" (early formed) wood is less dense than "summer" (late formed) wood. [Fig. II.3.10a-c, .20c(7)] (ES)
.075 Guttation: (see Hydathode, II.3.082).
.076 Haplocheilic: (see Stoma, II.3.190).
.077 Heartwood: dead, dried out, older wood (xylem), often infiltrated with oils, gums, resins, and tannins, and vessels blocked and more or less filled up with tyloses (II.3.207). (ES)
.078 Helical (thickening): (see Tracheid, II.3.202).
.079 Heterocellular: (see Rays, II.3.152).
.080 Histogen theory: a theory regarding the origin of tissues in the stem of a vascular plant in which three primordial layers, dermatogen, periblem, and plerome, give rise to a uniseriate external layer (epidermis), L , cortex, L , and central core (pith and primary vascular tissues), L , respectively; of limited morphological value but useful topographically. (Fig. II.3.14a) (E & M)
.081 Homocellular: (see Rays, II.3.152).
.082 Hydathode: cells or glandular structures that discharge water from the interior to the surface of a leaf in the process known as

II.3. MORPHOLOGY and ANATOMY

guttation. (Fig. II.3.15a) (ES)
.083 Hyperplasia: multiplication of callus cells above normal growth in wound healing. (ES)
.084 Hypertrophy: enlargement of intact (live) cells in response to wounding as part of the healing process. (ES)
.085 Hypodermis: the layers of cells immediately beneath the epidermis; often specialized and resembling an endodermis and containing suberin; absent in some stems; (see Exodermis, II.3.059). [Fig. II.3.13a(2)] (ES)
.086 Hypophysis: cell or cells in the developing embryo that result from transverse division of the next adjoining suspensor cell (i.e., the suspensor cell nearest the embryo proper) and give rise to the root tip. [Fig. II.2.3.3.03f-i(5)] (ES)
.087 Hypotrophy: asymmetric growth of radial organs with increased development of the lower surface when occupying a horizontal or descending position. (ES)
.088 Idioblast: a cell differing markedly from others of the same tissue in form, structure, or content; e.g., a crystal idioblast (see Parenchyma, II.3.125). (ES)
.089 Included (phloem): (see Phloem positions, II.3.136).
.090 Initial: (see Axial parenchyma, II.3.014).
.091 Intercalary meristem: an interpolated growing zone (a residual meristem derived from the original growing point or primary meristem), as in the basal portion of internodes or of petioles in monocots or the fascicular cambium in stems or pericycle in roots of dicots. [Fig. II.3.18a(2)] (E & M, ES)
.092 Interfascicular cambium: cambium (see Secondary growth of vascular system II.3.173).
.093 Internal (phloem): (see Phloem positions, II.3.136).
.094 Internode: the portion of a stem between nodes in which cell division and enlargement may progress acropetally or basipetally and the elongation of successive internodes may occur stepwise or overlap. [Fig. II.3.18a(3)] (ES)
.095 Interxylary (phloem): (see Phloem positions, II.3.136).
.096 Lateral meristem: (see Meristems, II.3.108).
.097 Lateral root: of endogenous origin, commonly originating in the pericycle, usually opposite a xylem pole in dicots. (ES)
.098 Laticifer: latex vessel or a series of latex vessels, which may be nonarticulated, articulated, or jointed, being then compound, and branching, as in Euphorbia, or anastomosing, as in Hevea; a characteristic feature of many tropical families such as Moraceae, Caricaceae, and Sapotaceae. (Fig. II.3.16a,b) (ES)
.099 Leaf gap: the parenchymatous region adaxial from the diverging leaf trace within the vascular cylinder; four main types of node occur in dicots: (1) unilacunar with one trace; (2) unilacunar with two traces; (3) trilacunar with three gaps and three traces, one medial and two lateral; and (4) multilacunar with several to many gaps and traces to the leaf. (Fig. II.3.05) (E & M)
.100 Leaf trace: the portion of the vascular system at each node that is deflected into the leaf attached at that node; the lower part of the vascular supply to the leaf and the arrangement of the trace or traces are related to the phyllotaxis. (Fig. II.3.05) (E & M)
.101 Lenticel: lenticular masses of loose cells protruding through fissures in the periderm (II.3.128) on stems and roots; their main function is gaseous exchange; usually arise beneath individual stomata but some do so independently; three types are found in dicots: (1) filling cells suberized, as in Magnolia and

II.3. MORPHOLOGY and ANATOMY

<u>Pyrus</u>; (2) closing layer of suberized cells formed at the end of the season, as in <u>Quercus</u>; and (3) loose strata alternating with narrow compact suberized strata, as in <u>Prunus</u>. (Fig. II.3.12b) (E & M)

.102 Libriform (fiber): (see Fibers, II.3.063).
.103 Lithocyst: a special enlarged cell in the epidermis which contains cystoliths (II.3.039). (Fig. II.3.08e) (ES)
.104 Lysigeny: formation of a space or canal by breaking down (i.e., dissolution) adjoining cells, as in oil glands of citrus. [Fig. II.3.12a(4)] (ES)
.105 Macrosclereid: (see Sclereids, II.3.170).
.106 Marginal: (see Axial parenchyma, II.3.014).
.107 Meristemoids: small, few-celled secondary meristems that differentiate into stomata, hairs, leaf primordia, and pith rays. (B & C)
.108 Meristems: perpetually young tissues primarily concerned with the formation of new cells; living cells may resume meristematic activities when properly stimulated by <u>dedifferentiation</u> (loss of previously developed characteristics) or <u>redifferentiation</u> (development of new characteristics); several different types: (1) <u>apical</u> at apices of main and lateral shoots (synonymous with shoot apex and root apex), subterminal in leaves; (2) <u>lateral</u>, parallel with sides of organ, e.g., vascular cambium and cork cambium; (3) <u>protomeristem</u>, initiating cells and most recent derivatives, including the protoderm, which differentiate into the epidermal system, i.e., undifferentiated epidermis regardless of origin; procambium (provascular tissue that produces primary vascular tissues); and ground meristem, the precursor of the fundamental or ground tissue system; (4) <u>intercalary meristem</u>, actively growing primary tissue somewhat removed from apical meristem, present in internodes and leaf sheaths of many monocots; and (5) <u>pulvini</u>, thickened regions in a leaf sheath or stem with meristematic activity. [Fig. II.3.18, .19, .21b(2)] (ES)
.109 Mesarch (xylem): (see Xylem differentiation, II.3.222).
.110 Mesogene: (see Stoma, II.3.190).
.111 Mesophyll: the ground tissue of a leaf that consists of palisade parenchyma and spongy parenchyma-containing chloroplasts (the latter with fewer chloroplasts than the palisade cells); the palisade cells commonly on the top (adaxial) side of the leaf and the spongy mesophyll, especially with many intercellular spaces, on the bottom (abaxial) side. [Fig. II.3.21a(4),(5)] (ES)
.112 Metaphloem: a primary phloem that matures after the growth in length of surrounding tissues is completed; usually contains none of the fibers that are present in protophloem (II.3.146). (ES)
.113 Metaxylem: a primary xylem that appears after the protoxylem and has spiral, reticulated and pitted secondary walls; matures after shoot elongation has ended. (Fig. II.3.22c-f) (ES)
.114 Multiple epidermis: the epidermis and the underlying hypodermis form a multiple-layered or multiseriate epidermis of 2 to 16 layers; the <u>velamen</u> on many epiphytic orchid roots is a multiple epidermis or <u>rhizodermis</u>. [Fig. II.3.13a(1+2), 4.08b(1)] (ES)
.115 Multiple perforations: (see Vessel member, II.3.217).
.116 Nectary: loosely packed cells that secrete (and can also absorb) sugar solution; may be floral or extrafloral (i.e., on vegetative parts of a plant). [Fig. II.3.15b, 2.2.24e(1)] (ES)
.117 Nonstoried wood: nonstratified wood; initials not arranged in horizontal tiers and ends overlap, as in gymnosperms and

II.3. MORPHOLOGY and ANATOMY

primitive dicots. (Fig. II.3.07a) (ES)
.118 Nonstratified wood: (see Nonstoried wood, II.3.117).
.119 Obturator: protuberance of the carpel wall through which pollen tubes grow to reach the ovule in certain plants, e.g., Casuarina equisetifolia, where the pollen tubes enter the ovule through the chalaza rather than the micropyle (the normal route). (Fig. II.3.09b) (EA)
.120 Ontogeny: the complete developmental history of an organism from egg (spore, bud, etc.) to adult stage. (ES)
.121 Osmophor: a special gland that produces fragrant volatiles, that may serve as pollinator attractants (or in some cases as protectants against predators). (Fig. II.3.15c,d) (ES)
.122 Osteosclereid: (see Sclereids, II.3.170).
.123 Paracytic: (see Stoma, II.3.190).
.124 Paratracheal: (see Axial parenchyma, II.3.014).
.125 Parenchyma: continuous tissues in the cortex of stem and roots and leaf mesophyll with vertical strands and rays in vascular tissues; of primary origin in the cortex, stem pith, and leaf, primary or secondary in vascular tissues; characteristically living cells capable of growth and division, often polyhedral in shape but may be stellate or much elongated; commonly with primary walls but may have secondary thickening; concerned with photosynthesis, storage, wound healing, and origin of adventitious structures; may be specialized as secretory or excretory cells. Several distinct types: (1) chlorenchyma, photosynthetic parenchyma; (2) storage parenchyma, with starch, proteins, fat or oil globules in vacuoles or in the cytoplasm; (3) idioblasts, special vesicles, sacs, or tubes with tannin; (4) water storage tissue; (5) aerenchyma, parenchyma with large and abundant intercellular spaces between the cells. [Fig. II.3.04, .21a(4,5)] (ES)
.126 Periblem: (see Histogen theory, II.3.080).
.127 Pericycle: the outermost layers of the phloem in most seed plants; part of the fundamental tissue, composed of parenchyma cells; may not always be clearly differentiated; commonly uniseriate in angiosperm roots, although some like palms, Morus, and Castanea have more than one layer. [Fig. II.3.03c, .11a(2), b(3)] (ES)
.128 Periderm: a dermal tissue of three layers: the cork tissue or phellem, cork cambium or phellogen, and phelloderm; phellogen arises in the epidermis, cortex, phloem, and root pericycle and produces phellem to the outside, phelloderm to the inside (may be absent); cork cells are commmonly tabular, compactly arranged, lack protoplasts at maturity, and have suberized walls; phelloderm usually consists of parenchyma cells. (Fig. II.3. 20a,b) (E & M)
.129 Perigene: (see Stoma, II.3.190).
.130 Perivascular (fiber): (see Fibers, II.3.063).
.131 Phellem: cork tissue (see Periderm, II.3.128).
.132 Phelloderm: the inner layer of the periderm (II.3.128).
.133 Phellogen: cork cambium (see Periderm, II.3.128).
.134 Phelloid: nonsuberized cells within the phellem (II.3.131).
.135 Phloem: a complex tissue, primary or secondary in origin, concerned with the support, conduction, and storage of food; the principal conducting cells are sieve cells and sieve-tube members (sieve tubes, which are enucleate at maturity) associated with parenchymatous companion cells; other phloem parenchyma are present in vertical files and as rays in secondary phloem;

II.3. MORPHOLOGY and ANATOMY

supporting cells are fibers and sclereids [Fig. II.3.11a(3), 24e-g, .26] (ES)
- .136 Phloem positions: found in several positions in relation to the xylem; <u>external</u> (abaxial, away from the axis, as in most angiosperms) or <u>internal</u> (adaxial), the latter also termed <u>intraxylary</u>; <u>included</u>, when phloem is included in secondary xylem; <u>concentric,</u> when layers of phloem alternate with those of xylem; and <u>foraminate</u>, when phloem strands are surrounded by xylem. (Fig. II.3.27, .28) (ES)
- .137 Phloic: (see Fiber, II.3.063).
- .138 Pitted: (see Tracheid, II.3.202).
- .139 Plerome: (see Histogen theory, II.3.080).
- .140 Polyderm: a special periderm (II.3.128), with suberized and nonsuberized (food storage) cells in alternate layers. (ES)
- .141 Primary tracheary element thickening: (see Tracheids, II.3.202).
- .142 Primary vascular system: the vascular system (i.e., xylem, phloem, and associated cambium) derived from the apical meristem. (Fig. II.3.01A,B,.02B,C,D) (ES)
- .143 Primary xylem: the first formed xylem in shoots and roots and initiated from the apical meristem; (see Protoxylem, II.3.147, and Metaxylem, II.3.113). (ES)
- .144 Procambium: the meristematic tissue from which the cambium is developed; it is unclear whether development is acropetal or basipetal or whether the procambium is even connected with the apical meristem; the procambium for development of adventitious shoots may be discontinuous with that of older shoots; (see Meristems, II.3.108). (Fig. II.3.01.A, B-1, .02C) (ES)
- .145 Protomeristem: (see Meristems, II.3.108).
- .146 Protophloem: a primary phloem, the first differentiated, with sieve elements but may lack companion cells (see also Metaphloem, II.3.112). (ES)
- .147 Protoxylem: a type of primary xylem, which may mature beyond the region of major elongation in roots but shows the beginning of vascular differentiation with annual and spiral, sometimes reticular thickening. [Fig. II.3.22a,b] (ES)
- .148 Provascular tissue: (see Meristems, II.3.108).
- .149 Pulvini: (see Meristems, II.3.108).
- .150 Radial: (see Vascular bundle arrangement, II.3.210).
- .151 Raphide: (see Crystals, II.3.037).
- .152 Ray cells: fusiform cells, either procumbent (oriented radially) or upright (oriented vertically); may be <u>uniseriate</u> (one cell wide), <u>multiseriate</u> (few to many cells wide), which usually tapers toward the upper and lower margins, or <u>biseriate</u> (two cells wide); (see Rays, II.3.154). (Fig. II.3.07) (ES)
- .153 Ray initial: (see Cambium, II.3.026).
- .154 Rays: found in two forms: <u>homocellular</u>, either procumbent or upright cells only, or <u>heterocellular,</u> both types of cells present in wood of angiosperms as radial files of parenchyma cells. (Fig. II.3.07) (ES)
- .155 Reaction wood: two forms, compression wood on the lower sides of branches and stems of conifers, tension wood on upper sides of branches and stems of dicots. (ES)
- .156 Redifferentiation: (see Meristems, II.3.108).
- .157 Resin duct: schizogenous intercellular spaces formed by separation of parenchyma cells in secondary xylem of gymnosperms and some angiosperms. [Fig. II.3.16c(1)] (ES)
- .158 Reticulate perforation: (see Vessel member, II.3.217).

II.3. MORPHOLOGY and ANATOMY

.159 Rhizodermis: (see Multiple epidermis, II.3.114).
.160 Rhytidome: the outer bark, consisting of the periderm (II.3.128) and tissues isolated by it. (Fig. II.3.20c,d) (ES)
.161 Ring porous wood: a type of wood in angiosperms in which the vessels are of unequal diameter; the largest are found in the early wood (see Diffuse porous wood, II.3.047). (Fig. II.3.10a) (ES)
.162 Root hairs: highly vacuolate, rarely branched and direct lateral extensions of the originating cells in the epidermis; develop acropetally (i.e., toward the apex) behind the zone of most active cell division; short-lived, the life span being a matter of days; range in length from mere bumps to structures 50 to 100 m or more in length; often absent, or apparently so, under field conditions on tropical plants [although they can be induced, as on species of Citrus and relatives, under proper conditions (Soule)]. [Fig. II.3.13a(1), b, c] (ES, E & M)
.163 Root nodule: proliferation of cortical cells induced by certain bacteria, e.g., Rhizobium. (ES)
.164 Sapwood: the most recently formed and active portion of the xylem (wood) with water-conducting elements still functional. (ES)
.165 Scalariform perforation: (see Vessel member, II.3.217).
.166 Scalariform thickening: (see Tracheid, II.3.202).
.167 Scanty: (see Axial parenchyma, II.3.014).
.168 Schizogeny: formation of an intercellular space or canal by splitting or delamination of adjacent cell walls. (ES)
.169 Schizolysigenous: formation of a space by a combination of splitting apart and dissolution of adjacent cells; (see Schizogeny, II.3. 168, and Lysigeny, II.3.104). [Fig. II.3.16c(1)] (ES)
.170 Sclereids: widely distributed; several types: brachysclereids or stone cells, roughly isodiametric; macrosclereids, elongated and rodlike (in epidermal layer of legume seeds); osteosclereids, bone-shaped (in leaves of many dicots, seed coats); astrosclereids, star-shaped (in dicot leaves); filiform, long, slender, fiberlike; and trichosclereids, branched, thin-walled, like hairs. (Fig. II.3.24a-d, .25) (ES, E & M)
.171 Sclerenchyma: strengthening elements of mature plant parts; thick secondary, often lignified walls and no protoplasts when mature; may develop in all primary and secondary parts; two forms, sclereids, which are short, sometimes elongated or branched, and fibers, which are long, slender cells; (see Fibers, II.3.063; Sclereids, II.3.170). (E & M, ES)
.172 Secondary growth in monocots: produced by resumption of cambial activity in parenchyma outside the vascular bundles which adds more vascular bundles to the stem; these and the associated parenchyma are seriated radially. (ES)
.173 Secondary growth of vascular system: requires cambium (lateral meristem), of which there is none if all the procambial cells are differentiated into primary vascular tissues; fascicular cambium is produced if some procambium remains with these bands interconnected with bands of interfascicular cambium from interfascicular parenchyma; patterns of secondary growth in gymnosperm and dicot stems are of three general types: (1) primary vascular tissues form an almost continuous cylinder and secondary tissues the same, as in Tilia; (2) primary vascular system of separate strands, secondary as a continuous cylinder, as in Prunus; (3) primary vascular system produces strands, whereas interfascicular cambium produces only ray parenchyma; therefore secondary tissues are also strands, as in Vitis and other liane stems. (Fig. II.3.06) (E & M)

II.3. MORPHOLOGY and ANATOMY

.174 Secondary meristem: meristematic tissues that arise anew from parenchyma in permanent tissues; e.g., cork cambium (phellogen) or interfascicular cambium. (Fig. II.3.20a, b, .06) (E & M)
.175 Secondary phloem: a complex tissue with sieve elements (sieve tubes), companion cells, phloem parenchyma, phloem fibers, and rays; growth increments are less distinct than those in the xylem; may have sclereids, secretory elemente, laticifers, various idioblasts, and crystals (especially in sclerified parenchyma strands) present, and rays may become dilated in some plants. (ES)
.176 Secondary xylem: xylem formed by cambium from fusiform (axial) and ray (horizontal) initials. (ES)
.177 Secretion: separation of a substance from a protoplast (see Excretion, II.3.057). (ES)
.178 Secretory canals: secretory channels found in axial parenchyma and ray systems of angiosperms, produced by schizogeny (splitting of cells) or lysigeny (dissolution of cells) or both; frequently without a differentiated epithelium (lining). [Fig. II.3.16a, b, c(a)] (ES)
.179 Secretory cells: found in two locations, on plant surfaces and internally; the former consist of glandular epidermal cells, hairs, and various glands (e.g., floral and extrafloral nectaries), certain hydathodes, and digestive glands (e.g., in carnivorous plants); the latter consist of secretory cells, intercellular cavities or canals lined with secretory cells (resin ducts, oil ducts), secretory cavities (oil cavities) from disintegration of secretory cells, and laticifers, which may be single cells (nonarticulated laticifers) or usually much branched or a series of cells (articulated laticifers) that contains latex and commonly multinucleate. (ES)
.180 Secretory space: a cavity or canal formed by schizogeny or lysigeny, the former being lined with an epithelium. (ES)
.181 Sieve area: specialized primary pit fields in sieve tubes (II.3.183). (ES)
.182 Sieve plate: the wall part of a sieve element (tube) that bears sieve areas; they are highly specialized and can be _simple_, with a single sieve area, or _compound_, with scalariform, reticulate, or other arrangements of the sieve area. (Fig. II.3.26a,d) (ES)
.183 Sieve tube: long series of sieve elements, a major component of the phloem. (Fig. II.3.26a-c) (ES)
.184 Simple: (see Vessel member, II.3.217).
.185 Slime: viscous, mainly proteinaceous, material that originates in the cytoplasm of a sieve element as discrete _slime bodies_ and may accumulate to form a _slime plug_. (Fig. II.3.26b,d) (ES)
.186 Sphaerite: (see Crystals, II.3.037).
.187 Spicules: minute needlelike crystals of oxalic acid found in cells of certain plants (see Crystals, II.3.037) (RH 58)
.188 Starch sheath: (see Endodermis, II.3.052). [Fig. II.3.16c(3)]
.189 Stele: the unit in the stem and root that combines the vascular tissues and the associated fundamental tissue; several types: (1) protostele; (2) siphonostele; (3) ectophloic; (4) amphiphloic; (5) dictyostele; (6) eustele; and (7) atactostele. (Fig. II.3.27) (ES, HI)
.189a. Actinostele: xylem and phloem tissues arranged in a radial pattern, as in many roots.
.189b. Amphiphloic stele: (see Siphonostele, II.3.189h).
.189c. Atactostele: composed of separate vascular bundles that are not

II.3. MORPHOLOGY and ANATOMY 204

.189d.
.189e.
.189f.
.189g.
.189h.

.190

.191
.192
.193
.194
.195
.196
.197
.198
.199
.200
.201
.202

distributed uniformly throughout the cylinder but are less numerous in the center than toward the periphery, as in monocot stems.

.189d. Dictyostele: an ectophloic stele that is dissected from the intrusion of wide rays or parenchymatous ground tissue (i.e., pith).
.189e. Ectophloic stele: (see Siphonostele, II.3.189h).
.189f. Eustele: collateral or bicollateral strands, with leaf gaps and interfascicular regions not clearly differentiated; arranged in a circle.
.189g. Protostele: a star-shaped stele with isolated xylem tissue that occupies the points, phloem tissue, the hollow between each point, and pith in the center, an early stage of stele development.
.189h. Siphonostele: two configurations, the ectaphloic, in which the center is pith with rings of xylem and phloem surrounding it, and the amphiphloic, in which the central pith is surrounded by rings of phloem, xylem, and phloem. (Fig. II.3.27c,d)
.190 Stoma (pl. stomata): apertures in the epidermis bound by two guard cells. Two main types in gymnosperms: (1) haplocheilic (simple lipped, with subsidiary cells perigene, i.e., not arising from a primary mother cell), and (2) syndetocheilic (compound lipped, with subsidiary cells mesogene, i.e., arising from a primary mother cell). Four main types in dicots: (1) anomocytic, irregularly celled, no subsidiary cells present; (2) anisocytic, unequal-celled, three subsidiary cells, one distinctly smaller than the other two around the stoma; (3) paracytic, parallel-celled, one or more subsidiary cells on either side of the stoma, their common walls being perpendicular to the long axis of the stoma; (4) actinocytic, variant, subsidiary cells arranged along radii of a circle. Four types of stoma complexes in monocots: two have four or more subsidiary cells (Rhoeo, Commelina), one has two (Gramineae), and one has none (Allium). (Fig. II.3.29, .30) (ES)
.191 Storage parenchyma: (see Parenchyma, II.3.125).
.192 Storied wood: stratified wood with fusiform initials in horizontal tiers; found in highly specialized dicots. (Fig. II.3.07b) (ES)
.193 Stratified wood: (see Storied wood, II.3.192).
.194 Styloid: (see Crystals, II.3.037).
.195 Suberization: process by which the cut surface forms a protective, corky layer, especially in conditions of high temperature and high humidity. (ES)
.196 Suberize: to heal or form a corky protective layer over a cut or wounded surface. (ES)
.197 Syndetocheilic: (see Stoma, II.3.190).
.198 Synplasm: cellular space.
.199 Tap root: the main or primary root in most dicots and gymnosperms; lateral (branch) roots appear in acropetal sequence; secondary growth is present in perennials; conductors of food and water as well as organs for storage and anchorage; feeder roots are the short, often fragile, and short-lived absorptive branches; (see II.1.3.016). (ES)
.200 Terminal: (see Axial parenchyma, II.3.014).
.201 Trace: individual vascular bundle to a leaf, branch, or flower. (ES)
.202 Tracheid: long, imperforate cells with only pit pairs on common walls; a tracheary element (water-conducting); primary tracheary

II.3. MORPHOLOGY and ANATOMY

elements have several types of thickening: <u>annular</u> (rings), <u>helical</u> (spiral), <u>scalariform</u> (elongated <u>pits</u>), <u>pitted</u> (elliptical pits, opposite or alternate and bordered), and <u>reticulate</u> (netted pattern, as in ferns, and some gymnosperms). (Fig. II.3. 22a-f, .23c,d) (ES)

.203 Transition region: the connection between the stem and root through the hypocotyl, a zone of considerable interest because of the different configurations of the vascular tissues of the two components, especially in the early stages of primary growth. (ES)

.204 Trichomes: hairs; numerous types: unicellular branched or unbranched; multicellular single-layered, several layers, dendroid (with treelike branching), stellate (branching in one plane), or a scale hair, being discoid = peltate; all may be glandular, with a stalk and a uni- or multicellular head; (see Hairs, II.5.6). (Fig. II.3.31) (E & M)

.205 Trichosclereid: (see Sclereids, II.3.170).

.206 Tunica-corpus theory: a later theory than the histogen of the origin of tissues in the stem of a vascular plant, in which there are two zones, the central part or <u>corpus</u> and the outer enveloping layer (or layers) or <u>tunica</u>; like the histogen theory, it has limited morphological value but is useful topographically. (Fig. II.3.14b) (E & M)

.207 Tylose: gummy or resinous secretion that forms in vessels and effectively blocks the passage of water. (Fig. II.3.10d) (ES)

.208 Tylosoid (cells): enlarging epithelial cells that eventually close a resin duct (II.3.157). (ES)

.209 Vascular bundle: a discrete group of xylem and phloem tissues, denoted as <u>open</u> if a cambium is present (i.e., additional xylem or phloem <u>cells</u> can be proliferated) or <u>closed</u> if a cambium is absent (in which case there would be no further increase in the number of cells that make up the bundle); (see Vascular bundle arrangement, II.3.210). (Fig. II.3.28) (ES)

.210 Vascular bundle arrangement: configuration of xylem and phloem tissues in stems, leaves, and roots; several distinct forms, such as <u>radial</u> with alternating xylem and phloem, <u>collateral</u> with phloem to the outside and xylem to the inside, <u>bicollateral</u> with phloem to the outside, xylem in the center and phloem to the inside, <u>amphicribal</u> with a cylinder of xylem surrounded by phloem, and <u>amphivasal</u> with a cylinder of phloem surrounded by xylem; the last two types are concentric, the former in monocots and dicots, the latter in ferns; (see also Stele, II.3.189). (Fig. II.3.28) (ES)

.211 Vascular bundles of leaves: the veins, which may be reticulate, parallel, or dichotomous (forked) (see Venation, II.1.2.121). (ES)

.212 Vascular strand: (see Vascular bundle, II.3.209).

.213 Vascular tissues: water- and food-conducting and storage elements and accessary cells, including a meristematic zone in many cases; formed in a cylinder in dicots and scattered bundles in monocots (exceptions in both). (Fig. II.3.06d-f) (ES)

.214 Vasicentric: (see Axial parenchyma, II.3.014).

.215 Velamen: a multiseriate epidermis, composed of compactly arranged nonliving cells with spirally, reticulately, or pitted thickening of their walls; an exodermis lies beneath it; found in aerial roots of tropical <u>Orchidaceae</u>, epiphytic <u>Araceae</u> (aroids), and some terrestrial monocots; apparently not an absorptive tissue;

II.3. MORPHOLOGY and ANATOMY

(see Multiple epidermis, II.3.114). (ES)
- .216 Vessel: a water-conducting element in the xylem (see Vessel member, II.3.217).
- .217 Vessel member: individual cells that make up a vessel; perforated, more or less drum-shaped with a perforation plate mainly in end walls; several types: simple, with one perforation; or multiple, with several perforations, like scalariform, parallel perforations; reticulate, net pattern, or ephedroid, approximately circular as in Ephedra; a tracheary element (water-conducting). (Fig. II.3.22g-i) (ES)
- .218 Water-storage tissue: (see Parenchyma, II.3.125).
- .219 Wide ray: the broadening of the older outer parts of phloem rays by tangential growth so that they appear to be wedge-shaped in transverse section (Rays, II.3.152). [Fig. II.3.09a(1)] (ES)
- .220 Xylary fiber: (see Fibers, II.3.063).
- .221 Xylem: a structurally and functionally complex tissue concerned with the conduction of water, storage of food, and support; primary or secondary in origin; the principal water-conducting cells are tracheids and vessel members (joined together as vessels); storage is in parenchyma cells arranged in vertical files and in secondary xylem in rays; mechanical (strengthening) cells include fibers and sclereids. (ES)
- .222 Xylem differentiation: three patterns occur in stems; endarch, from the center (meaning nearest the pith) out; mesarch, both ways from the middle; and exarch, from the outside toward the center; the last two are the most primitive; the exarch form is found in roots. (Fig. II.3.27; .28) (ES)

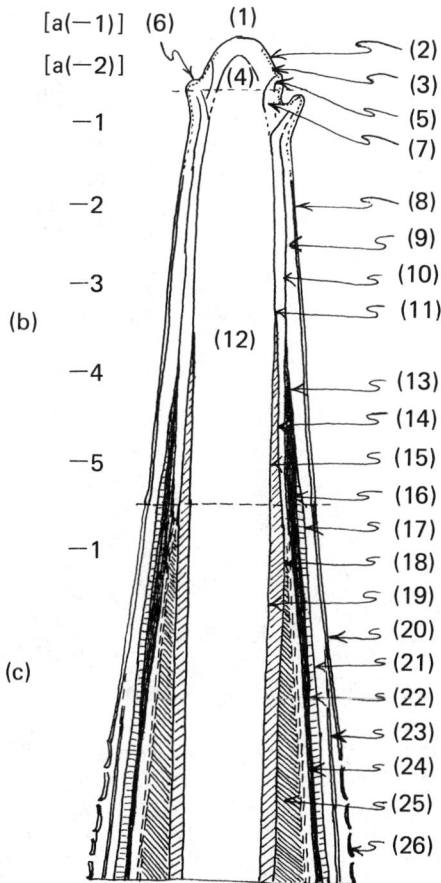

Fig. II.3.01. Anatomy: stem tissues in a woody dicotyledon (ls, diagrammatic): A-1, embryonic zone (0-0.02mm) and A-2, zone of determination (0.02-0.04mm): (1) promeristem, (2) protoderm, (3) protocortex, (4) protopith, (5) residual meristem, (6) leaf primordium. (B) Differentiation zone (0.04-25 mm) with beginning of (-1), procambial strand formation, (-2) phloem formation, (-3) xylem formation, (-4) primary phloem fiber formation, and (-5) end of primary tissue formation: (7) procambial strand, (8) epidermis, (9) primary cortex, (10) protophloem, (11) procambium, (12) pith, (13) primary (proto-) phloem, (14) cambium, (15) protoxylem, (16) proto- and metaphloem, (17) primary phloem fibers; (c) secondary thickening, beginning at (-1) with (18) cambium, (19) proto- and metaxylem (primary), (20) ruptured epidermis, (21) cortex, (22) primary phloem fibers, (23) cork cambium, (24) secondary phloem, (25) secondary xylem, and (26) cork. Adapted from Bell and Coombe, 1976; see Acknowledgments and Selected References.

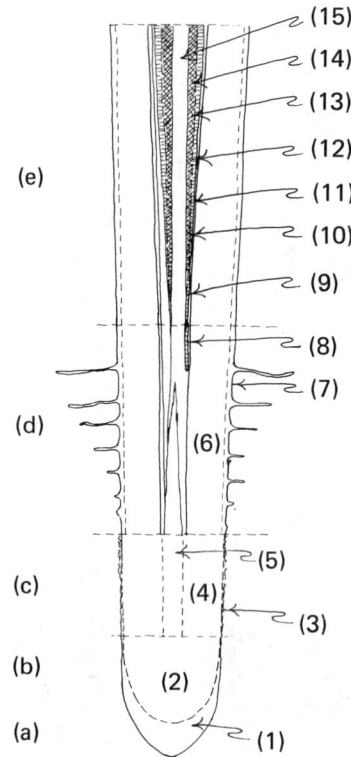

Fig. II.3.02. Anatomy: primary and secondary tissues in a young dicotyledon root (ls, diagrammatic): (a) rootcap (1); (b) cell division zone, with (2) protomeristem; (c) cell enlargement zone, with (3) protoderm, (4) ground meristem, (5) procambium; (d) primary tissue and root hair zone, with (6) cortex, (7) epidermis, and (8) primary phloem; (e) secondary tissue zone, with (9) endodermis, (10) pericycle, (11) primary phloem, (12) secondary phloem, (13) cambium, (14) secondary xylem and (15) primary xylem. Redrawn from Hill et al., 1960; see Acknowledgments and Selected References.

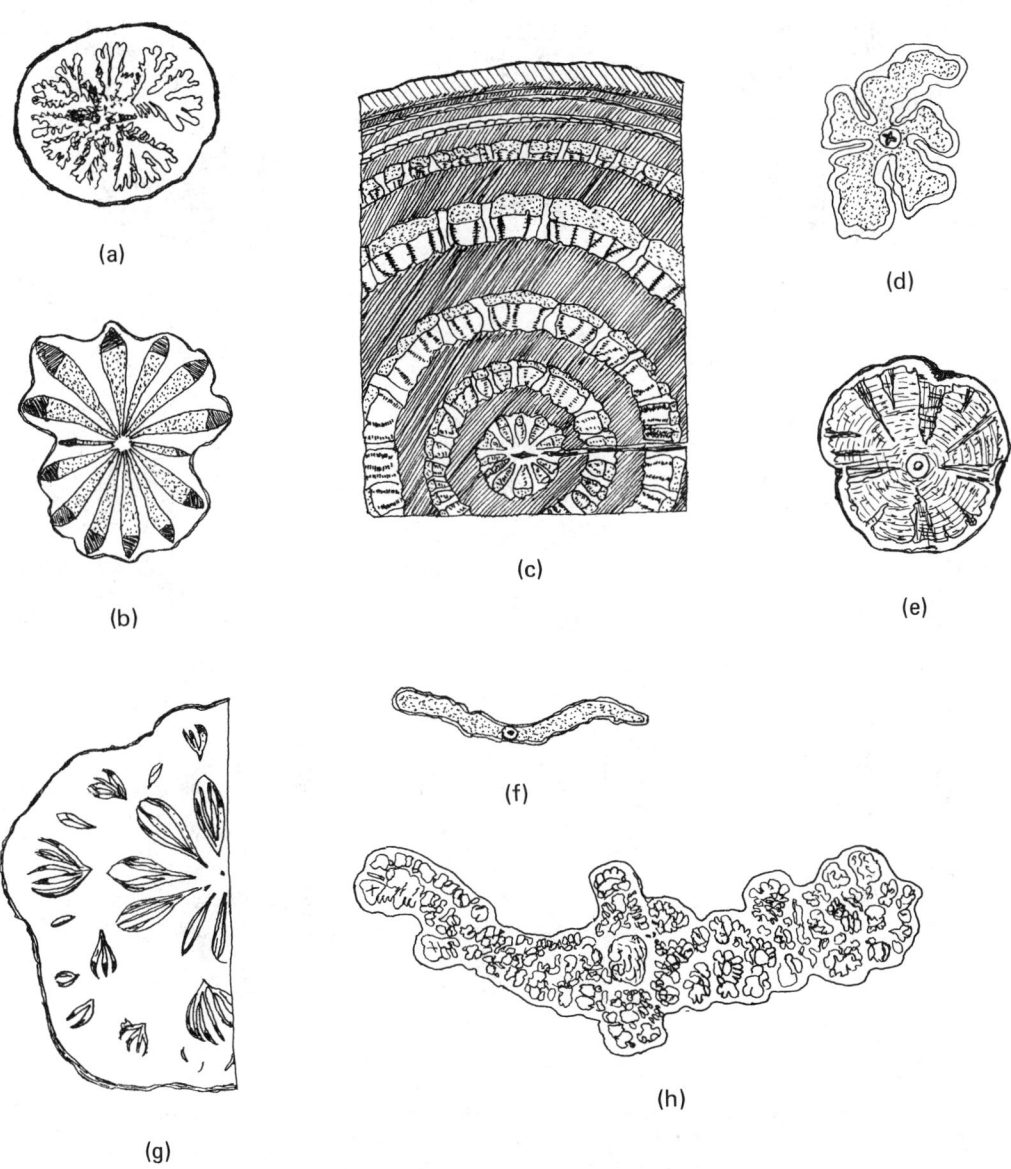

Fig. II.3.03. Anatomy: anomalous secondary growth in liane stems and beet root (ts): (a) Bignonia sp.; (b) Begonia fruticosa; (c) Beta vulgaris with alternate layers of vascular bundles (phloem stippled, xylem in radial rows) and proliferated pericycle (hatched); (d) Bauhinia rubiginosa; (e) Bignoniaceae; (f) Bauhinia sp.; (g) Piper fluminense; (h) Bauhinia langsdorffiana [xylem stippled, phloem dark in (b) and (g)]. Redrawn from Eames and MacDaniels, 1947; see Acknowledgments and Selected References.

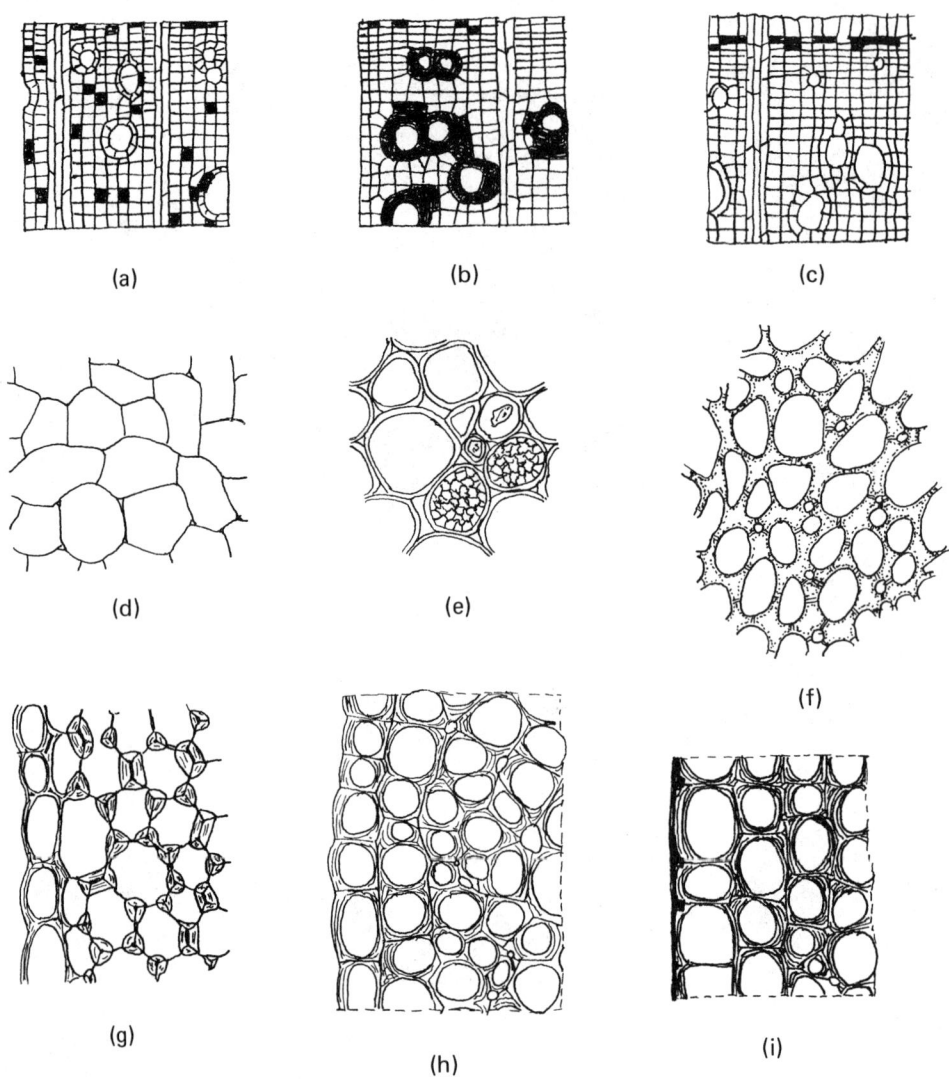

Fig. II.3.04. Anatomy: parenchyma, aerenchyma, and collenchyma: (a)-(c) wood parenchyma, (a) diffuse; (b) vasicentric; (c) terminal; (d)-(e) parenchyma: (d) Zea pith; (e) Castanea dentata twig pith with thick-walled lignified cells that contain starch grains or crystals; (f) aerenchyma in Canna leaf, with (1) large intercellular spaces; (g)-(i) collenchyma (epidermis at left): (g) Lactuca, lacunar; (h) Cucurbita, angular; (i) Sambucus, lamellar (thickening mainly in tangential walls). (a)-(e) Redrawn from Eames and MacDaniels, 1947; (f)-(i) redrawn from Esau, 1965; see Acknowledgments and Selected References.

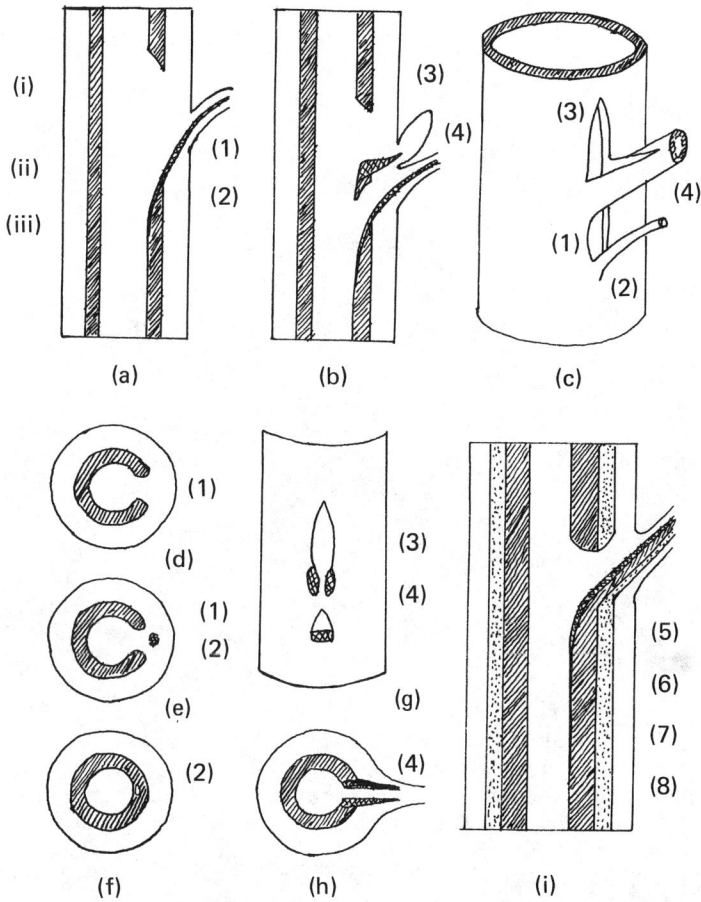

Fig. II.3.05. Anatomy: leaf and branch traces and gaps (mainly ls and ts, diagrammatic): (a) Longitudinal section of a node through (1) a leaf trace and (2) gap; (b) same but with branch (3) trace and (4) gap also present; (c) view of vascular cylinder showing (2) leaf and (4) branch traces and their (1,3) associated gaps; (d), (e), (f) at levels (i), (ii), (iii), respectively; (g) face view of (c); (h) ts of (g); (i) stem section with (5) protoxylem, (6) pith, (7) metaxylem, and (8) phloem indicated. Redrawn from Eames and MacDaniels, 1947; see Acknowledgments and Selected References.

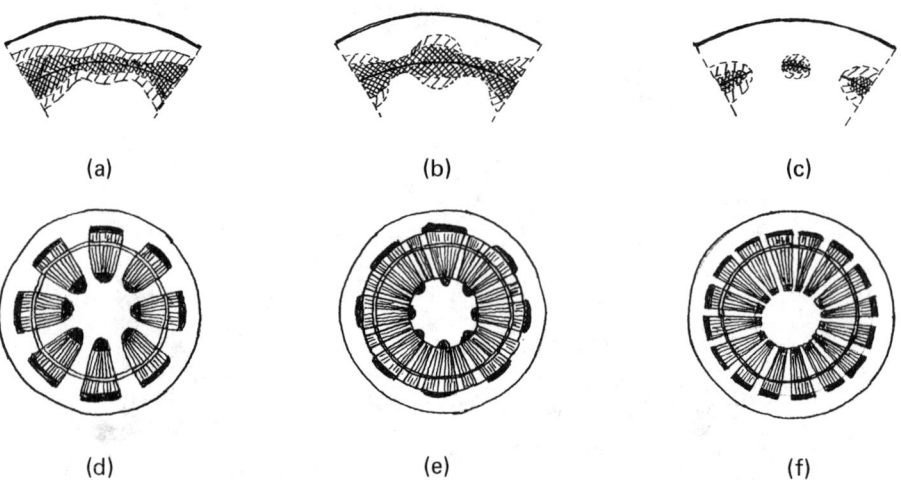

Fig. II.3.06. Anatomy: cambium cylinder formation (a)-(c) and types of thickening in dicotyledon stems (d)-(f) (ts, diagrammatic): (a) complete primary (hatched), secondary (crosshatched), and cambial cylinders; (b) secondary tissues only complete the vascular cylinder; (c) primary cylinder of discrete strands; (d) Aristolochia; (e) Ricinus; (f) Tilia. (a)-(c) Redrawn from Eames and MacDaniels, 1947; (d)-(f) redrawn from Bell and Coombe, 1976; see Acknowledgments and Selected References.

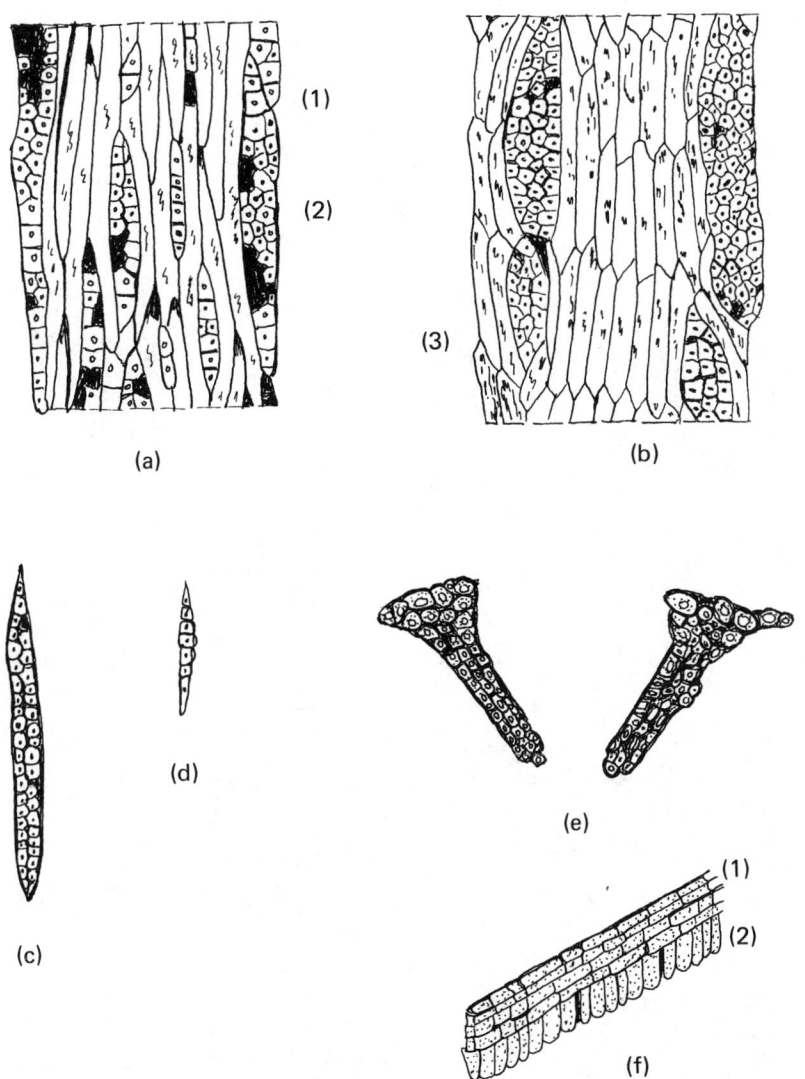

Fig. II.3.07. Anatomy: cambium initials and ray cells: (a) Juglans (tangential ls, X155), with (1) long fusiform initials (nonstoried cambium) and (2) heterocellular multiseriate rays; (b) Robinia (tangential long. sect., X155), with (3) short fusiform initials (storied cambium); (c)-(f) diagram of Liriodendron: (c) biseriate and (d) uniseriate heterocellular rays (tangential ls); (e) wide rays (ts); (f) biseriate ray (radial s), with (1) procumbent and (2) upright cells. (a)-(d), (f), Redrawn from Esau, 1965; (e) redrawn from Hill et al., 1960; see Acknowledgments and Selected References.

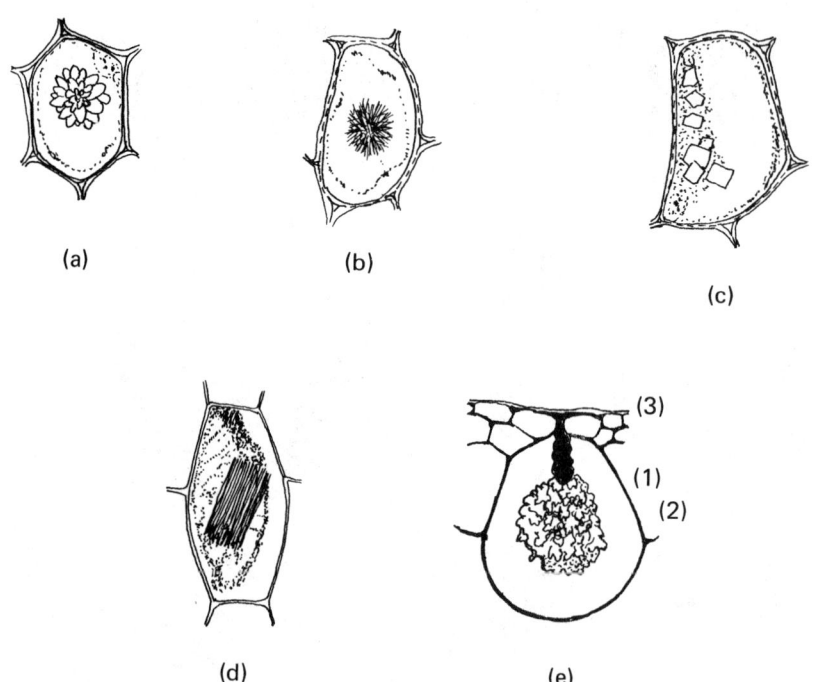

Fig. II.3.08. Anatomy: crystals, cystolith, druses, and lithocyst: (a)-(d) crystals, (a), (b) druses from Gnetum gnemon cortex; (c) prismatic and rhombohedral crystals from Gnetum indicum cortex [(a)-(c) X 800]; (d) raphides from Vitis vinifera leaf (X625); (e) lithocyst (1), with (2) mature cystolith in a Ficus elastica leaf with (3) multiple epidermis (X235). Redrawn from Esau, 1965; see Acknowledgments and Selected References.

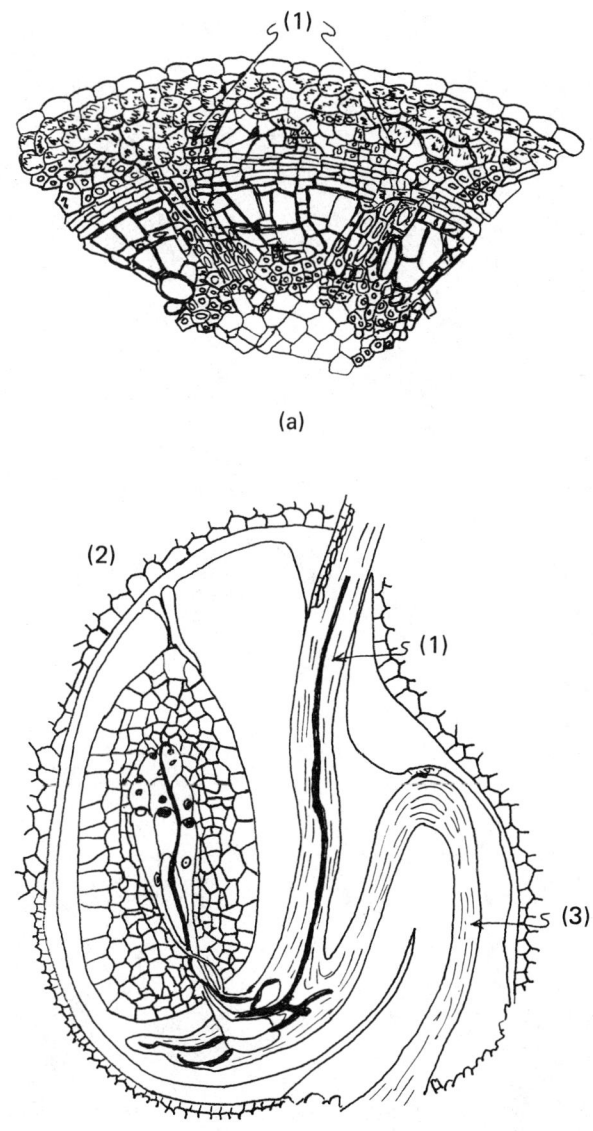

Fig. II.3.09. Anatomy: dilatation and obturator: (a) dilatation as typified by growth of (1) wide (phloem) rays in a Liriodendron stem; (b) obturator (1) in Casuarina equisetifolia, with (2) micropyle and (3) funicle. (a) Adapted from Hill et al., 1960; (b) redrawn from Eames, 1961; see Acknowledgments and Selected References.

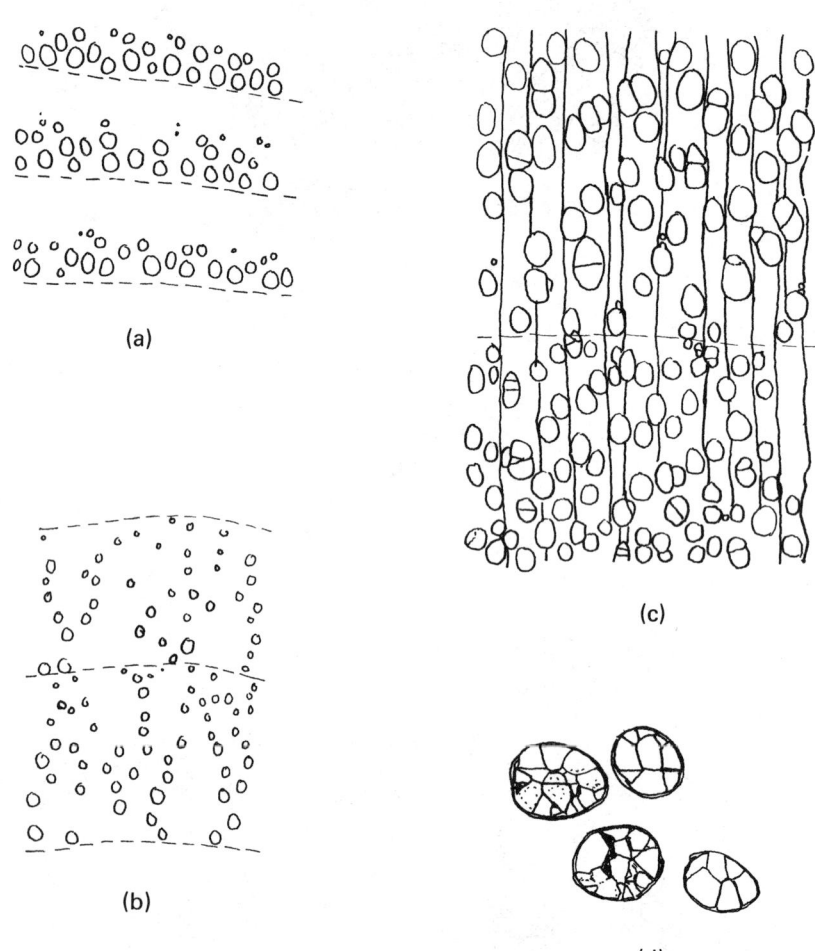

Fig. II.3.10. Anatomy: (a)-(c) vessel distribution in wood: (a) ring porous (Quercus bicolor); (b) semi-ring porous (Q. virginiana); (c) diffuse porous (Salix nigra) (X10, ts); (d) tyloses in Salix nigra vessels (X35, ts). Adapted from Esau, 1965; see Acknowledgments and Selected References.

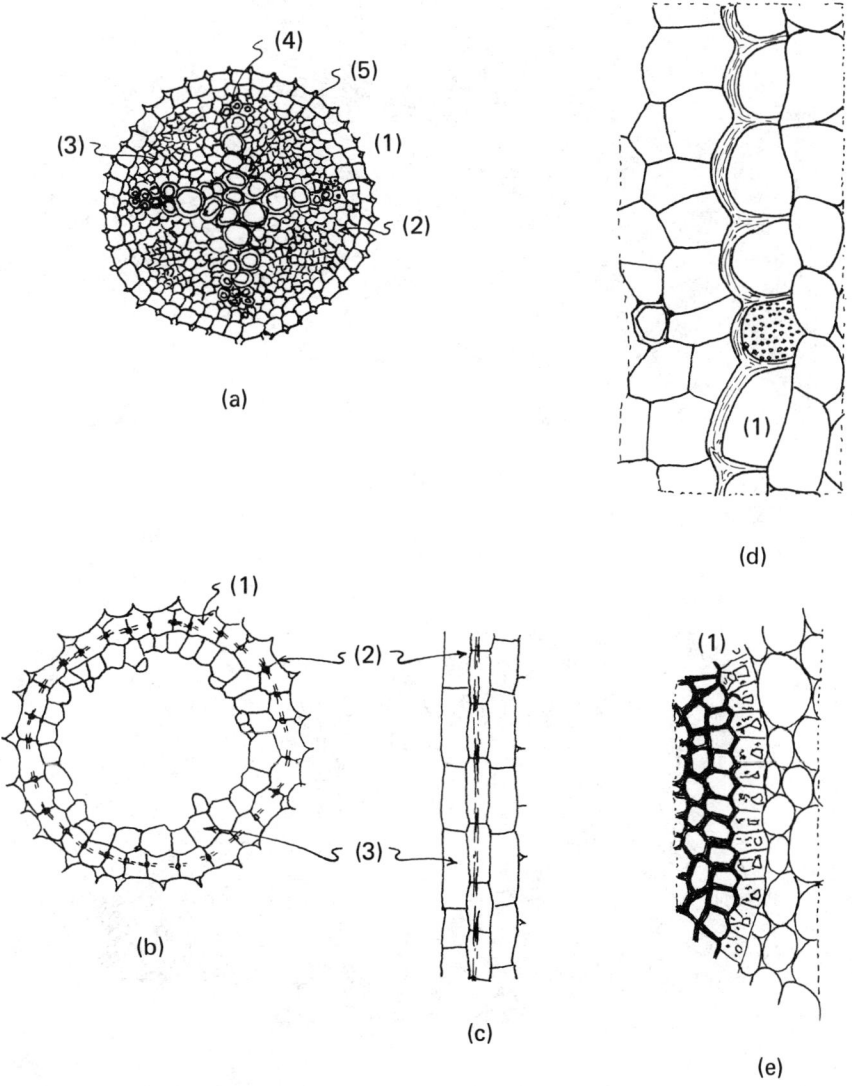

Fig. II.3.11. Anatomy: endodermis and pericycle in roots: (a) Ranunculus acris (ts), with (1) endodermis, (2) pericycle, (3) phloem, (4) protoxylem, and (5) metaxylem; (b) Convolvulus arvensis (ts, X225), with (1) endodermis, (2) casparian strip, and (3) pericycle; (c) radial s of (b) (X225); (d) Musa sapientum; (e) Smilax rotundifolia (ts), with (1) thick-walled type of endodermis. (a) Redrawn from Foster and Gifford, 1974; (b), (c) redrawn from Esau, 1965; (d), (e) redrawn from Eames and MacDaniels, 1947; see Acknowledgments and Selected References.

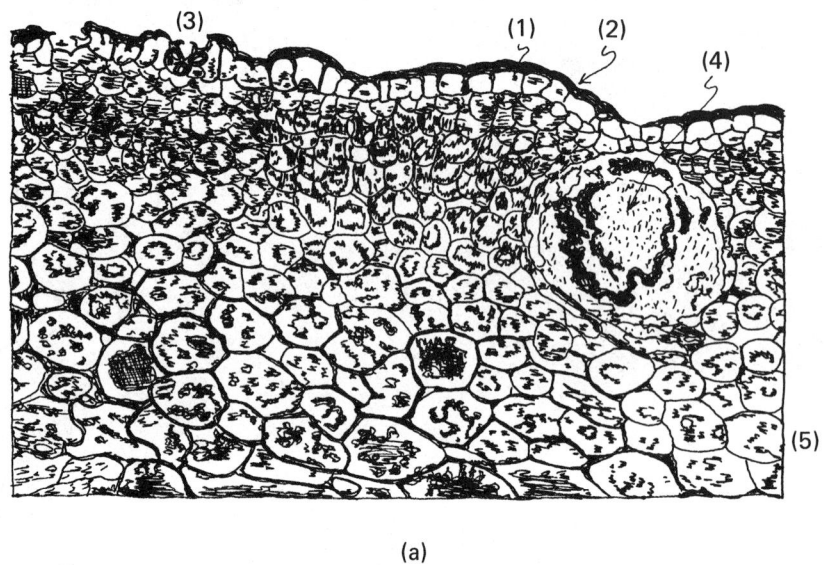

(a)

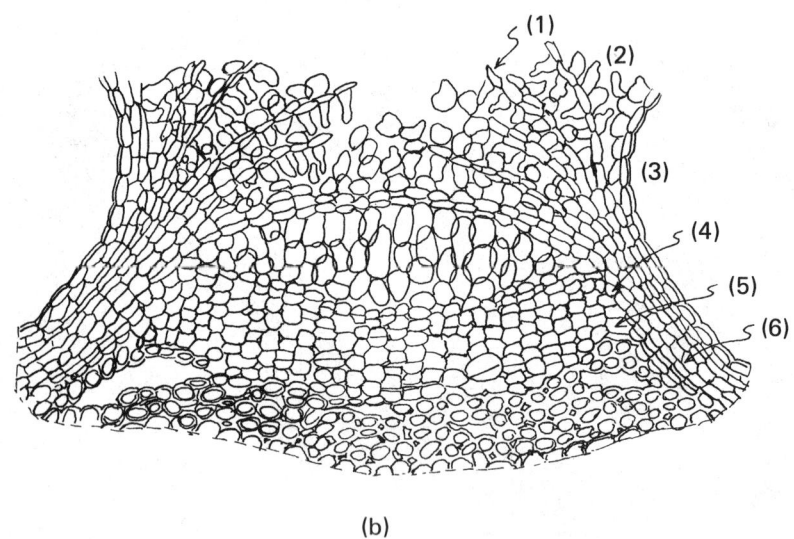

(b)

Fig. II.3.12. Anatomy: external stem features: (a) six-month-old Troyer citrange stem (ts, X250), near stem-root transition area, with (1) epidermis, (2) heavy cuticle, (3) stoma, (4) large lysigenous oil gland (ca. 100= m diameter), and (5) underlying cortex with numerous starch grains especially in the 6 to 8 cell layers nearest the epidermis; (b) lenticel in Prunus avium stem, with (1) closing cells, (2) complementary cells, (3) epidermis, (4) phellogen, (5) phelloderm, (6) phellem. (b) Redrawn from Eames and MacDaniels, 1947; see Acknowledgments and Selected References.

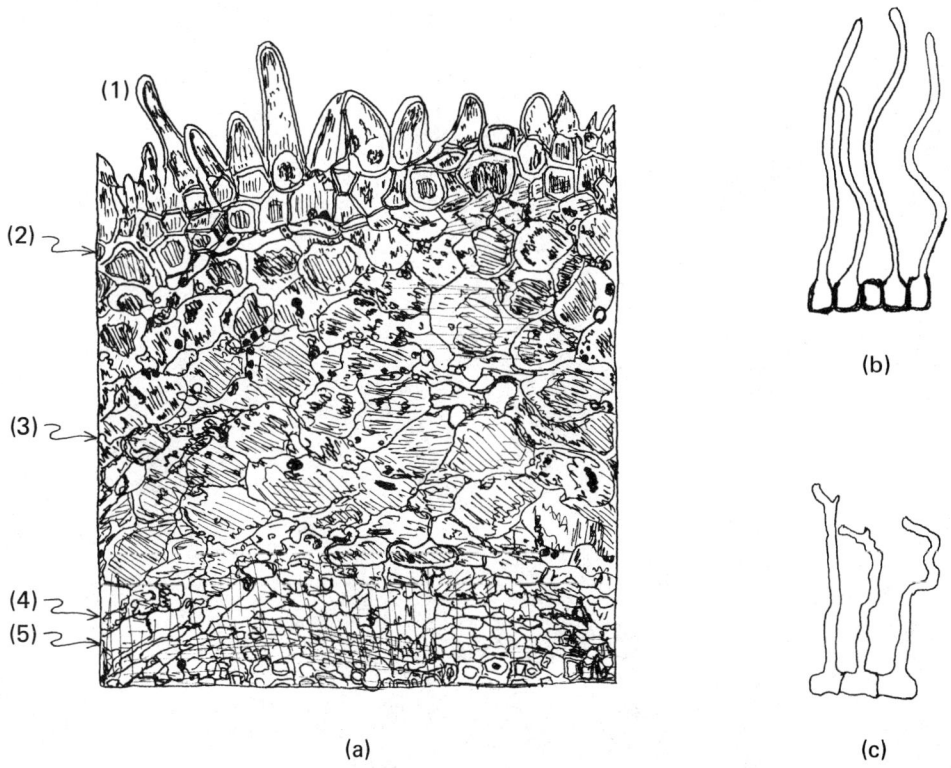

Fig. II.3.13. Anatomy. epidermis, root hairs: (a) four-month-old 'Orlando' tangelo seedling (ts, X250) with (1) root hairs of various lengths from about 5 to 50 m, (2) underlying suberized hypodermis cells, (3) cortex, (4) endodermis, and (5) pericycle (phloem, cambium, and secondary xylem are also present inward from the pericycle); (b), (c) root hairs grown in (b) water or moist air and (c) moist soil. (b), (c) Redrawn from Eames and MacDaniels, 1947; see Acknowledgments and Selected References.

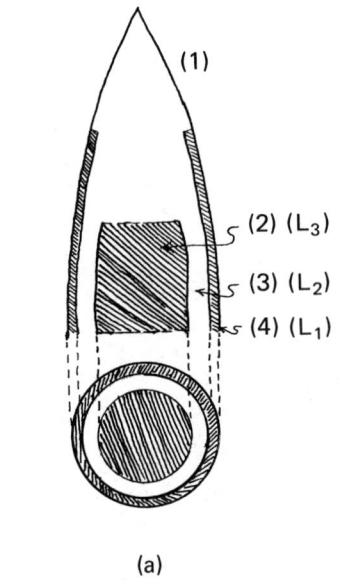

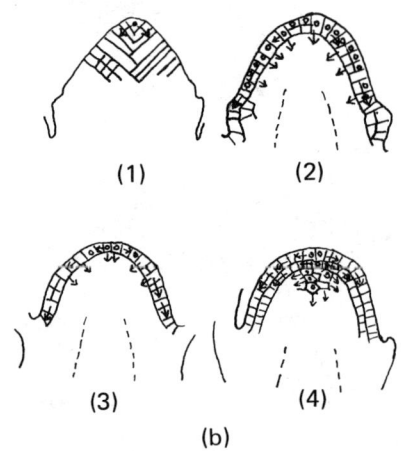

Fig. II.3.14. Anatomy: theories of meristem differentiation (diagrammatic): (a) histogen theory, with undifferentiated promeristem (1) which develops as (2) dermatogen (L_1), (3) periblem (L_2), and (4) plerome (L_3) layers; (b) tunica-corpus regions may develop as (1) a solitary initial with anticlinal division, (2) many superficial initials with anticlinal and periclinal divisions, (3) several superficial initials with anticlinal and periclinal divisions [the dermatogen (L_1) layers are largely anticlinal], (4) three-tiered initials, the two outer ones with anticlinal divisions only and the innermost (corpus) with division in all planes. Redrawn from Eames and MacDaniels, 1947; see Acknowledgments and Selected References.

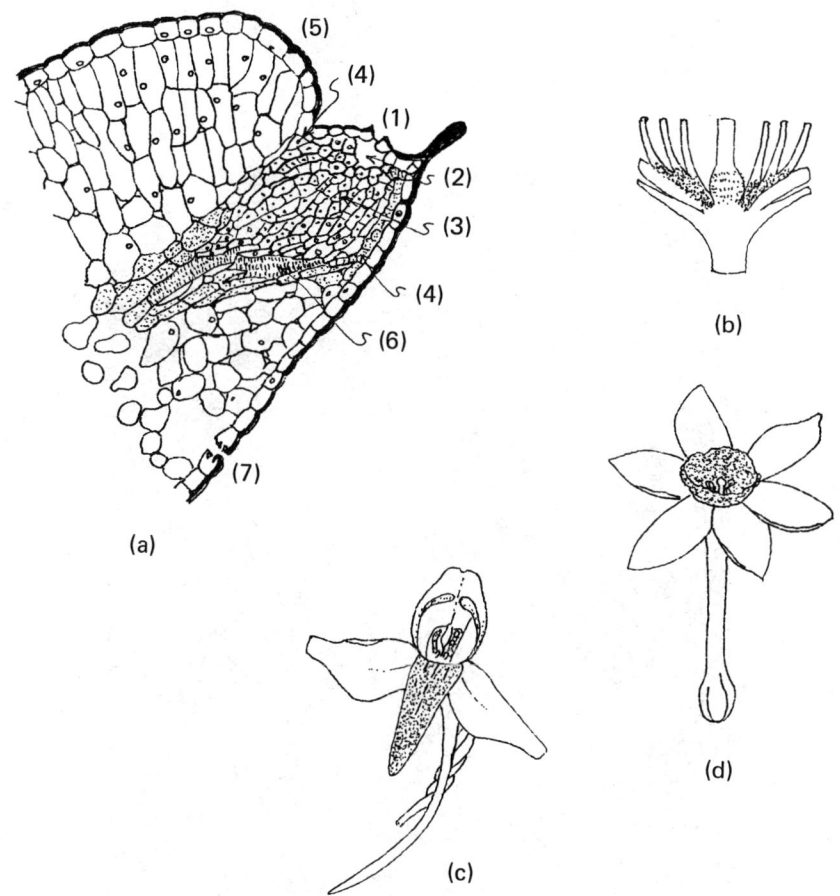

Fig. II.3.15. Anatomy: hydathode, nectary, osmphor: (a) hydathode of Saxifraga lingulata leaf (ls), with (1) pore, (2) water cavity, (3) epithem, (4) sheath, (5) cuticle, (6) xylem, (7) stoma; (b) nectary at base of stamens in Thea (=Camellia) flower; (c), (d) osmophors in (c) Platanthera bifolia and (d) Narcissus jonquilla. Redrawn from Esau, 1965; see Acknowledgments and Selected References.

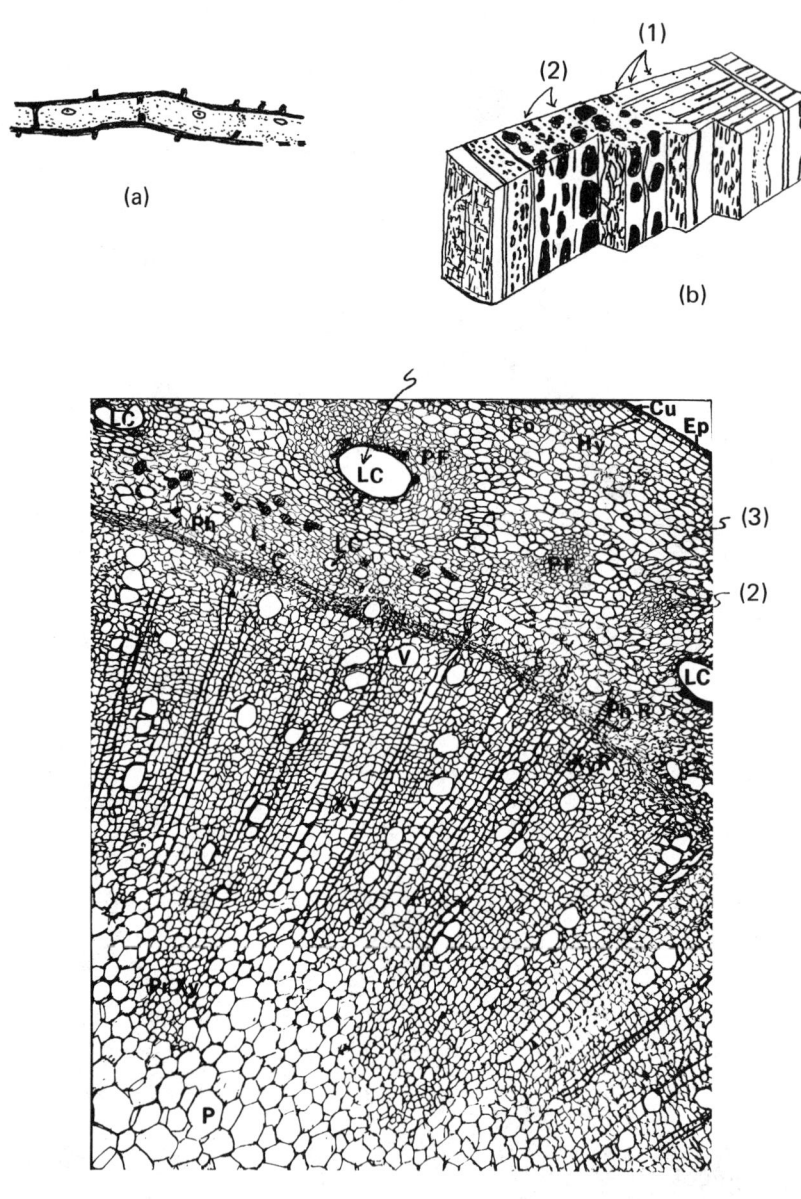

Fig. II.3.16. Anatomy: laticifers and resin canals: (a), (b) latex vessels, (a) origin of laticifer in a Hevea brasiliensis stem that shows successive stages in cell crosswall dissolution from left to right; (b) diagram of bark, with (1) rows of latex vessel cylinders and (2) sclereids (stone cells); (c) resin canals in a two-month-old 'Turpentine' mango (Mangifera indica) seedling (ts), with (1) schizolysigenous resin canals more or less surrounded by (2), a ring of "pericyclic fibers" and (3), a "starch sheath" several layers below the epidermis. (a), (b) Redrawn from Dijkman, 1951; see Acknowledgments and Selected References.

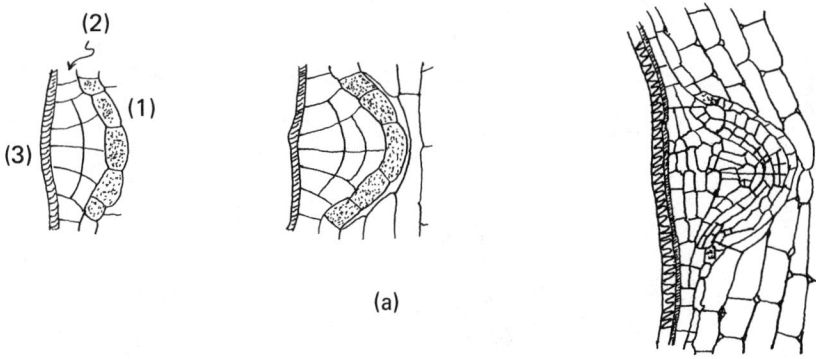

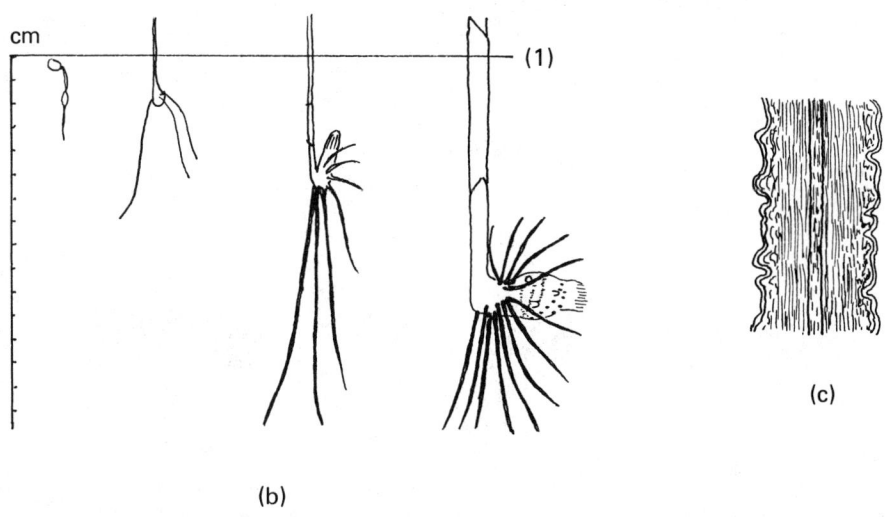

Fig. II.3.17. Anatomy: (a) lateral root development (Hypericum, radial s) that shows the origin from pericyclic cells and stretching of cells in parent root as it enlarges [(1) endodermis, (2) pericycle, (3) protoxylem]; (b), (c) contractile roots: (b) Arum maculatum showing germination and successive stages of tuber being pulled downward [(1) soil surface]; (c) contracted root of Lilium martagon (ls). (a) Redrawn from Eames and MacDaniels, 1947; (b), (c) redrawn from Bell and Coombe, 1976; see Acknowledgments and Selected References.

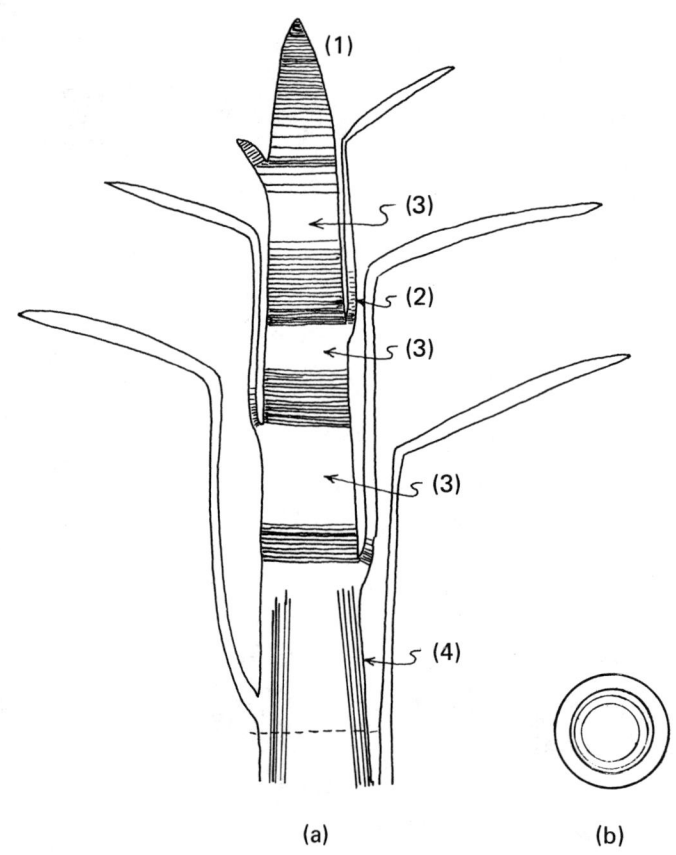

Fig. II.3.18. Anatomy: meristem positions (diagrammatic): (a) longitudinal view, with (1) apical, (2) intercalary meristems, (3) internodal region, and (4) lateral meristems; (b) transverse section of (a) at level of dashed line. Redrawn from Eames and MacDaniels, 1947; see Acknowledgments and Selected References.

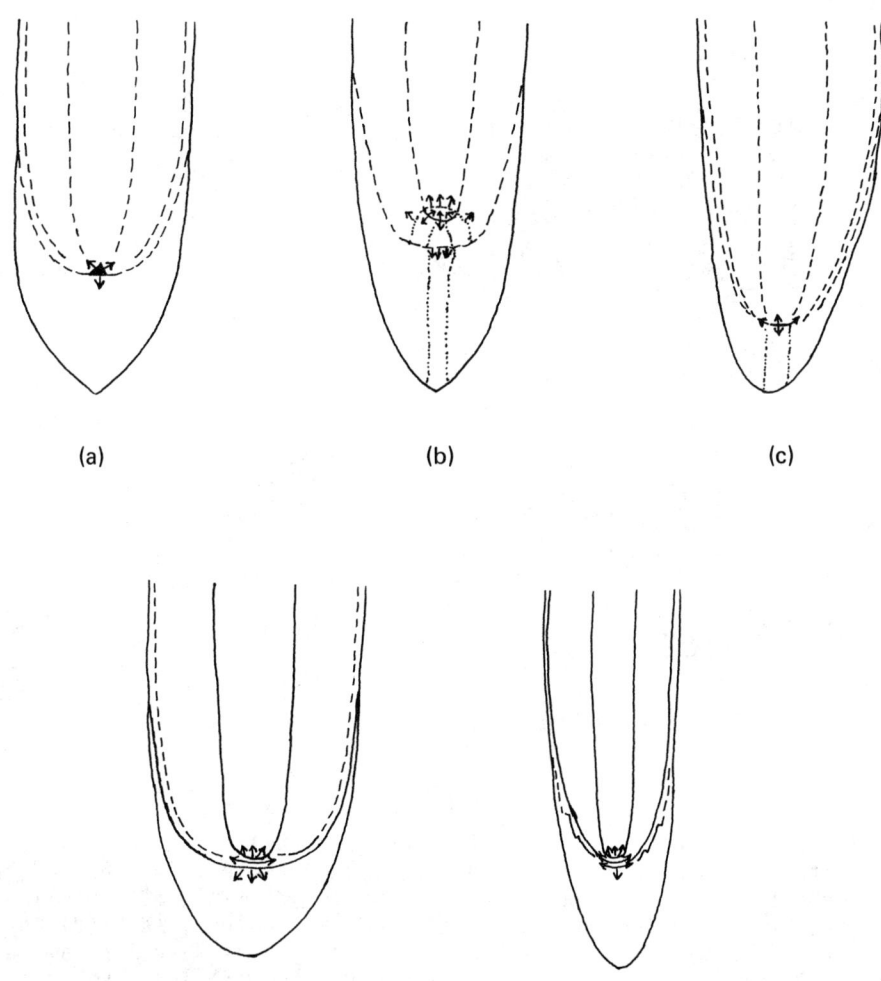

Fig. II.3.19. Anatomy: root apical meristems (ls): (a) Adiantum, with single apical cell (black triangle); (b) Pseudotsuga, with initial zone as black arc; (c) Allium, with poorly organized zone; (d) Zea (monocot), with three tiers of initials; (e) Nicotiana (dicot) with three tiers of initials. Redrawn from Esau, 1965; see Acknowledgments and Selected References.

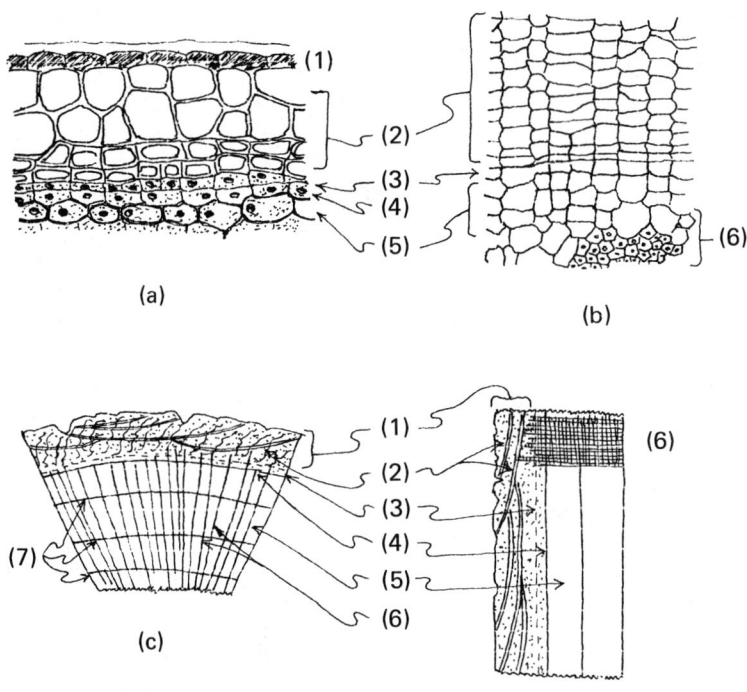

Fig. II.3.20. Anatomy. (a), (b) periderm (ts): (a) <u>Populus deltoides</u> twig, with (1) epidermis, (2) phellem, (3) phellogen, (4) phelloderm, and (5) cortex; (b) <u>Salix alba</u> var. <u>vitellina</u>, with (2) phellem, (3) phellogen, (4) phelloderm, and (6) secondary phloem; (c), (d) rhytidome (1), with (2) periderm, (3) active phloem, (4) vascular cambium, (5) xylem, (6) rays, and (7) growth rings [(c) ts, (d) ls]. [Note that the term "bark" (II.3.016) is a colloquial synonym for rhytidome]. (a), (b) Redrawn from Eames and MacDaniels, 1947; (c), (d) redrawn from Esau, 1965; see Acknowledgments and Selected References.

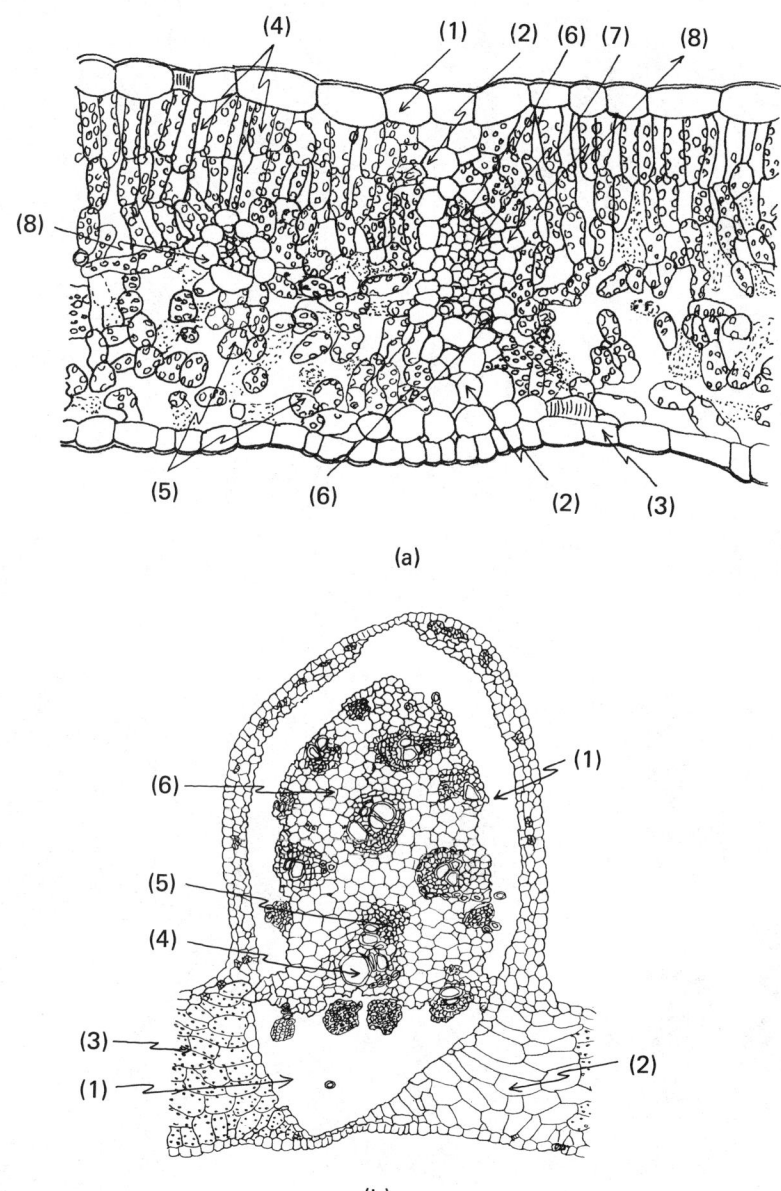

Fig. II.3.21. Anatomy: leaves: (a) internal structure of Pyrus communis leaf (ts), with (1) upper epidermis, (2) bundle sheath extension, (3) lower epidermis, (4) palisade parenchyma, (5) spongy parenchyma, (6) fibers, (7) xylem, (8) bundle sheath, (9) phloem; (b) main vein of Pinanga kuhlii leaflet near midrib (ts, X80), with (1) massive fiber sheath, (2) pulvinar tissue at junction with blade, (3) parenchyma with chloroplasts, (4) xylem, (5) phloem, (6) parenchyma. (a) Redrawn from Esau, 1965; (b) adapted from Corner, 1966; see Acknowledgments and Selected References.

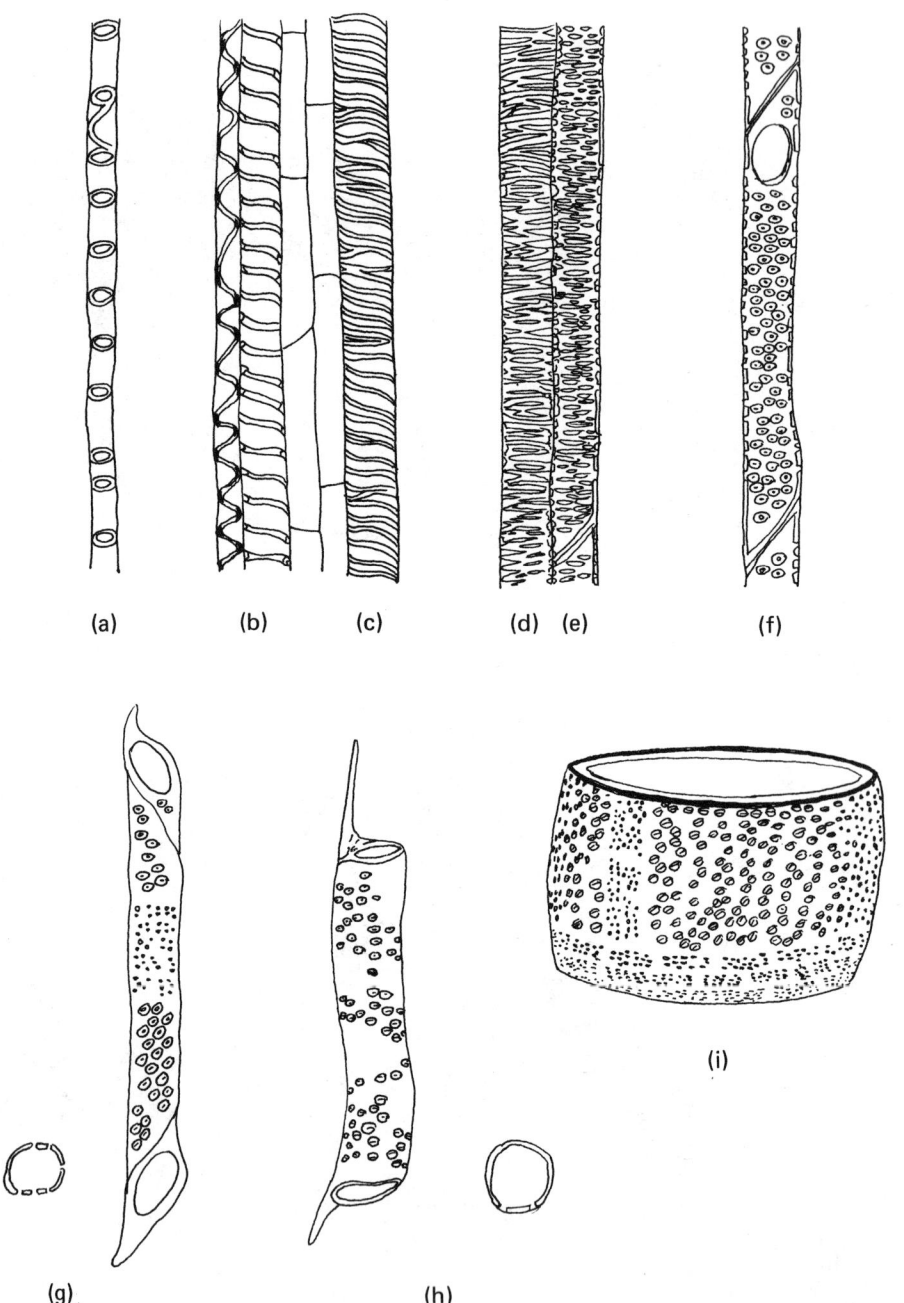

Fig. II.3.22. Anatomy: protoxylem, metaxylem, and vessel elements: (a), (b) protoxylem, (a) annular; (b) spiral; (c)-(f) metaxylem: (c) spiral [with parenchyma cells between it and (b)]; (d) scalariform; (e) scalariform-reticulated; (f) pitted vessel. (a)-(f) from Lobelia; (g)-(i) vessel elements: (g) Quercus alba, with ts at left; (h) Malus pumila with ts at right; (i) Quercus alba. Redrawn from Eames and MacDaniels, 1947; see Acknowledgments and Selected References.

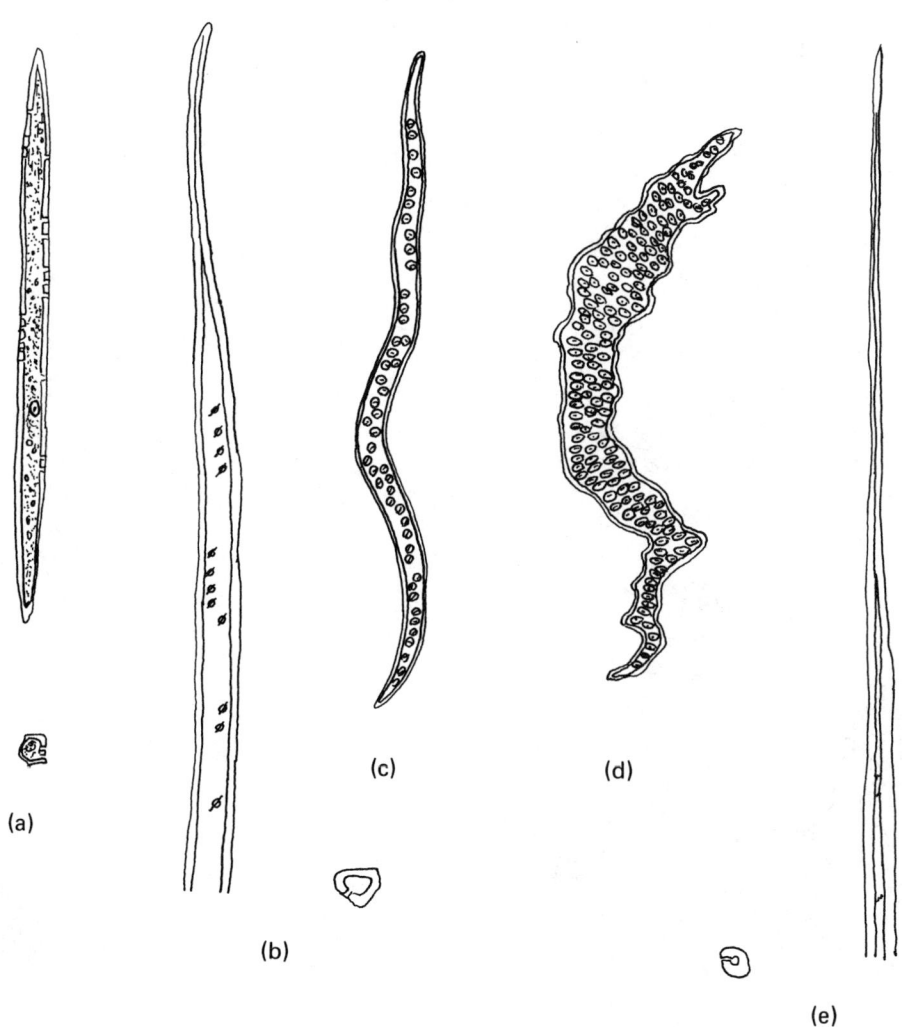

Fig. II.3.23. Anatomy: tracheids, fiber-tracheids, and fibers: (a) "substitute fiber" or "fusiform wood parenchyma cell" from Sassafras variifolium (ls, with ts below); (b) fiber-tracheid from Malus pumila (ls, with ts at right); (c), (d) tracheids from Quercus alba, (c) normal; (d) flattened and distorted from spring wood; (e) libriform fiber from Q. alba (ls, with ts at left). Redrawn from Eames and MacDaniels, 1947; see Acknowledgments and Selected References.

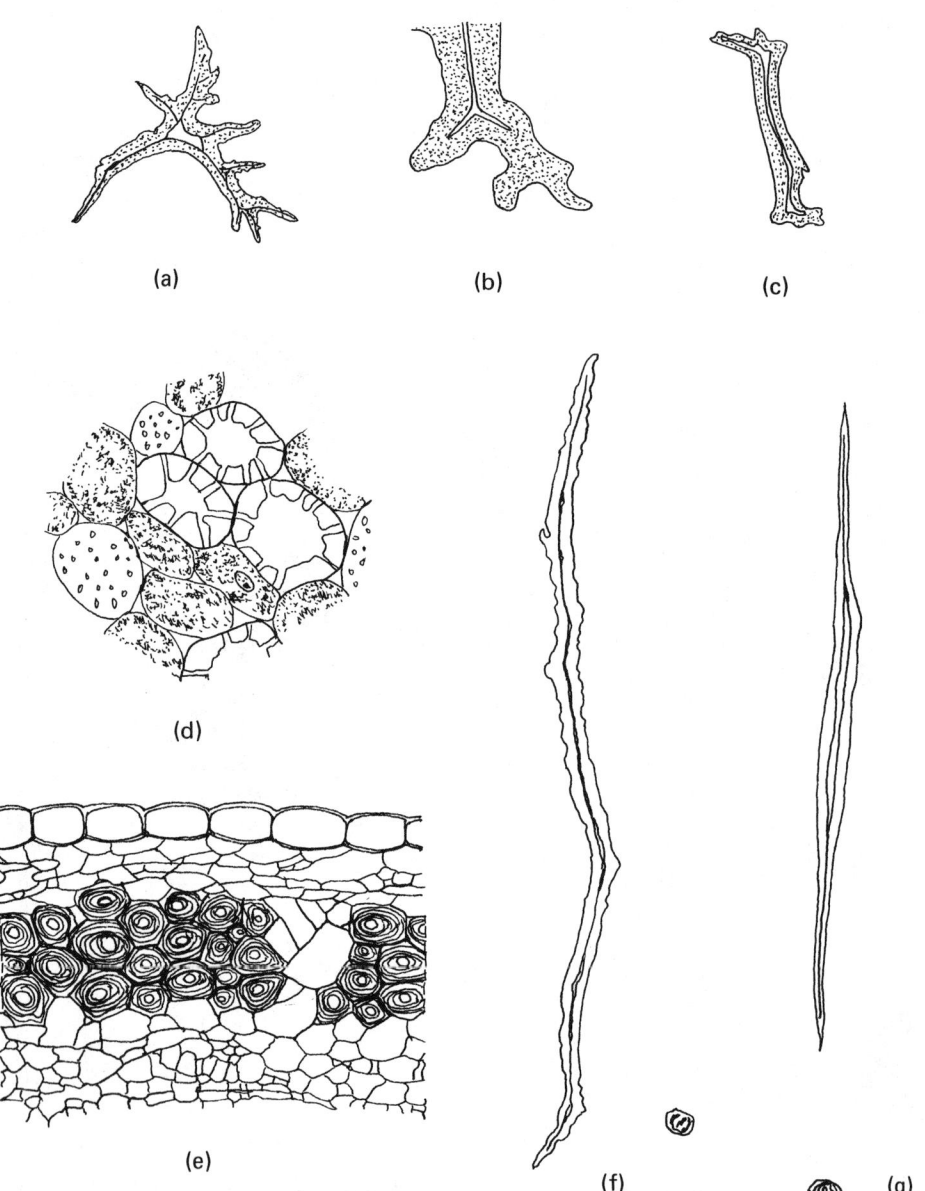

Fig. II.3.24. Anatomy: sclereids and phloem fibers: (a)-(c) leaf sclereids, (a) branched form in Trochodendron; (b) portion of sclereid; (c) columnar in Mouriria [(a), (c) X115, (b) X333]; (d) sclereids and parenchyma cells in Dracaena fragrans stem cortex, e. g., phloem fibers; (e) in primary phloem of Linum usitatissimum stem; (f) Malus pumila; (g) Robinia pseudacacia. [(a)-(c) ls; (d)-(e) ts; (f)-(g) ls, with ts at right and below)]. (a)-(c), (e) Redrawn from Esau, 1965; (d), (f), (g) redrawn from Eames and MacDaniels, 1947; see Acknowledgments and Selected References.

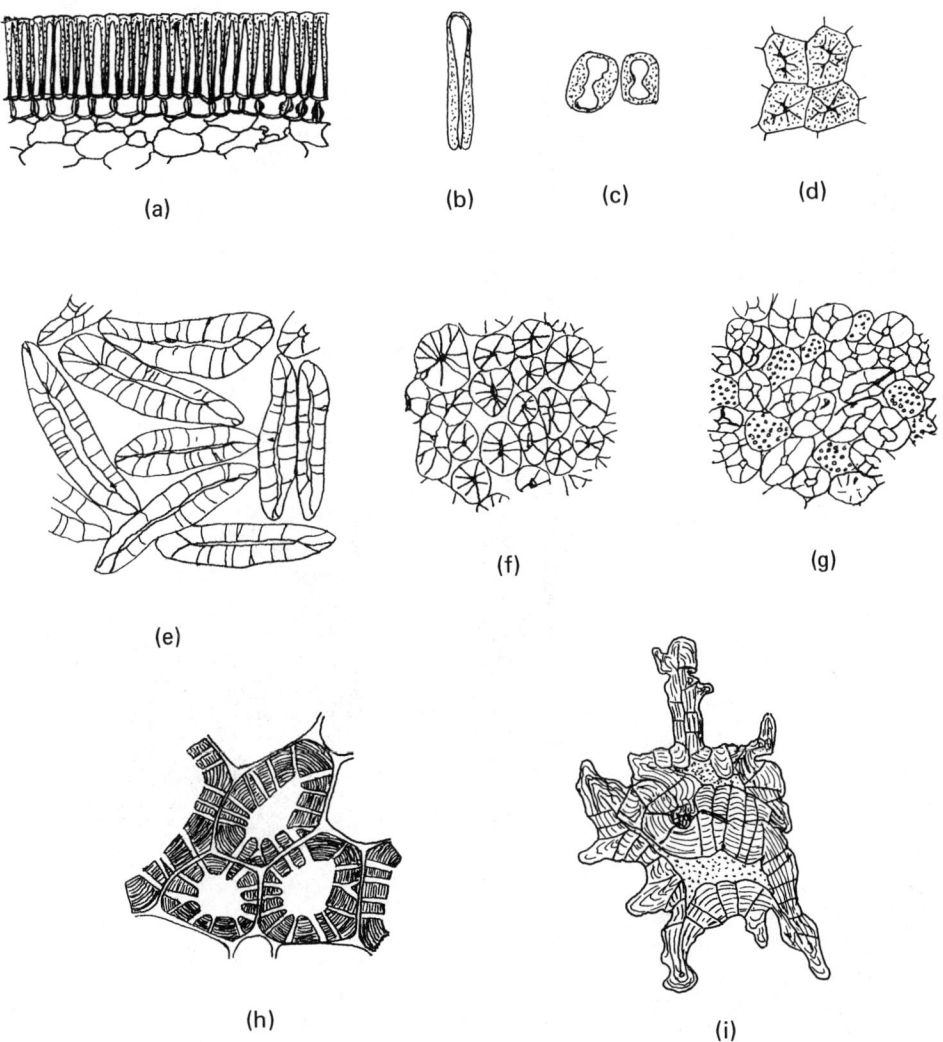

Fig. II.3.25. Anatomy: sclereids: (a)-(d) Phaseolus, (a) epidermis with solid layer of macrosclereids; (b), (c) individual epidermal and subepidermal sclereids, respectively; [(a)-(c) ts]; (d) face view; (e) Cocos nucifera, elongate sclereids (ls) in endocarp (shell); (f) same, ts; (g) "grit cells" from Pyrus communis pericarp; (h) Cydonia fruit; (i) irregular sclereids from Tsuga phloem [(g)-(i) ts]. (a)-(d), (h) Redrawn from Esau, 1965; (e)-(g), (i) redrawn from Eames and MacDaniels, 1947; see Acknowledgments and Selected References.

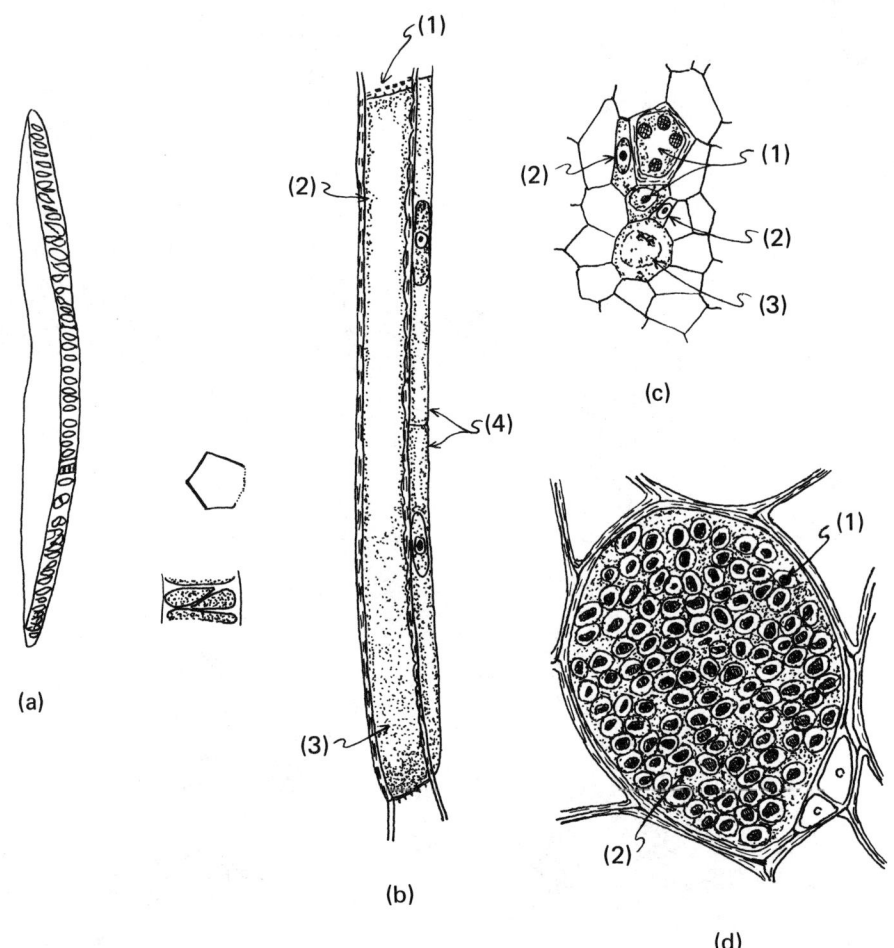

Fig. II.3.26. Anatomy: sieve cells and sieve tube elements: (a) <u>Malus pumila</u> side view with ts and detail of sieve plate at right; (b) mature sieve element of <u>Cucurbita</u> (ls), with (1) sieve plate, (2) parietal cytoplasm, (3) slime (denser below), (4) companion cells; (c) sieve-tube member of <u>Cucurbita</u> (ts), with (1) sieve-tube members, (2) companion cell, (3) cell before division; (d) sieve plate (<u>Cucurbita</u>), with (1) slime and (2) callose; [(b), (c) X730, (d) X590]. (a) Redrawn from Eames and MacDaniels, 1947; (b)-(d) redrawn from Esau, 1965; see Acknowledgments and Selected References.

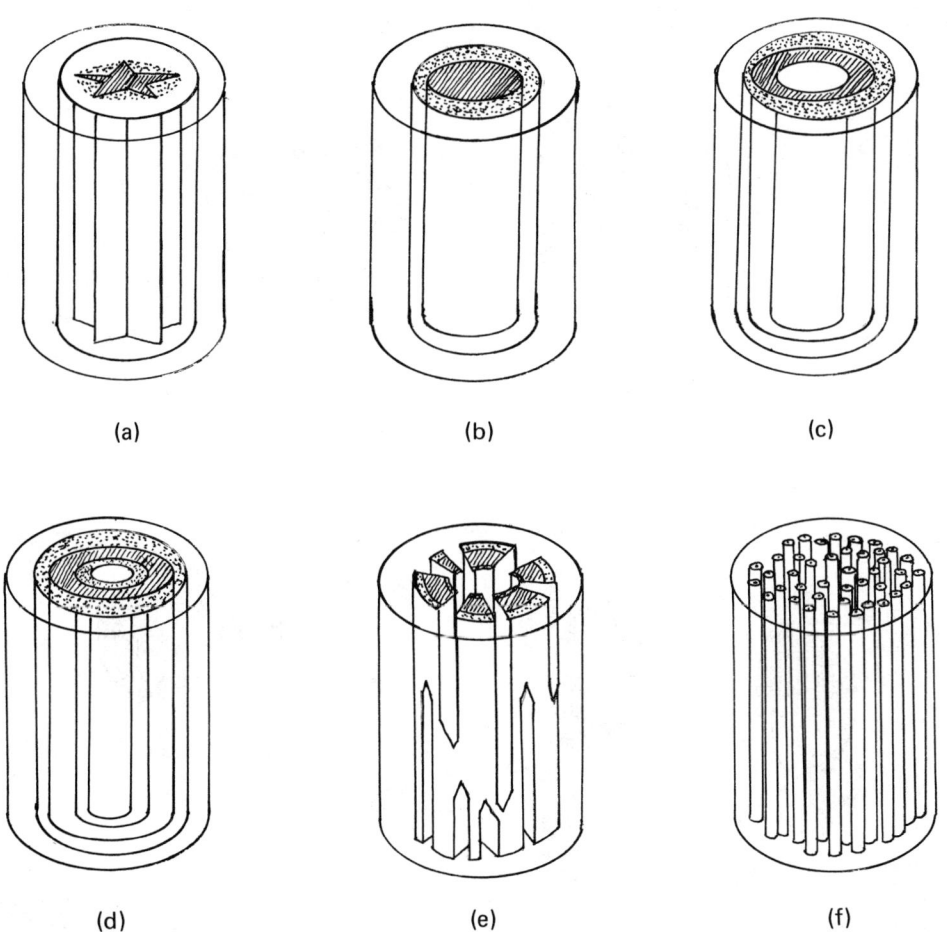

Fig. II.3.27. Anatomy: steles (diagrammatic): (a) protostele = actinostele; (b) protostele; (c) ectophloic siphonostele; (d) amphiphloic siphonostele; (e) dictyostele; (f) atactostele (xylem hatched, phloem stippled). Redrawn from Hill et al., 1960; see Acknowledgments and Selected References.

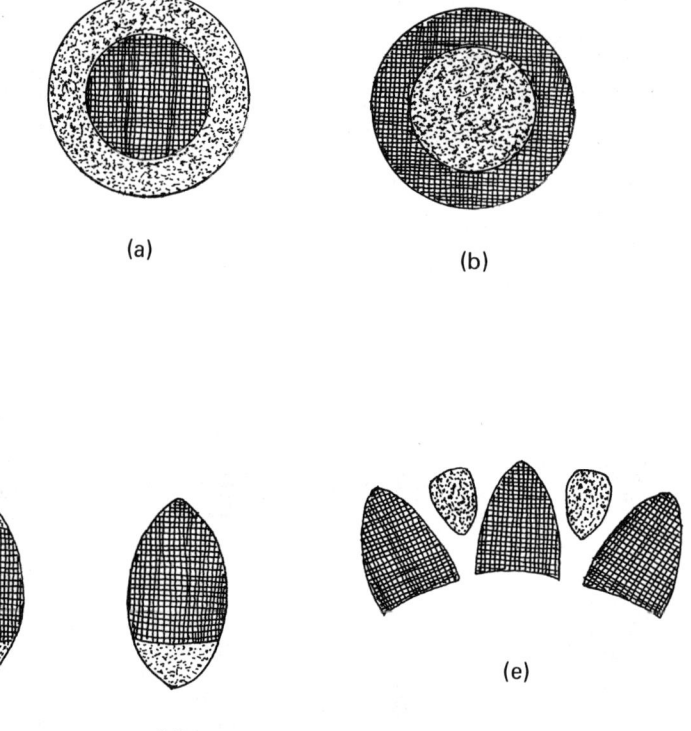

Fig. II.3.28. Anatomy: arrangement of xylem (crosshatched) and phloem (stippled) in primary tissues: (a) amphicribal; (b) amphivasal; (c) bicollateral; (d) collateral; (e) radial. Redrawn from Hill et al., 1960; see Acknowledgments and Selected References.

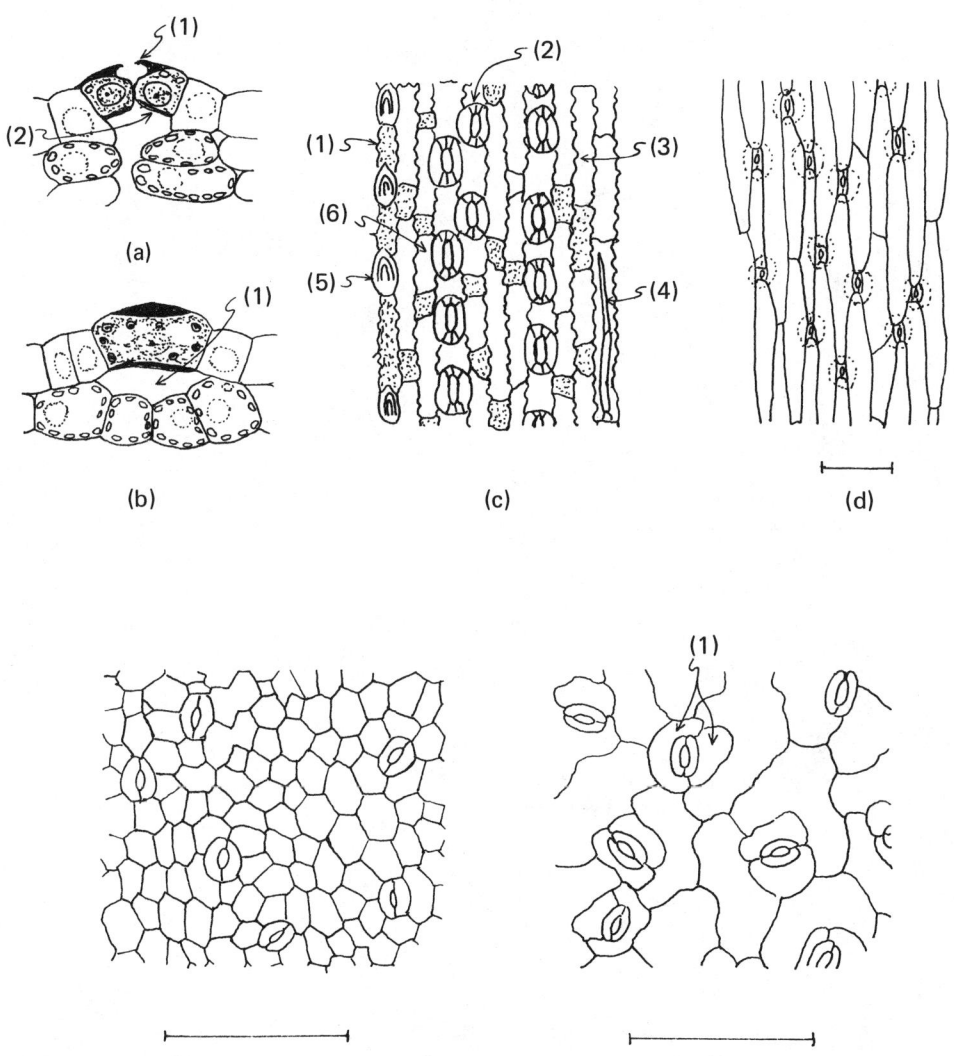

Fig. II.3.29. Anatomy: stomata: (a), (b) <u>Nicotiana</u> leaf stoma, (a) mature guard cells, with (1) upper and (2) lower ledges and unevenly thickened walls; (b) guard cell cut parallel to its long axis and perpendicular to leaf surface, with (1) substomatal chamber; (c)-(f) stomatal arrangements, (c) sugar cane, with (1) cork cells, (2) stoma, (3) long cell, (4) hair, (5) bristle, (6) short cell; (d) <u>Iris</u> with sunken cells in longitudinal rows; (e) <u>Vitis</u> with dispersed stomata; (f) <u>Vigna</u> paracytic, with subsidiary cells (lines = 100 m). Redrawn from Esau, 1965; see Acknowledgments and Selected References.

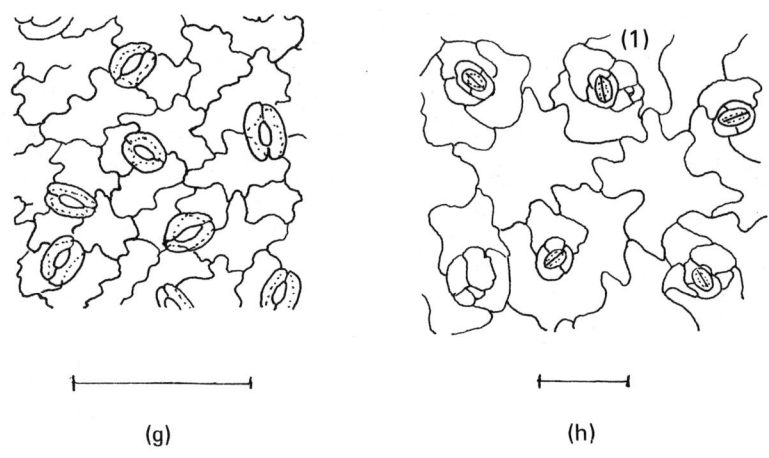

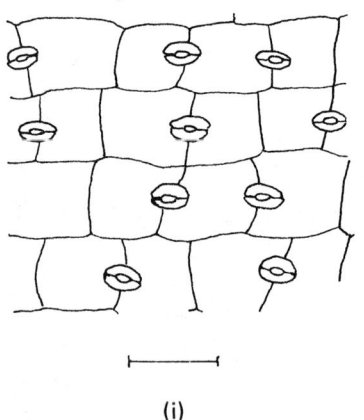

Fig. II.3.30. Anatomy: stomata: (g) Capsicum, raised stomata; (h) Sedum, variant of anisocytic [(1) subsidiary cells]; (i) Dianthus, diacytic (lines = 100 m). Redrawn from Easu, 1965; see Acknowledgments and Selected References.

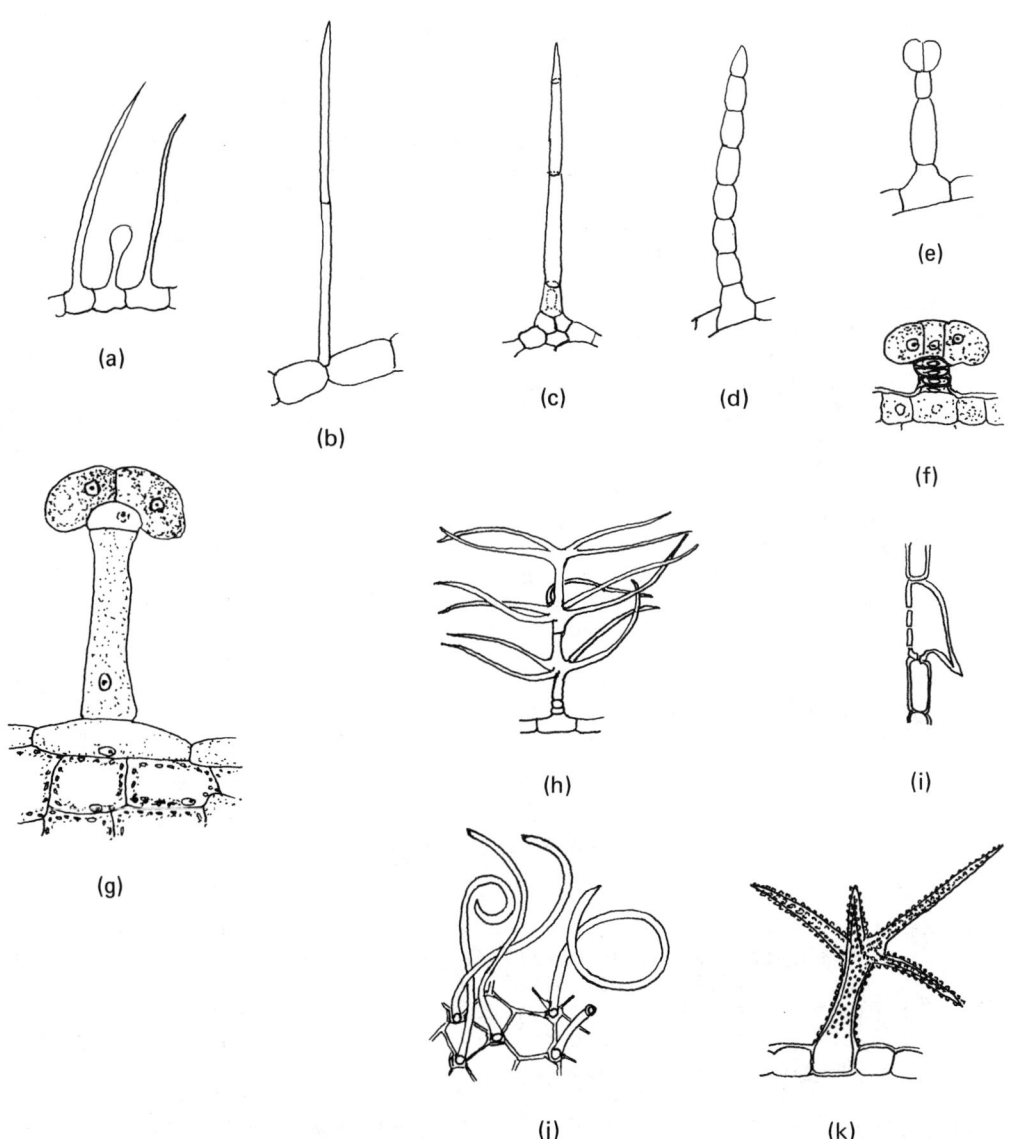

Fig. II.3.31. Anatomy: trichomes (hairs), secretory cells: (a) from Heliotropium calyx; (b) from Epigaea corolla; (c) from Cucumis leaf; (d) from Coreopsis leaf; (e) from Phryma corolla; (f) glandular (arrow) hair from Gaylussacia baccata ovary; (g) glandular (arrow) hair from Pinguicula leaf; (h) from young Platanus leaf; (i) from Avena leaf; (j) from Rubus fruit; (k) from Aubrieta stem. Redrawn from Eames and MacDaniels, 1947; see Acknowledgments and Selected References.

II.4.1. MORPHOLOGY and ANATOMY

II.4. Modifications

II.4.1. Life forms

.001 Chamaephyte: a plant in which the resting buds are located 10 to 50 cm above the soil, as in prostrate or creeping forms, heaths, cushion plants, and extreme desert plants. (Fig. II.4.01a) (B & C)
.002 Cryptophyte: a plant in which the resting buds are located below the soil surface, as on rhizomes or in bulbs. (Fig. II.4.01b) (B & C)
.003 Hemicryptophyte: a plant with its resting buds located at or close to the soil surface, as in tufted plants (e.g., many grasses), biennial and perennial rosette plants, and perennial stoloniferous herbs. (Fig. II.4.01c) (B & C)
.004 Perennating bud: an apical shoot meristem that is protected by dead leaves for winter survival of a biennial. (B & C)
.005 Phanerophyte: a plant in which the resting buds are located more than 50 cm above the soil surface, as in the larger woody trees and shrubs, most climbers, and many upright herbs and epiphytes of the humid tropics. (Fig. II.4.02a) (B & C)
.006 Therophyte: annual plant, whose embryos are protected from adverse conditions within resistant seeds. (Fig. II.4.02b) (B & C)
.007 Tropophyte: plant adapted to a seasonal climate. (Fig. II.4.01a,b) (B & C)
.008 Turion: scaly, often thick and fleshy, detached winter bud, by means of which water plants survive the winter; also the term commonly used to describe the production of adventitious shoot buds, as in Rubus. (TA)

II.4.2. Habitats

.000 Habitat: the type of locality and environmental conditions in which a plant occurs and is manifested in genotypic or phenotypic modifications (II.4.3. and II.4.4).
.001 Air plant: (see Epiphytes, II.4.2.006).
.002 Aquatic: growing in water; living in or frequenting the margins of water. (Fig. II.4.02c) (TA)
.003 Brackish: said of water with a high concentration of dissolved substances, usually somewhat salty as in streams subject to tide action. (TA)
.004 Catena: a chain or series of things connected together; a concept first utilized by Milne in Africa to describe the series of soil types developed at the top, middle, and base of a slope from the same parent rock; that is, the main factors that influence weathering processes are topography and rainfall (including infiltration, runoff, and downhill drainage) which in turn, affect the natural vegetation with a succession of plant associations from the crest to the valley, lake, or floodplain below; the latter are reflected in the choice of crops suitable for culture, depending on where the farm is located. (Fig. II.4.03) (L & M)
.005 Emersed, emergent: raised above and out of the water. (TA)

II.4.2. MORPHOLOGY and ANATOMY

.006 Epiphytes: air plants; plants that grow on others; a few are saprophytes, semiparasites, or parasites (II.4.2.026); two distinct groups in size and form: microepiphytes include mosses, liverworts, lichens, and algae found on leaves, smaller branches, and parts of the trunk or larger branches too smooth, too small, or too shaded and moist for larger plants; macroepiphytes are vascular plants. The former are the only group in temperate and boreal regions; both are a characteristic feature of tropical and warm subtropical forests. Most macroephiphytes are herbaceous perennials; they include many ferns but a limited number of angiosperm families (notably Orchidaceae, Bromeliaceae, Cactaceae, Asclepiadaceae, and Rubiaceae) and a few shrubs (e.g., Vaccinium spp.) and trees (e.g., Schlefflera spp.). Epiphytes have many xeromorphic characteristics, such as numerous adaptations for storage and/or absorption of water and retention of organic matter (e.g., rosetted leaves of bromeliads, pseudobulbs, and the velamen of some orchids, absorptive hairs, and specialized leaf forms). (Fig. II.4.02d) (L & J, B & C)

.007 Floating: floating on the surface of water. (Fig. II.4.02c; .05a) (TA)

.008 Geophyte, geophytic: plant growing in the ground; terrestrial. (Fig. II.4.02a). (H 69)

.009 Glycophyte: a plant that is relatively sensitive to a high concentration of salt; the reverse of a halophyte (II.4.2.010). (Fig. II.4.02a) (B & C)

.010 Halophytes: plants adapted to or tolerant of salt conditions found in two main situations, around salt lakes and pans in arid regions and along the seacoast; typical adaptations include succulent stems and leaves, thick cuticles, hairy stem and leaf surfaces, reduction of leaf surface (curled lamina), sunken stomata, and, in some plants, special excretory glands that exude a concentrated salt solution. Two main types of seashore vegetation are mangrove swamps dominated by mangroves in the tropics and warm subtropics and salt marshes dominated by low herbage in areas in which killing frosts limit growth of the former. Transitional zones from mangrove swamps or salt marshes to mesic types of vegetation may be gradual or abrupt, according to the slope of the land. (Fig. II.4.04a,b) (DV)

.011 Hammock: hardwood and palm forests usually dominated by broadleaved evergreen trees and limited to relatively small areas, growing on high upland to seasonally flooded soils, and containing a great diversity of temperate to tropical species; often a climax forest developed after a series or succession of other stages of vegetation. (Davis, 1943, lists 14 different types in southern Florida alone.) (DV)

.012 Heliophilous: attracted by sunlight; said of heliophytes, plants growing in, tolerating, or requiring full sunlight.

.013 Heliophobe: shade-loving; the opposite of heliophyte or heliophilous.

.014 Helophytes: marsh plants (see Hydrophytes, II.4.2.017; Hygrophytes, II.4.2.019). (Fig. II.4.04a, .04c, .06a)

.015 Hemiepiphytes: plants whose seeds lodge in bark crevices in trees, germinate and extend their stems and roots downward until they reach the ground and become terrestrial trees or shrubs, as in species of Fagraea, Schlefflera, Vaccinium, and Ficus (the last often becoming a "strangler"). (Fig. II.4.04d) (v.ST)

.016 Heterotroph: (see Parasites, II.4.2.026).

II.4.2. MORPHOLOGY and ANATOMY

.017 Hydrophyte, hydrophytic: water-loving; a plant growing in a naturally wet or marshy situation or is tolerant of such conditions. (Fig. II.4.02c, .05a) (TA)
.018 Hydrosere: a plant succession of aquatic and riparian forms in especially nutrient-rich (eutrophic) lakes. (TA)
.019 Hygrophyte: a plant growing in a damp habitat; i.e., a swamp plant. (Fig. II.4.04c)
.020 Immersed: growing under water. (TA)
.021 Macroepiphytes: vascular epiphytes; e.g., ferns and angiosperms. (Fig. II.4.02d, .04d, .10d) (L & J)
.022 Mesophyte, mesophytic: plants growing in a moist or humid situation that is neither excessively wet or dry, as most cultivated crops. (Fig. II.4.01b,c, .02a,b, .05b,c)
.023 Microepiphytes: epiphytic algae, lichens, or bryophytes on branches, limbs, or trunks (epixylous forms) and leaves (epiphyllous forms). (L & J)
.024 Myrmecophyte: plant that provides shelter or food or both to ants that live symbiotically with it. (Fig. II.4.05d) (CR 66, v.ST)
.025 Obligate halophyte: a plant, such as iodinebush, sampfire, and shadscale, that will grow only where the level of salts in the soil is high; (see Halophyte, II.4.2.010). (B & C)
.026 Parasites (heterotroph), semiparasites (semiheterotrophs): plants that obtain all or part of their nutrients from others; parasitic plants have leaves mostly reduced to scales and are devoid of chlorophyll; semiparasitic ones have leaves with chlorophyll; most are terrestrial herbs, undershrubs, shrubs, or small trees but a number are epiphytic (e.g., mistletoes); dodder (Convolvulaceae) and cassytha (Lauraceae) are twiners; (plants with mycorrhizae are technically semiparasitic). (Fig. II.4.05c,f) (HE)
.027 Perhumid: completely or thoroughly humid (virtually saturated atmosphere), as in a tropical rain forest. (L & J)
.028 Salt pan: areas of bare soil around depressions in salt flats (areas of salty soil rarely wet; i.e., not a marsh) in which the salt concentration may be too high for growth of halophytes. (DV)
.029 Saprophytes, Epiphytic: (see Epiphytes, II.4.2.006).
.030 Saprophytes: plants living on dead or decayed matter, such as fungi and certain orchids (e.g., Corallorhiza, Epigonium, and Noettia). (B & C)
.031 Semiheterotroph: (see Parasites, II.4.2.026).
.032 Semiparasite: (see Parasites, II.4.2.026).
.033 Slough: wet, deep mud or mire; a sluggish channel (i.e., slow-moving current of water); a wet or marshy place. (DV)
.034 Steppe: any large tract of arid land characterized by xerophilous (xerophytic) vegetation, found mostly in regions of extreme temperature range and loess (unstratified aeolian deposit, usually calcareous) soil.
.035 Submerged, submersed: growing under water. (TA)
.036 Swale: moist meadowy area lower than the surrounding land. (DV)
.037 Xerophyte: a plant that escapes, resists, avoids, or endures drought; the first three types show resistance by existing as dormant seeds during the dry season, resisting drought by storing water in succulent tissues, or avoiding drought by various anatomical modifications (mainly a deep root system); the last exhibit hardiness by nonsucculent perennials that lose large quantities of water during a drought but are not killed. (Fig. II.4.05e) (S & R, in part)

II.4.3. MORPHOLOGY and ANATOMY

II.4.3. Genotypic modifications

.000 Genotypic modification: a permanent alteration, thus hereditary, in form or structure of plants acquired in the course of adaptation over a long period of evolution; (see Phenotypic modifications, II.4.4.).

.001 Adaptation to seasonal climates: alterations in form or structure to enable plants to survive adverse conditions of cold, alternate cold, and heat, or desiccation by the development of protective coverings for buds, various types of storage organs, nanism, and the like. (HOT, B & C)

.002 Adaptations for gaining light: climbing plants (i.e., vines, lianes, scandent shrubs, and hemiepiphytes) have evolved numerous adaptations, such as twining, aerial roots, tendrils, cirrhose leaf tips, and various hooks, hairs, prickles, thorns, and whips. (CR 66, v.ST)

.003 Aerenchyma: air-conducting tissue in roots or stems, found in many plants in or near water or areas subject to poor aeration; a secondary parenchymatous respiratory tissue or modified periderm (cortical tissue). (Fig. II.3.04f) (ES)

.004 Aerial roots: postively haptotropic (= thigmotropic) adventitious roots, as in Vanilla, black pepper, Hedera spp. and creeping fig. (Fig. II.4.02d, .06b) (B & C)

.005 Antibiosis: antagonistic association between two organisms to the detriment of one; the opposite of symbiosis (II.4.3.067). (RB)

.006 Ascidiform: pitcher- or flask-shaped, an organ or appendage found in some epiphytes. (G 62)

.007 Brachytic: dwarfing (dwarfness) in plants characterized by the shortening of internodes only. (B & C)

.008 Buttress: enlargement of the base of trunks of emergent tropical trees that range from a small spur or swelling to massive structures, partly root, partly stem, reaching as high as 10 m up the stem and much further horizontally, thin and flat to thick, twisted, or anastomose; also found on many subtropical trees (e.g., Taxodium); (see Pneumorhizae, II.4.3.039). (Fig. II.4.10a,b) (L & J)

.009 Carnivorous plants: members of about 450 species in six families with various contrivances for detaining and capturing insects and other small animals for food, like pitfalls (passive traps), lobster pots, flypaper traps (passive or active traps), steel traps (active), and mouse traps. (Fig. II.4.06a) (LL)

.010 Cirrhose leaf tips: tendrillike extensions of the leaf, as in Gloriosa lilies and Flagellariaceae. (Fig. II.4.06d) (WE)

.011 Cirriferous: bearing a cirrus (II.4.3.012). (Fig. II.4.09e) (CR 66)

.012 Cirrus: the prolonged end of a leaf axis (rachis) modified as a whip used in climbing, as in certain Calamus spp. (rattans). (Fig. II.4.09e) (CR 66)

.013 Cladophyll: reduction of leaves to small scales or spines with branches becoming leaflike, as in Asparagus, Ruscus; (see also Phylloclad, II.4.3.036). (Fig. II.1.1.3.06a) (L 51)

.014 Clinging: vines growing upward and supported by (1) twining about an axis, (2) tendrils, (3) cirrhous leaf tips, (4) twisting petioles, or (5) aerial roots. Examples of (1) are morning glory, bryonia, and wisteria; of (2), grape and passion fruits; of (3), gloriosa lily; and of (5), black pepper, monstera, and vanilla; certain palms (rattans) climb by (1) or (4) as aided by hooks and spines. (Fig. II.4.06b-d, .09d-g) (B & C)

II.4.3. MORPHOLOGY and ANATOMY

.015 Crown: more or less swollen bases of stems at the point of junction with the roots, as in daylily and peony, and capable of separation from the original (mother) plant; (see V.1.4.024). (H & K)

.016 "Cushion" plant: the form of plants at high latitudes (i.e., above about 50° N or S) or altitude (above 3000-4000 m); becoming lower in stature and more or less ball-like in shape, with smaller leaves and shorter internodes or rosetted in many herbaceous perennials, as the climate becomes more rigorous. (Fig. II.4.01a) (B & C, v.ST)

.017 "Drip-tip": leaves with long (caudate) tips that are usually curved downward and serve as a means of rapid removal of excess moisture; characteristically found on plants in forest undergrowth, either on those that are inhabitants of the understory or on emergent seedlings of the upper story as they pass through this zone, in which there is a permanent or temporary excess of moisture (and usually heavy shade as well): noted on some Ficus spp. used as foliage plants and F. religiosa, as a street tree. (Fig. II.4.07, .08a) (L & J, RI)

.018 Ephemerals: many wild herbaceous annuals and perennials remain quiescent until a rain occurs; they then burst into growth and complete their seed-to-seed cycle (annuals) or flower, produce seed, and form new resting buds (perennials) within the space of a few weeks. (L 51)

.019 Epiphyllous: growing on leaves, like certain microepiphytes (II.4.2.023) (L & J)

.020 Epixylous: growing on branches, limbs, or trunks, like certain microepiphytes (II.4.2.023). (L & J)

.021 Flagelliferous: bearing a flagellum (II.4.3.022). (Fig. II.4.09d,f) (CR 66)

.022 Flagellum: a sterile unbranched inflorescesce produced as a spine-bearing whip used in climbing, as in certain Calamus spp. (rattans). (Fig. II.4.09d,f) (CR 66)

.023 Growth movements: two main types as adaptations to seasonal climates, the extension of plagiotropic axes, as in the diagonally downward growth of Gloriosa lily tubers, and contractile roots, as in lilies, crocus, and gladiolus. (Fig. II.3.17b,c) (B & C)

.024 Haustorium: three different meanings: (1) a specialized outgrowth of the stem or root of a parasitic plant that serves as an organ for the absorption of nutrients from living cells of its host; (2) the cotyledon of certain monocots (e.g., palms) that expands into the seed cavity and serves as an organ for the absorption of nutrients from the endosperm; (3) absorbing organ of plant parasitic fungi; (see V.3.3.2.199). [Fig. II.4.05f(1)]

.025 Heteroblasty: the property of a plant that produces seeds with several different germination requirements; a mechanism by which the chances of survival of at least some of the progeny are increased. (H & K)

.026 High moisture-humidity conditions in tropical forests: two main adaptations, "drip-tip" leaves (i.e., leaves with cuspidate caudate or apices) on understory and seedlings of canopy plants and linear-lanceolate leaves on riverside trees. (Fig. II.4.07, .08a) (L & J, RI)

.027 Hooks, hairs, prickles, thorns, whips: hooklike side shoots (Solanum dulcamara), stiff hairs (Galium, Humulus), prickles (Rosa, Rubus), and thorns (Bougainvillea); hooks, spine-hooks (some Caesalpinia, Acacia, Smilax, etc.); hooks, spine-hooks, leaflet-whips, leaf-end (cirrous) whips, leaf-sheath (flagella)

II.4.3. MORPHOLOGY and ANATOMY

whips (rattans in Palmae); all of these are used for support and in many instances as protection from predators. [As Corner (1966) so forcefully inferred: Woe be the traveller, other than an elephant or rhinoceros, who unwittingly becomes entangled in a rattan while walking through a Malayan (Malayasian) jungle!.] (Fig. II.4.09) (CR 66)

.028 Hot zone plants: adaptations to reduce excessive light or transpiration in plants that grow in areas with seasonally high temperatures include small, narrow leaves (e.g., Olea and Myrtus) or leaflets (many legumes), nyctinastic drooping of leaves (Acacia), thick felty mats of white hairs (Leucadendron), thick-walled epidermal cells with a lacquerlike cuticle (Laurus), and pulvini at the bases of leaves of leaflets (Theobroma, other Sterculiaceae, and many palms and legumes). (Note that these adaptations are similar in many respects to those found in xerophytes, II.4.2. 037.) (Fig. II.4.08b) (B & C)

.029 Knee-root: small loops of root that leave the soil and reenter it a short distance away; (see Pneumorhizae, II.4.3.039). (L & J)

.030 Laminar rolling: temporary (some grasses) or permanent (certain Ericaceae) rolling of leaf blades of xerophytic plants as moisture is lost. (B & C)

.031 Leaf, leaflet, leaf-end, leafsheath whips: (see Hooks, hairs, prickles, thorns, whips, II.4.3.027).

.032 Leaf modifications: sclerophylls (e.g., Laurus, Myrtus, and Olea; small, evergreen, frequently coriaceous, low sap content), various types of toughening and thickening of the cuticle, cutinization of outer epidermal walls, multilayered epidermis, subepidermal sclerotic layers, sunken stomata; permanent (certain Ericaceae) or temporary (some grasses) rolling of lamina; vertical position of falcate leaves ("shadeless forests" of Eucalyptus spp.); unifacial leaves (Tamaricaceae, Cycas, etc.); substitution of phyllodes (leaflike petioles, Acacia); reduction to scales (many families, e.g., Cupressaceae and Casuarinaceae); fugacious (Euphorbiaceae and some Asclepiadaceae). (Fig. II.4.08b, II.1.2. 04) (B & C)

.033 Leaf succulent: a type of xerophytic plant with thick, fleshy leaves filled with water-holding cells. (Fig. II.1.2.07a) (B & C)

.034 Nanism: more circumscribed branches, shorter axes, and a reduction in number and size of leaves, the dwarfing becoming more pronounced as precipitation decreases, a situation analogous to that found in plants at higher latitudes or altitudes. (Fig. II.4.01a) (v.ST, B & C)

.035 Peg-root: small vertical, pencil-size roots to 30 cm arising from a horizontal subterranean root; common in both saltwater and freshwater swamps all over the tropics; (see Pneumorhizae, II.4.3. 039). (Fig. II.4.11e,f) (L & J)

.036 Phylloclad: branches becoming flattened and assuming the function of leaves, as in Opuntia, Phyllocactus, and Muehlenbeckia platyclados. (Fig. II.4.05e) (L 51)

.037 Phyllode: leaflike petiole, as in Acacia; an adaptation of many species found in xerophytic habitats. (Fig. II.1.2.10c) (L & M)

.038 Pneumatothode: similar to peg-roots but only the apex protrudes above the soil surface; a loose powdery cortex forms just below the tip. (Fig. II.4.10c) (HT, L & J)

.039 Pneumorhizae: literally, air-breathing roots; i.e., roots of tropical (and subtropical) plants modified under conditions of poor drainage (e.g., from underlying hardpan or waterlogged sites) to

II.4.3. MORPHOLOGY and ANATOMY

assist in gas exchange; many different configurations, such as aerial stilt-roots (e.g., Pandanus, certain palms, and trees on waterlogged sites), buttresses (characteristic of large emergent trees), knee-roots (Symphonia, Mitragyna spp.), serial knee-roots (Brugiera, Ceriops spp.), peg-roots (Avicennia, Sonneratia, Raphia, Ancistrophyllum, Voacanga, and Anthocleista), stilted peg-roots (Xylopia staridtii), root-knees (Taxodium), and pneumatothodes (Phoenix spp., Elaeis, Laguncularia racemosa). (Fig. II.4.10c,11) (L & J)

.040 Prickle: (see Hooks, hairs, prickles, thorns, whips, II.4.3.027).

.041 Protection of resting buds: several distinct means, such as scales for dormant buds of the temperate zone and many woody tropical plants and stipules (e.g., Moraceae and Magnoliaceae), leaf primordia (e.g., Viburnum spp.), and resinous exudations for a great number of plants in warm climates. (HOT)

.042 Protective devices (against predators): armature in the form of spines (thorns), as on honey locust, paniala (Flacourtia cataphracta), natal plum (Carissa grandiflora), many palms, many Prunus spp., and most citrus species; prickles, as on most bromeliads, roses, brambles, and hollies; stinging or irriating hairs, as on Urticaceae, many Ficus spp., many grasses; ant plants (myrmecophytes, i.e., plants harboring ant colonies). (Note that many climbing blants utilize various forms of armature as adaptations for gaining light; armature is also a characteristic feature of many woody xerophytes.) (Fig. II.4.09)

.043 Pseudobulb: (see V.1.4.058).

.044 Pungent: terminating gradually in a hard sharp point, like a spine or prickle. (Fig. II.4.09a) (G 62)

.045 Recurved: curved backward, as in prickles on pineapple or Corypha elata leaves. (Fig. II.4.09b) (G 62)

.046 Rhizome: (see V.1.4.060).

.047 Root-knee: a large projection from a horizontal subterranean root that results from local intense cambial activity on the top side of the latter; e.g., bald cypress (Taxodium) (see Pneumorhizae, II.4.3.039). (Fig. II.4.11d) (L & J)

.048 Rootstock: (= Crown, II.4.3.015).

.049 Root tuber (tuberous root): a root modified as a storage organ, as in dahlia, sweet potato, some terrestrial orchids, cassava, and Dioscorea spp. (Fig. II.4.13a) (G 62)

.050 Sclerophyll: a type of leaf in which there is excessive development of sclerenchyma (i.e., thickened and lignified cell walls), characteristic of many hot climate or xerophytic plants (e.g., Nerium, Myrtus, Laurus, and Olea). (Fig. II.4.08b) (B & C)

.051 Serial knee-roots: similar to knee-roots but curving in and out of the soil; (see Pneumorhizae, II.4.3.039). (Fig. II.4.11c) (L & J)

.052 "Shadeless forest": vertical position of falcate leaves of Eucalyptus spp. under conditions of bright sunshine. (B & C)

.053 "Sinker": outgrowth of the haustorium in parasitic plants that extends into host tissues. [Fig. II.4.05f(2)] (HE)

.054 Spine: a modified stem with a sharp point; found on many woody xerophytes and other plants as protection against grazing animals. (Fig. II.4.09) (G 62)

.055 Spine-hooks: (see Hooks, hairs, prickles, thorns, whips, II.4.3.027).

.056 Stem succulent: type of xerophytic plant with an enlarged stem filled with water-holding cells. (B & C, S & R)

.057 Stem tubers: several distinct types: derived from the hypocotyl, as in Cyclamen, some radishes, or garden beet; from the stem

II.4.3. MORPHOLOGY and ANATOMY

proper, as in kohlrabi; from plagiotropic subterranean stems (stolons), as in potato; or from orthotropic subterranean stems (corms), as in <u>Ranunculus</u> spp., <u>Colchium</u>, <u>Crocus</u>, <u>Caladium</u>, Jerusalem artichoke, and gladiolus. [Fig. II.4.12a,b, .13b,c, f(2)] (B & C)

.058 Stenophyllous: = narrow leaved, such as the foliage on many plants near a stream. (Fig. II.4.08a) (L & J)

.059 Stilted peg-roots: similar to peg-roots except reaching two m and branching; these stilts are in addition to the ordinary adventitiously formed stilt-roots; (see Pneumorhizae, II.4.3.039). (Fig. II.4.11e) (L & J)

.060 Stilt-roots: adventitious roots arising from the lower part of the trunk to form a network of aerial stilts and weak underground roots; characteristic of trees that grow on waterlogged sites; the taproot is poorly developed or absent; (see Pneumorhizae, II.4.3.039). (Fig. II.4.11a) (L & J)

.061 Stolon: a plagiotropic stem usually above but near the surface of the ground; sometimes forms roots spontaneously; sometimes subterranean, as in the tuber-terminated stems of potato (<u>Solanum tuberosum</u>). (Fig. II.4.12a,b.d) (B & C)

.062 Stoloniferous: a plant with leafless runners whose tips root (note that "stolon" is often used as a synonym for the slender grass type of rhizome). (Fig. II.4.12d) (H & K)

.063 Storage organs: several types of modified stem, root, or combinations thereof with reserves of starch, inulin, oil, etc., such as rootstocks (crowns), fleshy rhizomes, stem tubers, bulbs, root tubers, taproot, taproot plus hypocotyl, and taproot plus hypocotyl plus stem base. (Fig. II.4.12, .13) (B & C)

.064 "Strangler": a plant that is initially an epiphyte but sends down roots that reach the ground and whose ramifications eventually overgrow and strangle its host, like certain species of <u>Ficus</u>, <u>Schlefflera</u>, and <u>Clusia</u>. (Fig. II.4.10d) (L & J)

.065 Succulent: development of water-storage tissue and the consequent thick fleshy appearance of leaves, as in <u>Aloe</u>, <u>Sedum</u>, <u>Sempervivum</u>, <u>Lithops</u>, and <u>Agave</u>; stems, as in <u>Cactaceae</u>, some <u>Euphorbia</u> spp., some <u>Stapelia</u> spp., <u>Cissus</u>, <u>Kleinia</u>, and baobab or kapok (Bombacaceae); or roots, as in <u>Umbelliferae</u>, <u>Cucurbitaceae</u>, <u>Compositae</u>, <u>Asclepiadaceae</u>, <u>Pelargonium</u>, and <u>Oxalis</u>. (Figs. II.1.2.07a; II.4.05e, .10b) (B & C)

.066 Symbiont: an organism living in symbiosis. (RB)

.067 Symbiosis: intimate association of two living organisms (see also Mycorrhiza, V.3.3.2.264). (RB)

.067a. Antipathetic: parasitism.

.067b. Conjunctive: more or less complete union, as in lichens, mycorrhizal associations, and rhizobial associations.

.067c. Disjunctive: no actual union, as in the cases of ant-myrmecophyte associations or epiphytes.

.068 Taproot plus hycocotyl plus stem storage: enlargement of the taproot, hypocotyl, and the basal part of the stem as a storage organ, as in rutabaga and celeriac. (B & C, G 62)

.069 Taproot plus hypocotyl storage: enlarged taproot plus hypocotyl as a storage organ, as in mangold (forage beet) and garden radish. [Fig. II.4.02b, .13f(3)] (B & C, G 62)

.070 Taproot storage: enlarged taproot as a storage organ, as in carrot, parsnip, and sugarbeet. [Fig. II.4.13f(1)] (B & C, G 62)

.071 Tendrils: positively haptotropic and often nutational movements of modified shoot axes (<u>Vitis vinifera</u>, Virginia creeper), axillary

II.4.3. MORPHOLOGY and ANATOMY

unbranched side shoots (Passifloraceae), leaf midribs (Cucurbitaceae), leaf or leaflet lamina (Pisum, Lathyrus, Bignoniaceae, and Cucurbitaceae), petioles (Tropaeolum), main and secondary leaf rachises (Clematis), stipules (Smilax), or inflorescences, especially the rachis (Vitis spp.). (Fig. II.4.06c) (B & C)

.072 Thorn: (see Hooks, hairs, prickles, thorns, whips, II.4.3.027).
.073 Tuberous root: (see Root tuber, II.4.3.049).
.074 Twining: winding (nutational) movements of stems, as in members of Convolvulaceae, Apocynaceae, Araliaceae, Bignoniaceae, and Dioscoreaceae. (Fig. II.4.05c) (B & C)
.075 Whip: (see Hooks, hairs, prickles, thorns, whips, II.4.3.027).
.076 Xeromorphic: a plant with adaptations to limit water loss and/or water storage, as in many epiphytes (II.4.2.006).

II.4.4. Phenotypic modifications (v. ST)

.000 Phenotypic modifications, phenotypic morphoses: responses of individual plants to environmental conditions, such as climate, soil, exposure, altitude, temperature, wind, fire, and living organisms (including man). Changed characters are not inherited, although the manner in which a plant reacts to environmental conditions is. Only experimental breeding can determine whether a particular environmental condition has induced the segregation of a specialized race, hence a separate genotype. Modifications have been separated by van Steenis (1948, 1957) into the following categories of biotic, climatic, edaphic, and intrinsic factors.
.001 Adaptation: the reaction of plants to environmental conditions (i.e., soil, site, and climate); may be transitory (phenotypic) or permanent (genotypic) (see Adaptation, V.3.3.3.004).
.002 Biotic factors:
.002a. Anthropomorphosis: pioneer plants and savanna trees; revegetation of areas devastated by landslides and forest fires; includes many forest trees which appear as shrubs that flower and fruit when only a few centimeters high; trees in a forest often lack branches up to 20-30 m and have a more or less columnar crown, whereas the same species in a savanna, or in an open park, generally have branches from the ground up and a nearly globose crown.
.002b. Cecidiomorphosis: (see Zoomorphosis, II.4.4.002h).
.002c. Myrmecomorphosis: ant plants, as invaders of Myrmecodia and Hydnophytum tubers or hollowing out pith or leaf sheaths of Endospermum, Wightia, and a number of palms; a type of symbiosis in many instances because the ants repel other invaders. (Fig. II.4.05d)
.002d. Pascuomorphism: (see Zoomorphosis, II.4.4.002h).
.002e. Phytomorphosis: fungus and bacterial diseases, the former often causing malformations in many plants and the latter having a symbiotic effect on leaves of certain species of Pavetta, Psychotria, Ardisia, and Smilax or in nodules on some Rubiaceae.
.002f. Pseudonanism: (see Zoomorphosis, II.4.4.002h).
.002g. Pyromorphosis: influence of fire; stunting and early flowering; also development of a swollen half-buried "lignotuber" in many plants. (Fig. II.4.14)

II.4.4. MORPHOLOGY and ANATOMY

.002h.   Zoomorphosis: insect galls mimicking actual plant structures (cecidiomorphosis) and pseudonanism (i.e., low stature and early flowering) from browsing of animals (pascuomorphosis).
.003 Climatic factors:
.003a.   Anemomorphosis: wind forms; peculiar oblique condensed, one-sided habit, and smaller leaves.
.003b.   Horamorphosis: seasonal variation; "out-of-season" bloom (second flowering in plants that normally bloom only once a year) with flowers half normal size and paler color; dwarfed seedlings with atypical immature foliage; certain perennials and biennials that act as annuals in tropical regions.
.003c.   Hygromorphosis: influence of drought; reduction in peduncle length or leaf size and flowering of certain forest trees or bamboos.
.003d.   Hypselomorphosis: phenotypic effect of altitude; plants characterized by reduced foliage, roundish leaves, recurved margins, reduced petioles, and more compact habit.
.003e.   Photomorphosis: effects of light, manifested in two different ways: (1) epiphytes that withstand dry conditions but need rather high atmospheric humidity and light, hence grow in exposed locations on rocks, old lava streams, landslides, poor sandy soils, and the like, but are more rigid and condensed, often fastigiate or without a woody base (as in some _Vaccinium_ spp.); (2) shade forms that have larger, thinner leaves, longer internodes, slenderer habit, and sometimes marked variations in leaf coloration and form.
.004   "Cushion plant": a form characteristic of high altitude (or latitude) in which the shoots are profusely branched, contracted into rosettes, and packed together to form firm cushions which remain close to the ground or are hemispherical in shape; the high altitude form may not remain so when grown at a lower elevation, a phenotypic rather than a genotypic modification.
.005 Edaphic factors:
.005a.   Edaphomorphosis: fumarole (II.4.4.006) and solfatara (II.4.4.011); the former enables lowland plants to grow at high altitudes (i.e., 3000 m or more), although much reduced in size, with a condensed habit and diminutive leaves; the latter produces one-sided growth that will revert to normal if the vent shifts and the fumes no longer blow on the plants.
.005b.   Hydromorphosis: water and swamp plants; water plants develop conspicuously different land forms as the level recedes; many herbaceous swamp plants bloom only when the water level drops; many shrubs and trees develop stilt roots and others have a more or less prominently swollen base in deep swamps.
.006 Fumarole: a volcanic steam vent.
.007 Intrinsic factors:
.007a.   Ontogenomorphosis: precocious flowering (paedogenesis); dimorphous foliage, flowers, fruits, and seeds, some of which represent juvenile characteristics; dimorphism, however, is common to many mature tropical plants.
.007b.   Paedogenesis: (see Ontogenomorphosis, II.4.4.007a).
.007c.   Teratologomorphosis: teratological (monstrous) forms; invirescentia (green shoots arising from flowers).
.008   Invirescentia: teratological (monstrous) forms of inflorescences, etc., which arise from otherwise normal places as a result of environmental stress.

II.4.4. MORPHOLOGY and ANATOMY

.009 Lignotuber: thickened, partially subterranean lower stem and upper taproot (main root) which results from regular burning of the vegetation. (Fig. II.4.14b)
.010 Pseudonanism: dwarfing of plants by continued grazing by animals or by fire. (Fig. II.4.14b)
.011 Solfatara: a volcanic vent that emits heated poisonous gases.

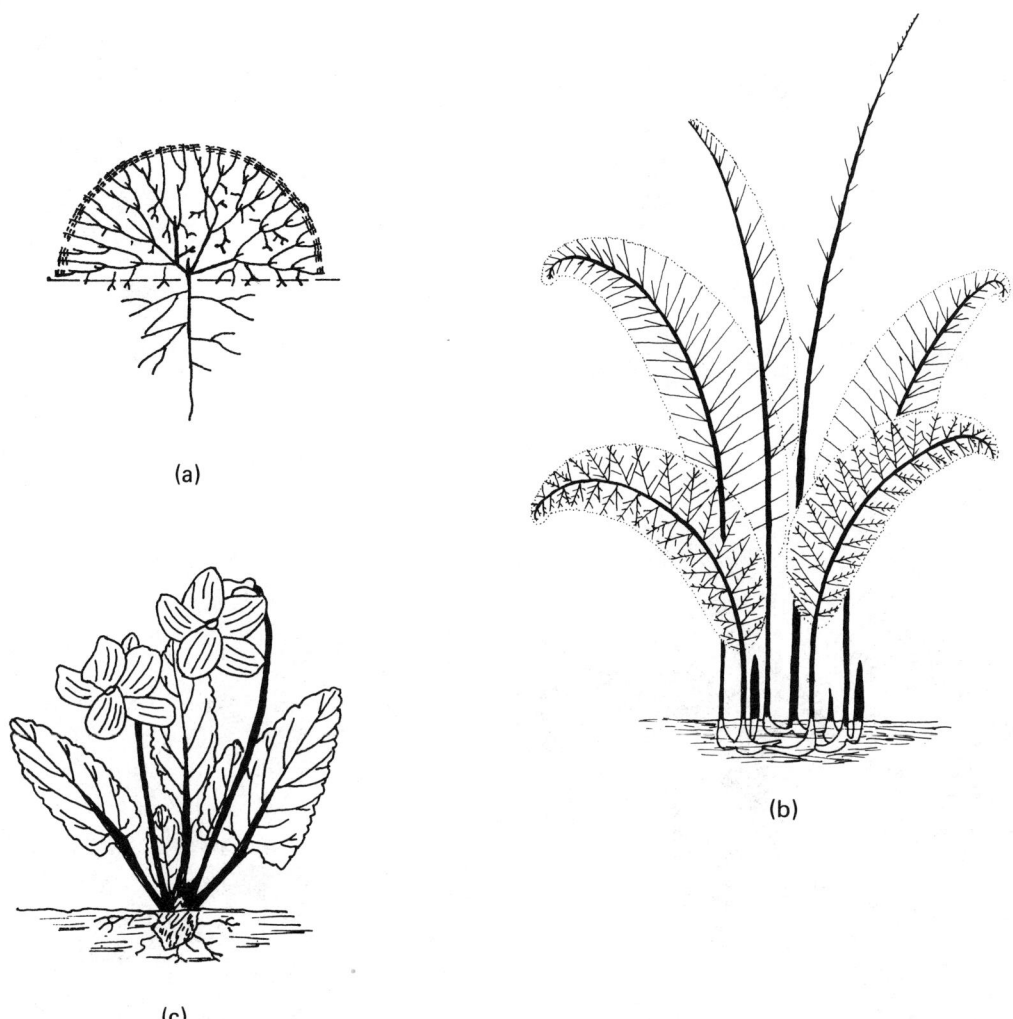

Fig. II.4.01. Life forms: (a) chamaephyte, tropophyte, cushion plant (<u>Azorella selago</u>), ls; (b) cryptophyte, bamboo; (c) hemicryptophyte, violet. (a) Redrawn from Bell and Coombe, 1976; (b) redrawn from Hallé et al., 1978; (c) redrawn from Gray, 1862; see Acknowledgments and Selected References.

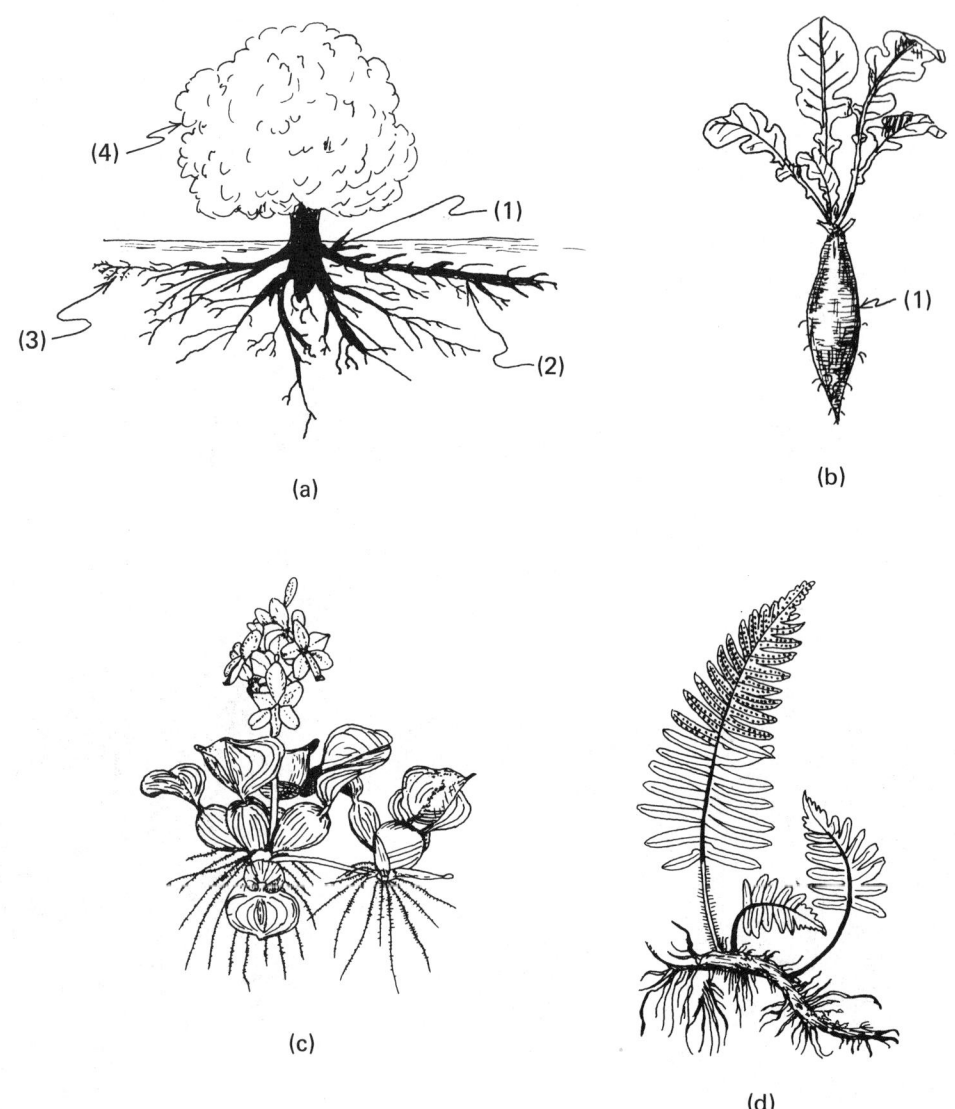

Fig. II.4.02. Life forms: (a) phanerophyte: citrus [(1)-(4) taproot, lateral root, feeder roots, and crown, respectively]; (b) therophyte: radish [(1) swollen storage root]; (c), (d) habitats: (c) aquatic, hydrophyte; water hyacinth (Eichhornia crassipes), a serious pest in lakes and waterways; (d) epiphyte, polypody fern (Polypodium). (a) Redrawn from Praloran, 1972; (b) redrawn from Gray, 1862; (c) redrawn from Lind and Morrison, 1974; (d) redrawn from Johnson 1633; see Acknowledgments and Selected References.

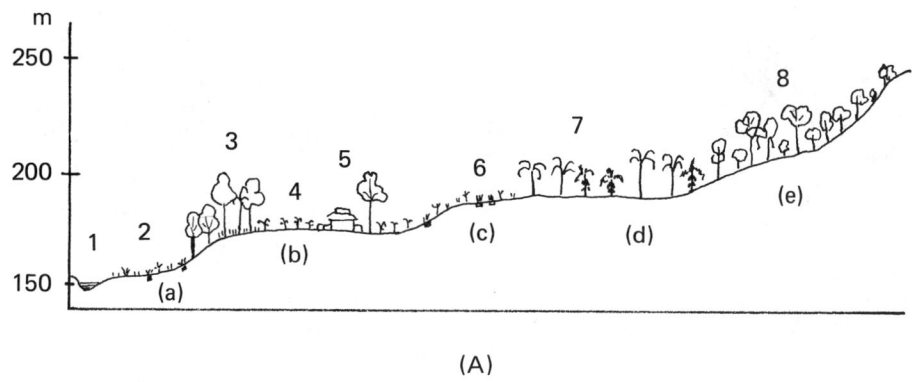

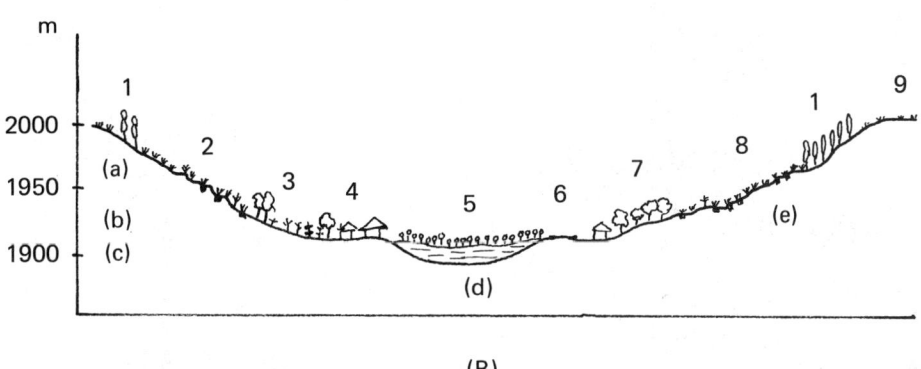

Fig. II.4.03. Catenas: (A) Kigezi, southwest Uganda, with (1) seasonal stream, (2) vegetables, sugar cane, sweet potatoes, (3) steep wooded slope (often rocky), (4) maize, (5) farmhouse, (6) okra, tomatoes, vegetables, (7) cacao, oil palm, (8) secondary forest; (a) valley bottom, (b) thin sandy loam, (c) stony ground, (d) terrace area with deep red loam, (e) thin loam. (B) "Huza hide of land," Kroba district, southeast Ghana, with (1) eucalyptus, (2) field crops cultivated on terraces, (3) tobacco, maize, (4) huts, (5) papyrus swamp with peat formation, (6) road, (7) wattle (Acacia) trees, beehives, (8) sorghum, maize, beans, pigs, etc. (e.g., fallow), (9) sparse grass pasture; (a) thin brown loamy soil; (b) deep red loam; (c) slate and phyllites; (d) vegetables, sweet potatoes, some fishponds in reclaimed swamp; (e) terraced slope with fields. Redrawn from Manshard, 1974; see Acknowledgments and Selected References.

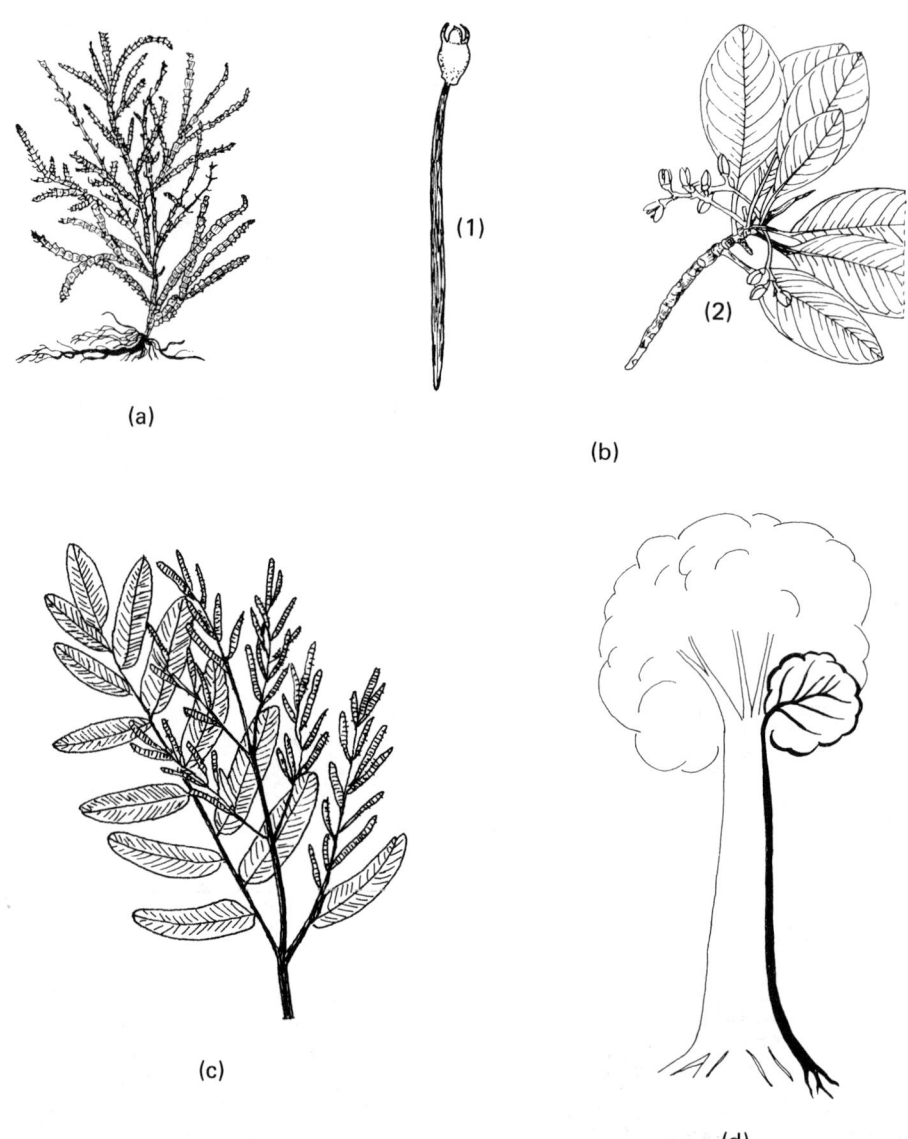

Fig. II.4.04. Habitats. (a), (b) halophytes, (a) saltwort (Salicornia); (b) mangrove (Rhizophora), with (1) germinated (viviparous) seed and (2) branch; (c) helophyte, hygrophyte, royal-fern (Osmunda regalis); (d) hemiepiphyte, Vaccinium laurifolium (arrow) at 1500 m., East Java. (a) Redrawn from Johnson, 1633; (b) redrawn from Lind and Morrison 1974; (c) redrawn from Britton and Brown, 1923; (d) redrawn from van Steenis, 1948; see Acknowledgments and Selected References.

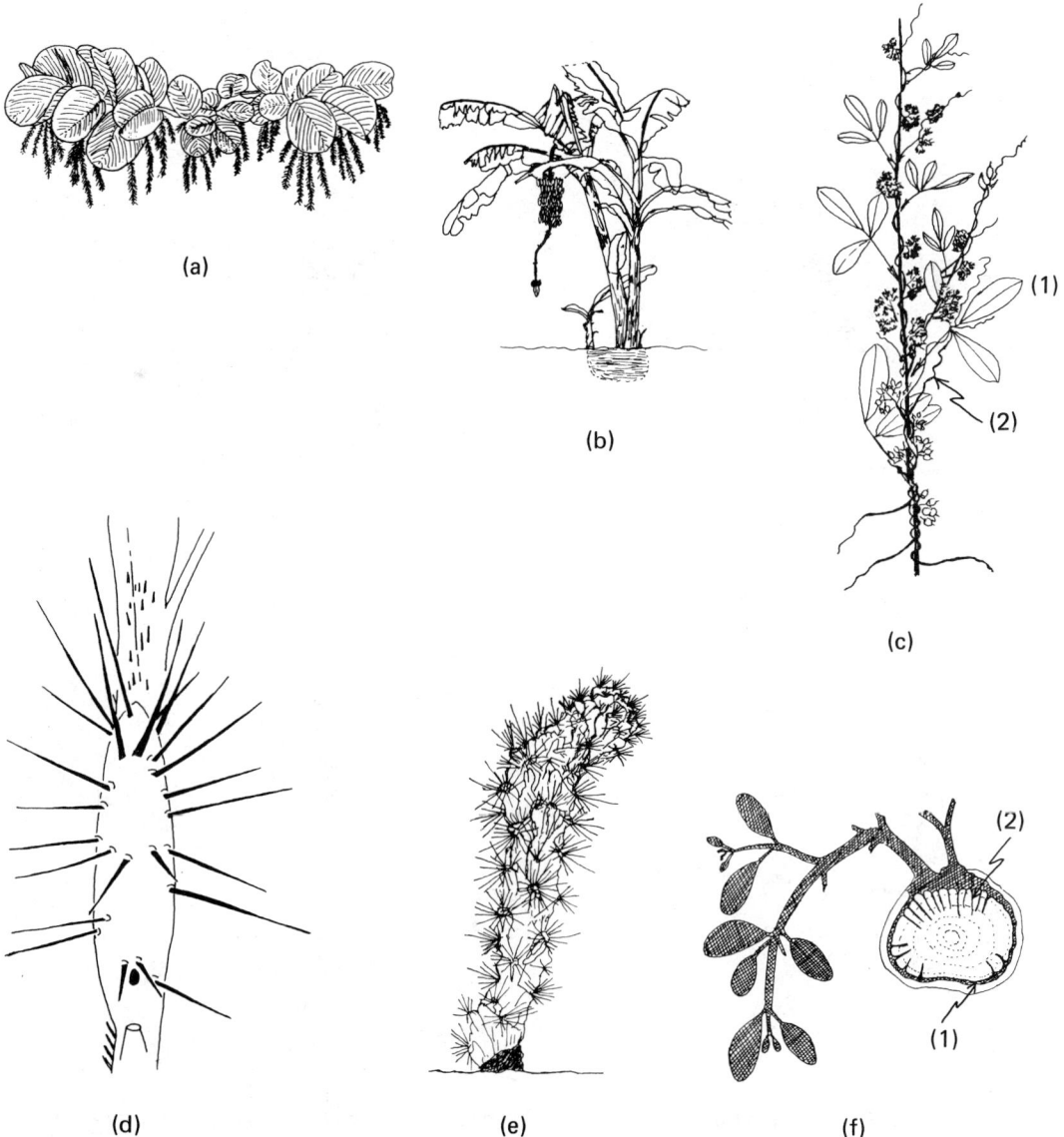

Fig. II.4.05. Habitats: (a) floating, hydrophyte; water fern (<u>Salvinia auriculata</u>; (b), (c) mesophytes, (b) 'Lacatan' banana (<u>Musa</u> sp.); (c) (1) alfalfa, with (2) dodder (parasitic vine); (d) myrmecophyte, <u>Korthalsia ochrea</u> (leaf sheath) with ant hole (arrow); (e) xerophyte, succulent, <u>Opuntia</u>; (f) parasite, mistletoe (epiphyte), with (1) haustorium and (2) "sinkers." (a) Redrawn from Lind and Morrison, 1974; (b) adapted from Tai, 1958; (c) redrawn from Muenscher, 1955; (d) redrawn from Corner, 1966; (e) redrawn from Bailey, 1928; (f) adapted from Heald, 1933; see Acknowledgments and Selected References.

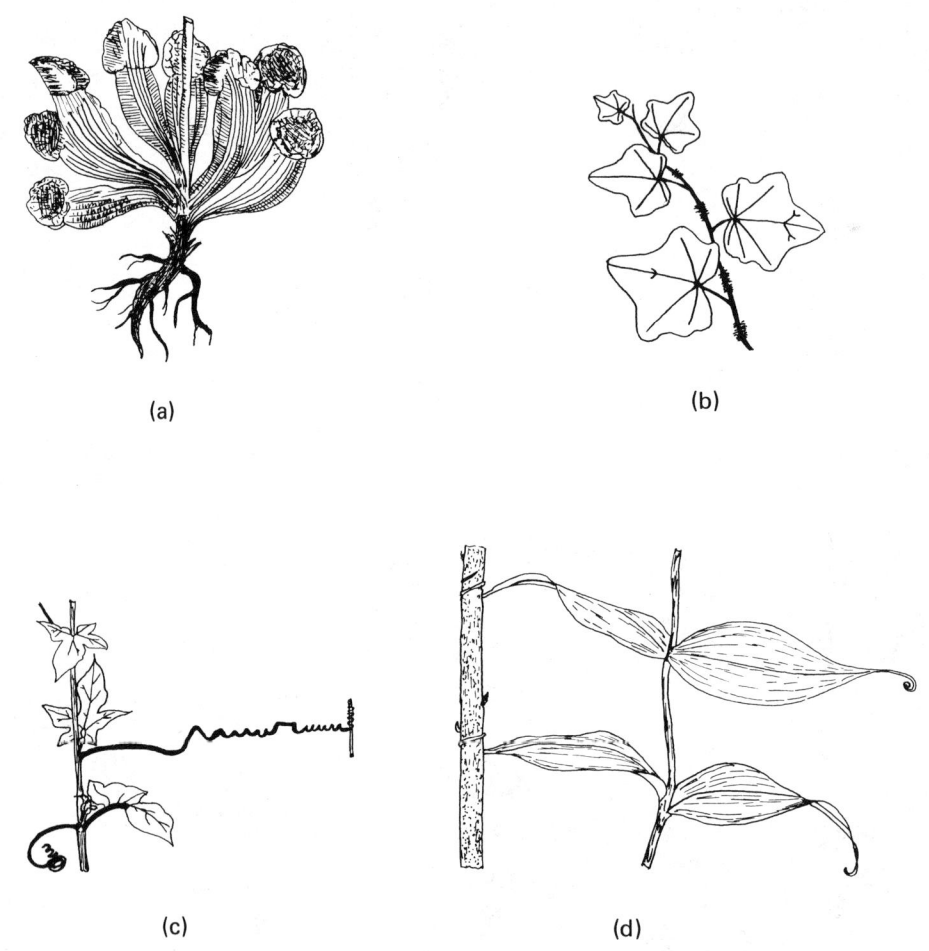

Fig. II.4.06. Modifications: (a) carnivorous, pitcher plant (<u>Sarracenia purpurea</u>, an early drawing of this American plant); (b)-(d) clinging, (b) <u>aerial</u> roots, English ivy; (c) tendrils, <u>Bryonia dioica</u>, with "inversion point" at arrow; (d) cirrhose leaf tips, <u>Gloriosa lily</u>. (a) Redrawn from Johnson, 1633; (b) redrawn from Watkins and Sheehan, 1975; (c) redrawn from Bell and Coombe, 1976; (d) adapted from West, 1957; see Acknowledgments and Selected References.

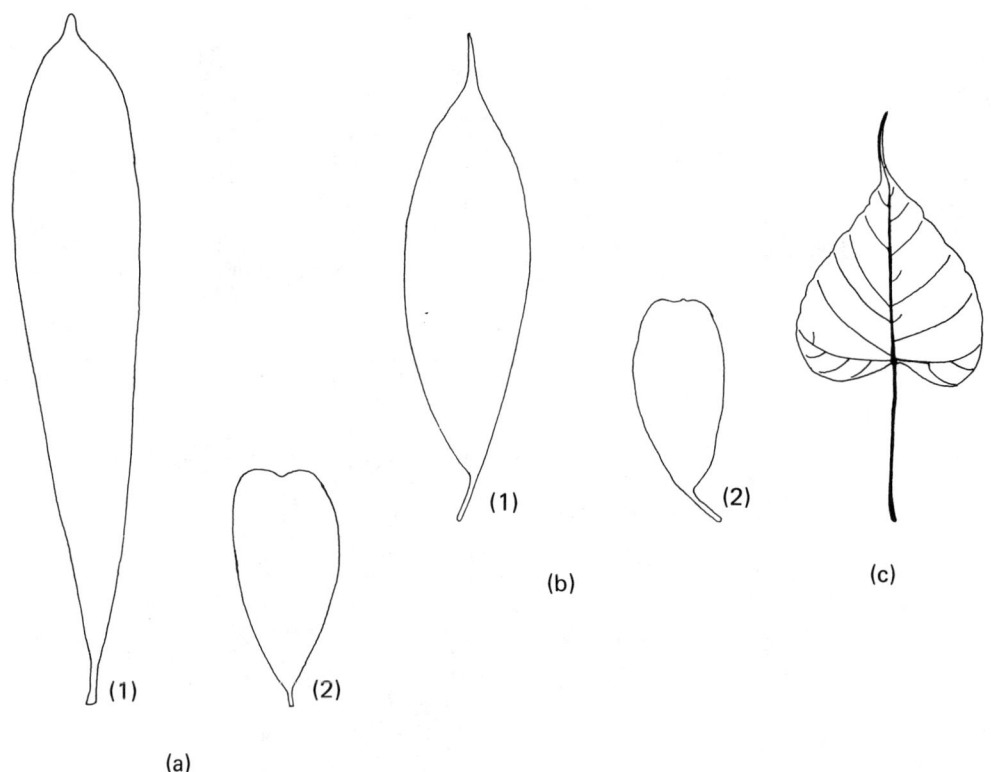

Fig. II.4.07. Modifications: drip-tips: (a) Lophira procera, Nigeria, (1) leaf from juvenile (4m) tree, (2) leaf from mature (30m) tree; (b) Catostemma fragrans, Guyana, (1) from young seedling, (2) from mature tree [(a), (b) about X 1/3]; (c) peepul or Bo tree (Ficus religiosa). (a), (b) Redrawn from Richards, 1952; (c) adapted from Watkins and Sheehan, 1975; see Acknowledgments and Selected References.

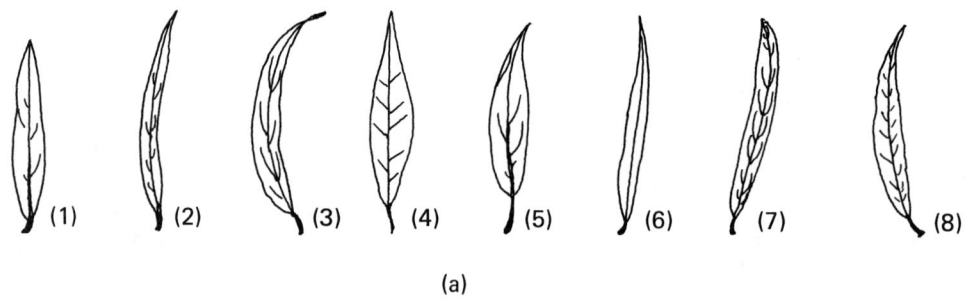

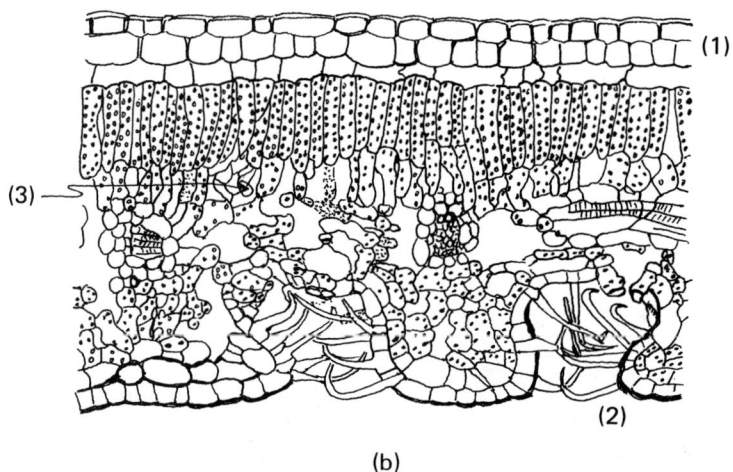

Fig. II.4.08. Modifications: (a) stenophyllous (stream-bank) leaves: (1) Neonauclea angustifolia (Rubiaceae), (2) Fagraea stenophylla (Loganiaceae), (3) Garcinia linearis (Guttiferae), (4) Syzygium neriifolium (Myrtaceae), (5) Saurauia angustifolia (Actinidiaceae), (6) Erycibe longifolia (Convolvulaceae), (7) Homonia riparia (Euphorbiaceae), (8) Psychotria acuminata (Rubiaceae); (b) Sclerophyllous (dry climate) leaf modifications: oleander (Nerium oleander), with (1) heavily cutinized multiple epidermis, (2) deep stomatal crypt lined with trichomes, (3) druse. (a) Redrawn from Merrill, 1945; (b) redrawn from Esau, 1965; see Acknowledgments and Selected References.

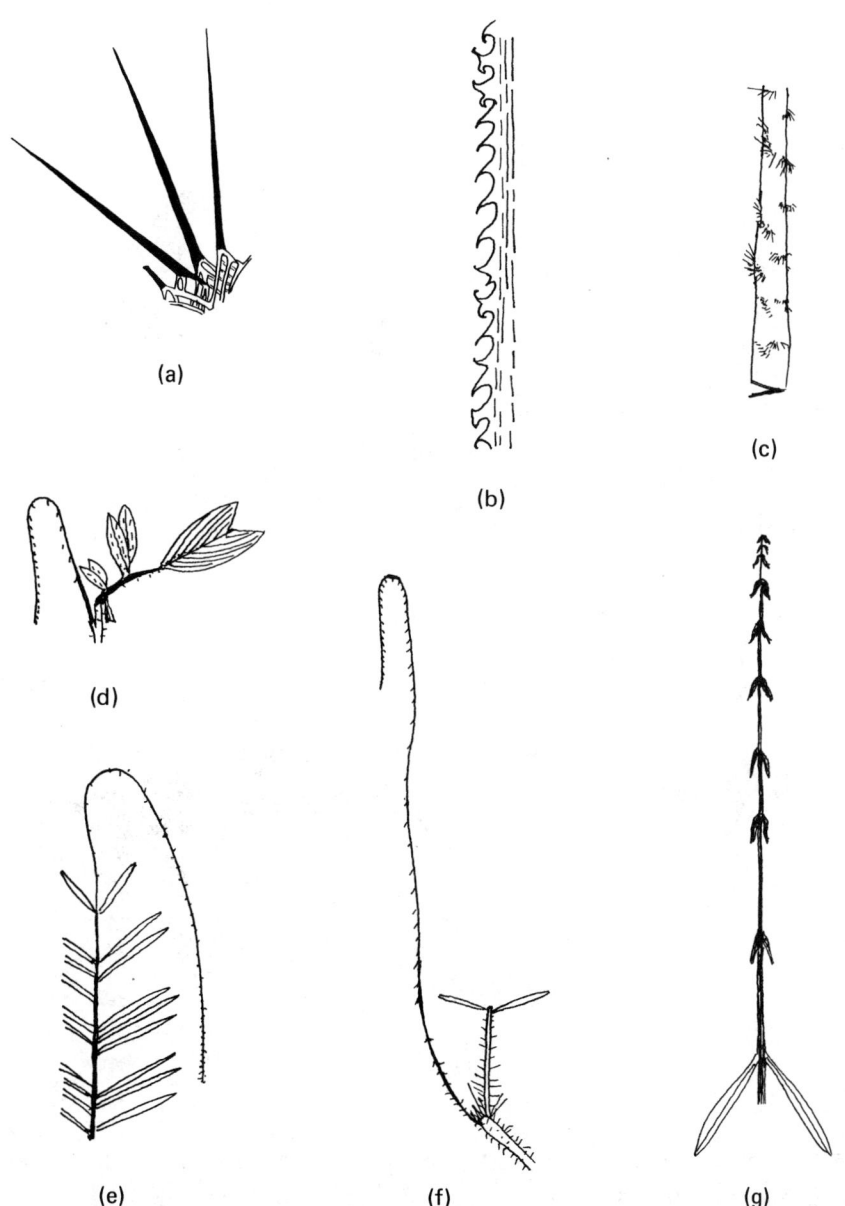

Fig. II.4.09. Modifications. hooks, etc., for climbing and protective devices: (a) fiber thorns of Trithrinax leaf sheath (X 1/3); (b) spiny edge of Corypha elata petiole (X 1/10); (c) combs of spines on Salacca leaf sheath (X 1/8); (d) flagelliferous whip from leaf sheath; (e) cirriferous (apical) leaf whip; (f) flagelliferous leaf sheath whip [(d)-(f) rattans, X 1/40]; (g) Desmoncus (rattan) with deflexed hooks. Redrawn from Corner, 1966; see Acknowledgments and Selected References.

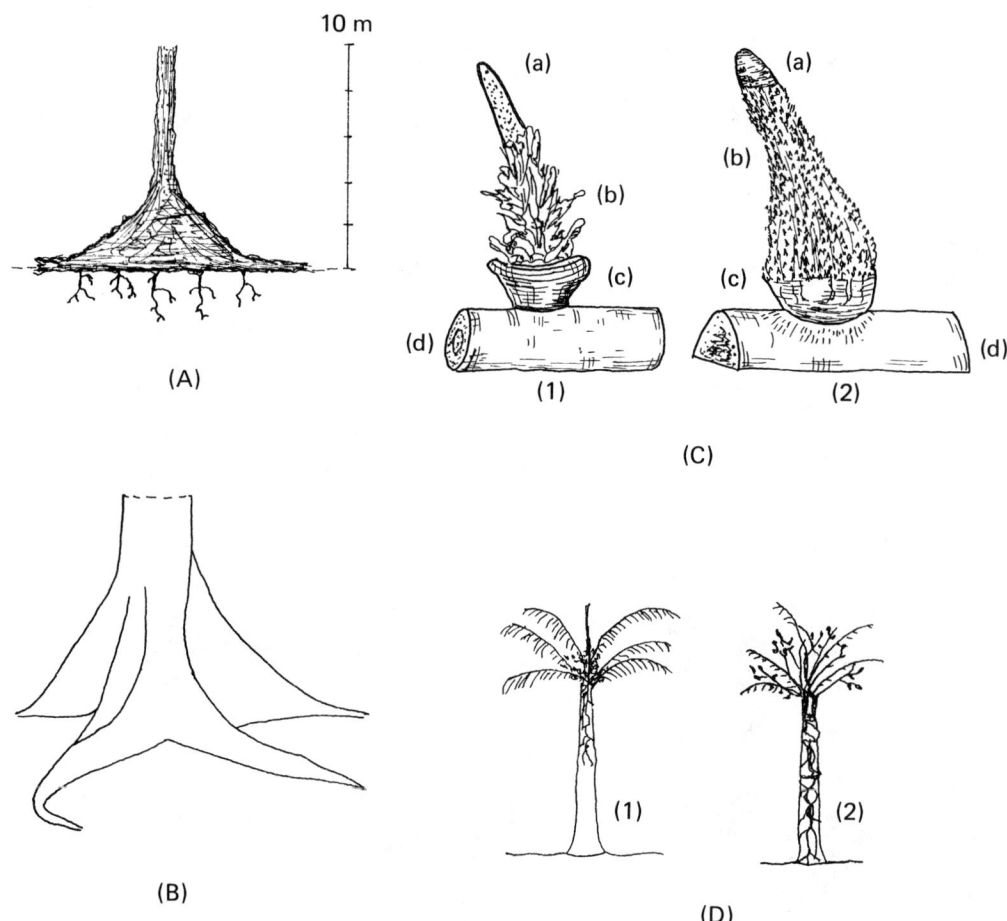

Fig. II.4.10. Modifications: (a), (b) buttresses, (a) <u>Piptideniastrum</u>; (b) <u>Ceiba</u>; (c) pneumatothodes on (1) young underground and (2) old aerial oil palm roots, [(a) stele, (b) cortex, (c) epidermis and hypodermis, (d) parent root]; (d) "strangler", <u>Ficus</u> sp. on oil palm, (1) early envelopment, (2) later stage with palm nearly dead. (a), (d) Redrawn from Longman and Jenik, 1974; (b) adapted from Watkins and Sheehan, 1975; (c) redrawn from Hartley, 1977; see Acknowledgments and Selected References.

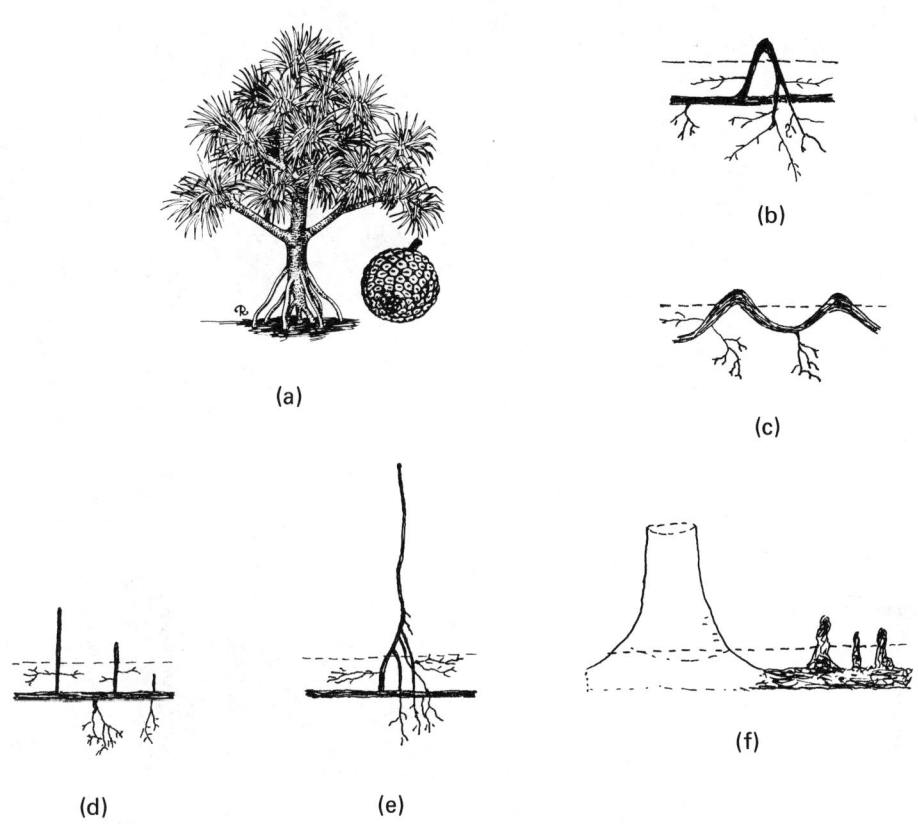

Fig. II.4.11. Modifications: pneumorhizae on hygrophytes: (a) aerial stilt roots, (1) Pandanus, (2) Uapaca sp.; (b) lateral knee-root; (c) serial knee-root; (d) root-knee [Taxodium, note very different structure compared with (b) or (c)]; (e) stilted peg-roots (Xylopia); (f) peg-roots (Avicennia). (Water level indicated by dashed line.) [a(1)] From Watkins and Sheehan, 1975; [a(2)-(f)] redrawn from Longman and Jenik, 1974; see Acknowledgments and Selected References.

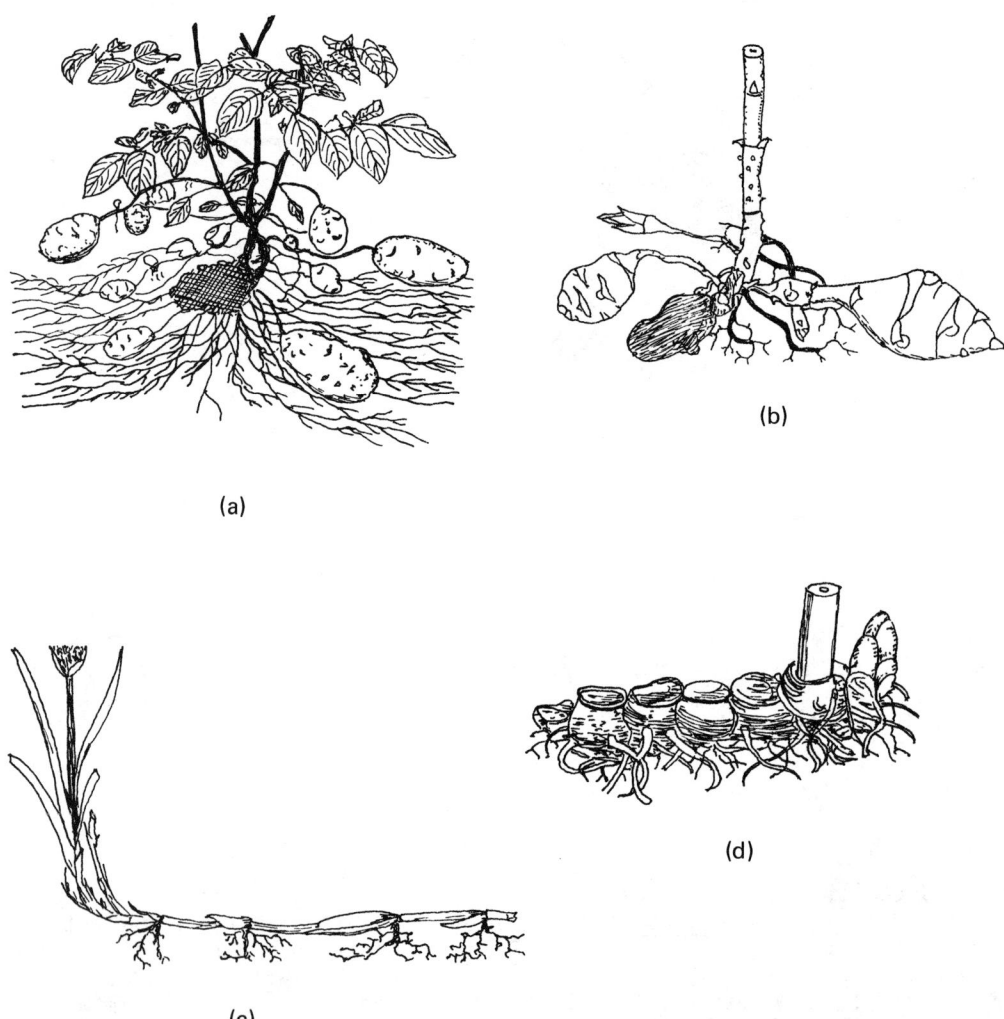

Fig. II.4.12. Storage organs: (a), (b) tubers, (a) potato (<u>Solanum tuberosum</u>); (b) Jerusalem artichoke (<u>Helianthus tuberosus</u>); (c) rhizome (fleshy), <u>Diphyllea cymosa</u>; (d) stolon, <u>Carex arenaria</u> [Note that (a) and (b) (see arrows) also have stolons.] (a) Redrawn from Bell and Coombe, 1976; (b)-(d) redrawn from Gray, 1862; see Acknowledgments and Selected References.

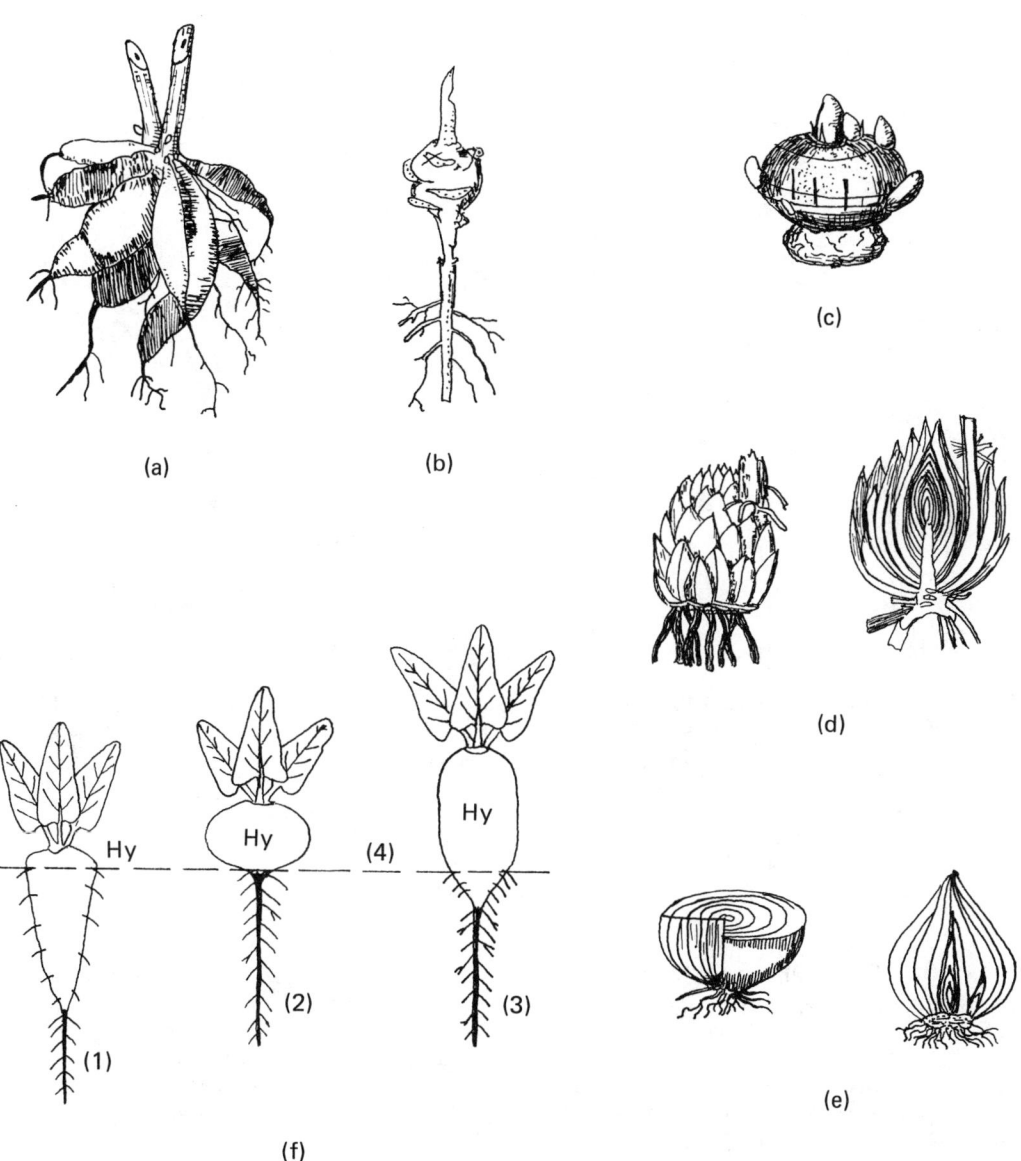

Fig. II.4.13. Storage organs: (a) root tubers (dahlia); (b) stem tuber (kohlrabi); (c) corm (crocus); (d) scaly bulb (lily) with ls at right; (e) tunicate bulb (onion), ts with ls at right; (f) storage "roots" as in (1) sugar beet (mostly root), (2) garden beet (mostly hypocotyl = Hy), (3) mangold (forage beet, enlarged taproot plus hypocotyl), (4) soil level. (a), (c)-(e) Redrawn from Gray, 1862; (b), (f) redrawn from Bell and Coombe, 1976; see Acknowledgments and Selected References.

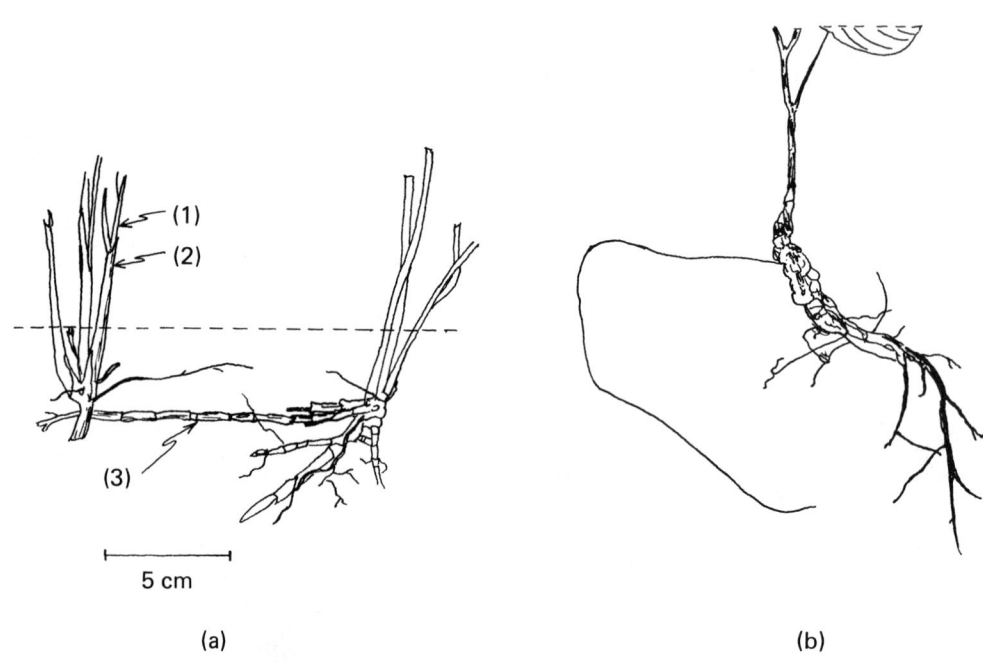

Fig. II.4.14. Phenotypic modifications: pyromorphosis: (a) regrowth after burning of Imperata cylindrica, the infamous cogon or alang-alang grass, with (1) burnt stem, (2) new shoot, and (3) food-storing rhizome; (b) lignotuber (arrow) of three-year-old Butea monosperma seedling developed on fireswept savannas in West Java (X 1/3). (a) Redrawn from Lind and Morrison, 1974; (b) redrawn from van Steenis, 1948; see Acknowledgments and Selected References.

II.5. General Morphological Terms

II.5.1. Color

.000 Color: the appearance of objects determined by the response of the eye to the wavelengths of light coming from these objects; the portion of the electromagnetic spectrum visible to the eye lies between approximately 400 and 700 nm, the ends intergrading into the ultraviolet (UV) below and infrared (IR) above this range; color as perceived by the eye depends in part on wavelength, in part on hue, tone and intensity (chroma), in part on psychological factors, such as illumination of the objects, and the age, emotion, and fatigue of the observer; it is then subjective unless standardized by reference to a color chart or fan, like those of Ridgway, Munsell, Maerz and Paul, and the Royal Horticultural Society, or color tubes used under specified conditions; more precise (objective) measurements for horticultural purposes may by obtained with instruments like the Hunter color difference meter and various types of spectrophotometer.
.000a. Chroma: (see Intensity, II.5.1.000c).
.000b. Hue: the attribute by which one color is distinguishable from another.
.000c. Intensity (chroma): the brilliance of a color, i.e., grayness, ranges from pure to neutral.
.000d. Tone, value: lightness or darkness of a color.
.001 Black, brown: (ST)
.001a. Black (niger, nigrescens): glistening, a little tinged with gray.
.001b. Brown (fuscus): tinged grayish or blackish.
.001c. Cinnamon (cinnamomeus): bright brown mixed with yellow and red.
.001d. Ferrugineus: (see Rusty, II.5.1.001l).
.001e. Fuligineus: (see Sooty, II.5.1.001m).
.001f. Fuscus: (see Brown, II.5.1.001b).
.001g. Hepaticus: (see Liver-colored, II.5.1.001i).
.001h. Ligneous (ligneus): yellowish brown.
.001i. Liver-colored (hepaticus): dull brown with a little yellow.
.001j. Niger, nigrescens: (see Black, II.5.1.001a).
.001k. Rufous (rufus): reddish brown.
.001l. Rusty (ferrugineus): light brown, with a little mixture of red.
.001m. Sooty (fuligineus): dirty (grayish) brown verging on black.
.002 Blue, violet: (ST)
.002a. Azureus: (see Sky blue, II.5.1.002i).
.002b. Blue (caeruleus): lighter and duller than indigo.
.002c. Caeruleus: (see Blue, II.5.1.002b).
.002d. Caesius: (see Lavender-colored, II.5.1.002f).
.002e. Cyaneus, cyano-: (see Prussian blue, II.5.1.002h).
.002f. Lavender-colored (caesius): pale blue with a slight mixture of gray.
.002g. Lilac (lilacinus): pale, dull violet mixed with a little white.
.002h. Prussian blue (cyaneus, cyano-): clear, bright blue.
.002i. Sky blue (azureus): light, pure, lively blue.
.002j. Violet (violaceus): pure blue stained with red.
.003 Gray (grey): a mixture of white and black ranging form nearly white to nearly black. (ST)
.003a Ash-gray (cinereus): mixture of pure white and pure black.

II.5.1. MORPHOLOGY and ANATOMY

.003b. Canescens: (see Hoary, II.5.1.003h).
.003c. Canus: (see Hoary, II.5.1.003h).
.003d. Cinereus: (see Ash-gray, II.5.1.003a).
.003e. Fumeus: (see Smoky, II.5.1.003n).
.003f. Grey: (= Gray, II.5.1.003).
.003g. Griseus: (see Pearl gray, II.5.1.003j).
.003h. Hoary (canus, canescens): grayish white caused by hairs overlying a green surface.
.003i. Lead-colored (plumbeus): gray bordering on blue with some metallic luster.
.003j. Pearl gray (griseus): pure gray verging to blue.
.003k. Plumbeus: (see Lead colored, II.5.1.003i).
.003l. Schistaceus: (see Slate gray, II.5.1.003m).
.003m. Slate-gray (schistaceus): gray bordering on blue.
.003n. Smoky (fumeus): gray changing to brown.
.004 Green: (ST)
.004a. Chloro-: (see Green, II.5.1.004d).
.004b. Elaio: (see Olive green, II.5.1.004e).
.004c. Glaucous, glaucescens: (see Sea green, II.5.1.004f).
.004d. Green (viridis, virescens, chloro-): clear green.
.004e. Olive green (olivaceus, elaio-): mixture of green and brown.
.004f. Sea green (glaucous, glaucescens): dull green passing into grayish blue.
.004g. Virescens: (see Green, II.5.1.004d).
.004h. Viridis: (see Green, II.5.1.004d).
.005 Red, purple: (ST)
.005a. Brick red (lateritus): dull scarlet with a mixture of yellow and gray.
.005b. Brown-red (rubiginosus): dull red with a slight mixture of brown.
.005c. Carmine (puniceus): pure red without admixture.
.005d. Carneus: (see Flesh-colored, II.5.1.005k).
.005e. Cinnabar (cinnabarinus): scarlet with a slight mixture of orange.
.005f. Coccineus: (see Scarlet, II.5.1.005v).
.005g. Coppery (cupreus): brownish red with a metallic luster.
.005h. Cupreus: (see Coppery, II.5.1.005g).
.005i. Erythro-: (see Red, II.5.1.005p).
.005j. Flame-colored (igneus): lively scarlet, fiery red.
.005k. Flesh-colored (carneus): paler than rosy.
.005l. Igneus: (see Flame-colored, II.5.1.005j).
.005m. Lateritus: (see Brick red, II.5.1.005a).
.005n. Puniceus: (see Carmine, II.5.1.005c).
.005o. Purple (purpureus): dull red with a slight amount of blue.
.005p. Red (ruber, erythro-): pure red.
.005q. Rosy (roseus): pale pure red.
.005r. Ruber: (see Red, II.5.1.005p).
.005s. Rubiginosus: (see Brown-red, II.5.1.005b).
.005t. Salmon-colored, (salmonicolor, salmonaceus): pink with a dash of yellow.
.005u. Sanguine (sanguineus): dull red passing into brownish black.
.005v. Scarlet (coccineus): carmine slightly tinged with yellow.
.005w. Vinaceus: (see Wine-colored, II.5.1.005x).
.005x. Wine-colored (vinaceus): deep red with some purple.
.006 Variegation, markings: (ST)
.006a. Albomaculatus: leaf variegation where the normal green is irregularly spotted with patches of paler green or white.

II.5.1. MORPHOLOGY and ANATOMY

- .006b. Banded (fasciatus): transverse stripes of one color crossing another.
- .006c. Blotched (maculatus): color in broad, irregular patches.
- .006d. Bordered (limbatus): when one color is surrounded by an edging of another.
- .006e. Dotted (punctatus): color in very small round spots.
- .006f. Edged (marginatus): when one color is surrounded by a very narrow rim of another.
- .006g. Fasciatus: (see Banded, II.5.1.006b).
- .006h. Guttatus: (see Spotted, II.5.1.006q).
- .006i. Limbatus: (see Bordered, II.5.1.006d).
- .006j. Maculatus: (see Blotched, II.5.1.006c).
- .006k. Marginatus: (see Edged, II.5.1.006f).
- .006l. Ocellated (ocellatus): a broad spot of one color has another spot of a different color within it.
- .006m. Painted (pictus): colors are disposed in streaks of unequal intensity.
- .006n. Pictus: (see Painted, II.5.1.006m).
- .006o. Punctatus: (see Dotted, II.5.1.006e).
- .006p. Sordid (sordidus): dull, dirty (usually yellowish but can apply to any color).
- .006q. Spotted (guttatus): color in small spots.
- .006r. Striped (vittatus): longitudinal stripes of one color crossing another.
- .006s. Variegated (variegatus): color disposed in various irregular sinuous spaces.
- .006t. Vittatus: (see Striped, II.5.1.006r).
- .006u. Zoned (zonatus): same as ocellated but the concentric bands are more numerous.
- .007 White, colorless: (ST)
- .007a. Albidus: (see Whitish, II.5.1.007o).
- .007b. Albus: (see White, II.5.1.007m).
- .007c. Aqueous (aqueus): water clear.
- .007d. Argenteus: (see Silvery, II.5.1.007k).
- .007e. Dealbatus: (see Whitened, II.5.1.007n).
- .007f. Hyaline (hyalinus): glassy.
- .007g. Lacteus: (see Milk-white, II.5.1.007h).
- .007h. Milk-white (lacteus): dull white to bluish.
- .007i. Niveus: (see Snow white, II.5.1.007l).
- .007j. Pellucid (pellucidus): clear.
- .007k. Silvery (argenteus): slightly bluish gray with a metallic luster.
- .007l. Snow white (niveus): pure white.
- .007m. White (albus): dead white.
- .007n. Whitened (dealbatus): slightly covered with white on a darker background.
- .007o. Whitish (albidus): slightly soiled white.
- .008 Yellow, orange: (ST)
- .008a. Apricot-colored (armeniacus): yellow with a perceptible mixture of red; dull orange.
- .008b. Armeniacus: (see Apricot-colored, II.5.1.008a).
- .008c. Aurantiacus: (see Orange-colored, II.5.1.008n).
- .008d. Aureus: (see Golden yellow, II.5.1.008j).
- .008e. Chryso-: (see Golden yellow, II.5.1.008j).
- .008f. Citrinus: (see Lemon-colored, II.5.1.008k).
- .008g. Ferrugineus: (see Rusty, II.5.1.008q).
- .008h. Flavus: (see Pale yellow, II.5.1.008o).

II.5.1. MORPHOLOGY and ANATOMY

.008i. Fulvus: (see Tawny, II.5.1.008s).
.008j. Golden yellow (aureus, chryso-): pure yellow, slightly grayish, but bright.
.008k. Lemon-colored (citrinus): pure yellow but not bright.
.008l. Lutescens: (see Pale yellow, II.5.1.008o).
.008m. Luteus: (see Yellow, II.5.1.008v).
.008n. Orange-colored (aurantiacus): redder than apricot-colored.
.008o. Pale yellow (flavus, lutescens): pure but paler than yellow.
.008p. Rubiginosus: (see Rusty, II.5.1.008q).
.008q. Rusty (rubiginosus, ferrugineus): reddish orange.
.008r. Sulfur-colored (sulfureus, sulphureus): pale lively yellow with a mixture of white.
.008s. Tawny (fulvus): dull yellow with a mixture of gray and brown.
.008t. Testaceous (testaceus): brownish yellow.
.008u. Xantho-: (see Yellow, II.5.1.008v).
.008v. Yellow (luteus, xantho-): gamboge yellow.

II.5.2. Direction and size

.001 Brachiate: when ramifications proceed from a common axis nearly at right angles alternately in opposite directions. (ST)
.002 Declining: (see Inclining, II.5.2.017).
.003 Deflexed: bent downward. (ST)
.004 Dependent, hanging: having a downward direction caused by its weight, like a branch, inflorescence, or fruit. (ST)
.005 Depressed: broad and dwarf as if growth were horizontal instead of vertical. (ST)
.006 Descending: having a direction gradually downward. (ST)
.007 Diffuse: spreading widely. (ST)
.008 Divaricate, straggling: turning off from anything irregularly but almost at a right angle, as the branches of many things. (ST)
.009 Drooping: inclining a little from the perpendicular so that the apex is directed toward the horizon. (ST)
.010 Dwarf: small, short, dense, compared with other taxa of the same genus or family; very small is the same, except that a general reduction in size is understood. (ST)
.011 Elongate, produced: lengthened, stretched out.
.012 Flaccid: limp, floppy, wilted.
.013 Flexuous: having a gently bending direction alternately inward and outward. (ST)
.014 Gigantic: tall but stout and well proportioned. (ST)
.015 Habit: growth form of a plant.
.016 Hanging: (see Dependent, II.5.2.004).
.017 Inclining, declining: same as reclining, except to a greater degree (also similar to deflexed). (ST)
.018 Inflexed, introflexed: suddenly bent downward. (ST)
.019 Introflexed: (see Inflexed, II.5.2.018).
.020 Inverted: having the apex of one thing in an opposite direction to that of another. (ST)
.021 Little: small in all parts but well proportioned; generally used in apposition to large. (ST)
.022 Lofty: same as tall but to a greater degree. (ST)
.023 Low: applied to a plant much smaller than other kindred forms. (ST)
.024 Oblique: when the margin points to the heavens, the apex to the horizon. (ST)

II.5.2. MORPHOLOGY and ANATOMY

.025 Obsolescent: vestigial, nearly extinct. (L 51)
.026 Obsolete: not evident, extinct; as the absence of an organ where it should normally appear. (L 51)
.027 Pendulous: hanging downward in consequence of the weakness of its support; essentially the same as dependent but the latter may straighten up when the weight is removed, whereas a pendulous body would not. (ST)
.028 Perpendicular, vertical: being at right angles with some other body. (Vertical usually means straight up.) (ST)
.029 Produced: (see Elongate, II.5.2.011).
.030 Reflexed: bent abruptly backward at more that a 90° angle. (ST)
.031 Retrorse: turned in a direction opposite to that of the apex of the body to which the part appertains. (ST)
.032 Rudimentary: vestigial, imperfectly developed. (ST)
.033 Size: the degree of development (in terms of dimensions) of a plant or organ. (ST)
.034 Straggling: (see Divaricate, II.5.2.008).
.035 Tall: said of plants whose height is greater than would be expected from their parts. (ST)
.036 Tortuous: having an irregular bending and turning direction. (ST)
.037 Vague: uncertain; having no particular direction, like many tendrils. (ST)
.038 Vertical: (see Perpendicular, II.5.2.028).

II.5.3. Duration

.000 Duration: the life span of a plant or organ. (Fig. II.5.01,.02)(G 62)
.001 Accrescent: becoming enlarged with age, as a fruiting calyx or corolla. (L 55)
.002 Annual: living but one year or one growing season. (Fig. II.5.01a) (ST)
.003 Aperiodical: plant that produces more than one crop a year or only one, but irregularly, rather than at a specific time of the year or over a span of years. (L & J)
.004 Aseasonal: strictly without season; said of plants that flower any time of the year. (Fig. II.5.01c) (L & J)
.005 Biennial: living two years or two growing seasons. (Fig. II.5.01b) (ST)
.006 Caducous: falling off very early. (Fig. II.5.02) (ST)
.007 Deciduous: finally falling off; also a plant that sheds its leaves annually (or all at one time and remains bare for some period); then used in apposition to evergreen. (Fig. II.5.02b) (ST)
.008 Ephemeral: diurnal; beginning and ending in a day, like individual flowers of some plants; short-lived flower or annual plant (e.g., chickweed and daylily). (L 51)
.009 Evergreen: remaining green throughout the year; a plant that sheds its leaves continuously throughout the year. (Fig. II.5.01c; .02c,d)
.010 Fugacious: falling off or perishing rapidly; ephemeral. (ST)
.011 Hapaxanthic, monocarpic: bearing fruit but once and then dying after fructification. Corner (1966) uses the former term for palms that flower and fruit only once and die, such as Corypha umbraculifera, and also for certain others (e.g., Caryota and Arenga) that flower and fruit in successively lower axils and die when the base is reached. (Fig. II.5.01c; .02d) (CR 66, ST)

II.5.3. MORPHOLOGY and ANATOMY

.012 Marcescent: withering or fading, not falling until the part that bears it is perfected but withering long before that time, like the corollas of heath or campanula. (ST)
.013 Monocarpic: (see Hapaxanthic, II.5.3.011).
.014 Perennial: lasting for several to many years; may be herbaceous or woody. (Fig. II.5.01c,d, .02b-d) (ST)
.015 Persistent: neither falling off nor withering until the part that bears it is perfected. (ST)
.016 Pleonanthic: flowering repeatedly during the plant's life, generally as axillary inflorescences. (Fig. II.5.02b,c) (HOT)
.017 Plietesial: requiring several to many years to reach the flowering state; then dying after fructification. (Fig. II.5.02d) (ST)
.018 Polycarpic: having the power of bearing fruit many times without perishing. (Fig. II.5.02b,c) (ST)
.019 Semelparous: equivalent to plietesial, as used in Burley and Styles (1976); a plant that reproduces once, usually after an extended vegetative period of several decades or centuries, like certain bamboos; synonymous with monocarpic (II.5.3.011), applied usually to short-lived forms like annuals or biennials but also to individual stems of certain perennial monocots (e.g., Musa and Zingiberaceae). (Fig. II.5.02d) (JZ)

II.5.4. Apex, base, margins

II.5.4.1. Apex (L 51, L 55; G 62, G 68; ST)

.001 Acuminate, taper-pointed: tapering gradually to a point; the sides are straight or convex and the angle of convergence is more than 90°. (Fig. II.5.03a) (ST)
.002 Acuminose: terminating gradually in a narrow flat end. (ST)
.003 Acute, sharp-pointed: terminating at once in a point, not abruptly but without tapering; with two almost straight lines that converge at an angle of less than 90°. (Fig. II.5.03b) (ST)
.004 Apiculate: terminating abruptly in a little point; differs from mucronate in not arising wholly from the midrib (costa). (Fig. II.5.03c) (ST)
.005 Aristate, awned: abruptly terminated (as a continuation of the midrib) in a hard, straight, subulate point of varying lengths. (Fig. II.5.03d) (ST)
.006 Awned: (see Aristate, II.5.4.1.005).
.007 Bitten off: (see Praemorse, II.5.4.1.016).
.008 Blunt: (see Obtuse, II.5.4.1.015).
.009 Caudate, tail-pointed: excessively acuminate with a long weak tail. (Fig. II.5.03e) (ST)
.010 Cirrhous: terminated with a spiral, or flexose, filiform appendage (which is an elongation of the midrib), as a cirrhose leaf on Gloriosa plants. (Fig. II.5.03f) (ST)
.011 Cuspidate: tapering gradually into a rigid point. (Fig. II.5.03g) (ST)
.012 Emarginate: having a notch at the end; indented with an acute sinus. (Fig. II.5.03h) (ST)
.013 Mucronate, mucronulate: abruptly terminated with a hard, short point; mucronulate is the diminutive. (Fig. II.5.03i) (ST)

II.5.4.1. MORPHOLOGY and ANATOMY

.014 Obcordate: deeply notched, the figure being reversely cordate. (Fig. II.5.03j) (ST)
.015 Obtuse, blunt: apex terminating gradually in a blunt end. (Fig. II.5.03k) (ST)
.016 Praemorse, bitten off: the same as truncate, except ragged and irregular as if gnawed or bitten off. (Fig. II.5.03l) (ST)
.017 Retuse: terminating in a round end with a rounded (small) sinus at the apex. (Fig. II.5.03m) (ST)
.018 Rotund: same as obtuse but like the arc of a circle. (Fig. II.5.03n) (ST)
.019 Sharp-pointed: (see Acute, II.5.4.1.003).
.020 Subtruncate: somewhat or slightly truncate. (Fig. II.5.03o)
.021 Tail-pointed: (see Caudate, II.5.4.1.009).
.022 Truncate: terminating abruptly as if a piece had been cut off. (Fig. II.5.03p) (ST)

II.5.4.2. Base   (L 51, L 55; G 62, G 68; ST)

.001 Acute: with the sides equally curved convexly to the base, the whole included in a 90° angle. (Fig. II.5.04a) (ST)
.002 Attenuate: with convex curved sides narrowed gradually and concavely to the base. (Fig. II.5.04b) (ST)
.003 Auriculate, eared: having at the base two rounded lobes that stand out from the rest of the figure (leaf blade) like little ears; a term also applied to a leaf with little ears. (Fig. II.5.04c) (ST)
.004 Blunt: (see Obtuse, II.5.4.2.010).
.005 Cordate: having two equal, more or less rounded lobes that form a deep sinus. (Fig. II.5.04d) (ST)
.006 Cuneate, wedge-shaped: differs from acute because the sides are straight and the included angle is either more or less than 90°. (Fig. II.5.04e) (ST)
.007 Halbert-shaped: (see Hastate, II.5.4.2.008).
.008 Hastate: having two equal, pointed, more or less triangular lobes pointed outward. (Fig. II.5.04f) (ST)
.009 Oblique: base is unequal-sided. (Fig. II.5.04g) (ST)
.010 Obtuse, blunt: base rounded enough to include a 90° angle. (Fig. II.5.04h) (ST)
.011 Rounded: rounded like the arc of a circle. (Fig. II.5.04i) (ST)
.012 Sagittate: having as the base two equal, pointed, more or less triangular lobes pointed downward. (Fig. II.5.04j) (ST)
.013 Subcordate: having two slight lobes and a shallow sinus. (Fig. II.5.04k) (ST)
.014 Subtruncate: (see II.5.4.1.020).
.015 Truncate: as if cut straight across. (Fig. II.5.04l) (ST)
.016 Wedge-shaped: (see Cuneate, II.5.4.2.006).

II.5.4.3. Margins   (L 51, L 55; G 62, G 68; ST)

.001 Angular: having several salient angles on the margin. (Fig. II.5.05a) (ST)
.002 Bicrenate: (see Crenate, II.5.4.3.009).
.003 Bidentate: (see Dentate, II.5.4.3.012).

II.5.4.3. MORPHOLOGY and ANATOMY

.004 Bifid: two-cleft.
.005 Biserrate: (see Serrate, II.5.4.3.032).
.006 Ciliate: fringed with a row of fine hairs. (Fig. II.5.05b) (ST)
.007 Cleft, split: divided nearly to the base into a determinate number of segments, with sinuses extending somewhat more than halfway from the margin to center; segments and sinuses are usually sharp and acute. (Fig. II.5.05c,d) (ST)
.008 Comb-shaped: (see Pectinate, II.5.4.3.028).
.009 Crenate, crenulate: having convex (rounded) teeth; the diminutive is crenulate; doubly crenate, when the teeth themselves are crenate, is bicrenate. (Fig. II.5.05e) (ST)
.010 Crispate, crisped, curled: having the margin divided and twisted with excessive irregularity; like parsley and certain types of lettuce and similar greens. (Fig. II.5.05f) (ST)
.011 Cut: (see Incised, II.5.4.3.021).
.012 Dentate, toothed, denticulate: having sharp teeth with concave or straight edges, usually pointed straight outward; the diminutive is denticulate; doubly dentate, when the teeth themselves are toothed, is bidentate. (Fig. II.5.05g,h) (ST)
.013 Dissected: divided into many slender segments. (Fig. II.5.05i) (ST)
.014 Divided: divisions separated clear to the midrib or base. (Fig. II.5.05j,k) (G 62)
.015 Doubly crenate = bicrenate: (see Crenate, II.5.4.3.009).
.016 Doubly dentate = bidentate: (see Dentate, II.5.4.3.012).
.017 Doubly serrate = biserrate: (see Serrate, II.5.4.3.032).
.018 Entire: strictly without any kind of marginal division; sometimes used in the sense of not pinnatifid or essentially without marginal division. (Fig. II.5.05l) (ST)
.019 Erose, gnawed: having the margin irregularly toothed as if bitten by an animal. (Fig. II.5.06a) (ST)
.020 Gnawed: (see Erose, II.5.4.3.019).
.021 Incised, cut: regularly divided with deep incisions. (Fig. II.5.06b,c) (ST)
.022 Lacerate, torn: irregularly divided with deep incisions. (Fig. II.5.06d) (ST)
.023 Laciniate, slashed: divided with deep, taper-pointed, cut incisions. (Fig. II.5.05i, .06e) (ST)
.024 Lobed: partly divided with a determinate number of segments. (Fig. II.5.06f,g) (ST)
.025 Palmate: (see II.1.2.071).
.026 Palmatifid: divided almost to the axis in segments whose midribs meet at a common point. (Fig. II.5.05d) (ST, G 62, G 68)
.027 Parted: divided into a determinate number of segments that extend near to the base of the part to which they belong. (Fig. II.5.06i,j) (ST)
.028 Pectinate, comb-shaped: the same as pinnatifid but the segments are numerous, close, and narrow like the teeth of a comb. (Fig. II.5.06h) (ST)
.029 Pedate: the same as palmate, except that the two lateral (lowest) lobes are themselves divided into smaller lobes, the midribs of which do not run directly to the same point as the rest. (Fig. II.1.2.16a, 5.06k) (ST, HS 69)
.030 Pinnatifid: divided almost to the axis in lateral segments, something like the side divisions of a feather. (Fig. II.5.05c) (ST, G 68)
.031 Repand: weakly sinuate. (Fig. II.5.07a) (ST)

II.5.4.3. MORPHOLOGY and ANATOMY

.032 Serrate: having more or less sharp, straight-edged teeth pointed toward the apex; the diminutive is serrulate; doubly serrate, when teeth are themselves serrate, is biserrate. (Fig. II.5.07c-e) (ST)
.033 Sinuate: having an uneven margin with alternate deep concavities and convexities (in and out). (Fig. II.5.07b) (ST)
.034 Sinuous: of a serpentine or wavy form; bending in and out.
.035 Sinus: space or recess between two lobes or division of a leaf or other expended organ. (Fig. II.5.061) (ST, G 62)
.036 Slashed: (see Laciniate, II.5.4.3.032).
.037 Split: (see Cleft, II.5.4.3.007).
.038 Squarrose-slashed: slashed with minor divisions at right angles to the others. (Fig. II.5.07f) (ST)
.039 Toothed: (see Dentate, II.5.4.3.012).
.040 Torn: (see Lacerate, II.5.4.3.022).
.041 Undulate, wavy: having an uneven, alternating convex and concave margin, i.e., up and down. (Compare with Sinuous, II.5.4.3.034.) (Fig. II.5.07g) (ST)

II.5.5. Shapes (L 51, L 55, G 62, G 68, ST)

.001 Acerose, needle-shaped: linear, rigid, tapering to a fine point from a narrow base. (Fig. II.5.08a) (ST)
.002 Acicular: needle-shaped, as a pine needle, slender, usually roundish in cross section and not flattened. (Fig. II.5.08b) (L 51)
.003 Arrow-shaped: (see Sagittate, II.5.5.052).
.004 Attenuate, tapering: gradually diminishing in breadth. (Fig. II.5.08c) (ST)
.005 Awl-shaped: (see Subulate, II.5.5.058).
.006 Band-shaped: long and narrow, with two margins parallel (length to breadth 12:1 or more). (Fig. II.5.08d) (ST)
.007 Cespitose, caespitose: forming dense patches or tufts. (Fig. II.5.08e) (ST)
.008 Circumscription: (see Outline, II.5.5.044).
.009 Columnar: having the shape and form of a column; a vertical body more or less circular in cross section and generally several times as tall as broad. (Fig. II.5.08f) (B-GH)
.010 Concave: shallowly hollowed; saucerlike. (Fig. II.5.09a) (G 62)
.011 Convex: rounded outward. (Ffig. II.5.09b) (G 62)
.012 Cordate, heart-shaped: having two round lobes at the base, the whole resembling a heart. (Fig. II.5.09c) (ST)
.013 Cristate: bearing a crestlike appendage. (Fig. II.5.08g) (ST, L 55)
.014 Cuneiform, wedge-shaped: inversely triangular with rounded angles, as a Ginkgo leaf or maidenhair fern. (Fig. II.5.09d) (ST)
.015 Deltoid: a solid with a triangular outline like the Greek delta ( ). (Fig. II.5.09e) (ST)
.016 Depauperate, starved: when some part is less perfectly developed than is usual in plants of the same family. (ST)
.017 Dorsiventral: having distinct dorsal and ventral surfaces, like most foliage leaves. (See Fig. II.1.2.04) (L 51)
.018 Eared: (see Auriculate, II.5.4.2.003).
.019 Elliptical, oval: elliptical, acute at each end (length to breadth 2:1 to 3:2, with sides curved equally from the middle). (Compare with ovate, which is egg-shaped.) (Fig. II.5.09f) (ST)

II.5.5. MORPHOLOGY and ANATOMY

.020   Ensiform, sword-shaped: lorate but quite straight, with the point acute. (Fig. II.5.09g) (ST)
.021   Entangled: when things are mixed in such an irregular manner they cannot be readily separated. (ST)
.022   Fasciated: when several contiguous parts grow unnaturally together into one. (Fig. II.5.09h) (ST)
.023   Fastigiate: when, e.g., all of the branches, are nearly parallel and usually vertical. (Fig. II.5.09i) (ST)
.024   Fiddle-shaped: (see Pandurate, II.5.5.047).
.025   Filiform: thin, cylindrical, as in dog-fennel (Eupatorium capillifolium). (Fig. II.5.10a) (MU)
.026   Hastate, halbert-shaped: abruptly enlarged at the base into two diverging lobes. (Fig. II.5.10b) (ST)
.027   Heart-shaped: (see Cordate, II.5.5.012).
.028   Hooked-back: (see Runcinate, II.5.5.051).
.029   Isobilateral: both surfaces alike, divisible into two similar halves, as in certain leaves. (Compare with dorsiventral, II.5.5.017.) (see Fig. II.1.2.04) (L 51)
.030   Kidney-shaped: (see Reniform, II.5.5.048).
.031   Lanceolate: lance-shaped; narrowly elliptical, tapering equally to each end (length to breadth 6:1 to 3:1). (Fig. II.5.10c) (ST)
.032   Ligule, ligulate, lorate, strap-shaped: an elongated, flattened, strap-shaped (i.e., narrow and moderately long with the two opposite margins parallel or essentially so) structure; ligulate: with or possessing a ligule; terms applied to several different organs: (1) leaves proper, as in some lilies (e.g., Liriope, Ophiopogon), then often denoted "lorate"; (2) the bractlike emergences (stipules) from the top of the leaf sheath at the base of the blade in monocots, especially grasses; (3) a corolla with strap-shaped petals (as in the Cichoreae, with all flowers ligulate, or only the marginal ray flowers in others) in the Compositae. (Compare with Ensiform, II.5.5.020.) [Fig. II.5.10d(1), (2), (3)] (ST, G 62, L 55)
.033   Linear: narrow, with the two opposite margins parallel. (Fig. II.5.10e) (ST)
.034   Lorate: (see Ligule, II.5.5.032).
.035   Lunate: crescent-shaped. (ST)
.036   Lyrate, lyre-shaped: same as pandurate, except that several sinuses occur on each side, gradually diminishing toward the base, as in Hieracium. (Fig. II.5.11a) (ST)
.037   Needle-shaped: (see Acerose, II.5.5.001).
.038   Obcordate: (see II.5.4.1.014). (Fig. II.5.11c)
.039   Oblanceolate: the reverse of lanceolate, the broadest half above the middle; the apex acute, obtuse, or otherwise. (Fig. II.5.11d) (G 68)
.040   Oblique: when the degree of inequality in the two sides is slight but the base is unequal, as in an elm leaf. (Fig. II.5.11e) (ST)
.041   Oblong: elliptical, obtuse at each end (length to breadth usually 2:1 to 3:2); sides almost parallel. (Fig. II.5.11f) (ST)
.042   Obovate: the reverse of ovate, with the broadest half above the middle and the narrower end toward the base. (Fig. II.5.11g,h) (G 68)
.043   Orbicular: perfectly circular. (Fig. II.5.11i) (ST)
.044   Outline, circumscription: figure represented by the margin of a leaf or fruit. (ST)
.045   Oval: (see Elliptical, II.5.5.019).

II.5.5. MORPHOLOGY and ANATOMY

.046 Ovate: oblong or elliptical but egg-shaped; broadest at the lower end (length to breadth 2:1 to 3:2; broadest below the middle). (Fig. II.5.11j). (ST)
.047 Pandurate, fiddle-shaped: obovate with a deep recess or sinus on each side. (Fig. II.5.11k) (ST)
.048 Reniform, kidney-shaped: cresent-shaped with the ends rounded. (Fig. II.5.11 l) (ST)
.049 Rhomboid: oval, a little angular in the middle. (Fig. II.5.11m) (ST)
.050 Roundish, rotund: orbicular, a little inclined to be oblong (length to breadth 6:5, broadest at middle). (Fig. II.5.11n) (ST)
.051 Runcinate, hooked-back: prominent teeth pointing toward the base, the apex usually more or less acute, as a dandelion leaf. (Fig. II.5.11b) (ST, G 62)
.052 Sagittate, arrow-shaped: gradually enlarged at the base into two straight lobes like the head of an arrow. (Fig. II.5.12a) (ST)
.053 Spatulate: oblong with the lower end much attenuated. (Fig. II.5.12b) (ST, G 68)
.054 Squarrose: when the parts spread out at right angles or thereabouts from a common axis. (Fig. II.5.12c) (ST)
.055 Starved: (see Depauperate, II.5.5.016).
.056 Strap-shaped: (see Ligule, II.5.5.032).
.057 Suborbicular: somewhat circular in outline.
.058 Subulate, awl-shaped: linear, narrow, tapering to a fine point from a narrow base, like juvenile leaves of <u>Juniperus</u>. (ST)
.059 Sword-shaped: (see Ensiform, II.5.5.020).
.060 Tapering: (see Attenuate, II.5.5.004).
.061 Triangular: having the figure of a triangle of any kind. (Fig. II.5.12e). (ST)
.062 Unequal, unequal-sided: when the two sides of a figure are not symmetrical as in a banana leaf. (Fig. II.5.12f) (ST)
.063 Wedge-shaped: (see Cuneiform, II.5.5.014).

II.5.6. Surface features

.000 Surface features: includes various types of hair and other outgrowth and surface markings, evenness, polish, and texture; combined here into a single group because of the imprecise fashion in which individual terms are used in the literature, as Lawrence (1951, 1955) and earlier writers have pointed out.
.001 Aciculated: marked with fine irregular streaks as if produced by the point of a needle. (Fig. II.5.13a) (ST)
.002 Aculeate, prickly: furnished with prickles. (Fig. II.5.13b) (ST)
.003 Alveolate, honeycombed: excavated in the manner of a section of honeycomb. (Fig. II.5.13d) (ST)
.004 Annulate, ringed: surrounded by elevated or depressed bands. (Fig. II.5.13c) (ST)
.005 Arachnoid, cobwebbed: covered with loose, white entangled hairs that resemble a spider web. (Fig. II.5.13e) (ST)
.006 Areolate: divided into a number of irregular squares or angular spaces. (Fig. II.5.13f) (ST)
.007 Asperous, rough, scabrous: covered with hard, short, rigid points; (see also Roughish). (Fig. II.5.13g,h) (ST)
.008 Barbed: (see Uncinate, II.5.6.144).

II.5.6. MORPHOLOGY and ANATOMY

.009 Bearded: having tufts of long weak hairs growing from different parts of the surface; also applied to bodies that bear long weak hairs in solitary tufts or parcels. (Fig. II.5.13i) (ST)
.010 Bloom: whitish, fine powdery covering of the surface, often waxy, as on some grapes (Vitis spp.), plums (Prunus spp.), and blueberries (Vaccinium spp.). (WS)
.011 Bossed: (see Umbonate, II.5.6.142).
.012 Bristle-pointed: (see Setose, II.5.6.123).
.013 Bristly: (see Echinate, II.5.6.036).
.014 Buckler-shaped: (see Scutate, II.5.6.120).
.015 Bullate: surface much puckered or blistered. (Fig. II.5.14a) (L 51)
.016 Caesious: like glaucous but greener. (ST)
.017 Callus: a hard prominence or protuberance; hardened or indurated as the point or serratures of certain leaves. (Fig. II.5.14b) (G 62)
.018 Candelabra: like a candlestick with several branches, as in certain types of hair. (Fig. II.5.14d) (ST)
.019 Canescent: gray-pubescent and hoary, or becoming so, as in certain brambles (Rubus). (Fig. II.5.14c) (ST)
.020 Cano-tomentose: midway between canescent and tomentose, as in certain brambles. (B-GH)
.021 Capillary, hair-shaped: same as filiform but more delicate. (Fig. II.5.14e) (ST)
.022 Capitate: with a knoblike head or tip, a term confined to cylindrical or terete bodies. (Fig. II.5.14f) (ST)
.023 Chaffy: (see Paleaceous, II.5.6.085).
.024 Ciliate: having fine hairs resembling an eyelash on or near the margin. (Fig. II.5.14g) (ST)
.025 Clavate, club-shaped: gradually thickening upward from a tapering base; shaped like a baseball bat. (Fig. II.5.14h) (ST)
.026 Club-shaped: (see Clavate, II.5.6.025).
.027 Cobwebbed: (see Arachnoid, II.5.6.005).
.028 Comose: bearing a tuft or tufts of hairs as in certain achenes of Compositae (e.g., sow thistle). (Fig. II.5.14i) (ST, G 62, 68)
.029 Crispy-hairy: hairs in short, close ringlets; undulated, frizzed. (Fig. II.5.14j) (ST)
.030 Dendritic, dendroid, treelike: divided at the top into a number of fine ramifications to resemble the head of a tree. (Fig. II.5.14k) (ST)
.031 Denuded: (see Naked, II.5.6.082).
.032 Dewy: (see Roridous, II.5.6.108).
.033 Dotted: (see Punctate, II.5.6.103; also II.5.1.006e).
.034 Downy: (see Pubescent, II.5.6.102).
.035 Dusty: (see Lentiginous, II.5.6.072).
.036 Echinate, bristly: furnished with numerous rigid hairs or straight prickles. (Fig. II.5.14 l) (ST)
.037 Farinose, mealy: covered with a sort of white, scurfy substance. (Fig. 5.15a) (ST)
.038 Feathery: (see Plumose, II.5.6.096).
.039 Felted-tomentose: covered with closely woven or matted woolly hairs, curled and appressed to the surface (more matted than tomentose). (Fig. II.5.15b) (L 51)
.040 Fenestrate: having numerous openings, irregularly reticulated. (Fig. II.5.15c) (ST)
.041 Fimbriate, fringed: having the margin bordered with long filiform processes thicker than hairs. (Fig. II.5.15d) (ST)
.042 Fissured: narrow openings, as in bark on trees. (Fig. II.5.15e) (G 62)

II.5.6. MORPHOLOGY and ANATOMY

.043 Flagelliform, whip-shaped: long, tapering, and supple. (Fig. II.5.15f) (ST)
.044 Floccose: covered with dense hairs that fall away in tufts. (Fig. II.5.15g) (ST)
.045 Foveolate: pitted with shallow distinct impressions. (Fig. II.5.15h) (ST)
.046 Fringed: (see Fimbriate, II.5.6.041).
.047 Frosted: (see Pruinose, II.5.6.100).
.048 Furrowed: (see Sulcate, II.5.6.137).
.049 Fusiform, spindle-shaped: thick, tapering to each end. (Fig. II.5.15i) (ST)
.050 Glabrate, glabrescent: becoming glabrous, although the former may indicate sparsely or minutely covered; thus the diminutive of glabrous. (ST)
.051 Glabrous, smooth: devoid of covering, the latter term also meaning free of any sort of unevenness. (ST, L 51)
.052 Glandular: covered with hairs that bear glands at the tips. (Fig. II.5.15j) (ST)
.053 Glandular-pubescent: (see Pubescent, II.5.6.102).
.054 Glaucous: covered with a fine bloom of whitish green color (glaucescent is slightly glaucous). (ST)
.055 Glittering: the same as laevigate, except the luster is a little broken from a slight irregularity of the surface. (ST)
.056 Glochidiate: barbed hairs disposed in tufts, as on cacti. (Fig. II.5.15k) (ST)
.057 Glutinous: (see Viscid, II.5.6.150).
.058 Greasy: (see Unctuous, II.5.6.145).
.059 Hair-pointed: (see Piliferous, II.5.6.092).
.060 Hair-shaped: (see Capillary, II.5.6.021).
.061 Hairy: (see Pilose, II.5.6.093).
.062 Hirsute, shaggy, villous, hirtuse: covered with long weak hairs. (Fig. II.5.15 l,m) (ST)
.063 Hispid: covered with long rigid hairs; the diminutive is hispidulus. (Fig. II.5.14n) (ST)
.064 Hoary, incanus: covered with short dense hairs placed so closely that the appearance of whiteness is given to the surface from which they grow. (ST)
.065 Honeycombed: (see Alveolate, II.5.6.003).
.066 Hooked: (see Uncinate, II.5.6.144).
.067 Incanus: (see Hoary, II.5.6.064).
.068 Knoblike: having an irregular roundish figure. (Fig. II.5.16a) (ST)
.069 Lacunose: having numerous large, deep depressions or excavations. (Fig. II.5.16b) (ST)
.070 Laevigate, polished: having the appearance of a polished substance. (ST)
.071 Lanate, woolly: covered with long, dense, curled, and matted hairs. (Fig. II.5.16c) (ST)
.072 Lentiginous, dusty: covered with minute dots as if dusted. (ST)
.073 Lepidote, leprous: covered with minute peltate scales. (Fig. II.5.16d,e) (ST)
.074 Lined, striate: marked with longitudinal lines. (Fig. II.5.16f) (ST)
.075 Loriform: strap-shaped hair with a long flexuous tip. (Fig. II.5.16g) (ST)
.076 Marginal: fixed on the edge of anything, like hairs (cilia) along the margin of azalea leaf. (ST)
.077 Mealy: (see Farinose, II.5.6.037).

II.5.6. MORPHOLOGY and ANATOMY

.078 Moniliform: cylindrical or terete and contracted at regular intervals; torulose, knotted, is nearly the same. (Fig. II.5.16h) (ST)
.079 Mucous, slimy: covered with a slimy excretion or with a coating readily soluble in water and becoming slimy when wetted. (ST)
.080 Multangulate: many pointed or cornered hair. (Fig. II.5.16j) (ST)
.081 Muricate: furnished with numerous short, hard excrescences which may be straight or curved. (Fig. II.5.16i) (ST)
.082 Naked, nude, denuded: the reverse of hairy, downy, or any similar term; not materially different from glabrous. (Fig. II.5.17a) (ST)
.083 Netted: (see Reticulate, II.5.6.106).
.084 Opaque: dull, the reverse of laevigate. (ST)
.085 Paleaceous, chaffy: covered with small, weak, erect membranous scales. (ST)
.086 Pannose: like felt or woolen cloth in texture; covered with a mass of woolly hairs. (Fig. II.5.17b) (ST)
.087 Papillose, pimpled: covered with minute tubercles or excrescences of uneven size and rather soft. (Fig. II.5.17c) (ST, HS 69)
.088 Pappus: tuft of delicate fibers or bristles at the tip of an achene, such as the feathery structure of a mature dandelion achene which is blown easily from the head and thus aids in its dispersal. (Fig. II.5.17d) (G 62)
.089 Patelliform: broad, round, thick; convex on the lower surface, concave on the upper. (Fig. II.5.17e) (ST)
.090 Peltate: (see Lepidote, II.5.6.073).
.091 Perulate: having buds covered with scales. (Fig. II.5.17f) (ST, G 62)
.092 Piliferous, hair-pointed: terminating in a very fine weak point. (Fig. II.5.17g) (ST)
.093 Pilose, hairy: covered with short, weak thin hairs. (Fig. II.5.17h) (ST)
.094 Pimpled: (see Papillose, II.5.6.087).
.095 Pitted: (see Scrobiculate, II.5.6.118).
.096 Plumose, feathery: consisting of long hairs that are themselves hairy. (Fig. II.5.17i) (LB)
.097 Polished: (see Laevigate, II.5.6.070).
.098 Powdery, pulverulent: covered with a fine bloom or powdery matter. (Fig. II.5.17j) (ST)
.099 Prickly: (see Aculeate, II.5.6.002).
.100 Pruinose, frosted: like dewy (roridous), except the appearance is more opaque as if the drops were congealed. (ST)
.101 Puberulous, puberulent: minutely pubescent; covered with exceedingly short, fine, rather dense straight hairs at right angles to the surface, scarcely perceptible to the naked eye except when viewed by transmitted light. (Fig. II.5.17k) (ST)
.102 Pubescent, downy: covered with short, weak, dense hairs; sometimes intermixed with glands or terminated by pinheadlike glands; then glandular-pubescent; also used as a general term meaning hairiness of any type. (ST; L 51, L 55; G 62)
.103 Punctate, dotted: covered with minute impressions as if made with point of a pin; may be colored (II.5.1.006e). (Fig. II.5.18a) (ST)
.104 Ramentaceous: bearing small, thin, loose, brownish scales on stems or leaves. (Fig. II.5.18b) (ST)
.105 Resinous: exuding droplets of resin. (Fig. II.5.18d) (G 62)
.106 Reticulate, netted: covered with a network of lines that project a little like those on a muskmelon. (Fig. II.5.18c) (ST)

II.5.6. MORPHOLOGY and ANATOMY

.107 Ringed: (see Annulate, II.5.6.004).
.108 Roridous, dewy: covered with little transparent elevations of parenchyma that have the appearance of fine drops of dew. (ST)
.109 Rough: (see Asperous, II.5.6.007).
.110 Roughish: (see Scabrous, II.5.6.113).
.111 Rugose: wrinkled, the elevations irregular; covered with reticulated lines with convex spaces between them. (Fig. II.5.18e) (ST)
.112 Scabrous: (see Asperous, II.5.6.007).
.113 Scabrous, scabrid, roughish: slightly covered with short, somewhat hard points. (Fig. II.5.18f) (ST)
.114 Scale: name given to many kinds of small, mostly dry, and appressed leaves or bracts. (G 62)
.115 Scale-shaped: (se Scutelliform, II.5.6.121).
.116 Scaly, squamate, squamose: covered with minute scales fixed at one end, as on leaves. (Fig. II.5.18g) (ST)
.117 Scarred: marked by the scars of bodies that have fallen off as leaf scars on a stem. (Fig. II.5.18h) (ST)
.118 Scrobiculate, pitted: having numerous small depressions or excavations. (Fig. II.5.18i,j) (ST)
.119 Scurfy: bearing small scales on the surface. (ST)
.120 Scutate, buckler-shaped: lens-shaped with an elevated rim. (Fig. II.5.18k) (ST)
.121 Scutelliform, scale-shaped: same as patelliform but oval. (Fig. II.5.19a) (ST)
.122 Sericeous: (see Silky, II.5.6.126).
.123 Setose, bristle-pointed: terminating gradually in a fine, sharp point. (Fig. II.5.19b) (ST)
.124 Shaggy: (see Hirsute, II.5.6.062).
.125 Shiny: having a smooth, even, polished surface. (ST)
.126 Silky, sericeous: covered with fine, close-pressed hairs silky to the touch. (Fig. II.5.19c) (ST)
.127 Slimy: (see Mucous, II.5.6.079).
.128 Smooth: (see Glabrous, II.5.6.051).
.129 Spiculate: covered with fine, fleshy erect points. (Fig. II.5.19d) (ST, L 55, ES)
.130 Spindle-shaped: (see Fusiform, II.5.6.049).
.131 Spiny, spinose, spinulose: furnished with spines; spinulose means furnished with small spines or scattered (few) spines. (See Fig. II.1.1.3.20) (ST)
.132 Squamate, squamose: (see Scaly, II.5.6.116).
.133 Stellate: divided into segments radiating from a common center, as a type of hair or scale. (Fig. II.5.19e) (ST)
.134 Stinging: (see Urent, II.5.6.146).
.135 Striate: (see Lined, II.5.6.074).
.136 Strigose: covered with sharp, appressed rigid hairs; sometimes considered equivalent to hispid. (Fig. II.5.19f) (ST)
.137 Sulcate, furrowed: marked with longitudinal channels. (Fig. II.5.20a) (ST)
.138 Tomentose: covered with dense, rather rigid, short hairs sensibly perceptible to the touch; the diminitive is tomentulose. (Fig. II.5.20b) (ST)
.139 Torulose: (see Moniliform, II.5.6.078).
.140 Treelike: (see Dendritic, II.5.6.030).
.141 Tuberculate, tubercled, verrucose: covered with little excrescences or warts, as the rind of some fruits. (Fig. II.5.20c) (ST)
.142 Umbonate, bossed: round, with a projecting point in the center. (Fig. II.5.20d) (ST)

II.5.6. MORPHOLOGY and ANATOMY

.143 Unarmed: destitute of any kind of spines or prickles. (ST)
.144 Uncinate, barbed, hooked: having hairs with tips bent back acutely or whose sides are provided with teeth or minute retrorse hooks. (Fig. II.5.20e) (ST)
.145 Unctuous, greasy: having a surface that feels greasy, although not actually so. (ST)
.146 Urent, stinging: covered with rigid, sharp-pointed, bristly hairs that emit an irritating fluid when touched. (Fig. II.5.20f) (RD)
.147 Velvety, velutinous: the same as tomentose but denser so that the surface resembles (and feels like) velvet. (Fig. II.5.20g) (ST)
.148 Verrucose: (see Tuberculate, II.5.6.141).
.149 Villous: (see Hirsute, II.5.6.062).
.150 Viscid, glutinous: covered with a gelatinous exudation. (ST)
.151 Whip-shaped: (see Flagelliform, II.5.6.043).
.152 Whitened: covered with an opaque white powder. (ST)
.153 Woolly: (see Lanate, II.5.6.071).

II.5.7. Texture (L 51, L 55, G 62, G 68, ST)

.001 Baccate: having a juicy, succulent texture, like many berrylike fruits.
.002 Bony: (see Osseous, II.5.7.026).
.003 Carneous: (see Fleshy, II.5.7.013).
.004 Cartilaginous: hard and tough. (ST)
.005 Chartaceous: (see Papery, II.5.7.027).
.006 Coated: (see Corticate, II.5.7.009).
.007 Coriaceous, leathery: having the consistency of leather. (ST)
.008 Corky: (see Suberous, II.5.7.031).
.009 Corticate, coated: harder externally than internally. (ST)
.010 Crustaceous: hard, thin, and brittle. (ST)
.011 Farinaceous, mealy: having the texture of flour in a mass. (ST)
.012 Fibrous: containing a great proportion of loose, woody fiber. (ST)
.013 Fleshy: firm, juicy, easily cut. (ST)
.014 Gelatinous: having the texture and appearance of jelly. (ST)
.015 Herbaceous: thin, green, and cellular; essentially the opposite of woody (ligneous). (ST)
.016 Horny: hard, brittle, fine in texture, but capable of being cut without difficulty. (ST)
.017 Indurate: hard. (ST)
.018 Lax, loose: of a soft cellular texture, like the pith of most plants. (ST)
.019 Leathery: (see Coriaceous, II.5.7.007).
.020 Ligneous, lignose, woody: having the texture of wood. (ST)
.021 Loose: (see Lax, II.5.7.018).
.022 Mealy: (see Farinaceous, II.5.7.011).
.023 Medullary, pithy: filled with spongy pith. (ST)
.024 Membranous, membranaceous: thin and semitransparent like a fine membrane. (ST)
.025 Oleaginous: fleshy in substance but filled with oil. (ST)
.026 Osseous, bony: like horny; brittle but cut with difficulty. (ST)
.027 Papery, chartaceous: having the consistency of writing paper and quite opaque. (ST)
.028 Pithy: (se Medullary, II.5.7.023).
.029 Scarious: having a thin, dry shriveled appearance. (ST)

II.5.7. MORPHOLOGY and ANATOMY

.030 Spongy: cellular, with the cellules filled with air and having the texture of a sponge. (ST)
.031 Suberous, corky: having the texture of cork. (ST)
.032 Succulent: cellular and juicy. (ST)
.033 Thick: something thicker than usual, for example, cotyledons; thickened. (ST)
.034 Waxy: having the color and texture of new wax. (ST)
.035 Woody: (see Ligneous, II.5.7.020).

II.6. Bryophytes, Pteridophytes, and Gymnosperms (L 51, L 55)

.001 Annulus: in ferns, the thick-walled ring that surrounds the sporangium and causes the rupture of the latter, when mature, by its contraction.
.002 Apophysis: swelling in the cone scale of certain conifers; also an enlargement or swelling of the stalk, or seta, at the base of the capsule in certain mosses.
.003 Archegonium: flask-shaped female sex organ in Bryophyta and Pteridophyta and some gymnosperms.
.004 Circinate, coiled: coiled inward from the tip; bent like the head of a crosier; the vernation of ferns.
.005 Club moss: plants of the genera Lycopodium or Selaginella. (Fig. II.6.01a)
.006 Coiled: (see Circinate, II.6.004).
.007 Conifer: any plant of the pine (Pinaceae) or other families in the Coniferae. (Fig. II.6.01c)
.008 Cycad: any plant of the family Cycadaceae, e.g, Zamia and Cycas. (Fig. II.6.01d)
.009 Dichotomous: having the divisions always in pairs, as in venation of Ginkgo and fern leaves.
.010 Fern: any plant of a large order, the Filicales, of the Pteridophyta. (Fig. II.6.01b)
.011 Gymnosperms (Gymnospermae): trees and shrubs, with usually evergreen needlelike or fernlike, sometimes broad-leaved foliage; vessels absent in wood (except Gnetales), resin canals often present, flowers reduced to pollen sacs and ovules usually arranged in strobili, unisexual monoecious or frequently dioecious, pollen windborne, except motile sperms in Cycadales and Ginkgo; seeds from naked ovules borne on megasporophylls (cone scale or carpel), cotyledons, two to several, rarely one; six orders, cycads (Cycadales), ginkgo (Ginkgoales), conifers (Coniferae), gnetum (Gnetales), ephedra (Ephedrales), and welwitschia (Welwitschiales).
.012 Heterosporous: bearing asexual spores of more than one kind.
.013 Indusium: in ferns, an outgrowth of the leaf that covers or invests the sorus in many species.
.014 Megasporophyll: a sporophyll (spore-bearing leaf) that develops only mega- (macro-) sporangia.
.015 Microsporophyll: a sporophyll that develops only microsporangia.
.016 Pteridophytes (Pteridophyta): ferns and allied groups, separated into five classes; horsetails (Equisetinae), clubmosses (Lycopodiinae), psilotas (Psilotinae), quillworts (Isoetinae), and ferns (Filicinae); characterized by vascular tissue in the sporophyte generation; true roots present in most; a clearly separated alternation of gametophytic and sporophytic generations, the latter dominant; egg cells borne in antheridia and

II.6. MORPHOLOGY and ANATOMY

motile sperm cells produced from antheridia and gametophyte minute, often lacking chlorophyll.
.017 Rhizophore: in Selaginella, one of the peculiar, leafless, dichotomous shoots that bear roots at the apex.
.018 Sorus: in ferns, one of the clusters of sporangia that form the so-called fruit dots on the fertile fronds.
.019 Spermatozoid: male gametes of bryophytes, pteridophytes, and certain gymnosperms.
.020 Sporocarp: a many celled body that produces spores or a capsulelike mass of sporangia, as in fern allies or mosses.
.021 Sporocyte: a spore mother cell of plants, usually denoted microsporocyte for the male and megasporocyte or macrosporocyte for the female.
.022 Sporophyll: spore-bearing leaf or modified leaf.
.023 Sporophyte: the spore-forming generation in the life cycle of plants; the asexual generation; normally diploid.
.024 Strobilus (strobile): conelike aggregation of sporophylls that bear sporangia, as in club mosses and gymnosperms.

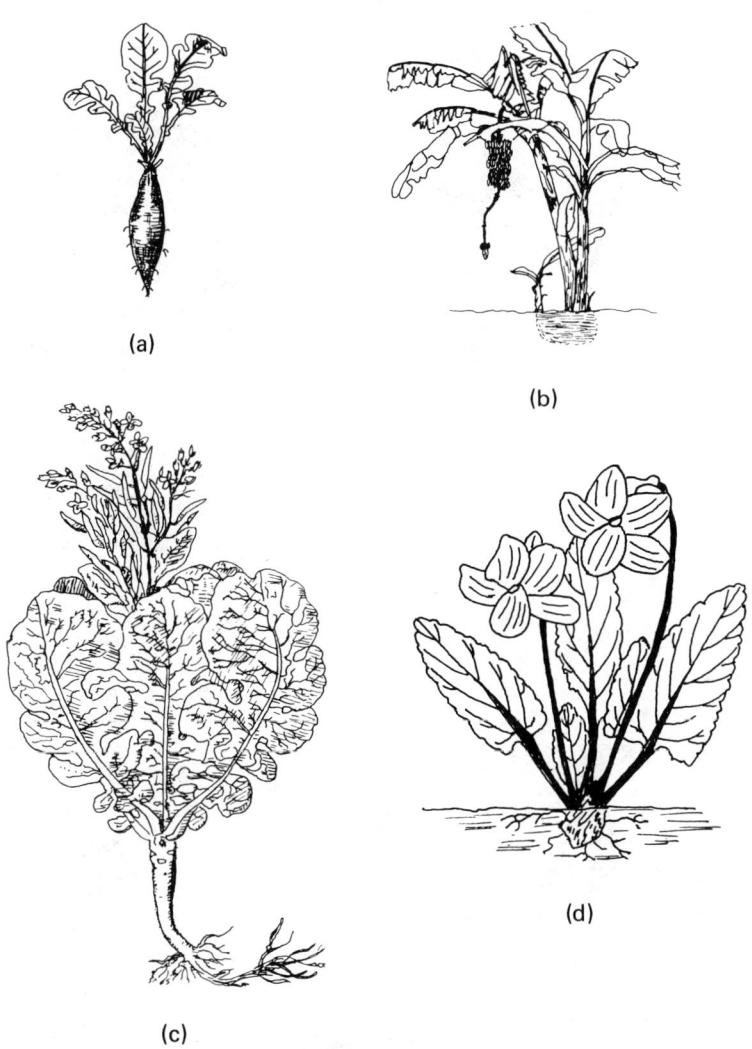

Fig. II.5.01. Duration: (a) annual, radish; (b) biennial, collard; (c) perennial, aseasonal, monocarpic (individual stems), semelparous, banana; (d) perennial, violet. (a), (d) Redrawn from Gray, 1862; (b) adapted from Tai, 1958; (c) redrawn from Johnson, 1633; see Acknowledgments and Selected References

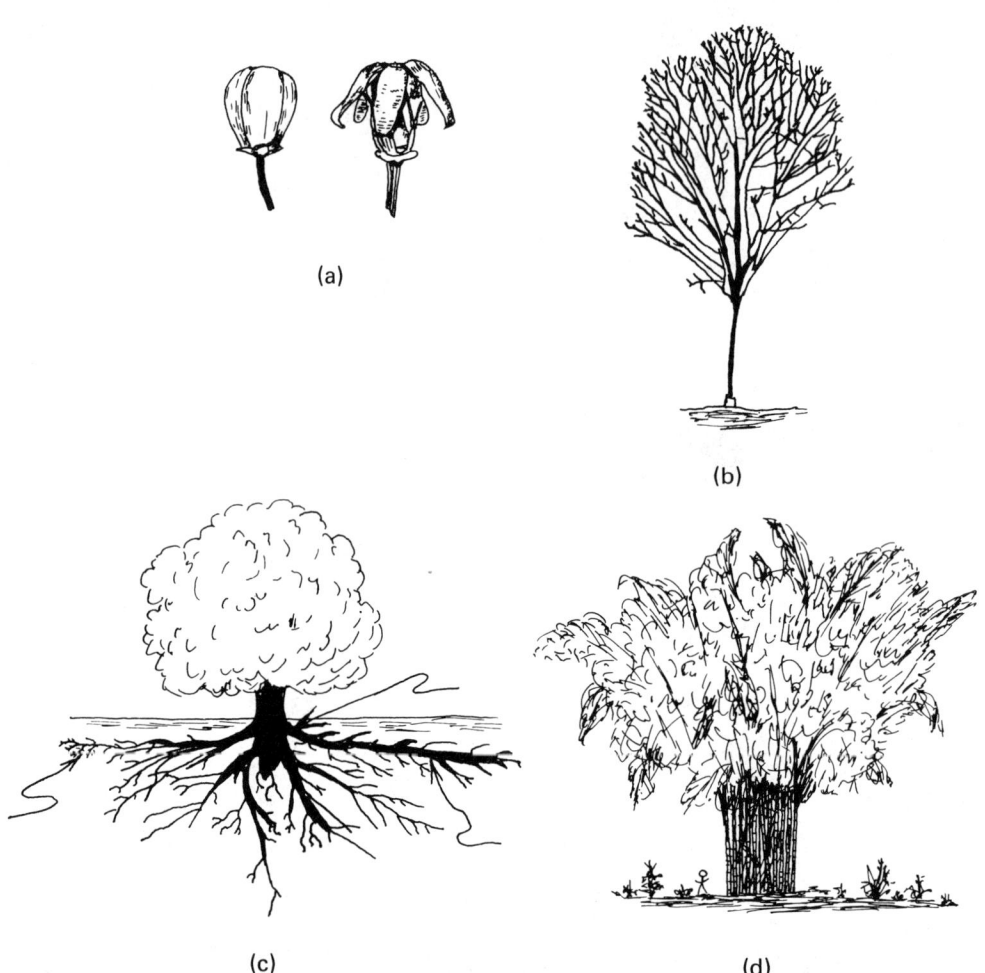

Fig. II.5.02. Duration: (a) caducous, petals of grape flower; (b) deciduous, perennial, pleonanthic, polycarpic, rubber (Hevea brasiliensis); (c) evergreen, perennial, pleonanthic, polycarpic, citrus; (d) evergreen, monocarpic, perennial, plietesial, semelparous, bamboo. (a) Redrawn from Bailey, 1928; (b) redrawn from Dijkman, 1951; (c) redrawn from Praloran, 1971; see Acknowledgments and Selected References.

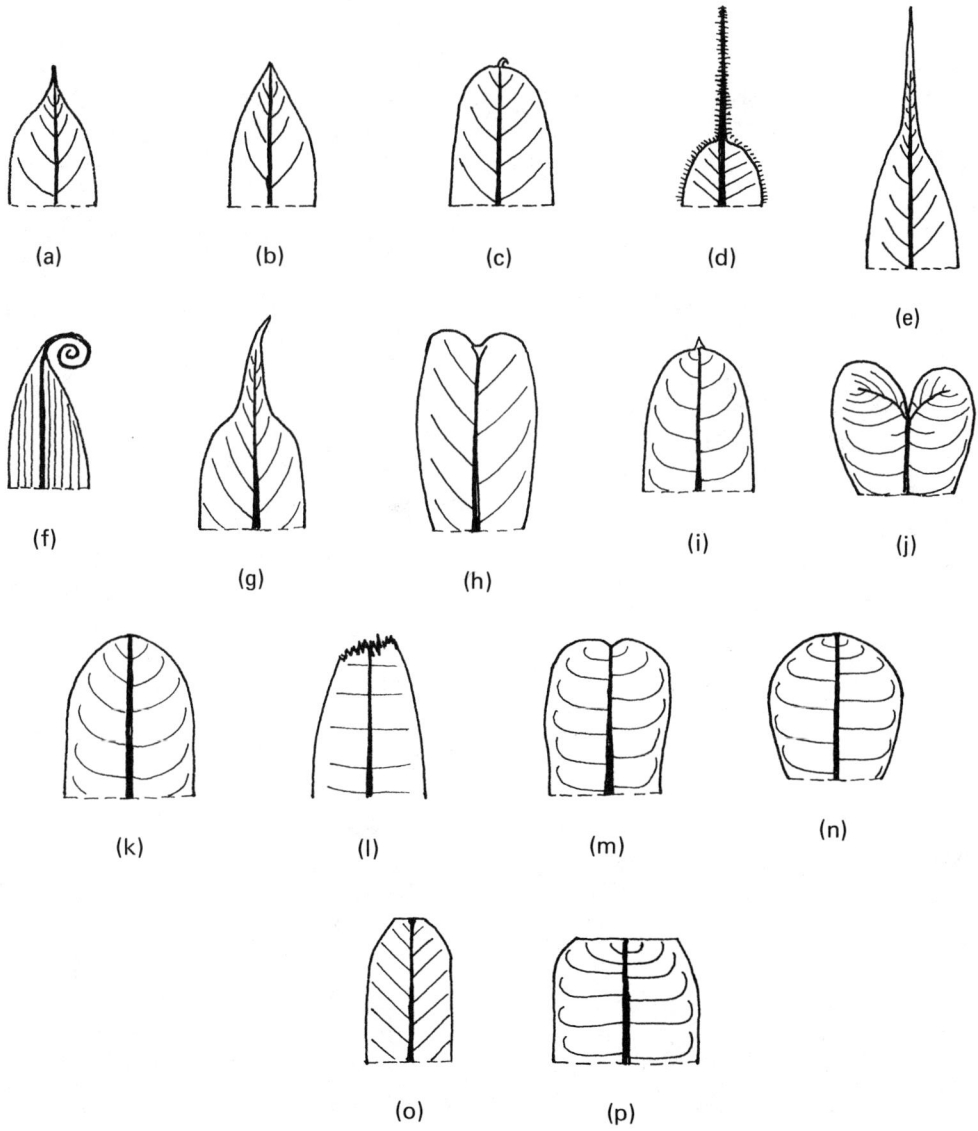

Fig. II.5.03. Apexes: (a) acuminate; (b) acute; (c) apiculate; (d) aristate; (e) caudate; (f) cirrhous; (g) cuspidate; (h) emarginate; (i) mucronate; (j) obcordate; (k) obtuse; (l) praemorse; (m) retuse; (n) rotund; (o) subtruncate; (p) truncate. (a)-(k), (m)-(p) Adapted from Lawrence, 1955; see Acknowledgments and Selected References.

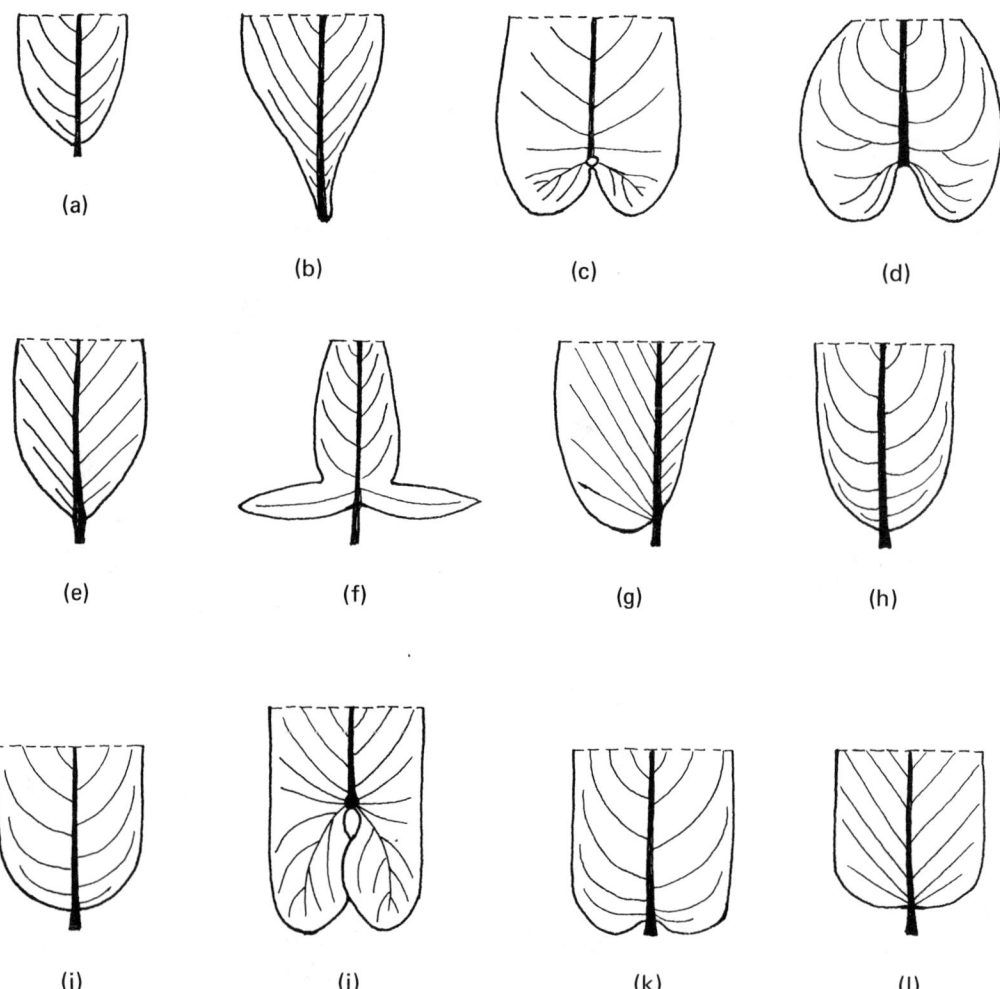

Fig. II.5.04. Bases: (a) acute; (b) attenuate; (c) auriculate; (d) cordate; (e) cuneate; (f) hastate; (g) oblique; (h) obtuse; (i) rounded; (j) sagittate; (k) subcordate; (l) truncate. Adapted from Lawrence, 1955; see Acknowledgments and Selected References.

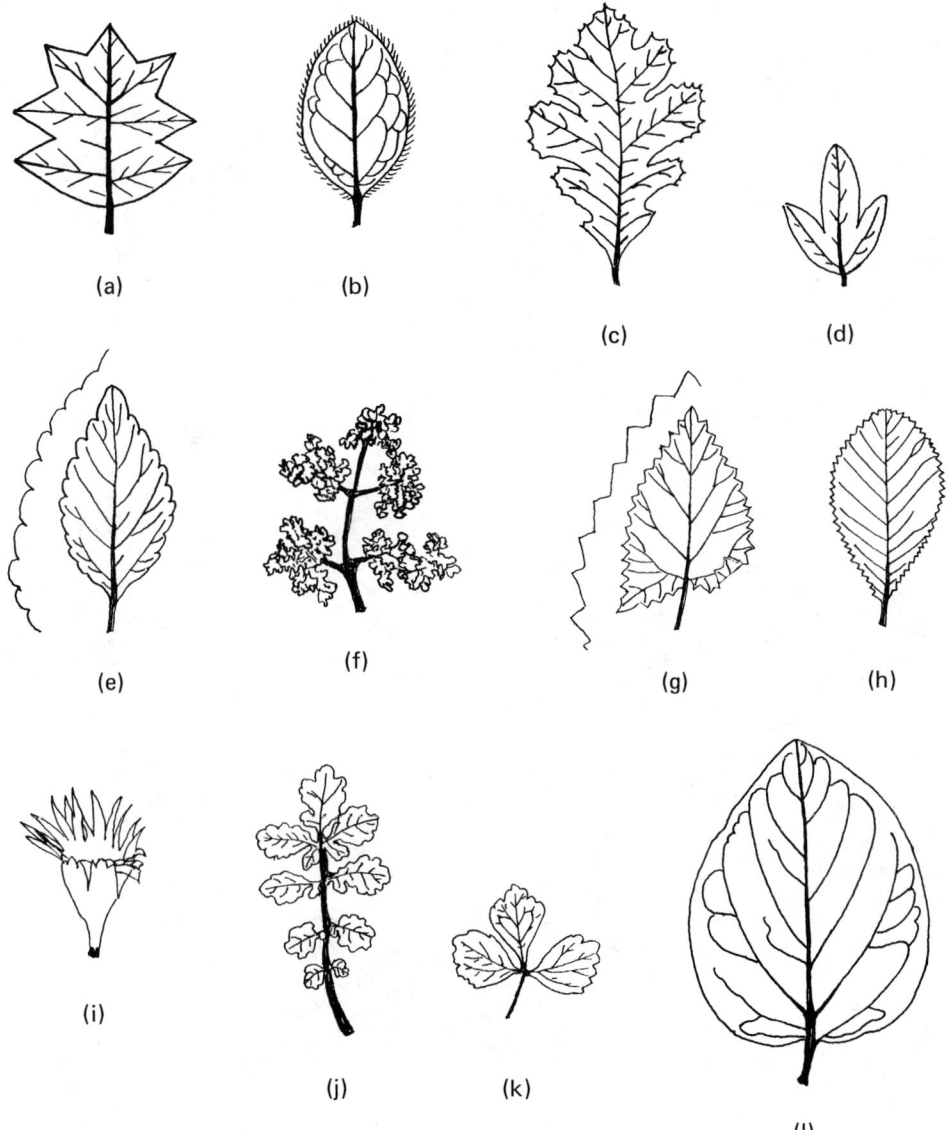

Fig. II.5.05. Margins: (a) angular; (b) ciliate; (c) cleft (pinnatifid); (d) cleft (palmatifid); (e) crenate; (f) crispate; (g) dentate; (h) denticulate; (i) dissected, laciniate; (j) divided (pinnately); (k) divided (palmately); (l) entire. (b), (e)-(i) Redrawn from Lawrence, 1955; (c), (d), (e) (outline), (g) (outline), (j), (k) Redrawn from Gray, 1862; see Acknowledgments and Selected References.

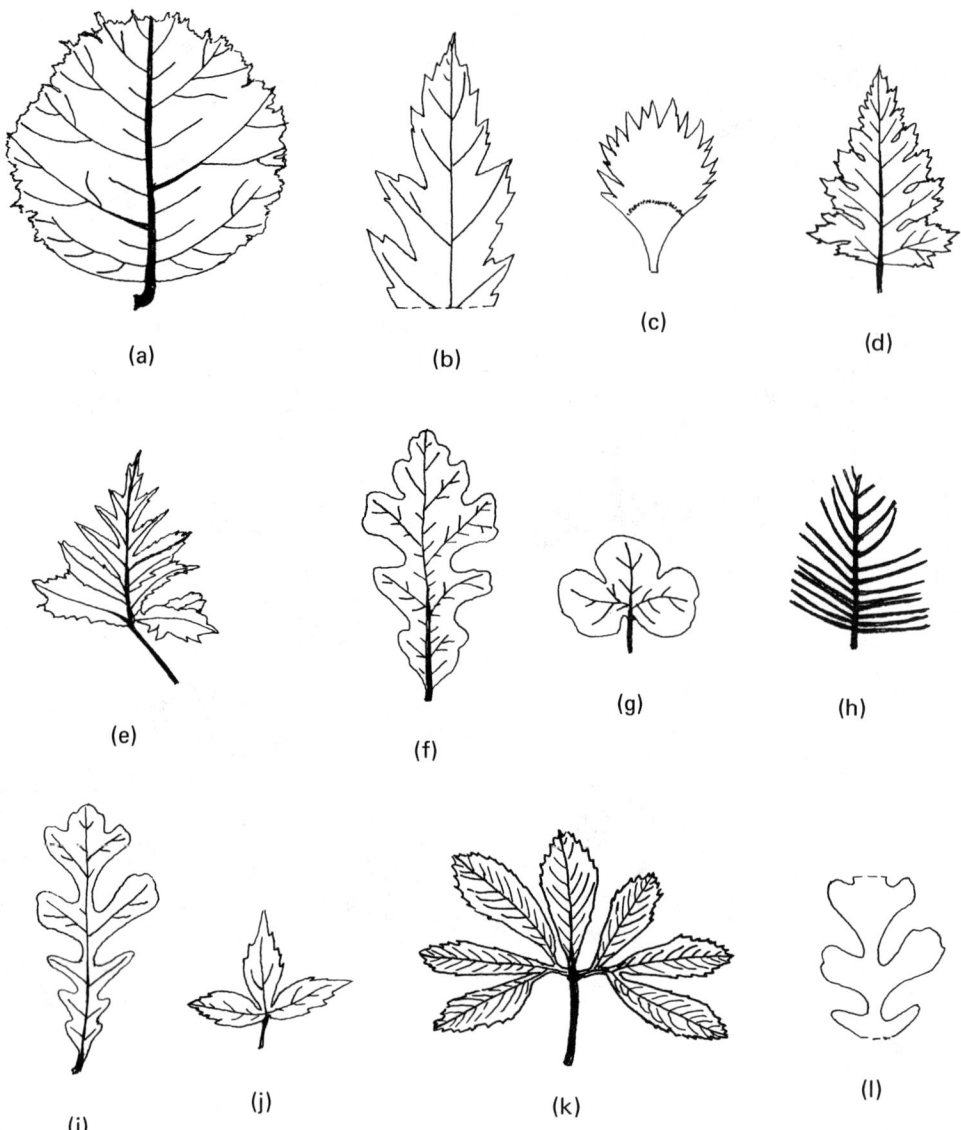

Fig. II.5.06. Margins: (a) erose; (b), (c) incised; (d) lacerate; (e) laciniate; (f) lobed (pinnatifid); (g) lobed (palmatifid); (h) pectinate; (i) parted (pinnately); (j) parted (palmately; (k) pedate; (l) sinus. (b), (f)-(l) Redrawn from Gray, 1862; (c), (d) redrawn from Lawrence, 1955; (e) redrawn from Hedrick, 1925; (k) redrawn from Hutchinson, 1969; see Acknowledgments and Selected References.

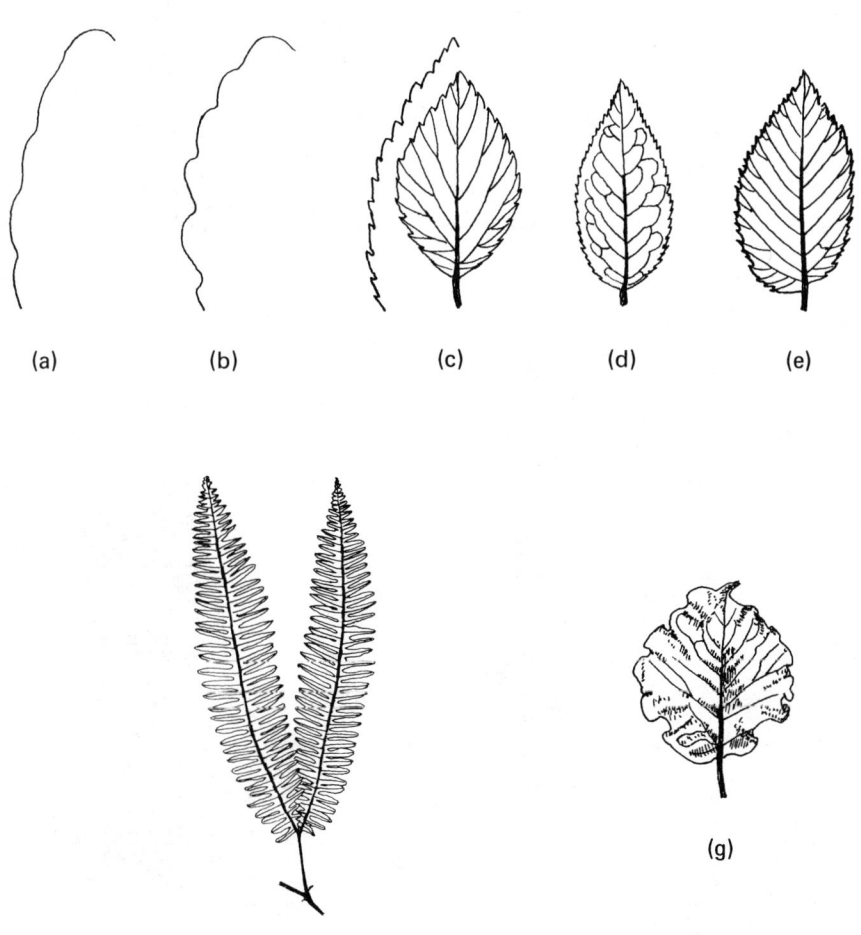

Fig. II.5.07. Margins: (a) repand; (b) sinuate; (c) serrate; (d) serrulate; (e) biserrate; (f) squarrose-slashed (Gleichenia dichotoma); (g) undulate. (a), (b), (c) Redrawn from Gray, 1862; (c), (d), (e), (g) redrawn from Lawrence, 1955; (f) redrawn from Schneider, 1893; see Acknowledgments and Selected References.

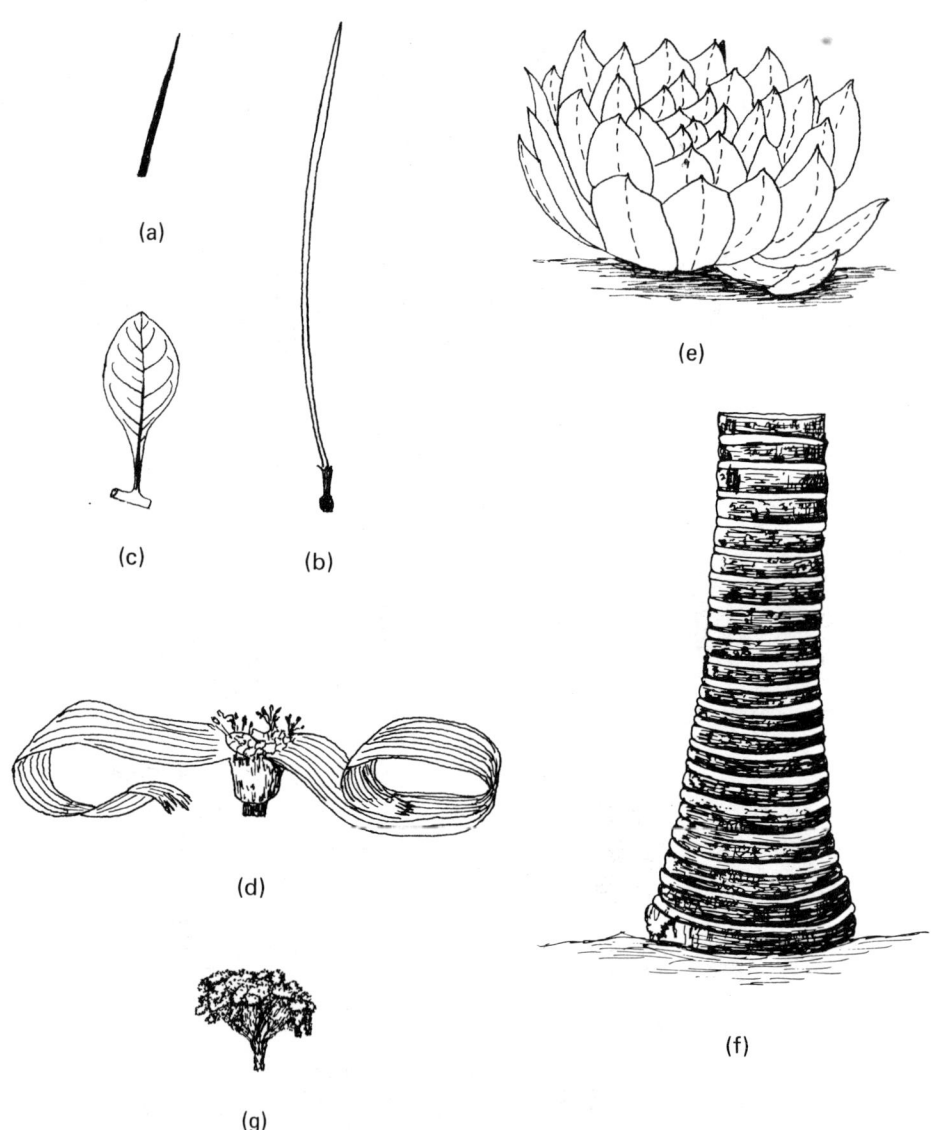

Fig. II.5.08. Shapes: (a) acerose; (b) acicular; (c) attenuate; (d) band-shaped (Welwitschia); (e) cespitose (Sempervivum), (f) columnar (Archontophoenix alexandrae var. beatricae); (g) cristate (Celosia). (c) Redrawn from Gray, 1868; (d) redrawn from Bell and Coombe, 1976; (e) redrawn from Bailey, 1928; (f) redrawn from Bailey, 1935; (g) redrawn from Lawrence, 1955; see Acknowledgments and Selected References.

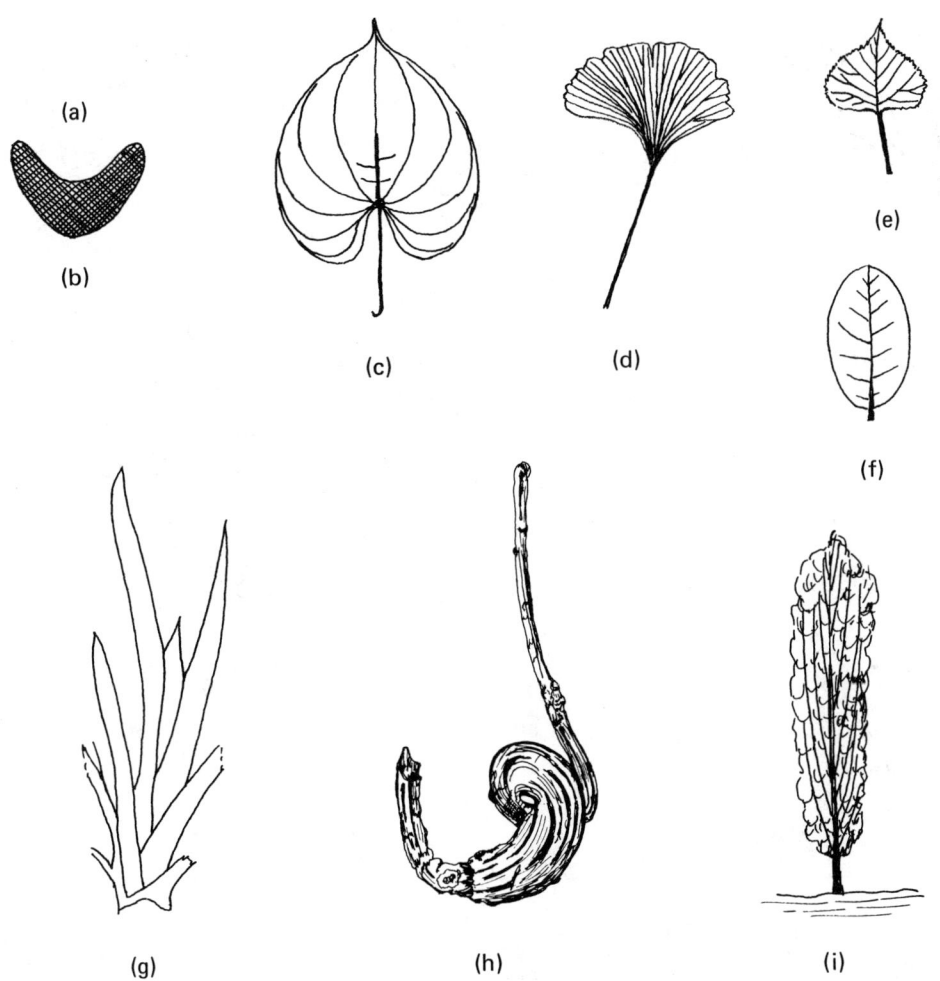

Fig. II.5.09. Shapes: (a) concave; (b) convex; (c) cordate (<u>Dioscorea bulbifera</u>); (d) cuneiform (<u>Ginkgo</u>); (e) deltoid (poplar); (f) elliptical; (g) ensiform (<u>Iris</u>); (h) fasciated (apple); (i) fastigiate (Lombardy poplar). (d) Redrawn from Bell and Coombe, 1976; (e) redrawn from Lawrence, 1955; (f), (g) redrawn from Gray, 1862; (h) redrawn from Heald, 1933; see Acknowledgments and Selected References.

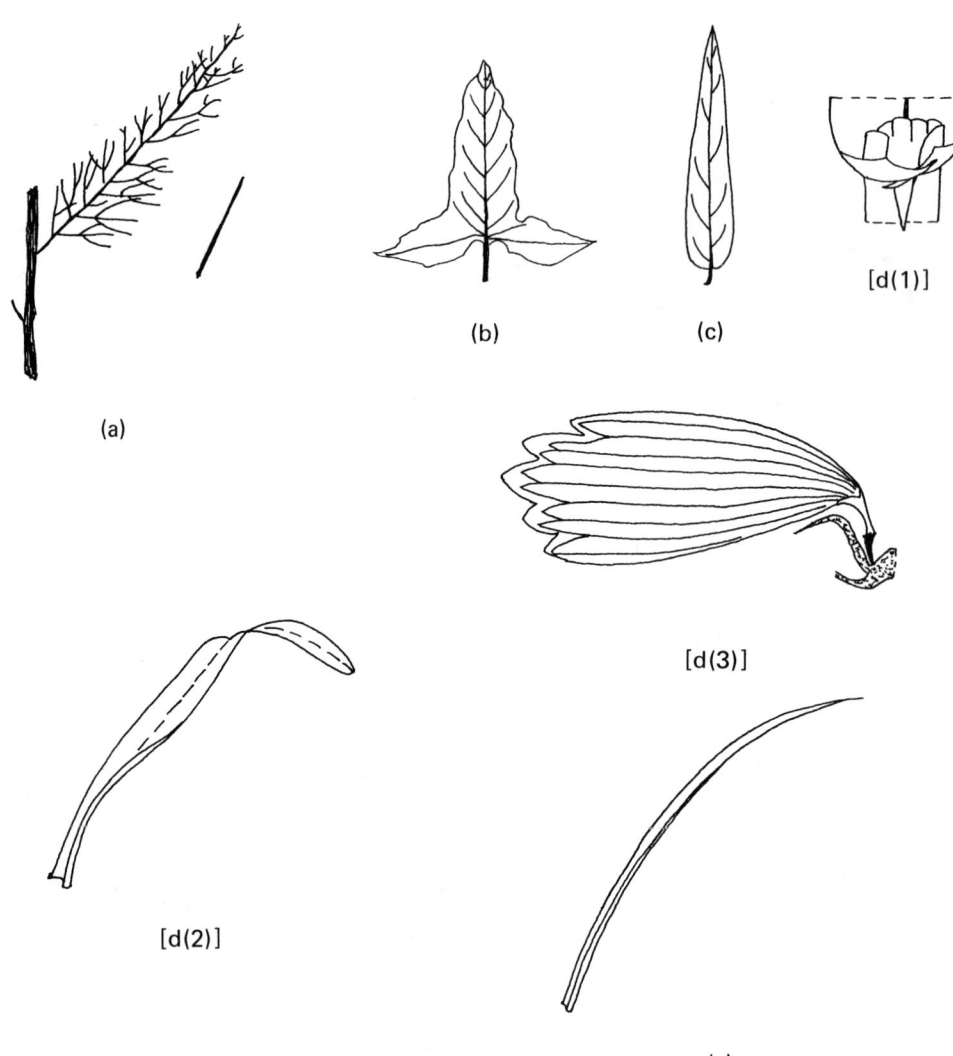

Fig. II.5.10. Shapes: (a) filiform (dog-fennel, detached leaf at right); (b) hastate; (c) lanceolate; [d(1)] ligule, ligulate, (2) lorate, (3) ligulate (composite flower); (e) linear. (a) Adapted from Muenscher, 1955; (b) after Lawrence, 1955; [d(1)] adapted from Robbins, 1924; [d(2)] (e) redrawn from Lawrence, 1955; [d(3)] redrawn from Gray, 1862; see Acknowledgments and Selected References.

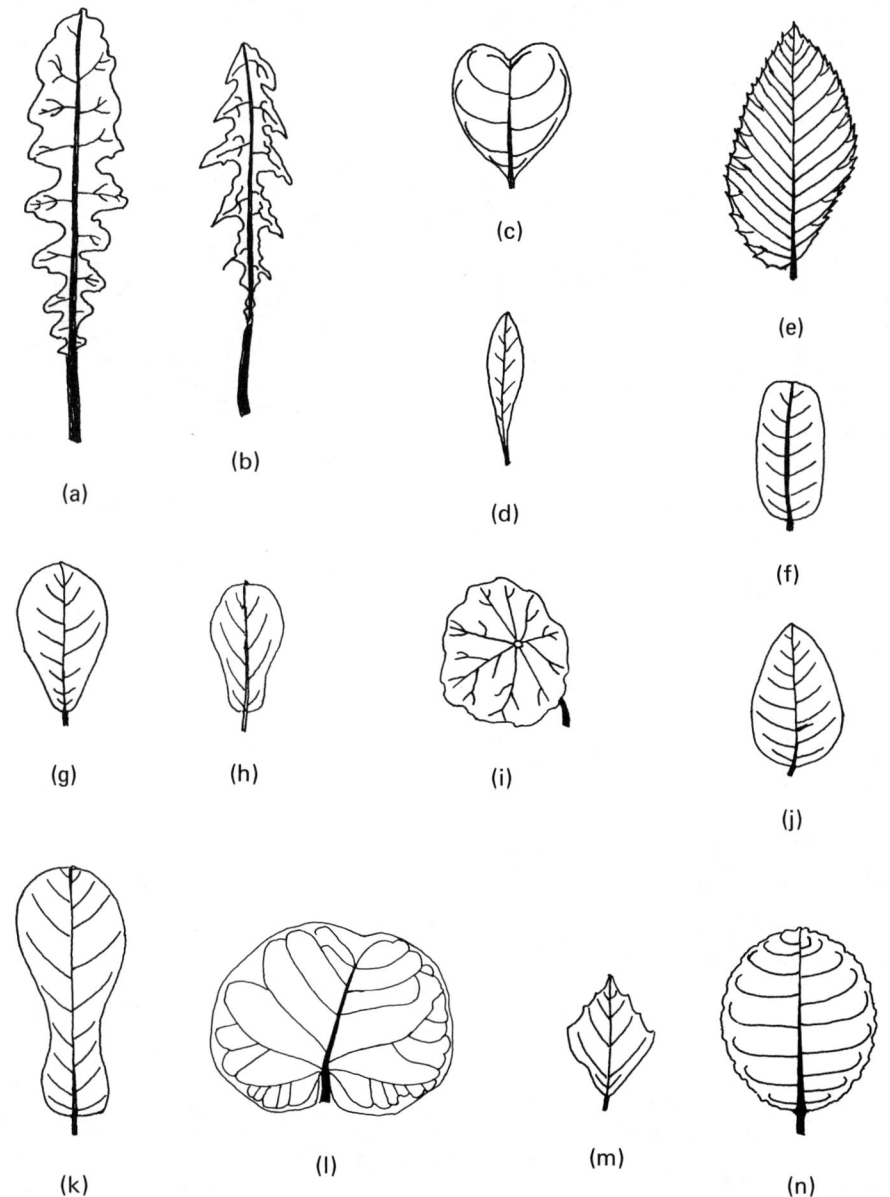

Fig. II.5.11. Shapes: (a) lyrate (<u>Hieracium</u>); (b) runcinate (dandelion); (c) obcordate; (d) oblanceolate; (e) oblique (elm); (f) oblong; (g), (h) obovate; (i) orbicular, peltate (nasturtium); (j) ovate; (k) pandurate; (l) reniform (seagrape); (m) rhombate; (n) roundish. (d), (f), (g), (j) Redrawn from Gray, 1868; (i) redrawn from Gray, 1862; (h), (k), (m) redrawn from Lawrence, 1955; see Acknowledgments and Selected References.

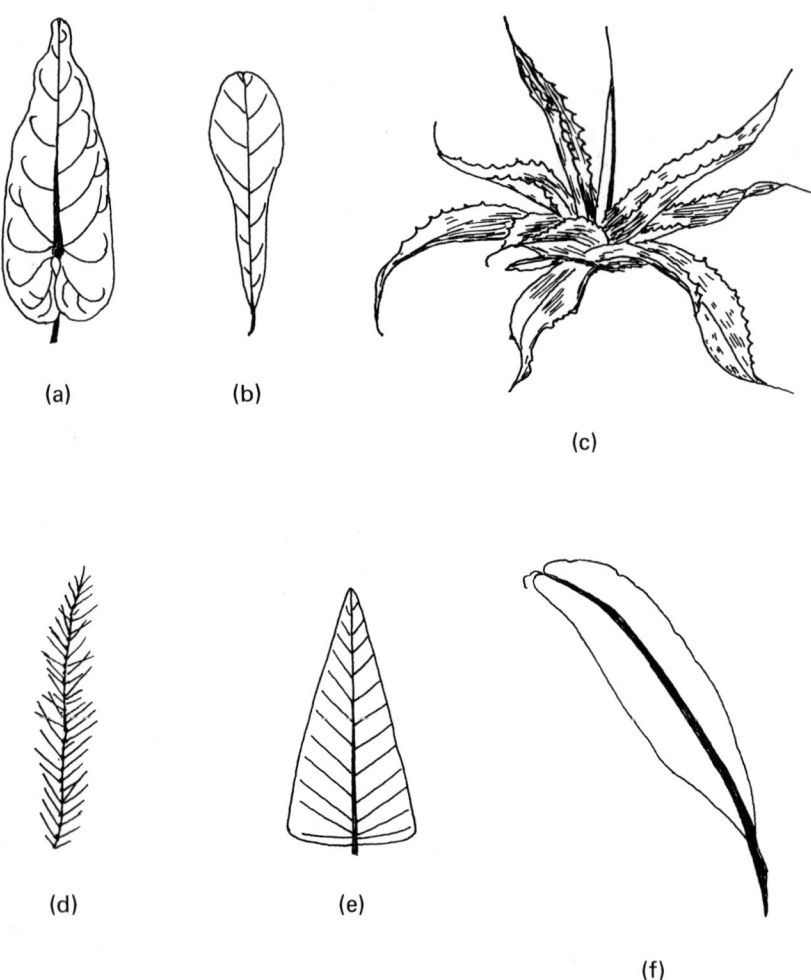

Fig. II.5.12. Shapes: (a) sagittate; (b) spatulate; (c) squarrose (Agave); (d) subulate (juvenile red cedar leaves); (e) triangular; (f) unequal (Musa sp. leaf). (a) Adapted from Lawrence, 1955; (b) redrawn from Gray, 1868; (c) redrawn from Gray, 1868; (f) adapted from Tai, 1958; see Acknowledgments and Selected References.

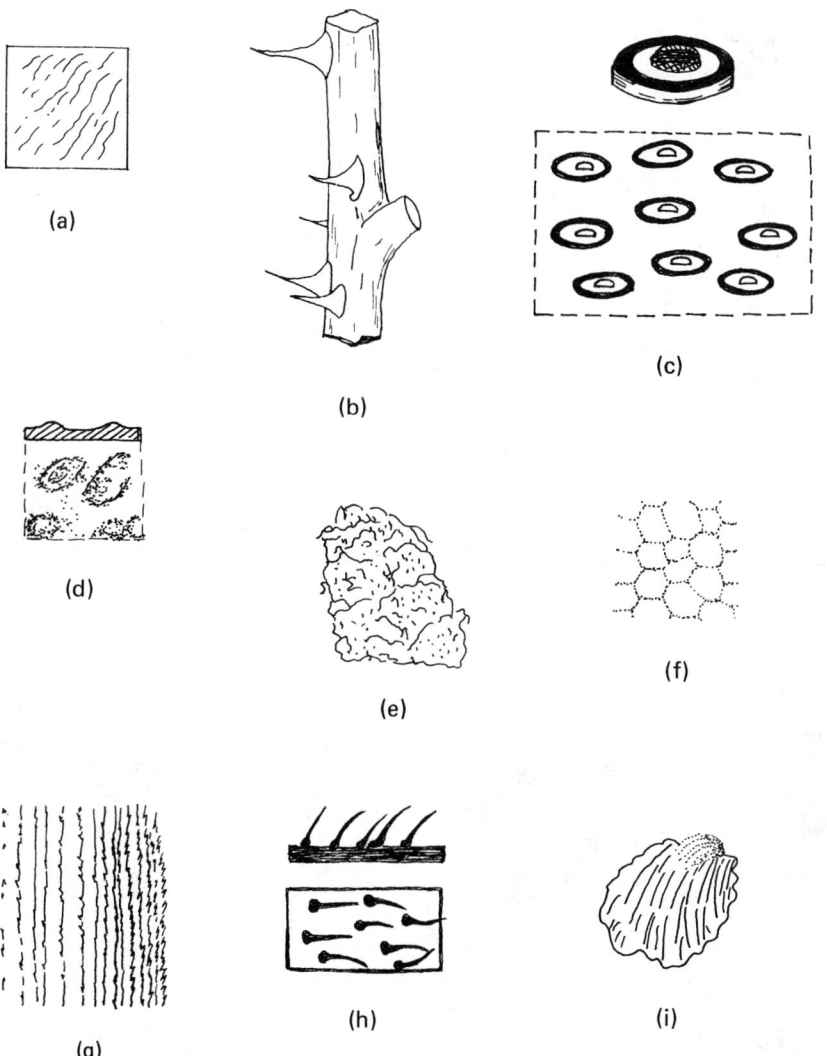

Fig. II.5.13. Surface features: (a) aciculated (diagrammatic); (b) aculeate; (c) annulate; (d) alveolate; (e) arachnoid; (f) areolate; (g), (h) asperous (scabrous); (i) bearded. (b) Redrawn from Bailey, 1941; (e), (g), (i) redrawn from Radford et al., 1974; (d), (f) redrawn from Stearn, 1966; (h) redrawn from Lawrence, 1951; see Acknowledgments and Selected References.

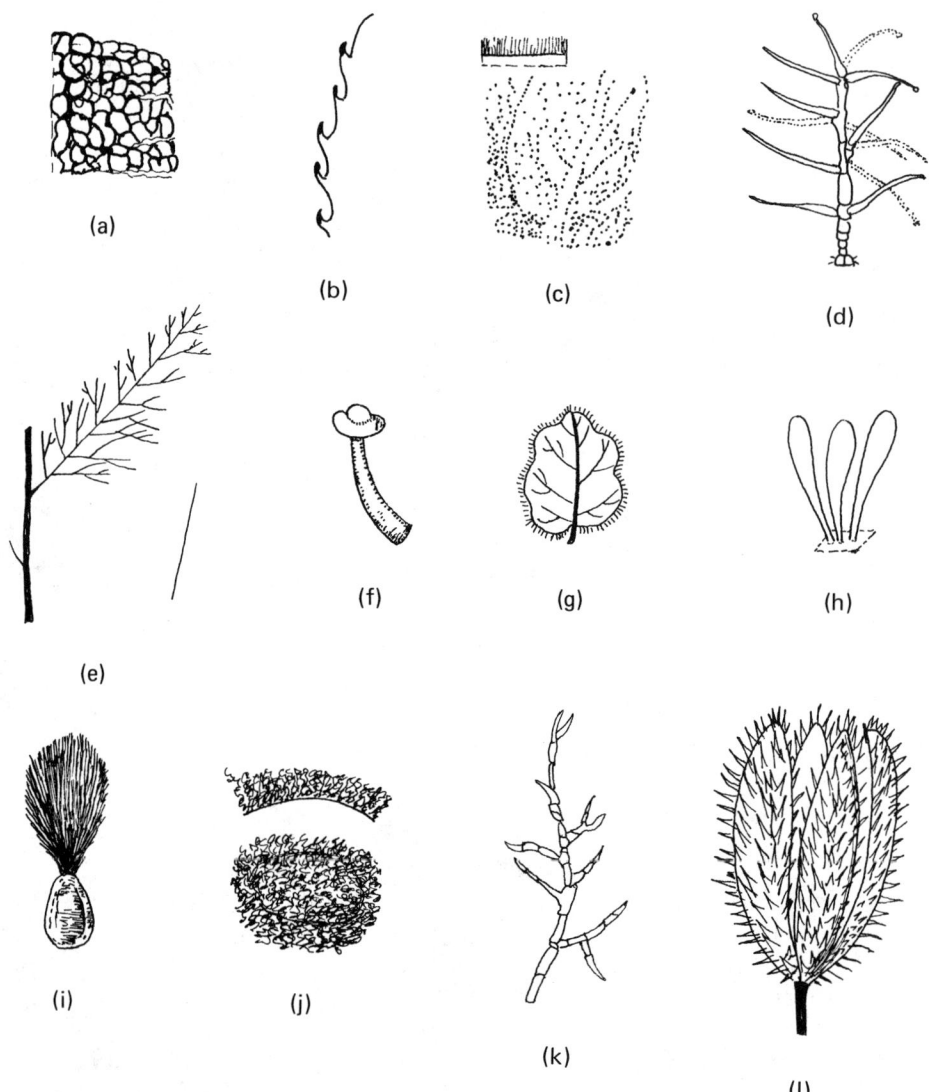

Fig. II.5.14. Surface features: (a) bullate; (b) callus (tips); (c) canescent; (d) candelabra (Platanus trichome); (e) capillary; (f) capitate; (g) ciliate; (h) clavate; (i) comose (milkweed); (j) crispy-hairy; (k) dendritic; (l) echinate (Clappertonia ficifolia). (a) Redrawn from Lawrence, 1951; (c), (g), (h), (k) redrawn from Radford et al., 1974; (d) redrawn from Esau, 1965; (f) redrawn from Lawrence, 1955; (i) redrawn from Gray, 1862; (j) adapted from U.S. Dept. Agr., 1961; (l) redrawn from Hutchinson, 1969; see Acknowledgments and Selected References.

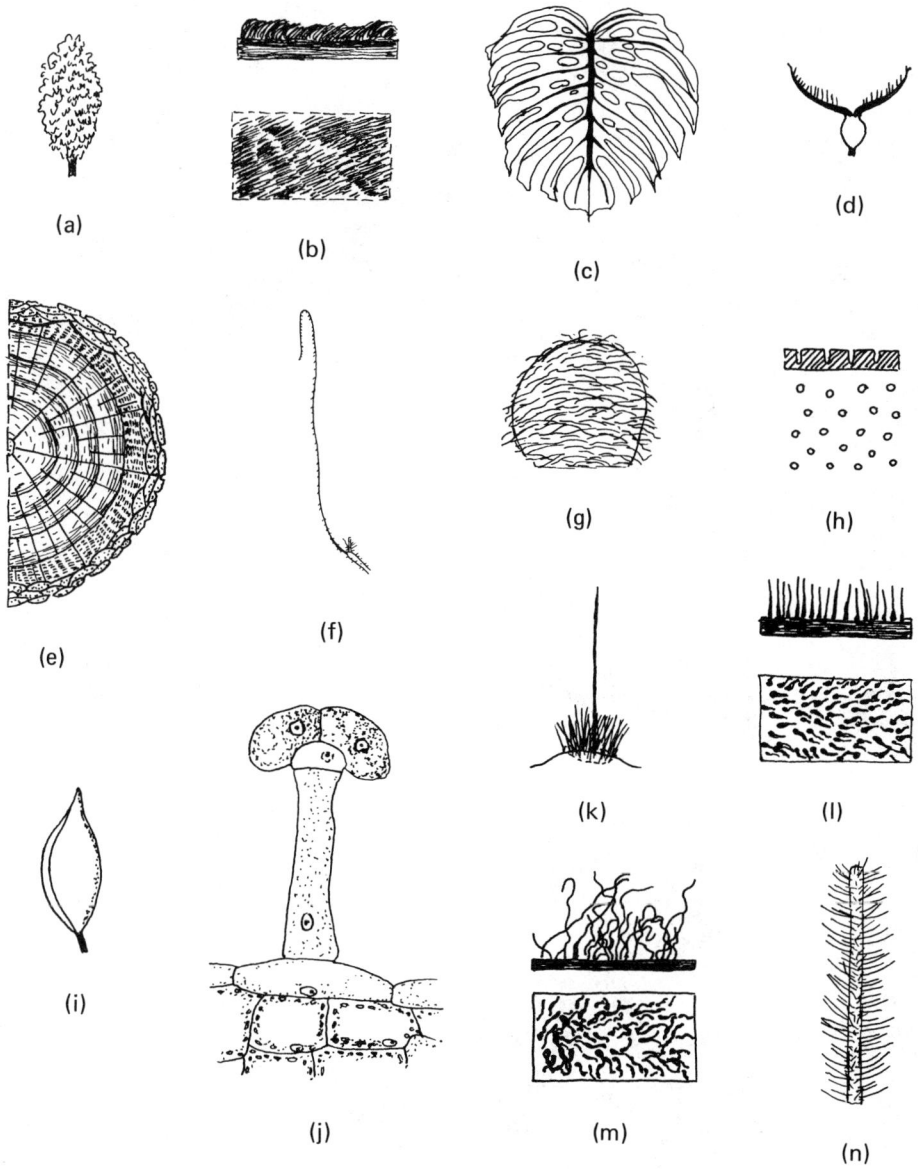

Fig. II.5.15. Surface features: (a) farinose; (b) felted-tomentose; (c) fenestrate (Monstera); (d) fimbriate; (e) fissured; (f) flagelliform; (g) floccose; (h) foveolate; (i) fusiform (Brachychiton populneum); (j) glandular; (k) glochidiate; (l) hirsute; (m) villous; (n) hispid. (a), (d), (g) Redrawn from Radford et al., 1974; (b), (l), (m), (n) redrawn from Lawrence, 1951; (k) adapted from Lawrence, 1955; (c) adapted from Watkins and Sheehan, 1975; (e), (j) redrawn from Eames and MacDaniels, 1947; (f) redrawn from Corner, 1966; (h) redrawn from Stearn, 1966; (i) redrawn from Hutchinson, 1969; see Acknowledgments and Selected References.

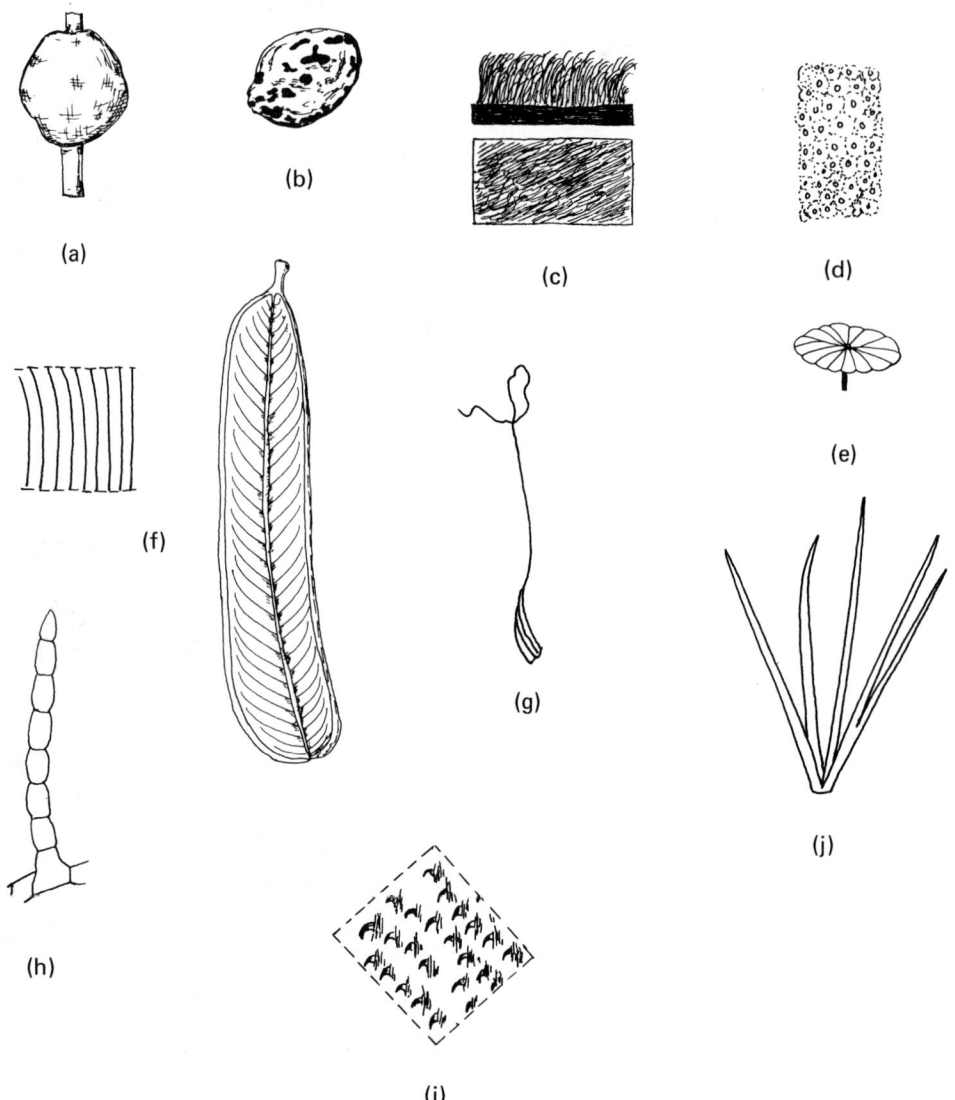

Fig II.5.16. Surface features: (a) knoblike; (b) lacunose (dried pea seed); (c) lanate; (d) lepidote, with (e) peltate scales; (f) lined (diagram, with Tetrapleura tetraptera pod at right); (g) loriform; (h) moniliform; (i) muricate (Annona muricata, X 1/4); (j) multangulate (trichome). (b) Adapted from U.S. Dept. Agr., 1961; (c) redrawn from Lawrence, 1951; (d), (e), (g), (j) redrawn from Radford et al., 1974; (f) pod redrawn from Hutchinson, 1969; (h) redrawn from Eames and MacDaniels, 1947; (i) adapted from Ochse, 1931; see Acknowledgments and Selected References.

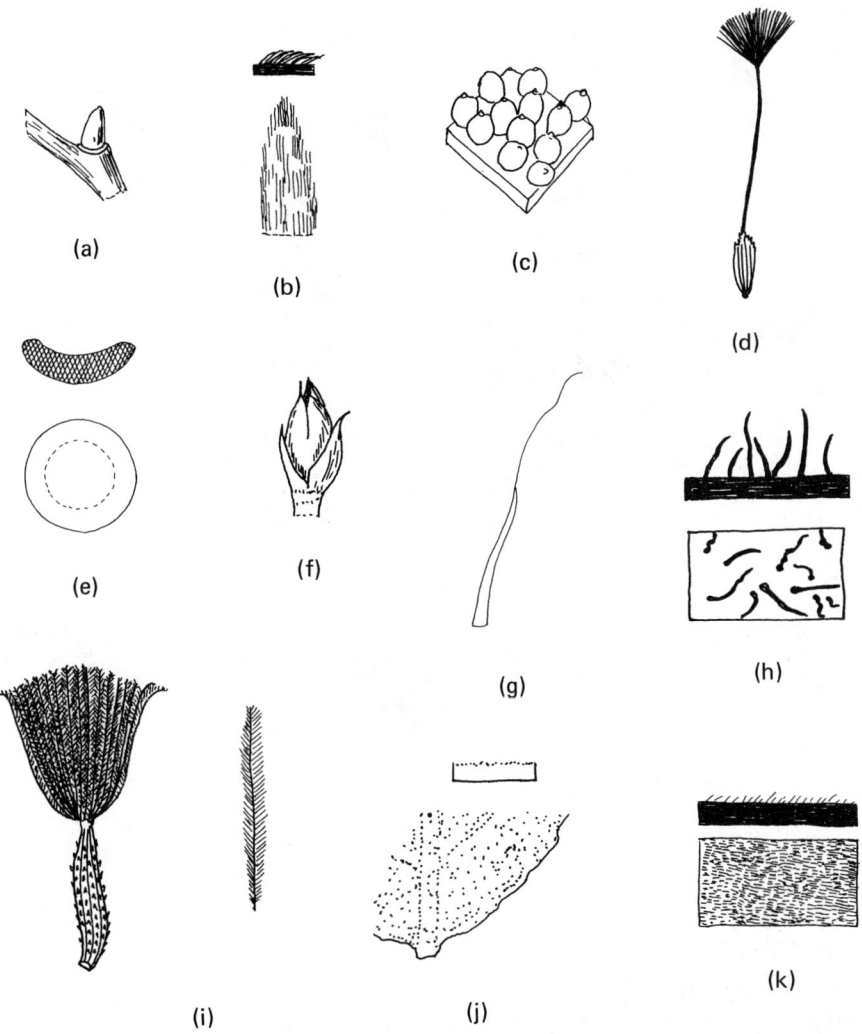

Fig. II.5.17. Surface features: (a) naked (Platanus bud); (b) pannose; (c) papillose; (d) pappus (dandelion); (e) patelliform (diagrammatic); (f) perulate (shagbark hickory); (g) piliferous (trichome); (h) pilose; (i) plumose (Thrincia hirta seed, with individual hair at right); (j) powdery; (k) puberulous. (a), (f) Redrawn from Gray, 1868; (b), (j) redrawn from Radford et al., 1974; (d) redrawn from Gray, 1862; (c), (h), (k) redrawn, (g) adapted from Lawrence, 1951; (i) adapted from Lubbock, 1894; see Acknowledgments and Selected References.

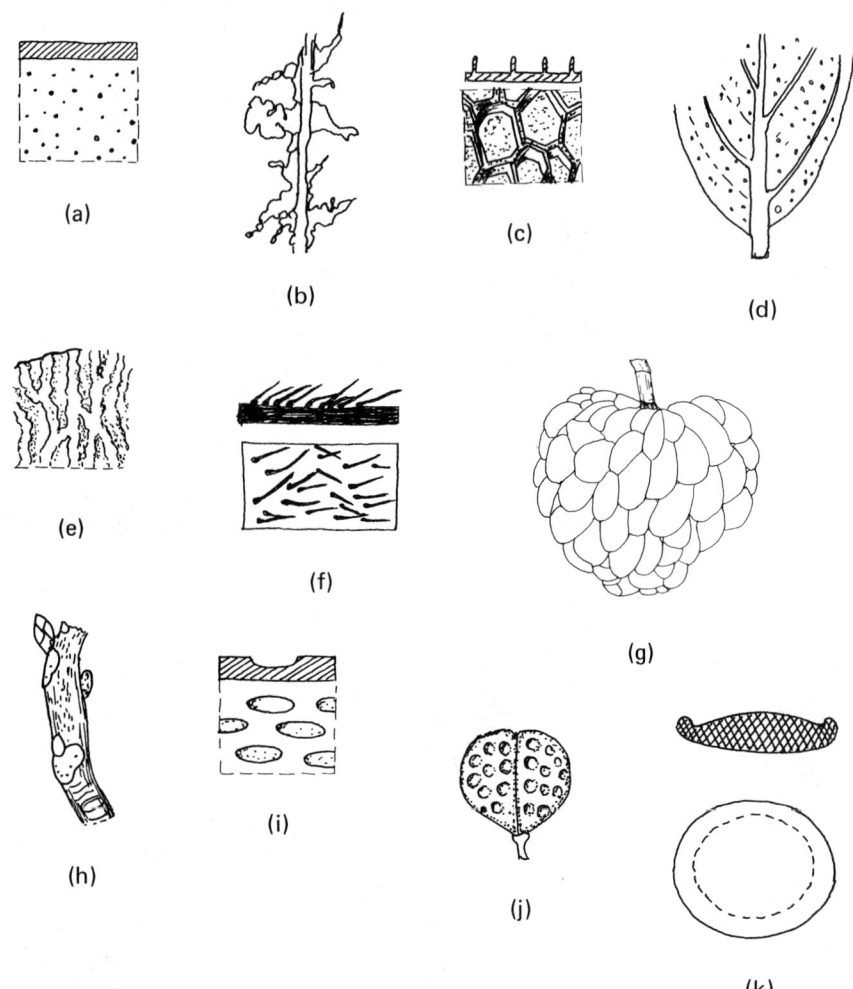

Fig. II.5.18. Surface features: (a) punctate; (b) ramentaceous; (c) reticulate; (d) resinous; (e) rugose; (f) scabrous; (g) squamose (Annona squamosa); (h) scarred; (i), (j) scrobiculate (Coronopus rytidocarpus); (k) scutate (diagrammatic). (a), (c), (e), (i) Redrawn from Stearn, 1966; (b), (d) redrawn from Radford et al., 1974; (f) redrawn from Lawrence, 1951; (g) adapted from Ochse, 1931; (h) redrawn from Gray, 1868; (j) redrawn from Hutchinson, 1969; see Acknowledgments and Selected References.

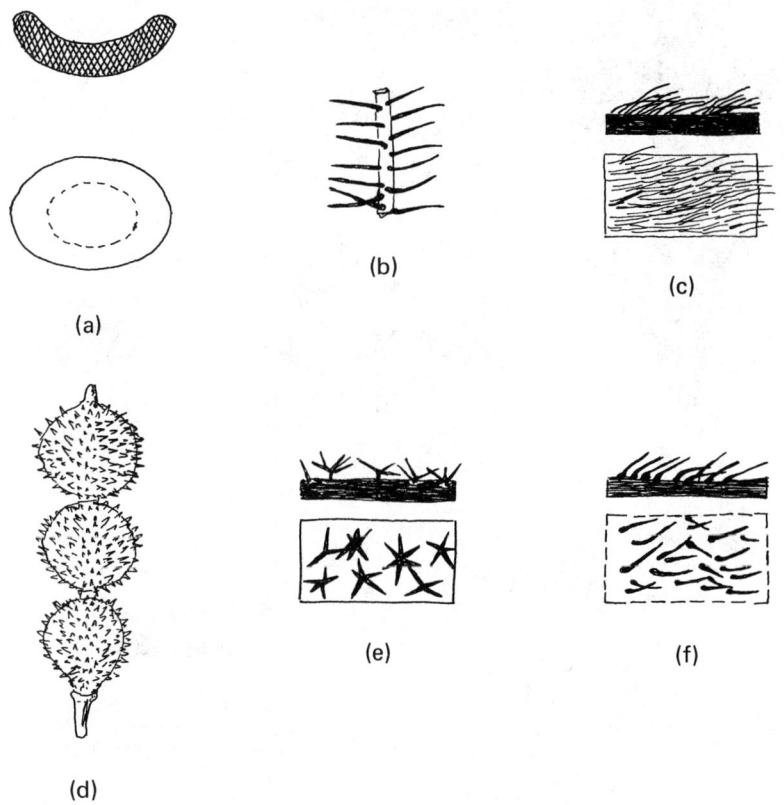

Fig. II.5.19. Surface features: (a) scutelliform (diagrammatic); (b) setose; (c) silky; (d) spiculate (Hedysarum coronarium); (e) stellate; (f) strigose. (b), (c), (e), (f) Redrawn from Lawrence, 1951; (d) redrawn from Hutchinson, 1969; see Acknowledgments and Selected References.

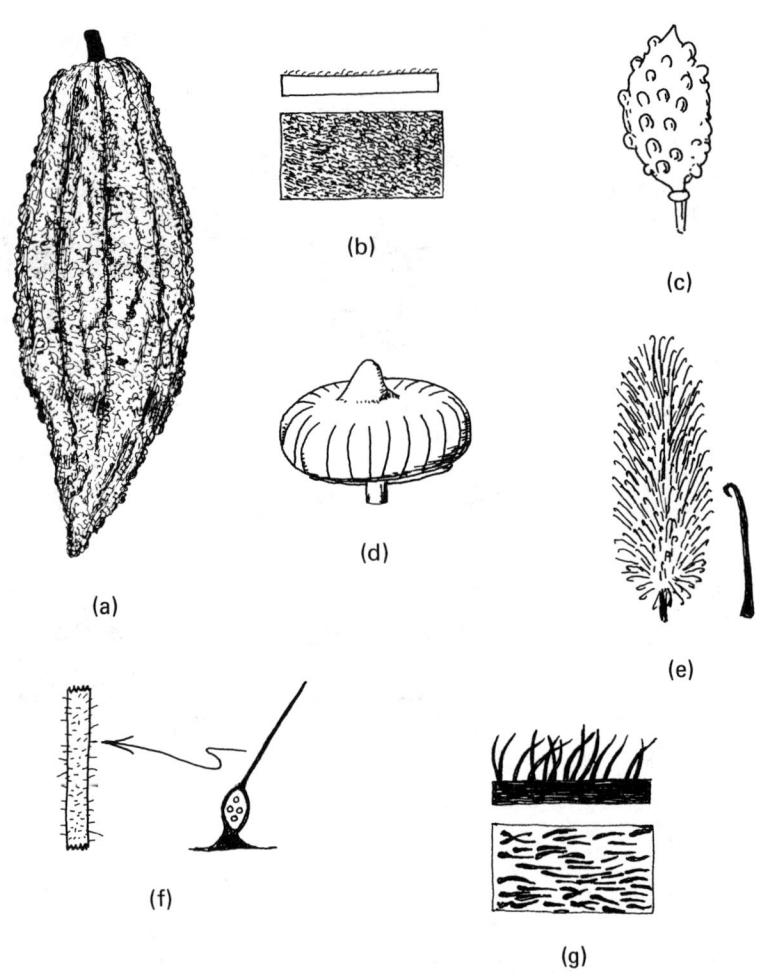

Fig. II.5.20. Surface features: (a) sulcate (<u>Theobroma cacao</u> "Cundiamor"); (b) tomentose; (c) tuberculate (<u>Trachylobium verrucosum</u>); (d) umbonate (<u>Potalia amara</u>); (e) uncinate (<u>Tourrettia lappacea</u>, with bristle at right); (f) urent (stinging nettle); (g) velvety. (a) Redrawn from Cuatrecasas, 1964; (b), (g) redrawn from Lawrence, 1951; (c)-(e) redrawn from Hutchinson, 1969; (f) adapted from Millspaugh, 1892; see Acknowledgments and Selected References.

Fig. II.6.01. Bryophytes, pteridophytes, and gymnosperms: (a) bryophyte, club moss (Selaginella); (b) pteridophyte, tree fern (Alsophila); (c), (d) gymnosperms: (c) conifer, China-fir (Cunninghamia); (d) Cycad: coontie (Zamia). (a) Redrawn from Bell and Coombe, 1976; (c) from, (b), (d) redrawn from Watkins and Sheehan, 1975; see Acknowledgments and Selected References.

III. HORTICULTURAL TAXONOMY and PLANT BREEDING

1. Taxonomy 302
   - 1.0. Horticultural Taxonomy 302
   - 1.1. Botanical categories, classification, nomenclature 302
   - 1.2. Biosystematics 307
2. Plant Breeding 311
   - 2.0. General 311
   - 2.1. Breeding and selection systems, gene pools 316
   - 2.2. Terms 317
     - Fig. III.2.2.01 327

III.1. Taxonomy

III.1.0. Horticultural taxonomy

.000 Taxonomy: the science of identification, nomenclature, and classification of objects.

.001 Horticultural taxonomy: the science of identification, nomenclature, and classification of cultivated plants, in which the underlying premises are (1) each plant has certain definable characteristics that serve to distinguish one from another, (2) each plant has one and only one valid scientific name, and (3) each plant bears a definite relationship to other members of the plant kingdom; complicating factors include gross morphological changes in nearly all cultigens, generally more luxuriant growth of cultigens, appearance of hybrids in all degrees of complexity, asexual perpetuation of monstrous or anomalous forms unknown in the wild, and numerous cases of overlapping characters and intermediate forms among cultigens.

III.1.1. Botanical categories, classification, and nomenclature

.001 Angiosperms (Angiospermae): flowering plants, characterized by the presence of vessels (lacking in a few families) in the wood, ovules enclosed in a megasporophyll (ovary), microsporangia (anthers) borne on a microsporophyll (filament), and a perianth of sepals and petals, either or both of which may be absent; two major subdivisions are dicotyledons and monocotyledons. (HS 73)

.002 Category: a level in a hierarchy; thus each plant from a taxonomic standpoint can belong to only one; a taxon such as sweet orange [Citrus sinensis (L.) Osbeck] belongs to the species category only but is a member of other taxa, such as the genus Citrus, subtribe Citrinae, tribe Citreae, subfamily Aurantioideae, and family Rutaceae. (L 51)

.003 Classification: the systematic arrangement or method of arrangement of plants (and animals) in groups or categories according to some definite plan or sequence; categories in common use are Division (Divisio), Class (Classis), Order (Ordo), Family (Familia), Genus

III.1.1. HORTICULTURAL TAXONOMY and PLANT BREEDING 303

(Genus), Species (Species), Variety (Varietas), and for cultivated plants, Cultivar; each species belongs to some genus that may or may not include other species, all agreeing in certain features characteristic of that genus; each genus is a member of a family that possesses certain family characters in common; each family is a member of an order and so on; categories of intermediate rank are also used and are named by adding the prefix sub- to the category immediately above. (L 51)

.004 Classification systems: the arrangement of plants in groups on the basis of their natural affinities (i.e., similarity of morphological characters) and presumed phylogeny (phylogenetic); those by Engler and Prantl (Die natuerlichen Pflanzenfamilien and Syllabus der Pflanzenfamilien) covers the entire plant kingdom and is widely used in Europe and the United States; others by Hutchinson (Families of Flowering Plants), used in the United Kingdom and Commonwealth countries, Takhtajan (Flowering Plants - Origin and Dispersal) in Russia, and Cronquist treat only flowering plants (i.e., angiosperms). (L 51)

.005 Common names: words in a local language used for wild or cultivated plants, especially the familiar or economically important, e.g., beet [=garden beet, sugarbeet or mangold (forage beet) Beta vulgaris L.] and sweet orange [Citrus sinensis (L.) Osbeck], and as such may be given legal status under the Federal Seed Act of 1939 as Amended as a "kind" (III.2.0.004). Common names, however, are not regulated under the International Code of Botanical Nomenclature or the International Code of Nomenclature of Cultivated Plants. Usage of common names varies tremendously around the world, particularly in the tropics, and is subject to confusion through lack of precision when the same word is applied to quite different plants; e.g., "corn" means Indian corn = maize (Zea mays) in the United States but "wheat" in the Old World; "sapote" is applied to fruit in at least three different families (Sapotaceae, Ebenaceae, and Rutaceae); and "pimento", "pimienta", "pimiento", and "pepper", of which the first is used for allspice in Jamaica and the others for black pepper (Piper nigrum) or Capsicum peppers. Conversely, different names may be used for the same plant, such as "bissy" = kola nut in Jamaica, avocado = alligator pear, palta (Peru), cupanda, zial, sikia, kulup, amo, and devora (South America), and sapodilla = naseberry, dilly, or chico. (L 51, OSDW)

.006 Diagnosis: a concise description of a species or group, giving its distinguishing characteristics. (L 51)

.007 Dicotyledons (Dicotyledoneae): plants woody or herbaceous, stems with vascular elements like a hollow cylinder or in bundles in a single circle (rarely scattered), leaves typically netted and veined in palmate or pinnate form, flowers basically with parts in fours or fives or multiples thereof, or numerous, embryos typically with two cotyledons. (HS 73).

.008 Family (Familia): a group of related plants that form a category ranking above a genus and below an order; usually includes several to many genera or, in an exceptional case, only one genus; names of families are derived from the genus designated as the type for the family and are plural substantives with the suffix -aceae, except for eight families for which there are alternative names (e.g., Palmae or Arecaceae, Umbelliferae or Apiaceae, Gramineae or Poaceae); a (presumably) natural grouping, usually separated from others of the same rank by characters inherent

III.1.1. HORTICULTURAL TAXONOMY and PLANT BREEDING 304

.009
.010
.011
.012
.013
.014
.015
.016
.017
.018

in reproductive structures, such as type of inflorescence, ovary position, type of placentation, number of carpels and pistils, type and position of ovules, embryology, special androecial conditions (e.g., monandry, diandry, and syngenesism), and disposition of the sexes (e.g., dioecism and monoecism). (L 51)

.009 "Fancy" name: (see Cultivar, III.2.0.003)

.010 Flora: an essentially monographic treatment or assemblage of plants of a given area, usually arranged in systematic fashion; e.g., G. Bentham and J. D. Hooker, 1862-1863. Genera plantarum. London. 3 vol.; or A. Engler and L. Diels, 1964. Syllabus der Pflanzenfamilien, 12th ed. Berlin, as world floras; or J. Hutchinson, 1948. British flowering plants. London. and M. L. Fernald, 1950. Gray's manual of botany, 8th ed., etc., as regional floras. (L 51)

.011 Form (Forma): a category ranking below a subspecies, used chiefly for certain minor variations (e.g., the yellow passion fruit, Passiflora edulis f. flavicarpa, whose fruit is yellow rather than purple as in P. edulis proper); a sporadic variant, equivalent to "variety" of some botanists but generally trivial such as corolla or fruit color or habit response. (L 51)

.012 Genus (Genus): a category of classification between a family and a species; a group of structurally or phylogenetically related species or consisting of an isolated species that exhibits unusually differential features (monotypic genus); distinctions between genera are sometimes empirical or arbitrary and liable to modification as knowledge advances; a category antedating binomial nomenclature, composed of plants with two or three characters of reproductive structures in common, although characters used for separation vary widely among different families. (L 51)

.013 Graft-chimaera: plants composed of tissues in intimate association from two different individuals; they originate by grafting and are not sexual hybrids; may be designated by formula or by a name with a plus (+) sign used instead of a X, as for a graft hybrid. (GL)

.014 Graft-hybrid: a sexual hybrid between two or more species or genera, which can be denoted by the botanical names of the parents connected by a multiplication sign (X) = formula, or a botanical name for an interspecific hybrid consisting of the generic name followed by a Latin collective epithet, the latter immediately preceded by X; or for an intergeneric hybrid, a "generic name" preceded by X and normally followed by a Latin collective epithet; a "generic" name of a multigeneric hybrid usually consists of a personal name with the suffix -ara. (GL)

.015 Grex (g.): swarm or flock, used in certain collective epithets to designate the nature of the unit. (GL)

.016 Identification: determination of a taxon as being identical with or similar to another and already known element or to an unknown element. (Note, however, complicating factors listed under Horticultural taxonomy, III.1.0.001.) (L 51)

.017 Index Kewensis: a complete world list of genera and species of spermatophytes; B.D. Jackson et al., comp. 1893-   . Index Kewensis plantarum phanarogamarum. 2(4) vol. Clarendon Press, Oxford. plus supplements to date (issued every five years); an invaluable reference source for published names of plants (including the author). (L 51)

.018 Index Londinensis: an index to illustrations of flowering plants, ferns, and fern allies, including genera, species, and varieties;

III.1.1. HORTICULTURAL TAXONOMY and PLANT BREEDING

O. Stapf et al., comp. 1920-   . Oxford; 6 vol. plus supplements. (L 51)

.019 International Code of Botanical Nomenclature: the rules for the use of names in Latin form for wild and cultivated plants: guiding principles are (1) botanical nomenclature is independent of zoological nomenclature; (2) the application of names of taxonomic groups is based on nomenclatural types; (3) each taxonomic group has only one correct name, the earliest (based on May 1, 1753, Linnaeus, Species Plantarum, ed.1, as the starting point) in accordance with the rules, except for nomina conservanda; (4) scientific names of taxonomic groups (taxa) are treated as Latin, regardless of their derivation; and (5) the rules are retroactive unless expressly limited. (L 51, SF)

.020 International Code of Nomenclature of Cultivated Plants: the rules for naming, etc., of cultivated plants: guiding principles are (1) importance of a precise, stable, internationally accepted system available for naming of cultivated plants; (2) use of botanical names in Latin form for cultivated and wild plants, except graft-chimaeras, governed by the International Code of Botanical Nomenclature; (3) promotion of uniformity, accuracy, and fixity in naming of agricultural, horticultural, and silvicultural cultivars (varieties), with specific exclusion of common names of genera and species; (4) importance of registration of cultivar names for nomenclatural stability; (5) strong urging that Articles and Recommendations of the Code be accepted and applied by those responsible for the legal use of cultivar names; and (6) Articles of Code are retroactive unless expressly limited. Cultivar names published on or after January 1, 1959, must follow this Code. (GL)

.021 Major taxonomic category: a group that does not form part of the plant name; i.e.,a tribe, family, or higher level; e.g., the tribe Citreae or family Rutaceae. (L 51)

.022 Minor taxonomic category: a group or groups that include levels that form part of the name, such as the genus and species plus infraspecific epithets if needed; e.g., Citrus sinensis cv. Hamlin. (L 51)

.023 Monocotyledons (Monocotyledoneae): plants herbaceous, less commonly woody; stems with scattered vascular bundles, not arranged in a single cylinder (minor exceptions), leaves usually parallel-veined with margins nearly always entire; flowers with parts three's or multiples thereof, rarely numerous; embryo with one cotyledon. (HS 73)

.024 Nomenclature: botanical nomenclature involves Latin names of species, genera, and other groups; the binomial system is mandatory under the International Code of Botanical Nomenclature, beginning with Linnaeus' Species Plantarum (1753) for botany; nomenclature of cultivated plants involves Latin names for species, genera, etc., but "fancy names" in a vernacular language for cultivars; the nomenclature of the latter per se is subject to the International Code of Nomenclature for Cultivated Plants. (L 51)

.025 Nomina familiarum conservanda: a conserved family name, specifically a name that would be rejected under strict application of the rules, especially that of priority, but is retained by a Botanical Congress; it is then conserved against all other names in the same rank based on the same type (nomenclatural synonyms) and against those names based on different types (taxonomic synonyms). (L 51)

III.1.1. HORTICULTURAL TAXONOMY and PLANT BREEDING

.026 **Nomina generica conservanda et rejicienda**: conserved and rejected names of genera (see nomina familiarum conservanda for details that are similar for genera, families, and intermediate taxa).
.027 Nothomorph (nm.): hybrid form, whether F_1, segregate, or backcross; the designation for different hybrid forms or groups of hybrid forms derived from the same parent species (including their infraspecific taxa) treated as belonging to a collective hybrid taxon of rank equivalent to species; the term "nothomorph" (abbrev. nm) is of rank equivalent to variety; such forms or groups are classed under the binary name assigned to the taxon and may be designated by an epithet preceded by this binary name and nm. (L 51)
.028 Order (Ordo): a category of classification that ranks above a family and below a class; the category "natural order" in older literature is equivalent to the modern "family"; order names (with some exceptions) have the suffix -ales and are usually derived from an included family (or occasionally from a family raised to order rank, such as Leguminales from Leguminosae; the latter is split into the Caesalpiniaceae, Mimosaceae, and Fabaceae). (L 51)
.029 Phylogeny: the evolution of a race or genetically related group of organisms, such as a species, family, or order. (L 51)
.030 Scientific name: a binary cognomen that consists of a generic name and specific epithet in Latin, plus the authority; rules for such are given in the International Code of Botanical Nomenclature in regard to validity, publication, and priority for wild and cultivated plants. (L 51)
.031 Species (Species; sp., pl. spp.): a category of classification lower than a genus or subgenus and above that of a subspecies or variety; a group of plants that possesses one or more characters in common distinguishing it from other groups and does or may interbreed and reproduce its characters in offspring, exhibiting only minor variations bridged over by differences ascribable to age, sex, polymorphism, individual peculiarity, or accident or selective breeding by man; species in a small genus will be separated by relatively large differences in characters, whereas species in a large genus are usually separated on the basis of very small differences that would be ignored in the former; interpretation of species limits depends largely on individual judgment; opinions among botanists differ widely in regard to polymorphic taxa or apomicts which have disparate morphological characters but hybridize freely; classical, or Linnaean, species are based on morphological (presumably constant) differences, biological species, or reproductive isolation. (L 51)
.032 Spermatophyta = Embryophyta Siphonogamia: a division that includes gymnosperms and angiosperms; now included in the Tracheophyta, vascular plants. (L 51)
.033 Subspecies (Subspecies; subsp.): a subdivision of a species commonly used to designate a group with more or less unstable characters and connected with some similar group by individuals with intermediate characters. (L 51)
.034 Taxon (plural: taxa): a general term applied to a taxonomic group of any rank or category. (L 51)
.035 Taxonomic types: (L 51, SF)
.035a. Nomenclatural (taxonomic) type: the constituent element of a taxon to which the name of the taxon is permanently attached, whether as a correct name or as a synonym; subdivided and defined taxonomically as follows:

III.1.1. HORTICULTURAL TAXONOMY and PLANT BREEDING

.035b. Cotype: second specimen from same plant from which the holotype was collected.
.035c. Holotype: the specimen or other element used by the author or designated by him as the nomenclatural type.
.035d. Isotype: any duplicate (part of a single gathering made by a collector at one time) of the holotype.
.035e. Lectotype: specimen or other element selected from the original source material to serve as a nomenclatural type when no holotype was designated at the time of publication or as long as it is missing.
.035f. Neotype: specimen or other element selected to serve as the nomenclatural type as long as all of the material on which the name of the taxon was based is missing.
.035g. Syntype: any one of two or more specimens cited by the author when no holotype was designated or any one of two or more specimens was simultaneously designated as the type.
.035h. Topotype: specimen collected from type locality (same station as type).
.036 <u>Tracheophyta</u>: the division of the plant kingdom that includes vascular plants (i.e., pteridophytes, gymnosperms, and angiosperms). (L 51)
.037 Variety (<u>Varietas</u>): a group of plants related by descent but distinguished from other similar groups by characters too inconsistent or too trivial to entitle it to recognition as a species or whose distinguishing characters are dependent on breeding controlled by man for their perpetuation; may be a morphological variant, geographical variant, a variant comingled with others of the same species, or a color or habit phase; now largely abandoned as a botanical category because of its indefiniteness and confusion with horticultural varieties; (see also Variety, III.2.0.005a). (L 51)

III.1.2. Biosystematics

.001 Acclimatization: the gradual process of the establishment of a plant in a new locality in which conditions differ markedly from those of its native habitat. (WF)
.002 Agrestial; agrestal: plant growing on arable land.
.003 Allopatric: geographically isolated and separate populations evolving into good species; (see Sympatric, III.1.2.058).
.004 Anthropogenous relic: a plant that will grow in a wild state only in areas in which it was formerly cultivated. (WF)
.005 Biotype: distinct physiological races or strains within morphological species; a population of individuals with identical genetic constitution; may be made up of homozygotes or heterozygotes, of which the former would be expected to breed true. (L 51)
.006 Cenospecies (coenospecies): a group of plants that represents one or more ecospecies of common evolutionary origin, so far as morphological, cytological, and experimental facts indicate; different cenospecies of the same comparium are usually separated by genetic barriers so absolute that hybrids between such cenospecies are sterile unless amphidiploidy occurs; (see Linneont, III.1.2.032). (v.ST, L 51)
.007 Clinal, clinal variation: more or less continuous morphological variation of a species (or genus or family) within an ecosystem (plant community). (v.ST)

III.1.2. HORTICULTURAL TAXONOMY and PLANT BREEDING

.008 Coenospecies: (see Cenospecies, III.1.2.006).

.009 Commiscuum (miscibility community): the total number of individuals connected genetically by miscibility (III.1.2.036). (v.ST)

.010 Comparium (hybridization community): a group composed of one or more cenospecies that are able to intercross; i.e., all of the individuals are connected by the possibility of hybridization. The hybrids must be viable but it is irrelevant whether they are capable of producing (functional) sexual organs. All commiscua between which a (sterile) hybrid is found belong to one comparium. Comparia are not miscible, thus are biosystemic units often comparable to a genus. (v.ST)

.011 Conspecific: of the same species; referring to different populations of an ecosystem.

.012 Convivium: each partial specific (i.e., subspecific) population or group of individuals that is more or less distinguishable (morphologically or ecologically) and maintains itself by isolation of whatever cause from other part(s) of the population but with which, however, it is completely miscible. (v.ST)

.013 Cultigen: plant or group of plants known only in cultivation; presumed to have originated under domestication. (v.ST)

.014 Cultispec: a cultigen worthy of being classified at the species level. (Name proposed by van Steenis, 1948, as the next higher category above cultivar.) (v.ST)

.015 Dipterocarp: a plant belonging to the Dipterocarpaceae, as in a forest or other ecosystem in which the dominant species belong to this family. (JG)

.016 Domesticate: to convert a wild plant into a cultivated crop by selection and adaptation.

.017 Ecology: the study of living organisms in relation to their environment and their effects on one another. (OS)

.018 Ecophene: a group of plants with similar genotypes that exhibit differences (i.e., have characteristic phenotypes) as a result of varied environments.

.019 Ecospecies: group of plants composed of one or more ecotypes within a cenospecies whose members are able to interchange their genes without detriment to the offspring (true only if environment does not favor one ecotype over another. (v.ST)

.020 Ecosystem: a complete plant community with biotic and abiotic (geological, edaphic, environmental, etc.) factors considered together as a unit.

.021 Ecotype: population distinguished by morphological and physiological characters, most frequently of a quantitative nature; interfertile with other ecotypes of the ecospecies but prevented from freely exchanging genes by ecological barriers for a phyletic unit adapted to a particular environment (but capable of producing fully fertile hybrids with other ecotypes of the same ecospecies if brought together). (v.ST)

.022 Endemic: native or confined naturally to a particular and usually restricted area or region. (L 55)

.023 Escape: cultivated plant that becomes naturalized in an area, with accompanying reversion to conditions of an independent existence. (WF)

.024 Exotic: introduced from another country.

.025 Genecological: study of genetic variations under ecological conditions, i.e., of plants as part of an ecosystem.

.026 Genospecies: species consisting of individuals with the same genotype.

III.1.2. HORTICULTURAL TAXONOMY and PLANT BREEDING

.027 Hybridization community: (see Comparium, III.1.2.010).
.028 Hybrid swarm: populations consisting of descendants of species hybrids, as at the border between geographical areas populated by these species. (v.ST)
.029 Indigen: native; a naturalized (thoroughly established) or adventive (naturalized, but unable to reproduce or compete in the new environment) plant. (L 55)
.030 Jordanon: basically a species (or a species concept) in which all forms, races, or other variations arising from environmental stress are accepted as separate entities of specific rank, the sole criterion being whether the differential characters, however produced, breed true. This concept denies (or ignores) the existence of polymorphism or that the same species can coexist in widely differing habitats or other nonuniform environments. It also ignores the fact that most, if not all, such geographical, altitudinal, and edaphic races comprised of such a species are able to interchange genes when they are grown together or deliberately crossed by hand pollination. (v.ST)
.031 Linnean species: (see Linneont, III.1.2.032).
.032 Linneont (= cenospecies): basically a species whose diagnosis (description) includes the entire range of variation (insofar as possible) found in the population it represents. Variations included are not only those observed in uniform habitats but also those induced by differing climatic, edaphic, or biotic environments, even though gross differences in morphological, particularly vegetative, but in addition peloric or other monstrous generative, characters thus may be represented. This concept accepts polymorphism as a normal consequence of the environment in which plants grow (as individuals and the total population); plants classified as Jordanons (i.e., ecological races; III.2.0.009) are distinguished here as taxa of infraspecific rank. (v.ST)
.033 Megatherm taxa: families, such as Dipterocarpaceae, and some genera that have a narrow range of adaptability with regard to low temperatures; thus species of these groups are seldom found above 1000 m in tropical regions and do not occur at higher latitudes. Certain genera, such as Calophyllum, Garcinia, Symplocos, and Pandanus, are typically tropical but do include some species that occur at high altitudes (e.g., above 1000 m), although they also are not as a rule found in temperate regions. (v.ST)
.034 Microtherm taxa: species of certain genera, such as Primula, Gentiana, Geranium, and Pedicularis, with a narrow range of ecological adaptation to constant high temperatures; thus they occur only at high altitudes (> 1000 m) in the tropics. (v.ST)
.035 Miscible, miscibility: the ability of two or more strongly disjunct (e.g., long isolated in time and space) taxa (species) to hybridize or be hybridized and produce fertile progeny. Eventually a complete series of intermediate forms, a panmictic population, may be formed in the progeny, either by crossing or selfing, by segregation, and recombination. These taxa (the parental forms) are part of a polymorphic population and are regarded as subspecies or varieties of a single linneont. (v.ST)
.036 Miscibility community: (see Commiscuum, III.1.2.009).
.037 Naturalization: the process of establishment of a plant brought into a region in which the conditions are similar to those whence it came. (WF)
.038 Naturalized: of foreign origin but established and reproducing itself as a native. (TA)

III.1.2. HORTICULTURAL TAXONOMY and PLANT BREEDING

.039 Neoendemic: recent new species.
.040 Neolithic: of or belonging to the late Stone Age.
.041 Neotropical (Neotropics): New World tropics or an organism from or inhabiting same; also applied to geological, edaphic, or other environmental features; (see Paleotropical, III.1.2.046).
.042 "Nomad" plants: ruderals (III.1.2.055), often arboreous (trees or shrubs), found in many locations disturbed by man's activities, like the edges of forests adjacent to clearings or openings in forests that have resulted from natural causes (e.g., windthrows and fire from lightning). (v.ST)
.043 Ochlospecies: complex reticulate variation without discontinuities throughout a species or part of one. (v.ST)
.044 Organogenetic: origin and development of organs in plants and animals.
.045 Ortet: (see III.2.0.006).
.046 Paleotropical (Palaeotropical): Old World tropics or an organism from or inhabiting same; also applied to geological, edaphic, and other environmental features; (see Neotropical, III.1.2.041).
.047 Panmictic unit: local population in which there is complete random mating (see Panmixis, III.1.2.048).
.048 Panmixis, panmixia, panmixy: a high degree of random mating within a breeding (wild) population; promiscuous interbreeding uninfluenced by selection.
.049 Paramorph, paramorphosis: genotypic variability within a specific (local) population that deviates from the average of the population as a whole. (v.ST)
.050 Phyletic: actual or presumed geneology of a species or higher category; the line of descent from earlier forms and its relation to other related taxa (this, however, is rarely known in practice at levels higher than species; therefore a source of much controversy among taxonomists). (HS 73)
.051 Pioneer species: plant forms first developing on a site after complete destruction of the previous flora, as by a landslide or volcanic action. (v.ST)
.052 Plant association: a major unit in a community organization characterized by essential uniformity in composition and structure, usually with two or more dominant species of a particular life form or habitat.
.053 Provenance: origin, source, or place where found or produced, as a cultivar or selection of a taxon.
.054 Rheophytes: plants with stenophyllous leaves (II.4.3.058). (v.ST)
.055 Ruderal: plant growing in rubbish or waste places. (v.ST)
.056 Segetal, segetal race: native, adventive, or ruderal species or infraspecific taxa that have become adapted as weed flora to the life cycles of various cereals and man's management of the latter. Initially the term applied to weeds in corn (= wheat) fields. They show a remarkable parallelism in their ecological and morphological characters with those of their cultigen hosts. Races developed in this manner commonly differ markedly from the wild population, often exhibiting traits unknown or rare in the latter. Segetal taxa may be closely related to the cultivated crop (e.g., sorghum) or belong to different genera or even families. (HN)
.057 Stropha: an individual of the species that is aberrant or unusual, probably equivalent to a cultivar (horticultural variety).
.058 Sympatric: same geographical distribution among plant species.
.059 Synecology: the structure, development, and distribution of communities in relation to their environments.

III.1.2. HORTICULTURAL TAXONOMY and PLANT BREEDING

.060 Syngameon: panmictic population; essentially equal to a linneont (III.1.2.032). (v.ST)
.061 Umbelliferous: pertaining to plants in the Umbelliferae (Apiaceae).
.062 Variation: the occurrence of differences among individuals of the same species or an equivalent taxon.
.063 Vicariism, vicarious species: a pair of species that is distinctly more closely related than to any other members of a group; they replace each other exclusively in an area, either latitudinally or altitudinally. (WF)
.064 Vicinism: natural crossing.

III.2. Plant Breeding

III.2.0. General

.001 Breeding: propagation of plants for the purpose of improvement by deliberate selfings or hybridizations and subsequent testing and selection for desired criteria and objectives.
.002 Clone: individuals derived by vegetative propagation or apomixis from a single original parent; (see Cultivar, III.2.0.003).
.003 Cultivar (abbrev. cv.): a term contracted from "cultivated variety" defined under the International Code of Nomenclature of Cultivated Plants (1969) (III.1.1.020) as "an assemblage of cultivated plants which is clearly distinguished by any characters (morphological, physiological, cytological, chemical, or others), and which, when reproduced (sexually or asexually), retains its distinguishing characters." This term is essentially different from the concept of botanical variety (III.1.1.037), which is always in the Latin form prescribed by the International Code of Botanical Nomenclature. These plants are named at three main levels: Genus, species, and cultivar, of which the first two are governed by the Botanical Code and the last is a "fancy name" in a modern language with capital initial letters and distinguished clearly from the botanical name or accepted common name by being enclosed in single quotation marks (e.g., 'Hamlin' sweet orange) or preceded by cv. [e.g., Citrus sinensis (L.) Osbeck cv. Hamlin]. Examples of cultivar categories distinguished under the Cultivated Plant Code (Art. II) follow:

   a. A clone or several closely similar clones, including distinguishable bud mutations derived from a clone. (Note, however, that neither a clone nor any other category listed is designated as a cultivar (= variety) until it has been released; i.e., when the name is validly published under prescribed rules.)
   b. One or more lines of normally self-pollinating individuals or inbred lines of normally cross-pollinating individuals.
   c. Cross-pollinated individuals that may show genetic differences but have one or more characters by which they can be differentiated from other cultivars of like or different origin.
   d. An assemblage of individuals reconstituted on each occasion by crossing, including single crosses, double crosses, three-way crosses, top crosses, and intervarietal hybrids. The primary difficulty in the foregoing categories, which refer to cultivated plants produced principally by systematic breeding and

release to growers, is the necessity for continual redefinition of guidelines for classifying populations (i.e., the establishment of precise criteria), particularly those of cultivated plants produced by sexual means such as agricultural (agronomic), vegetable, tree, shrub, and flower seeds. General practice for the latter includes the substitution of common for scientific names and variety for cultivar, variety being specifically defined as the exact equivalent of cultivar under the guidelines of the International Code of Nomenclature of Cultivated Plants; (see Federal Seed Act of 1938 as Amended and Regulations (1976), III.2.0.004; Guidelines for classifying cultivated plant populations (1978), III.2.0.005; and Plant patents, III.2.0.007). (GL)

.004 Federal Seed Act of 1938 as Amended and Regulations (1976): the U.S. statute governing aspects of seed production, handling, and sale.

.004a. Kind: means one or more related species or subspecies that, singly or collectively, is known by one common name; e.g., soybean, flax, carrot, and radish. (Art. I.11)

.004b. Variety: the name shall represent a subdivision of a kind that is characterized by growth, plant, fruit, seed, or other characters by which it can be differentiated from other sorts of the same kind. [Sec. 201.34(d)(1)]

.004c. Breeder seed: a class of certified seed directly controlled by the originating or sponsoring plant breeding institution, or person, or designee thereof, and is the source for the production of seed of the other classes of certified seed. [Sec. 201.2(bb)]

.004d. Foundation seed: a class of certified seed that is the progeny of Breeder or Foundation seed and is produced and handled under procedures established by the certifying agency . . . for the purpose of maintaining genetic purity and identity. [Sec. 201.2(cc)]

.004e. Registered seed: a class of certified seed that is the progeny of Breeder or Foundation seed and is produced and handled under procedures established by the certifying agency . . . for the purpose of maintaining genetic purity and identity. [Sec. 201.2(dd)]

.004f. Certified seed: a class of certified seed that is the progeny of Breeder, Foundation, or Registered seed . . . and is produced and handled under procedures established by the certifying agency . . . for the purpose of maintaining genetic purity and identity. [Sec. 201.2(ee)]

.004g. Off-type: a plant or seed that deviates in one or more characteristics from that described . . . as being usual for the strain or variety. [Sec. 201.2(ff)]

.004h. Inbred line: a relatively true-breeding strain that results from at least five successive generations of controlled fertilization or of backcrossing to a recurrent parent with selection, or its equivalent, for specific characteristics. [Sec. 201.2(gg)]

.004i. Single cross: the first generation hybrid between two inbred lines. [Sec. 201.2(hh)]

.004j. Foundation single cross: a single cross used in the production of a double, three-way, or top cross. [Sec. 201.2(ii)]

.004k. Double cross: the first generation hybrid between two single crosses. [Sec. 201.2(jj)]

III.2.0. HORTICULTURAL TAXONOMY and PLANT BREEDING

.004l. Top cross: the first generation hybrid variety. [Sec. 201.2(kk)]
.004m. Three-way cross: a first generation hybrid between a single cross and an inbred line. [Sec. 201.2(ll)]
.004n. Open-pollination: pollination that occurs naturally, as opposed to controlled pollination, by detasseling, cytoplasmic male sterility, self-incompatibility, or similar processes. [Sec. 201.2(mm)]
.004o. Purity: the name or names of the kind, type, or variety and the percentage or percentages thereof; the percentage of other agricultural or crop seed; the percentage of weed seeds, including noxious-weed seeds; the percentage of inert matter; the names of the noxious-weed seeds and the rate of occurrence of each. [Sec. 201.2(w)]
.005   Guidelines for classifying cultivated plant populations (1978): an appendix to the Federal Seed Program Review (1980) which gives more precise definitions of the various categories of cultivated varieties.
.005a. Variety (cultivar): means a subdivision of a kind (see III.2.0. 004a) that is <u>distinct</u> (in the sense that the variety can be differentiated by one or more identifiable morphological, physiological, or other characteristics from all other varieties of public knowledge), <u>uniform</u> (variations in essential and distinctive characters are describable), and <u>stable</u> (the variety will remain unchanged to a reasonable degree of reliability in its essential and distinctive characteristics and its uniformity when reproduced or reconstituted as required by the different categories of varieties). (Note that here "variety" and "cultivar" are considered exact equivalents in accordance with the International Code of Nomenclature of Cultivated Plants, 1969.)
.005a-1. Clonal varieties: consist of one clone or several closely similar clones propagated by asexual means, such as cuttings, tubers, corms, bulbs, rhizomes, divisions, grafts, or seed produced by obligate apomixis. Examples: 'Meyer' zoysiagrass, 'Elberta' peach, 'Russet Burbank' potato, 'Coastal' bermudagrass, 'Peace' rose, 'Iceberg' chrysanthemum. Examples of obligate apomicts: 'Troyer' citrange (rootstock), 'Higgins' buffelgrass.
.005a-2. Line varieties: one or more lines of self- or cross- fertilizing plants and single-line facultative apomicts with largely the same genetic background (a theoretical coefficient of parentage of 0.87 or higher and 95% apomixis for the single-line facultative apomicts, except in cases in which it is not possible to achieve 95% apomixis; a level as low as 80% may be classed as line varieties even though the variant plants present may differ in morphological characteristics) which are similar in essential and distinctive characteristics and are maintained or reproduced by controlled self- or sib-fertilization or line crossing of the plants (for self- or cross-fertilizing plants) and by close generation control (for single-line facultative apomicts). Examples of line varieties from normally self-fertilized crops: 'Gaines' wheat, 'Tendercrop' snap bean, 'Wayne' soybean; of line (inbred) varieties for normally cross-fertilized crops: 'MSU-713-5' gynoecious cucumber, 'WF9' corn, 'Nittany Lion Red' geranium, 'B2108A x B2108B' (line cross) onion, 'B2215C' onion; of single-line facultative apomicts

III.2.0. HORTICULTURAL TAXONOMY and PLANT BREEDING 314

(meet 95% apomixis requirement): 'Penstar' Kentucky bluegrass, 'Merion' Kentucky bluegrass; below 95% apomixis: Adelphi Kentucky bluegrass and Bristol Kentucky bluegrass.

.005a-3. Multiline varieties: two or more isogenic lines of normally self-fertilizing plants that are similar in most characteristics but differ in a limited number of describable physiological, morphological, or other essential or distinctive characteristics; derived by growing the component lines separately and compositing the lines to constitute the breeder class of seed. Examples: 'Multiline E-69' oat, 'Miramar-63' wheat.

.005a-4. Open-pollinated varieties of cross-fertilizing crops: consist of normally cross-fertilizing plants selected to a standard that may show variation but have one or more characteristics by which a variety can be differentiated from other varieties. Examples: 'Kenland' red clover, 'Nordan' crested wheatgrass, 'Yellow Bermuda' onion, 'Elbon' rye, 'Thumbelina' zinnia, 'Poinsett' cucumber, 'Golden Bantam' sweet corn.

.005a-5. Synthetic varieties:

.005a-5-a. First-generation synthetic varieties (Syn-1): first-generation progenies derived by intercrossing a specific set of clones or seed-propagated lines; these may include varieties of normally cross-fertilizing or self-fertilizing crops into which mechanisms have been introduced to maximize cross-fertilization such as male sterility or self- incompatibility. These varieties usually contain mixtures of seed that result from cross-, self-, and sib-fertilization; the variety consists of only the first-generation progenies after intercrossing and cannot be reproduced from seed of the first generation. Examples: 'Gahi' pearl millet, 'Vitagraze' rye, 'Tempo' alfalfa.

.005a-5-b. Advanced generation synthetic varieties (beyond Syn1): advanced generations derived from an initial intercrossing of a specific set of clones or seed-propagated lines; usually stable for only limited number of generations. Examples: 'Ranger' and 'Moapa' alfalfa, 'Saratoga' bromegrass, 'Pennlate' orchardgrass.

.005a-6. Hybrid varieties (F_1): first-generation (F_1) progenies from a cross, produced by controlling the pollination, between (1) two inbred lines, (2) two single crosses, (3) a single cross and an inbred line, (4) an inbred line or a single cross and an open-pollinated or a synthetic variety, or (5) two selected clones, seed lines, varieties, or species. A line cross between two closely related inbreds (theoretical coefficient of parentage at least 0.87) is considered equivalent to a line (inbred) variety; the hybrid variety cannot be reproduced from seed of the hybred generation. Examples of conventional hybrids: 'Hybrid-7' spinach, 'US13' hybrid corn, 'RS-610' hybrid grain sorghum, 'Moreton' hybrid tomato, 'Comanche' hybrid petunia; examples of varieties that contain substantial numbers of hybrid seeds: 'Market Prize' hybrid cabbage, 'Valley' hybrid sunflower, 'Picadilly' hybrid cucumber.

.005a-7. F_2 varieties: the next generation seed derived from the hybrid (F_1) generation; the variety cannot be perpetuated by growing additional generations. Examples: 'Foremost

III.2.0. HORTICULTURAL TAXONOMY and PLANT BREEDING 315

F_2' tomato, 'Market Pride' cantaloupe, 'Violet Blue' petunia, 'Seven-Eleven' pansy.

.005b. Composite-cross populations: a population generated by hybridizing more than two varieties and/or lines of normally self-fertilizing plants and propagating successive generations of the segregating population in bulk in specific environments so that natural selection is the principal force acting to produce genetic change; artificial selection may also be imposed on the population, the resulting population is expected to have a continuously changing genetic makeup; breeder seed is not maintained as originally released. Examples: 'Harlan' barley, 'Mezcla' lima bean.

.005c. Mixture (= blend): consists of seed of more than one kind or variety, each present in excess of 5% of the whole.

.005d. Variants and off-types in varieties:

.005d-1. Variants: (1) seeds or plants that are (a) distinct within the variety but occur naturally in it, (b) stable and predictable with a degree of reliability comparable to other varieties of the same kind, within recognized tolerances, when the variety is reproduced or reconstituted, and (c) were originally a part of the variety as released; variants are not to be considered off-types; (2) the Breeder should identify variants as a part of the variety description but the expected rate of occurrence of the variant need be stated only when the Breeder considers the variant to be an aid in identifying the variety; (3) tolerances in Table 4, Section 201.62, Part 201 of the Federal Seed Act should be applied to those variants that are described by the Breeder as useful in identification of that variety.

.005d-2. Off-type: means any seed or plant not a part of the variety in that it deviates in one or more characteristics of the variety as described and may include seeds or plants of other varieties; seeds or plants not necessarily of any variety; seeds or plants resulting from cross-pollination by other kinds or varieties; seeds or plants resulting from uncontrolled self-pollination during production of hybrid seed; and segregates from any of the above plants.

.006 Ortet: original ancestor of a vegetative clone. (WR)

.007 Plant patents: two statutes, the Plant Variety Protection Act of 1970 for sexually produced plants and the U.S. Patent Law of 1930 for asexually reproduced plants, provide for exclusive rights to propagation and collection of royalties for a 17-year period following the issuance of a U.S. Department of Agriculture certificate; general requirements under either law are a "novel variety" as defined therein; e.g., those for sexually produced plants are as follows:

.007a. Novel variety: may be represented by seed, transplant, or plants, subject to (1) "distinctness" in the sense that the variety clearly differs by one or more identifiable morphological, physiological, or other characteristics (which may include those evidenced by processing or product characteristics, e.g., milling and baking characteristics in the case of wheat) to which a difference in genealogy may contribute evidence, from all prior varieties of public knowledge at the date of determination; (2) "uniformity" in the sense that any variations are describable, predictable, and commercially acceptable; and (3) "stability" in the sense that the variety,

III.2.0. HORTICULTURAL TAXONOMY and PLANT BREEDING

when sexually reproduced or reconstituted, will remain unchanged with regard to its essential and distinctive characteristics with a reasonable degree of reliability commensurate with that of varieties of the same category in which the same breeding method is used. [Sec.41(a) P.V.P.A.]

.007b. Sexually reproduced: includes any production of a variety by seed. [Sec.41(f)]
.007c. Basic seed: means the seed planted to produce certified or commercial seed.
.008 Provenance: (see III.1.2.053).
.009 Race: a group or assemblage of organisms that exhibits general similarities but is not sufficiently distinct from other forms to constitute a species (e.g., the three races of avocado (Persea americana), Mexican, Guatemalan, and West Indian, of which the first is sometimes designated as a separate species, although it hybridizes freely with the other two); sometimes used for individuals of a particular geographical area, in this sense then equivalent to an ecotype (which is not a botanical term).
.010 Ramet: individual member of a clone. (WR)
.011 Selection: choice of a plant (or group of plants) that possesses a desired criterion (or criteria); note that the word is used in two senses; (1) it refers to the actual act of making a choice and (2) the plant or plants chosen (prior to being designated as a cultivar).
.012 Strain: a term incorrectly applied to selections of cultivar (= varieties); designation of a selection of a cultivar as a strain or equivalent term is not permitted under the International Code of Nomenclature of Cultivated Plants; any selection that shows sufficient variation from the parent cultivar to render it worthy of a name is to be regarded as a distinct cultivar; use of the term for a cultivar of hybrid origin, e.g., Lilium Olympic strain, is not recommended; such assemblages are more correctly termed "group". [Note that the Guidelines for Classifying Cultivated Plant Populations (1978) stipulate that variety and cultivar are considered exact equivalents; see III.2.0.005a.]

III.2.1. Breeding and selection systems, gene pools

.001 Breeding system: a particular mating system that involves a certain type or types of plant material, together with the necessary selection procedures; different strategies are used to improve self-pollinated crops to extract inbred pure lines and to improve cross-pollinated crops for population performance per se for the selection of superior heterozygous individuals to be vegetatively reproduced as clones or for the extraction of improved inbreds for use in hybrid production. (SM 79)
.001a. Clones: outbreeding mating system that involves perennial or quasi-annual vegetative material; homogeneous, heterozygous, isolated by selection of superior recombinants or transgressive segregates in the F_1, between heterozygous parental clones, and subsequently multiplied by vegetative propagation (monogenotypic); examples are potato, cassava, sweet potato, rubber, mango, avocado, apple, pear, banana, pineapple, strawberry, brambles, grape, peach, cherry, almond, citrus, date, Jerusalem artichoke, yams, black pepper (Piper), olive, fig, pistachio, and edible aroids.

III.2.1. HORTICULTURAL TAXONOMY and PLANT BREEDING

.001b. Hybrids: involve inbred lines with favorable combining ability of annual or biennial, or sometimes perennial, seed-propagating material; homogeneous, highly heterozygous, with inbred lines for use in hybrid production derived from continuous selfing of selected plants in cross-pollinated populations (verges on monogenotypic); examples are maize, onion, Brussels sprouts, kale, tomato, beets, cucurbits, black pepper (Piper), cloves, fig, radish, Chinese cabbage, and sunflower.

.001c. Inbred pure lines: involves inbreeding annual seed-propagated material; homogeneous, homozygous, isolated by selection of desired recombinants or segregates in F_2-F_7 generations of crosses between parental pure lines (generally monogenotypic, can be blended to form multilines); examples are tomato, lettuce, soybean, pea, cowpea, snap bean, field bean, Arabian coffee, pepper (Capsicum), eggplant, okra, lentil, and papaya ('Solo').

.001d. Open-pollinated populations: outbreeding mating system that involves annual, biennial, or perennial seed-propagated material; heterogeneous, heterozygous, constructed by changing gene-frequencies by selection (population improvement) or by making synthetics via parental lines or clones; verges on polygenotypic; examples are cacao, coconut, oil palm, maize, tea, onions, beet, carrot, cabbage, cauliflower, asparagus, and spinach.

.002 Gene pool system: three informal categories proposed by Harlan and de Wet (cited in Harlan, 1975) to provide a genetic perspective and focus for cultivated plants (HN):

.002a. Primary gene pool: corresponds to the traditional concept of biological species; crossing is easy, hybrids are generally fertile with good chromosome pairing, gene segregation is approximately normal, and gene transfer is generally simple amoung forms of this gene pool; almost always includes spontaneous races (wild and/or weedy, as well as cultivated races); the latter considered subspecies A and the former subspecies B.

.002b. Secondary gene pool: includes all biological species that will cross with the crop; approximates an experimentally defined cenospecies; gene transfer is possible; hybrids tend to be sterile, chromosomes pair poorly or not at all; some hybrids may be weak and difficult to bring to maturity and recovery of desired types in advanced generations may be difficult.

.002c. Tertiary gene pool: crosses can be made with the crop but hybrids tend to be anomalous, lethal, or completely sterile; gene transfer is either not possible with known techniques or extreme or radical measures are required (e.g., embryo culture, grafting or tissue culture to obtain hybrids, doubling chromosome number, or using bridging species to obtain some fertility); value chiefly informational, defines extreme outer limit of genetic distance potential; likely to be ill-defined until information is accumulated.

III.2.2. Terms

.001 Acquired character: modification impressed on an organism by environmental influences during development.

III.2.2. HORTICULTURAL TAXONOMY and PLANT BREEDING

.002 Additive genetic variance: that genetic variation (variance) which is due to additive effects of genes.
.003 Agameon: a species that contains only apomictic individuals.
.004 Allele = allel, allelic, allelomorph, allelomorphic: one of a pair or a series of factors that occur at the same locus on homologous chromosomes and is inherited in alternative pairs for this reason; one alternative form of a gene.
.005 Alloploidion: a species derived from allopolyploidy; individuals are usually highly variable and apomixis is not present.
.006 Allopolyploid: polyploid having chromosome sets from different sources such as different species; a polyploid that contains genetically different chromosome sets; e.g., from two or more species.
.007 Amphidiploid: plant possessing the somatic chromosomes of two species; the latter is definitely known in the case of an amphidiploid but not in that of an allopolyploid (III.2.2.006).
.008 Amphihaploid: plant having a haploid ($\underline{1n}$) set of chromosomes from each parent of an interspecific hybrid, thus being ($\underline{2x}$) rather than ($\underline{4x}$), as in the case of an amphidiploid (III.2.2.007).
.009 Amphipolyploid: polyploid that arises by the addition of the complete somatic chromosome sets of two or more known species.
.010 Androgenetic haploid: plant having chromosomes from the male parent only.
.011 Aneuploid: organism or cell having a chromosome number other than an exact multiple of the monoploid or basic number; hyperploid = higher, hypoploid = lower = nullisome.
.012 Anisogamete, anisogametic: a gamete of either of two kinds sexually differentiated in size or structure; the condition in which two such kinds of gametes are present.
.013 Anisogamous: gametes of unequal size.
.014 Anisogeny: production of ovules and pollen that exhibit a consistent difference in genetic constitution.
.015 Anisoploid: unequal ploidy levels (e.g., $\underline{2x}$ and $\underline{4x}$) among progeny of a polyploid cross.
.016 Apogameon: a species that contains both apomictic and nonapomictic individuals.
.017 Autopolyploid: polyploid that arises by the multiplication of the complete haploid genome of a single species.
.018 Backcross: cross of a hybrid to one of the parental types; the offspring of such a cross are referred to as the "backcross" (BC) generation or progeny; (see Test cross, III.2.2.175).
.019 Breeding cycle (breeding rotation): the shortest period between successive generations from germination of a seed to reproduction of the progeny; i.e., the seed-to-seed cycle.
.020 Breeding rotation: (see Breeding cycle, III.2.2.019).
.021 Bud sport: branch, flower, or fruit that differs genetically from the remainder of the plant (see also Chimera, II.2.2.024).
.022 Character (characteristic): expression of a gene or group of genes.
.023 Characteristic: (see Character, III.2.2.022).
.024 Chimera (chimaera): mixture of tissues of genetically different constitution in the same part of a plant; may result from mutation, irregular mitosis, somatic crossing-over, or artificial fusion of unlike tissues (e.g., a "graft hybrid"). (Fig. III.2.2.01) (N)
.024a. Cytochimera: chromosomal chimera; e.g., one having different chromosome numbers in the layers; similar but not necessarily identical to a mixoploid (III.2.2.024c).

III.2.2. HORTICULTURAL TAXONOMY and PLANT BREEDING

.024b. Mericlinal: periclinal form in which the outer layers of tissue occupy only a sector rather than completely enclosing the inner layer.
.024c. Mixoploid: chimera having different ploidy levels in the L_1, L_2, and L_3 layers, as in 2-4-4, 4-4-2, and the like.
.024d. Periclinal: concentric or parallel layers of genetically different tissues; several forms:
.024d-1. Epidermal mutant:
.024d-1-a. Diectochimera: a secondary change that involves the middle layer as well as the original outer one.
.024d-1-b. Ectochimera: the primary mutant involves the outer layer only.
.024d-2. Subepidermal mutants:
.024d-2-a. Endochimera: the primary mutant involves the inner layer only.
.024d-2-b. Mesochimera: the primary mutant involves the intermediate layer only.
.024d-2-c. Reversion: the primary endochimeral mutant is lost.
.024d-2-d. Solid: a secondary change from a mesochimera that involves the outer and intermediate layers.
.024e. Sectorial: genetically different tissues are situated side by side.
.025 Chromatid: longitudinal half-chromosome that appears between the early prophase and metaphase stages of somatic mitosis and between the diplotene (at least) and the second metaphase stage in meiotic mitosis.
.026 Chromosomes: microscopically small, dark-staining bodies visible in the nucleus of a cell at the time of nuclear division; the number in any species is usually constant; carriers of the genes, which are arranged in linear order.
.027 Cleistogameon: a species that reproduces in part, by cleistogamy; apomixis is not present.
.028 Colchiploid: colchicine-induced polyploid.
.029 Combining ability: "general combining ability" is the average progeny performance of a variety in a series of crosses; "specific combining ability" is the deviation from the performance predicted from general combining ability.
.030 Complementary genes: genes that are similar in phenotypic effect when present separately but react to produce new characters when they are combined; a 9:7 ratio results in the F_2 if two such genes are complementary for a dominant effect and a 15:1 ratio if they are complementary for a recessive effect; (see Duplicate factors, III.2.2.046).
.031 Crossing-over: the exchange of corresponding segments between chromatids of paired (homologous) chromosomes; it is a process inferred cytologically from new associations of parts of chromosomes, both of which may be observed in an exchange of factors and in combinations of factors differing from those that came in with the parents; the term "genetic cross-over" may be applied to these new gene combinations; (see Recombination, III.2.2.152).
.032 Cross-over unit: an exchange frequency of 1% between two pairs of linked genes.
.033 Cybrid: hybrid cytoplasm from protoplast fusion.
.034 Cytokinesis: the division of cytoplasm into cells.
.035 Cytoplasmic inheritance: transmission of hereditary traits from the maternal parents only by factors located in the cytoplasm.
.036 Deficiency: absence, "deletion", or inactivation of a segment of a chromosome.

III.2.2. HORTICULTURAL TAXONOMY and PLANT BREEDING

.037 Detassel: to remove the tassel (pollen-bearing organ) at the top of a corn plant before pollen is released.
.038 Diallele cross: the crossing in all possible combinations among a series of genotypes.
.039 Dihaploid: haploid (2$\underline{x}$) derived from a tetraploid (4$\underline{x}$).
.040 Diploid: having two chromosomes of each kind; somatic tissues of higher plants and animals are ordinarily diploid in chromosomal constitution in contrast to haploid (monoploid) gametes or the rare instances of haploid plants.
.041 Disjunction: separation; disjoined, as in the separation of homologous chromosomes during meiosis.
.042 Disome: (see Monosomic, III.2.2.113).
.043 Dominant: applied to a member of an allelomorphic (III.2.2.004) pair of characters with the quality of manifesting itself wholly or largely to the exclusion of the other member, or recessive.
.044 Double cross: the mating of two different sets of inbred lines to produce two different single crosses which are then mated, as in double-cross hybrid corn and the like.
.045 Drift: change in gene frequencies in small populations as a result of random variations.
.046 Duplicate factors (genes): different or independent factors with the same expression; the reverse of a multiple allelomorphic series in which changes in the same gene produce different effects.
.047 Duplication: occurrence of a segment more than once in the same chromosome or genome.
.048 Dysgenic: detrimental to hereditary qualities in a stock (e.g., a cultivar or population); biologically defective or deficient.
.049 Dysploid: a plant or species in which the chromosome number is more or less than the expected normal euploid number.
.050 Dysploidion: a species of morphologically similar members of a dysploid series; all members are sexually reproductive (i.e., apomixis is not present).
.051 Dysploidy: abnormal ploidy as in the appearance of diploid (2$\underline{x}$) or triploid (3$\underline{x}$) individuals in a normally tetraploid (4$\underline{x}$) population or of triploid and tetraploid ones in a normally diploid population.
.052 "Elite tree": plant of proved good combining ability.
.053 Embryogenetics: heredity and variation of embryos; the genetics of embryos.
.054 Epigenetic: a change in some morphological character as a result of localized influences different from the normal or usual pattern that occurs after development of an organism is initiated; a term used in connection with changes that result from plant tissue culture or animal embryological studies (used in apposition to Genetic, III.2.2.066).
.055 Epistasis: interallelic interaction; the suppression of the action of a gene or genes by a gene or genes not allelomorphic to those suppressed; suppressed genes are said to be "hypostatic"; the opposite of dominance which refers to the intraallelic action of members of an allelomorphic pair. An example is "piping", the leaf form typical of the "Maipure" group of pineapple (Ananas comosus), in which the upper and lower sides of the margin are folded over and fused and are completely spineless; the genotype is PPSS, although plants heterozygous for $\underline{P}$ are epistatic to the $\underline{S}$ or $\underline{s}$ alleles.

III.2.2. HORTICULTURAL TAXONOMY and PLANT BREEDING

.056 Euploid: an organism or cell having a chromosome number that is an exact multiple of the monoploid or haploid ($1n$) number; terms used for a euploid series are haploid ($1x$), diploid ($2x$), triploid ($3x$), tetraploid ($4x$), etc.
.057 Euploidion: a species sexually reproduced (i.e., apomixis is not present) and composed of segments with a common origin arranged in a euploid series; the segments are morpholoically separable but are similar and tend to intergrade.
.058 F_1: the first filial generation, the first generation of a given mating.
.059 F_2: the second filial generation, produced by crossing inter se or by self-pollination of the F_1.
.060 Fruit cycle: the period, or length of time, between fruit set and maturity.
.061 Fruiting cycle (fruiting rotation): the shortest period between successive generations of a plant; i.e., from propagule to fruit maturity (differs from breeding cycle in that the former may be reproduced vegetatively rather than from seed).
.062 Gene: the unit of inheritance located in the chromosome, which controls the development of a character by interacting with other genes and the environment.
.063 Gene flow: spread of genes by crossing.
.064 Gene frequency: the proportion in which alternative alleles of a gene occur in a population.
.065 General combining ability: (see Combining ability, III.2.2.029).
.066 Genetic: the normal or usual pattern of change in a morphological character that occurs after development of an organism is initiated.
.067 Genic sterility: a type of male sterility conditioned by nuclear genes; may be transmitted by either parent.
.068 Genome: a complete set of chromosomes, hence of genes, inherited as a unit from one parent; referred to as the basic number and represented by x.
.069 Genotype: the genetic constitution (gene makeup), expressed and latent, of an organism; individuals of the same genotype breed alike; contrast (this behavior) with phenotype (III.2.2.132).
.070 Germ plasm: the genetic material that provides the physical basis of heredity; also a collection of genotypes of an organism.
.071 Haploid: an organism or cell with only one complete set of chromosomes (i.e., $1n$); having half of one parent's chromosomes.
.072 Heredity: resemblance among individuals related by descent.
.073 Heritability: the proportion of variability that results from genetic causes; equivalent to total genetic variation, which is total variation less environmental variation; also that proportion of the variation of a population that is transmitted to progeny.
.074 Heterogameon: a species made up of races that if selfed, produce morphologically stable populations; apomixis is not present.
.075 Heterogenesis: alternation of generations, especially a unisexual-dioecious alternating with one or more parthenogenetic generations.
.076 Heterosis: hybrid vigor, usually expressed as markedly greater than either parent.
.077 Heterotypic division: (see Reductive division, III.2.2.154).
.078 Heterozygote: an organism with unlike members of any given pair or series of allelomorphs, consequently producing unlike gametes.
.079 Heterozygous: the condition in which homologous chromosomes of an individual possess different genes of the same allelomorphic series.

.080 Homogeneon: a genetically and morphologically homogeneous species in which apomixis is not present and all members are interfertile.
.081 Homologous: members of paired chromosomes in somatic cells; the former are similar in size, shape, and supposedly in function, one being derived from the male parent, the other from the female.
.082 Homotypic division: (see Reductive division, III.2.2.154; Fruit setting, Apomixis, IV.4.1.006).
.083 Homozygote: an individual whose chromosomes carry identical members of any given part of allelomorphs; the germ cells therefore are alike with respect to this locus and the individual will breed true.
.084 Homozygous: possessing identical genes with respect to any given pair or series of allelomorphs.
.085 Hybrid: the product of a cross between genetically unlike parents.
.086 Hybrid vigor: the situation in which the cross of two parents produces hybrids that show increased vigor in comparison to that of either parent.
.087 Hyperploid: (see Aneuploid, III.2.2.011).
.088 Hypoploid: (see Aneuploid, III.2.2.011).
.089 Hypostatic: (see Epistasis, III.2.2.055).
.090 Hypotriploid: a triploid ($3x$) lacking one or more chromosomes, as in instances in which $2n = 20$ instead of the expected 21 derived from a basic (x) number of 7.
.091 Ideotype: the ideal architectural plant type (cultivar).
.092 Inbreeding: the mating of individuals more closely related than individuals mating at random.
.093 Introgressive hybridization: hybridization followed by recrossing with the parental species in such a way that certain features of one species become transferred to the other species without impairment of taxonomic integrity.
.094 Inversion: a rearrangement of a group of genes in a chromosome in such a way that their linear order is reversed.
.095 Isogenetic lines: two or more lines differing at one locus only.
.096 Karyotype: the sum of the specific characters of a nucleus, such as chromosome number, size, and form.
.097 Land race: mixture of genotypes of an inbred crop locally adapted as a population; now largely superseded in most areas of crop culture but used and useful in less developed countries.
.098 Leptokurtic: flattopped, bell-shaped curve of frequency distribution.
.099 Lethal gene: a gene that renders inviable an organism or cell possessing it.
.100 Linkage: association of characters in inheritance as a result of the genes that determine them being physically located on the same chromosome, said characters or genes constituting a linkage group.
.101 Linkage group: (see Linkage, III.2.2.100).
.102 Locus: the fixed position of location of a gene on or in a chromosome.
.103 Maternal inheritance: inheritance from the female parent to offspring unaffected by inheritance from the male parent.
.104 Matroclinal: plant that exhibits certain characters inherited from the female parent as in certain banana (Musa) hybrids, which more nearly resemble the female parent rather than having intermediate characters.

III.2.2. HORTICULTURAL TAXONOMY and PLANT BREEDING

.105 Meiosis: reduction division; the process by which chromatin material (chromosomes) is reduced qualitatively and quantitatively to half the somatic number (i.e., from $2n$ to $1n$); it is completed in the two divisions, meiotic mitosis, that precede the formation of gametes (spores).
.106 Merogony: an individual with the egg cytoplasm from one parent and the egg nucleus from the other parent.
.107 Metaphase: the stage of cell division in which chromosomes are arranged in an equatorial plate or plane.
.108 Metric trait: a characteristic that shows continuous variation and may be determined by the interaction of polygenes and environments and in some cases by one of the monogenes when the environmental effect on the expression of the trait is large.
.109 Micton: a species of wide distribution, the result of hybridization of two or more species; all individuals are interfertile and have ancestral genotypes; apomixis is not present.
.110 Mitosis: the process by which a nucleus is divided into two daughter nuclei; the daughter nuclei are identical, qualitatively and quantitatively, in somatic mitosis, barring certain aberrations; spores in higher plants and gametes in animals are produced in two successive nuclear divisions in meiotic mitosis or meiosis (III.2.2.105).
.111 Modifier or modifying gene: a gene that affects the expression of another nonallelomorphic gene.
.112 Monoploid: (see Haploid, III.2.2.071).
.113 Monosomic, monosome: diploid organism that lacks one chromosome of its proper complement; "monosome" refers to a single chromosome, "disome," to two chromosomes of a kind, and "trisome" to three chromosomes of a kind.
.114 Multiple allele (allelomorph): a member or a series of more than two alternative forms of a gene.
.115 Mutation: a sudden variation that is inherited; the term is used loosely to include "point mutations" of a single gene and chromosomal changes.
.116 Nick: two different meanings: (1) a slight cut; (2) when two inbreds bloom at the same time so that they can pollinate one another and produce a good yield of seed.
.117 Nonallelic: two different genes that occupy different loci on the chromosomes of the genome.
.118 Offtype: plants in a seed field that deviate in one or more characteristics from those that are usual in the strain being grown.
.119 Oligogenic: a major gene, one or a few mendelian determinants.
.120 Operon: a set of structural genes controlled by an adjacent site on a strand of DNA; their activity is manifested in the controlled synthesis (transcription) of mRNA; may be catabolite-sensitive or -insensitive. (SB)
.121 Outbreeding: outcrossing; the random mating of individuals.
.122 Outcross: mating of a hybrid with a third parent; also an offtype plant resulting from pollen of a different sort contaminating a field of usually self-pollinated plants.
.123 Outcrossing: (see Outbreeding, III.2.2.121).
.124 Overdominanace: the heterozygote is superior to either homozygote.
.125 Pachytene: the double-thread stage of meiosis.
.126 Paleopolyploid (palaeopolyploid): early or primitive polyploid; e.g., one of ancient origin.
.127 Parageneon: a species in which apomixis is not present and with little morphological or genetic variation but contains some aberrant genotypes; all individuals are infertile.

III.2.2. HORTICULTURAL TAXONOMY and PLANT BREEDING

.128 Pedigree: a record of ancestry.
.129 Penetrance: the frequency of expression of a gene in individuals that carry it.
.130 Phenogen: a morphologically indistinguishable segment of a species which is intrafertile but intersterile.
.131 Phenon: a phenotypically homogeneous species whose individuals are sexually reproduced (i.e., apomixis is not present) but has intersterile segments.
.132 Phenotype: the observed character of an individual without reference to its genetic nature; individuals of the same phenotype look alike but may not breed alike (compare with Genotype, III.2.2.069).
.133 Phenotypic plasticity: the degree to which phenotypic expression is exhibited in a plant community under various environmental conditions.
.134 Plasmon: genetic elements contained in the cytoplasm.
.135 Plastome: inheritance of plastids by maternal cytoplasm of the egg.
.136 Pleiotrophy: the property of a gene by which it affects two or more characters.
.137 Point mutation: (see Mutation, III.2.2.115).
.138 Polygenes: genes whose effects are too slight to be identified individually.
.139 Polyhaploid: a haploid from a polyploid; i.e., a di-, tri-, or tetra-haploid.
.140 Polyploid: organism with more than two sets of a basic or monoploid number of chromosomes; e.g., triploid, pentaploid, hexaploid, heptaploid, and octaploid.
.141 Polyplotype: race or form of plant that differs from another in chromosome number, usually in ploidy. An increasing number of wild species have been ascertained to consist of two to several races that differ in ploidy level and are intersterile. Such species are better termed "species-complexes".
.142 Polysomaty: chromosome condition of a plant such as olive in which $2n = 55$ instead of the usual $2n = 2x = 46$ for the genus.
.143 Prepotency: the capacity of a parent to impress characters on its offspring so that they are more alike then usual.
.144 Promiscuous: heterogeneous or haphazard mixture.
.145 Prophase: the stages in mitosis or meiosis from the appearance of chromosomes to metaphase.
.146 Pure line: a strain of an organism that is comparatively pure genetically (homozygous) because of continued inbreeding or by other means.
.147 Putative: supposed, commonly thought or deemed, reputed, as a taxon of (presumed) hybrid origin.
.148 Qualitative character: variation is discontinuous.
.149 Quantitative character: variation is continuous.
.150 Random: chance without any type of discrimination.
.151 Recessive: applied to one member of an allelomorphic pair that lacks the ability to manifest itself wholly or in part when the other or dominant member is present.
.152 Recombination: observed new combinations of characters different from those exhibited by the parents; percentage of recombinations equals percentage of crossing-over only when the genes are relatively close together; cytological crossing-over refers to the process, whereas recombination or genetic crossing-over refers to the observed genetic result.
.153 Recurrent parent: the parent to which successive backcrosses are made.

III.2.2. HORTICULTURAL TAXONOMY and PLANT BREEDING 325

.154 Reductive division or heterotypic division: terms formerly applied to one of the two meiotic mitoses at which a particular author thought reduction and segregation occurred; contrasted with "homotypic division"; (see Fruit setting, Apomixis, IV.4.1.006)
.155 Restorer line: an inbred line that, when crossed on a male-sterile inbred, causes the resulting hybrid to be male fertile.
.156 Rheogameon: a species composed of segments with marked morphological divergence but gene exchange takes place between them; contiguous segments are interfertile; apomixis is not present.
.157 Sample: a finite series of observations taken from a population.
.158 Segregation: the parting of paternal from maternal chromosomes at meiosis and the consequent separation of differences as observed genetically in the offspring.
.159 Select-cross, select-crossing: similar to backcrossing except that a different recurrent parent is chosen each generation or each time a cross is made; used in tomato breeding (among others) for obtaining resistance to certain diseases, such as brown root rot (Pyraenochaeta lycopersici).
.160 Selective gameticide: a treatment that inactivates certain gametes, like one that produces male sterility but does not affect the female gametes.
.161 Semidwarf: dwarfed vegetative characteristics in a plant without dwarfing the reproductive characteristics (i.e., fruit size).
.162 Sex chromosome: chromosomes that are particularly connected with the determination of sex.
.163 Sex-limited: expression of a character in only one sex.
.164 Sex-linkage: association or linkage of a hereditary character with sex because its gene is on a sex-chromosome.
.165 Sib-mating: crossing of siblings or of two or more individuals of the same parentage (brother-sister mating).
.166 Sibs (siblings): progeny from the same parents.
.167 Sigmoid: S-shaped (growth curve).
.168 Somatic: referring to body tissues; having two sets of chromosomes, one set normally coming from the female parent, the other from the male; contrasted to germinal tissue (gametophyte) which will give rise to germ cells (gametes or spores).
.169 Species-complex: (see Polyplotype, III.2.2.141).
.170 Specific combining ability: (see Combining ability, III.2.2.029).
.171 Spiny tip: leaf form of the "Cayenne" group of pineapple (Ananas comosus) in which a few spines appear along the margin near the tip; the genotype is (pp)Ss, with P epistatic to S; (see Epistasis, III.2.2.055).
.172 Sport: (see Bud sport, III.2.2.021).
.173 Staminal carpellody: the transformation of stamens into fleshy carpellike structures during the early development of the flower, as in papaya.
.174 Synthetic: two distinct meanings: (1) an artificially produced material distinguished from that made by a living organism (although the precursor may come from a living organism); (2) an interbreeding population derived from the propagation of multiple hybrids.
.175 Telophase: the stage of mitosis after movement of chromosomes to the poles has ceased.
.176 Test cross: a cross made for the purpose of determining the genotype (hetero- or homozygosity of a given locus) of one of the parents; e.g., the unknown (either AA or Aa) crossed with a tester (aa, i.e., recessive for the particular gene) will have

all Aa or half Aa and half aa in the progeny according to whether the unknown is AA or Aa.
.177 Tetrad: the quadruple group of chromatids formed by the association of split homologous chromosomes to form bivalents during meiosis; also used with the same meaning as the term "quartet"; (see also Microspores, II.2.2.007t).
.178 Tetraploid: an organism whose cells contain four basic sets of chromosomes.
.179 Tetrasomic: an organism whose cells contain four chromosomes of one type; the rest of the chromosome complement is diploid (chromosome formula: $\underline{2n + 2}$); (see Trisomic, III.2.2.187)
.180 Totipotent: the potential of any living cell in a plant to reproduce the entire plant in tissue culture.
.181 Totipotentiality: the ability of a somatic cell to reproduce an entire plant somatically.
.182 Trait: a synonym of character with respect to function and performance.
.183 Transgressive segregation: the appearance in the F_2, or later generations of individuals, showing a more extreme development of character than either parent.
.184 Translocation: change in position of a segment of a chromosome to another part of the same chromosome or of a different chromosome.
.185 Triploid: an organism whose cells contain three basic sets of chromosomes.
.186 Trisome: (see Monosomic, III.2.2.113).
.187 Trisomic: an otherwise diploid organism that has an $\underline{extra}$ chromosome of one pair (chromosome formula: $\underline{2n + 1}$).
.188 Univalent: a chromosome unpaired at meiosis.
.189 Xenogenesis: fancied or supposed reproduction of an organism unlike the parents.
.190 Zygotene: a stage in the meiotic process when the threadlike chromosomes pair.

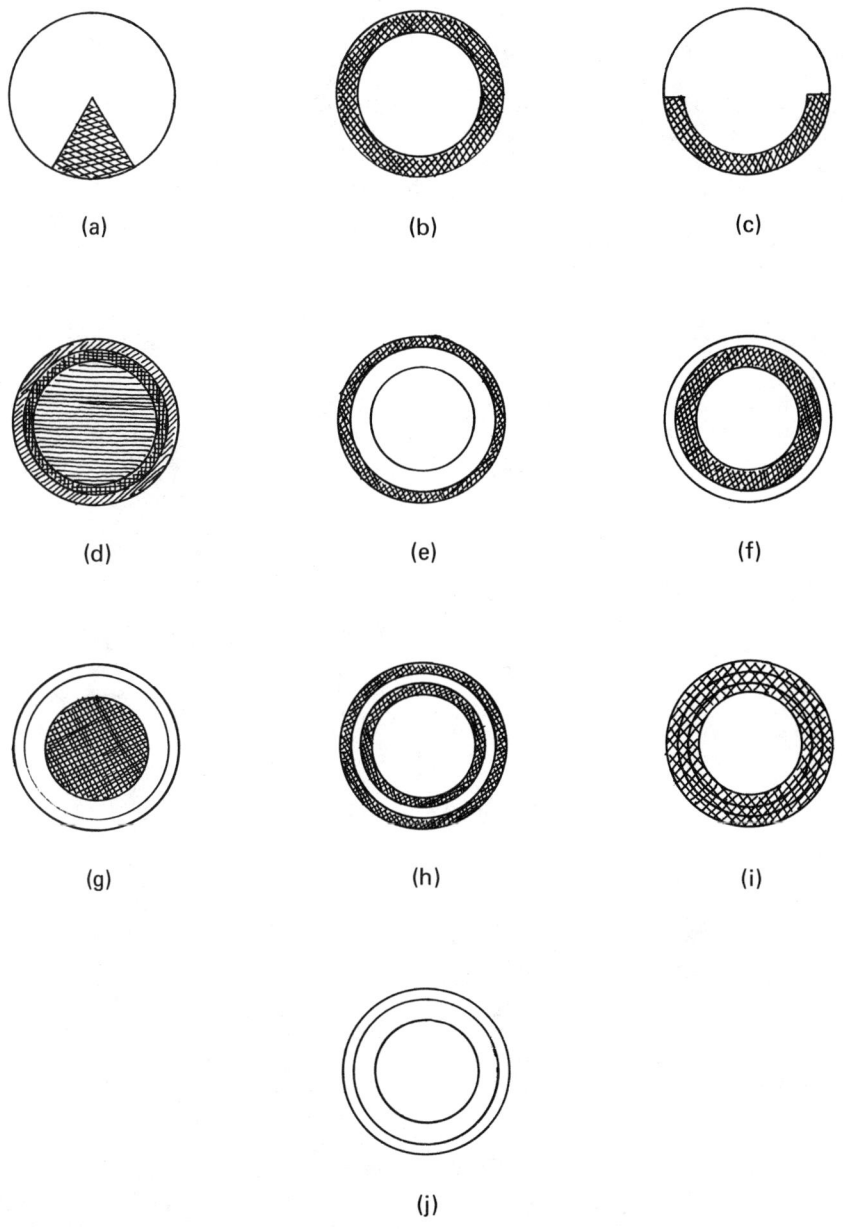

Fig. III.2.2.01. Chimeras: (a) sectorial; (b) periclinal; (c) mericlinal; (d) cytochimera, or mixoploid; (e) ectochimera; (f) mesochimera; (g) endochimera; (h) diectochimera; (i) solid; (j) reversion. Adapted from Sherman personal communication and Nybom, 1961; see Acknowledgments and Selected References.

IV. HORTICULTURAL PHYSIOLOGY and CROP ECOLOGY   328

0. General   328

1. Metabolic Sites and Processes   350

    1.1. Cell structure, organelles   350
    1.2. Major metabolic processes   352

2. Hormones and Growth Regulators, Other Metabolic Products 357

    2.1. Hormones and growth regulators   357
    2.2. Other metabolic products   359

3. Growth and Development   370

    3.1. Growth cycles, juvenility   370
    3.2. Photoperiodism, thermoperiodism   373
    3.3. Tropism, nastic movements, nutation   376

4. Fruit Setting   378

    4.1. Amphimixis (syngamy), apomixis   378
         Fig. IV.4.1.01-.03   382
    4.2. Carpel development   385
    4.3. Dichogamy, dicliny, hercogamy, heterostyly   385
    4.4. Pollination   387
    4.5. Pollinators, pollinator adaptations   391
    4.6. Polyembryony   392
    4.7. Dispersal   392
    4.8. Other terms   393

5. Water Relations, Cold, and Other Forms of Stress   396

    5.1. General terms   396
    5.2. Water relations (including stress aspects)   398
    5.3. Cold stress   407
    5.4. Other forms of stress   412
         5.4.1. High temperature (heat)   412
         5.4.2. Light Stress   413
         5.4.3. Salt (ionic), pressure, and flooding (waterlogging)
                stresses   414
         5.4.4. Atmospheric pollution and mechanical stresses 415

IV.0. General

.001 Absolute humidity: (see VI.5.025a).
.002 Absorbed dose (gray): (see Derived units with special names admitted for reasons of safeguarding human health, IV.0.260d).
.003 Absorbed dose index (gray): (see Derived units with special names admitted for reasons of safeguarding human health, IV.0.260d)
.004 Absorbed dose rate: (see Derived units expressed by special names, IV.0.260e).
.005 Acceleration: (see SI derived units, IV.0.260b).
.006 Activation energy: the input of the minimum quantity of kinetic

IV.0. HORTICULTURAL PHYSIOLOGY and CROP ECLOGY

energy required before a chemical reaction occurs. (AT)

.007 Activity (of a radionuclide) (bequerel): (see Derived units with special names admitted for reasons of safeguarding human health, IV.0.260d).

.008 Adaptability: the response of a plant to its environment, namely climatic, soil, and biotic conditions.

.009 Air flow rate: (see Air movement, IV.0.010).

.010 Air movement: the flow of air over or around plants or plant parts (e.g., detached fruits and leaves,) measured as (1) <u>air flow rate</u> or volume flow rate, the volume of air moving past <u>a given point</u> per unit time, expressed as cubic meters per second ($m^3 \cdot s^{-1}$), cubic millimeters per second ($mm^3 \cdot s^{-1}$), or cubic micrometers per second ($\mu m^3 \cdot s^{-1}$); (2) <u>wind speed</u> (wind velocity) expressed as kilometers per second ($km \cdot s^{-1}$), meters per second ($m \cdot s^{-1}$), millimeters per second ($mm \cdot s^{-1}$), or micrometers per second ($\mu m \cdot s^{-1}$). Note that the location of the anemometer or other instrument in relation to the plants, fruit, etc., should be stated. (SE 78, SE 79)

.011 Amount of substance: (see Mass concentration, IV.0.171a; SI base units, IV.0.260a).

.012 Ångström: (see Units in use temporarily with SI system, IV.0.260k).

.013 Angular acceleration (radian per second squared): (see SI derived units formed by using supplementary units, IV.0.260g).

.014 Angular velocity (radian per second): (see SI derived units formed by using supplementary units, IV.0.260g).

.015 Application rate: (see Mass, IV.0.171b).

.016 Area (A): preferred SI (MKS) units (IV.0.260b) are the square meter (m^2), square kilometer ($km^2 = 10^6 m^2$), square centimeter (cm^2 or $10^{-4} m^2$), square millimeter ($mm^2 = 10^{-6} m^2$), and square micrometer ($\mu m^2 = 10^{-12} m^2$) derived from it. The hectare ($10^4 m^2$) is not an SI unit but may still be used; one hectare equals 2.54 acres. The primary U.S. unit is the acre (A) of 43,560 square feet (ft^2) or 4046.73 square meters (m^2). (SE 79)

.017 Arrhenius plot: a graph that consists of the rates of a simple reaction expressed as the natural logarithm (ln), or 2.303 base$_{10}$ logarithm (log), plotted against the corresponding reciprocal absolute temperature (1/K). The line produced is straight, thus it is said to exhibit Arrhenius-type behavior if the changes in rate are exponential; the relationship is given by the equation $k_2 = \underline{A}_{exp}(-\underline{E}_a/RT)$, where $\underline{A}_{exp}$ (the preexponential factor, which is independent of temperature or nearly so) and the activation energy $\underline{E}_a$ are determined from a plot of ln k_2 against $1/\underline{T}$; i.e., ln $\underline{k}_2$ = ln $\underline{A}_{exp} - \underline{E}_a/RT$, where the intercept is ln $\underline{A}_{exp}$ and the slope of the line is $-\underline{E}_a/R$; $\underline{R}$ is the gas constant and $\underline{T}$ is the absolute temperature (K). (AT)

.018 Arrhenius-type behavior: (see Arrhenius plot, IV.0.017).

.019 Atmosphere (atm): a unit of pressure equal to 760 mm Hg at 0°C, 1.01325 bars, 101325 pascals, 1.01325×10^6 dynes $\cdot cm^{-2}$, $1.0332 \times 10^4 kg \cdot m^{-2}$, or g (gravitational acceleration) 9.80665 $m \cdot s^{-2}$; equivalent values at 15°C (288.16 K) are mass density 1.225040 kg $\cdot m^{-3}$ and specific weight 12.013284 $kg \cdot m^{-2} s^{-2}$. (Note that bar and dyne are non-SI units in use temporarily with the SI system, IV.0. 260k).

.020 Atmospheric moisture: water vapor in the air; (see Humidity, IV.5.025).

.021 Atto (a): (see SI unit prefixes, IV.0.260h).

.022 Average power: (see IV.0.157b-2-e-1).

IV.0. HORTICULTURAL PHYSIOLOGY and CROP ECOLOGY

.023 Avogadro constant (N_A): the number of particles [atoms, molecules, quanta (energy), photons (light), etc.] in the mass of a mole of substance (molal mass); equal to 6.022045×10^{23} (often rounded to 6.02×10^{23}) particles per mole (mol^{-1}). (Used in calculations where the number of particles per mole is involved.)
.024 Avogadro's number of photons (flat surface): (see Light terms and units, IV.0.157X-2).
.025 Avogadro's number of photons (point): (see Light terms and units, IV.0.157y-2).
.026 Bar: a unit of pressure equal to 750.06 mm Hg, 0.986 atm, 1×10^6 dynes $\cdot$ cm^{-2}, 1.01971×10^4 kg $\cdot$ m^{-2}, or 1×10^5 pascals. (see limits in use temporarily with the SI System, IV.0.260k).
.027 Barn: (see limits in use temporarily with the SI System, IV.0.260k).
.028 Base units SI: (see IV.0.260a).
.029 Bequerel: (see Activity, IV.0.260d).
.030 Biochemistry: branch of chemistry concerned with biological organisms and processes.
.031 British thermal unit (BTU): (see IV.5.005).
.032 Calorie: not an SI unit (see Other units, IV.0.260m), hence obsolete; defined in VI.5.006 because of its widespread use in the past.
.033 Candela (cd): (see Light terms and units, IV.0.157a; SI base units, IV.0.260a).
.034 Capacitance (farad): (see SI derived units with special names, IV.0.260c).
.035 Carbon dioxide (CO_2): a minor but biologically important component of air measured as millimoles per cubic meter (mmol $\cdot$ m^{-3}), by mass as grams per kilogram (g $\cdot$ kg^{-1}) or milligrams per kilogram (mg $\cdot$ k^{-1}), or by volume as cubic micrometers per cubic meter ($\mu m^3 \cdot m^{-3}$); mg$\cdot k^{-1}$ and $\mu m^3 \cdot m^{-3}$ are equivalent to ppm CO_2 by mass and volume, respectively, and are preferred units; typical CO_2 concentrations in air at sea level are 300 to 350 mg $\cdot$ kg^{-1} (mass) or $\mu m^3 \cdot m^{-3}$ (volume). (K & M)
.036 Carbon-14 (^{14}C): radioactive isotope of carbon. (S & R)
.037 Celsius temperature (°C): (see SI base units, IV.0.260a).
.038 Centi (c): (see SI unit prefixes, IV.0.260h).
.039 Chemiluminescence: a process by which products of oxidation reactions are generated in electronically excited states; therefore they glow (e.g., fireflies or old rotting stumps). (AL)
.040 Chromatography: a differential separation of mixtures of compounds or ions by the use of solvents or ionic absorbants; many different methods and techniques, such as paper, thin layer, and gas.
.041 Cis: (see Stereoisomers, IV.0.286).
.042 Colloid, colloidal system: the former is any substance in a state of fine subdivision or dispersion, with particles ranging between $10^{-2} \cdot$ mm and 1 nm in diameter; the latter is a complex in which a tremendously large interface is attained by the fine, heterogeneous dispersion of one or more substances in a second substance or material; each of the two phases may be a solid, liquid, or gas; materials in the colloidal state are distinguished especially by (1) a great effective surface or interface, (2) the capacity to fix and hold solids, gases, salts, and ions, and (3) catalysis, the tendency to hasten or retard chemical reactions; examples of colloidal systems are albumin, gelatin, starch, and soil clay minerals < 500 nm in diameter. (S & R)
.043 Concentration: (see Mass, IV.0.171c).

IV.O. HORTICULTURAL PHYSIOLOGY and CROP ECOLOGY

.044 Concentration (of amount of substance): (see SI derived units, IV.0.260c).
.045 Conductance (siemens): (see SI derived units with special names, IV.0.260c).
.046 Coulomb (C): (see SI derived units with special names, IV.0.260c).
.047 Coupled reactions: reactions that must be linked to other chemical reactions to expend the amount of energy for a reaction to take place. (LE 75)
.048 Crop ecology: the study of crop plants in relation to environment (i.e., climate, physiography, soil, and biotic factors) and their effects on one another.
.049 Curie (ci): (see Units in use temporarily with SI System, IV.0.260k).
.050 Current density: (see SI derived units, IV.0.260b).
.051 Dalton: equal to 1.0000 on the atomic mass scale; used for "molecular" weights of macromolecules, such as proteins (IV.2.2.041).
.052 Dark-field microscope: a compound (light) microscope in which a black plate is placed in the illuminating beam and what is seen is light scattered from the particles or tissues rather than the particles directly.
.053 Day (d): (see Units used with SI System, IV.0.260i).
.054 Deci (d): (see SI unit prefixes, IV.0.260h).
.055 Degree (°): (see Units used with SI system, IV.0.260i)
.056 Degrees Celsius (°C): see SI base units, IV.0.260a; Temperature, IV.0.295b).
.057 Degrees Fahrenheit (°F): (see Temperature, IV.0.295c).
.058 Deka (da): (see SI unit prefixes, IV.0.260h).
.059 Density, mass density (p): (see Mass, IV.0.171d; SI derived units, IV.0.260b)
.060 Derived SI units expressed in terms of base units: (IV.0.260b).
.061 Derived SI units expressed by means of special names: (IV.0.260c).
.062 Derived SI units with special names: (IV.0.260c).
.063 Dewpoint: (see VI.5.025b).
.064 Dialysis: the process of separating low molecular weight molecules from larger ones by a membrane that has pores large enough for only the smaller molecules to pass through, as in purification of a protein preparation by washing the contaminants through a cellophane membrane or tube. (S & R)
.065 Diastereomer: (see Stereoisomer, IV.0.286).
.066 Dose equivalent (sievert): (see Derived units with special names admitted for reasons of safeguarding human health, IV.0.260d).
.067 Dose equivalent index (sievert): (see Derived units with special names admitted for reasons of safeguarding human health, IV.0.260d).
.068 Dynamic viscosity (pascal second): (see SI derived units expressed by special names, IV.0.260e).
.069 Dyne (dyn): (see CGS units, IV.0.260l).
.070 Einstein (E): a number of quanta or photons equal to Avogadro's number (6.02×10^{23} atoms $\cdot$ mol^{-1}); can be calculated for any wavelength from $E = h\nu$, where E is the energy of a quantum or photon, ν is the frequency, and h is Planck's constant; equivalent to one mole per second per square meter (mol $\cdot$ s$^{-1} \cdot$ m^{-2}); the einstein is not a SI unit, the preferred designation is mol $\cdot$ s$^{-1} \cdot$ m^{-2}; (see Photosynthetically active radiation, IV.0.157m). (S & R, K & M)
.071 Electrical charge (coulomb): (see SI derived units with special names, IV.0.260c).

IV.0. HORTICULTURAL PHYSIOLOGY and CROP ECOLOGY

.072 Electrical charge density (coulomb per cubic meter): (see SI derived units expressed by special names, IV.0.260e)
.073 Electrical conductance (S): a reciprocal ohm (mho) now defined as siemens, which equals one mho; customary units are decisiemens per meter ($dS \cdot m^{-1}$), millisiemens per meter ($mS \cdot m^{-1}$), and microsiemens per meter ($\mu S \cdot m^{-1}$); used as a measure of the concentration of salts in soils and solution cultures. (G & B, K & M)
.074 Electric current (ampere): (see SI base units, IV.0.260a).
.075 Electric field strength (volt per meter): (see SI derived units expressed by special names, IV.0.260e).
.076 Electric potential (volt): (see SI derived units with special names, IV.5.260c).
.077 Electric resistance (ohm): (see SI derived units with special names, IV.0.260c).
.078 Electromotive force (volt): (see SI derived units with special names, IV.0.260c).
.079 Electron microscope: instruments of two major types [transmission electron microscope (TEM) and scanning electron microscope (SEM)] that utilize beams of electrons rather than light as the source of illumination. Primary components of the TEM are an electron source (gun) that accelerates particles toward an anode held at high positive potential (25-1000 kV), a series of electromagnetic condenser, objective, and projector lenses for image magnification, a fluorescent screen, and photographic apparatus. The SEM has an electron-optical column operated at 1-50 kV with additional electromagnetic lenses and a single or double deflection field to permit scanning at high resolution and depth of field (and consequent image clarity). Both instruments are operated under high vacuum. The TEM is used for sectioned specimens of 0.05-0.25 μm or less at 25-125kV and up to 5 μm at 500-1000 kV, whereas the SEM is used for solid specimens or thick sections; specimens must be specially prepared according to the type of biological material, with numerous procedures for both TEM and SEM work. (O & M)
.080 Electron spin resonance spectroscopy (ESR): detection of occurrence of molecules with unpaired electron spins by their characteristic behavior in a magnetic field. (LE 70)
.081 Electronvolt (eV): (see Subunits whose value in SI units is obtained experimentally, IV.0.260j).
.082 Electrophoresis: separation of proteins or other charged molecules in an electric field.
.083 Enantiomer: (see Stereoisomer, IV.0.286).
.084 Endergonic: (see Endothermic, IV.0.085).
.085 Endothermic (endergonic): a chemical reaction that absorbs energy (heat). (LE 70)
.086 Energy, work, quantity of heat (E): the capacity to do work; measured as joules (J) = newtons X meter = square meters X kilograms per second per second ($J = N \cdot m = m^2 \cdot kg \cdot s^{-2}$), of which the last is the SI (MKS) unit (see SI derived units with special names, IV.0.260c); note that power and radiant flux are also energy expressed as watts (W) = joules per second ($J \cdot s^{-1}$) = square meters X kilograms per second cubed ($m^2 \cdot kg \cdot s^{-3}$); (see also Light terms and units, IV.0.157). (G & B)
.087 Energy (flat surface): (see Light terms and units, IV.0.157x-1-a).
.088 Energy content (flat surface): (see Light terms and units, IV.0.157x-1).
.089 Energy content (point): (see Light terms and units, IV.0.157y-1).
.090 Energy density (joules per cubic meter): (see SI derived units ex-

IV.0. HORTICULTURAL PHYSIOLOGY and CROP ECOLOGY

pressed by special names, IV.0.260e).
.091 Energy flow rate (flat surface): (see Light terms and units, IV.0. 157x-1-b).
.092 Energy flow rate (point): (see Light terms and units, IV.0. 157y-1-a).
.093 Energy fluence (point): (see Light terms and units, IV.0.157y-1-b).
.094 Energy fluence rate (applied or absorbed)(point): (see Light terms and units, IV.0.157y-1-c).
.095 Energy flux (flat surface): (see Light terms and units, IV.0.157x-1-c).
.096 Energy sum: (see IV.0.157b-2-e-2).
.097 Enthalpy (Σ): (see VI.5.017).
.098 Entropy (S): basically the degree of disorder or randomness in a system, which is zero at 0 K (the Third Law of Thermodynamics, IV.0.154); e.g., the direction of a spontaneous change is derived from a state of low probability (i.e., greater disorder) to one of maximum probability (less disorder or a greater dispersal of energy); this also means that there is a certain quantity of unavailable energy or unavoidable waste when mechanical work is derived from the heat energy of a body, and the magnitude of this quantity is dependent on the quantity of heat in a body and its temperature multiplied by any lower temperature (the minimum available). Entropy per se cannot be calculated as an absolute value but rather as a change, i.e., ΔS, the difference in dispersal of energy when a system is changed from one state (set of conditions) to another, expressed as joules per kelvin per mol ($J \cdot K^{-1} \cdot mol^{-1}$). (AT, G & B)
.099 Entropy (joules per kelvin): (see SI derived units expressed by special names, IV.0.260e).
.100 Environment: a plant's surroundings, including air, soil, moisture, light, and heat.
.101 Erg (erg): (see CGS units, IV.0.5.260 l).
.102 ESR: (see Electron spin resonance spectroscopy, IV.0.080).
.103 Exa (E): (see SI unit prefixes, IV.0.260h).
.104 Exergonic: (see Exothermic, IV.0.031).
.105 Exothermic: a chemical reaction that releases energy (heat). (LE 70)
.106 Exposure (x- and γ- rays): (see Derived units expressed by means of special names, IV.0.260e).
.107 Farad (F): (see SI derived units with special names, IV.0.260c).
.108 Femto (f): (see SI unit prefixes, IV.0.260h).
.109 Fermi: (see Other units, IV.0.260m).
.110 Fluorescence microscope: a compound (light) microscope in which the source of illumination is a xenon or similar type of lamp and which causes certain plant tissues not otherwise easily distinguished (e.g., pollen tubes growing down a style) to glow (become fluorescent) after treatment with certain dyes (e.g., eosin or fluorescein).
.111 Force (newton): (see SI derived units with special names, IV.0.260c).
.112 Fraction (w): (see Mass, IV.0.171e).
.113 Free energy: energy in any particular chemical (or physical) system that is available to do useful work. (AT)
.114 Frequency (f): a term used in two different senses; (1) rotation (as a shaft, electron, etc.) as revolutions per second ($r \cdot s^{-1}$) and (2) the frequency of a periodic wave (e.g., light and/or

IV.0. HORTICULTURAL PHYSIOLOGY and CROP ECOLOGY

.115   sound), designated as the hertz (Hz) as one per second ($1 \cdot s^{-1}$); (see SI derived units with special names, IV.0.260c). (SE 79)
.115   Gal (gal): (see Units outside but used for limited time with SI system, IV.0.260k).
.116   Gamma (γ): (see Other units, IV.0.260m).
.117 γ(gamma): (see Other units, IV.0.260m).
.118   Gas constant (R): a constant whose numerical value is derived from the general equation for ideal gases, $R = PV/nT$ (or $V = nRT/P$), relating volume (V), pressure (P), and temperature (T) as degrees (K) for a given number of moles (n); $R = 8.20562 \; 10^{-2} m^3 \cdot atm \cdot kmol^{-1} \; K^{-1}$ at a standard volume (V_o) of $22.4136 \; m^3 \cdot kmol^{-1}$; $R = 82.05 (10^{-3} \cdot kg \cdot mol)^{-1} \times (10^{-6} m)^{-3} \; ^\circ X \; atm^{-1} \cdot K^{-1}$ or $0.08205 \; atm \cdot K \cdot (10^{-3} \cdot m^3)^{-1} \cdot (10^{-3} kg \; mol)^{-1}$, or $62.4 \; (10^{-3} kg \cdot mol)^{-1} \cdot mm \; Hg^{-1} \; (10^{-3} \cdot m^3)^{-1} \cdot K^{-1}$; or $8.31432 \; J \cdot K^{-1} \cdot mol^{-1}$; (see Arrhenius plot, IV.0. 017; Gibbs free energy, IV.0.121). (S & R)
.119   Gauss (Gs,G): (see CGS units, IV.0.260l).
.120   Gel: jellylike material formed by coagulation of a colloidal liquid.
.121   Gibbs free energy: the expression for free energy (i.e., the energy available for carrying out a chemical reaction or process, distinct from entropy (S), which is not available); $G = E + PV + TS$, where G is Gibbs free energy (J), E, internal energy, P barometric pressure (atm), V, volume (m^{-3}), T, absolute temperature (K) and S, entropy; free energy must decrease for a reaction to occur spontaneously and increase for a nonspontaneous reaction; calculated in practice as the change in free energy as $G = -RT \ln K_{eq}$, where ΔG is the change (J), R is the gas constant ($\overline{149.4} \; J \cdot mol^{-1}$), T is the absolute temperature (K), and $\ln K_{eq}$ is the natural logarithm of the equilibrium constant = activities of products/activities of reactants; (see Entropy, IV.0.098). (S & R)
.122   Giga (G): (see SI unit prefixes, IV.0.260g).
.123   Gravitational constant (G): $6.6720 \times 10^{-13} \; N \cdot m^2 \; kg^{-2} = 6.6720 \times 10^{-13} \cdot Pa \; kg^{-2} = 6.6720 \times 10^{-13} \; m^{-1} \cdot kg \cdot s^{-2}$ (see Gravity, IV.0.124).
.124   Gravity (g): the universal attraction that exists between all material bodies; the earth's gravitational attraction at sea level (45°32' N lat.) is $9.80665 \; m \cdot s^{-2}$ or $32.1740 \; ft \cdot s^{-2}$.
.125   Gray: (see Absorbed dose, IV.0.260d).
.126   Heat: a form of energy associated with and proportional to the molecular (or atomic motions) of a substance or body and which may be transferred from one body to another, (see also Thermal energy, VI.5.051).
.126a.   British thermal unit: (see VI.5.005).
.126b.   Calorie: (see VI.5.006)
.126c.   Capacity (joule per kelvin): (see SI derived units expressed by special names, IV.0.260e; Specific heat capacity, VI.5.051e)
.126d.   Enthalpy (Σ): (see VI.5.017).
.126e.   Flux: energy change per unit time, i.e., the rate of energy change; units are joules per second ($J \cdot s^{-1}$) or watts ($W = J \cdot s^{-1}$), the former used for work or quantity of heat and the latter for power or radiant flux. (HKS)
.126f.   Flux density: the rate of energy change per unit time per unit area; units are joules per second per square meter ($J \cdot s^{-1} \cdot m^{-2}$) or watts per square meter ($W \cdot m^{-2}$); (See SI derived units expressed by special names, IV.0.260e). (G & B) [Note that Flux and Flux density are not used or defined uniformly in the

IV.0. HORTICULTURAL PHYSIOLOGY and CROP ECOLOGY

literature and should not be used.) (HKS)
- .126g. Removal: (see VI.5.024).
- .126h. Specific heat capacity (C): (see Specific heat, VI.5.051e).
- .126i. Specific latent heat of liquefaction (fusion): (see Latent heat, VI.5.051c).
- .126j. Specific latent heat of vaporization (condensation): (see Latent heat, VI.5.051c).
- .126k. Surface convective coefficient of heat transfer (h): (see VI.5.048).
- .126l. Thermal conductivity (k) (watt per meter kelvin): (see SI derived units expressed by special names, IV.0.260d; VI.5.049).
- .126m. Thermal diffusivity (α): (see VI.5.050).
- .126n. Ton of refrigeration: (see VI.5.056).
- .127 Hectare (ha): (see Area, IV.0.016; Units in use temporarily with SI system, IV.0.260k).
- .128 Hecto (h): (see SI unit prefixes, IV.0.260h).
- .129 Henry (H): (see SI derived units with special names, IV.0.260c).
- .130 Hertz (hz): (see Frequency, IV.0.114; SI derived units with special names, IV.0.260c).
- .131 Hofmeister series: (see Lyotropic series, IV.0.166).
- .132 Homogenate: biological material finely divided and thoroughly mixed.
- .133 Hour (h): (see Units used with SI system, IV.0.260i).
- .134 Illuminance (E; lux): (see Light terms and units, IV.0.157c; SI derived units with special names, IV.0.260c).
- .135 Inductance (henry): (see SI derived units with special names, IV.0.260c).
- .136 Infrared spectrophotometer: an instrument used to measure spectra in the near-infrared (0.8 to 2.0 μm), infrared (2 to 16 μm), and far-infrared (16 to 300 μm) regions; the wavelength region of 2.5 to 16 μm is of greatest interest to organic chemists or biochemists because many bond deformations occur in this region; spectra are customarily plotted as light transmitted from 0 to 100%. (AL)
- .137 Inhibition: complete cessation of a process; "inhibition" is an absolute term (i.e., either yes or no); "partial inhibition" means (or should mean) a process is proceeding in normal fashion but the rate is reduced by the intervention of some condition (e.g., blockage of a reactant). (LE 70)
- .138 Irradiance (watt per square meter): (see Light terms and units, IV.0.157d; SI derived units expressed by special names, IV.0.260e).
- .139 Irradiation: (see Light terms and units, IV.0.157e).
- .140 Isoelectric pH, isoelectric point: the pH at which the solubility of a colloid (e.g., a protein) is lowest. (S & R)
- .141 Isopleth: a multivariate graph on which the relationship of one or between two (or more) variables is expressed as a function of two others (e.g., changes in leaf and air temperatures in a plant canopy expressed as a function of height above ground level and time of day), or data obtained from sample borings made to certain depths and distances apart in a soil catena are plotted and lines of equal value drawn to indicate the location of areas of low, medium or high percentages of base saturation in the overall profile.
- .142 Isotope: forms of an element with essentially identical chemical properties but differing atomic weights (i.e., the same number of protons but a different number of neutrons), as ^{14}C, ^{13}C, ^{12}C, etc.; some, such as ^{14}C are radioactive, and are denoted as radioisotopes or radionuclides.

IV.O. HORTICULTURAL PHYSIOLOGY and CROP ECOLOGY

.143 Joule (J): (see SI derived units with special names, IV.O.260c).
.144 Kelvin (K): (see SI derived units with special names, IV.O.260a; Thermodynamic temperature, IV.O.295d).
.145 Kerma: (see Derived units with special names admitted for reasons of safeguarding human health, IV.O.260d).
.146 Kinetic energy: energy resulting from the motion or activity of ions, atoms, or molecules; zero at 0 K. (S & R)
.147 Kilo (k): (see SI unit prefixes, IV.O.260h).
.148 Kilogram-force (kgf): (see Other units, IV.O.260m).
.149 Knot: (see Units in use temporarily with SI system, IV.O.260k).
.150 λ (lambda): (see Other units, IV.O.260m).
.151 Langley (ly): a measure of the total energy in sunlight or other light sources for plant growth, with a unit of one calorie g per square centimeter, equivalent to about 4.184 joules per cm^2 or 41840 $J \cdot m^{-2}$; the latter is an SI unit; subject to confusion as the calorie has several different units, hence values for number of joules.
.152 Langley per day: (see Light terms and units, IV.O.157b-2-e-3).
.153 Large calorie: (see Calorie, IV.5.006).
.154 Laws of Thermodynamics: (AT)
   First: Energy is neither created nor destroyed (energy may be converted into heat and conversely).
   Second: Energy conversions are never 100% efficient (it is always the higher state that loses energy or heat and the lower one that gains energy or heat).
   Third: The change in entropy of an isothermal process approaches zero as the temperature approaches 0 K and is zero at 0 K.
.155 Length (l): the standard SI unit is the meter (m); (see IV.O.260a); the ångström (Å), millimicron (mμ) and micron (μ) are absolete; the proper SI equivalents are 1Å = 10 nanometers (nm), 1mμ = 1nm (=10^{-9}m), and 1μ = 1 micrometer (1μm = 10^{-6}m).
.156 Light intensity: (see Radiant intensity, IV.O.157w).
.157 Light terms and units: it should be noted that there are substantial differences in opinion among photobiologists and other plant physiologists regarding the correct terminology and units of measure for light terms. Part of the confusion arises from the failure to distinguish clearly among different light sources. Photometric sources emit or sense radiation of wavelengths to which the human eye is sensitive. By contrast, radiometric sources emit or sense wavelengths to which plants are sensitive, including the proportion of radiation arriving vs. that intercepted at the plant. There is a concerted effort among scientists to report their data strictly in SI units. Certain of the latter, such as the lux and lumen, are valid, but considered obsolete principally because of the way the candela, on which both units are based, was defined prior to 1980 (see present definition in the 1981 edition for SI units). Use of a unit like "density" out of context (i.e., with terms not so defined in the SI) may be redundant, therefore erroneous. There is a large body of data in the older literature in which non-SI units, such as calorie, einstein, and langley, abound; this information should not be ignored, but rather converted where possible to more meaningful (i.e., current) terminology. Finally, it is incumbent on photobiologists to ensure that the terms used and data generated in their research are comprehensible to a wide spectrum of horticulturists, many of whom presently lack familiarity with SI

IV.O. HORTICULTURAL PHYSIOLOGY and CROP ECOLOGY

units and the metric system in general.

.157a. Candela (cd): luminous intensity of a source that emits monochromatic radiation of frequency 540×10^{12} Hz and radiant intensity 1/683 watt per steradian (sr); corresponds to 683 lumens (lm) per watt at approximately 555 nm wavelength; (see SI base units, IV.0.260a). (G & B)

.157b. Conversion units: (T & H)

.157b-1. Errors: uncertainties in interconversion of photometric (instruments for measurement of luminous intensity, luminous flux, illuminance or brightness, i.e., sources, by comparison with standards), radiometric (instruments to measure radiant energy, i.e., sinks), and quantum (photon) light units are attributable to measurement (variations among light sensors and time units) and conversion (lamp ageing, operating procedures, differences among lamps from different manufacturers, and lack of equivalence among sources of different spectral content); errors range from 10% up, with perhaps 25 to 30% error expected under non-ideal conditions (e.g., factors derived from broad-band measurements or gross averaging of daily or hourly values).

.157b-2. Formulas:

.157b-2-a. Mixed (photometric to radiometric): lux to watts per square meter (lx to $W \cdot m^{-2}$) is $[(lm \cdot m^{-2} lx^{-1}) \cdot (lx)(mW \cdot lm^{-1})(h \text{ of light})] [(10^3 \cdot mW^{-1} \cdot W^{-1})(24h)]^{-1}$; values in this conversion are dependent on the light source and waveband of interest; (see Tables 1 and 2 in T & H).

.157b-2-b. Mixed (photometric to quantum): lux to micromoles per second per square meter (lx to $\mu mol \cdot s^{-1} m^{-2}$) can be converted in either of two ways: (1) $lx \cdot constant^{-1} = \mu mol \cdot s^{-1} m^{-2}$ (see Table 3 in T & H for constant); or (2) convert lx to $W \cdot m^{-2}$ (see IV.0.157b-2-a) ignoring the time interval, then see Table 3 in T & H for $\mu mol \cdot s^{-1} m^{-2} \cdot (W \cdot m^{-2})^{-1}$.

.157b-2-c. Mixed (quantum to radiometric): conversion of $mol \cdot s^{-1} m^{-2}$ to $W \cdot m^{-2}$ entails knowledge of the number of photons per unit wavelength at each wavelength and energy per photon at each wavelength to calculate total energy; obtain the constant for a known light source from Table 3 in T & H), then $(\mu mol \cdot s^{-1} m^{-2}) \text{ constant}^{-1} = W \cdot m^{-2}$; or for a 24-h average $(W \cdot m^{-2} \cdot h \text{ irradiation}) \cdot 24h^{-1} = W \cdot m^{-2}$ (24h basis).

.157b-2-d. Photometric: 10.8 lux = 1 footcandle (= 1 lumen per square foot); i.e., 10 klx = 926 footcandles (U.S. unit).

.157b-2-e. Radiometric:

.157b-2-e-1. Average power: energy per unit time (e.g., per minute, hour, or day) reported as watts; can convert energy sum to a rate value by dividing the former by the time interval; for example, 1 kilowatt-hour per hour = 1 kilowatt or 1 joule per s = 1 watt therefore divide joules per hour by 3600 s to effect the conversion to watts.

.157b-2-e-2. Energy sum: energy summed over some time period (e.g. hours, or days) should be reported as kilowatt hours or joules, not $W \cdot m^{-2}$, which is a rate measurement. (The kilowatt-hour is not a SI unit).

.157b-2-e-3. Langley (Ly) per day: = 0.484 watt per square meter or

IV.O. HORTICULTURAL PHYSIOLOGY and CROP ECOLOGY 338

$207 \text{ Ly} \cdot \text{d}^{-1} = 100 \text{ W} \cdot \text{m}^{-2}$. (The langley is not an SI unit).

.157c. Illuminance (lux): lumen (lm) per square meter ($lx = lm \cdot m^{-2} = m^{-2} \cdot cd \cdot sr$); the candela, thus the lumen ($cd \cdot sr$) and lux (lx), has been redefined (G & B); (see Candela, IV.O.157a, .260a). (K & M)

.157d. Irradiance: energy intercepted per time and unit area, expressed as watts per square meter ($W \cdot m^{-2} = J \cdot s^{-1} \cdot m^{-2}$); wavelength region and time interval averaged must be specified (see SI derived units expressed by means of special names, IV.O.260e). (G & B, P & V)

.157e. Irradiation: strictly, the exposure of an object or body to ionizing radiation (i.e., a radioisotope), thus properly not a light term; sometimes used (incorrectly) instead of irradiance. (K & M)

.157f. Light intensity: (see Radiant intensity, IV.O.157w).

.157g. Luminance (L): luminous intensity from a source or the equivalent illuminance per unit area, expressed as candela per square meter ($cd \cdot m^{-2}$); (see Candela, IV.O.157a). (G & B, SE 79)

.157h. Luminous flux (luminous energy): the amount of luminous energy, expressed as lumens (= candelas per steradian, $cd \cdot sr^{-1}$). (G & B)

.157i. Luminous intensity: an SI base unit (see Candela, IV.O.157a).

.157j. Lux: (see Illuminance, IV.O.157c)

.157k. Photometric radiation: light of 380 to 780 nm wavelengths with cosine correction for the angle(s) of incidence; the unit is the lux (GCWG).

.157l. Photon flux density: [see Photosynthetically active radiation (PAR), IV.O.157m].

.157m. Photosynthetically active radiation (PAR): quantum (photon) flux density (light emitted) for 400- to 700 nm wavelengths with cosine correction; units are micromole per second per square meter ($\mu mol \cdot s^{-1} \cdot m^{-2}$) or microeinsteins per second per square meter ($\mu E \cdot s^{-1} m^{-2}$); irradiance (light arriving) with the same units may be substituted. (K & M, MO) [It should be pointed out that (1) density is used incorrectly because "flux" implies density (HKS); and (2) the einstein not only is not a SI unit, but is defined as (a) the quantity of radiant energy in Avogadro's number of photons or (b) Avogadro's number of photons. (T & H)]

.157n. Photosynthetic efficiency: (see IV.3.1.020).

.157o. Photosynthetic irradiance (PI): (see Photosynthetically active radiation, IV.O.157m).(K & M)

.157p. Photosynthetic photon density: measured as micromoles per square meter ($\mu mol \cdot m^{-2}$) or microeinsteins per square meter ($\mu E \cdot m^{-2}$); (see parenthetic comments under Photosynthetically active radiation, IV.O.157m).

.157q. Photosynthetic photon flux density (PPFD): same as quantum (photon) flux density (and subject to the same comments) (see Photosynthetically active radiation, IV.O.157m). (K & M)

.157r. Radian: the plane angle subtended by an arc equal to the radius; nondimensional (see IV.O.157ac).

.157s. Radiance: irradiance (IV.O.157d) per steradian; the unit is watt per square meter steradian ($W \cdot m^{-2} \cdot sr^{-1}$). (G & B)

.157t. Radiant energy: the energy associated with radiation, the wave lengths implied are those of the complete electromagnetic spectrum; equal to the mean of Planck's constant X frequency (h ν) X Avogadro's number of photons per mole (see IV.0.023); the unit is the joule ($J = m^2 \cdot kg \cdot s^{-2} mol^{-1}$).

.157u. Radiant flux (Q): the rate at which radiant energy is emitted or received; the unit is the watt ($W = J \cdot s^{-1}$). (G & B)

.157v. Radiant flux density (rfd): the rate (amount per unit time) at which radiant energy strikes a surface of unit area (square meter); the unit is watts per square meter ($W \cdot m^{-2}$); the converse of irradiance (i.e., radiant energy received) is that intercepted (see also comments under IV.0.157d and IV.O. 157m). (SE 78)

.157w. Radiant intensity: watt per steradian ($W \cdot sr^{-1}$).

.157x. Radiation arriving at a <u>flat surface</u>: acceptance angle is 180° or less and a cosine correction is applied for radiation arriving at angles not normal to the receiving surface: (HSK)

.157x-1. Energy content:
.157x-1-a. Energy applied (absorbed): joules per square meter ($J \cdot m^{-2}$).
.157x-1-b. Energy flow rate: joules per second ($J \cdot s^{-1}$) or watt (W).
.157x-1-c. Energy flux: watts per square meter ($W \cdot m^{-2}$).
.157x-1-d. Radiant energy: joules (J).
.157x-2. Photon content:
.157x-2-a. Avogadro's number of photons: (dimensionless).
.157x-2-b. Photon flow rate: number per second (s^{-1}) or moles per second ($mol \cdot s^{-1}$)
.157x-2-c. Photon flux: number per square meter per second ($m^2 \cdot s^{-1}$) or moles per square meter per second ($mol \cdot m^2 \cdot s^{-1}$).
.157x-2-d. Photons (number): (dimensionless).
.157x-2-e. Photons applied: number applied per square meter (m^{-2}) or mole per square meter ($mol \cdot m^{-2}$).

.157y. Radiation arriving at a <u>point</u>: detection has 360° spherical acceptance angle and measurements are the integral of the radiation from all directions: (HKS)

.157y-1. Energy content:
.157y-1-a. Energy flow rate: joules per second ($J \cdot s^{-1}$) or watts (W)
.157y-1-b. Energy fluence: amount of energy applied to an object as joules per square meter ($J \cdot m^{-2}$).
.157y-1-c. Energy fluence rate: energy applied (or absorbed, taking into account screening factors) as watts per square meter ($W \cdot m^{-2}$).
.157y-1-d. Radiant energy: joules (J).
.157y-2. Photon content:
.157y-2-a. Avogadro's number: (mol).
.157y-2-b. Photon flow rate: number per second (s^{-1}) or moles per second ($mol \cdot s^{-1}$).
.157y-2-c. Photon fluence: number of photons applied to an object or absorbed; the latter takes into account screening factors; number per square meter (m^{-2}) or moles per square meter ($mol \cdot m^{-2}$).
.157y-2-d. Photon fluence rate: number per square meter per second ($m^{-2} \cdot s^{-1}$) or moles per square meter per second ($mol \cdot m^{-2} \cdot s^{-1}$)

.157z. Spectral irradiance (spectral energy flux): energy intercepted from λ_1 to λ_2 (e.g., 250 to 850 nm) in less than 20-nm bandwidths with cosine correction; the unit is watts per square

IV.O. HORTICULTURAL PHYSIOLOGY and CROP ECOLOGY 340

        meter per second per nanometer [$W \cdot m^{-2} s^{-1} \cdot nm^{-1}(\lambda_1$ to λ_2 nm)]. (K & M)

.157aa. Spectral photon flux: quantum (photon) flux for λ_1 to λ_2 nm in less than 20-nm bandwidths with cosine correction; the unit is micromoles (quanta) per second per square meter per nanometer ($\lambda_1 - \lambda_2$ nm [$\mu mol \cdot s^{-1} m^{-2} nm^{-1}$ ($\lambda_1 - \lambda_2$ nm)]. (K & M)

.157ab. Speed of light in a vacuum (c): 2.9979246×10^8 m $\cdot s^{-1}$.

.157ac. Steradian (sr): the solid angle that encloses an area on a sphere equivalent to the square of the radius (IV.0.260f).

.157ad. Total irradiance: irradiance (IV.0.157c) with cosine correction. (K & M) (Note, however, the sources of error listed in IV.0.157b-1.).

.158 Liter (l,L): (see units used with the SI system, IV.0.260i).

.159 Lumen (lm): (see SI derived units with special names, IV.0.260c).

.160 Luminance (L): (see Light terms and units, IV.0.157g; SI derived units, IV.0.260b).

.161 Luminous energy (Q): (see Light terms and units, IV.0.157h).

.162 Luminous flux (Φ, lumen): (see Light terms and units, IV.0.157h; derived units with special names, IV.0.260c).

.163 Luminous intensity (I,cd): (see Light terms and units, IV.0.157i; SI base units, IV.0.260a).

.164 Lux (lx): (see SI derived units with special names, IV.0.260c).

.165 Lyophilize: freeze-drying the tissue under vacuum. (S & R)

.166 Lytropic (Hofmeister) series: tenacity of adsorption of cations to the surfaces of charged colloidal particles (i.e., to a clay mineral or organic matter in the soil); (most) $Al^{+3} > H^+ > Ba^{++} > Ca^{++} > Mg^{++} > K^+ > NH_4^+ > Na^+ > Li^+$ (least); or in soil, $H^+ > Ca^{++} > Mg^{++} > K^+ > NH_4^+ > Na^+$. (S & R)

.167 Macromolecule: a large molecule such as a protein or many polymers whose weight may range up to many thousands of daltons (IV.0.045). (S & R)

.168 Magnetic field strength: (see SI derived units, IV.0.260b).

.169 Magnetic flux (weber): (see SI derived units with special names, IV.0.260c).

.170 Magnetic flux density (tesla): (see SI derived units with special names, IV.0.260c).

.171 Mass (m): the SI basic unit (IV.0.260a) is the kilogram (kg); measured on a <u>beam</u> balance to be independent of gravity (see Weight, IV.0.325). (G & B)

.171a. Amount of substance (mol): the basic SI unit is the mole (mol), the amount of substance that contains as many elementary entities as there are unbound atoms in the ground state in 0.012 kilogram of carbon 12; elementary particles must be specified and may be atoms, molecules, ions, electrons, photons, other particles or specified groups of such particles. (Note that amount of substance, as defined, and mass are entirely different quantities; see IV.0.260a). (G & B)

.171b. Application rate: mass applied per unit area, expressed in kilograms per square meter (kg $\cdot m^{-2}$). (SE 79)

.171c. Concentration (of amount of substance): for solid nutrients is expressed as moles per cubic meter (mol $\cdot m^{-3}$), kilomole per cubic meter (kmol $\cdot m^{-3}$), millimoles per cubic meter (mmol m^{-3}), micromoles per cubic meter (μmol $\cdot m^{-3}$), or moles per kilogram (mol $\cdot kg^{-1}$), and for liquid nutrients, as micromoles per liter (μmol $\cdot L^{-1}$) or millimoles per liter (mmol $\cdot L^{-1}$); note that cubic decimeters (dm^{-3}) is preferred to liter as a

IV.0. HORTICULTURAL PHYSIOLOGY and CROP ECOLOGY 341

unit; (see SI derived units, IV.0.260b). (G & B, SE 79)
.171d. Density (ρ): mass per unit volume of substance; units are kilograms per cubic meter (kg $\cdot$ m^{-3}), grams per cubic meter (g$\cdot$ m^{-3}), milligrams per cubic meter (mg $\cdot$ m^{-3}), or micrograms per cubic meter (μg $\cdot$ m^{-3}); the term <u>mass density</u> is preferred to density. (SE 79)
.171e. Fraction (w): mass per mass, expressed as kilograms per kilogram (kg $\cdot$ kg^{-1}), grams per kilogram (g $\cdot$ kg^{-1}), milligrams per kilogram (mg $\cdot$ kg^{-1}), or micrograms per kilogram (μg $\cdot$ kg^{-1}).
.171f. Molality: gram-moles per kilogram of water equals 1000 X molal concentration [gram-moles per cubic decimeter (g mol $\cdot$ dm^{-3}) divided by total water concentration (grams per cubic decimeter, g $\cdot$ dm^{-3})] or [10^3 g-mol $\cdot$ dm^{-3} (g $\cdot$ dm^{-3})$^{-1}$]. (SE 79)
.172 Mass basis: (see Transpiration rate, IV.5.2.077b).
.173 Mass density: (see Mass, IV.0.171d; SI derived units, IV.0.260b).
.174 Mass spectrometer: an instrument for the determination of molecular composition by bombardment in the gas phase with an electron beam; components of the spectrometer are an ion source, magnetic or electric analyzer (effects separation essentially according to mass), and a detection-recording device. (AL)
.175 Maxwell (Mx): (see CGS units, IV.0.260l).
.176 Mean calorie: (see Calorie, VI.5.006).
.177 Mega (M): (see SI unit prefixes, IV.0.260h).
.178 Metric carat: (see Other units; IV.0.260m).
.179 Metric ton (tonne): (see Units used with the SI system, IV.0.260i).
.180 Micro (μ): (see SI unit prefixes, IV.0.260h).
.181 Micron (μ): (see Other units, IV.0.260m).
.182 Milli (m): (see SI unit prefixes, IV.0.260h).
.183 Minute: as time (min) or angle ('); (see Units used with SI system, IV.0.260i).
.184 Mole (mol): amount of substance (a basic SI unit); (see Mass concentration, IV.0.171b).
.185 Molality: (see Mass, IV.0.171f).
.186 Molar energy (joule per mole): (see SI derived units expressed by special names, IV.0.260e).
.187 Molar entropy (joule per mole kelvin): (see SI derived units expressed by special names, IV.0.260e).
.188 Molar heat capacity (joule per mole kelvin): (see SI derived units expressed by special names, IV.0.260e).
.189 Molar volume for ideal gas: $V_m = RT_o \cdot \rho_o^{-1}$, where V_m is the molal volume, R is the gas constant (IV.0.118), T_o = 273.15 K, and ρ_o = one atmosphere; V_o = 0.02241383 m^{-3} $\cdot$ mol^{-1}.
.190 Moment of force (meter newton): (see SI derived units expressed by special names, IV.0.260e).
.191 Nano (n): (see SI unit prefixes, IV.0.260h).
.192 Nautical mile: (see Units in use temporarily with SI system, IV.0.260k).
.193 Newton (N): kilogram per meter per second squared (m^{-1} $\cdot$ kg $\cdot$ s^{-2}) (see SI derived units with special names, IV.0.260c).
.194 Normal calorie: (see Calorie, VI.5.006).
.195 Nuclear magnetic resonance (NMR) spectrometer: an instrument for determining the number, nature, and environment of hydrogens in a molecule; the sample is placed in the field of an electromagnet and a radiofrequency field (i.e., an energy source) is applied by passing a current through a coil wound around the sample; the spectrum obtained is a plot of the induced voltage against the sweep (increase in strength) of the magnetic field; the area

IV.0. HORTICULTURAL PHYSIOLOGY and CROP ECOLOGY 342

under a peak of induced voltage is dependent on the number of H nuclei "flipping" (excitation from one orientation to another). (AL)
.196 Nuclide: (see Radioisotope, IV.0.248).
.197 Oersted (Oe): (see CGS units, IV.0.260l).
.198 Ohm (Ω): (see SI derived units with special names, IV.0.260c).
.199 Oxidation, oxidized form: the change in a compound from a less to a more positive state, i.e., by loss of one or more electrons to produce an ion or oxidized form. (AL)
.200 PAR: (see Light terms and units, IV.0.157m).
.201 Pascal (Pa): (see Pressure, IV.0.232; SI derived units with special names, IV.0.260c).
.202 Permeability (henry per meter): (see SI derived units expressed by special names, IV.0.260e).
.203 Permittivity (farad per meter): (see SI derived units expressed by special names, IV.0.260e).
.204 Peta (P): (see SI unit prefixes, IV.0.260g).
.205 Phase-contrast microscope: a compound (light) microscope that translates differences in phase of the light transmitted through, or reflected by, an object into differences in intensity of the image, thus providing contrast with the parts and with the background.
.206 Phot (ph): (see CGS units, IV.0.260l).
.207 Photometric radiation: (see Light terms and units, IV.0.157k).
.208 Photon content (flat surface): (see Light terms and units, IV.0.157x-2).
.209 Photon content (point): (see Light terms and units, IV.0.157y-2).
.210 Photon flow rate (flat surface): (see Light terms and units, IV.0.157x-2-b).
.211 Photon flow rate (point): (see Light terms and units, IV.0.157y-2-b).
.212 Photon fluence (applied or absorbed)(point): (see Light terms and units, IV.0.157y-2-c).
.213 Photon fluence rate (point): (see Light terms and units, IV.0.157y-2-d).
.214 Photon flux (flat surfaces): (see Light terms and units, IV.0.157x-2-c).
.215 Photons applied (flat surface): (see Light terms and units, IV.0.157x-2-e).
.216 Photons (flat surface): (see Light terms and units, IV.0.157x-2-d).
.217 Photosynthetically active radiation: (see Light terms and units, IV.0.157m).
.218 Photosynthetic efficiency: (see IV.3.1.020).
.219 Photosynthetic irradiance: (see Light terms and units, IV.0.157o).
.220 Photosynthetic photon density: (see Light terms and units, IV.0.157p).
.221 Photosynthetic photon flux density: (see Light terms and units, IV.0.157q).
.222 PI: (see Light terms and units, IV.0.157o).
.223 Pico (p): (see SI unit prefixes, IV.0.260h).
.224 Planck constant (h): equal to 6.626176×10^{-34} joules (J) per hertz (Hz); (see Frequency, IV.0.070).
.225 Plane angle: (see SI supplementary units; IV.0.260f).
.226 Poise (P): (see CGS units, IV.0.260l).
.227 Polymer, polymeric, polymerize: the first is any of two or more polymeric compounds, specifically one of higher molecular weight;

IV.0. HORTICULTURAL PHYSIOLOGY and CROP ECOLOGY

the second is compounds that consist of the same elements in the same proportions by weight but differ in molecular weight; the last is the change (by union of two or more molecules of the same kind) into another compound with the same elements in the same proportions but a higher molecular weight and different physical properties; examples of polymers are rubber, sterols, terpenes, and the phytol tail of chlorophyll, all of which are polymeric with respect to isoprene (IV.2.2.107). (AL, S & R)

.228 Potential difference (volt): (see SI derived units with special names, IV.0.260c).

.229 Potential energy: energy that is stored or inactive, like that released during a change of state (does not include entropy). (S & R)

.230 Power (P): an SI derived unit expressed as joules per second (J · s^{-1}), watts (W), or square meter X kilogram per second cubed (m^2 · kg · s^{-3}); watt is used for radiant flux (see SI derived units with special names, IV.0.260c). (G & B)

.231 PPFD: (see Light terms and units, IV.0.157q).

.232 Pressure (P), stress: an SI derived unit, the pascal (Pa) (see SI units with special names, IV.0.260c), which equals newtons per square meter (N · m^{-2}) or per meter X kilogram per second squared (m^{-1} · kg · s^{-2}), the last in MKS units; 1 pascal = 10^{-5} bar or 9.86293 · 10^{-6} atmosphere (atm). (G & B)

.233 Q_{10} (temperature coefficient): the increase in rate of a biological process or reaction with a 10°C rise in temperature; log Q_{10} = ($10/T_2 - T_1$) · log K_2/K_1, where T_1 and T_2 are the initial and final temperatures (°C) and K_1 and K_2 are the initial and final rates of the process or reaction. (S & R)

.234 Quantity of electricity (coulomb): (see SI derived units with special names, IV.0.260c).

.235 Quantity of heat (joule): (see SI derived units with special names, IV.0.260c).

.236 Rad [rad(rd)]: (see Limits outside but used for a limited time with SI system, IV.0.260k).

.237 Radian: (see SI supplementary units, IV.0.260f).

.238 Radiance: (see Light terms and units, IV.0.157s; SI derived units by using supplementary units, IV.0.260g).

.239 Radiant energy: (see Light terms and units, IV.0.157t).

.240 Radiant energy (flat surface): (see Light terms and units, IV.0.157x-1-d).

.241 Radiant energy (point): (see Light terms and units, IV.0.157y-1-d).

.242 Radiant flux (Q, watt): (see Light terms and units, IV.0.157u; Power IV.0.230; SI derived units with special names, IV.0.260c).

.243 Radiant flux density (rfd): (see Light terms and units, IV.0.157v).

.244 Radiant intensity (watt per steradian): (see SI derived units by using supplementary units, IV.0.260g).

.245 Radiation arriving at a flat surface: (see Light terms and units, IV.0.157x).

.246 Radiation arriving at a point: (see Light terms and units, IV.0.157y).

.247 Radiation terms: (see Light terms and units, IV.0.157c-y).

.248 Radioisotope (radionuclide): a radioactive isotope (e.g., ^{14}C), one which is undergoing the process of changing into another element by the emission of radiant energy; a widely used tool in biochemical or physiological research that involves movement or reactions of a given compound or constituent thereof. (S & R)

IV.O. HORTICULTURAL PHYSIOLOGY and CROP ECOLOGY

.249 Radionuclide: (see Radioisotope, IV.O.248).
.250 Reduction, reduced form: a change from a more to a less positive state, usually involving the gain of one or more electrons to give the reduced form of a compound. (LE 70)
.251 Relative humidity (RH): (see VI.5.025c).
.252 Rem: (see Limits in use temporarily with the SI system, IV.O.260k).
.253 Rfd: (see Light terms and units, IV.O.157v).
.254 Roentgen (R): (see Limits in use temporarily with the SI system, IV.O.260k).
.255 Rotation: (see Frequency, IV.O.114).
.256 Scanning electron microscope: (see Electron microscope, IV.O.079).
.257 Second ("): (see Units in use with the SI system, IV.O.260i).
.258 Siemens (S): (see Electrical conductance, IV.O.073; SI derived units with special names, IV.O.260c).
.259 Sievert: (see Dose equivalent, IV.O.260d).
.260 SI units: the International System of units based on the meter (m), kilogram (kg), and second (s) (MKS). (P & V, G & B)
.260a. Base units:

Quantity	Name (Symbol)
Amount of substance	mole (mol)
(Celsius temperature)	degrees Celsius (°C)
Electric current	ampere (A)
Length	meter (m)
Luminous intensity	candela (cd)
Mass	kilogram (kg)
Thermodynamic temperature	kelvin (K)
Time	second (s)

.260b. Derived units expressed in terms of base units:

Quantity	Name (Symbol)
Acceleration	meter per second squared ($m \cdot s^{-2}$)
Area	square meter (m^2)
Concentration (of amount of substance)	mole per cubic meter ($mol \cdot m^{-3}$)
Current density	ampere per square meter ($A \cdot m^{-2}$)
Density, mass density	kilogram per cubic meter ($kg \cdot m^{-3}$)
Luminance	candela per square meter ($cd \cdot m^{-2}$)
Magnetic field strength	ampere per meter ($A \cdot m^{-1}$)
Specific volume	cubic meter per kilogram ($m^{-3} \cdot kg^{-1}$)
Speed, velocity	meter per second ($m \cdot s^{-1}$)
Volume	cubic meter (m^3)
Wave number	1 per meter (m^{-1})

.260c. Derived units with special names:

Quantity	Name (Symbol)	Expressed in Other	SI base units
Capacitance	farad (F)	C/V	$m^{-2} \cdot kg^{-1} \cdot s^4 \cdot A^2$
Celsius temperature	degree celsius (°C)		K
Electric charge, quantity of electricity.	coulomb (C)		$s \cdot a$
Electric conductance	siemens (S)	A/V	$m^{-2} kg^{-1} \cdot s^3 \cdot A^2$

IV.0. HORTICULTURAL PHYSIOLOGY and CROP ECOLOGY

Quantity	Name (Symbol)		
Electric potential, potential difference, electromotive force	volt (V)	W/A	$m^2 \cdot kg \cdot s^{-3} \cdot A^{-1}$
Electric resistance	ohm (Ω)	V/A	$m^2 \cdot kg \cdot s^{-3} \cdot A^{-2}$
Energy, work, quantity of heat	joule (J)	N·m	$m^2 \cdot kg \cdot s^{-2}$
Force	newton (N)	---	$m \cdot kg \cdot s^{-2}$
Frequency	hertz (Hz)	---	s^{-1}
Illuminance	lux (lx)	---	$m^{-2} \cdot cd \cdot sr$
Inductance	henry (H)	Wb/A	$m^2 \cdot kg \cdot s^{-2} \cdot A^{-2}$
Luminous flux	lumen (lm)	---	$cd \cdot sr$
Magnetic flux	weber (Wb)	V·s	$m^2 \cdot kg \cdot s^{-2} \cdot A^{-1}$
Magnetic flux density	tesla (T)	Wb/m	$kg \cdot s^{-2} \cdot A^{-1}$
Power, radiant flux	watt (W)	J/s	$m^2 \cdot kg \cdot s^{-3}$
Pressure, stress	pascal (Pa)	N/m	$m^{-1} \cdot kg \cdot s^{-2}$

.260d. Derived units with special names admitted for reasons of safeguarding human health:

Quantity	Name (Symbol)	Expressed in Other SI based units	
Absorbed dose, specific energy imparted, kerma, absorbed dose index	gray (Gy)	J/kg	$m^2 \cdot s^{-2}$
Activity (of a radionuclide)	becquerel (Bq)	---	s^{-1}
Dose equivalent, dose equivalent index	sievert (Sv)	J/kg	$m^2 \cdot s^{-2}$

.260e. Derived units expressed by means of special names:

Quantity	Name (Symbol)	Expressed in SI base units
Absorbed dose rate	gray per second (Gy/s)	$m^2 \cdot s^{-3}$
Dynamic viscosity	pascal second (Pa·s)	$m^{-1} \cdot kg \cdot s^{-1}$
Electric charge density	coulomb per cubic meter (C/m^3)	$m^{-3} \cdot s \cdot A$
Electric field strength	volt per meter (V/m)	$m \cdot kg \cdot s^{-3} \cdot A^{-1}$
Electric flux density	coulomb per square meter (C/m^2)	$m^{-2} \cdot s \cdot A$
Energy density	joule per cubic meter (J/m^3)	$m^{-1} \cdot kg \cdot s^{-2}$
Exposure (x and γ rays)	coulomb per kilogram (C/kg)	$kg^{-1} \cdot s \cdot A$
Heat capacity, entropy	joule per kelvin (J/K)	$m^2 \cdot kg \cdot s^{-2} \cdot K^{-1}$
Heat flux density, irradiance	watt per square meter (W/m^2)	$kg \cdot s^{-3}$
Molar energy	joule per mole (J/mol)	$m^2 \cdot kg \cdot s^{-2} \cdot mol^{-1}$
Molar entropy, molar heat capacity	joule per mole kelvin (J/(mol · K))	$m^2 \cdot kg \cdot s^{-2} \cdot K^{-1} \cdot mol^{-1}$
Moment of force	meter newton (N · m)	$m^{-2} \cdot kg \cdot s^{-2}$
Permeability	henry per meter (H/m)	$m \cdot kg \cdot s^{-2} \cdot A^{-2}$
Permittivity	farad per meter (F/m)	$m^{-3} \cdot kg^{-1} \cdot s^4 \cdot A^2$

IV.O. HORTICULTURAL PHYSIOLOGY and CROP ECOLOGY

Specific energy	joule per kilogram (J/kg)	$m^2 \cdot s^{-2}$
Specific heat capacity, specific entropy	joule per kilogram kelvin [J/(kg·K)]	$m^2 \cdot s^{-2} \cdot K^{-1}$
Surface tension	newton per meter (N/m)	$kg \cdot s^{-2}$
Thermal conductivity	watt per meter kelvin [W/(m·K)]	$m \cdot kg \cdot s^{-3} \cdot K^{-1}$

.260f. Supplementary units:

Quantity	Name (Symbol)
Plane angle	radian (rad) = 180°/π (plane)
Solid angle	steradian (sr) = 180°/π (sphere)

.260g. Derived units formed by using supplementary units:

Quantity	Name (Symbol)
Angular acceleration	radian per second squared (rad·s^{-2})
Angular velocity	radian per second (rad·s^{-1})
Radiance	watt per square meter steradian (W·m^{-2}·sr^{-1})
Radiant intensity	watt per steradian (W·sr^{-1})

.260h. SI prefixes:

Factor	Prefixes	Symbols
10^{18}	exa	E
10^{15}	peta	P
10^{12}	tera	T
10^{9}	giga	G
10^{6}	mega	M
10^{3}	kilo	k
10^{2}	hecto	h
10^{1}	deka	da
10^{-1}	deci	d
10^{-2}	centi	c
10^{-3}	milli	m
10^{-6}	micro	μ
10^{-9}	nano	n
10^{-12}	pico	p
10^{-15}	femto	f
10^{-18}	atto	a

.260i. Units in use with the SI system:

Name	Symbol (Value in SI unit)
Day	d (1d = 24h = 86,400s)
Degree	° (1° = (π/180)rad)
Hour	h (1h = 60 min = 3,600s)
Liter	l, L (1L = 1dm^3 = 10^{-3}m^3)
Metric ton (tonne)	t (1t = 10^3kg)
Minute	min (1 min = 60s)
Minute	' (1' = (1/60)° = (π/10,800)rad)
Second	" (1" = (1/60)' = (π/648,000)rad)

IV.O. HORTICULTURAL PHYSIOLOGY and CROP ECOLOGY

.260j. Units whose value in SI units is obtained experimentally:

Name	Symbol (Approximate value)
Electron volt	eV ($1eV = 1.60219 \times 10^{-19}$ J approx.)
Unified atomic mass unit	u ($1u = 1.66053 \times 10^{-27}$ kg approx.)

.260k. Units in use temporarily with the SI system:

Name	Symbol	(Value in SI units)
Ångström	Å	($1Å = 0.1$ nm $= 10^{-10}$ m)
Are	a	($1a = 10^2 m^2 = dam^2$)
Bar	bar	($1bar = 0.1$ MPa $= 10^5$ Pa)
Barn	b	($1b = 100$ fm $= 10^{-28} m^2$)
Curie	Ci	($1Ci = 3.7 \times 10^{10}$ Bq)
Gal	Gal	($1Gal = 1cm \cdot s^{-2} = 10^{-2} \cdot m \cdot s^{-2}$)
Hectare	ha	($1ha = 1hm^2 = 10^4 m^2$)
Knot	---	(1 nautical mile h^{-1} = $(1,852/3,600) m \cdot s^{-1}$)
Nautical mile	---	(1 nautical mile = 1,852 m)
Rad	rad or (rd)	($1rad = 1cGy = 10^{-2} Gy$)
Rem	rem	($1rem = 1 cSv = 10^{-2}$ Sv)
Roentgen	R	($1R = 2.58 \cdot 10^{-4} C \cdot kg^{-1}$)

.260l. CGS units with special names (preferably not used with SI units):

Name	Symbol	(Value in SI units)
Dyne	dyn	($1dyn = 10^{-5}N$)
Erg	erg	($1erg = 10^{-7}J$)
Gauss	Gs, G	[($1Gs$ corresponds to 10^{-4}T)(tesla)]
Maxwell	Mx	[$1Mx$ corresponds to 10^{-8}Wb (weber)]
Oersted	Oe	[1 Oe corresponds to $(1000 \cdot 4\pi^{-1}) A \cdot m^{-1}$]
Phot	ph	($1ph = 10^4 lx$)
Poise	P	($1P = 1dyn \cdot s \cdot cm^{-2} = 0.1 Pa \cdot s$)
Stilb	sb	($1sb = 1cd \cdot cm^{-2} = 10^4 cd \cdot m^{-2}$)
Stokes	St	($1St = 1cm^2 \cdot s^{-1} = 10^4 m \cdot s^{-1}$)

.260m. Other units generally deprecated:

Name	Value in SI units
Calorie (cal)	---
Fermi	1fermi = 1 fm = 10^{-15}m
Gamma (γ)	$1\gamma = 1nT = 10^{-9}$T (Tesla)
γ (gamma)	$1\gamma = \mu g = 10^{-9}$kg
Kilogram-force (kgf)	1kgf = 9.80665N
(lambda)	$1\lambda = 1\mu L = 10^{-6}L = 10^{-9}m^3$
Metric carat	1 metric carat = 200mg = 2×10^{-4}kg
Micron (μ)	$1\mu = 1\mu m = 10^{-6}$m
Standard atmosphere (atm)	1atm = 101,325 Pa
Stere (st)	$1st = 1 m^3$
Torr	1torr = $(101,325 \cdot 760^{-1})$ Pa
X unit	---

.261 Sol: fluid mixture of a colloid and liquid; a liquid colloidal solution or suspension.

IV.0. HORTICULTURAL PHYSIOLOGY and CROP ECOLOGY

.262 Solid angle: (see SI supplementary units, IV.0.260f).
.263 Sonicate, sonification: the utilization of ultrasonic (ca 20 KHz) waves for disruption of cells and/or their organelles.
.264 Specific energy (joule per kilogram): (see SI derived units expressed by special names, IV.0.260e).
.265 Specific energy imparted (gray): (see Derived units with special names admitted for reasons of safeguarding human health, IV.0.260d)
.266 Specific entropy (joule per kilogram kelvin): (see SI derived units expressed by special names, IV.0.260e).
.267 Specific heat capacity (joule per kilogram kelvin): (see SI derived units expressed by special names, IV.0.260e; VI.5.051e).
.268 Specific humidity: (see VI.5.025a).
.269 Specific latent heat of condensation: (see VI.5.051c).
.270 Specific latent heat of fusion: (see VI.5.051c).
.271 Specific latent heat of liquefication: (see VI.5.051c).
.272 Specific latent heat of vaporization: (see VI.5.051c).
.273 Specific volume: (see SI derived units, IV.0.260b).
.274 Specific water potential: (see Water potential, VI.5.2.075).
.275 Spectral energy flux: (see Light terms and units, IV.0.157z).
.276 Spectral irradiance: (see Light terms and units, IV.0.157z).
.277 Spectral photon flux: (see Light terms and units, IV.0.157aa).
.278 Spectrometry: the measurement of the absorption or emission of light by a substance at specific wavelengths. (AL)
.279 Spectrophotometer: an instrument for measuring the intensity of colors at specific wavelengths or spectral regions. (AL)
.280 Speed: (see SI derived units, IV.0.260b).
.281 Speed of light in a vacuum: (see Light terms and units, IV.0.157ab).
.282 Standard atmosphere (atm): (see Other units, IV.0.260m).
.283 Stefan-Boltzmann law: the quantity of radiant energy (Q) emitted from an object is a function of the fourth power of the absolute temperature (T) of the emitting surface; $Q = \varepsilon \delta T^4$, where Q is the quantity of energy radiated in calories, ε is the emissivity (= about 0.98 for leaves), δ is the Stefan-Boltzmann constant, $8.132 \times 10^{-15} m^{-2} \cdot min^{-1} \cdot K^{-1}$, and T is the absolute temperature in K, (see VI.5.006 for definition of calorie in SI units). (AL)
.284 Steradian: (see SI supplementary units, IV.0.157ac, .260f).
.285 Stere (st): (see Other units, IV.0.260m).
.286 Stereoisomers: compounds with identical molecular formulas and order of atom linkage but differing in spatial arrangement of the atoms; those are mirror images of each other are <u>enantiomers</u> (optical isomers), those that are not are <u>diastereomers</u>, whose configuration is designated as <u>cis</u> (all on one side) or <u>trans</u> (not all on one side); all enantiomers are optically active, whereas only some diastereomers are optically active (but are nonmirror images); (see Structural isomers, IV.0.290). (AL)
.287 Stilb (sb): (see CGS units, IV.0.260l).
.288 Stokes (St): (see CGS units, IV.0.260l).
.289 Stress (pascal): (see SI derived units with special names, IV.0.260c).
.290 Structural isomers: compounds with identical molecular formulas but that differ from one another in the order in which their atoms are joined; (see Stereoisomers, IV.0.288). (AL)
.291 Succulent: a plant in which the tissues have an unusually high vacuole to cytoplasm ratio, thus very large cells; (see also Succulent, II.4.3.065). (S & R)
.292 Surface convective coefficient of heat transfer (h): (see VI.5.048).

IV.0. HORTICULTURAL PHYSIOLOGY and CROP ECOLOGY

.293 Surface tension (newton per meter): (see SI derived units expressed by special names, IV.0.260e).
.294 TEM: (see Electron microscope, IV.0.79).
.295 Temperature:
.295a. Coefficient: (see Q_{10}, IV.0.233).
.295b. Degrees Celsius (°C): defined as $t = T - T_o$, where T is the thermodynamic temperature and T_o = 273.15 kelvin (K) by definition, thus 1°C = 1 K; horticultural measurements of temperature should be sensed with a shielded and aspirated (> 3 m · s^{-1}) device (thermometer, thermocouple, thermistor, pyrometer, etc.) and the location in relation to the object, plant surface, etc., specified [i.e., kind of temperature, air, or leaf, device used, and location should be noted; 1°C = 5/9 (°F-32)]; (see IV.0.260a).
.295c. Degrees Fahrenheit (°F): equal to 1.8(°C) + 32; used in heating, refrigeration, air conditioning engineering, and other applications in the United States; conversions to degrees Celsius should be rounded to the nearest whole unit inasmuch as most thermometers are calibrated to the nearest °F and thermostats used in homes, cold storage units, etc., generally have a range of ± 2 or 3°F.
.295d. Thermodynamic: the kelvin (K), the unit of thermodynamic temperature is the fraction 1/273.16 of the thermodynamic temperature of the triple point of water; a SI base unit, see IV.0.260a). (G & B)
.296 Temperature coefficient: (see Q_{10}, IV.0.233).
.297 Tera (T): (see SI unit prefixes, IV.0.260h).
.298 Tesla (T): (see SI derived units with special names, IV.0.260c).
.299 Thermal conductivity (k, watt per meter kelvin): (see SI derived units expressed by special names, IV.0.260e; VI.5.049).
.300 Thermal diffusivity (d): (see VI.5.050).
.301 Thermodynamic temperature: (see SI base units, IV.0.260a, .295d).
.302 Thixotropy: the property exhibited by some gels of becoming liquid when shaken. (S&R)
.303 Time: (see SI base units, IV.0.260a).
.304 Tonne (t): (see Limits outside but used with SI system, IV.0.260i).
.305 Ton of refrigeration: (see VI.5.053).
.306 Torr: (see Other units, IV.0.260m).
.307 Total irradiance: (see Light terms and units, IV.0.157ad).
.308 Trans: (see Stereoisomer, IV.0.286)
.309 Transmission electron microscope: (see Electron microscope, IV.0.079).
.310 Transpiration rate: (see IV.5.2.077b).
.311 Ultracentrifuge: a high-speed centrifuge used for separations of proteins and other macromolecules. (S&R)
.312 Ultraviolet spectrometer: an instrument that measures electronic excitations obtained from compounds containing multiple bonds; similar to visible spectroscopy but concerned with wavelengths from about 50 to 400 nm, those below 190 nm are carried out in vacuum. (AL)
.313 Unified atomic mass unit (u): (see Units whose value in SI units is obtained experimentally, IV.0.260j).
.314 Vapor pressure: (see VI.5.025d).
.315 Velocity: (see SI derived units, IV.0.260b).
.316 Volt (V): (see SI derived units with special names, IV.0.260c).
.317 Volume (V): the preferred unit is the cubic meter (m^3) or cubic decimeter (dm^3 = $10^{-3} \cdot m^3$); the liter equals 1.000027dm^3 and is

IV.0. HORTICULTURAL PHYSIOLOGY and CROP ECOLOGY

.317 (cont.) widely used but is not a recommended unit, especially for precise measurements, (see SI derived units, IV.0.260b). (G & B, SE79)
.318 Volume basis: (see Transpiration rate, IV.5.2.074b-2).
.319 Volume fraction (θ): the units are cubic meter per cubic meter ($m^3 \cdot m^{-3}$) or cubic millimeter per cubic meter ($mm^3 \cdot m^{-3}$); compare with mass fraction (IV.0.171e).
.320 Volumetric water potential: (see Water potential, IV.5.2.075).
.321 Water:
.321a. Atmospheric moisture: (see Humidity, VI.5.025).
.321b. Density: (see Mass density, IV.0.171c).
.321c. Movement: (see IV.5.2.074).
.321d. Potential: (see IV.5.2.075).
.322 Watt (W): (see SI derived units with special names, IV.0.260c).
.323 Wave number: (see SI derived units, IV.0.260b).
.324 Weber (wb): (see SI derived units with special names, IV.0.260c).
.325 Weight: mass (IV.0.171) times the acceleration of gravity, i.e., the mass times the force with which a body is attracted toward the center of the earth; the unit is mass (kilograms) times the gravitational constant (IV.0.123), of which the latter will vary according to location; measured in practice on a spring balance. (SE 79)
.326 Wind speed: (see Air movement, IV.0.010).
.327 Work (joule): (see SI derived units with special names, IV.0.260c).
.328 X-unit: (see Other units, IV.0.260m).

IV.1. Metabolic Sites and Processes

IV.1.1. Cell structure, organelles

.001 Amyloplast: (see Leucoplast, IV.1.1.012).
.002 Chloroplast: a type of plastid, a double membrane bound organelle peculiar to higher plant cells, in which the photosynthetic apparatus is localized; e.g., in palisade tissues and spongy mesophyll of leaves and epidermal or subepidermal cells of petioles, rachises, stems, and other aerial organs characterized by their green color (which, however, may be more or less masked by other constituents in the plastids); the outer membrane is single, continuous, and rather fragile, whereas the inner membrane is also continuous but folded in a highly complex manner; (see Chromoplast, IV.1.1.003e). (S & R)
.002a. Granum: a stack of membrane-bound disclike or flattened vesicles formed in certain regions of the chloroplast.
.002b. Intergranal lamellae: longer lamellae that connect one granum to another and extend throughout the chloroplast matrix.
.002c. Lamella: a system of membrane-bound channels; part of the inner membrane of the chloroplast.
.002d. Stroma: chloroplast matrix in which granal and intergranal lamellae are imbedded.
.002e. Thylakoid: an individual unit of short lamella that collectively is comprised of a granum (IV.1.1.002a); pigments present include largely chlorophylls a and b, plus carotenes (pure hydrocarbons) and xanthophylls (oxygen-containing), of which some of the latter may also be found in the chloroplast envelope (i.e., outside the thylakoid); (see Chloroplast, IV.1.1.002). (S & R)

IV.1.1. HORTICULTURAL PHYSIOLOGY and CROP ECOLOGY

.003 Chromatin: (see Nucleus, IV.1.1.018).
.004 Chromoplast: pigmented membrane-bound plastids, other than chloroplasts; their function is not clear in many cases but they impart red, orange, or yellow coloration to many fruits, vegetables, and flowers. (S & R)
.005 Cytoplasm: the visibly least differentiated part of the protoplasm that encloses all of the other components. (S & R)
.006 Dictyosomes: flattened stacks of platelike membrane-bound sacs; small vesicles will frequently bud off one side and migrate within the cell; function in cell-wall synthesis in plants, more often play a role in secretory processes in animals; collectively referred to as the Golgi apparatus.
.007 Elaioplast: (see Leucoplast, IV.1.1.012). (S & R)
.008 Endoplasmic reticulum: a network of membranes within the cytoplasm of a cell; contains many enzymes and also functions in the transport of substances within the cell; may be called "rough endoplasmic reticulum" or "smooth endoplasmic reticulum", depending on the presence or absence of ribosomes IV.1.1.025). (S & R)
.009 Glyoxisome (glyoxysome): small membrane-bound organelle that contains enzymes necessary to carry out β-oxidation and the glyoxlate cycle; identical morphologically to a peroxisome but found in seeds that store fats. (S & R)
.010 Golgi apparatus: collective term for dictyosomes (IV.1.1.006). (S & R)
.011 Karyolymph: (see Nucleus, IV.1.1.018).
.012 Leucoplast: colorless membrane-bound plastid that functions primarily as a storage structure; one that contains starch is called an amyloplast (as in potato tubers) and one with oil droplets, an elaioplast (as in corn kernels); the interior is otherwise structureless and may occur in above-ground or subterranean plant parts. (S & R)
.013 Lipid body: (see Oleosome, IV.1.1.019).
.014 Microbodies: membrane-bound organelles of two types, peroxisomes (IV.1.1.021) and glyoxisomes (IV.1.1.009). (S & R)
.015 Microtubules: some tubules within the cytoplasm about 25 nm in diameter and variable in length; function in the separation of chromosomes during nuclear division (mitosis); also in cell-wall formation. (S & R)
.016 Mitochondrion: double membrane-bound organelle; inner membrane folded inward to form cristi (convoluted folds); site of cellular respiration, ATP production; contains respiratory enzymes and its own DNA. (S & R)
.017 Nucleolus: a viscous and semisolid body, denser than the karyolymph, that frequently contains vacuoles and crystallike bodies; its chief function is to synthesize RNA; located in the nucleus (IV.1.1.018). (S & R)
.018 Nucleus: double membrane-bound, with pores 40 nm in diameter at various intervals; inclusions are the matrix or karyolymph (nuclear sap), a reticulum composed of chromatin, which becomes aggregated into chromosomes, and a nucleolus or nucleoli (the number present being a species characteristic). (S & R)
.019 Oleosome (sphaerosome, lipid body): specialized bodies for storage of fats within a cell. (S & R)
.020 Organelle: a structure with a specialized function within a cell. (S & R)
.021 Peroxisome (peroxysome): small membrane-bound organelle that contains several oxidative enzymes; exists almost exclusively in

IV.1.1. HORTICULTURAL PHYSIOLOGY and CROP ECOLOGY

photosynthetic tissues (i.e., leaves) and often appears in direct contact with a chloroplast. (S & R)
.022 Plasmalemma: outer cell membrane 10 nm thick. (S & R)
.023 Plasmodesmata: minute pores in the cytoplasmic membrane. (S & R)
.024 Protoplast: all living cellular constituents, including the plasmalemma. (S & R)
.025 Ribosomes: sites of protein synthesis within cells; not bounded by a membrane; may be associated with the endoplasmic reticulum or found free in the cytoplasm; composed partly of RNA and are sites in which messenger RNA which carries genetic information from the nucleus is used to organize transfer RNA, bringing amino acids for protein synthesis from the cytoplasm. (S & R)
.026 Rough endoplasmic reticulum: (see Endoplasmic reticulum, IV.1.1.008).
.027 Smooth endoplasmic reticulum: (see Endoplasmic reticulum, IV.1.1.008).
.028 Sphaerosome: (see Oleosome, IV.1.1.019).
.029 Tonoplast: the vacuolar membrane. (S & R)
.030 Vacuole: large organelle in a mature cell that occupies 90% or more of the cell volume; surrounded by a membrane (tonoplast); contains a wide variety of substances, depending on the tissue, such as crystals, resins, pigments, latex, oil, tannins, certain enzymes, organic acids, and sugars (in some storage tissues). (S & R)

IV.1.2. Major metabolic processes

.001 Absorption spectrum: (see Photosynthesis, IV.1.2.036a).
.002 Action spectrum: (see Photosynthesis, IV.1.2.036b).
.003 Ammonification: (see Nitrogen cycle, IV.1.2.028a).
.004 Anaplerotic reactions: special enzymatic mechanisms by which tricarboxylic acid cycle intermediates, such as α-ketoglutarate, succinate, and oxaloacetate, removed from the Krebs cycle to serve as precursors of amino acids, can be replenished; at least one type of anaplerotic reaction, the glyoxylate cycle (IV.1.1.009), occurs in plants and many microorganisms. (LE 70)
.005 ATP: (see Electron transport (Cytochrome system), IV.1.2.040c).
.006 Calvin cycle: (see Photosynthesis, IV.1.2.036c).
.007 Carbon dioxide compensation point: the CO_2 concentration in which photosynthetic fixation just balances respiratory and photorespiratory loss, being 50-100 ppm for C-3 and 0-5 ppm CO_2 for C-4 plants (see Light compensation point, IV.5.4.2.002). (S & R)
.008 Carbon dioxide (CO_2) fixation: three different pathways for initial fixation of CO_2 in photosynthesis have been discovered. (S & R)
.008a. C-3 plants: species that fix CO_2 largely into 3-phosphoglyceric acid: found in all gymnosperms, pteridophytes, bryophytes, and algae, many monocots (e.g., oat, barley, rice, wheat, and Italian ryegrass), and dicots (e.g., beet, spinach, sunflower, lettuce, peanut, soybean, bean, cotton, and carrot).
.008b. C-4 plants: species that have 4-C acids (e.g., malate and aspartate) as primary CO_2 fixation products. Decarboxylation of these acids, unlike that in C-3 plants, leads to the production of sucrose, starch, etc., in vascular-bundle sheath cells rather than in the leaf mesophyll (chloroplast-containing) cells. This arrangement of one or occasionally two distinct

IV.1.2. HORTICULTURAL PHYSIOLOGY and CROP ECOLOGY

layers of tightly packed, often thick-walled, photosynthetic cells (the bundle sheath), which almost always surround the leaf vascular bundles (veins) and often contain far more chloroplasts, mitochondria, and other organelles, and smaller central vacuoles, compared with C-3 plants, is denoted Kranz anatomy. There is also little or no photorespiration; all of these features together constitute the "C-4 syndrome": they have been found so far in 485 spp., 117 genera, and 13 families, mostly monocots (e.g., Cyperus esculentus L., bermudagrass, crabgrass, bahiagrass, sugar cane, and maize) and a few dicots (e.g., globe amaranth, pigweed, and common purslane); some genera have both C-4 and C-3 species. (Note that C-3 is the "normal" situation, the interest in C-4 plants being that they are at least twice as efficient as C-3 in the conversion of solar energy to dry-matter production under conditions of high light intensity and temperatures 25 to 35°C).

.008c. Crassulacean acid metabolism (CAM) plants: species with stomata that open primarily at night and CO_2 fixation into organic acids, especially malic; predominantly xerophytic or epiphytic plants in 18 families characterized by a poorly developed palisade layer; leaf or stem cells mostly spongy mesophyll and bundle sheath cells are present, as in C-4 plants, but similar to mesophyll cells; CAM favored by hot days, high light intensities, cool nights, and dry soils; however, CAM metabolism may revert to that of C-3 (i.e., daylight CO_2 fixation) under certain growing conditions (e.g., after a rainstorm, higher night temperatures, or different day length).

.009 C-4 syndrome: (see Carbon dioxide fixation, IV.1.2.008b).
.010 Chlorophyll: (see Photosynthesis, IV.1.2.036d).
.011 Citric acid cycle: (see Krebs cycle, IV.1.2.040b).
.012 Cyanide-resistant respiration: (see Respiration, IV.1.2.040d).
.013 Cytochromes: (see Photosynthesis, IV.1.2.036e).
.014 Cytochrome system: [see Electron transport (cytochrome system), IV.1.2.040c].
.015 Denitrification: (see Nitrogen cycle, IV.1.2.028b).
.016 Electron transport system: [see Electron transport (cytochrome system) - oxidative respiration, IV.1.2.040c].
.017 Etioplast: plastids present in dark-grown seedlings, in which chlorophyll is not developed but carotenoids are, in the absence of light. (S & R)
.018 Fermentation: (see Respiration, IV.1.2.040e).
.019 Gluconeogenesis: the process by which phosphoenol pyruvate (PEP) generated from pyruvate is converted by a series of enzyme-mediated reactions to glucose-6-phosphate, which is a precursor in the formation of storage polymers (e.g., starch), monosaccharides other than glucose, disaccharides, and structural polysaccharides (e.g., cellulose, cell-wall polysaccharides, and fructosans) and free glucose (in animals). (LE 70)
.020 Glucose: [see Electron transport (cytochrome system), IV.1.2.040c].
.021 Glycolysis: (see Respiration, IV.1.2.040a).
.022 Hexose phosphate shunt: (see Pentose phosphate pathway, IV.1.2.040g).
.023 Hill reaction: (see Photosynthesis, IV.1.2.036f).
.024 Kranz anatomy: (see C-4 plants, CO_2 fixation, IV.1.2.008b).
.025 Krebs cycle: (see Respiration, IV.1.2.040b).
.026 Leaf area index (LAI): the ratio of leaf area (one surface only) of a crop to the ground area on which it grows. (KR)

IV.1.2. HORTICULTURAL PHYSIOLOGY and CROP ECOLOGY

.027 Nitrification: (see Nitrogen cycle, IV.1.2.028c).
.028 Nitrogen cycle: the overall process that involves fixation of atmospheric N_2, interconversions of nitrogen compounds in plants, animals, industry (including engines), water, volcanic emissions, and the like, and their return to the atmosphere.
.028a. Ammonification: conversion of organic nitrogen to ammonium ions (NH_4^+) by soil organisms.
.028b. Dentrification: the reduction of nitrates to nitrous oxide (N_2O) and free N_2, which takes place most rapidly in soils not too acid (above about pH 5) under conditions of poor aeration and in the presence of an active microbial population (i.e., in warm, wet soils well supplied with decomposable organic matter or by Nitrosomonas to N_2O when the temperature and P supply are low but the pH is high); the bacteria responsible are predominantly facultative anaerobes. (RU)
.028c. Nitrification: the production of nitrate (NO_3^-) from ammonia (NH_4^+), a two-stage biological process in which Nitrosomonas and related genera of bacteria oxidize ammonia and produce nitrite (NO_2^-) via hydroxylamine and Nitrobacter oxidize nitrite and produce nitrate; these two groups are the only autotrophs that have proved to carry out these oxidations, although a number of heterotrophs are suspected; Aspergillus flavus and possibly other fungi will produce nitrate from organic N sources but are, as a rule, active only on high N substrates. (RU)
.029 Nitrogen fixation: the process by which N_2 is reduced to ammonium ion, principally by free-living soil bacteria or blue-green algae living in symbiotic association with fungi (i.e., in lichens), pteridophytes, or bryophytes, and bacteria associated symbiotically with roots, especially legumes; other minor pathways include atmospheric fixation (to NH_4^+ and NO_3^-), volcanic activity, salt spray, and the deposition of salts by rainfall. (S & R)
.030 Operon: (see III.2.2.116).
.031 Oxidative phosphorylation: [see Electron transport (cytochrome system), IV.1.2.040c].
.032 Pasteur effect: (see Respiration, IV.1.2.040f).
.033 Pentose phosphate pathway (PPP): (see IV.1.2.040g).
.034 Phosphogluconate pathway: (see Pentose phosphate pathway, IV.1.2.040g).
.035 Photorespiration (Warburg effect): inhibition of photosynthesis by O_2 from much more rapid respiration under illumination in C-3 than in C-4 plants; i.e., inhibition of photosynthesis occurs from the faster release of CO_2 by respiration in the light. (S & R)
.036 Photosynthesis: the synthesis of organic compounds, specifically the formation of carbohydrates from the oxygen of water and the carbon of carbon dioxide effected with light as the energy source, in chlorophyll-containing tissues of plants (e.g., vascular plants, bryophytes, and certain algae and bacteria). (S & R)
.036a. Absorption spectrum: relative absorbance of various wavelengths of light of a compound such as chlorophyll _a_ or _b_ measured with a spectrophotometer. (S & R)
.036b. Action spectrum: effects of different wavelengths on the rate of photosynthesis and other photobiological processes, as shown in a graph of rate as a function of wavelengths. (S & R)
.036c. Calvin cycle: the series of reactions in the photosynthetic

IV.1.2. HORTICULTURAL PHYSIOLOGY and CROP ECOLOGY

process in which carbon dioxide and water are converted to hexose phosphates, thence to starch within the chloroplasts and to sucrose and other polysaccharides outside the plastids; a cyclic process in which the net efficiency of the reduction of CO_2 to hexose, compared with the energy input of eight photons (quanta of light at 600 nm), is about 30%; now generally referred to as the "reductive pentose phosphate pathway." (SY, S & R, KK)

.036d. Chlorophyll: a porphyrin that consists of four substituted pyrrole rings (one of which is partially reduced), with the four central N atoms bonded coordinately to a Mg^{++} ion and a phytol ($C_{20}H_{39}$) side chain; there are four forms, of which a and b are found in green plants, c, in brown algae, diatoms, and dinoflagellates, and d in red algae. (S & R, LE 70)

.036e. Cytochromes: iron-containing proteins (i.e., porphyrins), three of which exist in chloroplasts and function in the electron transport system involved in noncyclic photosynthetic phosphorylation. (S & R)

.036f. Hill reaction: the light-driven reaction that results in the splitting of water in the absence of CO_2 fixation (see Photosystems, IV.1.2.036g) by which the role of light and the sequence of reactions in the release of O_2 from water was elucidated. (S & R)

.036g. Photosystems: two main photosystems: I, which absorbs wavelengths above about 680 nm and also below, and II, which can absorb wavelengths shorter than 680 nm only; reaction centers are P700 for I and P680 for II; the photosystems act as electron transporters that use light as the source of energy in a series of reactions that results in the splitting of water to release oxygen. (S & R)

.037 Photosynthetic end products: include literally all of the compounds synthesized in a plant, whether formed as immediate products of photosynthesis proper (e.g., sugar mono- and diphosphates) or in numerous subsequent metabolic reactions via a host of different pathways.

.038 Photosynthetic rates: the quantitiy of CO_2 consumed and O_2 produced in photosynthesis is the gross rate, from which the amounts of CO_2 evolved and O_2 consumed in respiration and photorespiration must be subtracted to determine the net rate; the latter represents the gain in dry weight of the plant; C-4 plants have the highest and CAM plants, the lowest net photosynthetic rates; the range is 0 to about 70 mg CO_2/dm^2 leaf area per hour, measured by CO_2 exchange; rates are affected by the age and genetic constitution of the plants and environmental factors (e.g., light, temperature, and nutrients). (S & R)

.039 Reductive pentose phosphate pathway: (see Calvin cycle, IV.1.2.036c).

.040 Respiration (cellular basis): collectively the anabolic (synthetic) processes in plants to sequester or transfer energy and provide intermediates for numerous products; carbohydrates or products thereof are used as substrates in a series of reactions that consists of three major phases: glycolysis, Krebs (citric or tricarboxylic acid) cycle, and electron transport (cytochrome system)-oxidative phosphorylation. (S & R)

.040a. Glycolysis: a series of reactions, each of which is catalyzed by an enzyme, in which glucose, glucose-1-phosphate, or fructose released by hydrolysis of starch or other reserved polysaccharides is converted to pyruvic acid in the cytoplasm outside any organelle; an aerobic process dependent on the presence of O_2.

.040b. Krebs (citric or tricarboxylic acid) cycle: the second major phase in respiration which is carried out in the mitochondria and consists of the initial loss of CO_2 from pyruvate, a combination of the remaining 2-carbon acetate unit with coenzyme A (sulfur-containing) to form acetyl CoA, the removal of electrons from organic acid intermediates, and the release of two additional CO_2 molecules in the cycle proper; the cycle thus serves in part as a means of energy transfer and in part to form carbon skeletons that can be synthesized to certain amino acids; no O_2 is absorbed in any part of the cycle.

.040c. Electron transport (cytochrome system)-oxidative phosphorylation: the last major phase in respiration in which electrons are transferred by a series of intermediate compounds (i.e., several cytochromes, a quinone, and a riboflavin-containing protein) ultimately to combine with O_2 to form water and produce ATP (adenosine triphosphate); this is an oxidative phosphorylation in which the overall efficiency in sequestering energy in ATP in relation to that in glucose (the substrate in glycolysis) is about 40%; the remainder is lost as heat.

.040d. Cyanide-resistant respiration: the type of respiration that can continue in the presence of certain ions, such as cyanide (CN^-), azide (N_3^-), or carbon monoxide (CO), which combine with the iron in cytochrome oxidase and strongly inhibit aerobic respiration in most plants; mitochondria in CN^--resistant plants have an alternate electron transport pathway that leads mainly to the production of heat, not ATP, unless electron transport is fast enough to produce both (in which case there will be a rapid drain on reserve foods); examples are aroids that utilize the heat produced in volatilization of certain pheromones to attract pollinators (carrion flies and beetles).

.040e. Fermentation: a series of reactions that occurs under anaerobic conditions and involves the reduction of pyruvate (derived from glycolysis, IV.1.2.040a) to ethyl alcohol and CO_2 (or in animals to lactic acid and CO_2).

.040f. Pasteur effect: the acceleration of sugar utilization in respiration under conditions of low O_2.

.040g. Pentose phosphate pathway (PPP): also known as the phosphogluconate pathway and hexose phosphate shunt, an alternative respiratory (oxidative) pathway, a cyclic process that involves five-carbon sugar phosphates as intermediates for obtaining energy from oxidation of sugars into CO_2 and water; provides reactants such as ribose-5-phosphate used in nucleotide and nucleic acid synthesis and erythrose-4-phosphate for the synthesis of lignins and other aromatic compounds; like glycolysis, it is carried out in the cytoplasm outside any organelle but also occurs in chloroplasts; this process is the reverse of the Calvin cycle (IV.1.2.036c); i.e., sugar phosphates are oxidized in chloroplasts in the dark but inactivated in the light.

.041 Respiratory quotient (RQ): the volume of CO_2 released compared with the volume of O_2 taken up in respiration. (S & R)

.042 Shikimic acid pathway: a process in which erythrose 4-phosphate (a tetrose phosphate) is converted in several steps by way of shi-

IV.1.2. HORTICULTURAL PHYSIOLOGY and CROP ECOLOGY

kimic acid to anthranilic acid, thence to tryptophan or to prephenic acid, and then to phenylpyruvic acid, the precursor of phenylalanine. (S & R)

.043 Solarization: the light-dependent absorption of O_2 and the release of CO_2 that causes bleaching of chloroplast pigments; observed when plants are moved from shade to sunny conditions; the older leaves are particularly affected unless the plants are "hardened" by a gradual increase in light intensity over a period of time. (S & R)

.044 Tricarboxylic acid cycle: (see Krebs cycle, IV.1.2.040b).

.045 Warburg effect: (see Photorespiration, IV.1.2.035).

IV.2. Hormones and Growth Regulators, Other Metabolic Products

IV.2.1. Hormones and growth regulators

.001 Hormone: an organic compound synthesized in one part of a plant and translocated to another or used in situ at low concentrations (usually 1 μM or less), where it causes a physiological response; sugars, amino acids, organic acids, and other metabolites necessary for growth and development are not hormones since they are usually present at concentrations of 1 to 50 μM, nor are inorganic ions (e.g., K^+) because these are not synthesized in plants, or any hormone or hormonelike organic compound synthesized in organisms other than plants or by a chemist; now thought to consist of five groups: auxins, gibberellins, cytokinins, growth inhibitors, and ethylene; (see also Physiological sprays, V.3.4.000a). (S & R)

.002 Plant growth regulator: any hormone (e.g., auxin, gibberelliins, cytokinins, growth inhibitors, or ethylene) or hormonelike organic compound synthesized in organisms other than plants or by a chemist; an exogenous hormone; (see also Physiological sprays, V.3.4.000a). (S & R)

.003 ABA: (see Growth inhibitors, IV.2.1.037).

.004 Abscisic acid: (see Growth inhibitors, IV.2.1.037).

.005 Abscisin: (see Growth inhibitors, IV.2.1.037).

.006 (+)-Abscisyl-β-D-glucopyraniside: (see Growth inhibitors, IV.2.1.037).

.007 Alar: (see Growth inhibitors, IV.2.1.037).

.008 Ammonium (5-hydroxycarvacryl) trimethylchloride piperidine carboxylate: (see Growth inhibitors, IV.2.1.037).

.009 Amo-1618: (see Growth inhibitors, IV.2.1.037).

.010 Ancymidol: (see Growth inhibitors, IV.2.1.037).

.011 A-Rest: (see Growth inhibitors, IV.2.1.037).

.012 Auxins: the hormone indoleacetic acid (IAA); also indoleacetaldehyde, indolepyruvic acid, indoleacetonitrile, and indole ethanol, which are converted to IAA; growth regulators: naphthaleneacetic acid (NAA), naphthylacetamide (NAAm), β-napthoxyacetic acid (BNOA), indolebutyric acid (IBA), 2,4-dichlorophenoxyacetic acid (2,4-D), 2,4,5-triiodobenzoic acid (TIBA, an auxin synergist). IAA is produced in many locations, especially in actively growing shoots, young leaves, and developing embryos; it can control rate of cell enlargement, induce or retard abscission of young or mature fruit, prevent leaf abscission, stimulate ethylene synthesis

IV.2.1. HORTICULTURAL PHYSIOLOGY and CROP ECOLOGY

in fruit, rooting of stem cuttings, and cambial division, and inhibit lateral bud development; some growth regulators are also used as herbicides (e.g., 2,4-D, 2,4,5-T, 2,4,5-TP etc.). (WS, S & R)

.013 BA: 6-benzylamino purine = benzyladenine (see Cytokinins, IV.2.1.025).
.014 Benzyladenine: (see Cytokinins, IV.2.1.025).
.015 6-Benzylamino purine: (see Cytokinins, (IV.2.1.025).
.016 6-Benzylamino-9-(2-tetrahydropyranyl)-9H-purine: (see Cytokinins, IV.2.1.025).
.017 B-9: (see Growth inhibitors, IV.2.1.037).
.018 B-995: (see Growth inhibitors, IV.2.1.037).
.019 BNOA: β-naphthoxyacetic acid, (see Auxins, IV.2.1.012).
.020 BOH: (see Growth inhibitors, IV.2.1.037).
.021 CCC: (see Growth inhibitors, IV.2.1.037).
.022 2-Chloroethylphosphonic acid: (see Ethylene, IV.2.1.033).
.023 (2-Chlorethyl)trimethylammonium chloride (CCC, cycocel): (see Growth inhibitors, IV.2.1.037).
.024 Cycocel: (see Growth inhibitors, IV.2.1.037).
.025 Cytokinins: N^6-substituted adenine derivatives, such as zeatin, zeatin riboside, isopentyladenine, and dihydrozeatin, found in plants; synthetic compounds: kinetin, 6-benzylamino purine (benzyladenine, BA), 6(benzylamino)-9-(2-tetrahydropyranyl)-9H-purine (PBA); promote cell division, regulate nucleic acids (DNA and RNA), apical dominance and branching, bud initiation, enhance seed germination to some extent, influence transport of nutrients and metabolites, prevent abscission and senescence of flowers and fruits, prevent leaf senescence, and inhibit root initiation. (WS, S & R)
.026 Dichlorobenzyltributylphosphonium chloride (Phosfon-D): (see Growth inhibitors, IV.2.1.037).
.027 2,4-D: 2,4-dichlorophenoxyacetic acid, (see Auxins, IV.2.1.012).
.028 2,4-Dichlorophenoxyacetic acid: (see Auxins, IV.2.1.012).
.029 1,2-Dihydro-3,6-pyridazinedione (MH, maleic hydrazide): (see Growth inhibitors, IV.2.1.037).
.030 Dihydrozeatin: (see Cytokinins, IV.2.1.025).
.031 Dormin: (see Growth inhibitors, IV.2.1.037).
.032 Ethephon: 2-chloroethylphosphonic acid (see Ethylene, IV.2.1.033).
.033 Ethylene: an ubiquitous, volatile hormone; 2-chloroethylphosphonic acid (ethephon) an ethylene-releasing compound; increases respiration rate, hastens fruit ripening, stimulates root initiation in cuttings, promotes abscission, stimulates floral initiation, inhibits elongation, breaks rest in buds and seeds, causes gumming in some species, and inhibits lateral bud development; synthesized in plant parts under stress (or injured); induced in plants (e.g., for floral initiation) by the application of auxins (e.g., NAA and 2,4-D) or certain growth inhibitors (e.g., β-hydroxyethylhydrazine). (WS, S & R)
.034 GA_3, GA_{4-7}: (see Gibberellins, IV.2.1.036).
.035 Gibberellic acid: GA (see Gibberellins, IV.2.1.036).
.036 Gibberellins: natural products of <u>Gibberella fujikuroi</u> (fungus) or higher plants; about 50 have been identified; GA_3 and GA_{4-7} are commercially available; promote seed germination, cell elongation (and cell division) of intact plants, aid in breaking rest in dormant buds, prevent flower initiation, retard coloring and maturity of fruit, and induce parthenocarpic fruit set. (WS, S & R)

IV.2.1. HORTICULTURAL PHYSIOLOGY and CROP ECOLOGY

.037 Growth inhibitors: abscisic acid (ABA, abscisin, dormin), a natural growth inhibitor; inhibits growth of shoots, blocks action of auxin, GA, and cytokinins, sometimes stimulates fruit and leaf abscission and senescence, can substitute for chilling or in some instances (e.g., grapefruit) delay chilling injury, delays bud break, and inhibits germination; several other natural growth inhibitors are phaseic acid, 2-trans-phaseic acid, (+)-abscisyl-β-D-glucopyranoside, and theaspirone; synthetic inhibitors are succinic acid-2,2-dimethyl hydrazide (SADH, B-995), B-9, Alar), (2-chloroethyl)trimethylammonium chloride (CCC, Cycocel), β-hydroxyethylhydrazine (BOH), 1,2-dihydro-3,6-pyridazinedione (MH, maleic hydrazide), 2,4-dichlorobenzyltributylphosphonium chloride (Phosfon-D), ammonium (5-hydroxycarvacryl) trimethyl chloride piperidine carboxylate (Amo-1618), a-cyclopropyl-a-(p-methoxyphenyl)-5-pyrimidinemethanol (A-Rest = ancymidol). (WS, S & R)
.038 β-Hydroxyethylhydrazine (BOH): (see Growth inhibitors, IV.2.1.037).
.039 IAA: (see Auxins, IV.2.1.012).
.040 IBA: indolebutyric acid, (see Auxins, IV.2.1.012).
.041 Indoleacetaldehyde: (see Auxins, IV.2.1.012).
.042 Indoleacetic acid: (see Auxins, IV.2.1.012).
.043 Indoleacetonitrile: (see Auxins, IV.2.1.012).
.044 Indolebutyric acid: (see Auxins, IV.2.1.012).
.045 Indole ethanol: (see Auxins, IV.2.1.012).
.046 Indolepyruvic acid: (see Auxins, IV.2.1.012).
.047 Isopentenyl adenine: (see Cytokinins, IV.2.1.025).
.048 Kinetin: (see Cytokinins, IV.2.1.025).
.049 NAA: naphthaleneacetic acid (see Auxins, IV.2.1.012).
.050 NAAm: naphthylacetamide (see Auxins, IV.2.1.012).
.051 Naphthaleneacetic acid: (see Auxins, IV.2.1.012).
.052 β-Naphthoxyacetic acid: (see Auxins, IV.2.1.012).
.053 PBA: 6(benzylamino)-9-(2-tetrahydropyranyl)-9H-purine; (see Cytokinins, IV.2.1.025).
.054 Phaseic acid: (see Growth inhibitors, IV.2.1.037).
.055 2-trans-Phaseic acid: (see Growth inhibitors, IV.2.1.037).
.056 Phosfon-D: (see Growth inhibitors, IV.2.1.037).
.057 SADH: (see Growth inhibitors, IV.2.1.037).
.058 Succinic acid-2,2-dimethyl hydrazide (SADH, B995, B9, Alar): (see Growth inhibitors, IV.2.1.037).
.059 Theaspirone: (see Growth inhibitors, IV.2.1.037).
.060 TIBA: 2,3,5-triodobenzoic acid (see Auxins, IV.2.1.012).
.061 2,4,5-Trichlorophenoxyacetic acid (2,4,5-T): (see Auxins, IV.2.1.012).
.062 2,4,5-Trichlorophenoxypropionic acid (2,4,5-TP): (see Auxins, IV.2.1.012).
.063 2,3,5-Triiodobenzoic acid: (see Auxins, IV.2.1.012).
.064 Zeatin: (see Cytokinins, IV.2.1.025).
.065 Zeatin riboside: (see Cytokinins, IV.2.1.025).

IV.2.2. Other metabolic products

.001 Abscisic acid: (see Isoprenoids, IV.2.2.107).
.002 Acephalin: (see Phospholipid, IV.2.2.159).
.003 Adenine: (see DNA, IV.2.2.056; Purine bases, IV.2.2.173).
.004 Adenosine: (see Nucleoside, IV.2.2.141).
.005 Adenosine, deoxy-: (see Nucleoside, IV.2.2.141).

IV.2.2. HORTICULTURAL PHYSIOLOGY and CROP ECOLOGY

.006 Adenosine-5'-diphosphate: (see Nucleotide, IV.2.2.142).
.007 Adenosine-5'-monophosphate: (see Nucleotide, IV.2.2.142).
.008 Adenosine-5'-triphosphate: (see Nucleotide, IV.2.2.142).
.009 ADP: (see Adenosine-5'-diphosphate, IV.2.2.142).
.010 Aldonic acids: (see Sugar acids, IV.2.2.032c).
.011 Aldose: (see Monosaccharides, IV.2.2.031a).
.012 Alkaloids: heterocyclic, basic, nitrogenous compounds of plant origin, of which more then 3000 are known; apparent by-products of other metabolic pathways; many are highly toxic (e.g., strychnine and colchicine), stimulating (e.g., caffeine and theobromine), or drugs (e.g., quinine, morphine, and atropine). (S & R)
.013 Allelochemic: substance produced by one organism that influences another. (S & R)
.014 Allelopathy, allelopathic: production by plants of an allelochemic that is harmful to other plants, such as juglone from walnut and terpenes from soft chaparral, which inhibit germination of seeds or roots of other plants. (S & R)
.015 Allosteric enzymes: (see Enzymes, IV.2.2.058a).
.016 Amino acids: compounds with the general formula $RCHNH_2COOH$, of which $-NH_2$ is the amino group, $-COOH$ is the carboxyl group, and $R-$ is a side chain that can vary from a simple H (e.g., glycine) to various aliphatic (e.g., valine, leucine and lysine), aromatic (e.g., tyrosine and phenylalanine), or heterocyclic (e.g., proline, tryptophan, histidine) groups; 20 different forms are known, each of which has different properties; the fundamental building blocks for proteins, also occurring free. (S & R)
.017 Amino sugars: (see Carbohydrate derivatives, IV.2.2.032f).
.018 AMP: (see Adenosine-5'-monophosphate, IV.2.2.142).
.019 Amylopectin: (see Starch, IV.2.2.031c-a-1).
.020 Amylose: (see Starch, IV.2.2.031c-a-1).
.021 Anthocyanins: colored pigments, mainly present as glycosides: responsible for the red, purple, or blue of plant parts; one of the flavonoids (IV.2.2.068; see also Phenolics, IV.2.2.154). (S & R)
.022 Anthranilic acid: (see Shikimic acid pathway, IV.1.2.042).
.023 Arabinans: (see Storage polysaccharides, IV.2.2.031c-a-3).
.024 D-Arabinose: (see Cell-wall polysaccharides, IV.2.2.038; Hemicelluloses, IV.2.2.031c-b-2; Monosaccharides, IV.2.2.031a; Polysaccharides, IV.2.2.031c).
.025 Arginine: (see Proteins, IV.2.2.172).
.026 Ascorbic acid: (see Sugar acids, IV.2.2.032c).
.027 ATP: (see Adenosine-5'-triphosphate, IV.2.2.142).
.028 Atropine: (see Alkaloids, IV.2.2.012).
.029 Betalains: red and yellow pigments found only in the <u>Caryophyllales</u> (e.g., garden beet); not related to anthocyanins and do not occur together in the same plant; also do not undergo extensive color changes with pH; can be hydrolyzed into a sugar and a colored portion like most other flavonoids (IV.2.2.068). (S & R)
.030 Caffeine: (see Alkaloids, IV.2.2.012).
.031 Carbohydrates (saccharides): polyhydroxy aldehydes (terminate with a -CHO group), ketones and their derivatives (with a carbonyl group, RC=O, anywhere except at a terminal carbon), or compounds that yield only polyhydroxy aldehydes or ketones on hydrolysis; the principal constituent of all plant parts except seeds (which often contain substantial quantities of fats and proteins); classified in three groups as monosaccharides, oligosaccharides with 2 to 10 monosaccharide units joined in glycosidic linkage, and polysaccharides with long chains, which may be linear or

branched, of monosaccharide units. (LE 70)

.031a. Monosaccharides: (also known as simple or reducing sugars from their reaction with Benedict's or Fehling's solution); unbranched compounds, with the formula $(CH_2O)_n$, where $\underline{n}$=3 or a larger number, as aldoses (aldehydes) or ketoses (ketones) in two series ranging from trioses to octoses based, respectively, on glyceraldehyde and dihydroxyacetone; all except dihydroxyacetone contain one or more asymmetric C atoms and exist as stereoisomers in D (dextrorotary) or L (levorotary) forms; hexoses are by far the most abundant; e.g., glucose (the most important as the primordial or parent monosaccharide from which all of the others are derived), fructose, galactose, mannose, L-fucose, L-rhamnose, and L-sorbose, and form six-or five-membered ring structures (then denoted as pyranoses or furanoses, respectively; the former is the more stable configuration for aldohexoses and the latter is typical for ketohexoses); important aldopentoses include ribose, arabinose, and xylose as components of nucleic acids; an important ketopentose biologically is D-ribulose (the keto counterpart of ribose); derivatives of trioses and heptoses are intermediates in carbohydrate metabolism; (see also Gluconeogenesis, IV.1.2.020).

.031b. Oligosaccharides: the most common disaccharides include sucrose (glucose + fructose), which is abundant in plants, maltose (glucose + glucose), an intermediate product (repeating unit) of starch degradation, lactose (glucose + galactose) found in milk, cellobiose (glucose + glucose), and trehalose (glucose + glucose), of which the first and last are nonreducing sugars; raffinose (fructose + glucose + galactose) in sugar beets and other angiosperms and melezitose (glucose + fructose + glucose) in some conifers are examples of trisaccharides found in nature; linkages in oligosaccharides are α (1-4) for maltose, β (1-4) in cellobiose, and β (1-6) in gentiobiose; this type of linkage is denoted glycosidic because one unit acts as a sugar in hemiacetal form and the other, as an alcohol; (see also Carbohydrate (monosaccharide) derivatives, IV.2.2.032).

.031c. Polysaccharides (glycans): long linear or branched chains of monosaccharides (i.e., glycosides), principally as D-glucose but also D-mannose, D-fructose, D- and L-galactose, D-xylose, and D-arabinose; important in plants as storage (e.g., starch, inulin, mannans, and arabinans) or structural (e.g., cellulose and hemicelluloses) compounds.

.031c-a. Storage polysaccharides:

.031c-a-1. Starch: the most abundant storage polysaccharide in plants; usually deposited in the form of large granules which also contain proteins and certain enzymes; occurs in two forms: α-amylose, which consists of a long unbranched chain with D-glucose units bound in α (1-4) linkages and varies in molecular weight from a few to 500,000, and amylopectin, which is highly branched, with backbone linkage α (1-4) and branch linkages α (1-6); branches average about 12 units long and occur about every twelfth unit along the backbone; the whole has a molecular weight up to a million.

.031c-a-2. Inulin: found in Jerusalem artichoke and a few other plants; consists of D-fructose residues in β (2-1) linkage.

IV.2.2. HORTICULTURAL PHYSIOLOGY and CROP ECOLOGY

- .031c-a-3. Mannans, arabinans: mannose polysaccharides found in bacteria, yeast, molds, and higher plants, and arabinose polysaccharides, in plants, respectively.
- .031c-b. Structural polysaccharides:
- .031c-b-1. Cellulose: represents more than 50% of the organic carbon in the biosphere; the simplest and most abundant structural and cell-wall polysaccharide in plants; consists of 300 to as many as 14,000+ D-glucose residues organized into bundles of 80 parallel chains, or microfibrils, which are crosslinked with H bonds.
- .031c-b-2. Hemicelluloses: D-xylans (polymers of D-xylose) in β(1-4) linkage, with side-chains of arabinose and other sugars; provides part of the matrix that cements cellulose fibrils together around a plant cell.
- .032 Carbohydrate (monosaccharide) derivatives: monosaccharides that have undergone any of several types of reaction to convert them into other compounds; they include glycosides (reaction of the aldehydic carbon atom of an aldohexose with an alcohol), sugar alcohols (produced by reduction of the carbonyl group), sugar acids (oxidation at the aldehydic or the terminal C atom), phosphoric acid esters (replacement of an H at the 1 or 6 position or both, with phosphate), deoxy (desoxy) sugars (conversion of one or more CHOH groups into CH_2), and amino sugars (replacement of OH at the 2 position by an amino group). (LE 70)
- .032a. Glycosides: include, in addition to oligosaccharides and polysaccharides (see Carbohydrates, IV.2.2.031), pectins and other cell-wall polysaccharides (IV.2.2.038), and nucleosides (IV.2.2.141); a diverse but important group in plant metabolism.
- .032b. Sugar alcohols: include naturally occurring compounds such as sorbitol (from D-glucose), mannitol (from D-mannose), glycerol (an important component of lipids; see Fat, oil, IV.2.2.065), and inositol (a phospholipid, see IV.2.2.159).
- .032c. Sugar acids: two main groups, the aldonic acids (e.g., D-gluconic acid, ascorbic acid from D-glucose) from oxidation at the aldehydic C atom, and uronic acids (e.g., D-glucuronic acid from D-glucose, D-galacturonic acid, D-mannuronic acid) from oxidation at the C atom bearing the primary OH group.
- .032d. Phosphoric acid esters: important intermediates in carbohydrate metabolism.
- .032e. Deoxy (desoxy) sugars: the sugar component of nucleosides (IV.2.2.141), thus important in plant metabolism.
- .032f. Amino sugars: D-glucosamine and D-galactosamine are widely distributed in animal tissues as components of polysaccharides, chitin, and cartilage.
- .033 Carotene: (see Carotenoids, IV.2.2.034).
- .034 Carotenoids: yellow to reddish compound with conjugated (alternating single and double) bonds; two kinds, carotenes, which are hydrocarbons, and xanthophylls, which contain oxygen, are hydrophobic (water-hating) and not water-soluble; (see Isoprenoids, IV.2.2.107). (S & R)
- .035 Catalase: (see Porphyrins, IV.2.2.168).
- .036 Cellobiose: (see Oligosaccharides, IV.2.2.031b).
- .037 Cellulose: (see Structural polysaccharides, IV.2.2.031c-b-1).
- .038 Cell-wall polysaccharides: polymers of three hexoses, D-glucose, D-mannose, and D-galactose, two pentoses, D-xylose and L-arabinose, and five sugar derivatives, D-glucuronic acid, D-galacturonic acid, L-rhamnose (6-deoxy-mannose), L-fucose (6-deoxy- galac-

IV.2.2. HORTICULTURAL PHYSIOLOGY and CROP ECOLOGY

tose), and the methyl ester of D-galacturonic acid; pectin is a polymer that contains D-galacturonate; (see Gluconeogenesis, IV.1.2.019). (S & R, LE 70)

.039 Choline: (see Phospholipid, IV.2.2.159).
.040 Chlorophylls: (see Porphyrins, IV.2.2.168).
.041 Chromosomes: (see Proteins, IV.2.2.172).
.042 Coenzyme: (see Enzymes, IV.2.2.058b).
.043 Coenzyme A (CoA): (see Enzymes, IV.2.2.058c).
.044 Colchicine: an alkaloid produced by the autumn crocus; now widely used for chromosome doubling; (see also Alkaloids, IV.2.2.012).
.045 Cotyledons: (see Proteins, IV.2.2.172).
.046 Cuticle: a noncellular waxy covering that slows water loss and provides protection against some pathogens or minor mechanical damage; found on all aerial parts of a plant; (see Cuticle, II.3.038). (S & R)
.047 Cutin: (see Cuticle, II.3.038).
.048 Cyclic AMP: (see Nucleotide, IV.12.2.142).
.049 Cytochromes: (see Porphyrins, IV.2.2.168; Proteins, IV.2.2.172).
.050 Cytosine: (see DNA, IV.2.2.056; Nucleoside, IV.2.2.141; Pyrimidine base, IV.2.2.175).
.051 Deoxy (=desoxy) sugars, etc. (except DNA): alphabetized by the structural unit, not the substituent.
.052 Deoxyribonucleic acid: (see DNA, IV.2.2.056).
.053 Deoxyribonucleoside: (see DNA, IV.2.2.056; Nucleoside, IV.2.2.141).
.054 Dihydroxyacetone: (see Monosaccharides, IV.2.2.031a).
.055 Disaccharides: (see Oligosaccharides, IV.2.2.031b); Gluconeogenesis, IV.1.2.019).
.056 DNA (deoxyribonucleic acid): a long threadlike macromolecule coiled in a double helix along which genetic information is encoded in a precise sequence of base triplets that together constitute the fundamental unit of each gene on a chromosome; this molecule contains the template for myriad reactions and the components thereof in a cell. It is made up of a large number of deoxyribonucleotides with two main parts, a backbone of deoxyriboses linked by phosphodiester bonds (the linkage connects a 3' hydroxyl of sugar to a 5' hydroxyl of an adjacent one) and a variable part with respect to four kinds of base, adenine (A), guanine (G), thymine (T), and cytosine (C). Two polynucleotide chains run in opposite directions (i.e., one has 3'-5' phosphodiester bonds and the other, 5'-3' bonds), coiled in a double helix; the chains are linked by hydrogen bonds between pairs of bases, A always with T and G always with C; DNA is found almost entirely in the nucleus of cells. (SY)
.057 Endosperm: (see Proteins, IV.2.2.172).
.058 Enzymes: large organic compounds composed of protein alone or with a smaller, organic nonprotein portion (prosthetic group) attached; their function is to act as a catalyst and thereby greatly accelerate the rates of specific chemical reactions in plant or animal cells; several thousand have been identified; as catalysts they take part in reactions as substrates with specific binding sites for reactants but are not consumed themselves; (see also Hemes, IV.2.2.168; Peptide chains, IV.2.2.150). (LE 70)
.058a. Allosteric enzymes: those that combine with and respond, negatively or positively, to small molecules as part of the regulatory mechanism to control reaction rates (particularly with respect to inhibition or activation of the first enzyme of a series) in a metabolic pathway.

.058b.   Coenzyme: another organic compound, metal ion, or both, required for activity of a particular enzyme or enzyme system; many vitamins synthesized by plants act as enzymes or prosthetic groups.

.058c.   Coenzyme A (CoA): a derivative of pantothenate, one of the water soluble B complex of vitamins; (see Nucleotide, IV.2.2.142; Coenzyme, IV.2.2.058b).

.058d.   Isozymes (isoenzymes): multiple, distinct molecular forms of an enzyme that differ in net electric charge, hence are separable by gel electrophoresis of cell extracts. Isozymes usually have similar particle weights but differ with respect to their constituent polypeptide chains and kinetic characteristics, even though they catalyze the same overall reaction. They are important to the investigation of the molecular basis for cellular differentiation and morphogenesis.

.058e.   Kinetics of enzyme-catalyzed reactions: the overall rate and its relation to substrate concentration and product formed may be determined from the following equations which represent a steady-state condition:

.058e-1. Michaelis-Menten constant: $\underline{K_m} = k_2 + k_3/k_1 = [S]([E]-[ES])/[ES]$ where $\underline{K_m}$ is a "lumped" constant; k_1, k_2, and k_3 are specific rate constants [k_1 for E+S → ES, k_2 for E+S ← ES, and k_3 for ES → (E + P)]; [E] is the total enzyme concentration, [S] is the substrate concentration, [ES] is the concentration of the enzyme-substrate complex, and P is the product.

.058e-2. Michaelis-Menten equation: defines the quantitative relation between the initial enzyme reaction rate (v) and substrate concentration [S] if both V_{max} (the maximum velocity) and $\underline{K_m}$ (Michaelis-Menten constant) are known; $v = V_{max}[S]/K_m + [S]$.

.058e-3. Lineweaver-Burk plot: a double reciprocal plot of an enzyme-catalyzed reaction that provides a more accurate estimate of V_{max} than can be obtained from a plot of v vs [S]: $1/v = K_m/V_{max} \cdot 1/V_{max}$, where v is the initial rate of the enzymatic reaction, K_m is the Michaelis-Menten constant, V_{max} is the maximum velocity of the reaction, and [S] is the substrate concentration.

.058f.   Prosthetic group: a smaller, organic nonprotein radical (a fundamental constituent of a compound with a charge) usually tightly attached to the protein part of an enzyme by covalent bonds and generally essential to catalytic activity.

.058g.   Taxonomic classification of enzymes: the international classification divides enzymes into six groups, as oxidoreductases (oxidation-reduction reactions), transferases (transfer of functional groups), hydrolases (hydrolysis reactions), lyases (addition to double bonds), isomerases (isomerization reactions), and ligases (formation of bonds with ATP cleavage), each having several subcategories listed by code number and type of reaction catalyzed; trivial names are often substituted for systematic names for day-to-day reference and the latter are used in publications.

.059 Epicuticular waxes: (see Cuticle, II.3.038; Cuticle, IV.2.2.046).
.060 D-Erythrose: precursor of erythrose-4-phosphate. (IV.1.2.042).
.061 Erythrose-4-phosphate: (see Shikimic acid pathway, IV.1.2.042).
.062 Ethanolamine: (see Phospholipid, IV.2.2.159).
.063 FAD^+: (see Nucleotide, IV.2.2.142; Coenzyme, IV.2.2.058b).

IV.2.2. HORTICULTURAL PHYSIOLOGY and CROP ECOLOGY

.064 Farnesol: (see Isoprenoids, IV.2.2.107).
.065 Fat, oil: long-chain fatty acids esterified to glycerol (i.e., a triglyceride); fats are solid at room temperature, oils are liquid; may be saturated or unsaturated (i.e., contain one or more double bonds). (S & R)
.066 Fatty acids: a class of compound that contains a long hydrocarbon chain and terminal carboxyl (-COOH) group; those found in plants usually contain an even number of C atoms; the number varies typically between 12 and 24; may be saturated (i.e., all single bonds) or contain one or more double bonds; the latter when present are commonly in the cis configuration. (S & R)
.067 Flavin adenine dinucleotide (oxidized form): FAD^+; (see Nucleotide, IV.2.2.142; Coenzyme, IV.2.2.058b).
.068 Flavonoids: 15-C compounds generally distributed throughout the plant kingdom; usually colored by numerous conjugated (alternating single and double) bonds and often present in plant tissues as glycosides; three main groups: anthocyanins (IV.2.2.020), flavonols, and flavones (IV.2.2.069); many accumulate in the central vacuole, although apparently they are often synthesized in the chloroplasts. (S & R)
.069 Flavonols, flavones: yellowish or ivory-colored pigments closely related chemically to anthocyanins, which are also flavonoids; often contribute to the color of flowers or attract insects by their absorption of ultraviolet radiation; also widely distributed in leaves, especially in chloroplasts, in which they apparently deter would-be feeders and also protect against excessive ultraviolet radiation. (S & R)
.070 Fructosans: (see Gluconeogenesis, IV.1.2.019).
.071 D-Fructose: (see Inulin, IV.2.2.031c-a-2; Monosaccharides, IV.2.2.031a; Polysaccharides, IV.2.2.031c).
.072 L-Fucose: (see Cell-wall polysaccharides, IV.2.2.038; Monosaccharides, IV.2.2.031a).
.073 Furanoses: (see Monosaccharides, IV.2.2.031a).
.074 D-Galactosamine: (see Amino sugars, IV.2.2.032f).
.075 D-Galactose: (see Cell-wall polysaccharides, IV.2.2.038; Monosaccharides, IV.2.2.031a; Polysaccharides, IV.2.2.031c).
.076 L-Galactose: (see Polysaccharides, IV.2.2.031c).
.077 Galactose, 6-deoxy-: (see Cell-wall polysaccharides, IV.2.2.038).
.078 D-Galacturonate: (see Cell-wall polysaccharides, IV.2.2.038).
.079 D-Galacturonic acid: (see Cell-wall polysaccharides, IV.2.2.038; Sugar acids, IV.2.2.032c).
.080 D-Galacturonic acid methyl ester: (see Cell-wall polysaccharides, IV.2.2.038).
.081 Gentiobiose: (see Oligosaccharides, IV.2.2.031b).
.082 Gibberellins: (see Isoprenoids, IV.2.2.107).
.083 D-Gluconic acid: (see Sugar acids, IV.2.2.032c).
.084 D-Glucosamine: (see Amino sugars, IV.2.2.032f).
.085 D-Glucose: (see Cellulose, IV.2.2.031c-b-1; Cell-wall polysaccharides, IV.2.2.038); Monosaccharides, IV.2.2.031a; Oligosaccharides, IV.2.2.031b; Polysaccharides, IV.2.2.031c).
.086 Glucose-6-phosphate: (see Gluconeogenesis, IV.1.2.019).
.087 N-Glucosides: (see N-Glycosides, IV.2.2.094).
.088 D-Glucuronic acid: (see Cell-wall polysaccharides, IV.2.2.038; Sugar acids, IV.2.2.032c).
.089 Glycans: (see Polysaccharides, IV.2.2.031c).
.090 Glyceraldehyde: (see Monosaccharides, IV.2.2.031a).
.091 Glycerol: (see Sugar alcohols, IV.2.2.032b).

IV.2.2. HORTICULTURAL PHYSIOLOGY and CROP ECOLOGY

.092 **Glycine:** (see Amino acids, IV.2.2.016).
.093 **Glycolipid:** a lipid (IV.2.2.116) with a sugar esterified to the third hydroxyl group instead of phosphate; constitutes about 80% of the lipid fraction of chloroplasts in higher plants. (S & R)
.094 **Glycosides:** compounds formed from the linkage (with loss of a molecule of water) of a sugar and an alcohol, of which the alcohol can be another sugar; (see Carbohydrate derivatives, IV.2.2.032a). (LE 70)
.095 **Glycosidic linkage:** reaction of the pyranose or furanose forms of monosaccharides (hemiacetals, a ring in which oxygen is linked to adjacent C atoms) to form acetals from the reaction of an alcohol which has a reactive OH group and an aldehyde group; acetal formation in sugars involves the carbonyl group (present in aldehyde or ketone form) of one sugar and the alcohol group (OH) of another; the linkage is usually from the number 1 C (aldehyde end) to the number 4 C or 6 C (alcohol end), depending on the sugar; (see Carbohydrates, IV.2.2.031; Carbohydrate derivatives, IV.2.2.032). (LE 70)
.096 **Guanine:** (see DNA, IV.2.2.056; Nucleoside, IV.2.2.141; Purine bases, IV.2.2.173).
.097 **Hemes:** (see Porphyrins, IV.2.2.168).
.098 **Hemicelluloses:** (see Structural polysaccharides, IV.2.2.031c-b-2).
.099 **Hemoglobin:** (see Porphyrins, IV.2.2.168).
.100 **Hexoses:** (see Monosaccharides, IV.2.2.031a).
.101 **Histidine:** (see Amino acids, IV.2.2.016).
.102 **Histones:** (see Proteins, IV.2.2.172).
.103 **Inositol:** (see Sugar alcohols, IV.2.2.032b; Phospholipid, IV.2.2.159).
.104 **Inulin:** (see Storage polysaccharides, IV.2.2.031c-a-2).
.105 **Isocoumarin:** (see Phytoalexin, IV.2.2.162).
.106 **Isoprene:** (see Isoprenoid, IV.2.2.107).
.107 **Isoprenoids:** compounds consisting of dimers, trimers, or polymers of isoprene $[(CH_3)_2-C=CH-CH_2]$ units; include gibberellins, abscisic acid, farnesol, sterols, carotenoids, turpentine (and other terpenes), rubber, and the phytol tail of chlorophyll. (S & R)
.108 **Isozymes (isoenzymes):** (see Enzymes, IV.2.2.058d).
.109 **Jerusalem artichoke:** (see Inulin, IV.2.2.031c-a-2).
.110 **Ketoses:** (see Monosaccharides, IV.2.2.031a).
.111 **Kinetics of enzyme-catalyzed reactions:** (see Enzymes, IV.2.2.058e).
.112 **Lactose:** (see Oligosaccharides, IV.2.2.031b).
.113 **Leucine:** (see Amino acids, IV.2.2.016).
.114 **Lignin:** strengthening material that occurs with cellulose and other polysaccharides in the middle lamella and primary and secondary cellwalls of higher plants; (see Phenolics, IV.2.2.154). (S & R)
.115 **Lineweaver-Burk plot:** [see Enzyme kinetics, IV.2.2.058e(3)].
.116 **Lipids:** universal constituents of plant membranes; similar to triglycerides, except only two fatty acids (usually unsaturated) are present; the third hydroxyl group of glycerol is connected to any of several possible constituents. (S & R)
.117 **Lysine:** (see Amino acids, IV.2.2.016; Proteins, IV.2.2.172).
.118 **Maltose:** (see Oligosaccharides, IV.2.2.031b).
.119 **Mannans:** (see Storage polysaccharides, IV.2.2.031c-a-3).
.120 **Mannitol:** (see Sugar alcohols, IV.2.2.032b).
.121 **D-Mannose:** (see Cell-wall polysaccharides, IV.2.2.038; Monosaccharides, IV.2.2.031a; Polysaccharides, IV.2.2.031c).
.122 **Mannose, 6-deoxy-:** (see Cell-wall polysaccharides, IV.2.2.038).

IV.2.2. HORTICULTURAL PHYSIOLOGY and CROP ECOLOGY

.123 D-Mannuronic acid: (see Sugar acids, IV.2.2.032c).
.124 Melezitose: (see Oligosaccharides, IV.2.2.031b).
.125 Messenger RNA: (see mRNA, IV.2.2.188).
.126 Michaelis-Menten constant: (see Enzyme kinetics, IV.2.2.058e).
.127 Michaelis-Menten equation: (see Enzyme kinetics, IV.2.2.058e).
.128 Microfibrils: (see Cellulose, IV.2.2.031c-b-1).
.129 Monosaccharides: (see Carbohydrates, IV.2.2.031a).
.130 Morphine: (see Alkaloids, IV.2.2.012).
.131 Myoglobin: (see Porphyrins, IV.2.2.168).
.132 NAD^+: (see Coenzyme, IV.2.2.058b; Nucleotide, IV.2.2.142).
.133 NADH: (see Nucleotide, IV.2.2.142).
.134 NADPH: (see Nucleotide, IV.2.2.142).
.135 Nicotinamide adenine dinucleotide (oxidized form): NAD^+ ; (see Coenzyme, IV.2.2.058b; Nucleotide, IV.2.2.142).
.136 Nicotinamide adenine dinucleotide (reduced form): NADH; (see Nucleotide, IV.2.2.142).
.137 Nicotinamide adenine dinucleotide phosphate (oxidized form): $NADP^+$; (see Nucleotides, IV.2.2.142).
.138 Nicotinamide adenine dinucleotide phosphate (reduced form): NADPH; (see Nucleotide, IV.2.2.142).
.139 Nonreducing sugars: [see Oligosaccharides, IV.2.2.031b)].
.140 Nucleic acids: compounds characterized by the presence of a 5-carbon sugar (d-ribose or deoxy-D-ribose), phosphoric acid, and a nitrogenous base (derivatives of pyrimidine or purine); the sugar and base are often linked as a N-glycoside or nucleoside; often macromolecules, as of DNA or RNA and various individual nucleotides, which are usually esterified because the phosphate group is ionized at physiological pH; widely distributed in nucleic material and in other cell organelles in which biosyntheses occur; any of a number of compounds that act as metabolic regulators or intermediates in such activities. (LE 70)
.141 Nucleoside: N-glycoside (see IV.2.2.140) of a pyrimidine or purine base; i.e., D-ribose or 2-deoxy-D-ribose is linked with the C-1 carbon atom of the pentose to the N-1 of the pyrimidine or N-9 of the purine base; those with D-ribose form ribonucleosides and with 2-deoxy-D-ribose, deoxyribonucleosides; those formed from adenine (base) and D-ribose (sugar) are termed adenosine or from adenine plus deoxyribose, deoxyadenosine; other bases commonly incorporated include uracil (replaces thymine, one of the four bases in DNA, in RNA), thymine, cytosine, and guanine; their main functions are to serve as the backbone of the double helix in DNA and furnish suitable sites for the bonds between pairs of complementary bases on each strand and to act as substrates for a wide variety of metabolic regulators; (see Nucleic acids, IV.2.2.140). (LE 70)
.142 Nucleotide: a compound that consists of the phosphate ester of a nucleoside (IV.2.2.141), commonly through the OH group attached to the C-5 carbon of the pentose sugar portion; these are metabolic regulators, being activated precursors of DNA and RNA and their derivatives, that is activated intermediates in many biosyntheses; ribonucleotides, such as adenosine-5'-monophosphate (AMP) adenosine-5'-diphosphate (ADP), and adenosine-5'- triphosphate (ATP), have many functions; cyclic AMP mediates the action of many hormones and ATP is the universal energy currency in biological systems; adenine nucleotides are also components of three major coenzymes, NAD^+, FAD^+, and CoA; NADH and NADPH are used in respiration processes (IV.1.2.040); $NADP^+$ is used in the pentose

IV.2.2. HORTICULTURAL PHYSIOLOGY and CROP ECOLOGY 368

phosphate pathway (IV.1.2.040g) as an electron acceptor in the dehydrogenation of glucose-6-phosphate. (LE 70)
.143 Octoses: (see Monosaccharides, IV.2.2.031a).
.144 Oil: (see Fat, IV.2.2.065).
.145 Oligosaccharides: (see Carbohydrates, IV.2.2.031b).
.146 Pectin: (see Cell-wall polysaccharides, IV.2.2.038).
.147 Pentoses: (see Monosaccharides, IV.2.2.031a; Nucleotide, IV.2.2.142).
.148 PEP: (see Gluconeogenesis, IV.1.2.019).
.149 Peptide bond: (see Peptide chain, IV.2.2.150).
.150 Peptide chain: polymers that consist of amino acids and amides, each linked by a peptide bond; the latter involves the carboxyl end of one amino acid joined to the amino group of the next, with the loss of a molecule of water in the process; the mechanism of forming a peptide chain involves various kinds of RNA and enzymes which proceed to add amino acids, one at a time; (see Proteins, IV.2.2.172). (S & R)
.151 Perisperm: (see Proteins, IV.2.2.172).
.152 Peroxidase: (see Porphyrins, IV.2.2.168).
.153 Phaseolin: (see Phytoalexin, IV.2.2.162).
.154 Phenolics: compounds with benzene rings and various attached substituent groups, such as hydroxyl, carboxyl, or methoxyl, and other nonaromatic or ring structures; include aromatic amino acids, simple phenols, polyphenols, and derivatives like phytoalexins, coumarins, lignin, anthocyanins, flavonols, flavones (the last three collectively flavonoids), which arise from phosphoenol pyruvate (PEP), from glycolysis, and erythrose-4-phosphate from the pentose phosphate respiratory cycle via the shikimic acid pathway (IV.1.2.042) and subsequent reactions. (S & R)
.155 Phenylalanine: (see Amino acids, IV.2.2.016; Shikmic acid pathway, IV.1.2.042).
.156 Phenylpyruvic acid: (see Shikimic acid pathway, IV.1.2.042).
.157 Phosphodiester bond: (see DNA, IV.2.2.056).
.158 Phosphoenol pyruvate (PEP): (see Gluconeogenesis, IV.1.2.019; Phenolics, IV.2.2.154).
.159 Phospholipid: a lipid (IV.2.2.116) with a phosphate group esterified to the third hydroxyl group of glycerol; often other ions are esterified to the phosphate, e.g., choline (a lecithin), glycerol, inositol, ethanolamine, and acephalin. (S & R)
.160 Phosphoric acid esters: see Carbohydrate derivatives, IV.2.2.032d).
.161 Photosynthesis: (see Proteins, IV.2.2.172).
.162 Phytoalexin: a toxic substance, such as certain coumarin derivatives and various antifungal compounds, synthesized by plants in response to parasite invasion or infection by certain fungi; e.g., pisatin, phaseolin, orchinol, trifolirhizin, and isocoumarin from pea and bean pods, orchid tubers, red clover, and carrot root. (S & R)
.163 Phytol: (see Isoprenoids, IV.2.2.107).
.164 Pisatin: (see Phytoalexin, IV.2.2.162).
.165 Polyhydroxy aldehydes: (see Carbohydrates, IV.2.2.031).
.166 Polyhydroxy ketones: (see Carbohydrates, IV.2.2.031).
.167 Polysaccharides: (see Carbohydrates, IV.2.2.031c).
.168 Porphyrins: macrocyclic compounds that consist basically of four substituted pyrroles joined in a ring by methylene bridges and with the four central N atoms bonded coordinately to a Mg^{++} ion, as in chlorophylls, or to an iron ion, as in the hemes; the latter form the prosthetic group of proteins and enzymes such as

IV.2.2. HORTICULTURAL PHYSIOLOGY and CROP ECOLOGY   369

   myoglobin, hemoglobin, catalase, peroxidase, and cytochromes; (see also Chlorophyll, IV.1.2.036d; Cytochromes, IV.1.2.036e; Electron transport (cytochrome system), IV.1.2.040c; Enzymes, IV.2.2.058). (LE 70)
.169 Prephenic acid: (see Shikimic acid pathway, IV.1.2.042).
.170 Proline: (see Amino acids, IV.2.2.016).
.171 Prosthetic group: (see Enzymes, IV.2.2.058f).
.172 Proteins: N-containing compounds comprised of one or more peptide chains (IV.2.2.150) of tens to hundreds of amino acids that range in molecular weight from about 10,000 to more than 500,000 g/mol; every known enzyme has a protein as a major component; many contain only protein; some proteins have no catalytic function in microtubules or membranes; some, such as cytochromes, are involved strictly in electron transport during photosynthesis and respiration; chromosomes contain five major histones (positively charged proteins rich in arginine or lysine) held on DNA by ionic bonds and that are important in controlling the structure and genetic activity of the chromosomes; some proteins stored in endosperm, perisperm, cotyledons, or a pistillate gymnosperm gametophyte act as a reservoir for amino acids to be utilized by the seedling on germination. (S & R)
.173 Purine bases: two-ring heterocyclic compounds with four N atoms (one of which is in an amino group); adenine and guanine are common constituents of nucleosides (IV.2.2.141) and nucleotides (IV.2.2.142). (LE 70)
.174 Pyranoses: (see Monosaccharides, IV.2.2.031a).
.175 Pyrimidine bases: single-ring heterocyclic compounds with two N atoms; cytosine, thymine and uracil are common constituents of nucleosides (IV.2.2.141) and nucleotides (IV.2.2.142). (LE 70)
.176 Quinine: (see Alkaloids, IV.2.2.012).
.177 Raffinose: (see Oligosaccharides, IV.2.2.031b).
.178 Reducing sugars: (see Monosaccharides, IV.2.2.031a).
.179 Respiration: (see Proteins, IV.2.2.172).
.180 L-Rhamnose: (see Cell-wall polysaccharides, IV.2.2.038; Monosaccharides, IV.2.2.031a).
.181 Ribonucleic acid: (see RNA, IV.2.2.188).
.182 Ribonucleoside: (see Nucleoside, IV.2.2.141).
.183 Ribonucleotide: (see Nucleotide, IV.2.2.142).
.184 D-Ribose: (see Monosaccharides, IV.2.2.031a; Nucleic acids, IV.2.2.140; Nucleoside, IV.2.2.141; Nucleotide, IV.2.2.142).
.185 2-deoxy-D-Ribose: (see Nucleoside, IV.2.2.141; Nucleic acids, IV.2.2.140).
.186 Ribosomal RNA: (see rRNA, IV.2.2.188).
.187 D-Ribulose: (see Monosaccharides, IV.2.2.031a).
.188 RNA (ribonucleic acid): a long, unbranched macromolecule that consists of 75 to many thousand nucleotides joined by 3'-5' phosphodiester bonds; three types: messenger RNA (mRNA) is the template for protein synthesis; transfer RNA (tRNA) carries amino acids in activated form to the ribosome for peptide-bond formation in a sequence determined by mRNA; ribosomal RNA (rRNA) is the major component of ribosomes and is single-stranded as a rule; (see also Peptide chains, IV.2.2.150). (LE 70)
.189 Rubber: (see Isoprenoids, IV.2.2.107).
.190 Saccharides: (see Carbohydrates, IV.2.2.031).
.191 Sorbitol: (see Sugar alcohols, IV.2.2.032b).
.192 L-Sorbose: (see Monosaccharides, IV.2.2.031a).
.193 Starch: (see Storage polysaccharides, IV.2.2.031c-a-1).

IV.2.2. HORTICULTURAL PHYSIOLOGY and CROP ECOLOGY

.194 Stereoisomers: (see Monosaccharides, IV.2.2.031a).
.195 Sterols: (see Isoprenoids, IV.2.2.107).
.196 Storage polysaccharides: (see Polysaccharides, IV.2.2.031c-a).
.197 Structural polysaccharides: (see Polysaccharides, IV.2.2.031c-b)
.198 Strychnine: (see Alkaloids, IV.2.2.012).
.199 Sucrose: (see Oligosaccharides, IV.2.2.031b).
.200 Sugar acids: (see Carbohydrate derivatives, IV.2.2.032c).
.201 Sugar alcohols: (see Carbohydrate derivatives, IV.2.2.032b).
.202 Sugars, deoxy(desoxy): (see Carbohydrate derivatives, IV.2.2.032e).
.203 Taxonomic classification of enzymes: (see Enzymes, IV.2.2.058g).
.204 Terpenes: (see Isoprenoids, IV.2.2.107).
.205 Theobromine: (see Alkaloids, IV.2.2.012).
.206 Thymine: (see DNA IV.2.2.056; Nucleoside, IV.2.2.141; Pyrimidine bases, IV.2.2.175).
.207 Transfer RNA: (see tRNA, IV.2.2.188).
.208 Trehalose: (see Oligosaccharides, IV.2.2.031b).
.209 Trifolirhizin: (see Phytoalexin, IV.2.2.162).
.210 Triglyceride: an ester that consists of a molecule of glycerol, which has three reactive hydroxyl (-OH) groups, and three long-chain fatty acids; (see also Lipids, IV.2.2.116). (S & R)
.211 Trioses: (see Monosaccharides, IV.2.2.031a).
.212 Tryptophan: (see Amino acids, IV.2.2.016; Shikimic acid pathway, IV.1.2.042).
.213 Turpentine: (see Isoprenoids, IV.2.2.107).
.214 Tyrosine: (see Amino acids, IV.2.2.016).
.215 Uracil: (see Nucleoside, IV.2.2.141; Pyrimidine bases, IV.2.2.175).
.216 Uronic acids: (see Sugar acids, IV.2.2.032c).
.217 Valine: (see Amino acids, IV.2.2.016).

IV.3. Growth and Development

IV.3.1. Growth cycles, juvenility

.001 Afterripening: changes that take place within a seed (or bud) during dormancy, such as maturation of the embryo. (S & R)
.002 Aitonomic: interruptions in growth activity and cessation imposed by the environment; e.g., temporary dormancy caused by a dry period or cold snap; (see Autonomic, IV.3.1.003). (LA)
.003 Autonomic: innate (endogenous) rhymthicity of alternations between developmental activity and slowing or cessation of growth having its origin in the genotype (i.e., hereditary); (see Aitonomic, IV.3.1.002). (LA)
.004 Auxanometer: an instrument for determining and measuring growth that consists essentially of a lever with a long arm and a short arm, the latter attached to the plant. (MO)
.005 Cyclophysis: characters of a plant determined by age or progressive ontogenetic processes (usually decrease but can be increases in module lengths); (see also Topophysis, IV.3.1.014i). (WR)
.006 Dendrograph: an instrument for the automatic recording of diameter growth of trees. (AV)
.007 Development: the series of processes from the initiation of growth to death of a plant or plant parts (WT); the process of growth and differentiation (e.g., that of diameter growth of trees). (LA)

IV.3.1. HORTICULTURAL PHYSIOLOGY and CROP ECOLOGY

.008 Differentiation: the process by which cells become specialized. (S & R)
.009 Dormancy: a term used in several senses to denote the inability of a living plant or part to grow even under otherwise favorable environmental conditions; (see Quiescence, IV.3.1.021). (S & R, KK)
.010 Dormant: the condition of a seed when it fails to germinate or other parts to grow as a result of internal conditions, even though all external conditions (i.e., temperature and moisture) are suitable; equivalent to rest, as used by pomologists. (S & R)
.011 Endogenous winter dormancy (dormancy of temperate zone plants): cessation of growth from endogenous causes; a gradual process of readjustment to the winter state; three phases: (a) postdormancy, (b) predormancy, and (c) true dormancy. (LA)
.011a. Postdormancy: treatment with heat and additional light causes buds to develop once true dormancy has been broken; the postdormant period usually ends in February, after which environmental factors alone determine the time of sprouting.
.011b. Predormancy: shortening days after the summer solstice combined with decreasing temperatures induce the conclusion of the growth period and the transition to the dormant stage in some woody plants, the critical day length being around 12 to 15 hours and very long days (encountered in midsummer at high latitudes, e.g., 40° or more) that prevent the onset of winter dormancy; geophytes respond to long days and apparently have two built-in rest periods, one at the time of greatest winter cold, the other during the summer drought; decreasing day/night temperatures, usually at the critical level a little below 10°C, with gradual falling; decreasing temperatures are apparently more effective than short days.
.011c. True dormancy: complete dormancy in November and December in northern latitudes, in which the plant can no longer be induced to sprout by warming or a lengthened photoperiod, an important factor in their cold resistance; nontemperate zone species, such as citrus which lack deep winter dormancy, suffer injury from cold waves after mild weather in midwinter; still lower temperatures accelerate the transition to postdormancy; note, however, that many woody plants have a distinct cold requirement (usually measured as hours below a specified threshold temperature) that must be satisfied before they will resume normal growth.
.012 Growth: the irreversible increase in physical attributes (characteristics of a developing plant or plant part) (WT); in higher plants usually means the irreversible increase in plant or organ size measured as fresh or dry weight, of which the latter is preferred because fresh weight or volume increases may reverse as a result of water loss alone; growth may also be expressed as an increase in cell numbers as in populations of bacteria or algae; however, an increase in cell numbers in whole (higher) plants is not necessarily accompanied by increased dry weight or volume (done via a DNA assay); the parameter used to measure growth can vary, depending on the plant, tissue, or stage of development. (S & R)
.013 Growth curve: the curve produced when growth measured as weight, volume, etc., is plotted against time; it has a sigmoid shape for annuals and individual organs of annuals and perennials and for fruits, such as apple, pear, tomato, banana, strawberry, date, cucumber, citrus, avocado, pineapple, walnut, and pecan; it

IV.3.1. HORTICULTURAL PHYSIOLOGY and CROP ECOLOGY

has a double sigmoid shape for stone fruits, seeded grapes, blueberry, fig, pistachio, currant, olive, and raspberry. (WS)

.014 Juvenility: refers to the vegetative state of plants before they begin to flower; most often used with reference to woody plants, in which there are many manifestations with respect to morphological (e.g., growth habit, phyllotaxis, leaf shape, anatomy) and physiological (e.g., rooting, flowering ability, pigmentation, leaf abscission) characteristics associated with the period before they reach the adult (flowering and fruiting) stage.

.014a. Heteroblastic development: the abrupt change in a plant from the juvenile to the adult stage; (see Homoblastic development, IV.3.1.014b). (HOT)

.014b. Homoblastic development: the gradual change in a plant from the juvenile to the adult stage. (HOT)

.014c. Jarovization (= jarovisation): the process in which seeds are given chilling treatment after the radicle has begun to elongate (i.e., chilling treatment of partially germinated seeds); a means of shortening the vegetative period; (see also Vernalization, IV.3.1.014j).

.014d. "Mentor method": the grafting of a branch from an adult tree onto a seedling to stimulate flowering of the latter.

.014e. Neoteny: period of juvenility indefinitely prolonged; sometimes used in the sense of small size, as by Corner (1966) with reference to certain species of palms growing in moist or shaded locations, compared with more open and drier situations. (HOT, CR 66)

.014f. Paedomorphosis: the protraction of juvenile characteristics in the secondary xylem of dicotyledons that are not truly woody, manifested by a gradual, instead of a sudden, change in tracheary elements; (see also Acrotony, II.1.1.3.003; Basitony, II.1.1.3.014). (HOT)

.014g. Sphaeroblast (= prosphaeroblast): a meristemic mass on the trunk or branches of an apple tree (and others, e.g., maple and walnut) which gives rise to adventitious shoots or roots; seen most frequently on East Malling or Malling Merton rootstocks propagated in the southeastern United States, south of Missouri, or in northern Kentucky [e.g., observed from northwestern Arkansas (Fayette) and Kentucky southward; in some lots, 15% or more of the young trees are affected]; (see Burr knot, II.1.1.3.020). (W.B. Sherman, 1982, personal communication)

.014h. Topophysis: the long persistence of age effects after propagation; <u>cyclophysis</u> refers to the process if the propagule is a cutting or budling; <u>periphysis</u> refers to the process in which effects due to position on the plant are transmitted by budding or grafting. (WR)

.014i. Totipotency: the ability of a parenchyma cell to become meristematic and develop into an entire new plant. (S & R)

.014j. Vernalization: cold treatment for six weeks or more of partially germinated seed or other plant parts of biennials to induce flowering the same year as sown; a process reversible by exposure to high temperature, as in onion sets (see also IV.3.2.021e). (S & R)

.015 Maximum photosynthetic efficiency: (see Photosynthetic efficiency, IV.3.1.020).

.016 Morphogenesis: the developmental history of organisms or of their parts; equivalent to development (IV.3.1.007). (S & R)

IV.3.1. HORTICULTURAL PHYSIOLOGY and CROP ECOLOGY

.017 Periphysis: see Topophysis, IV.3.1.014h).

.018 Phenology: the study of the cycle of shoot growth, blooming, bearing of fruit, as well as other biological processes with periodicity, as they are affected by climatic changes during the year. (LA)

.019 Phenopause: externally visible changes in the course of a plant's life cycle; correlated in the first half of the year (or season) with certain temperature and photoperiod thresholds for opening buds, sprouting, onset of flowers, etc., and influenced in the latter half of the year (season) by all environmental conditions that delay or accelerate processes of maturation and ageing. (LA)

.020 Photosynthetic efficiency: the ratio of energy stored by the assimilation of CO_2 to the radiant energy absorbed by the photosynthetic system of a crop; neither part of the ratio, however, can be measured directly. Maximum photosynthetic efficiency is the maximum of $\mu(dN/dI)$, where μ = energy stored photochemically per g of CO_2 reduced (=10.8 J/g), N = net assimilation rate of CO_2 [g/(cm^2) · (sec)], and I = incident light flux (W/m^2); calculated theoretically as the product of (g mol wt of CO_2 = 44 g/mol) x (einsteins/unit of energy of radiation = 4.6 X 10^6 E/J) x (quantum yield of photochemical process = mol of CO_2 assimilated per einstein absorbed radiation)x(fraction of incident energy absorbed by chloroplasts; assumed equal to total absorbed fraction ca. 0.85) or about 18.6%, from which value photorespiration must be deducted from the net rate of photosynthesis to give true maximum photosynthetic efficiency. (MO)

.021 Quiescence: the condition of a seed or other plant part that prevents growth unless all external conditions (i.e., temperature and moisture) are favorable, i.e., are in the proper range. (S & R)

.022 Rest: (see Dormant, IV.3.1.010).

.023 Sphaeroblast: (see IV.3.1.014g).

.024 Viable: capable of living, growing, and developing; commonly used with reference to stored seeds. (S & R)

IV.3.2. Photoperiodism, thermoperiodism

.001 Action spectrum: the wavelength or wavelengths of light at which certain physiological processes (i.e., flowering, seed germination, and leaf expansion) are promoted or inhibited. (S & R)

.002 Energy budget equation: the relationships of factors affecting the energy, hence the thermal balance of plants; expressed as $Q_I + Q_M + Q_p + Q_{soil} + Q_H + Q_E = 0$, where Q_I is the radiation balance (see Net radiation balance, IV.3.2.007), Q_M is the energy turnover in metabolic processes (generally 1 to 2%), Q_p is the heat stored in the phytomass, Q_{soil} is the heat stored in the soil, and Q_H (sensible heat exchange) and Q_E (latent heat exchange) are the exchanges of energy with the surroundings by heat conduction and convection (Q_H) and evaporation or condensation (Q_E). (LA)

.003 Etiolation: the development of plants or plant parts in the absence of light, characterized by elongated shoots, small leaves, and lack of chlorophyll. (H & K, S & R)

.004 Eurythermic: plants that can thrive within a wide range of temperatures; e.g., terrestrial vascular plants; (see Stenothermic, IV.3.2.019). (LA)

IV.3.2. HORTICULTURAL POHYSIOLOGY and CROP ECOLOGY

.005 Fluorescence: short-delay emission, i.e., within 10^{-4} to 10^{-5} s, after absorption of the original photon, of light at a longer wavelength from an atom, ion, or molecule in an excited stage when it loses its excitation energy; (see Phosphorescence, IV. 3.2.008). (S & R)

.006 Irradiance: radiation (usually meaning light) received by an object. (S & R)

.007 Net radiation balance (Q_I):
$Q_I = \bar{I}_s + \bar{I}_l = (I_d + I_i - I_r) + I_a - I_g$, where $\bar{I}_s$ is the short wavelength (0.3 - 3 m) balance, $\bar{I}_l$ is the long wavelength thermal radiation balance, I_d is the direct solar radiation (i.e., direct sunlight), I_i is diffuse sky and cloud light (diffuse radiation from the sky and clouds), I_r is reflected shortwave radiation (short wavelength light reflected from leaves, etc.), I_a is reradiation at thermal wavelengths from the atmosphere, and I_g is the long wavelength (3-100 μm) thermal radiation from ground and plants. (S & R)

.008 Phosphorescence: the same as fluorescence (IV.3.2.005) but with a delay of 10^{-4} to 10 s or more. (S & R)

.009 Photoblastic: germination of seeds; positive when germination is promoted by red light but inhibited by far-red and negative when germination takes place in the dark. (S & R)

.010 Photochemical: pertaining to a chemical reaction activated by light. (S & R)

.011 Photodormancy: requirement of light for seeds to germinate (i.e., such seeds will not germinate unless exposed to light); may be induced by exposure to high temperature (e.g., lettuce); alternating low and high temperatures can substitute for the light requirement (e.g., Kentucky bluegrass), as can exogenous gibberellins or cytokinins. (S & R)

.012 Photomorphogenesis: control of morphogenesis (growth and differentiation of individual cells into recognizable tissues and organs) by light. (S & R)

.013 Photoperiod, photoperiodism: response of plants to day length, manifested most often in respect to flowering or seed germination; photoperiodism is the process; photoperiod, the duration of light, including civil twilight, or darkness, whether natural or artificial. (S & R)

.013a. Ambiphotoperiodic plant: flowering quantitatively inhibited by intermediate daylengths, e.g., tarweed and hooked bristlegrass; plants that flower with long or short but not intermediate daylengths.

.013b. Critical daylength: period (number of hours) in which a plant with a qualitative (absolute) daylength requirement below or above which it will not flower, depending on whether it is a long-day or short-day plant; note, however, that nightlength may be more important in determining a response than daylength per se.

.013c. Day-neutral plant: no apparent response to daylengths but flowering may be promoted by high, low, or alternating temperatures; e.g., cucumber, globe amaranth, sunflower, Jerusalem artichoke, rice, kidney bean, garden pea, and maize.

.013d. Florigen: the name for a hypothetical hormone said to be responsible for the initiation of flowering.

.013e. Induction cycle: the diurnal period without interruption required to elicit a photoperiodic response (i.e., a short night in the case of a long-day plant or a long night in the case of a short-day plant); (see Long-day plant, IV.3.2.013g).

.013f. Intermediate-day plant: flowers when days are neither too short nor too long; e.g., autumn coleus and sugar cane.

.013g. Long-day plant: one that flowers in response to long days (short nights; daylight approx. 11-16 h); may require only a single induction cycle (e.g., dill, white mustard, and spinach) or several cycles, being qualitative (e.g., oats, radish, and black-eyed susan); require vernalization (IV.3.2.021e), e.g., winter oats, sugar beet, winter wheat, and garden beet (low temperature); or quantitative [e.g., barley, winter rye, spring wheat; garden beets, lettuce, petunia (high temperature)].

.013h. Long-short-day plant: requires an alternation of long followed by short days, e.g., aloe, kalanchoe, and night-blooming jasmine (at 23°C).

.013i. Night interruption (night break): the inhibition of flowering in short-day plants and promotion in long-day plants by a light break during the night; a phytochrome-mediated response, with short-day plants generally more sensitive to the occurrence of the light break, its duration, and wavelength (red being more effective than far-red, whereas a mixture of the two is more effective with long-day plants).

.013j. Qualitative (absolute) response: inhibition of flowering (i.e., absolute response) by exposure to the wrong daylength or interruption of a photoperiod; [see Quantitative (facultative) response, IV. 3.2.013k].

.013k. Quantitative (facultative) response: progressive rather than absolute inhibition of flowering by a certain daylength or interruption of a photoperiod [see Qualitative (absolute) response, IV.3.2.013j].

.013l. Short-day plant: one that flowers in response to short days (or long nights; daylength is approximately 8 to 12 h); may require only a single inductive cycle (e.g., cocklebur, goosefoot, and duckweed) or several cycles for induction, being qualitative [e.g., chrysanthemum var., kalanchoe, maize, strawberry (at high temperature), and poinsettia (at high temperature)] or quantitative [e.g., chrysanthemum var., strawberry (at low temperature), sunflower, sugar cane, and onion (may require vernalization)].

.013m. Short-long-day plant: requires an alternation of short followed by long days, e.g., white clover, or orchardgrass, and Kentucky bluegrass (require vernalization or short days for induction and long days for development of the inflorescence).

.014 Photoreaction, photoreactive: reaction initiated or hastened by light.

.015 Photosynthetically active radiation (PAR): (see IV.0.157m; Photosynthetic efficiency, IV.3.1.020).

.016 Photothermal: pertaining to the combined effects of light and temperature.

.017 Phytochrome: light-absorbing pigment in plant tissues that control morphogenesis; two forms, red-absorbing, blue (P_r), with the maximum at about 667 nm and far-red-absorbing, greenish (P_{fr}) at about 727nm, with interconversion from one to the other under low irradiance levels and responses by red light almost always nullified by an intermediate or subsequent exposure to far-red light. (S & R)

.018 Q_I: (see Net radiation balance, IV.3.2.007).

.019 Stenothermic: plants that thrive only within a narrow but sometimes extreme temperature range, such as aquatic plants, particularly thallophytes (i.e., algae, bacteria, fungi, and lichens). (LA)
.020 Thermal induction: change in growth and development of plants brought about by a given temperature exposure. (S & R)
.021 Thermoperiodism: promotion of growth or development or both by alternating day and night temperatures; some plants, however, e.g., cocklebur, sugar beet, wheat, oats, bean, and pea, grow as well at an optimum constant temperature as they do when day and night temperatures vary. (S & R)
.021a. Cardinal temperatures: ranges about the genetically fixed norm within which the optimal, minimal, and maximal temperatures for the plant can shift as the plant adapts to environmental conditions. (S & R)
.021b. Devernalization: reversal of vernalization by immediate exposure to high temperatures (i.e., 15 to 30°C within four or five days or so, depending on the plant). (S & R)
.021c. Thermoperiod: the period of exposure of a plant to a particular temperature; specifically, that period characteristic of the diurnal alternation of day and night temperatures when both period and temperature are at or near the optimum for the induction of certain activities, such as growth or flowering. (B & K)
.021d. Thermoperiodicity: the sum of the responses of an organism to appropriately fluctuating temperatures. (AV)
.021e. Vernalization: the low-temperature promotion of flowering, which may be quantitative (facultative; e.g., result in faster flowering) or qualitative (absolute; e.g., flowering depends in an absolute sense on cold); the response may be inductive when flowering occurs subsequent to cold treatment, often being strongly promoted by long days, inhibited if plants are not exposed to cold, or noninductive, when flowering occurs during cold treatment, as with Brussels sprouts. (S & R)

IV.3.3. Tropism, nastic movements, nutation

.001 Circumnutation: (see Nutation, IV.3.3.004).
.002 Heliotrope: plant that turns toward sunlight.
.003 Nastic movements: movements of plant organs that take place in a direction independent of that of the stimulus, in contrast to tropisms in which the stimulus is directional; several forms. (B & C)
.003a. Chemonasty: movements in response to chemical stimulation (as in some carnivorous plants).
.003b. Haptonasty (thigmonasty): movements in reponse to touch; (e.g., tentacles of Drosera spp.).
.003c. Nyctinasty: the rhythmic movement of organs (e.g., flowers or leaves) in a diurnal cycle.
.003d. Photonasty: induced by changes in light intensity (as in opening and closing movements of developing leaves or flowers).
.003e. Seismonasty: movements induced by changes in turgor pressure (as in petals, stamens, stigmas, or leaves of certain plants); differs from ordinary wilting in occurring suddenly.
.003f. Thermonasty: induced directly or indirectly by heat (as in the opening or closing of flowers when brought from a warm to a

IV.3.3. HORTICULTURAL PHYSIOLOGY and CROP ECOLOGY

cold room or vice versa).
- .003g. Thigmonasty: (see Haptonasty, IV.3.3.003b).
- .004 Nutation: autonomic movements caused by growth irregularities not resulting from tropic or nastic stimuli, as in circumnutation, the circular pattern of "searching" movements of young tendrils. (B & C)
- .005 Phototaxis: movement of plant or organelle (e.g., chloroplasts) in response to light.
- .006 Seismomorphogenesis: (see Seismotropism, IV.3.3.008r).
- .007 Thigmomorphogenesis: (see Haptotropism, IV.3.3.008i).
- .008 Tropism: a movement that can be induced in various plant organs, such as stem, roots, and leaves, and takes place in a direction related to that of the stimulus. (B & C)
- .008a. Aerotropism: movement of a plant or organ in response to air from a given direction.
- .008b. Anisotropic: a plant or organ that assumes different positions in response to external stimuli (e.g., changes in light intensity).
- .008c. Autotropism: the tendency for an organ or plant to resume its original position once the stimulus is removed.
- .008d. Chemotropism: movement of a plant or organ in response to a chemical stimulus, being positive toward a higher concentration and negative toward a lower concentration.
- .008e. Diageotropic: the tendency of a plant part to assume a position at right angles to the direction of gravity.
- .008f. Diaphototropism: (see Plagiophototropism, IV.3.3.008p).
- .008g. Galvanotropism: movement of a plant or organ in response to an electric field from a given direction.
- .008h. Geotropism: the reaction of a plant or organ to gravity, being negatively geotropic if the growth or curvature is away from the center of the earth, positively geotropic if the growth of curvature is toward the center of the earth, as in main roots, or plagiotropic (diageotropic, transverse) if horizontal or obliquely downward.
- .008i. Haptotropism (thigmotropism): sensitivity to touch, as in tendrils and some other organs.
- .008j. Heliotropism: (see Phototropism, IV.3.3.008n).
- .008j-1. Hydrotropism: sensitivity to water, like the leaves of certain houseplants.
- .008k. Klinostat: an apparatus to determine the effect of gravity on a plant or plant part.
- .008l. Negatively geotropic: (see Geotropism, IV.3.3.008h).
- .008m. Orthophototropism: movement toward light (positive) or away from light (negative).
- .008n. Phototropism (heliotropism): curvature induced by unilateral light, positive if the organ bends toward the light or negative if it bends away from the light.
- .008o. Plagiogeotropism: (see Geotropism, IV.3.3.008h).
- .008p. Plagiophototropism (diaphototropism): movement perpendicular to the direction of the light stimulus, as of aerial roots or tendrils.
- .008q. Plagiotropic: (see Geotropism, IV.3.3.008h.).
- .008r. Seismotropism (seismomorphogenesis): effect of shaking on a plant.
- .008s Thermotropism: movement of a plant or organ in response to heat from a given direction.
- .008t. Thigmotropism: (see Haptotropism, IV.3.3.008i).

.008u. Transverse: (see Geotropism, IV.3.3.008h.).
.008v. Traumatotropism: movement of a plant in response to wounding applied in a given direction.

IV.4. Fruit Setting

IV.4.1. Amphimixis (syngamy), apomixis

.001 Agamospermy: (see Apomixis, IV.4.1.006).
.002 Aitionomic: the ability to develop parthenocarpic fruit only in response to a stimulus external to the ovary.
.003 Amphimixis: (see Syngamy, IV.4.1.026).
.004 Androgenesis: development in which the embryo contains only paternal chromosomes.
.005 Apoarchespory: (see Sporogenesis, IV.4.1.006b-6).
.006 Apomixis: reproduction of a plant without any form of sexual union, being obligate if by apomixis only or facultative if by apomixis or normal sexual reproduction. Basic subdivisions are agamospermy, reproduction by means of seeds, whether involving morphological alternation of generations (gametophytic apomixis) or without morphological alternation of generations (sporophytic apomixis or adventitious embryony), and vegetative reproduction, not by means of seeds. It should be pointed out that apomixis (agamospermy) is neither an evolutionary aberration (a so-called morphological "dead end") nor rare, having been found so far in more than 300 genera in 80 families. (KH) (Fig. IV.4. 1.01-.03.)
.006a. Apomictic processes (Sherman): (Fig. IV.4.1.01,.02)
.006a-1. Adventitious embryony: cells of the nucellus or integuments form embryos and continue the sporophytic generation. (Fig. IV.4.1.02b)
.006a-2. Apogamy: the formation of an embryo from a synergid or an antipodal (but not from the polar nuclei). The process usually requires the presence of the egg nucleus which may develop simultaneously into an embryo. [Fig. IV.4.1.02(a)(6)]
.006a-3. Apospory: the cell functioning as an embryo sac mother cell is derived from nonarchesporial tissue in the sporophyte, such as the nucellus or chalazal region of the ovule (the megaspore mother cell is normally derived from archesporial tissue). [Fig. IV.4.1.02(a)(4)]
.006a-4. Archesporial (recurrent) agamospermy: the megaspore mother cell produces a diploid gametophyte by diplospory and the gametophyte develops into a sporophyte by parthenogenesis, pseudogamy, or apogamy. (Fig.IV.4.1.02a)
.006a-5. Diplospory: the process by which one or more cells in the archesporial tissue (somatic chromosome number = $2n$) functions as a spore (embryo-sac mother cell directly to form a gametophyte, which is diploid ($2n$) because of the failure of normal meiosis. [Fig. IV 4.1.02a(3)]
.006a-6. Induced parthenogenesis: (see Parthenogenesis, IV.4.1.006a-10).
.006a-7. Monoploid (nonrecurrent) apomixis: a haploid ($1n$ gametophyte is produced by normal meiosis and develops into a haploid sporophyte by parthenogenesis, pseudogamy, or apogamy.

IV.4.1. HORTICULTURAL PHYSIOLOGY and CROP ECOLOGY

.006a-8. Natural vegetative reproduction: bulbs, bulbils, corms, tubers, layers, runners, pseudobulbs, and the like. [Fig. IV.4.1.01(12)]

.006a-9. Nonrecurrent apomixis: (see Monoploid apomixis, IV.4.1.006a-7).

.006a-10. Parthenogenesis: autonomous development of an embryo from the egg without the stimulus of pollination; "induced parthenogenesis" involves development of an embryo from the egg only, with the stimulus of pollination, whether the pollen tubes actually reach the embryo sac or not, as in Rubus; "pseudoparthenogenesis" involves the development of an embryo from the egg under the stimulus of foreign pollen. [Fig. IV.4.1.02a(5)]

.006a-11. Pseudogamy: parthenogenetic development of the female gamete (egg) which requires the stimulus of pollination but not the union of the gametes (fertilization); thus the embryo contains only nuclei derived from the female parent. Fertilization of the endosperm nuclei may be required for continued development of the embryo, as in citrus. [Fig. IV.4.1.02a(5)]

.006a-12. Pseudoparthenogenesis: (see Parthenogenesis, IV.4.1.006a-10).

.006a-13. Recurrent agamospermy: (see Archesporial agamospermy, IV.4.1.006a-4; Somatic agamospermy, IV.4.1.006a-14).

.006a-14. Somatic (recurrent) agamospermy: a cell from the nucellus undergoes somatic apospory to become a diploid ($2n$) gametophyte which develops into a diploid sporophyte (embryo) by parthenogenesis, pseudogamy, or apogamy. (Fig. IV.4.1.02a)

.006b. Apomictic processes (Khokhlov): (Fig. IV.4.1.03) (KH, SV)

.006b-1. Androgenesis: normal pollination, growth of the pollen tube, and separation of the pollen nuclei occur, but the egg nucleus degenerates prior to fusion, whereas the pollen nucleus which has entered the egg remains and divides as the embryo develops, as in hybrids of Nicotiana tabacum with N. langsdorfii and N. digluta. [Fig. IV.4.1.03(2)]

.006b-2. Apogamy: the formation of an embryo from a synergid (Liliaceae, orchids, etc.) or an antipodal (Allium, Eragrotis, and Alangium), but not from the polar (central) nuclei; the process usually requires the presence of the egg nucleus, which may develop simultaneously into an embryo; only mature embryos of the synergid type have been found so far. [Fig. IV.4.1.03(10,11)]

.006b-3. Parthenogenesis: autonomous development of embryos from the egg without the stimulus of pollination, as in Taraxacum.

.006b-3-a. Induced: development of an embryo from the egg cell requires the stimulus of pollination, whether the pollen tubes actually reach the embryo sac or do not, as in Potentilla, Rubus, Rudbeckia, Ranunculus, and Poa.

.006b-3-b. Pseudo: development of an embryo from the egg under the stimulus of foreign pollen.

.006b-4. Pseudogamy: normal pollination, growth of the pollen tube, and separation of the pollen nuclei occur but the pollen nucleus degenerates before or after entering the egg prior to fusion with the egg nucleus; the latter divides so that the embryo contains only nuclei derived from the female parent (the reverse of androgenesis, in which the embryo contains only male nuclei); fusion of the second pollen nucleus with the polar nuclei and subsequent development of

IV.4.1. HORTICULTURAL PHYSIOLOGY and CROP ECOLOGY

.006b-5.
: endosperm is an essential condition in the process, as in Potentilla. [Fig. IV.4.1.03(9)]

.006b-5. Semigamy (semigamic apomixis): normal pollination, growth of the pollen tube, and separation of the pollen nuclei to travel to the polar (endosperm) nuclei and egg occur but fusion does not take place in the latter; rather a binucleate zygote develops with independent divisions of egg and pollen nuclei so that the embryo has a mixture of cells with male derived and female derived nuclei. [Fig. IV.4.1.03(8)]

.006b-6. Sporogenesis: the development of a gametophyte from the megaspore mother cell (megasporocyte); this process consists of two divisions, heterotypic and homotypic; further development of the gametophyte involves three mitotic divisions except in the Eragrostis type of apospory and the Chloris and Panicum types of apoarchespory (with more than one megasporocyte, of which only one is functional) in which the embryo sac contains only four instead of the usual eight nuclei; also more than one embryo sac may be formed, as in Poa, in which embryo sacs with 2n and 1n nuclei occur together.

.006b-6-a. Aneuspory: sporogenesis that involves abnormal heterotypic (meiotic) and homotypic (mitotic divisions); four types, of which three have unreduced (2n) nuclei and the other is reduced (1n) but doubles later.

.006b-6-a-1. Apohomotypic: the heterotypic division is normal, i.e., two cells with 1n nuclei are formed, but the homotypic division is replaced by a prolonged resting stage from which an embryo sac is formed later, along with a doubling of the chromosome number to 2n in the nuclei, as in Datura. [Fig. IV.4.1.03(3)]

.006b-6-a-2. Mitotic: the heterotypic division is eliminated but the homotypic (mitotic) division does occur, giving a gametophyte with unreduced (2n) nuclei from one of the two cells of the latter. [Fig. IV.4.1.03(6)]

.006-6-a-3. Pseudohomotypic: the heterotypic division is abnormal and results in two cells with unreduced (2n) nuclei that do not undergo homotypic division; a gametophyte develops from one cell; the other degenerates. [Fig. IV.4.1. 03(5)]

.006b-6-a-4. Semiheterotypic: the heterotypic division is abnormal and results in only one cell with a 2n nucleus, but the homotypic division is normal; a gametophyte with unreduced (2n) nuclei is produced from one of the two cells from the homotypic division. [Fig. IV.4.1.03(4)]

.006b-6-b. Apospory: the megaspore mother-cell (megasporocyte) functions directly as the gametophyte (embryo sac), which may be generative or somatic in origin. [Fig. IV.4.1.03(7)].

.006b-6-c. Euspory: typical (normal) sporogenesis involves reduction of chromosome numbers to 1n in the heterotypic division, production of four cells in the homotypic division, degeneration of three of the latter, and the development of a single embryo sac with 1n nuclei from the fourth cell.

.007 Autonomic: the ability to set fruit parthenocarpically without the

IV.4.1. HORTICULTURAL PHYSIOLOGY and CROP ECOLOGY

stimulus of pollination.
.008 Centromere: the specialized chromosome or region of the chromosome to which the spindle fiber appears to be attached when the chromosomes are separating during the anaphase stage of cell division.
.009 Chromonemata: plural of chromonema, chromosome thread; subdivisions of chromatids (III.2.2.025); forerunners of chromatids in the succeeding cell division.
.010 Crossing-over: the interchange of corresponding segments of genetic material between two homologous chromosomes.
.011 Crossover unit: a measurement of the degree of linkage or crossover value between two genes expressed as a percentage.
.012 Cytoplasmic male sterility: type of male sterility conditioned by the cytoplasm rather than by nuclear genes and transmitted only by the maternal parent.
.013 Eugamy: (see Syngamy, IV.4.1.026).
.014 Fertilization: the process of fecundation which consists of the union of one of the male gametes (sperm) contained in the pollen grain with the female gamete (egg) in the ovule; the resulting zygote is $2n$; endosperm (II.2.3.3.015) results from fusion of the other male gamete in the pollen grain with the two polar nuclei so that this tissue has a $3n$ chromosome complement.
.015 Heterofertilization: double fertilization in which sperm from different pollen tubes fertilize the egg and polar (primary endosperm) nuclei.
.016 Heterotypic: the first meiotic division of a germ cell (i.e., meiosis, III.2.2.105).
.017 Homotypic: mitotic division of a germ cell (i.e., mitosis, III.2.2.110).
.018 Interkinesis: an abbreviated resting stage between the first and second meiotic divisions; analogous to premitotic interphase, although no DNA is replicated.
.019 Karyokinesis: division of the nuclear material during cell division.
.020 Oomorphogenesis: evolution of the egg.
.021 Parthenogenesis: (see IV.4.1.006a-10, IV.4.1.006b-3).
.022 Plastogamy: cytoplasmic fusion without nuclear fusion.
.023 Plastomere: the cytoplasmic portion of a sperm cell.
.024 Proembryo: the young embryo in the first stages of development.
.025 Pseudogamy: (see Apomixis, IV.4.1.006a-11; IV.4.1.006b-4).
.026 Syngamy (amphimixis, eugamy): the normal sexual process in which the two pollen nuclei (male gametes) fuse with the egg (female gamete) and polar nuclei, respectively. The asexual or spore-forming generation in the life cycle of plants is normally diploid. Normal morphological alternation of generations begins with the formation of the megaspore mother cell (and microspore mother cells to produce pollen) from tissues of the parent sporophyte. The megaspore mother cell undergoes two divisions (one meiotic and one mitotic) to produce four cells, three of which disintegrate in the development of the gametophyte ($1n$). The latter then undergoes three successive mitotic divisions to produce eight cells (nuclei), of which one is the egg (female gamete) in the embryo sac. Pollination, penetration of the pollen tubes into the embryo sac, and fusion of one pollen nucleus with the egg (fertilization), the other with two polar nuclei, give rise to the embryo and endosperm, respectively; the embryo is the filial sporophyte which in its turn will develop into a mature plant and repeat the cycle. (Fig. IV.4.1.01)
.027 Vegetative reproduction: (see Apomixis, IV.4.1.006).

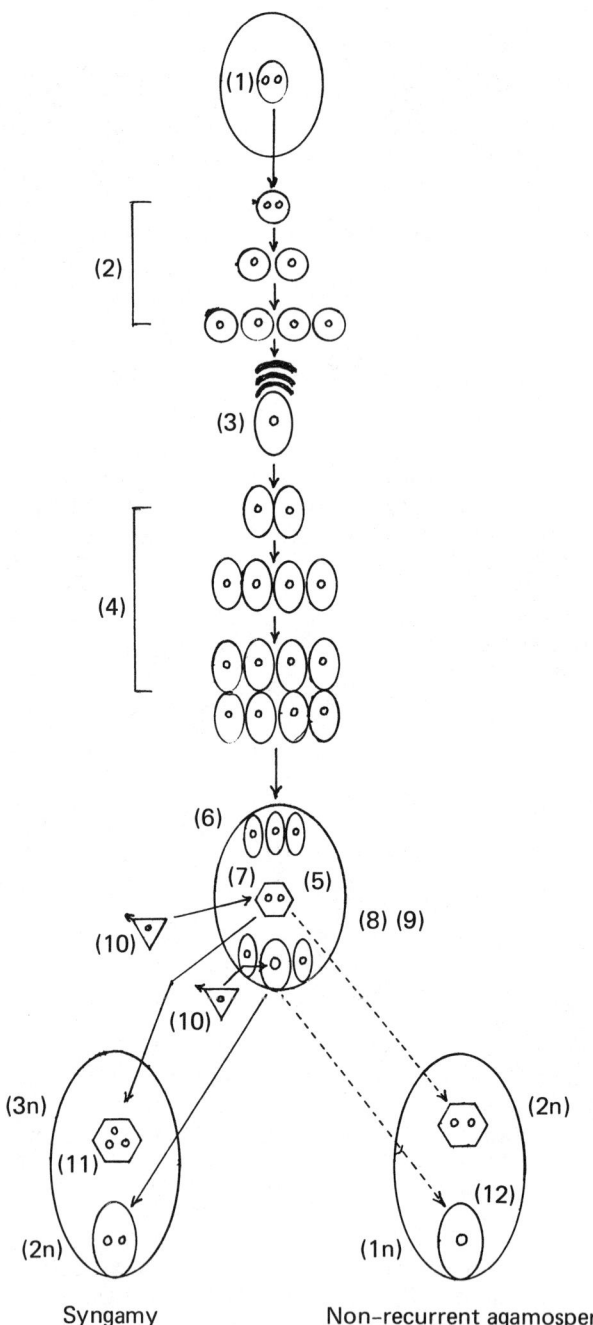

Fig. IV.4.1.01. Syngamy and nonrecurrent (gametophytic) agamospermy: (1) archesporial cell in nucellus, (2) heterotypic (meiotic) divisions, (3) megaspore mother cell (three degenerated cells above), (4) homotypic (mitotic) divisions, (5) embryo sac = gametophyte with (6) three antipodals, (7) two polar nuclei, (8) two synergids and (9) egg, (10) pollen nucleus, (11) normal embryo, (12) haploid embryo. Adapted from Soule and Sherman, 1978; see Acknowledgments and Selected References.

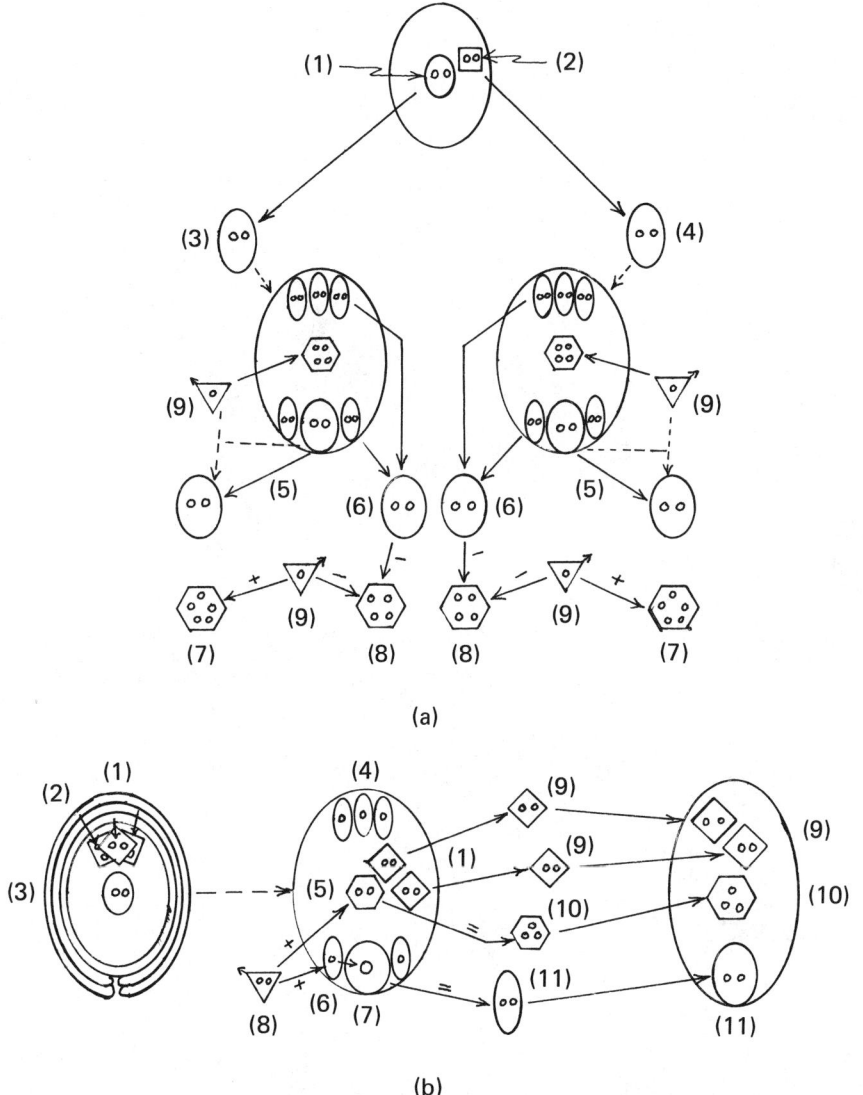

Fig. IV.4.1.02. Recurrent (gametophytic) and sporophytic agamospermy: (a). recurrent (gametophytic) agamospermy, (1) archesporial cell, (2) somatic cell, (3) gametophyte developed by diplospory, (4) gametophyte developed by apospory, (5) embryo developed from egg by parthenogenesis or pseudogamy, (6) embryo developed from synergid or antipodal by apogamy, (7) fertilized polar nuclei (by pseudogamy or apogamy), (8) nonfertilized polar nuclei; (b) sporophytic agamospermy whereby embryos develop adventitiously from (1) nucellar or (2) integumental cells concurrently with normal syngamy (the sexual embryo may be suppressed later, however), (3) archesporial cell, (4) antipodals, (5) embryo sac, (6) synergids, (7) egg, (8) pollen, (9) nucellar embryos, (10) fertilized polar nuclei, (11) fertilized egg = sexual embryo. Adapted from Soule and Sherman, 1978; see Acknowledgments and Selected References.

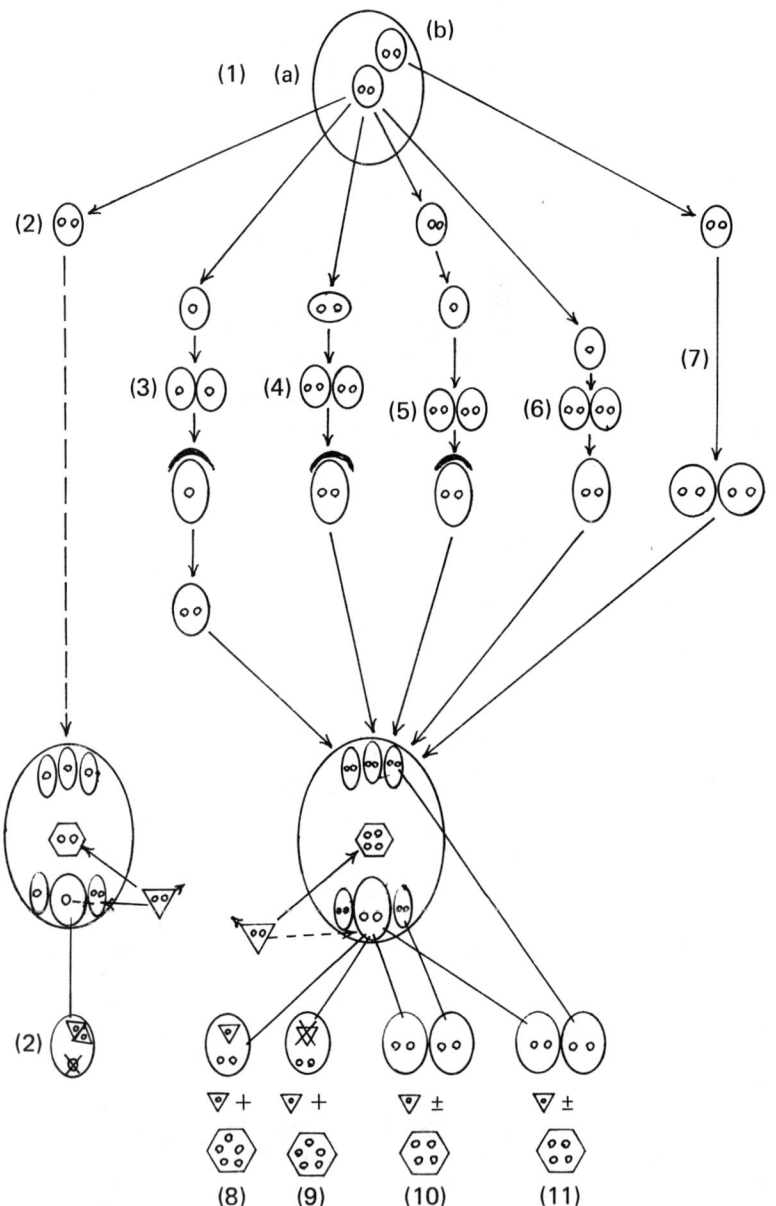

Fig. IV.4.1.03. Additional types of gametophytic agamospermy (apomixis): (1) nucellus with (a) archesporial and (b) somatic cells, (2) normal development of gametophyte but degeneration of egg prior to fertilization = androgenesis; (3) apohomotypic, (4) semiheterotypic, (5) pseudohomotypic, (6) mitotic aneuspory, (7) apospory leading to (8) semigamy, (9) pseudogamy, (10) and (11) apogamy. Adapted from Solntseva, 1976; see Acknowledgments and Selected References.

IV.4.2. Carpel development (GP-G)

.000 Carpel development: the process that involves formation of a fruit and seeds from a gynoecium and its parts, in the case of a simple fruit, or gynoecia and their parts, in the case of aggregate, accessory, or multiple (compound) fruits. Terms used to distinguish among the various influences of pollination or its lack on fruit set and seed formation.
.001 Automatic parthenocarpy: fruit without seeds are formed in the absence of pollination, but with pollination the following applies:
.001a. Complete: fruit without seeds are formed in the Cavendish types of banana which are both male-sterile and female-sterile.
.001b. Strong I: fruit without seeds are formed unless the pollenizer is a wild diploid (AA, AB, or BB) type for 'Gros Michel' bananas (AAA).
.001c. Strong II: fruit with rarely a few seeds may be formed in 'Tahiti' lime (like 'Gros Michel', incompletely but usually highly female-sterile).
.001d. Weak: fruit with an occasional seed may be found as a result of incomplete female sterility, as in 'Washington' navel sweet orange or 'Marsh' grapefruit.
.002 Conditional parthenocarpy: no fruit is formed without pollination but fruit without seeds develops if pollen of Ampelopsis is used for certain Vitis vinifera cultivars.
.003 Spermocarpy: fruit and seed are formed if pollination and fertilization take place, but neither develops if there is no pollination.
.004 Stenospermocarpy: neither fruit nor seed develops in the absence of pollination, but the following types occur with pollination: (a) normal fruit and aborted seeds occur in Vitis ('Sultanina'), (b) abnormal fruit ("cukes") and aborted seeds in avocado, (c) normal fruit and seeds with nucellar embryos as a result of single fertilization (fusion nucleus only) in (polyembryonic) mango, and (d) normal fruit and seeds with the formation of zygotic embryos which are usually crowded out by nucellar embryos in nonparthenocarpic polyembryonic citrus.
.005 "Transitional" parthenocarpy: initiation of fruit and adventive embryos occur in the absence of pollination, but fruit and seed maturation take place when flowers are pollinated, as in Eugenia jambos.

IV.4.3. Dichogamy, dicliny, hercogamy, heterostyly

.001 Androdioecious: staminate flowers on one plant and bisexual flowers on another plant of the same species. (L '55)
.002 Dichogamy: prevention of natural self-pollination in an individual flower by separation of pollen dehiscence and stigma receptivity in time (i.e., the organs are not functional simultaneously). (F & P)
.002a. Metandry: (see Protogyny, IV.4.3.002c).
.002b. Protandry (proterandry): "complete protandry" is the condition in which pollen is shed before the stigma is receptive in the same flower, as in Annona spp.; "incomplete protandry" refers to the condition in which pollen shedding begins before the stigma is receptive but continues afterward so that self-pol-

lination can take place, as in some Passiflora spp. (F & P)
.002c. Protogyny (proterogyny): the condition in which the stigma is receptive before pollen is shed from stamens on the same flower. (F & P)
.002d. Second-order dichogamy: the condition in which individual (monoclinous) flowers in an inflorescence are effectively unisexual by separation in time of pollen dehiscence and stigma receptivity but the two stages are not synchronized, thus allowing allogamy (cross pollination) but not autogamy (F & P). An analogous situation is one in which (a) where staminate and pistillate (diclinous) flowers develop in the same inflorescence but protandry is complete as in tall coconuts, thus necessitating cross-pollination between palms; however, protandry is incomplete in dwarf coconuts and self-pollination does occur; or (b) oil palm in which inflorescences have all staminate or all pistillate flowers but are produced in alternating cycles of a few to several months, thus preventing natural pollination to the extent that hand pollination is used in some areas (e.g., Sumatra) to obtain proper fruit set.
.002e. Synchronous dichogamy: (see Synchronous protogyny, IV.4.3.002f).
.002f. Synchronous protogyny (synchronous dichogamy): the condition in which all of the flowers opening at the same time on the same seedling or all trees of the same cultivar are first functionally pistillate and later become functionally staminate, as in avocado.
.003 Dimorphic: styles of two lengths, long and short, in the flowers, as in the Primulaceae, Rubiaceae, Boraginaceae, Oleaceae, Plumbaginaceae, Oxalidaceae, Polygonaceae, Turneraceae, Labiatae, Verbenaceae, and Gentianaceae. (F & P)
.004 Dioecious: the condition in which unisexual flowers, staminate and pistillate, are borne on separate plants, as in date, nutmeg, and many Flacourtiaceae; certain crops have flowers that are functionally dioecious, as in allspice. (L'55)
.005 Effective bloom period: the length of time a plant is in conspicuous bloom.
.006 Hercogamy (herkogamy): spatial separation of anthers and stigma in a flower; second-order hercogamy may occur in certain inflorescences in which protandry or protogyny creates a zone of functionally unisexual flowers. (F & P)
.007 Heterogamy: androecium and gynoecium mature in a flower at different times; synonymous with dichogamy (IV.4.3.002). (F & P)
.008 Heterogeny: (see Trimorphism, IV.4.3.019).
.009 Heterostyly: existence of styles of different length (i.e., long styles and short stamens or the reverse = "thrum" and "pin") in flowers, which prevents self-pollination; (see also Trimorphism, IV.4.3.019). (F & P)
.010 Homogamy: androecium and gynoecium mature in a flower at the same time; the opposite of dichogamy (IV.4.3.002). (F & P)
.011 Homostylic dimorphism: behavior as to fruit set is like that of plants with dimorphic heterostyled flowers but style length is the same, as in Armeria. (F & P)
.012 Illegitimate pollination: pollination of the same form of heterostyled flowers (e.g., stigma and anthers are not in the same relative position; i.e., thrum X thrum or pin X pin), whether on the same or different plants. (F & P)
.013 Legitimate pollination: pollination of different forms of heterostyled flowers where stigma and anthers are in the same relative

IV.4.3. HORTICULTURAL PHYSIOLOGY and CROP ECOLOGY

position, (i.e., thrum and pin), as in normal cross-pollination. (F & P)
.014 Monoecious: staminate and pistillate flowers borne on the same plant in one of three configurations: (a) flowers in the same inflorescence; e.g., tung, coconut, and rubber; (b) flowers in separate inflorescences produced at the same time; e.g., maize and Artocarpus; (c) flowers in separate inflorescences produced sequentially; e.g., oil palm. [Note that (c) exhibits second order dichogamy and (b) may also.]
.015 Polygamodioecious: the condition in which plants are functionally dioecious but there are a few flowers of the opposite sex or a few bisexual flowers appear on all plants during the bloom period. (L '55)
.016 Polygamomoneocious: the condition in which the inflorescences are basically monoecious but a few bisexual flowers appear on all plants during the bloom period. (L '55)
.017 Polygamous: the condition in which an inflorescence contains unisexual (usually staminate) and bisexual flowers, as in mango, cashew, and kola. (L '55)
.018 Pseudohermaphrodite: functional unisexualism in a plant that bears apparently perfect flowers, as in allspice, a condition quite different from and not to be confused with dichogamy, hercogamy, and the like. (F & P)
.019 Trimorphism (tristyly): styles of three lengths, long, medium and short, among the flowers, as in Lythraceae, Linaceae, Oxalidaceae, Pontederiaceae, and Connaraceae; = heterogeny. (F & P)
.020 Trioecious: individual plants of the same species (or cultivar) that bear staminate, pistillate, or perfect flowers, as in Carica papaya.
.021 Tristyly: (see Trimorphism, IV.4.3.019).

IV.4.4. Pollination

.000 Pollination: the transfer of pollen from an anther to a stigma, in which there are three phases: (a) the release of pollen from an anther, (b) transfer of pollen to a stigma, and (c) successful placement of pollen on the receptive surface of the stigma, followed by germination of the pollen grains (growth of the pollen tubes and their entry into the ovule are considered part of the last phase). (F & P)
.001 Allophilic blossom: one accessible for pollination by any visitor, there being no morphological adaptations unique to a specific pollinator. (F & P)
.002 Anthesis: flowering; strictly speaking, the time of flower expansion when pollination takes place. (F & P)
.003 Attractants: pollen as food, odor, nectar, perianth color, etc. (F & P)
.004 Blossom types (exclusive of those closed during anthesis): (F & P)
.004a. Inconspicuous: Betulaceae, Urticaceae, Zostera, etc.; plants with abiotic pollination plus mignonette, grapes, and certain night-pollinated blossoms characterized by small corollas and dull color.
.004b. Conspicuous: individual flowers of inflorescence are larger; several types that intergrade or overlap.
.004b-1. Bell-or funnel-shaped: Campanula (bell), Calystegia, Zantede-

schia (funnel); narrow or broad rim serves as a lighting platform and with a central column of anthers and gynoecium.

.004b-2. Dish- to bowl-shaped: Caltha, Rosa, Tulipa silvestris representative of former; composite heads (Chrysanthemum), umbels (Chenopodium, Viburnum), Magnolia, and Passiflora of latter; more diffuse and lower anthers and gynoecium than in bell or funnel; pollination of mess-and-soil type (IV.4.5.002j).

.004b-3. Flag: androecium and gynoecium on lower side of pollination unit; e.g., Papilionaceae (Fabaceae), Corydalis, Pelargonium, Dictamnus (primitive), and Pedilanthus.

.004b-4. Gullet: Labiatae, Scrophulariaceae, basitonic orchids (Ophyrideae) in which androecium and gynoecium are restricted to upper side of pollination unit; Iris (three units), Acanthus (parts of calyx), Centrosema (flag inverted), Ocymum (gullet reversed, approaching flag type), Mimetes hartogii (inflorescence functions as gullet); landing place is lower lip; also visited by hovering animals.

.004b-5. Head- or brush-shaped: Thalictrum, Salix, Phyteuma, Eupatorium cannabinum, and many Proteaceae; visited by fluttering pollinators or those that alight.

.004b-6. Tube: location central as in a corolla or to one side as in a spur; some tubes lack a rim (hovering visitors) or have a rim (visitors can alight); sometimes combined with other types, as in orchids.

.005 Breba: the first (spring) crop of syconia on the edible fig tree; (see Caprification, IV.4.4.007). (CN)

.006 Bud (pollination): artificial or natural pollination of a flower while still in the unopened or bud stage. (F & P)

.007 Caprification: the process of pollination of the fig flower, in which the blastophaga wasp transfers pollen from the wild fig (caprifig) to the stigmatic surfaces of the long-styled pistillate flowers in the syconia of the edible fig; caprification is required for the Smyrna and main (second) crop of San Pedro types, whereas fruit set is parthenocarpic in the common type; both staminate and short-styled pistillate flowers are present in caprifig syconia but protogyny precludes self-pollination; wasps overwinter as larvae in the "mamme" crop on the caprifig trees, emerge as adults, fly to the "profichi" crop, which develops on the preceding year's or older wood, and also to the "breba" (first) crop on edible fig trees, where they effect pollination but deposit their eggs in the "profichi" figs; these eggs hatch, emerge as adults, fly to the main (second) crop of edible figs to effect pollination, but return to lay their eggs in the "mammoni" syconia of the caprifig; adults emerge and lay eggs in the "mamme" syconia to complete the cycle. (CN)

.008 Caprifig: the wild fig in which the blastophaga wasp lives; (see Caprification, IV.4.4.007). (CN)

.009 Carrion flower: emits a carrion odor to attract certain pollinators like carrion flies and beetles. (F & P)

.010 Chalazogamy: penetration of the pollen tube through the chalazal end of the ovule. (EA)

.011 Chasmogamous: opening of the perianth at maturity for the purpose of pollination; the opposite of cleistogamous (IV.4.4.012). (F & P)

.012 Cleistogamous, cleistogamy: self-pollination or self-pollinating

IV.4.4. HORTICULTURAL PHYSIOLOGY and CROP ECOLOGY

    nature of closed flowers. (F & P)
.013 Deceptive device: texture, color, form, and/or smell utilized in the production and display of attractants but of no advantage to pollinators. (F & P)
.014 Depth effect: the attraction of more advanced pollinators such as bumblebees to deeper types of flowers. (F & P)
.015 Diurnal: opening during the day and closing at night, like many flowers; active by day like many animals; (see Nocturnal, IV.4.4.038; Crepuscular, IV.4.5.002c).
.016 Effective pollination period; (see IV.4.8.021).
.017 Elaiophore: plant that secretes oil or other fatty substances from special glands in or near the blossom which acts as an attractant for pollinators. (F & P)
.018 Emasculate: artificial removal of predehiscent (immature) stamens, often including petals and sepals prior to hand pollination.
.019 Euanthium: the flower as a unit for pollination. (F & P)
.020 Euphilic blossom: inflorescence (pseudanthium), flower or part of a flower (meranthium), in which pollination is restricted to highly specialized, sometimes unique agents. (F & P)
.021 Flagelliflory: flowers borne on long dangling stalks, as in Adansonia, Parkia, Marcgravia, Kigelia, Musa, and Eperua, thus adapting them to certain visitors. (F & P)
.022 Geitonogamy: pollination by means of pollen coming from another flower on the same seedling or clone. (F & P)
.023 Guiding structure: ribs, hairs, etc., which serve to make pollinators follow a certain route when visiting a flower. (F & P)
.024 Haplomorphic: flower with open bowllike shape. (F & P)
.025 Haptogamy: a form of geitonogamy (IV.4.4.022) in which the flowers touch one another. (F & P)
.026 Hemiphilic: flower adapted to use by visitors of an intermediate degree of specialization. (F & P)
.027 Heterandry (heteranthery): differentiation within an androecium between "feeding" anthers to furnish pollen for consumption and fertilization ("pollination") anthers, as in Lagerstroemia indica and Verbascum thapsus; Cassia also has a third type as a landing place for insect visitors. (F & P)
.028 Hypocraterimorphous: flower of the dish type with a funnel. (F & P)
.029 Intraovarian: a type of artificial pollination in which pollen is inserted directly into an ovary rather than on the stigma. (F & P)
.030 Light window: whitish or thin areas in an otherwise dark-colored corolla, as in Gentiana acaulis, whose function is to let in light attracting insects so they may see while crawling around the interior of the flower. (F & P)
.031 Main crop (figs): the second crop of syconia which is borne on wood of the current season in the edible fig; (see Caprification, IV.-4.4.007). (CN)
.032 Mamme: the syconia of the caprifig in which the blastophaga wasp overwinters; (see Caprification, IV.4.4.007). (CN)
.033 Mammoni: the second (summer) crop of syconia on the caprifig; (see Caprification, IV.4.4.007). (CN)
.034 Meranthium: (see Euphilic blossom, IV.4.4.020).
.035 Monophilic blossom: regular pollination is by representatives of a single species of visitor. (F & P)
.036 Nectar: sweet liquid secreted by the nectaries of a plant, often a means of attracting pollinators. (ES)
.037 Nectariferous: having nectaries or secreting nectar. (ES)

IV.4.4. HORTICULTURAL PHYSIOLOGY and CROP ECOLOGY

.038 Nocturnal: night activity; said of flowers opening or animal activity at night.
.039 Olfactoric: attraction of pollinators by scent. (F & P)
.040 Oligophilic blossom: regular pollination is by representatives of one major taxon of visitor. (F & P)
.041 Penduliflory: (see Flagelliflory, IV.4.4.021).
.042 Perfume blossoms: those in which the primary attractant, odor (which may often be highly offensive to man), is alone and not a lead to a more rewarding (from the pollinator's viewpoint) attractant. (F & P)
.043 Pioneer pollen: apple trees pollinated twice with a three-day interval between pollinations, with the result that the pollen applied the second time accounts for twice as many zygotes as the pollen applied first; the pollen applied first is denoted "pioneer pollen" because it seems to make it easier for the second pollen to function.
.044 Pleomorphic: an actinomorphic flower with a few axes of symmetry recognizable by more highly evolved pollinators. (F & P)
.045 Pollenizer: the cultivar used to furnish pollen.
.046 Pollen tube: the structure produced by the pollen grain when germinated on the stigma and which grows down the style (if one is present) and normally into the embryo sac via the micropyle; three nuclei are present, the pollen tube nucleus and two sperm; the latter pass through the pollen tube and one unites with the egg in the process of fertilization; the other combines with the polar nuclei in the process of triple fusion. (ES)
.047 Polyphilic: flowers pollinated by many different taxa of visitors. (F & P)
.048 Porogamy: penetration of the pollen tube through the micropyle, the usual avenue of entry into the ovule; (see also Chalazogamy, IV.4.4.010). (EA)
.049 Profichi: the first (spring) crop of syconia on a caprifig tree in which the blastophaga wasp lays its eggs; (see Caprification, IV.4.4.007). (CN)
.050 Pseudanthium: inflorescence; (see Euphilic blossom, IV.4.4.020).
.051 Recognition pollen: cultivar A is normally self-incompatible; therefore a mixture of pollen from it and killed pollen of cultivar B is used as a source in an attempt to force cultivar A to "recognize" cultivar B's pollen and accept its own pollen to effect fruit set.
.052 Self-pollination: strictly the transfer of pollen from the anthers of a flower of a cultivar (or seedling) to the stigma of the same flower. (F & P)
.053 Semitrap flower: pollinators get in through a conspicuous opening, but the interior of the flower is constructed to prevent them from leaving by any but a certain route, during the course of which they pick up or have pollen thrown on them. (F & P)
.054 Stereomorphic: flower with depth effect (IV.4.4.014). (F & P)
.055 Trap flower: similar to a semitrap flower, except that emergence of the pollinator is delayed until the flower changes configuration (if the flower does not the insect is trapped and usually dies). (F & P)
.056 Window opening: circular or other-shaped openings in a flower usually designed to force a visitor to follow a prescribed route, often into a semitrap, in the process of pollination. (F & P)

IV.4.5. Pollinators, pollinator adaptations (F & P)

.001 Pollinators (pollination agency): the means by which pollen is transferred from an anther to a stigma.
.001a. Abiotic pollinators: wind (anemophily), water (hydrophily) on the surface (ephydrophily) or under the surface (hyphydrophily), and gravity (geitonogamy).
.001b. Anemophily: (see Abiotic pollinators, IV.4.5.001a).
.001c. Biotic: pollination by animals, such as insects (entomophily), including beetles (cantharophily), dung and carrion beetles (coprophily, coprocantharophily), flies (myophily), dung and carrion flies (sapromyophily), bees (melittophily), small bees (micromelittophily), ants (myrmecophily), butterflies (psychophily), moths (phalaenophily), snails and slugs (malacophily), birds (ornithophily), and bats (chiropterophily).
.001d. Blastophaga wasp: Blastophaga psenes, the pollinator for edible fig (Ficus carica L.); (see Caprification, IV.4.4.007).
.001e. Cantharophily: (see Biotic pollinators, IV.4.5.001c).
.001f. Chiropterophily: (see Biotic pollinators, IV.4.5.001c).
.001g. Coprocanthrophily: (see Biotic pollinators, IV.4.5.001c).
.001h. Coprophily: (see Biotic pollinators, IV.4.5.001c).
.001i. Entomophily: (see Biotic pollinators, IV.4.5.001c).
.001j. Ephydrophily: (see Abiotic pollinators, IV.4.5.001a).
.001k. Geitonogamy: (see Abiotic pollinators, IV.4.5.001a).
.001l. Hydrophily: (see Abiotic pollinators, IV.4.5.001a).
.001m. Hyphydrophily: (see Abiotic pollinators, IV.4.5.001a).
.001n. Malacophily: (see Biotic pollinators, IV.4.5.001c).
.001o. Melittophily: (see Biotic pollinators, IV.4.5.001c).
.001p. Micromelittophily: (see Biotic pollinators, IV.4.5.001c).
.001q. Myophily: (see Biotic pollinators, IV.4.5.001c).
.001r. Myrmecophily: (see Biotic pollinators, IV.4.5.001c).
.001s. Ornithophily: (see Biotic pollinators, IV.4.5.001c).
.001t. Phalaenophily: (see Biotic pollinators, IV.4.5.001c).
.001u. Psychophily: (see Biotic pollinators, IV.4.5.001c).
.001v. Sapromyophily: (see Biotic pollinators, IV.4.5.001c).
.002 Pollinator adaptations:
.002a. Anthophilous: animal that feeds on or lives among flowers.
.002b. Constancy: individual quality in a polytropic (IV.4.5.002p) animal that is physiologically, physically, and ethologically able to and does visit any of a number of species; but the individual restricts its visits to a single plant species over a longer or shorter period.
.002c. Crepuscular: active in the twilight, like certain birds or insects.
.002d. Dystropic: unadapted or counteradapted visitation of an animal, frequently destructive but sometimes effecting pollination.
.002e. Eutropic: fully adapted blossom visitors.
.002f. Fidelity: behavior of an individual visitor (insect) to a flower.
.002g. Gastrilegic: insects (leaf-cutter bees) that collect pollen ventrally.
.002h. Hemitropic: visitor of an intermediate degree of specialization to flowers imperfectly adapted for its utilization.
.002i. Melliferous: producing or bearing honey.
.002j. Mess-and-soil: visitors (insects) trample around in feeding and accidentally pick up pollen, as in dish- to- bowl-shaped flow-

IV.4.5. HORTICULTURAL PHYSIOLOGY and CROP ECOLOGY

.002k.
ers, or have pollen thrown on them, as in Cytisus or Arum.

.002k. Monotropic (monolectic): pollinator that visits only one or some closely related species of plant.

.002l. Nectarivorous: an animal that feeds on nectar produced by plants (or certain animals on plants) with consequent pollination in many instances.

.002m. Nototribic: pollen carried or distributed by a flower, on the back of an insect (see Sternotribic, IV.4.5.002q).

.002n. Oligophagous: animal that eats only a few specific kinds of food like an insect that feeds on only one or a few species of plant.

.002o. Podolegid: insects that collect pollen on their feet.

.002p. Polytropic (polylectic): insects, etc., that visit many different taxa of plants.

.002q. Sternotribic: pollen carried, or distributed by a flower, on the underside of an insect.

.002r. Univoltine: having but one generation or brood per year; commonly applied to insects.

IV.4.6. Polyembryony

.000 Polyembryony: strictly, the presence of more than one embryo in an ovule or seed; embryos may be produced by amphimictic or apomictic processes or a combination.

.001 Gametophytic: two or more embryos formed from cells of the embryo sac or other nonsomatic tissue.

.001a. Cleavage (simple): division of a zygote or young embryo to give embryos of sexual origin.

.001a-1. Euploid: cleavage that involves different ploidy levels, including cases in which monosomic ($2n - 1$), disomic ($2n - 2$), etc., conditions are involved.

.001b. Compound: two or more embryos that result from simultaneous development of a zygote and an apogamic embryo with or without cleavage.

.001b-1. Euploid: compound but involving different levels of ploidy or monosomic, etc., conditions.

.001c. False: single embryos of zygotic (or apogamic) origin are developed but each comes from a separate embryo sac within an ovule, as in certain Prunus.

.001d. Multiple: several embryos, including those developed by apogamy, are produced from a single megaspore mother-cell.

.002 Complex: the presence of two or more embryos of recognizable gametophytic and sporophytic origin.

.003 Monoembryonic, monoembryony: the presence of only one embryo in an ovule or seed.

.004 Sporophytic: embryos are developed from somatic cells (e.g., the nucellus or integuments) by adventitious embryony.

IV.4.7. Fruit dispersal (PJ)

.001 Fruit dispersal agencies:
.001a Anemochory: wind.
.001b. Autochory: plant.

.001c. Barochory: weight.
.001d. Chiropterochory: bats.
.001e. Hydrochory: water.
.001f. Ichthyochory: fish.
.001g. Mammaliochory: mammals.
.001h. Ornithochory: birds.
.001i. Saurochory: reptiles.
.002   Diaspore: the plant or plant part to be spread; ranges among spermatophytes from spores, embryos, and seeds to simple, aggregate, spurious, or multiple fruits, seeds in whole plants or parts thereof, or vegetative pieces of the old plant.
.003   Dyszoochory: agency that destroys diaspores in feeding.
.004   Endozoochory: agency that swallows diaspores and destroys or voids them.
.005   Epizoochory: agency that transports diaspores on the outside, as in feathers or fur or clinging to feet.
.006   Frugiverous: animal that feeds mainly or exclusively on fruit; usually applied to birds, bats, and the like.
.007   Synzoochory: agency that transports diaspores in the beak or mouth.

IV.4.8. Other terms

.001   Aberration: nontypical form or function; some abnormality of an organism or part of a biological process.
.002   Aborted: (see Abortive, IV.4.8.003).
.003   Abortive: defective; barren; imperfectly developed; = aborted, as in failure of seeds to develop normally in otherwise sound fruit.
.004   Allogamy: (see Cross-pollination, IV.4.8.017).
.005   Androgentic: genetic behavior specific to the androecium (e.g., male sterility in a flower).
.006   Asexual: not involving germ cells or fusion of nuclei; said of reproduction or of an individual using this mode of reproduction.
.007   Autogamy: strictly, pollination within the same flower. (F & P)
.008   Barren: unproductive.
.009   Blanking: floret sterility in a head (e.g., a composite).
.010   Bolt: said of a plant that develops a flower stalk and seeds prematurely.
.011   Compatible: ability of pollen tubes to grow through a style and reach the ovules in time to effect fertilization.
.012   Coulure: failure of grape blossoms to set, resulting in premature drop; (cf. Millerandage, IV.4.8.041). (WS)
.013   Cross-compatible: pollen of one cultivar (A) is capable of growing in a style and fertilizing ovules in another cultivar (B); however, the reciprocal may not be cross-compatible.
.014   Cross-fertile: one cultivar (A) is used as a pollenizer for another cultivar (B); the latter produces a commercial crop with viable seeds.
.015   Cross-fruitful: one cultivar (A) is used as a pollenizer for another cultivar (B); the latter produces a commercial crop.
.016   Cross-incompatible: the reverse of cross-compatible (IV.4.8.013); cultivar (A) produces functional gametes (sperm) but the pollen tubes apparently grow too slowly in cultivar (B)'s style to effect fertilization (although this is only one of many causes that result in failure to produce viable seeds).
.017   Cross-pollination: the transfer of pollen from the anthers of a

flower of one seedling, clone, or cultivar to the stigma of another flower; subdivided into two categories; geitonogamy (same plant) and xenogamy (different plants); synonymous with allogamy. (F & P)

.018 Cross-sterile: the reverse of cross-fertile (IV.4.8.014); cultivar (A) is used as a pollenizer for cultivar (B) but the latter fails to produce fruit with viable seed.

.019 Cross-unfruitful: the reverse of cross-fruitful (IV.4.8.015); cultivar (A) is used as a pollenizer for cultivar (B) but the latter fails to produce a commercial crop. (Failure may not be complete, in which case the condition may be denoted "partially cross-unfruitful" or "partially cross-fruitful," according to the departure from the usual situation for the cultivars involved).

.020 Degree-day: a heat unit that represents one degree of temperature above a given mean daily-base value; used to determine how close a crop is to maturity or some physiological stage of development.

.021 Effective pollination period: embryo-sac longevity minus the time required to complete pollination and pollen-tube growth. (WS)

.022 Endosperm incompatibility: failure of the embryo to develop as a consequence of the failure of the second pollen nucleus to unite with the polar nuclei and produce endosperm (in other words, incompatibility here is secondary rather than primary).

.023 Fecundity: ability of flowers to produce germinatable seed in large numbers.

.024 Fertile: said of pollen-bearing anthers or of seed-bearing fruits.

.025 Fertility: the ability to produce functional gametes.

.026 Floral initiation: the first discernible change from a vegetative bud to a floral primordium. (WS)

.027 Fruitful: plant or cultivar that sets and matures a commercial crop of fruit.

.028 Fruit set: persistence and development of an ovary and/or adjacent tissue following anthesis. (WS)

.029 Impotence: sterility by reason of the inability to produce functional gametes of one or both sexes.

.030 Incompatible, incompatibility: the inability to produce sexual seed because of some physiological reason; e.g., the inability of viable pollen to grow rapidly enough in the style for the pollen tube to reach the ovule and release its sperm in time to effect fertilization; (see Cytoplasmic male sterility, IV.4.1.012; Endosperm incompatibility, IV.4.8.022; Nuclear incompatibility, IV.4.8. 0042).

.031 Infertile, infertility: general inability to produce viable gametes.

.032 Intercompatible: pollen produced by a cultivar of a combination is capable of growing rapidly enough to fertilize egg cells in ovules of the other cultivar.

.033 Interfertile: cultivars A and B both produce fruit with viable seed when pollinated by each other.

.034 Interfruitful: cultivars A and B both produce commercial crops when pollinated by each other.

.035 Interincompatible: cultivars A and B are unfruitful when pollinated by each other because of some physiological reason; e.g., pollen tubes of each cultivar grow too slowly in the pistil of the other to effect fertilization.

.036 Intersterile: neither cultivar A used as a pollenizer for cultivar B nor the reciprocal will produce fruit with viable seed; implies failure of either one to produce viable gametes.

IV.4.8. HORTICULTURAL PHYSIOLOGY and CROP ECOLOGY

.037 Interunfruitful: cultivars A and B both fail to produce commercial crops when pollinated by each other.
.038 June drop: the final postbloom shedding of fruit, often occurring in May or June. (WS)
.039 Latent bud: a usually concealed (i.e., undeveloped) bud that remains dormant for an indefinite period. (WS)
.040 Metaxenia: the direct effect of pollen on the parts of the seed and fruit outside the embryo sac; (see Xenia, IV.4.8.057).
.041 Millerandage: a condition in grapes in which the ovary persists but the seeds remain small or do not attain usual size. (WS)
.042 Nuclear incompatibility: failure of the egg and pollen nuclei to unite even though the latter has entered the egg, thus resulting in a form of apomixis (IV.4.1.006) if an embryo develops.
.043 Perennation: lasting of fruit long after the usual season of maturity.
.044 Precocious: exceptionally early in development, as in flowering; prematurely developed.
.045 Self-compatible a cultivar that produces pollen capable of germination, growing down the style (if present) and entering the ovule to effect fertilization on the same plant or another of the same cultivar.
.046 Self-fertile: the ability of a cultivar to produce fruit with viable seeds following self-pollination.
.047 Self-fruitful: a cultivar that sets and matures a commercial crop of fruit, whether accomplished by self-pollination or parthenocarpy.
.048 Self-incompatible: a cultivar that produces functional gametes; i.e., viable pollen and a functional ovary but self-unfruitful because of some physiological process; such a cultivar may, however, serve as an effective pollenizer for another cultivar and may also be self-fruitful by parthenocarpic fruit set.
.049 Self-sterile: the inability of a cultivar to produce fruit with viable seeds when self-pollinated because of morphologically nonfunctional male or female gametes; certain cultivars may, however, still be self-fruitful via parthenocarpy.
.050 Self-unfruitful: a cultivar that is unable to set a commercial crop of fruit when self-pollinated and unable to do so parthenocarpically.
.051 Serotinous: flowering or fruiting late in the growing season or late in seed dispersal (opposite of Precocious, IV.4.8.044).
.052 Somoplasm: protoplasm other than that of a gamete.
.053 Sterility: morphological nonfunction of gametes, male, female, or both.
.054 Unfruitful: plant or cultivar that fails to set or mature a commercial crop of fruit.
.055 Viable: (see IV.3.1.024).
.056 Viability: the ability to live, grow, and develop; the length of time that a seed or pollen grain retains its capacity to grow and develop in a normal manner.
.057 Xenia: the immediate effect of pollen on the endosperm as a result of double fertilization (union of sperm with the egg and two polar nuclei) in the seed parent; (see also Metaxenia, IV.4.8.040).
.058 Xenogamy: a plant that receives pollen from another plant; a form of cross-pollination. (F & P)
.059 Zygote: the cell produced by the union of two cells (gametes), a fertilized egg; also the individual developing from this cell.

IV.5.1. Water Relations, Cold, and Other Forms of Stress

IV.5.1. General terms

.001 **Avoidance:** a specialized type of adaptation of certain plants (i. e., about 450 species of 135 genera in 82 families) whose taproots, lateral roots, and/or adventitious roots are contractile; these organs undergo 10 to 70% shortening by passive contraction mainly of the cortical cells, thus pulling the growing points of many seedlings, bulbs, rhizomes, etc., into the soil or nearer the soil surface; such plants avoid adverse environmental conditions (e.g., cold, drought, and excess heat) by changing location in a spatial sense prior to the development of the stress condition; thus they differ from those plants that can escape or tolerate such conditions by appropriate adjustments in their life cycles (escape) or high or low plant water status (tolerance); most plants, excepting those with contractile roots, are immobile; hence those that cannot escape by completion of a shortened life cycle must be able to tolerate a stress if they are to endure; (see Tolerance, IV.5.1.019; Drought tolerance, IV.5.2.020). (TU)

.002 **Elastic, elasticity:** capable of recovering size or shape after deformation (e.g., by loss of turgor in a cell) or, in the case of a liquid, of resisting compression (e.g., as in water in a plant's xylem vessel); elasticity is the ability to recover size or shape after deformation.

.003 **Elastic resistance:** the ratio of an applied stress to the resulting strain from which a plant will recover without injury, as measured under standard conditions (synonymous with modulus of elasticity = Young's modulus of elasticity, a term widely used in engineering); as applied to plants, e.g., the ratio of the temperature of a plant's surroundings to that of the plant (or organ) at which a certain quantified reversible change is detectable or the ratio of water stress (bars) applied to that of the internal stress from which the plant will recover without injury; (compare with Plastic resistance, IV.5.1.012).

.004 **Elastic strain** (temporary stress reaction): a reversible chemical or physical change (i.e., a temporary, reversible response to plant stress); may, however lead to injury or death if maintained for a long enough period (e.g., a sustained water stress resulting from a prolonged drought).

.005 **Modulus of elasticity** (Young's modulus of elasticity): the standard or norm for any parameter or value measured under unit conditions; used especially in situations in which a stress or strain is involved; e.g., the firmness of cell walls expressed in bars (or mPa).

.006 **Modulus of elasticity:** (see Elastic resistance, IV.5.1.003).
.007 **Modulus of plasticity:** (see Plastic resistance, IV.5.1.012).
.008 **Pascal (Pa):** (see IV.0.232).
.009 **Permanent stress reaction:** (see Plastic strain, IV.5.1.013).
.010 **Plastic:** capable of being deformed continuously and permanently in any direction without rupture when placed under a stress that exceeds the yield value or, in a biological system, an irreversible chemical change; an example of a physical change is a cell deformed from loss of water to the point at which it could not

IV.5.1. HORTICULTURAL PHYSIOLOGY and CROP ECOLOGY

regain its former shape; an example of a chemical change is plasmolysis of cell contents.

.011 Plasticity: the ability to retain a shape attained by pressure deformation, as in the situation in which citrus fruit packed in a container retain their deformed shape instead of regaining their original spherical or oblate configuration.

.012 Plastic resistance (modulus of plasticity): the intensity of an external factor that produces a standard irreversible physical or chemical change in which the organism is being exposed to the stress for a standard time (e.g., the temperature required to kill a plant or plant part expressed as a certain percentage, as LD_{50}, or LD_{100}, when the material is subjected to a specified temperature regime for a given period); differs from elastic resistance (IV.5.1.003) in its objective which is to determine the level of the external factor (temperature, moisture level, etc.) at which an irreversible change occurs under standard conditions.

.013 Plastic strain (permanent stress reaction): ordinarily an irreversible physical or chemical change that increases with the stress and produces more and more injury as the stress persists; repairable by an active expenditure of metabolic energy (e.g., a portion of a plant is seriously injured by a freeze but may be replaced by growth of shoots within the injured area or from uninjured buds below it).

.014 Resistance adaptation: the capability of a plant to change gradually to decrease or prevent a plastic (irreversible) strain (e.g., one that results from cold, drought, excess water, or heat) when subjected to a stress; may be stable, hence genotypic (hereditary) or unstable, hence phenotypic and dependent on the developmental stage of the plant and its exposure to environmental factors.

.015 Resistance groups: classified by Turner (1979) into three major categories with respect to drought [a meterological term meaning a period without significant rainfall, i.e., any month with < 60 mm as defined by Mohr and van Baren (1954)] as drought escape, drought tolerance with high tissue water potential, and drought tolerance with low tissue water potential; the last two are divided into subcategories according to whether productive processes are reduced; (see Tolerance, IV.5.1.019). (TU)

.016 Strain: changes in cell form, size, or constituents induced or caused by an environmental factor; may be elastic, thus reversible or plastic, irreversible, depending on the magnitude and/or duration of the external conditions; measured in units appropriate to the environmental factor.

.017 Stress: (1) the force applied per unit area (engineering and phycal sciences); (2) any factor that disturbs the normal functioning of an organism (biology), e.g., one whose degree, as in the case of a plant water stress or deficit, depends on the extent to which water potential and cell turgor are below their optimum values. (T & K - KR)

.018 Stress resistance: there are four possible mechanisms: (a) stress exclusion (the capability of a plant to avoid a given stress by escape, i.e., a short life cycle, or true avoidance, i.e., having contractile roots); (b) stress survival (the capability of a plant to tolerate an internal stress, e.g., drought tolerance with low tissue water potential); (c) strain prevention (the capability of a plant to tolerate a plastic or elastic strain,

IV.5.1. HORTICULTURAL PHYSIOLOGY and CROP ECOLOGY

e.g., by maintenance of water uptake as in drought tolerance with high tissue water potential); and (d) strain repair (the capability of a plant to recover from a strain; e.g., regrowth of a plant killed back by cold). (TU)

.019 Tolerance: the capability of a plant to adjust (adapt) to and survive a stress imposed on it by given environmental conditions by changes in metabolic or growth process (e.g., osmotic adjustment to maintain cell turgor at low water potential); the degree of adaptability possible will depend to a large extent on whether the plant can continue to make the necessary adjustments until the condition that causes the stress is alleviated or the stress exceeds the limits of the plant's capability to adjust and death ensues; equilibrium in the true sense (i.e., a steady-state situation) is unlikely in a stress situation because adjustments must be continuous if the plant is to remain alive, i.e., can tolerate the stress. (TU)

IV.5.2. Water relations (including stress aspects)

.001 Acclimation: (1) the nonheritable (temporary phenotypic, thus readily reversable) modification of characters caused by exposure of plants to changes in environmental conditions, such as cooler, warmer, or drier weather (Kramer 1980). (2) the process whereby plants become inured to adverse conditions, such as soil pH that is too high or too low for a particular crop, the presence of potentially toxic ions (e.g., Al, B, S, Se, Te, Na, Cu, and Co), cold, drought, and flooding, by gradual exposure to reduced water supplies, increased internal aeration (as in a stagnant water situation), increased light intensity, low N-high K fertilization (plus adequate essential elements), reduction of metabolic activities (as temperatures decrease) leading to slower growth, and preferential absorption or sequestration of certain ions (by the application of soil amendments). Note, however, definition (1) refers to natural events, whereas (2) does or may involve man's intervention; and thus it becomes essentially synonymous with hardening (IV.5.2.030); also the degree of acclimation (or hardening) possible will depend on the crop as well as the particular regime under which a crop is grown. (S & R, T & K-KR)

.002 Adaptation: (1) the evolutionary origin of one or more characters; (2) as used in a stress sense, heritable (genotypic) modifications in structure or function that increase the probability of an organism's survival, and perhaps also reproduction in a given environment; (2) depends on a plant acquiring an optimum combination of characters that maximizes advantageous effects and minimizes deleterious ones (i.e., a combination beneficial from a horticultural standpoint). Such character expression may be phenotypically constant, unchanged over a wide range of environments (e.g., the production of water storage tissue and thick cuticle in succulents), or plastic that is responsive to different environments, like sun vs. shade leaves or the switch from C-3 to CAM metabolism in some succulents subjected to water stress. [Many scientists from Darwin on have pointed out that continual selection in cultivated crops greatly enhances the appearance and subsequent heritability of favorable adaptations (to the grower, that is) if they can be induced]. (T & K-KR)

IV.5.2. HORTICULTURAL PHYSIOLOGY and CROP ECOLOGY

.003 Adhesion: attraction between unlike molecules, as in water and other compounds; (see Cohesion, IV.5.2.006). (S & R)
.004 Antitranspirant: (see Drought protectants, IV.5.2.017).
.005 Cell-water potential: the potential for water to do work by moving from one energy level to a lower one; i.e., movement of water from a cell into its surroundings (a decrease in turgor that may result in plasmolysis and consequent death if carried too far) or vice versa (an increase in turgor that may cause the cell to burst if too great); osmotic potentials of most plants range from about -4 to -20 bars (mPa); those of halophytes, from -50 to -80 bars; calculated from the equation: $\Psi cell = \Psi s + \Psi p + \Psi m$, where Ψ cell is the cell-water potential, Ψm matric forces, Ψp pressure, and Ψs is the sum of the contributions of solutes; the last term is usually disregarded; note that Ψ cell is negative, which means that the cell is under tension; Ψ cell can also be expressed as $\Psi cell = P - \pi$, where P is the hydrostatic pressure (i.e., that exerted by the weight of a vertical column of water) and π is the osmotic pressure (bars, atm., or mPa). (KR)
.006 Cohesion: attraction of like molecules, like those of water which because of hydrogen bonding gives this substance unusually high tensile strength and surface tension (the reason why continuous columns of water can exist in plants such as rattans more than 200 m long). (KR)
.007 Critical water potential: two points on a water stress curve for a plant, one in which the stomata begin to close (also called the threshold water potential) and the second in which stomatal closure is complete, the critical level.
.008 Dehydration avoidance: (see Drought tolerance, IV.5.2.020).
.009 Dehydration (strain) tolerance measurement: the critical relative water content (i.e., at the drought killing point) can be determined in herbaceous plants, such as tomato, and used with water saturation deficit and relative water content (measurements of dehydration strain and avoidance, respectively) to give qualitative information regarding the kind of stress resistance developed by the plant tested.
.010 Diffusion pressure deficit (DPD): an obsolete term formerly used as a measure of the tendency of water to diffuse from a cell or vice versa; DPD = OP - TP, where OP is the osmotic pressure and TP, the turgor pressure within a cell; (see Cell-water potential, IV.5.2.005). (KR)
.011 Direct elastic dehydration strain: survival of a plant under water stress is dependent on prevention of equilibrium with its environment; the transfer process is always slow (compared with heat) and requires days, weeks, or months before equilibrium with the water potential of the environment is reached.
.012 DPD: (see Diffusion pressure deficit, IV.5.1.010).
.013 Drought: the absence of precipitation for a period long enough to cause depletion of soil moisture and damage to plants, in other words, a meteorological and environmental condition of sufficient duration to produce a plant water deficit or stress; may be permanent as in desert areas, seasonal as in areas with well-defined wet and dry periods, or unpredictable as in many humid climates. (T & K-KR)
.014 Drought avoidance: (see Avoidance, IV.5.1.001).
.015 Drought escape (avoidance): a term applied to desert ephemerals or some plants in areas with well defined wet and dry seasons in which they complete their life cycles, or at least their repro-

ductive cycles (i.e., mature early) before the onset of the next dry season and escape (avoid) being seriously stressed. The general mechanism is one of rapid phenological development and plasticity; e.g., plants growing as a result of winter rains usually have a rosette form and C-3 anatomy, but those after summer rains, an erect habit and Kranz anatomy typical of C-4 plants; an intermediate habit is particularly suited to wide developmental plasticity, a significant factor in the production of grains and other annuals in areas with limited or irregular rainfall. (T & K-KR,TU)

.016 Drought hardening: increased tolerance to water stress by manipulation of environmental conditions (including water stress), nutrition, and the application of drought protectants; (see IV.5.2.017).

.017 Drought protectants: including growth inhibitors (CCC, Phosfon-D), sugars, soluble proteins, polypeptides, and antitranspirants which act as physical barriers to moisture loss from leaves or induce alterations in cell permeability and other relevant properties by physiological or biochemical changes.

.018 Drought resistance measurement: the attempt to quantify the mechanisms involved in the several types of drought resistance (i.e., drought escape, drought tolerance at high water potentials, and drought tolerance at low water potentials); many different methods have been or are used, such as measurement of xerophytic characters or individual factors in drought resistance (both indirect and usually poorly correlated with field experience), yield, osmotic concentration, heat stability of chlorophylls. "bound water", sulfhydryl content, survival time in the field or drought chambers, and efficiency of water utilization; the primary problems in all of these methods have been to avoid wholesale confounding (in a statistical sense) and to reconcile criteria among growers, breeders, and physiologists in setting up and carrying out experiments.

.019 Drought stress tolerance: a measure of the water-potential stress in terms of water potential required for 50% of the plants tested to die from lack of moisture; expressed as $T_D = \Psi_o - \Psi\,p50$, or simply $-\Psi\,p50$, where T_D is the drought tolerance, and Ψ_o and $-\Psi p50$, are, respectively the water potentials at saturation and of the plants in the steady state at the point at which 50% are killed; this information is useful in determining how long a drought can be tolerated for a given crop and still have half the plants alive, as in a dry year when irrigation is not available.

.020 Drought tolerance: the capability of plants to survive drought specifically adaptations that enhance their power to withstand drought-induced stress; classified into two main groups: (1) dehydration postponement by means of morphological or physiological modifications that reduce transpiration or increase water absorption (e.g., a thick cuticle, leaf rolling, responsive stomata, deep root systems, and water storage); (2) dehydration tolerance, the degree of hydration without permanent injury in terms of yield or survival (varies widely depending on the stage of development, duration of the stress, and the kind of plant). (T & K-KR)

.021 Drought tolerance mechanisms: (see Drought tolerance, IV.5.2.020).

.022 Drought (water) stress: involves several different types of injury, such as (a) primary direct drought injury from a plastic (membrane) strain (see IV.5.2.042); (b) primary indirect-drought

IV.5.2. HORTICULTURAL PHYSIOLOGY and CROP ECOLOGY

injury from elastic or plastic metabolic strains plus elastic turgor loss or growth strains (see IV.5.2.044); and (c) secondary water-stress injury from a stress-induced strain; (see IV.5.2.056).

.023 Elastic growth strain: cell dehydration is elastic and completely reversible up to a point beyond which it is plastic, irreversible, and injurious; may be indirectly injurious by limiting the ability of the plant to send out new roots; two groups of plants, stenohydric (grow only within a narrow range of water stresses) and euryhydric (grow within a wide range of water stresses); those in the former group are highly sensitive to soil moisture stress; growth is often slowed drastically by stresses of only a few bars (also true for euryhydric forms but with a somewhat greater range).

.024 Enzyme inactivation: (see Primary indirect drought injury, IV.5.2.044).

.025 Euryhydric: (see Elastic growth strain, IV.5.2.023).

.026 Euryhydric species: plants that can tolerate, i.e., grow and adjust to, a wide range of drought stress, as in those that can maintain turgor and endure low tissue-water potential without reduction in their productive processes.

.027 Evaporation: loss of moisture from surfaces other than plants, e.g., from soil or free water (lake, pond, ocean, or pan); that from a free water surface is calculated as $E = C_{water} - C_{air}/r_{air} \cong 0.622 \ P_{air}/p \times e_{water} - e_{air}/r_{air}$; (see Transpiration, IV.5.2.065, for explanation of terms). (KR)

.028 Evapotranspiration: the combined loss of water by evaporation from the soil and transpiration from plant surfaces; measured by the water-balance, energy-balance, aerodynamic, or vapor-flow methods or estimated by empirical formulas from meteorological data, Penman's formula, or modifications thereof. (KR)

.029 Field capacity (FC): the water content of a soil after the drainage of gravitational water is essentially complete and the water content is momentarily stable; an arbitrary point on a water potential curve, often taken as 0.3 bar, but this value, like any other measurement of soil moisture, depends on soil structural characteristics (e.g., soil type and bulk density) and is recognized, in spite of its imprecise nature, as the effective upper limit of readily available water for plant use. (KR)

.030 Hardening: the equivalent of acclimation (IV.5.2.001) in depending on phenotypic modifications but differs in that hardening is deliberately induced to inure plants to otherwise intolerable conditions, as the hardening off of cuttings, layers, seedlings, and the like prior to transplanting. (T & K-KR)

.031 Homoiohydric plants: a large central vacuole is the common characteristic, the protoplasm being less affected by external conditions because of water stored in the vacuole; a protective cuticle slows down evaporation, stomata are in crypts often lined with hairs that reduce transpiration, and an extensive root system enables steady activity to take place in cells, despite sudden changes in humidity; thus they possess drought tolerance and may also show a considerable degree of drought avoidance, even though most are "water spenders" (see IV.5.2.080). (LA)

.032 Leaf water loss resistance: a measure of how rapidly moisture can be removed from a leaf, based on intrinsic properties of the leaf itself: $1/r_l = 1/r_c + 1/r_s$, where r_l is the total leaf resistance (s/cm), r_c is the cuticular resistance, and r_s is the total stomatal resistance which is the sum of r_m, r_c, and r_p, the resistance in mesophyll cell walls, intercellular spaces, and the stomatal pores, respectively; stomatal aperture is the main factor that affects stomatal resistance. (KR)

.033 **Moisture equivalent:** the specific capacity of a soil to retain capillary water; measured as the weight of water retained under a force of 1000X gravity expressed as the % of dry weight of soil; a measure of the water-storage capacity of the soil; approximately equal to field capacity (IV.5.2.029); obsolete. (KR)

.034 **Osmoregulation (turgor regulation):** the regulation, either by changes in solute concentration or turgor pressure, of osmotic potential within a cell by the addition or removal of solutes from a solution until the intracellular osmotic potential is approximately equal to that of the medium surrounding the cell; used to define the change in osmotic potential in higher plants in response to salinity. (see Osmotic adjustment, IV.5.2.036). (T & K-T & J)

.035 **Osmosis:** diffusion of fluid through a semipermeable membrane. (S & R)

.036 **Osmotic adjustment:** the lowering of osmotic potential in higher plants arising from the net accumulation of solutes (i.e., a net solute increase) in response to a water deficit; (see Osmoregulation IV.5.2.034). The degree of adjustment is measured as the change in π at a particular water potential or content, usually for comparative purposes at full ($\Psi = 0$, $P = \pi$) or zero ($\Psi = \pi$, $P = 0$) turgor. Factors affecting adjustment include the rate of water-deficit development (a function of soil volume occupied by roots, hydraulic conductivities of soil and plant, stomatal conductance, leaf area, and atmospheric demand for water), degree of stress (slower as the water deficit increases), environmental conditions (rate of drying, direct influence of temperature and light), and differences in cultivars; also may be transient or limited, especially with container-grown plants, and then may not fully maintain physiological and morphological processes. (T & K-T & J)

.037 **Osmotic potential:** expressed as $\pi = \pi_o V_o/V$, where V is the osmotic volume and π_o and V_o are the osmotic potential and volume, respectively, at some reference value, such as full turgor or zero turgor, with the assumption of passive solute concentration (i.e., cells act as perfect osmometers); active solute accumulation in the cell will result in a greater lowering of osmotic potential than predicted by the above equation (which is a form of the general equation for osmotic pressure (IV.5.2.038), used with special reference to water stress situations). (T & K-T & J)

.038 **Osmotic pressure (osmotic potential):** the pressure of a solution resulting from a substance or substances dissolved in it; the general equation is $\pi V = n_s RT$, where π is the osmotic pressure in bars (IV.0.026), atmospheres (IV.0.019), or pascals (IV.0.232), V is the volume of the solvent in liters, n_s is the moles of solvent, R is the gas constant (IV.0.118) and T is the absolute temperature (K); $\pi = 22.7$ bars (or 22.4 atm) at 273 K for one mole of solute in one liter of solvent (Van't Hoff equation for dilute solutions of nondissociating substances). Measured in practice by freezing point depression where $\pi : 22.7 = T : 1.86$, or relative vapor pressure, $e/e_o = n_w/(n_w - n_s)$, where e_o is the vapor pressure of water, n_w is moles of pure water, and n_s is moles of solute. (KR)

IV.5.2. HORTICULTURAL PHYSIOLOGY and CROP ECOLOGY

.039 Permanent wilting point (PWP): effectively, the lower end of the range of readily available water for most plants; the soil moisture content at which leaves on a plant wilt and do not recover when placed in a humid atmosphere; the usual test plant is a dwarf sunflower in a container, particularly when a comparison of PWP values among several soils is desired; or plants of a given species or cultivar may be used to determine the PWP; equivalent to approximately -15 bars but, like FC (IV.5.2.029), it is an arbitrary point on a water potential curve; the actual <u>quantity</u> of water remaining in a soil is dependent on structural characteristics, environmental factors, and the crop, whose osmotic potential (IV.5.2.037), thus the point at which permanent wilting will occur, may vary markedly from that of a standard test plant. (KR)

.040 Poikilohydric plants: match their water content with the humidity of their surroundings; found among thallophytes, dry habitat mosses, certain vascular cryptogams (some epiphytic ferns), and very few angiosperms; also pollen grains and embryos in seeds. (LA)

.041 Pressure chamber: a device for the measurement of water potential, especially of leaves or stems, by pressure equilibration. (KR)

.042 Primary direct drought injury: the injury is labeled direct if it occurs too rapidly to be metabolic or involves a dehydration strain severe enough to cause metabolic injury in spite of ample reserves and protoplasmic proteins in affected cells; (see Drought (water) stress, IV.5.2.022).

.043 Primary direct drought tolerance: (see Drought tolerance, IV.5.2.020).

.044 Primary indirect drought injury: several indirect effects of water loss: (a) <u>starvation</u> by increased respiration, reduced or inhibited <u>photosynthesis</u>, and reduced translocation of photosynthates which cause depletion of reserves; drought injury, however, usually occurs when reserves are ample and starvation is more likely to be due to the prevailing high temperature; (b) <u>protein breakdown</u>, some does occur, with particular accumulation of <u>proline</u> in water-stressed leaves (some <u>de novo</u>) and amides; (c) <u>enzyme inactivation</u>, inactivation of some enzymes but increases in others during drought stress; (d) <u>RNA decrease</u>, some destruction or polymeric-monomeric changes in mRNA and arrest of protein synthesis; (see Drought (water) stress, IV.5.2.022).

.045 Primary indirect drought tolerance: [see Drought (water) stress, IV.5.2.022)].

.046 Proline: (see Primary indirect drought injury, IV.5.2.044).

.047 Protein breakdown: (see Primary indirect drought injury, IV.5.2.044).

.048 Pseudohardening: the condition in which a plant will increase its capacity to endure a drought while maintaining a high tissue-water status but not increase a plant's capability to endure a low tissue-water potential.

.049 Readily available water: the amount of water in a soil presumably available to plants between field capacity and the permanent wilting point; said amount is widely variable, depending on soil structural characteristics and environmental factors. (KR)

.050 Relative drought index: = $WSD_{act.}/WSD_{crit.} \times 100$, where $WSD_{act.}$ and $WSD_{crit.}$ are the actual and critical water saturation deficits, respectively. (LA)

.051 Relative water content: (see Water saturation deficit, IV.5.2.078).

.052 RNA decrease: (see Primary indirect drought injury, IV.5.2.044).

.053 Saturation index: $H = T/P \times 100$, where H is the index of saturation or relative turgor, T is the actual turgor pressure, and P is the osmotic pressure or maximum potential turgor pressure; a means of comparing the relative turgidity (i.e., water content) of different plants or crops.
.054 Schollander pressure chamber: (see Pressure chamber, IV.5.2.041).
.055 Secondary drought-induced stress: [see Drought (water) stress, IV.5.2.022].
.056 Secondary water stress injury: reduced plant growth that results from an induced, or secondary, stress, such as reduced availability (uptake) of phosphorus, apparently from decreased translocation (the primary injury); [see Drought (water) stress, IV.5.2.022].
.057 Soil-water potential: essentially the tenacity with which the soil solution is held by the particles in the soil; $\Psi_{soil} = \Psi_m + \Psi_p + \Psi_s + \Psi_g$, where Ψ_m is the matric potential (forces associated with the structure and characteristics of the soil matrix), Ψ_p is the pressure potential, Ψ_s is the osmotic (solute) potential, and Ψ_g is the gravitational potential. (KR)
.058 Solutes: (see Drought protectants, IV.5.2.017).
.059 Specific survival time: indicates how long, with a given evaporative power of the air, that the leaves of a plant can remain undamaged after their stomata have closed as a result of a lack of water; expressed as $S_{st} = W_{av}/E_c$, where S_{st} is the specific survival time measured in hours or days, W_{av} is available water, and E_c is cuticular transpiration. (LA)
.060 Starvation: (see Primary indirect drought injury, IV.5.2.044).
.061 Stenohydric: (see Elastic growth strain, IV.5.2.023).
.062 Stenohydric species: plants that suffer impairment of their vital functions, even with slight increases in osmotic pressure.
.063 Threshold water potential: (see Critical water potential, IV.5.2.007).
.064 Total drought resistance: overall drought resistance is a combination of stress tolerance (T_d) and avoidance (A_{d50}); it is also equivalent to $R_d = -\Psi_{e50}$, where R_d is drought resistance and $-\Psi_{e50}$ is the water stress just sufficient to cause 50% mortality: $T_d \times A_{d50} = -\Psi_{p50} \times \Psi_{e50}/\Psi_{p50} = -\Psi_{e50}$; $R_d = -\Psi_{e50}$; therefore $R = T_d \times A_{d50}$.
.065 Transpiration: the loss of moisture from plants in the form of water vapor; measured by any of several methods, such as phytometer (periodic weighing of plants in containers), cut shoots, water vapor loss, and velocity of sap flow; a common method of rate measurement is $T = C_{leaf} - C_{air}/r_{leaf} - r_{air} \cong 0.622\, p_{air}/p \times e_{leaf} - e_{air}/r_{leaf} - r_{air}$, where T is the rate of transpiration in $g/cm^2 \cdot s$, C_{leaf} and C_{air} are the water vapor concentrations (g/cm^3) on evaporating surfaces within the leaf and in bulk air, r_{leaf} is the additional diffusive resistance within the leaf, r_{air} is the surface boundary layer resistance (s/cm), p_{air} is air density, p is atmospheric pressure (mm Hg), and e_{leaf} and e_{air} are vapor pressures (mm Hg) at the evaporative surfaces and in bulk air. (Note that Δe is the vapor pressure deficit.)
.066 Transpiration coefficient: (see Water requirement, IV.5.2.077).
.067 Transpiration efficiency: dry matter produced per kilogram of water lost; similar to water requirement, IV.5.2.077).
.068 Transpiration ratio: the net proportion of water vapor lost compared with net CO_2 uptake per unit time; thus the amount of water required per unit of dry weight increase; ranges for C-3, C-4,

IV.5.2. HORTICULTURAL PHYSIOLOGY and CROP ECOLOGY

and CAM plants are 450 to 950, 250 to 350, and 50 to 55 g of water per g of dry weight increase (seasonal basis), respectively. (S & R)

.069 Turgor: the normal distension of plant cells that results from internal pressure exerted against the cell walls; e.g., when water is absorbed. (LA)

.070 Turgor regulation: (see Osmoregulation, IV.5.2.034).

.071 Water balance: the difference between the rates of water intake and loss in a plant. (LA)

.072 Water capacitance: the capability of plants to store and release water, which may be a substantial quantity, especially in trees.

.073 Water content: the quantity of water in a plant, organ (i.e., leaf or fruit), or soil; determined on a dry weight, fresh-weight, or relative water-content basis; dry weight is the most precise, provided due precautions are taken not to use too low (incomplete removal of water) or too high (volatilization of certain inorganic elements or caramelization of organic compounds, e.g., carbohydrates) a temperature. (KR)

.074 Water movement:

.074a. Energy flux density (LE): the rate of energy change per unit time per unit area; units are joules per kilogram X kilograms per second per square meter ($J \cdot kg^{-1} \cdot kg\ s^{-1} \cdot m^{-2} = J \cdot s^{-1} \cdot m^{-2} = W \cdot m^{-2}$). (SE79)

.074b. Transpiration rate: (water vapor flux per one-sided leaf area):

.074b-1. Mass basis: the unit is kilograms per square meter per second ($kg \cdot m^{-2} \cdot s^{-1}$).

.074b-2. Volume basis: the unit is cubic meters per square meter per second ($m^3 \cdot m^{-2} \cdot s^{-1}$) or meters per second ($m \cdot s^{-1}$). (SE79)

.074b-3. Water vapor diffusion resistance: measured as seconds per meter ($s \cdot m^{-1}$).

.074b-4. Water vapor flux (rate of change): measured as kilograms per second ($kg \cdot s^{-1}$), grams per second ($g \cdot s^{-1}$), milligrams per second ($mg \cdot s^{-1}$) or micrograms per second ($\mu g \cdot s^{-1}$).

.074b-5. Water vapor flux density (rate of change per unit area) (E): kilograms, grams, milligrams or micrograms per second per square meter ($kg \cdot s^{-1} \cdot m^{-2}$, $g \cdot s^{-1} \cdot m^{-2}$, $mg \cdot s^{-1} \cdot m^{-2}$, or $\mu g \cdot s^{-1} \cdot m^{-2}$, respectively).

.075 Water potential:
Specific: the unit of a mass of water moved to a reference state = joules per kilogram ($J \cdot kg^{-1}$).
Volumetric: the energy needed to move a unit volume of water from the system to the reference position (i.e., pure free water at the same temperature as the system and at one atmosphere pressure); equals water density [$\rho_w(T)$] times specific water potential (note that ρ is temperature dependent); units are joules per cubic meter ($J \cdot m^{-3}$), or newtons per square meter ($N \cdot m^{-2}$) or pascals (Pa).

.076 Water relations: the study of the properties of water in relation to physiological and biochemical processes in plants or their parts. (S & R)

.077 Water requirement (transpiration coefficient): water lost per unit of dry weight produced; wide range among crop plants.

.078 Water saturation deficit (wsd), relative water content (rwc): both terms refer to the proportion of water in a plant compared with the quantity its tissues could contain in a saturated condition: wsd = saturation wt-natural wt/saturation wt-dry wt X 100; rwc = natural wt-dry wt/saturation wt-dry wt X 100, where saturation weight is the weight of the plant or plant part at maximum turgor pressure, natural weight is fresh weight at any given moment, and dry weight is the weight of tissues after drying to a constant value under standard conditions.

.079 Water savers: (1) plants that have the capability of enduring a drought while maintaining high tissue-water potential by the reduction of water loss; xerophytes, succulents, and some sclerophyllous plants accomplish this in short periods of stomatal opening, low rates of cuticular transpiration, reduction in leaf area, shallow-spreading root systems, and by CAM or C-4 carbon dioxide fixation pathways; (2) plants that have the capability of enduring a drought or low tissue-water potential by maintaining turgor (e.g., by osmotic adjustment), increased cell elasticity, and decreased cell size, plus increased protoplasmic tolerance to desiccation; these plants also have many of the same characteristics as those in (1), plus increased root density and depth (higher root-to-top ratio); (see Water spenders, IV.5.2.080). (TU)

.080 Water spenders: plants that endure a drought while maintaining high tissue-water potential by more efficient water uptake and utilization, e.g., by having a higher proportion of conducting-to-nonconducting tissues, high root-to-top ratio (i.e., deeper and more extensive root system), and increased liquid-phase conductance, dew absorption, increased cuticle development, sunken stomata, and higher trichome density (compared with nondrought conditions) in common with water savers; this group and (1) under water savers are not mutually exclusive as mechanisms for the reduction of water loss, and more efficient means of water uptake and use will often occur or develop simultaneously in the same plant; mechanisms for (2) under water savers are those for survival under extreme conditions. (TU)

.081 Water stress: (see Drought (water) stress, IV.5.2.022).

.082 Water turnover, water turnover rate: the former is the milligrams of water evaporated per gram of water content; the latter is the percentage of water present in the leaf that is lost per unit time (minute, hour, day). (LA)

.083 Water uptake by roots: an equation used to determine the quantity of water absorbed by a root system, thus permitting comparisons of the efficiency of the process among different plants or crops and of water uptake in relation to transpiration of dry matter accumulation in a given plant or among crops: $W_{abs} - A \times \Psi_{soil} - \Psi_{root} / \Sigma$, where W_{abs} is the amount of water absorbed by the roots in unit time, A is the exchange surface area, Ψ_{soil} and Ψ_{root} are the water potentials of the solid solution and roots, respectively, and Σ is the sum of the resistance to water transport in the soil and from the soil solution into the roots. (LA)

.084 Water-use efficiency (WUE): a measure of the dry matter produced by a plant per amount of water used; calculated as (1) the ratio of CO_2 fixed in photosynthesis to water lost by transpiration, as WUE= g CO_2 /g H_2O or mol CO_2/mol H_2O or g CO_2/g H_2O X mb VPD (of which the last component of the last equation reflects variations in relative humidity plus the transpiration rate); (2) the ratio of dry matter produced (crop yield) to water used in transpiration and evaporation, as WUE = dry matter or crop yield/evapotranspiration; the usual range in values for (1) is 0 to 1, with C-3 plants 0.3 to 0.4 and C-4 plants 0.8 to 0.9. (T & K-T & J)

.085 Xeromorphic structure: quantitative characteristics that may aid drought tolerance: (a) hereditary (i.e., genotypic) or (b) varying with the environment (i.e., phenotypic). Examples of (a) include phyllodes (Acacia), phylloclads (Opuntia, Phyllocactus), cladophylls (Asparagus, Ruscus), sclerophylls (Laurus, Myrtus, Olea, scale leaves (Casuarinaceae), succulent leaves (Sedum, Lithops, and Agave) and stems (Cactaceae and some Euphorbiaceae), or roots (Umbelliferae, Compositae, and Cucurbitaceae); many of these plants are more or less modified and heavily armed with spines (e.g., Cactaceae and Acacia), which are also protective devices against predators. Examples of (b) include laminar rolling (some grasses), vertical orientation of falcate leaves (Eucalyptus spp.), the tendency for lower leaves to be more mesomorphic (from higher humidity) than the upper ones and lower water content in the latter, greater sugar accumulation, higher cell sap concentration, the capability of drawing water from lower leaves (or from fruit), generally higher amino acid content, greater transpiration rates per unit area (among C-3 or C-4 or CAM plants), and lower cuticular transpiration and accelerated senescence of lower leaves when exposed to drought conditions; (see Epiphytes, II.4.2.006; Leaf modifications, II. 4.3.032; Succulent, II.4.3.065; Xerophyte, II.4.2.037). (L 51, B & C, TU)

.086 Xeromorphy: (see Drought escape, IV.5.2.015; Drought tolerance, IV. 5.2.020).

IV.5.3. Cold stress

.001 Adiabatic lapse rate: the rate of cooling as dry air rises, or about -1°C/100 m (the actual value being dependent on the barometric pressure, temperature, latitude, time of year, and time of day when readings are made). (S & R)

.002 Artificial freezing: testing of plant hardiness in freezing chambers, often measured as the frost-killing point, which may be incipient (the temperature that just begins to cause injury), ultimate 100% kill, or the 50% killing point; basic requirements for killing-point determinations are (a) inoculation of plants to ensure freezing (b) standard rate of cooling, (c) a single freeze used for standard length of time, (d) standard rate of thawing, and (e) standard postthawing conditions.

.003 Black heart: results from killing living cells of the wood in apple trees as a result of a single midwinter freeze; pathogen invasion and subsequent rotting may not become evident until two or three years later unless the tree is seriously weakened and blows over in a wind storm. (CS)

.004 Capacity adaptation: a measure of the organism's avoidance of temporary stress reaction.

.005 Chilling injury: basically, exposure of plants or plant parts to temperatures well above freezing; three types of injury: (a) direct, in which injury occurs rapidly, i.e., within a few hours, as in ripe bananas, synonymous with cold shock; (b) indirect, in which appearance of injury requires days or weeks of exposure as a result of starvation, protein breakdown, respiratory upset, toxins, or biochemical lesions; (c) secondary stress, in which desiccation of tissues occurs; (see Chilling resistance, IV.5.3.007).

.006 Chilling requirement: the necessity for temperate zone (e.g., apple, pear, and peach) and some subtropical (e.g., lychee, Japanese persimmon, and kiwi fruit) woody plants and winter annuals and biennials to undergo a cold period if their vegetative and floral buds are to open and grow normally in the spring; chilling is measured in terms of hours below a certain threshold temperature of about 7°C but above 0°C for temperate zone crops; the accumulation is hour for hour up to 7°, then diminishes to zero at about 15°C; the chilling requirement of plants, such as peaches or rabbiteye blueberry (Vaccinium ashei), bred or grown in warm, temperate, or cool, subtropical areas is apparently satisfied at a higher threshold temperature (i.e., near 10°) and the limit for zero accumulation is also higher; moreover, the chilling requirement of certain cultivars, e.g., 'Rome Beauty' apple (1200 h) and low-chilling (300 h or less) peaches, may be substituted for by any treatment, such as defoliation or a dry period, which will induce dormancy even under tropical conditions where temperatures never fall below 20°C.

.007 Chilling resistance: the ability of plants to withstand chilling; plants from temperate climates; psychrophiles are fully resistant and plants from tropical or subtropical climates show varying degrees of sensitivity.

.008 Chilling-resistant plants: three groups: (a) permanently chilling-resistant; possess membrane lipids with a low solidification temperature (i.e., a high degree of unsaturation), thus do not require hardening treatment; (b) chilling-sensitive plants that suffer direct or rapid chilling injury (cold shock); possess membrane lipids with a high solidification temperature (high degree of saturation); hardening would induce resistance only if it decreases the degree of saturation in the membrane lipids; (c) chilling-sensitive plants that suffer indirect (slow) injury; possess membrane lipids with a low solidification temperature (highly unsaturated); would survive brief chilling but later injury from peroxidation of lipids; hardening would induce resistance by preventing peroxidation at chilling temperatures.

.009 Cold acclimation: (see Frost hardening, IV.5.3.028).

.010 Cold hardening: (see Frost hardening, IV.5.3.028).

.011 Cold shock: (see Chilling injury, IV.5.3.005).

.012 Conditioning: a gradual lowering of temperature toward that critical for chilling injury; can provide only limited protection from cold shock and chilling.

.013 Cryobiology: the biology of freezing temperatures.

.014 Cryoprotectant: substances such as glycerol or dimethylsulfoxide (DMSO) that prevent injury to cells during freezing; work for tissue cultures of some plants but are not successful for normal plant tissues unless with some form of prehardening, precooling, and/or slow reduction of temperature.

.015 Dehardening, deacclimation: loss of hardening if a plant's rest period is completed or if growth is resumed; a serious problem with some crops that have relatively short rest periods; (see Frost hardening, IV.5.3.028).

.016 Direct chilling injury: (see Chilling injury, IV.5.3.005).

.017 Direct stress injury: (see Stress injury, IV.5.3.047).

.018 Double freezing point: living tissues of plants exhibit two distinct freezing points; one designated as the freezing point due to extracellular ice formation and the other, the lower one, the undercooling point due to intracellular ice formation (i.e., freezing of the cell sap) under artificial conditions; there are

several explanations of these "exotherms".
- .019 Eutectic point: the lowest temperature at which a solution of given composition will freeze when cooled or the solid will melt when warmed.
- .020 Freeze dehydration: (see Freezing injury, IV.5.3.023).
- .021 Freeze desiccation: (see Freezing injury, IV.5.3.023).
- .022 Freeze smothering: (see Freezing injury, IV.5.3.023).
- .023 Freezing injury: several distinct types: (a) <u>primary direct freezing injury</u> occurs only as a result of intracellular (intraprotoplasmal) freezing; the degree and extent of injury depend on the physiological state of the plant (i.e., hardened or unhardened), rates of cooling and warming, length of time frozen, repeated freezing and thawing, and postthawing treatment; (b) <u>secondary freezing injury</u> due to stresses induced by the primary stress: (1) <u>freeze smothering</u>, gas stress imposed by the ice barrier to gas diffusion, (2) <u>freeze desiccation</u>, freeze-induced water stress that leads to net evaporation from the plant, (3) <u>freeze dehydration</u>, freeze-induced water stress that leads to exomosis of water from cells to ice centers in the plant and can occur during freezing, while frozen, during thawing, or after thawing; there is no evidence of <u>primary indirect freezing injury</u>.
- .024 Freezing injury estimation: several methods, electrical conductivity as a measure of cell sap electrolytes, release of amino acids and other ninhydrin-reacting substances, conductance ratio of high and low frequency currents, vital staining of sections with neutral red, TTC (2,3,5-triphenyltetrazolium chloride) staining (widely used for seed testing, V.1.3.070), sugar content, impedance readings, and sulfhydryl content.
- .025 Freezing stress: injury produced by temperatures low enough to cause ice formation; can be intercellular (extracellular) or intracellular; the latter leads to death of the cell or cells in which it occurs and the former, only if water cannot be reabsorbed into the cells when thawed or if ice crystals have injured them; water diffuses out of the cells during slow cooling to prevent their contents from freezing until the temperature drops low enough or if the cooling rate is so rapid that it produces ice crystals small enough to penetrate the lipid membrane and act as nucleating agents.
- .026 Freezing-stress tolerance: an increase in tolerance of freezing stress is due to an increase in (a) avoidance or (b) tolerance of the dehydration strain; thus the stress tolerance (i.e., frost-killing temperature) can be calculated from the freezing-point lowering if the tolerance of the dehydration strain is known; freezing-point lowering (fpl) of plant sap ranges from 0.5 to 4°C and dehydration tolerance (dt) from 50 or below up to 96%; thus the range for frost-killing temperatures is -1° for 0.5° fpl and 50% dt to -70°C for 3° fpl and 96% dt.
- .027 Freezing tolerance: the primary form of freezing resistance in plants is tolerance of extracellular freezing; measured as winter hardiness or artificial freezing (to determine the frost-killing temperature, IV.5.2.026); inverse relationship between freezing tolerance and developmental stage; (see Frost hardening, IV.5.3.028).
- .028 Frost hardening, cold hardening, cold acclimation: seasonal changes in freezing tolerance vary from nearly none in new spring growth to a maximum in midwinter; some tender plants (e.g., tropicals) never develop frost resistance; low temperature itself is incap-

able of inducing hardening; hardening with alternating low and high temperatures is effective only if light is supplied during the high temperature period and consists of two or more stages at successively lower temperatures for maximum effect; hardening requires a reduction in growth in order to take place (i.e., growing plants won't harden); withholding water tends to increase cold tolerance, provided its effect is not offset by other conditions that promote growth; the annual curve for hardening and later dehardening is associated with day length as well as temperature; lengthening nights and lower temperatures improve hardening in the fall and shorter nights and increasing temperatures result in loss of hardening in late winter and spring months; induction of rest also plays a part in the fall and this may be as important as photoperiod; subtropical and tropical plants will acclimate to some extent (freeze at temperatures a few degrees lower) as a result of cold (below 20 or 10°C, depending on the plant) or lack of soil moisture which causes dormancy; this condition, however, with some exceptions (e.g., kumquat), can be readily reversed, regardless of the time of year.

.029 Frost plasmolysis: freeze-killed cells, specifically from the loss of water and consequent cell wall and protoplast contraction during extracellular ice formation and the inability of the protoplast to reabsorb the lost water when the extracellular ice has melted.

.030 Frost-resistance stable: plants whose hardiness fails to change much with external conditions.

.031 Hardening: (see Hardiness, IV.5.3.032).

.032 Hardiness: resistance to freezing, drought, or heat, etc.; the process of hardening is such that the tolerance to a given stress is increased; (see Pseudohardening, IV.5.3.041).

.033 Heat vs. chilling tolerance: kinds of indirect stress injury produced by chilling-sensitive and heat-intolerant plants are identical but tolerances to chilling and heat are mutually exclusive from the standpoint of lipids (high unsaturation in chilling-tolerant, low in thermotolerant) and proteins (those most easily denatured at high temperature give greatest stability at low temperature); (see Heat versus freezing tolerance, IV.5.3.034).

.034 Heat vs. freezing tolerance: similarities and differences include (a) development of freeze tolerance by hardening at low temperatures also results in direct-type thermotolerance because the hardened plant can prevent irreversible aggregation by avoidance or tolerance due to high hydrophobicity of proteins; (b) low-temperature hardening is slow, high (heat shock) hardening is rapid; (c) greater thermostability of many enzymes in heat-tolerant than in heat-sensitive plants but no pattern of cryostability in freeze-tolerant and less tolerant plants; (d) repair is a far more important factor in thermotolerance because of the more rapid metabolic rates at postheating than postthawing temperatures.

.035 Homoiotherm: an organism that does or can regulate its temperature without regard to that of its environment; e.g., a warm-blooded animal.

.036 Indirect injury: (see Chilling injury, IV.5.3.005).

.037 Indirect stress injury: (see Stress injury, IV.5.3.047).

.038 Poikilotherm: the opposite of homoiotherm; plants in general lack the capability of regulating their temperatures independently of their environment, which means that they can tolerate only moderately low or high temperatures, except in those instances in which contractile roots pull the growing point into the soil or nearer the soil surface and/or the plants have a rosette form as a means of avoidance (see Avoidance, IV.5.1.001); most poikilotherms also exhibit high-temperature tolerance because leaves can survive daytime temperatures in the 40 to 50°C range or even higher in some instances; it is noted, however, that fruit on citrus trees will not tolerate these temperatures if exposed but will if borne inside the canopy (e.g., 'Lisbon' instead of 'Eureka' lemon is grown on the Yuma mesa and 'Marsh' grapefruit in Iran, the latter being a phenotypic modification; Samedi and Cochran 1975).

.039 Primary direct freezing injury: (see Freezing injury, IV.5.3.023).

.040 Primary indirect freezing injury: (see Freezing injury, IV.5.3.023).

.041 Pseudohardening: the increased resistance to stress (e.g., freezing or drought) of a plant by changes in anatomical structure that lead to an increase in avoidance rather than tolerance; (see Hardiness, IV.5.3.032).

.042 Psychrophile: organism thriving at relatively low temperatures.

.043 Resistance adaptation: a measure of the organism's tolerance of a temporary stress.

.044 Secondary freezing injury: (see Freezing injury, IV.5.3.023).

.045 Secondary stress: (see Chilling injury, IV.5.3.005).

.046 Secondary stress injury: (see Stress injury, IV.5.3.047).

.047 Stress injury: the action of stress on a plant is manifested in several ways: (a) <u>direct stress injury</u>, induced by a direct plastic strain (e.g., freezing by exposure to a sudden low-temperature stress); (b) <u>indirect stress injury</u>, conversion of an elastic strain maintained for a long enough time to a plastic strain (e.g., chilling stress); (c) <u>secondary stress injury</u>, one stress (e.g., high temperature) may not be injurious but induces a second stress (e.g., water deficit) which may injure the plant; both stresses require a more or less lengthy period of exposure.

.048 Sunscald: thawing and subsequent rapid freezing of a tree trunk or limb exposed to bright sunlight; the freezing in this instance is mainly intracellular instead of extracellular.

.049 Supercooling (undercooling): cooling of a liquid or hydrated tissue below its freezing point without solidification; undercooling refers to supercooling below the expected (or known) freezing point; freezing requires not only a low enough temperature but also the presence of nuclei around which ice crystals can form; thus supercooling can occur only in situations in which nuclei (dust, certain bacteria, crystals of ice, or other substances) are absent or under any condition that may cause them to form; e.g., a citrus fruit may be exposed to a temperature well below its normal freezing point of -1 to -2°C and will remain unfrozen in a still atmosphere unless it is bumped when it immediately becomes rock hard (i.e., is frozen solid).

.050 Ultrarapid cooling (and warming): cooling and warming rates in excess of about 10,000°C/sec.

.051 Undercooling: (see Supercooling, IV.5.3.049).

.052 Undercooling point: (see Double freezing point, IV.5.3.018).

.053 Water of hydration: water molecules bound on ions and surfaces of protein molecules and polysaccharides by association with polar groups (hydroxyl, carboxyl, and amino); the tightness of the binding depends on the distance from the polar group; the closer

to the latter, the stronger the bonding force; amounts to 5 to 10% of the total cell water.
.054 Winter hardiness: the ability of plants to survive the rigors of winter.

IV.5.4. Other forms of stress

IV.5.4.1. High temperature (heat)

.000 High-temperature stress: for mesophiles (organisms that grow and develop at temperatures of about 10 to 30°) any temperature above about 35°C; the heat-killing temperature varies inversely with exposure time and as an exponential relationship (i.e., an Arrhenius plot results in a straight-line relationship between log heat killing rate and the reciprocal of the absolute temperature).
.001 Biochemical lesions: (see Heat injury, IV.5.4.1.005).
.002 Decomposition injury: (see Heat injury, IV.5.4.1.005).
.003 Direct heat injury: (see Heat injury, IV.5.4.1.005).
.004 Heat avoidance: heat-resistant plants may avoid excessive (i.e., lethal) temperatures by insulation (thicker bark on older trees), lowered respiration (e.g., aroid inflorescences), absorption of radiant energy by reflection, transmissivity or absorption by protective layers, and transpirational cooling (see Antitranspirants, IV.5.2.003); difficult to measure in practice from numerous variables involved; (see Poikilotherm, IV.5.3.038)
.005 Heat injury: several types are recognized: (a) direct injuries as in (1) membrane injury that results from direct strains such as protein denaturation and lipid mobility and (2) decomposition injury from chemical decomposition; (b) indirect injuries as in (1) starvation injury from direct kinetic strain (changes in metabolic reaction rates) or indirect strains (excess respiration over photosynthesis and loss of reserves), (2) biochemical lesions from direct kinetic strain or indirect strains (decreased net synthesis of essential coenzymes and disruption of the metabolic processes dependent on coenzymes), (3) toxicity injury from direct kinetic strain or indirect strains (decreased removal of toxic intermediates and disruption of metabolism by toxins), and (4) protein loss injury from protein denaturation (direct strain) or of protein synthesizing enzymes, inactivation of enzymes, and protein breakdown without resynthesis (indirect strains).
.006 Heat tolerance (thermotolerance): the ability of a plant to withstand heat; measured by many different methods; e.g., determination of the heat-killing temperature of plants or plant parts in a heat chamber (essential to maintain 100% relative humidity) or water bath (cytoplasmic streaming, conductivity, among others); some degree of hardening (increased heat tolerance) is made possible by brief but severe heat shocks, dry cultivation, increased light (?), and/or low mineral nutrition (not fully investigated).
.007 Membrane injury: (see Heat injury, IV.5.4.1.005).
.008 Protein-loss injury: (see Heat injury, IV.5.4.1.005).
.009 Starvation injury: (see Heat injury, IV.5.4.1.005).
.010 Thermotolerance: (see Heat tolerance, IV.5.4.1.006).
.011 Toxicity injury: (see Heat injury, IV.5.4.1.005).

IV.5.4.2. Light stress

.000 Light stress: a type of radiation stress, that caused specifically by near IR, visible, and UV wavelengths; (see also Ionizing radiation resistance, IV.5.4.2.001).

.001 Ionizing radiation resistance: resistance to X-rays, gamma rays (photons), alpha rays (positive, He nuclei), and beta rays (negative, electrons) by avoidance, such as small nuclear volume, water mainly in nonliving parts of cells, or impermeability to O_2; tolerance as avoidance of radiation strain, avoidance of ions, radicals, and their products by natural scavengers, tolerance of ions, etc., by protoplasmic substances, or tolerance of radiation strain, repair (metabolic reduction, etc.).

.002 Light-compensation point: the level of illumination required for photosynthesis to balance respiration and photorespiration (i.e., net assimilation of CO_2 must be sufficient to compensate for the respiratory and photorespiratory CO_2 evolution); the level is markedly dependent on temperature and is lower for shade than sun plants; (see Carbon dioxide compensation point, IV.1.2.007).

.003 Light-saturation point: the level of illumination at which the rate of photosynthesis reaches a plateau; i.e., no further increase will occur if the light intensity is increased.

.004 Photochemical strain: a rise in chemical activity as a direct effect of radiation; thus a primary direct injury.

.005 Photodynamic effect: an injurious, often lethal, effect of a photochemical strain induced by a photosensitizer; used in practice to kill bacteria.

.006 Photoinactivation: shutting down (inactivation) of some metabolic process (e.g., components of an enzyme system) by light, especially blue and near-UV; like photoinhibition, a primary indirect injury.

.007 Photoinhibition: reduction or cessation of a plant process, such as stem elongation, in the presence of light; a plastic strain that requires a certain exposure time and is not instantly reversible on removal of the stress; light appears to inhibit growth by upsetting the balance of growth regulators present in a plant part; a primary indirect injury.

.008 Photoreactivation: tolerance to UV light by a repair mechanism whose effect is to overcome inactivation of UV by short wavelength light (i.e., near UV or blue).

.009 Photoreversion: tolerance to UV light by a process that involves reversion or reversal of the effects of certain wavelengths of UV light by longer ones.

.010 Photosensitizer: light-absorbing compound, such as erythrosin, fluorescein, and rose bengal, as introduced materials or pigments in the plant itself (rarely, however, in higher plants), which increase the effect of a given level of irradiation.

.011 Radiation-induced stress: stress induced by high light intensity, such as sunscald or drought injury due to the evaporative effect of sun energy; both are secondary rather than primary in nature.

.012 Ultraviolet (UV) absorption: water absorbs UV much more intensely than visible radiation; anthocyanin and possibly other related but colorless pigments are responsible for the major absorption (hence protection) in plants; these pigments are associated with UV resistance by avoidance.

.013 Ultraviolet (UV) radiation: preferential absorption of UV light of

IV.5.4.2. HORTICULTURAL PHYSIOLOGY and CROP ECOLOGY

different wave lengths; that in the UV-B range (280-315 nm) is particularly injurious to various plant processes [e.g., photoinactivation of proteins, cyclic and noncyclic phosphorylation, growth inhibition (primary indirect injury) or death of epidermal cells (primary direct injury)].

.014 UV resistance mechanisms: several types, including avoidance, repair by photoreactivation, presence of higher amino acid content in light-resistant sun leaves; inactivation of enzymes or growth inhibition involves sulfhydryl groups; also oxidation and peroxidation processes.

IV.5.4.3. Salt (ionic), pressure, and flooding (waterlogging) stresses

.000 Salt (ionic) stress: two general types, chronic and acute; chronic stress, result from the gradual increase in elements essential for plant growth and development; acute stress results from certain elements (e.g., Hg, Cd, Co, Ni, and Cr) that are toxic in microamounts (e.g., at concentrations of 5 ppm or less) or others (essential elements like Cu, Zn, Mn, and B) that may be toxic singly or in combinations at higher concentrations; toxic (i.e., injurious) levels are pH-dependent in most instances (e.g., Al toxicity), and often induce deficiencies of other elements (e.g., excess Cu, Mn, and/or Zn are often manifested as Fe deficiency or excess As, as B deficiency).

.001 Calcicole (calciphile, calcitrophic): plants that can grow and develop on calcareous soils; may require high levels of Ca for proper growth and development; (see Calcifuge, IV.5.4.3.002).

.002 Calcifuge (calciphobe): plants whose growth is inhibited or exhibit injury or die when grown on calcareous soils; growth, etc., of these plants may then be influenced more by secondary effects of bicarbonate ions (HCO_3^-), pH (reduced solubility of essential ions), or mineral deficiency; (see Calcicole, IV.5.4.3.001).

.003 Calciphile: (see Calcicole, IV.5.4.3.001).
.004 Calciphobe: (see Calcifuge, IV.5.4.3.002).
.005 Calcitrophic: (see Calcicole, IV.5.4.3.001).
.006 Flooding (waterlogging): the partial or complete submersion of roots in water; flooding is ordinarily a temporary situation caused by heavy rains (e.g., from a hurricane) or excessive irrigation; waterlogging is a more or less permanent situation that results from poor internal drainage, a soil underlain with an impervious layer to give a perched water table, and/or land adjacent to a low, continuously wet area (e.g., swamp, marsh, pond, lake, or river) that may overflow periodically; flooding may result in little damage to roots, provided the water is moving and can be removed before the dissolved oxygen is depleted or there is substantial leaching of essential nutrients. A stagnant flooded or waterlogged situation results in the depletion of oxygen, which, with carbon dioxide respired from the roots, quickly produces an anaerobic condition that decreases water uptake, reduces transport of solutes, fermentation by soil bacteria, and the accumulation of reduced (hence more soluble) forms of manganese, iron, and the like, as well as sulfides, cyanide, etc., to toxic levels, all of which can kill or seriously impair normal functioning of the roots within a period of no more than 12 to 48 h. The sensitivity of crops to flooding or waterlogging varies

IV.5.4.3. HORTICULTURAL PHYSIOLOGY and CROP ECOLOGY

widely; most mesophytes are relatively intolerant, whereas swamp or marsh plants or mesophytes subject to periodic inundation may tolerate such conditions for many weeks or months, often by the development of some form of pneumorhizae (II.4.3.039) or aerenchyma tissue (II.4.3.003) to provide adequate aeration; these modifications may be hereditary (i.e., genotypic) or induced (i.e., phenotypic).

.007 Physiological drought: a droughtlike stress that occurs when the salt content (i.e., osmotic potential) of the soil water increases to a point at which absorption of water and essential ions into the roots is restricted; there is a wide range in salt tolerance among crops; many fruits (e.g., apple, pear, stone fruits, citrus, and avocado) are salt-sensitive, i.e.., growth is seriously restricted at or near 4 mmhos/cm salt concentration in soil water; fig, olive, grape, maize, rice, sorghum, and alfalfa are moderately tolerant (i.e., will tolerate up to 8 mmhos/cm); and date palm, sugar beet, cotton, and bermudagrass are tolerant (i.e., will tolerate somewhat > 8 mmhos/cm). (RU).

.008 Pressure stresses: symmetrical pressures, such as hydrostatic or gas, are ordinarily not injurious, whereas asymmetrical pressures, such as wind stress per se or wind-induced evaporative stress are artificial shearing processes, may cause actual mechanical damage (i.e., limb breakage or uprooting of plants by high winds), distortion of growth from constant wind, or crushing of tissue (e.g., cells under a cover slip on a slide). (GSK)

.009 Salt injury: may be primary, e.g., a toxicity or growth inhibition from excess salt (principally NaCl), secondary from osmotic effects, like a change in ionic balance or a decrease in water potential within the plant cells, or nutritional (i.e., adverse effects on K, Mg, and Ca levels).

.010 Salt injury resistance (or tolerance): several factors: (a) toxicity increases with temperature, (b) resistance is greater in the shade than light, (c) presence of Ca^{++} reduces injurious effects of Na^+, and (d) protection with growth regulators (e.g., IAA, kinetin, B995); avoidance can be (1) passive exclusion, (2) active extrusion, or (3) dilution after entry; wide variation in toxicity of various salts among crops; NaCl or $NaHCO_3$ in general are most toxic and $MgSO_4$ or $MgCl_2$, least toxic; hardening is possible only by culture in saline media or presoaking of seeds (not always beneficial).

IV.5.4.4. Atmospheric pollution and mechanical stresses

.001 Atmospheric pollution: a wide range of substances, which includes nitrogen oxides, ozone, sulfur compounds (e.g., H_2S, SO_2, various sulfides), fluorides, and other gases or particulates from industrial, urban, or natural sources (e.g., terpene derivatives from forest trees, particularly conifers), may injure plants directly or induce injurious secondary effects by disturbance (or inhibition in some cases) of various metabolic processes and induction of other stresses such as water stress and sensitivity to cold stress; time of exposure, repeated exposure, temperature, and other environmental factors affect pollution resistance (i.e., tolerance) and the possibility of isolating plant responses to a pollutant from other types of stress.

IV.5.4.3. HORTICULTURAL PHYSIOLOGY and CROP ECOLOGY

.002   Mechanical stress: various types of girdling, ringing, scoring, notching, twisting, bending, spreading, wind sway, and sleeving are utilized in horticultural crops to produce dwarf plants, induce flower-bud formation, promote fruiting, force vegetative buds into growth, produce strong sturdy trees, and control plant size; pruning, e.g., heading or thinning, per se dwarfs plants, and is a type of mechanical stress that some plants (e.g., some Myrtaceae) will not tolerate; graftage by budding, grafting, or approach grafting to produce compound trees is also a type of mechanical stress that is beneficial horticulturally. (GSK)

.002a. Girdling: two types: (1) a process by which a ring of bark is removed from the trunk or branch of a tree or a ligature is tied tightly as a means of dwarfing and to induce flower-bud formation and fruiting; (2) a ring of bark and wood is removed from the trunk of a tree; the object in this instance is to kill the plant. (GSK, TK)

.002b. Notching: a shallow cut into the wood just above a bud to force it into vegetative growth or just below a bud to promote fruit bud initiation. (TK)

.002c. Pleach: (see V.3.4.058).

.002d. Pollard: (see V.3.4.062).

.002e. Ringing (scoring): a form of girdling in which a narrow cut is made with a pruning knife or similar instrument around the circumference of a trunk or branch; differs from girdling proper in that no bark is removed.

.002f. Scoring: (see Ringing, IV.5.4.4.002e.). (GSK, TK)

.002g. Twisting, bending, and spreading: interweaving and/or forcible curvature of branches with appropriate weighting down or tieing to produce dwarfed plants, various forms of topiary work, or espaliers; also to induce growth of dormant lateral buds and to promote flower-bud formation and fruiting. (TK)

V.1. PROPAGATION, NURSERY HANDLING, SOILS, CROP PRODUCTION

1. Propagation and Nursery Handling — 417
    1.0. General — 417
    1.1. Nursery facilities, structures, equipment — 418
    1.2. Materials and supplies — 427
    1.3. Generative (seed) propagation — 432
        Fig. V.1.3.01 — 440
    1.4. Vegetative (asexual) propagation — 441
        Fig. V.1.4.01-.17 — 453
    1.5. Micropropagation — 470
        Fig. V.1.5.01 — 471

2. Soils — 472
        Fig. V.2.01 — 483

3. Crop Production (Fruits, Vegetables, and Woody Ornamentals) — 484
    3.0. General — 484
    3.1. Site Preparation — 484
        Fig. V.3.1.01 — 488
    3.2. Mineral nutrition — 490
    3.3. Pests, pest management — 497
        3.3.0. General — 497
        3.3.1. Pests - Animals — 498
        3.3.2. Pests - Plants — 509
        3.3.3. Pest management — 531
    3.4. Plant and crop-size control — 539
        Fig. V.3.4.01-.12 — 545
    3.5. Water regulation — 557

4. Crop Production (Commercial Floriculture) — 559
    4.1. Culture and shipment to market — 559
    4.2. Pests — 565
        4.2.1. Parasitic diseases, including animals — 565
        4.2.2. Physiological disorders (nonparasitic diseases) — 566

V.1. Propagation and Nursery Handling

V.1.0. General

.000 Propagation: the multiplication of plants by seeds [generative (sexual) propagation], numerous types of vegetative material (vegetative propagation), or *in vitro* culture (micropropagation); an ancient practice dating from the dawn of agriculture carried out in a nursery or directly in the field (e.g., many vegetables); partially a science, partially an art, in which success in many instances is highly dependent on the ingenuity of the propagator within the limitations of available material and environmental conditions; includes propagation per se and subsequent culture of plants to the stage at which they may be transplanted to permanent locations. (H & K)

V.1.0. PROPAGATION, NURSERY HANDLING, SOILS, CROP PRODUCTION

.001 Balled-and-burlapped plant: one dug from a field with a ball of earth around the roots, with a burlap wrap (jute cloth), or wire mesh plus burlap in the case of large trees, fastened securely around the roots; plants are often root-pruned beforehand, especially in the case of large or difficult to move specimens.

.002 Bare-rooted plant: one dug from the field, the soil shaken from the roots, and the top pruned heavily to match the size of the root system; used mainly for plants grown in light (i.e., sandy) soils, such as citrus in Florida.

.003 Medium (pl. media): the substrate in or on which plants or plant parts are grown or cultured; many different components and mixes; [see Propagation media (V.1.2.011)]. (H & K)

.004 Plant patents: (see III.2.0.007).

.005 Propagule: any part of a plant that may be used for sexual or asexual propagation.

.006 Rogue, roguing: the process of culling out or removing variations from the standard type. (H & K)

.007 Root-pruning: the practice of cutting off roots in a circle some distance from the stem with a sharp spade some months before the plant is to be dug and moved; used mainly for large or difficult to-move specimens (e.g., many trees). (H & K)

.008 Stock, stock plant: source of seeds or vegetative propagation material; also plants propagated or grown in a nursery; the lower portion of a plant propagated by graftage is properly termed a rootstock (V.1.4.028s). (H & K)

.009 Win: several meanings applicable to horticulture: (1) to gather or harvest a crop; (2) to extract peat, rock, or dirt from a pit or quarry; (3) to dry peat, seed, or turf by exposure to air.

.010 Winning: several meanings applicable to horticulture: (1) the drying of peat before it is ground and baled or bagged for shipment; (2) the drying and later extraction of seed from other plant parts; (3) the culture and harvesting of a crop.

V.1.1. Nursery facilities, structures, equipment

.000 Nursery facilities, structures, equipment: constitute the land, buildings, facilities, and mechanized and hand equipment required for nursery operations, including the propagation of seeds and vegetative materials, transplanting into containers or field rows, and growing on, with consequent maintenance (i.e., watering, fertilizing, pest control, and culling), until plants are ready for sale or, in the case of particular vegetables and some ornamental crops, seeds or other types of material (e.g., bulbs and corms), are sown or set directly in their permanent locations, which may be a field, greenhouse bench, or flower garden. (H & K)

.001 Bell jar: (see Propagating cases, V.1.1.046).

.002 Bottom heat: heating cables or pads or occasionally steam or hot water pipes or hot air tubes arranged in the bottom of a green house bench or ground bed to provide additional warmth to accelerate rooting of cuttings and the like.

.003 Bucket: cylindrical or inverted truncated cone-shaped metal (or plastic) container of 2- to 14-L capacity for carrying water or other materials.

.004 Capillary bench: (see Irrigation, V.1.1.017f).

V.1.1. PROPAGATION, NURSERY HANDLING, SOILS, CROP PRODUCTION

.005 Capillary mat: (see Irrigation, V.1.1.017f).
.006 Cold frame: a ground bed of variable dimensions but usually only 1 m (3 ft) or so wide and 1 to several m long, with low wood or masonry sides and sloping hinged cover of window sash, framed plastic panel(s), etc.; plants are grown directly in the medium or containers plunged or set on top of the soil or gravel that forms the floor of the frame; the purpose is to give the plants some protection while they are being hardened or grown on before being set out; often used in conjunction with a greenhouse or less frequently with a hot bed, by home or small commercial growers. (H & K)
.007 Container fillers: machines for filling flats, plant bands in flats, various types of pot, etc.; consist of a hopper for soil mix, feed mechanism, and conveyors for empty containers, filling station, and filled containers; may be more or less highly automated; many different versions and designs custom built or assembled from components available in industrial supply houses.
.008 Cooling: (see Greenhouse, V.1.1.017i).
.009 Coverings: (see Greenhouse, V.1.1.017d).
.010 Dibble: a garden hand tool for planting bulbs and the like; a modified trowel in which the blade part is a tapered cylinder with a conical end and curved or straight handle.
.011 Disk-harrow: a harrow (V.1.1.020) with disks about 40 to 45 cm in diameter (which may be flat, dished, or dished and scalloped) mounted at an angle to the line of travel; the angle is adjustable from straight (essentially cutting) to offset for more or less churning; usually consists of two sections with the disks angled toward the center; tandem disk-harrows are used to level fields after plowing.
.012 Disk plow: (see Plows, V.1.1.044).
.013 Dolly: a two or four wheeled flat-bed cart for transporting containers, packages, etc.; many different sizes and designs, depending on nursery requirements.
.014 Forklift, forklift truck: small motorized truck with hydraulic lift mechanism and broad flat tines for transporting pallets.
.015 Framing: (see V.1.1.017c).
.016 Glasshouse: (see Greenhouse, V.1.1.017).
.017 Greenhouse (glasshouse, plastic greenhouse): an enclosed structure for the propagation and container or bed culture of plants. (JR)
.017a. <u>Location</u>: relevant factors include (1) local building and zoning codes; (2) availability and initial cost of land; (3) land characteristics (e.g., local accessibility, elevation, topography, and drainage); (4) sources of irrigation (availability, legal constraints on usage, and chemical and biological quality); (5) temperature maxima and minima and wind patterns (land elevation, topography, slope exposure, and wind currents); (6) house orientation (normally north to south); (7) availability of labor; (8) transportation facilities (commercial carriers, airports, and highway systems); (9) accessibility to technical information, equipment dealers, maintenance services, horticultural supply sources, and utility services.
.017b <u>Types of structure</u>: several distinct types include (1) open, with posts and covering of polyethylene or Saran cloth; (2) ridge and gutter, as (a) individual (free-standing) houses, (b) ridge and furrow houses (several units without inside walls), or (c) "Dutch-type" houses which differ in width,

rafter lengths, and single-pane ridge-to-eave glazing; roofs may be low (1 to 2:12), medium (4 to 6:12), or high (6 to 7:12) pitch, usually even span; (3) quonset or vault design (bowed roof); (4) sawtooth (usually multiple with roof spans sloped in the same direction); and (5) inflatable polyethylene; dimensions may be 2.5-3 to 12 m or more wide, 4-6 to 30 m or more long, and 2 to about 4.5 m high at the ridge and 2 to 3 m high at the eaves.

.017c. Framing: usual materials are prefabricated aluminum alloy, precast steel, and wrought and malleable iron; standard weight galvanized (also wrought iron) pipe; wood, e.g., redwood, heart cypress, western cedar, heart pitch pine (all pressure treated with a preservative); temporary frames of PVC pipe, mesh, welded wire mesh, bamboo, and wire cables.

.017d. Coverings: plastic films (e.g., UV-resistant polyethylene and vinyls); polyester rigid fiberglass (flat or corrugated; coated with polyvinyl fluoride to retard photooxidation by UV radiation); glass (best but expensive; double-strength grade is recommended for greenhouse use).

.017e. Internal design: depends on intended type of use (i.e., plants for stock plants, pots, and/or baskets and intended plant size); suggested regular bench heights are 75 to 90 cm (30 to 36 in.) to the top of the sides, with the top bench in a double bench system at 120 to 150 cm (4 to 5 ft); widths of 120 to 150 cm for stock-plant production, 180 cm (6 ft) for potted plants, and for side benches (reachable from only one side) half-width; lengths about 15 m (50 ft); aisle widths of 45 to 90 or 120 cm, side aisles often being narrower than the center one, which may be wide enough to accommodate wheeled vehicles; may use pyramid racks, i.e., A-frame style with steps for potted plants, especially cascade or hanging types; benches may be movable (i.e., mounted on rollers to slide over a pipe framework; common in Europe); benches may be made with a wood, corrugated or flat asbestos-cement, plastic or metal bottom, with flat sides 15 to 20 cm high and set on pipe or other metal stringers (longitudinal supports) on concrete, wood, or steel pipe piers; may be equipped with facilities for automatic irrigation (see Irrigation, V.1.1.017f), shade frames for photoperiod control (see Shade, V.1.1.017g), and bottom heat (V.1.1.002).

.017f. Irrigation: several systems for watering (including fertilizers and/or pesticides) plants on greenhouse benches or in beds: (1) overhead sprinklers; (2) hoses with a "waterbreaker" (nozzle with fine holes in one or more screens); (3) capillary benches (i.e., watertight benches with sand or other suitable medium for subirrigation; may be intermittent level or constant level); (4) capillary mats (flat or slightly inclined benches covered with a watertight film with synthetic fiber (3 to 5 mm in diameter) mats laid on top; particularly well adapted to small pots); (5) "Viaflo" (twin-wall) tubes; (LA), a system that dispenses water to a medium at a low rate through tubes; (6) self-subirrigating containers made with a reservoir large enough to hold sufficient water for one to several weeks and a wick or layer of medium that acts as a capillary column to convey moisture to the plant root zone (now used widely for foliage or other plants in offices, etc.); (7) dew hoses, drip hoses, and "soaker" hoses; and (8)

V.1.1. PROPAGATION, NURSERY HANDLING, SOILS, CROP PRODUCTION

.017g.   ooze headers (canvas or plastic hoses with fine holes that permit water to ooze out continuously).

.017g.   **Shade:** several common methods include (1) paint (oil-based for glass, water-based for fiberglass or soft plastics); (2) fabrics (e.g., polypropylene cloth), generally used inside a greenhouse; and (3) preshaded fiberglass.

.017h.   **Heating:** several types, such as steam, hot water, direct forced air, unit heaters (gas-or fuel oil-fired), or solar (water) heat, of which the first two and the last run through pipes; the forced-air system runs through polyethlene delivery tubes or directly into the house via a fanjet blower system.

.017i.   **Cooling:** five different systems: (1) ventilators manually operated or thermostatically activated motor driven; (2) fan and pad systems (standard, with extreme temperatures) in which air pulled into a house by large fans is cooled by the evaporation of water [involves a change of state from liquid to vapor in which 580 gcal (=2428J · kg^{-1}) are absorbed at 30°C] flowing over pads of wood wool, Spanish moss, etc.; the amount of cooling depends on atmospheric humidity, thus becomes progressively larger as the humidity is lower; a remarkably efficient system even in humid areas such as the East or Southeast; (3) fog (high-pressure mist) systems with thermostatic and humidistatic controls; (4) ground irrigation (i.e., under-bench sprinkling); (5) forced air, where cool air is drawn in as warm air is blown out of the greenhouse.

.018   Ground bed: consists of old window sash or sheets of plastic film (e.g., polyethylene, mylar, or fiberglass) laid over simple wood, bamboo, or pipe frames to cover a ground bed (i.e., a specially prepared area, often built up 15 to 30 cm above the ground and lined with boards or concrete blocks around the perimeter); differs from a greenhouse, hot bed, or cold frame in having a nearby source of water but no heating or cooling facilities and ventilation provided by propping up a sash or rolling up the sides of the plastic film cover; much less elaborate in construction and usually designed for temporary (i.e., seasonal) use; (see Cold frame, V.1.1.006; Hot bed, V.1.1.026).

.019   Grubbing hoe: (see Mattock, V.1.1.032).

.020   Harrow: a cultivation implement for pulverizing or smoothing soil and sometimes for mulching, covering seeds, or removing weeds; several versions, of which the disk-harrow and spring-tooth are the most common.

.021   Headhouse: a building attached and opening directly to the (short) side of one or more greenhouses as a service area which includes propagation and transplanting facilities, working space for other greenhouse-related activities, storage space for materials (e.g., fertilizers and pesticides), supplies (containers and labels) and tools, boiler room (for steam or hot water heat), and greenhouse controls.

.022   Heating: (see Greenhouse, V.1.1.017h).

.023   Hiller plow: (see Plows, V.1.1.044).

.024   Hoe: a hand tool used mainly for weeding; two main configurations, those with a broad thin blade set transversely to the handle and used with a chopping motion and those with a blade (which is more or less triangular) set at an angle so that it can slide along (scuffle hoe); many versions of the former, e.g., garden hoes, in which the blade extends from a thin metal neck at the end of the handle, "eye" hoes, which are of heavier construction and made with an eye in the blade for the handle to slip through, and grubbing hoes (see Mattock, V.1.1.032), etc.

.025 Hoses: rubber, plastic, or composition (i.e., fabric plus rubber or plastic), usually multiply, flexible tubes for garden, lawn, nursery, etc., watering by hand, or with various types of sprinkler; often used in lieu of permanent sprinkler installations for small numbers of plants, especially those that require unusual care in maintaining proper moisture levels; hoses for general garden or nursery use generally measure 1.6 cm (5/8 in.) or 1.9 cm (3/4 in.), whereas those for parks, parkways, or golf courses are often 2.5 cm (1 in.) or 3.8 cm (1 1/2 in.), with operating pressures of about 1.4 to 4.2 kg/cm (20 to 60 psi) or more.

.026 Hot bed: a ground bed with low wood or masonry sides and a sloping cover of window sash, framed plastic panel(s), etc., which may be closed or opened at will; thus essentially the same as a cold frame (V.1.1.006) but differing in that steam or hot water pipes, electric cables, fresh manure (older propagation texts suggest horse manure), or decaying vegetable matter are laid under the medium in the bed to provide a source of heat; plants may be grown directly or in containers plunged in the medium; suitable for small numbers of plants in lieu of a regular greenhouse. (H & K)

.027 Internal design: (see Greenhouse, V.1.1.017e).

.028 Irrigation: facilities for watering greenhouses, various types of ground beds, stock blocks, nursery blocks, etc., will utilize hoses, sprinklers, watering cans, syringes, and the like; (see Irrigation, V.1.1.017f; Irrigation systems, V.3.5.013).

.029 Lathhouse: (see Shade houses, V.1.1.059).

.030 Light box: a small container with a metal or plastic frame and glass or plastic sides and top (which may be removable) equipped with lights (usually fluorescent) in which small numbers of plants may be grown from seed or cuttings for home use; usually mounted on casters for ease of movement. (H & K)

.031 Location: (see V.1.1.017a).

.032 Mattock (grubbing hoe): a tool for grubbing out roots and the like in places where an axe or saw cannot be used (e.g., close to a stump or large tree); it has a short, heavy, slightly curved blade with an eyehole for the handle, which is generally much heavier and shorter (1 m or sometimes less) than that of a hoe (a mattock with a sharp blade can easily cut through a root 10 cm in diameter).

.033 Moldboard plow: (see Plows, V.1.1.044).

.034 Mowers, mowing machines: three basic versions: (a) reel-type, in which blades mounted spirally on a horizontal shaft wipe against a stationary bedknife, used for fine-bladed grasses in home lawns, parks, and golf-course fairways and greens; those for general use usually have 3 to 5 or rarely 6, and for precise cutting (as on a green), 7 to 9 blades; made as single units or in gangs of 3 to 9 units and in cutting swath widths of about 40 to 75 - 100 cm, the narrower ones are for home use and the wider, for fairways and other large areas; (b) <u>sickle-bar</u>, in which a series of flat, replaceable blades mounted along a horizontal bar moves in a reciprocal motion in close contact with a stationary frame with projecting fingers that hold grass, leaves, stems, etc., in place as the movable blade does the actual cutting; used for rough mowing of pastures which have herbaceous plants; now largely superseded by rotary mowers; (c) <u>rotary</u>, in which two, or occasionally four, straight or angled blades are mounted on a vertical shaft and physically shear off anything they hit with the cutting edge on the outer end of each blade; now widely used for mowing everything from home lawns to the rough along golf-course fairways or highway berms to thickets of brush; made in sizes of 40 to 60 cm for home use and up to 1.0 to 1.5 m, multiple shafted or in gangs, for other applications.

.035 Nursery block: areas under irrigation but seldom shaded for growing-on plants in containers or in open ground; those in the latter are sold as balled-and-burlapped or bare root trees, shrubs, or vines.

.036 Nursery equipment: general nursery equipment (i.e., not used for propagation) includes (a) garden or farm tractors with implements for plowing, disking, rotary tilling, mowing, brush cutting, seeding, hole digging, fertilizing, and herbicide application; (b) soil sterilizer; (c) soil shredder; (d) mechanical soil mixer; (e) mechanical soil screens; (f) container-filling machines; (g) sprayers and dusters from hand or knapsack types to large mist blowers or airblast sprayers; (h) hand or tractor-mounted seeders; (i) hand tools (e.g., spades and shovels); and (j) pruning tools; available in a wide range of sizes and designs, according to requirements of a given nursery.

.037 Pick: tool for digging or prying out stones or loosening hard soil; has a double-ended blade with a stout tapering point on one end and a relatively narrow, slightly curved flat-pointed blade on the other.

.038 Pitch fork: a tool with long, sharp, slightly curved sharp-pointed steel tines attached to a long handle for picking up leaves, grass, or other light material; ordinary garden pitch forks have about 4 to 6 tines, manure forks may have 12 to 20, and a hay fork, about 3.

.039 Plant chipper: differs from soil or plant shredders in its series of short, curved, sharp projections mounted on a steel drum which effectively reduce large limbs or logs to chips of a size determined by the spacing and size of the teeth; trailer-mounted with blower and chute to convey the chips into a dump truck; used mainly in connection with removal of large trees in urban areas.

.040 Plant shredder: similar to a soil shredder (V.1.1.064) but with heavier knives to tear limbs, leaves, and other plant debris into small pieces; usually trailer-mounted and equipped with a blower and chute to convey chopped up material into a dump truck or into a pile.

.041 Plastic bag: (see Propagating cases, V.1.1.046).
.042 Plastic greenhouse: (see Greenhouse, V.1.1.017).
.043 Plastic mesh house: (see Shade houses, V.1.1.059).
.044 Plows: two major types (and many variations); (a) <u>moldboard plow</u>, which has three main parts: (1) share or cutting edge with a point at the tip and a heel at the outside corner, (2) moldboard, the curved face that turns the furrow, and (3) landside which takes the side pressures when the furrow is turned; (b) <u>disk plow</u>, which consists of a concave bowl set at an angle to the shaft on which it rotates, sometimes reversible to turn soil in the same direction in each row; used for sticky or very hard soils; both types may be single or ganged (i.e., two or more plows attached to a tractor); the moldboard type may also be double (i.e., one share with right and left moldboards), without a moldboard (i.e., a subsoil plow to loosen soil), or a hiller

type (i.e., two plows with moldboards turned toward each other). (B - M)

.045 Potato fork: a tool with four or five heavy, flattened, slightly curved, pointed tines at a right angle to the handle; used for digging roots or tubers (e.g., potato or sweet potato); also made in a lighter version with cylindrical, sharp-pointed tines for general garden use.

.046 Propagating cases: several types, of which the most elaborate is the Wardian case, a miniature glasshouse widely used for transporting plants before the days of air shipments; others are a bell jar placed over a plant in a container, or a newly grafted plant or a cutting to be rooted in a plastic bag; the last is utilized in a quasi-commercial way for cacao and certain other tropicals. (H & K)

.047 Propagation and transplanting facilities: a shed or protected area in which seeds may be extracted and dried, seed flats prepared, plants transplanted into containers; a potting bench or benches of convenient height for sitting or standing, 90 to 120 cm wide, and as long as needed (allowing about 2 to 2.5 m per worker), with room for soil, containers (pots), and flats; also a large flat-bottomed sink with running water, screens of different mesh sizes, drying cabinet, refrigerator or cold storage room for seed handling, space for preparation of cuttings, equipment for washing containers, and other operations; other facilities are three to five bins, each holding several cubic meters of medium, with a flat, level surface on which mixing may be done located adjacent to the potting-transplanting center; medium-to-large-sized nurseries will have a soil shredder, mechanical mixer, mechanical screens, and soil sterilizing equipment (preferably steam). (JR)

.048 Propagation structures: an enclosure or structure designed to provide propagules (seeds and cuttings) with the most favorable environment possible for their successful multiplication and later growth and development; numerous types according to the quantity and type of plants being multiplied: ground beds, greenhouses, hot beds, cold frames, shade houses, light boxes, propagating cases, or individual flats or pots covered with a pane of glass or plastic bag or bell jar; greenhouses, shade houses, and ground beds, particularly, are used for commercial culture of floricultural and foliage crops and for growing on woody and herbaceous ornamentals for landscape planting, bedding plants, and the like. (H & K)

.049 Proportionator, proportioner: a type of pump system for a greenhouse or similar installation by which fertilizers, pesticides, etc., may be introduced into the irrigation water in known quantities; these range from simple manually controlled tank-pump devices to elaborate fully automated systems with pH and conductivity (i.e., salt concentration) controls. (JR)

.050 Rake: a hand tool for raking up leaves and other light debris or leveling planting beds, lawn areas, and the like; two main versions, (1) a springtooth rake, which has a series of flat, narrow tines of spring steel, with curved ends, with the tines fanning out from the point of attachment on the handle, and (2) a garden rake, which has teeth projecting at right angles from a metal (or rarely wood) frame attached to the end of the handle; the springtooth is used for raking light debris, the garden rake for leveling and other heavier tasks.

.051 Range (greenhouse range): a series of two or more greenhouses (V.1.

1.017), constructed as a single unit with or without walls between individual houses, or separate greenhouses which are often attached to a headhouse (V.1.1.021).

.052 Reel-type mowers: (see Mowers, V.1.1.034).

.053 Rooting structures: special structures, such as double-glass frames, mistbeds, high-humidity beds, or individual containers in which proper conditions of heat, moisture, light, and aeration may be maintained during the rooting period of cuttings other than those that are direct-set (V.1.4.022f).

.053a. Double-glass: the traditional method consists of a heavily shaded cutting bed under glass in a greenhouse; cuttings are inserted with few or no leaves attached; many species, including most tropicals, are considered difficult or impossible to root.

.053b. "Electric leaf": a sensing device used to control the intervals at which a mist bed is watered; many different types, some operating on a resistance principle, others gravimetric, to detect when humidity in the vicinity of leaves on the cuttings has dropped low enough to require water; actually preferred to an interval timer if properly adjusted because it will vary the intervals between waterings and according to light conditions (i.e., passing clouds); (see Mist bed, V.1.1.053e). (H & K)

.053c. High-humidity beds: the forerunner of mist beds V.1.1.053e), which require a dependable source of electricity and water and work on the same principle; the medium is enclosed in a wood, concrete, or metal box with a movable lid of glass sash or plastic film which is shaded with muslin or burlap laid on top of the lid or draped over a bamboo or lath frame above the lid; some beds have means of dripping water over the cloth to prevent overheating; cuttings are syringed two to four times a day; in another type, cuttings are watered thoroughly at the time of insertion, the bed lid is kept tightly closed, and rewatered every 7 to 10 days; hard-to-transplant species are rooted in individual bags set closely together in the bed; used commercially for large-scale production of clonal cacao, tea, and other crops in tropical countries. (HA)

.053d. Individual containers: a minature version of a high-humidity propagation bed in which individual polyethylene bags are used as rooting chambers; the cutting and medium are watered thoroughly at the beginning, the top of the bag is tied tightly, and opened every few weeks to add water (if needed) until rooting is completed; differs from the procedure for hard-to-transplant species described under high-humidity beds (V.1.1.053c) only in that the latter are in open bags; used by small growers where a relatively few plants are needed. (UQ)

.053e. Mist bed: continuous (less desirable from leaching of nutrients) or intermittent application of water as a fine mist over leaf surfaces of cuttings; the essential conditions are high humidity, the highest possible light intensity without burning leaves or overheating cuttings, and a medium with perfect drainage; typically, beds are 1.2 to 1.5 m wide, 10 to 20 cm deep, and any desired length, with loose, porous medium, such as any convenient mixture of peat, perlite, vermiculite, sand, ground coir (coconut husk), and bagasse (sugar cane fiber) that is reasonably inert and will not settle and pack, underlain with gravel; mist heads (many types available) are mounted 30 to 45 cm above the bed and the water supply is controlled by a solenoid valve activated by an interval timer or "electric leaf"; widely used for a vast number of crops. (The medium is used only once or twice and then discarded to prevent buildup of fungi and bacteria.) (H & K)

.054 Rose-nozzle: a circular or fan-shaped nozzle with small perforations, which may be attached to a watering can, syringe, or garden hose for sprinkling.

.055 Rotary mowers: (see Mowers, V.1.1.034).

.056 Rotary tiller: a self-propelled hand-guided or tractor-mounted machine which has L-shaped blades mounted on a rotating shaft; used for the preparation of ground seedbeds or similar applications in which complete mixing of the soil down to a depth of 25 cm or more is desired; also a good means of ridding an area of subterranean rhizomes (e.g., bermudagrass and maidencane) which wrap around the tines as the machine moves across the field, brickbats, and pieces of roots.

.057 Self-subirrigating containers: (see Irrigation, V.1.1.017f).

.058 Shade: (see V.1.1.017g).

.059 Shade houses: structures built for the purpose of hardening-off plants or growing those requiring partial shade (e.g., ferns and foliage plants) in containers, ground beds, or raised beds; several different forms: <u>lathhouses</u> (wood lath 1.5 X 5 cm spaced 5 to 10 cm apart), <u>plastic mesh houses</u> (Saran, polypropylene, etc., mesh to give 20 to 80% shade), <u>wire mesh house</u> (fence wire mesh with Spanish moss or similar material growing on it), or palm leaves or fern fronds laid on a wood, bamboo, or pipe framework; variable in dimensions from a few square meters to several hectares (e.g., ferneries); most of the larger, newer ones are of Saran or other plastic mesh construction. (JR, H & K)

.060 Shovel: a tool for moving soil or debris, digging up plants, leveling ground areas, and turning over soil for a garden; two main configurations, (1) round-point for moving or digging operations, which has a wide, thin, rounded blade tapering to a broad point, and (2) square-point for leveling, which has a flat blade with a square end and turned up sides; both are made in a wide range of sizes and have a long or a short handle, the latter with a handhold at the top.

.061 Sickle-bar mowers: (see Mowers, V.1.1.034).

.062 Soil mixer: a concrete mixer, which has a motor engine and a tilted rotating urceolate-shaped drum with narrow slanting blades on the inside; it ranges in size from small wheel or sled-mounted of 0.25 m^3 capacity or less to regular truck-mounted units that hold several m^3.

.063 Soil screen: two versions; a flat rectangular screen with a metal or wood frame mounted horizontally or tilted slightly and vibrated or shaken in a reciprocating motion, or a cylindrical (drum) screen which is rotated; the latter is amenable to continuous feed for large-scale operations.

.064 Soil shredder: a modified hammer mill with rotating knives on a horizontal shaft and stationary knives between them to provide a shearing action to pulverize soil, peat, etc., fed into the machine from a hopper; usually self-contained (i.e., with a motor or engine) and mounted on wheels.

.065 Soil pasteurization (soil sterilization): three main methods: steam (wet or aerated), heat (wood, fire, electrical), and chemical; steam is generally preferred; 180°F (82°C) for 30 minutes has

V.1.1. PROPAGATION, NURSERY HANDLING, SOILS, CROP PRODUCTION

been recommended for years, although 140°F (60°C) for 30 minutes will not kill so many beneficial microorganisms (nor weed seeds) and avoids toxicity [mainly nonpeat, high organic-matter mixes or soluble Mn from exposure to temperatures above 185°F (85°C)]; dry heat is generally not recommended as too expensive and likely to "burn" soil; fumigants (chemicals), such as formaldehyde (good fungicide), methyl bromide plus chloropicrin (kills most nematodes, insects, weed seeds, and some fungi; toxic and odorless, hence usually contains 1% chloropicrin for smell), vapam, and fungicidal drenches, e.g., dexon (controls Phytophthora, Pythium), terrachlor (inhibits Rhizoctonia), and benomyl (inhibits Rhizoctonia, Fusarium, and Verticillium); all fumigants require a 2-to 14-day waiting period (note that environmental regulations may necessitate use of other materials); toxic to some plants; all sterilization should be done on moist soil. (H & K)

.066 Soil sterilization: (see Soil pasteurization, V.1.1.065).
.067 Spade: similar to a shovel but with a heavier, long, slightly rounded blade for digging plants, root pruning, turning over soil in a garden; a digging spade usually has a longer, narrower, heavier blade than the regular garden spade; can have a long or short handle, the latter with a handhold at the top.
.068 Spading fork: similar to a spade (V.1.1.067) but with several heavy flat tines; used for turning over soil in a garden.
.069 Stock nursery: special area for stock plants that serve as sources of seed, cutting wood, and budwood for budding and grafting.
.070 Subsoil plow: (see Plows, V.1.1.044).
.071 Syringe: a rubber bulb with a short neck and nozzle (with fine holes) used for applying a fine mist over a plant.
.072 Tine: a slender, pointed projecting part, usually one of a set of two or more, as on a pitchfork or spading fork.
.073 Trowel: a garden hand tool for general transplanting of small seedlings, setting of bulbs, etc.; consists of a broad-to-narrow, concave blade and handle; many shapes and sizes.
.074 Types of structure: (see V.1.1.017b).
.075 Viaflow: (see Irrigation, V.1.1.017f).
.076 Wardian case: (see Propagating cases, V.1.1.046).
.077 Water breaker: hose nozzle with fine holes in one or more screens.
.078 Watering can: a metal or plastic cylindrical container with a side spout and nozzle and a bail for carrying; used for watering small plants or applying liquid fertilizer or pesticides to or around them.
.079 Weeder: a hand tool with clawlike metal fingers for weeding in a garden or flower bed.
.080 Wheelbarrow: a type of cart with metal or wood body (the latter often with detachable sides and flat bottom) with a wheel at one end and handles and legs at the other; used for hauling soil, containers, etc., in quantities one person can move.
.081 Wire mesh house: (see Shade houses, V.1.1.059).

V.1.2. Materials and supplies

.000 Materials and supplies: include media, containers, fertilizers, pesticides, and growth regulators as materials, and labels, stakes, twine, other types of tie, pencils, record books as supplies; the distinction is that the former are used directly in plant culture

V.1.2. PROPAGATION, NURSERY HANDLING, SOILS, CROP PRODUCTION

(i.e., required by plants per se or as aids to their growth and development) and the latter are disposable items.

.001 Antitranspirants: a compound designed to reduce transpiration when sprayed or painted on leaves of newly transplanted trees, shrubs, or vines or used as a dip for cuttings in lieu of misting; may interfere with photosynthesis and respiration if the coating is too thick or complete; commercial formulations are "WiltPruf," "Mobileaf," or "Vapor Gard." (H & K)

.002 Budding knife: a single-bladed knife with a bone, wood, or plastic handle; may be single piece or the folding type; has a straight cutting edge that curves up at the end to form a sharp point useful in prying up bark flaps; (see Grafting knife, V.1.2.007). (H & K)

.003 Containers: many different types for germinating seeds, rooting cuttings, or growing on seedlings or nursery plants; most of those now in use are designed for mechanized filling with potting medium (except fiber blocks and others that already have medium in them).

.003a. Asphalt-coated felt paper pots: inexpensive, sturdy, lightweight, easily nested containers available in a number of sizes, suitable for temporary use (e.g., for bedding plants and the like).

.003b. Cones: styrofoam, styrene-latex, or other plastic molded cones or tubes with flanges for insertion in a wire mesh frame or made up in multiples of 10, 20, or 100 are used widely for machine planting of vegetables (e.g., tomatoes and squash) and flowers; also beginning to be used for greenhouse-grown citrus seedlings that will be bench-budded later.

.003c. Flats: shallow wooden, plastic, or metal trays with drainage holes in the bottom and available in various sizes from those easily moved by hand to large ones with pallet runners that are handled by forklift truck.

.003d. Paraffined paper, styrofoam cups: cheap, lightweight containers with holes punched for drainage for growing on and transferring young plants.

.003e. Peat or fiber blocks: made of solid material, sometimes with a prepunched hole, for germinating seeds or rooting cuttings of chrysanthemums, poinsettias, etc.; sometimes compressed to a wafer that expands to usable size when wetted.

.003f. Plant bands: wood, plastic, or paper containers that are glued or stapled as an endless strip and scored to form a square when opened up; usually bottomless, they are placed in a flat or similar rack for support.

.003g. Polyethylene bags, tubes: inexpensive containers available in many sizes as bags with perforated bottoms or seamless tubes that can be cut to a desired length and the bottom folded under; more difficult to fill and handle than plastic or metal pots or cans but widely used, especially in tropical regions where labor costs are generally low; bag-filling machines have been developed in other areas.

.003h. Pots: clay, plastic, metal, concrete, ceramic, asphalt roofing felt, or fiber containers ranging in size from diminutive "thimble" pots to 45 to 60 cm in diameter; round or square, plastic, fiber or peat and sometimes made into "packs" of 8 or 12 together (widely used for bedding plants); clay pots have largely been superseded by those of other materials; several distinct forms, such as the regular type, in which the depth is equal to or greater than the diameter, azalea type, the depth of which is about two-thirds that of the diameter, the bulb type, which is about a quarter to a third as deep as broad, plus others of special uses, e.g., for orchids with diagonal side slits; fiber pots are biodegradable and set out without removal of the plant from the container.

.004 Double-knife: two budding knives or razor blades separated by a wood spacer and securely fastened as a unit for use in patch-budding. (H & K)

.005 Foliage-plant potting mixes: several special media for foliage plants are recommended: (a) Cornell: (1) foliage plant mix: two parts sphagnum peat, one part horticultural vermiculite, one part horticultural perlite (plus 4.75 kg ground limestone, 1.2 kg superphospate, 0.6 kg potassium nitrate, 1.2 kg fritted trace elements, 0.45 kg ferrous sulfate, and 1.5 kg of Osmocote (14-14-14) per cubic meter); (2) epiphytic plants: one part peat, one part perlite, one part Douglas fir bark (plus 4.15 kg ground limestone, 2.4 kg superphospate, 1.6 kg potassium nitrate, 1.2 kg fritted trace elements, 0.45 kg ferrous sulfate, and 1.5 kg Osmocote per cubic meter). (b) University of Hawaii: one part wood shavings: two parts cinders. (c) Texas A & M: one part sphagnum peat, one part perlite. (d) University of Florida: (1) raised bench stock on solid bottom: three parts peat, one part sand; on wire bottom: three parts peat, one part perlite; (2) propagation benches: 100% peat, three parts peat, one part perlite (or styrofoam if iron sensitive); (3) potted plants: (a) two parts peat, one part bark, one part shavings; (b) one part peat, one part bark; [(a) and (b) for plants in greenhouses.]; (c) three parts peat, one part sand (for larger containers, e.g., in shade houses). (Note: 1 kg/m^3 = 1.69 lb/yd^3.) (JR)

.006 Grafting compound: two principal types: (a) grafting wax, mixtures of beeswax, rosin, raw linseed oil or tallow, and lampblack or powdered charcoal for hot wax, which must be melted for use, or hand wax, which consists of beeswax, tallow, and rosin (e.g., 2:1:4 or 1:2:4 mixtures with linseed oil sometimes substituted for tallow in the latter; and (b) water-asphalt emulsion (likely to be washed off if rained on before thoroughly dry; otherwise quite satisfactory and widely used). (H & K)

.007 Grafting knife: a single-bladed knife with wood, bone, or plastic handle; may be straight (nonfolding) or folding; the blade is longer and heavier than that of a budding knife (V.1.2.002) and differs in the tip being curved down to the lower edge (i.e., the cutting edge is straight instead of curved at the end). (H & K)

.008 Grafting wax: (see Grafting compound, V.1.2.006).

.009 Labels: wood, plastic, paper, or metal, painted or unpainted, usually rectangular strips inserted in containers, tied or wrapped around a stem, or fastened to a stake at the head of a row or rows of plants with information written or embossed on them for the purpose of identification; may be temporary or permanent; many different sizes, forms, and colors.

.010 Pallet: a rectangular or square wooden platform with dimensions about 100 X 120 cm and heavy (7.5 X 10 cm) runners beneath; usually two-way (i.e., can be picked up from either end, sometimes four-way (can be picked up from both ends and both sides); designed for transporting containers stacked on them with a forklift truck (V.1.1.014).

.011 Propagation media: individual components used for propagation purposes are soil, sand, peat, sphagnum moss, vermiculite, perlite, compost, manure, bark, sawdust, and shavings, plus innumerable soil mixes, including the traditional U.C., John Innes, and Cornell, etc.; major factors in choice of media for a particular mix are the purpose (i.e., for seed or use in containers), good drainage, good aeration, light weight, free from pathogens or weeds (or easily capable of being made so), chemical inertness (i.e., lack of toxic substances), stability (chemically and physically), and cost; (see Win, V.0.009, Winning, V.0.010, also Foliage plant potting mixes, V.1.2.005). (H & K)

.011a. "Activo": (see Compost, V.1.2.011c).

.011b. "Artificial" soil mixes: numerous formulations; e.g., equal parts of shredded bark, peat moss, perlite, and sand.

.011c. Compost: essentially a heterogeneous mixture of vegetative remains, including kitchen refuse of all (nonmetal) sorts, formed in layers alternating with soil and turned over occasionally until the whole is thoroughly decomposed, or simply refuse alone; knowledgeable growers inoculate a new compost pile with material from an older pile or sprinkle one of the proprietary bacteria-containing preparations (e.g., "Activo", V.1.2.011a) over each layer to hasten breakdown.

.011d. Cornell "Peat-Lite" mixes: (see Special soil mixes, V.1.2.011n).

.011e. John Innes mixes: (see Special soil mixes, V.1.2.011n).

.011f. Manure: the product of animals such as cattle, horses, sheep, goats, and poultry, which has been used for a long time in traditional propagation mixtures after it has decomposed thoroughly; may burn plants from evolution of ammonia when fresh (particularly poultry manure); frequently loaded with weed seeds and various pathogens, necessitating sterilization of any mix in which it is used; originally used because of the organic matter content and nutrients (which in decomposed manures are usually low and unbalanced), now largely superseded because of its cost and other drawbacks.

.011g. Milled sphagnum: (see Sphagnum moss, V.1.2.011o).

.011h. Peat (peat moss): a mass of semicarbonized vegetative material formed by partial decomposition of plant tissues (e.g., mosses and sedges) in water, contains less than 10% sand or other matter and is usually highly acid; peat moss is derived from mosses; sedge peat or peat humus, the least decomposed type, is often preferred for horticultural uses.

.011i. Perlite: ground pumice treated with steam to expand the particles; used as a propagation medium; may be toxic to some plants; made in horticultural and industrial grades, of which only the former should be used for plant purposes.

.011j. Sand: soil particles ranging from about 0.02 mm (fine sand) to 2 mm (coarse sand), free draining and essentially inert if well washed (e.g., with acid) to remove contaminants; the heaviest of all common propagation media, about 1.8 m. tons per m^3.

.011k. Sawdust: (see Shredded bark, sawdust, wood shavings, V.1.2.011 l).

.011l. Shredded bark, sawdust, wood shavings: materials from redwood, cedar, fir, or pine serve the same purpose as peat moss but with a slower decomposition rate; may be toxic when fresh; sufficient nitrogen should be added to take care of the decomposition requirements of sawdust.

.011m. Soil: ordinary garden or field soil, variable in physical and chemical composition as well as pathogens and weed seeds;

often a component of traditional mixes which should (must) be sterilized before use.

.011n. Special soil mixes: mixes developed to permit use of uniform, generally available materials; they require no previous preparation and are easily duplicated; (1) U.C. mixes: based on fine sand and finely shredded peat moss; proportions vary from 100% sand to 100% peat moss with specific fertilizer formulations for each, depending on whether the mix is to be stored indefinitely or used within a week of preparation; (2) John Innes mixes: one for seeds (John Innes seed compost): two parts loam, one part peat moss, and one part clean sand by volume, plus 2 lb (900 g) superphosphate and 1 lb (450 g) ground limestone per yd ($0.765 m^3$); one for potting (John Innes potting compost): seven parts loam, three parts peat moss, two parts clean sand by volume, plus 5 lb (2.27 kg) John Innes base and 1 lb (450 g) ground limestone; (3) Cornell "Peat-Lite" mixes, three basic mixes utilizing peat plus vermiculite or perlite plus fertilizer mixes A-C, A and B for containers, C for germinating seeds; (see also Foliage plant mixes, V.1.2.005).

.011o. Sphagnum moss: mosses (Sphagnum spp.) found in ditches, along the shores of lakes, etc.; collected and used fresh or more often dried; widely used after moistening for air layerage, packing material, and the like; milled sphagnum (scrubbed through a 2-5 mm wire mesh screen) is an excellent medium for germinating seeds, often used as a thin layer on top of soil in a seed flat or bed (does not as a rule need to be sterilized).

.011p. Traditional soil mixes: numerous variations but generally mixtures of sand, soil, and peat moss (or leaf mold or well-rotted manure) in proportions of 1 or 2:1:1 (by volume) for rooted cuttings and young seedlings or 1:2:1 (by volume) for general container-grown nursery stock; variable in physical and chemical properties from one batch to another.

.011q. U.C. mixes: (see Special soil mixes, V.1.2.011n).

.011r. Vermiculite: a type of micaceous mineral expanded with heat and often used in media (but must not be compressed or compacted when wet); also a type of clay mineral (V.2.098).

.011s. Wood shavings: (see Shredded bark, sawdust, wood shavings, V.1.2.011).

.012 Slow-release fertilizers: granules of soluble fertilizers coated with some type of membrane, resin, sulfur, and the like; glass frits; ureaform (urea-formaldehyde); also slightly soluble materials such as granulated magnesium-ammonium phosphate; (see Mineral nutrition, V.3.2.). (H & K)

.013 Soluble fertilizers: high-analysis mixtures of readily available source materials (i.e., those with high solubility in water, such as biuret-free urea, ammonium nitrate, potassium nitrate, and monoammonium phosphate); many different analyses, a common one being 20-20-20; (see Mineral nutrition, V.3.2.). (JR, H & K)

.014 Stakes: slender straight metal rods or wood (e.g., pine, cypress, redwood, or bamboo) sticks used for training plants or for attachment of labels at the head of a nursery row; generally 60 to 100 cm long.

.015 "Starter" ("booster") solution: a dilute solution of a soluble fertilizer, such as 20-20-20, applied to young plants shortly before or after transplanting to aid in their rapid establishment; (see Mineral nutrition, V.3.2.). (JR, H & K)

V.1.2. PROPAGATION, NURSERY HANDLING, SOILS, CROP PRODUCTION 432

.016 Water-asphalt emulsion: (see Grafting compound, V.1.2.006).
.017 Wrapping materials: a wide range of materials, such as plastic film strips or sheets, rubber strips, raffia, waxed string, budding cloth tape, masking tape, adhesive clear plastic tape, and various proprietary wraps with fasteners, may be used successfully for graftage purposes. (H & K)

V.1.3. Generative (sexual) propagation

.000 Generative (sexual) propagation: multiplication of plants by means of seeds. (H & K)
.001 Air screen cleaner: (see Seed-cleaning equipment, V.1.2.057).
.002 Alternating temperatures: many seeds require a diurnal temperature alternation of at least 10°C to germinate properly. (H & K)
.003 Bench-rooting, bench-rooted: the condition in a seedling primary root where curvature is pronounced or, in some instances, makes a complete loop at about the point of emergence from the seed; this results from seed-coat resistance; bench-rooting may also occur when a seedling is allowed to grow too long in a shallow container before it is transplanted, in which case the tap root will have an angular bend where its downward growth was stopped, some lateral growth and then further downward growth after transplanting; both types greatly reduce seedling vigor; plants with one bend should be discarded and those with two bends should have the taproot pruned above the lower bend if they are not culled out. (Fig. V.1.2.01a)
.004 Biostimulants: growth regulatorlike compounds that stimulate growth and development; e.g., a mixture of α-keto acids, as for tomato plants, applied at the time of sowing.
.005 Breeders' seed: (see Seed production, V.1.3.070).
.006 Broadcast seeder: (see Cyclone seeder, V.1.3.015).
.007 Certified seed: (see Seed production, V.1.3.070).
.008 Coefficient of velocity: a measure of the precocity of seedling emergence; expressed as $V_s = (100 \times A_1 + A_2 + A_3 - A_n)/A_1 T_1 + A_2 T_2 + - A_n T_n)$, where A is the number of seedlings that have emerged and T is the number of days after sowing.
.009 Color separator: (see Seed-cleaning equipment, V.1.3.057).
.010 Combine: a machine for harvesting plants and threshing seeds of certain vegetables, such as lettuce, spinach, radish, and cabbage, many flowers, and grain crops; [see Seed extraction (dry fruits), V.1.3.063].
.011 Composite mixture: breeder seed obtained by mechanically combining seed from two or more strains; the mixture is increased in successive steps in a certified seed program and distributed as a synthetic variety; (see Seed production, III.2.004).
.012 Contra-inoculation: treatment of legume seeds with the "wrong" species of Rhizobium; e.g., the use of the species that will produce nodules on soybeans on alfalfa or vice versa, neither of which will then produce nodules; (see Inoculation, V.1.3.041).
.013 Cool-season seeds: those that germinate at relatively low temperatures; the maximum is about 25°C; (see Thermodormancy, V.1.3.083). (H & K)
.014 Cross-inoculation: treatment of seeds from two different species of legume with a given species of Rhizobium or soil from a field in which one or the other legume has been grown results in proper

V.1.3. PROPAGATION, NURSERY HANDLING, SOILS, CROP PRODUCTION

nodulation on the roots of both; the success of this cross-inoculation procedure means that both species belong to the same group; (see Contra-inoculation, V.1.3.012, and Inoculation, V.1.3.041).

.015 Cyclone seeder: a device in which seeds are dropped from a hopper onto a revolving disk that disperses them by centrifugal force over an area; may be hand-operated, for seeding a home lawn, or motor-driven for larger areas, such as a golf course; also used to dispense fertilizers and granular or pelletized pesticides; the many different sizes and configurations available include those that use an air blast in lieu of a revolving disk.

.016 Cylinder separator: (see Seed-cleaning equipment, V.1.3.057).

.017 Cytex: a product derived from seaweed and containing cytokinins and used as a growth regulator (bioregulator) for certain vegetable crops. (BN)

.018 Debearder: a machine for removing awns or beards (tufts of hairs) from grains; (see Seed-cleaning equipment, V.1.3.057).

.019 Decorticator: (see Seed-cleaning equipment, V.1.3.057).

.020 Draper: (see Seed-cleaning equipment, V.1.3.057).

.021 Drill: a machine for dribbling seeds from a hopper along a furrow; may be rolled manually, like the familiar Planet Jr. drill, or tractor-mounted; usually with two to several hoppers and feeding devices for sowing several rows simultaneously; many different forms, sizes, and configurations.

.022 Electronic separator: (see Seed-cleaning equipment, V.1.3.057).

.023 Electrostatic separator: (see Seed-cleaning equipment, V.1.3.057).

.024 Endocarp removal: hastens germination of seeds such as _Prunus_ species and mango because the bony (_Prunus_) or cartilaginous (mango) endocarp must rot or imbibe enough water to split open to allow the seedling to emerge; this does not, of course, involve inhibitors which may be present in the seedcoats; the impediment to germination with respect to the endocarp is one of mechanical resistance.

.025 Excised embryo: (see Germination test, V.1.3.031; Micropropagation, V.1.5.).

.026 Fermentation: usually not desirable in extraction and cleaning of seeds from fleshy fruits; standard commercial practice (and necessary) for tomato seeds (also pepper and eggplant), where a mixture of seeds and juice is allowed to ferment for about two days at 24 to 27°C or treated with 8.4 L of hydrochloric acid per metric ton for 15 to 30 minutes (with 0.8% acetic acid added if there is danger of bacterial canker); also part of the processing procedure for cacao, coffee, and tea (the last is an aerobic oxidation instead of true fermentation which involves the degradation of sugars under conditions of insufficient oxygen); (see Seed extraction (fleshy fruits), V.1.3.064). (H & K)

.027 Finishing machines: many different types, whose purpose is to effect final separation of pure seed from others; (see Seed-cleaning equipment, V.1.3.057).

.028 Fluid drilling (fluid sowing): seeds sown by using a fluid: (a) spraying a suspension of dry seeds in a gel over the soil surface; (b) "hydroseeding," in which imbibed but not germinated seeds are sown with a small quantity of water; (c) "flow-sow," in which dry seeds are sown in a gel; (d) "fluid drilling," in which pregerminated seeds are suspended in a protective gel; the last (d) (pregerminated seeds delivered to the soil with a small quantity of water) is considered fluid drilling proper. (G - D)

.029 Foundation seed: (see III.2.0.004d).
.030 Gel sowing: the practice of sowing seeds, which may be pregerminated, suspended in a fluid gel; additives, such as pesticides, growth regulators, and small amounts of fertilizer, may be included for improved emergence and growth. (G - D)
.031 Germination test: seeds placed under optimum environmental conditions of light and temperature (plus various pretreatments; e.g., soaking in water, exposure to a certain temperature, and soaking in or addition of KNO_3 to the substrate); several different techniques are used: trays with seeds between or on top of blotters, plastic or paraffined cardboard boxes, covered glass Petri dishes, rolled paper towels, flats with sterile sand medium, seeds on peat moss in a glass baking dish covered with a pane of glass, excised embryos (for seeds requiring a long after-ripening period), a tetrazolium test (soaking of seeds for up to 24 hours in 0.1 to 1.0% 2,3,5-triphenyltetrazolium chloride solution), or X-ray radiography (a standard procedure for grain crops); (see Seed testing, V.1.3.075). (AS)
.032 Hammermill: a machine equipped with swinging hammers mounted along a horizontal shaft; (see Seed-cleaning equipment, V.1.3.057).
.033 Hardening-off: the process of inuring a plant (e.g., a young seedling, a newly rooted cutting or air layer, or one grown under partial shade) to the conditions under which it will have to live in its permanent location; this entails a gradual reduction in moisture (atmospheric humidity and soil water) and an increase in light intensity up to the level that the plant will have to tolerate later and may require anywhere from a week or two to about six weeks; hardening-off is the most critical stage in successful production of rooted cuttings or air layers as salable plants; losses here far exceed those in the rooting process or transplanting later unless done with great care; in fact, aertain crops like cacao are deliberately rooted in individual containers to avoid losses during transplanting, which may be 30% or more, compared with the usual rooting percentage of more than 90. (Fig. V.1.3.01b,c) (H & K)
.034 Hard seed: seeds with seed-coats impervious to water or oxygen required for germination, as in many Leguminosae. (US-S)
.035 Huller: (see Seed-cleaning equipment, V.1.3.057).
.036 Hydroseeding: (see Fluid drilling, V.1.3.028).
.037 Imbibition: absorption of liquid; the taking up of water by a seed from a moist medium in preparation for germination.
.038 Impermeable: impenetrable; e.g., when a seed-coat allows no passage of water or gases.
.039 Incubation: analogous to stratification (V.1.3.081) but consists of elevated rather than lowered temperatures as a means of speeding up and improving percentage germination of many palm seeds (many take up to two to three years); the procedure for oil palm consists of incubation at 38 to 40°C for 70 to 80 days with semidry (14.5 to 16% moisture content for dura, 18 to 20% for tenera type) seeds in polyethylene bags, then ambient temperature (25 to 28°C) for about 15 days (germination averages about 80%, compared with 40% or less and six to eight months for nonincubated seeds). (HT)
.040 Indent disk separator: (see Seed-cleaning equipment, V.1.3.057).
.041 Inoculation: treatment of seeds with specific species of Rhizobium known to be required for proper nodulation of roots, hence nitrogen-fixing capacity, for a given species or group of species of

V.1.3. PROPAGATION, NURSERY HANDLING, SOILS, CROP PRODUCTION

legume; at least eight groups of the latter, each requiring a different species of Rhizobium, have been found so far; (see Contra-inoculation, V.1.3.012, and Cross-inoculation, V.1.3.014).

.042 Light requirement for germination: germination of seeds of certain plants, such as Phacelia, Nigelia, Amaranthus, and Phlox, is inhibited by light, but light is required for germination of celery, lettuce, many conifers, many herbaceous flower plants, avocado, and coconut, among others. (H & K)

.043 Magnetic separator: (see Seed-cleaning equipment, V.1.3.057).

.044 Pebble mill: a rotating drum with pebbles inside, used for crushing or cleaning material; may be set to rotate horizontally or obliquely to give a tumbling action; (see Seed-cleaning equipment, V.1.3.057).

.045 Pelletized seeds: a type of coating in which clay or some other inert material, with fertilizer and/or pesticide added, is used to form balls or pellets to facilitate mechanical sowing and give good initial growth of seedlings; used especially for small irregular-shaped vegetable seeds, such as pepper, cauliflower, parsley, carrot, onion, beet, melons, and corn. (US-S)

.046 Peristaltic fluid drill: a mechanical seeder designed with a pulse action (peristaltic) metering device for sowing seeds suspended in a fluid gel medium. (SW)

.047 Plug-mix: a combination of a special soil mixture, such as Cornell "Peat-Lite" soil mix, Magamp, and seeds (e.g., tomato), moistened with water, mixed thoroughly, and dispersed as small plugs in premade holes along a row; one of several methods used for machine sowing small seeds.

.048 Precleaners: (see Seed-cleaning equipment, V.1.3.057b).

.049 Pregerminated seeds: the practice of germinating seeds under controlled conditions prior to sowing in the field; used for tomatoes and other vegetables, as well as certain other crops, e.g., oil palm, which requires special treatment to produce a high rate of germination; (see Gel sowing, V.1.3.030).

.050 Presowing treatments: many different kinds of treatment or combinations thereof are given seeds prior to sowing; e.g., those that hasten germination, increase germination percentage, overcome various types of dormancy or mechanical resistance to seedling emergence, provide more uniform germination, facilitate mechanized sowing, enhance the growth and development of seedlings, and prevent loss from diseases, insects, and rodents. (H & K)

.051 Purity test: visual division of the working sample after weighing into pure seed, other crop seed, weed seed, and inert material, including seedlike structures, empty or broken seeds, chaff, soil, stones and other debris; each component is weighed separately after sorting; (see Seed testing, V.1.3.075). (AS)

.052 Registered seed: (see Seed production, III.2.0.004c-f).

.053 Row-seeder: (see Drill, V.1.3.021).

.054 Scalper: (see Seed-cleaning equipment, V.1.3.057b).

.055 Scarification: mechanical (abrasion, notching, and clipping) or chemical (soaking in concentrated acid) presowing treatment of seeds to render hard or impervious seed-coats permeable to water or gases. (US-S)

.056 Scarifier: (see Seed-cleaning equipment, V.1.3.057b).

.057 Seed-cleaning equipment: machinery to remove dirt, leaves, stems, and chaff and measure or sense the differences in seed size, shape, weight, surface area, specific gravity, color, electronic (electrostatic) properties, texture, stickiness, and pubescence;

basically four groups of machines: (a) <u>separators</u> (to remove non-seed debris); (b) <u>precleaners</u> that consist of (1) scalpers (removal of material coarse enough to be easily separated in a perforated metal, inclined-reel screen or flat, perforated, mechanical shaker screen), (2) hullers (removal of outer seed coat or husk, as in bermudagrass, bahiagrass, or lespedeza), (3) scarifiers (abrasion of hard seed coats, as on hairy indigo, crotalaria, asparagus or okra), or (4) high-speed threshers, hammermills, debearders, or tumbling pebble mills (removal of awns and beards); (c) <u>air screen cleaner</u> (removes the bulk of foreign material, i.e., <u>weed seeds</u>); (d) <u>finishing machines</u>, final separation by (1) specific gravity (via air flotation or a machine with a tilted, oscillating deck for bentgrass, corn, or soybeans by weight or size), (2) indent disk and cylinder separators (with pockets to retain short and reject long seeds), (3) pneumatic (pressure) and aspirator (vacuum) separators (sorts by resistance to air-flow), (4) velvet roll (sorts smooth from rough seeds, e.g., dodder from other seeds), (5) spiral or inclined drapers, timothy bumper mill (removes weed seeds), vibrators, horizontal disk separators (uses centrifugal force to eliminate smooth or round seeds) for sorting by shape, (6) electronic (electrostatic) separator (uses difference in electrical properties as in watercress), (7) magnetic (seeds coated with iron powder) and buckhorn machines (the latter for separation of buckhorn plantain from legumes) for sorting rough or sticky surfaced seeds from the smooth, and (8) color separators for sorting light from dark-colored seeds, e.g., peas and beans; also <u>decorticators</u>, a machine to remove the outer corky parts of seedballs (multiple seeds) of beets; (see Winning, V.0.013). (US-S)

.058 Seed coatings; seeds treated before sowing with fungicide, insecticide, fertilizer, or combinations (including <u>Rhizobium</u> spores for leguminous seeds, e.g., soybean); commercially available for a wide range of crops; (see also Pelletized seeds, V.1.3.045). (US-S)

.059 Seed cracking: systems used in lieu of mechanical or chemical scarification for certain seeds with a hard shell, e.g., tea (<u>Camellia sinensis</u>) and other members of the same genus; that for tea consists of soaking seeds overnight in running water, then exposing them to bright sunshine for a few hours to induce cracks to form, which greatly improve percentage and rapidity of germination; a mechanical cracker can also be used, e.g., for stone fruits.

.060 Seed dormancy: classified by Hartmann and Kester (1975) in four groups: I. Regulators in external seed covering: (a) hard seed impermeable to moisture or resistant to embryo expansion (as in walnut, olive, or stone fruits); (b) chemical inhibitors. II. Morphologically undeveloped (rudimentary) embryos, e.g., palms, orchids. III. Internal (endogenous) dormancy: (a) physiologically shallow, as in most freshly harvested seed, common among herbaceous plants, disappears with dry storage; (b) physiologically intermediate; where moist chilling stimulates germination, e.g., conifers and other woody plants; (c) physiologically deep, where seeds need prolonged moist chilling, common in trees and shrubs and some herbaceous temperate zone plants; (d) seeds requiring a warm, then a cold, moist period, e.g., lily species, <u>Viburnum</u>, peony; (e) seeds requiring a cold, moist, then warm period, e.g., native temperate zone perennials. IV. Double dormancy: both seed coat and embryo dormancy present, as in various woody tree and shrub species. (H & K)

.061 Seed drying: four main methods for drying seeds: (a) sun drying (seeds spread in thin layers on concrete or asphalt pavement or on mats or cloth spread on the ground in the sun or partial shade); (b) drum, column, or tunnel direct-heated or indirect heated artificial driers, with or without forced air circulation (i.e., with a blower); (c) vacuum drying (in containers subjected to a vacuum from an aspirator or steam ejector); and (d) IR drying (radiant-heat drying); the first two are widely used; (see Win, V.0.012). (US-S)

.062 Seeder: a device for dispensing seed or seed-containing mixes in some predetermined manner; two basic types: (a) row-seeders or drills (V.1.3.020), from which seeds, etc., are sown in separate lines; (b) broadcast or cyclone seeders (V.1.3.014), from which seeds, etc., are flung by centrifugal force or blown more or less evenly over an entire area.

.063 Seed extraction (dry fruits): general procedures: (a) hand or machine harvesting of entire plants, fruit, or seed clusters (mainly with a combine or, for sweet corn and popcorn, a two-row corn picker); (b) windrowing (curing and drying on the ground or, for shattering types, on canvas); (c) threshing (as part of combining or after windrowing and drying); (d) recleaning (see Seed-cleaning equipment, V.1.3.057); (e) treatment (see Presowing treatments, V.1.3.050); (f) storage (bulk) or after packaging (see Seed packaging, V.1.3.067, Seed storage, V.1.3.073); numerous variations among crops; many are handled by several different methods (e.g., lettuce and other composites). (US-S, H & K)

.064 Seed extraction (fleshy fruits): general procedures: (a) hand or machine harvesting; (b) seed extraction (i.e., fruit cut open or ground up, seeds scooped out, as in cucurbits, or mixed juice and seed fermented, as in tomato, pepper, and eggplant, or seeds plus adhering pulp separated from other refuse as in tree fruits); (c) thorough washing; (d) drying (thoroughness depends on sensitivity of the seeds); (e) storage (dry or cool-moist). (US-S, H & K)

.065 Seed fiber mats: (see Seed tape, V.1.3.074).

.066 Seeding (sowing) rate: the number of seeds needed per unit (given) area (W) is obtained from the following equation: $W = (A \times S)(D \times P \times G \times L)$, where A is the area ($m^2$ or ft^2), S is the number of living seedlings desired per m^2 (ft^2), D is the average number of seeds per kg (or lb), P is the average purity, G is the effective percentage of germinated seeds that will live. (Note that P and G are obtained from Seed testing, V.1.3.075.) (H & K)

.067 Seed packaging: seeds are packed in burlap, cloth, or paper bags (often multiwall or laminated with asphalt, plastic films, or aluminum foil) for moisture, insect, and rodent resistance; also cellophane, pliofilm, polyester, polyvinyl, aluminum foil, polyethylene, and combinations for bags and packets; metal containers; cardboard boxes, often with liners of bags inside; and fiber or cardboard (or fiberboard) drums, lined or unlined, with snap-on lids; a vast assortment of commercial wholesale and retail containers from small packets up to and including hopper cars (railroad) or bulk trucks. (US-S)

.068 Seed paper grids: (see Seed tape, V.1.3.074).

.069 Seed priming: pretreatment of seed (e.g., lettuce which germinates poorly at temperatures above 30°C) in 1% potassium phosphate solution for 6 to 9 hours in the light and then air drying to overcome thermodormancy and increase the percentage and rate of germination.

V.1.3. PROPAGATION, NURSERY HANDLING, SOILS, CROP PRODUCTION

.070 Seed production: four main categories for nongrower production of seed for commercial (and home) use; (see III.2.0.004c to .004f).

.071 Seed propagation: (see Generative propagation, V.1.3.000).

.072 Seed sizing: sorting of seeds, such as radish and pepper, into categories based on diameter or thickness; markedly improves uniformity and time of germination over those of unsized seeds, thus facilitating the use of mechanized harvesting equipment and providing maximum yield.

.073 Seed storage: the period of seed storage is basically dependent on seed viability, which ranges from only a few days for some tropical species to several years; seed moisture content, holding temperature, and adequate ventilation to prevent fermentation are important; bulk seeds are dry stored in bins or tanks, bagged seeds in warehouses with air temperatures near ambient; refrigerated storage (i.e., 4 to 10°C) and humidity from 45 to 70% are recommended for extended holding; seeds in sealed containers are already, or should be, at proper moisture content; cool storage usually extends viability; packages must be labeled to show the species, cultivar, percentage of live seed, purity, content of noxious weeds, and seed treatment, if any, and usually a date or code number for the lot, plus instructions for storage if conditions are critical for retention of viability (e.g., some seeds should not be put in cold storage). (US-S, H & K)

.074 Seed tape: a water-soluble plastic strip to which seeds are bonded at precise intervals so that the strip merely needs to be unrolled; now widely used for flowers and vegetables for home gardens; seeds on paper grids or fiber mats are available for sowing some grasses.

.075 Seed testing: sampling and testing of agricultural, vegetable, flower, fruit, and forest seed lots to ensure compliance with federal (seeds in interstate commerce or of foreign origin) and/or state (seeds in intrastate commerce) regulations regarding the labeling (name, cultivar, and origin), germination percentage and percentage of pure seed, other crop seed, weed seed, and inert material; procedures for seed tests are published by the Association of Official Seed Analysts, U.S. Department of Agriculture, and the International Seed Testing Association, among others; the first two are the primary guides in the various states. (AS)

.076 Seed wafering: individual seed enclosed in a small wafer, 6 mm thick and 2 mm in diameter, made of finish aggregate vermiculite plus activated charcoal; wafers are machine drilled.

.077 Select seed: (see III.2.0.004c-f).

.078 Separators: machines designed to remove nonseed debris (e.g., dirt, leaves, stems, and chaff) from seeds; (see Seed-cleaning equipment, V.1.3.057).

.079 Soaking: immersion of seeds in water (preferably with frequent changes) to soften the seed-coats or endocarp, in some instances, and leach out growth inhibitors; recommended for seeds moist stored for more than a few days and dry-stored seeds; reduces bench rooting (V.1.3.003) and hastens germination.

.080 Sterilization: immersion of seeds in hot water (43 to 50°C) for periods up to 30 minutes to kill fungi, bacteria, borers, etc.; other treatments are soaking in pesticide solutions, with subsequent surface drying, shaking in a container with dry pesticide solutions and subsequent surface drying, or shaking seeds in a container with dry pesticide (commonly done prior to extended storage). (H & K)

V.1.3. PROPAGATION, NURSERY HANDLING, SOILS, CROP PRODUCTION

.081 **Stratification**; storage of seeds at low temperature, usually 0 to 10°C, over a period of a month to a year or more, according to the species, to complete necessary after-ripening before germination; alternate layers of seed and suitable medium, such as peat, perlite, sawdust, sand, vermiculite, or their mixtures, to maintain moisture, are placed in a box, flat, or plastic film bag for the purpose; seeds of certain species may also require one or more periods of warm storage interspersed with stratification treatment. (H & K)

.082 **Tetrazolium test**: (see Germination test, V.1.3.031).

.083 **Thermodormancy**: the sensitivity of the seeds of certain plants, such as lettuce, celery, endive, many flower and some woody perennial to high temperatures; i.e., they require a relatively low temperature to germinate; (see Cool-season seeds, V.1.3.13). (H & K)

.084 **Threshing, threshing machine**: separation of seeds or seedlike fruits from other harvested plant parts may be done by hand or with a machine designed for the purpose in lieu of using a combine (V.1.3.010); [see Seed extraction (dry fruits), V.1.3.063].

.085 **Transplants**: seedlings grown in seedbeds or containers and used as planting material in the field; the term usually refers to seedlings but applies equally well to clonal plants. (H & K)

.086 **Velvet roll**: (see Seed-cleaning equipment, V.1.3.057).

.087 **Warm-season seeds**: the minimum germination temperature for asparagus, sweet corn, and tomato is 10°C and for beans, eggplant, pepper, and cucurbits, 15°C, whereas lima beans, cotton, soybeans, and sorghum will show chilling injury at 10 to 15°C; (see Cool-season seeds, V.1.3.013). (H & K)

.088 **Windrow, windrowing**: plant debris piled up in long rows, generally in row middles, where it will not interfere with planting operations, as part of clearing and cleaning a field and left to decay or be burned (see Site preparation, V.3.1.001) or as part of seed extraction (dry fruits, V.1.3.063) to allow the material to dry and cure before threshing. (US-S)

.089 **Working sample**: the sample on which a seed test is actually to be run; the amount required varies with the kind of seed and is specified in the Rules for Seed Testing (i.e., listed in the Rules of the Association of Official Seed Analysts and U.S. Department of Agriculture regulations); procedures for obtaining the working sample ensure a uniform representation of the entire lot being tested; (see Seed testing, V.1.3.075). (AS)

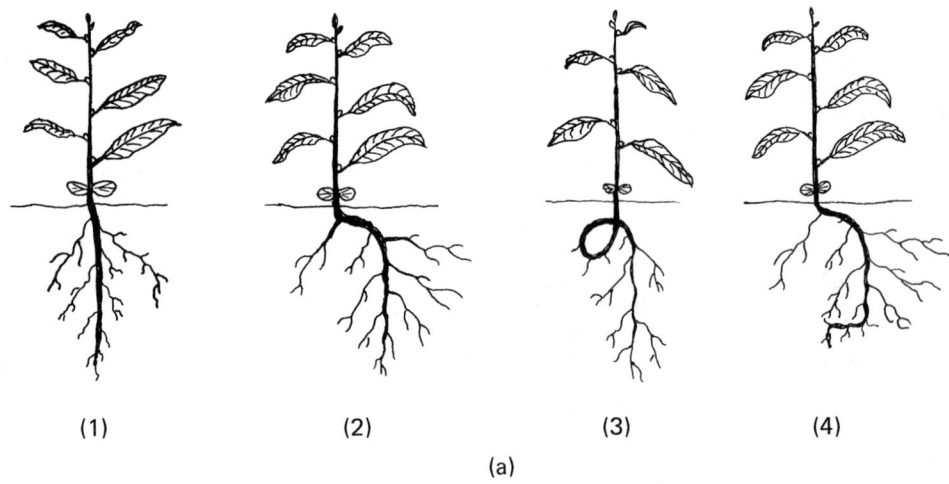

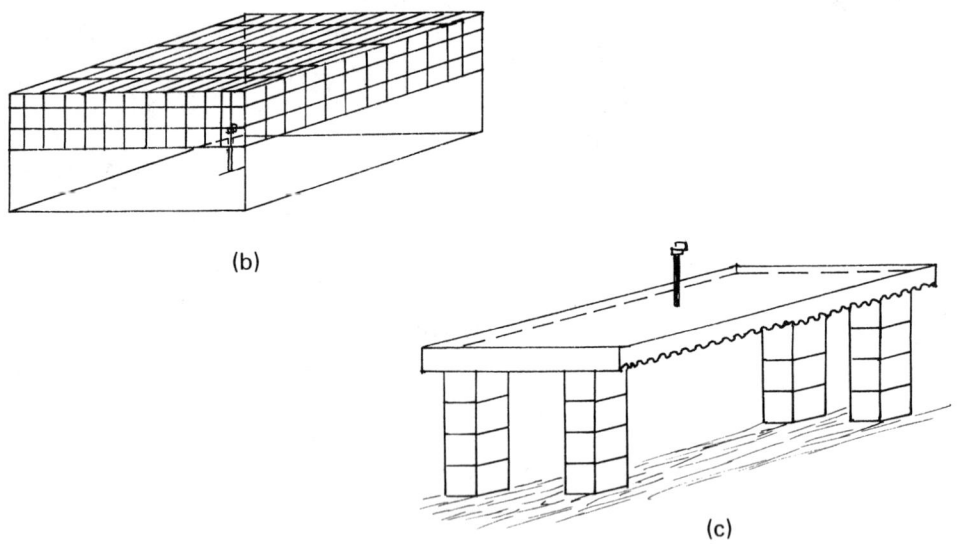

Fig. V.1.3.01. (a) Bench-rooting: (1) normal seedling, (2), (3) curvature from seedcoat resistance, (4) double curvature from seedcoat resistance and being grown too long in a shallow container; (b), (c) hardening-off; (b) shade house with mist irrigation (ground level), (c) mist bed (elevated).

V.1.4. PROPAGATION, NURSERY HANDLING, SOILS, CROP PRODUCTION

V.1.4. Vegetative (asexual) propagation

.000 Vegetative (asexual) propagation: multiplication or perpetuation of plants by asexual means (this does not include apomictic embryos; see Apomixis, IV.4.1.006); refers to methods involving plants on their own roots, such as runners, bulbils, suckers, layers, separations, divisions and cuttings, and plants on roots other than their own, graftage. (H & K)
.001 Air layerage: (see Layerage, V.1.4.036a).
.002 Antiseptics: (see Prerooting treatment, V.1.4.022a-3).
.003 Asexual propagation: (see Vegetative propagation, V.1.4.000).
.004 Bending: (see Graftage aftercare, V.1.4.028i).
.005 Bulb: (see Separation, V.1.4.071a).
.006 Bulbils: bublets formed on the aerial portion of a stem or inflorescence, as in Agave or Lilium spp. (Fig. V.1.4.16d). (G'62)
.007 Bulblet: minature bulb formed on scales or on underground parts of the stem; e.g., around the sides or base of an older bulb; (see Bulb, V.1.4.071a). (G'62)
.008 Callus: soft tissue composed of parenchyma cells produced in the process of wound healing in plants.
.009 Callusing: (see Prerooting treatments, V.1.4.022a-3).
.010 Certification (registration) program: a procedure for obtaining pathogen-free, high-quality, clonal planting material such as potato, strawberry, grapes, tree fruits, and nuts; usually conducted by state or regional agencies set up for that purpose or by individual growers under close supervision; may include scionwood and seed or clonal rootstock sources in crops produced by graftage; analogous to seed production programs; includes visual inspections, indexing, heat and/or chemical treatment, and culture in isolation as parts of the program prior to certification or registration; (see Indexing, V.3.3.2.221; Indicator plant, V.3.3.2.222). (H & K)
.011 Chinese layerage: (see Layerage, V.1.4.036a).
.012 Cion: (see Scion, V.1.4.067).
.013 Circumposition: (see Layerage, V.1.4.036a).
.014 Collection of cutting material: (see Cuttings, V.1.4.022a-1).
.015 Compatibility, congeniality: the condition of a plant that consists of two or more components (i.e., one produced by some form of graftage, thus on roots other than its own) and has a long, vigorous, economically productive life; anything less than this is denoted as incompatibility (uncongeniality) to some degree, which may range from slight to complete (i.e., the rootstock-scion combination does not unite at all or is short-lived); compatibility (congeniality) is generally good when the components belong to the same cultivar but becomes progressively less so as botanical affinity becomes more remote; the chances of successful union are reduced between species, or genera, and essentially zero between families; (see Incompatible, V.1.4.035). (H & K)
.016 Compound layerage: (see Layerage, V.1.4.036b-2).
.017 Congeniality: (see Compatibility, V.1.4.015).
.018 Corm: (see Separations, V.1.4.071b).
.019 Cormel: diminutive corms formed around the periphery of the parent corm (see Corm, IV.1.4.071b). (G 62)
.020 Crown: (see Divisions, V.1.4.024).
.021 Cutting preparation: (see Cuttings, V.1.4.022).
.022 Cuttings, cuttage: a form of vegetative propagation for plants on

their own roots; material is cut from the parent plant and induced to form roots and shoots under favorable environmental conditions; many different types. (H & K)

.022a. Cutting preparation: three principal stages, collection of material, preparation of individual cuttings, and prerooting treatments.

.022a-1. Collection of material: selected, healthy branches of the proper stage of maturity are cut from plants of the desired variety, immersed in water, or rolled in moist burlap or paper towels, placed in a polyethlene bag, and kept cool until used; any suitable branches of most plants may be taken, except those with dimorphic branches, i.e., orthotropic and irreversibly plagiotropic growth, of which only the former will produce a normal upright plant (e.g., Norfolk Island pine, coffee).

.022a-2. Preparation of cuttings: consists of removing unusable parts (i.e., tender tips, side branches, and broken ends), then trimming the remainder into convenient lengths, usually 10 to 25-30 cm, with a knife or sharp, double-bladed pruning shears; cuts are square (horizontal) or sometimes slanted; only sufficient leaves are removed (large ones may have to be cut in half) to enable insertion of the cutting into the rooting bed; hardwood stem or root cuttings must be marked (or cut) to indicate which end is to be inserted.

.022a-3. Prerooting treatments: four different treatments: (1) callusing, for hardwood or root cuttings only; (2) immersion in an antiseptic, such as dilute (1:10) Na or Ca hypochlorite solution, as the cuttings are made (sometimes omitted but apparently it does improve rooting of slow-to-root plants like tea); (3) quick dips of the basal end of cuttings in a growth regulator or mixture prepared as a concentrated solution or powder to induce quicker and heavier rooting; (4) wounding where vertical cuts are made or narrow strips of bark are stripped from the basal end of cuttings, usually before growth regulator treatment; optimum conditions for growth regulator treatment, the compound or mixture, concentration, and length of treatment vary widely with species of plant, wood maturity, and time of year.

.022b. Direct-set cuttings: two groups: (1) stem cuttings of sugar cane, black pepper, cottonwood, willows, and rhizomatous grasses, which are set directly in the field without callusing; (2) hardwood, truncheon, and piece or whole root cuttings, which are usually callused for a few weeks to months, then set directly in the orchard, fence row, or nursery block, or bench grafted and set out; (see Rooting structures, V.1.1.056).

.022c. Leaf: whole or parts of blades or blades with petioles attached, as for rex begonia, bryophyllum, sansevieria, and African violet. (Fig. V.1.4.01a-c)

.022d. Leaf-bud: similar to leaf cuttings, except a bud is present at the base of the petiole (i.e., a small piece of stem tissue), as in blackberry and hydrangea. (Fig. V.1.4.01d)

.022e. Root: whole or pieces of roots, as in red raspberry and horseradish. (Fig. V.1.4.01e,f)

.022f. Stem: several distinct forms.

.022f-1. Hardwood: leafless wood of deciduous (temperate zone), broad-leaved evergreens, or older wood of narrow-leaved evergreens (conifers), as in fig, grape, quince, rose, forsyth-

V.1.4. PROPAGATION, NURSERY HANDLING, SOILS, CROP PRODUCTION 443

.022f-2.   ia, and various conifers (e.g., junipers). (Fig. V.1.4.02b)
.022f-2. Herbaceous: soft, succulent growth of herbaceous plants, as in geranium, coleus, and chrysanthemum. (Fig.V.1.4.02a)
.022f-3. Semihardwood: leafy, firm, but not fully matured wood from broad-leaved evergreens or deciduous plants; several distinct forms include regular multinode (i.e., nodes at top and bottom, with or without others between), single-node (node at top only), "mallet" (single node with a short piece of stem that may be split lengthwise), and "heel" (multinode with a short piece of older wood at base), as in citrus, olive, camellia, and holly. (Fig. V.1.4.02c-e; .03a,b,d)
.022f-4. Softwood: similar to semihardwood, except that it consists of relatively soft, current-season, flush growth of woody plants; may be made with a heel; e.g., lilac, forsythia, and weigelia. (Fig. V.1.4.03c)
.022g. Truncheon: large branches cut or torn from a woody plant, as are olive and many species of tropicals ("living fences"). (Fig. V.1.4.03e)
.023 Dado graft: (see Whip, or tongue, graft, V.1.4.028k-9).
.024 Divisions: a form of vegetative propagation for plants grown on their own roots by means other than runners, suckers, layers, or cuttings; types of material include (1) crowns which require separation from the mother plant or from each other, as in everbearing strawberry, phlox, or day lily; (2) offsets which require separation from the mother plant, as in houseleek, pineapple, or date; and specialized vegetative structures, such as (3) rhizomes (e.g., canna, iris, many Zingiberaceae); (4) tubers (e. g., potato); (5) tuberous roots (e.g., sweet potato, dahlia, yams, and cassava); (6) pseudobulbs (e.g., many orchids), which may be planted whole or separated into several parts before planting, the usual practice with tubers or rhizomes (and occasionally with tuberous roots, most of which, however, require a piece of stem tissue with a good bud). (Fig. V.1.4.04,.05) (H & K)
.025 Forcing bud growth: (see Graftage aftercare, V.1.4.028i).
.026 Gootee: (see Layerage, V.1.4.036b).
.027 Gourmandizer: [see Graftage aftercare, V.1.4.028i(4)].
.028 Graftage: the general term for vegetative propagation of plants on roots other than their own; includes budding, grafting, approach grafting, inarching, and various forms of repair, such as bridge grafting and living braces. (H & K)
.028a. Approach grafting: the process by which a rootstock or an unrooted cutting growing in a container is united with a branch of a large plant (scion), with separation from the latter after the union is completed; used for plants difficult to propagate (e. g., allspice) or when large plants are desired, as for mango in Jamaica or India. (Fig. V.1.4.06)
.028b. Bench-graftage: literally, grafting of plants typically during late winter or early spring at a propagation bench instead of in the nursery; used widely among temperate-zone fruits such as apple, pear, and grape; also some ornamentals; rootstocks may be whole or piece roots or stems that have been stored and callused; grafting may be more or less mechanized (i.e., making cuts and wrapping); grafted plants may be stored at temperatures too low for growth to occur but high enough (usually 4 to 10°C) to enable the union to heal more or less completely and then at a lower temperature (2 to 4°C) to keep them dor-

mant until planting time in the spring; (see Graftage, V.1.4. 028).

.028c. Bridge grafting: a type of graftage to repair an established tree with an area girdled by deer or rabbits too high on the trunk for inarching (V.1.4.028m); the operation consists of making a series of T-cuts in the sound part of the trunk above (inverted-T) and below (upright-T) the girdled area, or alternatively, rectangular slots (inlay type); scionwood from the same tree or another of the same cultivar is cut slightly longer than the distance between the ends of the cuts; the ends of the individual sticks are trimmed on a slant to a thin edge (T-cuts) or a thicker one to fit in the slots (inlay cuts), slipped into place, secured with brads, and wrapped with grafting tape. (Fig. V.1.4.12a-f)

.028d. Buckhorning: heavy pruning of a tree damaged by cold or being prepared for topworking (V.1.4.028v) to leave stubs of large branches, which must be whitewashed, painted, or otherwise protected to prevent sunburn. (Fig. V.1.4.10a,b)

.028e. Bud and graft union healing: two main configurations, (1) the "butted end" type, i.e., all forms of grafting, chip budding, and inarching, in which the cambial layers must be in reasonable alignment and the bridge across the union is in the form of an arc; (2) the "cambial face" type, shield(T) budding, in which the scion is placed against a broad cambial surface and the cambial bridge is formed as an S-shaped curve from the scion to the bark flaps of the rootstock. Stages in the healing process are (a) precallus, suberization of cut surfaces to isolate unwounded cells and reduce loss of moisture (24 hours); (b) callus, proliferation of callus from both rootstock and scion (4 to 8 days for callus "bridge"); (c) cambial bridge, from rootstock to scion (10 to 15 days) and proliferation of vascular tissues (15 to 30 days); (d) healed union, complete cylinders of xylem and phloem (30 to 60 days). (Note: times given are averages for citrus in Israel and Florida and mango in Florida; de novo differentiation of tracheary or phloem sieve elements observed in tobacco and cotton apparently does not occur in woody perennials). (Fig. V.1.4.07)

.028f. Budding: the process of inserting a scion, which consists of a single bud on a shield, square, or rectangle of bark, with or without a small piece of wood attached, into a plant (=rootstock) with the intention that it will unite and grow there; the most common method is shield(T) budding used for stone and pome fruits, roses, citrus, and many other crops, chip budding (e.g., grape and mango), Forkert budding (e.g., mango and other fruit trees in the Far East), and patch, ring, and I budding for walnut and pecan. (Fig. V.1.4.08)

.028f-1. Annular bud: intermediate between patch (V.1.4.028f-6) and ring (V.1.4.028f-7) budding; i.e., a partial ring of bark is removed. (Fig. V.1.4.08i)

.028f-2. Chip bud: the single budeye version of veneer grafting (V.1.4.028k-8) in which a shallow vertical downward cut is made in the bark and outer layers of wood and a second one, horizontally angled (i.e., to form a "step"), is made at the bottom on the rootstock; the scion shield, which is cut with a square end and may or may not have a sliver of wood underneath (a shield with bark only is used in peach and some other plants to avoid transmission of virus carried in

V.1.4. PROPAGATION, NURSERY HANDLING, SOILS, CROP PRODUCTION

the xylem), is laid against the wound with the cambial layers aligned on at least one side; used when poor slippage or brittle bark prevents the use of a shield bud. (Fig. V.1.4. 08c,e)

.028f-3. Forkert bud: a scion shield with a bud eye, leaf scar, and some bark (with or without wood) is cut with a square end, as for a shield bud, and fitted against a cambial surface made on the rootstock by a horizontal cut and two downward cuts which produce a flap that is pulled back for insertion of the shield and replaced afterwards; widely used in the Far East for a great many plants but virtually unknown elsewhere. (Fig. V.1.4.08f)

.028f-4. Hanging bud: the upside down version of a chip bud (V.1.4.028 f-2). (Fig. V.1.4.08d)

.028f-5. I-bud: a scion patch similar to that used for patch budding (V.1.3.028f-6) is inserted under the flaps of bark produced by making a short vertical cut and two horizontal cuts, one at the top and one on the bottom, on the rootstock. (Fig. V.1.4.08j)

.028f-6. Patch bud: a square or rectangular patch of bark is removed from the rootstock and replaced with a similar patch, plus a bud, taken from a stick of scionwood; commonly done with a double-bladed knife; used for plants whose bark will not permit shield budding; e.g., pecan and walnut. (Fig. V.1.4. 08g)

.028f-7. Ring bud: similar to a patch bud (V.1.4.028f-6), except that a complete ring of bark is removed from the rootstock and replaced with a ring from the scion wood. (Fig. V.1.4.08h)

.028f-8. Shield(T) bud: two versions, upright and inverted, differing in the height on the rootstock at which the vertical cut is made (lower for inverted-T) and at which end the horizontal cut (crossbar of the T) is located; the most common type of graftage and one of the easiest, requiring only that the rootstock bark be slipping; the shield, which may be boat-shaped or square at one end and may consist of bark only or bark plus wood, is slipped under the flaps which are pried open to permit easy entry; the budeye may be covered, which is usual practice with citrus, or left uncovered. (Fig. V. 1.4.08a,b)

.028g. Cutting-graft: a scion inserted by splice, whip, or cleft grafting in a rootstock, which consists of a short piece of stem, wrapped with a rubber band or adhesive plastic tape and placed in a cutting bed for rooting and healing the union; cutting grafts can be of any convenient length; used commercially for roses, difficult-to-root conifers (on easily rooted rootstocks), and rhododendrons. (Fig. V.1.4.03d)

.028h. Double-working: graftage as a three-component system, rootstock, interstock, and scion, by a method of grafting or budding; commonly used to overcome incompatibility between scion and rootstock, as in 'Bartlett' pear on quince, obtain a certain degree of dwarfing in apples by the use of an interstock, such as EM IX or EM VII, provide a cold or disease-resistant trunk or crotch system, or provide a disease-resistant scion for a susceptible interstock (as in rubber grown in the Western Hemisphere); (see Graftage, V.1.4.028). (Fig. V.1.4.08k,09j)

.028i. Graftage aftercare: the five main operations are (1) unwrapping about two or three weeks after budding or six to eight weeks after grafting (or longer in some cases); (2) forcing bud growth, such as (a) lopping, in which the rootstock is cut part way through and the top laid over, (b) bending, in which the top of one rootstock is tied to the base of the next plant in the row, (c) topping, in which the rootstock is cut off in one or two stages, (d) nicking, in which a cut is made immediately above the scion, and (e) trimming, in which the rootstock is cut on a slant as close to the union as possible; (3) staking and tieing; (4) removal of unwanted sprouts (gourmandizers); and (5) training to force proper development of the main branch framework, which may be done by low (15 to 30 cm) or high (50 to 100 cm) heading. (Fig. V.1.4.11a-h)

.028j. Graftage height: no set height; low graftage (i.e., budding or grafting) is done at 5 to 15 cm and high graftage, at 20 to 60 cm or more above the soil level, depending on the presence of soil pathogens likely to be splashed up on the scion and custom (e.g., low budding of citrus in Florida vs. high budding in California).

.028k. Grafting: the process of inserting a scion, which consists of a piece of stem and two or more buds, in a plant (= rootstock) with the intention that it will unite and grow; several types and numerous methods, such as piece or whole root grafting of apple or pear by whip, or tongue, methods, crown grafting of Persian walnut by whip, or tongue, camellia, by cleft, narrow-leaved evergreens, by side grafting, and top grafting of various fruit trees by cleft, saw-kerf or notch, bark, side, veneer, and whip, or tongue, grafting.

.028k-1. Bark grafting: a type of graftage used for topworking a large cut-back limb or stump in which two to perhaps six vertical cuts are made through the bark around the perimeter; the flaps are gently pried open, scions whose bases have been trimmed to a thin wedge on one side are slipped into place under the bark, often anchored with brads, the wounds are wrapped, and cut surfaces are protected with grafting compounds. (Fig. V.1.4.10g)

.028k-2. Cleft graft (V-graft): two main versions, one involving young plants in containers or a nursery row and the other, large branches as in topworking and the like. The former consists of a rootstock with the top cut off square and a vertical cut to the desired depth made with a knife; a single scion, preferably of the same diameter as the rootstock, is cut as a two-sided wedge of the same length as the rootstock cut; the latter is pried open and the wedge is inserted. The cleft graft for topworking is done with a clefting iron driven in with a mallet to the desired depth of the cut; in the case of a branch with brittle wood or twisted grain, the cut may be made with a saw and trimmed smooth with a knife; two scions, with long tapering wedges at the basal end, are inserted into the cut, which is held open with the chisel point on the clefting iron, and adjusted to align their cambial layers with those of the stock; also, cut surfaces are covered with grafting compound and sometimes a heavy paper collar (painted white) is fastened around the end of the limb with moist, loose, light-colored material (sand or perlite) poured in to leave only the top few buds of the scion sticking out as protection from drying out during the healing process. (Fig.V.1.4.09c,d;10c)

V.1.4. PROPAGATION, NURSERY HANDLING, SOILS, CROP PRODUCTION

.028k-3. Inlay grafting: similar to bark grafting (V.1.4.028k-1), except that a rectangular slot is cut into the limb or stump instead of opening up bark flaps; the base of each scion is trimmed so that the cambial layers will be aligned with those of the rootstock and with a small step to rest on the flat end of the limb or stump; scions are held in position with one or two brads, wrapped securely, and all cut surfaces are protected with grafting compound. (Fig. V.1.4.10e;.12c,i)

.028k-4. Oblique cleft (saw-kerf) graft: a method of topworking large branches or trunks by sawing four to six V-shaped cuts around the perimeter of the cut off stub, smoothing them off with a knife, and then inserting scions whose basal ends have been tapered to fit snugly and have good alignment of the cambial layers; small brads may be used to hold the scions in place; cut surfaces are protected with grafting compound and the wounds are wrapped. (Fig. V.1.4.10d)

.028k-5. Saddle graft: upside down version of the V-graft. (Fig. V.1.4.09e)

.028k-6. Side graft: a cut slanted about 20 to 30° from the vertical is made in the rootstock; the basal end of the scion is trimmed as a 2-sided wedge, with the outer side shorter than the inner, to fit the rootstock cut; cambial layers are aligned on at least one side; used for brittle (i.e., twisted grain) or hard-wooded plants, as in many tropicals. (Fig. V.1.4.09h,i;10f)

.028k-7. Splice graft: consists of slanting cuts made at the base of the scion and top of the rootstock; components must have the cambial layers aligned on at least one side if their diameters differ; sometimes a short pin is inserted in the pith to help maintain alignment. (Fig. V.1.4.09a)

.028k-8. Veneer graft (side-veneer graft): similar to the side graft (V.1.4.028k-6), except that the main cut on the rootstock is made nearly vertical to remove a comparatively thin sliver of bark and wood with a second horizontal cut at an angle of 45° at the base of the main cut (i.e., identical to the cuts made for a chip bud, V.1.4.028f-2, but a longer and wider wound); the scion is trimmed as a one-sided wedge which is laid on the rootstock wound and the cambial layers aligned on at least one side; used for many tropicals. (Fig. V.1.4.09f,g)

.028k-9. Whip, or tongue, graft: the same as a splice graft (V.1.4.028k-7), except that a second vertical cut is made part way along each slanted cut so that the scion and rootstock are held firmly together when they are joined with cambial layers aligned on at least one side; various mechanized versions that use a pair of dado saws (rotary saw blades with spacers between them) may be done, as with grapes. (Fig. V.1.4.09b)

.028 l. Heteroplastic: grafting or transplanting of tissue between two unrelated plants, especially those of a different genus or species, as in many citrus budlings.

.028m. Inarching: the process by which several rootstocks (usually small seedlings, layers, or rooted cuttings) are planted adjacent to the base of a large tree and united to the latter above a poor union or girdled region. (Fig. V.1.4.12g-k)

.028n. Interstock: a stem piece inserted between a stock (rootstock) and scion in an operation called "double-working", which is done by grafting or budding simultaneously or successively. (Fig.V.1.4.09j)

.028o. Living braces: a type of repair graftage used to strengthen a narrow, weak, or partially split crotch in a valuable tree; horizontal cuts similar to those for a side graft are made on the side of each branch with a carpenter's chisel and mallet 25 to 50 cm above the crotch; the "scion" is cut from the same tree or another of the same cultivar; the ends, after the scion is cut to the proper length, are trimmed as two-sided wedges; the piece is inserted, checked for cambial alignment (on one side of the cuts only), fastened with brads, and wrapped securely; a loop of wire enclosed in a hose and with a turnbuckle fastened a short distance above the brace and tightened enough to relieve strain on the brace; it will remain until the wounds have completely healed. (Fig.V.1.4.013a-d)

.028p. Micrograftage: miniaturized budding (microbudding) or grafting (micrografting), which involves the use of single buds with little bark or wood or of diminutive scions (i.e., toothpick size) with several buds, respectively, inserted in young seedlings or clonal rootstocks a few weeks or months old; the former is used commercially for citrus and other fruits; the latter has long been used in virus transmission studies, in which very small scions may be inserted in tender tip growth of seedlings a few months old; the graftage process is more tedious and requires a razor blade or scalpel for the work, but wound healing is faster than with conventional methods. (Fig. V.1.4.11i-o)

.028q. Nicolin (Nicolieren) budding: a technique of double-budding to overcome localized incompatibility between a rootstock and scion; consists of a regular shield bud (with or without wood) as the scion and a second budless shield underneath it as the interstock; both shields are inserted into a regular T-cut on the rootstock and wrapped in normal fashion. (Fig. V.1.4.08k)

.028r. Nurse-root grafting: a type of graftage in which the real object is to have a stem cutting of a difficult-to-root plant on its own roots; this may be done by a root graft in which the union is buried so that scion rooting will take place (eventually, however, the tree must be dug up and the rootstock cut off, by an inverted root graft (the rootstock dies after the scion roots are formed), by using an incompatible rootstock that will gradually cease to function as the scion roots take over, or by girdling the scion just above the union with a tightly wound rubber budding strip that will disintegrate in a year or two. (Fig. V.1.4.13e-h)

.028s. Rootstock: the root system, the bottom part of a plant propagated by graftage.

.028t. Rootstock preparation: the two main phases for nursery graftage are (1) selection of seeds or vegetative material (for clonal rootstocks) from productive, disease-free plants, germination of seeds, rooting of cuttings or layers, lining out in nursery rows or transplanting to containers, continuous culling of bench-rooted, weak, or sickly plants, and adequate water, fertilizer, and protection from pests to maintain active vigorous growth; (2) trimming off side branches and thorns and wiping off soil from the stems; this is done immediately prior to the budding or grafting operation, which may be carried out when the rootstocks are large enough, i.e., anywhere from a few weeks to two or three years; rootstocks that lack good cambial activity may be given an application of quickly available N fertilizer, such as nitrate of soda or ammonium nitrate, about 10 to 14 days before graftage.

.028u. Scionwood preparation: ringing or scoring of branches behind the point at which they will be cut for bud or graftwood and/or cutting of leaves to short petiole stubs four to six weeks prior to graftage has been recommended for certain hard-to-propagate plants; most propagators agree, however, that selection of scionwood at the proper stage of development (i.e., full plump buds still dormant but ready to grow) from vigorous healthy trees, prompt removal of leaves, storage in a cool, moist location, immediate use (when possible), and, perhaps most important, a <u>sharp</u> knife will give good success in most instances.

.028v. Topworking: replacement of the original scion with a different one in an established tree, shrub, or vine; may be done by grafting into large cutback (buckhorned) branches or main trunk or budding into sprouts allowed to grow out from cutback branches; done to replace an uneconomic cultivar or a tree system damaged by cold or wind, but otherwise basically healthy (i.e., neither too old nor diseased). (Fig. V.1.4.10)

.028w. Wrapping: many different materials (V.1.2.017) may be used for wrapping newly inserted buds, grafts, or approach grafts; the primary objective is to hold the scion securely in place without crushing cells by wrapping too tightly or allowing movement by wrapping too loosely.

.029 Ground layerage: (see Layerage, V.1.4.036b).
.030 Growth regulators: (see Prerooting treatments, V.1.4.022a-3).
.031 Hardening-off: (see V.1.3.033).
.032 Hardwood cutting: (see Cuttings, V.1.4.022f-1).
.033 "Heel" cutting: (see Cuttings, V.1.4.022f-3).
.034 Herbaceous cutting: (see Cuttings, V.1.4.022f-2).
.035 Incompatible, incompatibility (uncongenial, uncongeniality): the condition of a plant produced by graftage in which symptoms of distress (e.g., overgrowth, of the union, poor or yellowed and weak growth, and twig dieback) appear in the nursery or may be delayed for 20 years or more; two main forms are (a) <u>localized</u>, in which a mutually compatible interstock is necessary, as in many pear cultivars on quince rootstock; (b) <u>translocated</u>, in which use of an interstock does not rectify the situation, as the result of latent virus, viroid, or other systematic disorder becoming manifest in a given combination; predictability of incompatibility in untested combinations is poor, necessitating actual trial of individual combinations to determine their short and long-term success. (ME, H & K)

.036 Layerage, layering, layers: a form of vegetative propagation for plants on their own roots which are difficult to multiply by other means; here roots are induced to form before removal of the material from the parent plant; several ground and air layerage methods are commonly used; can be done any time of year when root growth can be expected to take place. (Fig. V.1.4.14,.15) (H & K)

.036a. Air layerage (marcottage, Chinese layerage, circumposition, gootee): roots are induced to form on aerial stems by first removing a cylinder of bark or making an upward slanting cut that is wedged open; a ball of moist sphagnum moss, shredded coconut husk (coir), or similar material is placed around the wound and a protective wrap of polyethylene film, often with aluminum foil over it, is fastened securely to hold the medium in place and retain moisture, as in India rubber (and other Ficus spp.), lychee, pummelo, and nutmeg (in Grenada). (Fig. V.1.4.15)

.036b. Ground methods:
.036b-1. Mound (stool) layerage: a plant is cut back and soil, sawdust, peat, or other loose medium is mounded around the shoots as they grow out, as in gooseberry, apple. (Fig. V.1.4.14c-e)
.036b-2. Serpentine (compound) layerage: alternate sections of a vine or slender branch are buried and left exposed, as in grape and honeysuckle. (Fig. V.1.4.14f)
.036b-3. Simple layerage: individual low-hanging or slender branches of trees or shrubs are bent upward a short distance from the tip and maintained in firm contact with the soil; the underside of the bent portion may be nicked to promote faster rooting in some instances, as in honeysuckle, spiraea, and filbert. [Fig. V.1.4.14a,a(1)]
.036b-4. Tip layerage: ends of arching or trailing bramble (Rubus) canes are bent sharply upward close to the tip which is then partially buried until roots have formed. (Fig. V.1.4.14b)
.036b-5. Trench layerage: a form of mound layerage in which plants are established in the bottom of a shallow trench; the trench is gradually filled in with loose medium as shoots develop from pegged-down branches, as in apple, pear, and cherry. (Fig. V.1.4.14g-i)
.037 Leaf-bud cutting: (see Cuttings, V.1.4.022d).
.038 Leaf cutting: (see Cuttings, V.1.4.022c).
.039 Leptomorph rhizome: (see Rhizome, V.1.4.060a).
.040 Liner row: a row in the nursery in which liners (V.1.4.041) are grown on until they are large enough for budding or grafting; spacing in the row varies from about 15 to 50 cm or more, depending on how long the plants will remain, type of growth (i.e., spreading or upright), and presence of thorns; spacing between rows will be dictated largely by the cultivation equipment used and will usually vary from about 75 to 150 cm.
.041 Liners: seedlings, cuttings, or layers spaced closely in rows in a nursery for use later as rootstocks; may be transplanted from a seed flat, seedbed, cutting bench, or ground layerage bed or sometimes, in the case of seeds, directly seeded in the nursery row.
.042 "Living fences": (see Truncheon, V.1.4.022g).
.043 Localized incompatibility: (see Incompatible, V.1.4.035).
.044 Lopping: (see Graftage aftercare, V.1.4.028i).
.045 "Mallet" cutting: (see Cuttings, V.1.4.022f-3).
.046 Marcottage: (see Layerage, V.1.4.036a).
.047 Mesomorph rhizome: (see Rhizome, V.1.4.060b).
.048 Microbudding: (see Micrograftage, V.1.4.028p).
.049 Micrografting: (see Micrograftage, V.1.4.028p).
.050 Mound layerage: (see Layerage, V.1.4.036b-1).
.051 Nicking: (see Graftage aftercare, V.1.4.028i).
.052 Nicolieren budding: (see Nicolin budding, V.1.4.028q).
.053 Offsets: (see Divisions, V.1.4.024).
.054 Pachymorph rhizome: (see Rhizome, V.1.4.060c).

.055 Piece root grafting: (see Bench-graftage, V.1.4.028b).
.056 Preparation of cuttings: (see Cuttings, V.1.4.022a-2).
.057 Prerooting treatment: (see Cuttings, V.1.4.022a-3).
.058 Pseudobulb: thickened or bulbiform stems of certain epiphytic orchids that are solid and borne above ground; (see Divisions, V.1.4.024). (G'62)
.059 Registration program: (see Certification program, V.1.4.010).
.060 Rhizome: specialized plagiotropic stem in which the main axis grows horizontally just below or on the surface of the ground; three types: fleshy, in which the internodes are short and growth is determinate, as in Iris, or indeterminate, as in ginger; slender (fibrous) with long internodes and indeterminate growth, as in grasses (not a rootstock); and intermediate; i.e., pachymorph, leptomorph, and mesomorph, respectively; (see Divisions, V.1.4.024 and Stolon, II.4.3.061). Fig. V.1.4.04d,05a,b) (G'62, H & K)
.060a. Leptomorph rhizome: slender form with long internodes, as in lily-of-the-valley, many grasses, and some sedges (e.g., Carex arenaria).
.060b. Mesomorph rhizome: intermediate type between leptomorph and pachymorph.
.060c. Pachymorph rhizome: thick, fleshy, shortened form, as in iris, ginger, bamboo, sugar cane, and banana.
.061 Root cutting: (see Cuttings, V.1.4.022e).
.062 Root grafting: (see Bench-graftage, V.1.4.028b).
.063 Root sucker: (see Suckers, V.1.4.083).
.064 Runners: specialized stem that develops from a leaf axil at the crown of a plant, grows horizontally along the ground, and forms a new plant at one of the nodes, usually at or near the tip; as in strawberry and some ornamentals. (Fig. V.1.4.16b) (H & K)
.065 Saw-kerf graft: (see Oblique cleft graft, V.1.4.028k-4).
.066 Scaly bulb: (see Bulb, V.1.4.071a-1).
.067 Scion (cion): the plant being propagated vegetatively in graftage; the top part.
.068 "Seed" planting material of vegetative origin, such as stem cuttings for sugar cane, banana rhizomes, and cut up potatoes.
.069 Seedpiece: pieces cut from stem tissues for the purpose of vegetative propagation; frequently abbreviated to "seed" (V.1.4.068).
.070 Semihardwood cuttings: (see Cuttings, V.1.4.022f-3).
.071 Separations: (G'62, H & K)
.071a. Bulb: a modified orthotropic underground stem, very short, usually flattened; forms a basal plate, crowned by typically fleshy, more or less imbricated, nongreen, scalelike leaves and without distinct nodes or internodes; two main forms:
.071a-1. Scaly: leaves separate and lacking an enveloping, dry covering (tunic), as in lily (Lilium). (Fig. V.1.4.16c,d;.17)
.071a-2. Tunicate: leaves in concentric layers, of which the outermost are dry and membranous, as in onion (Allium), narcissus (Narcissus), amaryllis (Hippeastrum), hyacinth, and tulip. (Fig. V.1.4.16e)
.071b. Corm: a very short, thick, firm, fleshy orthotropic subterranean stem, usually broader than high and protected by a tunic; differs from a bulb (V.1.4.071a) in having distinct nodes and internodes; development is typically from the terminal bud; however, the lateral buds may be induced to grow if the apical bud is aborted or monocarpic or the corm is cut into pieces, as in gladiolus, crocus. Both bulbs and corms may be cut up, scored, scooped out, or separated to induce formation of additional plants. (Fig. V.1.4.16f)

.072 Serpentine layerage: (see Layerage, V.1.4.036b-2).
.073 Shield bud: (see V.1.4.028f-8).
.074 Side-veneer graft: (see Veneer graft, V.1.4.028k-8).
.075 Simple layerage: (see Layerage, V.1.4.036b-3).
.076 Single-node cutting: (see Cuttings, V.1.4.022f-3).
.077 Softwood cutting: (see Cuttings, V.1.4.022f-4).
.078 Sprout removal: (see Graftage aftercare, V.1.4.028i).
.079 Staking and tieing: (see Graftage aftercare, V.1.4.028i).
.080 Stem cutting: (see Cuttings, V.1.4.022f).
.081 Stolon: (see II.4.3.060).
.082 Stool layerage: (see Layerage, V.1.4.036b-1).
.083 Suckers: shoots that arise form adventitious buds on a root, either naturally or induced by wounding, as in red raspberry, blackberry, breadfruit, guava. (Fig. V.1.4.16a) (H & K)
.084 Tip layerage: (see Layerage, V.1.4.036b-4).
.085 Tongue-and-groove approach graft: (see Approach grafting, V.1.4.028a).
.086 Tongue graft: (see Whip graft, V.1.4.028k-9).
.087 Topping: (see Graftage aftercare, V.1.4.028i).
.088 Training: (see Graftage aftercare, V.1.4.028i).
.089 Translocated incompatibility: (see Incompatible, V.1.4.035).
.090 Trench layerage: (see Layerage, V.1.4.036b-5).
.091 Trimming: (see Graftage aftercare, V.1.4.028i).
.092 Truncheon: (see Cuttings, V.1.4.022g).
.093 Tuber: (see Divisions, V.1.4.024).
.094 Tuberous root: (see Divisions, V.1.4.024).
.095 Tunicate bulb: (see Bulb, V.1.4.071a-2).
.096 Uncongenial, uncongeniality: (see Incompatible, V.1.4.035).
.097 Unwrapping: (see Graftage aftercare, V.1.4.028i).
.098 V-graft: (see Cleft graft, V.1.4.028k-2).
.099 Wounding: (see Prerooting treatment, V.1.4.022a-3).

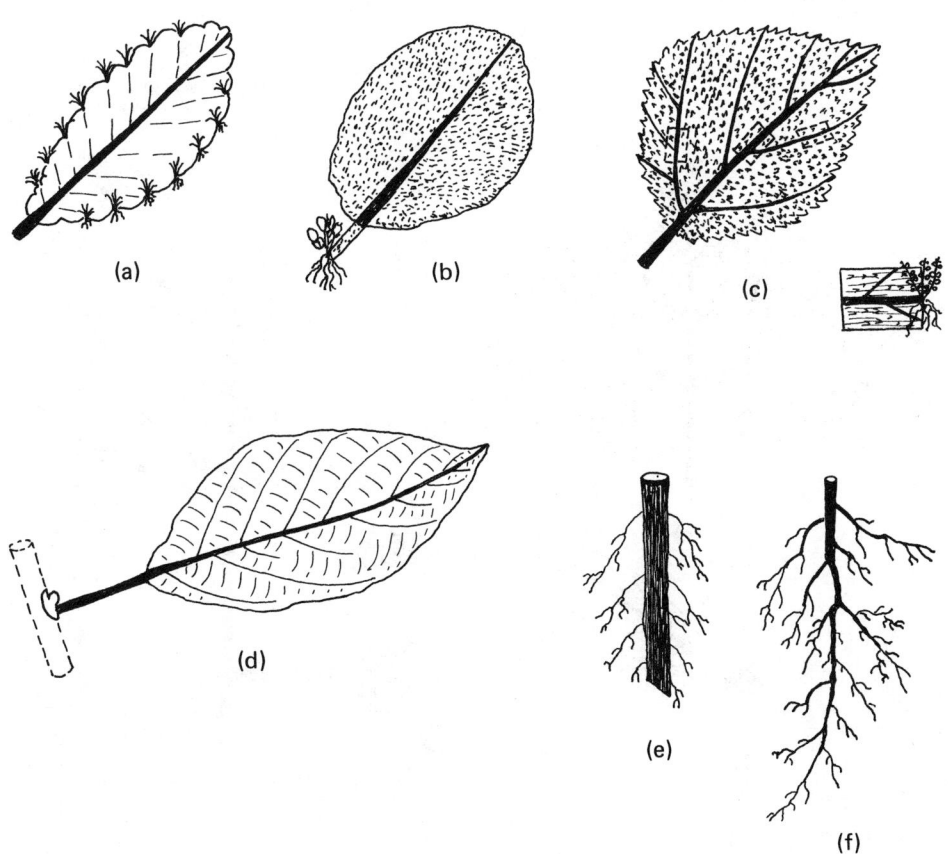

Fig. V.1.4.01. Leaf, leaf-bud and root cuttings: (a)-(c) leaf, (c) with enlargement of plantlet at right; (d) leaf-bud; (e) piece root; (f) whole root.

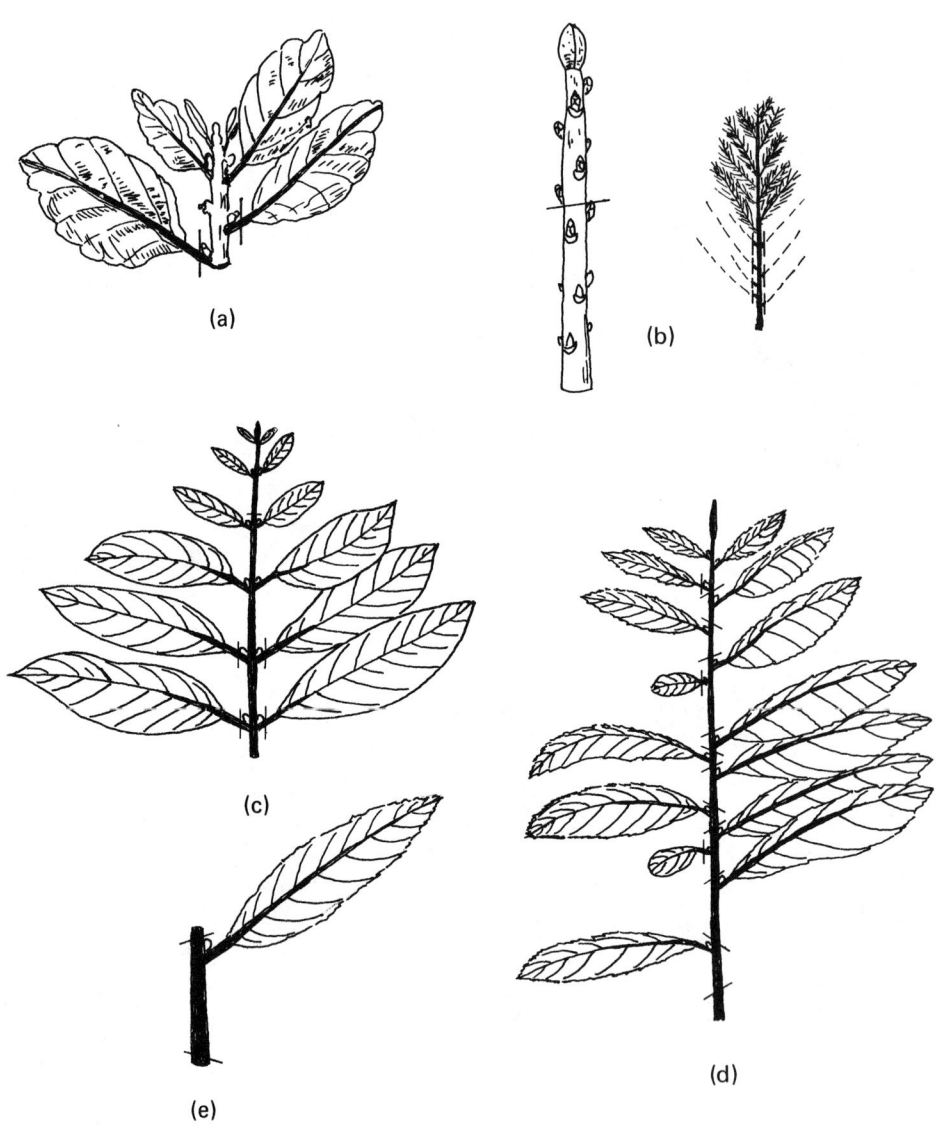

Fig. V.1.4.02. Stem cuttings: (a) herbaceous; (b) hardwood, leafless and leafy (conifer); (c)-(e) semihardwood, (c) multinode, (d), (e) single-node showing (d) cuts on a stem and (e) a single cutting.

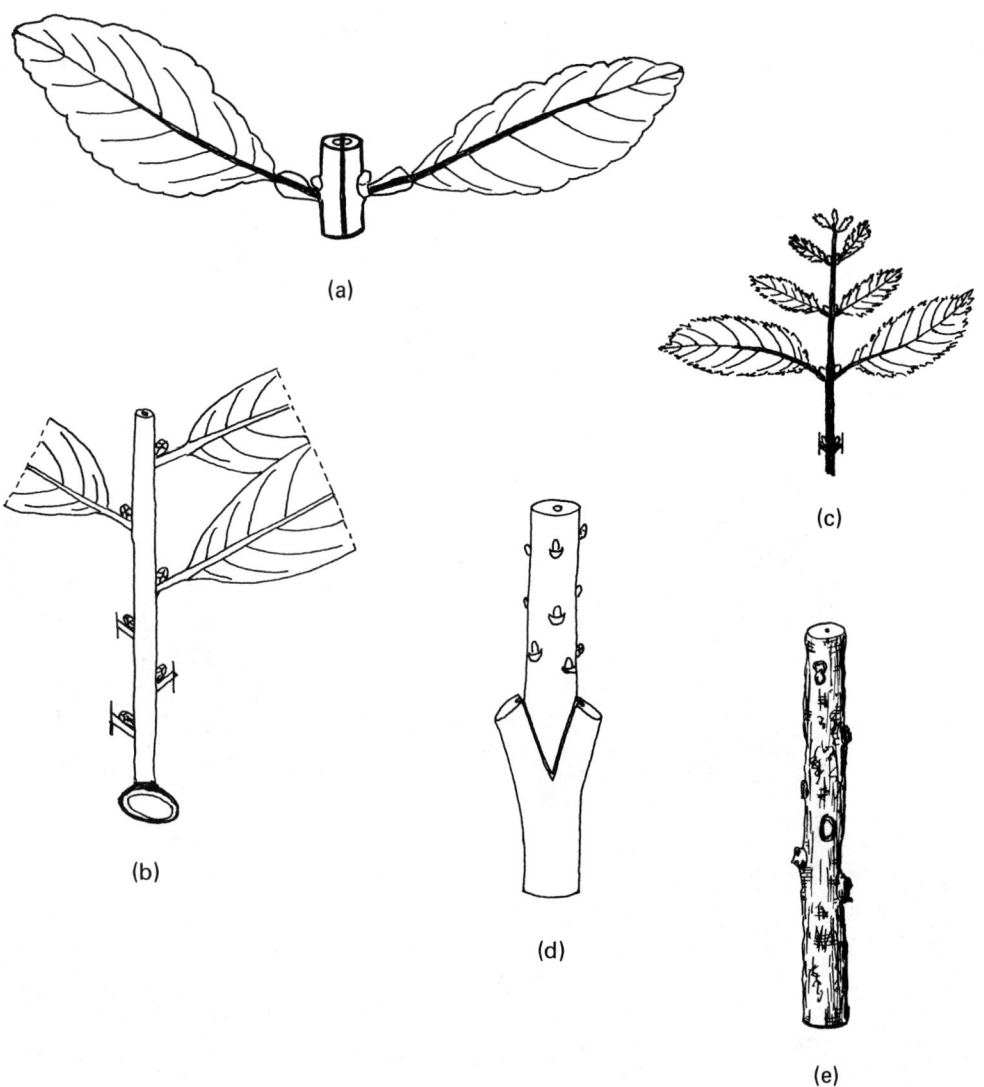

Fig. V.1.4.03. Cuttings (cont.): (a), (b) semihardwood: (a) mallet; (b) heel; (c) softwood; (d) cutting-graft (scion and rootstock both semihardwood); (e) truncheon.

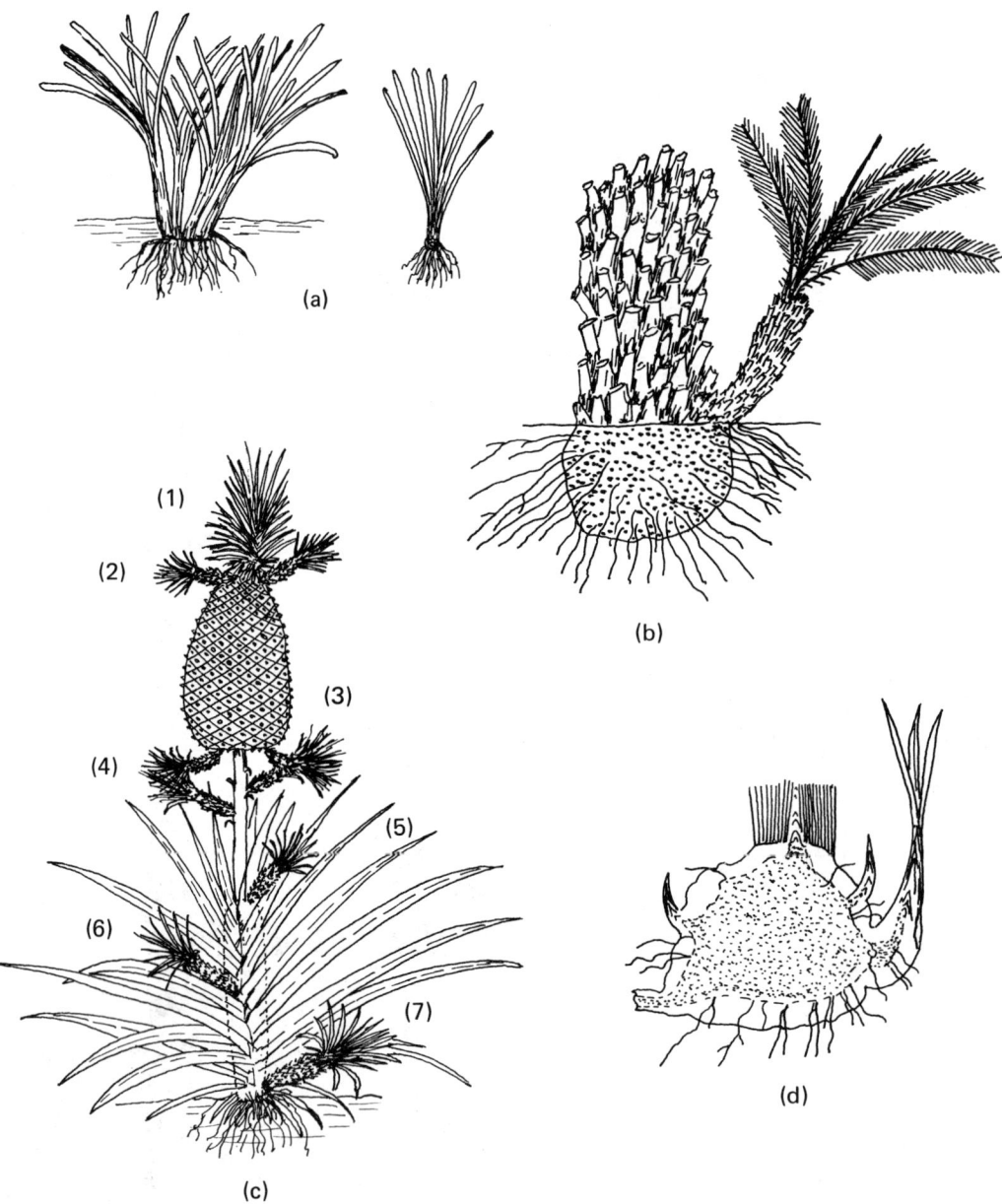

Fig. V.1.4.04. Divisions: (a) crowns (daylily); (b), (c) offsets, (b) date palm, (c) pineapple with (1) crown, (2) crown slips, (3) basal slips, (4) slips, (5) hapa, (6) sucker and (7) ratoon; (d) rhizome (t.s. of banana). (c) Redrawn from Py and Tisseau, 1965; (d) redrawn from Champion, 1963; see Acknowledgments and Selected References.

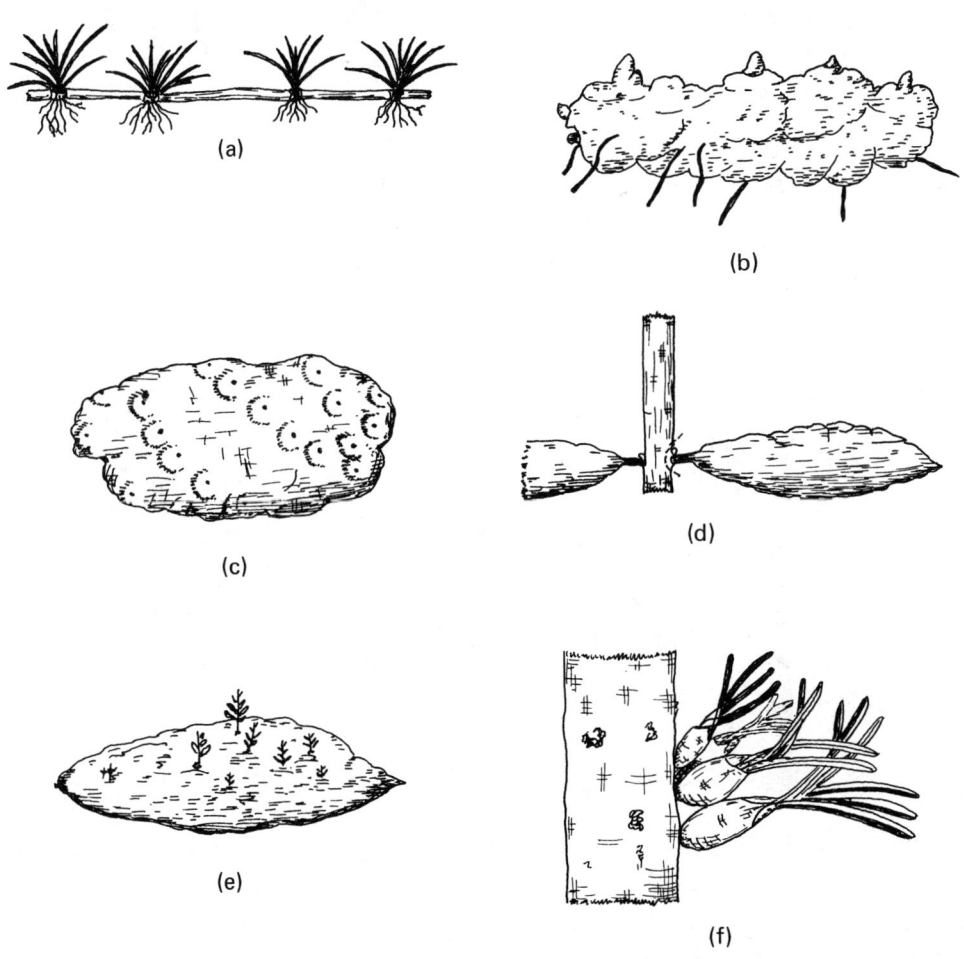

Fig. V.1.4.05. Divisions (cont.): (a) leptomorph type rhizome (="stolon"); (b) pachymorph type rhizome (ginger); (c) tuber (potato); (d), (e) tuberous roots, (d) dahlia; (e) sweet potato; (f) pseudobulbs (orchid). Redrawn from Soule and Young, 1975; see Acknowledgments and Selected References.

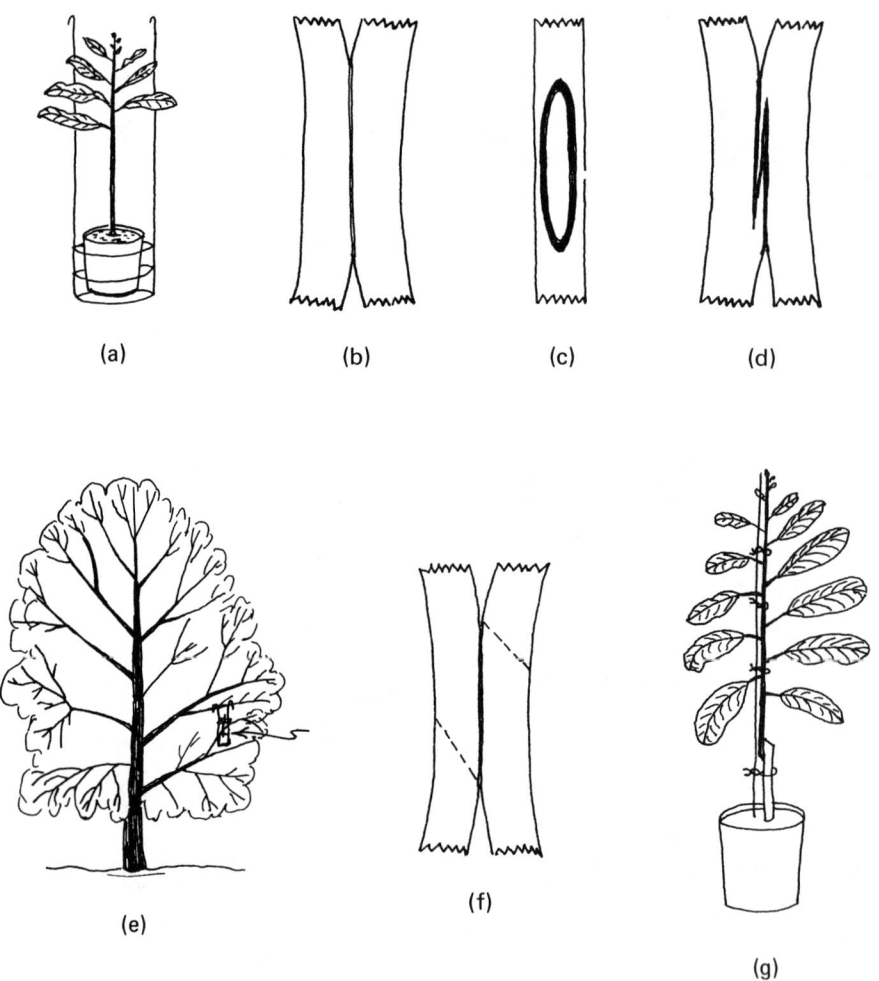

Fig. V.1.4.06. Approach grafting: (a) rootstock in container; (b) close-up of cuts made on rootstock and scion; (c) face view of a cut; (d) tongue type of cuts; (e) rootstock supported on tree after approach graft applied; (f) cuts made on scion (left) and rootstock (right) after union is complete; (g) young plant ready to be set out. Redrawn from Soule and Young, 1975; see Acknowledgments and Selected References.

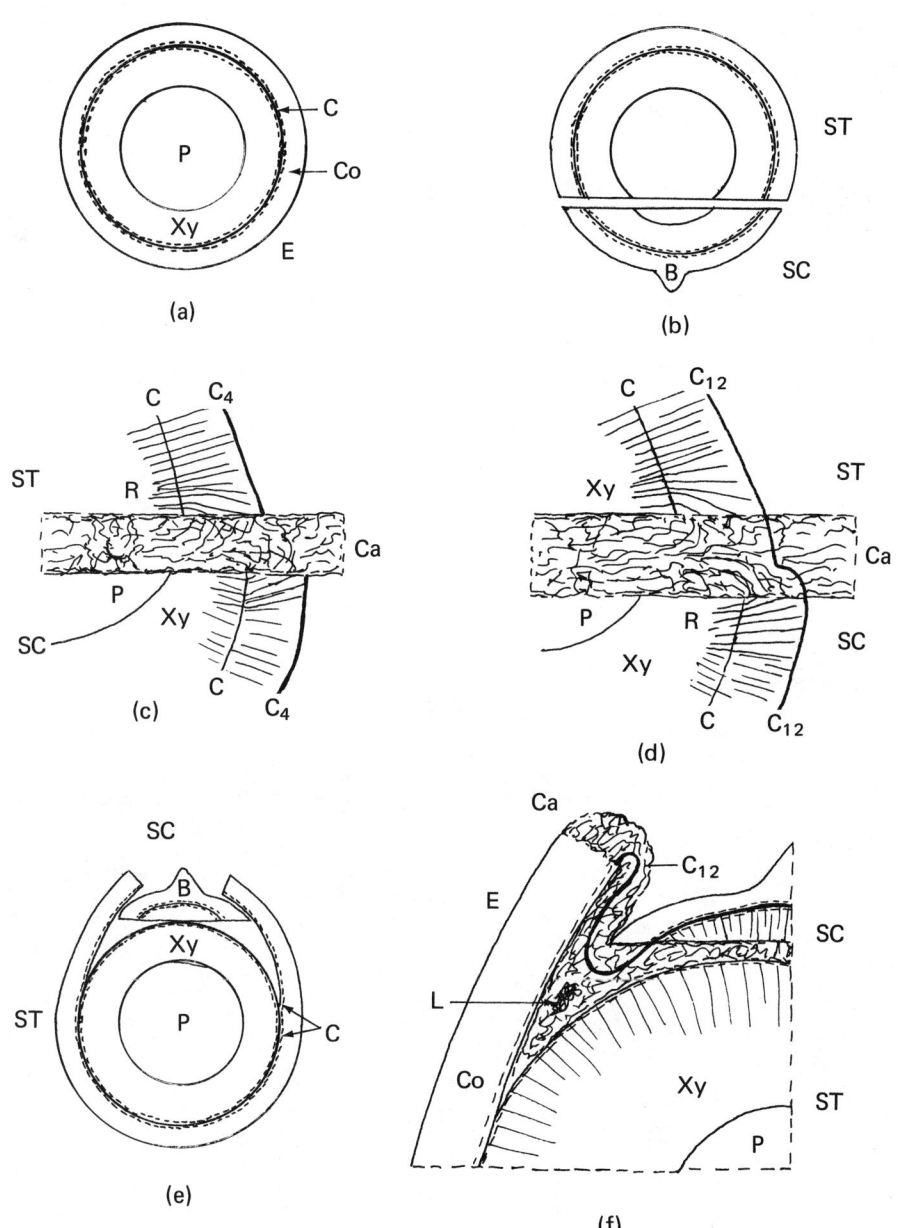

Fig. V.1.4.07. Bud (graft) union healing (median t.s.): (a)-(d) "butted end" form, (a) young seedling rootstock; (b) chip bud inserted; (c) callus bridge after 4-8 days (enlarged); (d) arched cambial bridge across callus after 12-14 days (enlarged); (e), (f) "cambial-face" form (shield budding), (e) immediately after budding; (f) completed S-shaped cambial bridge after 12-14 days. [ST = rootstock, SC = scion, P = pith, Xy = xylem, C = cambium, with most recently formed xylem and phloem adjacent (dashed lines), C_4, C_{12} = positions of cambium after 4-8 days and 12-14 days, Ca = callus, Co = cortex, R = rays (note bending near wound during healing), L = lignified area of isolated xylem in callus.] (e), (f) Adapted from Nanthachai, 1977; see Acknowledgments and Selected References.

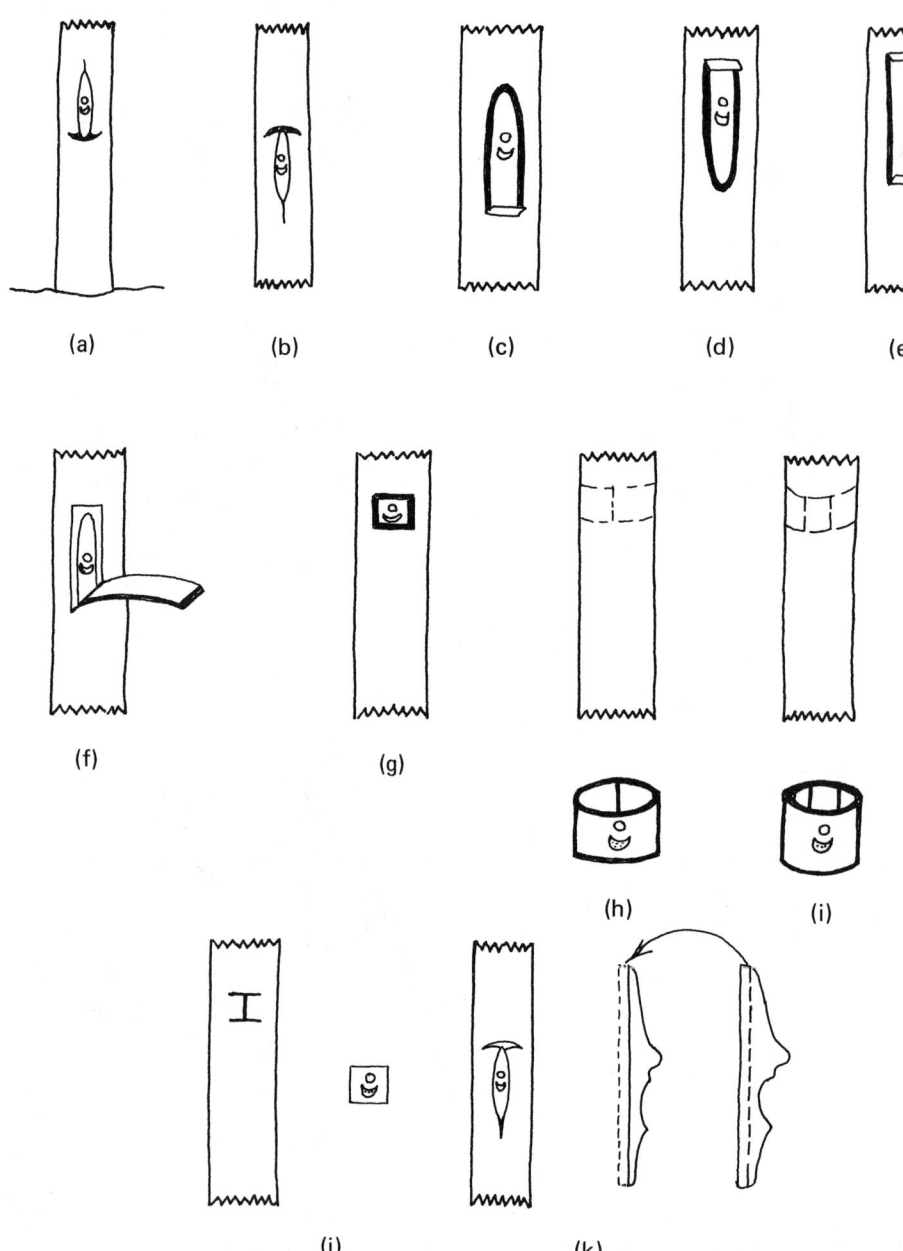

Fig. V.1.4.08. Budding: (a), (b) shield (T) bud, (a) upright T; (b) inverted T; (c) chip bud; (d) hanging bud; (e) "double" chip bud; (f) Forkert bud; (g) patch bud; (h) ring bud (with scion below); (i) annular bud (with scion below); (j) I-bud (with scion at right); (k) Nicolin bud (with details of scion and interstock at right). (a)-(j) Redrawn from Soule and Young, 1975; (k) adapted from Hartmann and Kester, 1975; see Acknowledgments and Selected References.

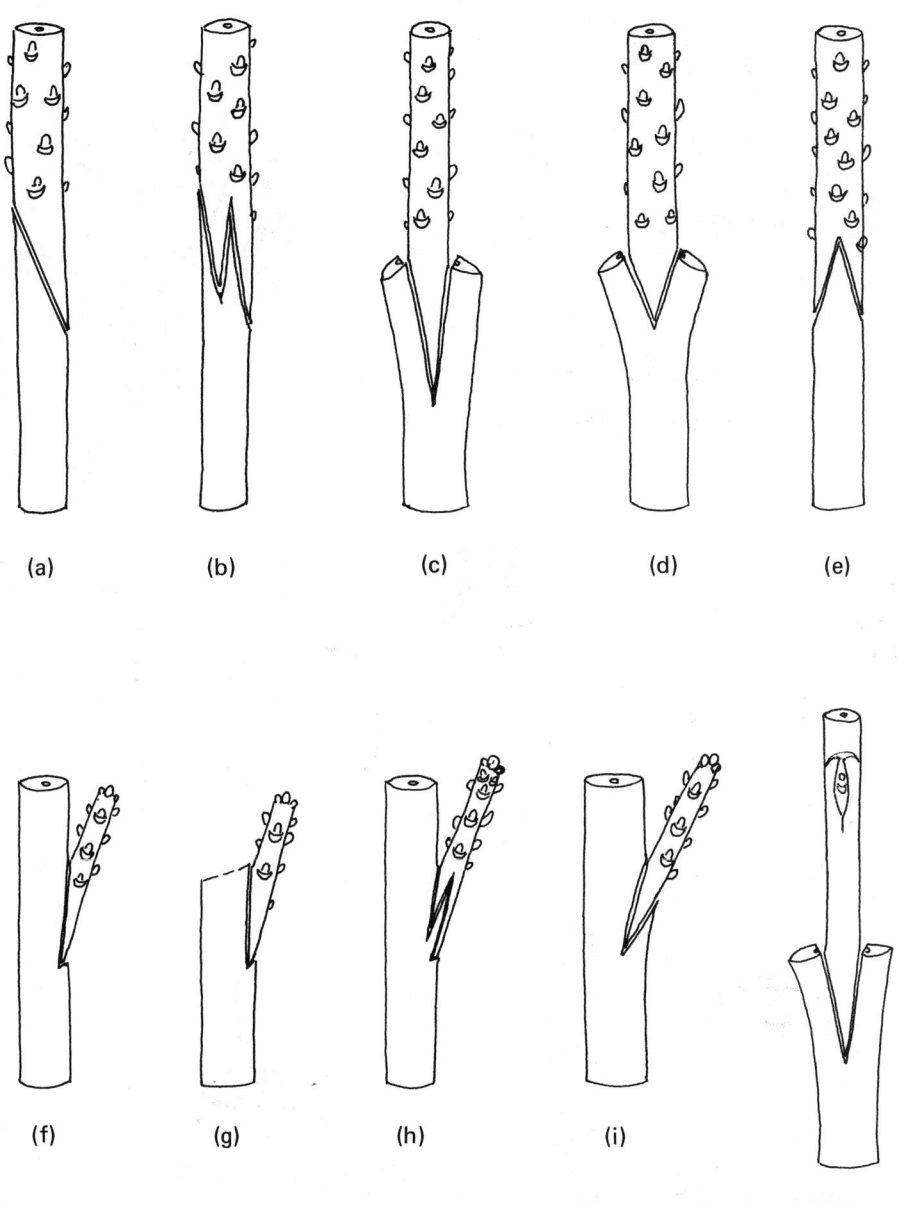

Fig. V.1.4.09. Nursery grafting: (a) splice graft; (b) whip (tongue) graft; (c) cleft graft; (d) V-graft (similar to cleft but with a shorter wedge); (e) saddle graft; (f), (g) veneer graft, (g) with rootstock cut off just above the graft at the time of propagation; (h) tongue type of side graft; (i) side graft; (j) double-working with cleft graft for interstock and upright-T bud above for the scion. Redrawn from Soule and Young, 1975; see Acknowledgments and Selected References.

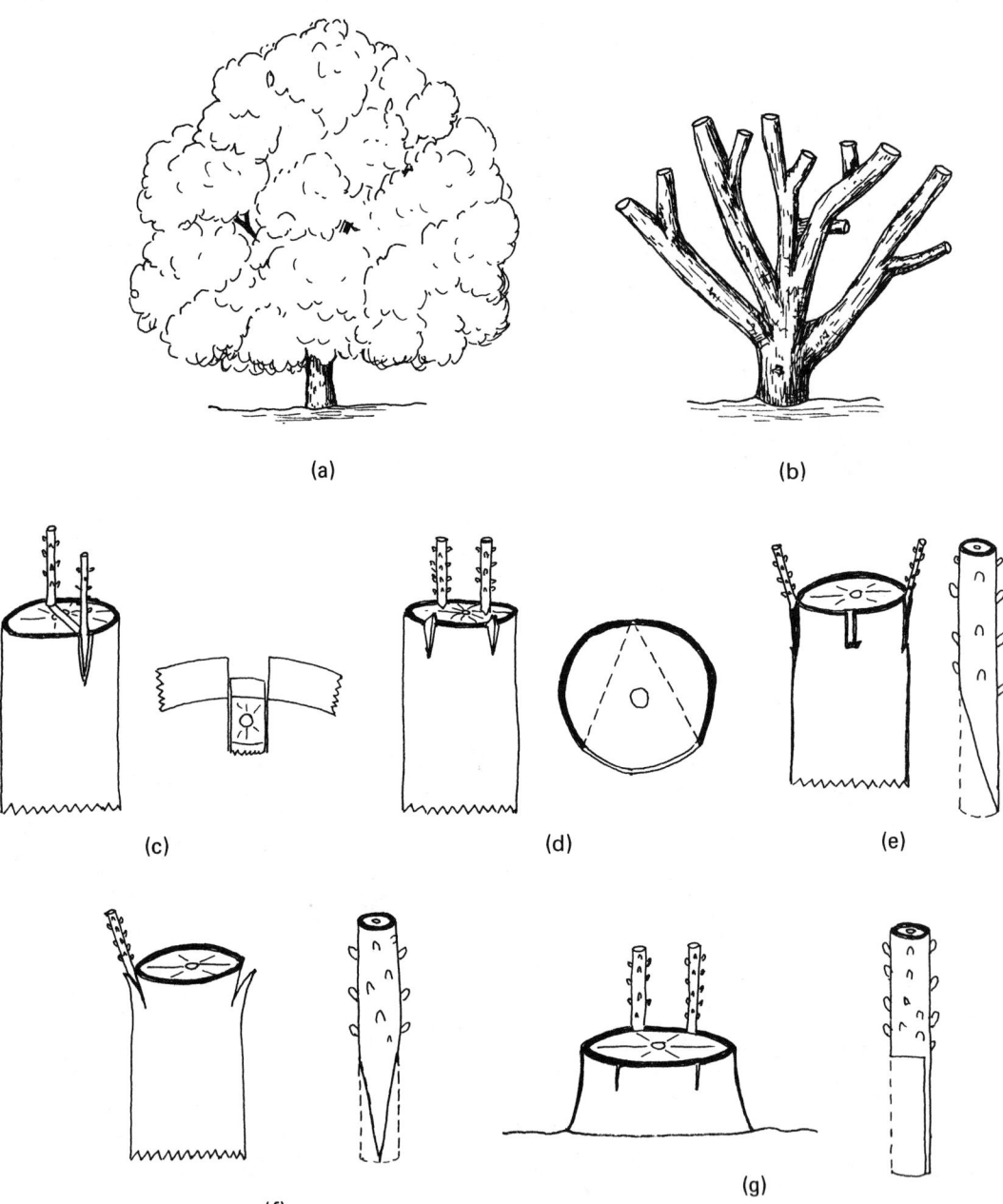

Fig. V.1.4.10. Topworking: (a) established tree; (b) buckhorned (= "hat-racked") tree; (c)-(g) types of grafting for topworking (details of scions at right); (c) cleft graft; (d) oblique cleft graft (cut usually made with a saw); (e) veneer or inlay graft; (f) side graft; (g) bark graft in stump. Redrawn from Soule and Young, 1975; see Acknowledgments and Selected references.

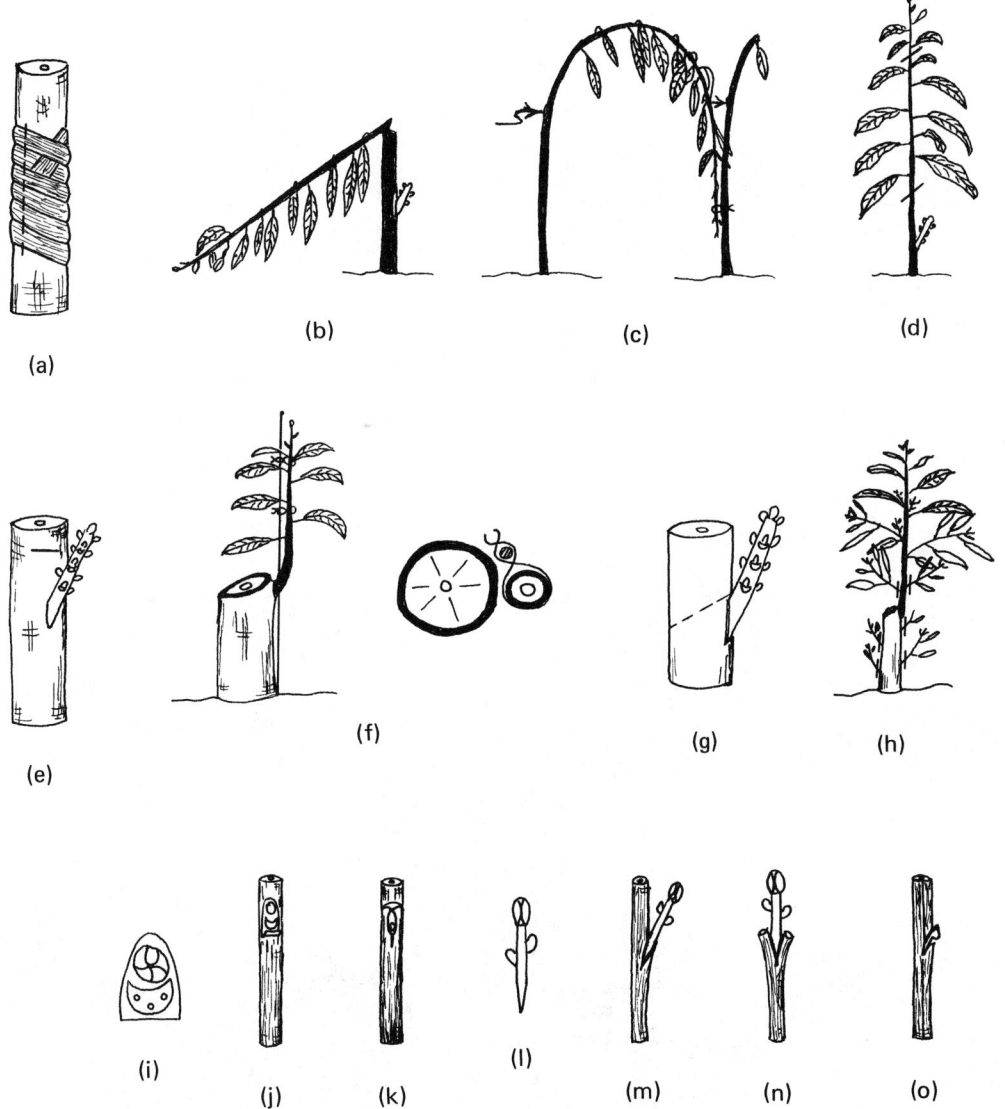

Fig. V.1.4.11. Graftage aftercare, micrograftage: (a)-(h) graftage aftercare, (a) unwrapping; (b)-(e) forcing scion growth, by (b) lopping; (c) bending; (d) topping; or (e) nicking; (f) staking and tieing (detail of proper tie at right); (g) cutting off rootstock immediately above scion; (h) sprout (gourmandizer) removal; (i)-(o) micrograftage, (i) bud shield (enlarged); (j) chip bud; (k) T bud; (l) scion as graft, base cut to a wedge; (m) side graft; (n) cleft (V) graft; (o) bud inserted as side graft (scions and rootstocks are 2-3 mm diameter necessitating use of a scalpel or razor blade and a steady hand). (a)-(h) Redrawn from Soule and Young, 1975; see Acknowledgments and Selected References.

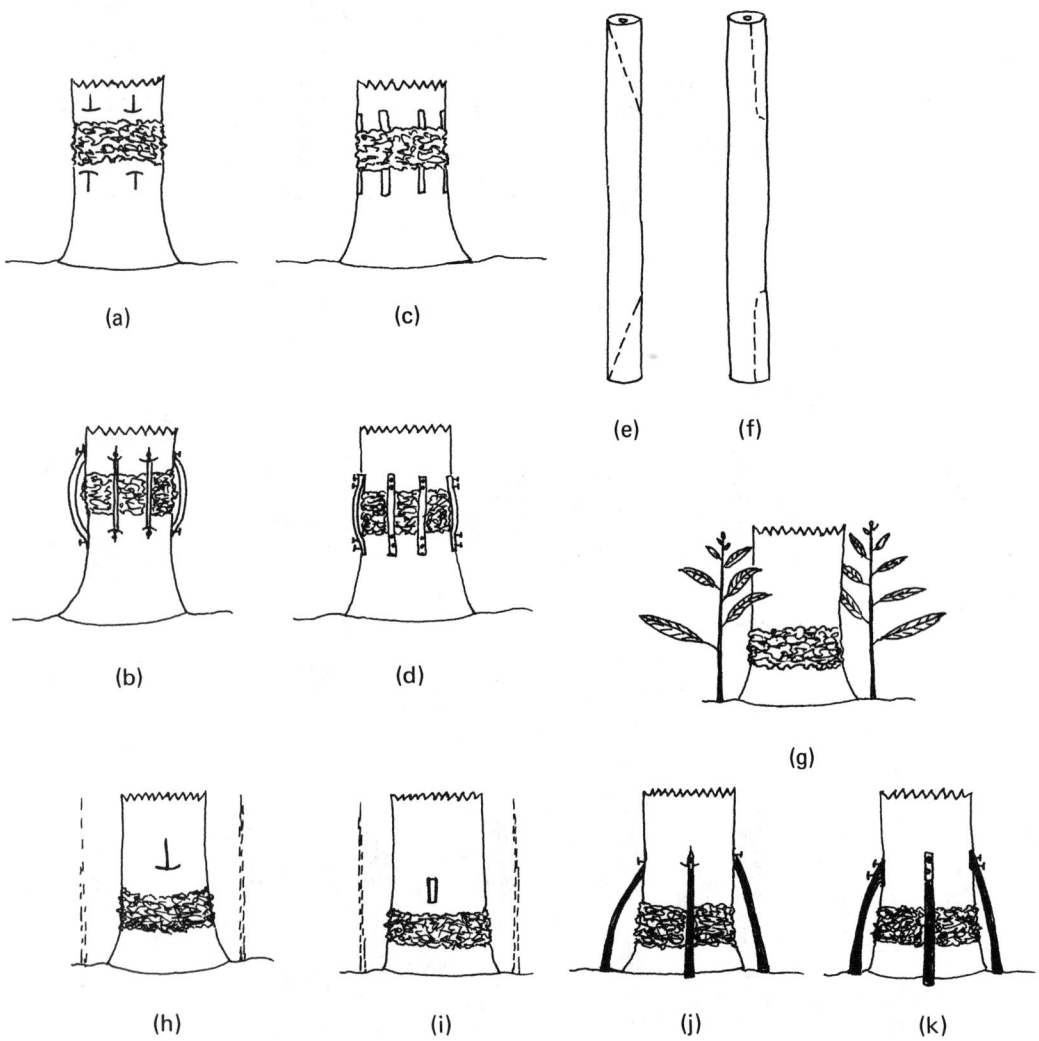

Fig. V.1.4.12. Bridge grafting; inarching: (a)-(f) bridge grafting (2 versions, T cut and inlay), (a) T cuts made above and below girdled area on trunk (too high for inarching); (b) scions in place and nailed in place with brads; (c) slots cut above and below girdled area; (d) scions nailed in place with brads; (e), (f) scions showing how ends are trimmed; (g)-(k) inarching, (g) young seedlings planted around base of tree; (h) T cuts made above girdled area; (i) inlay slots above girdled area; (j), (k) top end of seedlings trimmed as in (e) or (f) and nailed in place with brads. Redrawn from Soule and Young, 1975; see Acknowledgments and Selected References.

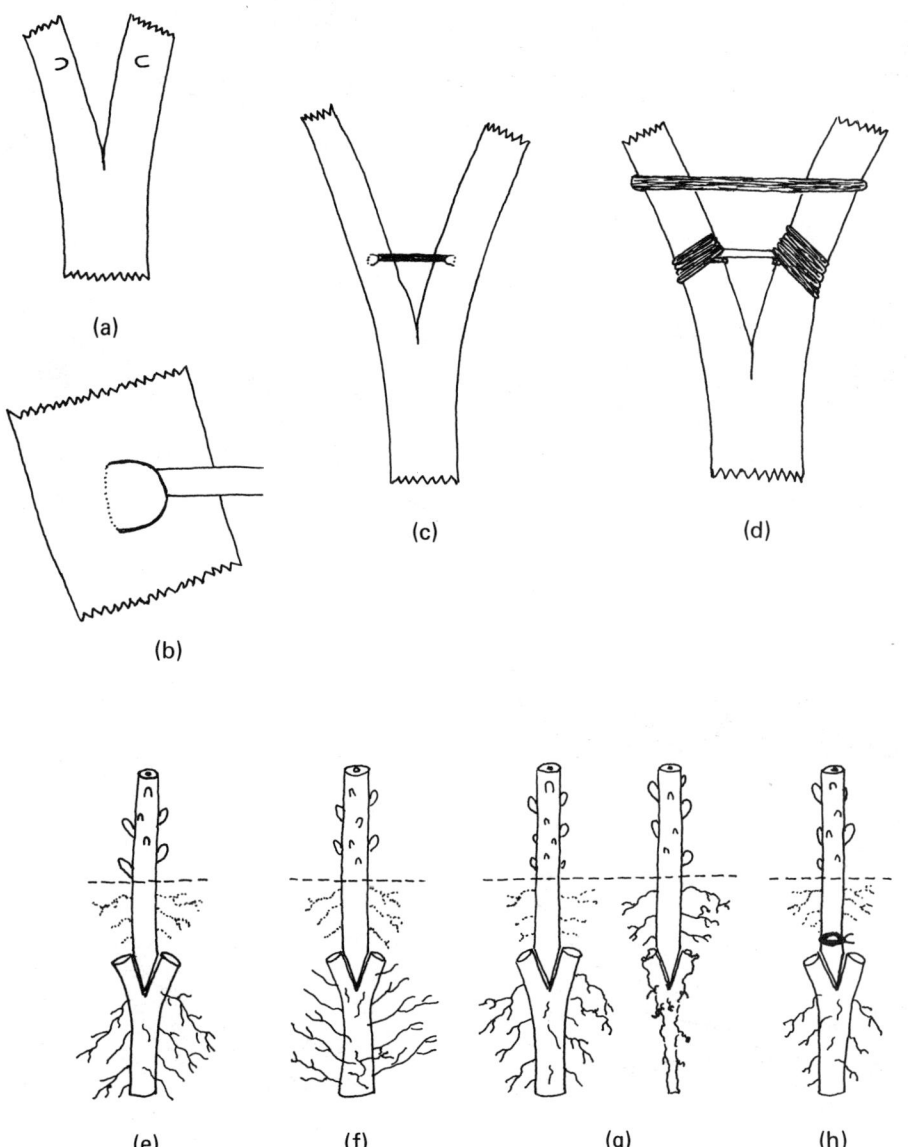

Fig. V.1.4.13. Repair graftage, nurse root grafting: (a)-(d) living brace (crotch repair), (a) horizontal cuts similar to a side graft; (b) enlargement of cut; (c) "scion" in place; (d) wounds wrapped securely with a loop of wire enclosed in hose above to relieve strain on the brace while it heals; (e)-(h) nurse root grafting, (e) root graft buried to allow scion rooting; (f) inverted root graft (rootstock dies after scion has rooted); (g) root graft with incompatible rootstock which soon dies (at right); (h) root graft with a tightly wound rubber strip just above the union (rootstock killed). (a)-(d) Redrawn from Soule and Young, 1975; (e)-(h) adapted from Hartmann and Kester, 1975; see Acknowledgments and Selected References.

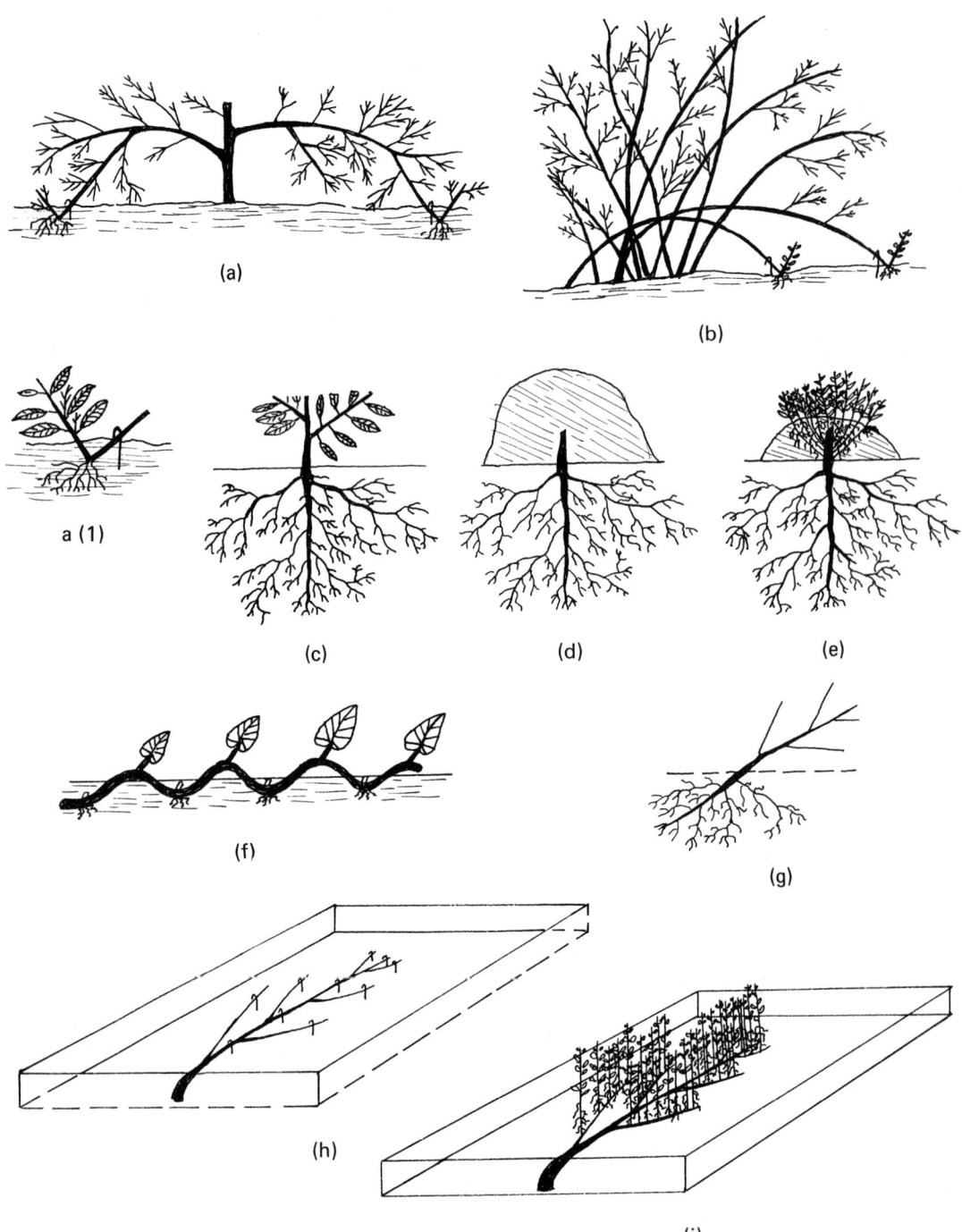

Fig. V.1.4.14. Layerage: (a) simple layering, enlargement at (1); (b) tip layering; (c)-(e) mound layering, (c) intact plant; (d) cut off and mounded with soil; (e) shoots with roots ready to be cut off; (f) serpentine layering; (g)-(i) trench layering, (g) plant set on slant; (h) pegged down in a shallow trench and covered with soil; (i) shoots with roots ready for removal. (c)-(e), (g)-(i) Adapted from Hartmann and Kester, 1975; see Acknowledgments and Selected References.

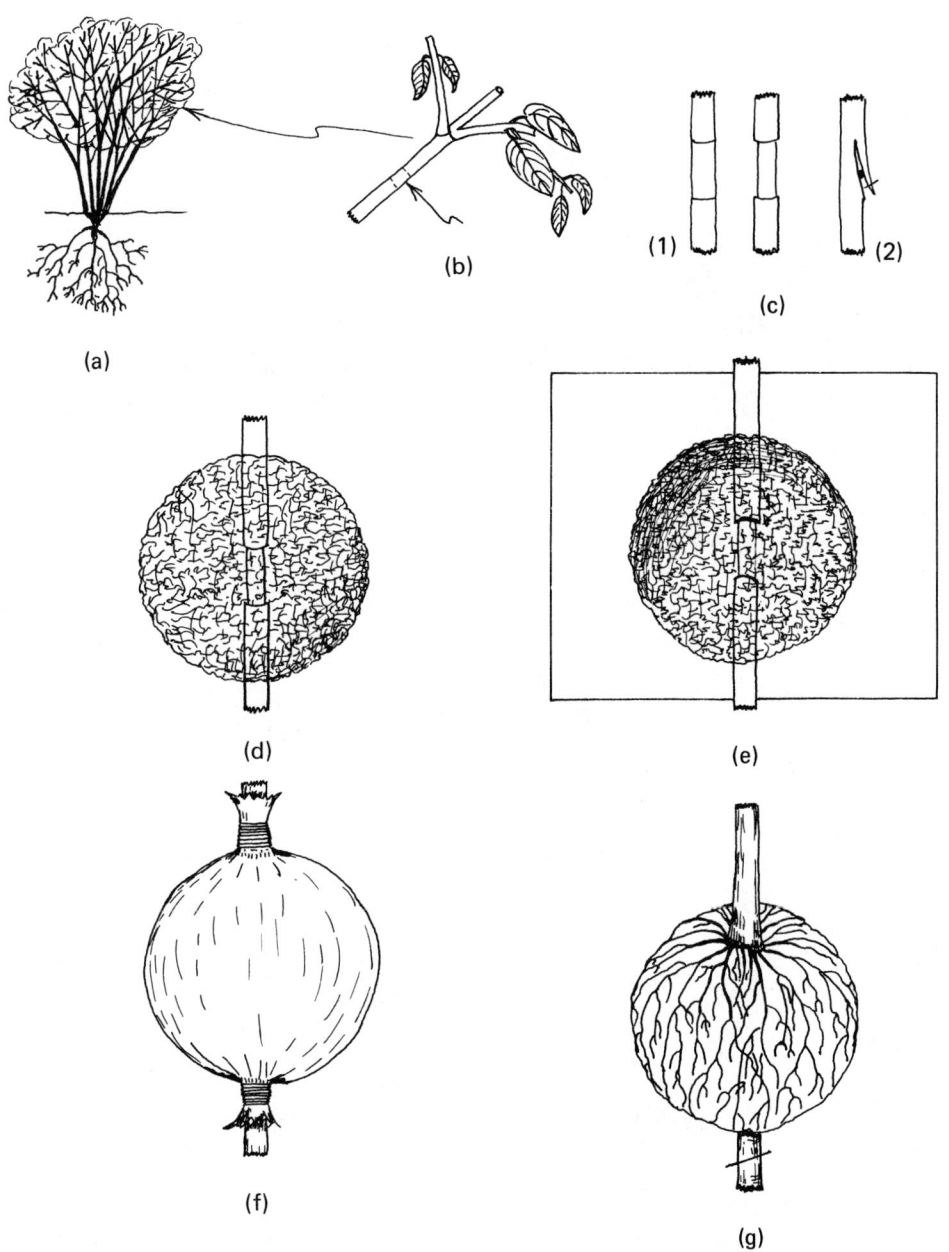

Fig. V.1.4.15. Air layering: (a) tree (shrub) to be air layered; (b) branch with area to be girdled or cut (arrow); (c) two versions, (1) cuts made top and bottom, then cylinder of bark removed or (2) slanting upward cut with tip cut off and used as a wedge; (d) ball of moist sphagnum moss or coir formed around wound; (e) sheet of polyethylene film wrapped around the ball; (f) tied securely (an outer foil wrap may also be applied); (g) a well-rooted layer ready for removal. Adapted from Ochse et al., 1961; see Acknowledgments and Selected References.

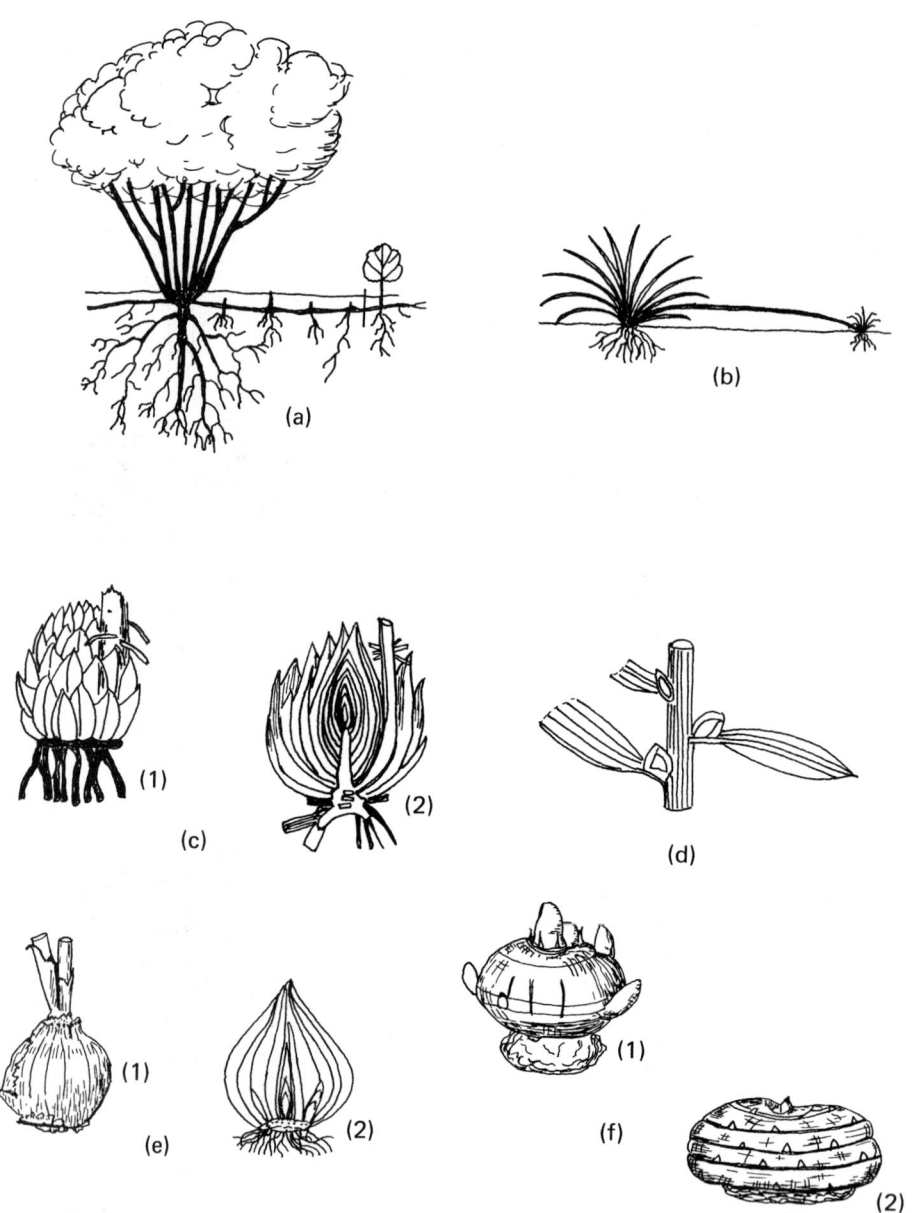

Fig.V.1.4.16. Root suckers, runners, separations, bulb, corms: (a) root sucker; (b) runner. (c)-(f) separations, (c)-(e) bulbs, (c) scaly (lily), (1) whole and (2) l.s.; (d) bulbil (aerial stem bulblet); (e) tunicate, (1) whole ("double-nosed" tulip), (2) l.s. (onion); (f) corms; (1) crocus, (2) gladiolus (both with tunic removed). (c), (d), [(e)(2)], [(f)(1)} Redrawn from Gray, 1862; [(e)(1)] redrawn from Lawrence, 1955; see Acknowledgments and Selected References.

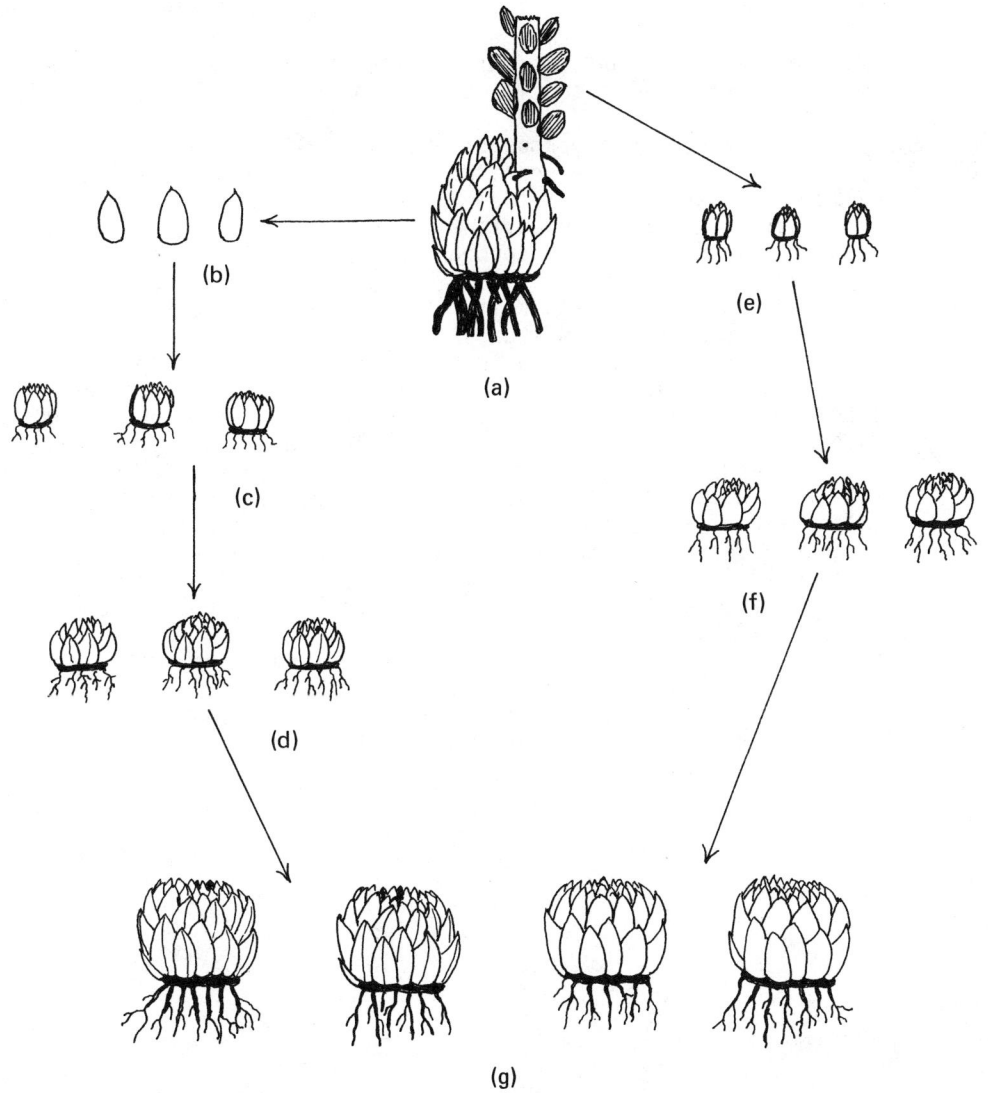

Fig.V.1.4.17. Propagation of commercial lily bulbs from scales and stem bulblets: (a) mother plant; (b) scales field planted to produce; (c) scale bublets the first year; (d) yearlings (1) the second year and (2) commercial bulbs the third year; (e) stem bulblets field planted to produce (f) yearlings the first year and (g) commercial bulbs the second year. Adapted from de Hertogh et al., 1971; see Acknowledgments and Selected References.

V.1.5. Micropropagation

.000 Micropropagation: miniaturized in vitro multiplication and/or regeneration of plant material under aseptic and controlled environmental conditions on specially prepared media that contain substances necessary for growth; used for three general types of tissue: excised embryos (embryo culture), shoot-tips ("meristem" culture or mericloning), and pieces of tissue that range from bits of stems, roots, seeds, fruits, anthers, and pollen grains to individual cells or cell parts (tissue culture). (Fig. V.1.5.01) (H & K)

.001 Embryo culture: (see Micropropagation, V.1.5.000).

.002 Embryoid: a structure resembling an embryo that is developed from callus in tissue culture; (see Micropropagation, V.1.5.000).

.003 Mericloning: (see Micropropagation, V.1.5.000).

.004 "Meristem" culture: (see Micropropagation, V.1.5.000).

.005 Micropropagation media: semisolid, with 0.5 to 1% agar, or liquid substrates that contain salts of N, P, K, Ca, Mg, S, Zn, Mn, Cu, B, Fe, and Mo, 2 to 4% sugars (sucrose or glucose), vitamins (e.g., thiamin, nicotinic acid, pyridoxine, inositol, panthothenic acid, and biotin), growth substances (e.g., naphthaleneacetic acid, 2,4-D, IAA, and kinetin), and organic complexes, such as coconut milk, yeast extract, or casein hydrolysate; the formulation depends on the kind of plant and purpose of the culture (i.e., to induce callus, shoots, and/or roots); used in test tubes, erlenmeyer flasks, bottles, or Petri dishes; liquid media require continuous shaking or rotation to provide necessary aeration; containers and media are autoclaved, with labile material run through a microbial filter and added later; (see Micropropagation, V.1.5.000).

.006 Serial divisions: separation of excised shoot-tip material that grows on culture medium at about monthly intervals to induce additional plantlets; (see Micropropagation, V.1.5.000).

.007 Tissue culture: (see Micropropagation, V.1.5.000).

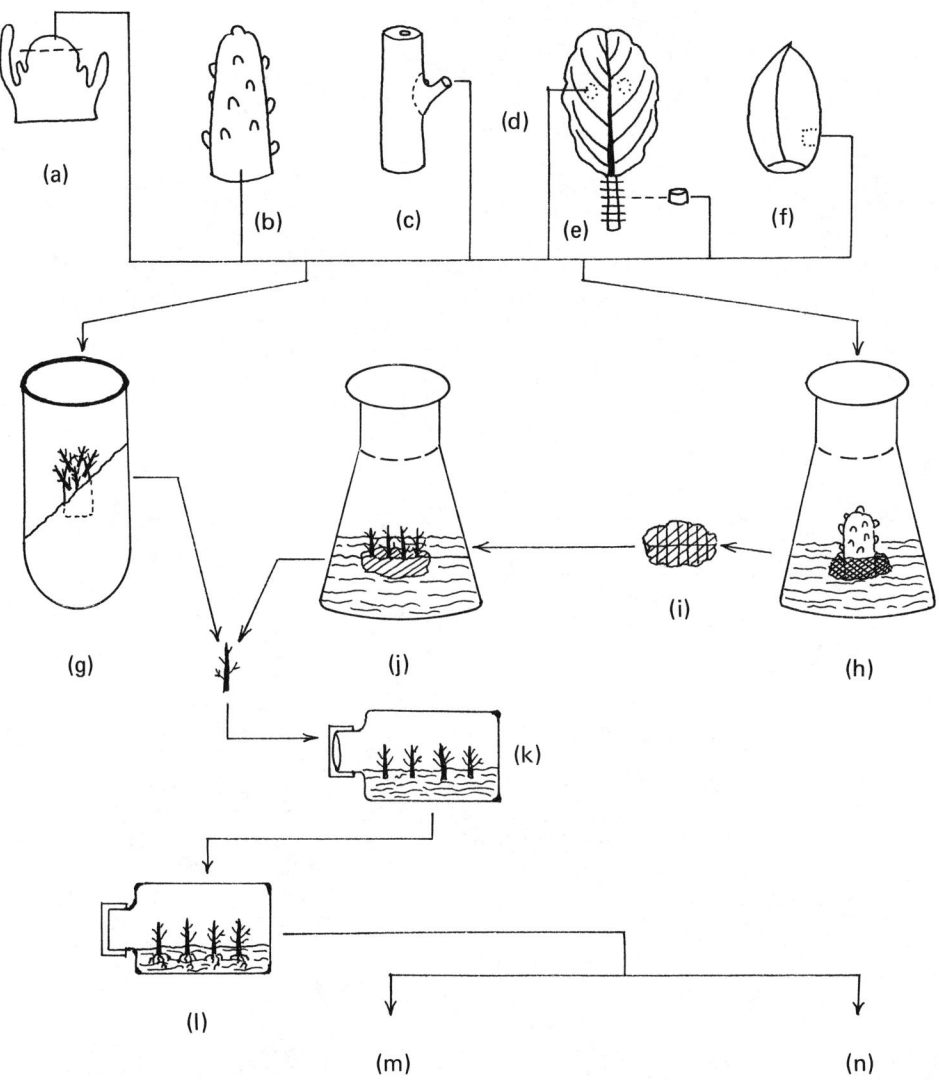

Fig. V.1.5.01. Micropropagation: (a)-(e) explant sources, (a) apical meristem; (b) shoot tip; (c) nodal section from stem; (d) leaf section; (e) petiole disks; (f) bulb scale section; (g)-(k) shoot formation, (g) shoots or adventitious buds; (h)-(j) shoots from callus; (k), (l) rooting in larger containers; (m), (n) hardening-off; (m) mist bed; (n) containers (pot or poly bag). Adapted from Young and Soule, 1978; see Acknowledgments and Selected References.

V.2. Soils

.000 Soils: originate from igneous, sedimentary, or metamorphic rocks which have been broken into minute fragments by chemical and physical processes of weathering under the combined influence of climate, physiography, vegetation, and time and thereby changed in whole or in part to a complex aggregation of primary and secondary minerals, organic matter, water, air, and biotic organisms; soil formation and its development into more or less discrete layers is a complex series of processes in which a major component is time on the order of many centuries.
.001 Aid plants: (see Soil management, V.2.110e).
.002 Albic: (see Soil horizons, V.2.107a).
.003 Alkalization: (see Solonization, V.2.120).
.004 Andic: (see Soil classification, Subgroup, V.2.096d).
.005 Angular blocky: (see Soil structure, V.2.115a).
.006 Arenic: (see Soil classification, Subgroup, V.2.096d).
.007 Argillans: (see Soil horizons, V.2.107b).
.008 Argillic horizon: (see Diagnostic horizons, V.2.028a).
.009 Attapulgite: (see Soil clay minerals, V.2.098).
.010 Azonal soil: skeletal and alluvial soils, i.e., those that lack horizons because of lack of time for zonation to develop, the manner of deposition (i.e., by wind or water), or both; (see Soil profile, V.2.112). (Y)
.011 Base saturation: the total exchangeable bases in a soil divided by the cation exchange capacity expressed as a percentage; (see Cation exchange, V.2.015).
.012 Calcic horizon: (see Diagnostic horizons, V.2.28b). (Y)
.013 Cambic: (see Soil horizons, V.2.107c).
.014 Carbon:nitrogen (C:N) ratio: the proportion of organic carbon to organic nitrogen in a soil; relatively constant for a given soil under a wide range of management conditions; a rough measure of humus content (rough because of the inherent errors in determining organic C or N). (L & B)
.015 Cation exchange capacity (CEC): the measure of negatively charged sites on the surface of clay or humus (organic matter) molecules; determined as the saturation CEC with ammonium acetate at pH 7.0 and expressed as milliequivalents (0.001 atomic or molecular wt/valence) per 100 g of dry soil; note, however, that the actual CEC may vary from the saturation CEC value, depending on soil pH. (RU)
.016 CEC: (see Cation exchange capacity, V.2.015).
.017 Class: (see Soil structure, V.2.115b).
.018 Clay loams: (see Soil textural classes, V.2.117g).
.019 Clay minerals: (see Soil clay minerals, V.2.098).
.020 Clays: (see Soil textural classes, V.2.117a).
.021 Coarse sand: (see Soil textural classes, V.2.117 1-4).
.022 Coarse sandy loam: (see Soil textural classes, V.2.117j-4).
.023 Coated: (see Soil horizons, V.2.107d).
.024 Columnar: (see Soil structure, V.2.115c).
.025 Cover crops: (see Soil management, V.2.110e).
.026 Crumb: (see Soil structure, V.2.115d).
.027 Cumulic: (see Soil classification, Subgroup, V.2.096d).
.028 Diagnostic horizons: layers of a soil profile used in classification; (see Soil horizons, V.2.107; Soil profile, V.2.112). (C & N)

V.2. PROPAGATION, NURSERY HANDLING, SOILS, CROP PRODUCTION

.028a.   Argillic horizon (subsurface): silicate clay accumulation.
.028b.   Calcic horizon: secondary calcium carbonate accumulation.
.028c.   Gypsic horizon (subsurface): secondary calcium sulfate accumulation.
.028d.   Histic epipedon (surface): very high in organic matter; wet during some parts of the year.
.028e.   Mollic epipedon (surface): thick, dark colored; high base saturation; strong structure.
.028f.   Ochric epipedon (surface): light colored; low organic content; too thin to be mollic, umbric, or histic.
.028g.   Salic horizon (subsurface): layer 15 cm (6 in.) or thicker, with secondary enrichment of salts.
.028h.   Spodic horizon (subsurface): illuvial accumulation of organic matter, iron, and aluminum.
.028i.   Umbric epipedon (surface): same as mollic, except low base saturation.
.029   Eluviation zone: (see Soil horizons, V.2.107e).
.030   Epipedon: (see Soil horizons, V.2.107f).
.031   Erosion, erosion controls: (see Soil management, V.2.110a).
.032   Family: (see Soil classification, V.2.096e).
.033   Fibrous clays: (see Soil clay minerals, V.2.098).
.034   Fine sand: (see Soil textural classes, V.2.117 l-2).
.035   Fine sandy loam: (see Soil textural classes V.2.117j-2).
.036   Fragipan: (see Soil horizons, V.2.107g).
.037   Gley: a soil with gray, blue-gray, or mottled horizons that result from impeded drainage and consequent reducing (i.e., anaerobic) conditions; the degree of mottling and development of blue-gray or gray horizons depends on the relative severity and length of time that the reducing conditions persist, the amount of drainage that takes place, and the content of accessible iron at temperatures high enough for microbial decomposition of organic matter under conditions of oxygen deficiency; anaerobic conditions result in the reduction of ferric iron to the ferrous form and insoluble manganese oxides to manganous forms; the reduced ions are much more mobile in the soil; gley or gleyed soils are common in marshy areas along the edge of lakes and rivers; (see Soil horizons, V.2.107; Soil profile, V.2.112). (Y)
.038   Grade: (see Soil structure, V.2.115e).
.039   Granular: (see Soil structure, V.2.115f).
.040   Great group: (see Soil classification, V.2.096c).
.041   Gypsic horizon: (see Diagnostic horizons, V.2.028c).
.042   Halloysite: (see Soil clay minerals, V.2.098c).
.043   Hardpan: hard, impervious layer in the soil; (see Soil horizons, V. 2.107; Soil profile, V.2.112).
.044   Histic epipedon: (see Diagnostic horizons, V.2.028d).
.045   Hydrous mica group: (see Soil clay minerals, V.2.098).
.046   Illite: (see Soil clay minerals, V.2.098).
.047   Illuviation zone: (see Soil horizons, V.2.107h).
.048   Kaolin: (see Soil clay minerals, V.2.098).
.049   Kaolinite group: (see Soil clay minerals, V.2.098).
.050   Limnic materials: (see Soil horizons, V.2.107i).
.051   Lithic: (see Soil classification, Subgroup V.2.096d).
.052   Lithic contact: (see Soil horizons, V.2.107j).
.053   Lithologic discontinuity: (see Soil horizons, V.2.107k).
.054   Loams: (see Soil textural classes, V.2.117i).
.055   Loamy coarse sand: (see Soil textural classes, V.2.117k-4).
.056   Loamy fine sand: (see Soil textural classes, V.2.117k-2).

V.2. PROPAGATION, NURSERY HANDLING, SOILS, CROP PRODUCTION

.057 Loamy sands: (see Soil textural classes, V.2.117k).
.058 Loamy very fine sand: (see Soil textural classes, V.2.117k-1).
.059 Massive: (see Soil structure, V.2.115g).
.060 **Mechanical aids**: (see Soil management, V.2.110c).
.061 Medium loamy sand: (see Soil textural classes, V.2.117k-3).
.062 Medium sand: (see Soil textural classes, V.2.117 l-3).
.063 Medium sandy loam: (see Soil textural classes, V.2.117j-3).
.064 Mollic epipedon: (see Diagnostic horizons, V.2.28e).
.065 Montmorillonite: (see Soil clay minerals, V.2.098).
.066 Montmorillonite group: (see Soil clay minerals, V.2.098).
.067 Ochric epipedon: (see Diagnostic horizons, V.2.28f).
.068 Order: (see Soil classification, V.2.096a).
.069 Organic layer designations: horizons in organic soils may be classed as _fibric_ (least decomposed), _hemic_ (intermediate by decomposed), or _sapric_ (highly decomposed); (see Organic soils, V.2.070: Soil horizons, V.2.107; Soil organic matter, V.2.111). (C & N)
.070 Organic soils: designated as _muck_ (well decomposed plant remains), _peat_ (raw undecomposed plant materials), _mucky peat_, and _peaty muck_ (materials intermediate in decomposition between muck and peat); (see Soil organic matter, V.2.111). (C & N)
.071 Orogeny: the process of mountain formation; (see Peneplanation, V.2.074).
.072 Paralithic contact: (see Soil horizons, V.2.107 l).
.073 Ped: individual lumps that form the body of the soil; (see Soil structure, V.2.115). (C & N)
.074 Peneplanation: the process of wearing land down to a low relief; the reverse of orogeny (V.2.071).
.075 Platy: (see Soil structure, V.2.115h).
.076 Plinthite: iron-rich, humus-poor mixtures of clay with quartz and other diluents that commonly occur as dark red mottles or reticulate patterns which harden irreversibly with repeated wetting and drying; (see Soil horizons, V.2.107). (C & N)
.077 Plow sole (plow pan): a hard layer of soil formed by continual plowing of a field at the same depth. (RU)
.078 Prismatic: (see Soil structure, V.2.115i).
.079 Pyrophyllite: (see Soil clay minerals, V.2.098).
.080 Quartzipsamments: (see Soil horizons, coated, V.2.107d).
.081 Regolith: (see Soil horizons, V.2.107m).
.082 Salic horizon: (see Diagnostic horizons, V.2.28g).
.083 Salinization: accumulation of soluble salts within a soil profile, whether by natural processes or as a result of irrigation, under conditions of low rainfall; the zone of maximum accumulation may occur at the surface or at some lower depth, depending on the level of the water table; (see Soil profile, V.2.112). (Y)
.084 Sands: (see Soil textural classes, V.2.117 l).
.085 Sandy clay loams: (see Soil textural classes, V.2.117h).
.086 Sandy clays: (see Soil horizons, V.2.117b).
.087 Sandy loams: (see Soil textural classes, V.2.117j).
.088 Sequum: (see Soil horizons, V.2.107n).
.089 Series: (see Soil classification, V.2.096f).
.090 Silt loams: (see Soil textural classes, V.2.117f).
.091 Silts: (see Soil textural classes, V.2.117e).
.092 Silty clay loams: (see Soil textural classes, V.2.117d).
.093 Silty clays: (see Soil textural classes, V.2.117c).
.094 Single grain: (see Soil structure, V.2.115j).
.095 Slickensides: polished or grooved surfaces on soil structural units

produced by one mass sliding past another; they are common in soils with swelling-type clays that undergo marked changes in moisture content; (see Soil structure, V.2.115). (C & N)

.096 Soil classification (U.S.): new system adopted by USDA in 1965; six categories from I (Order) to VI (Series); based on soil properties in the field that can be measured quantitatively and names with a definite connotative meaning that describes a major characteristic. (C & N)

.096a. Order: Category I, based on differences in measurable and visible soil horizon characteristics, whose presence or absence indicates the dominant active soil-forming process or the lack of such development; nine orders of mineral soil, one of organic soil, many of which fall into definite geographical ranges (i.e., importance of climate).

.096b. Suborder: Category II, based on developmental characteristics, such as accumulation of soluble materials in the profile, degree of gleying reflected by moisture conditions, presence or absence of B horizons, and/or broad differences in mineralogy and chemistry.

.096c. Great Group: Category III, based largely on the thickness and presence or absence of diagnostic horizons not used in Categories I or II; additional characteristics for the separation of groups are color, degree of base saturation, tonguing, irreversible hardening, and presence or absence of self-mulching clays.

.096d. Subgroup: Category IV, defined only in relation to Category III; either defined in terms of the central concept of the Great Group or in terms of intergrades to a category above IV; may also be in terms of extra grades (soil properties not characteristic of other Great Groups, Suborders, or Orders); e.g., <u>Andic</u> (high ash, pumice, or amorphous clay content), <u>Arenic</u> (50 to 100 cm of sandy materials over a B horizon), <u>Cumulic</u> (extra thick surface horizon that contain more organic matter than is ordinarily expected), and <u>Lithic</u> (bedrock near the surface).

.096e. Family: Category V, subdivided in subgroups largely on the basis of properties important to soil use and management; includes urban and industrial as well as agricultural uses; e.g., texture, mineralogy, reaction, temperature, permeability, horizon thickness, consistence, and coating.

.096f. Series: Category VI, somewhat closely related to family; consists of soils similar in all profile characteristics, except that surface soil texture may vary from place to place.

.097 Soil classification systems: USDA system (V.2.096) used in the United States; FAO-UNESCO, the only worldwide system; U.S.S.R.; Australia; several in Africa (CCTA, INEAC, ORSTOM, and Ghana); those in India, Sri Lanka, and other countries follow U.S. or French (ORSTOM) systems for the most part. (Y)

.098 Soil clay minerals: four main groups, characterized by two types of lattice: the 1:1 kaolinite group (a) consists of a sheet of silicon oxygen tetrahedra coalesced with a sheet of aluminum-hydroxyl octahedra; the 2:1 groups [(b) - (d)] have a sheet of tetrahedra on both sides of the sheet of octahedra; general formulas: (a) kaolinite group (1:1 clays) has low exchange capacity, $Si_2O_3O_2OHAl_2(OH)_3$, kaolin, halloysite; (b) hydrous mica group (2:1 clays), $Si_2O_3Al_2O_4(OH)_2SiO_3$, illite (moderate CEC), vermiculite (very high CEC); (c) montmorillionite group (2:1) clays, expanding lattice, high CEC, $Si_2O_3Al_2O_4(OH)_2Si_2O_3$, montmorillionite, pyrophyllite; (d) fibrous clays (2:1), $Si_2O_3Al_2O_4(OH)_2Si_2O_3$, attapulgite. (OSDW)

.099 Soil conditioners: materials used to stabilize or improve soil structural characteristics either by providing sources of reserve nutrients in sands or assisting in leaching out undesirable ions such as Na and thereby improving crumb structure of soils with appreciable clay content; two groups, natural and synthetic, the former including organic matter (i.e., green manures, peat, compost, etc.), gypsum (land plaster), and rock phosphate, among others, and the latter, linear organic polymers (e.g., "Krilium" = vinyl acetate maleic anhydride, hydrolyzed polyacryonitrile, various polyvinyl alcohols, etc.); synthetic conditioners have not been widely adopted primarily because of their cost and limited effectiveness, thus chief reliance has been placed on natural conditioners, which are often too expensive for large-scale use. (RU)

.100 Soil consistence: a combination of soil properties that determines its resistance and/or its ability to be molded or changed in shape; depends mainly on the forces of attraction between soil particles and varies with changes in moisture content: (1) when wet (at or slightly above field capacity), as stickiness (the quality of adhesion to other objects) or plasticity (the ability to change shape continuously under the influence of an applied force and to retain the impressed shape on removal of the stress); (2) when moist, as loose (noncoherent), friable (crushes easily under gentle to moderate pressure and coheres when pressed together), or firm (crushes under moderate pressure but resistance is distinctly noticeable); (3) when dry, as soft (weakly coherent and fragile) or hard (moderately resistant to pressure); and (4) cementation, hard brittle consistence caused by some substance other than clay minerals, as weakly cemented to indurated; (see Ped, V.2.073; Soil structure, V.2. 115). (C & N)

.101 Soil core drill: a hand or machine-driven drill that consists of a shaft and a steel tube with the bottom edge sharpened or cut as in saw teeth; used for collecting soil samples.

.102 Soil diffusivity: $Q = -D\, dm/dl$, where D is the diffusivity and dm/dl is the moisture content gradient in the soil; the relation between diffusivity and conductivity being $D = K\, ds/dm$, where ds/dm is the slope of the moisture-characteristic curve at suction $\underline{s}$; (see Soil hydraulic conductivity V.2.108). (RU)

.103 Soil erosion: (see Soil management, V.2.110a).
.104 Soil erosion controls: (see Soil management, V.2.110b).
.105 Soil erosion factors: (see Soil management, V.2.110c).
.106 Soil formation processes: a series of processes that leads to the formation of a soil, including physical and chemical (hydration, carbonation, hydrolysis, oxidation, reduction) weathering; translocation in solution (leaching, precipitation, chelation) or suspension (e.g., clays); clay mineral synthesis; lateral movement on the soil surface of the soil material or within the soil; and structure formation and biological processes (accumulation of dead material, humification, loss of organic matter, nitrification, and termite activity) from a parent material. (Y)

.107 Soil horizons: layers of soil profile, designated as 01, loose leaves and litter, largely undecomposed; 02, organic debris, partially decomposed; A1, well decomposed organic matter or humus; A2, light-colored horizon of maximum leaching or eluviation; A3,

	transitional from A to B but more like A; B1, transitional from A to B but more like B; B2, horizon of maximum accumulation of clay, iron, or organic matter or any combination of these materials; B3, transitional from B to C; C, parent material or any other unconsolidated geologic strata; (see Soil profile, V.2.112). (Fig.V.2.01a) (C & N)
.107a.	Albic horizon: surface or lower horizon with such thin sesquioxide and/or clay coatings on sand and silt that color is due to sand and silt particles.
.107b.	Argillans: translocated layers of clay particles (clay skins) oriented on ped surfaces or in pores.
.107c.	Cambic horizon: a horizon that occurs immediately below an epipedon (or at the surface if truncated) which has been altered by genetic soil development without mineral accumulation or extreme weathering.
.107d.	Coated: quartzipsamments (quartzic sands) with moisture equivalent of 2% or more (usually contain 5 to 15% silt, plus clay).
.107e.	Eluviation zone: soil horizons (A) from which materials have been removed (i.e., leached).
.107f.	Epipedon: the horizon that forms at the surface; includes the upper part of the soil that is darkened by organic matter, the upper eluvial horizons, or both.
.107g.	Fragipan: a loamy subsurface horizon with a low content of organic matter, high bulk density, seemingly cemented when dry and brittle when moist; dry fragments slake in water.
.107h.	Illuviation zone: soil horizons (B) that contain an accumulation of materials from overlying horizons (A).
.107i.	Limnic materials: includes organic or inorganic materials (1) deposited in water by precipitation or by the action of aquatic organisms such as algae or diatoms or (2) derived from underwater and floating aquatic plants subsequently modified by aquatic animals; e.g., layers of marl, coprogenous earth (sedimentary), or diatomaceous earth.
107j.	Lithic contact: boundary between soil and continuous, coherent underlying material (rock) with hardness of 3.0 (Moh's scale; i.e., as hard or harder than calcite).
.107k.	Lithologic discontinuity: a significant change in particle-size distribution or mineralogy that indicates a difference in the material from which the horizons have formed (master horizons are given roman numeral prefixes; e.g., II B2).
.107 l.	Paralithic contact: same as lithic contact (V.2.107j), but underlying material with hardness less than 3.0 (Moh's scale; i.e., softer than calcite).
.107m.	Regolith: the unconsolidated mantle of weathered rock and soil material on the earth's surface; the loose material above bedrock (which may or may not be the source of the regolith).
.107n.	Sequum: an illuvial (B) horizon with its overlying eluvial (A) horizon if one is present; if more than one sequum is present in a vertical sequence the lower ones are given A and B designations with one, two, or more prime accents, as needed (e.g., A'2, B'2, A"2, B"2).
.107o.	Solum: the upper part of a soil profile above the parent material in which the processes of soil formation are active; i.e., the A and B horizons.
.108	Soil hydraulic conductivity: expressed as $Q = -K\, ds/dl$, where Q is the water flux across a unit area perpendicular to the direction of flow, ds/dl is the suction gradient, and K is the hydraulic

(saturated soils) or capillary (unsaturated soils) conductivity; the latter is a constant at constant temperatures; K = kρg/n, where ρ is the density of water, g is the acceleration due to gravity, k is the soil permeability, and is the viscosity of the water; K may be expressed as a velocity (cm/s) or time (usually s), the latter if ds/dl is expressed as energy per unit mass. (RU)

.109 Soil lysimeter: a column of undisturbed soil entirely isolated from the land surface surrounding it and provided with effective and thorough drainage; may be set up as columns with masonry walls and perforated iron plate support or as tanks, cans, or jars for the study of percolation and leaching of soils under various cropping and fertilizing regimes. (L & B)

.110 Soil management: essentially a program to effect the transition from forest, savanna, grassland, etc., to productive cropland with minimum loss of essential nutrients or disturbance of physical structure and to maintain these conditions with measures to curtail losses from leaching and erosion, improve water relations, maintain soil organic matter, rehabilitate poor, badly eroded infertile soils, or reclaim waterlogged or saline soils so that mineral nutrition and other parts of the overall cultural program may be used to best advantage for optimum crop production. (OSDW)

.110a. Erosion controls: two main types, mechanical aids and vegetative covers (cover crops), based on three general principles: (1) brake (i.e., slow down) the force of rain hitting the soil and of water running along the surface, (2) absorb as much water into the soil as practicable, and (3) drain away the surplus safely.

.110b. Erosion factors: basically the runoff in humid climates, the portion of rain or melted snow or ice that is neither absorbed by the soil nor evaporated into the atmosphere; the proportion of water falling on an area that runs off and its destructive force are influenced by (1) the intensity, duration, frequency, and seasonal distribution of rainfall, (2) soil type, texture, structure, and moisture content, (3) depth to the ground-water table, (4) slope or inclination of the terrain, (5) type and kind of vegetative cover, and (6) the energy of a stream of water that varies as the square of its velocity but the capacity to transport soil particles (i.e., work performed) varies as the fifth power of its velocity.

.110c. Mechanical aids: several distinct types: (1) tillage across the slope rather than up and down; (2) planting along the contour, with or without buffer strips; (3) ridge, or broad based, terraces (slopes to 5 or perhaps 10%) in two general forms, Magnum and Nichols; (4) erosion dikes or narrow-based terraces (slopes to about 20%) run along the contour, with or without silt pits (short blind ditches to collect water and debris); (5) bench terraces (for slopes too steep for ridge terraces or erosion dikes) continuous or discontinuous (the latter for individual plants or for regular spacing on an uneven slope).

.110d. Soil erosion: two main agencies, water and wind, of which the first is the principal problem in humid climates and the second, in semiarid or arid climates; water erosion takes two forms; sheet, where thin layers are removed from a broad, nearly flat, inclined surface, and gully, where water flowing in channels down a slope forms rills, gullies, and finally rav-

V.2. PROPAGATION, NURSERY HANDLING, SOILS, CROP PRODUCTION 479

	ines; sheet erosion is the more insidious because all of the topsoil on a field may wash away before the damage is realized; gully erosion can be seen almost from the beginning of soil loss; the chief danger is the speed of channel formation; <u>wind erosion</u> is similar in its effects to sheet erosion and may be a serious menace in arid or semiarid regions and along the shores of bodies of water.
.110e.	Vegetative cover (cover crops, aid plants): permanent, mainly leguminous, cover of soil surfaces not occupied by crop plants; several distinct groups: (1) low covers (herbs, vines, creepers) in closed patterns (complete cover), rows for protection of bench and other terraces, drains, gutters, and the like; (2) medium covers (shrubs) in a regular pattern between rows of the main crop, in hedges, or outside the crop as a source of mulch and ravine wall protection; (3) tall covers or shade (trees) in regular patterns between rows of the crop, in rows, or for the protection of ravine walls and reforestation; (4) natural undergrowth; (see Vegetative cover advantages, Vegetative cover requirements; V.2.110e-1,.110e-2).
.110e-1.	Advantages: (1) beneficial influence on physical condition of topsoil and subsoil; (2) improvement of soil by N fixation; (3) temporary storage for nutrients; (4) higher humidity under a cover crop, which makes it possible to use slowly available, less readily leached fertilizer source materials; (5) reduction of erosion, crowding out of weeds, and slowed loss of nutrients from oxidation; (6) source of mulch material for the main crop; (7) protection from wind and shade provided by shrub or tree covers.
.110e-2.	Requirements: the following are required for the most economical utilization of mechanical aids and the crop mineral nutrition program: (1) easy multiplication, preferably by seed; (2) a root system that does not compete with the crop, yet has good soil-binding properties and does not require high fertility soils; (3) rapid growth and abundant leaf production in full sun and shade; (4) tolerance to pruning; (5) resistance to pests and drought; (6) ability to suppress weed growth; (7) easy eradication from a planting; (8) suitability for land reclamation; and (9) lack of unpleasant characteristics, such as spines, clinging tendrils, or hosts for vectors or pathogens.
.111	Soil organic matter: general term for undecayed and decaying plant and animal remains and humus in all degrees of degradation, in consequence of which the composition is not only variable but extremely complex; humus, an amorphous brown-to-black material, is fairly stable but can neither be separated from unhumified matter nor from the mineral constituents of a soil; the organic matter fraction, particularly humus, is highly colloidal with numerous charges; the equivalent cation exchange capacity ranges from 150 to 350-400, hence is immensely valuable in its manifold reactions with other soil constituents and as a reserve source of nutrients whose ions are adsorbed on it. (Y)
.112	Soil profile: a vertical section of a soil throughout all of its horizons (V.2.107) and extending into the parent material beneath; consists of organic debris, designated as O1 and O2, solum, as the zone of eluviation, A1, A2, and A3, zone of illuviation, B1, B2, and B3, and the parent material or any other unconsolidated geologic strata, C, with underlying bedrock, R, below

the profile proper; note that a soil profile may vary from one just developing to one fully mature or more or less eroded or, in some instances, may consist of multiple profiles (i.e., where successive volcanic depositions occur on top of an existing profile) then designated as a sequum (V.2.088). (Fig.V.2.01a) (C & N)

.113 Soil reaction: measured as pH in a soil-water suspension in dilutions ranging from a paste (1:1) to a 1:5 ratio.

.114 Soil separates: size categories for the mechanical composition of soils; U.S. system: gravel > 2.0 mm ($0.8m^2$/kg approx. surface area) very coarse sand 1.0-2.0 mm, coarse sand 0.5-1.0 mm ($2.3m^2$/kg), medium sand 0.25-0.5 mm ($22.6m^2$/kg), fine sand 0.1-0.25 mm, very fine sand 0.05-0.01 mm, coarse silt 0.02-0.05 mm, silt 0.005-0.02 mm ($226m^2$/kg), fine silt 0.002-0.005 mm, clay 0.0005-0.002 mm ($2260m^2$/kg), colloidal clay < 0.0005 mm. International system: gravel > 2mm, coarse sand 0.2-2.0 mm, fine sand 0.02-0.2 mm, silt 0.002-0.02 mm, clay < 0.002 mm; (see Soil textural classes, V.2. 117). (C & N)

.115 Soil structure: spatial arrangement of primary soil particles, components are type (shape), class (size), and grade (distinctness). (C & N)

.115a. Angular blocky: block or polyhedronlike with three dimensions of the same order of magnitude, arranged around a point, with the blocks or polyhedrons having plane or curved surfaces that are casts of the molds formed by the faces of the surrounding peds; (faces flattened; most vertices sharply angular).

.115b. Class: refers to size of aggregates, designated as very fine, fine, medium, coarse, or very coarse; used in connection with a set of standardized diagrams.

.115c. Columnar: the same as prismatic but with rounded caps.

.115d. Crumb: same as granular but with porous peds.

.115e. Grade: refers to the relative distinctness of aggregates as determined by their relative stability or durability and ease of separation from one another; characterized as weak (poorly formed undistinct peds barely observable in place), moderate (well formed distinct peds that are moderately durable and evident but not distinct in undisturbed soil), or strong (durable peds that adhere weakly to one another, withstand placement, and may become separated when soil is disturbed).

.115f. Granular: polyhedronlike or spheroidal with three dimensions of the same order of magnitude, arranged around a point with polyhedrons or spheroids with plane or curved surfaces that have slight or no accommodation to the faces of surrounding peds; relatively nonporous peds.

.115g. Massive: soil particles cling together in such huge masses with no definite orderly arrangement of natural lines of weakness so that their shape cannot be determined.

.115h. Platy: platelike peds with the vertical dimension limited and greatly less than the other two; arranged around a horizontal plane; faces mostly horizontal.

.115i. Prismatic: prismlike peds with two dimensions, the horizontal limited and considerably less than the vertical, arranged around a vertical line, with well defined vertical faces and angular vertices; without rounded caps.

.115j. Single grain: each soil particle is separate and distinct from all others; does not form aggregates; noncoherent.

.115k. Subangular blocky: same as angular blocky but with mixed rounded and flattened faces and with many <u>rounded</u> vertices.

.115 1. **Type:** refers to the shape of individual aggregates, noncoherent sands, and large coherent soil masses whose shape is unknown.
.116 **Soil texture:** the proportionate distribution of the different-sized mineral particles in a soil; does not include organic matter; (see Soil separates, V.2.114). (Fig. V.2.01b) (C & N)
.117 **Soil textural classes:** 12 major classes ranging from clays to sands for mineral soils; organic soils are designated separately. (Fig. V.2.01b) (C & N)
.117a. Clays: contain 0-45% sand, 0-40% silt, and 40-100% clay.
.117b. Sandy clays: contain 45-65% sand, 0-20% silt, and 35-55% clay.
.117c. Silty clays: contain 0-20% sand, 40-60% silt, and 40-60% clay.
.117d. Silty clay loam: contain 0-20% sand, 40-73% silt, and 27-40% clay.
.117e. Silts: contain 0-20% sand, 80-100% silt, and 0-12% clay.
.117f. Silt loams: contain 0-50% sand, 15-53% silt, and 0-27% clay.
.117g. Clay loams: contain 20-50% sand, 15-53% silt, and 27-40% clay.
.117h. Sandy clay loams: contain 45-80% sand, 0-28% silt, and 20-35% clay.
.117i. Loams: contain 23-52% sand, 28-50% silt, and 7-27% clay.
.117j. Sandy loams: contain 43-85% sand, 0-50% silt, and 0-20% clay; intermediate to sandy clay loams or loams and loamy sands; subdivided according to proportion of sand present.
.117j-1. Very fine sandy loam: contains > 30% very fine sand or > 40% fine and very fine sand, at least half of which is very fine sand and the combined very coarse and medium sands do not exceed 15%.
.117j-2. Fine sandy loam: contains >30% fine sand and <30% very fine sand or 15-30% very coarse, coarse, and medium sand.
.117j-3. Medium sandy loam: contains > 30% very coarse, coarse, and medium sand but < 25% very coarse sand and < 30% very fine or fine sand.
.117j-4. Coarse sandy loam: contains >25% very coarse and coarse sand and < 50% of any other single grade of sand.
.117k. Loamy sands: contain 70-90% sand, 0-30% silt, and 0-15% clay; subdivided as coarse, medium, fine, and very fine.
.117k-1. Loamy very fine sand: contains > 50% very fine sand.
.117k-2. Loamy fine sand: contains > 50% fine sand or < 25% very coarse, coarse, and medium sand and < 50% very fine sand.
.117k-3. Medium loamy sand: contains > 25% very coarse sand and < 50% of any other sand separate.
.117k-4. Loamy coarse sand: contains > 50% very coarse sand and < 50% of any other sand separate.
.117 l. Sands: loose and single-grained; must contain 85-100% sand, 0-15% silt, and 0-10% clay; subdivided as coarse, medium, fine, and very fine.
.117 l-1. Very fine sand: contains > 50% very fine sand.
.117 l-2. Fine sand: contains > 50% fine sand or < 25% very coarse, coarse, and medium sand and also < 50% very fine sand.
.117 l-3. Medium sand: contains > 25% very coarse, coarse, and medium sand and < 50% fine or very fine sand.
.117 l-4. Coarse sand: contains > 25% very coarse and coarse sand and < 50% of any other sand separate.
.118 **Soil types and phases:** no longer recognized as categories in the soil classification system (V.2.096); retained as names of individual kinds of soil and used as mapping units and for specialized soil research. (C & N)
.119 **Solodization:** leaching of solonized soils without Ca^{++} to replace

Na$^+$ may produce solodization when Na$^+$ is replaced by H$^+$, with the result that the soil becomes acid (often intensely so), the clay remains dispersed and intensely translocated, and the topsoil becomes sandy and bleached; (see Solonization, V.2.120). (Y)

.120 Solonization (alkalization): the process by which the soil exchange complex is appreciably saturated with sodium ions, being brought about by prolonged or repeated saturation with water in which sodium predominates over other cations (e.g., salty ground or irrigation water or sea-water flooding); the pH remains below 8.5 so long as free salts are present and their flocculating effect maintains soil structure and permeability but the pH rises to 8.5 to 10.0 if the free salts are leached (i.e., without concurrent replacement of Na$^+$ by Ca^{++}), whereupon the clay disperses with the collapse of the structure, clay translocates irreversibly to form an impermeable clay horizon, and humus disperses throughout the profile; (see Solodization, V.2.119). (Y)

.121 Solum: (see Soil horizons, V.2.107o).

.122 Spodic horizon: (see Diagnostic horizons, V.2.028h).

.123 Stokes law: the relationship between the velocity of sedimentation and intrinsic properties of the particles and fluid: $V = g(\Delta - p) d^2 / 18\eta$, where V is the settling velocity in cm/sec, g is the acceleration due to gravity (ca. 980 cm/sec), Δ is the density of the settling particle and p of water ($\Delta - p$ taken as 1.6), η is the viscosity of water (ca. 0.010 at 20°C), and d is the diameter of the spheres (commonly the equivalent diameter of soil particles); used for separation of particles < 0.05 mm for mechanical analysis; (see Soil separates, V.2.114). (RU)

.124 Subangular blocky: (see Soil structure, V.2.115k).

.125 Subgroup: (see Soil classification, V.2.096d).

.126 Suborder: (see Soil classification, V.2.096b).

.127 Subsoil: the lower part of the solum; i.e., the B horizons; (see Soil horizons, V.2.107). (C & N)

.128 Surface soil: soil ordinarily moved by tillage or its equivalent in uncultivated land; i.e., the upper part of the solum; topsoil; (see Soil horizons, V.2.107). (C & N)

.129 Topsoil: (see Surface soil, V.2.128).

.130 Truncated soil profile: one in which one or more of the upper layers have been removed, thus absent, from erosion (principally water but sometimes wind or ice); (see Soil profile, V.2.112).

.131 Type: (see Soil structure, V.2.115 l).

.132 Umbric epidedon: (see Diagnostic horizons, V.2.018i).

.133 Vegetative cover: (see Soil management, V.2.110e).

.134 Vegetative cover advantages: (see Vegetative cover, V.2.110e-1).

.135 Vegetative cover requirements: (see Vegetative cover, V.2.110e-2).

.136 Vermiculite: (see Soil clay minerals, V.2.098).

.137 Very fine sand: (see Soil textural classes, V.2.117j-1).

.138 Very fine sandy loam: (see Soil textural classes, V.2.117 l-1).

.139 Zonal soil: one with more or less well defined horizons; (see Soil profile, V.2.112). (C & N, L & B)

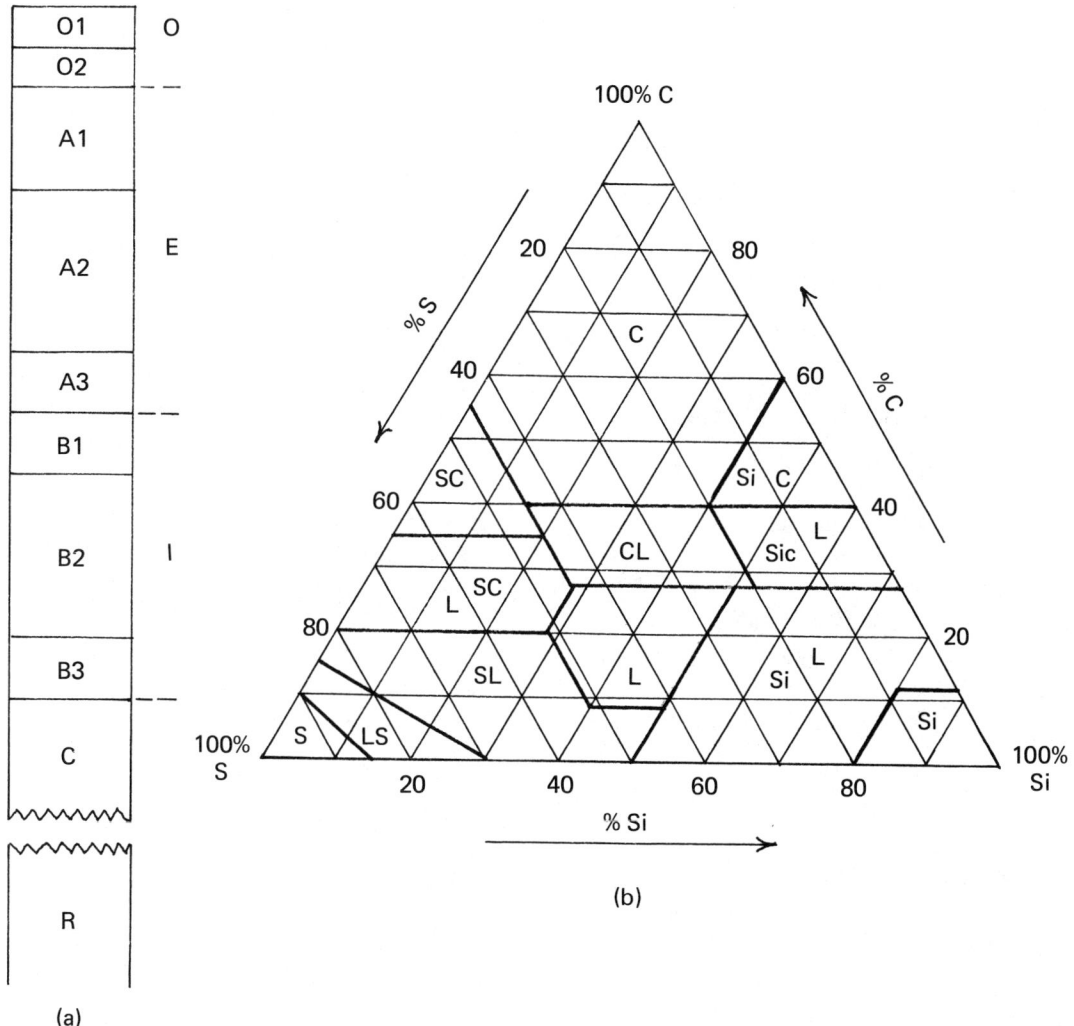

Fig. V.2.01. Soils: (a) hypothetical profile, O = organic debris, O1 = largely undecomposed loose leaves and litter, O2 - partially decomposed organic debris; E = eluviation zone, A1 = dark-colored horizon (mixture of mineral matter and well-decomposed organic matter or humus), A2 = light-colored horizon (maximum leaching), A3 - transitional (more like A); I = illuviation zone, B1 = transitional (more like B), B2 = maximum accumulation of clay, iron, and/or organic matter, B3 = transitional to C; C = parent material or other unconsolidated geologic strata; R = underlying bedrock; (b) Textural triangle, C = clay, S = sand, Si = silt; SC = sandy clay, SiC = silty clay, CL = clay loam, LSC = loamy sandy clay, SiCL = silty clay loam, LS = loamy sand, SL = sandy loam, L = loam, SiL = silty loam. Redrawn from Carlisle and NeSmith 1973; see Acknowledgments and Selected References.

V.3. Crop Production (Fruits, Vegetables, and Woody Ornamentals)

V.3.0. General

.000 Crop production: the site preparation, development of the planting, and cultural program phases of integrated crop management (V.3.0. 001); i.e., all of the operations up to disposal of the crop; (see Postharvest handling, VI).

.001 Integrated crop management (ICM): a systematic, planned approach to crop production, harvesting, and handling on a small or large scale in which sound horticultural principles and management practices are followed throughout; includes gathering information on crop requirements, crop problems, and environmental and economic site factors as a basis for the choice of a crop or crops and cultivars or selected seed sources for a given site, proper preparation of the site, development of the planting, a cultural program, and disposal of the crop to an appropriate market, based on information obtainable by systematic periodic inspections for specific site problems and planned utilization of labor, equipment, fertilizers, pesticides (including biological controls), irrigation and drainage, soil management practices, and harvesting, handling, and/or processing procedures for maximum long-term benefit to the crop or crops and the area in which the planting is located.

.002 Farm implements: (see Harrow, V.1.1.020; Plow, V.1.1.044, etc.).

.003 Farm and orchard equipment: includes wheeled or crawler-type tractors of many different sizes and styles, with points of attachment for implements used for cultivation, spraying, and other cultural operations; also many special-purpose machines, such as trenchers, backhoes, stump pullers, bulldozers, hay rakes, diggers, combines, and sprayers; (see Nursery facilities, structures, equipment, V.1.1.).

.004 Ley: arable land laid down with grass, often used as a fallow between grain or other annual crops; also a meadow or pasture (an English term little used in the United States). (RU)

V.3.1. Site preparation

.000 Site preparation: a series of operations to facilitate planting a crop; the entire sequence from natural cover or old crop land consists of (a) a survey to locate and stake site boundaries, road lines, block perimeters, main drainage and irrigation systems, and building sites; (b) clearing and cleaning done with bulldozers, road scrapers, draglines, backhoes, stump pullers, and the like, by hand labor (e.g., on steep slopes), or a combination; (c) laying out of blocks, plowing, leveling and cleaning the land, erection of terraces (if needed), installation of drainage and irrigation within blocks, construction of buildings and roads, and planting windbreaks and cover crops; (d) laying out planting sites (locations of individual plants) and rows or beds, construction of beds, and preplanting fertilizing, pesticide application, etc.; details will vary according to the existing cover and the crop or crops to be grown; much more intensive

V.3.1. PROPAGATION, NURSERY HANDLING, SOILS, CROP PRODUCTION

preparation is required for field-grown vegetables or herbaceous ornamentals.

.001 Aerator: a motorized implement with short (10 to 15 cm), closely set, hollow tubes mounted on a roller that removes small cores of soil as the machine moves along; made in several sizes for renovating golf-course greens, fairways, and the like; the usual practice is to topdress with prepared soil (greens) or sand (fairways) to improve aeration of compacted areas.

.002 Bed: (a) ridge of soil formed for planting row crops with furrows located on each side; (b) an area in which seedlings or transplants are grown for planting later in the field; may be level or built up.

.003 Blind cultivation: cultivation before the planted crop emerges.

.004 Colter (Coulter): a type of harrow (V.1.1.020) on which the regular circular or toothed dished disks are replaced with coulter blades (flat disks); each set of disks is set square (i.e., parallel instead of angled to the line being followed to make a series of narrow cuts across the field, usually in at least two directions (i.e., at right angles) for the purpose of renovating a large lawn, golf-course fairway, baseball park, or similar large area; it is particularly effective for cutting up and renovating stoloniferous grasses (those with subterranean stems), where the surface has become compacted sufficiently to impede their growth.

.005 Companion crop: crop grown with another to secure an earlier or larger return than from one crop alone.

.006 Cultivation: tillage of the soil around individual plants or rows of plants for the purpose of weed control and improvement of soil structure; gradually becoming less popular for many crops as growers realize the necessity of soil conservation and turn to sod (V.3.1.010) or stubble (V.3.1.013) culture and chemical weed controls (V.3.3.001).

.007 Gradient mulch system: a system of mulching in which the entire bed (i.e., sides as well as top) is covered with plastic film in combination with a reservoir of nutrients at the soil-bed surface and a constant water table (depth 40 to 45 cm for annual crops such as tomato) below. (GE)

.008 Planting systems: arrangement of crops in the field; based on four main considerations: the nature of the crop, land factors (i.e., topography, exposure, and soil type), cultural and harvesting operations, and grower preference; many different patterns. (Fig. V.3.1.01)

.008a. Avenue, hedgerow: patterns in which plants are spaced closely in rows; the avenue type consists of two to six or eight closely spaced rows with a wide middle between the rows, as in pineapple (usually two rows) or vegetables and often on beds (i.e., crop rows are elevated above the row middles for better drainage); the hedgerow has single rows of closely spaced plants with a wide middle between them, as in tea, rubber, cinchona, many vegetable crops, and vines on trellises.

.008b. Contour planting: a modification of one of the regular patterns used on sloping terrain; i.e., rows or groups of rows are laid out along the contour with a slight slope to provide good drainage with minimum erosion.

.008c. Double-setting: the practice of spacing plants (e.g., fruit crops) in a row at about half the usual distance or adding additional rows to a conventional square or rectangular system; used to provide better use of the land during the

V.3.1. PROPAGATION, NURSERY HANDLING, SOILS, CROP PRODUCTION

early years of long-lived crops but does necessitate removal of the extra plants or pruning to keep them in bounds when they begin to crowd; note that double-setting refers to interplanting of the same crop or cultivar, a quite different situation than interplanting (V.3.1.008f) proper.

.008d. Equilateral triangle: (see Hexagonal, V.3.1.008h).
.008e. Five-spot: (see Quincunx, V.3.1.008i).
.008f. Hedgerow: (see Avenue, V.3.1.008a).
.008g. Hexagonal (septuple or equilateral trangle): a pattern in which plants in alternate rows are spaced in the form of an equilateral triangle; the distance between rows is $a = \sqrt{3}/2$ (or 0.866a), where a is the length of one side of the triangle; allows maximum use of the ground with about 15% more plants per unit area, compared with a square pattern of the same dimensions; widely used in the tropics.
.008h. Interplanting: the practice of growing more than one crop, usually in alternate rows or beds (groups of rows); i.e., fruit trees with vegetables or short-cycle fruits; widely used in the tropics and indeed recommended in some situations to reduce pest problems (e.g., interplanting of cacao and robusta coffee in West Africa to reduce the spread of swollen shoot viruses on the cacao); most successful when the intercrop is short-lived or rows are spaced farther apart than normal.
.008i. Quincunx (Five-spot): a double pattern in which the main pattern is square or rectangular, with an additional plant in the center; used for interspersing one of smaller size or longer life; permits 78% more plants than the main square or rectangular pattern.
.008j. Rectangular (square): plants set out in straight rows (i.e., lined up at right angles to one another), with spacing between them in a row somewhat less than that between rows or rarely the reverse (rectangular) or the same distance between plants and between rows (square); a popular system on level or slightly rolling terrain, the easiest to lay out and manage with respect to cultural and harvesting operations.
.008k. Septuple: (see Hexagonal, V.3.1.008h).
.008l. Square: (see Rectangular, V.3.1.008j).
.008m. Triangular: a pattern in which plants in alternate rows are offset half the space between plants in a row; the distance between the rows is the same or more than that in a row; thus a series of isosceles triangles (two sides equal, instead of three as in an equilateral triangle) is formed; easier to lay out than the hexagonal pattern but results in 9% fewer plants than the equivalent square or rectangle.
.009 Renovation: the process of restoring the productivity of plants in solid stands or rows by pruning (trees, shrubs, vines), cultivation, fertilization, and reseeding.
.010 Sod culture: a cultural system in which the natural cover or one planted for that purpose is maintained by periodic mowing (or hand slashing) rather than cultivation; now used in many orchards and groves for long-lived woody perennials.
.011 Soil auger: a power-driven truck or tractor-mounted drill with a large helical screw bit used for digging planting holes in an orchard or grove.
.012 Spiker: similar to an aerator (V.3.1.002), except that the roller is studded with stout sharp-pointed spikes, the purpose of which is to punch holes through a sod or stubble cover to improve aera-

V.3.1. PROPAGATION, NURSERY HANDLING, SOILS, CROP PRODUCTION

tion and drainage.

.013 Stubble culture: a cultural system used primarily for annual crops (e.g., grains) in which the new crop is drilled or set out (vegetative material such as "seed" or corms) directly in a field on which plants of the preceding crop were cut off close to ground level rather than removed or incorporated into the soil as in the preparation of a seed bed.

.014 Tilth: the physical condition of the soil in its relation to plant growth; includes adequate aeration, sufficient rainfall, ready infiltration of rainfall, and a mellow, friable, easily handled soil (i.e., favorable consistency); a dynamic soil condition that must be renewed in the course of cropping; (see also Soil management, V.2.110). (RU)

.015 Windrow: (see Windrow, V.1.3.088).

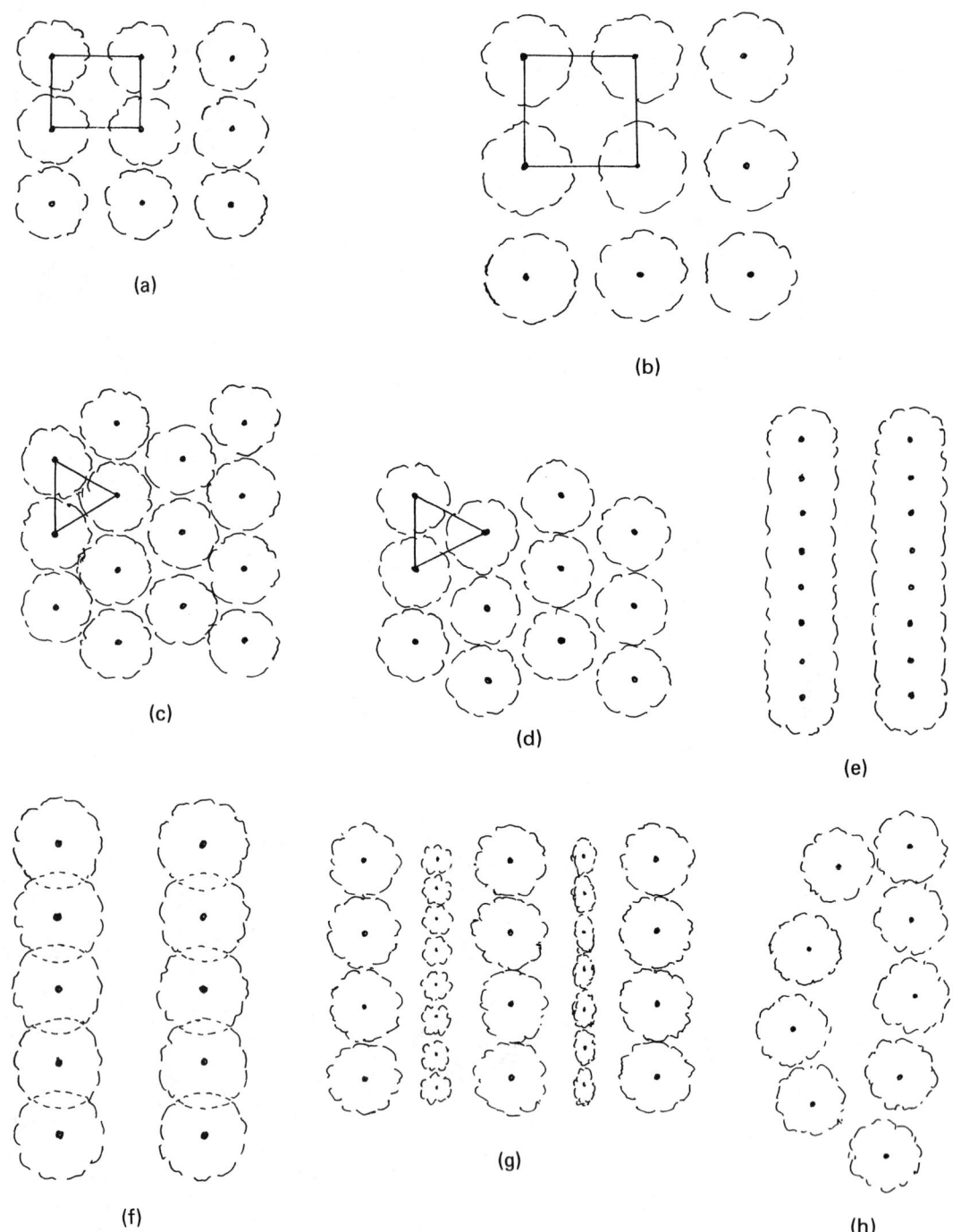

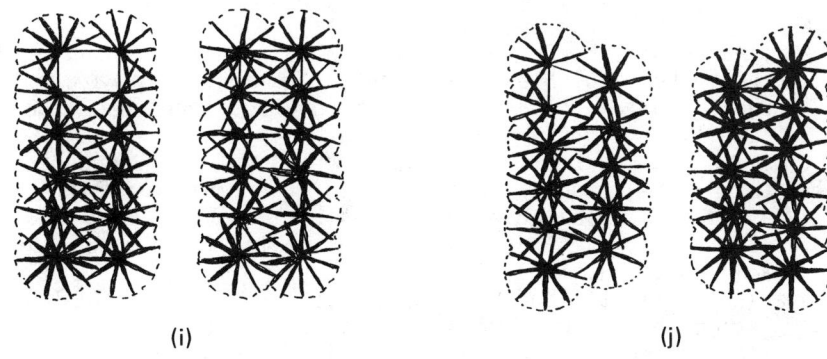

Fig. V.3.1.01 Planting systems: (a) rectangular; (b) square; (c) hexagonal (equilateral triangle, septuple); (d) triangular (isosceles triangle); (e) hedgerow; (f) double set (equivalent to an avenue system); (g) interset (interplanting); (h) contour planting; (i) double-row avenue (square set); (j) double-row avenue (triangular set); (k) quincunx (five spot).

V.3.2. Mineral Nutrition

.000 Mineral nutrition: the combined program, which includes fertilizers, nutritional sprays, organic materials (cover crops, leafy mulch materials), and soil amendments, designed to furnish plants with essential elements in quantities and at times to provide optimum yields and crop quality, including desired growth and development of landscape and garden ornamentals; based essentially on crop needs plus noncrop losses or use, i.e., total requirements include nutrients removed in harvesting, losses by erosion, leaching, oxidation and/or volatilization, utilization by cover crops, weeds and soil microorganisms, and immobilization on soil mineral colloids and organic matter.

.001 "Analysis": "as applied to fertilizers, designates the percentage composition of the product expressed in those terms that the law requires and permits." (Association of American Plant Food Control Officials). (FC)

.002 ATA: (see Chelating agents, V.3.2.012).

.003 Available phosphoric acid: available phosphoric anhydride (P_2O_5), the water soluble plus (neutral) ammonium-citrate-soluble phosphoric oxide in ammonium phosphate, superphosphate, or mixed fertilizers; determined by washing a weighed sample with water and then neutral ammonium citrate solution, the difference between the amount remaining in the residue and the total being available; available P_2O_5 in basic slag, fused phosphates, and rock phosphates is determined as the part soluble in 2% citric acid solution. (FC)

.004 Available plant food: that which is in a form available for assimilation by growing plants or capable of being converted to such a form in the soil during the growing season. (FC)

.005 Available potash: that which is soluble in water or ammonium oxalate solution. (FC)

.006 Biuret (carbamyl urea): a compound ($H_2NCONHCONH_2 \cdot H_2O$) formed by thermal decomposition of urea (H_2NCONH_2); highly phytotoxic to many crops, thus "biuret-free" urea is always specified for fertilizer use. (FC)

.007 Buffer: any salt that tends to preserve the original pH of a solution on addition of an acid or base, as in maintaining the pH of a culture solution in the desired range, despite the removal of ions by plant roots. (HW)

.008 Carbamyl urea: (see Biuret, V.3.2.006).

.009 Carbon dioxide fertilization: increased photosynthesis, hence growth with above-atmospheric CO_2 concentration of 1000 to 2500 ppm during daylight hours; needs a tightly closed greenhouse, good light, and adequate heat; effective for snapdragons, carnations, and chrysanthemums; does not work with all plants. (H & K)

.010 CDTA: (see Chelating agents, V.3.2.012).

.011 CDU: (see Controlled-release fertilizer, V.3.2.015).

.012 Chelating agents, chelate: a compound or group with two reactive valences that form a complex with and thus bind a metallic ion; examples of chelating agents are ethylenediaminetetraacetate (EDTA), N-hydroxyethylethylenediaminetriacetic acid (versenol, HEEDTA), N,N'-dihydroxyethylethylenediaminediacetic acid (versendiol, DHEEDDA), ethylenediamindi(o-hydroxyphenyl)-acetate (EDDHA), glycine (G), nitrilotriacetic acid (ATA), diethylenetriaminepentaacetic acid (DTPA), and cyclohexanetetraacetic acid

V.3.2. PROPAGATION, NURSERY HANDLING, SOILS, CROP PRODUCTION

(CDTA); examples of chelates are Fe-EDTA, Cu-EDTA, Zn-EDTA, and Mn-EDTA. (HW)

.013 Complete fertilizer: one that contains nitrogen (N), phosphorus (P), and potassium (K), plus, in some instances, certain minor elements and magnesium (Mg). (FC)

.014 Concentrated fertilizer: a mixed fertilizer that contains at least 30% N-P-K. (FC)

.015 Controlled-release fertilizer: one in which one or more nutrients have limited solubility in the soil solution, thus becoming available to the plant over a controlled period; e.g., ureaform, IBDU (isobutylidene diurea), CDU (crontonylidene urea), osmocote, SCU (sulfur-coated urea). (FC)

.016 Critical relative humidity: the humidity at which a given fertilizer salt or mixture begins to absorb moisture at 30°C; e.g., 59% for ammonium nitrate, 73% for urea, 79% for sulfate of ammonia, 96% for sulfate of potash. (FC)

.017 Crop logging: a sophisticated system of monitoring crop growth and development in relation to yield of 'Cayenne' pineapple in Hawaii; involves periodic measurements of environmental factors, foliar analyses for major and certain minor elements (e.g., Zn, Fe), leaf color (for N needs), and plant weights. (SN)

.018 Crotonylidene urea: (see Controlled-release fertilizer, V.3.2.015).

.019 Cyclohexanetetracetic acid: (see Chelating agents, V.3.2.012).

.020 Deliquescent: the property of a substance, such as some fertilizer materials, to absorb moisture from the air and become liquid. (FC)

.021 Determination of crop needs: four main methods: (a) pot and field experiments; (b) soil analyses; (c) diagnosis of deficiency or toxicity symptoms; and (d) foliar anaylsis.

.022 DHEEDDA: (see Chelating agents, V.3.2.012).

.023 Diethylenetriaminepentaacetic acid: (see Chelating agents, V.3.2.012).

.024 N,N'-Dihydroxyethylethylenediaminediacetic acid: (see Chelating agents, V.3.2.012).

.025 DTPA: (see Chelating agents, V.3.2.012).

.026 EDDHA: (see Chelating agents, V.3.2.012).

.027 EDTA: (see Chelating agents, V.3.2.012).

.028 Essential elements: elements necessary for plant growth and development; C, H, and O are derived from carbon dioxide and water and N, P, K, Ca, Mg, S, Cl, Fe, Mn, Cu, Zn, B, and Mo, from mineral or organic sources; the latter are divided into three groups on the basis of normal requirements as major (N, P, K), secondary (Ca, Mg, S, Cl) and minor (trace, micro) elements (Fe, Mn, Cu, Zn, B, and Mo); several others, e.g., Na, Si, Al, and Co occur generally in plants but are not essential, although Na may substitute in part for K and Si is a common strengthening agent. (HW)

.029 Ethylenediamindi(o-hydroxylphenyl)-acetate: (see Chelating agents, V.3.2.012).

.030 Ethylenediaminetetraacetate: (see Chelating agents V.3.2.012).

.031 Fertigation: inclusion of fertilizer with irrigation water applied in high or low-volume systems.

.032 Fertilization: the application of nutrients to plants.

.033 Fertilizer analysis: the content of an element or elements contained in a fertilizer expressed as a percentage of each; e.g., a 16-4-16-4 fertilizer would have 16% N, 4% P_2O_5, 16% K_2O, and 4% MgO, equivalent to 16% N, 1.75% P, 11% K, and 2.4% Mg; [see Fertilizer analysis (guaranteed), V.3.2.034]. (FC)

.034 Fertilizer analysis (guaranteed): the official (guaranteed) analysis of a fertilizer (individual source material or mixture) that will show the total percentage of each nutrient; the form [e.g., for N, the percentage of nitrate N, ammoniacal N, water-soluble organic N, and water-insoluble N; for P, the percentage available of P_2O_5 and insoluble P_2O_5; for K, the percentage of water-soluble K_2O; the percentage of Cl, if present; and for secondary plant foods, the percentage of total and water-soluble if more than 1% (or other nutrients like Al, Cu, Mn, and Fe as the element if less than 1%] and the source materials; this information is printed on a tag or directly on each bag. (FC)

.035 Fertilizer calculations: determination of total mineral nutrition requirements in amounts per season or year, based on (a) the ratio among essential elements (usually expressed in relation to N = 1), and (b) the quantity of fertilizer to be applied, determined from soil and plant analyses, estimates of noncrop utilization, and crop age, size, or yields; the total (X kg or g of ratio per season or year) is translated into actual amounts of source materials, then broken down to the individual applications of mixed fertilizers, separate source materials, or their combination, which will be made during the season or year; crop requirements, availability of source materials, cost, and grower preference are the basic factors that determine the type of program.

.036 Fertilizer conversion factors: 1 unit P_2O_5 = 0.4364 unit P, 1 unit K_2O = 0.8302 unit K; other nutrients are expressed as the oxide but may be converted to the element equivalent (i.e., Ca and Mg). (Note that many scientific journals, including those of the American Society for Horticultural Science require that fertilizer be stated as the element). (FC)

.037 Fertilizers: a major component, along with nutritional sprays and soil amendments of the mineral nutrition program in supplying the necessary quantities of essential elements to plants based on actual crop records (or estimated yields according to plant age and size), soil analyses, and foliar analyses (if available); factors include (a) the ratio of other elements to nitrogen (i.e., the proportion of P, K, etc., needed in relation to N = 1), (b) total amount of N,P,K, etc., to be applied per plant or per unit area, (c) number and timing of applications, (d) source materials, (e) calculation of amounts per application, (f) forms of application (i.e., complete NPK mixtures with or without minors, separate materials, or combinations), and (g) methods of application (e.g., solid material in bags or bulk or liquid in conjunction with irrigation, pesticide applications, or separately).

.038 Fertilizer source materials: individual compounds that contain essential elements for plant growth and development; two main groups: inorganic and organic, of which the former constitutes the higher analysis forms of one or two elements and the latter, lower analysis forms and usually several elements that are generally less readily available (hence have longer lasting effect); (see Source materials, V.3.2.084-.088). (FC)

.039 Fertilizer spreader: a device for dispensing fertilizers (and sometimes pesticides or other materials) in some predetermined manner according to the crop; many different forms, ranging from a person flinging material around or along plants in a row to large mechanized spreaders that dispense material broadcast over an area, along rows, or the sides of beds, on the soil surface or below it, as part of row preparation of a seeding operation.

.040 Foliar analyses: chemical analyses that consist of rapid colorimetric, spectrophotometric, and other methods for essential elements in leaves or other plant parts; useful as a basis for the adjustment of a nutrition program because the analyses reflect the current status in the crop rather than indirect soil analyses or "after the fact" pot or field experiments. (FC)

.041 G: (see Chelating agents, V.3.2.012).
.042 Glycine: (see Chelating agents, V.3.2.012).
.043 Granule, granular: particles of fertilizer in regular or irregular shapes, are designated as "uniform" if 95% of the product remains on a sieve within the range of 0.84 to 2.38 mm (i.e., 20 to 8 mesh); the term "pellet" or "pelletization" usually refers to seed coatings (V.1.3.058). (FC)
.044 Gypsum: $CaSO_4 \cdot 2H_2O$, a widely used conditioner for land on which peanuts and sometimes other legumes are grown, for treating alkali soils in arid regions, for soil reclamation, and to increase permeability of soil under irrigation; it is neutral in reaction, hence is not used for pH control. (FC)
.045 HEEDTA: (see Chelating agents, V.3.2.012).
.046 High-analysis fertilizer: a fertilizer in which the combined nitrogen, phosphate, and potash content is equal to or exceeds 20%; e.g., 20-20-20, 16-0-16, 33-0-0, 0-0-60. (FC)
.047 N-Hydroxyethylethylenediaminetriacetic acid: (see Chelating agents, V.3.2.012).
.048 Hygroscopic: the property of a substance, such as many fertilizer materials, to absorb moisture from the air and form a solid mass that must be broken up (very gently in the case of ammonium nitrate, an excellent high explosive when dry, exposed to an open flame or spark, or given a sharp blow) to be usable as a granular product. (FC)
.049 IBDU: (see Controlled-release fertilizer, V.3.2.015).
.050 Ion-exchange materials: substances, such as synthetic resins, zeolites (hydrous silicates), and bentonite or other clays, used for demineralizing (a) water and (b) sand cultures to modify the availability or combination of nutrients; both involve adsorption of cations and anions, in the former case to remove them from water and in the latter to add certain cations and/or anions in a desired ratio or concentration to the culture solution. (HW)
.051 Isobutylidene diurea: (see Controlled-release fertilizer, V.3.2.015)
.052 Land plaster: (see Gypsum, V.3.2.044).
.053 Limestone: a material incorporated in the soil for the purpose of raising pH; two main forms, calcic ($CaCO_3$) and dolomitic ($CaCO_3$-$MgCO_3$), of which the latter is often preferred because of its magnesium content which adds to the existing soil reserves of that element. (FC)
.054 Liquid fertilizers: term applied to anhydrous ammonia (NH^3), aqua ammonia (NH_4OH), various nitrogen solutions (aqueous solutions of urea and ammonium nitrate, or combination with or without ammonia added), and liquid mixed fertilizers as true solutions, suspensions (solids in liquids), or slurries (require constant agitation to keep solids suspended). (FC)
.055 Low-analysis fertilizer: a fertilizer mixture in which the combined nitrogen, phosphate, and potash content is less than 20%; e.g., 4-6-8, 4-7-5, 6-6-6. (FC)
.056 Macronutrient elements: those essential for plant growth and development and present in quantities exceeding 1 ppm, as in N,S,P,Ca,Mg,K (also Na which may substitute in part for K). (HW)

.057 Major elements: (see Essential elements, V.3.2.028).
.058 Micronutrient elements: those essential for plant growth and development and present in concentrations of 1 ppm or less, as in Cu, Mn, Zn, B, Mo, Fe, Co (the last is required when cobalamin, vitamin B_{12}, or related compounds are essential metabolites); (see Essential elements, V.3.2.028). (HW)
.059 Micronutrient spray: (see Nutritional spray, V.3.2.067).
.060 Minor elements: (see Essential elements, V.3.2.028).
.061 Minor element spray: (see Nutritional spray, V.3.2.067).
.062 Mixed fertilizer: (see Complete fertilizer, V.3.2.013).
.063 Mixed goods: (see Complete fertilizer, V.3.2.013).
.064 Nitrilotriacetic acid: (see Chelating agents, V.3.2.012).
.065 Nutrient availability: basically a function of five main factors; (a) the compound or form in which a particular essential element is applied; (b) the relative concentration of ions or molecules in the soil solution; (c) ionic solubility; (d) pH; (e) the form or forms of ions absorbed by a particular crop plant, all of which are interrelated; e.g., the availability of N sources varies from very rapid (ammonium nitrate), rapid (ammonium sulfate or urea), moderate (cottonseed meal or Peruvian guano), slow (processed tankage), to very slow (raw bone meal); most nutrients, except Mo, are most readily available for plant uptake at pH 5.5 to 6.5, with decreased availability (or toxicity in some instances) above or below this range (Mo is unusual in being more readily available above pH 7); relative concentrations of Ca or Mg vs K are important because too much of one will depress the availability of the other two.
.066 Nutrient deficiency: lack of an essential element to the extent that plant growth is slowed or visual symptoms appear, such as distinctive chlorotic patterns, overall fading, development of atypical color on leaves, noticeable reduction in growth, dying of twigs, production of malformed or multiple twigs, gumming of branches or fruit, reduction in fruit size, premature fruit drop; masking of one deficiency by another, restricted mobility of certain ions, and toxicity (e.g., heavy metals in acid soils) which induces a deficiency in another element occur frequently under field conditions.
.067 Nutritional spray: specifically, any plant nutrient applied in the form of a dilute spray; however, it generally refers to the application of Cu, Zn, Mn, B, and Mo as individual compounds or mixtures (particularly Cu, Zn, and Mn) as a spray; iron as ferrous sulfate may be sprayed but is usually applied in chelate form to the ground; nitrogen, alone of the major or secondary elements, is commonly applied in spray form, usually as urea or ammonium nitrate, both of which are quite soluble in water; e.g., to banana and pineapple, among others.
.068 Organic fertilizer: fertilizer made from plant or animal by-products or refuse; e.g., castor pomace, cottonseed meal, hull meal, garbage tankage, and tobacco stems from plant sources, and hoof meal, Peruvian guano, activated sludge, bone meal, and fish scrap from animal sources.
.069 Organic materials: residues from cover crops and leafy mulch material around plants, which have the combined role of maintaining soil physical structure and acting as a reserve source of nutrients; contributions of nutrients by organic materials may be substantial, in particular when leguminous plants are involved; thus the requirement for chemical fertilizers is reduced.

V.3.2. PROPAGATION, NURSERY HANDLING, SOILS, CROP PRODUCTION

.070 Osmocote: a coated slow-release fertilizer (e.g., 14-14-14, 18-9-9, 36-0-0) as granules or prills enclosed in a water-permeable resin film; (see Controlled-release fertilizer, V.3.2.015). (FC)

.071 Plant analysis: consists of subsamples for moisture determination and chemical analyses, either dry or wet ashing, and separate tests for total N (microkjeldahl), total P and S, total Cl, total Ca, Mg, K, and Na, total Cu and Zn, total Mo, total Fe and Mn, total Al, total B, etc., by standard methods; (see Foliar analysis, V.3.2.040). (CP-H)

.072 Plant sampling: similar to soil sampling (V.3.2.082) in the sense that enough plants or plant parts must be collected to furnish a representative sample but differing in that greater care is used in cleaning (especially for minor element determinations), drying at about 50°C, and grinding or crushing; samples for leaf analysis as a guide to fertilizer recommendations necessitate establishing sampling units (a certain number of plants), i.e., where to take leaves in relation to their development and the number of leaves (or leaflets) to be collected. (CP-H)

.073 Potential micronutrients: those elements, such as Al, Ga, Si, I, or V, that are present in tissues of many plants but whose essential nature has not yet been established. (HW)

.074 Prill: spherical pellets or granules made by passing a molten material or solution through cool air; (see Granule, V.3.2.043). (FC)

.075 Rock phosphate: basically calcium fluorphosphate of sedimentary origin plus more or less apatite of igneous origin; the former is porous and lower in density (i.e., about two-thirds bone phosphate); [see Source materials (Phosphate), V.3.2.087]. (FC)

.076 Sand culture: a type of solution culture in which sand or a similar medium is used as a substrate; (see Solution culture, V.3.2.083). (HW)

.077 SCU: (see Controlled-release fertilizer, V.3.2.015).

.078 Secondary elements: (see Essential elements, V.3.2.028).

.079 Soil amendment: a material applied to and thoroughly mixed into the soil for adjustment of pH to provide the most effective and efficient use of fertilizer by a given crop; customary materials are limestone (V.3.2.053) to raise pH or sulfur (V.3.2.089) to lower it; (see also Source materials, V.3.2.084-.088).

.080 Soil analysis: may consist of (a) fusion analysis (moisture determination, loss on ignition, fusion with Na_2CO_3, analyses for total Si,Fe,Al,Ti,Ca,Mg, etc.) to characterize soils, clays, and minerals or (b) determination of water-soluble constituents, cation exchange capacity, exchangeable cations, pH, carbonate, organic carbon, and N,P,K,Ca,Mg, etc.; the number of detailed analyses (e.g., of individual elements) depends on the information desired; (see Plant analysis, V.3.2.071). (CP-H)

.081 Soil conditioners: (see V.2.099).

.082 Soil sampling: the procedure for obtaining material for a soil analysis; (a) taking and mixing a series of cores from an area; (b) subsampling the original sample one or more times; (c) air drying, grinding, sieving, mixing, and storing, of which the first (a) is by far the most critical, i.e., the success of the analysis as a guide to fertilizer recommendations, etc., depends on securing a <u>representative</u> sample; (see Plant sampling, V.3.2.072). (CP-H)

.083 Solution culture: two meanings, (a) the culture of individual or small groups of plants in containers in solutions of nutrients to determine those essential for growth and deficiency, optimum and toxic levels of nutrients, including the proper pH range (i.e., solution cultures for research purposes); (b) commercial or home hydroponic culture of vegetables, fruit, etc., in tanks, thus differing from (a) in numbers of plants grown and in having solutions made up on the basis of research recommendations (rather than as the basis for such); both are usually carried out in an environmentally controlled greenhouse or other shelter, the degree of control of water purity, freedom of nutrients from trace contaminants, aeration, etc., being more rigorous as a general rule for research solution cultures; (see also Sand culture, V.3.2.076). (HW)

.084 Source materials (Magnesium): include, among others, magnesium sulfate (30% MgO, 24% S), sea water-magnesia (93% MgO), emjeo (27% MgO, 22% S), sulfate of potash-magnesia (18% MgO, 25% K_2O), sulpo-mag (18% MgO, 22% K_2O), etc. (FC)

.085 Source materials (Micronutrients): include borax (36% B_2O_3), copper sulfate (30% CuO), iron sulfate (33% Fe_2O_3), manganese sulfate (30% MnO), tec-man-gan (37% MnO), zinc sulfate (45% ZnO), sodium molybdate (58% MoO); neutral forms of copper, zinc, and manganese are available; they and various organic forms (thiocarbamates) are widely used in nutritional sprays or as pesticides; iron, copper, manganese, and zinc chelates are also available; (see Chelating agents, V.3.2.012). (FC)

.086 Source materials (Nitrogen): include anhydrous ammonia (83% N), sulfate of ammonia (20.5% N, 23% S), ammonium sulfate-nitrate (26% N, 11% S), ammonium phosphate (several forms, 11-21% N, 46-48% P_2O_5), ammonium nitrate (33.5% N; plus lime, 20.5% N, 23% Ca; solutions, 16-21% N), calcium cyanamide (20.6% N), calcium nitrate (15.5% N, 36% Ca), nitrate of soda (16% N), nitrate of soda-potash (15% N, 14% K_2O), cottonseed meal (ca 7% N), castor pomace (5.4% N), fish scrap (8% N, 6% P_2O_5), guano (12% N, 11% P_2O_5, 2.4% K_2O), urea (45-46% N), processed tankage (5-10% N), and bone meal (3.3-4.1% N). (FC)

.087 Source materials (Phosphate): include superphosphate (20% P_2O_5, 23% CaO), triple superphosphate (47% P_2O_5, 20% CaO, 2% S), phosphate rock (33% P_2O_5, 44% CaO), basic slag (8% P_2O_5, 40% CaO). (FC)

.088 Source materials (Potash): include muriate of potash (60% K_2O, 44% Cl), sulfate of potash (48% K_2O, 18% S), potassium nitrate (44% K_2O, 13% N), sulfate of potash-magnesia (25% K_2O, 18% MgO), sulpo-mag (22% K_2O, 18% MgO), and cotton bur ash (40% K_2O, 10% CaO). (FC)

.089 Sulfur: an essential element and a material incorporated into the soil to lower pH; certain sulfur-containing compounds, such as sulfate of ammonia or aluminum sulfate, are used for the same purpose, particularly around acid-loving plants such as azaleas and rhododendrons; (see Soil amendment, V.3.2.079). (FC)

.090 Sulfur-coated urea: (see Controlled-release fertilizer, V.3.2.015).

.091 Superphosphate: three main forms, (a) "regular" or "single," with about 20% P_2O_5 (= 8.73% P), made by acidulation of rock phosphate with sulfuric acid (thus containing 9% S plus 23% Ca); (b) "enriched," with 22 to 40% P_2O_5 (= 9.6 to 17.45% P), made by treatment of rock phosphate with a mixture of sulfuric and phosphoric acids: (c) "concentrated" or "triple" ("treble"), with 45 to 47% P_2O_5 (= 19.6 to 21% P), made by treatment primarily with phosphoric acid (thus containing only about 2% S plus about 20% Ca); [see Source materials (Phosphate), V.3.2.087]. (FC)

.092 Unit: a fertilizer term meaning 20 lb (9.07 kg) per short ton (907.

V.3.2. PROPAGATION, NURSERY HANDLING, SOILS, CROP PRODUCTION 497

18 kg) or 1%; the same in metric measure would be 10 kg per tonne; e. g., a 16-2-16-4 would have 16 units of N, 2 units of P_2O_5 (=0.87 units P), 16 units of K_2O (= 13.28 units of K), and 4 units of MgO (=2.41 units of Mg). (FC)

.093 Ureaform: a synthetic insoluble nitrogenous material made by reaction of urea and formaldehyde; (see Controlled-release fertilizer, V.3.2.015). (FC)
.094 Versendiol: (see Chelating agents, V.3.2.012).
.095 Versenol: (see Chelating agents, V.3.2.012).

V.3.3. Pests, pest management

V.3.3.0. Pests - General

.000 Pests: Organisms that provide harmful competition for space, light, and nutrients or attack a crop plant; include weeds and soil borne or inhabiting, systemic, production, and postharvest disorders.
.001 Weeds: broad-leaved annuals, annual grasses, perennial grasses, and other perennials that compete for space, light, and nutrients; in some instances are parasitic (e.g., dodder, mistletoe) or hosts for pests (e.g., insects, mites, pathogens) that attack crop plants; a highly diverse group; (see Weed, I.0.003). (TM)
.002 Soil-borne and inhabiting disorders: four main groups: (a) diseases caused by fungi and bacteria, including various root and foot rots, damping-off, and wilts; (b) insects, such as beetles, borers, grubs, mole crickets, ants, and termites; (c) rodents and other burrowing animals; and (d) nematodes, numerous parasitic species taken as a group that rank worldwide as one of the major limiting factors in horticultural crop production.
.003 Production disorders: a long list of insects, mites, rodents, and other animals, fungi, and bacteria that attack above-ground parts of plants.
.004 Systemic disorders: by definition those diseases that not only affect the entire plant system but also can be isolated from or transmitted by any part of the plant; includes viruses, viroids, mycoplasmas, and rickettsias; transmission is by plant tissue, insects, nematodes, pollen, in or on seeds, or mechanically (e. g., tools), etc.; a difficult group to work with because of the variability and slowness of symptom expression, presence of symptomless carriers, lack of suitable indicator plants for indexing, and inability in many instances to culture the organism outside living host tissue.
.005 Postharvest disorders: synonymous with "Market diseases" (VI.9.), those disorders that affect the marketability or usability of fresh fruits, vegetables, and other horticultural products; encompass three principal groups of disorders; (a) pathological (disease caused by fungi, bacteria, and other plant parasites; see Pests-Plants, V.3.3.2); (b) physiological (attributable to environmental conditions; see Nonparasitic diseases, V.3.3.2.270 and Postharvest physiological disorders, VI.9.3); (c) animals (principally insects and rodents; see Pests-Animals, V.3.3.1).

V.3.3.1. PROPAGATION, NURSERY HANDLING, SOILS, CROP PRODUCTION   498

V.3.3.1. Pests - Animals

.001 Alfalfa butterfly: (see Lepidoptera, V.3.3.1.132g).
.002 Ambrosia beetle: (see Coleoptera, V.3.3.1.132a).
.003 Ambush bug: (see Hemiptera, V.3.3.1.132c).
.004 Anguina spp.: (see Nematodes attacking above-ground plant parts, V.3.3.1.200a).
.005 Anthomyiids: (see Diptera, V.3.3.1.132b).
.006 Ants: (see Hymenoptera, V.3.3.1.132e).
.007 Aphelenchoides spp.: (see Nematodes attacking above-ground plant parts, V.3.3.1.200a).
.008 Aphid: (see Homoptera, V.3.3.1.132d).
.009 Arabis mosaic vector: (see Xiphinema spp., V.3.3.1.200c).
.010 Arachnids: invertebrate animals belonging to the phylum Arthropoda and class Arachnida, which include spiders, mites, ticks, harvestmen (daddy longlegs), scorpions, and tarantulas; characterized by two body regions, cephalothorax and abdomen, four pairs of legs (except some mites), no antennae, and respiration by means of tracheae (tubules), book lungs (contain numerous thin folds of membrane arranged like the leaves of a book), or diffusion through the body wall. (P & D)
.011 Armored scales: (see Homoptera, V.3.3.1.132d).
.012 Armyworm: (see Lepidoptera, V.3.3.1.132g).
.013 Assassin bug: (see Hemiptera, V.3.3.1.132c).
.014 Awl nematodes: (see Dolichodorus spp., V.3.3.1.200c).
.015 Bagworm moths: (see Lepidoptera, V.3.3.1.132g).
.016 Bark beetles: (see bark and ambrosia beetles, Coleoptera, V.3.3.1.132a).
.017 Bats: pollinators, fruit eaters, and seed dispersers for many species of tropical plants; sometimes destructive and then become pests. (F & P)
.018 Bean (lady) beetle: (see Coleoptera, V.3.3.1.132a).
.019 Bedbug: (see Hemiptera, V.3.3.1.132c).
.020 Bee flies: (see Diptera, V.3.3.1.132b).
.021 Bees: (see Hymenoptera, V.3.3.1.132e).
.022 Beet cyst nematode: (see Heterodera schachtii, V.3.3.1.200g).
.023 Beet leaf miner: (see anthomyiids, Diptera, V.3.3.1.132b).
.024 Beetles: (see Coleoptera, V.3.3.1.132a).
.025 Belonolaimus spp.: (see Migratory ectoparasitic nematodes attacking roots, V.3.3.1.200c).
.026 Birds: serious pests of ripening grains and crops with ripe fruit on them; fruit is consumed whole if small (e.g., holly berries by migrating cedar waxwings) or have holes pecked in them if large; predators of innumerable insects, hence greatly more beneficial than harmful overall; also pollinators of many tropical species of flowers. (P & D)
.027 Black flies: (see Diptera, V.3.3.1.132b).
.028 Blister beetles: (see Coleoptera, V.3.3.1.132a).
.029 Blow flies: (see Diptera, V.3.3.1.132b).
.030 Bollworm: (see Lepidoptera, V.3.3.1.132g).
.031 Braconid wasps: (see Hymenoptera, V.3.3.1.132e).
.032 "Brown ring" of bulbs: (see Foliar nematodes, V.3.3.1.200a).
.033 Brown-tail tussock moth: (see Lepidoptera, V.3.3.1.132g).
.034 Bugs: (see Hemiptera, V.3.3.1.132c).
.035 Bumble bees: (see Hymenoptera, V.3.3.1.132e).
.036 Burrowing nematodes: (see Radopholus similis, etc., V.3.3.1.200f).

V.3.3.1. PROPAGATION, NURSERY HANDLING, SOILS, CROP PRODUCTION 499

.037 Butterflies: (see Lepidoptera, V.3.3.1.132g).
.038 Cabbage butterfly: (see Lepidoptera, V.3.3.1.132g).
.039 Cabbage looper: (see Lepidoptera, V.3.3.1.132g).
.040 Cabbage maggot: (see anthomyiids, Diptera, V.3.3.1.132b).
.041 Cacopaurus spp.: (see Sedentary ectoparasitic nematodes attacking roots V.3.3.3.200b).
.042 Carpenter bees: (see Hymenoptera, V.3.3.1.132e).
.043 Carrion beetles: (see Coleoptera, V.3.3.1.132a).
.044 Caterpillar: larval form of moths, skippers, or butterflies; (see Lepidoptera, V.3.3.1.132g).
.045 Cave "cricket": (see Orthoptera, V.3.3.1.132h).
.046 Celery butterfly: (see Lepidoptera, V.3.3.1.132g).
.047 Chalcid wasps: (see Hymenoptera, V.3.3.1.132e).
.048 Chinch bug: (see Hemiptera, V.3.3.1.132c).
.049 Chrysanthemum midge: (see gall gnats, Diptera, V.3.3.1.132b).
.050 Cicada: (see Homoptera, V.3.3.1.132d).
.051 Cicada-killer: (see digger wasps, Hymenoptera, V.3.3.1.132e).
.052 Citrus nematode: (see Tylenchulus semipenetrans, V.3.3.1.200e).
.053 Citrus rust mite: (see Mites, V.3.3.1.172).
.054 Clear-winged moth: (see Lepidoptera, V.3.3.1.132g).
.055 Click beetles: (see Coleoptera, V.3.3.1.132a).
.056 Clothes moths: (see Lepidoptera, V.3.3.1.132g).
.057 Coconut weevil: (see Rhynchophorus palmarum, V.3.3.1.200a).
.058 Coffee lesion nematode: (see Pratylenchus coffeae, V.3.3.1.200a).
.059 Coreid: (see Hemiptera, V.3.3.1.132c).
.060 Corn earworm: (see Lepidoptera, V.3.3.1.132g).
.061 Crane flies: (see Diptera, V.3.3.1.132b).
.062 Crickets: (see Orthoptera, V.3.3.1.132h).
.063 Criconema spp.: (see Sedentary ectoparasitic nematodes attacking roots, V.3.3.1.200b).
.064 Criconemoides spp.: (see Sedentary ectoparasitic nematodes attacking roots, V.3.3.1.200b).
.065 Cucumber beetles: (see leaf beetles, Coleoptera, V.3.3.132a).
.066 Curculios: (see weevils, Coleoptera, V.3.3.1.132a).
.067 "Curly tip" of fig: (see Xiphinema, V.3.3.1.200c).
.068 Cutworm: (see Lepidoptera, V.3.3.1.132g).
.069 Cyclamen mite: (see Mites, V.3.3.1.172).
.070 Cyst nematodes: (see Heterodera spp., V.3.3.1.200e. V.3.3.1.200g).
.071 Dagger nematodes: (see Xiphinema spp., V.3.3.1.200c).
.072 Damsel bug: (see Hemiptera, V.3.3.1.132c).
.073 Darkling beetles: (see Coleoptera, V.3.3.1.132a).
.074 Deer: (see Large animals, V.3.3.1.140).
.075 Deer flies: (see Diptera, V.3.3.1.132b).
.076 Digger wasps: (see Hymenoptera, V.3.3.1.132e).
.077 Ditylenchus radicicola: (see Nematodes attacking above-ground plant parts, V.3.3.1.200a).
.078 Ditylenchus spp.: (see Nematodes attacking above-ground plant parts V.3.3.1.200a: Sedentary endoparasitic nematodes attacking roots, V.3.3.1.200g).
.079 Dolichodorus spp.: (see Migratory ectoparasitic nematodes attacking roots, V.3.3.1.200c).
.080 Dung beetles: (see scarab beetles, Coleoptera, V.3.3.1.132a).
.081 Eelworm: (see Nematode, V.3.3.1.183).
.082 Elephants: (see Large animals, V.3.3.1.140).
.083 Elm bark beetles: (see bark and ambrosia beetles, Coleoptera, V.3.3.1.132).
.084 Elytron (elytra): opaque, hardened front wing of beetles; (see Coleoptera, V.3.3.1.132a).

V.3.3.1. PROPAGATION, NURSERY HANDLING, SOILS, CROP PRODUCTION

.085 European corn borer: (see Lepidoptera, V.3.3.1.132g).
.086 False wireworm: larva of some darkling beetles; (see Coleoptera, V.3.3.1.132a).
.087 Field cricket: (see Orthoptera, V.3.3.1.132h).
.088 Fireflies: (see Coleoptera, V.3.3.132a).
.089 Flea beetles: (see leaf beetles, Coleoptera, V.3.3.1.132a).
.090 Flies: (see Diptera, V.3.3.1.132b).
.091 Flower bug: (see Hemiptera, V.3.3.1.132c).
.092 Foliar nematodes: (see Nematodes attacking above-ground plant parts V.3.3.1.200a).
.093 Four-footed butterfly: (see Lepidoptera, V.3.3.1.132g).
.094 Fruit flies: (see Diptera, V.3.3.1.132b).
.095 Fruitworm: (see Lepidoptera, V.3.3.1.132g).
.096 Gall gnats: (see Diptera, V.3.3.1.132b).
.097 Gall wasps: (see Hymenoptera, V.3.3.1.132e).
.098 Garden webworm: (see Lepidoptera, V.3.3.1.132g).
.099 Giant silkworm moth: (see Lepidoptera, V.3.3.1.132g).
.100 Giant snail: (see Snails, V.3.3.1.251).
.101 "Glowworm": larva of fireflies; (see Coleoptera, V.3.3.1.132a).
.102 Gophers: (see Rodents, V.3.3.1.224).
.103 Gossamer winged butterfly: (see Lepidoptera, V.3.3.1.132g).
.104 Grape fanleaf vector: (see _Xiphinema_ spp., V.3.3.1.200c).
.105 Grasshoppers: (see Orthoptera, V.3.3.1.132h).
.106 Grey garden slug: (see Slugs, V.3.3.1.250).
.107 Ground beetles: (see Coleoptera, V.3.3.1.132a).
.108 Grub: specifically, the larval form of certain higher (e.g., scarab) beetles (see Coleoptera, V.3.3.1.132a); often applied indiscriminately to all larval forms of beetles in general, plus larvae of other insects that are grublike.
.109 Gypsy moth: (see Lepidoptera, V.3.3.1.132g).
.110 Halteres: club-shaped, apparently sensory modified wings on flies; (see Diptera, V.3.3.1.132b).
.111 Hares: (see Rodents, V.3.3.1.224).
.112 Harvestmen: (see Arachnids, V.3.3.1.010).
.113 Hawk moth: (see Lepidoptera, V.3.3.1.132g).
.114 _Helicotylenchus_ spp.: (see Migratory semiendoparasitic nematodes attacking roots; Active, migratory endoparasitic nematodes attacking roots, V.3.3.1.200d,f).
.115 _Hemicriconemoides_ spp.: (see Sedentary ectoparasitic nematodes attacking roots, V.3.3.1.200b).
.116 _Hemicycliophora_ spp.: (see Sedentary ectoparasitic nematodes attacking roots, V.3.3.1.200b).
.117 Hessian fly: (see gall gnats, Diptera, V.3.3.1.132b).
.118 _Heterodera schachtii_: (see _Heterodera_ spp., V.3.3.1.200g).
.119 _Heterodera_ spp.: (see Sedentary semiendoparasitic nematodes attacking roots, V.3.3.1.200e; Sedentary endoparasitic nematodes attacking roots, V.3.3.1.200g).
.120 Hippopotami: (see Large animals, V.3.3.1.140).
.121 Honey bees: (see Hymenoptera, V.3.3.1.132e).
.122 _Hoplolaimus_ spp.: (see Migratory semiedoparasitic nematodes attacking roots, V.3.3.1.200d).
.123 Hornets: (see social wasps, Hymenoptera, V.3.3.1.132e).
.124 Horn fly: (see muscids, Diptera, V.3.3.1.132b).
.125 Hornworm: (see Lepidoptera, V.3.3.1.132g).
.126 Horse bot flies: (see Diptera, V.3.3.1.132b).
.127 Horse flies: (see Diptera, V.3.3.1.132b).
.128 House fly: (see muscids, Diptera, V.3.3.1.132b).
.129 Hypermetamorphosis: abnormal or unusual metamorphosis, as in blis-

V.3.3.1. PROPAGATION, NURSERY HANDLING, SOILS, CROP PRODUCTION   501

ter beetles where there are two or more distinct types of larvae; generally active and sedentary forms; (see Coleoptera, V.3.3.1. 132a).
.130 <u>Hypsoperine</u> spp.: (see Sedentary endoparasitic nematodes attacking roots, V.3.3.1.200g).
.131 <u>Ichneumon wasps</u>: (see Hymenoptera, V.3.3.1.132e).
.132 Insects: invertebrate animals belonging to the phylum Arthropoda and class Insecta (Hexapoda) in which the adult is characterized by one pair of antennae, three body regions, the head, thorax, and abdomen, three pairs of true legs, usually two pairs of wings, and tracheal respiration (i.e., via tubules); the largest group in the animal kingdom with about a million species described so far. (P & D)
.132a. Coleoptera: beetles, the largest order (approx. 40% of all species) of insects; include tiger beetles (predatory as larvae and adults), ground beetles (nocturnal, predatory), lady beetles (very small, 1 cm or less; only bean and squash beetles are "black sheep," others predators); fireflies ("lightning-bugs," predaceous, larva = "glowworm"), soldier beetles (predaceous), rove beetles (predaceous), carrion beetles (scavengers), skin beetles (scavengers, also household pests), darkling beetles (some stored-products pests, some larvae called false wireworms), pea and bean beetles (stored-legume pests), scarabs (May beetle, Japanese beetle, rose chafer, pests, others scavengers, e.g., dung beetles), click beetles (larvae = wireworms, important pests on roots and tubers), leaf beetles (rootworms, flea beetles, and cucumber beetles feed on foliage, some larvae on roots), long-horned beetles (larvae = roundheaded borers; adults feed on foliage and pollen, larvae tunnel heartwood; up to 10 cm long), metallic wood borers (larvae feed under bark or in dead or dying trees), blister beetles (foliage feeders as adults, larvae predatory; hyper-metamorphosis, with two or more distinct types of larva, generally active and sedentary forms), weevils (= snoutbeetles or curculios; many serious pests, can be up to 10 cm long; grubs commonly feed internally, adults externally), bark and ambrosia beetles (small with stubby snouts; many important pests of forest, shade, and fruit trees, e.g., shot-hole borers, peach bark beetle, and elm bark beetle, tunnel deep into heartwood and feed on cultivated fungi in galleries). Characterized by dimorphic wings (front ones hardened, opaque = elytra, rear ones membranous, folded beneath elytra), chewing mouthparts, complete metamorphosis and antennae of various types; larvae of certain higher forms (e.g., scarabs) = grubs (see V.3.3.1. 108), well-developed heads and jaws, usually three pairs of legs, no prolegs (weevil larvae entirely footless); pupae usually naked and appendages free.
.132b. Diptera: flies; include crane flies, midges, black flies, mosquitoes, gall gnats (some predaceous, some important pests; wheat midge, chryanthemum midge, hessian fly), horse and deer flies, bee flies (parasitic; larvae feed on caterpillars and grasshoppers), robber flies (predaceous), syrphid flies (important predators of aphids), fruit flies (many pests, e.g., Mediterranean, oriental, Mexican, melon flies as maggots), horse bot flies, warble flies, tachinid flies (parasitic on many pests), blow flies (screw-worm), muscids (tsetse, horn, house flies), anthomyiids (several important pests, e.g., cabbage maggot,

seed corn maggot, onion maggot, and beet leaf miner). Characterized by two wings or none, a pair of halteres (club-shaped, apparently sensory organs), complete metamorphosis, and a variety of antennal types and mouth parts; larvae = maggots, footless, head and mouthparts greatly reduced.

.132c. Hemiptera: the true bugs, i.e., leaf or plant bugs (small, many very destructive pests), ambush bugs (predators), assassin bugs (medium-large, to 5 cm, predatory), flower bugs (small, predaceous on mites, scales, aphids, leafhoppers, and eggs), damsel bugs (predaceous), lace bugs (several important pests), lygaeids (chinch bugs, otherwise predators), coreids (squash bug, plus some other pests), stink bugs (many predators, some pests); also bedbugs. Some wingless, mostly winged, piercing-sucking mouthparts, gradual metamorphosis.

.132d. Homoptera: cicadas (4-6 cm long, 17-year "locust" most destructive species), treehoppers (feed and oviposit in woody plants), planthoppers, spittlebugs (few pests but many hosts), leafhoppers (up to 1.3 cm, mostly small, many pests), aphids (=plant lice, small, nearly ubiquitous pests, important virus vectors), psyllids (=jumping plant lice, few important pests, e.g., those on pear and potato), whiteflies (several pests, mostly nuisances), scale insects (armored or hard and unarmored or soft; many species and numerous hosts), mealybugs (pests of many crops; virus vectors). Piercing-sucking mouthparts attached to base of head, winged and wingless forms (e.g., scales and mealybugs), male scales with one pair of wings; gradual to near complete metamorphosis.

.132e. Hymenoptera: bees, wasps, and ants, including bees (pollinators, honey and beeswax producers, e.g., honey bees, bumble bees, leaf-cutter bees, and carpenter bees), digger wasps (thread-waisted wasps, cicada-killers, mud-daubers, miners; many predatory, a few parasitic on homopterous insects), social wasps (hornets, yellow jackets, spider wasps, paper wasps), ants (large group, some pests, many predators), sawflies (defoliation and leaf mining by caterpillarlike larvae), gall wasps (some parasitic, mainly gall formers), ichneumon and braconid wasps (important parasites, particularly ichneumons), chalcid wasps (small to minute, some pests, mostly parasitic), proctotrupids (parasitic). Characterized by four membranous wings, legless, grublike, except sawflies [caterpillarlike with six to eight pairs of prolegs without crochets (hooked hairs)]; many species with highly developed social behavior.

.132f. Isoptera: termites; an important group in tropical and warm temperate areas of the world as pests primarily of wood and wood products but also of many living plants. Characterized by chewing mouth parts (their principal food is cellulose), gradual metamorphosis, and social with a well defined caste system of workers, soldiers, and reproductive forms (king and queen), of which only the last is winged and capable of forming new colonies.

.132g. Lepidoptera: moths, skippers, and butterflies, i.e., swallowtail (few pests, e.g., orange dog, celery butterfly), white and yellow butterflies (cabbage and alfalfa butterflies two common pests), gossamer winged butterflies, four-footed butterflies (few pests), skippers, giant silkworm moths (wingspread 25 cm or more; larvae feed on trees and shrubs), royal moths, sphinx or hawk moths (larvae = hornworms, important pests of tomato, grape, tobacco, catalpa, etc.), noctuids (many pests, e.g., cutworms, armyworms, bollworms, fruitworms, corn earworm, cabbage looper), tiger moths (larvae = "woolly bears," few pests), measuring worms (larvae of some species voracious feeders on woody plant foliage), tussock moths (gypsy, browntail, white-marked tussock moths important, among others), prominents (yellow-necked, walnut, red-humped caterpillars feed on woody plant foliage), clear winged moths (larvae are borers; many destructive species), bagworm moths (larvae build cases and feed on foliage), tent caterpillars (foliage feeders), tineids (clothes moths), leaf rollers (several important species), pyralids (many destructive species, e.g., European corn borer, sod webworms, pickleworm, melonworm, garden webworm, and meal moths). Dense covering of tiny scales and hairs on wings and bodies, mouthparts a long flexible tube for sucking or siphoning, palpi often prominent, various types of antennae, wings (two pairs) membranous; metamorphosis complete, larvae = caterpillars [elongated cylindrical, wormlike with three pairs of thoracic legs and fleshy abdominal prolegs (normally five pairs, fewer in loopers)], often elaborate silken cocoons for pupation.

.132h. Orthoptera: walkingsticks, praying mantids, roaches, long-horned grasshoppers (e.g., katydids, meadow grasshoppers, cave and Mormon "crickets"), short-horned grasshoppers or locusts, and crickets (e.g., mole crickets, field crickets and tree crickets). Characterized by strong mandibulate mouthparts, and gradual or simple metamorphosis; short-horned grasshoppers or locusts, Mormon cricket and some walkingsticks and crickets (e.g., mole and field crickets) are important plant pests.

.132i. Thysanoptera: (name means "fringed wings") include thrips, with many species destructive to plants and often occurring in vast numbers; adults are generally less than 2 mm long, with four wings, nymphs are wingless, or wings when present very narrow with greatly reduced venation and marginal setae (hairs), rasping-sucking mouthparts, and a gradual metamorphosis (with an incipient pupal stage).

.133 Instar: stages of insect development between moltings, as in scale insects, among others.
.134 Japanese beetles: (see scarab beetle, Coleoptera, V.3.3.1.132a).
.135 Jumping plant-lice: (=psyllids; see Homoptera, V.3.3.1.132d).
.136 Katydid: (see Orthoptera, V.3.3.1.132h).
.137 Lace bug: (see Hemiptera, V.3.3.1.132c).
.138 Lady beetles: (see Coleoptera, V.3.3.1.132a).
.139 Lance nematodes: (see Hoplolaimus spp., V.3.3.1.200d).
.140 Large animals: deer in many countries, elephants, hippopotami in Africa, and monkeys may browse anything edible when natural forage becomes scarce. (Elephants are also known to step deliberately on newly set coffee seedlings.)
.141 Larva: the wormlike, immature form of certain insects; some are denoted caterpillars (butterflies and moths), grubs (some higher beetles), or maggots (flies). (P & D)
.142 Leaf beetles: (see Coleoptera, V.3.3.1.132a).
.143 Leaf bug: (see Hemiptera, V.3.3.1.132c).
.144 Leafcutter bees: (see Hymenoptera, V.3.3.1.132e).
.145 "Leaf drop" of figs: (see Paratylenchus spp., V.3.3.1.200b).
.146 Leaf galls: (see Anguina spp., V.3.3.1.200a).
.147 Leafhopper: (see Homoptera, V.3.3.1.132d).

V.3.3.1. PROPAGATION, NURSERY HANDLING, SOILS, CROP PRODUCTION

.148 Leaf rollers: (see Lepidoptera, V.3.3.1.132g).
.149 Leptomorph: (see Rodents, V.3.3.1.224).
.150 "Lightningbug": (see Coleoptera, V.3.3.1.132a).
.151 Locust: (see Orthoptera, V.3.3.1.132h).
.152 Long-horned beetles: (see Coleoptera, V.3.3.1.132a).
.153 Long-horned grasshopper: (see Orthoptera, V.3.3.1.132h).
.154 Longidorus spp.: (see Migratory ectoparasitic nematodes attacking roots, V.3.3.1.200c).
.155 Lygaeid: (see Hemiptera, V.3.3.1.132c).
.156 Maggot: larval form of a fly; (see Diptera, V.3.3.1.132b).
.157 May beetle: (see Scarab beetles, Coleoptera, V.3.3.1.132a).
.158 Meadow grasshopper: (see Orthoptera, V.3.3.1.132h).
.159 Meal moths: (see Lepidoptera, V.3.3.1.132g).
.160 Mealybugs: (see Homoptera, V.3.3.1.132d).
.161 Measuring worms: (see Lepidoptera, V.3.3.1.132g).
.162 Mediterranean fruit fly: (see fruit flies, Diptera, V.3.3.1.132b).
.163 Meloidodera spp.: (see Sedentary endoparasitic nematodes attacking roots, V.3.3.1.200g).
.164 Meloidogyne spp.: (see Sedentary semiendoparasitic nematodes attacking roots, V.3.3.1.200e; Sedentary endoparasitic nematodes attacking roots, V.3.3.1.200g).
.165 Melon fruit flies: (see Diptera, V.3.3.1.132b).
.166 Melonworm: (see Lepidoptera, V.3.3.1.132g).
.167 Metallic wood borers: (see Coleoptera, V.3.3.1.132a).
.168 Mexican fruit fly: (see fruit flies, Diptera, V.3.3.1.132b).
.169 Mice: (see Rodents, V.3.3.1.224).
.170 Midges: (see Diptera, V.3.3.1.132b).
.171 Miner: (see digger wasps, Hymenoptera, V.3.3.1.132e).
.172 Mites: animals belonging to the order Acarina of the class Arachnida; certain of them, such as the several spider mites, which have oval bodies and four pairs of legs as adults but three pairs in the first postembryonic instar (hexapod larva) and the citrus rust mite, cyclamen mite, and other similar species, which have elongated bodies with two pairs of legs and are very small (ca. 0.1 mm in length), are serious pests of more than 200 host plants; various species of typhlodromid mites are important predators. (P & D)
.173 Mole cricket: (see Orthoptera, V.3.3.1.132h).
.174 Monkeys: (see Large animals, V.3.3.1.140).
.175 Mormon "cricket": (see Orthoptera, V.3.3.132h).
.176 Mosquitoes: (see Diptera, V.3.3.1.132b).
.177 Moths: (see Lepidoptera, V.3.3.1.132g).
.178 Mud-daubers: (see digger wasps, Hymenoptera, V.3.3.1.132e).
.179 Muscids: (see Diptera, V.3.3.1.132b).
.180 Nacobbus spp.: (see Sedentary endoparasitic nematodes attacking roots, V.3.3.1.200g).
.181 Needle nematodes: (see Longidorus spp., V.3.3.1.200c).
.182 Nema: (see Nematode V.3.3.1.183).
.183 Nematode: nema, eelworm; a round- or thread-worm of the phylum Nematoda (in some classifications a class of the phylum Aschelminthes); they are triploblastic, bilaterally symmetrical, unsegmented, invertebrate animals; many are free-living, whereas others parasitize animals or plants; (see Plant-parasitic nematodes, V.3.3.1.200). (RB, CMI)
.184 Neotylenchus abulbosus: (see Nematodes attacking above-ground plant parts, V.3.3.1.200a).
.185 Noctuid: (see Lepidoptera, V.3.3.1.132g).

V.3.3.1. PROPAGATION, NURSERY HANDLING, SOILS, CROP PRODUCTION

.186 Nothanguina cecidoplastes, N. phyllobia: (see Nematodes attacking above-ground plant parts, V.3.3.1.200a).
.187 Nothotylenchus acris: (see Nematodes attacking above-ground plant parts, V.3.3.1.200a).
.188 Nymph: immature form of certain insects (e.g., grasshoppers, cockroaches, true bugs) whose growing young resemble the parents in body form. (P & D)
.189 Onion maggot: (see anthomyiids, Diptera, V.3.3.1.132b).
.190 Orange dog: (see Lepidoptera, V.3.3.1.132g).
.191 Oriental fruit fly: (see fruit flies, Diptera, V.3.3.1.132b).
.192 Paper wasps: (see social wasps, Hymenoptera, V.3.3.1.132e).
.193 Paranguina spp.: (see Nematodes attacking above-ground plant parts, V.3.3.1.200a).
.194 Paratylenchus spp.: (see Sedentary ectoparasitic nematodes attacking roots, V.3.3.1.200b).
.195 Peach bark beetle: (see bark and ambrosia beetles, Coleoptera, V.3.3.1.132a).
.196 Phytophagous: feeding on plants; herbivorous, as many moths.
.197 Pickleworm: (see Lepidoptera, V.3.3.1.132g).
.198 Plant bug: (see Hemiptera, V.3.3.1.132c).
.199 Plant lice: (see Homoptera, V.3.3.1.132d).
.200 Plant parasitic nematodes: mostly minute vermiform animals that average about 1 mm in length, with a range in size of 0.3 (Paratylenchus spp.) to 10.0 mm (Longidorus spp.); some (e.g., gall-forming, root-knot nematodes, Meloidogyne spp.) extrude a gelatinous "egg-mass" or "egg-sac," or in cyst-forming nematodes (Heterodera) the eggs are retained to a greater or lesser extent in the tough resistant cyst (dead body wall of swollen mature female); include numerous pests of major economic importance and many others less troublesome; classified here in several groups according to the plant parts attacked (i.e., above-ground organs or roots) and whether the nematodes are ectoparasitic (live outside the host), semiendoparasitic (partly inside), endoparasitic (inside the host), sedentary, or migratory. (CMI, DU)
.200a. Nematodes attacking above-ground plant parts: Anguina spp. (stem leaf, or seed galls on cereals); foliar nematodes [Aphelenchoides spp. (attack leaves and buds of ferns, strawberries, chrysanthemums, many foliage ornamentals); "brown ring" of bulbs]; Ditylenchus spp. (many "races," more than 400 host species in 40 families; leaf, stem, and bulb endoparasites on bulb crops, beets, onion, oats, rye, and rice flowers; root galls by D. radicicola); Neotylenchus abulbosus; Nothanguina phyllobia, N. cecidioplastes (galls, hypertrophy or distortion); Nothotylenchus acris, Paranguina spp.; red ring (Rhadinaphelenchus cocophilus) on coconut, oil palm, and areca palm; [vector is coconut weevil (Rhynchophorus palmarum)]; exceptionally root-knot nematodes (Meloidogyne spp.), coffee lesion nematode (Pratylenchus coffeae), and spiral nematode (Scutellonema brachyurum).
.200b. Sedentary ectoparasitic nematodes attacking roots: Cacopaurus spp.; ring nematodes Criconema spp.; ring nematodes (Criconemoides spp.; grape, walnut, and grasses); sheathoid nematodes (Hemicriconemoides spp., spiny larva; grasses); sheath nematodes (Hemicycliophora spp.; grasses); Paratylenchus spp. (pineapple, "leaf drop" of figs, celery roots).
.200c. Migratory ectoparasitic nematodes attacking roots: sting nematodes (Belonolaimus spp.; cotton, soybean, cowpea, grasses,

V.3.3.1. PROPAGATION, NURSERY HANDLING, SOILS, CROP PRODUCTION

sweet potato, peanut, maize, and celery); awl nematodes (Dolichodorus spp.; celery roots, maize, beans, tomato, fruit trees, and strawberries); stubby-root nematodes (Trichodorus spp.; oats, cabbage, turnip, tomato, and lettuce; also virus vectors, e.g., potato corky ringspot); needle nematodes (Longidorus spp., cosmopolitan; decline of mint, fig, grasses, citrus; virus vectors; terminal root galls on many plants); dagger nematodes (Xiphinema spp., cosmopolitan; terminal galling and "curly tip" of fig, rose, and peanut; very extensive host range; also virus vectors of grape fanleaf, tomato and tobacco ring spots, and arabis mosaic).

.200d. Migratory semiendoparasitic nematodes attacking roots: spiral nematodes (Helicotylenchus spp., cosmopolitan; wide host range, grasses, coffee, cloves, maize, tobacco, rice, and banana); lance nematodes (Hoplolaimus spp.; cotton, grasses, cereals, maize, sugar cane, and apple); reniform nematodes (Rotylenchus spp.; pea, cowpea, tomato, soybean, carrot, sweet potato, tobacco, tea, cloves, and lemon; several hundred hosts); spiral nematodes (Scutellonema bradys; yams); stunt nematodes (Tylenchorhynchus spp.; tobacco, sugar cane, sweet potato, and rice).

.200e. Sedentary semiendoparasitic nematodes attacking roots: reniform nematodes (Rotylenchus spp.; several hundred hosts); Sphaeronema spp.; Trophotylenchus spp.; citrus nematode (Tylenchulus semipenetrans, cosmopolitan, 80 species of citrus affected); some root-knot nematodes (Meloidogyne spp.; root galls); cyst nematodes (some Heterodera spp.).

.200f. Active, migratory endoparasitic (female vermiform) nematodes attacking roots: root-lesion nematodes (Pratylenchus spp., cosmopolitan; pineapple, cowpea, date palm, potato tubers, coffee, banana, abaca, pea, bean, cabbage, cauliflower, tomato, apple, maize, tea, peanut, citrus, cherry, peach, narcissus, strawberry, mint, tobacco, and cereals); Pratylenchoides spp.; burrowing nematodes [Radopholus similis, R. spp., cosmopolitan in tropical-warm temperate areas; 150 hosts, including coffee, tea, sweet potato, pineapple, citrus (400 cvs.), sugar cane, banana, and pepper]; Rotylenchoides spp.; spiral nematodes (Helicotylenchus spp., cosmopolitan; wide host range).

.200g. Sedentary endoparasitic (female usually saccate) nematodes attacking roots: cyst nematodes (Heterodera spp., mainly temperate areas, cosmopolitan; wide host range, e.g., H. schachtii, beet cyst nematode with more than 100 hosts, potato, crucifers, legumes, cereals, sugar beet, and carrots); root-knot nematodes (Meloidogyne spp., cosmopolitan; immense host list; nearly all crops in warmer areas, greenhouse crops in cooler areas); Meloidodera spp.; Hypsoperine spp.; Nacobbus spp. (sugar beet plus about 40 others, including crucifers, lettuce cucumber, tomato, carrot, and pea); Ditylenchus spp. (D. radicicola severe root galling on cereals and grasses; other spp.; also bulb parasites).

.201 Potato corky ringspot vector: (see Trichodorus spp., V.3.3.1.200c).

.202 Pratylenchoides spp.: (see Active, migratory endoparasitic nematodes attacking roots, V.3.3.1.200f).

.203 Pratylenchus coffeae: (see Nematodes attacking above-ground plant parts, V.3.3.1.200a; Active, migratory endoparasitic nematodes attacking roots, V.3.3.1.200f).

.204 Pratylenchus spp.: (see Active, migratory endoparasitic nematodes

V.3.3.1. PROPAGATION, NURSERY HANDLING, SOILS, CROP PRODUCTION

attacking roots, V.3.3.1.200f).
.205 Praying mantid: (see Orthoptera, V.3.3.1.133h).
.206 Predaceous: (see Predatory, V.3.3.1.207).
.207 Predator, predatory: living by preying on animals, e.g., one insect preying on another; = predaceous.
.208 Proctotrupids: (see Hymenoptera, V.3.3.1.132e).
.209 Prominents: (see Lepidoptera, V.3.3.1.132g).
.210 Psyllid: (see Homoptera, V.3.3.1.132d).
.211 Pupa: developmental stage between a larva and an adult insect in those groups characterized by complete metamorphosis; often an elaborate silken cocoon in Lepidoptera (V.3.3.1.132g), usually naked in Coleoptera, V.3.3.1.132a). (P & D)
.212 Pyralids: (see Lepidoptera, V.3.3.1.132g).
.213 Rabbits: (see Rodents, V.3.3.1.224).
.214 Radopholus similis, R. spp.: (see Active, migratory endoparasitic nematodes attacking roots, V.3.3.1.200f).
.215 Rats: (see Rodents, V.3.3.1.224).
.216 Red-humped caterpillar: (see Lepidoptera, V.3.3.1.132g).
.217 Red ring: (see Rhadinaphelenchus cocophilus, V.3.3.1.200a).
.218 Reniform nematodes: (see Rotylenchus spp., V.3.3.1.200d, V.3.3.1.200e).
.219 Rhadinaphelenchus cocophilus: (see Nematodes attacking above-ground parts, V.3.3.1.200a).
.220 Rhynchophorus palmarum: (see Red ring, V.3.3.1.200a).
.221 Ring nematodes: (see Criconemoides spp., V.3.3.1.200b).
.222 Roaches: (see Orthoptera, V.3.3.1.132h).
.223 Robber flies: (see Diptera, V.3.3.1.132b).
.224 Rodents and leptomorphs: animals such as mice, rats, gophers, squirrels, rabbits and hares (leptomorphs) that gnaw bark or roots, bite into fruit, seeds, etc.; thus serious pests of a wide range of crops in the field or orchard and of stored products. (P & D)
.225 Root-knot nematodes: (see Meloidogyne spp., Nematodes attacking above-ground plant parts, V.3.3.1.200a; Sedentary semiendoparasitic nematodes attacking roots, V.3.3.1.200e; Sedentary endoparasitic nematodes attacking roots, V.3.3.1.200g).
.226 Root-lesion nematodes: (see Pratylenchus spp., V.3.3.1.200f).
.227 Rootworms: larvae of leaf beetles (Coleoptera, V.3.3.1.132a).
.228 Rose chafer: (see scarab beetles, Coleoptera, V.3.3.1.132a).
.229 Rotylenchoides spp.: (see Active, migratory endoparasitic nematodes attacking roots, V.3.3.1.200f).
.230 Rotylenchus spp.: (see Migratory semiendoparasitic nematodes attacking roots, V.3.3.1.200d; Sedentary semiendoparasitic nematodes attacking roots, V.3.3.1.200e).
.231 Roundheaded borers: larvae of long-horned beetles, (see Coleoptera, V.3.3.1.132a).
.232 Rove beetles: (see Coleoptera, V.3.3.1.132a).
.233 Royal moth: (see Lepidoptera, V.3.3.1.132g).
.234 Sawflies: (see Hymenoptera, V.3.3.1.132e).
.235 Scale insects: (see Homoptera, V.3.3.1.132d).
.236 Scarabs: (see Coleoptera, V.3.3.1.132a).
.237 **Scorpions:** (see Arachnids, V.3.3.1.010).
.238 Screw-worm: (see blow flies, Diptera, V.3.3.1.132b).
.239 Scutellonema brachyurum: (see Nematodes attacking above-ground plant parts, V.3.3.1.200a).
.240 Scutellonema bradys: (see Migratory semiendoparasitic nematodes attacking roots, V.3.3.1.200d).
.241 Seed corn maggot: (see anthomyiids, Diptera, V.3.3.1.132b).

V.3.3.1. PROPAGATION, NURSERY HANDLING, SOILS, CROP PRODUCTION

.242 Seed galls: (see Anguina spp., V.3.3.1.200a).
.243 Seventeen-year locust: (see Homoptera, V.3.3.1.132d).
.244 Sheath nematodes: (see Hemicycliophora spp., V.3.3.1.200b).
.245 Sheathoid nematodes: (see Hemicriconemoides spp., V.3.3.1.200b).
.246 Short-horned grasshopper: (see Orthoptera, V.3.3.1.132h).
.247 Shot-hole borers: (see bark and ambrosia beetles, Coleoptera, V.3.3.1.132a).
.248 Skin beetles: (see Coleoptera, V.3.3.1.132a).
.249 Skippers: (see Lepidoptera, V.3.3.1.132g).
.250 Slugs: more than 30 species of terrestrial pulmonate (i.e., with a lung) gastropods (univalve mollusks) are recorded in the United States; those of most importance as pests of greenhouse and garden plants are the spotted (Limax maximums), tawny (L. flavus), and gray (Deroceras agreste) garden slugs, which have an elongated fusiform shape, no shell, and are up to 20, 10, and 2.5 cm long, respectively; they devour young seedlings and succulent parts of plants and leave a trail of mucus where they crawl. (P & D)
.251 Snails: several hundred species of terrestrial gastropods (i.e., univalve mollusks) with a spiral or helical shell; mostly small and often highly decorative; colored from nearly white to brown to black and ornamented with contrasting stripes or mottling; about six or eight in the United States are pests of greenhouse and garden plants and occasionally orchard trees; the most serious worldwide is the giant snail (Achatina fulica), a monster with a shell 15 to 20 cm or more long and a proportionately large appetite. (P & D, OSDW)
.252 Snout beetles: (see weevils, Coleoptera, V.3.3.1.132a).
.253 Social wasps: (see Hymenoptera, V.3.3.1.132e).
.254 Sod webworm: (see Lepidoptera, V.3.3.1.132g).
.255 Soft scales: (see Homoptera, V.3.3.1.132d).
.256 Soldier beetles: (see Coleoptera, V.3.3.1.132a).
.257 Sorghum midge: (see gall gnats, Diptera, V.3.3.1.132b).
.258 Sowbug: a diminutive animal belonging to the class Crustacea; multi-segmented and elliptical in outline, with seven pairs of legs; a minor greenhouse pest. (P & D)
.259 Sphaeronema spp.: (see Sedentary semiendoparasitic nematodes attacking roots, V.3.3.1.200e).
.260 Sphinx moth: (see Lepidoptera, V.3.3.1.132g).
.261 Spider mites: (see Mites, V.3.3.1.173).
.262 Spiders: (see Arachnids, V.3.3.1.010).
.263 Spider wasps: (see social wasps, Hymenoptera, V.3.3.1.132e).
.264 Spiral nematode: (see Scutellonema brachyurum, V.3.3.1.200a).
.265 Spiral nematodes: (see Helicotylenchus spp., V.3.3.1.200d; Scutellonema bradys, V.3.3.1.201d).
.266 Spiral nematodes: (see Helicotylenchus spp., V.3.3.1.200f).
.267 Spittlebug: (see Homoptera, V.3.3.1.132d).
.268 Spotted garden slug: (see Slugs, V.3.3.1.250).
.269 Squash bug: (see Hemiptera, V.3.3.1.132c).
.270 Squash (lady) beetle: (see Coleoptera, V.3.3.1.132a).
.271 Squirrels: (see Rodents, V.3.3.1.224).
.272 Stable fly: (see muscids, Diptera, V.3.3.1.132b).
.273 Stem galls: (see Anguina spp., V.3.3.1.200a).
.274 Sting nematodes: (see Belonolaimus spp., V.3.3.1.200c).
.275 Stink bug: (see Hemiptera, V.3.3.1.132c).
.276 Stubby-root nematodes: (see Trichodorus spp., V.3.3.1.200c).
.277 Stunt nematodes: (see Tylenchorhynchus spp., V.3.3.1.200d).

V.3.3.1. PROPAGATION, NURSERY HANDLING, SOILS, CROP PRODUCTION

.278 Swallowtail butterflies: (see Lepidoptera, V.3.3.1.132g).
.279 Syrphid flies: (see Diptera, V.3.3.1.132b).
.280 Tachanid flies: (see Diptera, V.3.3.1.132b).
.281 Tarantulas: (see Arachnids, V.3.3.1.010).
.282 Tawny garden slug: (see Slugs, V.3.3.1.250).
.283 Tent caterpillar: (see Lepidoptera, V.3.3.1.132g).
.284 Termites: (see Isoptera, V.3.3.1.132f).
.285 Thread-waisted wasps: (see digger wasps, Hymenoptera, V.3.3.1.132e).
.286 Thrips: (see Thysanoptera, V.3.3.1.132i).
.287 Ticks: (see Arachnids, V.3.3.1.010).
.288 Tiger beetles: (see Coleoptera, V.3.3.1.132a).
.289 Tiger moth: (see Lepidoptera, V.3.3.1.132g).
.290 Tineid: (see Lepidoptera, V.3.3.1.132g).
.291 Tobacco ring spot vector: (see Xiphinema, V.3.3.1.200c).
.292 Tomato ring spot vector: (see Xiphinema, V.3.3.1.200c).
.293 Tree cricket: (see Orthoptera, V.3.3.1.132h).
.294 Treehopper: (see Homoptera, V.3.3.1.132d).
.295 Trichodorus spp.: (see Migratory ectoparasitic semiendoparasitic nematodes attacking roots, V.3.3.1.200c).
.296 Trophotylenchus spp.: (see Sedentary semiendoparasitic nematodes attacking roots, V.3.3.1.200e).
.297 Tsetse fly: (see muscids, Diptera, V.3.3.1.132b).
.298 Tussock moths: (see Lepidoptera, V.3.3.1.132g).
.299 Tylenchorhynchus spp.: (see Migratory semiendoparasitic nematodes attacking roots, V.3.3.1.200d).
.300 Tylenchulus semipenetrans: (see Sedentary semiendoparasitic nematodes attacking roots, V.3.3.1.200e).
.301 Typhlodromids: predaceous mites belonging to the genus Typhlodromus; (see Mites, V.3.3.1.172).
.302 Vector: an agent of dissemination or inoculation (e.g., an insect that transmits a virus or nematode).
.303 Walking stick: (see Orthoptera, V.3.3.1.132h).
.304 Walnut caterpillar: (see Lepidoptera, V.3.3.1.132g).
.305 Warble flies: (see Diptera, V.3.3.1.132b).
.306 Wasps: (see Hymenoptera, V.3.3.1.132e).
.307 Weevils: (see Coleoptera, V.3.3.1.132a).
.308 Wheat midge: (see gall gnats, Diptera, V.3.3.1.132g).
.309 White and yellow butterflies: (see Lepidoptera, V.3.3.1.132g).
.310 White flies: (see Homoptera, V.3.3.1.132d).
.311 White-marked tussock moth: (see Lepidoptera, V.3.3.1.132g).
.312 Wireworms: larvae of click beetles; (see Coleoptera, V.3.3.1.132a).
.313 "Woolly bear": (see Lepidoptera, V.3.3.1.132g).
.314 Xiphinema spp.: (see Migratory ectoparasitic nematodes attacking roots, V.3.3.1.200c).

V.3.3.2. Pests - Plants

.001 Actinomycetes: organisms characterized by fine hyphae, usually less than 1.0 m in diameter, that readily break into fragments resembling bacterial cells; classified as bacteria by some, as fungi by others; one pathogen (e.g., Streptomyces scabies, the causal agent of common potato scab). (R, RB)
.002 Aecidium (pl. aecidia): a cup-shaped aecium that contains a peridium (membrane of sterile cells). (RB)

V.3.3.2. PROPAGATION, NURSERY HANDLING, SOILS, CROP PRODUCTION

.003 Aecium (pl. aecia): the fruiting body of certain rust fungi (V.3.3.2.181f) produced immediately after the pycnium (V.3.3.2.349); the aecium produces dicaryotic (pair of haploid nuclei) spores (aeciospores) in chains; spores may infect the suscept in which they are formed but usually infect an unrelated alternate suscept; aecia may be the aecidium, cacoma, peridermium, or roestelium type. (RB)

.004 Aerobic: living or active only in the air, like many microorganisms; may be obligate or facultative; (see Anaerobic, V.3.3.2.013). (CMI)

.005 Aerogenic: derived from the air (i.e., aerobic), as in the fermentative metabolism of carbohydrates by the pectolytic, soft-rot bacterium (Erwinia carotovora); see Anaerogenic V.3.3.2.014). (CMI)

.006 Agent of inoculation: transports inoculum from its source to or into the infection court; e.g., wind, splashing rain, running water, insects, man, and other animals. (R)

.007 Agrobacterium: (see Bacterial diseases, V.3.3.2.040).

.008 Alga (pl. algae): aquatic plants or occupying damp habitats for the most part; unicellular or multicellular, with filamentous thalli; distinguished from fungi by the presence of chlorophyll; rarely pathogenic, an exception being Cephaleuros virescens (= C. mycoidea), an algal spot that attacks leaves of some fruit trees (e.g., citrus). (RB)

.009 Alternaria solani: (see Deuteromycetes, V.3.3.2.181g).

.010 Alternate host: one of two species of plants required as a host by heteroecious rust fungi for the completion of their developmental cycles. (RB)

.011 Alternative host: a plant other than the main host that is fed on by the parasite; not required for completion of the developmental cycle of the parasite. (RB)

.012 Amphiospore: a uredinospore (V.3.3.2.441) with thickened walls and capable of hibernating. (HE)

.013 Anaerobic: living or active only in the absence of air, like many microorganisms; may be obligate or facultative; (see Aerobic, V.3.3.2.004). (CMI)

.014 Anaerogenic: not derived from the air, as in the fermentative metabolism of carbohydrates by nonpectolytic species of Erwinia, including E. salicis and E. amylovora, which cause dry necroses, galls, or wilts, and the pectolytic soft-rot bacterium E. aroideae; (see Aerogenic, V.3.3.2.005). (CMI)

.015 Angular leaf spot of cotton: (see Bacterial diseases, V.3.3.2.040).

.016 Antagonistic symbiosis: parasitism; one organism of an association benefits at the expense of the other. (RB)

.017 Antheridium: fingerlike male sexual organ of the oomycetes (V.3.3.2.181d). (RB)

.018 Anthocyanescence, anthocyanescent: reddish-purple color in tissues that are normally green; often an plesionecrotic symptom of plant disease caused by an infectious agent (e.g., a virus), usually appearing in the margins around holonecrotic spots in green leaves. (RB)

.019 Anthracnose: common name of plant diseases usually characterized by the development of ulcerous lesions on stems, leaves, and fruit; often caused by certain imperfect fungi that produce conidia in acervuli (shallow, saucer-shaped structures) (e.g., Colletotrichum, Gloeosporium, Kabatiella); (see Fungus diseases, V.3.3.2.181e, V.3.3.2.181g). (RB)

.020 Anthracnose of cucurbits: (see Deuteromycetes, V.3.3.2.181g).

.021 Antibiosis: an association between two organisms that is detrimental to the vital activities of one of them. (RB)
.022 Antibiotic: a chemical substance produced by a microorganism; can inhibit the growth of other microorganisms or destroy them (CMI)
.023 Apothecium (pl. apothecia): sessile or stalked saucer-shaped fruiting structure of Ascomycetes of the series Discomycetes; the inner surface is lined with a hymenium (regular layer) that consists of asci and paraphyses (sterile cells); (see Fungus diseases, V.3.3.2.181e). (RB)
.024 Apple blotch: (see Deuteromycetes, V.3.3.2.181g).
.025 Apple canker: (see Basidiomycetes, V.3.3.2.181f).
.026 Apple fruit spot: (see Deuteromycetes, V.3.3.2.181g).
.027 Apple rust: (see Basidiomycetes, V.3.3.2.181f).
.028 Apple scab: (see Ascomycetes, V.3.3.2.181e).
.029 Appressorium: a bulbous or lobed swelling of a hyphal tip often held to the surface of the substrate by a gelatinous secretion; suscept tissue is penetrated by a slender peg that forms at the center of the appressorium.
.030 Armillariella mellea: (see Basidiomycetes, V.3.3.2.181f). (RB)
.031 Ascomycetes: class of fungi characterized by endogenous sexual production of spores (ascospores) in the organ of meiosis (ascus, a saclike cell); (see Fungus diseases, V.3.3.2.183e). (RB)
.032 Attenuation: lessening of the capacity of a parasitic organism or virus to cause disease; reduction of its virulence. (CMI)
.033 Autoecious: completion of the life cycle on one host (especially of rusts). (RB)
.034 Autotroph, autotrophic: organism capable of self-nourishment, as some bacteria (e.g., Nitrosomonas, Nitrobacter) and green (chlorophyllous) plants.) (RB)
.035 Avocado root rot: (see Oomycetes, V.3.3.2.181d).
.036 Axenic: culture in the absence of living bacteria; pure culture. (CMI)
.037 Axeny: inhospitality, "passive" as opposed to "active" resistance of a plant to a pathogen. (CMI)
.038 Bacillus: (see Bacterial diseases, V.3.3.2.040).
.039 Bacteria: small (0.5-20 µm), motile, or nonmotile microorganisms that contain no chlorophyll, with a single chromosome (prokaryotic, i.e., not enclosed in a membrane-bound nucleus) and sometimes a plasmid (V.3.3.2.320) present as in Agrobacterium tumefaciens; reproduce by simple cell division. (R)
.040 Bacterial diseases: plant-pathogenic bacteria occur in six genera: Erwinia (rots, wilts, blights), Bacillus (rots, spore-formers), Corynebacterium (wilts, blights), Agrobacterium (crown gall, nonself-limiting galls), Xanthomonas (wilts, leaf spots), and Pseudomonas (rots, wilts, leaf spots); all are rod-shaped (i.e., bacilli), usually actively motile, aerobic, or facultatively anaerobic (some Erwinia spp. are unusual in aerogenic or anaerogenic fermentative metabolism of carbohydrates), mostly Gram-negative (not stained with crystal violet); most species invade water-conducting elements of vascular tissues and/or parenchymatous tissues, which produce vascular wilts (e.g., in solanaceous, musaceous, and other hosts attacked by Pseudomonas solanacearum, cucurbits by Bacillus tracheiphilus, etc.) in the former and soft rot (e.g., of carrot by Erwinia carotovora), fire blight (E. atroseptica), angular leaf spot, bacterial blight of cotton (Xanthomonas campestris pv. malvacearum), citrus canker (X. campestris pv. citri), potato ring rot (Corynebacterium michiganense

pv. sependonicum) in the latter; a few produce tubercles, tumors, or galls or induce adventitious shoots or roots (i.e., are hyperplastic) like crown gall (Agrobacterium tumefaciens), olive tubercle (Pseudomonas syringae pv. savastonoi), etc.; many of these disorders are of great economic importance, especially in humid warm temperate, subtropical, and tropical areas; many of these disorders also produce symptoms similar to those of fungi but are characterized, thus distinguishable, by sliminess and oozing of gum from affected internal tissues, usually a foul odor, and highly contagious (e.g., Moko, Xanthomonas campestris pv. celebense, in Central America or citrus canker in the United States). (R)

.041 Bacterial blight of cotton: (see Bacterial diseases, V.3.3.2.040).
.042 Bacterial wilt of banana: (see Bacterial diseases, V.3.3.2.040).
.043 Bacterial wilt of potato: (see Bacterial diseases, V.3.3.2.040).
.044 Bacteriophage (phage): a virus that replicates in bacteria.
.045 Banana wilt: (see Deuteromycetes, V.3.3.2.181g).
.046 Basidiomycetes: class of fungi characterized by exogenous production of spores (basidiospores) on club-shaped organs of meiosis (basidia); (see Fungus diseases, V.3.3.2.181f). (RB)
.047 Basidiospore: sexually produced exogenous spore of a basidiomycete (V.3.3.2.181f); one of the several forms of spores produced by rust fungi. (RB)
.048 Basidium (pl. basidia): typically club-shaped organ of meiosis of basidiomycetes (V.3.3.2.181f); basidiospores (usually four) are borne externally on stalks (sterigmata) that extend from the apex of the basidium. (RB)
.049 Bean anthracnose: (see Deuteromycetes, V.3.3.2.181g).
.050 Bioassay: quantitative estimation of biologically active substances by the extent of their actions under standardized conditions on living organisms. (CMI)
.051 Biotype: a population of individuals with identical genetic but varying physiological characters: (1) a subdivision of a physiologic race; subrace; (2) (of bacteria), a subdivision of a species, subspecies, or serotype distinguished by some special physiological character. (CMI, RB)
.052 Bitter rot of apple: (see Ascomycetes, V.3.3.2.181e).
.053 Blackleg of potato: (see Bacterial diseases, V.3.3.2.040).
.054 Black molds: (see Zygomycetes, V.3.3.2.181c).
.055 Black rot of grape: (see Ascomycetes, V.3.3.2.181e).
.056 Black rot of strawberries, Rubus, Ribes, cucurbits, beets, peas, and crucifers: (see Ascomycetes, V.3.3.2.181e).
.057 Black shank of tobacco: (see Oomycetes, V.3.3.2.181d).
.058 Black spot of roses: (see Ascomycetes, V.3.3.2.181e).
.059 Blight: a nonrestricted necrotic symptom characterized by the ultimate death of tissues throughout an organ, such as leaf or flower. (RB)
.060 Blight of stone fruits: (see Deuteromycetes, V.3.3.2.181g).
.061 Blight of strawberry, Rubus, Ribes, cucurbits, beets, peas, and crucifers: (see Ascomycetes, V.3.3.2.181e).
.062 Blister of ferns, alder, poplar, birch, and cherry: (see Ascomycetes, V.3.3.2.181e).
.063 Blossom blast of hibiscus: (see Zygomycetes, V.3.3.2.181c).
.064 Blossom blast of squash: (see Zygomycetes, V.3.3.2.181c).
.065 Blotch: necrotic spot with fibrillose margins of visible mycelial strands. (RB)
.066 Blue mold of apple: (see Deuteromycetes, V.3.3.2.181g).

V.3.3.2. PROPAGATION, NURSERY HANDLING, SOILS, CROP PRODUCTION 513

.067 Blue mold of citrus: (see Deuteromycetes, V.3.3.2.181g).
.068 Blue mold of tobacco: (see Oomycetes, V.3.3.2.181d).
.069 Blue molds: (see Deuteromycetes, V.3.3.2.181g; Oomycetes, V.3.3.2.181d).
.070 Botrytis cinerea: (see Deuteromycetes, V.3.3.1.181g).
.071 Bremia lactucae: (see Oomycetes, V.3.3.2.181d).
.072 Brown root rot: (see Basidiomycetes, V.3.3.2.181f).
.073 Brown rot of citrus, deciduous fruits: (see Oomycetes, V.3.3.2.181d).
.074 Brown rot of pome and stone fruits: (see Deuteromycetes, V.3.3.2.181g).
.075 Brown rot of potato: (see Bacterial diseases, V.3.3.2.040).
.076 Brown rot of stone fruits, pome fruits: (see Ascomycetes, V.3.3.2.181e).
.077 Brown spot of corn: (see Chytridiomycetes, V.3.3.2.181a).
.078 Buckeye rot of tomato fruits: (see Oomycetes, V.3.3.2.181d).
.079 Cacao (Ceratostomella) wilt: (see Ascomycetes, V.3.3.2.181e).
.080 Cacoma: an aecium (V.3.3.2.003) surrounded by fungal filaments but has no peridium. (RB)
.081 Cane blight of rose, Rubus, apple: (see Ascomycetes, V.3.3.2.181e).
.082 Canker: a necrotic symptom of disease in woody plant parts; the necrosis is sharply limited and restricted to a definite area of cortical tissue and malformed bark that is surrounded by callus. (RB)
.083 Canker: (see Basidiomycetes, V.3.3.2.181f).
.084 Capnodium citri: (see Ascomycetes, V.3.3.2.181e).
.085 Carnation bud rot: (see Deuteromycetes, V.3.3.2.181g)
.086 Carnation rust: (see Basidiomycetes, V.3.3.2.181f).
.087 Carrier: an organism that bears an infectious agent internally but shows no marked symptoms of the disease caused by that carrier; e.g., a symptomless carrier of a virus disease or a vector. (RB)
.088 Caulimovirus: a group of viruses characterized by virions that consist of DNA rather than RNA; cause mosaic-mottle types of disease in their hosts, e.g., strawberry, cauliflower, dahlia, and carnation; (see Virus, V.3.3.2.461) (SP)
.089 Ceratocystis fimbriata: (see Ascomycetes, V.3.3.2.181e).
.090 Ceratocystis paradoxa: (see Ascomycetes, V.3.3.2.181e).
.091 Ceratostomella paradoxa: (see Ascomycetes, V.3.3.2.181e).
.092 Cercospora musae: the imperfect state of Mycosphaerella musicola; (see Ascomycetes, V.3.3.2.181e).
.093 Chestnut blight: (see Ascomycetes, V.3.3.2.181e).
.094 Chlamydospore: the thick-walled asexual spore that develops directly from hyphal cells. (RB)
.095 Chlorosis, chlorotic: hypoplastic symptom of plant disease characterized by a deficiency of chlorophyll that is due to its failure to develop fully. (RB)
.096 Choanephora: (see Zygomycetes, V.3.3.2.181c).
.097 Citrus canker: (see Bacterial diseases, V.3.3.2.040).
.098 Citrus knot: (see Deuteromycetes, V.3.3.2.181g).
.099 Cladosporium fulvum: (see Deuteromycetes, V.3.3.2.181g).
.100 Claviceps purpurea: (see Ascomycetes, V.3.3.2.181e).
.101 Cleistothecium (pl. cleistothecia): a completely enclosed ascigerous (ascusbearing, i.e., perfect stage) fruiting body of certain fungi (e.g., powdery mildews); (see Ascomycetes, V.3.3.2.181e). (RB)
.102 Clitocybe tabescens: (see Basidiomycetes, V.3.3.2.181f).
.103 Clone: a group of organisms decended by mitosis from a common ancestor. (CMI)

V.3.3.2. PROPAGATION, NURSERY HANDLING, SOILS, CROP PRODUCTION 514

.104 Club root of crucifers: (see Plasmodiophoromycetes, V.3.3.2.181b).
.105 Coenocytic: an adjective that describes thalli, especially those of archimycetes and phycomycetes, which consist of multinucleate masses of protoplasm. (RB)
.106 Coffee rust: (see Basidiomycetes, V.3.3.2.181f).
.107 Collectotrichum gloeosporioides: imperfect state of Glomerella cingulata; (see Ascomycetes, V.3.3.2.181e).
.108 Collectotrichum lagenarium: (see Deuteromycetes, V.3.3.2.181g).
.109 Collectotrichum lindemuthianum: (see Deuteromycetes, V.3.3.2.181g).
.110 Commensualism: a symbiosis in which one organism is benefitted but the second is neither benefitted nor harmed. (RB)
.111 Common smut of corn: (see Basidiomycetes, V.3.3.2.181f).
.112 Conidium (pl. conidia): an asexual fungal spore borne exogenously on a specialized hypha (conidiophore). (RB)
.113 Corynebacterium spp.: (see Bacterial diseases, V.3.3.2.040).
.114 Coryneum beijerinckii: (see Deuteromycetes, V.3.3.2.181g).
.115 Cotton wilt: (see Deuteromycetes, V.3.3.2.181g).
.116 Cottony rot of lettuce, celery, other vegetables, lemon: (see Ascomycetes, V.3.3.2.181e).
.117 Cronartium ribicola: (see Basidiomycetes, V.3.3.2.181f).
.118 Crown and trunk canker of deciduous fruits, black walnut: (see Oomycetes, V.3.3.2.181d).
.119 Crown gall: (see Bacterial diseases, V.3.3.2.040).
.120 Cucurbit wilt: (see Bacterial diseases, V.3.3.2.040).
.121 Culture: growth of an organism for the purpose of experiment, especially on laboratory media (culture media); often used in the sense of isolate or strain. (CMI)
.122 Dahlia blight: (see Zygomycetes, V.3.3.2.181c).
.123 Damping-off: a disease or necrotic symptom of disease in seedlings in which the seedling stem is decayed near the soil line and the seedling topples; damping-off pathogens may also prevent seed germination and kill the sprout before it emerges from the soil; (see Basidiomycetes, V.3.3.2.181f). (RB)
.124 Damping-off of seedlings: (see Basidiomycetes, V.3.3.2.181f).
.125 Diaporthe batatis: (see Ascomycetes, V.3.3.2.181e).
.126 Diaporthe citri: (see Ascomycetes, V.3.3.2.181e).
.127 Dicaryon (dikaryon): a pair of haploid nuclei that occur in a cell; the nuclei undergo simultaneous division (conjugate division) on formation of each new cell; typified in the diploid phase (dicaryon phase) of basidiomycetes (V.3.3.2.181f). (RB)
.128 Dieback: a necrotic sympton of disease in which death of shoot tissues begins at the tip and progresses back toward the main stem. (RB)
.129 Dieback of peach, other stone fruits: (see Ascomycetes, V.3.3.2.181e).
.130 Diplocarpon rosae: (see Ascomycetes, V.3.3.2.181e).
.131 Diplodia foot rot, dieback, stem-end rot of citrus, etc.: (see Deuteromycetes, V.3.3.2.181g).
.132 Diplodia natalensis: (see Deuteromycetes, V.3.3.2.181g).
.133 Disease: one or more harmful physiological processes caused by continuous irritation of a plant by a primary causal agent; results in morbific cellular activity and is expressed in characteristic pathological responses called symptoms; (see Bacterial diseases, V.3.3.2.040; Fungus diseases, V.3.3.2.181). (R)
.133a. Causative agents of disease: several distinct groups that show signs of plant pathogens, thus provide objective evidence of causative agents of disease:(1) viroids and viruses (ultrami-

croscopic particles); (2) prokaryotic organisms, including mycoplasmalike organisms (MLOs), rickettsialike organisms (RLOs) bacteria, actinomycetes (single cells); (3) fungi and algae (spores, fruiting, and vegetative bodies); (4) protozoans; (5) nematodes (eggs, immature forms, adults); (6) insects and mites (eggs, immature forms, adults); (7) seed plants (vegetative and reproductive structures; parasites, semiparasites).

.133b. Disease symptoms: three main groups: (1) cytological (cytoplasmic and nuclear inclusions); (2) histological (phloem necrosis); (3) morphological, as in (a) necrosis (spots, ring spots, etchings, tip blights, etc.), (b) hypoplastic (chlorosis, mosaic, and dwarfing), (c) hyperplastic (tumefaction, enation, oedema, anthocyanescence, and phyllody).

.133c. Metabolic classification of diseases: physiological manifestations of symptoms may be divided into seven groups: (1) storage disorders (e.g., rots, as brown rot of stone fruits); (2) digestion of tissues (e.g., damping-off of seedlings); (3) absorption and gummulation (e.g., mushroom root rots of trees) (4) destruction of meristems (e.g., club root of crucifers, loose smut of oats, root-knot nematodes); (5) interference with water conduction (e.g., fusarium-induced wilts); (6) interference with photosynthesis (e.g., bacterial blights of beans, leaf blights of corn, late blight of potatoes, powdery mildew of cucurbits, stem rust of cereals); (7) interference with translocation (e.g., cucumber mosaic). (Note that viruses, viroids, and mycoplasmalike organisms may be included under one or more of the above headings.)

.133d. Physiological effects of infections: three main groups: (1) translocation (systemic viruses move and replicate in phloem); (2) photosynthesis (chlorosis, reduced photosynthetic rates); (3) growth processes (enations, "shoe-string" leaves, tumefactions, witches' broom).

.133e. Transmission: two groups: (1) from one generation to the next, e.g., in vegetative organs (most common method) or by seed (relatively uncommon, sometimes important); (2) from one plant to another, as by (a) insects and mites, (i) stylet-borne (mosaic viruses; aphid vectors), (ii) circulative (yellows viruses; leafhopper vectors), or (iii) by white flies, mealybugs, mites, weevils and other beetles, flies, (b) graftage, (c) dodder (Cuscuta spp.), (d) mechanical methods, (e) nematodes, or (f) fungi.

.134 Dispersal (dissemination): spread of a pathogen or other pest in an area of its geographical range. (RB)
.135 Dothidella ulei: (see Ascomycetes, V.3.3.2.181e).
.136 Downy mildew of crucifers, beets, and rose: (see Oomycetes, V.3.3.2.181d).
.137 Downy mildew of cucurbits: (see Oomycetes, V.3.3.2.181d).
.138 Downy mildew of grape: (see Oomycetes, V.3.3.2.181d).
.139 Downy mildew of grasses: (see Oomycetes, V.3.3.2.181d).
.140 Downy mildew of lettuce: (see Oomycetes, V.3.3.2.181d).
.141 Downy mildews: (see Oomycetes, V.3.3.2.181d).
.142 Drop of lettuce, celery, other vegetables, lemons: (see Ascomycetes, V.3.3.2.181e).
.143 Dry heart rot of strawberry: Rubus, Ribes, cucurbits, beets, peas, crucifers; (see Ascomycetes, V.3.3.2.181e).
.144 Dry rot of sweet potato: (see Ascomycetes, V.3.3.2.181e).
.145 Early blight of potato: (see Deuteromycetes, V.3.3.2.181g).

.146 Ectoparasite: parasite living on the outside of its host (e.g., stubby-root and dagger nematodes); (see Endoparasite, V.3.3.2.151). (RB)
.147 Ectosymbiosis: symbiosis in which one member (microsymbiote) develops on the outside of the other member; (see Endosymbiosis, V.3.3.2.152). (RB)
.148 Ectotrophic: refers to a mycorrhiza in which the mycelium forms an external covering on the root (e.g., pine); (see Mycorrhiza, V.3.3.2.264). (RB)
.149 Edema: (see Oedema, V.3.3.2.273).
.150 Elsinoe fawcettii: (see Ascomycetes, V.3.3.2.181e).
.151 Endoparasite: parasite living within its host; (see Ectoparasite, V.3.3.2.146). (RB)
.152 Endosymbiosis: symbiosis in which one member (microsymbiote) lives within the other; (see Ectosymbiosis, V.3.3.2.147). (RB)
.153 Endothia parasitica: (see Ascomycetes, V.3.3.2.181e).
.154 Endotrophic: refers to a mycorrhiza in which the mycelium grows within the cortical cells of the root (e.g., in orchids); (see Mycorrhiza, V.3.3.2.264). (RB)
.155 Entyloma dahliae: (see Basidiomycetes, V.3.3.2.181f).
.156 Enzyme-linked immunosorbent assay: a rapid method for the detection of virus diseases that involve samples of pedicel bark and vascular fruit calyx tissue exposed to antibodies prepared from antiserum of a particular virus (e.g., tristeza); (see Virus, V.3.3.2.461).
.157 Epidemic: outbreak of a disease in a high percentage of a population. (RB)
.158 Epinasty: downward curling of a leaf blade due to cell growth on the upper side of a petiole being more rapid than that on the lower side or a twisted or misshapen stem; often a hyperplastic symptom of plant disease (e.g., the disease caused by potato virus Y in Physalis floridana). (RB)
.159 Epiphytology: the study of epiphytotics, with special reference to occurrence and severity of a plant disease, influenced by such environmental factors as moisture, temperature, wind, and nature of the soil; analogous to "epidemiology." (RB)
.160 Epiphytotic: occurrence of a plant disease in abundant proportions in an extensive area. (RB)
.161 Epiphytotic disease (epiphytotic): a disease suddenly, widely, and destructively affecting plants in a locality. (CMI)
.162 Ergot: (see Ascomycetes, V.3.3.2.181e).
.163 Erwinia: (see Bacterial diseases, V.3.3.2.040).
.164 Escape: failure of inherently susceptible plants to become diseased, even though disease is prevalent. (RB)
.165 Etiology: the study of cause; the phase of plant pathology that deals with the casual agent and its relations with the susceptible plant. (RB)
.166 Eukaryote: organism in which the chromosomes are located in a membrane-bound nucleus in each cell. (R)
.167 Exobasidium azalea, E. spp.: (see Basidiomycetes, V.3.3.2.181f).
.168 Facultative parasite: an organism that is normally saprophytic but can live as a parasite. (RB)
.169 Facultative saprophyte: an organism that is normally parasitic but can live saprophytically. (RB)
.170 Fasciation: hyperplastic symptom characterized by a fusing (and flattening) of such plant organs as stems. (RB)
.171 Fasciculation: hyperplastic symptom characterized by a clustering

V.3.3.2. PROPAGATION, NURSERY HANDLING, SOILS, CROP PRODUCTION

of plant organs such as shoots into a structures as witches' broom. (RB)
.172 Field immune: (of plants) not being infected by a disease in the field, although susceptible under experimental conditions. (RB)
.173 Fig rust: (see Basidiomycetes, V.3.3.2.181f).
.174 Fire blight: (see Bacterial diseases, V.3.3.2.040).
.175 Fomes spp.: (see Basidiomycetes, V.3.3.2.181f).
.176 Foot and root rot of citrus, deciduous fruits, black walnut: (see Oomycetes, V.3.3.2.181d).
.177 Forma specialis (f.sp.): special form; a biotype (or group of biotypes) of a species of pathogen that differs from others in its ability to infect selected genera or species of susceptible plants; (see Physiologic race, V.3.3.2.309). (RB)
.178 Fruit rot of squash: (see Zygomycetes, V.3.3.2.181c).
.179 Fungi: nonchlorophyllous plants whose vegetative bodies (thalli) are threadlike (hyphae) or, in some groups, amoeboid; saprophytic or parasitic, the latter including many important pathogens, perhaps 10,000 in all, of which several hundred are of great economic importance as pests of cultivated fruits, vegetables, and ornamentals. (R)
.179a. Taxonomy of fungi, lichens, algae, myxomycetes and bacteria: rules of the International Code of Botanical Nomenclature govern names of taxa, ranks, publication of names, etc.; note, however, that there is now a separate code for bacteria (see V.3.3.2.179b); beginning dates for lichens, algae (some exceptions), myxomycetes, and bacteria are May 1, 1753 (as for Spermatophyta and Pteridophyta), and for fungi, Uredinales (rusts), Ustilaginales (smuts), and Gasteromyces, December 31, 1801, and Fungi Cateri January 1, 1821; endings are -mycota, -mycotina, -mycetes, and -mycetidae for divisions, subdivisions, classes, and subclasses, respectively, of fungi; those of lower ranks are the same as for spermatophytes; the nomenclature of special forms (formae speciales) within a species, e.g., parasitic fungi, is not governed by the Code; it is permissible under the Code for a fungus to have two names, one for the state bearing asexual spores ("imperfect state") and one for the state bearing sexual spores ("perfect state"), the latter taking precedence over the former (e.g., Ceratocystis paradoxa for pineapple heart rot, the imperfect state of which is named Thielaviopsis paradoxa; Thanatephorus cucumeris for damping-off of seedlings, the imperfect state being Rhizoctonia solani). (DY, R)
.179b. International Code of Nomenclature of Bacteria: originally established as the International Bacteriological Code of Nomenclature in 1947 (Buchanan et al., 1948); reestablished in 1975 under the present name and revised in 1976. Provides that after January 1, 1980, priority of publication shall date from January 1, 1980, and all prior published names on the Approved Lists of Bacterial Names of the International Committee on Systematic Biology shall be treated as though they had been published for the first time on that date; those names validly published before January 1, 1980, but not included on the Approved Lists will have no further standing in nomenclature, thus are available for reuse in naming new taxa (quoted from Dye et al., 1980). (DY)
.179b-1. Pathovar: contraction of "patho(genic) var(iety)"; refers to a strain or set of strains with the same or similar charac-

teristics, differentiated at the infrasubspecific level from other strains of the same species or subspecies on the basis of distinctive pathogenicity to one or more plant hosts (Dye et al., 1980); established by the International Society of Plant Pathology Committee on Taxonomy of Pathogenic Bacteria; analogous to "cultivar" used for cultivated plants. (DY)

.180 Fungi imperfecti: a form class erected to contain those fungi whose sexual reproduction is unknown; (see Deuteromycetes, V.3.3.2. 181g). (RB)

.181 Fungus diseases: important pathogens are noted in seven classes of fungus. (R, HE)

.181a. Chytridiomycetes: mostly obligate, very primitive parasites; characterized by intracellular (i.e., develop within rather than between host cells) thalli bounded by cell walls; the fungus body consists of fine fibrils and enlarged portions that occupy host cells; colored resting zoospores (V.3.3.2. 476) formed from enlarged portions, set free by disintegration of host cells and germinate to form uniflagellate swarmspores; example: brown spot of corn (Physoderma zeae-maydis).

.181b. Plasmodiophoromycetes: primitive fungi related to slime molds; the vegetative body is a plasmodium (V.3.3.2.322) that forms spherical, smooth-walled, uninucleate, resting spores, which when set free germinate to form amoeboid, biflagellate swarmspores (holocarpic reproduction); plasmogamy (V.3.3.2.323) and karyogamy (V.3.3.2.232) are not much separated in space and time (i.e., development is rapid); new infections result from swarmspores (zoospores); example: clubroot of crucifers (Plasmodiophora brassicae).

.181c. Zygomycetes: black molds and allies; much branched, multinucleate, nonseptate mycelium, hyphae frequently inflated or restricted rather than uniform; asexual spore formation as spores in sporangia (spore cases) or conidia in conidiophores, sexual reproduction by zygospores; mostly saprophytes or weak parasites; examples: leak of potato, soft rot and ring rot of sweet potato, boll rot of cotton, vegetable rot (Rhizopus nigricans and other spp.), blossom blast, fruit rot of squash, hibiscus, dahlia blight (Choanephora spp.).

.181d. Oomycetes: downy mildews and allies; mostly obligate parasites with well developed, nonseptate (coenocytic) mycelium which is inter- or intracellular and diploid; asexual reproduction is by conidia (zoosporangia) with an infection hypha as the first product of germination; sexual reproduction is by unequal and dissimilar gametes, a large (female) oogonium that produces a uni- or multinucleate gamete, and a smaller antheridium that gives rises to several male gametes (meiosis occurs in the gametangia, not in the germinating oospore); examples: damping- off, stem rot (Pythium debaryanum); late blight of potato (Phytophthora infestans); brown rot, pod rot, foot and root rot, crown and trunk canker of citrus, deciduous fruits, black walnut, cacao, buckeye rot of tomato fruits (Phytophthora citrophthora, P. parasitica, and/or P. palmivora), avocado root rot (P. cinnamomi), black shank of tobacco (P. nicotianae), omnivorous (damping-off, blight, crown rot, fruit rot) disease (P. cactorum); downy mildews of grape (Plasmopara viticola), cucurbits (Peronoplasmopara cubensis), grasses (Sclerospora macrospora), lettuce (Bremia lactucae), and crucifers, beet, rose (Peronospora spp.), blue mold of tobacco (P. hyoscamyi), etc.

.181e. Ascomycetes: powdery mildews and some anthracnoses, scabs, blights, etc.; a diverse group of obligate parasites characterized by separate mycelia (haploid) with conidia, septate mycelia with antheridia (haploid) and conidia, septate mycelia with antheridia (haploid; V.3.3.2.017), and ascogonia and trichogynes (haploid; V.3.3.2.437) with plasmogamy (V.3.3.2.323), or binucleate nuclei (each haploid) which produce asci with karyogamy (diploid; V.3.3.2.232) and meiosis (reduction division) to yield ascospores (haploid; V.3.3.2.031); perfect (sexual) and imperfect states are known [however, the fungus is classified under Deuteromycetes (Imperfect Fungi) if the perfect state is not known, as often happens; then the anomaly of having different species of a genus listed in two separate classes]; examples are listed in four groups according to type of fruiting body: (1) none: peach leaf curl (Taphrina deformans), pear leaf blister (T. bullata), blister, leaf curl, witches' broom of ferns, poplar, birch, alder, cherry (T. spp.); (2) apothecia (V.3.3.2.022): brown rot of stone fruits, pome fruits (Sclerotinia spp.), cottony rot, drop, of lettuce, celery, other vegetables, lemon (S. sclerotiorum), etc.; (3) perithecia (V.3.3.2.298): ergot (Claviceps purpurea), South American leaf blight of rubber (Dothidella ulei), heart rot of pineapple, sugar cane [Ceratocystis (= Cerastomella) paradoxa], cacao (Ceratostomella) wilt (Ceratocystis fimbriata), dieback of peach, other stone fruits (Valsa leucostoma), bitter rot, ripe rot of apple (Glomerella cingulata), root rots (Rosellinia spp.), dry rot of sweet potato (Diaporthe batatis), melanose, phomopsis stem-end rot of citrus (D. citri), chesnut blight (Endothia parasitica), black rot of grape (Guignardia bidwellii), apple scab (Venturia inaequalis), sour orange scab (Elsinoe fawcettii), leaf spot (blight, ring spot, black rot, dry heart rot) of strawberry, Rubus, Ribes, cucurbits, beets, peas, crucifers (Mycosphaerella spp.), sigatoka (leaf spot) of banana (M. musicola = Cercospora musae), cane blight of rose, Rubus, and apple (Leptosphaeria coniothyrium; (4) cleistothecia (V.3.3.2.101): powdery mildews of grape (Uncinula necator), cucurbits (Erysiphe cichoracearum), composites (Sphaerotheca fuliginea), apple (Podosphaeria leucotricha); black spot of roses (Diplocarpon rosae), root rot of tobacco, beans, peas, etc. (Thielavia basicola), and sooty mold of citrus (Capnodium citri).

.181f. Basidiomycetes: smuts; rusts; damping-off, root rots; galls; mushrooms, etc.; a diverse group of parasites characterized by basidia (V.3.3.2.048) as organs of karyogamy and meiosis, basidiospores, and several other types of spore as inocula and hyphal fusion by plasmogamy; (1) smuts: obligate parasites with a stage that develops independently of the host, spore masses generally black, often breaking into fine, dustlike powder, also sporidia (V.3.3.2.402); examples: common smut of corn (Ustilago zeae), loose smut of oats (U. avenae), onion smut (Urocystis cepulae), white smut of dahlia (Entyloma dahliae); (2) rusts: ubiquitous parasites that attack nearly all groups of economic plants; like smuts, many rusts have a sporulating stage independent of the host and in certain species the development of heteroecism (separation of spore forms on two separate, unrelated hosts; typical rusts have polymorphic spores with a succession of five types during the life cycle (pycniospores, V.3.3.2.348; aeciospores, V.3.3.2.003; urediniospores, V.3.3.2.442; teliospores, V.3.3.2.423; and basidiospores, V.3.3.2.048); some species, however, may omit one or more spore forms (i.e., be short-cycle); examples: stem rust of grains (Puccinia graminis, an enormous number of physiological races), apple rust (Gymnosporangium juniperi-virginianae), white-pine blister rust (Cronartium ribicola), grape rust (Phakospora vitis), fig rust (Physopella fici), coffee rust (Hemileia vastatrix), carnation rust (Uromyces caryophyllatus), snapdragon rust (Puccinia antirrhini), etc.; (3) other basidiomycetes: (a) no fruiting bodies; e.g., damping-off of seedlings (Thanatephorus cucumeris, perfect stage of Rhizoctonia solani); (b) gall-formers: galls of ericaceous plants (Exobasidium azalea, etc.); (c) mushrooms, shelf fungi, fairy clubs, e.g., apple and pear canker (Septobasidium pedicellatum), seedling diseases of garden crops (Typhula spp.), white, brown root, and heart rot (Fomes spp.), red root rot (Ganoderma pseudoferreum), mushroom root rots (Clitocybe tabescens, Armillariella mellea), and witches' broom of cacao (Marasmius perniciosus).

.181g. Deuteromycetes (Imperfect Fungi): a heterogeneous and diverse group of fungi whose only known state is asexual (i.e., they have lost the ability to produce the perfect state or are those for which it has not yet been found), including many important parasites; characterized by spores borne on conidia, directly from hyphal cells (chlamydospores), sclerotia present in some instances as structures for survival; and the several types of fruiting body are listed in the following examples: (1) None: blue mold of citrus (Penicillium italicum), blue mold of apple (P. expansum), green mold (P. digitatum), wilt, yellows of cabbage (Fusarium conglutinans), tomato wilt (F. lycopersici), early blight of potato (Alternaria solani), potato wilt, blight (F. oxysporum), banana wilt, Panama disease (F. oxysporum f. sp. cubense), cotton wilt (F. oxysporum f. sp. vasinfectum). (2) Sporodochia: brown rot of pome and stone fruits (Monilina fructicola), gray mold rot of fruits, lettuce black rot, leaf blight (Botrytis cinerea), Texas root rot of cotton (Phymatotrichum omnivorum), carnation bud rot (Sporotrichum anthophilum), potato wilt (Verticillium alboatrum), tomato leaf mold (Cladosporium fulvum), etc.; acervuli: examples: bean anthracnose (Colletotrichum lindemuthianum), anthracnose of cucurbits (C. lagenarium), blight of stone fruits (Coryneum beijerinckii). (3) Pycnidia: apple blotch (Phyllosticta solitaria), apple fruit spot (Phoma pomi), tomato fruit rot (P. destructiva), citrus knot (Sphaeropsis tumefaciens), diplodia foot rot, dieback, stem-end rot of citrus, etc. (Diplodia natalensis), gladiolus hard rot (Septoria gladioli), etc.

.182 Fusarium conglutinans: (see Deuteromycetes, V.3.3.2.181g).
.183 Fusarium lycopersici: (see Deuteromycetes, V.3.3.2.181g).
.184 Fusarium oxysporum: (see Deuteromycetes, V.3.3.2.181g).
.185 Fusarium oxysporum f. sp. cubense: (see Deuteromycetes, V.3.3.2.181g).
.186 Fusarium oxysporum f. sp. vasinfectum: (see Deuteromycetes, V.3.3.2.181g).
.187 Gall: tumefaction or tumor; hyperplastic symptom of plant disease

characterized by local swelling or outgrowth of tissue composed of unorganized cells; (see Tumefaction, V.3.3.2.438). (RB)
.188 Galls of ericaceous plants: (see Basidiomycetes, V.3.3.2.181f).
.189 Ganoderma pseudoferreum: (see Basidiomycetes, V.3.3.2.181f).
.190 Gladiolus hard rot: (see Deuteromycetes, V.3.3.2.181g).
.191 Glomerella cingulata: (see Ascomycetes, V.3.3.2.181e).
.192 Grape rust: (see Basidiomycetes, V.3.3.2.181f).
.193 Gray mold rot of fruits: (see Deuteromycetes, V.3.3.2.181g).
.194 Green mold: (see Deuteromycetes, V.3.3.2.181g).
.195 Guignardia bidwellii: (see Ascomycetes, V.3.3.2.181e).
.196 Gummosis: a plant disease with the secretion of gum as a well-marked symptom.
.197 Gymnosporangium juniperi-virginianae: (see Basidiomycetes, V.3.3.2.181f).
.198 Hadromycosis: (see Tracheomycosis, V.3.3.2.435).
.199 Haustorium: absorbing organ of certain parasites; those of plant-parasitic fungi enter plant cells and invaginate the cytoplast as they enlarge into simple or branched structures for absorption of food. (RB)
.200 Heart rot of sugar cane: (see Ascomycetes, V.3.3.2.181e).
.201 Heart rot of pineapple: (see Ascomycetes, V.3.3.2.181e).
.202 Heart rots: (see Basidiomycetes, V.3.3.2.181f).
.203 Hemileia vastatrix: (see Basidiomycetes, V.3.3.2.181f).
.204 Heteroecious: of rust fungi (Uredinales, see Basidiomycetes, V.3.3.2.181f), which requires more than one host for a complete developmental cycle. (RB)
.205 Heterotopy: hyperplastic symptom in which an organ develops in a position other than its normal one (e.g., the development of ears in the tassels of corn plants). (RB)
.206 Heterotroph, heterotrophic: organism obtains nourishment from an outside source, as a parasite or saprophyte; (see Ascomycetes, V.3.3.2.181e). (RB)
.207 Holocarpic reproduction: in fungi, reproduction in which the entire fungal body (thallus) is segmented into spores. (RB)
.208 Holonecrotic, holonecrosis: dead tissue; (see Plesionecrosis, V.3.3.2.326). (RB)
.209 Host: a living organism from which a parasite derives its sustenance. (RB)
.210 Hydrosis, hydrotic: necrotic symptom of disease characterized by water-soaking of tissues. (RB)
.211 Hyperparasite: organism that parasitizes another parasite. (RB)
.212 Hyperplasia: overgrowth that results from an abnormal increase in the numbers of cells; (compare with Hypertrophy, V.3.3.2.214; Hypoplasia, V.3.3.2.215). (RB)
.213 Hypersensitive: giving violent local reactions to attack by a pathogen; the prompt death of tissues around the points of entry prevents further spread of infection. (CMI)
.214 Hypertrophy: overgrowth that results from an abnormal increase in the size of cells; (compare with Hyperplasia, V.3.3.2.212). (RB)
.215 Hypoplasia: underdevelopment that results from an abnormal paucity of cells, as in dwarfing; (see Hyperplasia, V.3.3.2.212). (RB)
.216 Immune: exempt from infection. (CMI)
.217 Immunity: in plants the ability to remain free from disease because of inherent structural or functional properties; (see Resistance, V.3.3.2.354). (RB)
.218 Imperfect fungi: (see Deuteromycetes, V.3.3.2.181g).
.219 Imperfect state (stage): the vegetative state in which asexual

spores (such as conidia), or no spores, are produced. (RB, CMI)
.220 Incubation: of a pathogen, its development, growth, and penetration of a plant prior to infection; or its activity within plant tissues subsequent to penetration and up to the appearance of symptoms, signs of disease, or both. (RB)
.221 Indexing: process of determining the presence of disease in a plant by transferring inoculum from the plant to another in which diagnostic symptoms develop; the second plant is termed a "test plant" or "indicator plant" (V.3.3.2.222). (RB)
.222 Indicator plant: one that reacts to certain viruses or environmental factors with specific symptoms and is used for identification of the viruses or the environmental factors. (CMI)
.223 Infect, infection: of a parasite or pathogen; to enter and grow, or to replicate within plant tissues; the relationship is simply parasitic if the association is noninjurious; the relationship is pathogenic as well as parasitic if it is injurious. (RB)
.224 Infectious: refers to a pathogen that can be transmitted from one suscept to another by an external agent; also refers to any disease whose pathogen can be so transmitted. (RB)
.225 Infest, infestation: to be present in numbers; infestation does not imply disease and is not to be confused with infection. (RB)
.226 Inoculum: the portion of a pathogen that may be transferred to an infection court (e.g., a fungal spore, a bacterial cell). (RB)
.227 Intumescence: hyperplastic symptom characterized by blisterlike swelling (due to water excess) on the surface of plant organs. (RB)
.228 In vitro: refers to biological processes that are allowed to occur (or are carried out) in isolation from the whole organism. (RB)
.229 In vivo: refers to biological processes that occur within the living organism. (RB)
.230 Isogamy: the condition in which gametes are morphologically similar. (RB)
.231 Isolate: (1) (verb) to separate a microorganism from host or substrate and establish it in pure culture; (2) (noun) a single spore or pure culture and the subcultures derived from it; also applicable to viruses. (CMI)
.232 Karyogamy (caryogamy): fusion of two haploid nuclei. (RB)
.233 Klendusity: special kind of disease escape, in which a susceptible plant avoids disease because of an intrinsic property of the plant itself that greatly reduces the chances of its being inoculated, even though there may be an abundance of inoculum in the area. (RB)
.234 Late blight of potato: (see Oomycetes, V.3.3.2.181d).
.235 Leaf curl of ferns, alder, birch, cherry, poplar: (see Ascomycetes, V.3.3.2.181e).
.236 Leaf spot of banana: (see Ascomycetes, V.3.3.2.181e).
.237 Leaf spot of strawberry, Rubus, Ribes, cucurbits, beets, peas, crucifers: (see Ascomycetes, V.3.3.2.181e).
.238 Leak of potato: (see Zygomycetes, V.3.3.2.181c).
.239 Leptosphaeria coniothyrium: (see Ascomycetes, V.3.3.2.181e).
.240 Lesion: localized area of disease tissue (e.g., a spot, a scab). (RB)
.241 Lettuce black rot: (see Deuteromycetes, V.3.3.2.181g).
.242 Lettuce leaf blight: (see Deuteromycetes, V.3.3.2.181g).
.243 Life cycle (life history): the complete succession of changes undergone by an organism during its life; a new cycle occurs when an identical succession of changes is initiated. (RB)

V.3.3.2. PROPAGATION, NURSERY HANDLING, SOILS, CROP PRODUCTION

.244 Loose smut of oats: (see Basidiomycetes, V.3.3.2.181f).
.245 Lyophilization: long-term preservation of viable spores, etc., by quick freezing at low temperatures and desiccation under high vacuum. (CMI)
.246 Marasmius perniciosus: (see Basidiomycetes, V.3.3.2.181f).
.247 Masked virus: one carried by a plant that does not show symptoms of its presence. (CMI)
.248 Mechanical inoculation: of plant viruses, a method of experimentally transmitting the pathogen from plant to plant; juice from a diseased plant is rubbed on test-plant leaves that usually have been dusted with carborundum or some other abrasive material. (RB)
.249 Melanose of citrus: (see Ascomycetes, V.3.3.2.181e).
.250 Metaplasia: changed condition of a structure or organ; a hyperplastic class of symptoms characterized by overdevelopment other than that due to hypertrophy or hyperplasia (e.g., abnormal starch accumulation, virescence). (RB)
.251 Microbe: microscopic organism. (RB)
.252 Microorgaanism: microscopic organism. (RB)
.253 Mildew: a plant disease in which the pathogen is a growth on the host surface; a powdery mildew is caused by one of the Erysiphaceae (see Ascomycetes, V.3.3.2.181e), a downy mildew, by one of the Peronosporaceae (see Oomycetes, V.3.3.2.181d), a dark (sooty) mildew, by one of the Meliolaceae or Capnodiaceae; (see Ascomycetes, V.3.3.2.181e). (CMI)
.254 Mold (mould): a microfungus with a well-marked mycelium or spore mass of varying color; superficial fungal growth on various substrates. (RB, CMI)
.255 Monilinia fructicola: (see Deuteromycetes, V.3.3.2.181g).
.256 Mosaic: symptom of certain viral diseases of plants; a patchy pattern of green and light green or yellow in leaves or fruit that are normally green. (RB)
.257 Mottle: arrangement of spots or confluent blotches of color, often symptomatic of virus diseases. (CMI)
.258 Mummy: a dried shriveled fruit (e.g., a peach fruit infected with Monilinia fructicola); (see Deuteromycetes, V.3.3.2.181g). (RB)
.259 Mushroom (toadstool): fleshy fruiting body of a fungus, especially of a basidiomycete (V.3.3.2.181f) of the family Agaricaceae. (RB)
.260 Mushroom root rots: (see Basidiomycetes, V.3.3.2.181f).
.261 Mutualistic symbiosis (mutualism): symbiosis beneficial to both members of the association. (RB)
.262 Mycoplasma: microscopic organisms now classed as viruses. (R)
.263 Mycoplasmalike organisms (MLOs): "bacteria without cell walls," of which about 70-odd plant pathogens are known (e.g., dwarf disease of mulberry trees; also similar to "yellows" viruses); now classed as viruses. (R)
.264 Mycorrhiza: literally, "fungus root," an association of fungal mycelium with the roots of a higher plant, from which association the latter may derive benefit; the relation may be one of mutualistic symbiosis; may be ectotrophic, on the surface of the roots; endotrophic, within the roots; pseudotrophic, in which the fungus is parasitic; or tolypophagous, in which the fungus is killed and digested by the host. (CMI)
.265 Mycosphaerella musicola: (see Ascomycetes, V.3.3.2.181e).
.266 Mycosphaerella spp.: (see Ascomycetes, V.3.3.2.181e).
.267 Mycotrophic: refers to green plants with mycorrhizae. (RB)
.268 Myxomycetes (Mycetozoa): the slime molds, a class of fungi charac-

terized by ameboid vegetative protoplasts, plasmodia, and by brightly colored spore-bearing capillitia (knobs). (RB)
.269 Necrosis: death of plant cells, especially when tissues are darkening, a common symptom of fungus infection. (CMI)
.270 Nonparasitic diseases: disorders in which no primary parasite (fungus, bacterium, virus, RLO, MLO, insect, mite, slug, snail, etc.) is involved; brought about by abnormal conditions of temperature (e.g., freezing, chilling injury), light (sunburn, sunscald, etc.), atmospheric impurities (acid rain, pollutant damage), soil moisture disturbances, nutritional imbalances (e.g., relative proportions of certain elements to others, deficiencies, and toxicities), lightning injury (actually quite common in fields and orchards), high winds (blow downs), or effects of toxic chemicals (i.e., sprays or dusts, especially herbicides), among others; many of those that occur in the field (garden or orchard) are causal agents themselves (thus directly related) of or contributory agents (i.e., indirectly related) to numerous physiological disorders associated with harvesting and subsequent handling of fresh produce; (see Market diseases, VI.9.3). (S & G)
.271 Obligate: refers to an organism that is restricted to a particular set of environmental conditions without which it cannot survive; an obligate parasite is an organism that can function only by parasitizing another organism; (see Facultative parasite, V.3.3.2.168, Facultative saprophyte, V.3.3.2.169). (RB)
.272 Obligate parasite: capable of living only as a parasite. (CMI)
.273 Oedema (edema): intumescence or blister formation due to an increase in intercellular water, as in leaves. (RB)
.274 Olive tubercle: (see Bacterial diseases, V.3.3.2.040).
.275 Omnivorous: [of plant (or animal) parasites], attacking a number of different hosts. (CMI)
.276 Omnivorous disease: (see Oomycetes, V.3.3.2.181d).
.277 Onion smut: (see Basidiomycetes, V.3.3.2.181f).
.278 Oogonium: in fungi, the female reproductive structure of oomycetes such as Pythium spp. and Phytophthora spp. (V.3.3.2.181d). (RB)
.279 Oomycetes: a class of fungi (the "algal fungi") characterized by coenocytic mycelia (thalli that consist of multinucleate masses of protoplasm) and sexual reproduction by unlike cells; (see also V.3.3.2.181d and .477). (R)
.280 Oospore: a spore formed after the fertilization of a large, passive female cell (oogonium) by a small, active male cell. (RB)
.281 Panama disease: (see Deuteromycetes, V.3.3.2.181g).
.282 Parasexual cycle: a mechanism by which recombination of hereditary properties is based on mitosis. (CMI)
.283 Parasite: an organism that obtains nourishment from cells of another living organism, the host, while contributing nothing to the host's survival. (RB)
.284 Pathogen: casual agent of disease in a plant; the suscept. (RB)
.285 Pathogenesis: that portion of the life cycle of a pathogen during which it becomes, and continues to be, associated with its suscept. (RB)
.286 Pathogenicity: the ability of a pathogen to cause disease. (RB)
.287 Peach leaf curl: (see Ascomycetes, V.3.3.2.181e).
.288 Pear canker: (see Basidiomycetes, V.3.3.2.181f).
.289 Pear leaf blister: (see Ascomycetes, V.3.3.2.181e).
.290 Penicillium digitatum: (see Deuteromycetes, V.3.3.2.181g).
.291 Penicillium expansum: (see Deuteromycetes, V.3.3.2.181g).
.292 Penicillium italicum: (see Deuteromycetes, V.3.3.2.181g).

V.3.3.2. PROPAGATION, NURSERY HANDLING, SOILS, CROP PRODUCTION 525

.293 Perfect state(stage): of a fungus; that part of the life cycle during which spores are produced sexually, in contrast to the imperfect state that refers to the remainder of the life cycle (or in some fungi is the only state known; known as Fungi Imperfecti = Deuteromycetes). (RB)
.294 Peridermium (pl. peridermia): an aecium, of the sort found in Cronartium, Melampsora, and Coleosporium, with a peridium that extends prominently beyond the chains of aeciopores; (see Basidiomycetes, V.3.3.2.181f and Roestelium, V.3.3.2.366). (RB)
.295 Perithecium (pl. perithecia): a spherical or flask-shaped, thick-walled fruiting body of ascomycetes (V.3.3.2.181e) consists of an internal hymenium of asci and paraphyses (sterile filaments) and an ostiole through which ascospores are discharged. (RB)
.296 Peronoplasmopara cubensis: (see Oomycetes, V.3.3.2.181d).
.297 Peronospora hyoscamyi: (see Oomycetes, V.3.3.2.181d).
.298 Peronospora spp.: (see Oomycetes, V.3.3.2.181d).
.299 Persistent virus: one that remains infective within its insect vector for a long period. (CMI)
.300 Phakospora vitis: (see Basidiomycetes, V.3.3.2.181f).
.301 Phoma destructiva: (see Deuteromycetes, V.3.3.2.181g).
.302 Phoma pomi: (see Deuteromycetes, V.3.3.2.181g).
.303 Phomopsis citri: imperfect state of Diaporthe citri; (see Ascomycetes, V.3.3.2.181e).
.304 Phomopsis stem-end rot of citrus: (see Ascomycetes, V.3.3.2.181e).
.305 Phycomycetes: a class of fungi that formerly contained oomycetes (V.3.3.2.279) and zygomycetes (V.3.3.2.477). (R)
.306 Phyllody: a change of floral leaves (petals) to foliage leaves.
.307 Phyllosticta solitaria: (see Deuteromycetes, V.3.3.2.181g).
.308 Phymatotrichum omnivorum: (see Deuteromycetes, V.3.3.2.181g).
.309 Physiologic race: a biotype (i.e., one of a group of forms alike in morphology but unlike in certain cultural, physiological, biochemical, pathological, or other characters) or groups of closely related biotypes that differ from other biotypes of that species in the ability to infect particular varieties of the susceptible plant species; (see Forma specialis, V.3.3.2.177). (RB)
.310 Physoderma zeae-maydis: (see Chytridiomycetes, V.3.3.2.181a).
.311 Physopella fici: (see Basidiomycetes, V.3.3.2.181f).
.312 Phytoalexin: in hypersensitive hosts, a metabolic by-product, toxic to both parasite and host, that is formed only in response to attack by the parasite. (RB)
.313 Phytophthora cactorum: (see Oomycetes, V.3.3.2.181d).
.314 Phytophthora cinnamomi: (see Oomycetes, V.3.3.2.181d).
.315 Phytophthora citrophthora: (see Oomycetes, V.3.3.2.181d).
.316 Phytophthora infestans: (see Oomycetes, V.3.3.2.181d).
.317 Phytophthora nicotianae: (see Oomycetes, V.3.3.2.181d).
.318 Phytophthora palmivora: (see Oomycetes, V.3.3.2.181d).
.319 Phytophthora parasitica: (see Oomycetes, V.3.3.2.181d).
.320 Plasmid: an extrachromosomal genetic element, i.e., an extra loop of DNA, found in certain bacteria; e.g., Agrobacterium tumefaciens, the casual agent of crown gall (V.3.3.2.040). (R)
.321 Plasmodiophora brassicae: (see Plasmodiophoromycetes, V.3.3.2.181b).
.322 Plasmodium: a multinucleate mass of naked protoplasm formed by the union of a number of amoebalike organisms; the vegetative body of a slime mold formed by the union of myxamoebae developed from spores. (R)
.323 Plasmogamy: the initiation of the diploid phase in the life cycles

- .324 Plasmopara viticola: (see Oomycetes, V.3.3.2.181d).
- .325 Pleomorphism (polymorphism): the occurrence of several forms in the life cycle (e.g., many rust fungi are pleomorphic in that they produce as many as five different spore forms in their complete life cycle). (RB)
- .326 Plesionecrosis: symptom exhibited by tissues not yet dead but in the process of dying (e.g., wilting). (RB)
- .327 Podosphaeria leucotricha: (see Ascomycetes, V.3.3.2.181e).
- .328 Pod rot of cacao: (see Oomycetes, V.3.3.2.181d).
- .329 Potato fusarium blight: (see Deuteromycetes, V.3.3.2.181g).
- .330 Potato fusarium wilt: (see Deuteromycetes, V.3.3.2.181g).
- .331 Potato ring rot: (see Bacterial diseases, V.3.3.2.040).
- .332 Potato wilt: (see Deuteromycetes, V.3.3.2.181g).
- .333 Powdery mildew of apples: (see Ascomycetes, V.3.3.2.181e).
- .334 Powdery mildew of cucurbits, composites: (see Ascomycetes, V.3.3.2.181e).
- .335 Powdery mildew of grape: (see Ascomycetes, V.3.3.2.181e).
- .336 Powdery mildews: (see Ascomycetes, V.3.3.2.181e).
- .337 Primary cycle: the first cycle of a pathogen initiated after a period of pathological inactivity, usually after rest or seasonal inactivity (dormancy). (RB)
- .338 Primary symptom: in a virus disease the first to appear in which case more than one type of symptom may be produced. (CMI)
- .339 Prokaryote, prokaryotic organisms: primitive single-celled plants with a single chromosome that is not within a membrane-bound nucleus, in contrast to eukaryotes in which the chromosomes are within a membrane-bound nucleus in each cell; pathogenic prokaryotes include mycoplasmalike organisms (MLOs), rickettsialike organisms (RLOs), bacteria, and actinomycetes. (R)
- .340 Prolepsis: a hyperplastic symptom of disease in which organs appear before the natural time (e.g., the sprouting of shoots from adventitious buds after disease has impaired the metabolism of the organ in question). (RB)
- .341 Proliferation: rapid and repeated production of new cells, tissues, or organs; specifically, a hyperplastic symptom of plant disease in which organs continue to develop after they have reached the point beyond which they normally do not grow. (RB)
- .342 Pseudomonas spp.: (see Bacterial diseases, V.3.3.2.040).
- .343 Pseudotrophic: (see Mycorrhiza, V.3.3.2.264).
- .344 Puccinia antirrhini: (see Basidiomycetes, V.3.3.2.181f).
- .345 Puccinia graminis: (see Basidiomycetes, V.3.3.2.181f).
- .346 Pustule: a pimplelike, eruptive fruiting structure, such as the uredinium of a rust fungus (V.3.3.2.181f). (RB)
- .347 Pycnidium: hard-walled, flask-shaped, fungal fruiting body that contains conidia (see Basidiomycetes, V.3.3.2.181f). (RB)
- .348 Pycniospore (pl.pycnia): (see Pycnium, V.3.3.2.345).
- .349 Pycnium (pl. pycnia): a spermagonium; a flask-shaped fruiting body of a rust fungus; contains haploid pycniospores (spermatia) and filaments (receptive hyphae) that extend through the ostiole (pore); a pycnium and its contents are of one mating type. (RB)
- .350 Pythium debaryanum: (see Oomycetes, V.3.3.2.181d).
- .351 Race: (1) a genetically and, as a rule, geographically distinct mating group within a species; (2) physiological race. (CMI)
- .352 Range: of a plant pathogen; the geographical region or regions in

V.3.3.2. PROPAGATION, NURSERY HANDLING, SOILS, CROP PRODUCTION

which it is known to occur. (RB)
.353 Red root rots: (see Basidiomycetes, V.3.3.2.181f).
.354 Resistance: the ability or power of a plant to remain relatively unaffected by a disease because of its inherent genetic and physiological or structural characteristics. (RB)
.355 Response: the change produced in an organism by a stimulus. (RB)
.356 Resting spore: fungal spore, usually thick-walled, that can remain viable in a dormant condition for an extended period (examples: chlamydospores, oospores, amphiospores, and teliospores). (RB)
.357 Rhizoctonia solani: imperfect state of Thanatephorus cucumeris; (see Basidiomycetes, V.3.3.2.181f).
.358 Rhizomorph: compact strand of hyphae formed by the longitudinal joining of hyphal strands into a bundle; it has a hard outer covering, grows from the apex, and serves as a survival organ and in the transport of food materials within the thallus. (RB)
.359 Rhizopus nigricans: (see Zygomycetes, V.3.3.2.181c).
.360 Rhizopus spp.: (see Zygomycetes, V.3.3.2.181c).
.361 Rickettsialike organisms (RLO's): obligately parasitic, small bacteria with cell walls usually scalloped; most invade xylem tissues (e.g., Pierce's disease of grape) but a few invade phloem tissues (e.g., club leaf of clover). (R)
.362 Ring rot of sweet potato: (see Zygomycetes, V.3.3.2.181c).
.363 Ring spot: a circular area of chlorosis; the center remains green, symptomatic of many virus infections. (CMI)
.364 Ring spot of strawberry, Rubus, Ribes, cucurbits, beets, peas, crucifers: (see Ascomycetes, V.3.3.2.181e).
.365 Ripe rot of apple: (see Ascomycetes, V.3.3.2.181e).
.366 Roestelium (pl. roestelia): in Gymnosporangium spp. an aecium with a strongly developed peridium (wall) extending beyond the chains of aeciospores; (see Peridermium, V.3.3.2.294). (RB)
.367 Root rot of tobacco, beans, peas, etc.: (see Ascomycetes, V.3.3.2.181e).
.368 Root rots: (see Ascomycetes, V.3.3.2.181e).
.369 Rosellina spp.: (see Ascomycetes, V.3.3.2.181e).
.370 Russeting: hyperplastic symptom of disease in which brownish, roughened areas form on the skin of fruit or tubers because of excessive cork-cell production. (RB)
.371 Rust: disease caused by a rust fungus (order Uredinales); also, the fungus itself. (RB)
.372 Rusts: (see Basidiomycetes, V.3.3.2.181f).
.373 Saltation: mutation occurring in the asexual stage of fungal growth, especially in culture (i.e., in an isolate known to be a pure genotype); dissociation. (RB)
.374 Saprogenesis: survival; that phase of the life cycle of a pathogen during which it is not actively causing disease in a living suscept. (RB)
.375 Sarcody: a hyperplastic symptom in which swellings occur above and below portions or organs that are tightly encircled; e.g., a stem might be "choked" by a twining vine. (RB)
.376 Scab: hyperplastic symptom characterized by hard, crusty lesions formed by excessive cork formation. (RB)
.377 Sclerospora macrospora: (see Oomycetes, V.3.3.2.181d).
.378 Sclerotinia sclerotiorum: (see Ascomycetes, V.3.3.2.181e).
.379 Sclerotinia spp.: (see Ascomycetes, V.3.3.2.181e).
.380 Sclerotium: hard, compact mass of fungal tissue that consists of an outer sclerotized rind and an inner parenchymatous medulla (central part of organ); capable of surviving long periods of adverse

environmental conditions. (RB)
.381 Secondary cycle: a cycle initiated by inoculum from a primary (or another secondary) cycle without an interposed resting or dormant period for the pathogen. (RB)
.382 Secondary symptom: in a virus disease one following the primary symptom when more than one type of symptom is produced. (CMI)
.383 Seedling disease of garden crops: (see Basidiomycetes, V.3.3.2.181f).
.384 Sensitive: reacting with severe symptoms to the attack of a given pathogen. (CMI)
.385 Sensitivity: the tendency of an organism attacked by a disease to show more or less strong symptoms. (CMI)
.386 Septobasidium pedicellatum: (see Basidiomycetes, V.3.3.2.181f).
.387 Septoria gladioli: (see Deuteromycetes, V.3.3.2.181g).
.388 Shot hole: symptom characterized by the dropping out of roundish fragments of leaf tissue attacked by certain leaf-spotting organisms. (RB)
.389 Sigatoka disease of banana: (see Ascomycetes, V.3.3.2.181e).
.390 Smut: disease caused by smut fungi (order Ustilaginales); also, the fungus itself. (RB)
.391 Smuts: (see Basidiomycetes, V.3.3.2.181f).
.392 Snapdragon rust: (see Basidiomycetes, V.3.3.2.181f).
.393 Soft rot of carrots: (see Bacterial diseases, V.3.3.2.040).
.394 Soft rots: (see Zygomycetes, V.3.3.2.181c).
.395 Sooty mold: (1) see dark mildew; (2) Capnodiaceae growing on honeydew excreted by aphids; (see Ascomycetes, V.3.3.2.181e). (CMI)
.396 Sooty mold of citrus: (see Ascomycetes, V.3.3.2.181e).
.397 Sour orange scab: (see Ascomycetes, V.3.3.2.181e).
.398 South American leaf blight of rubber: (see Ascomycetes, V.3.3.2.181e).
.399 Sphaeropsis tumefaciens: (see Deuteromycetes, V.3.3.2.181g).
.400 Sphaerotheca fuliginea: (see Ascomycetes, V.3.3.2.181e).
.401 Spore: of fungi; a one-to-many-celled reproductive unit that becomes detached from the parent and can germinate to give rise to a new individual. (RB)
.402 Sporidium (pl. sporidia): a small spore produced on a promycelium, as in the smuts and rusts; (see Basidiomycetes, V.3.3.2.181f). (RB)
.403 Sporodochium (pl. sporodochia): of fungi; an asexual fruiting body in which conidiospores develop over the surface of an erumpent, cushion-shaped fungal structure. (RB)
.404 Sporotrichum anthophilum: (see Deuteromycetes, v.3.3.2.181g).
.405 Spot: symptom of disease characterized by a limited necrotic area, as on leaves, flowers, and stems. (RB)
.406 State: (1) (of fungi) one phase of a pleomorphic fungus; e.g., the imperfect state characterized by asexual spores, the perfect state characterized by sexual spores; (2) (of bacteria) the name given to the rough, smooth, mucoid, and similar variants that arise in culture. (CMI)
.407 Stem rot: (see Oomycetes, V.3.3.2.181d).
.408 Stem rust of grains: (see Basidiomycetes, V.3.3.2.181f).
.409 Sterile: unable to reproduce sexually; also taken by many to mean free from living microorganisms. (RB)
.410 Sterilized: free from living microorganisms; axenic. (RB)
.411 Stimulus: an environmental change capable of inciting a change in the activities of an organism without itself providing energy for the new activities; (see Response, V.3.3.2.355). (RB)

V.3.3.2. PROPAGATION, NURSERY HANDLING, SOILS, CROP PRODUCTION

.412 Strain: (many meanings) (1) a group of similar isolates; race; form; (2) the descendants of a single isolation in pure culture; isolate; (3) (of bacteria) a cultivar; (4) (of plant viruses) a group of viruses with most of its antigens in common with another group (strain). (CMI)
.413 Suppression: hypoplastic symptom charactrized by the failure of plant organs or substances to develop. (RB)
.414 Suscept: any plant or species of plant that is susceptible to disease; an abbreviated term denoting "susceptible plant" or "susceptible species." (RB)
.415 Symbiont (symbiote): one member of a symbiotic relationship. (RB)
.416 Symbiosis: a vital association of two dissimilar organisms, often used in the restricted sense to mean mutualistic symbiosis. (RB)
.417 Symptom: a visible expression by a suscept of a pathological condition. (RB)
.418 Symptomology: the study of symptoms of disease and signs of pathogens for the purpose of diagnosis. (RB)
.419 Systemic: (1) (of a plant pathogen) occurring throughout a plant; (2) (of a chemical) absorbed into the plant through roots or foliage. (CMI)
.420 Taphrina bullata: (see Ascomycetes, V.3.3.2.181e).
.421 Taphrina deformans: (see Ascomycetes, V.3.3.2.181e).
.422 Taphrina spp.: (see Ascomycetes, V.3.3.2.181e).
.423 Teliospore, teleutospore: a spore of the rust fungi, usually a resting spore that germinates to produce a promycelium (analogous to a basidium, V.3.3.2.048) in which meiosis occurs. (RB)
.424 Texas root rot of cotton: (see Deuteromycetes, V.3.3.2.181g).
.425 Thanatephorus cucumeris: (see Basidiomycetes, V.3.3.2.181f).
.426 Thermal death point: the high temperature at which death of an organism occurs after a specified length of time, usually 10 minutes. (RB)
.427 Thielavia basicola: (see Ascomycetes, V.3.3.2.181e).
.428 Thielaviopsis basicola: imperfect state of Ceratocystis paradoxa; (see Ascomycetes, V.3.3.2.181e).
.429 Tolerant: able to endure infection by a particular pathogen, without showing severe disease, or giving little reaction to the effect of other factors (e.g., a virus tolerant of heat). (CMI)
.430 Tolypophagous: (see Mycorrhiza, V.3.3.2.264).
.431 Tomato fruit rot: (see Deuteromycetes, V.3.3.2.181g).
.432 Tomato leaf mold: (see Deuteromycetes, V.3.3.2.181g).
.433 Tomato wilt: (see Deuteromycetes, V.3.3.2.181g).
.434 Toxin: a toxic compound produced by a microorganism. (CMI)
.435 Tracheomycosis: a wilt disease in which the invading organism is mainly confined to the xylem = hadromycosis. (CMI)
.436 Transmission: the dissemination of pathogens and the inoculation of suscepts. (RB)
.437 Trichogyne: in some algae, lichens, and fungi a projection from the female sex organ that receives the male gamete or nuclei before fertilization (karyogamy). (RB)
.438 Tumefaction: a plant tumor or gall. (RB)
.439 Typhula spp.: (see Basidiomycetes, V.3.3.2.181f).
.440 Uncinula necator: (see Ascomycetes, V.3.3.2.181e).
.441 Urediniospore: (see Uredinium, V.3.3.2.442).
.442 Uredinium (pl. uredinia); uredium, uredosorus: in a rust fungus a group of spore-bearing filaments crowded together on which masses of red "summer spores" (urediniospores) are formed. (RB)
.443 Urocystis cepulae: (see Basidiomycetes, V.3.3.2.181f).

.444 Uromyces caryophyllatus: (see Basidiomycetes, V.3.3.2.181f).
.445 Ustilago avenae: (see Basidiomycetes, V.3.3.2.181f).
.446 Ustilago zeae: (see Basidiomycetes, V.3.3.2.181f).
.447 Valsa leucostoma: (see Ascomycetes, V.3.3.2.181e).
.448 Vascular wilt of solanaceous crops: (see Basidiomycetes, V.3.3.2.181f).
.449 Vector: (see V.3.3.1.302).
.450 Vegetable rot: (see Zygomycetes, V.3.3.2.181c).
.451 Vein banding: a symptom of virus-infected leaves in which veinal tissue is lighter green than that of healthy plants. (CMI)
.452 Vein clearing: a symptom of virus-infected leaves in which tissue is lighter than that of healthy plants. (RB)
.453 Venturia inaequalis: (see Ascomycetes, V.3.3.2.181e).
.454 Verticullium alboatrum: (see Basidiomycetes, V.3.3.2.181f).
.455 Virescence: a symptom in which green pigmentation occurs in plant tissues not normally green. (RB)
.456 Virion: an individual virus particle that may be occluded (be produced with a protein envelope) or nonoccluded (acquires a protein envelope later; e.g., from the cell membrane). (MZ)
.457 Viroid: a viruslike particle that lacks the protein coat around the core of nucleic acid; a causal agent of viruslike disorders such as exocortis of citrus.
.458 Virulence: relative ability to cause disease; a measure of pathogenicity. (RB)
.459 Virulent: strongly pathogenic. (CMI)
.460 Viruliferous: containing a virus; of an insect vector, containing virus and capable of introducing it into a suscept. (RB)
.461 Virus: an ultramicroscopic (one dimension less than 200 nm), obligately parasitic and infectious pathogen of disease; consists of a core of nucleic acid (usually RNA but occasionally DNA, as in caulimoviruses, V.3.3.2.088) in a protein coat; several hundred plant pathogens are known. (RB)
.462 Virus cross protection: infection with a mild strain of virus to protect against a severe strain; e.g., tristeza citrus virus in Brazil.
.463 Vivotoxin: a substance produced by the pathogen, its host, or both, which operates in the causation of disease but is not itself the primary agent; (see also Toxin, V.3.3.2.434). (CMI)
.464 White-pine blister rust: (see Basidiomycetes, V.3.3.2.181f).
.465 White root rot: (see Basidiomycetes, V.3.3.2.181f).
.466 White smut of dahlia: (see Basidiomycetes, V.3.3.2.181f).
.467 Wilt of cabbage: (see Deuteromycetes, V.3.3.2.181g).
.468 Wilting: plant disease, a plesionecrotic symptom characterized by loss of turgor, that results in drooping leaves, stems, and flowers. (RB)
.469 Witches' broom: a hyperplastic symptom due to proleptic development of many weak shoots from adventitious buds in a stem. (RB)
.470 Witches' broom of ferns, alder, birch, cherry, poplar: (see Ascomycetes, V.3.3.2.181e).
.471 Witches' broom of cacao: (see Basidiomycetes, V.3.3.2.181f).
.472 Xanthomonas: (see Bacterial disease, V.3.3.2.040).
.473 Yellowing: a plesionecrotic symptom characterized by the turning yellow of plant tissues that once were green; chloranemia is the necrotic symptom of yellowing, a loss of chlorophyll. (RB)
.474 Yellows: applied to plant diseases of which yellowing is a conspicuous symptom; e.g., peach yellows virus, cabbage yellows (Fusarium conglutinans). (CMI)

.475 Yellows of cabbage: (see Deuteromycetes, V.3.3.2.181g).
.476 Zoospore: a swarm-spore, a naked spore (motile by means of one to many flagella) produced within a sporangium. (RB)
.477 Zygomycetes: a class of fungi (the "algal fungi") characterized by coenocytic mycelia (thalli consists of multinucleate masses of protoplasm) and sexual reproduction by the union of two like sex cells; (see also V.3.2.2.181c and .279). (R)

V.3.3.3. Pest management

.000 Pest management: a key element in crop production and management and other components, such as mineral nutrition, plant size, crop control (training, pruning, growth regulators), and water regulation (plus replacements and rehabilitation for long-lived perennial crops); includes (1) methods of exclusion, via quarantines and importation of plant material after thorough inspection, fumigation, and an isolation period when needed; (2) eradication, the complete elimination of a pest; (3) breeding and selection of resistant or tolerant cultivars or seedling lines (i.e., budwood registration programs and use of certified or registered seeds, trees, and other plants); (4) biological control (native and imported predators); (5) sanitation, and (6) pesticides, (i.e., mulches and herbicides for weeds, fumigants for soil-borne pests, and fungicides, insecticides, and miticides for specific pests on specific crops, with appropriate inspection to ensure their most efficient use); (see Integrated pest management, V.3.3.3.055).
.001 Acaricide: (see Miticide, V.3.3.3.058).
.002 Acid equivalent (ae): theoretical yield of parent acid from the active ingredient content of a formulation. (TM)
.003 Active ingredient (ai): chemical(s) in a formulated product principally responsible for the herbicidal effects and shown as such on herbicide labels. (TM)
.004 Adaptation: from an evolutionary standpoint, a characteristic of a living organism that improves its chances for survival in the environment of its habitat; change brought about in a population of an organism as a result of exposure to a particular set of environmental conditions; the change enables the organism to adjust to the environmental conditions in question. (RB)
.005 Adjuvant: substance added to a chemical spray formulation to enhance the action of the main ingredient. (TM)
.006 Aerosol: fine liquid particles suspended in air, which may be produced by liquified gas formulations released by capillary or expansion-chamber nozzles, steam, or air atomization of liquid, spinning discs and rotors, forcing liquid under high pressure through atomizing nozzles, heat vaporization, or a combination of these methods. (UF)
.007 Antagonism: opposing action of different chemicals such that the action of one is impaired or the total effect is less than that of one component used separately. (TM)
.008 Attractant: (see Pheromone, V.3.3.3.073).
.009 Bait: a pesticide formulation in which the insecticide or rodenticide is mixed with a carrier-diluent, such as corn grits, wheat bran, sugar, peanut hulls, or other material, the carrier being attractive to the pest; the usual concentration of active ingredient ranges from 0.5 to 5%. (UF)

.010 Band treatment: applied to a continuous restricted area such as on or along a crop row. (TM)
.011 Basal herbicide treatment: applied to encircle the stem of a plant above and at the ground such that foliage contact is minimal; a term used mostly to describe treatment of woody plants. (TM)
.012 Biological control: utilization of natural or imported predators and parasites to keep pests from causing economic damage to plants or plant products; a vital part of an overall pest management program. (Note that biological control does not entail eradication of a given pest because a minimum level of the latter must be present for predators or parasites to survive.); (see also Biorational pesticides, V.3.3.3.013).
.013 Biorational pesticides: viruses, bacteria, protozoa, fungi, naturally occurring biochemicals, or insect growth regulators, etc., which may be used for their pesticidal action on other organisms (e.g., viral or bacterial infection of parasitic insects). (CO)
.013a. Cytoplasmic polyhedrosis virus: a viral disease of insects, mainly larvae of certain lepidoptera, characterized by the formation of polyhedral (or spherical) inclusion bodies in the cytoplasm of midgut epithelial cells. (MZ)
.013b. "Elcar": commercially available nucleopolyhedrosis virus (NPV) insecticide designed to control Heliothis insects. (CO)
.013c. Entomopox virus: viral infection of insects characterized by oval- or spindle-shaped inclusion bodies that contain virus particles and located in the cytoplasm of infected cells; used to control imported cabbage worm. (GR)
.013d. Granulosis: a virus disease of certain insects characterized by the presence of minute granular inclusion capsules in infected cells; used to control citrus red mite. (MZ)
.013e. Nonoccluded virus: characterized by virus particles that are not surrounded by a protein matrix; used to control citrus red mite. (MZ)
.013f. Nucleopolyhedrosis virus (NPV): a viral disease of insects, mainly larvae of certain lepidoptera (moths, butterflies) and hymenoptera (bees, wasps, and ants); characterized by the formation of polyhedral (rod-shaped) inclusion bodies in the nuclei of infected cells; multiply in the epidermis, tracheal matrix, fat body, and blood cells of lepidopteral larvae and in the midgut of hymenopteral larvae; used to control corn ear worm (Heliothis zae), gypsy moth, European spruce sawfly, cabbage looper, Douglas-fir tussock moth, coconut palm beetle, rhinoceros beetle, and, experimentally, spruce budworm. (MZ)
.014 Broadcast treatment: applied over an entire area. (TM)
.015 Brush control: herbicidal control of woody plants, such as brambles, sprout clumps, shrubs, trees, and vines. (TM)
.016 Calibration: determination of the exact amount of pesticide applied per plant or per unit area by actual experiment under conditions identical with those for commercial application; major variables include discharge rate (gallons or liters per minute), tractor speed (miles or kilometers per hour), number of nozzles, swath width (feet or meters) or tree spacing (feet or meters), and spray concentration (dilute or 6X, 8X, etc.); especially critical for herbicide applications in which overlapping of swaths may cause injury but also essential for the proper use of equipment for other pesticides. (HOBJ)
.017 Carrier: gas, liquid, or solid used to dilute, propel, or suspend a pesticide during its application. (TM)

V.3.3.3. PROPAGATION, NURSERY HANDLING, SOILS, CROP PRODUCTION

.018 Compatible: mixable in the pesticide formulation or spray tank for application in the same carrier without undesirably altering the separate effects of the components. (TM)
.019 Compatibility: several forms of incompatibility in mixtures of pesticides; (a) toxicity or length of residual effectiveness is reduced; (b) causes injury to plants (i.e., phytotoxic) as a mixture but individual components used alone are safe; (c) loss of effectiveness when left standing in a tank too long. (UF)
.020 Concentration: amount of active ingredient or pesticide equivalent in a quantity of diluent, usually expressed as a percentage. (TM)
.021 Contact herbicide: one that is phytotoxic by contact with plant tissue rather than as a result of translocation. (TM)
.022 Control: reduction of crop losses. (RB)
.023 Coverage: distribution of a pesticide over a continuous area like the leaves of a plant. (CMI)
.024 Defoliant: chemical or method of treatment that causes only the leaves of a plant to fall off or abscise; the fruit remains attached. (TM)
.025 Desiccant: any substance or mixture of substances used to accelerate the drying of plant tissue. (TM)
.026 Diluent: any essentially inert gas, liquid, or solid used to reduce the concentration of the active ingredient in a formulation; usually water. (TM)
.027 Directed herbicide application: precise application to a specific area or plant organ, such as a row or bed or the lower leaves and stems. (TM)
.028 Disinfest: to kill pathogens that have not yet initiated disease but occur in or on such inanimate objects as soil, tools, etc., or occur on the surface of such plant parts as seeds. (RB)
.029 Dust: pesticide formulation into a fine powder with a carrier-diluent, such as talc, clay, calcium carbonate, pyrophyllite, diatomaceous earth, or sulfur. (UF)
.030 Duster: a device for the application of a pesticide in powdered form; many different forms include hand-operated units (hand-held or knapsack), tractor-drawn units, and aircraft.
.031 Early postemergence: a pesticide (e.g., a herbicide) applied during the cotyledonary growth phase of the crop or weed seedlings. (TM)
.032 Emergence treatment: pesticide applied during the visible emerging phase of the specified crop or weed. (TM)
.033 Emulsifiable concentrate (ec): a product made by dissolving a technical grade pesticide in a solvent like xylene and adding an emulsifier like Triton X-155; usually contains 20 to 75% actual insecticide; mixed and diluted with water to form an emulsion spray. (TM)
.034 Emulsifier: surface-active substance that promotes the suspension of one liquid in another. (TM)
.035 Emulsion: suspension of one liquid as minute globules in another (e.g., oil dispersed in water). (TM)
.036 Entomopox virus: (see Biorational pesticides, V.3.3.3.013c).
.037 Eradication: a principle of plant-disease prevention characterized by the destruction or removal of a pathogen already established in a given area; (applies also to animal pests). (RB)
.038 Exclusion: the principle of plant-disease prevention in which the pathogen is prevented from entering a given region (e.g., by seed disinfestation or quarantine); (also applicable to animal pests). (RB)
.039 Fertilizer-pesticide mixture: a carefully blended (for uniformity)

formulation that contains fertilizer and insecticide or fungicide for control of certain pests; used for the same purposes as separate fertilizer and pesticide applications, which are often preferred because of difficulties in blending and stability and flow characteristics of a prepared mixture. (UF)

.040 Filler: diluent in powder form. (CMI)
.041 Foliar application: application of a herbicide to the foliage of plants. (TM)
.042 Formulation: (a) a pesticide or fertilizer preparation supplied by a manufacturer; (b) the process carried out by manufacturers in the preparation of pesticide-fertilizer mixtures. (TM)
.043 Fumigant: pesticides that are usually liquids at low temperatures or under pressure and will volatilize to form a gas when heated or released from pressure; may be applied to airtight enclosures or to the soil, the latter often being covered with a vapor-impermeable barrier to retain the gas for the required period; certain fumigants with low volatility may be formulated as granules which may be applied to the soil and covered in some manner (e.g., disked into the soil). (UF)
.044 Fungicide: a substance that kills or inhibits fungus spores or mycelium; <u>eradicant</u> <u>fungicide</u>: (a) a fungicide applied to a substrate in which the fungus is already present; (b) a fungicide used in disease control after infection has been established; <u>protective</u> <u>fungicide</u>: a fungicide used as a protectant. (CMI)
.045 Fungistatic: able to prevent the growth or development of fungus spores without killing them or to prevent fungus growth without being fungicidal. (CMI)
.046 Granule, granular: dry formulation of pesticide and/or other components in discrete particles generally less than 10 mm ; the carrier is clay or some other essentially inert material; usually formulated for ground (or aerial) application. (FC)
.047 Granulosis: (see Biorational pesticides, V.3.3.3.013d).
.048 Grazing: use of animals for weed control; many disadvantages, such as likelihood of crop damage from browsing, soil compaction by the hooves of animals (i.e., horses, cattle, sheep, or goats), likelihood of erosion from overgrazing, and the limited quantity of nutrients contained in animal manure; geese, however, have proved successful (provided that plantings are well fenced and predators kept away) for ornamentals in containers, pecans, and the like.
.049 Herbicides: a popular means of killing weeds by the use of a chemical or mixture of chemicals applied to plant surfaces, the soil, or both, in areas in which labor is scarce (or expensive); many different formulations, whose effectiveness in any given situation depends on the crop, composition of the weed flora, time of year, rainfall, soil type and dosage, and coverage for best control; continued use results in a gradual shift in weed-flora composition to a higher proportion of more resistant species, such as grasses and sedges which replace the broad-leaved forms.
.050 Humectant: material with a high water-attracting capacity.
.051 Insecticide: a substance that kills insects, usually being most effective when applied to a juvenile stage (i.e., egg or larva) rather than to the adults (imagines).
.052 Inspection (scouting): the process of examining plants or samples thereof in a field, orchard, grove, or nursery carefully and thoroughly to determine which pests (specific diseases, insects, mites, etc.) are present, their numbers, stage of development,

V.3.3.3. PROPAGATION, NURSERY HANDLING, SOILS, CROP PRODUCTION

and potential or actual damage to the crop or crops; this is a necessary adjunct to integrated pest management; however, it is standard practice in citrus and many other fruits to make periodic inspections that will form the basis for specific recommendations for the spray program with respect to timing, materials, coverage, and sequence (i.e., the order in which blocks or parts of blocks will be treated); experience and weather conditions will dictate how frequently inspections for a given pest or group of pests should be made.

.053 Integrated pest management (IPM): "...the selection, integration and implementation of pest control based on predicted economic, ecological and sociological consequences; seeks maximum use of naturally occurring pest controls, including weather, disease agents, predators, and parasites; uses in addition various biological, physical and chemical control and habitat modification techniques; artificial controls are imposed only as required to keep a pest from surpassing intolerable population levels predetermined from accurate assessments of the pest damage potential and the ecological, sociological and economic costs of the control measures. The presence of a pest species does not necessarily justify action for its control, and in fact tolerable infestations may be desirable, providing for important beneficial insects, for example." (Definition as two italicized paragraphs in S. L. Poe, 1981. An overview of integrated pest management. HortScience 16: 501; see Integrated crop management, V.3.0.001.)

.054 Invert emulsion: suspension of minute water droplets in a continuous oil phase. (TM)

.055 Late postemergence: a pesticide (e.g., a herbicide) applied after the specified crop or weeds are well established. (TM)

.056 Layby application: a pesticide applied with or after the last cultivation of a crop. (TM)

.057 Mist spraying: method in which concentrated spray is atomized into a high-velocity air stream, the air acting as diluent and carrier. (CMI)

.058 Miticide (=Acaricide): a substance that kills mites.

.059 Mulches: material laid around individual plants or along rows; many different types of organic mulch, such as grasses, leguminous and nonleguminous broad-leaved plants, coconut husks, and cut up banana pseudostems, and synthetic mulches, such as paper, asphalt roofing felt, and plastic film, of which the last as strips or sheets has proved most successful (cheaper and much longer lived as a rule); now often used for annual crops in conjuction with preplanting herbicides.

.060 Mutagen: compound having the property to induce mutations. (TM)

.061 Nematocide: a substance that kills nematodes.

.062 Nonoccluded virus: (see Biorational pesticides, V.3.3.3.013e).

.063 Nonselective herbicide: a chemical that is generally phytotoxic without regard to species; may be a function of dosage, method of application, etc. (TM)

.064 Noxious weed: weed specified by law as being especially undesirable, troublesome, and difficult to control; the definition varies according to legal interpretations. (TM)

.065 NPV: (see Biorational pesticides, V.3.3.3.013f).

.066 Nucleopolyhedrosis virus: (see Biorational pesticides, V.3.3.3.013f).

.067 Overtop application: pesticide applied over the top of transplanted plants or seedlings (plants grown in place) by airplane or raised spray boom of ground units; a broadcast application above the plant canopy. (TM)

.068 **Pellet, pelletize, pelleting:** dry formulation of pesticide and other components in discrete, more or less spherical particles, usually larger than 10 mm^3; often used for seeds, particularly those of irregular shape. (TM)

.069 **Persistent herbicide:** a herbicide that, when applied at the recommended rate, will harm susceptible crops planted in normal rotation after harvesting the treated crop or that will interfere with regrowth of native vegetation in noncrop sites. (TM)

.070 **Pesticide:** material that when properly applied (i.e., at the proper dosage, timing, method of application, and coverage for a particular organism or group), will kill or prevent the growth and spread of a pest or group of pests; a general term that includes insecticides, miticides, rodenticides, fungicides, herbicides, and nematocides.

.071 **Pesticide application:** sprays, dusts, poison baits, etc., used to control specific pests; important considerations are the proper material, timing, dosage, and coverage of plant parts to effect the desired control, use of "stickers" to prolong the effectiveness of materials (residual effectiveness), correct handling with respect to compatibility and phytotoxicity and precautions with regard to human hazards during handling, mixing, and application, and the required postspraying or dusting period prior to cultural operations and/or harvest. Proper use of pesticides requires not only careful attention to label recommendations but also prior inspection to ascertain when the least quantity can be applied for maximum effectiveness and minimum damage to biological controls. (Note that pesticide use in nurseries is similar, except perhaps of greater reliance on spot spraying or dusting.)

.072 **Pesticide equivalent:** the possible or theoretical yield of pesticidal compound from the active ingredient in a formulation (applicable especially to compounds not derived from acids).

.073 **Pheromone:** chemical(s) secreted to the outside by an individual and received by a second individual <u>of the same species</u>, in which it releases a specific reaction; e.g., a definite behavior or developmental process. (Original definition from P. Karlson and A. Butenandt, 1959. Pheromones (ectohormones) in insects. Annu. Rev. Entomol. 4: 39-68; courtesy of Dr. J. A. Coffelt, U. S. Dept. Agr. Insect Attractants Laboratory, Gainesville, FL.)

.074 **Photobiodegradable mulch:** plastic mulch film derived from a polymer that will totally degrade within two years from the combined action of light and soil organisms. (CA)

.075 **Phytoncide:** a chemical substance produced by higher green plants that can inhibit the growth of microorganisms. (CMI)

.076 **Phytosanitation:** any measure involving the removal or destruction of infected plant material likely to form a source of reinfection by plant disease. (CMI)

.077 **Phytotoxin:** a toxin toxic to a plant. (CMI)

.078 **Postemergence:** application made after emergence of the specified weed or planted crop. (TM)

.079 **Preemergence:** application made prior to emergence of the specified weed or planted crop. (TM)

.080 **Preemergence incorporated:** applied after seeding and incorporated into the soil above the seed. (TM)

.081 **Preplanting application:** a pesticide applied on the soil surface prior to seeding or transplanting. (TM)

.082 Protectant: a substance that protects an organism against infection by a pathogen. (CMI)
.083 Protection: as a principle of plant-disease control, the placement of a barrier between suscept and pathogen (e.g., the use of protective chemical sprays or dusts). (RB)
.084 Quarantine: (a) control of import and export of plants to prevent spread of diseases and pests; (b) holding of imported plants in isolation to ensure their freedom from diseases and pests. (CMI)
.085 Rate: amount of pesticide equivalent (or herbicide acid equivalent) applied per unit area or other treatment unit. (TM)
.086 Residual action: the continued action of a material, such as the effect of an insecticide on insects that come in contact with it for some time after its application. (FC)
.087 Rodenticide: a substance that kills mice, rats, and other rodents.
.088 Sanitation: the periodic removal of dead, diseased, or badly infested branches or plants and decayed or infested fruit, mowing of natural or planted cover around and under plants, and pruning of low hanging or dense clusters of branches to reduce or eliminate breeding sites for pests and provide good air circulation (particularly important for reducing disease problems).
.089 Scalicide: a substance that kills scale insects.
.090 Scout, scouting: to determine by appropriate sampling the actual level of a pest population and its damage to a crop at any given time; a necessary adjunct to integrated pest management (V.3.3.3.053) or inspection (V.3.3.3.052).
.091 Selective herbicide: a chemical that is more toxic to some plant species than to others; may be a function of dosage or mode of application. (TM)
.092 Smokes: clouds of pesticide particles (usually less than $0.1 \mu m$) produced by heat; similar to aerosols (V.3.3.3.006) in uses and properties. (UF)
.093 Soil applications: a pesticide applied to the soil rather than to vegetation. (TM)
.094 Soil injection: placement of the pesticide (herbicide, fumigant, insecticide, etc.) beneath the soil surface with a minimum of mixing or stirring of the soil, by using an injection blade, knife, or tine. (TM)
.095 Soil layered: placement of the pesticide (e.g., herbicide) or fertilizer beneath the soil surface under a lifted or tilled layer of soil. (TM)
.096 Soluble solid: dry pesticide formulation that is soluble in the carrier liquid; (see Suspension, V.3.3.3.107). (TM)
.097 Solution: a solid, liquid or gas dissolved in a solvent; e.g., in pest control a technical-grade insecticide or other pesticide dissolved in a solvent such as petroleum oil, xylene, or acetone; used frequently as residual sprays in households. (UF)
.098 Spray and dust program: includes four main components: (a) periodic inspections (V.3.3.3.052), (b) sanitation (V.3.3.3.088), (c) biological control (V.3.3.3.012), (d) pesticide application (V.3.3.3.071).
.099 Spray drift: movement of spray away from the intended area of application. (TM)
.100 Sprayer: a device for the application of a pesticide in liquid form; many different forms include simple hand-held and -pumped units (e.g., trombone sprayer), water hose-attached units, knapsack sprayers (with hand pumps), truck-mounted or tractor-hauled, high-pressure sprayers with handguns, tractor-drawn or self-pro-

pelled row-crop and broadcast and air-blast or air-carrier sprayers, and aircraft (fixed wing and helicopter) sprayers may be high-volume (for dilute applications), low-volume airblast or air-carriers, or mist blowers (ultralow volume sprays usually as aerosols).

.101 Spread: uniformity and completeness with which a pesticide deposit covers a continuous surface. (CMI)
.102 Spreader: a substance added to a spray to assist in its even distribution over the target. (CMI)
.103 Sticker: material added to a pesticide to increase tenacity; a substance added to a spray to make it adhere to the target. (CMI)
.104 Supplement: material added to a pesticide to improve some physical or chemical property; adjuvant; auxiliary. (CMI)
.105 Surfactant: material (i.e., detergent, adjuvant, or other surface-active agent) that favors or improves the emulsifying, dispersing, spreading, wetting, or other surface-modifying properties of a liquid used as a pesticide. (TM)
.106 Susceptibility: the lack of capacity of a plant to tolerate pesticide treatment without injury. (TM)
.107 Suspension: finely divided solid particles dispersed in a liquid, gas, or solid. (TM)
.108 Synergism, synergistic: the mutual interaction of two substances which results in an effect greater than the additional effects of the two when used separately; e.g., the increased pesticidal value of certain mixtures of pesticides or of pesticides and nontoxic materials. (TM)
.109 Systemic: (see Systemic, V.3.3.2.419).
.110 Tenacity: property of a pesticide deposit to resist removal by weathering. (CMI)
.111 Tolerance: (a) of diseased plants, the ability to endure disease without serious injury or crop loss; (b) the capacity to withstand herbicide (or other pesticide) treatment without marked deviation from normal growth or function; (c) the legal maximum residue allowed for a pesticide in or on plant tissues. (TM)
.112 Topical application: pesticide treatment of a localized surface site such as a single leaf blade, petiole, or growing point. (TM)
.113 Translocated herbicide: herbicide that is moved within the plant, being either phloem mobile or xylem mobile, although the term is frequently used in a restricted sense to refer to one that is phloem mobile. (TM)
.114 Ultralow volume concentrate: formulations of certain pesticides applied undiluted by aerial sprays, mist blowers (aerosol sprayers), and the like. (UF)
.115 Vapor drift: movement of chemical vapors from the area of application.
.116 Volatile herbicide: herbicide with a sufficiently high vapor pressure that, when applied at normal rates and temperatures, its vapors may cause irreversible injury to desirable plants away from the site of application; such injury is often difficult to distinguish from that of spray drift. (TM)
.117 Volunteer plant: an unwanted plant (i.e., weed) growing from seed (or vegetative material) that remains on a field from a preceding crop.
.118 Water-dispersible slurry: a two-phase concentrate of solid particles suspended in a liquid that is capable of suspension in water.
.119 Weed control: four general systems: (a) mulches, (b) cultivation,

V.3.3.3. PROPAGATION, NURSERY HANDLING, SOILS, CROP PRODUCTION

(c) herbicides, (d) grazing by animals used to control or eradicate weeds that compete with the crop for water, nutrients, and light; the process of limiting or killing weeds for economic, esthetic, public health, or other reasons. (TM)

.120 Weed eradication: the elimination from a site of all live plant parts, including seeds, of a weed. (TM)

.121 Weed killer: (see Herbicide, V.3.3.3.049).

.122 Weeds: classified from the standpoint of chemical control (i.e., their reaction to limited selectivity shown by most herbicides), as broad-leaved, annual and perennial grasses and other perennials. (TM)

.123 Weed science: identification, classification, and nomenclature of weeds and measures for their control. (TM)

.124 Wettable powder (wp): a pesticide in powder form blended with an inert dust and wetting agent such as Triton X-100; forms a suspension in water which must be kept agitated. (TM)

.125 Wetting agent: substance that serves to reduce interfacial (surface) tension and in spray solutions or suspensions helps to make better contact with treated surfaces. (TM)

V.3.4. Plant and crop-size control

.000 Plant and crop size control: the production and maintenance of the most efficient plant form, dimensions, and flowering or fruitfulness for the desired objective (i.e., esthetic value of ornamentals, and optimum fruiting of trees, shrubs, or vines); several different methods include (a) dwarfing rootstocks and interstocks; (b) inoculation with certain viruses or viroids (e.g., exocortis for citrus in Australia and ornamental citrus in California); (c) bending, twisting, girdling, and/or starvation (e.g., Bonzai, V.3.4.004; Topiary, V.3.4.074; or Espalier, V.3.4.025 for ornamentals); (d) pruning, including training (see Pruning, V.3.4.000b); and (e) use of growth regulators for fruit set and thinning (see Physiological sprays, V.3.4.000a); one or both of the last two are used for a wide range of crops. (Fig. V.3.4.01)

.000a. Physiological sprays: the application of growth regulators to modify plant growth and development processes; e.g., to induce floral initiation [i.e., with ethylene generators (ethephon) or auxins on many crops, B-9 on azaleas], chemical thinning (auxins, ethephon), fruit set (auxins, GA. SADH, CCC), to improve fruit shape and size (GA), prevent premature ripening (GA), prevent preharvest drop (auxins), reduce acidity in grapefruit (arsenate), induce fruit abscission (ethephon, cyclohexamide), and retard growth (Cycocel, CCC for poinsettias); concentrations of individual compounds and timing with respect to plant development and environmental conditions are critical in eliciting the desired response; (see Growth regulators, IV. 2.1.). (WS)

.000b. Pruning: the mechanical removal of vegetative and/or flowering and fruiting growth from a plant to regulate its size and crop load; a dwarfing process that nevertheless is beneficial for optimal crop production; two main types: (a) <u>thinning</u>, the more or less complete removal of individual branches, and (b) <u>heading</u>, trimming off the ends of branches; (a) is done to thin out dense growth and eliminate weak, thin, unproductive

V.3.4. PROPAGATION, NURSERY HANDLING, SOILS, CROP PRODUCTION

wood and improve light distribution; pruning encompasses two main stages: (1) _training_ and (2) _maintenance pruning_, the former to construct the future main branch framework of the tree, shrub, or vine and the latter to maintain proper plant size and cropping in addition to sanitary pruning (removal of dead, diseased, or infested branches, cross-overs, and the like); numerous systems have been devised for groups of crops or individual crops (or in some cases individual cultivars) (V.3.4.067) and carried out with a wide diversity of equipment (V.3.4.063). (Fig.V.3.4.02a-c; .03) (WS, J)

.001 Agobio system: (see Coffee, V.3.4.017).
.002 Bayonet: (see Coffee, V.3.4.017).
.003 Bending (festoons): a practice in which vigorous stems in young trees are induced to form floral buds by bending pairs of shoots and fastening them together to form roughly circular festoons. (Fig. V.3.4.12e) (TK, BU)
.004 Bonzai: a form of extreme dwarfing of trees developed in Japan; these plants are often only 30 to 60 cm tall when 200 years old or more. (Fig. V.3.4.12a) (TK)
.005 Bourgeon: a developing shoot. (SI)
.006 Bourse (cluster base): the thickened swollen base of an inflorescence in apple or pear; persistent and may bear floral or vegetative buds. (Fig. V.3.4.02e) (SI)
.007 Bourse bud: a floral or vegetative bud arising on a bourse. (Fig. V.3.4.02e,f) (SI)
.008 Brambles (_Rubus_ spp.): several training and pruning systems; trailing species of blackberry and other brambles are trained on upright or crossarm trellises, the former with two or three horizontal wires and the latter with a wire at the end of each crossarm and one or two horizontal wires below, upright blackberry and black raspberry as individual bushes or on trellises; pruning consists of continual removal of second-year canes that have fruited and thinning out new (one-year-old) canes to encourage better flowering and fruiting. (Fig. V.3.4.05) (WS, J)
.009 Brazilian system: (see Coffee, V.3.4.017).
.010 Brindille: thin lateral shoot about 7.5 to 30 cm long, usually terminating in a floral bud. (Fig. V.3.4.02h) (SI)
.011 Bucksaw: a saw for "bucking" large limbs into firewood; the blade has coarse teeth, as in a regular crosscut log saw, with a looped (wide U-shaped) handle to which the ends of the blade are fastened; used to remove large limbs (i.e., 5 to 15 cm) for topworking and the like; (see Pruning equipment, V.3.4.063).
.012 Candelabra palmette: (see Palmette, V.3.4.055).
.013 Candelabra system: (see Coffee, V.3.4.017).
.014 Central leader: (see Fruit trees, V.3.4.029).
.015 Chain saw: motor-driven saw with cutting teeth on an endless chain; used for the same general purposes as a bucksaw (V.3.4.011; Pruning equipment, V.3.4.063).
.016 Cluster base: (see Bourse, V.3.4.006).
.017 Coffee: at least 11 different pruning systems are used in various parts of the world for Arabian, robusta, and Liberian coffee: (a) _topping_ (grown as a single stem after top cut off with free development of laterals which are thinned occasionally); (b) _single-stem_ (top pinched off head high, also side shoots, fruiting wood on laterals which are thinned occasionally); (c) _modified single-stem_ (bayonet) [similar to (b) but a double-stem rotation system with two to four harvests between renewals]; (d) _vertical or_

V.3.4. PROPAGATION, NURSERY HANDLING, SOILS, CROP PRODUCTION

upright multiple-stem (East Indian) (young trees pinched at 75 cm, three to five good verticals selected then renewed in rotation every year for two to five years); (e) inclined or leaning multiple-stem (seedlings headed at 75 cm; several shoots are allowed to develop, then pulled over, and replaced every three to six years); (f) Hawaiian system (intensive upright stem rotation of four to six years); (g) umbrella (Colombian) system (seedling cut off at head height; side branches from nodes near top are encouraged with frequent thinning of laterals); (h) candelabra (Costa Rican) system (series of pairs of shoots allowed to develop, headed at 30 cm, then 100 cm and 200 cm to give uprights, which are rotated as bearing declines); (i) agobio (Guatemalan) system (top bent over at 45° angle, three to four good uprights selected and bent over in turn, with continual rotation of oldest vegetative shoots); (j) "poda" (Brazilian) system (four to five seedlings planted deep in holes, bent outward, pruned about every four years to remove dead wood); (k) Hawaiian stump system [as for (f) except individual rows are cut back to the stump in a rotation of four to six years.] (Fig. V.3.4.06,.07,.08,.09a,b) (OSDW, G & F)

.018 Colombian: (see Coffee, V.3.4.017).
.019 Cone: a variation of the pyramid form (V.3.4.068), usually about as wide as tall. (TK)
.020 Contre-espalier: plants grown on lattices, trellises, etc., not attached to a wall. (Fig. V.3.4.11). (TK)
.021 Cordon: a single-dimension tree that consists of spurs close to and along the stem; may be trained to be horizontal, oblique, vertical, serpentine, or spiral, and single (one-armed) or double. (Fig. V.3.4.11b,c,e) (TK)
.022 Costa Rican system: (see Coffee, V.3.4.017).
.023 Dard: an eye or bud surrounded by a rosette of small leaves; may develop into a shoot or fruiting spur. (Fig. V.3.4.02d,g) (SI)
.024 Dwarf pyramid: (see Pyramid, V.3.4.068).
.025 East Indian system: see Coffee, V.3.4.017).
.026 Espalier: specifically, plants with their scaffold branches fastened directly to a wall or on a lattice or trellis attached to a wall. (Fig. V.3.4.011a,b,e) (TK)
.027 Fan system: (see Grapes, V.3.4.032).
.028 Festoons: (see Bending, V.3.4.003).
.029 Fruit trees: trained in the nursery or shortly after planting: (a) high-headed (unbranched newly set tree cut back to 0.75 to 1.0 m); (b) low-headed (cut back to 0.4 to 0.6 m) as (1) a central leader, side branches removed initially, then three or four wide-angled side branches are allowed to develop from the trunk and the central leader is maintained; (2) modified central leader, similar to the central leader, except that the main shoot is headed back and the uppermost branch of the three or four side branches is allowed to grow to become the modified leader; (3) open center, in which the main leader is suppressed and the side branches form the principal framework of the tree; (1) and (2) are widely used for temperate zone, subtropical, and tropical fruit and nut trees, except peach, Japanese plums, almond, and apricot which are usually grown as (3). (Fig. V.3.4.03) (WS, J)
.030 Geneva double-curtain system: (see Grapes, V.3.4.032).
.031 Goblet: a tree trained in the form of a vase or goblet. (Fig. V.3.4.11f) (TK)
.032 Grapes (and many other vines): several distinct pruning systems

V.3.4. PROPAGATION, NURSERY HANDLING, SOILS, CROP PRODUCTION

(and modifications thereof) are carried out in the field after training the vines; all but one (head system) is also used for passion fruits, Chinese gooseberry, and the like: (a) <u>six-cane Kniffen system</u>, in which pairs of laterals are trained onto each of three horizontal wires; (b) <u>fan system</u>, in which several upright vines from the main stem are spread out and trained up a wood trellis or series of horizontal wires; (c) <u>horizontal-arm spur system</u>, in which two laterals are trained along the bottom wire or wood slat of a trellis and a series of spaced out uprights is trained and fastened to the higher wires or slats; (d) <u>Geneva double-curtain system</u>, with two horizontal wires 170 cm above the ground and 120 cm apart on which fruiting laterals are supported with a center wire 130 cm above the ground to support the main stems; and (e) <u>head system</u>, used primarily for vinifera (European) cultivars in California, with a main trunk with spur pruning; (see Renewal spur, V.3.4. 069). (Fig. V.3.4.10) (WS, J)

.033 Guatemalan system: (see Coffee, V.3.4.017).
.034 Hawaiian stump system: see Coffee, V.3.4.017).
.035 Hawaiian system: (see Coffee, V.3.4.017).
.036 Hawk-billed knife: a pruning knife (V.3.4.064; see Pruning equipment, V.3.4.063).
.037 Heading: (see Pruning, V.3.4.000b).
.038 Head system: (see Grapes, V.3.4.032).
.039 Hedger: (see Pruning equipment, V.3.4.063).
.040 Hedging machine: a device for the mechanized removal of growth from the sides of trees, shrubs, or vines; two principal types, sickle bar (heavy-duty version of a regular sickle-bar mower, used for relatively small trees) and circular saws (multiple saws, each 25 to 75 cm or more in diameter, used for hedging large trees up to 6 or 8 m or more in height); usually self-propelled and adjustable for cut angle; (see Pruning equipment, V.3.4.063).
.041 Hedge shears: a type of shears with flat blades and angled handles for trimming light growth on hedges or tufts of grass; may be hand-operated or electric; the latter is usually the sickle-bar type (i.e., equipped with a reciprocating blade with projecting cutting edges on the movable blade and similar ears on the stationary blade to provide multiple cutting surfaces); (see Pruning equipment, V.3.4.063).
.042 High-headed: (see Fruit trees, V.3.4.029).
.043 Horizontal-arm spur system: (see Grapes, V.3.4.032).
.044 Inclined multiple-stem: (see Coffee, V.3.4.017).
.045 Kniffen system: (see Grapes, V.3.4.032).
.046 Leader: the terminal branch of a tree (i.e., the extension of the original shoot) or the principal, usually terminal, shoot of a lateral branch; (see Pruning, V.3.4.000b). (WS)
.047 Leaning multiple-stem: (see Coffee, V.3.4.017).
.048 Lopping shears: pruning shears with long handles and usually heavier blades (than those of hand shears) for cutting off limbs up to about 5 cm in diameter; (see Pruning equipment, V.3.4.063).
.049 Lorette system: a special type of pruning developed in France and done only in the summer; used to induce early fruiting of cordon, U or double-U palmettes, and winged pyramid forms of not overly vigorous pome and stone fruit cultivars. (SI)
.050 Low-headed: (see Fruit trees, V.3.4.029).
.051 Maintenance pruning: (see Pruning, V.3.4.000b; Pruning systems, V. 3.4.067).
.052 Modified central leader: (see Fruit trees, V.3.4.029).

.053 Modified single-stem: (see Coffee, V.3.4.017).
.054 Open center: (see Fruit trees, V.3.4.029).
.055 Palmette: a special tree form that differs from the cordon in having lateral scaffold branches; four main groups, horizontal, oblique, candelabra (branches trimmed horizontally then vertically), Verrier (arms diminished in lateral or vertical extension with each tier above the base), and fan (branches radiate from a short main trunk). (Fig. V.3.4.11a,d,f,g) (TK)
.056 Pillar tree: pruning system developed in England for modified free-standing cordons; involves a standardized method for regular renewal of new wood. (Fig. V.3.4.04a-f) (TK)
.057 Pinching: removal of tender tips from a plant, usually with the fingers, to induce branching, as in azaleas, chrysanthemums, and other ornamentals in pots or garden beds (similar to "skiffing" in tea culture, V.3.4.073).
.058 Pleach: to braid and intertwine young woody stems. (Fig. V.3.4.12d).
.059 "Poda" system: (see Coffee, V.3.4.017).
.060 Pole saw: a pruning saw mounted on the end of a wood or metal pole for cutting off branches otherwise inaccessible from the ground; an electric or air-driven pole saw differs by usually having a circular blade; (see Pruning equipment, V.3.4.063).
.061 Pole shears: pruning shears mounted at the end of a wood or metal pole with a rope for operating the movable blade; used for cutting off small branches (i.e., to about 2.5 cm in diameter) a few meters high in a tree; may be hand-operated or air- or hydraulic-driven; (see Pruning equipment V.3.4.063).
.062 Pollard: perennial pruning of branches at or near the same point, which results in a distinctive, thick, bushy appearance of the trees. (Fig. V.3.4.12c)
.063 Pruning equipment: devices for effecting the removal of vegetative and/or fruiting growth from a plant; two main groups: (a) hand-held or -operated tools, such as pruning knives (hawk-billed knives), pruning shears (secateurs), pruning saws, bucksaws, chain saws, topping shears, hedge shears, pole shears, and pole saws plus electric, air-driven, or hydraulic hedge trimmers, shears, and saws; (b) machines, such as hedgers and toppers in which cutters may have reciprocating (i.e., sickle-bar type) or rotary (saw type) motion; many different forms and sizes of hand-held (or -operated) or mechanized pruning equipment.
.064 Pruning knife: a single-bladed knife, usually with at least a slight concave curve in the blade (i.e., hawk-billed) and a heavy handle; may be a solid or folding type; (see Pruning equipment, V.3.4.063).
.065 Pruning saw: a saw with fairly coarse teeth set to cut on the return stroke (i.e., toward the person holding it); usually made with a curved (concave) blade; used for limbs too large to be cut off with pruning or lopping shears; (see Pruning equipment, V.3.4.063).
.066 Pruning shears(secateurs): heavy forms of scissors of two general types: anvil in which there is only one cutting edge that is closed against a stationary blade (anvil), and scissors, in which both blades are movable and one or both sharpened for cutting; blades may be straight or, more often, curved; (see Pruning equipment, V.3.4.063).
.067 Pruning systems: specialized training and maintenance pruning regimes for particular crops or groups of crops; (see Fruit trees,

V.3.4.029; Brambles, V.3.4.008; Grapes, V.3.4.032; Coffee, V.3.4.017; Tea, V.3.4.073).

.068 Pyramid: free-standing tree with the greatest width at the base and tapering regularly up to the top; always taller than wide; many versions, of which the dwarf pyramid is produced commercially; the winged pyramids have cordonlike scaffold branches arranged in tiers and trained at an angle of 30 to 45° from the horizontal. (Fig. V.3.4.12d) (TK)

.069 Renewal spur: a short stem with a single or only a few buds that will grow out and replace old fruiting laterals on a grape vine; (see Grapes, V.3.4.032).

.070 Secateur: (see Pruning shears, V.3.4.066).

.071 Single-stem: (see Coffee, V.3.4.017).

.072 Spindle bush: a modified dwarf pyramid form in which fruit are borne on short branches; lateral shoots are tied down in a horizontal position. (Fig. V.3.4.04g) (TK, RHS)

.073 Tea: grown usually as flat-topped hedges from seedlings or rooted cuttings; training consists of three successive prunings at about 15-, 30-, and 40- cm heights at roughly yearly intervals, beginning about the second year; the object is to build a broad structure of framework branches to support the largest possible number of young shoots, with about three rounds of skiffing (light pruning to even up the harvesting surface) prior to bringing the bushes into production in the fourth year); maintenance or "production" pruning may be done annually when bushes are dormant in winter or about every other year with year-round growth; literally dozens of programs, some as stepwise cuts 5 cm above previous ones beginning at 45 cm and reaching 90 cm (this to 110 cm is the maximum height for economic harvesting) in about 16 years, some as alternate low and high heading (e.g., successively at 50, 58, 54, 62 cm), and some as three headings at the same level, then three more 5 cm higher and so on, certain of the cycles operating for 35 to nearly 60 years before returning to the original pruning height of 45 to 50 cm (well cared for tea plants have an economic life of a century or more). (Fig. V.3.4.09c-e)

.074 Thinning: (see Pruning, V.3.4.000b).

.075 Topiary: the art of training, by continual cutting and trimming, trees or shrubs into odd or ornamental shapes; also a garden characterized by such work. (Fig. V.3.4.12b) (SC, GO)

.076 Topper (topping machine): a device for the mechanized removal of growth from the tops of trees, shrubs, or vines; two principal types, sickle-bar (heavy-duty versions of a regular sickle-bar mower, used for light topping on relatively short plants) and circular saws (multiple saws, each 25 to 75 cm or more in diameter, used for large trees); (see Pruning equipment, V.3.4.063).

.077 Topping: (see Coffee, V.3.4.017).

.078 Topping machine: (see Pruning equipment, V.3.4.063).

.079 Training: (see Pruning equipment, V.3.4.063).

.080 Umbrella: (see Coffee, V.3.4.017).

.081 Upright multiple-stem: (see Coffee, V.3.4.017).

.082 Verrier palmette: (see Palmette, V.3.4.055).

.083 Vertical multiple-stem: (see Coffee, V.3.4.017).

.084 Winged pyramid: (see Pyramid, V.3.4.068).

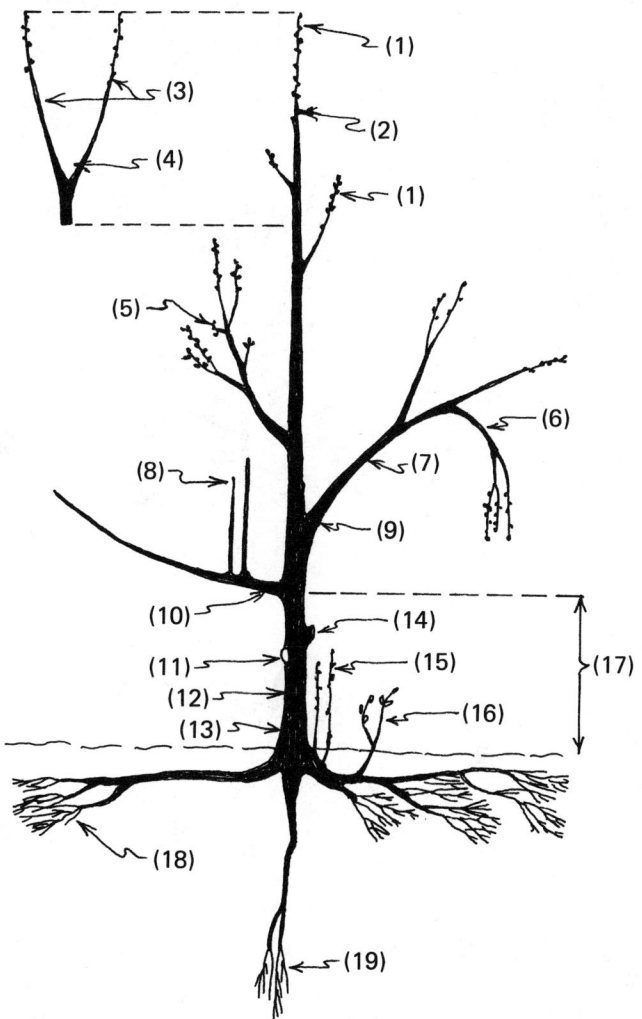

Fig. V.3.4.01. Plant framework in relation to pruning terminology: (1) terminal; (2) central leader; (3) double leader; (4) weak crotch; (5) spurs; (6) hanger (drooping branch); (7) scaffold branch; (8) watersprouts; (9) weak (narrow angle) crotch; (10) strong (wide angle) crotch; (11) properly pruned stub; (12) trunk; (13) crown; (14) improperly pruned stub; (15) crown sucker; (16) root sucker; (17) heading height; (18) lateral roots; (19) tap root. Adapted from Christopher, 1957; see Acknowledgments and Selected References.

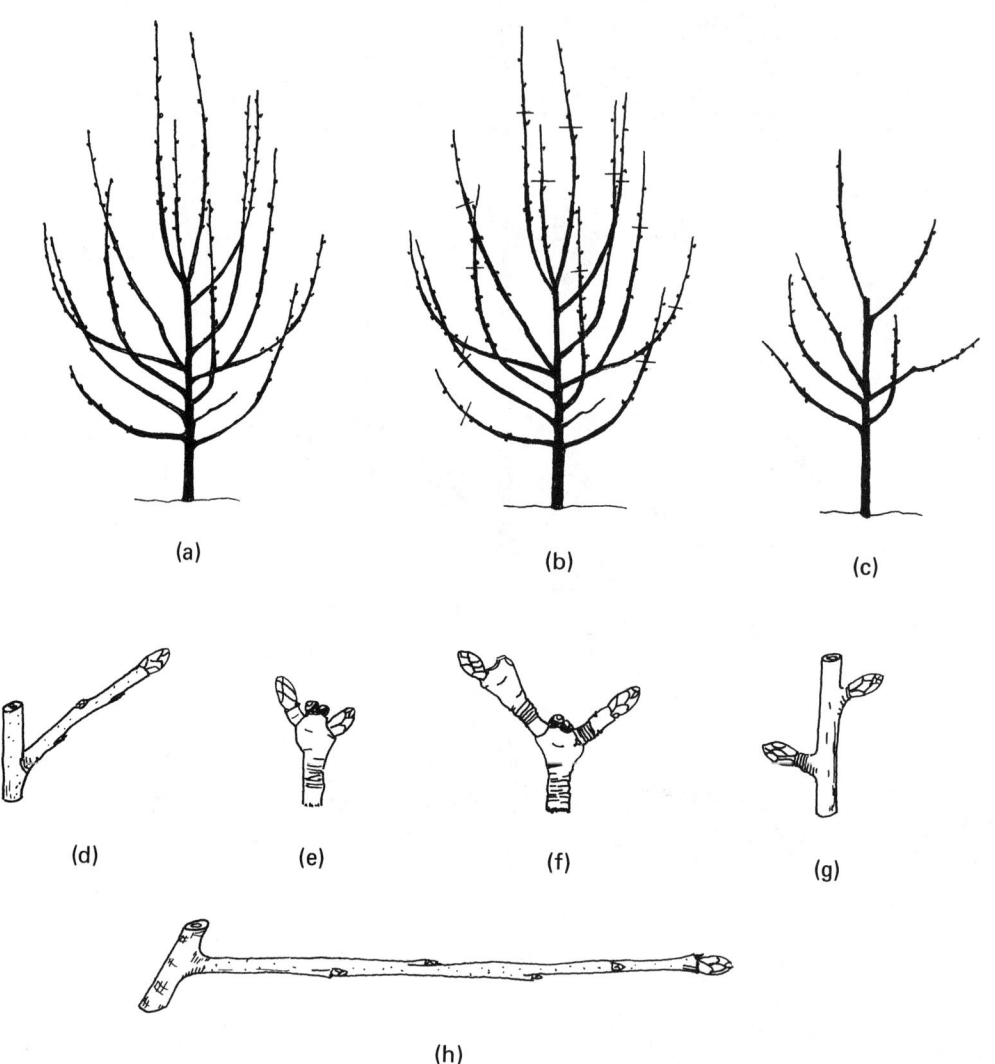

Fig. V.3.4.02. Basic types of pruning, special terms: (a)-(c) basic types of pruning, (a) intact tree; (b) heading back; (c) thinning; (d)-(h) special terms, (d) dard developed into a shoot with terminal fruit bud; (e) bourse with 2 bourse buds; (f) three-year-old spur with a bourse bud (left) and short bourse shoot; (g) 2 dards; (h) brindille with terminal fruit bud. (a)-(c) Adapted from Tukey, 1964; (d)-(h) adapted from Scientific Horticulture, 1955; see Acknowledgments and Selected References.

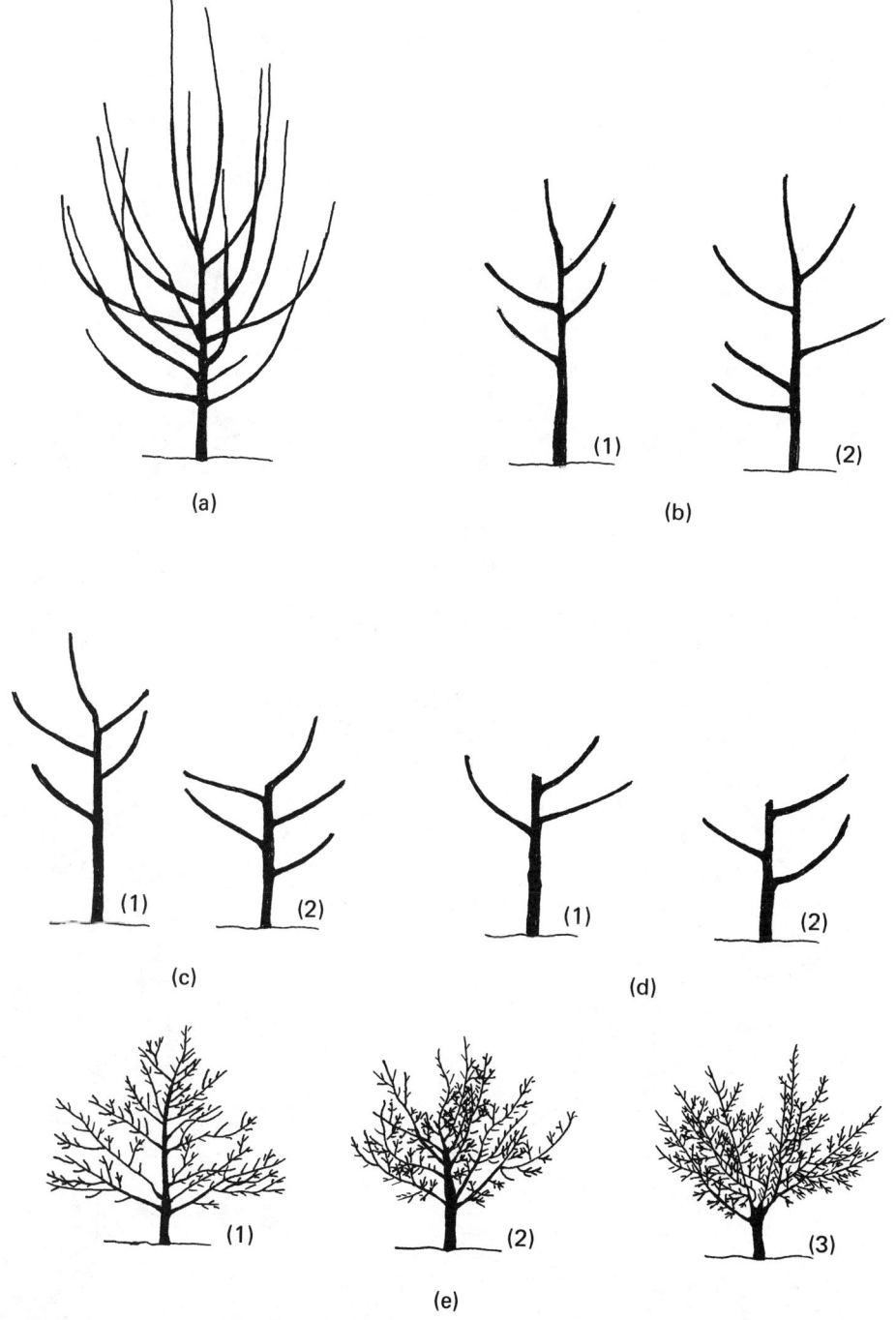

Fig. V.3.4.03. Leader training and head heights: (a) two-year-old tree (apple) at planting; (b) thinning to produce central leader, (1) high, (2) low headed; (c) modified central leader, (1) high, (2) low headed; (d) open center, (1) high, (2) low headed; (e) trained tree, (1) central leader, (2) modified central leader, (3) open center. (a), (e) Adapted from Tukey, 1964; (b)-(d) adapted from Janick, 1972; see Acknowledgments and Selected References.

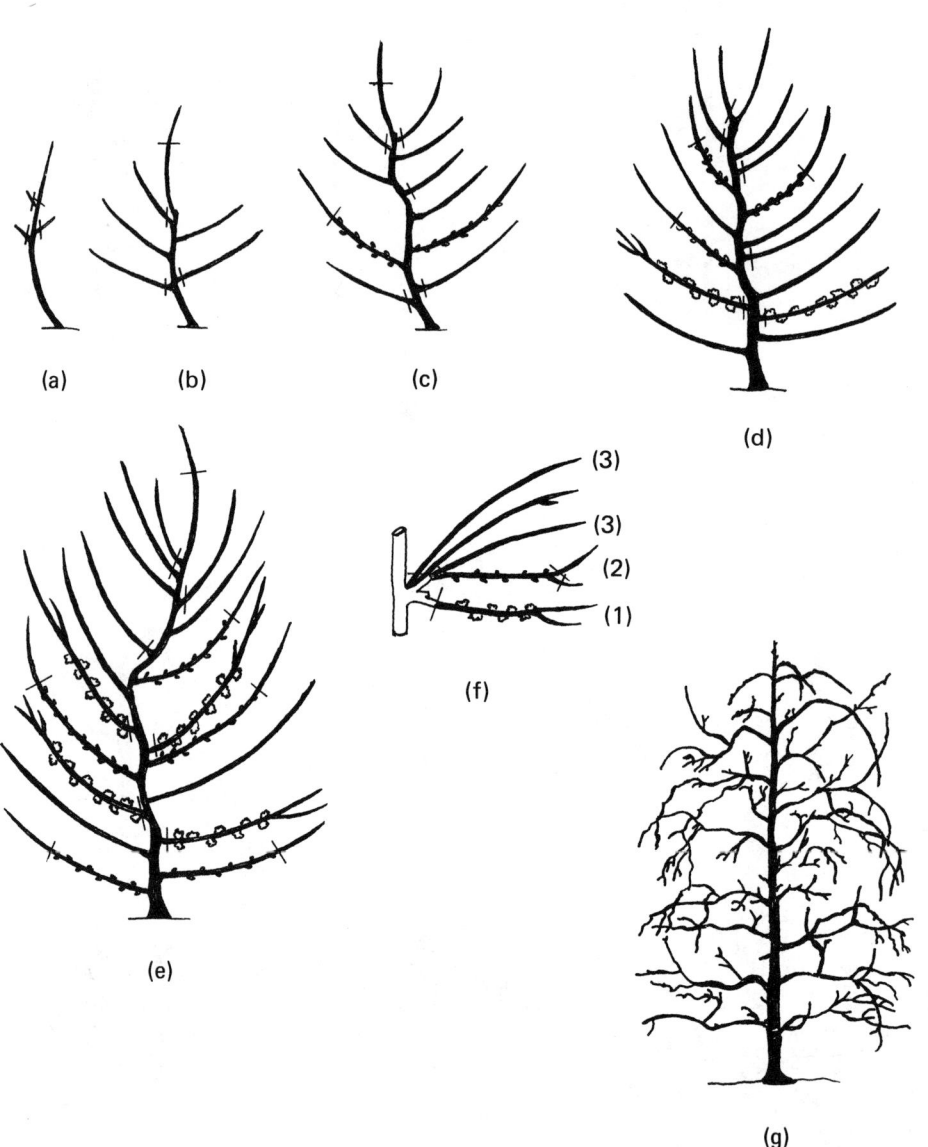

Fig. V.3.4.04. Apple pruning: (a)-(f) pillar trees, (a) newly planted whip headed at 75 cm; (b) end of first season; (c) end of second season (fruit buds on some branches); (d) end of third season; (e) end of fourth season; (f) mature tree, (1) three-year laterals, (2) two-year laterals, (3) one-year laterals (about 20 to 25 left per tree (- indicates where cut is made); (g) spindle bush. (a)-(f) Redrawn from Janick, 1972; (g) adapted from Royal Horticultural Society, 1950; see Acknowledgments and Selected References.

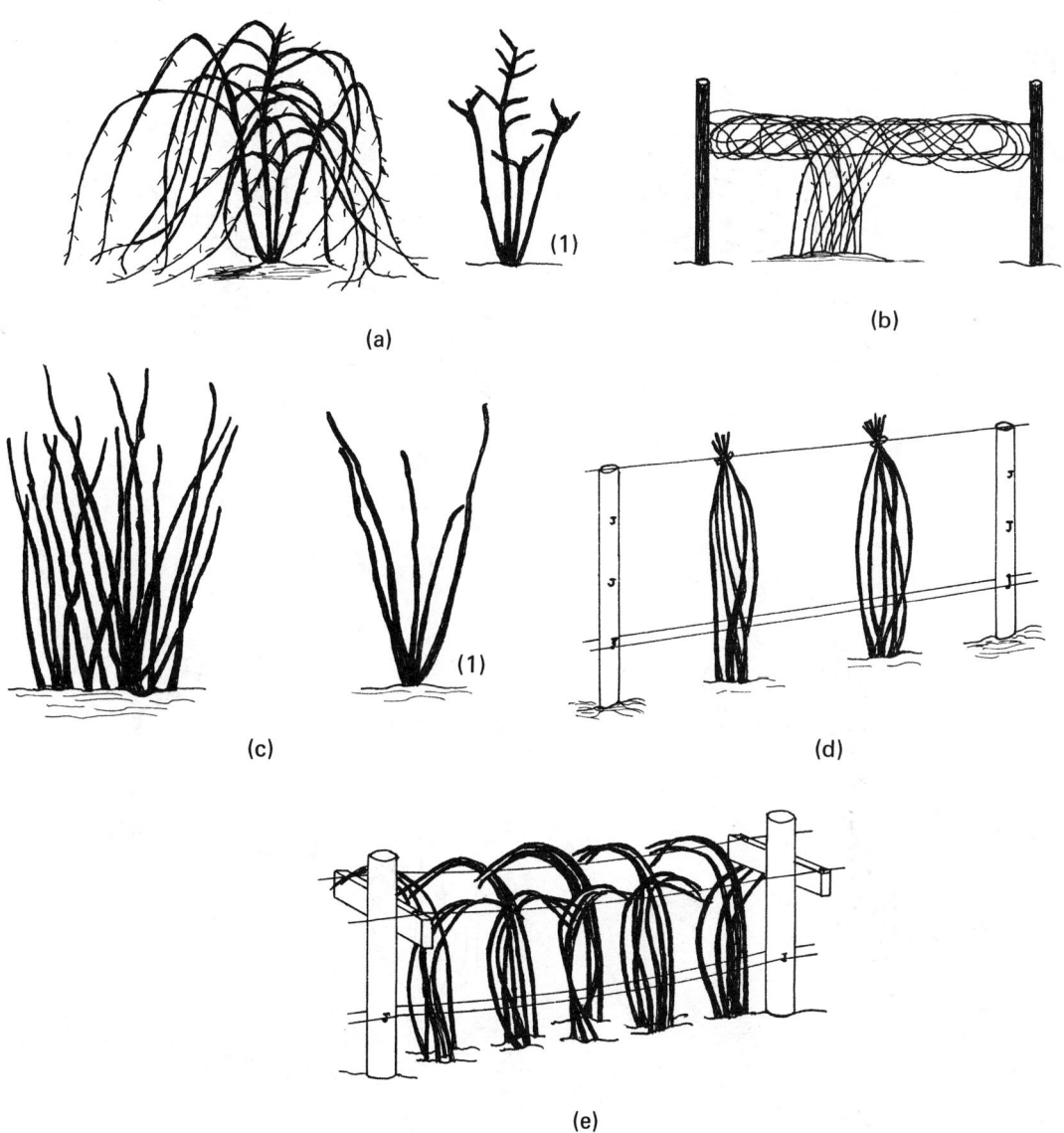

Fig. V.3.4.05. Pruning-training systems for brambles: (a) upright blackberry, black raspberry, (1) after pruning; (b) trailing blackberry (dewberry); (c) red raspberry, (1) thinned and pruned; (d) three-wire trellis; (e) crossarm system for red raspberry. Redrawn from Westwood, 1978; see Acknowledgments and Selected References.

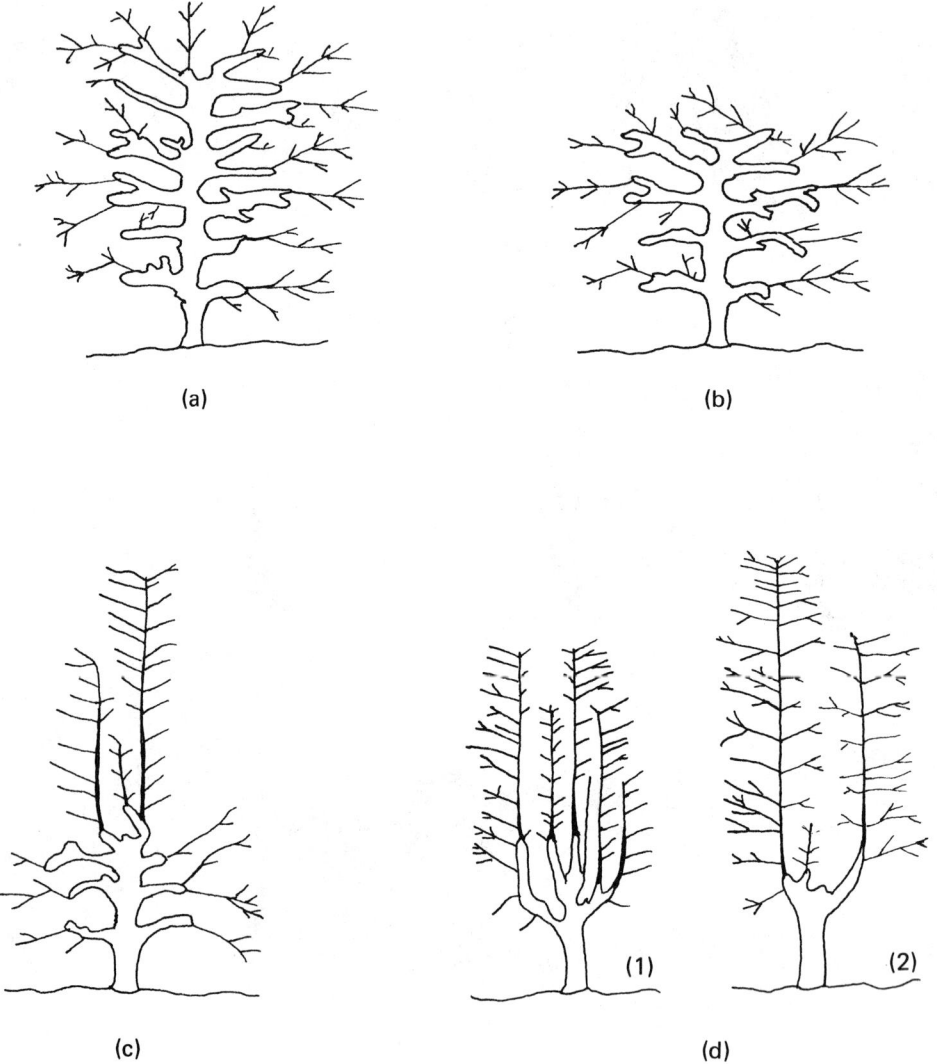

Fig. V.3.4.06. Coffee pruning: (a) topping; (b) single-stem; (c) modified single-stem or bayonet (modified topping); (d) vertical or upright multiple-stem (multiple vertical or tall vertical) trained as (1) two uprights of same age or (2) one upright. Adapted from Goto and Fukunaga, 1956; see Acknowledgments and Selected References.

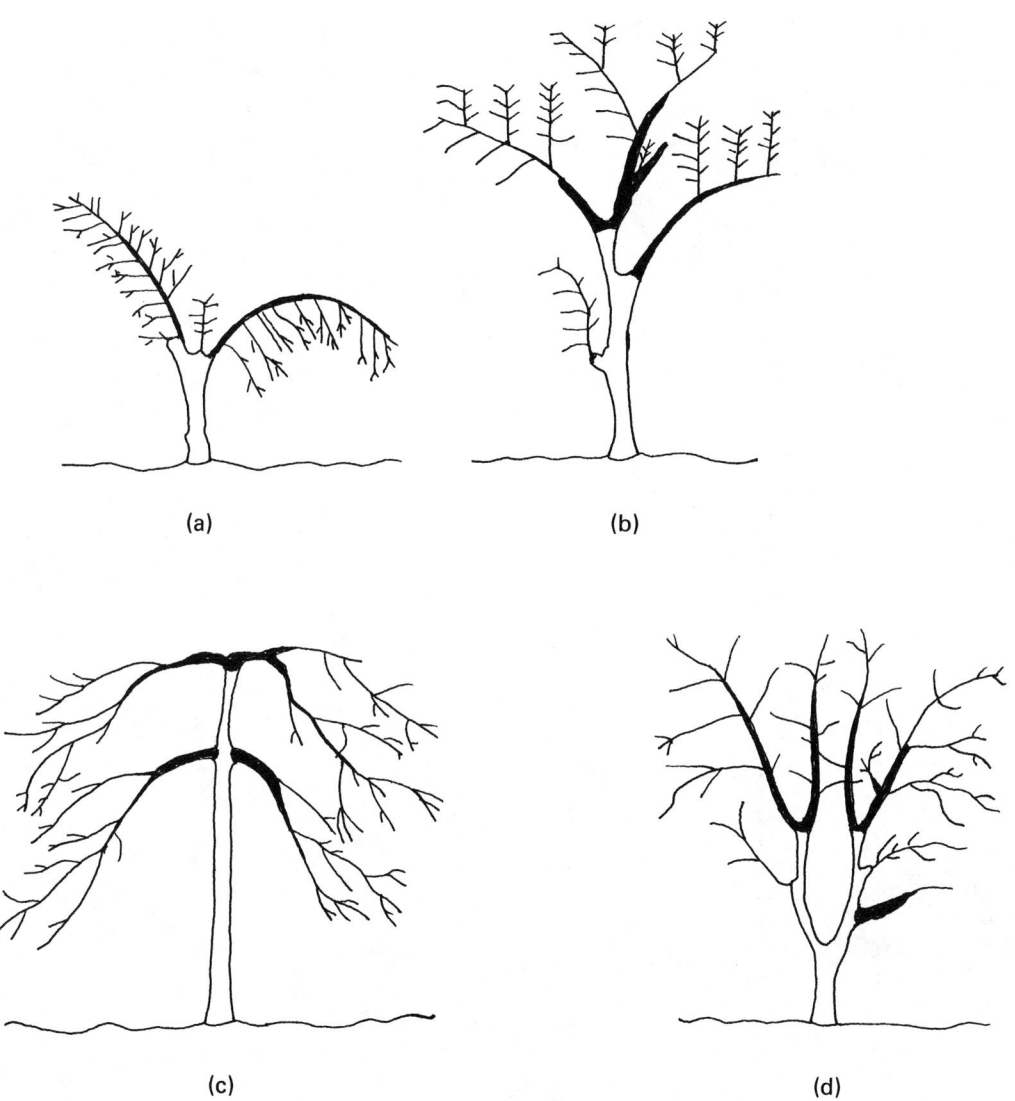

Fig. V.3.4.07. Coffee pruning (cont.): (a), (b) inclined (leaning) multiple-stem (leaning vertical or secondary vertical); (c) umbrella (Colombian) system; (d) candelabra (Costa Rican) system. (a), (b) Redrawn from Goto and Fukunaga, 1956; (c), (d) redrawn from Ochse et al., 1961; see Acknowledgments and Selected References.

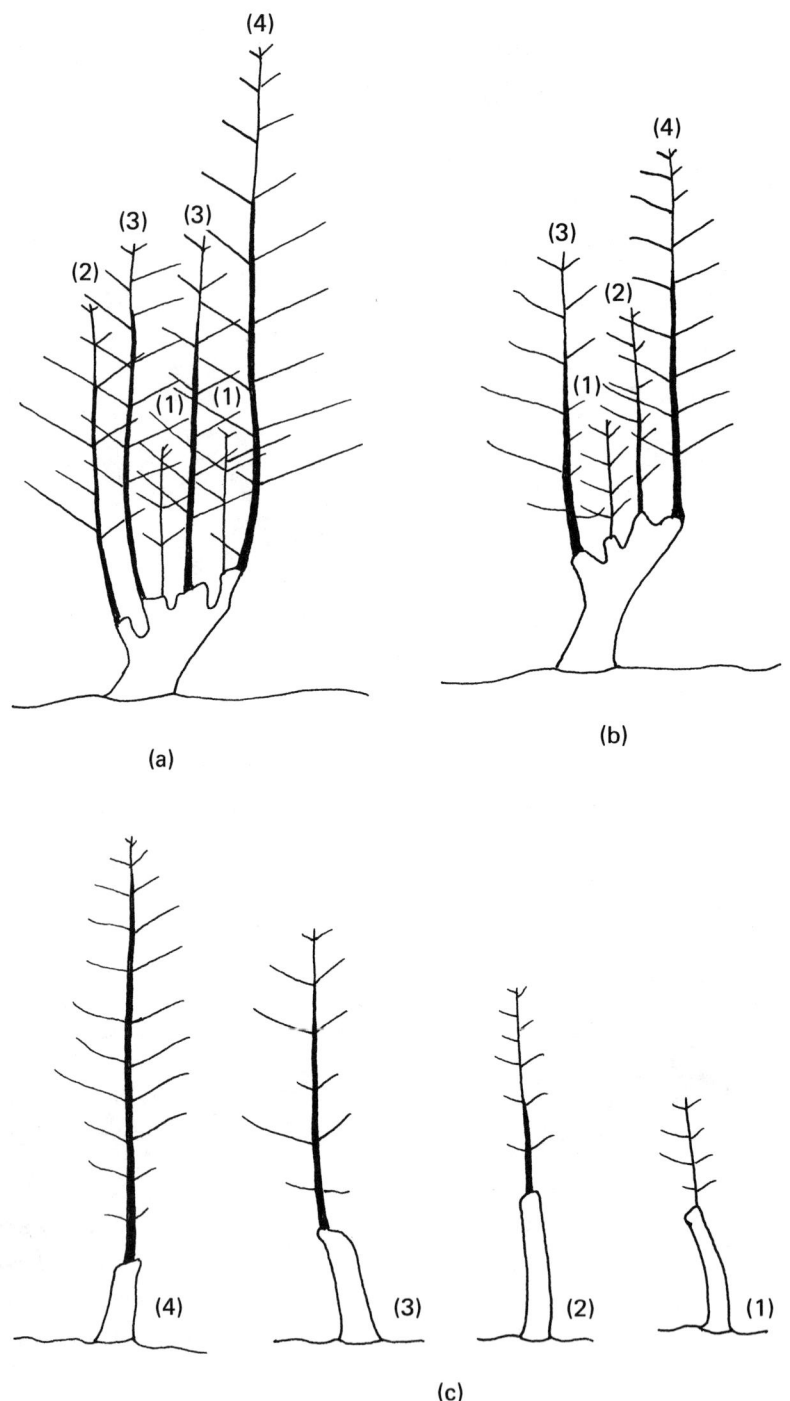

Fig. V.3.4.08. Coffee pruning (cont.): Hawaiian system, (a), (b) original version; (a) with six and (b) with four verticals; (c) new stump version [numbers indicate year of cycle; (1) and (2) are vegetative and (3) and (4), fruiting]. (a), (b) Redrawn from Goto and Fukunaga, 1956; see Acknowledgments and Selected References.

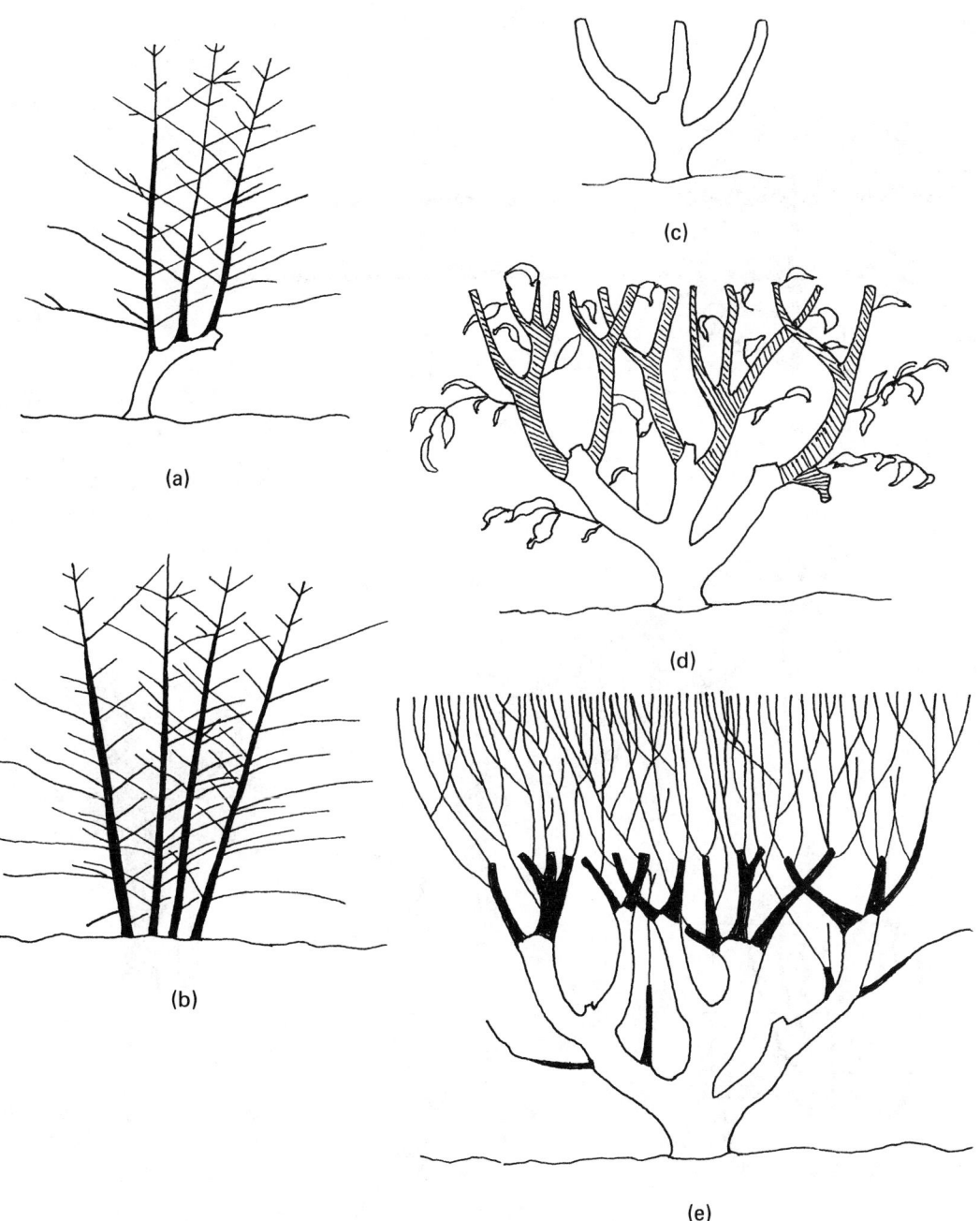

Fig. V.3.4.09. Coffee pruning (cont.); tea training: (a) agobio (Guatemalan) system; (b) "poda" (Brazilian) system. Tea training: (c) first pruning at 15 cm; (d) second pruning at 30 cm; (e) third pruning at 40 cm. (a) Redrawn from Ochse et al., 1961; (c)-(e) redrawn from TRI East Africa, 1967; see Acknowledgments and Selected References.

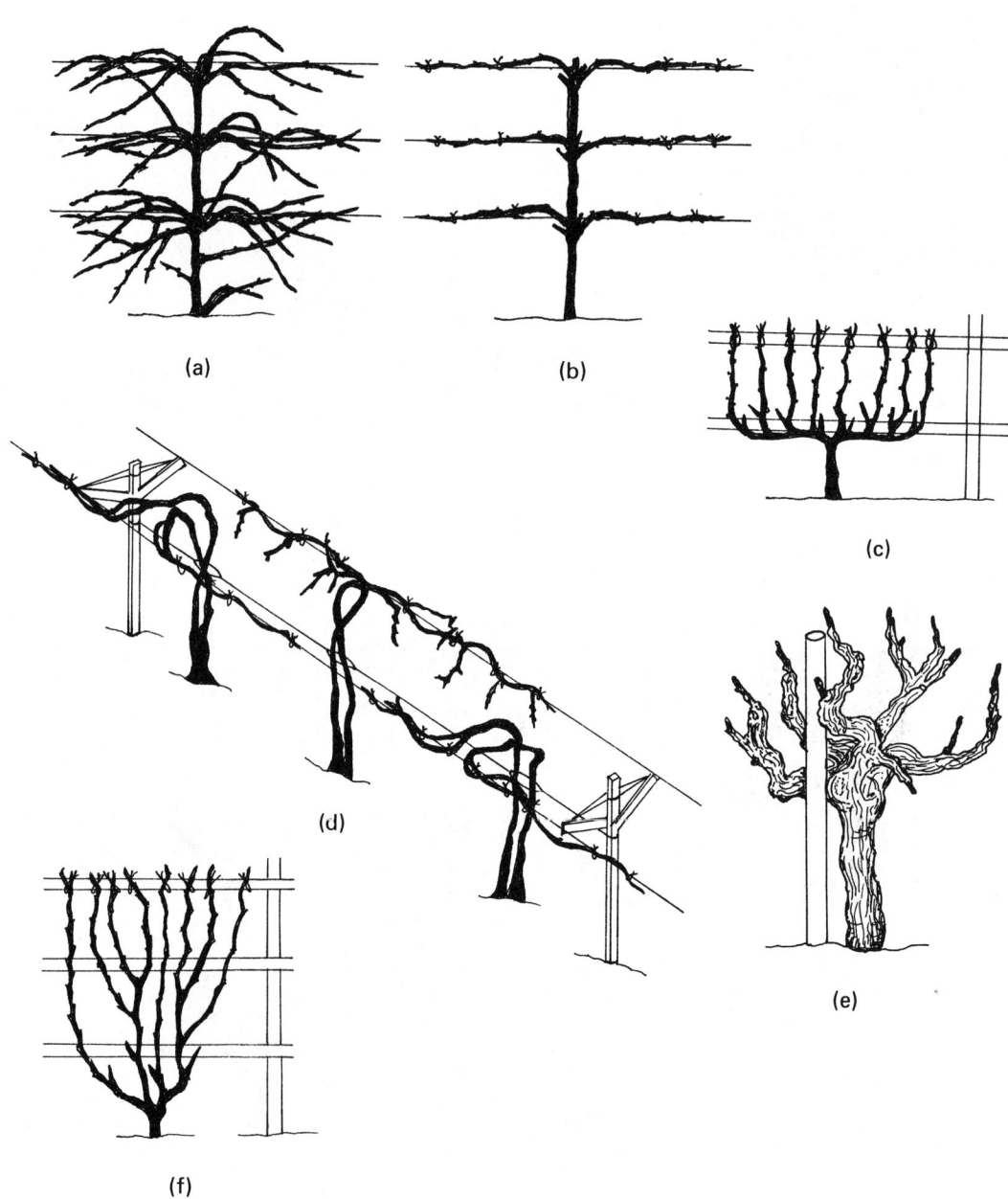

Fig. V.3.4.10. Grape training: (a) unpruned; (b) six-cane Kniffen system; (c) horizontal-arm spur system; (d) Geneva double-curtain system; (e) head system with spurs (vinifera cultivars in California); (f) fan system. Redrawn from Westwood, 1978; see Acknowledgments and Selected References.

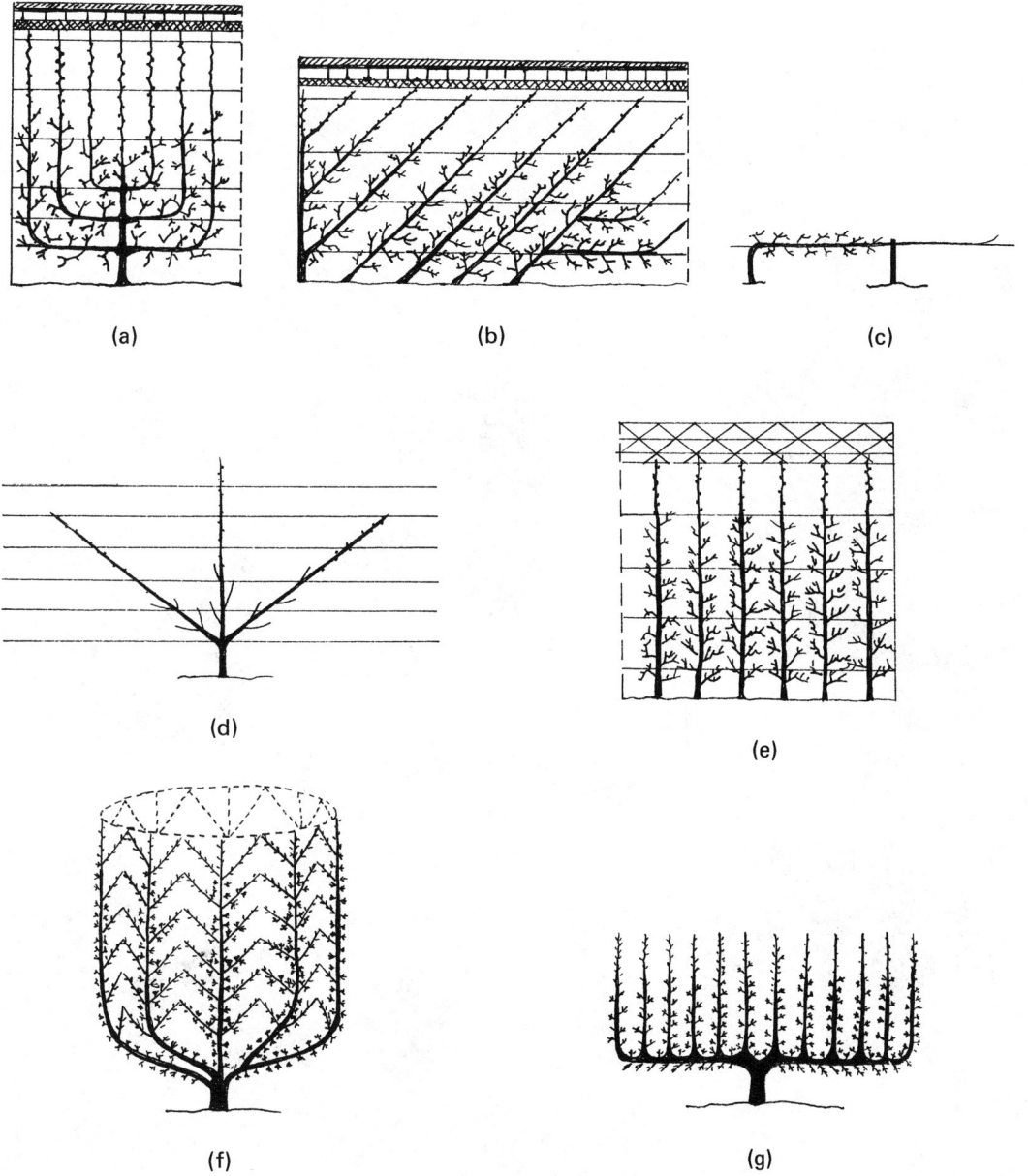

Fig. V.3.4.11. Espaliers, contre-espaliers (trellises), and free-standing forms of dwarfed fruit trees: (a) double-U palmette; (b) oblique cordons; (c) single-arm horizontal cordon; (d) Verrier palmette with three branches; (e) vertical cordons; (f) goblet with vertical branches; (g) candelabra with vertical arms [(a), (b), (e) espaliers; (c), (d) contre-espaliers; (f), (g) free standing]. (a)-(e) Redrawn from Tukey, 1964; (f) after Du Breuil, 1876; (g) after Lucas, 1909; see Acknowledgments and Selected References.

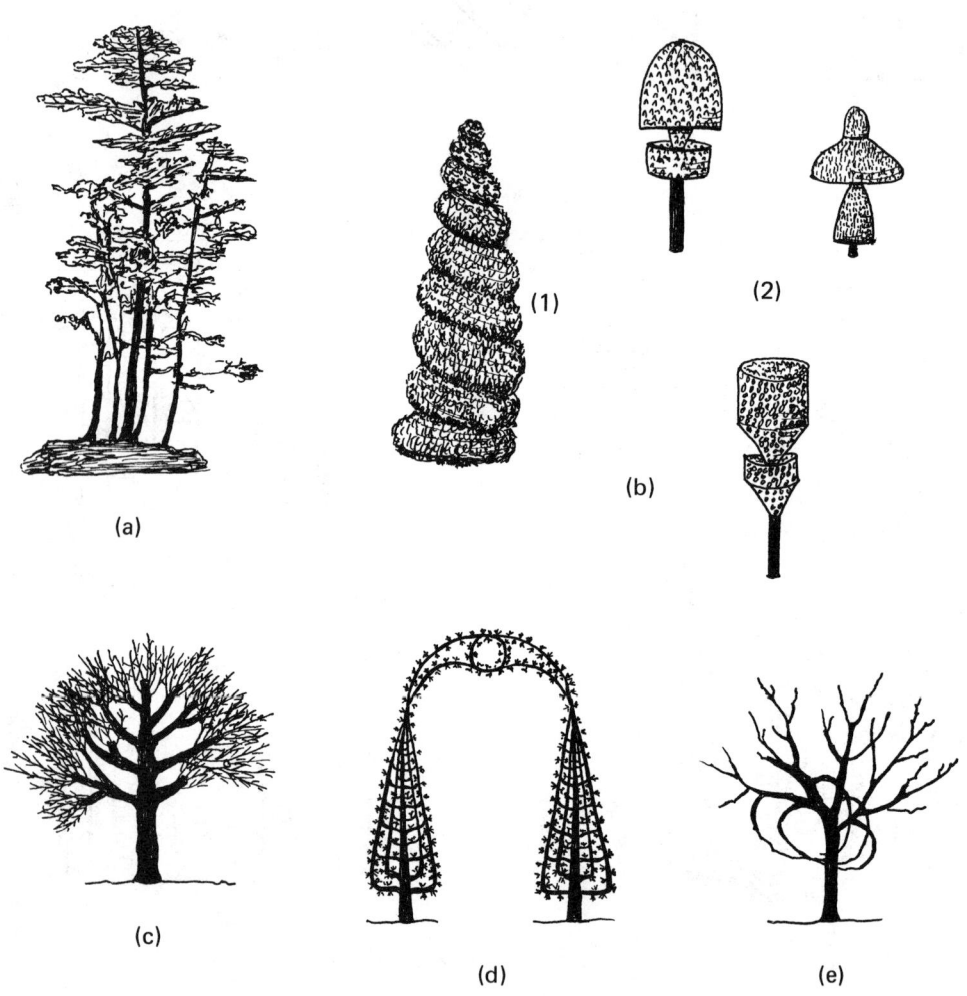

Fig. V.3.4.12. Dwarfed trees and art forms: (a) bonzai (pine trees); (b) some topiary forms in (1) United States and (2) English gardens; (c) pollarding (ls); (d) pleaching (archway of two winged pyramids with intertwined terminal arms); (e) bending or festooning (practiced occasionally in commercial orchards). (a) Redrawn from Tukey, 1964; [b.(1)] redrawn from Schuler, 1967; [b. (2)] redrawn from Gothein, 1928; (d) after Lucas, 1909; (e) adapted from Bush, 1943; see Acknowledgments and Selected References.

V.3.5. Water regulation

.000 Water regulation: the part of the cultural program that deals with the provision of proper soil-moisture conditions for the crop or crops; includes irrigation and drainage, the former is comprised of ways of supplementing precipitation and the latter, of prompt safe removal of excess water.

.001 Absorption blocks: porous ceramic, fiberglass, or other types of small block; e.g., Bouyoucos blocks, used for determining soil moisture content as a function of electrical resistance; now superseded by soil tensiometers largely because of the lack of consistent (i.e., reproducible) readings from a number of causes (slow response, tendency to decompose, disintegrate, or acquire salt deposits, and limited range of accuracy according to moisture content); (see Irrigation timing, V.3.5.013). (S & H)

.002 Drainage: removal of excess water from a field; two main types: (a) surface ditches, which consist of short tertiary ditches or shallow furrows and secondary or lateral and primary or main open ditches; the number, spacing, length, grade, depth, and slope of walls are determined by soil type and topography; have a graduated fall of about 0.5 to 1% from their beginning to the outfall; (b) subsoil drains, which may be made with a subsoil mole or pipe, are used on heavy soils when lateral movement of water is slight.

.003 Irrigation: supplemental water applied when precipitation and other sources of soil moisture are not adequate for crop requirements; may be seasonal or year-round, depending on the crop and the quantity and distribution of rainfall during the growing season; essential in arid or semiarid regions.

.004 Basin: (see Irrigation systems, V.3.5.012).
.005 Crown: (see Irrigation systems, V.3.5.012).
.006 Ditch: (see Drainage, V.3.5.002; Irrigation systems, V.3.5.012).
.007 Drip: (see Irrigation systems, V.3.5.012).
.008 Fixed: (see Irrigation systems, V.3.5.012).
.009 Flood: (see Irrigation systems, V.3.5.012).
.010 Furrow: (see Irrigation systems, V.3.5.012).
.011 Irrigation efficiency: the cost of a unit volume of water delivered to the soil surface, measured in terms of energy output per unit of crop harvested. (HAR)

.012 Irrigation systems: numerous types classifiable in two main groups and several subgroups: (a) surface, (1) flood or crown, in basins as a single-tree or multiple-tree type or furrow (ditch), as single or multiple ditches per row middle; (2) portable perforated pipe; (3) overhead sprinklers, as in portable or self-propelled volume guns, center-pivot or self-propelled traveling guns, permanent-set sprinklers; (4) under-tree sprinklers; (5) low-pressure, low-volume microsprinklers (under-tree); (6) drip (trickle); (7) perforated hose (seepers); (8) pop-up and (9) fixed (non-pop up) sprinklers (for home lawns, baseball fields, and parks); (b) subsurface, (1) mole drains (seepage) for field use, and (2) subsurface flood (for greenhouse benches and the like), tile drains for field or greenhouse use); (note that some of the subsurface types do double duty for irrigation and drainage). (T)

.013 Irrigation timing: several methods of determining when irrigation water should be applied, basically according to a calendar schedule or to need, as judged by visible symptoms (e.g., wilting of

- .014 Irrigation water measurement: the amount applied per application depends on soil storage capacity in the rooting area, infiltration rate (movement of water into the soil), and permeability (movement of water within the soil), all of which are functions of soil structural characteristics, and rate of application (i.e., centimeters or inches per hour), which is determined by pipe and nozzle sizes and water pressure for a given irrigation system. (T)
- .015 Microsprinkler: (see Irrigation systems, V.3.5.012).
- .016 Mole drains: (see Drainage, V.3.5.002; Irrigation systems, V.3.5.012).
- .017 Overhead sprinkler: (see Irrigation systems, V.3.5.012).
- .018 Perforated hose: (see Irrigation systems, V.3.5.012).
- .019 Perforated pipe: (see Irrigation systems, V.3.5.012).
- .020 Permanent-set sprinklers: (see Irrigation systems, V.3.5.012).
- .021 Pop-up sprinkler: (see Irrigation systems, V.3.5.012).
- .022 Riparian rights: the rights of a person who owns land with, containing, or bordering on a watercourse or other body of water in or to its banks, bed, or waters; and to have access to and have reasonable use, subject to state and/or federal restrictions, of same. (HC)
- .023 Seep irrigation: a type of furrow irrigation in which excess water is collected in drainage ditches that flow into catch basins, whence water is pumped via a pipe line into the reservoir (lake) serving as the source of irrigation water; distribution and return flow lines are PVC (plastic) pipe; (see Irrigation systems, V.3.5.012). (PV)
- .024 Soil neutron probe: an instrument for measuring soil moisture as a function of neutron absorption by water molecules (specifically, hydrogen ions); consists of a shielded radioisotope source in a probe, which is lowered into a hole in the soil made with a core drill, and neutron counter; requires careful calibration, hence is best suited for research measurements; (see Irrigation timing, V.3.5.013). (T)
- .025 Soil tensiometer: a device for measuring soil moisture; consists of a porous ceramic bulb connected to a suction gauge; (see Irrigation timing, V.3.5.013). (T)
- .026 Subsurface flood: (see Irrigation systems, V.3.5.012).
- .027 Surface: (see Irrigation systems, V.3.5.012).
- .028 Tile drains: (see Drainage, V.3.5.002; Irrigation systems, V.3.5.012).
- .029 Traveling gun: (see Irrigation systems, V.3.5.012).
- .030 Under-tree sprinkler: (see Irrigation systems, V.3.5.012).
- .031 Volume gun: (see Irrigation systems, V.3.5.012).
- .032 Water quality: suitability of water in a chemical sense for irrigation is dependent on the analysis of total dissolved salts and concentrations of chlorides, sodium, and boron and other toxic elements with respect to crop sensitivity, nature of the soil, type of irrigation management, and the climate; suitability in the physical and biotic senses is dependent on the size range of solid particles in the water, the likelihood of precipitation of dissolved salts (e.g., carbonates or sulfides), and build-up of

V.3.5. PROPAGATION, NURSERY HANDLING, SOILS, CROP PRODUCTION

bacterial or algal slime in pipe lines; these factors are of greater importance in low-volume, low-pressure systems with small nozzle orifices, as in drip or microsprinkler systems; (see Irrigation systems, V.3.5.012). (HC)

.033 Water rights: riparian rights (V.3.5.022); a right to the use of water, as for irrigation, originally acquired by appropriation and perfected by beneficial use or derived by ownership of riparian land; in the United States. These rights, if acquired by appropriation, may rest in the company making the diversion or in the individual to whose land the water is delivered, depending on the statutes and court decisions of the state concerned. (HC)

.034 Water supply: may come, for irrigation purposes, from an irrigation district (set up under state law), a mutual water company (water-user cooperatives), a lake, privately owned wells, or a private reservoir; irrigation districts may issue water permits to users or members of a mutual water company may have vested water rights; growing water shortages in many states have resulted in tighter restrictions on water permit allocations and on water rights, drilling of wells or use of lakes or reservoirs on private land for irrigation purposes; (see Water rights, V.3.5.033; Riparian rights, V.3.5.022). (HC)

V.4. Crop Production (Commercial Floriculture)

V.4.1. Culture and shipment to market

.001 After-lighting: an interrupted light program, now rarely used, to open up or increase the size of standard chrysanthemum blooms; typically consists of 12 short, then 10 long days, and back to short days.

.002 Artificial long days: interruption of the dark period or the extension of natural daylength to prevent flower bud initiation of short-day plants such as chrysanthemums or poinsettia. (LR)

.003 Basal plate: perennial, shortened modified stem, of a bulb, which has a growing point and interjoined scales and roots; (see Scaly bulb, V.4.1.104). (HR)

.004 Basal root: adventitious roots that develop from the basal plate; (see Scaly bulb, V.4.1.104). (HR)

.005 Bedding plants: (see Commercial floriculture, I.2.002a).

.006 Black cloth shading: used to reduce photoperiod and control height of Easter lilies, chrysanthemums, poinsettias, and the like. (LR)

.007 Blackout system: a means of covering plants with black sateen cloth or polyethylene film to shorten the photoperiod and promote flowering in short-day plants, such as Chrysanthemum morifolium. (LR)

.008 Blind shoots: stems that remain vegetative under conditions that normally stimulate inflorescence formation. (LR)

.009 Break: a new lateral shoot, often developed following removal of apical dominance by pinching. (LR)

.010 Bulb dealer: individual or firm that stores, ships, sells, and distributes bulbs. (HR)

.011 Bulb forcer: individual or firm that produces flowers or container-grown plants from bulbs. (HR)

.012 Bulb forcing: flowering of a bulb or plant by modifying growing conditions that simulate those required in nature; with spring-flowering bulbs, the natural seasons are reproduced by using a rooting room or outdoor rooting bed and a greenhouse. There are two basic techniques: (a) standard forcing, in which programming substitutes for the summer, fall, and winter and a greenhouse phase substitutes for spring; adaptable to all bulbs (i.e., tulip hyacinth, daffodil, grape hyacinth, and crocus) except Dutch iris; (b) special precooling, which has two phases: programming that substitutes for summer and winter and greenhouse that substitutes for fall and spring; tulips and Dutch iris are forced as cut flowers by special precooling. (HR)

.013 Bulb grower: individual or firm that propagates bulbs for wholesale purposes. (HR)

.014 Bulbing: (see Filling, V.1.045).

.015 Bulb jobber: (see Bulb dealer, V.4.1.010).

.016 Bulblet: small bulb produced on the underground portion of the stem around the base of an older bulb or by scale propagation; called "bulblet" until grown independently for a year; (see Stem bublet, V.4.1.115; Scale bulblet, V.4.102). (HR)

.017 Bulbous plants: (see Commercial floriculture, I.2.002b).

.018 Bulb pan: half-sized (half-depth) pot used for forcing spring-flowering bulbs. (HR)

.019 Bulb precooling (PC): dry or cold-moist (e.g., lilies) storage of bulbs at temperatures between 2 and 9°C (35 to 48°F). (HR)

.020 Bulb producer: (see Bulb grower, V.4.1.013).

.021 Bulb size: measured as the transverse circumference in centimeters or inches. (HR)

.022 Bulb tray: a wooden flat with a woven wire mesh bottom or a plastic flat used for storage of bulbous crops. (HR)

.023 Bullhead: a globose flower bud that usually results in a malformed flower. (LR)

.024 Bypassing shoot: vegetative shoot that develops immediately below the flower bud and often occurs on azaleas (analogous in some respects to a watersprout, which azaleas in open ground also produce, usually nearer the base of the plant). (LR)

.025 Case-cooled bulbs: bulbs, especially lilies, given low-temperature treatment while still in the shipping container. (LR)

.026 Commercial bulb: one of a size suitable for commercial forcing.

.027 Commercial floriculture: (see Ornamental horticulture, I.2.000).

.028 Container forcing: (see Standard forcing, V.4.1.114).

.029 Controlled temperature (lily bulb) forcing (CTF): procedure for forcing 'Ace' and 'Nellie White' lilies in which nonprecooled bulbs are potted, placed in a controlled temperature area at 15.6 to 17.2°C for two to four weeks, then cooled at 1.7 to 4.4°C for six to seven weeks prior to being placed in the greenhouse. (HR)

.030 Conventional forcing: (see Standard forcing, V.4.1.114).

.031 Crown bud: chrysanthemum inflorescence formed under adverse conditions, such as improper daylength; buds may abort. (LR)

.032 Cut flowers: (see Commercial floriculture, I.2.002c).

.033 Cyathium: inflorescence of plants such as poinsettia (in which the showy parts, as in Bougainvillea spp., are actually bracts); relatively inconspicuous but bears pistils, stamens, and nectaries (glands). (LR)

.034 Cyclic lighting: intermittent illumination during the dark period to simulate a long day and prevent flowering of short-day plants; (see Artificial long days, V.4.1.002). (LR)

.035 Daughter bulb: scales and leaves initiated by, and developing below

V.4.1. PROPAGATION, NURSERY HANDLING, SOILS, CROP PRODUCTION

and around, the new daughter apex which arises from a bud in the axil of a scale subtending the old (mother) bulb axis. (HR)
.036 Devernalization: negation of a vernalization stimulus by temperatures above a critical level. (HR)
.037 Direct forcing: (see Special precooling, V.4.1.111).
.038 Disbud, disbudding: the removal of lateral flower buds on stems of plants such as carnation or chrysanthemum to permit the remaining buds to develop into larger flowers. (LR)
.039 Dormancy: a physiological state of a healthy bulb characterized by a temporary delay in the sprouting or elongation of the daughter stem axis. (HR)
.040 Double nose bulb: a commercial bulb with two daughter bulbs and stem axes with a common basal plate; e.g., lily or Narcissus. (HR)
.041 Dry storage: storage of bulbs prior to planting; also used occasionally to refer to storage of flowers after harvesting (meaning in this sense flowers not stored in water). (HR)
.042 EC_e : electrical conductance measured as decisiemens per meter; (see Mho, V.4.1.071). (LR)
.043 Eskimo tulip: bulbs that have been fully rooted at 9°C and subsequently frozen at -3 to -5°C until removed eight to ten months later; used for fall flowering of tulips. (HR)
.044 Eye: a lateral bud, as on a rose stem. (LR)
.045 Filling: enlargement of the scales associated with storage of food reserves in the bulb; thought to be the bulbing response in lilies. (HR)
.046 "5° Forcing": (see Special precooling, V.4.1.111).
.047 Flower bud abortion: cessation of floral bud development at any stage. (HR)
.048 Flower differentiation: complete morphological development of the floral organs following initiation. (HR)
.049 Flower grower: (see Bulb forcer, V.4.1.011).
.050 Flower induction: an unobservable, preparatory step that occurs prior to visible flower-bud initiation. (HR)
.051 Flower initiation: visible organization of flower primordia at the stem apex. (HR)
.052 Foliage plants: (see Commercial floriculture, I.2.001d).
.053 Foliage plant selection: several connotations: (a) plants for use in breeding new cultivars; (b) plants chosen by a grower as sources of propagation material (e.g., seeds and cutting wood); (c) plants for use in specific indoor locations; the evaluation and selection of appropriate specimens from measurements and estimates of year-around light intensity, quality, and duration, high and low temperatures, relative humidity, water quality, expected pedestrian traffic patterns and location of heating, and cooling or ventilation systems. (JR)
.054 Forcing: acceleration of flowering of many plants by manipulation of environmental conditions (e.g., Bulb forcing, V.4.1.012). (HR)
.055 Fumigun: a large injection device for hand-injection of a fumigant into a medium (e.g., soil or soil mix). (LR)
.056 Geotropic bending: upward curvature (negative geotropic bending) of the tips of spikes, such as snapdragon or gladiolus, when held horizontally [stems of many plants with normally upright stems (e.g., palms) will show this curvature but it usually takes place slowly, i.e., over a period of weeks rather than a few hours, as in many herbaceous flowers]. (LR)
.057 Gooseneck, gooseneck stage of narcissus: the stage when the unopen-

ed flower bud of a daffodil curves at about right angles to the scape on which it is borne, the curvature occurring in the basal portion of the flower proper just above the point of articulation with the scape; flowers harvested earlier than this stage will not bend over or open properly. (HR)

.058 Grassy growth: excessive and noticeable production of axillary branches; e.g., on snapdragon stems. (LR)

.059 Greenhouse phase (forcing): that portion of forcing which encompasses the time from placing the plants in the greenhouse until flowering. (HR)

.060 Green pruning: pruning of actively growing rose bushes without benefit of a dormancy period. (LR)

.061 Hanging baskets: (see Commercial floriculture, I.2.002e).

.062 Hardening: conditioning of pot plants by using a cool (10 to 13°C = 50 to 55°F) growing regime just prior to flowering; also conditioning of cut flowers in water after they have been dry-stored and prior to sale or use in floral arrangements. (HR)

.063 Heat delay: a delay in flower-bud initiation as a result of abnormally high temperature. (LR)

.064 Juvenility: the early vegetative stage of growth when a bulb can readily increase in size but will not respond to flower induction treatments. (HR)

.065 Leaf counting: a procedure used to time Easter lily flowering according to the number of leaves on the plants. (LR)

.066 Light flux: the product of light intensity times the duration of light. (LR)

.067 Long-day treatment (LDT): use of light (10:00 p.m. to 2:00 a.m.) at the time of shoot emergence to promote rapid flowering of Easter lilies. (HR)

.068 Marketing phase (bulbous plants): movement of the plants and/or flowers from the forcing facilities to the wholesaler and/or the retailer at the proper stage of development so that the consumer receives the maximum possible enjoyment; may take place in some instances when the pots are removed from rooting areas (e.g., "Indoor spring gardens"); in other cases the plant and/or flowers are sold when they are just beginning to color (bud stage). (HR)

.069 Maturity: the measure of the capacity of a healthy daughter stem axis of a bulb to sprout without delay and to respond to flower-inducing treatments. (HR)

.070 Mat watering (capillary watering): irrigation of plants in containers by capillarity; mats are composed of fabric, cellulose, or other water-absorbing materials. (LR)

.071 Mho: reciprocal ohms, a measure of electrical conductance of water or soil solution; expressed as millimhos (0.001 mho) per centimeter = 1 decisiemen per meter; (see Electric conductance, IV. 0.073). (LR)

.072 Mother blocking: technique by which bulblets of selected mother bulbs are grown in units to establish true-to-name foundation stock and to obtain improved disease control. (HR)

.073 Mother bulb: that portion of the lily bulb that is currently flowering and producing a daughter bulb in the axil of a scale subtending the mother axis; the old mother scales encompass the new daughter axis. (HR)

.074 Multibranched plant: a plant with several shoots and flowers originating from a single cutting by pinching; a term often used to distinguish between a pot with a single bushy (i.e., multi-stemmed) plant versus one with several rooted cuttings. (LR)

V.4.1. PROPAGATION, NURSERY HANDLING, SOILS, CROP PRODUCTION

.075 Natural cooling: a technique in which nonprecooled commercial bulbs are planted immediately on arrival and grown under cool natural conditions but with frost protection prior to being placed in the greenhouse. (HR)
.076 Nonprecooled (NP) bulbs: bulbs which have received no precooling temperatures less than 13°C (55°F). (HR)
.077 Normal forcing: (see Standard forcing, V.4.1.114).
.078 NP: (see Nonprecooled bulbs, V.4.1.076).
.079 Osmocote: (see V.3.2.071).
.080 Panning: transplanting or potting rooted cuttings or bulbs. (LR)
.081 PC: (see Bulb precooling, V.4.1.019); Precooling, V.4.1.088).
.082 Perched water table: concentration of moisture at the bottom of a growing container; also a water table usually relatively near the surface in a soil profile containing an impermeable layer that prevents downward movement of the water. (LR)
.083 Physiological disorders: abnormal growth and development brought about by suboptimal environmental conditions; [see Physiological disorders (nonparasitic diseases), V.4.2.2]. (HR)
.084 Pinch: removal of a shoot apex to overcome apical dominance and promote lateral branching; usually done by hand when growth is tender. (LR)
.085 Planting back: a bulb that did not achieve commercial size during a normal production period and has been returned to the field. (HR)
.086 Planting stock of bulbous plants: bublets and yearlings of lily. (HR)
.087 Pot plants: (see Commercial floriculture, I.2.002f).
.088 Precooling (PC): a dry cold or cold-moist treatment prior to planting which induces rapid shoot elongation and flowering; bulbs are usually packed in moist peat. (HR)
.089 Preheating: the use of a 21°C temperature prior to planting nonprecooled bulbs or to a precooling period. (HR)
.090 PR: (see Prepared hyacinths, V.4.1.091).
.091 Prepared (PR) hyacinths: hyacinths that have been harvested earlier than normal and given special temperature treatments by the bulb grower or exporter so that they may be forced very early in the season. (HR)
.092 Production phase: all aspects of bulb production that lead to the sale of forcing-sized bulbs; may take one to three years, depending on the species. (HR)
.093 Programming phase: the portion of the forcing program in which the bulbs are induced to flower; methods used are precooling (V.4.1.088), natural cooling (V.4.1.075), and controlled temperature forcing (V.4.0.029) from the time they are harvested until they are placed under greenhouse conditions. (HR)
.094 Pruning: removal of vegetative shoots, or the shaping of plants, such as azaleas, by trimming with hand shears. (LR)
.095 Quantitative long-day plant: a plant that is not inhibited from flowering by short-day treatment but hastened by long-day treatment. (LR)
.096 Regular (RG) hyacinths: hyacinth bulbs that have been harvested at normal times and given temperature treatments for medium and late forcing periods. (HR)
.097 Response group: a classification of cultivars on environmental-based response; e.g., for chrysanthemums the number of the response group indicates the weeks required to flower after the start of short days; for snapdragons it indicates the time of year that the cultivar should be in flower. (LR)

.098 RG: (see Regular hyacinths, V.4.1.096).
.099 Rogueing (roguing): the selective removal of undesirable or diseased bulbs (see also V.1.0.006). (HR)
.100 Rooting bed: an outdoor area used to root and satisfy the cold requirement of bulbs. (HR)
.101 Rooting room: a controlled-temperature facility used to root and satisfy the cold requirement of bulbs. (HR)
.102 Scale bulblet: small bulb produced by the scaling method of production. (HR)
.103 Scaling: a technique for the propagation of foundation stock, wherein bulb scales are removed and planted; produces numerous bulblets from a single mother bulb (Fig. V.1.4.17) (HR)
.104 Scaly bulb: a bulb whose modified leaves are comprised of thick, fleshy, imbricated nongreen scales, as in lilies. (HR)
.105 Self-branching: the condition in which axillary buds may initiate growth without pinching. (LR)
.106 Shattering of snapdragon: abscission of snapdragon florets caused by unfavorable environmental conditions (e.g., ethylene) or pollination. (LR)
.107 Slabs: foliage plants, e.g., vining types of Philodendron and other Araceae, grown in containers and trained up a flat slab of rough sawn wood with the bark on one side (usually bald cypress or similar rot-resistant wood), which may be 30-90 cm (12 to 36 in.) tall, depending on the container size (15 to 30 cm or more). (JR)
.108 Slab-side of carnations: a carnation flower whose petals elongate and expand first on one side of the calyx; a lopsided flower. (LR)
.109 Slipping stage of carnations: the time when the inflorescence emerges from the leaf sheaths. (LR)
.110 Solubridge: an instrument used to measure electrical conductance; (see Mho, V.4.1.071). (LR)
.111 Special precooling (SPC) for tulips: a special technique used to program tulips for cut-flower forcing; the entire cold requirement is given as a dry cold treatment and rooting takes place during the greenhouse phase of forcing; also known as "5°(C) forcing" and "direct forcing"; this technique is also used for Dutch iris. (HR)
.112 Spring-flowering bulbs: a broad classification given to those bulbous plants that are planted outside in the fall, overwintered under low temperatures, and flower in the spring months. (LR)
.113 Stage "G" of tulip, etc.: term used to indicate gynoecium formation in tulip and other species of Liliaceae. (LR, HR)
.114 Standard forcing: a technique used to program spring-flowering bulbs in which all or most of the cold requirement is given after planting the bulbs; rooting takes place during the programming phase of forcing; also known as container, normal, or conventional forcing. (LR)
.115 Stem bulblet: a small bulb produced at an underground stem node; a source of planting stock. (Fig. V.1.4.17) (HR)
.116 Stem root: adventitious root produced on the underground portion of the stem in distinction to one from the basal plate. (HR)
.117 Summer sprouting of lily: premature growth of the daughter axis before the normal scale complement and bulb size are achieved. (HR)
.118 Tepal: (see II.2.2.061f).
.119 Totem pole plants: foliage plants, usually vining types of Philodendron and other Araceae, grown in containers and trained up a length of young sapling (bald or pond cypress poles of about 5 to

V.4.1. PROPAGATION, NURSERY HANDLING, SOILS, CROP PRODUCTION

7.5 cm = 2 to 3 in. in diameter sawn into lengths of 30 to 120 cm = 1 to 4 ft, with care to leave the bark intact). (JR)
.120 Tunic: dry, papery scales (modified leaves) that surround the fleshy organs of a bulb or corm; e.g., of an amaryllis, narcissus, or tulip bulb or a gladiolus corm. (G'62)
.121 Vernalization: cold-moist treatment applied to a bulb or plant to induce or hasten the development of the capacity for flowering. (LR)
.122 Yearlings: lily planting stock at the end of the first growing season from a bulblet; can be produced from stem or scale bulblets. (Fig. V.1.4.17) (HR)
.123 Year-round flowering: the control of day length and temperature to produce flowering plants throughout the year; often used to describe azalea and chrysanthemum programs. (LR)

V.4.2. Pests

V.4.2.1. Parasitic diseases, including animals

.001 Basal rot of lilies: chocolate-colored rot of lilies caused by Fusarium oxysporum f.sp. lilii; usually confined to warm climates; symptoms are stunting, yellowing, and premature dying, with eventual rotting of the entire bulb. (HR)
.002 Botrytis spp.: "fire" disease of lilies and other bulbs, especially tulips; attacks leaves or flowers, with spots, yellowing, and shriveling of leaves; leads to eventual total necrosis of aboveground parts; species that attack plants in the field and greenhouse are different. (HR)
.003 Bulb flies: larvae of three species of Diptera are important bulb pests; control is by hot-water or chemical treatment. (RE)
.004 Cucumber mosaic virus: aphid-borne virus with many strains and hosts; often produces no distinct symptoms in lilies; (see Fleck, V.4.2.1.009). (HR)
.005 Curl-stripe: white striping and twisting of leaves; appears shortly after emergence and is commonly associated with crooked terminal growth; possibly caused by one or more viruses. (HR)
.006 "Fire" disease of lilies: (see Botrytis spp., VI.4.2.1.002).
.007 Fire of bulbs: (see Botrytis spp., VI.4.2.1.002).
.008 "Fleck": (see Virus diseases of bulbs, V.4.2.1.022).
.009 Fleck: disease caused by a combination of cucumber mosaic virus (V.4.2.1.005) and lily symptomless virus (V.4.2.1.013); characterized by small white and brown spots and streaks on leaves and often deformed flowers. (HR)
.010 Fusarium oxysporum f. sp. lilii: (see Basal rot, V.4.2.1.001).
.011 Gummosis of bulbs: production of gum in tulip, Narcissus, hyacinth, and grape hyacinth caused by ethylene gas (which can be produced in quantity by Fusarium (hence the admonition to keep infected bulbs away from healthy ones and provide adequate ventilation of bulb storage to prevent accumulation of volatiles; also certain fruits, e.g., apples, are excellent sources of ethylene). (HR)
.012 Lily symptomless virus: an aphid-borne virus that alone produces no distinct symptoms in lilies. (HR)
.013 Mosaic of lily: yellowing patterns in the leaves (of lilies) produced by a virus related to the tulip-breaking virus. (HR)

.014 Nematodes (eel worms) attacking bulbous plants: Ditylenchus dipsaci attacks stems of narcissus and tulip and Pratylenchus spp., their roots and those of other bulbs; control is by hot-water or chemical treatment; (see also Plant parasitic nematodes, V.3.3.1.200). (RE)

.015 Penicillium rot of lilies, hyacinths and other bulbs: diseases caused by Penicillium spp., typified by the presence of blue-gray mold, large grayish-brown lesions on the outer scales, and black lesions on the bases of the inner scales, often near the point where the old stem was attached; rotting of bulbs may be rapid during storage. (HR)

.016 Pythium spp. on bulbs: one of several soil-borne fungi that can infect bulbs; in forcing, generally most serious as a bulb rot of special precooled tulips. (HR)

.017 Root base rot of lilies: a disease of the basal plate and roots involving a complex of organisms, possibly including nematodes and mites, Fusarium oxysporum, F. solani, and Cylindrocarpon radicicola; symptoms include root rotting, discoloration of root (vascular) traces into the lower basal plate, and, in advanced stages, rotting of the scale bases; severely infected bulbs have low root-producing potential and may fail to emerge during greenhouse forcing. (HR)

.018 Root rot: a condition of lily roots that may be caused by several organisms, either singly or in combination, such as Pythium, Rhizoctonia. and Fusarium spp.; nematodes and mites may also be involved. (HR)

.019 Rosette: a virus disease that causes down-curling of the leaves and shortened internodes with a yellowing and flattening [i.e., rosetting of the plants; (see also Virus diseases of bulbs, V.4.2.1.022)]. (HR)

.020 Sours of bulbs: a soft rot of bulbs caused by Fusarium; characterized by a sour smell and evolution of ethylene, which has a number of adverse physiological effects (gum formation, improper flower development), especially on tulips. (RE)

.021 Twist: sickle-shaped leaves that appear in one or two whorls in the midportion of the stem; a species of Pseudomonas has been implicated but the exact cause is unknown. (HR)

.022 Virus diseases of bulbs: numerous viruses known in tulip, Narcissus, and other bulbous plants include tulip-breaking virus, cucumber mosaic, yellow stripe, and white streak transmitted by aphids; tobacco necrosis, by fungi; tobacco rattle, tobacco ringspot, arabis mosaic, raspberry ringspot, tomato black ring, and strawberry latent ringspot, by nematodes; narcissus mosaic and possibly others, by mechanical means; several, such as cucumber mosaic and tobacco rattle, cause "fleck" (= corky fleck) in tulip; stunting and rosette symptoms are also common symptoms in lily, tulip, and others. (RE)

.023 Yellow disease: bacterial rot of hyacinth caused by Xanthomonas campestris pv. hyacinthi. (RE)

V.4.2.2. Physiological disorders (nonparasitic diseases)

.001 Antholyse (antholysis): retrograde metamorphosis, i.e., a metamorphosis of floral organs in which they become more or less foliaceous. (HR)

V.4.2.1. PROPAGATION, NURSERY HANDLING, SOILS, CROP PRODUCTION

.002 Blindstoken: a treatment for increasing the weight of tulip bulbs by killing the flower before field planting. (RE)

.003 Bullnose (bull-nose): physiological disorder of daffodils (Narcissus) characterized by failure of the flower to open properly after it reaches the gooseneck stage of development. (HR)

.004 Chalking of tulip: hard, dry, white spots resembling chalk in stored tulips; a physiological disorder generally resulting from poor handling techniques. (RE)

.005 Double stem: condition produced when a single bulb produces two separate scapes; often occurs in hyacinths. (HR)

.006 "Dropper" ("sinker") of tulip: a hollow, stolonlike (occasionally branched) structure which usually grows vertically downward or nearly so (rarely horizontal) from a mature tulip bulb and contains a daughter bulb near its tip, so that next year's bulb is at a lower level than its mother bulb; conditions favorable for dropper formation include shallow planting of the mother bulb, wet soil, low water table, high soil carbon-dioxide content, and a soil pH near neutrality; it occurs regularly in some species and cultivars but seems unrelated to horticultural groups, earliness of flowering, soluble materials in the bulb, or number of roots. (RE)

.007 Fasciation: fusion of two scapes to form a single floral stalk; occurs frequently with hyacinths. (HR)

.008 Feathering: a condition in which all or part of the inner scales developing around a daughter axis elongate faster than they are filled; may project out of the top of the bulb and will appear papery; can occur during the period from late spring to harvest; often confused with summer sprouting (V.4.1.117). (RE)

.009 Flat stem: (see Fasciation, V.4.2.2.007).

.010 Flower (bud) blasting: the failure of a bulb to produce a marketable flower after floral initiation; caused mainly by high temperatures before planting (e.g., "heating in transit"); (see also Kernrot, V.4.2.2.015; Flower blindness, V.4.2.2.011). (LR)

.011 Flower blindness of bulbs: failure of a bulb to produce any floral parts; characterized by complete lack of flower parts at the tip of the scape where the flower normally occurs; (compare with Flower blasting, V.4.2.2.010). (HR)

.012 Flower bud abortion: cessation of floral bud development at any stage. (LR)

.013 Flower bud abscission: dropping of flower buds at any stage of development. (HR)

.014 Hard base of tulip: a condition found particularly in forced, especially 5°C, tulips, when the tunic does not disintegrate sufficiently to permit normal emergence of roots; aggravated by early lifting and heat treatment; curable by cutting through the tunic just above the normal root-emergence area. (RE)

.015 Kernrot (stem rot): a physiological disorder of tulip characterized by the whole stem above the top leaf, including the flower, being necrosed and reduced in size; the disorder has a complex origin, being commonest when ventilation is poor, and some bulbs are infected with Fusarium oxysporum; ethylene is implicated but not directly responsible; other factors are forcing bulbs that are too small, forcing too early or at too high (or with widely fluctuating) temperatures, and poor rooting, plus feeding of mites on anthers that leaves wounds contaminated by fungal and bacterial spores whose growth leads to kernrot. (RE)

.016 Knuckling of tulip: a condition in some tulip cultivars, especially

V.4.2.2. PROPAGATION, NURSERY HANDLING, SOILS, CROP PRODUCTION

when large bulbs (more than 13 cm) are being forced, in which the shoot emerges in a doubled-over position with the leaf tip retained in the bulb; the cause is unknown. (RE)

.017 Leaf scorch: a physiological disorder manifested as sickle-shaped necrotic areas that develop along the margins and tips of leaves. (HR)

.018 Sinker: (see "Dropper" of tulip, V.4.2.2.006).

.019 "Spitting" of hyacinth: abscission of the entire scape and inflorescence from the basal plate during forcing. (HR)

.020 Stem topple of tulip: a physiological disorder characterized by the collapse of a small portion of the internode of the scape located just underneath the flower; can occur either just before or during flowering. (LR)

.021 Tied-leaf: a condition in tulips in which the demarcation between the last leaf and the perianth is unclear (i.e., the last leaf is more or less petaloid and partly fused to the perianth), thus preventing elongation of the stem on one side just below the flower so that the latter is frequently bent at an angle instead of being vertical on the peduncle; apparently caused (or influenced) by high temperature that induces slightly premature flower initiation (i.e., before the last leaf is finally formed). (RE)

VI.1. POSTHARVEST HANDLING and MARKETING

1. Maturity — 569
       Fig. VI.1.01,02 — 573
2. Harvesting — 575
3. Packinghouse — 577
4. Containers — 585
5. Refrigeration, Storage — 588
6. Carriers (Loading, Routing, Services) — 594
7. Laws, Rules, and Regulations — 597
   7.1. Perishable Agricultural Commodities Act (PACA) — 597
   7.2. Other laws, rules, and regulations — 609
8. Destination Markets, Business Terminology — 616
9. Market Diseases of Fruits and Vegetables — 626
   9.1. Pathological (parasitic) disorders — 626
   9.2. Animal pests that cause market disorders — 659
   9.3. Physiological market disorders — 663
10. Processing — 674

VI.1. Maturity

.001 Ageing: any increment of time which may (or may not) be accompanied by physiological change; (see Senescence, VI.1.016). (WT)

.002 Climacteric period: the period in the development of some plant parts that involves a series of biochemical changes associated with the natural respiratory rise and autocatalytic production of ethylene. The climacteric period consists of the preclimacteric, preclimacteric minimum, climacteric rise, climacteric peak, and postclimacteric phases. (WT) [see Maturity (climacteric fruits), VI.1. 006; Maturity (non-climacteric fruits), VI.1.007]. (Fig. VI.1.01)

.003 Firmness ("pressure") tester: a spring-loaded device with several interchangeable tips used for determining the flesh or rind firmness of fruits or vegetables as a guide to their stage of maturity or ripening; used for a number of deciduous fruits and also certain citrus (e.g., limes) to ascertain whether they have lost enough turgidity to permit safe harvesting (turgid fruit are highly sensitive to handling which causes rupturing of oil cells in the rind); note that it is force rather than pressure that is actually measured; therefore such determinations should be reported in $\underline{force}$ units as newtons ($N = m \cdot kg \cdot s^{-2}$); values such as pound-force ($\overline{lbf} = 0.2248N$) or kilogram-force ($kgf = 0.1002N$) should not be used (see IV.0.260m).

.004 Horticultural maturity: the state of development when a plant or plant part possesses the prerequisites for utilization by consumers for a particular purpose (WT); the stage at which growth or development is optimal for a particular use (may be at an optimum point of development or consumption or for processing at harvest or may ripen to acceptable quality after harvest). Three classes of horticultural maturity are (a) harvested physiologically immature: green cucumbers, green tomatoes, summer squash, gooseberries, cherries for brining; (b) harvested firm mature but ripened later: pears, winter apples, fresh prunes, plums, apricots, peaches; (c) harvested when ripe: berries, cherries, slicing tomatoes, nuts, prunes for canning or drying, fruit for roadside markets. (Fig. VI.1.02) (WS)

.005 Maturation: the stage of development leading to the attainment of physiological or horticultural maturity (WT); the final stage of fruit development prior to ripening; includes physical changes, such as decrease in firmness, changes in texture, decrease in skin chlorophylls, and increases in carotenoids, xanthophylls, and anthocyanins (the first two for ground color, the last for blush or overcolor), and internal chemical and physiological changes, such as a decrease in starch (some fruits), increases in sugars, other soluble solids, soluble pectin, decrease in acidity, and, for some, a decrease in respiratory activity. (Fig. VI.1.02) (WS)

.006 Maturity (climacteric fruits): physiological maturity, maturity in the ordinary sense, is the attainment, after full development, of the stage just prior to the start of ripening (see also Horticultural maturity, VI.1.004). (Fig. VI.1.01) (WS)

.007 Maturity (nonclimacteric fruits): fruits that do not contain starch, inulin, oil, or other sources of energy on reaching maturation and in consequence do not undergo ripening (and the concomitant rise in respiratory activity); therefore they may be harvested when they become edible; there is no clearly defined stage of maturation or maturity and the quality of such fruit (with minor exceptions such as lemons which are harvested immature and cured) is at its highest point at the time it is removed from the plant; examples are citrus fruits, pineapple, and a number of other tropical fruits. (S & G)

.008 Maturity index: factors measured to indicate horticultural maturity include flesh firmness, skin and/or flesh color, sugar content, total soluble solids, total (titratable) acid, soluble solids-acid ratio, juice content, chlorophyll content, carotenoid content, number of days from full term, and heat-unit accumulation (degree days). (WS)

.008a. °Brix: a measure of the total soluble solids in fruit and vegetable juices; calibrated in terms of pure sucrose, i.e., 1° Brix = 1% sucrose; originally adopted for processed products, now used interchangeably with total soluble solids for fresh fruits; measured with a Brix hydrometer (specific gravity) or a refractometer equipped with a sucrose scale (refractive index). (S & G)

.008b. Color break: the natural change in ground color of a fruit (excludes sunburn and insect damage among others) whereby the chlorophylls diminish and other pigments (i.e., carotenoids and anthocyanins) become evident. (FDOC)

.008c. Hydrometer: an instrument for measuring specific gravity; used for determining total soluble solids or °Brix in fruit juices, then calibrated in terms of percentage of pure sucrose, us-

VI.1. POSTHARVEST HANDLING and MARKETING

ually at 20°C, although some may be at 17.5°C; differs from a refractometer by requiring a sample of several hundred milliliters; thus is used for determining total soluble solids or Brix values of composite samples (e.g., in citrus, 20 oranges or 10 grapefruit of a given size). (S & G)

.008d. Refractometer: an instrument used for measuring the refractive index of liquids, such as juices or oil content of horticultural fruits; two main types, the large Abbé for precise determinations (i.e., refractive indices to four decimal places) and the hand model for field use; the former may have an auxiliary juice scale calibrated in °Brix (or % total soluble solids) = % pure sucrose in addition to the regular scale; the latter normally has only a single scale calibrated in °Brix (for sucrose or total soluble solids determinations); usually calibrated for 20°, although some models may be 25 or 28°C; requires a sample consisting of only a drop or two of liquid. (S & G)

.008e. Total soluble solids: the sugars, organic acids, and other soluble components in the juice of fruits; measured in terms of percentage of pure sucrose as the major constituent, either by specific gravity with a hydrometer or by refractive index with a refractometer; both instruments are equipped with an appropriate scale calibrated for a given temperature (usually 20°C); note that a correction for acid may be necessary if the juice contains more than about 2% (e.g., when determining total soluble solids in a lemon, lime, or other acid fruit). (FDOC, S & G)

.008f. Total soluble solids: total (titratable) acid ratio: essentially the ratio of sugars to acid in fruit juice, hence sweetness; a widely used measure for maturity of citrus fruits, e.g., in California and other areas with cool dry climates, where it is often the only standard. (S & G)

.008g. Total (titratable) acid: the organic acids in the juice of fruits, usually measured by titration with standard alkali, using phenolphthalein or another suitable indicator; calibration of the standard alkali (NaOH) will vary according to the principal acid present; that for citrus, pineapple, mango, and a number of other fruits is in terms of percentage anhydrous citric acid, but malic acid for apples and tartaric acid for grapes; true total acid is rarely measured because it requires a double titration (first with standard HCl to displace acid groups on various organic compounds, then with standard alkali). (FDOC, S & G)

.009 Maturity standards: legal (or industry) requirements for color break and certain internal qualities, such as juice content (measured as percentage by weight or volume, gallons per box, cc per fruit, etc.), percentage sugars, total soluble solids (as °Brix = % pure sucrose), total (titratable) acid (as percentage anhydrous citric, malic, etc.), total soluble solids: total acid ratio, and the like; limits are set by legislative action; therefore they are completely arbitrary but must be within the limits set by nature to be workable; (note that standards for grades of fresh fruits or vegetables often include maturity standards). (FDOC)

.010 Maturity test: determination by chemical or physical means of the actual content or level of a given quality, such as moisture, firmness, color break, juice, and total soluble solids; may be official or unofficial (e.g., determination of total soluble solids by hydrometer is official, that by refractometer, unofficial for Florida citrus fruit); methods used for official tests must be (a) simple (i.e., do not require expensive equipment or highly skilled personnel), (b) quick, and (c) reliable (i.e., results reflect the actual content or level within a reasonable limit of error when a test is performed). (S & G)

.011 **Maturity test sample:** usually a composite of a certain number or weight of fruit or vegetable determined by prior testing to be adequate to provide an accurate reflection of the content or level of internal qualities of the lot; usually consists of a separate sample for each size rather than one of random sizes. (S & G)

.012 **Physiological maturity:** the stage of development when a plant or plant part will continue ontogeny even if detached. (see Horticultural maturity, VI.1.003). (Fig. VI.1.02) (WT)

.013 **Ripening:** (1) the composite of the processes that occur from the latter stages of growth and development through the early stages of senescence and that results in characteristic and/or food quality, as evidenced by changes in composition, color, texture, or other sensory attributes (WT); (2) the transformation of physiologically mature fruit from an unfavorable state of firmness, texture, color, flavor, and aroma to a more favorable state for consumption; may occur before harvest, as in berries, stone fruits, nuts, figs, and grapes, or largely or entirely after harvest, as in pear, quince, late apples, avocado, and persimmons; may be more or less synonymous with maturity in nonclimacteric fruits (or vegetables), such as citrus, which do not ripen after harvest. (Fig. VI.1.02) (WS)

.014 **Sample:** a certain number or weight of objects of a lot (e.g., of harvested fruit or vegetables) or a population (e.g., fruit on a single tree, a block of trees, or an entire grove or orchard or a field of vegetables); the key factor is that the sample be representative of the lot or population. (S & G)

.015 **Sampling:** the choice of a certain number or weight of objects to constitute a sample for purposes of testing for maturity and/or grade; a sample for maturity may be obtained directly from trees, usually at a convenient height at the cardinal points, or from containers after harvesting and always of a given size (prior tests have shown that this is necessary in most instances); a sample for grade may be obtained from packing bins or packed containers (thus sized) and will consist of a number large enough to ensure that it is representative of the lot (usual practice is to take samples from each size in a lot and inspect them separately for compliance with standards for grade); (note that sampling for unofficial and official purposes is essentially the same, the primary difference being that the former is done by the grower or packinghouse personnel and the latter, by government inspectors). (S & G, FDOC)

.016 **Senescence:** those processes that follow physiological maturity or horticultural maturity and lead to death of tissue; (see Ageing, VI.1.001). (Fig.VI.1.02) (WT)

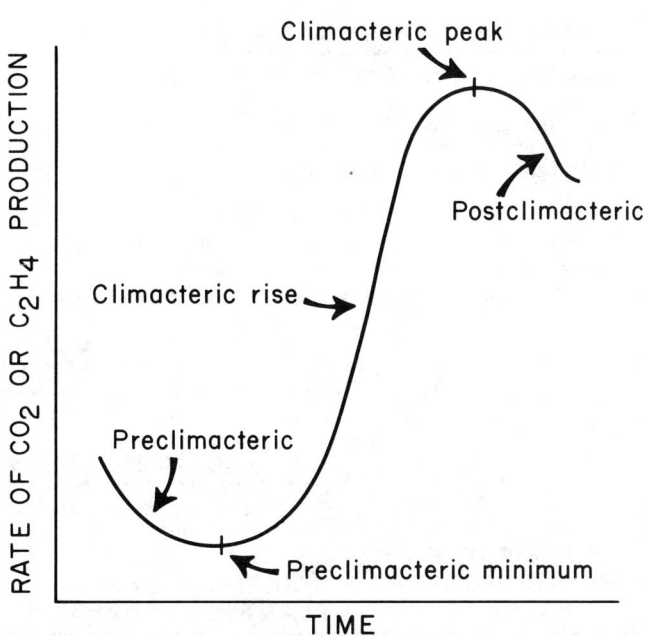

Fig.VI.1.01. Phases of the climacteric period. From Watada et al., 1984; see Acknowledgments and Selected References.

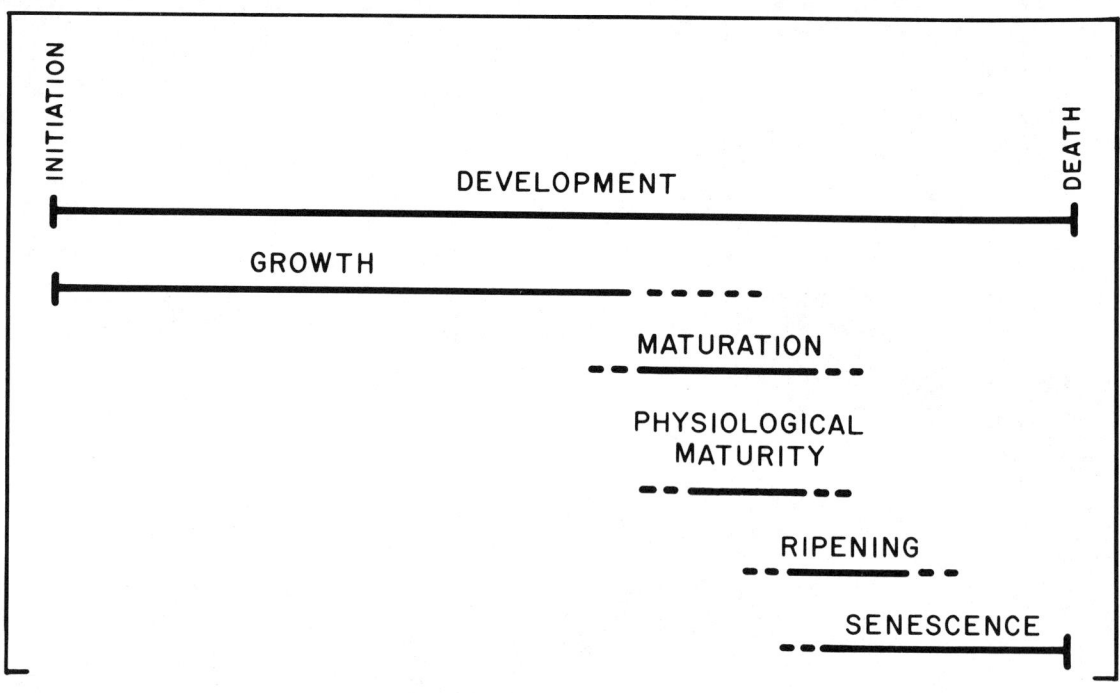
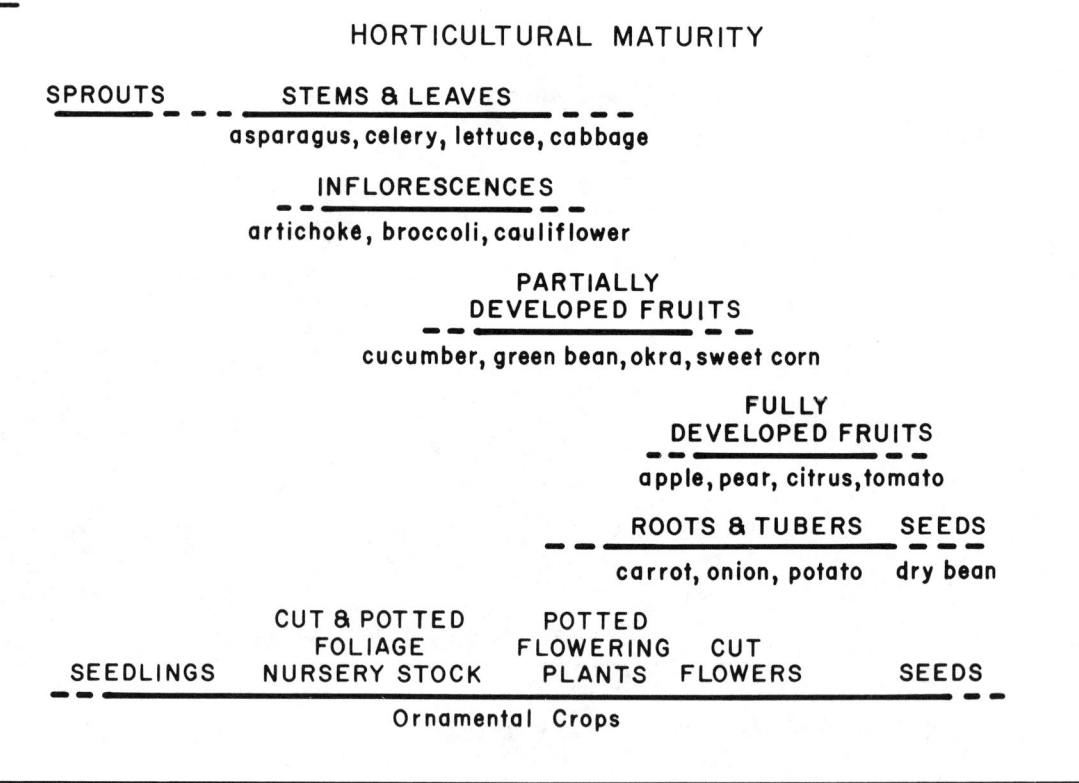

Fig.VI.1.02. Stages of development and senescence based on physiological processes and usage of horticultural crops. From Watada et al., 1984; see Acknowledgments and Selected References.

VI.2. POSTHARVEST HANDLING and MARKETING

VI.2. Harvesting

.000 Harvesting: the gathering of fruit, the cutting of stems with flowers, the digging of bulbs, tubers, tuberous roots, and the like at the proper stage of development for use in fresh or processed form as fruits or vegetables or for their esthetic value as floral arrangements or other decorative purposes; the first of a series of operations to effect the transfer of fruits, vegetables, cut flowers, or foliage from the orchard, field, greenhouse, or garden to the consumer or processor efficiently, economically, and with minimum wastage and/or deterioration in quality.
.001 Air blast shaker: (see Harvesting methods, VI.2.022b).
.002 Augers: (see Harvesting methods, VI.2.022b).
.003 Boom shaker: (see Harvesting methods, VI.2.022b).
.004 Bush harvester: (see Harvesting methods, VI.2.022b).
.005 Clean picking: (see Harvesting types, VI.2.023).
.006 Clipping: (see Harvesting methods, VI.2.022a).
.007 Corn picker: (see Harvesting methods, VI.2.022a).
.008 Cross-row conveyor central collection: (see Harvesting methods, VI.2.022c).
.009 Cutting: (see Harvesting methods, VI.2.022a).
.010 Digger: (see Harvesting methods, VI.2.022b).
.011 Digging: (see Harvesting methods, VI.2.022a,b).
.012 Field-to-packinghouse handling: fruits, vegetables, cut flowers, or foliage are harvested, placed in field boxes, crates, baskets, hampers, pallet boxes, or bins, and loaded by hand, forklift, or hydraulic lift onto small or large trucks, flatbed semitrailers, bulk trailers or semitrailers (processed fruit and vegetables), or stacked for hauling with a straddle carrier for transportation to the packinghouse; may involve separate in-field or in-orchard to roadside loading and reloading at roadside or direct packing into shipping containers in the field; or some mechanically harvested vegetables may be loaded directly into a truck as part of the harvesting operation.
.013 Field wrapping: packaging or wrapping vegetables in the field at the time of harvest. (RS)
.014 Flatbed truck with boom-type lift: a hydraulic lifter that picks up loaded pallet boxes by the top (i.e., by clamps), stacks them two (or sometimes more) high on the same truck or a flatbed semitrailer, or dumps them into a bulk trailer.
.015 Flexible hooks: (see Harvesting methods, VI.2.022b).
.016 Foliage shaker: (see Harvesting methods, VI.2.022b).
.017 Forking: (see Harvesting methods, VI.2.022a).
.018 Forklift tractor: a tractor equipped with a mast and forks at both ends for transporting, loading, and unloading pallet boxes at roadside near an orchard or grove or at a packinghouse; now often superseded by industrial forklift trucks for nongrove or field transporting.
.019 Hand harvesting: (see Harvesting methods, VI.2.022a).
.020 Handling systems: the movement of a product from one place to another in some predetermined organized manner which includes a number of unit operations (VI.3.079) linked together by conveyors (VI.3.019); may be simple, as in the movement of produce from the field (or orchard) to a roadside stand, where it may be sold directly from field containers, or highly complex and involving several systems linked together, as in movements of large-volume,

VI.2. POSTHARVEST HANDLING and MARKETING

commercial products, such as citrus, apples, lettuce, and bananas, with distinct field-to-packinghouse, packinghouse, transportation-to-market destination, and market (terminal and in-store) handling procedures; (many of the operations are similar but may be done in many different ways).

.021 Hand scoop: (see Harvesting methods, VI.2.022).

.022 Harvesting methods: numerous ways in which harvesting may be accomplished, classifiable in two primary groups: (a) those involving hand labor with or without tools as aids, and (b) those involving machines, with the only labor being the driver; also a third intermediate group, denoted as (c) semimechanized, in which machines are used as collecting devices or picking aids along with laborers who do the actual harvesting.

.022a. Hand harvesting: includes (a) physical removal of fruit or whole plants, including some vegetables, often with a combined twist and pull (pulling); (b) clipping with shears of many different styles, often blunt nosed with slightly curved blades (clipping); (c) cutting with a short hand-held curved knive, as with many flowers, vegetables, and foliage; (d) "snatching" fruit, using a hook on the end of a long pole; (e) cutting with a curved knife and bag to catch the fruit on the end of a pole, as often used for avocado, mango, or other large soft fruits; (f) cutting with a curved knife and rag on the end of a pole (the rag soaked with disinfectant to swab the cut on the plant) as for cacao; (g) cutting off an entire bunch with a machete, as for banana and plantain, with a preliminary cut in the pseudostem to cause it to topple over in the case of tall plants; (h) cutting off a bunch of oil palm fruit with a chisel-pointed steel rod (or a short machete); (i) snapping off or a simple bending and breaking of a pineapple peduncle close to the fruit; (j) plucking of tender tea stems with two partially grown leaves and a bud; (k) digging, forking, or plowing out tubers, tuberous roots, bulbs, and the like with a spade, digging fork, potato fork, or hand plow; (l) scooping blueberries or cranberries with a multitined hand scoop. (WS, OSDW, S & G)

.022b. Mechanized harvesting: (a) various types of digging machine for potato, carrot, radish, and other root and tuber crops; (b) corn pickers; (c) vine or bush harvesters for tomato, beans, etc.; (d) mechanical rakes, scoops, or sweeps for small fruits (e.g., blueberry, cranberry, currants, gooseberry) and nuts; (e) trunk, limb, or foliage shakers for prunes, apple, cherry, nuts, grape, and citrus (experimental), such as pole, cable, boom, oscillating air blast, and water blast shakers with or without catch frames and fruit collectors; (f) rotating rollers and cones, flexible hooks, vacuum cups, rubber spindles, rubber augers, stripper fingers, such as experimental machines for citrus for processing; grape cluster harvester; (g) tea harvester (sicklebar type with rubber fingers and stationary loops, developed in Russia); (h) mowers with catch frames for seed harvesting. (HD, FD, WH)

.022c. Semimechanized harvesting: (a) mobile (usually hydraulic) platform or basket as harvesting aids for tall plants, such as date, papaya, and cherry; (b) cross-row conveyors with central collection-grading-packing station ("mule train") for row crops like pineapple, lettuce, and tomato among others; (c) hand-held pneumatic power shaker for fruit and nut crops.

VI.2. POSTHARVEST HANDLING and MARKETING

.023 Harvesting types: two forms: (a) selective (spot-picking), when certain fruit, are harvested from an orchard (grove), field, or garden on the basis of color, size, or maturity; this operation is repeated two or three to several times; and (b) nonselective (clean picking), when all the fruit (except those of obviously different blooms, as may occur in citrus, or stages of ripening, as in coffee or black pepper and many vegetables) are harvested at one time. (S & G, OSDW)
.024 Limb shaker: (see Harvesting methods, VI.2.022b).
.025 Mechanical rake: (see Harvesting methods, VI.2.022b).
.026 Mechanical scoop: (see Harvesting methods, VI.2.022b).
.027 Mechanical sweep: (see Harvesting methods, VI.2.022b).
.028 Mechanized harvesting: (see Harvesting methods, VI.2.022b).
.029 Mobile (hydraulic platform): (see Harvesting methods, Vi.2.022c).
.030 Mower: (see Harvesting methods, VI.2.022b).
.031 "Mule train": a semimechanized harvesting system that consists of a central collection unit (truck or truck and trailer, often with facilities for grading and packing), which travels down a service road, with two projecting wings that consist of conveyor belts suspended over the rows of the crop; workers in each row cut, snap, or pull off the fruit or vegetable from individual plants and place them on the conveyor; originally devised for tomato and lettuce harvesting; now also used for pineapple and other low-growing row crops.
.032 Nonselective harvesting: (see Harvesting types, VI.2.023).
.033 Plucking: (see Harvesting methods, VI.2.022a).
.034 Pulling: (see Harvesting methods, VI.2.022a).
.035 Rotating cones: (see Harvesting methods, VI.2.022b).
.036 Rotating rollers: (see Harvesting methods, VI.2.022b).
.037 Rubber spindles: (see Harvesting methods, VI.2.022b).
.038 Selective harvesting: (see Harvesting types, VI.2.023).
.039 Semimechanized harvesting: (see Harvesting methods, VI.2.022c).
.040 Snapping: (see Harvesting methods, VI.2.022a).
.041 "Snatching": (see Harvesting methods, VI.2.022a).
.042 Spot-picking: (see Harvesting types, VI.2.023).
.043 Stripper fingers: (see Harvesting methods, VI.2.022b).
.044 Tea harvester: (see Harvesting methods, VI.2.022b).
.045 Trunk shaker: (see Harvesting methods, VI.2.022b).
.046 Vacuum cups: (see Harvesting methods, VI.2.022b).
.047 Vine harvester: (see Harvesting methods, VI.2.022b).
.048 Water blast shaker: (see Harvesting methods, VI.2.022b).

VI.3. Packinghouse

.000 Packinghouse (packing shed): a building to which fruits, vegetables or other products (e.g., cut flowers and bedding and foliage plants) are brought to ready them for shipment to market, including necessary cleaning, disinfecting, sorting, sizing, packing, and loading; no two packinghouses have the same mix or volume of produce, house size, or layout, with all degrees of operation efficiency among them; three basic types: commercial packinghouses, mail-order and gift houses, and roadside stands; many operations formerly carried out in some vegetable and fruit packinghouses are now often done in the field (e.g., field sorting and packing leafy vegetables and pineapples for fresh consump-

VI.3. POSTHARVEST HANDLING and MARKETING

tion. (RS, S & G)
.001 Bag filling: (see Packing operations, VI.3.053).
.002 Bassine: (see Brushes, VI.3.007; Washing, VI.3.081).
.003 Belt-and-roll sizer: (see Sizing, VI.3.064a).
.004 Benlate: (see Fungicide application, VI.3.033).
.005 Benomyl: (see Fungicide application, VI.3.033).
.006 Box (crate) nailer: a machine for putting together wood boxes and/or nailing on the cover.
.007 Brushes: cylindrical transverse-mounted brushes with natural or synthetic fibers inserted as rows of tufts fastened to a wood or metal core; common natural materials are bassine (palmyra palm) for some vegetables, palmetto, tampico (sisal) and horse hair, the first three for washing, the last for other operations; synthetic materials are Saran or nylon, made in various grades of stiffness; brush diameters range from 15 to 45 cm (6 to 18 in.) and lengths of about 75 to 240 cm (30 to 96 in.); used for many purposes in a packinghouse, such as wipe-offs. (RS, S & G)
.008 Bulk filling: (see Packing operations, VI.3.053b).
.009 Carton filling: (see Packing operations, VI.3.053c).
.010 Carton stapler: a heavy-duty, hand-held or mechanical stapler for fastening cartons together prior to filling or for closing the lids after filling.
.011 Central sizer: (see Sizing, VI.3.064a).
.012 Clamp-lift truck: a lift truck equipped with adjustable clamps for handling several stacks of field boxes (8, 12, or 16) as a unit load.
.013 Clamp truck (hand-clamp truck): a kind of hand cart that consists of a strong braced frame terminating in a pair of handles at one end and supported on a pair of small, heavy, broad-rimmed wheels; equipped with adjustable metal jaws for clamping field boxes or similar field containers, which are usually handled in stacks of four or five.
.014 Closing hammer: a hand-held device for closing and fastening the lids of wirebound boxes securely.
.015 Closing machine: a machine for closing and fastening securely the lids of wirebound boxes; note that a separate machine is required for each type of box which has a different number or spacing between the wires.
.016 Color add: a cosmetic treatment that consists of the application of a dye (e.g., F.D.&C Citrus Red No. 2) to the surface of oranges, 'Temple' orange and tangelos in Florida, and certain vegetables (e.g., sweet potatoes); neither legal nor required on citrus fruits in other states; a time-dependent operation with dye solution applied as a drench or dip under specified conditions of time and temperature and must be followed by a water rinse and water elimination. (FDOC)
.017 Coloring: (see Degreening, VI.3.023).
.018 Consumer packs: (see Packing operations, VI.3.053d).
.019 Conveyor: any mechanism for transporting a product from one place to another; may involve physical handling of a product in some type of container or a mechanical device, such as a road vehicle (truck, semitrailer, trailer), rail car or belt, slat, bucket or roller types of conveyor or elevator, or forklift truck.
.020 Count fill: (see Packing operations, VI.3.053c).
.021 2,4-D: (see Fungicide application, VI.3.033).
.022 Defuzzer: a machine in a peach packingline that removes the fuzz (epicarpal hairs) from the fruit.

VI.3. POSTHARVEST HANDLING and MARKETING

.023 Degreening (coloring, gassing, quailing): acceleration of the natural process of color change, whereby chlorophylls disappear and other pigments (i.e., carotenoids) become evident, by the use of ethylene or an ethylene-producing compound, high humidity, and continuous air circulation and ventilation according to the commodity; done routinely in citrus packinghouses, especially in Florida in the early part of the season before the fruit have attained good color on the tree; also used by some shippers to ripen tomatoes and by many wholesalers and retailers to ripen tomatoes and bananas in terminal markets or distribution centers; not to be confused with color-add (VI.3.016). (FDOC, S & G)

.024 Dolly: a small, heavy rectangular frame supported on four small wheels, two of which swivel, for moving heavy objects; usually used indoors but may be outside if it is equipped with rubber tires.

.025 Drier-polisher: a machine with fans and heaters used alone or in combination with a water eliminator to remove all surface moisture from fruit or other produce which are to be treated with a solvent-type wax; originally designed to polish fruit (or vegetables) to give them a proper shine, now unnecessary because present-day "waxes" are self-polishing; proper design with minimum brushing (preferably none) and removal of warm moist air are essential. (S & G)

.026 Dry washing: (see Washing, VI.3.081).

.027 Dumping: the initial stage of the packingline onto which fruit, vegetables, etc., are emptied dry from field containers (a) via (1) a trash eliminator onto a conveyor belt with (2) surge control, (3) presizing, and (4) pregrading, or (b) wet via a trash eliminator into a tank of water (e.g., many deciduous fruits, bananas). (S & G, WS)

.027a-1. Trash elimination, trash eliminator: a series of parallel bars, sloping belts, vacuum brushes, pintle brushes, and the like, to remove loose debris (i.e., soil, twigs, and leaves) from fruit or vegetables as they are dumped from field containers onto the packingline.

.027a-2. Surge control: a system for evening out the flow of produce being fed to a packingline from a dumper; may be twin belts (one fast, one slow), return-flow belts with a traveling shear (reversible direction belts with a movable bar to guide the flow of fruit), or a short, wide, slow-moving belt.

.027a-3. Presizing: removal of products too small or too large to be shipped under Marketing Agreement Regulations or market requirements by a sizer set to sort out the shippable produce and eliminate the others; a common system is a longitudinal belt-and-roll sizer with three rollers per belt.

.027a-4. Pregrading: removal of rots, splits, or other obviously damaged products to prevent them from contaminating wash water and equipment on the packingline; grading proper for color defects, etc., is a separate operation done after the fruits or vegetables have been washed and/or trimmed.

.028 Electronic sorting, electronic sorter: mechanical (photoelectric) sorting for color and/or defects; originally devised for sorting lemons and tomatoes for color; major drawbacks are cost ($10,000 to $25,000), low fruit volume handled per machine (600 to 1200 fruit per minute = 190 to 380 bu/h, or roughly 10% or less of the capacity of single citrus packinglines), and the difficulty of

VI.3. POSTHARVEST HANDLING and MARKETING

sorting for defects in humid areas. (S & G)

.029 Flick-bar: a series of metal bars mounted on a sliding frame under soft (horsehair grade) brushes in a water eliminator or under those in a water-emulsion wax applicator; the bars are spaced and adjusted for height so that each barely touches the bottom of a brush and throws off excess water when the brushes are revolving, in the water eliminator while fruit, etc., are running through the machine, and in the wax applicator during the rinsing operation to clean the brushes and brush bed when switching over from a nonfungicidal wax to fungicidal wax or vice versa; the bars are slid out of contact when either machine is started and during the waxing operation. (S & G)

.030 "Flip-over": (see Waxes, VI.3.085c,d).

.031 Fruit and vegetable packinglines: essential operations for fruits and vegetables not field sorted and packed are (a) dumping; (b) washing, dry cleaning, or trimming; (c) grading, (d) sizing, and (e) packing or packaging; optional operations: (f) color add (cosmetic), (g) waxing (cosmetic), and (h) fungicide application (may be mandatory).

.032 Fungicidal waxes: (see Waxes, VI.3.085c,d).

.033 Fungicide application: commonly used postharvest fungicides, such as sodium-o-phenylphenate (SOPP), thiabendazole (TBZ), or benomyl (benlate), may be applied to unwashed fruits or vegetables (a) before washing (TBZ, benomyl), (b) during washing (SOPP), (c) following water elimination (TBZ, benomyl), (d) in solvent-type "waxes" (benomyl), or water-emulsion waxes (SOPP, TBZ, benomyl, of which the last is unstable and must be freshly prepared); diphenyl is used as impregnated pads in citrus cartons, often in conjunction with SOPP); 2,4-D is applied as a preharvest spray or postharvest dip to retain the calyces (buttons) on lemons in California. (S & G)

.034 Gassing: (see Degreening, VI.3.023).

.035 Gift packs: (see Packing operations, VI.3.053e).

.036 Glue sealer: (see Packing operations, VI.3.053).

.037 Grader: a person, usually a woman, who sorts fruit, etc., on a packingline.

.038 Grade table: consists of two basic types of conveyor: wooden rollers or canvas belts; important points in design and operation include good lighting (cool, white fluorescent lamps to provide about 1900 to 2000 lumens/m), minimum lifting, capability of varying the product flow rate and/or the number of graders according to the proportion of product to be sorted out, positioning of least experienced graders upstream, environmental control (air conditioning and noise deadening), and independent variation of forward belt (roller) travel and speed of rotation (e.g., by use of "reverse-roll" = powered rollers, so that the product rotates with the direction of travel instead of the opposite, as in conventional rollers). (S & G)

.039 Grading: sorting of a product to eliminate those items that do not meet certain standards of color, firmness, shape, size, freedom from unhealed cuts, bruises, or damage, etc., for a given grade; may be done by hand or some mechanical means (e.g., electronic sorting); differs from grade inspection (VI.8.038c) in being carried out by packinghouse personnel; also the classification of plants and flowers for market, based on size and quality. (S & G, LR)

.040 Hand-clamp truck: (see Clamp truck, VI.3.013).

VI.3. POSTHARVEST HANDLING and MARKETING

.041 Hand truck: a kind of hand barrow or hand cart that consists of a strong, braced frame terminating in a pair of handles at one end and supported on a pair of small, heavy broad-rimmed wheels; used for moving stacks of containers in a packinghouse, warehouse, or store. (S & G)
.042 Horsehair: (see Brushes, VI.3.007).
.043 Lift truck: a small motorized four-wheel industrial truck equipped with a fixed boom-type hydraulic lift with tines for handling pallet boxes or palletized containers; stacks may be two to eight to ten high, depending on boom height; used in packinghouses, storages, and warehouses.
.044 Longitudinal belt-and-roll sizer: (see Sizing, VI.3.064).
.045 Nylon: (see Brushes, VI.3.007; Washing, VI.3.081).
.046 Order assembly: an area in a packinghouse in which different kinds of produce, grades, sizes, and containers are sorted into orders for shipment; conveyors are used to reduce handling and palletizing containers by hand or mechanically; located between the end of the packingline and the loading area. (S & G)
.047 Pack, package, packaging, packing:
.047a. Pack: to fill a receptacle with something, compactly and securely, as for shipment; the act, process, or manner of packing.
.047b. Package: a container, i.e., a box, case, barrel, or crate.
.047c. Packaging: refers to the placement of produce, etc., in consumer containers; e.g., for retail display and sale.
.047d. Packing: as used in a postharvest (horticultural) sense synonymous with pack as defined above; may be carried out with fruits, vegetables, etc., direct from the sizer or sorting table or from storage (e.g., apples and pears) located at the shipping point (packinghouse); (see Packing operations. VI.3.053).
.048 Package: (see Pack, VI.3.047b).
.049 Packaging: (see Pack, VI.3.047c).
.050 Packer: strictly, a person or machine that places a product into a container; also one who operates a packinghouse. (RS)
.051 Packing: (see Pack, VI.3.047d).
.052 Packinghouse layout: three major working areas: prepackingline (unloading, temporary storage, and, for citrus fruits, degreening), packingline, and postpacking line (precooling and/or storage, order assembly, palletizing, loading), arranged to provide linear flow of the product from the point of unloading at one side of the house to the point of loading at the other with adequate space for each of the necessary operations for the particular product or group of products; note that a wide variation exists in the number, type, and scale of operations, depending on whether the produce is fruits, vegetables, cut flowers, or foliage or potted plants, etc., and within each of these groups. (S & G)
.053 Packing operations: filling of containers with a given kind or specified mixture of produce by hand, semimechanized, or mechanized (automated) procedures. (S & G)
.053a. Bag filling: small numbers of plastic film (polyethyléne), plastic mesh, or cloth mesh bags may be filled by hand; large quantities are filled by using (a) semiautomatic machines, where the packer holds the mouth of the bag up to a spout that delivers the proper number or weight of fruit, etc., or (b) fully automatic machines, that cut, open, fill, and close bags in rolls or endless tubes; bags are closed with staples or twist ties with hand-operated or mechanical staplers or tiers; plas-

tic mesh and film bags are shipped in bagmaster containers which are packed by hand or as the final operation of the fully automatic bagging machine; cloth mesh and large plastic bags for potatoes, onions, and the like are generally shipped loose.

.053b. Bulk filling: an intermediate operation in which sized fruit, etc., is delivered directly from the sizer or by a conveyor into a bulk container, usually a pallet box, for shipment or storage (e.g., apples, pears, California and Arizona citrus, some vegetables); the latter are packed in smaller containers (e.g., cartons) when shipped.

.053c. Carton filling: automated packing of fiberboard cartons by (a) volume fill, where fruit, etc., are poured into the container and vibrated to settle the pack, or (b) count fill, where a definite amount of fruit is poured in; cartons are filled upside down and run through a glue sealer for closing; widely used for citrus.

.053d. Consumer packs: packaging of individual pieces of produce (as heads of lettuce) in film or in multiples in tubes, trays, and the like with stretch film overwraps; usually done at a terminal warehouse or supermarket warehouse or store rather than at the shipping point, mainly for flexibility in the type of packaging and the necessity of repackaging containers with decayed product. (RS)

.053e. Gift packs: a type of hand packing differing mainly in the wide assortment of containers used and inclusion of more than one kind of fruit, plus jellies, nuts, etc.

.053f. Place packing: placement by hand, or with mechanical aids, of fruit, vegetables, etc., in a shipping container in some regular pattern or arrangement; containers may be wirebound or wooden boxes or crates, fiberboard cartons, wood or fiberboard lugs, bags, trays (tray packs), or individual cells (cell packs). (S & G, RS)

.054 Packout: the proportion (percentage) of produce that is shippable of that harvested; also the total amount of a particular crop that is packed. (S & G, RS)

.055 Palmetto: (see Brushes, VI.3.007; Washing, VI.3.081).
.056 Perforated belt sizer: (see Sizing, VI.3.064c).
.057 Place packing: (see Packing operations, VI.3.053f).
.058 Pressure sealer: a machine for sealing the lid of packed cartons; differs from a glue sealer in that the flaps already have glue on them.
.059 Quailing: (see Degreening, VI.3.023).
.060 Reverse roll: (see Grade table, VI.3.038).
.061 Saran: (see Brushes, VI.3.007; Washing, VI.3.081).
.062 Shear (wipe-off): a device for directing or deflecting the flow of produce safely from one conveyor to another conveyor, grade table, packing bin, etc.; mounted at an angle (commonly 45° or thereabouts) to the first one or to reverse the flow of produce on the same conveyor; may be a flat or concave metal plate, an adjustable arm or plate, or an old horsehair brush (or one of similar softness) on a spindle mounted in the angle. (S & G)
.063 Sisal: (see Brushes, VI.3.007, Washing, VI.3.081).
.064 Sizing: sorting by diameter (most fruits, some vegetables), weight (some fruits), or length (some vegetables, cut flowers, or foliage) to enhance eye appeal and for ease of packing; several distinct types of sizer. (S & G, RS)

VI.3. POSTHARVEST HANDLING and MARKETING

.064a. Belt-and-roll: consists of five to eight cylindrical or bottle-shaped rollers set in a line with a belt inclined at about 30°; two configurations, longitudinal, where a single sizer line is mounted over a row of bins, and central, where several lines of sizers are combined into one machine and fruit (or vegetables) of a given size drop onto a single collection belt: sizes <u>one</u> fruit per roller at a time, regardless of configuration; widely used for citrus and avocados.

.064b. Miscellaneous: includes sizing by length, as determined by marks or stops on a table or moving belt or by eye (done by packers of many crops, including gift houses).

.064c. Perforated belt: requires a separate belt with holes in it for each size; thus adjustment and as many small machines as sizes are needed; used for tomatoes and some other products (capacity is too low for most citrus packinghouses).

.064d. Transverse: consists of pairs of rollers that diverge (i.e., the space between them widens) as they move along the machine; sizes as many fruit simultaneously as the machine width will allow, with fruit touching the rollers at four points instead of only two for belt-and-roll sizers; used for prolate or spherical-shaped fruit.

.064e. Weight: fruit ride along in small, spring-loaded cups with a trapdoor that releases when the cup reaches a predetermined location along the sizer; very gentle; used for apples and other deciduous fruits, avocado, etc., handled in relatively low volume (i.e., up to 200 to 400 bu/h).

.065 Slab wax: (see Waxes, VI.3.085).
.066 Soak tank: (see Washing, VI.3.081).
.067 Sodium-o-phenylphenate: (see Fungicide application, VI.3.033).
.068 Solvent-type waxes: (see Waxes, VI.3.085).
.069 SOPP: (see Fungicide application, VI.3.033).
.070 Source packaging: (see Source wrapping, VI.3.071).
.071 Source wrapping (source packaging): wrapping or packaging a product at the shipping point in the field or packinghouse; most commonly used for lettuce, celery, carrots, apples, potatoes, and radishes. (RS)
.072 Stamping: citrus fruits and some other produce (e.g., bananas, melons) must be individually stamped or have a pressure-sensitive label pasted on them to indicate their origin (i.e., state or country) or brand or grade designation; usually done before sizing. (S & G)
.073 Sudsing: (see Washing, VI.3.081).
.074 Tampico: (see Brushes, VI.3.007; Washing, VI.3.081).
.075 TBZ: (see Fungicide application, VI.3.033).
.076 Thiobendazole: (see Fungicide application, VI.3.033).
.077 Transverse sizer: (see Sizing, VI.3.064d).
.078 Trimming: removal of outer leaves or tops, shortening of stems, and the like in lieu of washing and/or to prepare the produce for grading and packing.
.079 Unit operation: a single, distinguishable step in a handling system; e.g., harvesting (removal of a product from its growing place), loading containers onto a truck, dumping, washing, grading, sizing, packing shipping containers, and loading.
.080 Volume fill: (see Packing operations, VI.3.053c).
.081 Washing: removal of dirt, spray, or fertilizer residues, scales, and the like from the product; may be a wet process, which consists of wetting with a series of coarse spray nozzles or a

VI.3. POSTHARVEST HANDLING and MARKETING

"soak" tank (not recommended for citrus and certain other fruits but used with deciduous fruits and bananas to avoid bruises), sudsing (application of foamed soap or detergent), washing (revolving brushes, bassine for vegetables, palmetto, sisal, nylon, or Saran fiber for fruits and some vegetables), rinsing and water elimination (sponge-rubber sectioned rollers or soft fiber brushes with wringer rolls or "flick bars" beneath, or brass or aluminum rollers); "dry washing" consists of wiping fruit or vegetables with a rag to remove surface dirt, as done with mango, avocado, cucumbers, etc., in small operations. (S & G, RS)

.082 Water elimination: (see Washing, VI.3.081).
.083 Water eliminator: (see Drier-polisher, VI.3.025).
.084 Water-emulsion waxes: (see Waxes, VI.3.085).
.085 Waxes: food-grade synthetic resins (e.g., coumarone indene and epolene) or true waxes (long-chain hydrocarbons), such as paraffin (retards moisture loss), carnauba (imparts gloss), and microcrystalline wax; four different types of wax and methods of application. (S & G)

.085a. Fungicidal waxes: solvent-type or water-emulsion "waxes" with a fungicide (e.g., sodium-o-phenylphenate, thiabendazole, or benomyl), the last sometimes added with a metering pump as needed rather than as an integral part of the formulation; differences in regulations on fungicides for fruit, etc., going to export versus domestic markets, plus the necessity for thorough cleaning of equipment before and after their use, have led to waxing and fungicide application as separate operations.

.085b. Slab wax: mixture of paraffin and carnauba in slabs under and in contact with revolving brushes which transfer the wax to the product (small packinglines, especially gift houses).

.085c. Solvent-type "waxes": coumarone indene or other resins dissolved in an organic solvent and applied as a fine mist onto the product as it passes through the applicator; drying is done on a slat conveyor with a "flip-over" (two short conveyors with a short drop in the middle to allow the product to turn over and dry on the bottom side).

.085d. Water-emulsion waxes: microcrystalline or other waxes or synthetic resins emulsified in water and applied as a foam, dip, spray, or brush wipe; the commonest method is spraying with "wigwag" nozzles (i.e., nozzles attached to an endless chain) mounted over a bed of slowly rotating brushes; drying is done on a "wax-setter," a pair of conveyors with a "flip-over" (VI.3.085c) near the middle.

.086 Waxing: done for two reasons, to replace the natural wax on a product removed during washing and thereby reduce loss of moisture and to impart a high gloss to the surface; the first is essential, the second is cosmetic; many different materials, mixtures and formulations are used for waxing fruits and vegetables (e.g., cucumbers); (see Waxes, VI.3.085). (S & G, RS)

.087 Wax-setter: (see Waxes, VI.3.085d).
.088 Weight sizer: (see Sizing, VI.3.064e).

VI.4. POSTHARVEST HANDLING and MARKETING

VI.4. Containers

.000 Container: general term for a receptacle of any type used in harvesting, handling, transportation, storage, or marketing of horticultural products; several general categories, each with many types, e.g., containers used in connection with harvesting proper and transportation from the field, storage containers, shipping containers, and market (consumer) containers. (BL68, RS)

.000a. Harvesting (field): picking bags, baskets, crates, hampers, flats, bulb trays, field boxes, pallet boxes (pallet bins), bins; all of these are used as harvesting containers; picking bags, baskets, and crates are usually emptied into larger containers (e.g., field boxes, pallet boxes, or bins) for transportation to the packinghouse or processing plant, although baskets, and crates may be stacked on a truck or trailer and hauled to a nearby packinghouse or, in some instances, taken to a truck or loaded onto a conveyor leading to a truck, where the produce is sorted and packed directly into shipping containers (e.g., lettuce, melons); (see Handling systems, VI.2. 021).

.000b. Market (consumer): plastic (mesh, perforated film) or cloth bags, overwraps, stretchwrap trays and baskets; fiberboard cartons are often used for displaying loose fruit or vegetables.

.000c. Shipping: fiberboard cartons, wooden boxes, wirebound boxes, and combinations; bulk bins, pallet boxes; tray packs (many types in cartons); cloth and plastic (mesh and perforated) bags in bagmaster cartons; wooden baskets and hampers; fiberboard, plastic, or metal flats; fiberboard lugs; loads are assembled on pallets or slip sheets (i.e., palletized) for commercial shipments. (BL68)

.000d. Storage: pallet boxes, bags, flats, bulb trays, and hampers; most storage of fruits and some vegetables involve pallet boxes, sometimes bags (e.g., potatoes and certain other root or tuber crops), or hampers; flats and bulb trays are used for bulbous crops (e.g., gladiolus, amaryllis, and lily), where good aeration is essential for proper curing (many bulbs, etc., may be mixed with peat when they need to remain moist); containers that have no pallets will usually be palletized for easier handling.

.001 Bag(s): perforated or nonperforated polyethylene, plastic mesh, cloth mesh, or burlap containers for fruits (e.g., apples, plums, peaches, and citrus fruit) and vegetables (e.g., onions and potatoes); closed with "twist 'ems" or staples; a consumer-type container usually holding larger quantities than a tray pack; also a shipping container for onions, potatoes, and the like. (RS)

.002 Bagmaster carton: fiberboard, wirebound or wooden container for shipment of polyethlene or plastic mesh bags; usually about a fourth larger than a regular carton which holds the same volume or weight of produce. (S & G, FDOC)

.003 Baskets: several distinct types: (a) woven wood (veneer), round containers with flat or rounded wood bottoms, sloping sides, wire handles, and a woven wood bulged lid with ears, made in 1-, 1/2-, and 1/4-bushel sizes; used for harvesting (open), shipment (lidded and strapped), or marketing (open) of some fruits and vegetables; now mainly used for roadside sales; (b) woven wood (veneer), rectangular containers with rounded bottoms and wire bails, made in 1-, 1/2-, and 1/4-bushel sizes, open topped; used mainly for roadside sales; (c) berry baskets of light wood veneer or plastic in quart, pint, and half-pint sizes for strawberries, blueberries, blackberries, raspberries, and the like, or small quantities of larger fruits (e.g., tomatoes or peaches); packed for shipment in covered wood, fiberboard, or plastic flats; sold singly in grocery stores or at roadside stands; (d) fancy woven reed or wood baskets of many different shapes, sizes, and designs for gift packages of fruit and other products; also for plant containers.

.004 Bin: a square or rectangular (sometimes cylindrical), shallow (i.e. 30 to 60 cm deep) metal, wooden, or plastic box used for hauling fruit from a grove (orchard) to a packinghouse (fresh fruit) or cannery (processed fruit); variable in width and length but often 1.3 to 1.5 m or more wide and about double that in length; often stacked several high for transportation with a straddle-carrier; now largely superseded by pallet boxes. (S & G)

.005 Bulb tray: (see V.4.1.022).

.006 Bulk bin: three versions: (a) cannery bin which is a tall rectangular bin with slatted sides, sloping bottom, and a capacity of about 300 to 600 bushels or more, used for temporary storage of fruit for processing; (b) fresh-fruit bin which is similar to a cannery bin except that it has a zigzag arrangement of baffles that supports ciderpress cloth, with a capacity of 160 to 240 bushels for temporary storage or degreening of fruit (citrus) at a packinghouse; now largely obsolete; (c) a shipping container for transportation of certain fruits and vegetables in bulk; essentially the same as a pallet box. (RS, S & G)

.007 Cardboard: strictly, a single or multilayer, usually thin, paper product that differs from fiberboard (VI.4.010) in that it lacks a center corrugated layer or layers and is used for construction of light containers for flowers and the like; also applied to fiberboard (corrugated) containers presumably to distinguish them from industrial pressed fiberboard.

.008 Carton: fiberboard or combination fiberboard and wood or veneer, rectangular or square container for shipping produce, plants, etc.; it may be constructed with internal fiberboard or plastic trays which may have molded pockets (tray pack); many different sizes and types of construction (i.e., one piece, telescoping); may be made of wax-impregnated fiberboard for shipment of bananas and the like. (FDOC, BL68)

.009 Crate: a slatted rectangular or square wooden box, usually deeper than a lug (VI.4.014); also a colloquial term applied to many different types of wooden container used for harvesting or shipment of fresh produce (e.g., field crates and shipping crates).

.010 Fiberboard: a multilayer kraft paper product used for manufacture of containers; many different types, weights of material, and configurations; commonly constructed with solid outer layers bonded to an inner corrugated layer or layers; sometimes termed "cardboard" (VI.4.007) to avoid confusion with the pressed fiberboard sheets used for industrial or home construction. (FDOC, BL68)

.011 Flat: a shallow rectangular container usually with a slotted bottom and solid sides made of wood, metal, plastic, or fiberboard; those of wood, metal, or plastic are used for the propagation of seedlings and other nursery purposes (e.g., sale of bedding plants); those of plastic or fiberboard also for shipping boxes or cartons (e.g., strawberries, other berries, and tomatoes in

VI.4. POSTHARVEST HANDLING and MARKETING

baskets packed in a single layer or tomatoes often in two layers); sometimes used as harvesting containers when they will usually be much larger (i.e., 60 to 75 wide X 100 to 120 long X 10 to 15 cm deep). (RS, BL68)

.012 Hamper: conical basket with a wooden bottom, woven or solid veneer sides, and wire handles, used without a cover for harvesting or with a cover for storage or shipment; also a fiberboard container, 33 X 33 X 107 to 130 cm, used for shipment of gladiolus; holds 15 to 24 ten-spike bunches wrapped in kraft paper or polyethylene and must be transported in a vertical position. (LR, BL68)

.013 Hanging basket: a container, usually plastic, used for bedding plants, foliage plants, poinsettias, and other plant materials; usually suspended from supports in a greenhouse and to contain plants in the home or garden. (LR)

.014 Lug: originally a shallow wooden box, now wooden, plastic, or more commonly fiberboard or combination wood and fiberboard (for shipment of soft produce items such as grapes, tomatoes and cherries) also used for harvesting and storage. (RS, BL68)

.015 Mesh: an open-weave material, usually plastic; allows maximum visibility and ventilation; used mostly for bags and for wrapping pallet loads. (RS)

.016 Netting: (see Mesh, VI.4.015).

.017 Pallet: a wooden frame with slotted bottom and heavy runners usually 120 cm (44 in.) square or 100 X 120 cm (40 X 44 in.) on which boxes or cartons are stacked for storage, loading, and shipment; made for handling with a forklift; an integral part of nearly all modern handling systems; present industry efforts are aimed at standardizing produce shipments on 100 X 120 cm pallets. (RS)

.018 Pallet box (pallet bin): wood, plastic, or metal box with solid or slotted sides, slotted bottom and runners beneath, usually 60 to 80 cm (30 to 32 in.) high and 120 cm (44 in.) square, often made with a reinforced top rim so that it can be moved with a hydraulic clamp-type lift instead of a forklift for harvesting operations; has largely superseded other, smaller containers for fruit handling and/or storage in most areas; capacity is about 16 to 20 bushels (400 to 500 kg); usually dumped from the top (i.e., rotated on a dumper to let its contents roll out onto a trash eliminator or into water); sometimes constructed with a trap door on one side for unloading; note that efforts have been made to change the dimensions to 100 X 120 cm (40 X 44 in.) to conform with pallets used in terminal markets. (S & G)

.019 Papier maché: a pressed paper product made of paper pulp mixed with size, glue, rosin, clay, or the like and used for trays.

.020 Picking bag: a cloth (canvas) bag with shoulder strap and bottom flap that can be released for emptying; several sizes ranging from 1/4 bushel to a bushel or more. (S & G)

.021 Shrink film: a clear plastic film that conforms to the object or product it covers when heat is applied; now widely used for consumer packages. (RS)

.022 Slip sheet: a form of pallet; a thin piece of material made of fiberboard or plastic that requires a specially equipped forklift to move a unitized load; growing in popularity because it is light weight and easily disposable. (RS)

.023 Stretchwrap, overwrap: plastic film used to protect individual fruit or vegetables or those in trays; films are heat-expandable

VI.4. POSTHARVEST HANDLING and MARKETING

.024   (stretchwrap) or self-adhering (overwrap) applied to fit tightly around the produce.

.024   Tray: a small container for fruits and vegetables; usually made of pulp (papier maché) or plastic; such containers are usually designed for consumer packs; larger trays are used for shipping. (RS)

.025   Tray pack: small shallow plastic or papier maché container in which the produce is protected from injury by an overwrap, as in a consumer container; also a type of shipping container; (see Carton, VI.4.008).

.026   Wirebound box: shipping container made of wood veneer and reinforced ends with wires wrapped around with closing loops; many different sizes and shapes; used for many different products; now superseded by cartons for many applications. (BL68, FDOC)

VI.5. Refrigeration, Storage

.000   Refrigeration: a general term for any type of cooling.
.001   Absolute humidity: (see Humidity, VI.5.025a).
.002   Air precooling: (see Precooling, VI.5.035a).
.003   Air storage: (see Storage, VI.5.047a).
.004   "Body" icing: finely chopped ice packed inside a container and mixed with the produce to absorb heat in transit. (JM)
.005   British thermal unit (BTU): the quantity of energy required to heat one pound of water at its maximum density one degree Fahrenheit (°F), i.e., from 39° to 40°F (3.89° to 4.44°C); equal to 0.252 calories kilogram (cal kg) (obsolete) or 1054.35 joules (=1054.35 $m^2 \cdot kg \cdot s^{-2}$); a widely used term in heating, refrigeration and air conditioning calculations; common units are BTU per pound (=1054.35J) and BTU per hour (=1054.35J h^{-1}). (ASHRAE)
.006   Calorie (cal): the quantity of heat necessary to raise the temperature of one gram of water from 3.5° to 4.5°C; a <u>mean</u> calorie is 1/100 the quantity of heat necessary to raise one gram of water from 0° to 100°C; a <u>normal</u> calorie involves a temperature change frdom 14.5° to 15.5°C; the <u>large</u> (food) calorie is equal to 1000 small calories; 1 cal g = 4.184J or 3.9683 X 10^{-3} BTU, 1 cal g (mean) = 4.19002J or 3.974 X 10^{-3} BTU, 1 cal g (15°C = normal) = 4.1858J or 3.97 X 10^{-3} BTU, [not an SI unit (see IV.0.260m), hence obsolete.]
.007   Carbon dioxide treatment (CO_2): use of high (10 to 40%) levels of CO_2 for short periods of time (2 to 16 days) prior to storage or during shipment. (SK,I)
.008   CA storage: (see Storage, VI.5.047b).
.009   Ceiling jet precooling room: (see Precooling, VI.5.035a-1).
.010   Conduction: (see Thermal energy transfer, VI.5.052a).
.011   Controlled atmosphere storage: (see Storage, VI.5.047b).
.012   Convection: (see Thermal energy transfer, VI.5.052b).
.013   Conventional forced-air precooling room: (see Precooling, VI.5.035a-3).
.014   Cooling coefficient (C): the rate of cooling divided by the temperature difference between the object and the surroundings: $C = TR_{ma} \cdot (\theta \cdot \Delta t_m)^{-1}$, where TR_{ma} is the mass-average (object) temperature reduction in degrees, θ is time in hours, and Δt_m is the logarithmic mean temperature difference. (GU, SYB)
.015   Dewpoint: (see Humidity, VI.5.025b).

VI.5. POSTHARVEST HANDLING and MARKETING

.016 Dry-pack storage: the storage of cut flowers in vaporproof containers with the stems not in water, usually at 0°C. (LR)

.017 Enthalpy (Σ): the total heat content of an object or substance, measured as BTU per pound (=1054.35J) or gram calories per gram (g cal$\cdot$g^{-1} = 4.184J) as the difference between any temperature and 32°F (0°C); used in heating, refrigeration and air conditioning calculations; $\Sigma = (t_1 - t_o) \, W \cdot c_p$, where t_1 is any temperature, t_o = 32°F or 0°C, W = mass (as lb or kg) and c_p is the specific heat. (ASHRAE)

.018 Field heat: (see Thermal energy, VI.5.051a).

.019 Forced-air precooler: (see Precooling, VI.5.035a-4).

.020 Gas storage: (see Controlled atmosphere storage, VI.5.047b).

.021 Half-cooling time: the period required to reduce the initial temperature difference between an object (i.e., a load of produce) and its surroundings by one-half; calculated as $Z = \ln 1/2/C$, where C is the cooling coefficient $(t_1-t) \cdot [\theta^{-1} \cdot (t-t_o)_{av}]^{-1}$, where t is the temperature of the object at time θ, t_1 is the initial temperature of the object, t_o is the temperature of the surroundings, and $(t - t_o)_{av}$ is the average temperature difference (i.e., the logarithmic mean temperature difference). Note that 2Z represents the three-quarters cooling time and 3Z, the seven-eighths cooling time. (GU)

.022 Heat: (see Thermal energy, VI.5.051).

.023 Heat of respiration: (see Thermal energy, VI.5.051b).

.024 Heat removal: the amount of heat extracted per unit mass cooled; two equations:

.024a. Theoretical: $(Qw) = hA\Delta t_m \cdot W^{-1}$, where h is the surface convective coefficient of heat transfer (VI.5.048), A is the surface area of the object per unit mass (ft$^2 \cdot$lb^{-1} or m$^2 \cdot$kg^{-1}), Δt_m is the logarithmic mean temperature difference (°F or °C) (VI.5.030), and W is the total mass (lb or kg) of the object being cooled. (JK)

.024b. Experimental: $(Qf) = c_p \Delta t_{ma} \cdot W^{-1}$, where c_p is the specific heat (e.g., 0.9 for oranges), Δt_{ma} is the mass average temperature reduction (°F or °C) (VI.5.030), and W is the total mass (lb or kg) of the objects being cooled. (SYB)

.025 Humidity (atmospheric moisture): water vapor in the air; measured or expressed in several ways: (S & G)

.025a. Absolute (specific) humidity: the measure of the mass of water in a given weight of air, usually expressed as grams per kilogram of dry air (=0.1428g$\cdot$kg^{-1}) or grams per cubic meter (g$\cdot$m^{-3}) in metric units or grains per pound of dry air (gr$\cdot$lb^{-1}) in U. S. units; a far superior measure to relative humidity since it shows the actual quantity of water present for any given temperature (e.g., 4.28g$\cdot$kg^{-1} at 6°C and 95% RH but 25.7g$\cdot$kg^{-1} at 29.5°C and 95% RH, a six-fold difference).

.025b. Dewpoint: the temperature to which moist air must be lowered (at constant pressure) to initiate condensation (100% RH).

.025c. Relative humidity (RH): the ratio, expressed as percent (%), between the quantity of water vapor present and the maximum possible at the temperature and barometric pressure (with correction for partial pressure of the water vapor); relative humidities <u>can only be compared</u> at the same temperature and barometric pressure; often measured as the difference in temperatures of a dry-bulb and wet-bulb pair of thermometers (calibrated to 1° or less) rotated at a speed of at least 3m$\cdot$s^{-1} until equilibrium is reached and RH read from a table or psychrometric chart.

VI.5. POSTHARVEST HANDLING and MARKETING

.025d. Vapor pressure: the partial pressure of the water component of any dry air plus water vapor combination with correction to a standard barometric pressure; directly correlated with absolute humidity at the same temperature and barometric pressure.

.026 Hydrocooling: (see Precooling, VI.5.035b).

.027 Hypobaric storage: (see Storage, VI.5.047c).

.028 Latent heat: (see Thermal energy, VI.5.051c).

.029 Logarithmic mean temperature (T_m): $T_m = t_{a1} - t_{a2} \cdot [\ln(t_{a1} \cdot t_{a2}^{-1})]^{-1}$, where t_{a1} and t_{a2} are initial and final medium temperatures, respectively. (GU, SYB)

.030 Logarithmic mean temperature difference (ΔT_m): $\Delta T_m = [(t_{ma1} - t_{a1}) - (t_{ma2} - t_{a2})] \cdot [\ln(t_{ma1} - t_{a1}) \cdot (t_{ma2} - t_{a2})^{-1}]^{-1} = (\Delta t_1 - \Delta t_2) \cdot [\ln(\Delta t_1 \cdot \Delta t_2^{-1})]^{-1}$, where t_{ma} = mass-average fruit temperature, t_a = medium temperature, and subscripts 1 and 2 denote respective cooling times. (SYB)

.031 Low pressure storage: (see Storage, VI.5.047c).

.032 Mass-average temperature (t_{ma}): the temperature of an object at the center of mass at a specified time and under adiabatic (stable, i.e., no removal or addition of heat) conditions, $\Delta t_{ma} = t_c - ma \cdot (t_c - t_s)$, where t_c and t_s are the center and surface temperatures (assuming a linear distribution) and ma is the center of mass in terms of the radius ratio; the last is 0.7937r for a homogeneous sphere and 0.7071r for a homogeneous cylinder but must be determined experimentally for a nonhomogeneous object or one that does not approximate a sphere or cylinder. (SYB)

.033 Modified atmosphere storage: (see Storage, VI.5.047d).

.034 Portable precoolers: (see Precooling, VI.5.035a-5).

.035 Precooling: rapid cooling of produce prior to transportation of storage; there are two basic types, air precooling and hydrocooling, of which the former has several versions. (GU, SYB)

.035a. Air precooling:

.035a-1. Ceiling jet cooling room: cold room equipped with air jet nozzles directed vertically downward, the floor being marked so that containers can be stacked to have channels lined up with the nozzles (half-cooling time approximately six h).

.035a-2. Cold room: similar to a room used for produce storage, where the rate of cooling is a few degrees per hour (half-cooling time approximately 10 h for grapes in lidded lugs).

.035a-3. Conventional forced-air cooling room: cooling with an air-delivery system sufficient to provide a small pressure difference and thereby cause the cold air to be pulled or pushed through a stack of containers.

.035a-4. Forced-air precooler: a specially designed chamber in which stacks of containers are set on top of perforated floor grates over troughs through which cold air is blown to provide a small pressure differential across the containers; usually constructed so that each trough can be operated independently, thus allowing simultaneous loading, precooling, and unloading; the most efficient method for produce that cannot be hydrocooled or vacuum-cooled (half-cooling time about 1.5 h). (SYB)

.035a-5. Portable precoolers: large, 10- to 20-ton refrigeration units mounted on a truck or dolly for precooling containers in a semitrailer or rail car (half-cooling time approximately 3 h); now rarely used.

.035a-6. Rail-car cooling: a system in which the flow of air through

VI.5. POSTHARVEST HANDLING and MARKETING

the load is reversed and the mechanical refrigeration unit (or ice bunkers) is used to cool the produce overnight while the car is still at dockside (i.e., not in transit, with power for the fan motors supplied from the packinghouse); now rarely used.

.035b. Hydrocooling: cooling with water at 1 to 2°C (34 to 36°F) flooded over loose or packaged (wooden containers) produce on a conveyor (generally); cooling time varies from about 20 to 45 min; used for broccoli, carrots, cauliflower, celery and peaches; should not be used with chilling-sensitive produce; not recommended for citrus and some other crops because of inordinately high decay when warmed. (S & G, RT)

.036 Precooling formulas: equations for computation of precooling times, where temperatures are in °F or °C (if all in the same units); M = cooling medium, S = initial product, and F = final product; cooling times (hours or minutes, if all in the same units) are given by H to half-cool, V to seven-eighths cool, and T to temperature F (final product); log = $base_{10}$ logarithms, ln = $base_e$ natural logarithms: (a) $T = 3.32 H \log (S - M)/(F - M)$; $H = 0.301 T/\log (S - M)/(F - M)$; (b) $T = 1.44 H \ln (S - M)/(F - M)$; $H = 0.693 T/\ln(S - M)/(F - M)$; (c) $T = 1.107 V \log (S - M)/(F - M)$; $V = 0.903 T/\log (S - M)/(F - M)$; (d) $T = 0.481 V \ln (S - M)/(F - M)$; $V = 2.08 T/\ln (S - M)/(F - M)$; note: Use of (a), (b), (c), or (d) is suggested, depending on whether H or V and "log" or "ln" are more familiar and convenient; all give the same results; H is probably more familiar; V is usually closer to commercial precooling practice; (from R.F. Kasmire 9/2/76 sheet for cooler operators in California).

.037 Psychrometric chart: a graph with lines that relate dry bulb temperature, wet bulb temperature, percent relative humidity, absolute (specific) humidity, vapor pressure, and enthalpy at a given barometric pressure (and usually with an equation for conversion of other barometric pressures to that of the graph); used in lieu of tables for humidity or heat content calculations, especially when interconversion of different measures of humidity is desired; units may be U.S. (i.e., °F, grains/lb, psi) or metric (i.e., °C, g/kg, J/g, g/m^2). (S & G)

.038 Radiation: (see Thermal energy transfer, VI.5.052c).
.039 Rail-car cooling: (see Precooling, VI.5.035a-6).
.040 Relative humidity: (see Humidity, VI.5.025c).
.041 Self-controlled atmosphere: (see Storage, VI.5.047e).
.042 Sensible heat: (see Thermal energy, VI.5.051d).
.043 Specific gravity: the ratio of the mass of a solid object to the mass of an equal volume of water at 4°C or some other specified temperature (e.g., 20°C); applies also to gases, in which case the standard is the mass of dry air at a specified temperature and barometric pressure. (ASHRAE)
.044 Specific heat (c_p): (see Thermal energy, VI.5.051e).
.045 Specific humidity: (see Humidity, VI.5.025a).
.046 Sprouting (potatoes): shoots growing from the eyes on potatoes; usually appear after several months in storage and when storage temperatures rise; inhibitors control early sprouting. (RS)
.047 Storage: the practice of holding fresh produce (fruits, vegetables, etc.) for the purposes of extending the market season and to avoid market gluts; usually involves some form of environmental control (i.e., of temperature, humidity, air flow, and air quality) in rooms designed specifically for maintaining conditions

suitable for a particular commodity or group of commodities; numerous studies with a wide range of fruits and vegetables and flowers and commercial experience emphasize the necessity for harvesting at the proper stage of maturity, freedom from diseases or blemishes, and careful handling if storage, especially for long periods (e.g., several months) is to be successful; common types of storage include the following: (L & H)

.047a. Air storage: regulated temperature and relative humidity with continuous air flow and ventilation, the last to remove volatiles, particularly ethylene; the CO_2 level should be about 0.003 to 1% and O_2, 21% (i.e., the level in normal air).

.047b. Controlled atmosphere (CA) storage: regulated CO_2 and O_2 (i.e., at respective levels higher than and lower than those in normal air); general practice is 2 to 3%, O_2, and 2 to 5%, CO_2, although proper temperature, O_2 and CO_2 levels vary with the cultivar, region, and season. (SK, I)

.047c. Low-pressure (LP) or hypobaric storage: use of very low barometric pressures. (LR, SK, I)

.047d. Modified atmosphere storage (MA): holding produce under conditions of atmosphere modified by package, overwrap, box liner, or pallet cover. (SK, I)

.047e. Self-controlled atmosphere (SCA): a type of MA in which sealed polyethylene bags of a defined thickness with fixed ratios of fruit weight to surface area are used, the theory being that a correct and beneficial mixture of CO_2 and O_2 will be maintained around the fruit, provided the right film and temperatures are used. (SK, I)

.048 Surface convective coefficient of heat transfer(h): the equation for spheres is $hDk^{-1} = 0.33 (\underline{v_m} D \rho \cdot \mu^{-1})^{0.6} X (\mu c_p \cdot k^{-1})^{1/3}$, where D is the object (e.g., fruit) diameter (ft or m), k is thermal conductivity [BTU per (hr)(ft)(°F) or $J \cdot (s \cdot m \cdot °C)^{-1}$], v_m is the mean air velocity through the voids around the objects (ft $\cdot$ m^{-1} or m$\cdot$s^{-1}), D is fruit density (lb$\cdot$ft^{-2} or kg$\cdot$m^{-3}), μ is dynamic viscosity [lb$\cdot$(h$\cdot$ft)$^{-1}$ or kg$\cdot$(s$\cdot$m)$^{-1}$], and c_p is specific heat [BTU$\cdot$(lb$\cdot$°F)$^{-1}$ or J$\cdot$(kg$\cdot$°C)$^{-1}$]; the units for h are BTU $\cdot$ [h$\cdot$(ft^{-2}) °F]$^{-1}$ or J$\cdot$(m$^2 \cdot$ s$\cdot$°C)$^{-1}$; used in refrigeration engineering.

.049 Thermal conductivity (k): the rate of heat (BTU or J) movement through an object or substance per unit time (h or s) per unit area (ft^2 or m^2) perpendicular to the direction of heat flow per degree (°F or °C) per unit length (ft or m) of path; the original equation can be simplified as heat per (unit time, unit length, degree); units are BTU$\cdot$(h$\cdot$ft$\cdot$°F)$^{-1}$ or J$\cdot$(s$\cdot$m$\cdot$°C)$^{-1}$; this is the basic equation for thermal conduction in the steady state. (MA)

.050 Thermal diffusivity (α): thermal conductivity (k, VI.5.049) per unit density (ρ as lb$\cdot$ft^{-3} or kg$\cdot$m^{-3}) times specific heat (c_p as BTU$\cdot$(lb$\cdot$°F)$^{-1}$ or J$\cdot$(kg$\cdot$°C)$^{-1}$; that is $\alpha = k \cdot (\rho c_p)^{-1}$, may be expressed as [BTU$\cdot$(hr$\cdot$ft$\cdot$°F)$^{-1}$]$\cdot$[(lb$\cdot$ft^{-3})$\cdot$(BTU$\cdot$(lb$\cdot$°F)$^{-1}$] or [J$\cdot$(m$\cdot$s$\cdot$°C)$^{-1}$]$\cdot$[(kg$\cdot$m^{-3})$\cdot$(J$\cdot$kg$\cdot$°C^{-1})$^{-1}$]; this is the basic equation for thermal conduction in the unsteady state; used in refrigeration engineering. (MA)

.051 Thermal energy (heat): several types are of concern in handling produce, measured as BTU's (US) or joules (metric). (S & G)

.051a. "Field" heat: the quantity of thermal energy removed in cooling or heating an object from one temperature to another, being the product of the temperature difference, mass, and heat

VI.5. POSTHARVEST HANDLING and MARKETING

capacity (specific heat).

.051b. Heat of respiration: thermal energy released by produce (and animals) in connection with metabolic processes (e.g., the conversion of sugars to carbon dioxide and water), usually expressed as BTU's per lb or per ton per hour or day or as joules per gram (or kilogram) per hour or day. (AY, RT)

.051c. Latent (specific latent) heat: thermal energy released or absorbed from an object when a change of state occurs that from solid to liquid (specific latent heat of liquefaction) or from liquid to solid (specific latent heat of fusion, equal to the former with reversed sign) for a unit mass at the same temperature; the unit of energy is joules per kilogram ($J \cdot kg^{-1}$) the value for 1 kg of water at 0° being 333.5J. Specific latent heat of vaporization (condensation) is the thermal energy required to vaporize (or condense with reversed sign) a unit mass at the same temperature; values for 1 kg of water are 2493.7J, 2447.6J, and 2257.7J at 0°, 20° and 100°C, respectively. Latent heat is important in orchard heating, evaporative cooling of greenhouses, and vacuum cooling of vegetables.

.051d. Sensible heat: thermal energy, including both field heat and heat of respiration, removed in cooling an object from one temperature to another; (note that heat of respiration is time-dependent). (AY, RT)

.051e. Specific heat capacity (c_p): the energy required to raise (or lower) the temperature of a unit mass (lb or kg) one degree (°F or K); units are BTU per pound per degree F ($BTU \cdot lb^{-1} \cdot °F^{-1}$), long used in heating, refrigeration and air conditioning engineering calculations, or joules per kilogram per degree kelvin ($J \cdot kg^{-1} \cdot K^{-1}$), now recommended in engineering and other fields. (ASHRAE)

.052 Thermal energy transfer: movement of thermal energy from one point to another may be effected by any of three different processes:

.052a. Conduction: transfer of thermal energy from one molecule to another, as in a solid or liquid. (ASHRAE)

.052b. Convection: transfer of thermal energy by active motion of molecules in a medium, which may be a gas, liquid, or solid; the process is termed "natural" if produced solely by differences in temperature or "forced," i.e., when molecules are moved with a fan or pump. (ASHRAE)

.052c. Radiation: the process by which energy is propagated through space, as stated by the Stefan-Boltzman law: $E = k(T^4 - T_o^4)$, where T is the absolute temperature (K) of the body, T_o is the absolute temperature of the surroundings, k is a constant, and E is the energy transferred in unit time by a black body. (S & R)

.053 Ton of refrigeration: equals 288,000 BTU or 12,000 BTU per hour, the latter equal to 37,971 kg of ice melted per hour; total equivalent energy is 303.65 megajoules (MJ) and 12.65 MJ respectively; used in refrigeration and air conditioning engineering. (Note that BTU's rather than tons of refrigeration are now used in calculations, the latter term being obsolete.)

.054 "Top" icing: finely chopped ice scattered over the top of containers in a load to absorb heat in transit. (AY, RT)

.055 Vacuum cooling: evaporative cooling of produce, a process that utilizes the heat of vaporization of water (see VI.5.051c) in the change of state from liquid to vapor and low pressure (typically 3 to 4 mmHg, the vapor pressure at 0°C being 4.6 mm = 0.18 in.)

to effect extremely rapid cooling; equipment consists of a large autoclave (sealed chamber with stout walls) with vacuum equipment (multiple effect steam jet or equivalent), and refrigeration to condense the water vapor and allow its removal from the system; used widely and most efficiently for produce with a high surface to volume ration, such as leafy vegetables, for which typical cooling times range from 10 to 20 minutes. (RT, F & R, RS)

.056 Vapor pressure: (see Humidity, VI.5.025d).

VI.6. Carriers (Loading, Routing, Services)

.000 Carrier: two meanings: (a) the company or agent that is responsible for transporting produce; (b) the actual truck, rail car, ship, or cargo plane in which the produce is transported.

.001 Bill of lading: similar to a manifest (VI.6.020), except that it is made up in contract form for transportation of the produce and signed by the carrier's agent who thereby acknowledges receipt; a negotiable instrument (meaning it can be sold or used as collateral) if drawn up in order form.

.002 Break bulk: the practice of restacking containers (i.e., tearing down palletized loads) at dockside prior to loading on shipboard (bear in mind steamship loads are normally 100,000 to 200,000 containers or more per trip).

.003 Bulk shipment: shipment of loose fresh produce by rail car or truck, not in containers; e.g., potatoes or watermelons. (RS)

.004 Common carrier: truck or trucking company that is licensed and regulated by the Interstate Commerce Commission, as opposed to exempt haulers (VI.6.006), who are truckers not regulated by the ICC. (RS)

.005 Dead head: when a truck has to run empty (e.g., when an over-the-road trucker hauls produce one way but is unable to get a return load). (RS)

.006 Exempt hauler (exempt trucker): truckers who haul fresh fruits and vegetables that are exempt from regulations under the Interstate Commerce Commission. (RS)

.007 Exempt load: a load of produce hauled by an exempt (unregulated) trucker. (RS)

.008 Exempt trucker: (see Exempt hauler, VI.6.006).

.009 Incompatible commodities: required transit temperatures, emission of volatiles (especially ethylene) harmful to other products, and whether products give off or absorb objectionable odors determine whether perishable food products can be shipped (or stored) together; (see Mixed loads, VI.6.022).

.010 Independent trucker: a person who owns and operates his own truck and commonly hauls produce over-the-road. (RS)

.011 Intermodal: a system in which more than one mode of transportation is used to ship a product from the shipper to the receiver; truck trailers on rail flat cars (TOFC) are the most common combination in the produce industry; (note that export shipments by boat are always intermodal; i.e., a ship plus a truck or rail car at both ends). (AY, RS)

.012 Invoice: an itemized, written list of produce shipped or conveyed to a buyer, with quantities, value or prices, and applicable charges; unlike a manifest, it does not necessarily accompany the shipment.

VI.6. POSTHARVEST HANDLING and MARKETING

.013 LCL: less than carload lot, a term in common use to define a price quoted by a shipper for wholesale lots of merchandise in less-than-carload lots. (RS, BL69)
.014 Less than carlot: (see LCL, VI.6.013).
.015 Less than truckload: (see LTL, VI.6.019).
.016 Loading: the movement of packed containers from the packinghouse floor into a carrier, which may be a semitrailer for over-the-road hauling, a semitrailer for railroad hauling (trailer on flat car, piggyback), a refrigerated rail car for overland, or a ship for export shipments; a skilled operation that requires placement by hand of containers lifted from a conveyor or by forklift of loads on pallets or slip sheets to prevent shifting and to provide adequate ventilation during transit; (see Loading pattern, VI.6.017). (S & G, BL69)
.017 Loading pattern: a precise arrangement of containers with the necessary bracing to prevent shifting during transit; loading patterns for containers shipped in rail cars or trailers on flat cars are described in the "Container and Loading Rules Tariff" and "Trailer on Flat Car Fruit and Vegetable Tariff" under Interstate Commerce Commission Regulations; U.S. Department of Agriculture research workers have devised several loading patterns for truck shipments, e.g., airflow and chimneystack; recommended patterns for ships roughly follow those for other carriers, the primary problems being hold sizes and depths (usually several times the dimensions of other carriers) and adequate refrigeration facilities; (see Loading pattern choice, VI.6.018). (AY, RT, BL68, BE)
.018 Loading pattern choice: the arrangement of containers in a load is determined or limited by type, size, and shape of the containers themselves, plus (a) interior dimensions of the vehicle, car, or ship hold, (b) weight or density of the commodity, (c) need for proper weight distribution to provide maximum permissible axle loadings, (d) specific refrigeration and ventilation requirements of the commodity being shipped, and (e) whether the load is made up of mixed commodities and container sizes.
.019 LTL: less than truckload shipment. (RS, AY, BL68)
.020 Manifest: a list (invoice) of a shipment that specifies grade, size, number, product, shipper, buyer, etc., of each container; it is not negotiable because it is not drawn in order form; accompanies the shipment with a certificate of inspection, both of which are required for acceptance of a commercial shipment by a carrier.
.021 Middleman: (see VI.8.066).
.022 Mixed loads: several meanings with reference to truck shipments: (a) two or more types of fruit and/or vegetable in a long-distance shipment, (b) a load from a grocery distribution center, with meats, dairy products, produce, and dry goods all in the same space, (c) a load with two or more types of container (e.g., fiberboard cartons and bushel baskets); (see Incompatible commodities, VI.6.009). (RS)
.023 MUM: modularization, unitization, and metrification; a system recommended by the United Fresh Fruit and Vegetable Association and the Palletization and Productivity Committee for unitized forklift handling that can be used for all produce commodities; involves use of the metric-sized 100 X 120 cm pallet as the standard unitized shipping pallet or an equivalent-sized slip sheet and changes in shipping carton and crate sizes (e.g., 60 X 50, 60 X 40, 50 X 40, 50 X 30, and 40 X 30 cm) to unitize efficiently a

VI.6. POSTHARVEST HANDLING and MARKETING

        100 X 120 cm unit base for ease of loading highway and TOFC vans. (K)

.024 Palletization: the system of shipping fruits and vegetables on pallets. (RS)

.025 Piggyback: (see Trailer on flat car, VI.6.033).

.026 Plan III: the piggyback (trailer on flat car, VI.6.033) system in which the railroads furnish the flat car and the shipper or co-operatives furnish the trailers; the railroads provide ramp-to-ramp service and shippers provide terminal service; the most common system used in the produce industry; examples are trailers owned by MARTRAC (UPS), GELCO, Co-op, CORNUCOPIA, or Inter-modal, which are used for hauling "fresh" perishables eastbound from western states (e.g., California) and dry freight westbound. (RS, K)

.027 Protective services: (see Services, VI.6.029).

.028 Routing: the art of working out the most feasible route from shipping point to a given destination; mainly for rail shipments (cars or piggyback trailers) in which several alternative routes or carriers may be involved in reaching a certain destination (e.g., New York, Chicago, or other large city in which there may be a dozen or more terminals). (BL69)

.029 Services: two sets, recommended and provided for perishable produce:

.029a. <u>Recommended protective services</u>: for truck and rail shipments, include the recommended transit temperature, desired relative humidity, highest freezing point, and top-ice and/or package ice preferable for each commodity, as given in U.S. Dept. Agr. Handbk. 105 for the protection of perishable foods during transport by motortruck and U.S. Dept. Agr. Handbk. 195 for the protection of rail shipments of fruits and vegetables. (RT, JM)

.029b. <u>Provided protective services for truck shipments</u>: not standardized in the sense that specific rules for cooling, heating, ventilation, etc., must be met; hence it is essential that shipper and trucker know the requirements of the specific commodity and meet them within the capabilities of the truck's equipment; failure to do so is responsible for most of the problems encountered in transit. (AY)

.029c. <u>Provided protective services for rail car and trailer on flat-car shipments</u>: standard ventilation, standard refrigeration, initial and subsequent reicing, and special heater protective service, and the like are spelled out in the "Perishable Protective Tariff" and "Trailer on Flat Car Fruit and Vegetable Tariff," along with rules on other services such as delivery, loading, and unloading of trailers. (JM, BE)

.030 Statement: an abstract of an account, usually submitted on a periodic basis (e.g., monthly), as a summary of invoices issued, plus any prior balance or credit due.

.031 Straight load: a shipment consisting of just one product. (RS)

.032 TOFC: (see Trailer on flat car, VI.6.033).

.033 Trailer on flat car (TOFC): intermodal method of hauling a full trailer load of fresh produce on a railroad flat car; also known as piggyback. (BE, RS)

.034 Transportation problems: deterioration of fresh produce during transportation to market may be (a) product related, (b) packinghouse related, or (c) driver related, but usually results from a combination of all three; typical causes are (1) dirty trailer

VI.6.1. POSTHARVEST HANDLING and MARKETING

(i.e., not cleaned after previous load, a source of contamination), (2) improper precooling (product temperature too high when loaded), (3) broken trailer insulation (increases cooling or heating load), (4) faulty thermostat (results in a too high or low trailer temperature), (5) improper loading pattern (prevents proper air circulation thus buildup of volatiles and CO_2), (6) load shifting (improper bracing blocks proper air flow), (7) broken containers (from load shifting or someone climbing on them), (8) road heat (reflection of heat through trailer floor changes temperature distribution pattern), and (9) inadequate inspection during transit (failure of driver to check load periodically). (AY)

.035 Unitization: the process of standardizing containers and pallets so that more efficient loads can be achieved; standardizing to conform to a distributor's warehouse slot. (RS)

.036 Van: two different meanings: (a) a semitrailer with a completely enclosed body used for over-the-road or piggyback hauling of produce; (b) a small vehicle of about 1/2-to-1-ton capacity with a body enclosing the engine and cargo and passenger space and used for business or recreational purposes. (K)

.037 Van modular containers: intermodal (VI.6.011) refrigerated containers in 10-, 20-, and 40- ft lengths (equivalent to 1/4, 1/2, and full trailer, respectively); can be transported by rail, truck, ship, or cargo plane; some have movable bulkheads to provide separate compartments for incompatible loads.

.038 Way bill: a document issued with every shipment of railroad freight; lists the details regarding shipment (including services), route, and charges.

VI.7. Laws, Rules, and Regulations

VI.7.1. Perishable Agricultural Commodities Act (PACA)

.000 Perishable Agricultural Commodities Act of 1930, as Amended (PACA) (1982): "an act to suppress unfair and fraudulent practices in the marketing of perishable agricultural commodities in interstate and foreign commerce"; i.e., "you must not steal and you must not lie"; spells out in the Act and its regulations the rights and responsibilities that the produce industry feels are needed to put everyone on an equal basis; administered and enforced by the U.S. Department of Agriculture; industry-sponsored and industry-financed; (excerpted from Perishable Agricultural Commodities Act of 1930 as Amended. 1982; see Acknowledgments and Selected References).

.001 Acceptance: (a) any act by the consignee that signifies acceptance of a shipment, including diversion or unloading; (b) any act by the consignee inconsistent with the consignor's ownership but signifies acceptance only if the act is wrongful against the latter and is ratified by him; (c) failure of the consignee to give notice of rejection to the consignor within a reasonable time; [PACA 46.2 (dd)].

.002 "Accommodation advance" ("regular advance"): used in connection with an advance of money or credit against anticipated net proceeds to be realized from the sale of consigned produce; means

VI.7.1. POSTHARVEST HANDLING and MARKETING

that the consignor has received an advance of money or credit and that if the consigned produce does not sell for enough to cover the cost of transportation and handling, including the customary or agreed commission and the advance made, the consignor must return to the person making the advance a sum equal to the deficit sustained.

.003 Account promptly: means rendering to the principal a true and correct accounting: (a) for buying or brokerage transactions, within 24 hours after date of shipment; (b) consignment or joint-account transactions, within 10 days after date of final sale with respect to each shipment and for a grower's agent or shipper, within five days after payment has been made (cooperative associations who have seasonal pools are exempt); (c) consignment or joint account transaction, within 10 days after date of receipt of payment of a carrier claim filed; [PACA sec. 46.2 (z)].

.004 AF: (see FOB acceptance final, VI.7.1.021).

.005 Broker: any person engaged in the business of negotiating sales and purchases of any perishable agricultural commodity in interstate or foreign commerce for or on behalf of the vendor or purchaser, except that no person shall be deemed to be a "broker" if such person is an independent agent negotiating sales for and on behalf of the vendor and if the only sales of such commodities negotiated by such person are sales of frozen fruits and vegetables with an invoice value not in excess of $200,000 in any calendar year; [PACA, sec. (7)].

.006 Brokerage fee: a fee charged for effecting a sale or making a valid and binding contract; may be charged to only one of the parties (seller and buyer) of the contract unless the parties make a prior agreement to split the fee; employment of another broker or selling agent, including auction companies, is not permitted without specific prior approval; accounting must be true and accurate and full payment prompt when the broker acts as the seller's agent; [PACA, sec. 46.28 (b)].

.007 Broker operations: the function of a broker is to negotiate, for or on behalf of others, valid and binding contracts; several types of operation in two general groups: (a) usual: consists of negotiation of the purchase and sale of produce of one or several commodities; broker acts as agent of the buyer or seller but not both parties; seller generally invoices the buyer unless a specific agreement is made in which the seller invoices the broker, who, in turn, invoices the buyer and collects and remits to the seller; broker may distribute to pool buyers or may be authorized by the seller to dispose of the produce for the seller's account as negotiated or as consignment sales; (b) buying broker acts as the buyer's representative in negotiating purchases at shipping points, terminal markets, or intermediate points; (PACA, sec. 46.27).

.008 CAC: (see CAF, VI.7.1.009).

.009 CAF, CAC, CIF: mean "cost and freight," "cost and charges," and "cost, insurance, and freight," respectively; CAF sales shall be deemed to be the same as FOB sales, except that the selling price shall include the correct freight charges to destination; CAC sales shall be deemed to be the same as FOB sales, except that the selling price includes the correct freight and refrigeration or heater charges to destination; CIF sales shall be deemed to be the same as FOB sales, except that the selling price includes insurance and the correct freight and refrigeration or heater

charges to destination.
.010 Carload (carlot, car): when used in offers, quotations, or contracts, in which the quantity is not more definitely specified and in the absence of well-established trade custom or standard of size of a "carload," "carlot," or "car" of the produce in question, it means not less than the minimum quantity required by the carrier's tariff applicable to the movement and not more than 10% in excess of such minimum tariff requirements, except that when the carrier's tariffs provide alternative rates and minimum the buyer shall state which tariff minimum must be observed, and, in event of failure to do so, the shipper may exercise discretion, in no case, however, exceeding the higher alternative minimum quantity provided by the tariff, with only the variations therefrom permitted by this paragraph; [PACA, sec. 46.43 (w)].
.011 Cash sale: means that the buyer is required to pay the seller within 24 hours after his acceptance of the shipment; [PACA sec. 46.43 (gg)].
.012 CIF: (see CAF, VI.7.1.009).
.013 Commission merchant: any person engaged in the business of receiving in interstate or foreign commerce any perishable agricultural commodity for sale on commission or for or on behalf of another; [PACA sec. 1 (5)].
.014 Complaint: (a) any person complaining of any violation of any provision of section 2 by any commission merchant, dealer, or broker may, at any time within nine months after the cause of action accrues, apply to the Secretary by petition, which shall briefly state the facts, whereupon if, in the opinion of the Secretary, the facts contained therein warrant such action a copy of the complaint made shall be forwarded by the Secretary to the commission merchant, dealer, or broker, who shall be called on to satisfy the complaint, or to answer it in writing, within a reasonable time to be prescribed by the Secretary. (b) Any officer or agency of any state or territory having jurisdiction over commission merchants, dealers, or brokers in such state or territory and any employee of the United States Department of Agriculture or any interested person may file, in accordance with rules and regulations of the Secretary, a complaint of any violation of any provision of this Act by any commission merchant, dealer, or broker and may request an investigation of the complaint by the Secretary; [PACA sec. 6 (a), (b)].
.015 Dealer: any person engaged in the business of buying or selling produce in wholesale or jobbing quantities in commerce; includes (a) jobbers, distributors, and other wholesalers; (b) retailers, when the invoice cost of all produce purchases exceeds $200,000 during a calendar year (means all purchases of fresh and frozen fruits and vegetables without regard to quantity or whether intrastate, interstate, or foreign commerce); (c) growers who market produce grown by others (excludes produce for processing or canning in the state where grown, but not if the canned or processed product is frozen or packed in ice or consists of cherries in brine); [PACA sec. 46.2 (m)].
.016 "Delivered" ("delivered sale"): means that the produce is to be delivered by the seller on board car or truck, or on dock if delivered by boat, at the market in which the buyer is located or at such other market agreed on, free of any and all charges for transportation or protective service. The seller assumes all risks of loss and damage in transit not caused by the buyer;

e.g., a sale of "U.S. No. 1 potatoes delivered Chicago" means that the potatoes, when tendered for delivery at Chicago, shall meet all the requirements of the U.S. No. 1 grade in quality and condition.

.017 Employ, employment: any affiliation of any person with the business operations of a licensee, with or without compensation; includes ownership or self-employment; [PACA sec. 1 (10)].

.018 FAS steamer: means that the produce is to be delivered free alongside the steamer, in suitable shipping condition (VI.7.1.070, VI.7.1.071), in accordance with the terms of the contract, and that the buyer assumes all responsibility and risk of damage thereafter; [PACA sec. 46.43 (O)].

.019 FOB: for example, FOB Winter Haven, FL, or FOB Salinas, CA; means that the produce quoted or sold is to be placed free on board the boat, car, or other agency of the through land transportation at the shipping point, in suitable shipping condition (VI.7.1.070, VI.7.1.071), and that the buyer assumes all risk of damage and delay in transit not caused by the seller, irrespective of how the shipment is billed. The buyer shall have the right of inspection at destination before the goods are paid for to determine whether the produce shipped complied with the terms of the contract at time of shipment, subject to the provisions covering suitable shipping condition; [PACA sec. 46.43 (i)].

.020 FOB acceptance (shipping point acceptance): means that the buyer accepts the produce at shipping point and has no right of rejection. The buyer has recourse against the seller if the produce was not in suitable shipping condition (see VI.7.1.070; VI.7.1.071) or has recourse for a material breach of contract, provided that the shipment is not rejected. The buyer's remedy under this method of purchase is by recovery of damages from the seller and not by rejection; [PACA sec. 46.43 (1)].

.021 FOB acceptance final (AF, shipping point acceptance final): means that the buyer accepts the produce at shipping point and has no right of rejection. Suitable shipping conditions do not apply under this trade term. The buyer has recourse for a material breach of contract, provided that the shipment is not rejected. The buyer's remedy under this type of contract is by recovery of damages from the seller and not by rejection of the shipment; (PACA sec. 46.43 (m)].

.022 FOB inspection and acceptance arrival: means that the produce quoted or sold is to be placed by the seller free on board car or other agency of through transportation at shipping point, the cost of transportation to be borne by the buyer, but the seller is to assume all risk of loss and damage in transit not caused by the buyer, who has the right to inspect the goods on arrival and to reject them if, on inspection, they fail to meet the specifications of the contract of sale at destination. The buyer may not reject without reasonable cause. Such a sale is FOB only in price and is on a delivered basis in regard to grade, quality, and condition; [PACA sec. 46.43 (dd)].

.023 FOB sale at delivered price: means the same as FOB, except that transportation charges from shipping point to destination shall be borne by the seller; i.e., the sale, is FOB in grade, quality, and condition and delivered at price.

.024 FOB steamer: means that the produce is to be placed free on board a steamer at the shipping point, in suitable shipping condition (see VI.7.1.070; VI.7.1.071), in accordance with the terms of the

contract and that the buyer assumes all responsibility and risk of damage thereafter; [PACA sec. 46.43 (n)].
.025 Fresh fruits and fresh vegetables: include all produce in fresh form generally considered perishable, whether packed in ice or held in common or cold storage, but does not include those perishable fruits and vegetables that have been manufactured into articles of food of a different kind or character. The effects of the following operations shall not be considered as changing a commodity into a food of a different kind or character: water or steam blanching, chipping, color adding, curing, cutting, dicing, drying for the removal of surface moisture; fumigating, gassing, heating for insect control, ripening, and coloring; removal of seed, pits, stems, calyx, husk, pods, rind, skin, peel, etc.; polishing, precooling, refrigerating, shredding, slicing, trimming, washing with or without chemicals; waxing, adding sugar or other sweetening agents; adding ascorbic acid or other agents used to retard oxidation; mixing several kinds of sliced, chopped, or diced fruits or vegetables for packaging in any type of container; or comparable methods of preparation; [PACA sec. 46.2 (u)].
.026 Frozen fruits and vegetables: include all fresh fruits and fresh vegetables as defined (VI.7.1.025), when such produce is in frozen form; [PACA sec. 46.2 (v)].
.027 Full payment promptly: normally means within 10 days (five days for grower's agents or shippers), unless other arrangements made by agreement (e.g., partial payments during the shipping season of a crop or a closed or advise bill of lading that requires cash on delivery); [PACA sec. 46.2 (aa)].
.028 Good delivery: means in connection with FOB contracts of purchase and sale (unless otherwise agreed to between the contracting parties) that the commodity must meet the requirements of the contract at the time of loading or sale and, if the shipment is handled under normal transportation service and conditions, meet the following additional requirements on delivery at the contract destination: provides for an additional Good Delivery tolerance over the destination tolerance for the applicable grade in the U.S. Standards for Grades of Lettuce;)PACA sec. 46.44).
.029 Grower: any person who raises produce for marketing; [PACA sec. 46.2 (p)].
.030 Grower's agent: any person operating at shipping point who sells or distributes produce in commerce for or on behalf of growers or others and whose operations may include planting, harvesting, grading, packing, and furnishing containers, supplies, or other services. (RS)
.031 Growers' agents: sell and distribute produce for or on behalf of growers and others and, in addition, may perform a wide variety of services, such as financing, planting, harvesting, grading, packaging, and furnishing labor, seed, containers, and other supplies or services; several types of contract from being limited only to sales to blanket authority to market and distribute the produce, using their discretion with regard to the best methods, can sell, consign, or ship on joint account, use the services of brokers or sell through terminal market auctions, grant credit, make adjustments in the invoice price, handle claims, or abandon shipments without consulting the growers; some agents have an agreement with growers to pool the produce and render accountings on the basis of average or prorated prices after

VI.7.1. POSTHARVEST HANDLING and MARKETING

deducting prorated expenses; some contracts specify certain fixed charges (harvesting, grading, packing, etc.), plus a selling fee, thus reducing record requirements. The important thing is to have a contract in which everything is spelled out; (PACA sec. 46.30, 46.32).

.032 Guaranteed advance: use in connection with an advance payment on consigned produce; means that the person making the advance guarantees that the net proceeds to the consignor shall at least equal the amount advanced and that the consignor cannot be held liable for any deficit that results from the sale of produce if the deficit is not occasioned by or contributed to by an act of the consignor; [PACA sec. 46.43 (aa)].

.033 Immediate shipment: (see "Tomorrow's shipment, VI.7.1.074).

.034 In transit, roller, rolling car: means that the produce referred to is in the possession of the transportation company and under movement from the shipping point when the quotation is made, that the car is moving over a route in line of haul between the point of origin and the market in which delivery is to be made, and that it has been moving since the date of shipment, without any delay attributable to the shipper or his agent. Unless otherwise specifically agreed, if a roller, rolling car, or a car in transit is sold FOB shipping point, the buyer shall be deemed to have assumed only the lowest all-rail freight charges applicable for the shipment between the point of origin and the contract destination specified by the parties, with other charges that would have accrued if the car had been shipped direct to the contract destination: it is also provided that the buyer is not liable for payment for protective services if the seller does not inform the buyer of the kind and extent of such services ordered from the carrier. [PACA sec. 46.43 (q)].

.035 Investigation: the Secretary of Agriculture will investigate a complaint, if reasonable grounds are apparent, and afford the person concerned an opportunity for a hearing (depositions or verified statements of fact may be used in support of the complaint and respondent's answer in lieu of a hearing if the amount claimed as damages does not exceed $3000); if the complainant is a nonresident of the United States, a bond for double the amount of the claim is required before any formal action can be taken (bond may be waived if there is an agreement of reciprocity between the countries).

.036 Joint account partners' duties: (see Receiving market commission merchant's duties, VI.7.1.050).

.037 Joint account-split above: means that the receiving joint partner will pay promptly the agreed cost of the shipment to the other joint partner. After disposition of the produce, the parties will divide equally the profits of the shipment after deduction of the cost of the shipment and proper expenses from the gross proceeds. The receiving joint partner will pay all expenses and cannot recover any loss that results from the joint venture; [PACA sec. 46.43 (hh)].

.038 Joint account transaction: a produce transaction in commerce in which two or more persons participate under a limited joint venture agreement in which they agree to share in a prescribed manner the costs, profits, or losses that result from the transaction.

.039 Liability: (a) if any commission merchant, dealer, or broker violates any provision of section 2 he shall be liable to the person

VI.7.1. POSTHARVEST HANDLING and MARKETING

or persons injured thereby for the full amount of damages sustained in consequence of such violation; (b) such liability may be enforced (1) by complaint to the Secretary as hereinafter provided or (2) by suit in any court of competent jurisdiction; however, this section shall not in any way abridge or alter the remedies now existing at common law or by statute and the provisions of this Act are in addition to such remedies.

.040 License required: (a) no person shall at any time carry on the business of a commission merchant, dealer, or broker without a license that is valid and effective at that time; (b) separate licenses are required for each person, and more than one trade name may be used by the same person only after such trade names have been approved in writing by the director; (c) joint account arrangements between two or more licensees are not considered to result in separate firms and therefore do not require separate licenses; (PACA sec. 46.3).

.041 License termination, suspension, revocation, cancellation: a license can be suspended if a trade name has been disapproved, automatically terminated on the date of discharge in bankruptcy, automatically cancelled if the bond posted when the license was issued is terminated without approval, automatically suspended if notice has been given of an increased bond that has not been posted, suspended or revoked for violations of sections dealing with unfair conduct, false or misleading statements, failure to keep proper records, or to permit their inspection, or terminated if the license fee for renewal is not paid within 60 days after the termination date; (PACA sec. 46.9).

.042 Official inspection certificate: issued after inspection by USDA or other licensed inspectors (i.e., state inspectors licensed to determine the class, quality and/or condition of any lot of any perishable agricultural commodity offered for interstate or foreign shipment); official inspection certificates for fresh fruits and vegetables shall be received as primafacie evidence of the truth of the statements contained therein; [PACA sec. 14 (a)].

.043 Perishable agricultural commodity: fresh fruits and vegetables of every kind and character, whether frozen or packed in ice, and cherries in brine (i.e., packed in a water solution containing SO or other bleaching agent of sufficient strength to preserve the product with or without the use of hardening agents); [PACA sec. 1 (4)].

.044 Price arrival: in the absence of a contrary specific understanding this means that the produce is shipped direct to the customer or to an agent of the consignor, for the benefit of the customer; the price is subject to agreement between the customer and the consignor on arrival of the produce at the customer's destination, with sufficient time permitted for inspection.

.045 Prompt shipment: means that the conditions of the offer, order, or confirmation will be met if the shipment is under billing by the transportation company in time to move on a transportation facility scheduled to leave not more than 72 hours later than allowed under "today's shipment" (VI.7.1.073); [PACA sec. 46.43 (d)].

.046 Purchase after inspection: means a purchase of produce after inspection or opportunity for inspection by the buyer or his agent. Under this term the buyer has no right of rejection and waives all warranties regarding quality or condition, except those expressly made by the seller; [PACA sec. 46.43 (ff)].

.047 Quick shipment: means that the conditions of the offer, order, or confirmation will be met if the shipment is under billing by the transportation company in time to move on a transportation facility scheduled to leave not more than 48 hours later than allowed under "today's shipment." (VI.7.1.073); (PACA sec. 46.43).

.048 Reasonable time: (a) frozen fruits and vegetables, rail shipments 48 hours; truck shipments 12 hours after notice of arrival and accessible for inspection; (b) fresh fruits and vegetables, 24 and 8 hours, respectively; delay permitted because of adverse weather, inability to obtain Federal inspection, or for shipments arriving on nonwork days or after or near the close of regular business hours.

.049 Receiving market commission merchant: any person operating on a receiving market who is engaged in the business of receiving produce in commerce for sale, on commission, for or on behalf of another; [PACA sec. 46.2 (r)].

.050 Receiving market commission merchant's duties: required to exercise reasonable care and diligence in disposing of the produce promptly and in a fair and reasonable manner; cannot employ another person or firm, including auction companies, sell consigned produce outside the market area, or average or pool sales without permission of the consignor; must keep complete and detailed records to cover produce received, sales, quantities lost, dates and cost of repacking or conditioning, unloading, handling, freight, demurrage or auction charges, and any other expenses, preferably with a definite prior agreement on the amount of the commission and other charges that will be assessed; (PACA sec. 46.29).

.051 Regular advance: (see Accommodation advance, VI.7.1.002).

.052 Rejection without reasonable cause: (a) refusing or failing without legal justification to accept produce within a reasonable time; (b) informing the seller, shipper or his agent that produce, complying with contract, will not be accepted; (c) indicating an intention not to accept produce by an act or failure to act inconsistent with the contract; or (d) any rejection following an act of acceptance; [PACA sec. 46.2 (bb)].

.053 Reparation order: an order by the Secretary of Agriculture that fixes the amount of reparation plus reasonable fees and expenses after he has determined that the commission merchant, dealer, or broker has violated any provision related to unfair conduct; both the complainant and the respondent have the right of appeal to a U.S. district court, the former if the reparation is not paid and the latter if he disputes the order; however, the licensee must have made an appeal or payment in full within five days after expiration of the period for compliance; otherwise his license will be suspended automatically until he does pay the amount owed, plus interest (suspension is also automatic 30 days after dismissal of an appeal or 10 days after court stay).

.054 Responsibly connected: affiliated or connected with a commission merchant, dealer, or broker as (a) partner in a partnership or (b) officer, director, or holder of more than 10% of the outstanding stock of a corporation or association; [PACA sec. 1 (9)].

.055 Roller: (see In transit, VI.7.1.034).

.056 Rolling acceptance: means that the buyer accepts at the time of purchase produce that is in the custody of the transportation company and under movement from the shipping point, under the terms and conditions described in "In transit" (VI.7.1.034) and "Tramp car" (VI.7.1.076), except that the buyer assumes full responsibility for transportation of the goods from time of purchase, has no recourse against the seller because of any change in condition after time of purchase, unless the goods at the time of sale were not in suitable shipping condition, and has no right of rejection on arrival. The buyer's remedy under this method of purchase is by recovery of damages from the shipper and not by rejection of the shipment; by agreement between the parties, however, the purchase may be made subject to inspection at any specified point while the car is rolling or in transit, and the point at which the buyer will assume transportation charges may be specified without affecting the time of acceptance of the commodity; [PACA sec. 46.43 (s)].

.057 Rolling acceptance final: means the same as "Rolling acceptance", except that the buyer has no recourse against the seller because of any change in condition of the produce in transit; the buyer has recourse against the seller for any material breach of the contract providing the shipment is not rejected; the buyer's remedy under this type of contract is recovery of damages from the seller and not by rejection; [PACA sec. 46.43 (t)].

.058 Rolling car: (see In transit, VI.7.1.034).

.059 Shipper: purchases produce from growers in his own name; distributes the produce in commerce by selling, consigning, or joint accounting the shipments and assumes any loss or profits resulting from these operations; also may handle produce on joint account with growers or others; duties vary with the particular contract; must keep full, accurate records; (PACA sec. 46.30, 46.31). (RS)

.060 Shipping point acceptance: (see FOB acceptance, VI.7.1.020).

.061 Shipping point acceptance final: (see FOB acceptance final, VI.7.1.021).

.062 Shipping point inspection: means that the seller is required to obtain federal or federal-state inspection, or other private inspection mutually agreed on, to show the compliance of the lot sold with the quality, condition, and grade specifications of the contract, and that the seller assumes the risk incident to incorrect certification; [PACA sec. 46.43 (x)].

.063 Shipping-point inspection final: or inspection final following the name of the state or point, as California inspection final, means that the seller is required to obtain federal or federal-state inspection, or other private inspection mutually agreed on, to show the compliance of the lot sold with the quality, condition, and grade specifications of the contract, and that the buyer assumes the risk incident to incorrect certification and is without recourse against the seller because of quality, condition, and grade; [PACA sec. 46.43 (y)].

.064 Shipment as soon as possible [shipment as soon as car (truck) can be secured]: means that the shipper is uncertain when the shipment can be made but expects to make it within a reasonable time and will make it as soon as possible; in any case, when these words are used, the buyer shall, at any time after seven days from the date the order is given, have the right to cancel the order or contract of sale if notice of his decision to cancel shall have been received by the shipper before shipment has been made; [PACA sec. 46.43 (h)].

.065 Shipment first part of week (shipment early part of week): means that the produce referred to shall be under billing on Monday or Tuesday of the week specified in time to be picked up by a train

VI.7.1. POSTHARVEST HANDLING and MARKETING

scheduled to move these days' loadings from the shipping point; when used in connection with shipments by truck, this term shall mean that the goods shall be loaded and shall actually start from loading point to destination before midnight on Tuesday of the week specified; [PACA sec. 46.43 (e)].

.066 Shipment last of week (shipment latter part of week): means that the produce referred to shall be under billing by the transportation company in time to move on a transportation facility schedul%d to leave on Friday or Saturday of the week specified; when used in connection with shipments by truck this term shall mean that the goods shall be loaded and shall actually start from loading point to destination before midnight on Saturday of the week specified; [PACA sec. 46.43 (g)].

.067 Shipment middle of week: means that the produce referred to shall be under billing by the transportation company in time to move on a transportation facility scheduled to leave Wednesday or Thursday of the week specified; when used in connection with shipments by truck this term shall mean that the goods shall be loaded and shall actually start from loading point to destination before midnight on Thursday of the week specified; [PACA sec. 46.43 (f)].

.068 Shipment on a specified date: (see Today's shipment, VI.7.1.073).

.069 Subject approval government inspection: means that the seller is required to obtain federal or federal-state inspection, or other private inspection mutually agreed on, and to communicate correctly, by wire or other agreed means, the statements on the certificate regarding quality, condition, and grade, and other essential information, whereupon the buyer, on approval thereof, will be deemed to have accepted the produce without recourse against the seller because of quality, condition, and grade; [PACA sec. 46.43 (z)].

.070 Suitable shipping condition (direct shipments): means that the commodity, at time of billing, is in a condition, if the shipment is handled under normal transportation service and conditions, to ensure delivery without abnormal deterioration at the contract destination agreed on between the parties; however, it will be considered abnormally deteriorated if a "Good delivery standard" for a commodity is set forth (VI.7.1.028) and that commodity at the contract destination contains deterioration in excess of any tolerance provided therein; the seller has no responsibility for any deterioration in transit if there is no contract destination agreed on between the parties; [PACA sec. 46.43 (j)].

.071 Suitable shipping condition (reconsigned roller or tramp cars): means that the commodity, at time of sale, meets the requirements of this phrase as defined in VI.7.1.070 relating to direct shipments; [PACA sec. 46.43 (k)].

.072 Suspension and revocation of license: the license of any commission merchant dealer, or broker who (a) has violated any of the unfair conduct provisions or (b) has been found guilty in Federal court of having violated section 14(b) (tampering in any way with any certificate of inspection) may be suspended for up to 90 days (also if proper accords, records, and memoranda relating to all transactions are not kept) or revoked if the violation is flagrant or repeated (or if false or misleading statements were made in the license application); no licensee can employ any person, or any person who is or has been responsibly connected with any person, (1) whose license has been revoked, (2) is in a period of suspension, or (3) against whom there is an unpaid reparation

VI.7.1. POSTHARVEST HANDLING and MARKETING

award issued within two years (anyone who operates or tries to operate without a valid license may be fined $500 plus $25 per day and the Secretary may also obtain a court injunction to restrain further continuance); (PACA sec. 8); (this Act has big teeth!).

.073 "Today's shipment" (shipment on a specified date): means in connection with shipments by rail, that the goods referred to shall be under billing by the transportation company on the date the order is given or on the date specified in time to be picked up by a train scheduled to move that day's loadings from the shipping point; when used in connection with shipments by boat this term shall mean that the goods shall be placed alongside the boat and be under billing in time to be loaded and shipped on a boat scheduled to leave before midnight of the date specified; when used in connection with shipments by truck this term shall mean that the goods shall be loaded and shall actually start from loading point to destination before midnight of the date specified; [PACA sec. 46.43 (a)].

.074 "Tomorrow's shipment" (immediate shipment): means that the shipment referred to shall be under billing by the transportation company in time to move on a transportation facility scheduled to leave not more than 24 hours later than allowed under "Today's shipment" (VI.7.1.073); [PACA sec. 46.43 (b)].

.075 "Track sale" ("sale on track"): (a) means a sale of produce on track after transit and after inspection or opportunity for inspection by the buyer, or his agent, who shall be considered to have waived any right to reject the commodity purchased on receipt by him or his duly authorized representative from the seller or his duly authorized representative of the bill of lading, delivery order, or other document enabling him to obtain the goods from the carrier; (b) the above definition shall not be construed as depriving the buyer of a right to reparation when the unloading of the car demonstrates that a part of the lading which was not accessible to inspection was of a quality or condition materially inferior to the portion that was accessible to inspection; but notice of intention to file a claim for reparation must be given the seller within 24 hours after receipt by the buyer of the delivery order or bill of lading; (c) if the seller gives the date of arrival when quoting the price the buyer shall, in the absence of any written memorandum of sale to the contrary, assume all charges that accrue on the shipment from the date of arrival; if the seller fails to furnish the date of arrival when quoting price the buyer may, in the absence of any written memorandum of sale that includes the date of arrival or written statement that specifies who shall assume the charges that have accrued after arrival, assume that the shipment arrived at the point of sale on the day on which the purchase was made and shall be liable only for such charges that would properly attach to a shipment arriving on the date the purchase was made; [PACA sec. 46.43 (u)].

.076 Tramp car (tramp car sale): means that the produce has left the shipping point under a bill of lading issued prior to the day on which the quotation is made and has moved, or is moving, over a route out of line of haul with the market in which it is to be delivered, or in which it is being offered or quoted, or has been moving over a route in line of haul between the point of origin and the market in which it is to be delivered, or in which it is being offered or quoted but has been delayed in transit by the seller or has been held by the transportation company at diversion or other points en route awaiting instructions from the shipper and by such holding or delay has missed scheduled movement between points of shipment and the market in which it is to be delivered as the result of the transaction in question; unless otherwise specifically agreed, if a "tramp car" is sold FOB shipping point or a "tramp car sale" is made FOB shipping point the buyer shall be deemed to assume only the lowest authorized all-rail freight charges applicable for the shipment between the point of origin and the contract destination agreed on between the parties, with such other charges that would have accrued if the car had been shipped direct to the contract destination provided that the buyer is not liable for payment for protective services if the seller does not inform him of the kind and extent of such services ordered from the carrier; [PACA sec. 46.43 (r)].

.077 Unfair conduct: the following are unlawful in or in connection with any transaction in interstate or foreign commerce: (a) for any commission merchant, dealer, or broker to engage in or use any unfair, unreasonable, discriminatory, or deceptive practice in connection with the weighing and counting or in any way determining the quantity of any perishable agricultural commodity received, bought, sold, shipped, or handled in interstate or foreign commerce; (b) for any dealer to reject or fail to deliver in accordance with the terms of the contract without reasonable cause any perishable agricultural commodity bought or sold or contracted to be bought, sold, or consigned in interstate or foreign commerce by such dealer; (c) for any commission merchant to discard, dump, or destroy without reasonable cause any perishable agricultural commodity received by such commission merchant in interstate or foreign commerce; (d) for any commission merchant, dealer, or broker to make, for a fraudulent purpose, any false or misleading statement in connection with any transaction involving any perishable agricultural commodity received in interstate or foreign commerce by such commission merchant or bought, sold, or contracted to be bought, sold, or consigned in such commerce by such dealer or the purchase or sale of which in such commerce is negotiated by such broker; or to fail or refuse truly and correctly to account and make full payment promptly in respect of any transaction in any such commodity to the person with whom such transaction is had; or to fail, without reasonable cause, to perform any specification or duty, express or implied, arising out of any undertaking in connection with any such transaction; (e) for any commission merchant, dealer, or broker to misrepresent by work, act, mark, stencil, label, statement, or deed the character, kind, grade, quality, quantity, size, pack, weight, condition, degree of maturity, or state, country, or region of origin of any perishable agricultural commodity received, shipped, sold, or offered to be sold in interstate or foreign commerce; (f) for any commission merchant, dealer, or broker, for a fraudulent purpose, to remove, alter, or tamper with any card, stencil, stamp, tag, or other notice placed on any container or railroad car containing any perishable agricultural commodity if such card, stencil, stamp, tag, or other notice contains a certificate or statement under authority of any federal or state inspector or in compliance with any federal or state law or regulation of the grade or quality of the commodity contained in such container or railroad car or the state or country in which such commodity was produced; (g) for any commission merchant, dealer, or broker, without the consent of an inspector, to make, cause, or permit to be made any change by way of substitution or otherwise in the contents of a load or lot of any perishable agricultural commodity after it has been officially inspected for grading and certification, but this shall not prohibit resorting and discarding inferior produce; (PACA sec.2).

VI.7.2. Other laws, rules, and regulations

.001 Active ingredient (def.): (see Federal Insecticide, Fungicide & Rodenticide Act of 1947; VI.7.2.017a).

.002 Adulterated (def.): (see Federal Insecticide, Fungicide & Rodenticide Act of 1947; VI.7.2.017b).

.003 Adulterated food: (see Federal Food, Drug & Cosmetic Act, as Amended; VI.7.2.016a).

.004 Agricultural Marketing Agreement Act of 1937, as Amended: the enabling legislation for federal-handler (growers and shippers) marketing agreements and marketing orders for milk, fruits (including pecans and walnuts but not including apples and not including fruits other than olives for canning), tobacco, vegetables (not including vegetables other than asparagus for canning), soybeans, and naval stores; basically, a mechanism for controlling the total quantity of a commodity marketed or transported in a given period by limitations on grade, size, or quality, types of container, and handler quotas (prorata shipments); also the control and disposition of surpluses or the establishment of reserve pools for a commodity.

.004a. Marketing Agreement: a formal agreement between the U.S. Secretary of Agriculture and handlers of a commodity specified under the Agricultural Marketing Agreement Act of 1937, as amended, which becomes effective after handlers representing not less than 50% of the volume (80% of the volume for California citrus fruits) have signed the agreement. (RS, S & G)

.004b. Marketing Order: an order issued by the Secretary of Agriculture which defines the production areas, commodity (including species and/or cultivars); the administrative body, composition, and duties; expenses and assessments; regulations of shipments (i.e., by grade, size, or both, and prorata shipment by handlers); exemptions, inspection, and certification; marketing policy; at least two-thirds of the producers (three-fourths in the case of California citrus fruits) must favor or have approved the order before it becomes effective; each Marketing Order (and Marketing Agreement) must be reapproved each year by at least 50% of the handlers. (RS, S & G)

.004c. Marketing Regulation: a rule issued by the U.S. Secretary of Agriculture on the recommendation of the Administrative Committee which specifies the minimum size or grade or other restriction placed on shipment of a commodity during a particular period; note, however, that the Secretary may refuse to issue a regulation if it conflicts with a provision of the Agricultural Marketing Agreement Act or the Marketing Agreement (usually happens only when handlers wish to restrict grades or sizes when a commodity is in short supply and the price is above

VI.7.2. POSTHARVEST HANDLING and MARKETING

"parity"). (RS, S & G)

.005 Antidote (def.): (see Federal Insecticide, Fungicide & Rodenticide Act of 1947; VI.7.2.017c).

.006 Color additive: (see Federal Food, Drug & Cosmetic Act, as Amended; VI.7.2.017b).

.007 Coloring of economic poisons (def.): (see Federal Insecticide, Fungicide & Rodenticide Act of 1947; VI.7.2.017d).

.008 Copeland Act of 1938: (see Federal Food, Drug & Cosmetic Act, as Amended; VI.7.2.016).

.009 Defoliant (def.): (see Federal Insecticide, Fungicide & Rodenticide Act of 1947; VI.7.2.017e).

.010 Delaney Amendment: (see Federal Food, Drug & Cosmetic Act, as Amended; VI.7.2.016c).

.011 Desiccant (def.): (see Federal Insecticide, Fungicide & Rodenticide Act of 1947; VI.7.2.017f).

.012 Device (def.): (see Federal Insecticide, Fungicide & Rodenticide Act of 1947; VI.7.2.017g).

.013 Economic poison (def.): (see Federal Insecticide, Fungicide & Rodenticide Act of 1947; VI.7.2.017h).

.014 Environmental Protection Agency (EPA): a federal agency established in 1970 with the specific responsibilities of (a) issuing regulations to set standards for pollutants of all types (e.g., emissions from various industrial, home, municipal, farm, or other sources; dumping of wastes into lakes, groundwater sources, streams, rivers, or oceans; and disposal of wastes) on a national basis; and (b) enforcement of these standards; as in the case of other federal acts (e.g., Food, Drug & Cosmetic Act, Occupational Safety and Health Act, Insecticide, Fungicide and Rodenticide and Nematocide, Plant Regulator, Defoliant and Desiccant Acts, and Federal Environmental Pesticide Control Act of 1972), states have enacted counterpart legislation, which may be more restrictive (e.g., California's laws on pollution and occupational safety and health).

.015 EPA: (see Environmental Protection Agency, VI.7.2.014).

.016 Federal Food, Drug & Cosmetic Act, as Amended: the Copeland Act of 1938 (which superseded the old Pure Food Laws of 1906) and subsequent amendments (e.g., Miller Act of 1954, Harris Act of 1958, Haley Bill of 1960, Color Additive Amendments of 1960, including the Delaney Amendment) which define food (including the establishment of definitions and standards of identity for avocados, cantaloupes, citrus fruits, and melons, relating only to maturity and the effects of freezing), drugs, cosmetics, labeling, misbranding, color additives, and procedures for establishing tolerances for pesticides (banned unless cleared), color additives, etc., as well as penalties (e.g., seizure of produce for misbranding, with the right to a jury trial).

.016a. Adulterated food: a food is deemed adulterated; (a) (1) if it bears or contains any poisonous or deleterious substance that may render it injurious to health (but not considered adulterated if it is not an added substance and if the quantity present does not ordinarily render it injurious to health); (2) if it bears or contains any added poisonous or added deleterious, unsafe substance; (3) if it consists in whole or in part of any filthy, putrid, or decomposed substance or if it is otherwise unfit for food; or (4) if it has been prepared, packed, or held in unsanitary conditions in which it may have become contaminated with filth or rendered injurious to health

VI.7.2. POSTHARVEST HANDLING and MARKETING

.016b.
...or (5) if its container is composed in whole or in part of any poisonous or deleterious substance that may render the contents injurious to health; (b)...(3) if damage or inferiority has been concealed in any manner or (4) if any substance has been added or mixed or packed therewith to increase its weight or bulk or reduce its quality or strength, or make it appear better or of greater value than it is; (c) if it bears or contains color other than one from a batch certified in accordance with regulations;...(FD&C Act, Chap. IV. Sec. 402; clauses relating to animal products deleted).

.016b. Color additive: a dye, pigment, or other substance synthesized, extracted, isolated, or otherwise derived, with or without intermediate or final change of identity, from a vegetable, animal, or mineral source and when added or applied to a food is capable (alone or by reaction with other substances) of imparting color thereto.

.016c. Delaney amendment: a paragraph in the Color Additive Amendment of 1960 to the Food, Drug & Cosmetic Act, which stipulates that a color additive is deemed unsafe and cannot be listed for any use, whether ingested or not, if it is found to induce cancer in man or animal in the course of tests "appropriate" for the evaluation of the safety of the additive.

.016d. Food and Drug Administration: an agency of the Department of Health and Human Welfare that administers the Food, Drug & Cosmetic Act and publishes regulations under and interpretations of the Act.

.016e. Misbranded food: a food is deemed misbranded: (a) if its labeling is false or misleading in any particular; (b)...offered for sale under the name of another food; (c)...is an imitation of another food unless its label bears,...the word "imitation" and...the name of the food imitated; (d) its container is made, formed or filled to be misleading; (e) in package form unless it bears a label that contains (1) the name and place of business of the manufacturer, packer, or distributor and (2) an accurate statement of the quantity of the contents in terms of weight, measure, or numerical count (reasonable variations permitted and exemptions of small packages established by regulations); (f) any word, statement, or other information required by or under authority of this act to appear on the label or labeling is not prominently placed thereon with such conspicuousness (compared with other words, statements, designs, or devices in the labeling) and in such terms that will render it likely to be read and understood by the ordinary individual under customary conditions of purchase and use; (g) purports to be or is represented as a food for which a definition and standard of identity has been prescribed by regulation...unless (1) it conforms to such definition and standard and (2) its label bears the name of the food specified in the definition and standard and, insofar as required by such regulations, the common names of optional ingredients (other than spices, flavoring and coloring) present in such food; (h) purports to be or is represented as (1) a food for which a standard of quality has been prescribed by regulations...and its quality falls below this standard unless its label bears ...a statement that it does so, or (2) a food for which a standard or standards of fill of container...and it falls below the standard of fill applicable thereto unless its label bears...a

VI.7.2. POSTHARVEST HANDLING and MARKETING

statement that it does so;....(FD&C Act. Chap. IV, Sec. 403); this is the section of the Act under which most legal action is initiated; clauses relating to animal products deleted.

.016f. Pesticide chemical: any substance which, alone, in chemical combination, or in formulation with one or more substances, is an "economic poison" within the meaning of the Federal Insecticide, Fungicide and Rodenticide Act (of 1947) as now in force or as hereafter amended (e.g., the Federal Nematocide, Plant Regulator, Defoliant and Desiccant Act of 1959 and Federal Environmental Pesticide Control Act of 1972), and which is used in the production, storage, or transportation of raw agricultural commodities. (Note that any "pesticide" is banned unless cleared, i.e., a tolerance has been established or it has been exempted from a tolerance); (from Miller Bill amending FD&C. Act, Public Law 85-518, 1954).

.016g. Raw agricultural commodity: any food in its raw or natural state, including all fruits that are washed, colored, or otherwise treated in the unpeeled form prior to marketing. (From Miller Bill amending FD&C Act, Public Law 83-518, 1954).

.017 Federal Insecticide, Fungicide and Rodenticide Act of 1947; Nematocide, Plant Regulator, Defoliant and Desiccant Amendment of 1959; and Federal Environmental Pesticide Control Act of 1972: the 1947 Act and 1959 Amendment defined and specified uses, registration, labeling, misbranding, and enforcement of provisions relating to pesticides, with administration under the Department of Agriculture prior to 1970, at which time authority was transferred to the Environmental Protection Agency (EPA). The 1972 Act was a substantial revision of the 1947 Act and 1959 Amendment, wherein provisions of the latter (i.e., use, registration, and regulation of pesticides) were extended to <u>intrastate</u> commerce; authority was given EPA to establish tolerances, classify pesticides for general or restricted use (with regulations set up by which appropriate state agencies would certify pesticide applicators for the latter group), register pesticide manufacturing plants, develop procedures and regulations for storage or disposal of pesticide containers, issue experimental use permits, conduct research on pesticides and alternatives, and monitor pesticide use and presence in the environment; (note that enforcement of tolerances on pesticides set up by EPA remains under the Food and Drug Administration of the Department of Health and Human Welfare, with problems of worker health and safety under the Occupational Safety and Health Agency of the Department of Labor; local problems of pesticide use, pollution and waste disposal, etc., are enforced by state counterparts of EPA under regulations that may actually be more stringent than federal standards).

.017a. Active ingredient: (1) in the case of an economic poison other than a plant regulator, defoliant or desiccant an ingredient that will prevent, destroy, repel, or mitigate insects, nematodes, fungi, rodents, weeds, or other pests; (2) in the case of a plant regulator an ingredient that, by physiological action, will accelerate or retard the rate of growth or rate of maturation or otherwise alter the behavior of ornamental or crop plants or the produce thereof; (3) in the case of a defoliant an ingredient that will cause the leaves of foliage to drop from a plant; (4) in the case of a desiccant an ingredient that will artificially accelerate the drying of plant tissue

.017b. Adulterated: applies to any economic poison if its strength or purity falls below the professed standard or quality expressed on its labeling or under which it is sold, if any substance has been substituted wholly or in part for the article, or if any valuable constituent of the article has been wholly or in part abstracted.

.017c. Antidote: a practical immediate treatment of poisoning that includes first-aid treatment.

.017d. Coloring of economic poisons: unlawful to distribute, sell, offer for sale, receive, deliver, or offer to deliver in the original unbroken package to any other person economic poisons, commonly known as standard lead arsenate, basic lead arsenate, calcium arsenate, magnesium arsenate, zinc arsenite, sodium fluoride, sodium fluosilicate, and barium fluosilicate unless they have been distinctly colored or discolored..., or any other white-powder economic poison that the Secretary ...shall, by regulation, require to be distinctly colored or discolored....

.017e. Defoliant: any substance or mixture of substances intended for causing the leaves or foliage to drop from a plant, with or without causing abscission.

.017f. Desiccant: any substance or mixture of substances intended for artificially accelerating the drying of plant tissue.

.017g. Device: any instrument or contrivance intended for trapping, destroying, repelling, or mitigating insects or rodents or destroying, repelling, or mitigating fungi, nematodes, or other pests designated by the Secretary, but not including equipment used for the application of economic poisons when sold separately therefrom.

.017h. Economic poison: (1) any substance or mixture of substances intended for preventing, destroying, repelling, or mitigating any insects, rodents, nematodes, fungi, weeds, and other forms of plant or animal life or viruses, except viruses on or in living man or other living animals, which the Secretary of Agriculture shall declare to be a pest, and (2) any substance or mixture of substances for use as a plant regulator, defoliant, or desiccant; (note that any economic poison must be registered, be in the registrant's or manufacturer's unbroken immediate container, and the label must bear the skull and crossbones, the work "poison" prominently in red, and a statement of an antidote.)

.017i. Fungi: all nonchlorophyll-bearing thallophytes (i.e., all nonchlorophyll-bearing plants of a lower order than mosses and liverworts), e.g., rusts, smuts, mildews, molds, yeasts, and bacteria, except those on or in living man or other animals.

.017j. Fungicide: any substance or mixture of substances intended for preventing, destroying, repelling, or mitigating any fungi.

.017k. Herbicide: any substance or mixture of substances intended for preventing, destroying, repelling, or mitigating any weed.

.017 l. Inert ingredient: one that is not active.

.017m. Ingredient statement: (1) a statement of the name and percentage of each ingredient, with the total percentage of the inert ingredients, in the economic poison; (2) a statement of the name of each active ingredient, with the name of each and total percentage of the inert ingredients, if any, in the economic poison (except option 1 shall apply if the preparation is highly toxic to man); in addition to (1) or (2), in case the economic poison contains arsenic in any form, a statement of the percentages of total and water soluble arsenic, each calculated as elemental arsenic.

.017n. Insect: any of the numerous small invertebrate animals generally having a more or less obviously segmented body, for the most part belonging to the class Insecta, and comprised of six-legged, usually winged forms; e.g., beetles, bugs, bees, and flies; also belonging to other allied classes of arthropods whose members are wingless and usually have more than six legs; e.g., spiders, mites, ticks, centipedes, and wood lice; [note: (1) all insects belong to the class Insecta: (2) spiders, mites, and ticks belong to the Arachnida, centipedes, to the Chilopoda, and wood lice, to the Crustacea or the order I-soptera (termites) of Insecta. Courtesy Dr. James A. Coffelt].

.017o. Insecticide: any substance or mixture of substances intended for preventing, destroying, repelling, or mitigating any insects that may be present in any environment whatsoever.

.017p. Misbranded: applies (1) to any economic poison or device, if its labeling bears any statement, design, or graphic representation relative thereto or to its ingredients, that is false or misleading in any particular; (2) to any economic poison...(a) if it is an imitation of or is offered for sale under the name of another economic poison; (b) if its labeling bears any reference to registration under this Act; (c) if the labeling accompanying it does not contain directions that are necessary for use and if complied with are adequate for the protection of the public; (d) if the label does not contain a warning or caution statement that may be necessary and if complied with adequately to prevent injury to living man and other vertebrate animals, vegetation, and useful invertebrate animals; (e) if the label does not bear an ingredient statement on that part of the immediate container and on the outside container or wrapper, if there be one, through which the ingredient statement on the immediate container cannot be clearly read, of the retail package that is presented or displayed under customary conditions of purchase: provided that the Secretary may permit the ingredient statement to appear prominently on some other part of the container, if the size or form of the container makes it impracticable to place it on the part of the retail package that is presented or displayed under customary conditions of purchase; (f) if any work, statement, or other information required by or under authority of this Act to appear on the label or labeling is not prominently placed there...; or (g) if, in the case of an insecticide, nematocide, fungicide, or herbicide, when used as directed or in accordance with commonly recognized practice it shall be injurious to living man or other vertebrate animals or vegetation, except weeds, to which it is applied, or to the person applying such economic poison; or (h) if in the case of a plant regulator, defoliant, or desiccant when used as directed it shall be injurious to living man or other vertebrate animals or vegetation to which it is applied, or to the person applying such economic poison; provided that physical or physiological effects on plants or parts thereof shall not be deemed to be injurious when this is the purpose for which the plant regulator, defoliant, or desiccant was applied in accordance with the label claims and recommendations.

VI.7.2. POSTHARVEST HANDLING and MARKETING

.017q. Nematocide: any substance or mixture of substances intended for preventing, destroying, repelling, or mitigating nematodes.
.017r. Nematode: invertebrate animals of the phylum Nemathelminthes and class Nematoda; i.e., unsegmented round worms with elongated, fusiform, or saclike bodies covered with cuticle and inhabiting soil, water, plants, or plant parts; may also be called nemas or eelworms.
.017s. Plant regulator: any substance or mixture of substances, intended by physiological action, for accelerating or retarding the rate of growth or rate of maturation or for otherwise altering the behavior of ornamental or crop plants or the produce thereof but shall not to the extent that they are intended as plant nutrients, trace elements, nutritional chemicals, plant inoculants, and soil amendments.
.017t. Rodenticide: any substance or mixture of substances intended for preventing, destroying, repelling, or mitigating rodents or any other vertebrate animal that the Secretary shall declare a pest.
.017u. Weed: any plant that grows where it is not wanted.
.018 Food and Drug Administration: (see Federal Food, Drug & Cosmetic Act as Amended; VI.7.2.016d).
.019 Fungi (def.): (see Federal Insecticide, Fungicide & Rodenticide Act of 1947; VI.7.2.017i).
.020 Fungicide (def.): (see Federal Insecticide, Fungicide & Rodenticide Act of 1947; VI.7.2.017j).
.021 Haley Bill of 1960: (see Federal Food, Drug & Cosmetic Act, as Amended; VI.7.2.016).
.022 Harris Act of 1958: (see Federal Food, Drug & Cosmetic Act, as Amended VI.7.2.016).
.023 Herbicide (def.): (see Federal Insecticide, Fungicide & Rodenticide Act of 1947; VI.7.2.017k).
.024 Inert ingredient (def.): (see Federal Insecticide, Fungicide & Rodenticide Act of 1947; VI.7.2.017 l).
.025 Ingredient statement (def.): (see Federal Insecticide, Fungicide & Rodenticide Act of 1947; VI.7.2.017m).
.026 Insect (def.): (see Federal Insecticide, Fungicide & Rodenticide Act of 1947; VI.7.2.017n).
.027 Insecticide (def.): (see Federal Insecticide, Fungicide & Rodenticide Act of 1947; VI.7.2.017o).
.028 Marketing Agreement: (see Agricultural Marketing Agreement Act of 1937, as Amended; VI.7.2.004a).
.029 Marketing Order: (see Agricultural Marketing Agreement Act of 1937, as Amended; VI.7.2.004b).
.030 Marketing Regulation: (see Agricultural Marketing Agreement Act of 1937, as Amended; VI.7.2.004c).
.031 Miller Act of 1954: (see Federal Food, Drug & Cosmetic Act, as Amended; VI.7.2.016).
.032 Misbranded (def.): (see Federal Insecticide, Fungicide & Rodenticide Act of 1947; VI.7.2.017p).
.033 Misbranded food: (see Federal Food, Drug & Cosmetic Act, as Amended; VI.7.2.016e).
.034 Nematocide (def.): (see Federal Insecticide, Fungicide & Rodenticide Act of 1947; VI.7.2.017q).
.035 Nematocide, Plant Regulator, Defoliant and Desiccant Amendment of 1959: (see Federal Insecticide, Fungicide & Rodenticide Act of 1947; VI.7.2.017).
.036 Nematode (def.): (see Federal Insecticide, Fungicide & Rodenticide

Act of 1947; VI.7.2.017r).
- .037 Occupational Safety and Health Act of 1970: designed "to assure so far as possible every working man and woman in the nation safe and healthful working conditions and to preserve our human resources..."; covers every employer in a business with one or more employees, except those under other federal laws (e.g., the Federal Coal Mine Health & Safety Act and Federal Metal and Nonmetallic Safety Act), with separate provisions for federal, state, and local government employees; (see Occupational Safety and Health Administration, VI.7.2.038).
- .038 Occupational Safety and Health Administration (OSHA): agency in the U.S. Department of Labor responsible for administering the Act; has specific congressional mandate (a) to encourage employers and employees to reduce hazards in the workplace and start or improve existing safety and health programs; (b) to establish employer and employee responsibilities; (c) to set mandatory job safety and health standards; (d) to provide an effective enforcement program (by inspections that include checking of required records, citations, and penalties); (e) to encourage states to administer and enforce their own occupational safety and health programs to be at least as effective as the federal program; and (f) to provide for reporting procedures on job injuries, illnesses, and fatalities; inspections made by Compliance Safety and Health Officers and Industrial Hygienists to ensure compliance with regulations on safety or health standards; citations with proposed penalties issued by an area director with right of appeal to the Occupational Safety & Health Review Commission (and U.S. Circuit Courts); Part 1910, Chap. XVII of Title 29 of the Code of Federal Regulations contains occupational safety and health standards, most of which apply to some phase of horticulture.
- .039 OSHA: (see Occupational Safety & Health Administration, VI.7.2.037).
- .040 Pesticide chemical: (see Federal Food, Drug & Cosmetic Act, as Amended; VI.7.2.016f).
- .041 Plant regulator (def.): (see Federal Insecticide, Fungicide & Rodenticide Act of 1947; VI.7.2.017s).
- .042 Raw agricultural commodity: (see Federal Food, Drug & Cosmetic Act as Amended; VI.7.2.016g).
- .043 Rodenticide (def.): (see Federal Insecticide, Fungicide & Rodenticide Act of 1947; VI.7.2.017t).
- .044 Weed (def.): (see Federal Insecticide, Fungicide & Rodenticide Act of 1947; VI.7.2.017u).

VI.8. Destination Markets, Business Terminology (RS)

- .001 Auction: a central location to which growers bring (or send) their produce to be sold to the highest bidder.
- .002 Bird dog: at least two definitions: (a) an independent fruit buyer who purchases a crop "on the tree" for sale to a packinghouse or processor; e.g., citrus fruit in Florida; (b) an on-the-ground inspector hired by a broker, buyer, or buying organization to inspect produce in the field and inform the client of its quality and other factors.
- .003 Blemish: (see Grade, VI.8.038a).
- .004 Board of directors: a body whose individuals are elected by the

VI.8. POSTHARVEST HANDLING and MARKETING

owners of an enterprise to act as their representatives; it is responsible for managing the affairs of the enterprise and, in turn, is accountable to the owners (i.e., stockholders, of a corporation or grower members of a cooperative association); its primary function is to formulate policies for the guidance of the executive (officers) and lower levels.

.005 Brand: a label or trademark to identify a shipper's product; "any word, name, symbol, or device or any combination thereof adopted and used by a manufacturer or merchant to identify his goods and distinguish them from those manufactured or sold by others" (Trademark Act of 1946 definition); many produce shippers have several brands of the same item that represent different degrees of quality and are registered with the appropriate state agency.

.006 Breaker: a tomato just beginning to turn color from a solid green; e.g., most tomatoes harvested in western Mexico.

.007 Bulk display: a retail display of unpackaged fruits or vegetables.

.008 Buyer (on-the-ground): person hired by a retail or wholesale organization to be located in the growing area and to inspect and buy produce for that particular organization.

.009 Buyer (retail): a person at headquarters level who buys fresh produce for a retail or wholesale organization.

.010 Carlot, carlot equivalent: a commonly used measure of shipping volume; one rail car or truck load of fresh produce or the equivalent thereof.

.011 Cartage firm: a company that specializes in delivering produce from a terminal market company to the firm's customers.

.012 Centralized packaging: produce items packaged in a centralized location (e.g., a warehouse that serves several chain stores), as opposed to packaging in the back room of a retail supermarket.

.013 Chain, chain store: a large number of supermarkets owned by a single company and using a single name or trademark (e.g., A & P, Winn-Dixie, and Kroger).

.014 Code dating: the practice of marking a code on an item to indicate its salable life.

.015 Commission merchant: any person operating in a receiving market who is engaged in the business of receiving produce in commerce for sale for or on behalf of another.

.016 Condition: defined by the U.S. Department of Agriculture Agricultural Marketing Service (1981) as "shall be deemed to include the stage of maturity, decay, freezing injury, shriveling, flabbiness or any other deterioration that may have occurred or progressed since the product was harvested and may continue to progress"; terms used in connection with "condition" in USDA Market News reports (MNS): (a) good: does not justify any price reduction because of condition factors; (b) fair: having a slight degree of off-condition factors that may warrant a small price reduction, compared with "good"; shelf-life for flowers is questionable; (c) ordinary: having a greater degree of off-condition factors that may warrant a substantial price reduction, compared with "good"; shelf-life for flowers is considered rather short; (d) poor: so badly off-condition to warrant heavy price reduction; very short shelf-life for flowers that must be moved immediately or dumped; (e) holdovers: refers (presumably) to merchandise that has been on the market longer than normal but remains near its original condition; prices are discounted to clear up supplies or because probable shelf-life is reduced (not specific and to be used rarely).

VI.8. POSTHARVEST HANDLING and MARKETING

.017 Consignment sale: a sale in which the produce has no set price when it is shipped; a broker tries to sell the produce for "the best price he can get" and will take his brokerage fee from that price.

.018 Consumer cooperative: consumers organized as a nonprofit corporation to operate food markets to serve member families or the general public.

.019 Convenience store: a small compact self-service store open long hours and featuring a limited line of brands and sizes.

.020 Cooperative: a wholesale warehouse operation that is owned by individual stores supplied by that warehouse; each store has stock in the company and receives dividends at the end of each year.

.021 Co-op store: retail store owned by a consumer cooperative and operated by professional managers.

.022 Cost plus: a system of wholesale selling in which merchandise is billed to the retailer at cost, plus a percentage markup or dollar charge, for services rendered; calculated at the end of the order at a fixed rate or billed by product groups in varying amounts.

.023 Count: the number of items in a container corresponding with the size of the produce; the smaller the number, the larger is the size.

.024 Cross-merchandising: a company that specializes in delivering produce from a terminal market company to the firm's customers.

.025 Dark storage: term used to describe the time that foliage plants remain in darkness in shipping containers during transit or storage. (LR)

.026 Deal: a period of time in which growers and shippers in a certain area market their crops (e.g., when Salinas, California, firms begin shipping vegetables) or the national marketing of a single commodity (e.g., apples).

.027 Demand: represents the immediate or current desire for a commodity, coupled with the ability and willingness of the buyer to pay for it; should be considered in relationship to what might be considered normal for the season (U.S. Dept. Agr. Market News Service definition, 1981); terms (MNS): (a) exceeds supply or offerings; demands substantially greater than supply or offerings; (b) very good: well above average for seasonally normal offerings; (c) good: better than average and trading more active than normal; (d) moderate to fairly good: average buyer interest and trading; (e) fairly light: buyer interest and trading slightly below average; (f) light: below average; (g) very light: few buyers interested in trading; (h) almost no demand: stagnant condition on the market with little interest and few or no sales.

.028 Direct buying: when a retail or wholesale organization purchases fruits and vegetables directly from the shipper without using any firm (e.g., broker) in the middle of the system.

.029 Discoloration: (see Grade, VI.8.038b).

.030 Display case: cases, usually refrigerated, installed in the selling area of a retail market.

.031 Distribution (retail): percentage of produce sales to total supermarket sales.

.032 Distributor: firm at the receiving end of the marketing system, usually a wholesaler, which supplies produce to retail and/or foodservice outlets and to jobbers.

.033 EDP: electronic (computer) data processing.

.034 Estimated inventory: a retail practice of arriving at the total cash value of a store inventory by multiplying the sum item count by the retail price and subtracting the estimated gross-profit margin.

.035 Ethnic produce: fruits and vegetables preferred by certain races, nations, or ethnic groups.

.036 Floral preservative: chemical (many proprietary compounds and/or mixtures) added to water to extend the vase life of cut flowers. (LR)

.037 General manager: the person who is responsible for the supervision of the several operating departments (production, packinghouse, processing plant, sales manager, feed and fertilizer plant, etc.) in an organization that grows, handles, and ships produce.

.038 Grade: a measure of produce quality defined by federal and state departments of agriculture; most commodities have two to four grades, plus subclasses in some instances, of which only the top one or two will normally be marketed; terminology is not consistent in designation of grades, the top grade for apples being U.S. Extra Fancy (followed by U.S. Fancy and U.S. No. 1) and for potatoes, U.S. No. 1; furthermore, the top grade is often not the one commonly marketed (e.g., U.S. No. 1 for citrus fruits) and standards for the grade marketed sometimes also differ among states of origin (U.S. No. 1 standards for California oranges are much more restrictive than those of Florida oranges); note that inspections for grade are made first at the shipping point but may also be made at destination. (US - GN)

.038a. Blemish: a disfiguration of the surface of a fruit, vegetable, etc., caused by mites, insects, diseases, nutritional problems, chemicals, careless handling, weather, or other conditions; a grade-lowering factor that detracts from external appearance and affects marketability of the product; may occur in the grove, orchard, field, greenhouse, packinghouse, shipment, or market handling. (S & G)

.038b. Discoloration: superficial light-brown, smooth lesions on the surface of a fruit; e.g., rust mite of citrus; also any color that is lighter or darker than normal for any reason; may be a separate category under standards for grade (e.g., subclasses of grades for Florida citrus fruits), or considered as a blemish. (S & G)

.038c. Grade inspection (fresh produce): involves checking for decay, bruises, unhealed cuts, various categories of damage (blemishes), firmness, etc., to ensure that standards for a grade of fresh produce have been met, including maturity standards, and compliance with marketing agreement and other regulations; carried out by federal-state inspectors.

.038d. Grade inspection (processed products): carried out by federal inspectors to ensure compliance with standards for grade and federal and state food and drug regulations. (FDOC, US-SG)

.038e. Grade standard: specified limits for varietal characteristics, size, pack, color, form, maturity, texture, freedom from or free of injury from certain specified blemishes, etc., for a given commodity, with provisions for variations incidental to proper grading and handling (i.e., tolerances for defective specimens in an individual inspection sample and a lot). (US-SG)

.038f. Tolerance: several different connotations: (a) with respect to grading of produce or a product, the allowance for human or mechanical error (usually about 1% for decay and 5 to 10% for

VI.8. POSTHARVEST HANDLING and MARKETING

serious blemishes at shipping point); (b) the <u>maximum</u> level of a pesticide residue on or in produce postharvest; and (c) the <u>minimum</u> level of fungicide applied to Florida citrus fruit (Fla. Dept. of Citrus Official Rule 20-33), for shipping intrastate, interstate, or for export (some exceptions). (S & G, US-SG)

.039 Grade inspection (fresh produce): (see Grade, VI.8.038c).
.040 Grade inspection (processed products): (see Grade, VI.8.038d).
.041 Grade standard: (see Grade, VI.8.038e).
.042 Gross profit, gross profit margin, gross profit markup: the spread as cash or percentage points between the cost and selling price of an item; called "margin" if on the basis of cost; the first is used almost exclusively in the food retailing industry.
.043 Grove run: (see Tree run, VI.8.125).
.044 Harvesting foreman: the person actually in charge of harvesting operations and delivering the produce to the packinghouse.
.045 Holdover: (see Condition, VI.8.016e).
.046 House: a firm's facility on a terminal market.
.047 Impulse item: an item that, because of its color and attractiveness, causes customers to buy it, even though it is not on their shopping lists; many fresh fruits and vegetables, which includes strawberries, bananas, plums, cherries, and grapes, fall into this category.
.048 Independent store: store or supermarket owned by an individual not associated with a chain-store organization.
.049 In-store packaging: packaging of produce in a back room of a supermarket.
.050 Inventory: the quantity of goods or materials on hand.
.051 Inventory turns: the rate at which a product moves through a warehouse; expressed in turns per year.
.052 Jobber: normally a distributor who buys produce in small lots from a local wholesaler or distributer and resells to food service or retail firms.
.053 Labor: persons who carry out assigned tasks under instructions from a supervisor; several distinct categories, such as clerical (i.e., typists, accountants, and bookkeepers), skilled (mechanics, operators of certain complicated machinery, graders, and packers), semiskilled (mechanics helpers, tractor drivers, and plant diggers), and common (harvesters, grove, or field labor).
.054 Loss leader: an item in an advertisement or merchandised in-store that is priced substantially less than its competition; an item sold at less than cost.
.055 Lumper (swamper): a person on a terminal market or at a retail warehouse dock who unloads trucks or produce; is not employed by the terminal market or retail organization but works independently or belongs to a union; hired by truckers to unload the product.
.056 "Mad customer index": the number of bags or other small containers in a lot of produce that contain at least one concealed rot or other type of damage that renders the item unfit for consumption; derived from the expression on a housewife's face when she finds a rotten fruit or vegetable in a bag brought home from the grocery store; this means that it is the <u>number</u> of bags in a lot with a decayed product that is important rather than the percentage of decay, which actually may be quite low. (S & G)
.057 Management: the collective body of those who manage or direct an enterprise or interest; can be an individual who manages his own business (sole proprietorship), a group of partners (partner-

VI.8. POSTHARVESTING HANDLING and MARKETING

ship), or a corporation; also the persons who have overall supervision of day-to-day operations in an organization, the main function of the general and departmental managers.

.058 **Markdown:** a reduction in price of an item below its regular shelf price during the period of a sale; a reduced price to move merchandise for which a shrinkage record is kept; a controlled (i. e., deliberate) reduction of price.

.059 **Market:** represents the price level in buying and selling at which a commodity is traded (U.S. Market News Service term).

.060 **Market (facility):** an area in which several receiving firms are located in proximity, such as a terminal market (VI.8.118) or state farmers market (VI.8.113).

.061 **Market share:** a retailer's percentage of the total sale of a city or other geographical area.

.062 **Markup:** the difference between the warehouse cost and the ultimate retail price of an item, expressed as a percentage of warehouse cost; e.g., the markup is 100% if the cost of an item shipped to the retail warehouse is 50 cents and the same item is sold in the retail store for $1.00.

.063 **Mature:** a fully developed produce item but not necessarily ripened; e.g., mature green tomatoes.

.064 **Merchandise mix (product mix):** the retail sale of products with varying margins to achieve a specific margin objective; also called mix.

.065 **Merchandising:** the art of selling fruits and vegetables in retail supermarkets, including advertising and display techniques.

.066 **Middleman:** an individual or firm that buys produce with the expectation of reselling it for a profit; usually refers to a wholesaler, distributor, or terminal market operator.

.067 **Mix-and-match:** offering the customer a choice of several different or similar items at a total price (e.g., six pieces of fruit for one dollar).

.068 **Multideck:** a display fixture with shelves one above the other; usually a refrigerator case.

.069 **Multiple pricing:** pricing a single item in units of more than one (e.g., three for 50 cents).

.070 **Net profit:** the difference between gross profit and the cost of doing business; usually reported two ways by a corporation; e.g., before and after taxes.

.071 **Net sales:** gross sales minus adjustments and returns; the final sales figure.

.072 **Nutrient labeling:** the practice of listing nutrient ingredients for produce items in supermarkets [mandated by the Food & Drug Administration but suspended under a court injunction now (July 1983) made permanent]; good idea but is impractical for fresh fruits and vegetables (too variable in the amount of nutrients present).

.073 **Offering:** the volume of product voluntarily made available for sale; characterized as "light," "moderate," or "heavy;" often prefixed with modifiers "very" or "fairly."

.074 **Officers:** persons, such as the president, vice-presidents for certain broad areas of responsibility, secretary, treasurer, and controller, who constitute the executive level of an organization, and are elected by the board of directors, recommend policies to the latter, and are responsible for carrying them out (i.e., overall objectives, goals, long-range planning, and other items established by the board).

.075 **Orchard run:** (see Tree run, VI.8.125).

VI.8. POSTHARVEST HANDLING and MARKETING

.076 PACA: (see Perishable Agricultural Commodities Act, VI.7.1).
.077 Packinghouse foreman: the person actually in charge of operations from the time produce is delivered to the packinghouse until it is loaded on a carrier for shipment.
.078 Packinghouse manager: the person who is responsible for harvesting and handling fresh produce, including transportation from the field, packinghouse operations, order assembly, and loading for shipment on a carrier; works in liaison with the sales manager.
.079 Palatability: literally, the taste of a fruit or vegetable at the time of consumption; a subjective measure of acceptability, which includes appearance, flavor, aroma, sweetness (tartness), bitterness, and astringency; objectively determinable only by use of taste panels that involve large numbers (30 to 50) of people or other statistically acceptable tests.
.080 Peak: the time in which the biggest volume of a certain item is being shipped, as in peak volume.
.081 Perishables: foods that deteriorate from exposure to environmental changes; foods that require refrigeration or special handling in transit and at the point of sale; more specifically those that require careful, i.e., special, handling in harvesting, packing, loading, shipping, and other stages up to and including the consumer's home.
.082 "Pick-your-own": (see You-pick operation, VI.8.138).
.083 Place pack: hand arrangement of produce in a store to make a better display; (see also Packing operations, VI.3.053).
.084 Point of purchase/point of sale material (POP, POS): interchangeable terms used to describe in-store advertising/merchandising material; point of purchase refers to the place of contact between consumer-purchasers, the point of final sale; point of sale refers to the location of a product in the selling area of a retail store or display.
.085 Pool: a separate account set up by a grower cooperative or processing plant for a commodity (can be a separate cultivar, a group of cultivars, or all cultivars of a given crop and a certain grade or grades) to cover a stipulated period (day, week, month, crop season, or year) during which the fruits or vegetables from participating growers will be marketed and individual growers will be paid on the basis of the average sales return, less expenses per unit; there may be only one pool in operation at a time or several when different cultivars or crops are being marketed simultaneously; separate pools are established, as a rule, for fresh and processed produce, largely because of the greater lag time when a processing plant's products are sold, compared with that for fresh produce; a typical pool system for fresh fruit or vegetables might be an early season pool often of short duration (i.e., a week or two), a midseason pool of several weeks to months when the great bulk of the harvest is made, and a late season pool of a few weeks, particularly if the historical market pattern shows high prices at the beginning and end of a season, with a lower average in midseason. (MK)
.086 Pooling: a type of marketing system used by grower cooperative associations in which a cooperative can handle and pay for members' produce; it can involve the commingling of the products from many producers, the combining of sales returns and operating expenses, and the prorating of net returns among members in proportion to the volume of business each transacts in the cooperative over a certain period of time; most cooperatives stipulate in their

VI.8. POSTHARVESTING HANDLING and MARKETING

contact with a grower that it shall handle the grower's production and will treat all of the produce as one lot in a given pool rather than several lots sold on the basis of individual ownership; the contract will also stipulate when the produce is commingled, that is, whether it will be done prior to running the produce over the packingline or on the basis of marketable produce (=packout) of a stipulated grade or grades. Pooling is (a) a means of spreading market risks, (b) makes planning of an effective marketing program by controlling the time, place, and form in which the produce is sold, and (c) makes it easier for a cooperative to obtain needed finances, both in working capital in the form of a stipulated per unit retain and outside loans; most pooling systems are operated on a seasonal or annual basis; thus it is customary for the cooperative to make an advance payment or payments as produce is received and pay the remainder less retain at the conclusion of the season or year. (MK)

.087 POP: (see Point of purchase/point of sale material, VI.8.084).
.088 POS: (see Point of purchase/point of sale material, VI.8.084).
.089 Prepack, prepackaging: the practice of packaging, labeling, and price marking produce in advance rather than at the time of sale; may be done before the product reaches the supermarket or in a back room of the store.
.090 Production manager: the person who is responsible for all phases of crop production, including clearing of land, preparation for planting, planting, and the various phases of the cultural program up to but usually not including harvesting; will have one to several foremen to supervise individual work crews.
.091 Product mix: a term common to the foliage plant industry (and others) that describes the ratio of different plants produced by individual producers or by the industry annually. (JR)
.092 Product mix: (see Merchandise mix, VI.8.064).
.093 Pull date: the date after which perishable products such as fresh produce should not be sold.
.094 Quality: term to measure appearance and worth of a product; includes size, color, shape, texture, cleanliness, freedom from defects and other more permanent physical properties of a product that affect its market value (USDA, AMS); qualifying terms for quality (USDA Market News, 1981): (a) fine: better than good; superior in appearance, color, and other quality factors; (b) good: stock that has a high degree of merchantability with a small percentage of defects; includes U.S. No. 1, generally, 85% U.S. No. 1, or better, on some commodities (e.g., lettuce); (c) fair: higher percentage of defects than good; roughly 75% U.S. No. 1 with some leeway in either direction; (d) ordinary: fairly high percentage of defects; roughly 50 to 65% U.S. No. 1; (e) poor: high percentage of defects, with low degree of sale potential except to "low-priced" trade; > 50% grade defects.
.095 Reach and frequency: reach is the percentage of all television households in the market area or "universe," which are exposed at least once within a given period; frequency is the average number of times each of those TV households has been exposed within that same period.
.096 Receiver: anyone, whether a retail chain, cooperative, voluntary, wholesale, or terminal market operator, who receives produce shipments from growing areas.
.097 Reefer: a refrigerated rail car, truck trailer, or marine container van; also a trade term used to designate controlled-temperature

VI.8. POSTHARVEST HANDLING and MARKETING

containers used for transporting bulbs. (RS, LR)

.098 Retail: often used to denote the retail price of a product; a retail outlet of any sort is one from which sales are made directly to the consumer.

.099 Retain: a specified (i.e., part of the grower's contract) sum per unit of produce (e.g., bushel, box, hundredweight) that a growers' cooperative holds back (i.e., retains) when growers are paid for their produce after it has been received and marketed under a pool, the money so derived being used as working capital for the cooperative; the usual system is to roll over the retain after a certain number of years (e.g., 8 to 10); the retain subtracted 8 or 10 years earlier is returned to the grower, provided the cooperative is solvent; (see Pool, VI.8.085; Pooling, VI.8.086).

.100 Roller: a car load or truck load of produce that has been shipped but not yet sold; usually occurs when FOB prices are low and the shipper hopes the price will improve by the time the product reaches a market.

.101 Sales agent: usually handles marketing of a product for a grower or group of growers.

.102 Sales manager: the person in a packinghouse or processing plant who has the responsibility for procuring orders and arranging for harvesting and necessary handling of frozen produce according to availability and delivery date; a clerk under his supervision will handle details of carrier procurement, routing, services, etc., if there is no separate traffic manager (as in a large packinghouse or processing plant).

.103 Scanner: an electronic checking mechanism that reads the Universal Product Code (UPC) and translates it into an item description and the price; if it is connected to an electronic cash register, it also records the sale.

.104 Seasonal item: a product with more consumer acceptance at one time of year than another (e.g., lemons); an item with an annual seasonal sales peak.

.105 Service wholesaler: a wholesale organization that supplies complete purchasing and merchandising services for a group of independent stores or for small chains.

.106 Shipping clerk: the person in the packinghouse (shipper's) sales department who arranges the details of procurement, routing, services required, and necessary papers for truck, piggyback (trailer on flat car), rail car, or steamer shipments of produce with the local truck broker or individual truckers, railroad freight agent, or shipping agent for the steamship company.

.107 Shorts: the practice of retail chains or other firms that usually buy direct from shippers but will fill the gap by buying from terminal market operations or other wholesalers when they end up short of a product; thus they are buying "short."

.108 Shrinkage: loss of product in the marketing chain; may include such factors as concealed damage during shipment, loss in handling and displaying the product, and trimming, theft, and damage by customers at retail; soft fruits are highly perishable in the summer, thus highly susceptible to shrinkage; lettuce and tomatoes are also high-shrink items.

.109 Space allocation: shelf space provided for an item or family group of produce items in a retail store and usually based on the weekly unit sales rate.

.110 Space management: regulation of the space occupied by items in the produce department; movement records are used to determine space

VI.8. POSTHARVESTING HANDLING and MARKETING

allocation.
.111 Special display: space given to the promotion of an item in addition to regular display space.
.112 Staple items: refers in the produce department to items that are sold year around, such as lettuce and potatoes; these are items that usually are not so heavily promoted as seasonal items.
.113 State farmers market: a state-owned market to which farmers can come to sell their produce; states with the most farmers' markets are Florida, Georgia, North Carolina, and Alabama.
.114 Store-door delivery: direct delivery from the shipper to the retail store, which bypasses the wholesale or retail warehouse; e.g., watermelons.
.115 Street: refers to the terminal market as a rule.
.116 Supervisors: persons who have direct responsibility for the performance of certain designated operations; includes the general manager, department managers, and foremen of an organization, of whom the last have the dual function of supervising the work of others as well as working themselves.
.117 Swamper: (see Lumper, VI.8.055).
.118 Terminal market: a central location at which several wholesalers, distributers, and/or jobbers are grouped together; the market can be owned by the state, city, or private companies.
.119 Terminal packaging: packaging done by a firm at a terminal market.
.120 Tie-in display: when two different types of produce item, such as lettuce and tomatoes, are displayed together; or when a produce item, such as bananas, is displayed with nonproduce items, such as vanilla wafers.
.121 Tonnage items: large volume items in a produce department, usually available year-around; e.g., potatoes, bananas, lettuce, apples, and onions.
.122 Traffic manager: the person in a large packinghouse or sales organization (e.g., Blue Goose, Sealdsweet, or Sunkist), who arranges the details of procurement and routing of shipments; (see Shipping clerk, VI.8.106).
.123 Traffic pattern: the route consumers take when shopping in a produce department.
.124 Tree fruit: any fruit that is grown on trees but usually refers to soft fruits such as peaches, plums, nectarines, and pears.
.125 Tree (orchard, grove) run: ungraded produce straight from the plant; includes fruit or vegetables sold on a roadside stand or as part of a "you-pick" operation.
.126 Truck lot: truck load of product.
.127 Tube tomatoes: a container with a polyethylene window that holds three or four tomatoes.
.128 Turnover: the rate of rotation and/or replacement of inventory; normally computed by dividing the cost of goods sold by average inventory at cost.
.129 Turns: the number of times a product completes a cycle of moving through a warehouse or retail store; the number of times a warehouse is rotated completely.
.130 Unit pricing: the individual price appearing on a shelf tag expressed in cents per unit of volume or weight; e.g., price per pound.
.131 Universal product code (UPC): a product coding system designed to allow simpler and more accurate product identification as goods move from the shipper to the wholesaler to the retailer and to allow the use of scanner-equipped checkstands that, in turn,

VI.8. POSTHARVEST HANDLING and MARKETING

speed customer checkout operations.
.132 Vase life: the longevity of cut flowers.
.133 Vine ripe: a tomato term originally meaning "turning color to nearly red"; now means "breaking to bright red", as used in the produce industry.
.134 Volume: "in volume" means that a certain shipping area is into heavy shipments following a period when shipments were light.
.135 Voluntary: a grocery wholesaler who sponsors a voluntary organization of independent chains or stores; generally stores in such a group adopt a common name and engage in joint advertising.
.136 Warehouse store: a bare bones, no frills, low-prices store in which products are displayed directly in their cartons and items are boxed and bagged by the customers.
.137 Work order: written instructions that specify when, where, how, and with what labor, equipment, materials, etc., a given task is to be performed; can be for a field (production) crew, harvesting foreman's crew, packinghouse foreman, and maintenance foreman.
.138 You-pick operation: a type of direct sale in which an individual customer does the actual harvesting of produce in the field, orchard, grove, or vineyard; used by small growers for a wide range of crops, especially those that are highly perishable and/or labor intensive.

VI.9. Market Disorders of Fruits and Vegetables

.000 Market disorders: those disorders that affect marketable or usable products (i.e., fruit, vegetables, and cut flowers) and may originate in the field and orchard (i.e., preharvest) or at some subsequent stage of harvesting and handling, including storage; a large, highly diverse group of parasitic and nonparasitic disorders encompassing those that develop in the orchard, grove, or field prior to harvest or at some subsequent stage; divided here into three main groups according to whether the causal agencies are (a) pathological (fungi, bacteria, etc.), (b) animal parasites (insects, mites, etc.) or (c) physiological (nonparasitic); (see also Pests-Plants, V.3.3.2; Pests-Animals, V.3.3.1; Nonparasitic diseases, V.3.3.2.270).

VI.9.1. Pathological (parasitic) disorders

.000 Pathological (parasitic) market diseases: includes diseases caused by fungi, bacteria, viruses and viruslike organisms which attack fruits and vegetables; those for bulbous crops, etc. are listed under Crop Production (Commercial Floriculture), V.4.).
.001 Albugo candida: (see White rust of radish, VI.9.1.566).
.002 Albugo occidentalis: (see White rust of spinach, VI.9.1.567).
.003 Algal spot (Cephaleuros virescens): superficial mosslike, nearly circular, slightly raised spots up to about 8 mm in diameter. (SHJ, RB))
.004 Alternaria brassicae: (see Alternaria leaf spot of cabbage, VI.9.1.007; Alternaria root rot and leaf spot on rutabaga and turnip, VI.9.1.010; Brown rot on cauliflower, VI.9.1.125).
.005 Alternaria citri: (see Alternaria rot of lemons, VI.9.1.014).

VI.9.1. POSTHARVEST HANDLING and MARKETING

.006 Alternaria herculea: (see Alternaria root rot and leaf spot of rutabaga and turnip, VI.9.1.010)

.007 Alternaria leaf spot of cabbage, cauliflower, broccoli (Alternaria brassicae, A. oleracea): transit and storage decay, secondary decay, and blemishes that require heavy trimming are main market losses; first spots are grayish-brown to black, about 1.5 mm in diameter, mainly on older outer leaves; old spots to 2.5 cm or more, A. brassicae, with concentric zonation; those of A. oleracea not zonate; often growth of dark brown to black mold is evident. (R & S)

.008 Alternaria oleracea: (see Alternaria leaf spot of cabbage, VI.9.1. 007).

.009 Alternaria porri: (see Purple blotch of onions, VI.9.1.426).

.010 Alternaria root rot and leaf spot on rutabaga and turnip (Alternaria brassicae; A. herculea): serious decay of turnip roots in field; lesions, usually circular, dark brown, firm, with characteristic concentric light and dark brown rings and gray to grayish-brown surface mycelia; not moist or soft or bad smelling as with bacterial soft rot. (R & S)

.011 Alternaria rot of apple (Alternaria tenuis): more or less round, brown-to-black lesions around wound; fruit rot becomes spongy with black streaked flesh in advanced stages. (PCM)

.012 Alternaria rot of cherries, peaches, nectarines, apricots, plums (Alternaria sp.): dark brown-to-black lesions on cherries and plums, black-to-greenish-black on peaches, nectarines, and apricots; firm, moist, smooth (cherries) or slightly sunken; may be covered with olive green spores; rotted tissue cone-shaped, easily lifted out. (HSK)

.013 Alternaria rot of fig (Alternaria tenuis): surface spotting and rot of mainly ripe fruit. (SHJ)

.014 Alternaria rot of lemons, black rot of oranges (Alternaria citri): stem-end rot on lemons that appears when the calyx dies, but an internal black rot on oranges provides little or no external indication of its presence; the latter also occurs in grapefruit and tangerines but is ordinarily less prevalent. (SHJ, S & G)

.015 Alternaria rot of muskmelon (Alternaria tenuis): circular or oval brown-to-black sunken lesions up to 6.5 cm in diameter with fairly regular definite and slightly watersoaked edges and frequently alternating light and dark concentric rings, later black with deep grayish-olive fluffy mat of fungus on honeyball melons, often only on rind, sometimes penetrates flesh with decayed tissues that lift out easily; external lesions on cantaloupes more or less definite, with dark sporulating mycelia. (R & S)

.016 Alternaria rot of squash (Alternaria tenuis): tan, smooth lesions, 6 to 25 mm in diameter, often with greenish-black mold over the surface; yellowish-to-black, fairly dry, spongy decay pockets 13 mm or more deep in hard-shell cultivars. (R & S)

.017 Alternaria rot of tomato, pepper, eggplant (Alternaria tenuis): weak pathogen; enters through wounds or chilling injury lesions; small circular, grayish-green, watersoaked spots (pepper); then brown-to-black in pepper and eggplant, or without a definite margin, flattened or slightly sunken, firm, extending into flesh that is a dark-brown-to-black, dry corelike mass of decayed tissues (tomato) or spongy and tannish (eggplant). (MCW)

.018 Alternaria solani: (see Early blight rot of tomato, VI.9.1.210).

.019 Alternaria tenuis: (see Alternaria rot of apple, VI.9.1.011; Alternaria rot of fig, VI.9.1.013; Alternaria rot of squash, VI.9.1.

VI.9.1. POSTHARVEST HANDLING and MARKETING   628

016; Alternaria rot of tomato, pepper, and eggplant, VI.9.1.017).
.020 Alternaria tomato: (see Nailhead spot of tomato, VI.9.1.325).
.021 Alternaria sp.: (see Alternaria rot of cherries, peaches, nectarines, apricots, plums, VI.9.1.012; Core rot of apple, VI.9.1.171, Heart rot of pomegranate, VI.9.1.286).
.022 Angular leaf spot: (see Bacterial spot of cucumber, VI.9.1.069).
.023 Anthracnose of avocado, etc. (Colletotrichum gloeosporioides): a weak parasite ubiquitous in subtropical and tropical areas; forms small, light-brown, circular lesions that darken to greenish black, sometimes 2.5 to 5 cm in diameter, and form a fairly firm mass that can be easily cut from a fruit; enters through lenticels (e.g., avocado or mango) or wounds. (SHJ, R & L)
.024 Anthracnose of banana (Gloeosporium musarum): primarily a wound parasite that develops as a finger-stalk rot, finger spotting, or tip rot as the fruit softens; probably the most common market disease; Cavendish types are particularly susceptible. (SHJ)
.025 Anthracnose of citrus (Colletotrichum gloeosporioides): a secondary invader of most citrus; severe only on 'Robinson' tangerine. (S & G, SHJ)
.026 Anthracnose of garden pea (Colletotrichum pisi): appears on pods as circular spots with pale centers and dark-brown margins; yellow to salmon-colored wet spore masses develop under moist conditions. (SMF)
.027 Anthracnose of mango (Colletotrichum gloeosporioides): the most important market disease; fruit are susceptible from bloom time to maturity; early latent infections or wound invasions develop as black lesions as the fruit matures and softens. (SHJ)
.028 Anthracnose of muskmelons (Colletotrichum lagenarium): seen occasionally on honeydew melons; circular or oval spots about 6 to 20 mm in diameter, shallow at first, soon sharply sunken, with salmon or pink-colored spore masses over the surface later dark and penetrating deeply in to the flesh. (R & S)
.029 Anthracnose of papaya (Colletotrichum gloeosporioides): wound parasite on fruit and vegetative parts of plant; fruit most susceptible as they become mature. (SHJ)
.030 Anthracnose (bitter rot) of peaches, nectarines, apricots (Glomerella cingulata): same organism that causes bitter rot of apples (VI. 9.1.079) with similar symptoms. (HSK)
.031 Anthracnose of pepper (Colletotrichum gloeosporioides): small, water-soaked spots, later sordid green to greenish brown, slightly sunken circular spots with definite margins and covered with raised black pustules that may exude creamy-to pink spore masses (sometimes blackening with age); wound parasite. (MCW)
.032 Anthracnose of snap beans (Colletotrichum lindemuthianum): minute oval or circular greenish-brown specks, later becoming brick or rust-red-to-black around the borders; centers are sunken and dark and up to 13 mm in diameter; centers covered under humid conditions with flesh-colored spores that may form a slimy mass but are gray, brown, or black under dry conditions; seed-borne. (SMF)
.033 Anthracnose of tomato (Colletotrichum coccodes): primarily a disease of ripe fruit; small, circular, slightly sunken, water soaked spots, later darker, more depressed, to 13 mm diameter, sometimes reveals concentric ring markings; surface covered with cream-to-salmon-pink spore masses; may penetrate deeply into the flesh (vine-ripened fruit). (MCW)
.034 Anthracnose of turnip, radish, cabbage, Chinese cabbage, collards (Colletotrichum higginsianum): causes grayish-to-light-tan spots

VI.9.1. POSTHARVEST HANDLING and MARKETING

with regular depressed margins in stored turnips; usually followed by bacterial soft rot. (R & S)

.035 Anthracnose of watermelon (<u>Colletotrichum lagenarium</u>): slightly elevated pimples with yellow, translucent centers; nailhead (small, circular, raised welts) lesions on mature fruit, dark green, later brown, sunken, and cracking; slowly advancing dry rot of rind, often black, corky lesions around the bottom surface in contact with the soil. (R & S)

.036 Apple-cedar rust (<u>Gymnosporangium juniperi-virginianae</u>): grayish yellow-to-yellow spots 3 to 20 mm or more into the flesh; red cedar is the alternate host for the rust. (PCM)

.037 Apple scab (<u>Venturia inaequalis</u>): one of the most important apple diseases in areas of summer rains; infections and consequent formation of scab lesions may occur at any stage of fruit development, during transit or in storage; lesions begin as small, irregularly circular, raised, dark spots, then olive green to dark brown and velvety, becoming brown and corky and up to 2.5 cm or more across; fruit are often misshapen; also attacks leaves and twigs. (PCM)

.038 <u>Ascochyta pinodella</u>, <u>A. pinodes</u>, <u>A. pisi</u>: (see Pod spot of garden pea, VI.9.1.405).

.039 <u>Aspergillus alliaceus</u>: (see Aspergillus bulb rot of garlic and onions, VI.9.1.041).

.040 Aspergillus black-mold rot of fig (<u>Aspergillus niger</u>): dull white-to-pinkish surface and flesh-color spotting with production of black spores internally; a major cause of spoilage of fresh figs. (SHJ)

.041 Aspergillus bulb rot of garlic and onions (<u>Aspergillus alliaceus</u>): slight swelling and darkening of tissues, becoming water-soaked, soft, with a sharp demarcation line, later shrinking and brownish, with white mycelial mat between scales, and eventually white sclerotia with yellowish-brown spore cluster; not in the United States. (SMF)

.042 <u>Aspergillus niger</u>: (see Aspergillus black-mold rot of fig, VI.9.1.040; Aspergillus rot of citrus, VI.9.1.043; Black mold rot of cherries, peaches, nectarines, apricots, VI.9.1.083; Black mold rot of onions and garlic, VI.9.1.084; Heart rot of pomegranate, VI.9.1.286).

.043 Aspergillus rot of citrus (<u>Aspergillus niger</u>, A. spp.): minor decay of citrus fruit held at high temperatures; spreads by contact. (SHJ)

.044 <u>Aspergillus</u> sp.: (see Core rot of apple: VI.9.1.171).

.045 <u>Aspergillus</u> spp.: (see Aspergillus rot of citrus, VI.9.1.043; Black rot of corn, VI.9.1.090; Heart rot of pomegranate, VI.9.1.286).

.046 Aster yellows: viruslike disorder that produces twisted, dwarfed, curled heart leaves and soft heads in lettuce; may also have internal brown leaf spotting. (RFS)

.047 Avocado scab (<u>Sphaceloma perseae</u>): important disease in warm, humid areas; causes raised, more or less circular dark-to-purplish-brown corky lesions on the rind; wide variation in resistance among cultivars (e.g., 'Lula' quite susceptible, 'Waldin' and 'Pollock' quite resistant); also occurs on foliage and twigs. (R & L, SHJ)

.048 Bacterial blight of celery (<u>Pseudomonas syringae</u> pv. <u>apii</u>): numerous irregular-to-circular yellow, later rusty-brown spots with a yellow halo on the leaflets. (SMF)

.049 Bacterial blight of garden pea (<u>Pseudomonas syringae</u> pv. <u>pisi</u>):

causes water-soaked, irregular, slightly sunken spots on pods; may spread to and discolor seeds; spots may enlarge during transit. (SMF)

.050 Bacterial blights (common, fuscous, and halo) of lima beans (Xanthomonas campestris pv. phaseoli, Pseudomonas syringae pv. phaseolicola): also affect garden and field beans; small watersoaked areas that gradually enlarge, may coalesce, may have distinct zoning and narrow reddish-brown or brick-red ban around lesion, and may occur along sutures of pods; seeds often infected, becoming discolored, shriveled, and/or rotted; exudate light-cream or silver colored in halo blight, yellow in common and fuscous blights; halo blight lesions more watersoaked, circular, and remain the grayish-green instead of yellow and reddish-brown typical of the other two. (SMF)

.051 Bacterial brown rot of parsnip (Pseudomonas marginalis pv. pastinacae): transit, storage, and market disease. (SMF)

.052 Bacterial canker of tomato (Corynebacterium michiganense pv. michiganense): spots on mature-green or ripening fruit 1.5 to 5 mm in diameter, superficial, white-to-light brown; slightly raised later with dark-brown center and a white halo, never soft; may be infected internally from bacteria in vascular system; also causes seedling wilt and cankers on older plants. (MCW)

.053 Bacterial fruitlet rot: (see Brown rot of pineapple, VI.9.1.127).

.054 Bacterial leaf spot of cauliflower, cabbage, broccoli (Pseudomonas syringae pv. maculicola: minor disease that disfigures cabbage leaves, broccoli stems, and florets and cauliflower leaves and curds; lesions are small, water-soaked spots, sometimes gray-to brown (cauliflower) or purplish-gray-to-black (broccoli). (R & S)

.055 Bacterial ring rot of potato (Corynebacterium michiganense pv. sepedonicum): attacks foliage and tubers; the latter show light yellow discoloration of vascular tissues which break down and exude cheesy ooze when squeezed; now largely eradicated from many areas by use of certified ring rot-free seed. (CMI)

.056 Bacterial root rots of horseradish (Erwinia carotovora, etc.): some discoloration, some soft rot, some dry rot; enters through injuries. (R & S)

.057 Bacterial soft rot of asparagus (Erwinia carotovora): develops in soft, tender tips or at cut ends of spears; a soft, slippery, watery rot with a disagreeable odor. (SMF)

.058 Bacterial soft rot of cabbage and other crucifers (Erwinia carotovora, etc.): secondary invader following black rot, clubroot, downy mildew, or alternaria leaf spot; water-soaked or greasy spots on leaves or white-to-grayish soft areas at base of heads and putrid odor with extensive decay on cabbage and brussels sprouts; yellow-to-yellowish-brown areas becoming more watery; somewhat sunken on cauliflower florets; dark-green water-soaked or greasy lesions on broccoli. (R & S)

.059 Bacterial soft rot of carrot, parsnip (Erwinia carotovora): roots may be infected anywhere but decay frequently develops at the crown and proceeds down the core; lesions are grayish-brown; affected tissues appear water-soaked, soft, and watery; later slimy with a putrid odor. (SMF)

.060 Bacterial soft rot of celery (Erwinia carotovora): starts as small, soft, water-soaked spots that may occur anywhere on the stalk; quickly destroys leaflet tissues; decayed tissues on petioles are wet, mushy, light-brown, and slough away easily from healthy tissue; usually no offensive odor as in tomatoes and many other

VI.9.1. POSTHARVEST HANDLING and MARKETING

vegetables. (SMF)

.061 Bacterial soft rot of lettuce (slime) (Pseudomonas marginalis pv. marginalis, P. cichorii, P. viridilivida, Xanthomonas campestris pv. vitians, Erwinia carotovora): water-soaked spots, often discolored, becoming soft, slimy, mushy, and wet; often a secondary invader. (RFS)

.062 Bacterial soft rot of muskmelons, cucumbers, squash, pumpkins (Erwinia aroidea, E. carotovora): water-soaking and pronounced flesh softening at infection site; fruit becomes soft, mushy, and eventually shells with broken down tissues; offensive odor. (R & S)

.063 Bacterial soft rot of onions (Erwinia carotovora): invade neck tissues as plants approach maturity; affect scales singly; affected tissues glassy or water-soaked; pale yellowish-to-light-brown, becoming soft with a watery foul-smelling liquid forced out if the bulb is pressed; bulbs with mechanical injuries, sunscald, or bruises are particularly susceptible. (SMF)

.064 Bacterial soft rot of potato (Erwinia carotovora): invades tubers through wounds.

.065 Bacterial soft rot of snap beans (Erwinia carotovora): may follow bacterial blight or a fungus rot or enter through weakened tissues; causes a slimy, somewhat watery rot with a putrid odor. (SMF)

.066 Bacterial soft rot of spinach (Erwinia carotovora, etc.): watersoaked, muddy green, or greasy; tissues rapidly become soft, disintegrate, and develop a putrid odor. (RFS)

.067 Bacterial soft rot of tomato, pepper, eggplant (Erwinia carotovora): primarily a decay during transit and ripening (green tomato fruit more susceptible) but enters earlier through wounds; water-soaked spots rapidly enlarge, with affected tissues very soft and watery, easy rupturing of skin, and often foul odor (tomato) from secondary bacteria. (MCW)

.068 Bacterial speck of tomato (Pseudomonas syringae pv. tomato): superficial, dark brown, slightly raised specks 0.6 to 1.3 mm in diameter; also affects leaves, petioles, and stems. (MCW)

.069 Bacterial spot (angular leaf spot) of cucumber (Pseudomonas syringae pv. lachrymans): occurs on leaves and fruit; also affects honeydew melons and squash; minute, circular, water-soaked areas, which later dry out, crack and become chalky white with sunken centers; often a gummy exudate on lesions; brown lesions in maturing fruit with discoloration along vascular system to seeds. (R & S)

.070 Bacterial spot of honeydew melons (Pseudomonas syringae pv. lachrymans): slightly sunken, circular to oblong, water-soaked, greenish-tan spots 3 to 6 mm in diameter coalescing to form large brown or black areas. (R & S)

.071 Bacterial spot of lima beans (Pseudomonas syringae pv. syringae): nearly circular, small brown lesions with a water-soaked halo and cream-colored exudate; may cause reddish discoloration of seeds, also spongy excrescences from inner pod walls; spots on pods smaller than those of bacterial blight. (SMF)

.072 Bacterial spot of peaches, nectarines, apricots (Xanthomonas campestris pv. pruni): appears as faint brown, circular spots 0.5 to 1 mm in diameter on maturing peach fruit, later darkened and slightly sunken with water-soaked margins, cracking with a greenish halo and yellow gummy exudate; most prevalent in humid areas. (HSK)

.073 Bacterial spot of tomato and pepper (Xanthomonas campestris pv.

VI.9.1. POSTHARVEST HANDLING and MARKETING

vesicatoria): small, dark, slightly raised dots, sometimes with a narrow water-soaked border, later brownish-black elevated, scablike areas 3 to 6 mm in diameter with feathered or irregular margins; remain superficial. (MCW)

.074 Bacterial wilt (Stewart's disease, Stewart's wilt) of corn (Erwinia stewartii): a vascular type of wilt that attacks sweet (most susceptible) dent and popcorn; now largely controlled by resistant hybrids; infected from "seeds" and through flea beetle wounds.

.075 Bacterial wilt of snap beans (Corynebacterium flaccumfaciens pv. flaccumfaciens): vascular disease, thus infects pods and seeds if plants survive; infection from seeds; may have dark-green, water-soaked, elongated spots along the suture, where seeds are attached; seriously affected pods tend to wither; may enter the hilum to form yellow masses under the seedcoat or small, yellow, crustlike deposits on the surface of the seeds. (SMF)

.076 Bacterial zonate spot of cabbage, other crucifers and cucurbits (Pseudomonas cichorii): round-to-irregular lesions, 1.5 to 13 mm in diameter, buff-to-light brown, becoming darker brown later and zoned; infected areas firm and pliable, only slightly soft; enters through wounds. (R & S)

.077 Beet scab (Streptomyces scabies): superficial lesions like those on potato but more protruding and rounded. (RFS)

.078 Big vein: virus or viruslike disease of lettuce that causes vein clearing, wrinkling, thickening, and puckering of leaf tissue and smaller, less firm coarse, ribby, unattractive heads. (RFS)

.079 Bitter rot of apple, pear, peach, quince, cherry (Glomerella cingulata): brown-to-black discrete spots from pinpoints to the whole side of a fruit, firm with underlying affected flesh, cone shaped, brown, and moist; may occur in the orchard or later. (PCM, HSF)

.080 Bitter rot of peaches, nectarines, apricots: (see Anthracnose of peaches, etc., VI.9.1.030).

.081 Blackleg of potato (Erwinia atroseptica): serious bacterial disease that affects foliage and tubers and causes a soft black decay at the stem end of the latter; transmitted mechanically by contaminated tools and through wounds. (CMI)

.082 Black mold of carrot (Thielaviopsis basicola): black, irregularly scattered spots 6 to 20 cm, often with dark gray discoloration to 1.5 mm depth in underlying tissues. (SMF)

.083 Black mold rot of cherries, peaches, nectarines, apricots (Aspergillus niger): watery, mushy rot on cherries; small, tan slightly sunken spots with wrinkling in concentric circles as spots enlarge; flesh below brown and mushy; easily scooped out on peaches, etc.; spores yellow, then sooty black. (HSF)

.084 Black mold rot of onions and garlic (Aspergillus niger): main symptoms are the presence of black, powdery, spore masses on the outside scale or between that and the next inner one; tends to follow the veins; may have semiwatery, sunken, discolored areas or dry, papery, sometimes highly colored affected tissues. (SMF)

.085 Black pit of citrus (Pseudomonas syringae pv. syringae): brown-to-black sunken pits or lesions, 6 to 12 mm in diameter penetrating the albedo; lemon fruit is most susceptible; orange and grapefruit are more affected by citrus blast (disease of leaves and twigs); found mainly in cool regions as a wound parasite. (SHJ)

.086 Black rot of apple, pear, and quince (Physalospora obtusa): single, firm, brown spot becoming dark brown-to-black and enveloping entire fruit; sometimes pycnidia present. (PCM)

VI.9.1. POSTHARVEST HANDLING and MARKETING

.087 Black rot of beet (Phoma betae): occurs mainly on topped beets in storage; starts at the root tip, crown, or in wounds; surface lesions become black, slightly sunken, and dark-brown-to-black, with a sharp line of demarcation internally; initially brown and water-soaked, later black, somewhat granular and dry, with spongy cavities lined with mycelia; grayish-white mycelia on older surface lesions but no pycnidia. (RFS)

.088 Black rot of cabbage, cauliflower, other crucifers (Xanthomonas campestris pv. campestris): serious disease that opens the way for secondary invaders; vascular disease, veins appearing black or dark-shadowed when overlain by flesh on cabbage; quick wilting and blighted appearance of leaves on cauliflower; black staining pronounced in kohlrabi; small brown spots plus discoloration in Chinese cabbage; root center and vascular ring in rutabaga; no odor unless from a secondary invader. (R & S)

.089 Black rot of carrot (Stemphylium radicinum): may occur anywhere on the fleshy root; lesions are circular to irregular with sharp margins, usually slightly depressed, shallow, with greenish-black-to-black, dry mealy tissues; may penetrate deeper and be soft and wet under moist conditions; mainly a rot of topped, stored carrots. (SMF)

.090 Black rot of corn (Aspergillus spp.): causes a powdery black rot of kernels.

.091 Black rot (Gummy stem blight) of cucumber (Mycosphaerella citrullina): shows as inconspicuous, water-soaked spots with patches of white mycelia; later gradual darkening when black fruiting bodies appear. (R & S)

.092 Black rot (Gummy stem blight) of squash (Mycosphaerella citrullina): water-soaked, greenish-brown lesions that become firm and black in acorn types; hard, dry lesions dotted with minute pycnidia on stored squash. (R & S)

.093 Black rot of oranges: (see Alternaria rot of lemons, VI.9.1.014).

.094 Black rot of pineapple: (see Brown rot of pineapple, VI.9.1.127).

.095 Black rot of sweet potatoes (Endoconidiophora fimbriata): most serious market disease next to rhizopus soft rot (VI.9.1.457); circular, brown, slightly sunken superficial spots about 6 mm in diameter, becoming 13 to 50 mm; black-to-greenish-black with small long-necked perithecia; firm, shallow rot; diseased tissues very bitter. (RFS)

.096 Black rot of watermelon (Mycosphaerella citrullina): greenish-tan-to-black spots 1.3 to 6.5 cm in diameter; black centers with pycnidia (small black specks); larger lesions often have brownish-black concentric zones; eventually involves flesh. (R & S)

.097 Black rot (water blister, soft rot, water rot) of pineapple (Ceratocystis paradoxa = Thielaviopsis paradoxa): wound parasite entering through the cut stem or injuries on the side of the fruit; produces soft brown, water-soaked softening of tissues; later turns black, mainly internally; typical fermented odor; serious disease of ripe fruit, particularly 'Cayenne'. (SHJ)

.098 Black scurf, stem, and tuber rot of potato (Rhizoctonia solani): resembles dried muck or dirt on the skin of affected tubers; may girdle the bases of stems of sprouts arising from affected seed pieces; also causes damping-off.

.099 Black spot: (see Brown rot of pineapple, VI.9.1.127; Scab of peaches, nectarines, apricots, plums, VI.9.1.470).

.100 Black spot of pear and quince (Fabraea maculata): black, roughly circular, slightly sunken spots 1.5 to 9 mm in diameter; often surrounded by a red ring; often numerous; may occur on any part of the fruit; also attacks foliage. (PCM)

.101 Blossom-end rot of apple (<u>Botrytis cinerea</u>): superficial infection that dies and leaves a dry, dark, dusty blemishlike lesion that may be invaded later by secondary fungi. (PCM)

.102 Blossom-end rot of squash (<u>Choanephora cucurbitarum</u>): dense growth of immature, white, spore-bearing structures; later brown, finally purplish-black on fading flowers; then works down into new fruit or later causes a rapidly progressing soft wet rot; profusion of black sporangia. (R & S)

.103 Blotch of apple (<u>Phyllosticta solitaria</u>): dark, blotchlike, hard, sunken lesions with fringed margins, 6 to 12 mm in diameter on fruit; also attacks leaves and twigs. (PCM)

.104 Blue-green rot of corn (<u>Penicillium</u> spp.): causes rotting of kernels. (WALK)

.105 Blue mold (<u>Penicillium italicum</u>), green mold (<u>P. digitatum</u>) of citrus: important postharvest diseases; attack fruit on the tree or in the packinghouse, transit, storage, and market; green mold is a wound parasite; blue is also but will spread by contact and will grow at temperatures too low for green, thus is more common in storage; green mold is an efficient ethylene producer and ubiquitous in citrus areas; some fungicide-resistant strains. (S & G, SHJ)

.106 Blue mold rot of apple, pear, quince (<u>Penicillium expansum</u>): soft, watery, sharply delineated spots, sometimes with blue-green spores; the internal appearance is glassy and the lesion may be readily scooped out; flesh has a musty taste; generally a wound parasite but can penetrate through lenticels. (PCM)

.107 Blue mold rot of cherries (<u>Penicillium expansum</u>): wound parasite that enters through cracks; circular, flat, light-brown, very soft, and watery spots that crack and show white mycelia and bluish-green spores over the surface. (HSK)

.108 Blue mold rot of fig (<u>Penicillium</u> sp.): found on fresh figs in the market. (SHJ)

.109 Blue mold rot of garlic and onions (<u>Penicillium</u> spp.): starts on garlic as light-yellowish lesions, changing to blue or bluish-green as spores form; rapid breakdown of tissues; individual cloves are soft, spongy, or powdery-day; finally completely broken down into greenish-tan or gray masses. (SMF)

.110 Blue mold rot of muskmelons (<u>Penicillium</u> spp.): found on cantaloupes, honeyball, and honeydew melons; usually restricted well defined spot, circular to oval, slightly depressed, darker color than normal, 1.3 to 5 cm in diameter; abundant bluish or greenish spore masses along skin cracks and over inner surface of flesh cavities. (R & S)

.111 Blue mold rot of peaches, nectarines, apricots, plums (<u>Penicillium</u> sp.): attacks overripe or injured fruit; small tan spots with moist, easily broken skin, tan, moist, fairly firm flesh and sporulation later as spots enlarge on peaches, nectarines, and apricots; that on plums is similar but enters through wounds; serious rot during storage, transit, and marketing, (HSK)

.112 Blue mold rot of pomegranate (<u>Penicillium expansum</u>, other <u>P</u>. spp.): wound parasite causes pockets of blue-or green-spored colonies of fungi; may be overgrown by gray mold or heart rot organisms. (SHJ)

.113 Blue mold rot of sweet potatoes (<u>Penicillium</u> sp.): soft wound decay, with characteristic white mycelia and blue-green tufts

- .114 Boil smut: (see Common smut of corn, VI.9.1.169).
- .115 Botryodiplodia fruit rot of banana (Botryodiplodia theobromae): a wound parasite that causes main-stalk and tip rots; the infection occurs in the field or after harvest and becomes evident later; water-soaked lesions turn blackish and are often covered with pycnidia. (SHJ)
- .116 Botryosphaeria rot of apple (Botryosphaeria ribis): small brown specks or spots, usually with red halos; often slightly sunken, 3 to 6 mm in diameter; rapidly envelop entire fruit, then become soft and somewhat bleached (= white rot); a disease of mature fruit. (PCM)
- .117 Botrytis allii: (see Gray mold of onions, VI.9.1.264).
- .118 Botrytis byssoidea: (see Gray mold of onions, VI.9.1.264).
- .119 Botrytis cinerea: (see Blossom-end rot of apple, VI.9.1.101; Botrytis rot of citrus, VI.9.1.120; Ghost spot of tomato, VI.9.1.255; Gray mold rot of apple, VI.9.1.265; Gray mold rot of asparagus, VI.9.1.266; Gray mold rot of cabbage, VI.9.1.267; Gray mold rot of carrot, VI.9.1.268; Gray mold rot of fig, VI.9.1.269; Gray mold rot of garden pea, VI.9.1.270; Gray mold rot of lettuce, VI.9.1.272; Gray mold rot of pomegranate, VI.9.1.273; Gray mold rot of snap beans, VI.9.1.275; Gray mold rot of stone fruits, VI.9.1.276; Grey mold rot of tomato, pepper, eggplant, VI.9.1.277).
- .120 Botrytis rot (Botrytis cinerea) of citrus: mainly a storage, transit, and market disease; common on lemons but will attack other citrus; produces characteristic gray spores; enters through wounds or uninjured rind. (SHJ)
- .121 Botrytis squamosa: (see Gray mold of onions, VI.9.1.264).
- .122 Botrytis sp.: (see Core rot of apple, VI.9.1.171; Gray mold rot of globe artichoke, VI.9.1.271; Gray mold rot of rhubarb, VI.9.1.274).
- .123 Bremia lactucae: (see Downy mildew of lettuce, VI.9.1.204).
- .124 Brooks' spot: (see Fruit spot of apple, VI.9.1.231).
- .125 Brown rot of cauliflower (Alternaria brassicae): same organism causes black leaf spot or black mold on other crucifers; causes browning of individual or groups of buds, light brown to chestnut brown to olivaceous as lesions age and enlarge to form numerous spots or large blotches; may be dry and firm or wet and soft. (R & S)
- .126 Brown rot of citrus (Phytophthora spp.): firm leathery brown rot on citrus, most important on lemons; spread by spores splashing onto low-hanging fruit in the grove, also by contact in storage or transit; most prevalent in rainy weather; several species of Phytophthora can cause brown rot (also foot and root rots), such as P. citrophthora, P. syringae, and P. hibernalis in California, P. parasitica in Florida, and P. palmivora and P. cinnamomi in tropical areas. (S & G, SHJ)
- .127 Brown rot of pineapple (Erwinia ananas, Fusarium moniliforme, Penicillium funiculosum, Pseudomonas ananas): bacteria (Erwinia or Pseudomonas) or fungi (Fusarium or Penicillium) that produce internal, brown, rotten spots in or between fruitlets of mature or ripe fruit; many different names (e.g., fruitlet core rot, fruitlet brown rot, fruitlet black rot, ripe fruit rot, marbled fruit, bacterial fruitlet rot, exogenous brown discoloration, black rot, black spot); bacterial invasions produce gumming;

(continued from previous: breaking through the epidermis of chilled or frozen roots; often a musty odor. (RFS))

VI.9.1. POSTHARVEST HANDLING and MARKETING

fungal do not. (SHJ)

.128 Brown rot of pome fruits (Monilinia fructicola, M. Laxa): wound parasite; lesions circular; medium brown; the entire fruit becomes black with a velvety sheen; a softer, paler colored rot than black rot (VI.9.1.086). (PCM)

.129 Brown rot of stone fruits (Monilinia fructicola, M. Laxa): important disease of cherries, peaches, nectarines and apricots, particularly in humid areas or wet seasons; not so serious on plums or prunes; typically water-soaked, glassy light-brown spots, darkening later, with margins fading into normal fruit color; develops in the orchard and packinghouse, in transit, marketing, or homes, and on immature fruit. (HSK)

.130 Brown spot of celery (Cephalosporium apii): causes irregular, tan-to-brown shallow lesions on the inside and outside of petioles and on leaflets; may coalesce to form areas 1.3 to 4.5 cm long, and curling or distortion of mature petioles; may be followed by bacterial soft rot as a secondary invader. (SMF)

.131 Brown spot of stone fruits: (see Coryneum blight of stone fruits, VI.9.1.176).

.132 Buckeye spot of tomato (Phytophthora spp.): attacks fruit near or in contact with soil; hard, brown-to-grayish-green, remaining moderately firm with an irregular, indefinite margin; water-soaked appearance and grayish-green color, often with zones of alternating shades of brown and grayish brown; chocolate brown on green fruit, but lacking the dark brown color and zoning on ripe fruit; caused by several species of Phytophthora; e.g., P. capsici, P. drechsleri, and P. parasitica. (MCW)

.133 Bull's eye rot of apple (Pezicula malicorticis): spots pale yellow or brown with a pale center (bull's eye), 13 to 25 mm in diameter, flat to sunken, fairly firm; also produces branch cankers. (PCM)

.134 California aster yellows on celery, anise, dill: virus or viruslike disease that causes shortening, chlorosis, bending and twisting of the inner petioles, premature blanching, brittleness and cracking of outer petioles, and yellowed leaflets. (SMF)

.135 Cephaleuros virescens: (see Algal spot, VI.9.1.003).
.136 Cephalosporium apii: (see Brown spot of celery, VI.9.1.130).
.137 Cephalothecium roseum: (see Pink mold rot of apple, pear, (VI.9.1.392); Pink mold rot of muskmelons and cucumbers (VI.9.1.393).
.138 Ceratocystis paradoxa: (see Black rot of pineapple, VI.9.1.097; Thielaviopsis stalk rot of banana, VI.9.1.542).
.139 Cercospora apii: (see Early blight of celery, VI.9.1.209).
.140 Cercospora beticola: (see Cercospora leaf spot of beet, VI.9.1.141).
.141 Cercospora leaf spot of beet (Cercospora beticola): small spots with definite reddish-brown-to-purple margins and ashy-gray-to-light-tan centers; bacteria sometimes causes soft rot. (RFS)
.142 Cercospora purpurea: (see Cercospora spot (blotch) of avocado, VI.9.1.143).
.143 Cercospora spot (blotch) of avocado (Cercospora purpurea): important disease of avocado in warm, humid areas; lesions are slightly sunken, brown to dark brown, sharply but irregularly outlined and 3 to 6 mm in diameter, with characteristic short, gray, spore-bearing tufts of fungus on the surface. (R & L, SHJ)
.144 Charcoal rot of honeydew melons [Macrophomina phaseoli (Sclerotium bataticola)]: gray-or-purplish-gray, water-soaked spots with thin wrinkled crust of minute, black, tightly packed sclerotia;

VI.9.1. POSTHARVEST HANDLING and MARKETING

underlying flesh water-soaked; otherwise normal colored but with pungent sour odor and taste. (R & S)

.145 Charcoal rot of pepper (Macrophomina phaseoli): external shriveling with internal discoloration and presence of sclerotia. (MCW)

.146 Charcoal rot of sweet potatoes (Macrophomina phaseoli): most often starts at the upper end of the sweet potato; light brown discoloration of surface and internal tissues with a sharp line of demarcation; later remains firm and dark brown and the skin shrivels; finally converts to a hard, dry, black mummy; at least three distinct color zones in affected tissues, the youngest being light brown and slightly spongy, the older ones grayish black, dry, and firm with small sclerotia within the tissues (none visible on the surface). (RFS)

.147 Choanephora cucurbitarum: (see Blossom end rot of squash, VI.9.1.102).

.148 Cladosporium carpophilum: (see Scab of peaches, nectarines, apricots, plums, VI.9.1.470).

.149 Cladosporium cucumerinum: (see Cladosporium rot of muskmelons, VI.9.1.152).

.150 Cladosporium herbarum: (see Cladosporium rot of stone fruits, VI.9.1.153; Cladosporium rot of tomato and pepper, VI.9.1.154; Cladosporium spot of fig, VI.9.1.155; Seed spotting of lima beans, VI.9.1.483).

.151 Cladosporium piscola: (see Pea scab, VI.9.1.336).

.152 Cladosporium rot of muskmelons (Cladosporium cucumerinum): occurs frequently on cantaloupes, honeyball, and honeydew melons; a green mold rot occurs at the stem scar of cantaloupes and honeyball melons, at cracks and other points, and forms a thin, superficial olive green, later black, mold with frayed, dirty looking netting on the former or as scattered spots on the latter; spots from specks of 2.5 cm or more, circular to oval, with fairly regular, distinct margins; smooth, shiny, black, often with a buff band, sometimes heavy sporulating mat over affected areas on honeydew melons. (R & S)

.153 Cladosporium rot of stone fruits (Cladosporium herbarum): wound parasite that forms a hard, dry, gray-to-black, cone-shaped lesion that can be easily removed from surrounding healthy flesh; rarely occurs on peaches, nectarines, or apricots. (HSK)

.154 Cladosporium rot of tomato and pepper (Cladosporium herbarum): small light tan spots becoming slightly sunken, 3 to 10 mm in diameter, with dark-brown-to-black centers and tan or light brown margins; in delayed shipments and ripening rooms, 13 mm or more in diameter, shiny black and smooth centers with a light brown border. (MCW)

.155 Cladosporium spot of fig (Cladosporium herbarum): olive-green-to-yellowish-olive spots that develop as the fruit mature. (SHJ)

.156 Clubroot of crucifers (Plasmodiophora brassicae): important for marketing turnip, rutabaga, and radish; enlargement of main and/or lateral roots; sordid gray or pale yellow, smooth and firm, becoming dark colored later; infected roots are unmarketable. (R & S)

.157 Cochliobolus heterostrophus: (see Southern corn leaf blight, VI.9.1.514).

.158 Colletotrichum capsici: (see Ripe rot of pepper, VI.9.1.467).

.159 Colletotrichum circinans: (see Smudge of onions, leeks, shallots, VI.9.1.492).

.160 Colletotrichum coccodes: (see Anthracnose of tomato, VI.9.1.033).

.161 Colletotrichum erumpens: (see Rhubarb anthracnose, VI.9.1.460).

.162 Colletotrichum gloesporioides: (see Anthracnose of avocado, VI.9.1.023; Anthracnose of citrus, VI.9.1.025; Anthracnose of mango, VI.9.1.027; Anthracnose of papaya, VI.9.1.029; Anthracnose of pepper, VI.9.1.031).
.163 Colletotrichum higginsianum: (see Anthracnose of turnip, etc., VI.9.1.034).
.164 Colletotrichum lagenarium: (see Anthracnose of muskmelons, VI.9.1.028; Anthracnose of watermelon, VI.9.1.035; Cucumber anthracnose, VI.9.1.183).
.165 Colletotrichum lindemuthianum: (see Anthracnose of snap beans, VI.9.1.032).
.166 Colletotrichum pisi: (see Anthracnose of garden pea, VI.9.1.026).
.167 Common bean mosaic of snap beans: virus disease characterized by the shiny or greasy appearance of pods and other symptoms similar to those of southern bean mosaic (VI.9.1.512); another strain (in Florida) produces long, fairly broad stripes of normal and affected tissues but no greasy or water-soaked appearance of the latter. (SMF)
.168 Common blight of lima beans: (see Bacterial blights of lima beans, VI.9.1.050).
.169 Common smut (boil smut) of corn (Ustilago maydis): produces dry, hard, warty galls or tumors up to 15 cm or more in diameter on ears, tassels, leaves, stalks, and brace roots; controlled by resistant hybrids. (CMI)
.170 Coniothyrium sp.: (see Core rot of apple, VI.9.1.171).
.171 Core rot (moldy core) of apple (Penicillium sp., Physalospora sp., Alternaria sp., etc.): growth of fungi in the core region causes rot; species of Penicillium, Physalospora, Alternaria, Fusarium, Rhizoctonia, Coniothyrium, Aspergillus, and Botrytis are also found, of which the first three are the most common. (PCM)
.172 Corticium centrifugum: (see Fisheye rot of apple, VI.9.1.223).
.173 Corynebacterium flaccumfaciens pv. flaccumfaciens: (see Bacterial wilt of snap beans, VI.9.1.075).
.174 Corynebacterium michiganense pv. michiganense: (see Bacterial canker or tomato, VI.9.1.052).
.175 Corynebacterium michiganense pv. sepedonicum: (see Bacterial ring rot of potato, VI.9.1.055).
.176 Coryneum blight of stone fruits (Coryneum carpophilum): causes small, dry, purplish-red, circular spots 3 to 6 mm in diameter; cream-colored in the center with dark-red-to-brown margins; later slightly sunken and rusty brown; primarily affects twigs, dormant buds, leaves, and blossoms. (HSK)
.177 Cottony leak of eggplant (Pythium aphanidermatum): bleaching, development of tan discoloration; slight wrinkling of skin with affected tissues beneath watery and light brown; abundant cottony white surface mold. (MCW)
.178 Cottony leak of snap beans (Pythium butleri plus other Pythium spp.): one of the most serious diseases after harvest; usually starts where pods were in contact with the soil; water-soaking in browned (abraded cuticular tissue) or normal tissue, then white, cottony mycelia; spreads by contact to form "nests" of decayed pods, which are soft and watery. (SMF)
.179 Cottony rot (Sclerotinia sclerotiorum) of lemons: rapid, spreading, contact decay chiefly of lemons; characterized by white cottony mycelia. (SHJ)
.180 Crater rot of carrot (Rhizoctonia carotae): first evident as small pitted spots with whitish mold; later enlarges into brown, sunken

craters 6 to 13 mm in diameter, 3 to 13 mm deep, and lined with white-to-cream colored mold; decayed tissues below surface lesions are light brown, usually firm and dry, but the fungus may spread over the entire root and cause a firm decay with high humidity; secondary decay may develop from other rots during prolonged storage. (SMF)

.181 Crater spot of celery (Rhizoctonia solani): appears as tan-to-brown, ovoid, sharply delimited, sunken lesions on the inner and outer surfaces of petioles in contact with the soil; often followed by bacterial soft rot. (SMF)

.182 Crown rot of banana: a decay caused by Botryodiplodia theobromae, Thielaviopsis paradoxa, Gloeosporium musarum, Deightoniella torulosa, Fusarium roseum, or Verticillium theobromae, individually or as mixtures of two or more species; affects the curved portion of the fingers (crown) of boxed bananas. (SHJ)

.183 Cucumber anthracnose (Colletotrichum lagenarium): causes malformed fruit and reduces yields by foliage and stem injury; more or less circular, water-soaked areas with abundant orange, pink, and slimy spores later. (R & S)

.184 Cucumber mosaic on celery: virus disease; extremely variable with a range of symptoms that includes various degrees of mottling, deformation, twisting, and necrosis on leaflets and petioles. (SMF)

.185 Deightoniella torulosa: (see Crown rot of banana, VI.9.1.182; Speckle of banana, VI.9.1.516).

.186 Diaporthe batatis: (see Sweet potato dry rot, VI.9.1.535).

.187 Diaporthe citri: perfect state of Phomopsis citri; (see Melanose of citrus, VI.9.1.310; Stem-end rots of citrus, VI.9.1.528).

.188 Diaporthe phaseolorum: (see Pod blight of lima beans, VI.9.1.401).

.189 Diaporthe phaseolarum var. sojae: (see Phomopsis rot of tomato, VI.9.1.359).

.190 Diplodia ear rot (dry rot) of corn (Diplodia zeae, D. macrospora): bleaching of husks; ears become grayish brown and shrunken if infected early; grayish-white mold on and between kernels at harvest if infected later or mold only on the cob and kernels discolored if infected late in the season (i.e., near harvest; usually begins at the butt of the ear and involves all of the kernels as it progresses upward); D. macrospora is found in warm, humid areas; also causes stalk rot. (CMI)

.191 Diplodia macrospora: (see Diplodia ear rot of corn, VI.9.1.190).

.192 Diplodia natalensis: (see Diplodia rot of muskmelons and cucumbers, VI.9.1.193; Diplodia rot of peaches, VI.9.1.194; Diplodia stain of onions, VI.9.1.195; Stem-end rots of avocado, VI.9.1.527; Stem-end rots of citrus, VI.9.1.528; Stem-end rot of mango, VI.9.1.524; Stem-end rot of papaya, VI.9.1.525; Stem-end rot of watermelon, VI.9.1.526).

.193 Diplodia rot of muskmelons and cucumbers (Diplodia natalensis): buff-colored, later black, lesion (usually one per fruit); decayed tissues spongy and somewhat softened with pungent sour odor and taste. (R & S)

.194 Diplodia rot of peaches (Diplodia natalensis): essentially a wound parasite that develops after harvest; also spreads by contact; spots and decayed flesh are light brown, somewhat watery and there is often a sour odor; surface growth is profuse, fine, cottony white, with gray-to-black fruiting bodies later. (HSK)

.195 Diplodia stain of onions (Diplodia natalensis): superficial disease; silvery-gray-to-black discoloration of outer scales around upper half of bulb; pycnidia form singly or in clusters of two or

.196 Diplodia tubericola: (see Java black rot of sweet potatoes, VI.9.1.296).
.197 Diplodia zeae: (see Diplodia ear rot of corn, VI.9.1.190).
.198 Distichlis stricta: (see Orange rust of spinach, VI.9.1.330).
.199 Dothiorella gregarea: (see Stem-end rot of papaya, VI.9.1.525).
.200 Dothiorella sp.: (see Stem-end rot of avocado, VI.9.1.527).
.201 Double-virus streak of tomato (Marmor tabaci plus M. dubium var. vulgare): combination of two virus diseases; fruit rough and misshapen, with small, raised, glassy, or brown and corky, irregularly shaped areas; caused by plants infected with tomato mosaic and later with latent potato virus. (MCW)
.202 Downy mildew of crucifers (Peronospora parasitica): chiefly on cabbage, cauliflower, radishes, and turnips; destructive on radishes and cabbage on the market and in storage; small pale greenish-yellow angular spots with fine white mildew on the underside of leaves as spots enlarge; may have grayish-black discoloration of stalk as fungus enters and no odor; brown mottling or streaks from crown into root in turnips; later blackened; brown-to-black epidermal discolorations; sometimes elongated streaks in radishes. (R & S)
.203 Downy mildew of garden pea (Peronospora pisi): slightly raised, irregular, yellowish then brown patches on pods, with white mycelia and oospores immediately below the surface blotches and white tufts of feltlike tissue on the inner surface of the pods. (SMF)
.204 Downy mildew of lettuce (Bremia lactucae): light-greenish-to-yellow areas up to 1.3 cm in diameter with fuzzy, whitish-gray mold on undersurfaces; later turn brown when secondary invaders (e.g., bacterial soft rot, gray mold rot) move in quickly; at least five physiologic races are known. (RFS)
.205 Downy mildew of lima beans (Phytophthora phaseoli): attacks all young tender parts of bush and pole cultivars; infected young pods wilt, shrivel, and gradually die; may have irregular patches of white, woolly, or feltlike mycelia and fruiting structures on developing pods; also purplish border to lesions; may infect seeds. (SMF)
.206 Downy mildew of parsley (Plasmopara nivea): field disease of parsley. (SMF)
.207 Downy mildew of spinach (Peronospora effusa): first appears as pale yellow, irregular spots without clearly defined margins on both leaf surface; whitish-gray-to-lilac or purple-gray mold on the undersurface under humid conditions; severely infected leaves may dry up, become water-soaked, brownish, and soon decay. (RFS)
.208 Dry rot: (see Diplodia ear rot of corn, VI.9.1.190).
.209 Early blight of celery (Cercospora apii): appears on leaves; consists of circular, pale yellow spots often with a gray cast; lesions run lengthwise on petioles; distinguished from late blight (VI.9.1.297) by the yellowish-gray color and absence of black fruiting bodies. (SMF)
.210 Early blight rot of tomato (Alternaria solani): spots are dark brown, leathery, somewhat sunken, usually less than 2.5 cm diameter, with concentric markings and seldom extending more than 1.3 cm deep. (MCW)
.211 Elsinoe australis: (see Sour orange scab, VI.9.1.505).
.212 Elsinoe fawcettii, E. fawcettii var. scabiosa: (see Sour orange scab, VI.9.1.505).
.213 Endoconidiophora fimbriata: (see Black rot of sweet potatoes, VI.9.1.095).

VI.9.1. POSTHARVEST HANDLING and MARKETING

.214 Endosepsis (soft rot) of fig (Fusarium moniliforme): internal rot from spores carried from infected caprifigs by the blastophaga wasp. (SHJ)
.215 Erwinia ananas: (see Brown rot of pineapple, VI.9.1.127).
.216 Erwinia aroidea: (see Bacterial soft rot of muskmelons, etc., VI.9.1.062; Potato seed piece decay, VI.9.1.407).
.217 Erwinia atroseptica: (see Blackleg of potato, VI.9.1.081).
.218 Erwinia carotovora: (see Bacterial root rots of horseradish, VI.9.1.056; Bacterial soft rot of asparagus, VI.9.1.057; Bacterial soft rot of cabbage, etc., VI.9.1.058; Bacterial soft rot of carrot, parsnip, VI.9.1.059; Bacterial soft rot of celery, VI.9.1.060; Bacterial soft rots of lettuce, VI.9.1.061; Bacterial soft rot of muskmelons, etc., VI.9.1.062; Bacterial soft rot of onions, VI.9.1.063; Bacterial soft rot of potato, VI.9.1.064; Bacterial soft rot of snap beans, VI.9.1.065; Bacterial soft rot of spinach, VI.9.1.066; and Bacterial soft rot of tomato, pepper, eggplant, VI.9.1.067).
.219 Erwinia stewartii: [see Bacterial wilt (Stewart's disease) of corn, VI.9.1.074].
.220 Erysiphe polygoni: (see Powdery mildew of snap beans, VI.9.1.410).
.221 Exogenous brown discoloration: (see Brown rot of pineapple, VI.9.1.127).
.222 Fabraea maculata: (see Black spot of pear and quince, VI.9.1.100).
.223 Fisheye rot of apple (Corticium centrifugum): a firm rot with cream-colored-to-brown spots with a pale center, up to 2.5 cm in diameter; tough, leathery skin and spongy or stringy tissues that do not separate readily from healthy tissues. (PCM)
.224 Flyspeck of pome fruits (Microthyriella rubi): small, sightly raised black specks. (PCM)
.225 Foot and crown rots: (see Phytophthora rots of rhubarb, VI.9.1.386).
.226 Foot rot of sweet potato (Plenodomus destruens): firm to spongy dark brown decay with evident shriveling and presence of dark brown pycnidia in the skin; most infections are at the proximal end or in wounds; important storage and market problems. (RFS)
.227 Freckles: (see Scab of peaches, nectarines, apricots, plums, VI.9.1.470).
.228 Fruitlet black rot, fruitlet brown rot, fruitlet core rot, fruitlet rot: (see Brown rot of pineapple, VI.9.1.127).
.229 Fruit rot: (see Heart rot of pomegranate, VI.9.1.286).
.230 Fruit rot of eggplant: (see Phomopsis rot of eggplant, VI.9.1.358).
.231 Fruit (Brook's) spot of apple (Mycospaerella pomi): scattered, highly colored spots 3 to 6 mm in diameter; deep red or black on red and dark-green on green or yellow areas in midsummer, becoming slightly sunken and flecked with black late in the season. (PCM)
.232 Fusarium brown rot (Fusarium spp.) of citrus: a minor decay of oranges and lemons similar to anthracnose or alternaria rot; primarily a wound parasite, often starting at the stem end. (SHJ)
.233 Fusarium bulbigenum batatas: (see Stem rot of sweet potatoes, VI.9.1.530).
.234 Fusarium bulb rot of onions (Fusarium zonatum forms): attacks uninjured bulbs or through bruises or insect injuries; roots rotted off, bulb base with white-to-pinkish mold and is dry or watery, depending on Fusarium cepae and other F. spp.; species progresses slowly as a storage rot. (SMF)
.235 Fusarium cepae: (see Fusarium bulb rot of onions, VI.9.1.234).

VI.9.1. POSTHARVEST HANDLING and MARKETING

.236 <u>Fusarium</u> <u>culmorum</u>: (see Fusarium rot of muskmelons, VI.9.1.247).
.237 <u>Fusarium</u> <u>graminium</u>: (see Fusarium rot of muskmelons, VI.9.1.247).
.238 <u>Fusarium</u> <u>moniliforme</u>: (see Endosepsis of fig, VI.9.1.214; Brown rot of pineapple, VI.9.1.127).
.239 <u>Fusarium</u> <u>moniliforme</u> var. <u>subglutinans</u>: (see Fusarium rot on muskmelons, VI.9.1.247).
.240 <u>Fusarium</u> <u>oxysporum</u>: (see Fusarium bulb rot of onions, VI.9.1.234).
.241 <u>Fusarium</u> <u>oxysporum</u> f. 1.: (see Surface rot of sweet potatoes, VI.9.1.534).
.242 <u>Fusarium</u> <u>oxysporum</u> f. 2.: (see Stem rot of sweet potatoes, VI.9.1.530).
.243 <u>Fusarium</u> (pink) kernel rot of corn (<u>Gibberella</u> <u>fujikuroi</u>, G. fujikuroi var. <u>subglutinans</u>): pale pink-to-lavender discoloration of kernel caps; enters mainly through wounds made by ear worms and borers. (CMI)
.244 <u>Fusarium</u> <u>roseum</u>: (see Crown rot of banana, VI.9.1.182).
.245 <u>Fusarium</u> rot of asparagus (<u>Fusarium</u> spp.): occurs on tips, bracts, and other parts of spears, first as a white, later slightly pinkish mold; tissues are water-soaked then yellow to brown, becoming soft and watery and covered with white fluffy mold; lesions may be 5 to 7.5 cm long on the sides of spears. (SMF)
.246 <u>Fusarium</u> rot of carrot (<u>Fusarium</u> spp.): a shallow, spongy, occasionally brown decay, with lesions at the crown or sides of roots, usually at the site of injuries; a sparse white mold may develop on the surface or in infected tissues; decay usually develops only after several months storage. (SMF)
.247 <u>Fusarium</u> rot of muskmelons (<u>Fusarium</u> <u>scirpi</u> var. <u>acuminatum</u>, F. culmorum, F. graminium, F. moniliforme var. subglutinans): common disease of cantaloupes, honeyball, honeydew, Persian, Spanish, casaba, and other melons; small scattered lesions coalesce later, with the development of white or pinkish white mycelia and spongy, diseased internal tissue that can be lifted out intact; spots on white-skinned cultivars brownish at first; may also have purplish-red discoloration produced by <u>Fusarium</u> scirpi var. acuminatum internally and an extremely bitter taste (other species do not produce the bitter flavor). (R & S)
.248 <u>Fusarium</u> rot of tomato (<u>Fusarium</u> spp.): weak pathogen causes slight water-soaking, sinking, softening, and wrinkling of affected tissues (vine-ripe); firm rot; internally a pale brown corelike mass that can be lifted out in a single piece (ripening room). (MCW)
.249 <u>Fusarium</u> <u>scirpi</u> var. <u>acuminatum</u>: (see Fusarium rot of muskmelons, VI.9.1.247).
.250 <u>Fusarium</u> <u>zonatum</u> forms: (see Fusarium bulb rot of onions, VI.9.1.234).
.251 <u>Fusarium</u> sp.: (see Core rot of apple, VI.9.1.171).
.252 <u>Fusarium</u> spp.: (see Fusarium brown rot of citrus, VI.9.1.232; Fusarium bulb rot of onions, VI.9.1.234; Fusarium rot of asparagus, VI.9.1.245; Fusarium rot of carrot, VI.9.1.246; Fusarium rot of tomato, VI.9.1.248; Sweet potato end rots, VI.9.1.536).
.253 Fuscous blight of lima beans (<u>Xanthomonas</u> <u>campestris</u> pv. <u>phaseoli</u> mutant (?): (see Bacterial blights of lima beans, VI.9.1.050).
.254 <u>Geotrichum</u> <u>candidum</u>: (see Sour rot of carrot, VI.9.1.506; Sour rot of citrus, VI.9.1.507; Sour rot of peaches, VI.9.1.508; Sour rot of pome fruits, VI.9.1.509; Sour rot of tomato, VI.9.1.510).
.255 Ghost spot of tomato (<u>Botrytis</u> <u>cinerea</u>): causes a white-to-yellowish ring surrounding a green center; 3 to 6 mm in diameter. (MCW)
.256 <u>Gibberella</u> <u>fujikuroi</u>:(see Fusarium kernel rot of corn, VI.9.1.243).

VI.9.1. POSTHARVEST HANDLING and MARKETING

.257 **Gibberella fujikuroi** var. **subglutinans**: (see Fusarium kernel rot of corn, VI.9.1.243).
.258 **Gibberella (red) ear rot of corn (Gibberella zeae)**: pink-to-brick-red mold discoloration of ears, husks, and kernels, usually beginning at the tip of the ear and progressing downward; also causes a stalk rot and seedling blight of corn and scab or head blight of small grains (e.g., barley and wheat). (CMI)
.259 **Gibberella zeae**: (see Gibberella ear rot of corn, VI.9.1.258).
.260 **Gloeodes pomigena**: (see Sooty blotch of apple, VI.9.1.503).
.261 **Gloeosporium musarum**: (see Anthracnose of banana, VI.9.1.024; Crown rot of banana, VI.9.1.182).
.262 **Glomerella cingulata**: (see Anthracnose of peaches, etc., VI.9.1.030; Bitter rot of apple, pear, peach, quince, cherry, VI.9.1.079).
.263 **Gray ear rot of corn (Physalospora zeae)**: causes bleached husks, tiny black specks throughout the pith of the cob; slate gray, light-weight ears and kernels with small black specks or stripes beneath the seed coat; also attacks leaves.
.264 **Gray mold (Neck rot) of onions, shallot, garlic (Botrytis allii, B. byssoidea, B. squamosa)**: typical gray mold rot that affects bulbs in transit, storage, and marketing; colored bulbs more resistant. (SMF)
.265 **Gray mold rot of apple, pear (Botrytis cinerea)**: a firm, pale, translucent area becoming pale brown to brown with paler edges and eventually enveloping the entire fruit; frequent in cold storage, usually from earlier infections. (PCM)
.266 **Gray mold rot of asparagus (Botrytis cinerea)**: causes a tip wilt in the field or decay after harvest; tissues water-soaked, with some white mold; later grayish-brown with smoky, granular spore masses. (SMF)
.267 **Gray mold rot of cabbage, cauliflower, broccoli, turnip, rutabaga (Botrytis cinerea)**: important as transit and storage disease; greenish-brown, water-soaked areas become grayish-brown with brown surface mold later on green leafy tissues; gray-to-light-brown and fairly moist in fleshy tissues of cauliflower and roots; no odor unless bacterial soft rot is also present. (R & S)
.268 **Gray mold rot of carrot (Botrytis cinerea)**: affected tissues are water-soaked spongy, and light brown, with lesions occurring anywhere, usually at injuries; whitish-gray, later grayish-brown; mold and granular spore masses develop on lesions; also sclerotia under humid conditions. (SMF)
.269 **Gray mold rot of fig (Botrytis cinerea)**: market disease chiefly of overripe fruit. (SHJ)
.270 **Gray mold rot of garden pea (Botrytis cinerea)**: starts as small, water-soaked, gradually enlarging spots that become grayish-buff color; pale gray mycelia may develop on the pod surface, later becoming brownish and covered with small granular clusters of spores. (SMF)
.271 **Gray mold rot of globe artichoke (Botrytis sp.)**: sometimes serious; lesions moist to wet, odorless, reddish-brown-to-brown, with definite, slightly water-soaked margins; commonly originates at cuts or splits. (RFS)
.272 **Gray mold rot of lettuce (Botrytis cinerea)**: tissues with water-soaked, grayish-green, or brownish areas that become soft and slimy with smoky gray mycelia and spore masses. (RFS)
.273 **Gray mold rot of pomegranate (Botrytis cinerea)**: light brown, firm decay, starting at the calyx; the rind is tough and leathery, the

VI.9.1. POSTHARVEST HANDLING and MARKETING

interior dark and disintegrated. (SHJ)
.274 Gray mold rot of rhubarb (Botrytis sp.): small red spots on the petiole and water-soaked brown areas at the petiole base or elsewhere; grayish, smoke-colored mycelia and grayish-brown, granular spore masses later; enters through wounds or apparently normal tissues. (RFS)
.275 Gray mold rot of snap beans (Botrytis cinerea): enters through weakened, dea, or wounded tissues; soft, watery, but not mushy; water-soaked or pale brown with smoky colored mycelia (moist conditions) and spore masses in advanced stages; not a serious disease. (SMF)
.276 Gray mold rot of stone fruits (Botrytis cinerea): lesions appear as a light brown spot on the skin, then covered with grayish spores and the underlying tissue becomes watery and darker brown; enters through wounds or by contact; affects mainly mature and stored fruit; numerous hosts among fruits, vegetables, and other crops. (HSK)
.277 Gray mold rot of tomato, pepper, eggplant (Botrytis cinerea): gray to grayish brown spots with regular, sharply defined margins that may be reddish-purple in eggplant; penetrates deeply into fruit (tomato) but is firm enough to be readily separated in eggplant. (MCW)
.278 Greasy pod: (see Common bean mosaic of snap beans, VI.9.1.167).
.279 Green mold of citrus: (see Blue mold of citrus, VI.9.1.105).
.280 Green mold rot: (see Cladosporium rot of muskmelons, VI.9.1.152; Cladosporium rot of stone fruits, VI.9.1.153).
.281 Green mold rot on plums: (see Alternaria rot of cherries, VI.9.1.012).
.282 Gummy stem blight: (see Black rot of cucumber, VI.9.1.086; Black rot of squash, VI.9.1.087).
.283 Gymnosporangium clavipes: (see Quince rust on apple, VI.9.1.437).
.284 Gymnosporangium juniperi-virginianae: (see Apple-ceder rust, VI.9.1.036).
.285 Halo blight of lima beans (Pseudomonas syringae pv. phaseolica): (see Bacterial blights of lima beans, VI.9.1.050).
.286 Heart rot (fruit rot) of pomegranate (Aspergillus niger, A. spp., Alternaria sp.): mass of blackened arils and slightly abnormal skin color. (SHJ)
.287 Helminthosporium carbonum: (see Southern corn leaf spot, VI.9.1.515).
.288 Helminthosporium carposaprum: (see Helminthosporium rot of tomato, VI.9.1.290).
.289 Helminthosporium maydis: (see Southern corn leaf blight, VI.9.1.514).
.290 Helminthosporium rot of tomato (Helminthosporium carposaprum): not known to be in the United States; lesions are similar to those of alternaria rot (VI.9.1.017). (MCW)
.291 Helminthosporium turcicum: (see Northern corn leaf blight, VI.9.1.329).
.292 Heterosporium leaf spot (pinhead rust) of spinach (Heterosporium variabile): weak pathogen that attacks weakened plants; leaf spots numerous, roughly circular, light brown, and slightly depressed, with sharply defined brown-to-purple borders; abundant layers of olive-green-to-black velvety layers of sporulating mold on both surfaces later; spots most abundant on older leaves. (RFS)
.293 Heterosporium variabile: (see Heterosporium leaf spot of spinach, VI.9.1.292).

VI.9.1. POSTHARVEST HANDLING and MARKETING

.294 Hormodendrum rot of corn (Hormodendrum spp.): causes a deep greenish-black rot on kernel caps; also spreads on and between kernels on ears in the field in wet, warm weather after the ears are mature.
.295 Itersonilia perplexans: (see Parsnip canker, VI.9.1.331).
.296 Java black rot of sweet potatoes (Diplodia tubericola): a slow-developing rot, most prevalent in the southern United States and tropics; brown and moderately firm in early stages; later the central part is light brown and the skin and underlying tissues become dark-brown-to-black from pycnidia which form domelike elevations on the surface; eventually a hard, dry, black mummy. (RFS)
.297 Late blight of celery (Septoria apiicola): few-to-numerous small yellowish or chlorotic, later brownish, grayish, or sometimes black spots on leaflets; may coalesce to form large areas of dead tissue; shiny black pycnidia on both dead and green tissues; spots are elongated, brown, without definite margins, and frequently coalesce on petioles, also with pycnidia; serious disease. (SMF)
.298 Late blight of potato (Phytophthora infestans): probably the most serious of all potato diseases; forms a reddish-brown-to-purplish dry rot in tubers or seed pieces from which the infection spreads to the leaves; several distinct races. (CMI)
.299 Late blight rot of tomato (Phytophthora infestans): brown-to-rusty-tan blotches without definite margins; firm, with a pebbly, roughened surface; occasionally water-soaked with a green margin and brown center, or as small irregularly shaped spots with numerous small sunken areas over much of the affected surface. (MCW)
.300 Latent potato virus: (see Marmor dubium var. vulgare, Double-virus streak of tomato, VI.9.1.201).
.301 Leaf curl of peaches and nectarines (Taphrina deformans): red blotches, sometimes depressed and lacy, or large, partly red, wartlike protuberances on fruit, which may be misshapen or cracked. (HSK)
.302 Leaf roll of potato: virus disease transmitted by aphids, tubers, or stalk grafts; reduces size of plants and size and set of tubers; characterized by upward rolling of leaves, leathery texture, chlorosis or lighter green color, and plants about half normal size.
.303 Leaf spot: (see Rhubarb stem spot, VI.9.1.461).
.304 Lethum australiense var. typicum: (see Spotted wilt of tomato, VI.9.1.522).
.305 Macrophomina phaseoli: (see Charcoal rot of honeydew melons, VI.9.1.144; Charcoal rot of pepper, VI.9.1.145; Charcoal rot of sweet potatoes, VI.9.1.146).
.306 Marbled fruit: (see Brown rot of pineapple, VI.9.1.127).
.307 Marmor dubium var. vulgare: (see Double-virus streak of tomato, VI.9.1.201).
.308 Marmor tabaci: (see Tomato mosaic, VI.9.1.543).
.309 Marmor tabaci plus M. dubium var. vulgare: (see Double-virus streak of tomato, VI.9.1.201).
.310 Melanose (Diaporthe citri) of citrus: several forms, such as speck type (individual small pustules), mudcake (large scablike patches of coalesced pustules), tear stain (where dew has carried spores down the sides of the fruit, most common on grapefruit), and star (pustules with starlike projections radiating out from them, caused by spraying with copper after an infection has occurred);

attacks leaves, young twigs, and fruit of all Citrus, particularly ripe fruit; overwinters in dead wood; hence a disease of older trees. (S & G, SHJ)
.311 Microthyriella rubi: (see Flyspeck, VI.9.1.224).
.312 Moldy core: (see Core rot of apple, VI.9.1.171).
.313 Monilinia fructicola: (see Brown rot of pome fruits, VI.9.1.128; Brown rot of stone fruits, VI.9.1.129).
.314 Monilinia laxa: (see Brown rot of pome fruits, VI.9.1.128; Brown rot of stone fruits, VI.9.1.129).
.315 Monilochaetes infuscans: (see Sweet potato scurf, VI.9.1.537).
.316 Mosaic of lettuce: virus disease that may produce a mild mottling, irregularly shaped heads, vein clearing, and flecking of leaves; progressive marginal browning and decay and shorter storage life. (RFS)
.317 Mottle necrosis (ring rot) of sweet potato (Pythium ultimum, P. scleroteichum): small, sunken, grayish-brown spots, and areas around attachment of secondary roots (natural infections); sometimes penetrates deeply; may be mottled-marbled, grayish-to-chocolate-brown with pockets; another type consists of bands or rings of grayish-brown surface decay, that soon becomes dry and sunken. (RFS)
.318 Mucor sp.: (see Soft rot of fig, VI.9.1.497).
.319 Mucor rot of sweet potatoes (Mucor racemosus): enters through wounds or dead rootlets; moist, clayish white, with distinct starchy odor; later firm to spongy and diseased tissues fibrous, stringy; similar to Rhizopus soft rot (VI.9.1.457) but develops below 10°C. (RFS)
.320 Mycosphaerella brassicicola: (see Ring spot of crucifers, VI.9.1.465).
.321 Mycosphaerella citrullina: (see Black rot of cucumber, VI.9.1.091; Black rot of squash, VI.9.1.092; Black rot of watermelon, VI.9.1.096).
.322 Mycosphaerella pinodes: perfect state of Ascochyta pinodes; (see Pod spot of garden pea, VI.9.1.405).
.323 Mycosphaerella pomi: (see Fruit spot of apple, VI.9.1.231).
.324 Myrothecium sp.: (see Ring rot of tomato, VI.9.1.464).
.325 Nailhead spot of tomato (Alternaria tomato): eliminated by use of resistant cultivars in the United States; spots are shallow, small, circular, slightly sunken, tan-to-brown-to-black, usually less than 6 mm in diameter, sometimes with a green halo. (MCW)
.326 Neck rot of onions: (see Gray mold of onions, VI.9.1.264).
.327 Nematospora phaseoli: (see Yeast spot of lima beans, VI.9.1.575).
.328 Nigrospora cob rot of corn (Nigrospora oryzae): shredding of cob usually from the butt end, with gray discoloration and disintegration in internal tissues; kernels with masses of black spores near their tips, light ears, and bleached, loose kernels.
.329 Northern corn leaf blight (Helminthosporium turcicum): dark, grayish-green, water-soaked, later greenish-tan spots, becoming spindle-shaped and up to 5 cm wide by 15 cm long; produces chaffy kernels in ears; starts on lower leaves, spreads, and kills the plant; now controlled mainly by resistant lines. (CMI)
.330 Orange rust of spinach (Puccinia aristidae): light yellow spots on underside of leaves, later conspicuous with light yellow to orange sori; spots may enlarge in transit and marketing; spreads from wild salt grass (Distichlis stricta) to spinach. (RFS)
.331 Parsnip canker (Itersonilia perplexans): a destructive disease of stored parsnips; slightly depressed rust-colored, later black,

.332 Parsnip scab (Streptomyces scabies): field disease of parsnips. (SMF)
.333 Peach blight: (see Coryneum blight of stone fruits, VI.9.1.176).
.334 Pear scab (Venturia pirina): attacks fruit at all stages (also leaves and young twigs); lesions become corked, russeted, and larger and rougher than on apples; fruit often badly misshapen. (PCM)
.335 Pear stony pit: a virus disease of 'Bosc', 'Winter Nelis', 'Anjou', 'Hardy', and 'Forelle'; 'Bartlett', a symptomless carrier; deep cone-shaped pits 6 mm or deeper into the flesh, with fused stone cells at the base of each and brown colored flesh around the stones. (PCM)
.336 Pea scab (Cladosporium pisicola): dark brown or black, irregular, raised, scablike spots, which may be numerous, on pods; small blisters or brown and black lesions that later become sunken and covered with greenish, powdery spores on seeds. (SMF)
.337 Pea streak: virus disease that produces spotted, purplish-gray or brown, pitted, seriously malformed pods. (SMF)
.338 Pellicularia rolfsii: (see Sclerotium rot of snap beans, VI.9.1.481).
.339 Penicillium digitatum: (see Blue mold of citrus, VI.9.1.105).
.340 Penicillium expansum: (see Blue mold rot of apple, pear, quince, VI.9.1.106; Blue mold rot of cherries, VI.9.1.107; Blue mold rot of pomegranate, VI.9.1.112).
.341 Penicillium funiculosum: (see Brown rot of pineapple, VI.9.1.127).
.342 Penicillium hirsutum: (see Penicillium root rot of horseradish, VI.9.1.233).
.343 Penicillium italicum: (see Blue mold of citrus, VI.9.1.105).
.344 Penicillium root rot of horseradish (Penicillium hirsutum): stored roots with decay similar to Rhizoctonia, except that greenish-blue spore masses are in abundance. (R & S)
.345 Penicillium sp.: (see Blue mold rot of fig, VI.9.1.108; Blue mold rot of peaches, etc., VI.9.1.111; Blue mold rot of sweet potatoes, VI.9.1.113; Core rot of apple, VI.9.1.171).
.346 Penicillium spp.: (see Blue-green rot of corn, VI.9.1.104; Blue mold rot of muskmelons, VI.9.1.110; Blue mold rot of garlic and onions, VI.9.1.109; Blue mold rot of pomegranate, VI.9.1.112).
.347 Peronospora effusa: (see Downy mildew of spinach, VI.9.1.207).
.348 Peronospora parasitica: (see Downy mildew of crucifers, VI.9.1.202).
.349 Peronospora pisi: (see Downy mildew of garden pea, VI.9.1.203).
.350 Pezicula malicorticis: (see Bull's eye rot of apple, VI.9.1.133).
.351 Phialophora malorum: (see Side rot of apple, VI.9.1.490).
.352 Phoma apiicola: (see Phoma root rot of celery, VI.9.1.355).
.353 Phoma betae: (see Black rot of beet, VI.9.1.087).
.354 Phoma destructiva: (see Phoma rot of tomato, VI.9.1.356).
.355 Phoma root rot of celery, celeriac (Phoma apiicola): confined mainly to crown tissues, which are bluish-green, later black and completely disintegrated in severe cases; black pycnidia are formed in abundance over the decayed areas. (SMF)
.356 Phoma rot of tomato (Phoma destructiva): transit and market decay that enters through stem scars, cracks, and other wounds; small, slightly sunken brown spots becoming circular or elliptical with brownish-to-black centers, leathery, firm and dotted with black pycnidia; penetrates deeply into the pulp. (MCW)

VI.9.1. POSTHARVEST HANDLING and MARKETING 648

.357 Phomopsis citri: (see Stem-end rots of citrus, VI.9.1.528; Stem-end rot of papaya, VI.9.1.525).
.358 Phomopsis rot of eggplant (Phomopsis vexans): small, roughly circular tan-to-light-brown, slightly depressed spots, becoming darker brown, somewhat zoned up to 5 to 7.5 cm diameter; internal decay is soft and spongy with light brown flesh discoloration. (MCW)
.359 Phomopsis rot of tomato (Diaporthe phaseolorum var. sojae): wound parasite; smooth, turgid lesions turning brown to nearly black; very firm, later shriveling and cracking and with a firm spongy core of decayed flesh permeated with mycelia. (MCW)
.360 Phomopsis sp.: (see Stem-end rots of avocado, VI.9.1.527).
.361 Phomopsis vexans: (see Phomopsis rot of eggplant, VI.9.1.358).
.362 Phyllosticta solitaria: (see Blotch of apple, VI.9.1.103).
.363 Phyllosticta straminella: (see Rhubarb stem spot, VI.9.1.461).
.364 Physalospora ear rot of corn (Physalospora zeicola): similar to gray ear rot (VI.9.1.263) but does not attack leaves.
.365 Physalospora obtusa: (see Black rot of apple, VI.9.1.086).
.366 Physalospora rhodina: perfect state of Diplodia natalensis (see Stem-end rots of citrus, VI.9.1.528).
.367 Physalospora sp.: (see Core rot of apple, VI.9.1.171).
.368 Physalospora zeae: (see Gray ear rot of corn, VI.9.1.263).
.369 Physalospora zeicola: (see Physalospora ear rot of corn, VI.9.1.364).
.370 Phytophthora cactorum: (see Phytophthora rot of apple, VI.9.1.380; Phytophthora rots of rhubarb, VI.9.1.386).
.371 Phytophthora capsici: (see Buckeye spot of tomato, VI.9.1.132; Phytophthora rot of pepper, VI.9.1.384).
.372 Phytophthora cinnamomi: (see Brown rot of citrus, VI.9.1.126).
.373 Phytophthora citrophthora: (see Brown rot of citrus, VI.9.1.126).
.374 Phytophthora drechsleri: (see Buckeye spot of tomato, VI.9.1.132).
.375 Phytophthora hibernalis: (see Brown rot of citrus, VI.9.1.126).
.376 Phytophthora infestans: (see Late blight rot of potato, VI.9.1.298); Late blight rot of tomato, VI.9.1.299).
.377 Phytophthora palmivora: (see Brown rot of citrus, VI.9.1.126).
.378 Phytophthora parasitica: (see Brown rot of citrus, VI.9.1.126; Buckeye spot of tomato, VI.9.1.132; Phytophthora rot of pepper, VI.9.1.384; Phytophthora rots of rhubarb, VI.9.1.386).
.379 Phytophthora phaseoli: (see Downy mildew of lima beans, VI.9.1.205).
.380 Phytophthora rot of apple, pear (Phytophthora cactorum): light brown, firm decay, later becoming somewhat spongy, more intensely colored, and deeply water-soaked; minor as a rule. (PCM)
.381 Phytophthora rot of asparagus (Phytophthora spp.): large water-soaked greenish or light brown lesions on side of spears above the base; may encircle the spears with tissue collapse; some appressed whitish-gray mold may develop in advanced stages; no odor unless bacteria present. (SMF)
.382 Phytophthora rot of eggplant (Phytophthora spp.): dark brown with lighter colored borders; penetrates deeply into flesh with discolored vascular bundles. (MCW)
.383 Phytophthora rot of muskmelons (Phytophthora spp.): irregular bleached area neither sunken nor well defined (although sometimes slightly sunken and reddish brown), flesh slightly softer and spongier than normal; blistering of epidermis; shriveling, wrinkling, folding, and flattening of rind as lesion enlarges; also wet, appressed, sordid white, felty mycelial growth; flesh

.384 Phytophthora rot of pepper (Phytophthora capsici, P. parasitica): migrates from plant to cause small, green, water-soaked spots that enlarge and elongate down the sides of the pod from the stem scar; often covered with superficial grayish-white-to-tan mold. (MCW)

.385 Phytophthora rot of tomato (Phytophthora sp.): occurs on mature green and ripening fruit; firm, water-soaked areas with little discoloration; later an appressed white mold on the surface which finally wrinkles. (MCW)

.386 Phytophthora rots (foot and crown rots) of rhubarb (Phytophthora parasitica, P. cactorum): watery, greenish-brown sunken lesions start at the petiole base and progress rapidly, causing brown decay; secondary bacterial infections develop quickly. (RFS)

.387 Phytophthora sp.: (see Phytophthora rot of tomato, VI.9.1.385).

.388 Phytophthora spp.: (see Buckeye spot of tomato, VI.9.1.132; Phytophthora rot of asparagus, VI.9.1.381; Phytophthora rot of eggplant, VI.9.1.382; Phytophthora rot of muskmelons, VI.9.1.383)

.389 Phytophthora syringae: (see Brown rot of citrus, VI.9.1.126).

.390 Pinhead rust: (see Heterosprium leaf spot of spinach, VI.9.1.292).

.391 Pink kernel rot: (see Fusarium kernel rot of corn, VI.9.1.243).

.392 Pink mold rot of apple, pear (Cephalothecium roseum): resembles fisheye and bull's eye rots except for irregularly shaped spots and white-to-pinkish mold growth. (PCM)

.393 Pink mold rot of muskmelons and cucumbers (Cephalothecium roseum): lesions as brown discoloration, spots 2.5 cm in diameter to a third of the surface, later covered with mycelia and abundant pink spore masses; extremely bitter flesh taste. (R & S)

.394 Piricularia grisea: (see Pitting of banana, VI.9.1.395).

.395 Pitting of banana (Piricularia grisea): a widespread disease, usually on Dwarf and Giant Cavendish fruit; also finger stalks and cushions; small dark-brown-to-black pits up to about 6 mm in diameter; infection occurs in field on young fruit and becomes evident during ripening. (SHJ)

.396 Plasmodiophora brassicae: (see Clubroot of crucifers, VI.9.1.156).

.397 Plasmopara nivea: (see Downy mildew of parsley, VI.9.1.206).

.398 Plenodomus destruens: (see Foot rot of sweet potato, VI.9.1.226).

.399 Pleospora lycopersici: (see Pleospora rot of tomato, VI.9.1.400).

.400 Pleospora rot of tomato (Pleospora lycopersici): in western United States and Mexico; lesions start as small, brown, oval spots near the stem scar or wounds and enlarge to 1.3 cm or more in diameter with black, pimplelike perithecia in the center; develops during transit. (MCW)

.401 Pod blight of lima beans (Diaporthe phaseolorum): rarely attacks young pods, lesions reddish-brown, circular or oval, usually around edges of pod, later studded with minute, gray, pimplelike protuberances that emerge as black pustules or pycnidia that may cover the entire pod; may damage the seeds which may show small brown spots or become discolored and shrunken. (SMF)

.402 Pod-distorting mosaic of snap beans: virus disease; a strain of bean yellow mosaic; affected pods are distorted and malformed, with irregular, rough, warty areas; mottling is not so pronounced as in pod mottle (VI.9.1.403) or greasy pod; (see Common bean mosaic of snap beans, VI.9.1.167). (SMF)

.403 Pod mottle of snap beans: virus disease; pods often malformed; affected areas are dark green, water-soaked, glassy, and somewhat sunken in contrast to normal tissues. (SMF)

VI.9.1. POSTHARVEST HANDLING and MARKETING

.404 Podosphaeria leucotricha: (see Powdery mildew of apple, VI.9.1. 408).
.405 Pod spot of garden pea (Ascochyta pisi, A. pinodes, A. pinodella). commonly seen on pods as more or less circular, sharply depressed spots, 3 to 6 mm in diameter, with a pale-tan-to-pinkish center and dark margin; other forms are usually not sunken, 3 mm or less in diameter, dark-brown-to-black and penetrate pods and seeds; may sporulate under moist conditions. (SMF)
.406 Potato scab (Streptomyces scabies): causes irregular-shaped or deep russet scab lesions on tubers, rendering them worthless if severe; now controlled mainly by use of resistant cultivars. (CMI)
.407 Potato seed piece decay (Erwinia aroidea): soft bacterial rot that invades seed pieces.
.408 Powdery mildew of apple, pear (Podosphaeria leucotricha): small grayish or feltlike patches, followed by stunting and lacy type russeting of fruit; also attacks twigs, foliage, and flowers. (PCM)
.409 Powdery mildew of peaches, nectarines, apricots (Sphaerotheca pannosa var. persicae): begins as light green or white spots covered with whitish mold; later reddish or pinkish and then dark brown, with dead surface tissue and hard underlying flesh. (HSK)
.410 Powdery mildew of snap and lima beans (Erysiphe polygoni): affected areas on pods are yellowish to reddish brown, often seriously twisted and malformed; sometimes spots of dead, somewhat sunken, discolored tissues; mild infections cause reddish-brown, speckled, superficial discolorations with indefinite margins. (SMF)
.411 Pox: (see Soil rot of sweet potatoes, VI.9.1.500).
.412 Pseudomonas ananas: (see Brown rot of pineapple, VI.9.1.127).
.413 Pseudomonas cepacia: (see Sour skin of onions, VI.9.1.511).
.414 Pseudomonas cichorii: (see Bacterial soft rots of lettuce, VI.9.1. 061; Bacterial zonate spot of cabbage, etc., VI.9.1.076).
.415 Pseudomonas marginalis pv. marginalis: (see Bacterial soft rots of lettuce, VI.9.1.061).
.416 Pseudomonas marginalis pv. pastinacae: (see Bacterial brown rot of parsnip, VI.9.1.051).
.417 Pseudomonas syringae pv. apii: (see Bacterial blight of celery, VI. 9.1.048).
.418 Pseudomonas syringae pv. lachrymans: (see Bacterial spot of cucumber, VI.9.1.069; Bacterial spot of honeydew melons, VI.9.1.070).
.419 Pseudomonas syringae pv. maculicola: (see Bacterial leaf spot of cauliflower, etc., VI.9.1.054)
.420 Pseudomonas syringae pv. phaseolicola: (see Bacterial blights of lima beans, VI.9.1.050).
.421 Pseudomonas syringae pv. pisi: (see Bacterial blight of garden pea, VI.9.1.049).
.422 Pseudomonas syringae pv. syringae: (see Bacterial spot of lima beans, VI.9.1.071; Black pit of citrus, VI.9.1.085).
.423 Pseudomonas syringae pv. tomato: (see Bacterial speck of tomato, VI.9.1.068).
.424 Pseudomonas viridilivida: (see Bacterial soft rots of lettuce, VI. 9.1.061).
.425 Puccinia aristidae: (see Orange rust of spinach, VI.9.1.330).
.426 Purple blotch of onions (Alternaria porri): affects stems, leaves, and bulbs; infection at necks of toppled bulbs or through wounds; first semiwatery; colored deep yellow turning to wine red and eventually dark brown to black. (SMF)
.427 Pustular spot: (see Coryneum blight of stone fruits, VI.9.1.176).

.428 Pyrenochaeta lycopersici: (see Brown root rot of tomato, VI.9.1.576).
.429 Pythium aphanidermatum: (see Cottony leak of eggplant, VI.9.1.177).
.430 Pythium butleri: (see Cottony leak of snap beans, VI.9.1.178).
.431 Pythium rot of tomato (Pythium sp.): causes an extensive, firm, water-soaked area that soon envelopes the entire fruit. (MCW)
.432 Pythium rot of watermelon (Pythium spp.): minor disease; water-soaked spots soon turning light, bluish, or chocolate brown, with flesh flaccid and marshy odor. (R & S)
.433 Pythium scleroteichum: (see Mottle necrosis of sweet potatoes, VI.9.1.317).
.434 Pythium sp.: (see Pythium rot of tomato, VI.9.1.431).
.435 Pythium spp.: (see Pythium rot of watermelon, VI.9.1.432).
.436 Pythium ultimum: (see Mottle necrosis of sweet potatoes, VI.9.1.317).
.437 Quince rust on apple (Gymnosporangium clavipes): produces dwarfing and distortion at the calyx end and internal discoloration. (PCM)
.438 Red ear rot: (see Gibberella ear rot of corn, VI.9.1.258).
.439 Rhizoctonia carotae: (see Crater rot of carrot, VI.9.1.180).
.440 Rhizoctonia ear rot of corn (Rhizoctonia solani): husks and kernels covered with pink mycelia in early stages, later becoming gray.
.441 Rhizoctonia head and root rots of cabbage, turnips, rutabagas, radishes (Rhizoctonia solani = Thanatephorus cucumeris): several strains of the fungus; moderately coarse, cream-colored-to-brown surface mycelia and irregularly shaped sclerotia; affected tissues light and semiwatery to dark brown or brownish black; head rot of cabbage has dark brown sunken lesions on the stem, midribs, and blades of head leaves; root rots enter through wounds with round light brown, later dark brown, slightly sunken lesions; internal tissues watery or spongy; sometimes concentric rings of light and dark brown on stored turnips. (R & S)
.442 Rhizoctonia pod rot: (see Soil rot of snapbeans, VI.9.1.499).
.443 Rhizoctonia root rot of horseradish (Rhizoctonia solani): light-yellow-to-grayish-tan, rather dry decay, with brownish-black sclerotia as infection progresses; infected tissues separate readily. (R & S)
.444 Rhizoctonia solani: see Black scurf, stem and tuber rot of potato, VI.9.1.098; see Crater spot of celery, VI.9.1.181; Rhizoctonia ear rot of corn, VI.9.1.440; Rhizoctonia head and root rots of cabbage, etc., VI.9.1.441; Rhizoctonia root rot of horseradish, VI.9.1.443; Soil rot of snap beans, VI.9.1.499; Soil rot of tomato, VI.9.1.501; Soil rot of watermelon, VI.9.1.502).
.445 Rhizoctonia sp.: (see Core rot of apple, VI.9.1.171).
.446 Rhizopus arrhizus: (see Rhizopus soft rot of carrot, VI.9.1.453; Rhizopus soft rot of cucurbits, VI.9.1.455).
.447 Rhizopus nigricans: (see Rhizopus soft rot of sweet potatoes, VI.9.1.457; Soft rot of fig, VI.9.1.497).
.448 Rhizopus oryzae: (see Rhizopus soft rot of carrot, VI.9.1.453).
.449 Rhizopus rot of apple and pear (Rhizopus stolonifer): dark colored, coarse mold with a gray-white and black appearance. (PCM)
.450 Rhizopus rot of avocado (Rhizopus stolonifer): primarily a decay of ripening fruit at room temperature; a weak wound parasite that forms soft, dark brown lesions with a coarse white-to-gray mycelia that bear black sporangia. (SHJ)
.451 Rhizopus rot of stone fruits (Rhizopus stolonifer): important postharvest disease, especially of overripe fruit; lesions begin as water-soaked, tender, tan-to-brown areas with rapid development

VI.9.1. POSTHARVEST HANDLING and MARKETING

of mycelia and jet black spores; causes "leaky" boxes from collapse of fruit under pressure. (HSK)

.452 Rhizopus rot of tomato, pepper, eggplant (Rhizopus stolonifer): causes large, water-soaked, soft lesions with no discoloration and rapidly envelops the entire fruit; coarse mycelia with grayish spore balls are present on the surface; brownish, soft, wet decay in eggplant. (MCW)

.453 Rhizopus soft rot of carrot (Rhizopus tritici, R. stolonifer, R. arrhizus, R. oryzae): brownish water-soaked lesions with soft watery tissues, distinguished from bacterial soft rot by the presence of coarse, white mycelia; like gray mold rot, a problem in common storage. (SMF)

.454 Rhizopus soft rot of crucifers (Rhizopus stolonifer): attacks wounded or bruised tissues, worse on more fleshy vegetables; light brown, soft, watery decay, sometimes with coarse stringy mycelia that bear white and black sporangia; usually no odor. (R & S)

.455 Rhizopus soft rot of cucurbits (Rhizopus stolonifer, R. arrhizus, etc.): common on all types of muskmelon, cucumber, squash, and pumpkin; also watermelon; causes softening and pronounced water-soaking of flesh, often no external evidence; slightly sour odor and taste; decayed flesh sharply delimited from healthy tissue. (R & S)

.456 Rhizopus soft rot of snap beans (Rhizopus stolonifer, R. tritici): soft watery rot under moist, warm conditions, which soon develops coarse, white, later gray, stringy mycelia that bear white sporangia that soon turn black; spread from diseased to sound beans, resulting in a nest of decayed pods. (SMF)

.457 Rhizopus soft rot of sweet potatoes (Rhizopus nigricans, R. tritici): most important storage, transit, and market disease; decay evident in less than 24 hours, usually entering through wounds; soft and watery, with yellowish-brown liquid, pleasant fermented or yeasty odor, becoming cinnamon-to-chocolate-brown, withered and firm, and a hard brown mummy (under dry conditions); side infections may develop as the ring rot stage, which often forms a dry groovelike ring after tissues shrivel and shrink. (RFS)

.458 Rhizopus stolonifer: (see Rhizopus rot of apple and pear, VI.9.1.449; Rhizopus rot of avocado, VI.9.1.450; Rhizopus rot of stone fruits, VI.9.1.451; Rhizopus rot of tomato, pepper, eggplant, VI.9.1.452; Rhizopus soft rot of carrot, VI.9.1.453; Rhizopus soft rot of crucifers, VI.9.1.454; Rhizopus soft rot of cucurbits, VI.9.1.455; Rhizopus soft rot of snap beans, VI.9.1.456).

.459 Rhizopus tritici: (Rhizopus soft rot of carrot, VI.9.1.453; Rhizopus soft rot of snap beans, VI.9.1.456; Rhizopus soft rot of sweet potatoes, VI.9.1.457).

.460 Rhubarb anthracnose (Colletotrichum erumpens): oval, soft, watery translucent spots; small black specks (acervuli) appear when lesions are about 1.3 cm in diameter; later, stalks are soft, rotten, and covered with acervuli; spread by contact, raindrops, or insects. (RFS)

.461 Rhubarb stem spot (leaf spot) (Phyllosticta straminella): causes small greenish-yellow spots, later tan with wine-red borders, and eventually ragged holes in leaves; small, oval-to-oblong, reddish-brown spots, becoming 1.3 cm or more long with age on stems; pycnidia (small black fruiting bodies) eventually appear on the dead tissues; infection can occur in healthy uninjured stems if moist. (RFS)

.462 Ring pox of apricots: a virus disease; water-soaked gray spots and

rings on and just under the skin, becoming reddish purple, later brown-to-reddish-brown, slightly raised and bumpy; sometimes forms larger irregular blotches. (HSK)
.463 Ring rot: (see Mottle necrosis of sweet potatoes, VI.9.1.317).
.464 Ring rot of tomato (Myrothecium sp.): large, conspicuous, charply defined circular-to-oval, flattened lesions, with affected tissues brown-to-black, deep penetrating, and removable readily as a single, firm, spongy lump easily separated from healthy flesh; spore masses develop in narrow concentric bands alternating with white mold over the lesions. (MCW)
.465 Ring spot of crucifers (Mycosphaerella brassicicola): important on cauliflower, less so for cabbage, brussels sprouts, kale, rutabagas, radishes, or white turnips; small, circular, dark-centered, with greenish-yellow, water-soaked borders, becoming light-to-grayish-brown with narrow olive-green or olive-gray borders; black pycnidia and perithecia in concentric rings around center or scattered. (R & S)
.466 Ripe fruit rot: (see Brown rot of pineapple, VI.9.1.127).
.467 Ripe rot of pepper (Colletotrichum capsici): appears on ripening capsules as yellowish-to-light-pink circular spots, 3 to 6 mm in diameter; becomes water-soaked, soft, with sordid brown-to-black centers and reddish-brown borders; pink spore masses on larger spots; a disease of pimentos in Georgia. (MCW)
.468 Russet ring of apple: a virus disease that produces rings of russeted skin tissue surrounding free areas. (PCM)
.469 Rust of peaches and plums (Tranzschelia discolor): primarily affects foliage and bark, occasionally fruit; water-soaked, dark green spots 3 to 6 mm in diameter, becoming sunken with deep yellow or orange centers; underlying flesh is tough and leathery and clings tightly to adjacent healthy tissue. (HSK)
.470 Scab (freckles, black spot) of peaches, nectarines, apricots, plum (Cladosporium carpophilum): circular, olive-green (yellow on plum) spots about 1.5 to 3 mm in diameter, with indistinct, later well-defined margins; later enlarge, coalesce, and turn dark olive or black, with a cork layer below them; misshaping of fruit occurs as they develop. (HSK)
.471 Sclerotinia intermedia: (see Watery soft rot of celery, VI.9.1.559).
.472 Sclerotinia minor: (see Watery soft rot of celery, VI.9.1.559; Watery soft rot of lettuce, VI.9.1.561
.473 Sclerotinia rot: (see Watery soft rot of cabbage, etc., VI.9.1.557).
.474 Sclerotinia sclerotiorum: (see Cottony rot of lemons, VI.9.1.179; Watery soft rot of asparagus, VI.9.1.556; Watery soft rot of cabbage, etc., VI.9.1.557; Watery soft rot of carrot, VI.9.1.558; Watery soft rot of celery, VI.9.1.559; Watery soft rot of lettuce, VI.9.1.561).
.475 Sclerotinia spp.: (see Watery soft rot of snap beans, VI.9.1.562).
.476 Sclerotiniose: (see Watery soft rot of lettuce, VI.9.1.561).
.477 Sclerotium bataticola: (see Charcoal rot of honeydew melons, VI.9.1.144).
.478 Sclerotium cepivorum: (see White rot of onions, garlic, shallot, Welsh onion, leek, VI.9.1.565).
.479 Sclerotium rolfsii: (see Sclerotium rot of citrus, VI.9.1.480; Sclerotium rot of snap beans, VI.9.1.481; Sclerotium rot of tomato, VI.9.1.482).
.480 Sclerotium rot of citrus (Sclerotium rolfsii): minor decay of fruit

that touches the ground. (SJH)
.481 Sclerotium rot (southern blight) of snap beans (Sclerotium rolfsii = Pellicularia rolfsii): watery soaked areas on pods; development of coarse white mycelia with spherical white, later brown, sclerotia; generally of minor importance after harvest. (SMF)
.482 Sclerotium rot (southern blight) of tomato (Sclerotium rolfsii): decayed areas yellowish, slightly sunken, cracking when about 2 cm in diameter; advances rapidly, penetrates deeply into affected tissues; mushy soft with characteristic odor; penetrates uninjured skin; sclerotia smooth, light brown; many hosts among vegetables, including watermelons and ornamentals. (MCW)
.483 Seed spotting of lima beans (Cladosporium herbarum): occurs during marketing from contamination during shelling; superficial brown spots, 1.5 to 3 mm diameter on seed coat, enlarging and becoming covered with olive-green fungus which has a granular appearance from the dense mass of spores. (SMF)
.484 Septoria apiicola: (see Late blight of celery, VI.9.1.297).
.485 Septoria blight of parsley (Septoria petroselini): field disease of parsley. (SMF)
.486 Septoria citri: (see Septoria spot of citrus, VI.9.1.488).
.487 Septoria petroselini: (see Septoria blight of parsley, VI.9.1.285).
.488 Septoria spot of citrus (Septoria citri): a minor disease of lemons, grapefruit, and 'Valencia' oranges; usually round, dark brown, sunken rind lesions, 10 to 12 mm in diameter, typically with a narrow green, later reddish-brown halo; infected fruit develop an off-flavor and fall from the tree. (SHJ)
.489 Shot hole: (see Coryneum blight of stone fruits, VI.9.1.176).
.490 Side rot of apple (Phialophora malorum): oval or slightly irregular, usually less than 2.5 cm in diameter; slightly sunken, brown-to-dark-brown, tender lesions which separate cleanly from healthy tissues; mainly a saprophyte. (PCM)
.491 Slime: (see Bacterial soft rot of lettuce, VI.9.1.061).
.492 Smudge of onions, leeks, shallots (Colletotrichum circinans): blemishes outer scales but can destroy bulbs; small dark-green-to-black dots, often grouped together to form concentric rings or blotches. (SMF)
.493 Smut of onions and leeks (Urocystis cepulae): causes poor yields, stunts growth of plants, and shrinks bulbs excessively in storage; dark-colored, slightly raised streaks or blisters filled with greenish-black powdery spore masses on leaves and young bulbs of green onions; lesions slightly raised; brown-to-black pustules on bulbs reaching trade channels; does not develop or spread in transit or storage. (SMF)
.494 Snap bean rust (Uromyces phaseoli var. typica): starts as small brown spots and later appears as rusty-brown pustules, 1.5 to 3 mm in diameter, with typical slits and feathering. (SMF)
.495 Soft rot: (see Black rot of pineapple, VI.9.1.097).
.496 Soft rot of fig: (see Endosepsis of fig, VI.9.1.214).
.497 Soft rot (watery soft rot) of fig (Rhizopus nigricans, Mucor sp.): disease of soft overripe fruit on the tree or in the market. (SHJ)
.498 Soilage: contamination of dirt, spores (e.g., from Penicillium spp.), leakage of fluid, etc., from one fruit to another in containers during storage or shipment; thus not only necessitates sorting of decayed from sound fruit but also cleaning of the latter before they can be marketed. (S & G)
.499 Soil rot (Rhizoctonia pod rot) of snap beans, lima beans (Rhizoc-

tonia solani): infects plants, including pods, in the field; starts as small rusty-brown-to-brown, irregular spots without definite margins and occurs singly or in groups; becomes soft and brown or dried out, roughened, and chocolate brown, often surrounded by rust-colored tissue; may show concentric zonation; surface mycelia may be stimulated and grow over and infect nearby healthy pods under moist conditions. (SMF)

.500 Soil rot (pox) of sweet potatoes (Streptomyces ipomoea): irregular-sized and shaped; superficial dry brown pits or pock marks, 3 to 25 mm in diameter, smooth or slightly sunken; the epidermis cracks and exposes a cavity below; usually firm and dry; roots often misshapen. (RFS)

.501 Soil rot of tomato (Rhizoctonia solani): small circular brown spots; those 6 mm or larger have concentric ring markings that become less evident as fruit ripens; then spots are moderately firm, reddish brown, with water-soaked borders. (MCW)

.502 Soil rot of watermelon (Rhizoctonia solani): fairly firm, sharply separated small yellow or tan spots becoming brown as they enlarge; cream-colored mold seldom seen; flesh rarely decayed. (R & S)

.503 Sooty blotch of apple and pear (Gloeodes pomigena): dark granular spots or smudges on fruit; superficial and can be rubbed off. (PCM)

.504 Souring of fig: a condition caused by various yeasts and bacteria that induce watery, colorless pulp and an alcoholic or sour pungent odor. (SHJ)

.505 Sour orange scab (Elsinoe fawcettii): a serious disease of lemons, 'Temple' orange, sour orange, and some tangelos; sweet oranges are resistant and grapefruit and tangerines, intermediate; attacks only young tissues of leaves, twigs, and fruit; produces irregular scabby areas or warty protuberances; sweet orange scab (E. australis) and Tryon's scab (E. fawcettii var. scabiosa) are not known in North America. (S & G, SHJ)

.506 Sour rot of carrot (Geotrichum candidum): invades wounds or weakened tissues; decayed tissues are colorless, watery, and have a decided vinegary odor; develops occasionally in bulk or prepackaged carrots that have not been precooled promptly. (SMF)

.507 Sour rot of citrus (Geotrichum candidum): now primarily a major disease because of rough harvesting and consequent dropping of fruit on the ground where the organism is prevalent; a messy, stinking deterioration from spores entering through minor wounds to form a water-soaked spot that rapidly enlarges; becomes dark buff-yellow and involves the entire fruit; resistant to currently legal fungicides. (S & G, SHJ)

.508 Sour rot of peaches (Geotrichum candidum): affects mainly ripe fruit that have been bruised or cut; surface lesions are depressed, dark brown with purple margins; internal lesions are wet, soft, dark brown with occasional cavities and vinegary odor. (HSK)

.509 Sour rot (Geotrichum candidum) of pome fruits: wound parasite; enters through wounds on edge of stem scar; dull, greasy, water-soaked-to-bleached, fairly firm until late; then a pickled appearance and sour odor; soft and water-soaked lesions on ripe or ripening fruit. (PCM)

.510 Sour rot of tomato (Geotrichum candidum): a wound parasite; dull, greasy, water-soaked-to-bleached appearance on green fruit, often extending in a sector over the shoulder or developing all over

from the stem scar; affected tissues remain fairly firm until late, then a pickled appearance and sour odor. (MCW)
.511 Sour skin of onions (Pseudomonas cepacia): attacks only certain of the outer fleshy scales which become slimy and yellow; bulbs may shrink, with outer dry scales slipping off readily but interior firm; secondary invaders may, in part, produce sour odor. (SMF)
.512 Southern bean mosaic of snap beans: virus disease that produces malformed pods with irregularly shaped, dark green, water-soaked areas, intermingled with normal gray-green areas on green-podded beans. (SMF)
.513 Southern blight: see Sclerotium rot of snap beans, VI.9.1.481; Sclerotium rot of tomato, VI.9.1.482).
.514 Southern corn leaf blight (Helminthosporium maydis = Cochliobolus heterostrophus): grayish-to-straw-colored spots up to 0.6 cm wide and 4 cm long; spores and spots light colored; reduces yields, weakens stalks, and damages fodder value of plants.
.515 Southern corn leaf spot (Helminthosporium carbonum): two known races; race 1 has circular or oval, yellowish-green, later reddish-tan, and covered with dark spores often arranged in concentric rings; spots up to 2.5 cm in diameter; infects sheaths, ears, and husks; affected kernels turn black; race 2 has tan or light brown angular spots up to 0.6 cm wide and 2.5 cm long; also causes blackening of ears.
.516 Speckle of banana (Deightoniella torulosa): widespread disease with lesions as round reddish-brown or black spots with a dark green, water-soaked halo, about 2 mm in diameter, on young green fruit in the field; affects only the peel. (SHJ)
.517 Sphaceloma perseae: (see Avocado scab, VI.9.1.047).
.518 Sphaerotheca pannosa var. persicae: (see Powdery mildew of peaches, nectarines, apricots, VI.9.1.409).
.519 Spotted wilt of celery: virus disease manifested as numerous small yellow spots that soon become necrotic on older leaflets; sunken brown patches of dead tissues appear on the petioles. (SMF)
.520 Spotted wilt of garden peas: virus disease characterized by irregular brown patterns with concentric markings on the pods. (SMF)
.521 Spotted wilt of lettuce: virus disease that produces brown, necrotic, pitted spots, usually most evident along basal portions of midribs, often affecting one side of the plant more than the other; external symptoms are less pronounced when plants are affected late in growth. (RFS)
.522 Spotted wilt of tomato (Lethum australiense var. typicum): virus disease affecting pepper, eggplant, other vegetable crops, many ornamental plants (e.g., dahlia); spots with concentric red and yellow bands, roughened appearance from raised center, about 1.3 cm in diameter; often surface mottling from uneven ripening; fruit infected in field or greenhouse; thrips main vector plus mechanical transmission. (MCW)
.523 Star crack of apple: a virus disease that produces corky, star-shaped cracks, mainly on the calyx end of fruit. (PCM)
.524 Stem-end rot of mango (Diplodia natalensis): a soft, watery, blackish, sour-smelling rot; occurs mainly on fruit harvested when immature or ripened in poorly ventilated rooms. (SHJ)
.525 Stem-end rot of papaya (Diplodia natalensis): similar to that of citrus, mango, and other tropical fruits; Phomopsis citri and Dothiorella gregarea may also cause stem-end rots. (SHJ)
.526 Stem-end rot of watermelon (Diplodia natalensis): progressive browning and shrivelling first of the stem and later, the fruit tis-

VI.9.1. POSTHARVEST HANDLING and MARKETING

sues; abundant dark gray mycelia and black pycnidia later. (R & S)

.527 Stem-end rots of avocado: primarily Diplodia natalensis, but also Phomopsis sp. and Dothiorella sp., mainly of immature fruit; the rot begins as a dark brown, firm ring around the stem end and rapidly invades the rind and flesh; affected fruit usually have a distinct off-flavor. (SHJ, R & L)

.528 Stem-end rots of citrus: principally Phomopsis citri = Diaporthe citri and Diplodia natalensis = Physalospora rhodina; primarily major diseases of citrus in warm, humid regions; spores enter calyx tissues or lodge beneath the calyx at the time of flowering and remain quiescent until fruit are harvested, or through any other wound in mature fruit; typically appear as water-soaked spots, then turn tan (Phomopsis) to blackish-brown (Diplodia); the former grows more slowly (optimum temperature 24°C) than the latter (optimum 30°C); thus found in the cooler months of the harvest season; both are serious problems for which prompt fungicidal treatment after harvest is essential. (S & G, SHJ)

.529 Stemphylium radicinum: (see Black rot of carrot, VI.9.1.089).

.530 Stem rot (wilt) of sweet potatoes (Fusarium bulbigenum batatas, F. oxysporum f. 2): a vascular disease that produces internal brownish-black discolorations but no external symptoms. (RFS)

.531 Stewart's disease, Stewart's wilt: (see Bacterial wilt of corn, VI.9.1.074).

.532 Streptomyces ipomoea: (see Soil rot of sweet potatoes, VI.9.1.500).

.533 Streptomyces scabies: (see Beet scab, VI.9.1.077; Parsnip scab, VI.9.1.332; Potato scab, VI.9.1.406).

.534 Surface rot of sweet potatoes (Fusarium oxysporum f. 1.): small, superficial, circular, light brown spots, enlarging, slightly sunken later, 6 to 8 mm in diameter, and seldom penetrating more than 6 mm or so into the flesh, even in advanced stages; no surface mold or fruiting bodies. (RFS)

.535 Sweet potato dry rot (Diaporthe batatis): dark brown, firm decay starting at the end of the sweet potato; broken down tissues lose water, wither, become black and hard, and the surface is soon covered with black pycnidia.

.536 Sweet potato end rots (Fusarium spp.): commonly invade through wounds or as secondary infections; many different symptoms, the most serious being a firm dry rot and withering of the ends of the sweet potato. (RFS)

.537 Sweet potato scurf (Monilochaetes infuscans): blemishes lower market value; skin-deep, small, grayish-brown spots and blotches, often running together and sometimes cracking when the root loses moisture rapidly and becomes worthless; most infection in the field but some under humid storage. (RFS)

.538 Taphrina deformans: (see Leaf curl of peaches and nectarines, VI.9.1.301).

.539 Thanatephorus cucumeris: perfect stage of Rhizoctonia solani; (see Crater spot of celery, VI.9.1.181; Rhizoctonia head and root rots of cabbage, etc., VI.9.1.441; Soil rot of snapbeans, VI.9.1.499; Soil rot of watermelon, VI.9.1.502).

.540 Thielaviopsis basicola: (see Black mold of carrot, VI.9.1.082).

.541 Thielaviopsis paradoxa: imperfect state of Ceratocystis paradoxa; (see Crown rot of banana, VI.9.1.182; Black rot of pineapple, VI.9.1.097).

.542 Thielaviopsis stalk rot of banana (Ceratocystis paradoxa = Thielaviopsis paradoxa): main stalk and finger stalk rot as a wound

VI.9.1. POSTHARVEST HANDLING and MARKETING

parasite; occurs in the plantation and after harvest; same disease as black rot of pineapple. (SHJ)
.543 Tomato mosaic (Marmor tabaci): virus disease; superficial cream-colored streaks more or less broken by dark green areas radiating from stem area; also affects peppers and eggplants. (MCW)
.544 Tranzschelia discolor: (see Rust of peaches and plums, VI.9.1.469).
.545 Trichoderma rot of citrus (Trichoderma viride): soft brown rot of lemons, grapefruit, or oranges with a coconutlike odor; enters through injuries or occasionally by contact. (SHJ)
.546 Urocystis cepulae: (see Smut of onions, VI.9.1.493).
.547 Uromyces phaseoli var. typica: (see Snap bean rust, VI.9.1.494).
.548 Ustilago maydis: (see Common smut of corn, VI.9.1.169).
.549 Venturia inaequalis: (see Apple scab, VI.9.1.037).
.550 Venturia pirina: (see Pear scab, VI.9.1.334).
.551 Verticillium alboatrum: (see Verticillium wilt of potato, VI.9.1.553).
.552 Verticillium theobromae: (see Crown rot of banana, VI.9.1.182).
.553 Verticillium wilt of potato (Verticillium alboatrum): causes wilting of tops and vascular discoloration of stems, tubers, and roots; fungus is harbored in tubers and persists in soil. (MRW)
.554 Water blister: (see Black rot of pineapple, VI.9.1.097).
.555 Water rot: (see Black rot of pineapple, VI.9.1.097).
.556 Watery soft rot of asparagus (Sclerotinia sclerotiorum): irregular, watery, odorless lesions that become covered with white-to-grayish appressed mold; hard black sclerotia in advanced stages. (SMF)
.557 Watery soft rot (Sclerotinia rot) of cabbage, rutabagas, turnips, etc. (Sclerotinia sclerotiorum): attacks a wide range of crops, including many crucifers; characteristic soft, watery decay often covered with white cottony mold; leak when badly decayed but no disagreeable odor; mainly found in transit, storage, or on the market. (R & S)
.558 Watery soft rot of carrot (Sclerotinia sclerotiorum): water-soaking and softening of decayed tissues, with white fluffy or appressed mold in advanced stages and later white, bluish, finally black sclerotia. (SMF)
.559 Watery soft rot of celery, anise, parsley, parsnip (Sclerotinia sclerotiorum, S. intermedia, S. minor): water-soaking and softening of decayed tissues which are generally light brown with a pinkish-brown border; no characteristic odor; white more or less appressed fungus growth in advanced stages and white, bluish, then black sclerotia developed on the surface or in cavities. (SMF)
.560 Watery soft rot of fig: (see Soft rot of fig, VI.9.1.497).
.561 Watery soft rot of lettuce (Sclerotinia sclerotiorum, S. minor): water-soaked, light or pinkish brown, with development of white cottony mold; heads turn into a wet, leaking mess; may have black sclerotia in advanced stages. (RFS)
.562 Watery soft rot of snap beans (Sclerotinia spp.): watery soft rot that enters through wounds or weakened or healthy tissues; first water-soaked, later tan-colored lesions of various shapes and sizes anywhere on pods; white cottony or somewhat appressed mold and sclerotia (white, later black, and hard) appear later; rapid development at room temperature. (SMF)
.563 Western celery mosaic: virus disease; produces green-to-light-green mottling or vein banding with cupping and distortion of leaflets; the petioles sometimes become flabby during transit or on the

VI.9.1. POSTHARVEST HANDLING and MARKETING

market. (SMF)
.564 White rot: (see Botryosphaeria rot of apple, VI.9.1.116).
.565 White rot of onions, garlic, shallot, Welsh onion, leek (Sclerotium cepivorum): field disease mainly, with symptoms like gray mold (VI.9.1.264) or fusarium bulb rot (VI.9.1.234) but whiter mycelia and sclerotia more spherical than on other bulb rots. (SMF)
.566 White rust of radish (Albugo candida): a problem only when mustard greens or radishes are marketed with their tops; white pustules that contain powdery spores damage appearance of leaves. (R & S)
.567 White rust of spinach (Albugo occidentalis): develops within leaf tissues and later produces numerous tiny sori filled with whitish spore masses; slightly indefinite yellowing of adjacent areas on both leaf surfaces; may have oospores (blackish) or sometimes mosaiclike discoloration of leaves. (RFS)
.568 Wild salt grass: (see Orange rust of spinach, VI.9.1.330).
.569 Wilt of sweet potatoes: (see Stem rot of sweet potatoes, VI.9.1.530).
.570 Xanthomonas campestris pv. campestris: (see Black rot of cabbage, etc., VI.9.1.088).
.571 Xanthomonas campestris pv. phaseoli: (see Bacterial blights of lima beans, VI.9.1.050).
.572 Xanthomonas campestris pv. pruni: (see Bacterial spot of peaches, etc., VI.9.1.072).
.573 Xanthomonas campestris pv. vesicatoria: (see Bacterial spot of tomato and pepper, VI.9.1.073).
.574 Xanthomonas campestris pv. vitians: (see Bacterial soft rot of lettuce, VI.9.1.061).
.575 Yeast spot of lima beans (Nematospora phaseoli): enters pods through insect injuries and attacks seeds; active infections have dark brown, irregular sunken areas, sometimes with rupturing of the seedcoat and grayish-brown, granular tissues. (SMF)
.576 Brown root rot of tomato (Pyrenochaeta lycopersici): a soil-infesting pathogen of tomato in the Netherlands, United Kingdom, United States, and Canada. (MCW, V & R)

VI.9.2. Animal pests that cause market disorders

(Note: All except a very few of the animal pests listed here actually damage produce prior to harvest; i.e., entry or feeding of insects and mites takes place in the orchard (grove) or field and causes blemishes or other grade-lowering defects, thus reducing the product's marketability.)

.001 Acrosternum hilare: (see Catface of stone fruits, VI.9.2.013).
.002 Anarsia lineatella: (see Fruit gumming of stone fruits, VI.9.2.035).
.003 Aonidiella aurantii, A. citrina: (see Armored scale injury on citrus, VI.9.2.007).
.004 Apple maggot injury (Rhagoletis pomonella): tunneling of flesh by maggots, often just below the skin in winding fashion (hence "railroad worm"); fruit often deformed. (PCM)
.005 Archips argyrospilus: (see Leaf roller injury of apple, VI.9.2.044)
.006 Argrotaenia velutinana: (see Leaf roller injury of apple, VI.9.2.044).

VI.9.2. POSTHARVEST HANDLING and MARKETING

.007 Armored scale injury on citrus: several species, e.g., California red scale (Aonidiella aurantii), Florida red scale (Chrysomphalus anonidum, chaff scale (Parlatoria pergandii), Glover scale (Lepidosaphes gloverii), purple scale (L. beckii), yellow scale (Aonidiella citrina), all of which are more or less thoroughly parasitized on fruit; Florida red scale degreens the rind beneath it, whereas Glover, chaff, and purple scales do not; the latter three leave green spots on colored fruit; wash off easily only if dead. (SHJ)

.008 Aspidiotus ancylus: (see Putnam scale on apple, VI.9.2.065).
.009 Aspidiotus forbesi: (see Forbes scale on apple, VI.9.2.031).
.010 Aspidiotus perniciosus: (see San José scale on apple, VI.9.2.073).
.011 Brown stinkbug: (see Catface of stone fruits, VI.9.2.013).
.012 California red scale: (see Armored scale injury on citrus, VI.9.2.007).
.013 Catface of stone fruits: distortion of fruit by feeding of insects such as the tarnished plant bug (Lygus lineolaris), lygus bug (L. hesperus), brown stinkbug (Euschistus servus), green stinkbug (Acrosternum hilare), Consperse stinkbug (Euschistus conspersus), Say stinkbug (Pitedia sayi), and plum curculio (Conotrachelus nenuphar), (HSK)
.014 Chaetanaphothrips orchidii: (see Thrips injury of citrus, VI.9.2.080).
.015 Chaff scale: (see Armored scale injury on citrus, VI.9.2.007).
.016 Chrysomphalus anonidum: (see Armored scale injury on citrus, VI.9.2.007).
.017 Citrus rust mite russeting (Phyllocoptruta oleivora): several types of discoloration of rind; early season injury is dull and silvery-smooth or rough (when severe); late season, a light brown, smooth to dark brown, somewhat shriveled or tear-stained (streaks) discoloration; prevalent in warm, humid areas; necessitates subclasses of U.S. Grades for Florida citrus. (S & G, SHJ)
.018 Codling moth injury of apple and pear (Laspeyresia pomonella): serious insect pest; damage by worms (caterpillars) burrowing into fruit, then leaving to cause holes for entry of fungi. (PCM)
.019 Conotrachelus nenuphar: (see Catface of stone fruits, VI.9.2.013; Plum curculio injury on apple, VI.9.2.063).
.020 Consperse stinkbug: (see Catface of stone fruits, VI.9.2.013).
.021 Corn earworm (cotton bollworm, tomato fruit worm) (Heliothis zea): larvae feed on silks and gradually penetrate to the kernels; serious pest on cultivars with loose husks, rendering ears unfit for marketing. (P & D)
.022 Cotton bollworm: (see Corn earworm, VI.9.2.021).
.023 Edema: (see Thrips injury on cabbage, VI.9.2.079).
.024 Eriophyes pyri: (see Pear leaf blister mite injury on apple and pear, VI.9.2.057).
.025 European apple sawfly injury (Hoplocampa testudinea): superficial feeding of larvae on very young apples causes broad (6 mm), slightly curved streaks with a brown, corky, checked, slightly sunken surface on mature fruit. (PCM)
.026 European corn borer (Pyrausta nubilalis): larvae tunnel and feed inside the stalk, ears, tassel, leaf midribs, and brace roots; serious pest of sweet corn, popcorn, and field corn (dent and flint): reduces market value of infested ears. (P & D)
.027 Euschistus conspersus: (see Catface of stone fruits, VI.9.2.013).
.028 Euschistus servus: (see Catface of stone fruits, VI.9.2.013).
.029 Florida red scale: (see Armored scale injury on citrus, VI.9.2.007).

VI.9.2. POSTHARVEST HANDLING and MARKETING

.030 Flower thrips (Frankliniella tritici): serious greenhouse pest of flowers; see Pansy spot of apple (VI.9.2.054). (P & D)
.031 Forbes scale on apple (Aspidiotus forbesi): causes injuries similar to San José scale (VI.9.2.073). (PCM)
.032 Frankliniella bispinosa: (see Thrips injury of citrus, VI.9.2.080).
.033 Frankliniella occidentalis: (see Thrips injury to stone fruits, VI.9.2.084).
.034 Frankliniella tritici: (see Flower thrips; VI.9.2.030; Pansy spot of apple, VI.9.2.054).
.035 Fruit gumming of stone fruits: larvae of the oriental fruit moth (Grapholitha molesta) and peach twig borer (Anarsia lineatella) cause gumming of fruit (as do certain bacterial and fungal diseases). (HSK)
.036 Fruit tree leaf roller: (see Leaf roller injury of apple, VI.9.2.044).
.037 Globe artichoke plum moth (Platyptilia carduidactyla): larvae damage buds in feeding, thus causing them to be malformed. (RFS)
.038 Glover scale: (see Armored scale injury on citrus, VI.9.2.007).
.039 Grapholitha molesta: (see Fruit gumming of stone fruits, VI.9.2.035; Oriental fruit moth injury on quince, VI.9.2.053).
.040 Green fruitworm injury: (see Leaf roller injury of apple, VI.9.2.044).
.041 Green stinkbug: (see Catface of stone fruits, VI.9.2.013).
.042 Hoplocampa testudinea: (see European apple sawfly injury, VI.9.2.025).
.043 Laspeyresia pomonella: (see Codling moth injury of apple and pear, VI.9.2.018).
.044 Leaf roller injury of apple: large depressed areas of russeted and corky tissues, often with consequent fruit distortion caused by feeding on small fruit; several species: fruit-tree leaf roller (Archips argyrospilus), green fruitworm (Lithophane antennata and related species), larvae of rusty tussock moth (Orgyia antiqua) and red-banded leaf roller (Argyrotaenia velutinana). (PCM)
.045 Lepidosaphes beckii, L. gloverii: (see Armored scale injury on citrus, VI.9.2.007).
.046 Lithophane antennata: (see Leaf roller injury of apple, VI.9.2.044).
.047 Lygus bug: (see Catface of stone fruits, VI.9.2.013).
.048 Lygus hesperus: (see Catface of stone fruits, VI.9.2.013).
.049 Lygus lineolaris: (see Catface of stone fruits, VI.9.2.013).
.050 Meloidogyne spp.: (see Root knot of carrot, parsnip, VI.9.2.070; Root knot on sweet potatoes, VI.9.2.071).
.051 Orgyia antiqua: (see Leaf foller injury of apple, VI.9.2.044).
.052 Oriental fruit moth: (see Fruit gumming of stone fruits, VI.9.2.035).
.053 Oriental fruit moth injury on quince (Grapholitha molesta): larvae often honeycomb fruit with tunnels; a major reason why commercial quince culture has been abandoned in the eastern United States. (PCM)
.054 Pansy spot of apple (Frankliniella tritici): lobed, often more than 13 mm across, white or greenish (on green or yellow cultivars) to light red or pink (red cultivars) spots, with greenish-brown corky center caused by ovipositing of flower thrips. (PCM)
.055 Parlatoria pergandii: (see Armored scale injury on citrus, VI.9.2.007).
.056 Peach twig borer: (see Fruit gumming of stone fruits, VI.9.2.035).
.057 Pear leaf blister mite injury on apple and pear (Eriophyes pyri):

VI.9.2. POSTHARVEST HANDLING and MARKETING

blisterlike lesions in early season, becoming irregular-shaped russet spots, often with deformation and cracking on fruit at harvest. (PCM)

.058 Pear thrips: (see Thrips injury to stone fruits, VI.9.2.081).
.059 Phyllocoptruta oleivora: (see Citrus rust mite russeting, VI.9.2.017).
.060 Pitedia sayi: (see Catface of stone fruits, VI.9.2.013).
.061 Platyptilia carduidactyla: (see Globe artichoke plume moth, VI.9.2.037).
.062 Plum curculio: (see Catface of stone fruits, VI.9.2.013).
.063 Plum curculio injury on apple (Conotrachelus nenuphar): injury ranges from small round, russeted scars to badly deformed ("catfaced") misshapen fruit or sometimes open punctures, caused by feeding and oviposition. (PCM)
.064 Purple scale: (see Armored scale injury on citrus, VI.9.2.007).
.065 Putnam scale on apple (Aspidiotus ancylus): causes injuries similar to San José scale (VI.9.2.073). (PCM)
.066 Pyrausta nubilalis: (see European corn borer, VI.9.2.026).
.067 Red-banded leaf roller: (see Leaf roller injury of apple, VI.9.2.044).
.068 Rhagoletis pomonella: (see Apple maggot injury, VI.9.2.004).
.069 Rice weevil (Sitophilus oryzae): adults and larvae injure kernels of corn ears, causing up to 90% damage when harvesting is delayed. (P & D)
.070 Root knot on carrot, parsnip (Meloidogyne spp.): early infestations may cause branching and poor development of taproots; later ones produce galls that give affected roots a rough, lumpy, or knotty appearance. (SMF)
.071 Root knot on sweet potatoes (Meloidogyne spp.): causes galls on rootlets or scablike abrasions, ringlike lesions, or cracks on fleshy roots; presence of nematodes indicated as small brown spots, 3 mm or less under the surface. (RFS)
.073 San José scale on apple and pear (Aspidiotus perniciosus): occurs wherever apples, other temperate zone fruits, and many woody ornamentals are grown; causes small (about 3 mm in diameter) red spots where the scales have fed; the scale proper is about 1.5 mm, gray-to-grayish brown, with a small elevated sordid yellow center. (PCM)
.074 Say stinkbug: (see Catface of stone fruits, VI.9.2.013).
.075 Scirtothrips citri: (see Thrips injury of citrus, VI.9.2.080).
.076 Sitophilus oryzae: (see Rice weevil, VI.9.2.069).
.077 Taeniothrips inconsequens: (see Thrips injury to stone fruits, VI.9.2.081).
.078 Tarnished plant bug: (see Catface of stone fruits, VI.9.2.013).
.079 Thrips injury (Edema) on cabbage (Thrips tabaci): insects feed in open spaces between head leaves, causing superficial white or yellowish pustules (edema), 1.5 to 3 mm in diameter; also causes scarified rusty patches on outer leaves. (R & S)
.080 Thrips injury on citrus: Scirtothrips citri in California; Frankliniella bispinosa and Chaetanaphothrips orchidii in Florida; characteristic russeting of the rind around the stem or stylar end or where fruit are in contact (orchid thrips); caused by feeding of adults or larvae on small fruit; usually classified as discoloration or scars in Florida; (see Citrus rust mite russeting, VI.9. 2.017). (SHJ)
.081 Thrips injury to stone fruits: stippling, mottling, russeting, or silvering of the skin and sometimes checking or pitting, caused

VI.9.2. POSTHARVEST HANDLING and MARKETING

by feeding of thrips (e.g., pear thrips, Taeniothrips inconsequens) or western flower thrips (Frankliniella occidentalis). (HSK)
.082 Thrips tabaci: (see Thrips injury on cabbage, VI.9.2.079).
.083 Tomato fruit worm: (see Corn earworm, VI.9.2.021).
.084 Western flower thrips: (see Thrips injury to stone fruits, VI.9.2.081).
.085 Yellow scale: (see Armored scale injury on citrus, VI.9.2.007).

VI.9.3. Physiological market disorders

.000 Physiological market disorders: disorders that affect produce at some stage but in which no primary parasite is involved; they are brought about by abnormal environmental conditions of some sort; classified into several general groups according to when they originate: (a) production-related as (1) nutrition-related (relative proportions among essential elements, deficiencies, toxicities), (2) weather-related (abnormal conditions of temperature, light, wind, soil moisture, atmospheric humidity), (3) maturity-related (i.e., stage of maturation at harvest), (4) atmospheric pollutants, toxic sprays or dusts; (b) harvest-induced (handling practices); (c) postharvest (disorders produced or induced by handling, transportation, and/or storage conditions with respect to temperature, humidity, CO_2 and O_2 levels, and presence of volatiles such as ethylene, aldehydes, and ketones as respiratory products).
.001 Ageing of citrus: a type of stem-end rind breakdown (VI.9.3.123) that occurs in storage and is usually accompanied by a stale or off-flavor in fruit held for a long period (e.g., eight to twelve weeks for oranges or four to six weeks for grapefruit). (SHJ)
.002 Albedo browning of lemons: a type of chilling injury, (VI.9.3.029). (SHJ)
.003 Ammonia injury of apples: darkening of tissues around lenticels or wounds or changes in pigmentation (i.e., red to blue-black or yellow to olive-green). (PCM)
.004 Ammoniation of citrus: (see Exanthema, VI.9.3.043).
.005 Apple scald (ordinary scald): superficial death and browning of affected tissues; the appearance, severity, and degree of browning are variable among and within cultivars; a storage disorder, with immature fruit and the greener side most affected; associated with accumulation of toxic levels of volatiles in skin tissues (thus inhibitors and/or good ventilation reduces incidence). (PCM)
.006 Arsenic toxicity: (see Boron deficiency of citrus, VI.9.3.017).
.007 Baldwin spot: (see Bitter pit, VI.9.3.008).
.008 Bitter pit (stippen, Baldwin spot) of apple: primarily a postharvest disorder but has its origin and sometimes develops on the tree before harvest; it is distinctly more prevalent and worse under conditions of water stress, heavy pruning, heavy N fertilization, calcium deficiency, and Ca:K:Mg ratio such as to increase water competition between leaves and fruit; starts internally as necrosis of cortical cells adjacent to the ends of the ultimate branches of vascular bundles; causes water soaking and premature skin coloration, the skin later developing small roundish pits, 1.5 to 3 mm in diameter and gray to black, mostly over the calyx

VI.9.3. POSTHARVEST HANDLING and MARKETING

half of the fruit; affected commercial cultivars include 'Baldwin', 'Northern Spy', 'Grimes Golden', 'Stayman', and 'Delicious'. (PCM)

.009 Black end of pears: a blackening of tissues surrounding the calyx opening accompanied by a pointed or peaked appearance; associated with 'Bartlett', 'Anjou', 'Winter Nelis', 'Comice', etc., on Japanese pear (Pyrus pyrifolia var. culta) rootstock. (PCM)

.010 Black heart of celery: brown water-soaked lesions along margins of young heart leaves, later becoming dry and black; caused by lack of calcium. (SMF)

.011 Black leaf speck of cabbage, Chinese cabbage, and cauliflower: small sharply sunken brown or black specks on leaves; occurs under refrigeration in transit and storage and in association with sudden sharp temperature drops in the fields; cause unknown. (R & S)

.012 Blossom-end clearing of grapefruit (stylar-end clearing, water log): a translucent area several centimeters in diameter at the stylar end as a result of rupture of albedo cells and adjacent juice vesicles; appears about 24 hours after mature, heavy, thin-skinned, seedless grapefruit have been dropped during handling. (S & G, SHJ)

.013 Blossom-end rot of pepper: similar to blossom-end rot of tomato (VI.9.3.014). (MCW)

.014 Blossom-end rot of tomato: small water-soaked spots, becoming dried out, light-to-dark-brown, sunken, and leathery; usually appears on one third-to-half-grown fruit; caused by faulty nutrition associated with irregular moisture supply and high temperature. (MCW)

.015 Blossom-end rot of watermelon: smooth, leathery, firm, dark green, or brown delimited areas, 2.5 to 7.5 cm in diameter around the point of blossom attachment; caused by faulty nutrition associated with irregular moisture supply and high temperature. (R & S)

.016 Blotchy ripening: (see Internal browning complex of tomato, VI.9.3.074).

.017 Boron deficiency of citrus: numerous symtoms, e.g., raised bumps on the surface of citrus fruit with gum pockets in the albedo beneath; identical with arsenic toxicity (on grapefruit) and with symptoms produced from some virus diseases. (SHJ)

.018 Boron deficiency of crucifers: browning necrosis in fleshy pith of stem and core, blisters and cracks on leaf midribs, and hollowing of stems; similar symptoms in rutabaga and turnip; discoloration of flower buds in broccoli (also bitter flavor); cabbage, cauliflower, rutabaga, and turnip most sensitive. (R & S)

.019 Brown core: (see Pithy brown core of pears, VI.9.2.099).

.020 Brown core (stem-cavity browning) of apples: browning and necrosis of flesh around the seed cavity and in the flesh underlying the stem cavity; a low-temperature storage disorder that develops when 'McIntosh' and some other cultivars grown in the Northeast are held for long periods. (PCM)

.021 Brown heart: (see Core breakdown of pears, VI.9.3.033).

.022 Browning of lychee: discoloration of the pericarp associated with desiccation of fruit after harvest; affected fruit sometimes develop an off-flavor. (SHJ)

.023 Brown staining of grapefruit and mandarins: a diffuse, irregular superficial discoloration of the peel; a type of chilling injury (VI.9.3.029). (SHJ)

.024 Brown wall: (see Internal browning complex of tomato, VI.9.3.074).

VI.9.3. POSTHARVEST HANDLING and MARKETING

.025 Bruises on apples: several fairly distinct forms, depending on where and when they occur; usually firm and dry, sometimes scarred or corky from preharvest injury; small, shallow, sometimes numerous when damaged during harvesting and packing; or variable in size, soft, water-soaked, and often discolored when the injury occurs on ripe fruit coming out of storage or during market handling. (PCM)

.026 Bruises on pears: several types, friction (caused by rubbing of fruit or on hard surfaces during packinghouse operation), impact (injury caused by dropping, as dry, corky, or pithy spots), or pressure (flattened areas with more or less discoloration and deterioration of tissues). (PCM)

.027 Bruising of stone fruits: mechanical injury to tissues caused by fruit striking sharp surfaces (cutting), compression in containers, impact on machinery during handling or shipment, and vibration rubbing or rolling against fruit or brushing; injured tissues are water-soaked; later turn brown and often develop off-flavors and/or decaying. (HSK)

.028 Carbon dioxide injury of apples: death and discoloration of skin tissues, sometimes extending into the flesh, 'Delicious' and 'Golden Delicious' held in 5% CO_2 at 0°C; also severe injury with 15% CO_2 and zero O_2 at 0°C (a severe combination CO_2 -O_2 type of injury). (PCM)

.029 Chilling injury: a complex disorder with a broad spectrum of symptoms that ranges from large, unsightly, sunken pits (e.g., in the rind of certain chilling-susceptible citrus fruits such as grapefruit) to surface dulling, darkening, or discoloration or watery areas and internal browning, water-soaking (watery breakdown) of tissues, uneven ripening, and loss of flavor or necrotic areas in leaves in which death ensues after prolonged exposure; occurs in chilling-sensitive plants at temperatures of about 0° for sweet potato to 13°C for banana, with many fruits, vegetables, and ornamentals in the 4 to 6°C range; products of most tropical and some subtropical plants show visual (external) and internal manifestations of the disorder which often appear only upon ripening, even though causal temperatures may have occurred weeks or months earlier, as in the case of banana; exposure to adverse conditions need be only a few hours, as with ripe bananas (and some ornamentals), or a period of from 10 days to several weeks or even months; undoubtedly the single most important factor in the shipment of a great many tropical fruits, vegetables, and ornamentals; (see also Chilling injury, IV.5.3.005). (GSK, PA)

.030 Cloud: (see Internal browning complex of tomato, VI.9.3.074).

.031 Concentric ring stipple of citrus: weather-induced rupturing of oil glands in the flavedo when turgid citrus fruit on well-watered trees are subjected to sudden cold; symptoms are discrete spots, large "scalded" areas, or the formation of concentric rings of small lesions when drops of dew distribute the peel oil; limes, lemons, and navel oranges are highly susceptible; (see Oleocellosis, VI.9.3.090). (SHJ)

.032 Controlled-atmosphere storage injury of apple: brown core, internal browning, or soft scald and soggy breakdown produced when holding a number of cultivars at -1 to 0°C in CA storage. (PCM)

.033 Core breakdown of pears: also known as internal breakdown, core rot, or brown heart; a softening and browning of the core tissues, often extending into the surrounding flesh; has a disagreeable odor; appearance in storage and degree of susceptibility de-

pends on fruit maturity; the more mature, the greater the susceptibility. (PCM)
.034 Core rot: (see Core breakdown of pears, VI.9.3.033).
.035 Cork of apples: boron-deficiency symptoms, as round-to-irregular, shallow surface spots or lesions in the core or core and flesh, the former cracking and becoming corky; also abnormally early ripening, abnormally flat, dull color, and development of internal breakdown when stored. (PCM)
.036 Cork spot of 'Anjou' pears: a bumpy, uneven appearance of the surface as fruit approach maturity, with large masses of brown or grayish necrotic cells underlying individual spots (and not visible externally so are detected only by cutting fruit); cause unknown. (PCM)
.037 Cork spot (York spot) of apples: similar in some respects but a different disorder than bitter pit (VI.9.3.008); develops in the orchard as slightly flattened, apparently water-soaked skin areas, with corky areas below surrounded by tougher than normal tissues, and often more than 6 mm in diameter; occurs on 'York Imperial', 'Delicious', 'Golden Delicious', 'Stayman', 'Grimes Golden', and 'Jonathan', among others; apparently boron-calcium related. (PCM)
.038 Cracked stem of celery: small white specks in petioles; affected tissues collapse, become light yellow, and later form a corky layer; transverse cracks appear as affected petioles elongate; basically boron deficiency as such or an imbalance among B and N, K, Ca, and Na. (SMF)
.039 Cracking of stone fruits: occurs in cherries from absorption of moisture by fruit and the tree, being worse at high temperature and with sugar content of 17 to 20-21%; peaches and plums sometimes crack from renewed growth when a rain follows a period of dry weather; cracks may occur on the sides but often along the suture in plums or prunes. (HSK)
.040 Creasing of citrus: grooves or furrows irregularly distributed on the fruit surface with the underlying albedo discontinuous or completely absent; primarily a preharvest condition of oranges and mandarins but found also postharvest on 'Robinson' tangerines; the exact cause is unknown but K is implicated. (S & G)
.041 Drought injury of stone fruits: irregular or circular patches in prunes later turning brown; sometimes the underlying flesh turns brown, becomes very firm, and often develops gum pockets; shriveling, mainly at the stem end, may also occur; often associated with B deficiency or previous injury by cold weather. (HSK)
.042 Endoxerosis (internal decline, wither tip) of lemons: apparently caused by water stress in arid growing areas; symptoms are the collapse of internal tissues within the stylar end, discoloration, and sometimes production of cavities; affected fruit are highly susceptible to alternaria decay. (SHJ)
.043 Exanthema (ammoniation) of citrus: black or brown raised lesions on citrus fruit caused by copper deficiency. (SHJ)
.044 Formaldehyde injury (halo spot) of apples: brown to black spots centered around lenticels on 'Golden Delicious', often with halos of lighter brown tissue; (now virtually eliminated by discontinuing use of ureaform resins in packing materials). (PCM)
.045 Freezing injury of cabbage, etc.: similar in appearance to injury to leaves of other crops; i.e., glassy appearance when frozen, water-soaked, and later discoloration (especially more sensitive tissues as vascular bundles); outer leaves more resistant to

freezing than inner leaves and stem; fleshy roots (e.g., turnips, radishes, rutabagas, and horseradish) often show only discoloration of vascular tissues. (R & S)

.046 Freezing injury of celery: in leaves, drying out and papery; in petioles, isolated sunken brown lesions, sometimes with a partial separation of the epidermis. (SMF)

.047 Freezing injury of citrus: typically a collapse and subsequent desiccation of affected internal tissues, with little or no external symptoms in fruit with a thick rind (e.g., grapefruit); internal freeze damage may grow progressively worse after packinghouse grading, particularly if fruit are several weeks in transit to export markets; may also occur in fruit during transit or storage if temperatures are allowed to drop below -1 to -4°C for a few hours. (S & G, SHJ)

.048 Freezing injury of globe artichoke: breaking, cracking, and blistering of the epidermis on exposed parts, as slight, or killing and blackening of the buds, as severe injury. (RFS)

.049 Freezing injury of onion: water-soaked, grayish-yellow fleshy scales, sometimes with opaque areas; caused by exposure to temperatures in transit or storage below about -1 to -3°C. (SMF)

.050 Freezing injury of pome and stone fruits: may occur in the orchard as fruit russeting but more often during handling and storage as water-soaked translucent spots on the skin or flesh tissues, later collapsing, browning, and becoming soft and mushy; caused by holding fruit below their freezing point, which ranges from -3° to slightly below 0°C, according to the kind of fruit and cultivar. (PCM)

.051 Freezing injury of sweet potato: varies from yellowish-brown discoloration of vascular tissues and yellowish-green water-soaked appearance of other tissues when slightly frozen to tissue collapse on thawing, with subsequent softening and eventual mummification when severely frozen. (RFS)

.052 Freezing injury of tomato: glassy water-soaked appearance of fruit with a sharp demarcation between severely frozen (i.e., held at or exposed to -1.3°C or lower) and uninjured tissue; light injury in the field may be manifested later as a pebbly surface with a slight-to-moderate yellowish color. (MCW)

.053 Frost and freezing injury of apples: early injury (i.e., bloom or early postbloom) as a partial or complete russeted girdle about midway between the stem and stylar ends; late (i.e., near maturity) on-tree or in-transit freezing may range from slight discoloration of the skin and/or brown vascular bundle strands (instead of normal greenish) to more or less severe temporary shriveling, with subsequent water-soaking and discoloration when thawed; also much more susceptible to bruising and with a shorter shelf-life, even though injury may appear minimal. (PCM)

.054 Fruit cracking of apples: cracking of the skin and flesh from absorption of enough water into the fruit to create too much internal pressure; occurs most frequently on 'Stayman', 'Wealthy', and 'York Imperial'. (PCM)

.055 Fruit tumor (waxy blister) of tomato: white-to-cream-colored irregular blisters, 3 to 6 mm in diameter and often more than 3 mm high, becoming light-to-dark-brown, depressed, and cracked as fruit ripen. (MCW)

.056 Granulation of citrus: a curious gelatinization of the juice within individual vesicles, thus severely decreasing extractable juice; two forms, of which one found in oranges (particularly 'Valencia'), tangerines, and mandarin hybrids (e.g., 'Temple') shows the disorder only at the stem end and is clearly related to advancing maturity, whereas the other in navel oranges tends to occur through the center of fruit and the jelled vesicles show sharply increased levels of Mg and Ca, the latter mainly in water-soluble form; occurs preharvest in the United States but postharvest (with 'Valencia' and navel oranges) in South Africa. (S & G)

.057 Gray wall: (see Internal browning complex of tomato, VI.9.3.074).

.058 Growth cracks of sweet potato: longitudinal or transverse splits and fissures caused by irregular or interrupted growth. (RFS)

.059 Growth cracks of tomato: ruptures of the skin and flesh radiating from the stem end, or, less frequently, as concentric cracks around the shoulder area; caused by absorption of soil moisture by vines with near-ripe fruit, following a period of dry weather. (MCW)

.060 Hail damage on citrus: characterized by a round or elliptical, deep, dark-colored indentation in the rind on the upper side of a citrus fruit caused by a glancing blow of hail stone, thus rendering it unfit for fresh shipment; a direct hit on a mature or nearly mature fruit causes it to burst and shatter. (S & G, SHJ))

.061 Hail injury of apples: more or less sunken areas, 6 to 12 mm in diameter, with fruit slightly misshapen (early damage) and the skin intact or torn but with brown, spongy, dry flesh beneath. (PCM)

.062 Hail injury of stone fruits: irregular sunken scars as small or deep punctures or shallow or deep gouges, with abnormal development of the injured area and often gum pockets in the flesh beneath; may heal over if fruit are young, often with russeting, browning, or light red pigmentation (peaches). (HSK)

.063 Halo spot: (see Formaldhyde injury of apples, VI.9.3.044).

.064 Heat injury of avocados: uneven softening or failure to soften normally, darkening of flesh, small brown spots on the skin, and off-flavor when held at 25 to 30°C, with scaldlike skin discoloration and rubbery flesh at 33°C. (SHJ)

.065 Heat injury of stone fruits: several manifestations: pit burn in apricots with premature flesh softening and later browning near the pit; internal browning of Italian and French prunes with translucent flesh which later browns, breaks down, and often develops cavities near the pit; Kelsey spot, surface burn with flesh browning below, usually on the stylar end of 'Kelsey' plum; also other cultivars, especially Japanese types; caused by temperatures above 38 to 40°C. (HSK)

.066 Hollow stem: (see Pithiness of celery, VI.9.3.098).

.067 Internal black spot of beet: boron deficiency.

.068 Internal breakdown: (see Core breakdown of pears, VI.9.3.033).

.069 Internal breakdown of apples: a breakdown and browning of the flesh on one side or around a bruise but may involve the entire fruit, sometimes with an outer shell of healthy flesh; occurs most often on large, overmature fruit; a sign of senescence (i.e., end of storage live). (PCM)

.070 Internal breakdown of pomegranate: a disorder of unknown but definitely physiological origin. (SHJ)

.071 Internal breakdown of stone fruits: a gradual discoloration and breaking down of tissues near the pit of fruit held in storage more than a few weeks. (HSK)

.072 Internal breakdown of sweet potatoes: internal tissues become pithy or dry and spongy, with cavities in the center of the root; most

VI.9.3. POSTHARVEST HANDLING and MARKETING

prevalent in storage kept too warm and too dry or where slight chilling injury has occurred. (RFS)

.073 Internal breakdowns of citrus and other fruits: include freeze injury (disruption of internal tissues from exposure to freezing temperatures), various forms of internal browning (e.g., apple, pear, and pineapple), granulation (solidification of vesicles at the stem end of citrus fruits), and watery breakdown. (S & G, SHJ)

.074 Internal browning complex of tomato: a group of similar disorders that includes internal browning, brown wall, vascular browning, gray wall (thin wall) of California, gray wall of Florida, blotchy ripening, and cloud caused by water imbalance, plus high temperatures and/or nutrient imbalance; characterized by gray-brownish discoloration of internal tissues in green fruit which become brown, extend to the surface, and form lesions that remain greenish or yellow in ripe fruit; may affect only the outer wall on the shoulders but may extend the entire length of the fruit and involve the placentas and central axis; fruit with blotchy ripening have hard green, waxy, or glassy areas in addition to other symptoms. (MCW)

.075 Internal browning of apples: a firm browning of the core area and surrounding tissues of fruit of certain cultivars (e.g., 'Yellow Newtown' and 'Rhode Island Greening'), stored at -1 to 0°C for some length of time; cool weather and lack of sunshine during the growing season are predisposing factors. (PCM)

.076 Internal browning of pineapple: a form of chilling injury produced when fruit are stored for more than a few days or shipped at 0 to 4°C. (SHJ)

.077 Internal browning of prunes: (see Heat injury of stone fruits, VI.9.3.065).

.078 Internal decline: (see Endoxerosis, VI.9.3.042).

.079 Jonathan spot of apples: light brown, irregular, or somewhat lobed and slightly sunken to deep brown, or black sharply sunken spots at the lenticels mainly on the blush side; occurs primarily postharvest (i.e., in storage, transit, or market) on 'Jonathan' and a number of other cultivars; cause unknown. (PCM)

.080 Kelsey spot: (see Heat injury of stone fruits, VI.9.3.065).

.081 Low-oxygen injury of apples: a cooked appearance of the skin and necrosis of flesh with development of a strong fermented odor in 'McIntosh' and 'Jonathan' stored at 3 or 0°C and zero 0 ; (see Carbon dioxide injury of apples, VI.9.3.028). (PCM)

.082 Low-temperature breakdown of cucumbers: (see Chilling injury, VI.9.3.029).

.083 Low-temperature breakdown of melons: (see Chilling injury, VI.9.3.029).

.084 Marginal browning of lettuce: yellowing followed by browning (necrosis) of margins of wrapper (outer) leaves of head lettuce and romaine; caused by adverse growing or improper transit or storage conditions; seen mostly as of transit origin and apparently senescence associated. (RFS)

.085 Mechanical injuries of tomato: wide variety of blemishes, scars, discolorations, and disfigurations, plus punctures, rim cuts, bruises, and compression deformation. (MCW)

.086 Membranous stain of lemons: a type of chilling injury (VI.9.3.029). (SHJ)

.087 Modified atmosphere injury of citrus: a type of scald that often covers a large portion of the fruit surface; brown, abruptly

VI.9.3. POSTHARVEST HANDLING and MARKETING

sunken areas or pebbly appearance on oranges (particularly 'Valencia') stored in an atmosphere with O_2 below 5% and CO_2 5% or higher. (SHJ, S & G)

.088 Oil-gland darkening of citrus: a type of chilling injury (VI.9.3.029). (SHJ)

.089 Oil spotting: (see Oleocellosis, VI.9.3.090).

.090 Oleocellosis (oil spotting) of citrus: necrosis of epidermis adjacent to oil glands ruptured by rough handling or harvesting turgid fruit; limes, lemons, and navel oranges are quite susceptible; terpenes in the oil glands are highly phytotoxic. (S & G, SHJ)

.091 Ordinary scald: (see Apple scald, VI.9.3.005).

.092 Ozone injury of apples: killing of cells, tissue collapse, browning, and formation of small pits at the lenticels with 3.25 ppm ozone introduced into a cold-storage room. (PCM)

.093 Pencil stripe of celery: narrow brown lines on petioles and petiolules; associated with high soil P. (SMF)

.094 Peteca of lemons: a type of pitting developed in storage; (see Chilling injury, VI.9.3.029). (S & G, SHJ)

.095 'Pineapple' orange pitting: peel pitting, which is particularly severe in 'Pineapple' but does appear on other cultivars; barely discernible at harvest but tends to develop markedly after harvest, particularly if fruit are held at low humidity; associated with crop load, with low K levels implicated, although there is no clear-cut relationship. (S & G)

.096 Pink rib of lettuce: faint-to-pronounced pinkish discoloration of midrib; cause unknown. (RFS)

.097 Pit burn of apricot: (see Heat injury of stone fruits, VI.9.3.065).

.098 Pithiness (hollow stem) of celery: central tissues become soft and spongy and either die and become pithy or collapse to produce a hollow stem; two forms, hereditary and physiological; the latter occurs if plant growth is arrested or plants are overmature before harvest or held too long in storage at 0°C. (SMF)

.099 Pithy brown core (brown core) of pears: pithy brown discolored areas in the core region, ranging from flecks between the carpels to the entire core, plus the surrounding flesh; considered a form of CO_2 injury to 'Anjou', 'Bartlett', and 'Bosc' pears. (PCM)

.100 Pitting of citrus: a typical form of chilling injury (VI.9.3.029) exhibited by chilling-sensitive grapefruit; depressed areas in the rind turn tan-to-brown, being circular and 6 to 12 mm in diameter when pits occur singly or form large irregular patches when they coalesce; also may occur on fruit on a tree following a prolonged period (two or three weeks) of cold weather around 4 to 5°C. (S & G, SHJ)

.101 Pressure bruise: common occurrence in bulk storages (e.g., apples and potatoes), usually when harvested under adverse weather conditions at that time. (RS)

.102 Puffiness of citrus: separation of the peel from the segments in mandarin-type cultivars caused by a combination of advancing maturity, vigorous trees (i.e., more prevalent and appears earlier on rough lemon or comparable rootstocks), and an irregular water supply; a late-season disorder as a rule. (S & G, SHJ)

.103 Puffiness of tomato: hollow, light-weight fruit, which may be more or less deformed, with large internal air spaces caused by abnormal pollination temperatures (i.e., below 15.5° or above 38°C), excessive soil moisture or drought, and unbalanced nutrition (high N). (MCW)

VI.9.3. POSTHARVEST HANDLING and MARKETING

.104 Rib discoloration of lettuce: creamy yellow or light brown, firm lesions about 6 mm wide usually less than 2.5 cm long and clearly visible on both surfaces; originates in the field but cause is unknown. (RFS)
.105 Rind staining of navel oranges: reddish-brown blemishes caused by handling fruit with physiologically overmature peel; now prevented in California by use of GA sprays to delay senescence. (S & G)
.106 Rumple of lemons: epidermal furrows similar to creasing (VI.9.3.040) but caused by necrosis of the albedo in those areas; strongly maturity-related, appearing late in the season and usually on lemons allowed to color on the tree. (S & G)
.107 Russeting of beans: basically a form of chilling injury (VI.9.3.029). (SMF)
.108 Russeting of stone fruits: abnormal roughening and scarring of normally smooth fruit surfaces; caused by frost injury, spray injury, insect damage, and/or rubbing of young fruit against twigs. (HSK)
.109 Russet spotting of head lettuce: few-to-numerous irregular 1.5 mm specks or spots 0.5 to several centimeters in diameter; light yellow that becomes pink, brown, olive brown, or dark brown; appear anywhere on leaf but most often near the base of the head; a general term for several leaf discolorations that originate in the field or in transit or storage. (RFS)
.110 Scab spot complex of carrot: one or more pockets of blackish necrotic tissues covered by normal ones; may also have sharp constrictions at the point of damages; lesions may be few or numerous, separate or coalesced, or may crack and become sunken; appear scablike and elongated tangentially; cause unknown. (SMF)
.111 Senescent scald of 'Golden Delicious' apple: similar to apple scald (VI.9.3.005) but more severe on late-harvested fruit (e.g., mature or past their prime) and on yellow fruit. (PCM)
.112 Senescent scald of pears: brown-to-black skin discoloration of 'Bartlett', 'Bosc', etc., that have turned color in storage and lost their capacity to ripen normally; worse on immature fruit; characteristic foul taste and odor even before onset of discoloration. (PCM)
.113 Shriveling of stone fruits: overall loss of moisture from handling and storage at too low humidity; becomes progressively more severe as the temperature is raised. (HSK)
.114 Skin checks on tomatoes: tiny cracks, usually over the shoulders in segments of circles around the stem scar; important mainly as blemishes; probably result from fluctuations of temperature and rain or dew. (MCW)
.115 Sloughing of red grapefruit: large, moist, chocolate-brown areas of the peel that slough off under light finger pressure; develops several days after shipment and occurs only in very early harvests; now seldom seen in Florida since maturity standards were raised. (S & G)
.116 Soft scald of apples: peculiar patches and ribbonlike areas of brown tissue on the surface with a sharp line of demarcation; caused apparently by abnormal respiratory activity induced by exposure of mature or late-picked fruit to near-freezing temperatures; affects many commercial cultivars. (PCM)
.117 Soggy breakdown of apples: light brown, becoming darker, discoloration in the cortex with affected tissues sharply delimited; generally moist and soggy, with a fermented taste; a storage disorder of fruit held at about -1.5 to 0°C and greatly increased

by delayed storage; affects 'Jonathan' and a few other cultivars. (PCM)
.118 Speckle of watermelon: numerous small, pale-to-orange-yellow, superficial, smooth, firm spots from about 1.5 to 6 mm in diameter in the rind; cause unknown. (R & S)
.119 Split pits of stone fruits: incomplete closure of the endocarp (stone) suture caused by abnormal growth of flesh during the pit-hardening stage in peaches and plums; weather conditions, small crop load, and high soil N are contributory factors. (HSK)
.120 Splitting and cracking of pomegranate: apparently caused by a sudden fluctuation in soil and air moisture content (i.e., by heavy rains, irrigation, or dry winds); worse on overmature fruit. (SHJ)
.121 Splitting of fig: cracking at the ostiole associated with cool, damp weather and turgid fruit in the orchard; also package pressure and rough handling during transit and in the market; globose or oblate fruit have a greater tendency to split than the elongated or pyriform. (SHJ)
.122 Stem-cavity browning of apples: (see Brown core of apples, VI.9.3.020).
.123 Stem-end rind breakdown of citrus: a disorder characterized by the collapse and subsequent necrosis of tissues around the stem end of citrus fruits but not immediately adjacent to it; initiated by an ill-defined imbalance in nutrition that involves N and K; its development is dependent on handling with care not to let the fruit dry out. (S & G)
.124 Stippen: (see Bitter pit, VI.9.3.008).
.125 Stylar-end clearing: (see Blossom-end clearing of grapefruit, VI.9.3.012).
.126 Sulfur dioxide injury on stone fruits: pitting and bleaching of cherries, peaches, nectarines, and plums stored or shipped with table grapes fumigated with SO_2. (HSK)
.127 Sunburn of citrus: a flat, pale, leathery area on the exposed side of a fruit caused by insolation; rarely occurs on the plant unless the fruit has an unusually thick pedicel that prevents it from hanging in a normal position (e.g., 'Honey' tangerine and 'Owari' satsuma) or a leaf that has provided shade is blown or moved away; all citrus cultivars and probably most other fruits are also subject to sunburn injury which can occur in only a few hours between harvest and delivery to a packinghouse. (SHJ)
.128 Sunburn of fig: dead, hard, tan-to-dark-brown bands around the ostiole or blotches or spots on the sides; most common on weak trees or those given inadequate irrigation. (SHJ)
.129 Sunburn of pineapple: rarely occurs on the plant unless the fruit topples over.
.130 Sunburn of pomegranate: (see Sunscald of pomegranate, VI.9.3.135).
.131 Sunburn, sunscald, delayed sunscald of apples: the first is a golden or bronze skin discoloration caused by exposure of one side of a fruit to intense sunlight on the tree; the second is a true burn (white or tan spots) caused by sudden exposure of shaded fruit to sunlight; the latter are brown, somewhat shriveled and sunken areas not evident when fruit were stored. (PCM)
.132 Sunscald of cabbage: water-soaked or blistered, irregularly shaped area which dries out quickly to become a bleached, papery, or parchmentlike island surrounded by healthy tissue; occurs only on leaves exposed on top of the head. (R & S)
.133 Sunscald of vegetables: actual burns produced on the surface of

VI.9.3. POSTHARVEST HANDLING and MARKETING

fruit or vegetables by droplets of water acting as lenses in bright sunshine.

.134 Sunscald of onion: bleached areas 2.5 to 5 cm in diameter, becoming soft and slippery, caused by exposure of immature uncured bulbs to bright sunshine in hot weather. (SMF)

.135 Sunscald (sunburn) of pomegranate: brown, slightly russeted, tough, leathery areas of the rind near the stem end; apparently caused by unequal insolation. (SHJ)

.136 Sunscald of snap beans: tiny reddish-brown spots that become diagonal streaks; in severe cases water-soaking, death, browning, and subsequent shriveling of affected tissues; more likely to occur when the soil is wet or humidity is high during periods of intense sunlight after several days of cloudy weather. (SMF)

.137 Sunscald of tomato and pepper: whitish, shiny, blistered areas on the sides or upper half, later becoming sunken, pale yellowish, and often wrinkled; caused by sudden exposure to sunlight; more serious in hot weather. (MCW)

.138 Superficial scald of 'Anjou' pear: brown skin discoloration but no sloughing or unpleasant odor as with senescent scald (VI.9.3.112); often develops post-cold storage. (PCM)

.139 Surface breakdowns of citrus: include wind scars (abrasion of young fruit by leaves), various forms of pitting (indentations), scalds (superficial necrosis of adjacent cells form ruptured oil cells), as in citrus, and superficial discoloration; may or may not involve underlying tissues except as means of entry for fungi and bacteria. (SHJ)

.140 Tip burn of cabbage: yellowing, browning, and gradual drying of leaf margins; often only the outer leaves are affected; occurs most frequently in soils low in K and high in N. (R & S)

.141 Tip burn of head lettuce: begins as small spots or flecks near the margin of outer head or inner wrapper leaves, which enlarge and coalesce to form an irregular brown, necrotic, wilted border up to 1.3 cm or more wide; cause unknown. (RFS)

.142 Translucent scales of onions: grayish, water-soaked appearance of one or more fleshy scales; cause unknown but high temperatures in the last few weeks before onions are stored or cooled are involved. (SMF)

.143 Uneven ripening: a common symptom of chilling injury (VI.9.3.029) in many tropical and some subtropical fruits and vegetables.

.144 Vascular browning: (see Internal browning complex of tomato, VI.9.3.074).

.145 Water core of apple: water-soaked regions in the flesh, commonly near the core but sometimes in other parts or in the whole of the fruit, becoming hard and glassy when severe; apparently an internal form of sunburn; particularly prevalent in regions of intense heat and sunshine, especially late in the season. (PCM)

.146 Water log: (see Blossom-end clearing of grapefruit, VI.9.3.012).

.147 Waterspot of citrus: epidermal cells ruptured by endosmosis; typically a problem of navel oranges grown under usually low humidity conditions, then subjected to a cool, wet period; oil sprays can accentuate the problem. (SHJ)

.148 Watery breakdown: (see Chilling injury, VI.9.3.029).

.149 Waxy blister: (see Fruit tumor, VI.9.3.055).

.150 Waxy breakdown of garlic: small, slightly sunken, light-yellow areas that become amber in clove flesh; cause unknown. (SMF)

.151 Wind scar of citrus: large silvery or tan blemishes on citrus fruit, which are actually produced during the first 12 weeks when

VI.9.3. POSTHARVEST HANDLING and MARKETING

still very small (less than 1 cm in diameter), from rubbing against a leaf or twig; fruit are virtually immune to such injury during later stages of development. (S & G)
.152 Wither tip: (see Endoxerosis of lemons, VI.9.3.043).
.153 York spot: (see Cork spot of apples, VI.9.3.037).
.154 Zebra skin of tangerines: the rupture of albedo and flavedo cells which produces a blackening of the peel over individual segments; caused by a sudden excess of soil water taken up by tangerine trees previously under a moderate-to-severe drought stress, then fruit are harvested before the cells have had time to lose their dangerously high turgor pressure (a matter of 5 to 7 days after the rain or irrigation). (S & G)

VI.10. Processing

.000 Processing: a general term for various means of preserving plant products for later consumption or use; canning of whole or parts of fruits or vegetables, concentration of juices, making of jellies, jams, pickles, chutneys, drying (dehydration) of whole or parts of fruits or vegetables, freeze drying, fermentation; includes the necessary steps to prepare coffee, tea, cacao, kola, coconut, oil palm, macadamia nut, cashew, rubber, and spices prior to shipment to market; excludes storage of fresh fruits and vegetables to extend their market availability and, according to the PACA definition, frozen fruits and vegetables.
.001 AVP evaporator: [see Plate evaporator, VI.10.010(d)].
.002 Barbeque: (see Kiln, VI.10.023).
.003 Beer: several meanings; (a) one of a class of alcoholic beverages brewed from malt (germinated barley or other grains) or malt substitute with the addition of hops to give a bitter taste; (b) an alcoholic beverage made from the fermentation of the roots or fruits, etc., of plants such as cassava, banana, yams, and the like in tropical regions; (c) artificially carbonated beverages resembling beer in color but usually more or less sweet, such as ginger ale and root beer.
.004 Brandy: a product obtained by distillation of almost any fermented fruit juice; refers to that made from wine if used without a modifier; e.g., peach brandy and apple jack.
.005 Can: a cylindrical metal (coated steel or aluminum), plastic, paper, or a combination of these materials, with a hermetically sealed lid (sometimes under vacuum or with an inert gas in the headspace); made in many different standard sizes and used for a wide assortment of products; also the act of preserving fruits or vegetables in sealed cans, jars, or bottles.
.006 Chilled juice: a nonfrozen juice product made directly from blended fresh juice, reconstituted frozen concentrate, or a mixture of the two, flash pasteurized, packaged in consumer-size containers, and held at or near 0°C; now home delivered or sold in grocery stores with dairy products or produce. (US-SG)
.007 Chutney: a spicy, highly seasoned condiment of Far Eastern origin, consisting of unripe, acid fruits cut up into chunks and blended with various spices, herbs, and vinegar into a thick mixture as a side dish for meats, fish, and the like. (BI)
.008 Concentration: the removal of water from fruit or vegetable juices; carried out commercially in multiple-effect evaporators under

high vacuum produced by steam ejectors to achieve removal of water at temperatures near or slightly below ambient; the usual practice is to overconcentrate, then add back fresh (single-strength) juice plus volatile constituents flashed off in the evaporation process, thus restoring essential fresh flavor to the product.

.009 Drying: the exposure of whole or parts of fruits or vegetables or other plant parts to heat to effect removal of sufficient moisture that the resulting product may be shipped or stored for later consumption; may be done by exposure to sunshine (considered as the best method for many products, e.g., fruits, coffee beans, cacao beans, and spices), in a kiln, as in coconut for copra, in a drying cabinet, as in many home fruits and vegetables, or in a drum drier. (OSDW)

.010 Evaporators: equipment in a processing plant used for removal of water from fruit or vegetable juices by evaporation; several different basic types, such as plate (AVP), low-pressure, or vacuum and flash evaporators, which are widely used for heat-sensitive fluids in which it is imperative to induce water vaporization without adverse effects on flavor and/or nutritionally important constituents: (a) flash evaporator: a system for "flashing off" water from a fruit or vegetable juice by (1) preheating to about 99°C, injecting clean steam to effect superheating, and passing the liquid down a single-pass heat exchanger where it flashes into a mixture of vapor and liquid; or (2) by heating to about 90°C, then flashing through a series of down-flow, single-pass evaporators with progressively lower pressures; juice can be concentrated to 65° Brix in about three minutes in an evaporator with seven stages and four effects, plus a flash cooler (known as the Thermally Accelerated Short-Time Evaporator = TASTE); (b) multiple-effect evaporator: a type of low-pressure (vacuum) evaporator in which juice is concentrated in a series of stages rather than in a single one; the hot vapor arising from the first stage is used to heat the second stage, the vapor from the second, for the third and so on, thus permitting use of heat energy from the steam more than once; this would not be the case in a single-effect evaporator; such a system may operate at 27°C (80°F) or lower; (c) low-pressure (vacuum) evaporator: a type of evaporator used for heat-sensitive fruit or vegetable juices in which high-pressure steam is injected into the evaporator and allowed to escape through a large ejector or "hog" jet to establish the initial vacuum; preheated juice is pumped or sucked into the heat exchanger, steam is admitted into the steam chest, boiling begins, and vapors are separated and removed continuously to maintain the vacuum; the latter is self-sustaining once the juice is boiling and vapor removal starts; (d) plate (AVP) evaporator: a type of double-effect evaporator in which a specially designed compact plate heat exchanger is used instead of the conventional evaporator with a tubular heat exchanger, thus permitting higher evaporating temperatures for each of the two effects; used in conjection with a flash cooler as the final stage when a concentrate without a cooked odor or taste (acceptable only for beverage base) is desired.

.011 Extractor (juice extractor): a machine used for extracting juice from fruit by pressure, cutting them in half, and reaming out the pulp, squeezing with mechanical fingers, and the like; may include some type of screen for separation of the juice from seeds, membranes, rind, and other nonjuice portions as an integral part of the extractor, the whole mass may be fed into a finisher, or large particles may be removed in the extraction procedure and smaller ones, in a finisher; (VI.10.014).

.012 Fermentation: the anaerobic process by which sugars in juice (e.g., of grapes) are converted to alcohol by yeasts to make wine; the further conversion into vinegar is an oxidation process of the alcohol that involves certain bacteria often present as contaminants; also the process by which cacao beans and adhering pulp, piled in a mass of at least 200 kg, heat up to about 50°C over a period of two to seven or eight days (depending on the type of cacao); the pulp liquefies and drains away and the beans undergo complex chemical and physical changes in the course of developing typical chocolate color, flavor, and aroma. (OSDW)

.013 "Fermentation": a strictly oxidative process that involves suitably withered and rolled or ground up tea leaves under conditions of moderate temperature and high humidity in which their constituents, notably polyphenols, undergo complex changes that produce the typical astringency, flavor, and aroma of the black tea of commerce. (OSDW)

.014 Finisher: a machine composed of a rotating screw inside a close fitting cylindrical or conical housing perforated as a screen; crude juice or puree from a pulper is fed via a hopper or external screw drive into the finisher which squeezes the juice through the perforations and the pulp, seeds, pieces of rind, etc., through the outlet end; the finisher screw is adjustable so that more or less pressure may be put on the mass of pulp according to the proportion of juice it is desired to extract and how much fine pulp is to be retained in the juice; finishers are commonly run in tandem, the first one to remove larger particles through larger perforations and less pressure and the second one with a fine mesh screen and more pressure to eliminate the smaller particles.

.015 Flash evaporator: [see Evaporator, VI.10.010(a)].

.016 Flash pasteurization: a type of pasteurization in which a thin film of juice is allowed to flow over a hot plate set at an angle; the period of contact is about 2 seconds, during which time the juice is heated to 82 to 85°C and cooled immediately thereafter; now widely used for concentrated and chilled juices; (see Pasteurization, VI.10.026).

.017 Freeze drying (lyophilizing): a process of drying in which a plant or plant part is frozen quickly and the moisture therein is removed by sublimation (direct evaporation of a solid without going through the liquid state) and high vacuum; originally developed as a laboratory procedure, now widely used commercially (e.g., freeze-dried coffee and many other products).

.018 Frozen concentrate: a juice product from which a substantial portion of the water has been removed by high-vacuum concentration (see Concentration, VI.10.008) and then frozen and packaged in consumer size or institutional containers or put up in drums for later use.

.019 Hot-pack concentrate: a juice product from which a substantial portion of the water has been removed by low-vacuum concentration rather than the high-vacuum process used for frozen concetrate; used mainly for manufacturing beverage base and the like.

.020 Jam: a thick, nearly solid, sweet flavorful product obtained by boiling whole or cut up fruit and sugar until the mass becomes

nearly or quite homogeneous; many kinds made from individual sorts of fruit or sometimes blended.
.021 Jelly: a product made in a manner similar to that for jam, except that fruit juice, strained to remove seeds, and sugar are boiled until the mixture will set when cooled; addition of pectin is often necessary to obtain proper solidification.
.022 Juice extractor: (see Extractor, VI.10.11).
.023 Kiln (barbeque): an enclosed structure with a grate on which halved coconut meats are exposed to heat from a low fire (dangerous) or current of hot air; many different versions. (OSDW)
.024 Low-pressure (vacuum) evaporator: [see Evaporator, VI.10.011(c)].
.025 Multiple-effect evaporator: [see Evaporator, VI.10.011(b)].
.026 Paste: a smooth food product made by evaporation or grinding, as in tomato or almond pastes.
.027 Pasteurization: the use of heat to kill bacteria and other harmful organisms in food products; the typical treatment for canned fruits, fruit juices, and vegetables is about 30 minutes at 55 to 60°C; (see Flash pasteurization, VI.10.016).
.028 Pickles: a product made from whole or chopped up vegetables, especially cucumbers, or unripe fruit mixed with brine or vinegar alone, or more often with various spices added.
.029 Pounds of juice: a common method of sale of citrus fruit (e.g., grapefruit) to be used for processing into single-strength (canned) juice. (S & G)
.030 Pounds-solids: the product of pounds of juice X percentage of total soluble solids, usually expressed on a per box (= 1 3/5 bushels or 90 lb = 40.8 kg for Florida oranges) basis; a common method of sale for fruit to be made into frozen concentrated orange juice. (S & G)
.031 Pulper: a machine that chops up whole fruits or vegetables and screens out the larger particles, including seeds if present, to form a puree that can be used directly or fed into a battery of finishers to obtain juice.
.032 Punch: an alcoholic or nonalcoholic beverage, variable in composition but usually based on wine or some sort of distilled liquor, fruit, soda, or fruit juices with sugar, lemon, and other ingredients added to form a flavorful combination.
.033 Puree: fruit or vegetable pulp strained through a sieve to remove the seeds, pieces of rind, etc.; may be cooked or not, depending on the product; ranges in viscosity from a thick juice to a mass that must be removed from the container with a spoon or spatula.
.034 Squash: a sweetened fruit juice drink, usually carbonated; thus similar to "soda pop" but sometimes of thicker consistency because of the presence of at least some pulp; one of a number of "thirst quenchers" developed in tropical regions, particularly British Commonwealth countries. (BI)
.035 TASTE: [see Flash evaporator, VI.10.010(a)].
.036 Thermally Accelerated Short-Time Evaporator: [see Flash evaporator, VI.10.010(a)].
.037 Vacuum evaporator: [see Evaporator, VI.10.011(C)].
.038 Vinegar: an acidulous liquid prepared by a two-stage process that consists first of the fermentation of a fruit juice and then oxidation of the alcohol produced to acetic acid; used for many purposes as a condiment or preservative.
.039 Wine: an alcoholic beverage made by fermenting fruit juice, most commonly from grapes or the sap of many other plants, notably palms, with or without additional sugar, and yeast, either natural or added from a culture. (OSDW)

SELECTED REFERENCES

Addicott, F. T. 1969. Ageing, senescence, and abscission in plants; Phytogerontology. HortScience 4:114-116.
Allinger, N. L., M. P. Cava, D. C. DeJongh, C. R. Johnson, N. A. Lebel, and C. L. Stevens. 1971. Organic chemistry. Worth Publishers, New York. (AL)
Alvim, P. de T. 1977. Cacao, p.279-313. In P. de T. Alvim and T. T. Kozlowski (Eds.). Ecophysiology of tropical crops. Academic Press, New York. (AV)
Alvim, P. de T. and T. T. Kozlowski, Eds. 1977. Ecophysiology of tropical crops. Academic Press, New York.
American Society of Heating, Refrigerating and Air Conditioning Engineers. 1981. 1981 Refrigeration fundamentals. Chap. 35. ASHRAE, Atlanta, GA. (ASHRAE)
Ashby, B. H. 1970. Protecting perishable foods during transport by motortruck. U. S. Dept. Agr., Agr. Handbk. 105 (revised). (AY)
Association of Official Seed Analysts. 1970. Rules for testing seeds. Proc. Assn. Official Seed Anal. 60 (No.2):1-116. (AS)
Atkins, P. W. 1978. Physical chemistry. W. H. Freeman & Co., New York. (AT)
Bailey, L. H. 1928. The standard cyclopedia of horticulture, 3rd ed., The Macmillan Co., New York, 6(3) vol. (B-M)
Bailey, L. H. 1934. 6. American palmettoes. Gent. Herb. III, fasc. VI:277. (B-GH)
Bailey, L. H. 1935. 8. The king palms of Australia-Archontophoenix. Gent. Herb. III, fasc. VI:397,405. (B-GH)
Bailey, L. H. 1936. 1. The arecastrums. 2. The butias. Gent. Herb. IV, fasc. I:10,27. (B-GH)
Bailey, L. H. 1941. Species Batorum. The genus Rubus in North America I. Gent. Herb. V, fasc. I:2,9,12,15. (B-GH)
Bailey, L. H. 1944a. Revision of the palmettoes. Gent. Herb. VI, fasc. VII:371,377. (B-GH)
Bailey, L. H. 1944b. 9. Rubus on Hispaniola. Gent. Herb. VI, fasc. VI:347. (B-GH)
Bailey, L. H. 1944c. 10. Rubi of Panama. Gent. Herb. VI, fasc. VI:360. (B-GH)
Bailey, L. H. 1945. Species Batorum. The genus Rubus in North America IX. Arguti. European species introduced. Gent. Herb. V, fasc. IX:649, 657,671,750,795,817. (B-GH)
Bailey, L. H. 1947. 4. Species studies in Rubus. Species Batorum-Addendum I. Gent. Herb. VIII, fasc. III:337. (B-GH)
Bailey, L. H. 1949. New and adopted names. 1. Discussions in botanical names of cultivated plants. Gent. Herb. VIII, fasc. I:42. (B-GH)
Bailey Hortorium Staff, L. H. (comp.). 1976. Hortus third. The Macmillan Co., New York.
Baker, K. F., Ed. 1957. The U. C. system for producing healthy containergrown plants. California Agr. Exp. Sta. Man. 23. (BA)

*Letters in parentheses are author abbreviations used in text; see also Acknowledgments.

SELECTED REFERENCES

Bartholomew, D. P. and S. B. Kadzimin. 1977. Pineapple. In P. de T. Alvim and T. T. Kozlowski (Eds.). Ecophysiology of tropical crops. Academic Press, New York. (B & K)

Ball, G. J., 1973. The Ball red book, 12th ed. George J. Ball, West Chicago, IL.

Bell, P. and D. Coombe, translators. 1976. Strasburger's textbook of botany (English ed.). Gustav Fischer Verlag, Stuttgart, and Longman Group, London. (B & C)

Berry, Z. C. 1972. Trailer on flat car fruit and vegetable tariff. Freight Tariff 910-J. Southern Freight Tariff Bureau, Atlanta, GA. (BE)

Block, W. E. 1968. Container and loading rules tariff. Freight tariff 823-E. Southern Freight Tariff Bureau, Atlanta, GA. (BL68)

Block, W. E. 1969. Florida citrus fruit tariff. Southeastern Freight Tariff Bureau, Atlanta, GA. (BL69)

Bowman, E. K., A. H. Spurlock, S. Hedden, and W. Grierson. 1971. Modernizing handling systems for Florida citrus from picking to packing line. U.S. Dept. Agr. Mkt. Res. Rept. 915.

Braudeau, Jean. 1969. Le cacaoyer. G.-P. Maisonneuve & Larose, Paris. (BR)

Britton, N. L. and H. C. Brown. 1923. An illustrated flora of the United States, Canada and the British possessions. New York Botanical Gardens, Bronx Park, New York. Vol. I, p.7. (B & B)

Bryan, H. H., P. H. Everett, L. N. Shaw, K. Pohronezny, V. H. Waddill. 1980. Cultural management for machine harvest of market tomatoes on mulched beds. Proc. Fla. State Hort. Soc. 93:227-230. (BN)

Buchanan, R. E., R. St. John-Brooks, and R. S. Breed. 1948. International bacteriological code of nomenclature. J. Bact. 55:287. (BH)

Burkill, I. H. 1966. A dictionary of the economic products of the Malay peninsula. Ministry of Agriculture and Co-operatives, Kuala Lumpur, Malaysia. (BI)

Burley, J. 1976. Genetic systems and genetic conservation of tropical pines. p.85-100. In J. Burley and B. T. Styles (Eds.). Tropical trees. Variation, breeding and conservation. Linnean Soc. (London) Symp. Ser. 2. (Academic Press, New York). (BU)

Burley, J. and B. T. Styles, Eds. 1976. Tropical trees. Variation, breeding and conservation. Linn. Soc. Symp. Ser. No. 2:1-243. (Linn. Soc. London) (Academic Press, New York).

Bush, C. S. 1973. Grades and standards for nursery plants, Part I, 3rd ed. Division of Plant Industry, Fla. Dep. Agr. & Consumer Serv., Gainesville, FL.

Bush, R. 1943. Tree fruit growing. Lawrence Pollinger, London. (BS)

Carlisle, V. W. and J. NeSmith. 1973. Florida soil identification handbook. University of Florida, Soil Science Department and U.S. Dept. of Agriculture, Soil Conservation Service, Gainesville, FL. (C & N)

Carnell, D. and K. Thompson. 1980. Experiences with photobiodegradable mulches in Florida. Proc. Fla. State Hort. Soc. 93:371. (CA)

Champion, J. 1963. Le Bananier. G.-P. Maisonneuve et Larose, Paris. (CH)

Chapman, G. P. 1964. Pollination and the yields of tropical crops: An appraisal. Euphytica 13(1964):187-197. (CP-G)

Chapman, H. D. and P. F. Pratt. 1961. Methods of analysis for soils, plants and water. Agricultural Sciences Division, University of California, Berkeley, CA. (CP-H)

Chesson, J. H. and P. F. Burkner. 1977. Citrus harvesting in California and Arizona. Proc. Int. Soc. Citriculture 2-397-402.

Childers, N. F. 1975. Modern fruit science, 6th ed. Horticultural Publications, New Brunswick, NJ. (CS)

Christie, J. R. 1959. Plant nematodes. Their bionomics and control. Fla. Agr. Exp. Sta., Gainesville, FL. 32611.

SELECTED REFERENCES

Christopher, E. P. 1957. The pruning manual. The Macmillan Co., New York. (CI)
Cohen, M. 1981. Beneficial effects of viruses for horticultural plants. p. 394-411. In Jules Janick (Ed.). Horticultural reviews, Vol. 3. AVI Publishing Co., Westport, CT. (CO)
Commonwealth Mycological Institute. 1968. Plant pathologist's pocketbook. Commonwealth Mycological Institute, Kew, England. (CMI)
Condit, I. J. 1947. The fig. Chronica Botanica Co., Waltham, MA. (John Wiley & Sons, New York). (CN)>
Cook, J. G. 1969. ABC of plant terms. Merrow Publishing Co., Watford, England.
Coppock, G. E. and S. L. Hedden. 1977. Citrus harvesting in Florida. Proc. Int. Soc. Citriculture 2:393-397.
Corner, E. J. H. 1966. The natural history of palms. University of California Press, Berkeley, CA. (CR66)
Corner, E. J. H. 1976. The seeds of dicotyledons. (2 vol.) Cambridge University Press, New York. (CR76)
Coste, R. 1968. Le cafeier. G.-P. Maisonneuve & Larose, Paris. (CT)
Coulter, J. M. and C. J. Chamberlain. 1903. Morphology of angiosperms. D. Appleton & Co., New York.
Crane, J. C. 1969. The role of hormones in fruit set and development. HortScience 4:108-111.
Crocker, T. E. and R. L. Phillips. 1975. Maintenance hedging and topping of Florida citrus. Fla. Coop. Extens. Serv. Cir. 388.
Cuatrecasas, J. 1964. Cacao and its allies. A taxonomic revision of the genus Theobroma. Contrib. U.S. Nat. Herb. 35(6):379-614. (CU)
Darwin, Charles. 1892. The various contrivances by which orchids are fertilized by insects, 2nd rev. ed. D. Appleton & Co., New York. (DR)
Davis, J. H., Jr. 1943. The natural features of southern Florida. Fla. Geological Bul. 25. Tallahassee, FL.
Davis, P. H. and V. H. Heywood. 1965. Principles of angiosperm taxonomy. D. Van Nostrand Co., New York.
Dijkman, M. J. 1951. Hevea. Thirty years of research in the Far East. University of Miami Press, Coral Gables, FL. (DK)
Dilley, D. R. 1969. Hormonal control of fruit ripening. HortScience 4:111-114.
Du Breuil, M. S. 1876. Culture des arbres et arbrisseaux. Paris. (DB)
Dunn, R. A., Ed. 1979. Florida nematode control guide. Department of Entomology & Nematology, University of Florida, Gainesville, FL. (DU)
Dye, D. W., J. F. Bradbury, M. Goto, A. C. Hayward, R. A. Elliott, and M. N. Schroth. 1980. International standards for naming of phytopathogenic bacteria and a list of pathovar names and pathotype strains. Rev. Plant Pathology 59(4):153-165. (DY)
Eames, A. J. 1961. Morphology of the angiosperms. McGraw-Hill Book Co., New York. (EA)
Eames, A. J. and L. H. MacDaniels. 1947. An introduction to plant anatomy, 2nd ed. McGraw-Hill Book Co., New York. (E & M)
Echlin, P. 1968. Pollen. Sci. Amer. 218(April):80-90. (EC)
Egler, F. E. 1948/49. The terminology of floral types. Chronica Bot. 14(4/6):169-173.
Entomology and Nematology Department, University of Florida, 1978. Florida insect control guide. Fla. Coop. Extens. Serv., Gainesville, FL.(UF)
Epstein, E. 1973. Roots. Sci. Amer. 228(May):48-58.
Esau, K. 1965. Plant anatomy, 2nd ed. John Wiley & Sons, New York. (ES)
Faegri, K. and L. van der Pijl. 1979. The principles of pollination ecology, 3rd ed. Pergamon Press, New York. (F & P)
Farm Chemicals. 1983. Plant food dictionary. Meister Publishing Co., Wil-

loughby, OH. (FC)
Florida Department of Agriculture. 1968. Certification seed law. Rules and regulations. Chap. 575, Florida Statutes. Fla. Dept. Agr., Tallahassee, FL. p. 2809-2810, 215-224B.
Florida Department of Agriculture and Consumer Services. 1970. Florida seed law. Rules and regulations. Chap. 578, Florida Statutes. Fla. Dept. Agr. & Consumer Services, Tallahassee, FL. p. 2880-2887, 211-214B.
Florida Department of Citrus. 1984. Florida citrus fruit laws (Chapter 601, Florida Statutes). Fla. Dept. Citrus, Lakeland, FL.
Florida Department of Citrus. 1984. Official rules affecting the Florida citrus industry. Fla. Dept. Citrus, Lakeland, FL. (FDOC)
Foster, A. S. and E. M. Gifford, Jr. 1974. Comparative morphology of vascular plants, 2nd ed. W. H. Freeman & Co., New York. (F & G)
Freeman, Brian. 1978. Cuticular waxes of developing leaves and fruits of citrus and blueberry: Ultrastructure and chemistry. Ph.D. dissertation, University of Florida, Gainesville, FL. (FR)
Fremond, Y., R. Ziller, and M. de L. de Lamothe. 1963. Le cocotier. Maisonneuve et Larose, Paris. (FZL)
Fridley, R. B. and P. A. Adrian. 1966. Mechanical harvesting equipment for deciduous tree fruits. Calif. Agr. Exp. Sta. Bul. 825. (FD)
Fridley, R. B., P. A. Adrian, L. L. Claypool, A. D. Rizzi, and S. J. Leonard. 1971. Mechanical harvesting of cling peaches. Calif. Agr. Expt. Sta. Bul. 851. (FD)
Friedman, B. A. and W. A. Radspinner. 1956. Vacuum-cooling fresh fruits and vegetables. U.S. Dept. Agr. AMS-107. (F & R)
Frost, H. B. and R. K. Soost. 1968. Seed reproduction: Development of gametes and embryos. Chap. 4. In W. Reuther, L. D. Batchelor and H. J. Webber (Eds.). The citrus industry, Vol. II. Division of Agricultural Sciences, University of California, Berkeley, CA.
Geraldson, C. M. 1950. Importance of water control for tomato production using the gradient mulch system. Proc. Fla. State Hort. Soc. 93:278-279. (GE)
Gilmour, J. S. L., F. R. Horne, E. L. Little, Jr., F. A. Stafleu, and R. H. Richens, Eds. 1969. International code of nomenclature of cultivated plants - 1969. International Bureau for Plant Taxonomy of the International Association for Plant Taxonomy, Utrecht, The Netherlands. (GL)
Goldman, D. T. and R. J. Bell, Eds. 1981. The international system of units (SI). U.S. Dept. Commerce, National Bur. Stds. NBS Special Pub. 330 (4th ed.). (G & B)
Gomez, R. E. 1971. Anatomical aspects of avocado stems and their relation to rooting. Ph.D. dissertation, Dept. of Fruit Crops, University of Florida, Gainesville, FL.
Gothein, M. L. 1928. A history of garden art. J. M. Dent & Sons, London. (GO)
Goto, Y. B. and E. T. Fukunaga. 1956a. Coffee. Care of the mature orchard. Hawaii Agr. Extens. Cir. 358. (G & F)
Goto, Y. B. and E. T. Fukunaga. 1956b. Coffee. Care of the young orchard. Hawaii Agr. Extens. Cir. 357. (G & F)
Granados, R. R. 1976. Infection and replication of insect pathogenic viruses in tissue cultures. Adv. in Virus Res. 20:189-236. Academic Press, New York. (GR)
Grant, Verne. 1951. The fertilization of flowers. Sci. Amer. 184:52-56.
Gray, Asa. 1862. Introduction to structural and systematic botany, 5th ed. Ivison, Phinney & Co., New York. (G62)
Gray, Asa. 1868. Gray's school and field book of botany. Ivison, Blakeman,

Taylor & Co., New York. (G68)

Gray, D. 1981. Fluid drilling of vegetable seeds. p. 1-27. In Jules Janick (Ed.). Horticultural reviews. Vol. 3. AVI Publishing Co., Westport, CT. (G-D)

Grierson, W. (1981). Physiological disorders of citrus fruits. 1981. Proc. Internat. Soc. Citriculture (Paper no. 622). 22 p.

Grierson, W., W. M. Miller, and W. F. Wardowski. 1978. Packingline machinery for Florida citrus packing houses. Fla. Agr. Exp. Sta. Bul. 803.

Grierson, W., J. Soule, and K. Kawada. 1982. 8. Beneficial effects of physiological stress. p. 247-271. In Jules Janick (Ed.). Horticultural reviews. Vol. 4. AVI Publishing Co., Westport, CT. (GSK)

Growth Chanber Working Group of the American Society for Horticultural Science. 1980. Guidelines for measuring and reporting the environment for plant studies. HortScience 15(6):719-720. (GCWG)

Guillou, R. 1960. Coolers for fruits and vegetables. Calif. Agr. Exp. Sta. Bul. 773. (GU)

Gurney, H. P. and J. Lurie. 1923. Charts for estimating temperature distributions in heating or cooling solid shapes. Indus. Eng. Chem. 15(11):1170-1172.

Haarer, A. E. 1964. Modern banana production. Leonard Hill, London.

Hallé, F., R. A. A. Oldeman and P. B. Tomlinson. 1978. Tropical trees and forests: An architectural analysis. Springer-Verlag, Berlin. (HOT)

Hardy, F., Ed. 1960. Cacao manual. Inter-American Institute of Agricultural Sciences, Turrialba, Costa Rica. (HY)

Harlan, J. R. 1975. Crops and man. American Society of Agronomy, Madison, WS. (HN)

Harrison, D. E. 1977. Energy management in irrigation. Fla. Coop. Extens. Serv. EC-12. (HAR)

Harrison, D. S. 1978. Irrigation systems for crop production in Florida. University of Florida, IFAS, Water Resources Council WRC-8. (HAR)

Harrison, D. S. and R. C. J. Koo. 1978. Irrigation methods and equipment for production of citrus in Florida. University of Florida, IFAS, Water Resources Council WRC-10. (HAR)

Harrison, D. S., J. R. Orsenigo, R. F. Brooks, and P. J. Jutras. 1967. Calibration of pesticide applicators. Fla. Agr. Extens. Serv. Circ. 275B. (HOBJ)

Hartley, C. W. S. 1977. The oil palm, 2nd ed. Longman Group, London. (HT)

Hartmann, H. T. and D. E. Kester. 1975. Plant propagation, 3rd ed. Prentice-Hall, Englewood Cliffs, NJ. (H & K)

Harvey, J. M., W. L. Smith, Jr., and J. Kaufman. 1972. Market diseases of stone fruits: Cherries, peaches, nectarines, apricots and plums. U.S. Dept. Agr., Agr. Handbk. 414. (HSK)

Hayward, H. E. 1938. The structure of economic plants. The Macmillan Co., New York. (HY)

Heald, F. D. 1933. Manual of plant diseases, 2nd ed. McGraw-Hill Book Co., New York. (HE)

Hedden, S. L., J. D. Whitney, and G. E. Coppock. 1977. Mechanical systems to harvest citrus fruit for juice processing plants in Florida. Proc. Int. Soc. Citriculture 2:418-423.

Hedrick, U. P. 1917. The peaches of New York. N.Y. Dept. Agr., Albany, NY. (HD)

Hedrick, U. P. 1921. The pears of New York. N.Y. Dept. Agr., Albany, NY. (HD)

Hedrick, U. P. 1925. The small fruits of New York. N.Y. Dept. Agr., Albany, NY. (HD)

Hertogh, A. A. De. 1975. A need for uniformity in terminology and data requirements for research on bulbous species. p. 39-44. In A. R. Rees and H. H. van der Borg (Eds.). Second international symposium on flower bulbs. Acta Hort. 47:1-446. (HR)

Hertogh, A. A. De. 1981. Holland bulb forcer's guide. Netherlands Flower-Bulb Institute, New York. (HR)

Hertogh, A. A. De, A. N. Roberts, N. W. Stuart, R. W. Langhans, R. H. Lawson, H. W. Wilkins, and D. C. Kiplinger. 1971. A guide to terminology for the Easter lily (Lilium longiflorum Thunb.). HortScience 6(2):121-123. (HR)

Hewitt, E. J. 1966. Sand and water culture methods used in the study of plant nutrition. Commonwealth Bur. Hort. & Plantation Crops Tech. Commun. 22 (revised 2nd ed.). (HW)

Hill, J. B., L. O. Overholts, H. W. Popp, and A. R. Grove, Jr. 1960. Botany, 3rd ed. McGraw-Hill Book Co., New York. (HI)

Hilltop Orchards and Nurseries. 1982. 1982 Catalog. Hilltop Orchards & Nurseries, Hartford, MI. (HON)

Holman, R. M. and W. W. Robbins. 1927. A textbook of general botany. John Wiley & Sons, New York.

Holmes, M. G., W. H. Klein and J. Sager. 1985. Photons, flux and some light on philology. HortScience 20(1):29-31. (HKS)

Horticultural Education Assn. Fruit Committee. 1955. A glossary of terms used in pruning fruit trees. Scientific Horticulture 11:67-74. (SI)

Hume, H. H. 1907. Citrus fruits and their culture, 2nd ed. Orange Judd Co., New York. (HM)

Hume, H. H. 1926. The cultivation of citrus fruits. The Macmillan Co., New York. (HM)

Hutchins, W. A. 1967. Irrigation water rights in California. Calif. Agr. Exp. Sta. Cir. 452. (HC)

Hutchinson, John. 1969. Evolution and phylogeny of flowering plants. Academic Press, London. (H69)

Hutchinson, John. 1973. The families of flowering plants, 3rd ed. Oxford University Press, London. (H73)

Isenberg, F. M. R. 1979. Controlled atmosphere storage of vegetables. In Jules Janick (Ed.). Horticultural reviews. vol. 1. AVI Publishing Co., Westport, CT. (I)

Jakob, M. and G. A. Hawkins. 1958. Elements of heat transfer, 3rd ed. John Wiley & Sons, New York. (JK)

Jamison, W. T. (agent). 1968. Perishable protective tariff. National Perishable Freight Committee, Chicago, IL. (JM)

Janick, J. 1972. Horticultural science, 2nd ed. W. H. Freeman & Co., New York. (J)

Janzen, D. H. 1976. Why do bamboos wait so long to flower? In J. Burley and B. T. Styles (Eds.). Tropical trees. Variation, breeding and conservation. p. 135-139. Linnean So. (London) Symp. Ser. 2. (Academic Press, New York.) (JZ)

Jaynes, R. A. 1979. Handbook of North American nut trees. Northern Nut Growers Assn., Hamden, CT. (JY)

Jensen, W. A. 1973. Fertilization in flowering plants. Bioscience 23:21-26. (JN)

Joiner, J. N., Ed. 1981. Foliage plant production. Prentice-Hall, Englewood Cliffs, NJ. 614 p. (JR)

Johnson, Thomas. 1633. John Gerard's the herbal or general history of plants (revised and enlarged). Adam Islip, Joice Norton & Richard Whitakers, London. (Unabridged ed. of original published by Dover Publications, New York, 1975.) (JH)

Jong, K. 1976. Cytology of the Dipterocarpaceae. p. 79-84. In J. Burley and B. T. Styles (Eds.). Tropical trees. Variation, breeding and conservation. Linnean Soc. (London) Symp. Ser. 2. (Academic Press,

New York.) (JG)

Khokhlov, S. S., Ed. 1976. Apomixis and breeding. Amerind Publishing Co. Pvt. Ltd., New Delhi. (Printed for Agr. Res. Serv., U.S. Dept. Agr. and National Science Foundation) (English translation from Russian.) U.S. Dept. Commerce, National Technical Information Service, Springfield, VA.(KH)

Kramer, P. J. 1969. Plant and water relationships: A modern synthesis. McGraw-Hill Book Co., New York. (KR)

Kramer, P. J. 1980. Drought stress and the origin of adaptation. In N. C. Turner and P. J. Kramer (Eds.). Adaptation of plants to water and high temperature stress. John Wiley & Sons, New York. (T & K - KR)

Kramer, P. J. 1983. Water relations of plants. Academic Press, New York. (KR)

Krizek, D. T. 1977. Reviewers comments. HortScience 12(5):438. (KZ)

Krizek, D. T. and J. C. McFarlane. 1983, 1984. Controlled-environment guidelines. HortScience 18(5):662-664; 19(1):7. (K & M).

Kucharek, T., G. Simons, and R. S. Mullin, Eds. 1978. Florida plant disease control guide. Plant Pathology Dept., University of Florida, Gainesville, FL. (KSM)

Larcher, W. 1980. Physiological plant ecology, 2nd ed. Springer-Verlag, Berlin. (LA)

Larson, R. A., Ed. 1980. Introduction to floriculture. Academic Press, New York. (LR)

Lawrence, G. H. M. 1951. Taxonomy of vascular plants. The Macmillan Co., New York. (L51)

Lawrence, G. H. M. 1955. An introduction to plant taxonomy. The Macmillan Co., New York. (L55)

Lehninger, A. L. 1970. Biochemistry. Worth Publishers, New York. (LE70)

Lehninger, A. L. 1975. Biochemistry, 2nd ed. Worth Publishers, New York. (LE75)

Levitt, J. 1972. Responses of plants to environmental stresses. Academic Press, New York.

Lind, E. M. and M. E. Morrison. 1974. East African vegetation. Longman Group, London. (L & M)

Little, E. L., Jr. and F. H. Wadsworth. 1964. Common trees of Puerto Rico and the Virgin Islands. U.S. Dept. Agr., Agr. Handbk. 249. (L & W)

Lloyd, F. E. 1942. The carnivorous plants. Chronica Botanica Co., Waltham, MA. (John Wiley & Sons, New York.) (LL)

Longman, K. A. and J. Jenik. 1974. Tropical forest and its environment. Longman Group, London. (L & J)

Lubbock, Sir John. 1894. Flowers, fruits, leaves. Macmillan & Co., London. (LB)

Lucas, Fr. 1909. De Lehre von Baumschnitt. Stuttgart. (LS)

Lyon, T. L. and H. O. Buckman. 1946. The nature and properties of soils, 4th ed. The Macmillan Co., New York. (L & B)

Lutz, J. M. and R. E. Hardenburg. 1968. The commercial storage of fruits, vegetables, and florist and nursery stocks, rev. U.S. Dept. Agr., Agr. Handbk. 66. (L & H)

Maerz, A. and M. R. Paul. 1930. A dictionary of color. McGraw-Hill Book Co., New York.

Magness, J. R., G. M. Markle and C. C. Compton. 1971. Food and feed crops of the United States. New Jersey Agr. Exp. Sta. Bul. 828. (MMC)

Maheshwari, P. 1950. An introduction to the embryology of angiosperms. McGraw-Hill Book Co., New York. (MH)

Manshard, W. 1974. Tropical agriculture. Bibliographisches Institut AG, Mannheim. (Longman Group, London.) (MS)

Markeson, C. B. 1959. Pooling and other grower payment methods as used by

local fruit, vegetable and tree nut cooperatives. U.S. Dept. Agr., Farmer Coop. Serv., Gen. Rept. 67. (MK)
Maynard, D. N. and O. A. Lorenz. 1979. 2. Controlled-release fertilizers for horticultural crops. p. 79-140. In J. Janick (Ed.). Horticultural Reviews, Vol. 1. AVI Publishing Co., Westport, CT. (M & L)
Mazzone, H. M. and G. H. Tignor. 1976. Insect viruses: Serological relationships. Advances in Virus Res. 20:237-270. (Academic Press, New York. (MZ)
McAdams, W. H. 1942. Heat transmission, 2nd ed. McGraw-Hill Book Co., New York. (MA)
McCain, A. H., R. D. Raabe, and Stephen Wilhelm. 1979. Plants resistant to verticillium wilt, rev. Calif. Coop. Extens. Serv. Leaflt. 2703.
McColloch, L. P., H. T. Cook, and W. R. Wright. 1968. Market diseases of tomatoes, peppers, and eggplants, rev. U.S. Dept. Agr., Agr. Handbk. 28. (MCW)
McGregor, S. E. 1976. Insect pollination for cultivated crop plants. U.S. Dept. Agr., Agr. Handbk. 496. (MG)
Merrill, E. D. 1945. Plants of the Pacific world. The Macmillan Co., New York. (MR)
Metcalf, C. L. and W. P. Flint. 1939. Destructive and useful insects, 2nd ed. McGraw-Hill Book Co., New York.
Millspaugh, C. F. 1892. American medicinal plants. J. C. Yorston & Co., Philadelphia. 2 vol. (Unabridged ed., Dover Publications, New York. 1974.) (ML)
Mitchell, C. A. and H. C. Dostal. 1977. Light intensity, foot-candles and lux are obsolete terms. HortScience 12(5):437. (M & D)
Mohr, E. C. J. and F. A. van Baren. 1954. Tropical soils. Uitgeverij W. van Hoeve, The Hague, The Netherlands.
Monteith, J. L. 1977. Climate. p. 1-27. In P. de T. Alvim and T. T. Kozlowski (Eds.). Ecophysiology of tropical crops. Academic Press, New York. (MO)
Mosse, B. 1962. Graft incompatibility in fruit trees. Commonwealth Bur. Hort. & Plantation Crops Tech. Commun. 28. (ME)
Muenscher, W. C. 1955. Weeds, 3rd ed. The Macmillan Co., New York. (MU)
Munsell, A. H. 1963. Munsell book of color, cabinet ed. Munsell Color Co., Baltimore, MD.
Nanthachai, P. 1976. Comparative studies of rootstock-scion combinations of citrus with special reference to anatomical structure. Ph.D. dissertation, Dept. of Fruit Crops, University of Florida, Gainesville, FL.
Nybom, Nils. 1961. Mutation and plant breeding. Nat. Acad. Sci. Nat. Res. Council Publ. 891:252-294. (N)
O'Brien, T. P. and M. E. McCully. 1981. The study of plant structure. Principles and selected methods. Termarcarphi Pty., Ltd., Wantirna, Victoria, Austrailia. (O & M)
Ochse, J. J. 1931a. Fruits and fruit culture in the Dutch East Indies. G. Kolff & Co., Djakarta. (OJ1)
Ochse, J. J. 1931b. Vegetables of the Dutch East Indies. Kementerian Pertanian, Djakarta. (OJ1)
Ochse, J. J., M. J. Soule, Jr., M. J. Dijkman, and C. Wehlburg. 1961. Tropical and subtropical agriculture. The Macmillan Co., New York. 2 vol. (OSDW)
Oosting, H. J. 1948. The study of plant communities. W. H. Freeman & Co., New York. (OS)
Osman, A. M. A., W. Reuther, and L. C. Erickson. 1974. Xenia and metaxenia studies in the date palm *Phoenix dactylifera* L. 51st. Date Growers

SELECTED REFERENCES

Institute 51:6-16.
Page, C. H. and P. Vigoureux, Eds. 1974. The international system of units (SI). U.S. Nat'l. Bur. Stand. Spec. Publ. 330. (P & V)
Pantastico, E. B. 1975. Postharvest physiology, handling and utilization of tropical and subtropical fruits and vegetables. AVI Publishing Co., Westport, CT. (PA)
Peairs, L. M. and R. H. Davidson. 1956. Insect pests of farm, garden, and orchard, 5th ed. John Wiley & Sons, New York. (P & D)
Phillips, R. L. 1980. Principles and practices for Florida citrus. Fla. Coop. Extens. Serv. Cir. 477.
Pierson, C. F., M. J. Ceponis, and L. P. McCulloch. 1971. Market diseases of apples, pears, and quinces. U.S. Dept. Agr., ARS, Agr. Handbk. 376. (PCM)
Pijl, L. van der 1972. Principles of dispersal in higher plants, 2nd ed. Springer-Verlag, Berlin, New York. (PJ)
Poe, S. L. 1981. An overview of integrated pest management. HortScience 16:501-506.
Praloran, J. C. 1971. Les agrumes. C.-P. Maisonneuve & Larose, Paris. (PR)
Prevatt, J. W., C. D. Stanley, and W. E. Waters. 1980. Evaluation of a water conveyance and recovery system for seep irrigation. Proc. Fla. State Hort. Soc. 93:252-256. (PV)
Proebsting, E. L., Jr. 1970. Relation of fall and winter temperatures to flower bud behavior and wood hardiness of deciduous fruit trees. HortScience 5:422-424.
Purseglove, J. W. 1968. Tropical crops. Dicotyledons. 2 vol. Longman Group, Harlow, UK. (PG68)
Purseglove, J. W. 1972. Tropical crops. Monocotyledons. 2 vol. Longman Group, Harlow, UK. (PG72)
Py, C. and M.-A. Tisseau. 1965. L'Ananas. Maisonneuve et Larose, Paris. (P & T)
Radford, A. E., W. C. Dickison, J. R. Massey, and C. R. Bell. 1974. Vascular plant systematics. Harper & Row, New York. (RD)
Ramsey, G. B. and M. A. Smith. 1961. Market diseases of cabbage, cauliflower, turnips, cucumbers, melons and related crops. U.S. Dept. Agr., Agr. Handbk. 184. (R & S)
Ramsey, G. B., B. A. Friedman, and M. A. Smith. 1959. Market diseases of beets, chicory, endive, escarole, globe artichokes, lettuce, rhubarb, spinach and sweet potatoes. U.S. Dept. Agr., Agr. Handbk. 155. (RFS)
Redit, W. H. 1969. Protection of rail shipments of fruits and vegetables. U.S. Dept. Agr., Agr. Handbk. 195 (revised). (RT)
Rees, A. R. 1972. The growth of bulbs. Academic Press, London. (RE)
Reuther, W., Ed. The citrus industry, rev. ed., Vol. III. Division of Agricultural Science, University of California, Berkeley, CA.
Richards, P. W. 1952. The tropical rain forest. Cambridge University Press, New York. (RI)
Ridgway, R. 1912. Color standards and nomenclature. A. Hoen & Co., Baltimore, MD. (distributor).
Risch, C. Ed. 1981. Produce glossary. p. A-5-A-12. In C. Risch (Ed.). The Packer's 1981 produce availability and merchandising guide. The Packer, Shawnee Mission, Kansas. (RS)
Roberts, D. A. 1978. Fundamentals of plant pest control. W. H. Freeman & Co., New York. (R)
Roberts, D. A. and C. W. Boothroyd. 1972, 1975. Fundamentals of plant pathology. W. H. Freeman & Co., New York. (RB)
Robbins, W. W. 1924. The botany of crop plants, 2nd ed. P. Blakiston's Son & Co., Wynnewood, PA. (RB24)

SELECTED REFERENCES

Rosengarten, F., Jr. 1969. The book of spices. Livingston Publishing Co., Wynnewood, PA.
Royal Horticultural Society. 1950. Fruit year book 1950. Royal Horticultural Society, London. (RHS)
Ruehle, G. D. 1958. Miscellaneous tropical and subtropical crops. Fla. Agr. Extens. Bul. 156A. (RH)
Ruehle, G. D. 1963. The Florida avocado industry. Florida Agr. Exp. Sta. Bul. 602. (RH)
Ruehle, G. D. and R. B. Ledin. 1960. Mango growing in Florida. Fla. Agr. Extens. Bul. 174. (R & L)
Russell, E. W. 1973. Soil conditions and plant growth, 10th. ed. Longman Group, Harlow, UK. (RU)
Sachs, M., D. J. Cantliffe, and J. T. Watkins. 1980. Germination of pepper at low temperatures after various pretreatments. Proc. Fla. State Hort. Soc. 93:258-260.
Sachs, R. M. and W. P. Hackett. 1969. Control of vegetative and reproductive development in seed plants. HortScience 4:103-107.
Salisbury, F. B. and C. W. Ross. 1978. Plant physiology, 2nd ed. Wadsworth Publishing Co., Belmont, CA. (S & R)
Sanford, W. G. 1962. Pineapple crop log - concept and development. Better Crops with Plant Food. p. 32-42. American Potash Institute, Washington, DC. (SN)
Sargent, C. S. 1905. Manual of the trees of North America. Houghton Mifflin Co., Boston, MA. (SA)
Savage, E. F. 1970. Cold injury as related to cultural management and possible protective devices for dormant peach trees. HortScience 5:425-428.
Savage, M. J. 1978. Modifications for guidelines in reporting studies in controlled environment chambers. HortScience 13(3):217-218. (SE78)
Savage, M. J. 1979. Use of the international system of units in the plant sciences. HortScience 14(4):492-495. (SE79)
Schneider, G. 1893, 1894. The book of choice ferns. L. Upcott Gill, London. Vol. II, p. 221; Vol. III, p. 36. (SR)
Schoen, J. F. and L. C. Shenberger. 1973. Growth performance tests for the presence of Rhizobia on preinoculated seed. Contrib. No. 30 to Handbook on Seed Testing. Association of Official Seed Analysts.
Schuler, S. 1967. America's great private gardens. The Macmillan Co., New York. (SC)
Science Kit & Boreal Laboratories. 1980. Turtox key card for pollen grains. Science Kit & Boreal Laboratories, Tonawanda, NY. (SKB)
Shaw, L. N., H. H. Bryan, and D. A. Nichols. 1980. An intermittent peristaltic fluid drill for vegetables. Proc. Fla. State Hort. Soc. 93: 256-257. (SW)
Sheehan, T. and M. Sheehan, 1979. Orchid genera illustrated. Van Nostrand Reinhold Co., New York. (SS)
Shepherd, R. J. 1976. DNA viruses of higher plants. Advances in Virus Res. 20:305-339. Academic Press, New York. (SP)
Simmonds, N. W. 1976. Evolution of crop plants. Longman Group, London. (SM76)
Simmonds, N. W. 1979. Principles of crop improvement. Longman Group, London. (SM79)
Smajstrala, A. G. and D. S. Harrison. (nd). Measurement of soil water for irrigation management. Florida Coop. Extens. Serv. Cir. 532. (S & H)
Small, J. K. 1938. Ferns of the southeastern United States. Science Press, Lancaster, PA. (SL)
Smith, K. M. 1972. A textbook of plant virus diseases. Academic Press, New York.

SELECTED REFERENCES

Smith, M. A., L. P. McColloch, and B. A. Friedman. 1966. Market diseases of asparagus, onions, beans, peas, carrots, celery and related vegetables. U.S. Dept. Agr., Agr. Handbk. 303. (SMF)

Smith, S. R. 1960. Doing business the PACA way...pointers on trading practices prescribed by the Perishable Agricultural Commodities Act. U.S. Dept. Agr., AMS-358.

Smock, R. M. 1979. Controlled atmosphere storage of fruits. p. 301-336. In Jules Janick (ED.). Horticultural Reviews, Vol. 1 AVI Publishing Co., Westport, CT. (SK)

Smoot, J. J., L. G. Houck, and J. R. Johnson. 1971. Market diseases of citrus and other tropical fruits. U.S. Dept. Agr., Agr. Handbk. 398 revised. 115 p. (SHJ)

Solntseva, M. P. 1976. Basis of embryological classification in angiosperms. p. 89-100. In S. S. Khoklov (Ed.). Apomixis and breeding. Amerind Publishing Co., Pvt. Ltd., New Delhi (for ARS U.S. Dept. Agriculture and National Science Foundation). U.S. Dept. Commerce, National Technical Information Service, Springfield, VA. (SV)

Soule, M. J., Jr. (James) 1949. Some ecological aspects of South Florida. Dept. of Botany, University of Miami, Coral Gables, FL.

Soule, James. 1974. Plant propagation laboratory manual (PLS311). Fruit Crops Dept., IFAS, University of Florida, Gainesville, FL.

Soule, James. 1979. Principles of tropical fruit culture. Fruit Crops Dept., IFAS, University of Florida, Gainesville, FL.

Soule, James. 1980a. Horticultural morphology. Fruit Crops Dept., IFAS, University of Florida, Gainesville, FL.

Soule, James. 1980b. Horticultural taxonomy. Fruit Crops Dept., IFAS, University of Florida, Gainesville, FL.

Soule, James. 1981. Rootstock-scion relationships. Fruit Crops Dept., IFAS, University of Florida, Gainesville, FL.

Soule, James and William Grierson. 1981. Citrus maturity and packinghouse procedures. Fruit Crops Dept., IFAS, University of Florida, Gainesville, FL. (S & G)

Soule, James and W. B. Sherman. 1978. Glossary for horticultural crops, rev. Fruit Crops Dept., IFAS, University of Florida, Gainesville, FL. (S & S)

Soule, J., G. E. Yost, and A. H. Bennett. 1966. Certain heat characteristics of oranges, grapefruit and tangelos during forced-air precooling. Trans. ASAE 9(3):355-358. (SYB)

Soule, J., G. E. Yost, and A. H. Bennett. 1969. Experimental forced-air precooling of Florida citrus fruit. U.S. Dept. Agr. Mkting. Res. Rept. 845. (SYB)

Soule, J. and M. J. Young. 1975, 1976. Plant propagation laboratory manual. Fruit Crops Dept., IFAS, University of Florida, Gainesville, FL. (S & Y)

Stafleu, F. A., Ed. 1971. International code of botanical nomenclature. A. Oosthoek's Uitgeversmaatschappij N.F., Utrecht, Netherlands. (SF)

Stearn, W. T. 1966. Botanical Latin. David & Charles Publishers, Newton Abbott, Devon, UK. (ST)

Steenis, C. G. G. J. van. 1948-1954. General considerations. p. XIII-LXIX. In C. G. G. J. van Steenis, (Gen. Ed.). Flora malesiana, Series I. Vol. 4. P. Noordhoff, Groningen, The Netherlands. (v.ST)

Steenis, C. G. G. J. van. 1955-1958. Specific and infraspecific limitations. p. CLXVII-CCXXXIV. In C. G. G. J. van Steenis (Gen. Ed.). Flora malesiana, Series I, Vol. 5. P. Noordhoff, Groningen, The Netherlands. (v.ST)

Stewart, J. K. and H. M. Cathey. 1963. Hydrocooling vegetables. U.S. Dept. Agr. Mkting. Res. Rept. 637.

SELECTED REFERENCES

Strickberger, M. W. 1976. Genetics, 2nd ed. The Macmillan Co., New York. (SB)
Stryer, L. 1975. Biochemistry. W. H. Freeman & Co., New York. (SY)
Sumner, H. R. and D. B. Churchill. 1977. Collecting and handling mechanically removed citrus fruit. Proc. Int. Soc. Citriculture 2:413-418.
Surre, C. and R. Ziller. 1963. Le palmier a huile. G.-P. Maisonneuve & Larose, Paris. (S & Z)
Tai, E. A. 1958. Timing the Lacatan banana crop. Jamaica Banana Board, Research Department, Kingston. (TI)
Tarver, D. P., J. A. Rodgers, M. J. Mahler, and R. L. Lager. 1978. Aquatic and wetlands plants of Florida. Bureau of Aquatic Plant Research and Control, Florida Dept. of Natural Resources, Tallahassee, FL. (TA)
Tea Research Institute of East Africa. 1967. Tea estate practice 1966. Tea Research Institute of East Africa, Kericho, Kenya. (TRI)
Teem, D. H., Ed. 1977. Florida weed control guide. Department of Agronomy, University of Florida, Gainesville, FL. (TM)
Thimijan, R. W. and R. D. Heins. 1983. Photometric, radiometric, and quantum light units of measure: A review of procedures for interconversion. HortScience 18(6):818-822. (T & H)
Thompson, H. G. 1949. Vegetable crops. McGraw-Hill Book Co., New York. (TP)
Tibbitts, T. W. 1977. Reviewer's comments. HortScience 12(5):437-438. (TB)
Tomlinson, P. B. and A. M. Gill. 1973. Growth habits of tropical trees: Some guiding principles. p. 129-143. In B. J. Meggers, E. S. Avensu, and W. D. Duckworth (Eds.). Tropical forest ecosystems in Africa and South America: A comparative review. Smithsonian Institution Press, Washington, DC. (T & G)
Tucker, D. P. H. 1978. Citrus irrigation management. Florida Coop. Extens. Serv. Cir. 444. (T)
Tukey, H. B. 1964. Dwarfed fruit trees. Cornell University Press, Ithaca, NY. (TK)
Turner, N. C. 1979. Drought resistance and adaptation to water deficits in crop plants. p. 343-372. In H. Mussel and R. C. Staples (Eds.). Stress physiology in crop plants. John Wiley & Sons, New York. (TU)
Turner, N. C. and M. M. Jones. 1980. Turgor maintenance by osmotic adjustment: A review. In N. C. Turner and P. J. Kramer (Eds.). Adaptation of plants to water and high temperature stress. John Wiley & Sons, New York. (T & K - T & J)
Underwood, L. S., L. L. Tieszen, A. B. Callahan, and C. E. Folk, Eds. 1979. Comparative mechanisms of cold adaptation. Academic Press, New York.
United Fresh Fruit and Vegetable Association. 1978. There is a better way (MUM = modularization, unitization, and metrification). United Fresh Fruit & Veg. Assn.
U.S. Congress. 1930. Perishable Agricultural Commodities Act [Public No. 325, 71st Congress (S108)] (original act). U.S. Government Printing Office, Washington, DC.
U.S. Congress. 1937. Agricultural Marketing Agreement Act of 1937 [Public No. 137, 75th Congress. Chapter 296-1st Session (H.R. 5722)]. U.S. Government Printing Office, Washington, DC.
U.S. Congress. 1938. The Copeland Act of 1938. (Public No. 717, 75th Congress) (Chapter 675-3rd Session) (Federal Food, Drug and Cosmetic Act). U.S. Government Printing Office, Washington, DC.
U.S. Congress. 1947. Federal Insecticide, Fungicide and Rodenticide Act of 1947 (PL 104; 80th Congress, Chapter 125-1st Session, H.R. 1237; June 25, 1947). U.S. Government Printing Office, Washington, DC.
U.S. Congress. 1958. Harris Act (Public Law 85-929; 85th Congress H.R.

13254, Sept. 6, 1958) (Food Additives Amendment of 1958). U.S. Government Printing Office, Washington, DC.
U.S. Congress. 1959. Nematocide, Plant Regulator, Defoliant and Desiccant Amendment of 1959 (Public Law 86-139, 86th Congress, H.R. 6436, August 7, 1959) (Amendment to Federal Insecticide, Fungicide and Rodenticide Act of 1947). U.S. Government Printing Office, Washington, DC.
U.S. Congress. 1960. Public Law 86-537; 86th Congress, H.R. 7480, June 29, 1960. (Haley Bill amending Federal Food, Drug and Cosmetic Act). U.S. Government Printing Office, Washington, DC.
U.S. Congress. 1960. Public Law 86-618; 86th Congress, S.2197, July 12, 1960. [Color Additive Amendments of 1960 (including Delaney Amendment)]. U.S. Government Printing Office, Washington, DC.
U.S. Congress. 1970. Plant Variety Protection Act. (Public Law 92:577, 91st Congress, S.3070. December 24, 1970.) U.S. Government Printing Office, Washington, DC.
U.S. Congress. 1972. Federal Environmental Pesticide Act. (December 15, 1972). U.S. Government Printing Office, Washington, DC.
U.S. Dept. Agriculture. 1896. Nut culture in the United States. U.S. Dept. Agr., Div. Pomology. (US-P)
U.S. Dept. Agriculture. 1949. Trees. U.S. Dept. Agr., Agr. Yearbk., 1949.
U.S. Dept. Agriculture. 1960. Grade names used in U.S. Standards for farm products, rev. U.S. Dept. Agr., Agr. Handbk. 157. (US-GN)
U.S. Dept. Agriculture. 1960. Perishable Agricultural Commodities Act and Regulations, rev. U.S. Dept. Agr., Agr. Marketing Serv., S.R.A.-AMS 121.
U.S. Dept. of Agriculture. 1961. Seeds. U.S. Dept. Agr., Agr. Yearbk. 1961. (US-S)
U.S. Dept. of Agriculture. 1968. Federal Seed Act of August 9, 1939, rev. U.S. Dept. Agr., Consumer & Marketing Serv.
U.S. Dept. of Agriculture. 1980. Executive summary report. Federal seed program review (Roy C. Creech, consultant). IV.A. Guidelines for categorizing varieties and other classes of plant populations (p.18-20); guidelines for classifying cultivated plant populations (p.77-81). U.S. Dept. of Agriculture, Washington, DC.
U.S. Dept. of Agriculture, Agricultural Marketing Service. 1976. Parts 201-202-Federal Seed Act Regulations. Interstate commerce, foreign commerce, and general regulations. U.S. Dept. Agr., AMS.
U.S. Dept. of Agriculture, Agricultural Marketing Service. 1980. The market news service on fruits, vegetables, ornamentals, and specialty crops. U.S. Dept. Agr., AMS, Mkting. Bul. 61 rev. (MNS)
U.S. Dept. of Agriculture. 1982. Perishable Agricultural Commodities Act and Regulations. U.S. Dept. of Agriculture. (PACA)
Urquhart, D. H. 1961. Cacao. John Wiley & Sons, New York. (UQ)
Volin, R. and L. Ramos. 1980. The association of brown root rot resistance with yield components and root weight among tomato selections grown in infested and non-infested soil in Dade County, Florida. Proc. Fla. State Hort. Soc. 93:224-226. (V & R)
Walker, D. R. 1970. Growth substances in dormant fruit buds and seeds. HortScience 5:414-417.
Watada, A. E., R. C. Herner, A. A. Kader, R. J. Romani, and G. L. Staby. 1984. Terminology for the description of developmental stages of horticultural crops. HortScience 19(1):20-21. (WT)
Watkins, J. V. 1969. Florida landscape plants. University of Florida Presses, Gainesville, FL. (W)
Watkins, J. V. and T. J. Sheehan. 1975. Florida landscape plants, rev. ed. University of Florida Presses, Gainesville, FL. (W & S)

SELECTED REFERENCES

Weaver, R. J. 1972. Plant growth substances in agriculture. W. H. Freeman & Co., New York.

Wehlburg, C., S. A. Alfieri, Jr., K. R. Langdon, and J. W. Kimbrough. 1975. Index of plant diseases in Florida. Florida Dept. of Agriculture & Consumer Services, Division of Plant Industry, Bul. 11. (WALK)

Weiser, C. J. 1970. Cold resistance and acclimation in woody plants. HortScience 5:403-410.

West, Erdman. 1957. Poisonous plants around the home. Florida Gr. Exp. Sta. Cir. S-100. (WE)

Westwood, M. N. 1970. Rootstock-scion relationship in hardiness of deciduous fruit trees. HortScience 5:418-421.

Westwood, M. N. 1978. Temperate-zone pomology. W. H. Freeman and Co., New York. (WS)

White, J. M. 1980. Cabbage yield, head weight, and size as affected by plant growing containers. Proc. Fla. State Hort. Soc. 93:266-267.

Whitney, J. D. and H. R. Sumner. 1977. Mechanical removal of fruit from citrus trees. Proc. Int. Soc. Citriculture 2:407-412. (WH)

Wilson, R. F. 1938. Horticultural colour chart. British Colour Council and Royal Horticultural Society, London. 2 vol.

Wilson, W. C., R. E. Holm, and R. K. Clark. 1977. Abscission chemicals-aid to citrus fruit removal. Proc. Int. Soc. Citriculture 1:404-406.

Wolfe, H. S. and S. J. Lynch. 1940. Papaya culture in Florida. Florida Agr. Exp. Sta. Bul. 350. (W & L)

Wright, J. W. 1976. Introduction to forest genetics. Academic Press, New York. (WR)

Wulff, E. V. 1943. An introduction to historical plant geography. Chronica Botanica Co., Waltham, MA. (John Wiley & Sons, New York.) (WF)

Young, A. 1976. Tropical soils and soil survey. Cambridge University Press, New York. (Y)

Young, M. J. and James Soule. 1978. Plant propagation laboratory manual (PLS 311). University of Florida, IFAS, Dept. of Fruits Crops, Gainesville, FL. (Y & S)

Young, R. 1970. Induction of dormancy and cold hardiness in citrus. HortScience 5:411-413.

Zielinski, Q. B. 1955. Modern systematic pomology. Wm. C. Brown Co., Dubuque, IA.

INDEX OF TERMS

A

A & P(*VI.8.013)617
ABA(*IV.2.1.027)357
Abaxial(II.2.2.001)118
Abbé refractometer(*VI.1.008d)571
Aberration(IV.4.8.001)393
Abdomen(*V.3.3.1.010,*.132)498,501
Abiotic pollinators(IV.4.5.001a)391
Abnormal:
  growth and development(*V.4.1.083) 563
    increase in cell numbers(*V.3.3.2. 212)521
    increase in cell sizes(*V.3.3.2.214) 521
    paucity of cells(*V.3.3.2.215)521
    starch accumulation(*V.3.3.2.250) 523
Aborted(*IV.4.8.003)393
  seeds(IV.4.2.004)385
Abortive(II.2.2.27d(1);IV.4.8.003)156, 393
Above-ground parts(*V.4.2.1.002)565
Abrasion(V.1.3.055)435
Abscisic acid(*IV.2.1.037;*2.2.107) 359,366
Abscisin(*IV.2.1.037)359
(+)-Abscisyl-β-D-glucopyranoside (*IV. 2.1.037)359
Abscission(II.1.2.001;.01a,b;*IV.2.1. 033)59,66,358
  of florets(*V.4.1.106)564
  zone[II.2.2.01(8,12,18)]131
Absolute:
  humidity(*VI.5.025a,*.037)589,591
  response(*IV.3.2.013j)375
  temperature(*IV.5.2.038)402
Absorbed dose:
  (gray)(*IV.0.260d)345
  index(*IV.0.260d)345
  rate(*IV.0.260e)345
Absorbing organ(V.3.3.2.199)521
Absorption:
    and gummulation(*V.3.3.2.133c) 515
  blocks(V.3.5.003,*.013)557
  spectrum(IV.1.2.036a)354
Absorptive hairs(*II.4.2.006)239
Acaricide(*V.3.3.3.058)535
Acarina(*V.3.3.1.172)504
Acaulescent(II.1.1.3.001,.01a)21,37
Acceleration(*IV.0.260b)344

Acceptability(*VI.8.079)622
Acceptance(VI.7.1.001)597
Accessibility(V.1.1.017a)149
Accessory:
  (bud)(II.1.1.1.001,.01a)19,28
  (fruit)(II.2.3.1.001,*.011,.01d
    .04i(2),.05d,.06a-c,.07;*IV.4.2.
    000)158,159,168,171,174,385
Acclimation(IV.5.2.001,*.030)398,401
Acclimatization(III.1.2.001)307
Accommodation advance(VI.7.1.002)597
Account promptly(VI.7.1.003)598
Accountant(*VI.8.053)620
Accrescent(*II.2.2.002;.5.3.001)118, 267
  (involucre)(*II.2.1.042,.10d,e)102, 113
Accrete(II.2.2.002,.15i)118,144
Accumbent(II.2.3.3.001,.03a)163,183
Acephalin(*IV.2.2.159)368
Acerose, needle-shaped(II.1.2.001;.5. 5.001;.5.08a)59,271,288
Acervuli(*V.3.2.2.019,.*181g)510, 520
Acetal(*IV.2.2.095)366
  formation(*IV.2.2.095)366
Acetic acid(*VI.10.038)677
Acetone(*V.3.3.3.097)537
Acetyl CoA(*IV.1.2.040b)356
Achatina fulica(*V.3.3.1.251)508
Achene(II.2.3.1.002,.01a-c,e-g;*.5.6. 028)158,168,274
Acicular(II.5.5.002;.5.08b)271,288
Aciculated(II.5.6.001;.5.13a)273.293
Acid:
  equivalent(ae)(V.3.3.3.002)531
  -loving plants(*V.3.2.089)496
  rain(*V.3.3.2.270)524
  soil(*V.2.119)481
Acidity(*VI.1.005)570
Acorn(II.2.3.1.04d)171
Acquired character(III.2.2.001)317
Acre(A)(*IV.0.016)329
Acropetal(*II.1.1.1.030;.1.1.3.002; .01b;*.3.144)20,22,37,201
Acrosternum hilare(*VI.9.2.013)660
Acrotony(II.1.1.3.003,.02a)22,38
Actinocytic(*II.3.190)204
Actinomorphic(II.2.2.061c-1,.20a-c, .21,.22;*IV.4.4.044)126,149-151,390
Actinomycetes(V.3.3.2.001,*.133a,

INDEX OF TERMS 694

*.339)509,514,526
Actinostele(II.3.189a,.27a)203,233
Action spectrum(IV.1.2.036b;.3.2.001) 355,373,374
Activated charcoal(*V.1.3.076)438
  sludge(*V.3.2.068)494
Activation energy(IV.0.006)328
Active:
  content(*V.3.3.3.002)531
  extrusion(*IV.5.4.3.010)415
  ingredient(a.i.)[*V.3.3.3.002,.003, *.072;VI.7.2.017a(def.)]531,536, 612
  migratory, endoparasitic(female vermiform) nematodes attacking roots (V.3.3.1.200f)506
Activity (of a radionuclide) (becquerel):(*IV.0.260d)345
Activo(*V.1.2.011c)430
Aculeate(II.5.6.002;.5.13b)273,293
Acuminate(II.5.4.1.001;.5.03a)268,283
Acuminose(II.5.4.1.002)268
Acute:
  (apex)(II.5.4.1.003;.5.03b)268,283
  (base)(II.5.4.2.001;.5.04a)269,284
  (stress)(*IV.5.4.3.000)414
Acylic(*II.1.2.104)63
Adaptability(IV.0.008)329
Adaptation(II.4.4.001;IV.5.1.001;.5.2. 002;V.3.3.3.004)246,396,398,531
  for gaining light(II.4.3.002)241
  to seasonal climates(II.4.3.001)241
Adaxial(II.2.2.003)118
Adding sugar, other sweetening agents (*VI.7.1.025)600
Additive genetic variance(III.2.2.002) 318
Adenine(IV.2.2.056,*.141,*.142,*.173) 363,367,369
Adenosine(*IV.2.2.141)367
  deoxy-(*IV.2.2.141)367
  -5-diphosphate(*IV.2.2.142)367
  -5-monophosphate(*IV.2.2.142)367
  -5-triphosphate(*IV.2.2.142)367
Adherent(II.2.2.004)118
Adhesion(IVv.5.2.002)399
Adhesive:
  clear plastic tape(*V.1.2.017)432
  plastic tape(*V.1.4.028g)445
Adiabatic:
  conditions(*V.5.032)590
  lapse rate(IV.5.3.001)407
Adjuvant(V.3.3.3.005,*.104)531,538
Adnate(II.2.2.005,.08c)118,137
ADP(*IV.2.2.142)367
Adpressed[II.2.2.006,.23a(3)]118,152
Adsorbed ions(*V.2.111)479

Adsorption of cations and anions(*V.3. 2.050)493
Adulterated:
  (def.)VI.7.2.017b)613
  food(def.)(VI.7.2.016a)610
Advance payment(*VI.8.086)622
  on consigned produce (*VI.7.032)601
Advanced generation synthetic varieties(III.2.0.005a-5b)314
Adventitious:
  buds(*V.1.4.083;.3.3.2.340)452,526
  embryony(IV.4.1.006a-1;*4.6.004)378, 392
  root(roots)[II.1.3.001,.*004,*.009, *.015,.01,.02c(6);*4.3.060;*V.3.3. 2.040;*4.1.004,*.116]94,96,97,245, 511,559,564
  shoot(s)[II.1.1.3.13a(2);*.3.144;*V. 3.3.2.040]49,204,511
  shoot buds(*II.4.1.008)238
  structures(*II.3.125)200
Adventive(*III.1.2.029)309
Adverse conditions(*IV.5.2.001)398
Advertising techniques(*VI.8.065)621
Aecidium (pl. aecidia)(*V.3.3.2.003) 510
Aeciospores(*V.3.3.2.003,*.181f,*.294, *.366)510,519,525,527
Aecium (pl. aecia)(V.3.3.2.003,*.294, *.366)510,525,527
Aeolian deposit(*II.4.2.034)240
Aeration(*V.3.1.001,*.014,*.083)485 .487,495
  increased internal(*IV.5.2.001)398
Aerator(V.3.1.001,*.012)485,486
Aerenchyma(*II.3.125,.04f;.4.3.003; *IV.5.4.3.006)200,210,241,414
Aerial roots(*II.1.3.002,.01a,e;*4.3 .002,.004,*.014;.4.02d,.06b;*IV.3 .3.008p)94,96,241,250,254,378
  sprays(*V.3.3.2.114)538
  stilt roots(*II.4.3.039;.4.11a)243, 259
Aerobic(V.3.3.2.004,*.040)510,511
  oxidation(*V.1.3.026)433
Aerodynamic(*IV.5.2.028)401
Aerogenic(V.3.3.2.005)510
Aerosol(V.3.3.3.006,*.092)531,537
  sprayers(*V.3.3.3.114)538
Aerotropism(IV.3.3.008a)377
Aestivation(*II.2.2.032)122
AF(*VI.7.1.021)600
A-frame style(*V.1.1.017e)420
Africa(*V.2.097)475
After-lighting(V.4.1.001)559
  -ripening(IV.3.1.001;*V.1.3.081)370, 439

INDEX OF TERMS

Agameon(III.2.2.003)318
Agamospermy(*IV.4.1.006)378
Agar(*V.1.5.005)470
Agaricaceae(*V.3.3.2.259)523
Ageing(VI.1.001,*.016)569,572
  of citrus(VI.9.3.001)663
Agent(*VI.6.000)594
  of inoculation(V.3.3.2.006)510
Aggregate(II.2.3.1.003,*.011,.06)158, 159,173
  fruit(*IV.4.2.000)385
Aggregated(II.2.1.001,.05a)100,108
Aggregates(*V.2.115b,e,j)480
  individual(*V.2.115l)481
Aging(*IV.3.1.019;see ageing)373
Agobio system(*V.3.4.017i,.09a)541,553
Agreement of reciprocity(*VI.7.1.035) 602
Agrestial; agrestal(III.1.2.002)307
Agricultural:
  Marketing Agreement Act of 1937, as Amended(VI.7.2.004,*.004a,*c)609
  seed lots(*V.1.3.075)438
  uses(*V.2.096e)475
Agrobacterium(*V.3.3.2.040)511
  tumefaciens(*V.3.3.2.039,*.040, *.320)511,525
Aid plants(*V.2.110e)479
Air(*V.2.000)472
  atomization(*V.3.3.3.006)531
  blast(*IV.3.3.008a;V.1.3.015)377, 433
  -blast shaker(*VI.2.022b)576
  -blast sprayer(*VI.3.3.3.100)537
  -carrier sprayer(*VI.3.3.3.100)537
  circulation(*V.3.3.3.088;*VI.3.023) 537,577
  conditioning(*VI.5.005,*.017,*.051e, *.053)588,593
  continuous circulation(VI.3.023)577
  density(*IV.5.2.065)404
  diluent and carrier(*V.3.3.3.057)535
  -driven hedge trimmers(*V.3.4.063) 543
  -driven pole saw(*V.3.4.060)543
  -driven pole shears(*V.3.4.061)543
  -driven saws(*V.3.4.063)543
  -driven shears(*V.3.4.063)543
  drying(*V.3.2.082)495
  evaporating surfaces(*IV.5.2.065)404
  flotation(*V.1.3.057)435
  flow (loading pattern)(*VI.6.017)595
  flow rate(IV.0.010)329
  high-velocity stream(*V.3.3.3.057) 535
  layer(*V.1.3.033)434
  layerage(*V.1.4.036,.036a,.15)449, 467
  movement(IV.0.010)329
  plants(*II.4.2.006)239
  precooling(VI.5.035,*.035a)590
  screen cleaner(*V.1.3.057)435
  storage(VI.5.047a)592
  temperature(*V.1.3.073)438
Aircraft(*V.3.3.3.030)533
  sprayers(*V.3.3.3.100)537
Airplane(*V.3.3.3.067)535
Airtight enclosures(*V.3.3.3.043)534
Aitionomic(IV.3.1.002;.4.1.002)370,378
Alabama(*VI.8.113)625
Alae[II.2.2.061c-2,.23h(2)]127,152
Alar(*IV.2.1.037)359
Alate(II.1.1.3.004;.1.2.004,.01d;.2.3. 2.001,.01c,d)22,59,66,161,175
  (testal outgrowths)(II.2.3.3.042a, .11d,e)166,192
Albedo[II.2.3.2.029a,.05d(3)]162,179
  browning of lemons(VI.9.3.002,*.029) 663,665
Albic horizon(V.2.107a)477
Albidus(II.5.1.007a)265
Albomaculatus(II.5.1.006a)264
Albugo:
  candida(*VI.9.1.566)659
  occidentalis(*VI.9.1.567)659
Albumen(II.2.3.3.002,*.015)163,164
Albumin(*IV.0.042)330
Albus(II.5.1.007m)265
Alcohol, alcohols(*II.3.038;*IV.2.2. 032b;VI.10.012,*.038)195,362,676, 677
Aldehydes(*II.3.038;*IV.2.2.031,.095; VI.9.3.000)195,361,366,663
Aldohexoses(*IV.2.2.031a,*.032)361,362
Aldonic acids(*IV.2.2.032c)362
Aldopentoses(*IV.2.2.031a)361
Aldoses(*IV.2.2.031a)361
-Ales(*III.1.1.028)306
Alfalfa butterfly(*V.3.3.1.132g)502
Alga, Algae(*II.4.2.006;*IV.1.2.029, *.036d,*3.2.019;V.2.107i;3.3.2. 008,*.133a,*.179a,*.437)239,354, 355,371,376,477,510,514,517,529
  (epiphytic)(*II.4.2.023)240
"Algal fungi"(*V.3.3.2.279,*.477)524, 531
Algal spot(*V.3.3.2.008;VI.9.1.003) 510,626
Aliform(*II.3.014)194
Alkali soils(*V.3.2.044)493
Alkalization(*V.2.120)482
Alkaloids(IV.2.2.012)360
Allele=allel;allelic;allelomorph;alle- lomorphic(III.2.2.004)318

INDEX OF TERMS

Allelochemic(IV.2.2.013)360
Allelopathy, allelopathic(IV.2.2.014) 360
Allogamy(*IV.4.3.002d;*.4.8.017) 386,393
Allopatric(III.1.2.003)307
Allophilic blossom(IV.4.4.001)387
Alloploidion(III.2.2.005)318
Allopolyyploid(III.2.2.006)318
Allorhizic(II.1.3.003)94
Allosteric enzymes(*IV.2.2.058a)363
Almost no demand(*VI.8.027)618
Alpha ray resistance(*IV.5.4.2.001) 413
Alternaria:
  brassicae(*VI.9.1.007,*.010,*.125) 626,634
  citri(*VI.9.1.014)627
  herculea(*VI.9.1.010)627
  oleracea(*VI.9.1.007)627
  porri(*VI.9.1.426)650
  solani(*V.3.3.2.181g;*VI.9.1.210) 520,640
  tenuis(*VI.9.1.011,*.013,*.015, *.016,*.017)627
  tomato(*VI.9.1.325)646
  sp.(*VI.9.1.012,*.171,*.286)627,638, 646
Alternaria:
  leaf spot of cabbage, cauliflower, broccoli etc.(VI.9.1.007)627
  root rot and leaf spot of rutabaga and turnip(VI.9.1.010)627
  rot of:
    apple(VI.9.1.011)627
    cherries, peaches, nectarines, apricots, plums(VI.9.1.012)627
    fig(VI.9.1.013)627
    lemons, black rot of oranges(VI.9.1.014)627
    muskmelon(VI.9.1.015)627
    squash(VI.9.1.016)627
    tomato, pepper, eggplant(VI.9.1.017)627
Alternate:
  (buds)(*II.1.1.1.002,.01b)19,28
  host(V.3.3.2.010)510
  (leaves)(II.1.2.005,.01e)59,**66**
  suscept(*V.3.3.2.003)510
Alternative(*II.1.1.2.001,.01a)20,34
  host(V.3.3.2.011)510
Alternating temperatures(V.1.3.002)432
Altitude(*II.4.4.003d)247
Aluminum (Al)(*V.2.028h;*.3.2.028, *.034,*.073)473,491,495
  alloy framing(*V.1.1.017c)420
  foil(*V.1.3.067;*1.4.036a)437,449

-hydroxyl octahedra(*2.097)475
phosphate (hydrous)(*II.3.037)195
rollers(*VI.3.081)583
sulfate(*VI.3.2.089)496
total(*V.3.2.071)495
Alveolate(II.5.6.003;.5.13d)273,293
Amaryllids(*II.2.2.022)121
Ambiphotoperiodic plants(IV.3.2.013a) 374
Ambrosia beetles(*V.3.3.1.132a)501
Ambush bug(*V.3.3.1.132c)502
Ament(*II.2.1.013)100
Amide(s)(*IV.2.2.150)368
Amino (polar groups)(*V.5.3.053)411
  acid content(*IV.5.2.085;.5.4.2.014) 406,414
  acids(*IV.2.1.001;.2.2.016,*.154, *.172)357,360,368,369
  acids release(*IV.5.3.024)409
  sugars(IV.2.2.032f)362
Ammonia(*V.1.2.011f)430
  injury of apples(VI.9.3.003)663
Ammoniacal N(*V.3.2.034)491
Ammoniation of citrus(*VI.9.3.043) 666
Ammonification(IV.1.2.018a)354
Ammonium:
  acetate(*V.2.015)472
  citrate(*V.3.2.003)490
  (5-hydroxycarvacryl)trimethylchloride piperidine carboxylate(*IV. 037)359
  nitrate(*V.1.2.013;*.1.4.028t;.3.2. 016,*.048,*.054;*.065,*.067,*.086) 431,448,491,493,494,496
  oxalate solution(*V.3.2.005)490
  phosphate(*V.3.2.003,*.086)490,496
  sulfate(*V.3.2.065)494
  sulfate-nitrate(*V.3.2.086)496
Amo(*III.1.1.005)303
Amo-1618(*IV.2.1.037)359
Amoeboid(*V.3.3.2.179)517
  biflagellate swarmspores(*V.3.3.2. 181b)518
  vegetative protoplasts(*V.3.3.2.268) 523
Amount of substance(*IV.0.171a,*.260a) 340,344
AMP(*IV.2.2.142)367
Ampere(*IV.0.260a)344
  per meter(*IV.0.260b)344
  per square meter(*IV.0.260b)344
Amphicribral(*II.3.2.210;.28a)205,234
Amphidiploid(III.2.2.007)318
Amphihaploid(III.2.2.008)318
Amphimictic(*IV.4.6.000)392
Amphimixis(*IV.4.1.026)381

INDEX OF TERMS

(syngamy) apomixis(IV.4.1)378-382
Amphiospore(V.3.3.2.012,*.356,*.442)
    510,527,529
Amphiphloic stele(II.3.189b,.27d)204,
    233
Amphipolyploid(III.2.2.009)318
Amphitropous[II.2.2.041t-4-k-1;.2.2.
    26c(2)]124,145
    (seed-form)(II.2.3.3.04b)185
Amphivasal(*II.3.210,.28b)205,234
Amplex(*II.1.1.2.012)21
Amplexicaul(*II.1.2.075)62
    stem-clasping(II.1.2.006,.02a)59,67
Amylopectin(*IV.2.2.031c-a-1)361
Amyloplast(*IV.1.1.012)351
Amylose(*IV.2.2.031c-a-1)361
Anaerobic(V.3.3.2.013)510
    condition(s)(*IV.5.4.3.006;*V.2.037)
    414,473
Anaerogenic(V.3.3.2.014)510
"Analysis"(V.3.2.001)490
Anapleurotic reactions(IV.1.2.004)352
Anarsia lineatella(*VI.9.2.035)661
Anastomose(II.2.121a,.22a)64,87
Anatomical structure changes(*IV.5.3.
    041)411
Anatomy(II.3.000,.01,.02)194-237,207,
    208
Anatropous:
    (ovule)(*II.2.2.041t-4-k-2,.016c(3)
    ;*.2.3.3.034,.06d)124,145,165,187
    (seed-form)(*II.2.3.3.012,.036a,
    .04c)163,165,185
Ancymidol(*IV.2.1.037)359
Andic(*V.2.096d)475
Androdioecious(IV.4.3.001)385
Androecium[II.2.1.01(2);2.2.007,.01a
    (6,7);*IV.4.4.027]104,118,130,389
Androgenesis(*IV.4.1.004)378
    (Khokhlov)[IV.4.1.006b-1-4,.03(2)]
    379,380,384
Androgenetic(IV.4.8.005)393
    haploid(III.2.2.010)318
Androgynophore[II.2.2.041a,.27a(1)]
    122,156
Anemochory(IV.4.7.001a)392
Anemomorphosis(II.4.4.003a)247
Anemophily(*IV.4.5.001a)391
Aneuploid(III.2.2.011)318
Aneuspory[IV.4.1.006b-6-a,.03(6)]380,
    384
(Angiospermae)(*III.1.1.001)302
Angiosperms(*I.1.000;*II.3.047,*.052,
    .136,.152,*.157,*.161,*.178;*.4.2.
    021;III.1.1.001;*IV.5.2.040)3,196,
    201,202,203,240,302,403
Angström(*IV.0.260k)347

Anguina spp.(*V.3.3.1.200a)505
Angular(II.1.2.007,.02b;.3.04h;5.4.3.
    001;.5.05a)59,67,210,269,285
    acceleration(*IV.0.260g)346
    blocky(V.2.115a)480
    leaf spot of cotton(*V.3.3.2.040)511
    leaf spot of cucumber(*VI.9.1.069)
    631
    velocity(*IV.0.260g)346
    vertices(*V.2.115a,i)480
Anhydrous ammonia(*V.3.2.054)493
Animal(s)(*IV.4.4.038,*.4.5.001c,
    *.002a,*b,*d,*l,*n)390,391,392
    hooves(*V.3.3.3.048)534
    kingdom(*V.3.3.1.132)501
    manure(*V.3.3.3.048)534
    other(*V.3.3.2.006)510
    parasites(*VI.9.000)626
    pests(*V.3.3.3.037,*.038)533
    remains(*V.2.111)479)
Animal pests causing market disorders
    (VI.9.2.)659-663
Anisocytic(*II.3.190,.30h)204,236
Anisogamete, anisogametic(III.2.2.012)
    318
Anisogamous(III.2.2.013)318
Anisogeny(III.2.2.014)318
Anisogynous(II.2.2.008,.03a)121,132
Anisomerous(II.2.2.009,.03b,.24c)121,
    132,153
Anisophyllous(II.1.2.008,.02c)59,67
Anisophylly(II.1.2.009,.02d)59,67
Anisoploid(III.2.2.015)318
Anisopterous(II.2.3.2.002,.01a,b)161,
    175
Anisotony(II.1.1.3.005,.02b)22,38
Anisotropic(IV.3.3.008b)377
Annual(*II.4.1.006;.5.3.002;.5.01a)
    238,267,281
    and perennial grasses(*V.3.3.3.122)
    539
    crops(*V.3.0.004;.3.1.013)484,487
    grasses(*V.3.3.0.001)497
    seasonal sales peak(*VI.8.104)624
Annular:
    bud(V.1.4.028f-1,.08i)444,460
    (disk)(II.2.2.03c)132
    (embryo)(II.2.3.3.003,.02b)163,183
    (thickening)(*II.3.201,.22a)204,228
Annulate:
    (disk)[II.2.2.27c(3)]156
    ringed(II.5.6.004;.5.13c)273,293
Annulus[II.2.2.010,.03c,.27c(3);.6.
    001]121,132,156,279
Anocymidol(*IV.2.1.037)359
Anomalous secondary growth(II.3.008,
    .03)194,209

INDEX OF TERMS

Anomocytic(*II.3.190)204
Ant(s)(*II.4.2.024;*IV.4.5.001c;*V.3.
    3.1.132e)240,392
  myrmecophyte association(*II.4.3.
    067c)245
  plants(*II.4.3.042;*.4.4.002c)244,
    246
Antagonism(V.3.3.3.007)531
Antagonistic symbiosis(V.3.3.2.016)
    510
Antennae(*V.3.3.1.132)501
Anterior(*II.2.2.007h-4,*.046)119,125
Anther(II.2.2.007a,.01a(6),.11a,b)118,
    130,140
  cap[II.2.2.007b,.24f(1),.25b(1)]118,
    153,154
  positions(II.2.2.07,.08)136,137
Antheridia(*V.3.3.2.181e)519
Antheridium(V.3.3.2.017)510
Anthers(*V.1.5.000)470
  feeding of mites(*V.4.2.2.015)567
Anthesis(IV.4.4.002;.4.8.028)387,394
Anthocyanescence, anthocyanescent(V.3.
    3.2.018,*.133b)510,515
Anthocyanins(IV.2.2.021,*.029,*.068,
    .069,.154;*.5.4.2.012;*VI.1.005,
    *.008b)360,365,368,414,570
Antholyse (antholysis)(V.4.2.2.001)566
Anthomyiids(*V.3.3.1.132b)501
Anthophilous(IV.4.5.002a)391
Anthracnose(V.3.3.2.019,*.181g)510,
    519,520
  of avocado(IV.9.1.023)628
    banana(VI.9.1.024)628
    citrus(VI.9.1.025)628
    cucurbits(*V.3.3.2.181g)520
    garden pea(VI.9.1.026)628
    mango(VI.9.1.027)628
    muskmelons(VI.9.1.028)628
    papaya(VI.9.1.029)628
    peaches, nectarines, apricots(VI.
      9.1.030)628
    pepper(VI.9.1.031)628
    snap beans(VI.9.1.032)628
    tomato(VI.9.1.033)628
    turnip, radish, cabbage, Chinese
      cabbage, collards(VI.9.1.034)628
    watermelon(VI.9.1.035)629
Anthranilic acid(*IV.1.2.042)357
Anthropogenous relic(III.1.2.004)307
Anthropomorphosis*II.4.4.002a)246
Antibiosis(II.4.3.005;V.3.3.2.021)241,
    511
Antibiotic(V.3.3.2.022)511
Antibodies(*V.3.3.2.156)516
Anticlinal (divisions)(II.3.14b)220
Anticous(*II.2.2.046)125

Antidote(VI.7.2.017c)613
Antigens(*V.3.3.2.412)529
Antipathetic(II.4.3.067a)245
Antipetalous(*II.2.2.007d-6)118
Antipodals[II.2.2.041t-4-b-1,.16b(12);
    IV.4.1.006a-2,.006b-2,.01(6),.02a
    (6),b(4)]124,145,378,379,382,383
Antiraphe(II.2.3.3.004,.04d)163.189
Antisepalous (stamens)(*II.2.2.007d-6)
    118
Antiseptic[*V.1.4.022a-3(2)]442
Antiserum(*V.3.3.2.156)516
Antitranspirant(s)(*IV.5.2.017;V.1.2.
    001)400,428
Ants(*IV.4.4.001c;*V.3.3.1.132e;*.3.3.
    3.013f)391,502,532
Anvil-type pruning shears(*V.3.4.066)
    543
Aonidella aurantii; A. citrina(*VI.9.
    2.007)658
Apatite(*V.3.2.075)495
Aperiodical(II.5.3.003)267
Aperturate(*II.2.2.007t-10)120
Apetalous(II.2.2.01e)130
Apex, base, margins(II.5.4)268-271
  terms(II.5.4.1.;.5.03)268-279,283
Aphelenchoides spp.(*V.3.3.1.200a)505
Aphid(*V.3.3.1.132d,*3.3.2.395;*.4.2.
    1.022)502,528,566
  borne virus(*V.4.2.1.004,*.012)565
  vectors(*V.3.3.2.133e)515
Apiaceae(*III.1.1.008,*1.2.061)303,311
Apical(*II.1.1.1.031;2.2.007w-1)20,120
  bud(*V.1.4.071b)451
  dominance(*IV.2.1.025;*V.4.1.084)
    358,363
  dominance removal(*V.4.1.009)559
  meristem[*II.3.108,.142,.143,.18a
    (1);V.1.5.01a]199,201,224,471
  (pendulous)(II.2.2.15d)144
  pore[II.2.2.10e(1)]139
  tubes(II.2.2.07h-1)119
Apicifixed(II.2.2.007w-1,.08b,d)120,
    137
Apiculate(II.2.2.08a,.09a;.5.4.1.004;
    .5.03e)132,138,268,283
Apoarchespory(*IV.4.1.006b-6)380
Apocarpous(II.2.2.041b)122
  (fruit)(II.2.3.2.003,.02a)161,176
  (gynoecium)(II.2.2.14a,.22b)143,151
Apogameon(III.2.2.016)318
Apogamic embryo(*IV.4.4.001b,*001c)392
Apogamy(IV.4.1.006a-2,4,7,14,.006b-2,
    .02(6,7),.03(10,11);*4.6.001d)378,
    379,383,384,392
Apohomotypic[IV.4.1.006b-6-a-1,.03(3)]
    380,384

Apomictic:(*IV.4.6.000)392
  embryo(*II.2.3.3.014;*V.1.4.000)163, 441
  processes(IV.4.1.006a,b)378,379
Apomixis(IV.4.1.006,.03;*4.8.042)378, 384,395
Apophysis(II.6.002)279
Apoplasm(II.3.011)194
Apospory[IV.4.1.006a-3,.006b-6-b, .02a(4),.03(7)]378,380,383,384
Apothecium(pl. apothecial)(V.3.3.2. 023,*.181e)511,519
Apotracheal(*II.3.014)194
Appearance(*VI.8.079)622
  of symptoms(V.3.3.2.220)522
Appendage(II.1.2.011,.02r)59,67
"Apple"[II.2.3.1.04i(2)]171
Apple blotch(*V.3.3.2.181g)520
  canker(*V.3.3.2.181f)519
  -cedar rust(V.3.3.2.181f;VI.9.1.036) 519,629
  fruit spot(*V.3.3.2.181g)520
  maggot injury(IV.9.2.004)659
  rust(*V.3.3.2.181f)520
  scab(*V.3.3.2.181e;VI.9.1.037)519, 619
  scald(VI.9.3.005)663
Application:
  number and timing(*V.3.2.037)492
  rate(IV.0.171b)340
Appressed(*II.2.2.006)118
Appressorium(V.3.3.2.029)511
Approach:
  grafting(*V.1.4.028,.028a,.06)443, 458
  grafts (wrapping)(*V.1.4.028w)449
Appropriation(*V.3.5.033)559
Apricot-colored(II.5.11.008a)265
Aqua ammonia(*V.3.2.054)493
Aquatic(s)(*I.2.009;II.4.2.002;.4.02c) 6,238,250
  animals(*V.2.107i)477
  organisms(*V.2.107i)477
  plants(*IV.3.2.019;V.2.107i;.3.3.2. 008)376,477,510
Aqueous (aqueus)(II.5.1.007c)265
Ara (suffix)(*III.1.1.014)304
Arabinans(*IV.2.2.031c-a-3)361,362
D-Arabinose(*V.2.2.031a,*c,*c-b-2, *.038)361,362
  polysaccharides(*IV.2.2.031c-a-3)362
Arabis mosaic(*V.3.3.1.200c;*.4.2.1. 022)506,566
  vector(*V.3.3.1.200c)506
Arable land(*V.3.0.004)484
Arachnida(*V.3.3.1.010,*.172;VI.7.2. 017n)498,504,614

Arachnids(V.3.3.1.010)498
Arachnoid(II.5.6.005;.5.13e)273,293
Arborescent(II.1.1.3.006)22
Arced(II.1.2.012)59
Archegonium(II.6.003)279
Archesporial:
  cell[IV.4.1.01(1),.02a(1),b(3), .03(1)(a)]382,383,384
  (recurrent) agamospermy(IV.4.1.006a- 4)378
  tissue(*IV.4.1.006a-5)378
Archesporium(II.2.2.007c)118
Archimycetes(*V.3.3.2.105)514
Archips argyrospilus(*VI.9.2.044)661
Area(IV.0.016)329
Area(A)(*IV.0.260b)344
Arecoid palms(*II.1.2.054)61
Areolate(II.5.6.006;.5.13f)273,293
Arenic(*V.2.096d)475
A-rest(*IV.2.1.037)359
Argenteus(II.5.1.007k)265
Argillans(V.2.107b)477
Argillic horizon(V.2.028a)473
Arginine(*IV.2.2.172)369
Argyrotaenia velutinana(*VI.9.2.044) 661
Arid regions(*V.3.2.044)493
Aril[*II.2.3.3.042a,.042b,.05a(2), .08a-c,.09b(2),d(2),.10a,c(1), .11b(5),d(2)]166,186,189,190,191,192
Arillate[II.2.3.3.042c,.10b(5)]166,191
  (seeds)(II.2.3.3.10)191
Arillode(II.2.3.3.042d)166
Arillostome[II.2.3.3.042e,.10c(3), d(3)]166,191
Aristate(II.5.4.1.005;.5.03d)268,283
Armed(II.1.1.3.007,.02c)22,38
Armeniacus(*II.5.1.008a)265
Armillariella mellea(V.3.3.2.181f)519
Armored scale(s)(*V.3.3.1.132d)502
  injury on citrus(VI.9.2.007)660
Armyworm(*V.3.3.1.132g)502
Aroma(*VI.8.079)622
Arrangement of the androecium in a flower(II.2.2.007d,.04-.06)118,133- 135
Arrhenius:
  plot(IV.0.017;.5.4.1.000)329,412
  -type behavior(*IV.0.017)329
Arrow-shaped(*II.5.5.052)273
Arsenate(*V.3.4.000a)539
Arsenic toxicity(*VI.9.3.017)664
Arthropoda(*V.3.3.1.010,*.132)498,501
Articulate(II.1.2.013,.03a)60,68
Articulated(II.1.2.014,.03a)60,68
  laticifers(*II.3.179)203
Artificial:

driers(*V.1.3.061)437
freezing(IV.5.3.002,*.027)407,409
long days(V.4.1.002)559
pollination(*IV.4.4.029)439
shearing(*IV.5.4.3.008)415
soil mixes(V.1.2.011b)430
Arundinaceous(II.1.1.3.008)22
Ascendent, ascending(II.1.1.3.009, .03a)22,39
Aschelminthes(*V.3.3.1.183)504
Asci(*V.3.3.2.023,*.295)511,525
Ascidiform(II.4.3.006)241
Ascigerous fruiting body(*V.3.3.2.101) 513
Ascochyta:
    pinodella(*VI.9.1.405)650
    pinodes(*VI.9.1.405)650
    pisi(*VI.9.1.405)650
Ascogonia(*V.3.3.2.181e)519
Ascomycetes(*V.3.3.2.023,*.181e,*.295) 511,519,525
Ascorbic acid(*IV.2.2.032c;*VI.7.1. 025)362,601
Ascospores(*V.3.3.2.031,*.181e,*.295) 511,519,525
Aseasonal(II.5.3.004;.5.01c)267,281
Asexual(IV.4.8.006)393
    fungal spore(*V.3.3.2.112)514
    propagation(*V.1.4.000)441
    spore (thick walled)(*V.3.3.2.094) 513
    spores(*V.3.3.2.179a,*.219)517,521
Ash(*II.1.1.3.067)25
    content(*V.2.096d)475
    -gray(II.5.1.003a)263
Ashing, wet or dry(*V.3.2.071)494
Aspergillus:
    alliaceus(*VI.9.1.041)629
    flavus(*IV.1.2.018c)354
    niger(*VI.9.1.040,*.043,*.083,*.084, *.286)629,632,644
    sp.(*VI.9.1.171,*.286)638,644
    spp.(*VI.9.1.043,*.090,*.286)629, 633,644
Aspergillus:
    black-mold rot of fig(VI.9.1.040)629
    bulb rot of garlic and onions(*VI.9. 1.041)629
    rot of citrus(VI.9.1.043)629
Asperous(II.5.6.007;.5.13g,h)273,293
Asphalt(*V.1.3.067)437
    -coated felt paper pots(V.1.2.003a) 428
    roofing felt(*V.3.3.3.059)535
    roofing felt pots(*V.1.2.003h)428
Aspidiotus:
    ancylus(*VI.9.2.065)662

forbesi(*VI.9.2.031)661
perniciosus(*VI.9.2.073)662
Aspirator separator(*V.1.3.057)436
Assassin bug(*V.3.3.1.132c)502
Association of Official Seed Analysts and Certifying Agencies (1976)(*III. 2.0.004c-f;*V.1.3.075)312,438
Aster yellows(VI.9.1.046)629
Astringency(*VI.8.079)622
Astrosclereid(*II.3.170)202
Asymmetrical pressures(*IV.5.4.3.008) 415
ATA(*V.3.2.012)490
Atactostele(*II.3.189c,.27f)203,233
Ategmic(II.2.3.3.005)163
Atmosphere(IV.0.019;*.5.2.038)329,402
Atmospheric:
    humidity(*V.1.3.033;*VI.9.3.000)434, 663
    impurities(*V.3.3.2.270)524
    moisture(IV.0.020;*VI.5.025)329,589
    nitrogen(*IV.1.2.028)354
    pollutants(*VI.9.3.000)663
    pollution(IV.5.4.4.001)415
    and mechanical stresses(IV.5.4.4) 415
    pressure(*IV.5.2.065)404
Atomic weight(*V.2.015)472
Atomizing nozzles(*V.3.3.3.006)531
ATP(*IV.1.1.016,*.1.2.040c;*.2.2.142) 351,356,367
Atropine(*IV.2.2.012)360
Attapulgite(*V.2.098)475
Attenuate(II.5.4.2.002;.5.5.004;.5. 04b,.08c)269,271,284,285
Attenuation(V.3.3.2.032)511
Atto(a)(*IV.0.260h)346
Attractant(s)(IV.4.4.003,*.013;*V.3.3. 3.073)387,536
Auction(IV.8.001)616
Augers(*VI.2.022b)576
Aurantiacus(*II.5.1.008n)266
Aurantioideae(*III.1.1.002)302
Aureus(II.5.1.008j)266
Auriculate(II.5.4.2.003;.5.04c)269,284
Australia(*V.2.097)475
Autochory(IV.4.7.001b)392
Autoclave(*VI.5.055)593
Autoclaved(*V.1.5.005)470
Autoecious(V.3.3.2.033)511
Autogamy(*IV.4.3.002d;.4.8.007)386,393
Automatic:
    (bagging) machines(*VI.3.053a)581
    irrigation(*V.1.1.017e)420
    parthenocarpy(IV.4.2.001)385
Autonomic(IV.3.1.003;.4.1.007)370,381
Autopolyploid(III.2.2.017)318

Autotroph, autotrophic(V.3.3.2.034)511
Autotropism(IV.3.3.008c)377
Auxanometer(IV.3.1.004)370
Auxiliary(*V.3.3.3.104)538
Auxins(IV.2.1.012,*.033;*V.3.4.000a)
  357,358,539
Availability of nutrients(*V.3.2.065)
  494
Available:
  phosphoric acid(V.3.2.003)490
  plant food(V.3.2.004)490
  potash(V.3.2.005)490
  year-around(VI.8.121)625
Avenue(V.3.1.008a,.01f,i,j)485,488-489
Average power(IV.0.157b-2-e-1)337
Avocado:
  root rot(*V.3.3.2.181d)518
  scab(VI.9.1.047)629
Avogadro:
  constant(IV.0.023)329
  number of photons (flat surface)(IV.
    0.157x-2-a)339
  number (point)(IV.0.157y-2-a)339
Avoidance(IV.5.1.001;*.5.2.015;*.5.3.
  038,*.041;*.5.4.2.001)396,400,410,
  411,413
  UV resistance(*IV.5.4.2.012)413
  UV mechanisms(*IV.5.4.2.014)414
AVP evaporator(*VI.10.010d)675
Awl:
  nematodes(*V.3.3.1.200c)505
  -shaped(II.5.5.058)273
Awned(*II.5.4.1.005)268
Awns(*V.1.3.018)433
Axenic(V.3.3.2.036,*.410)511,528
Axeny(V.3.3.2.037)511
Axial parenchyma (wood)(II.3.014,.04)
  194,210
Axil(II.1.1.3.010;*V.4.1.035)22,560
Axile(II.2.2.041t-2-a,.02(6,16),
  .14c,d)123,131,143
  -parietal(II.2.2.041t-2-b,.14d)123,
  143
Axillary(II.1.1.1.004,*.012,.01c;.2.1.
  065;.2.3.2.004,.03d)19,28,103,161,
  177
  buds(*V.4.1.105)564
Axis(II.1.1.3.011)22
Axle loading(*VI.6.018)595
Azalea type pots(*V.1.2.003h)428
Azide(*IV.1.2.040d)356
Azonal soil(V.2.010)472
Azureus(II.5.1.002i)263

B

BA(*IVv.2.1.025)358

Baccate(II.2.3.1.004;.5.7.001)158,278
Bacillus tracheiphilus(*V.3.3.2.040)
  511
Backcross(III.2.2.018)318
Backhoe(s)(*V.3.0.003;*V.3.1.000)484
Bacteria(*II.3.163;*IV.1.2.018b,*c,
  .029,.036,*.2.2.0.031c-a-3;*.3.
  1.012;*.3.2.019;*.5.3.049;.5.4.2.
  006;*V.1.3.080;.3.3.0.003,*.005,
  .3.3.2.001,*.034,*.036,*.040,
  *.179a,*b,*b-1,*.270,*.320,*.339,
  .361,.406,*.412;*.3.3.3.013;VI.
  7.2.017i;*.9.000;*.9.1.000)203,
  354,362,371,376,411,414,438,497,
  509,511,512,517,524,525,526,527,
  528,529,532,613,626
  priority of publication(*V.3.3.2.
    179b)517
Bacterial blight(s):
  of beans(*V.3.3.2.133c)515
  of celery(VI.9.1.048)629
  of cotton(*V.3.3.2.040)511
  of garden pea(VI.9.1.049)629
  of lima beans(VI.9.1.050)630
Bacterial brown rot of parsnip(VI.9.1.
  051)630
Bacterial canker of tomato(*V.1.3.026;
  VI.9.1.052)433,630
Bacterial cell(*V.3.3.2.226)522
Bacterial diseases(*II.4.4.002e;V.3.3.
  2.040)246,511
Bacterial fruitlet rot(*VI.9.1.127)635
Bacterial infection(*V.3.3.3.013)532
Bacterial leaf spot of cauliflower,
  cabbage, broccoli(VI.9.1.054)630
Bacterial ring rot of potato(VI.9.1.
  055)630
Bacterial root rots of horseradish(VI.
  9.1.056)630
Bacterial rot of hyacinth(*V.4.2.1.
  023)566
Bacterial soft rot(s) of:
  asparagus(VI.9.1.057)630
  cabbage and other crucifers(VI.9.1.
    058)630
  carrot, parsnip(VI.9.1.059)630
  celery(VI.9.1.060)630
  lettuce(VI.9.1.061)631
  muskmelons, cucumbers, squash, pump-
    kins(VI.9.1.062)631
  onions(VI.9.1.063)631
  potato(VI.9.1.064)631
  snap beans(VI.9.1.065)631
  spinach(VI.9.1.066)631
  tomato, pepper, eggplant(VI.9.1.067)
    631b
Bacterial speck of tomato(VI.9.1.068)

INDEX OF TERMS

631
Bacterial spot of:
  cucumber(VI.9.1.069)631
  honeydew melons(VI.9.1.070)631
  lima beans(VI.9.1.071)631
  peaches, nectarines, apricots(VI.9.1.072)631
  tomato and pepper(VI.9.1.073)631
Bacterial wilt of:
  banana(*V.3.3.2.040)511
  corn(VI.9.1.074)632
  potato(*V.3.3.2.040)511
  snap beans(VI.9.1.075)632
Bacterial zonate spot of cabbage, other crucifers and cucurbits(VI.9.1.076)632
Bacteriophage(phage)(V.3.3.2.044)512
Bag(s)(*V.1.3.067;*.3.2.037;*VI.4.000a,*b,*c,*d,.001;*.8.056)437,492 585,620
  filling(*VI.3.053a)581
  -filling machines(*V.1.2.003g)428
Bagasse(*V.1.1.053e)425
Bagmaster:
  carton(*VI.3.053a;*.4.000c,*.002) 581,585
  containers(*VI.3.053a)581
Bagworm moths(*V.3.3.1.132g)502
Bait(V.3.3.3.009)531
Baldwin spot(*VI.9.3.008)663
Balled-and burlapped plant(V.1.0.001,*.036)418,420
Bamboo(*V.1.1.017c;*.1.2.014)420,431
  frame(*V.1.1.053c,*.059)425,426
Banana pseudostems(*V.3.3.3.059)535
  wilt(*V.3.3.2.181g)520
Band:
  -shaped(II.5.5.006;.5.08d)271,288
  treatment(V.3.3.3.010)532
Banded(*II.3.014;.5.1.006b)194,265
Bar(*IV.0.019,.026,*.260k;*.5.2.038) 329,330,347,402
Barbed(II.5.6.144)278
Barbeque(*VI.10.023)677
Bare bones store(*VI.8.136)626
Bare root(*V.1.1.035)423
  -rooted plant(V.1.0.002)418
Bark(*II.3.016,.20c,d;*V.1.2.005d,*.011)192,266,429
  beetles(*V.3.3.1.132a)501
  cuts(*V.1.4.022a-3)442
  cylinder(*V.1.4.036a)449
  grafting(*V.1.4.028k-1,.10g)446,462
Barn(*IV.0.260k)347
Barochory(IV.4.7.001c)393
Barometric pressure(*VI.5.025c,*.037,*.043)589,591

Barren(IV.4.8.008)393
Basal(II.1.1.3.012)22
  cell(II.2.3.3.02b)183
  -erect(II.2.2.15c)144
  herbicide treatment(V.3.3.3.011)532
  ovules(II.2.2.041t-2-c,.15c)123,144
  plate(*V.1.4.071a;.4.1.003,*.116;*.4.2.1.017;*.4.2.2.019)451,559, 564,566,568
  ,common(*V.4.1.040)561
  root(V.4.1.004)559
  rot of lilies(V.4.2.1.001)565
  slips[V.1.4.04c(3)]456
  style[II.2.2.17(1)]146
Base:
  saturation(V.2.011,*.096c)472,475
    high(*V.2.028e)473
    low(*V.2.038i)473
  SI units(IV.0.260a)344
  (terms)(II.5.4.2)269
Baseball park(*V.3.1.004)485
Basic seed(III.2.0.007c)316
  slag(*V.3.2.003,.087)490,496
Basidiomycetes(V.3.3.2.046,.181f)512, 519
Basidiospore(s)(*V.3.3.2.046,.047,*.181f)512,519
Basidium(V.3.3.2.048,*.423)512,529
Basifixed(II.2.2.007w-3,.07a,h,.08a) 120,136,137
Basiflory[*II.2.1.004,.05d(1)]100,108
Basin(*V.3.5.012)557
Basipetal(*II.1.1.1.030;.1.1.3.013,.03b)20,22,39
Basitony (branching)(II.1.1.3.014,.03c)22,39
Baskets(*VI.2.012;*.4.000a,*b,.003*.011,*.012)575,585,586,587
  bushel(*VI.6.022)595
Bassine(*VI.3.007,*.081)578,583
Bast fibers(*II.3.063)197
Bats(*IV.4.5.001c;*.4.7.001d,*.006;V.3.3.1.017)391,393,498
Bayonet(*V.3.4.017,.06c)540,550
Beaked(*II.2.2.041w,*y)122,125
Bean anthracnose(*V.3.3.2.181g)520
Bean (lady) beetle(*V.3.3.1.132a)501
Bearded[II.5.6.009;.5.13(i)]274,293
Beards(*V.1.3.018)433
Becquerel(*IV.0.260d)345
Bed(V.3.1.002;*.3.3.3.027)485,533
  construction(*V.3.1.000)484
Bedbug(*V.3.3.1.132c)502
Bedding plants(*I.0.001;.2.001a;*V.1.1.048;*.1.2.003a;*VI.3.0000;*.4.011,*.013)3,5,424,428,577,586,587
Bedknife(*V.1.1.034)422

Bedrock(*V.2.096d,*.107m,*.112,.01a)
475,477,479,483
Beds(*V.3.1.008a;*.3.2.039)486,492
 alternate(*V.3.1.008h)486
Bee flies(*V.3.3.1.132b)501
Beech(*II.1.1.3.067)25
Beer(VI.10.003)674
Bees(*IV.4.5.001c,*.002g;*V.3.3.1.
 132e;*.3.3.3.013f;*VI.7.2.017n)391,
 502,614
Beet:
 cyst nematode(*V.3.3.1.200g)506
 internal black spot(VI.9.3.067)668
 leaf miner(*V.3.3.1.132e)501
 scab(VI.9.1.077)632
Beetles(*IV.1.2.040d;*.4.4.009;*.5.
 001c;*V.3.3.1.084,*.086,.132a,*.3.3.
 3.133d;*VI.7.1.017n)356,389,391,499,
 500,501,515,614
Belonolaimus spp.(*V.3.3.1.200c)505
Bell:
 jar(*V.1.1.046,*.048)424
 -or funnel-shaped(IV.4.4.004b-1)387
 -shaped(*II.2.2.061c-1-b)126
Belt(s)(*V.3.027)579
 -and-roll sizer(*VI.3.064a,*d)583
 conveyor(*VI.3.019)578
 return-flow(*VI.3.027a-2)579
 sloping(*VI.3.027a-1)579
 twin(*VI.3.027a-2)579
Bench:
 and other terraces(*V.2.110e)479
 -budded(*V.1.2.003b)428
 -graftage(*V.1.4.028b)443
 grafted(*V.1.4.022b)442
 heights(*V.1.1.017e)420
 -rooted(*V.1.4.028t)448
 -rooting(*V.1.3.003,*.079,.01a)432,
 438,440
 terraces(*V.2.110c)478
Bending(*IV.5.4.4.002,.002g;*V.1.4.
 028i,.11c;*.3.4.000,.003,.12e)416,
 446,463,539,540,556
Beneficial use(*V.3.5.033)559
Benlate(*VI.3.033)580
Benomyl(*VI.3.033,*.085a)580,584
Bentham and Hooker(*III.1.1.010)304
Bentonite(*V.3.2.050)493
Benzyladenine(*IV.2.1.025)358
6-Benzylamino purine(*IV.2.1.025)358
6-Benzylamino-9-(2-tetrahydropyranyl)-
 9H-purine(*IV.2.1.025)358
Berry(II.2.3.1.005,.05a)158,172
 baskets(*VI.4.003)585
Beta rays(*IV.5.2.001)414
 resistance(*IV.5.4.2.001)413
Betalains(IV.2.2.029)360

Beverage base(*VI.10.019)676
 crops(*I.0.001;.1.001)3,4
Beyond Syn-1(*III.2.0.005a-5-b)314
Bicarbonate ions(*IV.5.4.3.002)414
Bicollateral(*II.3.210,.28c)205,234
Bicrenate(*II.5.4.3.009)270
Bidentate(*II.5.4.3.012)270
Biennial(*II.3.026;*.4.1.004;*.4.4.
 003b;.5.3.005;.5.01b)195,238,247,
 267,281
Bifacial leaf(II.1.2.016,.04a)60,69
Bifid(II.5.4.3.004)270
Bifoliolate(II.1.2.017,.03b)60,68
Big vein(VI.9.1.078)632
Bijugate(*II.1.2.069,.15b)62,80
Bilabiate(II.2.2.061c-14a,.20e,.23g)
 127,149,152
Bilaterally symmetrical(*V.3.3.1.183)
 504
Bill of lading(VI.6.001,*.020)594,595
 advise(*VI.7.1.027)600
 closed(*VI.7.1.027)600
Bin(VI.4.000a,*c,.004)586
Binate(*II.1.2.017)60
Bins(*V.1.1.047)424
Bioassay(V.3.3.2.050)512
Biochemical lesions(*IV.5.3.005;*.5.4.
 1.005)407,411
Biochemistry(IV.0.030)330
Biological:
 control(*V.3.3.3.000,.012,*.053,
 *.098)531,532,535,537
 controls(*V.3.0.001)484
 minimum damage(*V.3.3.3.071)536
 processes(*V.2.106)476
Biologically active substance(*V.3.3.
 2.050)512
Biorational pesticides(V.3.3.3.013)532
Biostimulants(V.1.3.004)432
Biosystematics(III.1.2.)307
Biotic:
 factors(II.4.4.002)246
 organisms(*V.2.000)472
 (pollinators)(IV.4.5.001c)391
Biotin(*V.1.5.005)470
Biotype(III.1.2.005;V.3.3.2.051,*.177,
 *.309)307,512,517,525
Bipulvinate(II.1.2.18b)83
"Bird dog"(VI.8.002)616
Birds(*IV.4.5.001c,*.002c;*.4.7.001h,
 *.006;V.3.3.1.026)391,393,498
Biseriate rays, ray cells(II.3.152,
 *.153,.07c,f)195,201,213
Biserrate(*II.5.4.3.032;.5.07e)271,
 287
Bisexual[II.2.2.013;.2.3.2.03c(2);*IV.
 4.3.001,*.015,*.016,*.017]121,177,

385,387
Bissy(*III.1.1.005)303
Bitegmic(II.2.3.3.006)163
Bitten off(*II.5.4.1.016)269
Bitter:
  pit of apple(VI.9.3.008)663
  rot of apple(*V.3.3.2.181e)519
  rot of apple, pear, peach, quince, cherry(VI.9.1.079)632
  rot of peaches, nectarines, apricots(*VI.9.1.030)628
Bitterness(*VI.8.079)622
Biuret(V.3.2.006)490
  -free urea(*V.1.2.013;*.3.2.006)431,490
Black(II.5.1.001,.001a)263
  cloth shading*V.4.1.006)559
  end of pears(VI.9.3.009)664
  flies(*V.3.3.1.132b)501
  heart of celery(VI.9.3.010)664
  leaf speck of cabbage, Chinese cabbage, and cauliflower(VI.9.3.011)664
  mold of carrot(VI.9.1.082)632
    rot of cherries, peaches, nectarines, apricots(VI.9.1.083)632
    rot of onions and garlic(VI.9.1.084)632
  molds(*V.3.3.2.181c)518
  pit of citrus(VI.9.1.085)632
  rot of apples, pear, and quince(VI.9.1.086)632
    of beet(VI.9.1.087)633
    of cabbage, cauliflower, other crucifers(VI.9.1.088)633
    of carrot(VI.9.1.089)633
    of corn(VI.9.1.090)633
    of cucumber(VI.9.1.091)633
    of grape(*V.3.3.2.181e)519
    of oranges(*VI.9.1.014)627
    of pineapple(VI.9.1.097,*.127)633,635
    of squash(VI.9.1.092)633
    of strawberries, Rubus, Ribes, cucurbits, beets, peas, crucifers(*V.3.3.2.181e)519
    of sweet potatoes(VI.9.1.095)633
    of watermelon(VI.9.1.096)633
  sateen cloth(*V.4.1.007)449
  scurf, stem and tuber rot of potato(VI.9.1.098)633
  shank of tobacco(*V.3.3.2.181d)518
  spot:
    of peaches, nectarines, apricots, plums(*VI.9.1.470)653
    of pear and quince(VI.9.1.100)633
    of pineapple(*VI.9.1.127)635

    of roses(V.3.3.2.181e)519
  tea(*I.1.001a;*VI.10.013)4,676
Blackleg of potato(*V.3.3.2.040;VI.9.1.081)511,632
Blackout system(V.4.1.007)559
Bladdery(*II.2.3.2.017,.02c)161,176
Blade[II.1.2.020;.03c(1);.2.1.01(12)]60,68,104
Blanching(*VI.7.1.025)600
Blanking(IV.4.8.008)393
Blastophaga psenes(*IV.4.5.001d)391
Blastophaga wasp(*IV.4.4.007,*.008;.4.5.001d)388,391
Blemish(VI.8.038a,b)619
Blemishes:
  free of injury from(*VI.8.038e)619
  freedom from(*VI.8.038e)619
Blight(V.3.3.2.059)512
  disease(*V.3.3.2.181d)518
  of stone fruits(*V.3.3.2.181g)520
  of strawberry(*V.3.3.2.181e)519
Blights(*V.3.3.2.040,*.181e)511,519
Blind:
  cultivation(V.3.1.003)485
  shoots(V.4.1.008)559
Blindstoken(V.4.2.2.002)567
Blister:
  beetles(*V.3.3.1.129,*.132a)500,501
  of ferns, alder, poplar, birch, cherry(*V.3.3.2.181e)519
Blisterlike swelling(*V.3.3.2.227)522
Block(s)(*V.2.115a)580
  perimeters(*V.3.1.000)484
Bloom(II.5.6.010)274
Blossom:
  blast of hibiscus(*V.3.3.2.181c)518
  blast of squash(*V.3.3.2.181c)518
  -end clearing of grapefruit(VI.9.3.012)664
  -end rot of apple(VI.9.1.101)634
  of pepper(VI.9.3.013)664
  of squash(VI.9.1.102)634
  of tomato(VI.9.3.014)664
  of watermelon(VI.9.3.015)664
  types(IV.4.4.004)387
Blotch(V.3.3.2.065)512
  of apple(VI.9.1.103)634
  of avocado(*VI.9.1.143)636
Blotched(II.5.1.006c)265
Blotchy ripening of tomato(*VI.9.3.074)669
Blowdowns(*V.3.3.2.270)524
Blow flies(*V.3.3.1.132b)501
Blue(II.5.1.002b)263
  Goose(*VI.8.122)625
  -green rot of corn(VI.9.1.104)634
  mold of apple(*V.3.3.2.181g)520

INDEX OF TERMS 705

    of citrus(*V.3.3.2.181g;*VI.9.1.
      105)520,634
    of tobacco(*V.3.3.2.181d)519
    rot of apple, pear, quince(VI.9.1.
      106)634
      of cherries(VI.9.1.107)634
      of fig(VI.9.1.108)634
      of garlic and onions(VI.9.1.109)
        634
      of muskmelons(VI.9.1.110)634
      of peaches, nectarines, apri-
        cots, plums(VI.9.1.111)634
      of pomegranate(VI.9.1.112)634
      of sweet potatoes(VI.9.1.113)634
  molds(*V.3.3.2.181d,*.181g)518,520
Blunt (apex)(*II.5.4.1.015)269
  (base)(*II.5.4.2.010)269
B-9(*IV.2.1.037;*V.3.4.000a)359,539
B-995(*IV.2.1.037,*.5.4.3.010)359,415
BNOA(*IV.2.1.012)357
Board of directors(VI.8.004,*.074)616,
  621
Boat-shaped(II.2.1.023,.07b)100,110
"Body" icing(VI.5.004)588
BOH(*IV.2.1.037)359
Boil smut(*VI.9.1.169)638
Bole(II.1.1.3.015,.09b)22,45
Boll rot of cotton(*V.3.3.2.181c)518
Bollworm(*V.3.3.1.132g)502
Bolt(IV.4.8.010)393
Bombacaceae(*II.4.3.065)245
Bonding force(*IV.5.3.053)411
Bone:
  meal(*V.3.2.068,*.086)494,496
  meal, raw(*V.3.2.065)494
  phosphate(*V.3.2.075)495
Bony(II.5.7.026)278
Bonzai(V.3.4.004,.12a)540,556
Book lungs(*V.3.3.1.010)498
Bookkeeper(*VI.8.053)620
Boom shaker(VI.2.022b)576
"Booster" solution(*V.1.2.015)431
Borage(*I.3.006)7
Borax(*V.3.2.085)496
Bordered(II.5.1.006d)265
Borers(*V.1.3.080)438
  European corn(*V.3.3.1.132g)502
  metallic wood(*V.3.3.1.132a)501
Boron(B)(*V.3.2.028,*.058,*.067;*.3.5.
  032)491,494,558
  deficiency of citrus(VI.9.3.017)664
  of crucifers(VI.9.3.018)664
  total(*V.3.2.071)495
Bossed(II.5.6.142)277
Bostryx(II.2.1.006,.05b)100,108
Botanical affinity(*V.1.4.015)441
Botanical categories(III.1.1)302-307

Botryodiplodia fruit rot of banana(VI.
  9.1.115)635
Botryodiplodia theobromae(*VI.9.1.115,
  *.182)635,639
Botryosphaeria rot of apple (Botryosphaeria ribis)(V.9.1.116)635
Botrytis:
  allii(*VI.9.1.264)643
  byssoidea(*VI.9.1.264)643
  cinerea(*V.3.3.2.181g;*VI.9.1.101,
    .120,.255,*.265,*.266,*.267,
    .268,.269,*.270,*.272,*.273,
    .275,.276,*.277)520,634,635,643,
    644
  squamosa(*VI.9.1.264)643
  sp.(*VI.9.1.171,*.271,*.274)638,644
Botrytis rot of citrus(VI.9.1.120)635
Bottles(*V.1.5.005)470
Bottom heat(V.1.1.002,*.017e)418.420
"Bound water"(*IV.5.2.018)400
Bourgeon(V.3.4.005)540
Bourse(V.3.4.006,.02e,f)540,546
  bud(V.3.4.007,.02e,f)540,546
Bouyoucos blocks(*V.3.5.003)557
Bowed roof (greenhouse)(*V.1.1.017b)
  419
Box(*V.1.3.081;VI.4.004,*.009)439,
  585,586
  nailer(VI.3.006)578
Brachiate(II.5.2.001)266
Brachysclereid(*II.3.170)202
Brachytic(II.4.3.007)241
Brackish(II.4.2.003)238
Braconid wasps(*V.3.3.1.132e)502
Bract, bracteate[II.2.1.007,.01;.2.2.
  01g,.23i(3)]100,104,130,152
Bracteole(II.2.1.009,.01i)100,104
Bracts(*V.4.1.033)560
Brads(*V.1.4.028c,*k-1,*-3,*-4,*.028o)
  444,446,447,448
Brambles(V.3.4.008,.05)540,549
Branch(V.3.4.017,*.029,*.046,*.060,
  .061,.062)540,541,542,543
  framework(*V.1.4.028i)445
  gap(II.3.021,.05)194,211
  horizontal(*V.3.4.068)544
  lateral scaffold(*V.3.4.055)543
  short(*V.3.4.072)544
  system(*II.1.1.3.019,.04)22,40
  trace(II.3.021,*.201,.05)194,204,211
Branched(II.1.1.3.076;*.3.204)25,205
Branches(*II.3.056,*.201)196,204
  axillary(*V.4.1.058)562
Branching(*V.3.4.057)543
  by apposition (sympodial)(II.1.1.3.
    017,.03d)22,39
  by substitution (sympodial)(II.1.1.

3.018,.03e)22,39
Brand(VI.8.005)612
  designation(*VI.3.072)583
Brandy(VI.10.004)674
Brass rollers(*VI.3.081)583
Brazilian system(*V.3.4.017,.09b)541,
  553
Break(V.4.1.009)559
  bulk(VI.6.002)594
Breaker(VI.8.06)616
Breba(IV.4.4.005,*.007)388
Breeder seed(III.2.0.004c)312
Breeding(III.2.0.001)311
  and selection(*V.3.3.3.000)531
  and selection systems(III.2.1)316
  cycle (breeding rotation)(III.2.2.
    019)318
  sites(*V.3.3.3.088)537
  system(III.2.1.001)316
Bremia lactucae(*V.3.3.2.181d;*VI.9.1.
  204)518,640
Brick red(II.5.1.005a)264
Brickbats(*V.1.1.056)425
Bridge grafting(*V.1.4.028,*.028c,
  .12a-f)443,444,464
Brindille(*V.3.4.010,.02h)540,546
Bristle[II.3.29c(5)]235
  pointed(*II.5.6.123)277
Bristly(*II.5.6.036)274
British thermal unit(VI.5.005)588
Brittle wood(*V.1.4.028k-2)446
°Brix(*VI.1.008a,*c,*d)570,571
  hydrometer(*VI.1.008a)570
Broad-based terraces(*V.2.110c)478
Broadcast:
  application(*V.3.3.3.067)535
  seeder(*V.1.3.015,*.062)432,437
  treatment(V.3.3.3.014)532
Broad-leaved
  annuals(*V.3.3.0.001)497
  evergreens(*V.1.4.022f-1,*-3)442,443
  perennials(*V.3.3.3.122)539
  plants(*V.3.3.3.059)535
  (weeds)(*V.3.3.3.049)534
Broken:
  containers(*VI.6.034)596
  trailer insulation(*VI.6.034)596
Broker(VI.7.1.005,*.039,*.040,*.053,
  .054,.072,*.077)598,602,604,606,
  607
  operations(VI.7.1.007)598
Brokerage fee(VI.7.1.006)598
  transactions(*VI.7.1.003)597
Brook's spot(*VI.9.1.231)641
Broom-shaped(*II.2.1.048)102
Brown(II.5.1.001,.001b)263
  algae(*IV.1.2.036d)355

core of apples(VI.9.3.020)664
core of pears(*VI.9.3.099)670
heart(*VI.9.3.033)665
-red(II.5.1.005b)264
-ring of bulbs(*V.3.3.1.200a)505
root rot(*V.3.3.2.181f)519
root rot of tomato(*III.2.2.159;VI.
  9.1.576)325,659
rot of potato(*V.3.3.2.040)511
  of cauliflower(VI.9.1.125)635
  of citrus(*V.3.3.2.181d;VI.9.1.
    126)518,635
  of deciduous fruits(*V.3.3.2.181d)
    518
  of pineapple(VI.9.1.127)635
  of pome and stone fruits(*V.3.3.2.
    181g)520
  of pome fruits(VI.9.1.128)636
  of stone fruits(*V.3.3.2.133c;VI.
    9.1.129)515,636
  of stone fruits, pome fruits(*V.3.
    3.2.181e)519
spot of celery(VI.9.1.130)636
  of corn(*V.3.3.2.181a)518
  of stone fruits(*VI.9.1.176)638
staining of grapefruit and mandarins
  (VI.9.3.023)664
stinkbug(*VI.9.2.013)660
-tail tussock moth(*V.3.3.1.132g)502
wall of tomato(*VI.9.3.074)669
Browning of lychee(VI.9.3.022)665
Browsing(*V.3.3.3.048)534
Bruises(*VI.8.038c)619
  on apples(VI.9.3.025)665
  on pears(VI.9.3.026)665
Bruising of stone fruits(VI.9.3.027)
  665
Brush control(V.3.3.3.015)532
  -shaped(II.2.1.048)102
Brushes(VI.3.007,*.081)578,583
  soft fiber(*VI.3.081)583
Bryophytes(*IV.1.2.008a,*.029)352,354
BTU(*VI.5.005)588
Bucket(V.1.1.003;*VI.3.019)418,578
Buckeye:
  rot of tomato fruits(*V.3.3.2.181d)
    518
  spot of tomato(VI.9.1.132)636
Buckhorn machine(*V.1.3.057)436
Buckhorned(*V.1.4.028v)449
Buckhorning(V.1.4.028d,.10b)444,462
Buckler-shaped(*II.5.6.120)277
Bucksaw(V.3.4.011)540
Bud(s)(II.1.1.000;*IV.2.1.012,*.033,
  *.036)19,358
  and graft union healing(V.1.4.028e)
    403

blasting(*V.4.2.2.010)519
break(*IV.2.1.037)359
dormant(*IV.2.1.036)358
dormant lateral(*IV.5.4.4.002g)416
initiation(*IV.2.1.025)358
lateral development(*IV.2.1.033)358
newly inserted(*V.1.4.028w)449
pollination(*IV.4.4.006)389
protective covering(*II.4.3.002)241
rest breaking(*IV.2.1.033,*.036)358
shield(V.1.4.11i)463
sport(III.2.2.021)318
union healing(V.1.4.07)417
Budding(*IV.3.1.014h,*.5.4.4.002;*V.1.
4,028,*.028f,*h,*t,*.040,.08)372,
416,444,445,448,450,**460**
knife(V.1.2.002,*.007)428,429
Budeye(*V.1.4.028f-8)445
Budwood registration programs(*V.3.3.
3.000)531
Buffer(V.3.2.007)490
Bug(s)(*V.3.3.1.132c;*VI.7.2.017n)502,
614
Building:
and road construction(*V.3.1.000)484
and zoning codes(*V.1.1.017a)419
sites(*V.3.1.000)484
Build-up of bacterial or algal slime
(*V.3.5.032)558
Bulb, bulbs(*I.2.001b;*II.3.035,*.063;
.4.1.002,.13d,e;*IV.5.1.001;*V.
1.4.007,.071a,.16c-e,.17;*.4.1.
003,*.085,*.088,*.093,*.120,
.121;.4.2.2.002,*.004,*.005,
.006,.008,*.010,*.015)5,195,
196,238,261,397,441,451,**468,469**,
559,563,565,567
axis, old(*V.4.1.035)560
cold requirement(*V.4.1.100)564
contractile roots(*IV.5.2.001)396
crops(I.3.004)7
daughter(*V.4.1.040,*.073;*.4.2.2.
006)561,562,567
dealer(V.4.1.010)559
exporter(*V.4.1.091)563
"fire"(*V.4.2.1.002)565
flies(V.4.2.1.003)565
food reserve storage(*V.4.1.045)561
forcer(V.4.1.011)559
forcing(V.4.1.012)559
forcing size(*V.4.1.093)563
frozen(*V.4.1.043)561
grower(V.4.1.013,*.091)560,563
gummosis(V.4.2.1.011)565
harvesting(*VI.2.000)575
healthy(*V.4.1.029)561
infected(*V.4.2.1.011,*.017)565,566

jobber(*V.4.1.010)559
mature(*V.4.2.2.006)567
mother(*V.4.2.2.006)567
natural vegetative reproduction(*IV.
4.1.006a-8)379
nonprecooled commercial(*V.4.1.075,
*.079)563
pan(*V.4.1.018)560
pests(*V.4.2.1.003)565
precooling(V.4.1.019)560
temperatures(*V.4.1.076)563
producer(*V.4.1.013)560
production(*V.4.1.092)563
Pythium spp.(V.4.2.1.016)566
removal(*V.4.1.099)564
rot of special precooled tulips(V.4.
2.1.016)566
rotting(*V.4.2.1.015)566
scale section(V.1.5.01f)471
selected mother(*V.4.1.072,*.103)
562,564
size(V.1.021,*.117)560,564
small(*V.4.1.102,*.115;.4.2.2.015)
564,567
sours(V.4.2.1.020)566
storage(*V.4.2.1.015)566
ventilation(*V.4.2.1.011)565
tray(*V.4.1.022;*VI.4.000a,*d)560,
585
virus diseases(V.4.2.1.022)566
Bulbil(s)(*II.2.3.3.020a;*V.1.4.000,
.006,.16d)164,441
natural vegetative reproduction(*IV.
4.1.006a-8)379
Bulbing(*V.4.1.045)561
Bulblet(s)(*II.2.3.3.020a;*V.1.4.006,
.007;.4.1.016,*.072,*.086,*.103,
*.122)164,441,560,563,564,565
stem or scale(*V.4.2.1.122)565
Bulbous or lobed swelling(*V.3.3.2.
029)511
Bulbous plants(*I.0.001;.2.001b;*V.4.
1.112)3,5,564
Bulk(*V.3.2.037)492
bin(s)(*VI.4.000c,.006)585,586
density, high(*V.2.107g)477
display(VI.8.007)617
filling(*VI.3.053b)582
shipment(VI.6.003)594
trailer(*VI.2.012,*.014)575
trucks(*V.1.3.067)437
Bullate(II.5.6.015;.5.14a)274,**294**
Bulldozer(*V.3.0.003,*.3.1.000)484
Bullhead(V.4.1.023)560
Bullnose(V.4.2.2.003)567
Bull's eye rot of apple(VI.9.1.133)636
Bumble bees(*IV.4.4.014;*V.3.3.1.132e)

INDEX OF TERMS

389,502
Bundle caps(*II.3.063)197
  sheath[II.3.022,.21a(8),b(1)]194,227
Bundles(*II.2.2.007d-15,.04f)119,133
Bur[II.2.3.1.04g(1);.2.3.2.006,.02(b)]
  161,171,176
  knot(II.1.1.3.020,.05a;IV.3.1.014g)
  22,41,372
Burlap(*V.1.1.053c)425
  bags(*V.1.3.067)437
  containers(*VI.4.000d,.001)585
  wrap(*V.1.0.001)418
Burr (see Bur)
Burrowing animals(*V.3.3.0.002)497
  nematodes(*V.3.3.1.200f)506
Bush harvester(*VI.2.022b)576
Business terminology(*VI.8)616-626
"Butted end"(*V.1.4.028e,.07a-d)444,
  459
Butterflies(*IV.4.5.001c;*V.3.3.1.
  132g;*.3.3.3.013f)391,502,532
Buttress(II.4.3.008,*.039,.10a,b)241,
  243,258
Buyer:
  (on the ground)(VI.8.008)617
  (retail)(VI.8.009)617
Bypassing shoot(V.4.1.024)560

C

CA(*VI.5.047b)592
Cabbage:
  butterfly(*V.3.3.1.132g)502
  looper(*V.3.3.1.132g;.3.3.3.013f)
    502,532
  maggot(V.3.3.1.132b)501
  worm, imported(*V.3.3.3.013c)532
  yellows(*V.3.3.2.474)530
CAC(*VI.7.1.009)598
Cacao (Ceratostomella) wilt(*V.3.3.2.
  181e)519
Cacoma(*V.3.3.2.003,.080)510,513
Cacopaurus spp.(*V.3.3.1.200b)505
Caducous[II.2.2.27c(2);.5.3.006;.5.
  02a]156,267,282
Caeruleus(*II.5.1.002b)263
Caesious(II.5.6.016)274
Caesius(*II.5.1.002f)263
Caespitose(II.5.5.007)271
CAF(VI.7.1.009)598
Caffeine(*IV.2.2.012)360
Calcarate[II.2.2.061c-14-b,.23c(2),
  d(1)]127,157
Calcareous soils(*IV.5.4.3.001,*.002)
  414
Calceolate[II.2.2.061c-14-c,.24e(1)]
  127,152

Calcic horizon (subsurface)(V.2.028b)
  473
  limestone(*V.3.2.053)493
Calcicole(IV.5.4.3.001)414
Calcifuge(IV.5.4.3.002)414
Calciphile(*IV.5.4.3.001)414
Calciphobe(*IV.5.4.3.002)414
Calcite(*V.2.107j,*l)477
Calcitrophic(*IV.5.4.3.001)414
Calcium (Ca)(*V.3.2.028,*.056,*.065)
  491,493,494
  carbonate(*II.3.039;*V.3.3.3.029)
    196,533
  cyanamide(*V.3.2.086)496
  effects on Na (*IV.5.4.3.010)415
  fluorphosphate(*V.3.2.075)495
  hypochlorite solution(*V.1.4.022a-3)
    442
  nitrate(*V.3.2.086)496
  oxalate(*II.3.037)195
  total(*V.3.2.071,*.080)495
Calculated amounts per application(*V.
  3.2.037)492
Calendar schedule(*V.3.5.013)557
Calibration(V.3.3.3.016)532
California:
  aster yellows on celery, anise, dill
    (VI.9.1.134)636
  red scale(*VI.9.2.007)660
Callose[II.3.023,*.041,.26d(2)]194,196
  232
Callus(*II.3.083;.5.6.017;.5.14b;V.1.
  4.008,*.028e,.07c;*.1.5.005;*.3.3.
  2.082)198,274,294,441,444,459,470,
  513
Callused(*V.1.4.022b)442
Callusing(*V.1.4.022a-3,b)442
Calorie (cal)(*IV.0.260m;VI.5.006)347,
  588
Calvin cycle(IV.1.2.036c,*.040g)354,
  356
Calyptra[II.1.2.021,.03(d)]60,**68**
Calyptrate]II.2.2.27c(2)]156
Calyptrogen(II.3.025,.19c,d)195,225
Calyx[II.2.1.01(4);.2.2.061a,.01a(9);
  *V.4.1.108]104,126,130,564
CAM(IV.1.2.008c,*.038;*.5.2.002,*.068
  *.085)353,355,398,404,406
  CO_2 fixation(*IV.5.2.079)406
  metabolism(*IV.5.2.002)398
  plants(*IV.5.2.068)404
  xeromorphic structure(*IV.5.2.085)
    406
Cambial:
  activity(*V.1.4.028t)448
  alignment of layers(*V.1.4.028o)448
  bridge(*V.1.4.028e,.07d,f)444,**459**

face(*V.1.4.028e,.07e,f)444,459
layers(*V.1.4.028e,*f-2,*k-2,*-4,
 -6,-7,*-8,*-9)444,446,447
Cambic horizon(V.2.107c)477
Cambium[II.3.026,*.173,.01B(14),C(18),
 .02e(13),.06a-c,.07a,b]195,202,
 207,208,212,213
 cylinder formation(II.3.06a-c)212
 initials(II.3.07a,b)213
Campanulate(II.2.2.061c-1-b,.21e)126,
 151
Campylotropous:
 (ovule)[II.2.2.041t-4-k-3,.16c(4);
 .2.3.3.004,.036a-4,.02c]125,145,
 163,165,183
 (seed)(II.2.3.3.036a-1,a-2,.02c,
 .04d,.06c)165,185,187
Can(VI.10.005)674
Canaliculate[II.1.2.022;.05(a)]60,70
Candela (cd)(*IV.0.157,.157a,*.260a)
 336,337,344
 per square meter(*IV.0.260b)344
Candelabra(II.5.6.018;.5.14d;*V.3.4.
 055)274,294,543
 palmette(*V.3.4.055,.11g)543,555
 system(*V.3.4.017,.07d)541,551
Cane(*II.1.1.3.008)22
 blight of rose, Rubus, apple(*V.3.3.
 2.181e)519
Canescens(*II.5.1.003h)264
Canescent(II.5.6.019;.5.14c)274,294
Canker(V.3.3.2.082,*.181f)513,520
Cannery(*VI.4.004)586
 bin(*VI.4.006)586
Canning fruits and vegetables(*VI.10.
 000)674
Cano-tomentose(II.5.6.020)274
Cans(*V.2.109)478
Cantharophily(*IV.4.5.001c)391
Canus(*II.5.1.003h)265
Canvas belts(*VI.3.038)580
Capacitance(*IV.0.260c)344
Capacity adaptation(IV.5.3.004)407
Capillary(II.5.6.021;.5.14e)274,294
 bench(*V.1.1.017f)420
 mat(*V.1.1.017f)420
 nozzles(*V.3.3.3.006)431
 water(*IV.5.2.033)402
 watering(*V.4.1.070)562
Capillitia(*V.3.3.2.269)524
Capitate(II.2.2.18a;.5.6.022;.5.
 14f)147,274,294
Capitulum(*II.2.1.037)101
Capnodiaceae(*V.3.3.2.253,*.395)523,
 528
Capnodium citri(*V.3.3.2.181e)519
Caprification(IV.4.4.007)388

Caprifig(IV.4.4.008,*.049)388,390
Capsule(II.2.3.1.006;*.2.3.2.045;.2.3.
 1.02g-n,.03a-c;*.6.002)158,163,169,
 170,279
Car(*VI.7.1.010)599
Caramelization(*IV.5.2.073)405
Carbamyl urea(*V.3.2.006)490
Carbohydrate (monosaccharide) deriva-
 tives(IV.2.2.032)362
Carbohydrates(*II.3.023;*IV.1.2.040;
 .2.2.031;*.5.2.073)194,355,360,405
Carbon (C)(*V.3.2.018)491
 -14(IV.0.036)330
 monoxide(*IV.1.2.040d)356
 :nitrogen (C:N) ratio(V.2.013)472
Carbon dioxide (CO_2)(IV.0.035;*V.3.2.
 028;*VI.5.051b)330,491,493
 buildup(*VI.6.034)596
 compensation point(IV.1.2.007)352
 fertilization(V.3.2.009)490
 fixation(IV.1.2.008,*.036f)352,355
 fixed(*IV.5.2.084)406
 injury of apples(VI.9.3.028)665
 levels(*VI.9.3.000)663
 net assimilation(*IV.5.4.2.002)413
 treatment(VI.5.007)588
 uptake(*IV.5.2.068)404
Carbonate(*V.3.2.080)495
Carbonates(*V.3.5.032)558
Carbonation(*V.2.106)476
Carborundum(*V.3.3.2.248)523
Carboxyl(*IV.2.2.154;.5.3.053)368,411
Cardboard(VI.4.007,*.010)586
 boxes(*V.1.3.067)437
 drums(*VI.3.067)437
Cardinal temperatures(IV.3.2.021a)376
Careless handling(*VI.8.038a)619
Cargo plane(*VI.6.000)593
 van modular containers(*VI.6.037)597
Carina[II.2.2.061c-3,.23h(3)]127,152
Carinate[II.2.2.061c-14-d,.20d,.23h
 (3)]128,149,152
Carload(VI.7.1.010;*.8.100)599,624
Carlot(*VI.7.1.010;*VI.8.010)599,617
 equivalent(VI.8.010)617
Carmine(II.5.1.005c)264
Carnation:
 bud rot(*V.3.3.2.181g)520
 rust(*V.3.3.2.181f)520
Carnauba (wax)(*VI.3.085,*.085b)584
Carneous(II.5.7.013)278
Carneus(*II.5.1.005k)264
Carnivorous plants(*II.3.179;.4.3.009;
 .4.06a)203,241,254
Carotene(*IV.1.1.002e;*.2.2.034)351,
 362
Carotenoid(s)(IV.2.2.034,*.095;*VI.1.

005,*.008b;*.3.023)362,366,570,578
content(*IV.1.008)570
Carpel(II.2.2.041e)122
  development(IV.4.2.000)385
Carpenter bees(*V.3.3.1.132e)502
Carpophore[II.2.3.1.007,*.022,.03k(1)]
  158,159,170
Carrier(s)(V.3.3.2.087;*.3.3.3.009,
    .017;VI.6.000,*.016)513,531,532,
    594
  -diluent(*V.3.3.3.009,*.029)531,533
  liquid(*V.3.3.3.096)537
  procurement, routing, services(*VI.
    8.102)624
Carrion:
  beetles(*IV.1.2.040d;*.4.4.009;*.4.
    5.001c;*V.3.3.1.132a)356,388,391,
    501
  flies(*IV.1.2.040d;*.4.4.009;*.4.5.
    001c)356,388.391
  flower(IV.4.4.009)388
Cartage firm(IV.8.011)617
Cartilage(*IV.2.2.032f)362
Cartilaginous(II.5.7.004)278
Carton(s)(VI.4.008,*.026)586,588
  filling(VI.3.053c)582
  stapler(VI.3.010)578
Caruncle(II.2.3.3.042f)167
Caryophyllaceous(*II.2.2.061c-14-2,
    .23f)128,152
Caryopsis(II.2.3.1.008,.01k-o)159,168
Case-cooled bulbs(V.4.1.025)560
Casein hydrolysate(*V.1.5.005)470
Cash sale(VI.7.1.011)599
Casparian strip[*II.3.052,.11b(2)]196,
    217
Caste system(*V.3.3.1.132f)502
Castor pomace(*V.3.2.068,.086)494,496
CA storage(*VI.5.047b)592
Catalase(*IV.2.2.168)368
Cataphyll(II.1.2.023,.05b)60,70
Catch:
  basins(*V.3.5.023)558
  frame(*VI.2.022b)576
Category(III.1.1.002)302
Catena(II.4.2.004,.03;*IV.0.141)238,
    251,335
Caterpillar(s)(*V.3.3.1.132g,*.141;
    *VI.9.2.018)502,503,660
Caterpillarlike(*V.3.3.1.132e)502
"Cat faced"(*VI.9.2.063)662
Catface of stone fruits(VI.9.2.013)
    660
Cation exchange capacity(*V.2.011,
    .015,*.111;*.3.2.080)472,479,495
Cations(*V.2.120)482
Catkin(II.2.1.013,.05c)100,108

Cattle(*V.1.2.011f;*.3.3.3.048)430,534
Caudate(II.5.4.1.009;.5.03e)268,283
Caudex(II.1.1.3.021,.05b)23,41
Caudicle[II.2.2.007e,*.007t-9,.25e
    (2)(ii),.26d(2),e(2)]119,120,154,155
Cauliflorous(II.1.1.1.005;.01e;*.2.1.
    015)19,28,100
Cauliflory(II.2.1.015,.05d)100,108
Caulimovirus(es)(V.3.3.2.088,*.461)
    513,530
Cauline(II.1.1.3.022)23
Causal agent(V.3.3.2.165)516
Causative agents of disease(V.3.3.2.
    133a)514
Cause(*V.3.3.2.165)516
Cave "cricket"(*V.3.3.1.132h)503
CCC(IV.2.1.037;*.5.2.017;*V.3.4.000a)
    359,400,539
CCTA(*V.2.097)475
CDTA(*V.3.2.012)490
CDU(V.3.2.015)491
CEC(*V.2.015)472
Cecidiomorphosis(*II.4.4.002h)247
Cedar(*V.1.2.011)430
  (bark)(*V.1.2.011)430
  waxwings(*V.3.3.1.026)498
  western(*V.1.1.017c)420
Celery butterfly(*V.3.3.1.132g)502
Ceiling jet cooling room(VI.5.035a-1)
    590
Cell(*II.2.2.041p)123
  dehydration(*IV.5.2.023)401
  division(*IV.2.1.036;*.4.1.019)358,
    381
  division zone(II.3.02b)208
  elasticity(*IV.5.2.079)406
  elongation(*IV.2.1.036)358
  enlargement zone(II.3.02c)208
  packs(*VI.3.053f)582
  parts(*V.1.5.000)470
  permeability(*IV.5.2.017)400
  sap concentration(*IV.5.2.085)406
  sap electrolyte(*IV.5.3.024)409
  size(*IV.5.2.079)406
  turgor(*IV.5.1.017)397
  -wall polysaccharides(*IV.1.1.019;
    .2.2.031c-b-1,.032a,.038,*.114)
    353,362,366
  water(*IV.5.3.053)411
  -water potential(IV.5.2.005)399
  water potential decrease(*IV.5.4.3.
    009)415
Cellobiose(*IV.2.2.031b)361
Cellophane(*V.1.3.067)437
Cellulose(*II.3.038;*IV.1.2.019;*.2.2.
    031c,c-b-1,.114;*V.3.3.1.132f;.4.1.
    070)196,353,361,362,366,502,562

Celsius temperature(*IV.0.260a,c)344
Cementation(*V.2.100)476
Cenospecies(III.1.2.006;*.032)307,309
Centi(c)(*IV.0.260h)346
Centipedes(*VI.7.2.017n)613
Central:
  core(II.3.080)197
  leader[(V.3.4.029,.01(2),.03b,
    e(1)]541,545,547
  sizer(*VI.3.064a)582
  vacuole(*V.5.2.031)401
Centralized packaging(IV.8.012)617
Centric(*II.1.1.3.033)23
Centrifugal(II.2.2.014)121
Centripetal(II.2.2.015)121
Centromere(IV.4.1.008)381
Cephaleuros
  mycoidea(*V.3.3.2.008)510
  virescens(V.3.3.2.008;*VI.9.1.003)
    510,626
Cephalosporium apii(*VI.9.1.130)636
Cephalothecium roseum(*VI.9.1.392,
  *.393)649
Cephalothorax(*V.3.3.1.010)498
Ceramic:
  absorption blocks(*V.3.5.001)557
  porous bulb(*V.3.5.025)558
  pots(*V.1.2.003h)428
Ceratocystis:
  fimbriata(*V.3.3.2.181e)519
  paradoxa(*V.3.3.2.179a,*.181e;*VI.9.
    1.097,*.542)517,519,633,657
Ceratostomella:
  paradoxa(*V.3.3.2.181e)519
  wilt of cacao(*V.3.3.2.181e)519
Cercospora:
  apii(*VI.9.1.209)640
  beticola(*VI.9.1.141)636
  musae(*V.3.3.2.181e)519
  purpurea(*VI.9.1.143)636
Cercospora:
  blotch of avocado(*VI.9.1.143)636
  leaf spot of beet(*VI.9.1.141)636
  spot of avocado(VI.9.1.143)636
Certificate of inspection(*VI.6.020)
  595
Certification program(V.1.4.010)441
Certified seed(III.2.004f;*V.3.3.3.
  000)312,531
Cespitose(II.5.5.007;.5.08e)271,288
C-4 plants(*IV.1.2.007,.008b,*c,*.035,
  .038;.5.2.015,*.068,*.084,*.085)
  352,353,354,355,399,404,406
  CO fixation(*IV.5.2.079)406
  "syndrome"(*IV.1.2.008b)352
  transpiration ratio(*IV.5.2.068)404
  water use efficiency(*IV.5.2.084)406

xeromorphic structure(*IV.5.2.085)
  406
Chaetanaphothrips orchidi(*VI.9.2.080)
  662
Chaff[II.2.2.23i(3);*V.1.3.051]152,435
  scale(*VI.9.2.007)660
Chaffy(*II.5.6.085)276
Chain, chain store(*VI.8.012,.013)617
  saw(V.3.4.015)540
Chalaza[II.2.2.041t-4-a,.16b(3);.2.3.
  3.007,.04(g,h),.10a(5);*.3.119]124,
  145,163,185,191,200
Chalazal region(*IV.4.1.006a-3)378
Chalazogamy(IV.4.4.010)388
Chalcid wasps(*V.3.3.1.132e)502
Chalking of tulip(V.4.2.2.004)567
Chamaephyte[II.4.1.001;.4.01(a)]238,
  249
Channel formation(*V.2.110d)418
Channeled(*II.1.2.022)60
Character (characteristic)(III.2.2.
  022)318
Charcoal rot:
  of honeydew melons(VI.9.1.144)636
  of pepper(VI.9.1.145)637
  of sweet potatoes(VI.9.1.146)637
Chartaceous(*II.5.7.027)278
Chasmogamous(IV.4.4.011)388
Chelating agents, chelate(V.3.2.012,
  *.067)490,494
Chelation(*V.2.106)476
Chemical(s)(*V.1.1.065;*.3.3.3.049;
  *VI.8.038a)426,534,619
  analyses(*V.3.2.040,*.071)492,495
  change, irreversible(*IV.5.1.010,
    .012,.013)396,397
  change, reversible(*IV.5.1.004)396
  control(*V.3.3.3.053)535
  decomposition(*IV.5.4.1.005)412
  (scarification)(*V.1.3.055)435
  spray formulation(*V.3.3.3.005)531
  stimulation(*IV.3.3.003a,.008d)376,
    377
  substance(*V.3.3.3.075)536
  thinning(*V.3.4.000a)539
  treatment(V.1.4.010;*.4.2.1.003,
    *.014)441,565,566
  weed controls(*V.3.1.006)485
Chemiluminescence(IV.0.039)330
Chemistry(*V.2.096b)475
Chemonasty(*IV.3.3.003a)376
Chemotropism(IV.3.3.008d)377
Cherries in brine(*IV.7.1.015,*.043)
  599,603
Chesnut blight(*V.3.3.2.181e)519
Chico(*III.1.1.005)303
Chilled juice(VI.10.006)673

Chilling:
  injury(*IV.2.1.037;.5.3.005,*.012;
    V.1.3.087;.3.3.2.270;VI.9.3.029,
    .072,.100)359,407,408,439,524,
    665,668,670
  requirement(*I.1.016;IV.5.3.006)5,
    407
  resistance(IV.5.3.007)408
  -resistant plants(IV.5.3.008)408
  -sensitive plants(*IV.5.3.008,*.033)
    408,410
  -sensitive produce(*VI.5.035b)591
  -tolerant(*IV.5.3.033)410
Chilopoda(*VI.7.2.017n)613
Chimera (chimaera)(III.2.2.024,.01)
  318,327
Chimneystack (loading pattern)(*VI.6.
  017)595
Chinch bug(*V.3.3.1.132c)502
Chinese layerage(*V.1.4.036a)449
Chip budding(*V.1.4.028e,*.028f,f-2,
  *-4,.08c,.11i)444,445,460,463
Chipping(*VI.7.1.025)600
Chiropterochory(IV.4.7.001d)393
Chiropterophily(*IV.4.5.001c)391
Chitin(*IV.2.2.032f)362
Chlamydospore(s)(V.3.3.2.094,*.181g,
  *.356)513,520,527
Chloranemia(*V.3.3.2.473)530
Chlorenchyma(*II.3.125)200
Chlorides(*V.3.5.032)558
Chlorine (Cl)(*V.3.2.018)491
  total(*V.3.2.071)495
Chloro -(*II.5.1.004d)264
2-Chloroethylphosphonic acid(*IV.2.1.
  033)358
(2-Chloroethyl)trimethylammonium chlo-
  ride(*IV.2.1.037)359
Chlorophyll(s)(*IV.0.227;*.1.1.002e;
    .1.2.036a,d;.2.2.107,*.168;*.3.
    2.003;*.5.2.018;*V.3.3.2.008,
    *.473;*VI.3.023)343,351,354,355,
    366,368,373,400,510,530,578
  content(*VI.1.008)570
  deficiency(*V.3.3.2.095)513
  heat stability(*IV.5.2.018)400
Chlorophyllous(*V.3.3.2.034)511
Chloropicrin(*V.1.1.065)426
Chloroplast(*II.3.111;IV.1.1.002,
  .021;.1.2.036c,e,*.040g;*.2.2.
  069,*.093;*.3.3.005)199,350,352,
  355,356,365,366,377
  pigments(*IV.1.2.043)357
Chlorosis, chlorotic(V.3.3.2.095,
  .133b,d,.363)513,515,527
Choanephora:
  cucurbitarum(*VI.9.1.102)634

  spp.(*V.3.3.2.181c)518
Chocolate(*VI.10.012)676
Choline(*IV.2.2.159)368
Choricarpelly(II.2.3.2.007,.03a)161,
  177
Choripetalous(*II.2.2.061c-11)127
Chorisepalous(*II.2.2.061a-1)126
Chroma(*II.5.1.000c)263
Chromatid(III.2.2.025;*IV.4.1.009)319,
  381
Chromatin(IV.1.1.018)351
Chromatography(IV.0.040)330
Chromonema(*IV.4.1.009)381
Chromonemata(IV.4.1.009)381
Chromoplast(IV.1.1.004)351
Chromosome(s)(III.2.2.026;*IV.1.1.018;
  .2.2.172;.4.1.014)319,351,369,
  381
  homologous(*IV.4.1.010)381
  paternal(*IV.4.1.004)378
  somatic number(IV.4.1.006a-5)378
  thread(*IV.4.1.009)381
Chronic stress(*IV.5.4.3.000)414
Chrysanthemum midge(*V.3.3.1.132b)501
Chryso-(*II.5.1.008j)266
Chrysomphalus anonidum(*VI.9.2.007)660
Chupon[II.1.1.3.024,.05d(1),.13b(1)]
  23,41,49
Chutney(s)(*VI.10.000,.007)674
Chytridiomycetes(*V.3.3.2.181a)518
Cicada(*V.3.3.1.132d)502
  -killer(*V.3.3.1.132e)502
Cichoreae(*II.5.5.032)272
CIF(*VI.7.1.009)598
Ciliate(II.5.4.3.006;.5.6.024;.5.05b,
  .14g)270,274,285,294
Cincinnus[II.2.1.024a,.03a(8)]101,106
Cinders(*V.1.2.005)429
Cinereus(*II.5.1.003a)263
Cinnabar (cinnabarinus)(II.5.1.005e)
  264
Cinnamon (cinnamoneus)(II.5.1.001c)263
Cion(*V.1.4.067)451
Circinate[II.1.1.2.003,.01(b);.6.004]
  20,34,279
Circular saws(*V.3.4.040,*.063,*.075)
  442,443,444
Circulative(*V.3.3.2.133e)515
Circumnutation(*IV.3.3.004)377
Circumposition(*V.1.4.036a)449
Circumscissile(II.2.3.1.006a)158
Circumscription(*II.5.5.044)272
Cirrhose leaf tips[*II.4.3.002,.010,
  *.014;.4.06(d)]241,254
Cirrhous(II.5.4.010;.5.03f)268,283
Cirriferous[II.4.3.011;.4.09(e)]241,
  257

INDEX OF TERMS

Cirrus(II.4.3.012;.4.09e)241,257
Cis(*IV.0.286;*.2.2.066)348,365
Citreae(*III.1.1.002,*.021)302,305
Citric acid(*IV.1.009)571
    anhydrous(*VI.1.008g)571
    cycle(*IV.1.2.040b)356
Citrinae(*III.1.1.002)302
Citrinus(*II.5.1.008k)266
Citrus(*I.1.003)4
Citrus(*III.1.1.002,*.005,*.022)302,
    303,305
    canker(*V.3.3.2.040)511
    knot(*V.3.3.2.181g)520
    nematode(*V.3.3.2.200e)506
    red mite(*V.3.3.3.013d,e)532
    rust mite(*V.3.3.1.172)504
    -russeting(VI.9.2.017)660
City(*VI.8.118)625
Cladogenous(II.1.3.004)94
Cladophyll(II.1.1.3.025,.06a;.4.3.
    013;*IV.5.2.085)23,42,241,406
Cladosporium:
    carpophilum(*VI.9.1.470)653
    cucumerinum(*VI.9.1.152)637
    fulvum(*V.3.3.2.181g)520
    herbarum(*VI.9.1.153,*.154,*.155,
        *.483)637,654
    pisicola(*VI.9.1.336)647
Cladosporium:
    rot muskmelons(VI.9.1.152)637
        of stone fruits(IV.9.1.153)637
        of tomato and pepper(VI.9.1.154)
        637
    spot of fig(VI.9.1.155)637
Clambering(II.1.1.3.026,.06c)23,42
Clamp-lift truck(VI.3.012)578
Clamp truck (hand-clamp truck)(VI.3.
    013)578
Class(V.2.115b)480
Classification(III.1.1;.003)302-307,
    302
Classification systems(III.1.1.004)303
Classis(*III.1.1.003)302
Clavate(II.5.6.025;.5.14h)274,294
Clavate (stigma)(II.2.2.18b)147
Claviceps purpurea(*V.3.3.2.181e)519
Clawed(*II.2.2.061c-14-w)128
Clay(s)(*V.2.015,*.076,*.096d,*.106,
    .107,.107a,*b,*d,*.114,*.117,*a,
    .120;.3.2.050,*.080;*.3.3.3.029)
    472,474,475,476,480,481,482,493,
    495,533
    accumulation[V.2.107,.01a(B2)]476,
        483
    amorphous(*V.2.096d)475
    coatings(*V.2.107a)477
    content(*V.2.099)476

    expanding lattice(*V.2.098)475
    horizon(*V.2.120)482
    impermeable(*V.2.120)482
    loams(*V.2.117g,*h)481
    minerals(*IV.0.042,*.166;*V.1.2.
        011r;*.2.095,*.098,*.100)330,340,
        431,474,475
    mineral synthesis(*V.2.106)476
    molecules(*V.2.015)472
    particles(*V.2.107b)477
    pots(*V.1.2.003h)428
    sandy(*V.2.117b,*h)481
    self-mulching(*V.2.096c)475
    silty(*V.2.117c,*d)481
    skins(*V.2.107b)477
    swelling-type(*V.2.095)474
    translocation(*V.2.120)482
Clean picking(*VI.2.023)577
Cleaning*V.3.2.072)495
Cleanliness(*VI.8.094)623
Clear-winged moth(*V.3.3.1.132g)502
Cleavage (simple) (polyembryony)(IV.4.
    6.001a,*.001b)392
Cleft[II.5.4.3.007;.5.05(c,d)]270,284
    graft(*V.1.4.028k,k-2,.09c-d,.10c,
        .11n)446,461-463
    grafting(*V.1.4.028k,k-2)446
Clefting iron(*V.1.4.028k-2)446
Cleistogameon(III.2.2.027)319
Cleistogamous, cleistogamy(*IV.4.4.
    011,.012)388
Cleistothecium (pl. cleistothecia)(V.
    3.3.2.101,*.181e)513,519
Clerical(*VI.8.053)620
Click beetles(*V.3.3.132a)501
Climacteric:
    fruits(*VI.1.006)570
    peak(*VI.1.002)569
    period(*VI.1.002,.01)569,573
    rise(*VI.1.002)569
Climate(s)(*V.2.000;*.3.5.032)472,558
    arid(*V.2.110d)478
    humid(*V.2.110b,*d)478
    semiarid(*V.2.110d)478
Climatic factors(II.4.4.003)247
Climbers (most)(II.4.1.005)238
Climbing(*II.1.1.3.085)26
    palms(*II.0.022)11
    plants(*II.4.3.002)241
Clinal, clinal variation(III.1.2.007)
    307
Cling, cling-stone(II.2.3.2.029b)162
Clinging(II.2.2.004;.4.3.014;.4.06
    b-d,.09d-g)118,241,254,257
    tendrils(*V.2.110e-2)479
Clipping(*V.1.3.055;*VI.2.022a)435,576
Clitocybe tabescens(*V.3.3.2.181f)519

INDEX OF TERMS

Clonal:
  plants(*V.1.3.085)439
  rootstock(*V.1.4.028p,*s)448
  rootstock sources(*V.1.4.010)441
  varieties(III.2.0.005a-1)313
Clone(III.2.0.002,*.003a,*.005a-1,
  .006,.010;.2.1.001a;*IV.4.4.022;
  *.4.8.017;V.3.3.2.103)311,313,315,
  316,389,393,513
Closed:
  flowers(*IV.4.4.012)388
  patterns(*V.2.110e)479
Closing:
  cells[II.3.12b(1)]218
  hammer(VI.3.014)578
  machine(VI.3.015)578
Cloth:
  bags(*V.1.3.067;*VI.4.000b,*c,*.001,
  *.020)437,585,587
  budding (grafting) tape(*V.1.2.017)
  432
  mesh bags(*VI.3.053a;*.4.000b,*.001)
  581,585
Clothes moths(*V.3.3.1.132g)502
Cloud of tomato(*VI.9.3.074)669
Club:
  leaf of clover(*V.3.3.2.361)527
  moss(*II.1.1.3.053;.6.005,*.016)24,
  279
  root of crucifers(*V.3.3.2.133c,
  *.181b;VI.9.1.156)518,637
  - shaped(*II.5.6.025)274
<u>Clusia</u>(*II.4.3.064)245
Cluster base(*V.3.4.006)540
Clustered(*II.2.1.001)100
Clustering of shoots(*V.3.3.2.171)516
C:N ratio(*V.2.014)472
CoA(*IV.2.2.058c,*.142)364,367
Coalescent(II.2.2.061b)126
Coarse:
  sand(*V.2.114,.117l-4)480,481
  sandy loam(V.2.117j-4)481
  silt(*V.2.114)480
Coated(*II.5.7.009;V.2.107d)278,477
Cobalamin(*V.3.2.058)494
Cobalt (Co)(*V.3.2.018,*.058)491,494
Cobwebbed(*II.5.6.005)273
Coccineus(II.5.1.005r)264
Coccus[II.2.3.1.009,.02g(1)]159,169
Cochlear(II.1.1.2.004,.01c)20,34
Cochleate(II.2.2.041f,.15f)122,144
<u>Cochliobolus</u> <u>heterostrophus</u>(*VI.9.1.
  514)656
Coconut:
  husk(*V.1.1.053e)425
  husks(*V.3.3.3.059)535
  milk(*V.1.5.005)470

  palm beetle(*V.3.3.3.013f)532
  weevil(*V.3.3.1.200a)505
Cocoon(*V.3.3.1.132g)503
Code dating(VI.8.014)617
Code of Federal Regulations(*VI.7.2.
  038)615
Codling moth injury of apple and pear
  (VI.9.2.108)660
Coefficient of velocity(V.1.3.008)432
Coenocytic(V.3.3.2.105)514
  mycelia(*V.3.3.2.279,*.477)524,531
  mycelium(*V.3.3.2.181d)518
Coenospecies(*III.1.2.006)307
Coenzyme(IV.2.2.058b;*.5.4.1.005)364,
  413
  A (CoA)(*IV.1.2.040b;.2.2.058c,
  *.142)356,364,367
  metabolic processes(*IV.5.4.1.005)
  412
  net synthesis(*IV.5.4.1.005)412
Coffee(V.3.4.017)540
  lesion nematode(*V.3.3.1.200a)505
  (pruning)(V.3.4.017,.06,.07,.08,.09)
  540,550-553
  rust(*V.3.3.2.181f)519
Coflorescence[II.2.1.017,*.052,.06a
  (1)]100,103,109
Coherent(II.2.2.016,.05g)121,134
Cohesion(IV.5.2.006)399
Coiled(II.2.3.3.008,.02d;.6.004)163,
  183,279
Coir(*V.1.1.053e,*.1.4.036a)425,449
<u>Cola</u>(*II.2.3.1.017,*.2.3.3.014c)159,
  163
Colchicine(*IV.2.2.012,.044)360,363
Colchiploid(III.2.2.028)319
Cold(*IV.5.1.014;*.5.2.001)397,398
  acclimation(*IV.5.3.028)409
  dry treatment(*V.4.1.111)564
  frame(V.1.1.006,*.018,.048)419,421,
  424
  hardening(*V.5.3.028)409
  -moist treatment(*V.4.1.121)565
  requirement(*V.4.1.111,*.114)564
  -resistant crotch system(*V.1.4.
  028h)445
  -resistant trunk system(*V.1.4.
  028h)445
  room(VI.5.035a-2)590
  shock(*IV.5.3.005,*.008,*.012)407,
  408
  storage room(*V.1.1.047)424
  stress(IV.5.3.)404
  stress sensitivity(IV.5.4.4.001)415
  tolerance(*IV.5.3.028)409
  treatment(*IV.3.2.021e)376
Cole crops(I.3.005)7

INDEX OF TERMS

Coleoptera(V.3.3.1.132a)501
Coleoptile[II.2.3.3.014a,.01b(5)]163, 182
Coleorhiza[II.2.3.3.014b,.01b(9)]163, 182
Coleosporium(*V.3.3.2.294)525
Collateral(*II.3.210,.28d;*VI.6.001] 205,234,594
Collenchyma(*II.3.022,.030,*.067, .04g-i)194,195,197,210
Collection of (cutting) material(V.1. 4.22a-1)442
Colleter(II.3.031)195
Colletotrichum(*V.3.3.2.019)510
  capsici(*VI.9.1.467)653
  circinans(*VI.9.1.492)654
  coccodes(*VI.9.1.033)628
  erumpens(*VI.9.1.460)652
  gloeosporioides(*V.3.3.2.181e;*VI.9. 1.023,*.025,*.027,*.029,*.031)519 527,628
  higginsianum(*VI.9.1.034)628
  lagenarium(*V.3.3.2.181g;*VI.9.1. 028,*.035,*.183)520,628,629,639
  lindemuthianum(*V.3.3.2.181g;*VI.9. 1.032)520,628
  musarum(VI.9.1.024)618
  pisi(*VI.9.1.026)628
Colloid(IV.042,*.261)330,347
Colloidal(*V.2.111)479
  clay(*V.2.114)480
  system(*IV.0.042,*.261)330,347
Colombian (pruning system)(*V.3.4. 017,.07c)541,551
Color(II.5.1.000;*IV.4.4.013;*V.2. 096c;*VI.8.038e,*.094)263,389,475, 619,623
  add(VI.3.016,*.031)578,580
  adding(*VI.7.1.025)601
  additive(*VI.7.2.016,.016b)610,611
  Additive Amendments of 1960(*VI.7.2. 016)610
  break(VI.1.008b,*.009,*.010)570,571
  reddish-purple(*V.3.3.2.018)510
  separators(*V.1.3.057)436
Colorimetric methods(*V.3.2.040)492
Coloring(*VI.3.023)579
  of economic poisons(VI.7.2.017d)613
Colorless(II.5.1.007)265
Colpate, colpi(*II.2.2.007t-10)120
Colter(V.3.1.004)485
Column(II.2.2.017,.06f,.025b-d)121, 135,154
Columnar(II.1.1.3.028,.05c,.09b;.5.5. 009;.5.08f;V.2.115c)23,41,45,271, 288,480
Columns(*V.2.109)478

Coma(II.2.3.3.009;.2.3.1.01f,g)163,168
Combination fiberboard and wood or veneer(*VI.4.008)586
Combine(s)(V.1.3.010,*.063,*.085;*.3. 0.003)432,437,439,484
Combining ability(III.2.2.029)319
Comb-shaped(II.5.4.3.028)270
Commensualism(V.3.3.2.110)514
Commercial:
  bulb(V.4.1.026)560
  crop(*IV.4.8.014,*.015,*.019,*.034, *.037,*.047,*.050,*.054)393,394
  floriculture(*I.0.001,*.002;*.2.001) 3,5
  hydroponic culture(*V.3.2.083)495
  packinghouses(*VI.3.000)577
  retail containers(*V.1.3.067)437
  size(*V.4.1.085)563
  wholesale containers(*V.1.3.067)437
Commingled produce(*VI.8.086)622
Commiscuum(III.1.2.009)308
Commission merchant(VI.7.1.013,*.039, *.040,*.049,*.050,*.053,*.054,*.072, *.077;.8.015)599,602,603,604,606, 607,617
Commisure[II.2.3.1.010,*.022,.03k (2,6)]159,170
Commodity(*VI.8.085)622
Common(*V.8.053)620
  bean mosaic of snap beans(VI.9.1. 167)638
  blight of lima beans(*VI.9.1.050)630
  carrier(VI.6.004)594
  names(III.1.1.005)303
  smut of corn(*V.3.3.2.181f;.9.1.169) 519,638
Comose(II.5.6.028,.14k)274,294
Companion cells[II.3.032,*.175,.26b (4),c(2)]195,203,**232**
Companion crop(V.3.1.005)485
Company(*VI.6.000)594
Comparium(III.1.2.010)308
Compatibility(V.1.4.015;.3.3.3.019, *.071)441,533,536
Compatible(IV.4.8.011;V.3.3.3.018)393, 533
Complaint(VI.7.1.014,*.035)599,602
Complementary cells[II.3.12b(2)]218
  genes(III.2.2.030)319
Complete:
  cover(*V.2.110e)479
  fertilizer(V.3.2.013)491
  (flower)(II.2.2.018,.01a,b)121,130
  NPK mixtures(*V.3.2.037)492
  (parthenocarpy)(IV.4.2.001a)385
  protandry(*IV.4.3.002b)385
Complex(V.3.2.012)490

INDEX OF TERMS

(polyembryony)(IV.4.6.002)392
Compliance safety and health officers
  (*VI.7.2.038)616
Composite(*IV.4.8.009)393
  - cross populations(III.2.0.005b)315
  mixture(V.1.3.011)432
Compost(*V.1.2.010,.011c;*.2.099)429,
  430,476
Compound:
  corymb[II.2.1.02a(5)]105
  dichasium[*II.2.1.030,.03a(2).08c]
    101,106,111
  fruit(II.2.3.1.011,.06,.07)159,173,
    174
  inflorescence(II.2.1.02b)105
  layerage(*V.1.4.036b-2)450
  (leaf)(II.1.2.016,.05c)60,70
  or form applied(*V.3.2.065)494
  (polyembryony)(IV.4.6.001b)392
  (sieve plate)(*II.3.182)203
  trees(*IV.5.4.4.002)416
  umbel(II.2.1.03b)106
Compressed(II.2.2.019,.28a)121,157
Compression wood(*II.3.155)201
Computer(*VI.8.033)618
Concave(II.5.5.010;.5.09a)271,289
Concealed damage(*VI.8.108)624
Concentrated:
  fertilizer(V.3.2.014)491
  spray(*V.3.3.3.057)535
  superphosphate(*V.3.2.091)496
Concentration(IV.0.171c;V.3.3.3.020;
    VI.10.008)340,533,674
  (of amount of substance)(IV.0.171c,
    *.260b)340,344
Concentrations of individual compounds
  (*V.3.4.000a)539
Concentric:
  cracks around shoulder(*VI.9.3.059)
    668
  (phloem)(*II.3.136)201
  ring stipple of citrus(VI.9.3.031)
    665
Concrete:
  mixer(*V.1.1.062)426
  pots(*V.1.2.003h)428
Condiments, spices, and essential oils
  (I.3.006)7
Condition(VI.8.016)617
Conditional parthenocarpy(IV.4.2.002)
  386
Conditioner(*V.3.2.044)493
Conditioning(IV.5.3.012)408
Conductance(*IV.0.260c)344
  ratio(*IV.5.3.024)409
Conducting-to-nonconducting tissues
  (*IV.5.2.080)406

Conduction(VI.5.052a)593
Conduplicate(II.1.1.2.005,.01d;.2.2.
  17a)20,34,146
Cone(V.1.2.003b;*.3.4.067)428,541
Confluent(*II.2.2.016;*.3.014)121,194
Congeniality(*V.1.4.015)441
Congested(*II.2.1.001)100
Conglomerated(*II.2.1.001)100
Conical(II.2.3.2.008,.01d)171,176
Conidia(*V.3.3.2.019,*.181e,*.181g,
  .219,.347)510,519,520,521,526
Conidiophore(s)(*V.3.3.2.112,*.181c)
  514,518
Conidiospore(*V.3.3.2.403)528
Conidium (pl. conidia)(V.3.3.2.112)514
Conifer(II.6.007,.01c)279,301
Coniothyrium sp.(*VI.9.1.171)638
Conjugate division(*V.3.3.2.127)514
Conjunctive(II.4.3.067)245
Connate(II.2.2.020,.18k)121,147
  - perfoliate(II.1.2.027,.05d)60,70
Connective[II.2.2.007f,.08f,.11b(2)]
  119,137,140
Connivent(II.2.2.021,.09c)121,138
Conoidal(II.2.3.2.009,.02e)161,176
Conotrachelus nenuphar(*VI.9.2.013,
  *.063)600,662
Consignment:
  sale(VI.8.017)618
  transactions(*VI.7.1.003)597
Consignor(*VI.7.1.032)602
Conspecific(III.1.2.011)308
Consperse stinkbug(*VI.9.2.013)660
Conspicuous(IV.4.4.04b)387
Constancy(IV.4.5.002b)391
Consumer:
  containers(*VI.4.000b,*.001,*.025)
    585,588
  cooperative(VI.8.018)618
  packages (shrink film)(*VI.4.021)587
  packs(*VI.3.053d)582
  route(*VI.8.123)625
Consumer's home(*VI.8.081)622
Contabescence(II.2.2.007g,.05c)119,
  134
Contact herbicide(V.3.3.3.021)533
Container, containers(V.1.2.003;*.1.3.
  063,*.085;*.1.4.028a,*k-2,*t;*.1.
  5.005,*.4.1.107,*.119;*VI.3.046;
  .4.000;*.6.003)428,437,439,443,
  446,448,470,563,572,581,585,594
  and Loading Rules Tariff(*VI.6.017)
    595
  "body" icing(VI.5.004)588
  fillers(V.1.1.007)419
  filling machines(*V.1.1.036)423
  forcing(V.4.1.114)564

INDEX OF TERMS

   -grown nursery stock(*V.1.2.011p)431
   -grown plants(V.4.1.011)559
   loading pattern(*VI.6.017)595
   loading pattern choice(*VI.6.018)595
   manifest(*VI.6.020)595
   mixed sizes(*VI.6.018,*.022)595
   moistures(*V.4.1.082)563
   paper(*V.1.2.003f)428
   plants in(*V.4.1.070)562
   plastic(*V.1.2.003f)428
   pot or poly bag(V.1.5.01n)471
   refrigerated(*VI.6.037)597
   restocking(*VI.6.002)594
   size(*V.4.1.107)564
   "top" icing(*VI.5.054)593
   types (limitations)(*VI.7.2.004)608
   unitization(*VI.6.035)596
   washing facilities(*V.1.1.047)424
   wood(*V.1.2.003f)428
Continuous(II.1.2.028)60
   air flow(*VI.5.047a)592
   growth(II.1.1.3.029,.07a,b)23,43
Contorted(II.1.1.2.006,.01e)20,34
Contorto-duplicate(II.1.2.007)20
Contour planting(V.3.1.008b)485
Contract(*VI.8.086)622
   form(*VI.6.001)594
Contractile roots(II.3.17b,c;.4.3.023;*IV.5.1.001,*.018;*.5.3.038)223,242,396,397,410
Contracts(*VI.7.1.031)601
Contra-inoculation(V.1.3.012)432
Contre-espalier(V.3.4.020,.11c,d)541,555
Control(V.3.3.3.022)533
Controlled:
   atmosphere storage(VI.5.047b)592
   injury of apple(VI.9.3.032)665
   release fertilizer(V.3.2.015)491
Controlled-temperature area(*V.4.1.029)560
   containers for bulb transportation (*VI.8.097)623
   facility(*V.4.1.101)564
   forcing(*V.4.1.093)563
   (lily bulb) forcing(V.4.1.029)560
Controller(*VI.8.074)621
Convection(VI.5.052b)593
Convenience store(VI.8.019)618
Conventional forced air cooling room (IV.5.035a-3)590
   forcing(*V.4.1.114)564
Convergent(*II.2.2.021)121
Conversion units(IV.0.157b)337
Convex(II.5.5.011;.5.09b)271,289
Conveyor(*VI.2.010;*.3.019,*.046,*.053b,*.062)575,578,581,582

Convivium(III.1.2.012)308
Convolute(II.1.1.2.008,01f)20,34
Cool storage(*V.1.3.073;*.1.4.028u) 438,449
Cooling(*V.1.1.017;*VI.5.000)421,588
   coefficient(VI.5.014,*.021)588,589
Cool-season seeds(V.1.3.013)432
Co-op, cooperative(*VI.6.026;*.8.096) 596,623
   association(*VI.8.004)616
   associations with seasonal pools (*VI.7.1.003)598
Copeland Act of 1938(*VI.7.2.016)610
Copper (Cu)(*V.3.2.028,*.034,*.058,*.067)491,493,494
   chelate(*V.3.2.085)496
   neutral(*V.3.2.085)496
   sulfate(*V.3.2.085)496
   total(*V.3.2.071)495
Coppery(II.5.1.005g)264
Coppice(II.1.1.3.030)23
Coprocanthrophily(*IV.4.5.001c)391
Coprogenous earth(*V.2.107i)477
Coprophily(*IV.4.5.001c)391
Cordate(II.5.4.2.005;.5.5.012;.5.04d,.09c)269,271,284,289
Cordon(V.3.4.021,*.049,*.055,.11b,c,e) 541,542,543,555
   modified free-standing(*V.3.4.056) 543
Core:
   breakdown of pears(VI.9.3.033)665
   drill(*V.3.5.024)558
   rot of apples(VI.9.1.171)638
   rot of pears*VI.9.3.033)665
Coreid(*V.3.3.1.132c)502
Cores(*V.3.2.082)495
Coriaceous(II.5.7.007)278
Cork[II.3.01C(26)]207
   cambium[*II.3.108,*.174,.01C(23)] 199,203,207
   cells[II.3.29c(1)]235
   of apples(VI.9.3.035)666
   spot of 'Anjou' pears(VI.9.3.036)666
   spot of apple(VI.9.3.037)666
Corky(II.5.7.031)279
   fleck(*V.4.2.1.022)566
Corm(s)(*I.2.001b;*II.4.3.057;.4.13c; V.1.4.071b,.16f;*.3.1.013;*.4.1.120)5,244,261,451,468,487,565
   natural vegetative reproduction(*IV.4.1.006a-8)379
Cormel(*V.1.4.019,*.071b)441,451
Corn(maize)[II.2.1.14a,b;.2.3.1.01o(1,2);.2.3.3.01b]117,168,182
   caryopsis(II.2.3.1.008;.2.3.3.01b) 159,182

INDEX OF TERMS 718

detassel(III.2.2.037)320
earworm(*V.3.3.1.132g;*.3.3.3.013f;
   VI.9.2.021)502,532,660
grits(*V.3.3.3.009)531
picker(*V.1.3.063;*V.2.022b)437,576
plant organs(*V.3.3.2.205)521
Cornell:
   foliage plant mixes(*V.1.2.005)429
   "Peat-Lite" mixes(*V.1.2.011,*.011n)
      430,431
   "Peat-Lite" soil mix(*V.1.3.047)435
Corniculate[II.2.2.061c-14-g,.26a(4)]
   128,155
CORNUCOPIA(*VI.6.026)595
Corolla[II.2.1.01(3);.2.2.061c,.26a
   (2)]104,126,155
   dark-colored(*IV.4.4.030)389
Corona[II.2.2.022,.27a(2),b(1)]121,156
Coronate[II.2.2.061c-14-h,.26a(3),.27a
   (2),b(1)]128,155,156
Coronet(II.2.2.022)121
Corporation(*VI.8.004,*.057,*.070)616,
   620,621
Corrugate(*II.1.1.2.010,.01g,h)21,34
Corpus[II.3.14b(4)]220
Cortex(*II.3.063,*.067,*.073,*.080,
   01c(21),.02d(6),.12a(5),.13a(3),
   .20a(5),.24d;.4.10c(b)]197,207,208,
   218,219,226,230,258
Cortical:
   cells (root)(*V.3.3.2.154)516
   fiber(*II.3.063)197
   tissue(*V.3.3.2.082)513
Corticate(II.5.7.009)278
Corticium centrifugum(*VI.9.1.223)641
Corymb[II.2.1.020,.02a(4,5),.053]100
   105,108
Corymbose, corymbiform(II.2.1.021,
   .06b,.12b)100,109,115
Corynebacterium(*V.3.3.2.040)511
   flaccumfaciens pv. flaccumfaciens
      (*VI.9.1.075)632
   michiganense pv. michiganense(*VI.
      9.1.052)630
      pv. sepedonicum(*V.3.3.2.040;*VI.
         9.1.055)511,630
   spp.(*V.3.3.2.040)511
Coryneum:
   beijerinckii(*V.3.3.2.181g)520
   carpophilm(VI.9.1.176)638
Coryneum blight of stone fruits(VI.9.
   1.176)638
Cosmetics(*VI.7.2.016)610
Cost:
   and charges(*VI.7.1.009)598
   and freight(*VI.7.1.009)598
   , insurance, and freight(*VI.7.1.

009)598
Costapalmate(II.1.2.029,.06a)60,71
Costa Rican system(*V.3.4.017,.07d)
   541,551
Costate(II.1.2.121b,.22b)64,87
Cost-plus(VI.8.022)617
Cotton:
   bollworm(*VI.9.2.021)660
   bur ash(*V.3.2.088)496
   -seed(*I.1.008)4
   -seed meal(*V.3.2.065,*.068,*.086)
      494,496
   wilt(*V.3.3.2.181g)520
Cottony:
   leaf of eggplant(VI.9.1.177)638
   of snapbeans(VI.9.1.178)638
   rot of lemons(VI.9.1.179)638
      of lettuce, celery, other vege-
      tables, lemon(*V.3.3.2.181e)519
Cotyledon, cotyledons[II.2.3.3.014c,
   .01a(1),b(2),.03f-i(3),.07e(2);*.5.
   7.033;*.6.011;*IV.2.2.172]163,182,
   184,188,279,369
Cotyledonary growth phase(*V.3.3.3.
   031)533
Cotyloform(II.2.2.061c-1-c,.22a)126,
   151
Cotype(III.1.1.035b)307
Coulomb(*IV.0.260c)344
   per cubic meter(*IV.0.260e)345
   per kilogram(*IV.0.260e)345
   per square meter(*IV.0.260e)345
Coulter blades(*V.3.1.004)485
   (harrow)(*V.3.1.004)485
Coulure(IV.4.8.012)393
Coumarins(*IV.2.2.154,*.162)368
Coumarone indene(*VI.3.085,*.085c)584
Count(VI.8.023)618
   fill(*VI.3.053c)582
Coupled reactions(IV.0.047)331
Court decisions(*V.3.5.033)559
Cover crop(*V.2.110a,*e;*.3.1.000;*.3.
   2.000,*.069)478,479,484,490,494
   utilization(*V.3.2.000)490
Coverage(*V.3.3.3.023,*.049,*.052)533,
   534
Covering seeds(*V.1.1.020)421
Coverings(V.1.1.017d)420
Cracked stem of celery(VI.9.3.038)666
Cracking of stone fruits(VI.9.3.039)
   666
Crane flies(*V.3.3.1.132b)501
Crassulacean acid metabolism (CAM)(IV.
   1.2.008c)352
Crate(s)(*VI.2.012,*.4.000a,.009,
   *.014)575,585,586,587
   field(*VI.4.009)586

nailer(*VI.3.006)578
shipping(*VI.4.009)586
Crater:
  rot of carrot(VI.9.1.180)638
  spot of celery(VI.9.1.181)638
Crateriform(II.2.2.023,.22b)122,151
Crawler-type tractors(*V.3.0.003)484
Creasing of citrus(VI.9.3.040,*.106)
  666,667
Creepers(*V.2.110e)479
Creeping legumes(*I.2.011)6
Cremocarp(*II.2.3.1.032,.03i-k)160,170
Crenate, crenulate(II.5.4.3.009;.5.
  05e)270,285
Crepuscular(IV.4.5.002c)391
Crested(II.2.3.3.010)163
  (stigma)(II.2.2.18c)147
Crickets(*V.3.3.1.133h)503
Criconema spp.(V.3.3.1.200b)505
Criconemoides spp.(V.3.3.1.200b)505
Crispate, crisped(II.5.4.3.010;.5.
  05f)270,285
Crispy-hairy(II.5.6.029;.5.14j)274,294
Cristi(*IV.1.1.016)352
Cristate(II.5.5.013;.5.08g)271,288
Critical:
  daylength(IV.3.2.013b)374
  relative humidity(V.3.2.016)491
  water potential(IV.5.2.007)399
Crochets(*V.3.3.1.132e)502
Cronartium ribicola(*V.3.3.2.181f,
  *.294)519,525
Crop, crops(*V.3.1.008;*.3.3.3.032,
    .052,.055,*.056.*.069;*VI.8.
    085)485,533,534,535,536,622
  age, size or yields(*V.3.2.035)492
  control(*V.3.3.3.000)531
  damage(*V.3.3.3.048,*.090)534,537
  disposal(*V.3.0.001)484
  ecology(IV.0.048)331
  growth and development(*V.3.2.017)
    491
  land(*V.2.110)478
  logging(V.3.2.017)491
  losses(*V.3.3.3.022,*.111)533,538
  needs(*V.3.2.000;*.3.5.013)490,557
  pests(*V.3.3.1.,*.224)498-509,507
  plant(*V.3.2.065)494
  plants(*IV.5.2.077)405
  plants (tender growth)(*V.3.5.013)
    557
  ,preceding(*V.3.3.3.117)538
  problems(*V.3.0.001)484
  production(*V.2.110;*.3.0.000,*.001;
    *VI.8.090)478,484,623
  production and management(*V.3.3.3.
    000)531

Production (Commercial Floriculture)
    (V.4)559-568
  (Fruits, Vegetables and Woody Or-
    namentals(V.3)484-559
  quality(*V.3.2.000)490
  records(*V.3.2.037)492
  requirements(*V.3.2.035)492
  ,row(*V.3.3.3.010)532
  seed(*V.1.3.051)435
  sensitivity(*V.3.5.032)558
  water uptake(*IV.5.2.083)406
  yields(*V.3.2.000)490
Cross:
  -arm system(*V.3.4.05)549
  -compatible(IV.4.8.013)393
  -fertile(IV.4.8.014,*.018)393,394
  -fruitful(IV.4.8.015,*.019)393,394
  -incompatible(*IV.4.8.013,.016)393
  -inoculation(V.1.3.014)432
  -merchandising(VI.8.024)618
  over(*IV.4.1.011)381
  over unit(IV.4.1.011)381
  -pollination(IV.4.8.017)393
  -row conveyor-central collection
    (*VI.2.022c)576
  -sterile(IV.4.8.018)395
  -unfruitful(IV.4.8.019)395
Cross overs(*V.3.4.000b)539
Crosscut log saw(*V.3.4.011)540
Crossing-over(III.2.2.031;IV.4.1.010)
  319,381
Crotch repair(V.1.4.013a-d)465
Crotonylidene urea(*V.3.2.015)491
Crowded(II.1.2.031,.07a)60,72
Crown[II.1.1.3.031,.07c,.08a;*.2.2.
    022;.4.3.015,*.063;*V.1.4.024,
    *.064,.04a,c(1);V.3.4.01(13);.3.5.
    012]23,43,44,121,242,245,443,451,
    456,545,557
  and trunk canker of deciduous fruits
    -black walnut(*V.3.3.2.181d)518
  bud(V.4.1.031)560
  gall(*V.3.3.2.040,*.320)511,525
  grafting(*V.1.4.028k)446
  rot(*V.3.3.2.181d)518
    of banana(VI.9.1.182)639
  -shaft(*II.1.1.3.031)23
  slips[V.1.4.04c(2)]456
  sucker[V.3.4.01(15)]545
Crowned(II.2.2.025)122
Cruciform(II.2.2.061c-1-d,.21d)126,151
Crumb(V.2.115d)481
Crumpled(II.1.1.2.010,.01h)21,32
Crustacea(*V.3.3.1.258;*V.7.2.017n)
  508,614
Crustaceous(II.5.7.010)278
Cryobiology(IV.5.3.013)408

Cryoprotectant(IV.5.3.014)408
Cryostability(*IV.5.3.034)410
Cryptogams(*IV.5.2.040)404
Cryptophyte(II.4.1.002;.4.01b)238,249
Crystal violet(*V.3.3.2.040)511
Crystals(II.3.037,*.175,.08a-d;*IV.1.
  1.030)195,203,214,352
CTF(*V.4.1.029)560
C-3 plants(*IV.1.2.007,.008a,*b,*c,
  .035;.5.2.002,*.015,*.068,*.084,
  *.085)352,353,354,398,399,404,406
  anatomy(*IV.5.2.015)399
  metabolism(*IV.5.2.002)398
  transpiration ratio(*IV.5.2.068)404
  water use efficiency(*IV.5.2.084)406
  xeromorphic structure(*IV.5.2.085)
  406
Cubic:
  meter(*IV.0.260b)344
  meter per kilogram(*IV.0.260b)344
Cucullate(II.2.2.061c-14-i,.26)128,155
Cucurbit wilt(*V.3.3.2.040)511
Cucurbits(I.3.007)7
Cucumber:
  anthracnose(VI.9.1.183)639
  beetles(*V.3.3.1.132a)501
  mosaic(*V.3.3.2.133c)515
  mosaic on celery(VI.9.1.184)639
  mosaic virus(V.4.2.1.004,*.009,
    *.022)565,566
Cu-EDTA(*V.3.2.012)490
Culling(*V.1.4.028t)448
Culm(*II.1.1.3.021,.032,.08b)23,44
Cultigen(*III.1.0.001;.1.2.013)302,308
Cultispec(III.1.2.014)308
Cultivar, cultivars(*III.1.1.003,
  .024;.1.2.053,*.057;*.2.0.003,
  .005a,.011,*.012;*.2.2.091;*IV.
  4.8.013,*.014,*.015,*.016,*.017,
  .018,.019,*.032,*.033,*.034,
  .035,.036,*.037,*.045,*.046,
  .047,.048,*.049,*.050);*.5.2.036
  *.039;*V.1.3.073;*.1.4.015,
  *.028c,*o;*.3.0.001;*.3.1.008c;
  .3.3.3.000;.4.2.2.006,*.016;*VI.
  8.085)302,305,310,311,313,316,322,
  393,394,395,402,403,438,441,444,
  448,484,485,531,567,622
  breeding new(*V.4.1.053)561
  classification*V.4.1.097)563
  uneconomic(*V.1.4.028v)449
Cultivated:
  crops (most)(*II.4.2.022)240
  fruits, vegetables and ornamentals
    (*V.3.3.2.179)517
Cultivation(*V.3.0.003;.3.1.006,*.009,
  .010;.3.3.3.119)484,485,486,538

Cultural:
  operations(*V.3.0.003;.3.1.008,
    *.008j;.3.3.3.071)484,485,486,536
  program(*V.2.110;*.3.0.001)478,484
  system(*V.3.1.010,*.013)486,487
Culture(*V.1.5.005;.3.3.2.121,*.373)
  470,514,517
  and shipment to market(V.4.1.)559
  in isolation(V.1.4.010)441
  medium(*V.1.5.006)470
  solution(*V.3.2.007,*.050)490,493
Cumulic(*V.2.096c)475
Cuneate(II.5.4.2.006;.5.04e)269,284
Cuneiform, wedge-shaped(II.5.5.014;.5.
  09d)271,289
Cupola-shaped[II.2.2.061c-1-e,.27a(3)]
  126,156
Cup-shaped(II.2.2.061c-1-f,*.061c-1-g,
  .27(1)]126,156
Cupreus(*II.5.1.005g)264
Cupule(II.2.3.1.013,.04d)159,171
Curculios(*V.3.3.1.132a)501
Curie (ci)(*IV.0.260k)347
Curing(*VI.7.1.025)600
Curled(*II.5.4.3.010)270
Curl-stripe(V.4.2.1.005)565
"Curly-tip" of fig(V.3.3.1.200c)505
Current:
  density(*IV.0.260b)344
  -season flush growth(*V.1.4.022f-4)
    443
Curvative(II.1.1.2.011,.01i)21,34
Curved(*II.1.2.012)59
  (embryo)(II.2.3.3.011,.02e,f)163,183
Curve-ribbed(II.1.2.121c,.22c)64,87
Cuscuta spp.(*V.3.3.2.133e)515
Cushion plant(s)(*II.4.1.001;.4.3.016;
  .4.01a)238,242,249
Cushioned(*II.1.2.086)62
Cuspidate(II.5.4.1.011;.5.03g)268,283
Customer:
  bagged(*VI.8.136)626
  checkout operations(*VI.8.131)625
  damage(*VI.8.108)624
  harvesting(*VI.8.138)626
Cut(*II.5.4.3.021)270
  flower longevity(*VI.8.132)626
  flowers(*I.0.001;.2.001c;*V.4.1.012,
    .062;VI.2.000;.3.000)3,5,559,
    562,575,577
Cuticle[II.2.3.2.05d(2);.2.3.3.12e(4);
  .3.038,*.054,.12a(2),.15a(5);IV.2.
  2.046;*.5.2.002]179,193,195,196,
  218,221,363,398
  development(*IV.5.2.080)406
  lacquerlike(*II.4.3.028)243
  thick(*IV.5.2.002,*.020)398,400

INDEX OF TERMS

Cuticular resistance(*IV.5.2.032)401
 transpiration(*IV.5.2.059,*.079,
 *.085)404,406
Cutin(*II.3.038)195
Cuttage(*V.1.4.022)441
Cutting, cuttings, cuttage(*IV.2.1.
 012,*.033;.5.2.030;*V.1.1.053;*.1.
 3.033;*.1.4.000,.022,*.024,*.041,
 .01-.03;*.4.1.074;*VI.2.022a;*.7.
 1.025)358,401,418,425,434,441,443,
 450,453-455,562,576,601
 bed(*V.1.4.028g)445
 bench(*V.1.4.041)450
 deciduous(*V.1.4.022f-1,*-3)442,443
 direct-set(V.1.4.022b)442
 -graft(V.1.4.028g,.03d)445,455
 hardwood(*V.1.4.022a-3,*.022b)442
 hardwood stem(*V.1.4.022a-2,.022f-1)
 442
 herbaceous(V.1.4.022f-2)443
 leaf(V.1.4.022c)442
 leaf-bud(V.1.4.022d)442
 off (rootstock)(*V.1.4.028i,.11g)
 445,463
 piece root(*V.1.4.022b,*e)442
 preparation(V.1.4.022a)442
 preparation space(*V.1.1.047)424
 root(*V.1.4.022a-2,*-3,.022e)442
 rooting(*V.1.4.022a-3,*.028t)442,448
 semihardwood(V.1.4.022f-3)443
 softwood(V.1.4.022f-4)443
 stem(*V.1.4.002b)442
 truncheon(V.1.4.022g)443
 unrooted(*V.1.4.028a)443
 whole root(*V.1.4.022b,*e)442
Cutworm(*V.3.3.1.132g)502
Cyaneus, cyano-(*II.5.1.002h)263
Cyanide accumulation(*IV.5.4.3.006)414
 -resistant respiration(*IV.1.2.
 040d)356
Cyathiform[II.2.2.061c-1-g,.27b(1)]
 126,156
Cyathium(V.4.1.033)560
Cybrid(III.2.2.033)319
Cycad(II.6.008,.01d)279,301
Cyclamen mite(*V.3.3.1.172)504
Cyclic(II.2.2.026)122
 AMP(*IV.2.2.142)367
 lighting(V.4.1.034)560
 phosphorylation(*IV.5.4.2.013)413
Cyclohexamide(*V.3.4.000a)539
Cyclohexanetetraacetic acid(*V.3.2.
 012)490
Cyclone seeders(V.1.3.015,*062)433,437
Cyclophysis(IV.3.1.005,*.014h)370,372
a-Cyclopropyl-a-(p-methoxyphenyl)-5-
 pyrimidine methanol(*IV.2.1.037)359

Cycocel(*IV.2.1.037;*V.3.4.000a)359,
 539
Cylindraceous(II.2.2.061c-1-h,.23e)
 126,152
Cylinder separator(*V.1.3.057)436
Cylindrical[II.1.1.3.033,.08b(2)]23,44
 unifacial leaf(*II.1.2.119,.07a)64,
 69
Cylindrocarpon radicicola(*V.4.2.1.
 017)566
Cymba[II.2.1.022,.07a,.13b(2)]100,110,
 116
Cymbiform(II.2.1.023,.07b)100,110
Cyme[II.2.1.024,.03a(3),.06d,.07d,e,
 .09f]100,106,109,110,112
Cymiform(II.2.1.06c)109
Cymose(II.2.1.025,.06c)101,109
 paniculate(II.2.1.026,.04a)101,107
Cymule[II.2.1.027;.07c(2)]101,110
Cypress(*V.1.1.017c;.1.2.014;*.4.1.
 107,*.119)420,431,564
Cyst, cyst-forming nematodes(*V.3.3.1.
 200,*.200e,g)505,506
Cystolith[II.3.036,.08e(2)]196,214
Cytex(V.1.3.017)433
Cytochimera(III.2.2.024a,.01d)318,327
Cytochrome oxidase(*IV.1.2.040d)356
 system(IV.1.2.040c)356
Cytochromes(IV.1.2.036e,*.040;*.2.2.
 168,*.172)355,356,368,369
Cytokinesis(III.2.2.034)319
Cytokinins(IV.2.1.025;*V.1.3.017)358,
 433
Cytological symptoms(*V.3.3.2.133b)515
Cytoplasm(*II.3.125;IV.1.1.005;*.3.3.
 3.013a,*c)200,351,532
Cytoplasmic:
 inclusions(*V.3.3.2.133b)515
 inheritance(III.2.2.035)319
 male sterility(IV.4.1.012)381
 polyhedrosis virus(V.3.3.3.013e)532
Cytosine(*IV.2.2.056,*.141,*.175)363,
 367,369

D

2,4-D(*IV.2.1.012,*.033;*V.1.5.005;
 *VI.3.033)357,358,580
Daddy longlegs(*V.3.3.1.010)498
Dado:
 graft(*V.1.4.018k-9)447
 saws(*V.1.4.028k-9)447
Dagger nematodes(*V.3.3.1.200c;*.3.3.
 2.146)505,516
Dahlia blight(*V.3.3.2.181c)518
Dairy products(*VI.6.022)595
Dalton(IV.0.051)331

Damage(*VI.8.038c)619
Damsel bug(*V.3.3.1.133c)502
Damp habitats(*V.3.3.2.008)510
Damping-off(*V.3.3.0.002,*.3.3.2.123,
   *.181d,*f;*VI.9.1.098)514,518,519,
   632
   of seedlings(*V.3.3.2.133c,*.179a,
      *.181f)515,517,519
Dandelion(*I.3.016)8
Dangling stalks(*IV.4.4.021)389
Dard(V.3.4.023,.02d,g)541,546
Dark:
   -field microscope(IV.0.052)331
   interruption(*V.4.1.002)559
   mildew(*V.3.3.2.253)523
   period(*V.4.1.034)560
   red mottles(*V.2.076)474
   storage(IV.8.025)618
Darkling beetles(*V.3.3.1.132a)501
Daughter:
   axis(*V.4.1.117;*.4.2.2.008)564,567
   bulb(V.4.1.035)560
Day (d)(*IV.0.260i)346
   length(*IV.3.2.013;.5.4.028)374,409
   length control(*V.4.1.123)565
   length, improper(*V.4.1.031)560
   -neutral plant(IV.3.2.013c)374
Deacclimation(*IV.5.3.015)408
Dead:
   head(IV.6.005)594
   material accumulation(*V.2.106)476
   tissue(*V.3.3.2.208)521
Deal(VI.8.026)618
Dealbatus(II.5.1.007m)265
Dealer(VI.7.1.015,*.039,*.040,*.053,
   .054,.072,*.077)599,602,604,606,
   607
Debearder(V.1.3.018,*.057)433,436
Decay(*VI.8.038c,*.056)619,620
Deceptive device(IV.4.4.013)389
Deci (d)(*IV.0.260h)346
Deciduous(II.5.3.007;.5.02b)267,282
   crops(*I.1.016)5
Decisiemens(*V.4.1.071)562
Declining(*II.5.2.017)266
Decomposed, decomposition(*V.2.067,
   *.070)474
   injury(*IV.5.4.1.005)412
Decompound(II.1.2.035,.07b,.21b)60,72,
   86
Decorticator(*V.1.3.057)436
Decurrent(II.1.2.036,.07c)60,72
Decumbent(II.1.1.3.023,.09a)23,45
Decussate[II.1.1.1.006,.02f;.1.2.037,
   .07d;.2.2.12c(3)]19,29,60,72,141
Dedifferentiation(*II.3.108)199
Deer(*V.1.4.028c;*.3.3.1.140)444,503

flies(*V.3.3.1.132b)501
Deficiencies(*V.3.3.2.270;*VI.9.3.000)
   524,663
Deficiency(III.2.2.036)319
   masking(*V.3.2.066)494
Definitive callose(II.3.041)196
Deflexed(II.4.09g;5.2.003)257,266
Defoliant(V.3.3.3.024;VI.7.2.017e,*p)
   533,613,614
Defuzzer(VI.3.022)578
Degradation(*V.2.111)479
Degree(°)(*IV.0.260i)346
   -day(IV.4.8.020;*VI.1.008)395,570
Degreening(VI.3.023,*.052)578,581
Degrees:
   Celsius (°C)(*IV.0.260a,*c,.295b)
      344,349
   Fahrenheit(°F)(IV.0.295c)349
Dehardening(IV.5.3.015,*.028)408,409
Dehiscence(II.2.2.007h)119
Dehydration(*VI.10.000)674
   avoidance(*IV.5.3.026)409
   postponement(*V.5.2.020)400
   strain(*IV.5.2.042)403
   (strain) tolerance measurement(IV.5.
      2.009)399
   tolerance(*IV.5.2.020;*.5.3.026)400,
      409
Deightoniella torulosa(*VI.9.1.182,
   *.516)639,656
Deka (da)(*IV.0.260h)346
Delaney Amendment(*VI.7.2.016,.016c)
   610,611
Delayed sunscald of apples(VI.9.3.
   131)672
Deliquescent(II.1.1.3.035,.09c;V.3.2.
   020)23,45,491
"Delivered" ("delivered sale")(VI.7.1.
   016)599
Deltoid(II.5.5.015;.5.09e)271,289
Demand(VI.8.027)618
Dendritic(*II.3.204;.5.6.030;.5.14k)
   205,274,294
Dendrograph(IV.3.1.006)370
Dendroid(*II.5.6.030)274
Denitrification(IV.1.2.028b)354
De novo differentiation(*V.1.4.028e)
   444
Density(IV.0.171d,*.260b)341,344
Dentate, denticulate(II.5.4.3.012;.5.
   05g,h)270,285
Denuded(*II.5.6.082)276
Deoxy (=desoxy) sugars, etc. (except
   DNA)(IV.2.2.051)363
   sugars(IV.2.2.032e)362
Deoxyribonucleic acid(*IV.2.2.056)363
Deoxyribonucleoside(*IV.2.2.056,*.141)

363,367
Deoxyribonucleotide(*IV.2.2.056)363
Deoxyriboses(*IV.2.2.056)363
Department managers(*IV.8.116)625
Departmental managers(*VI.8.057)620
Depauperate(II.5.5.016)271
Dependent(II.5.2.004)266
Depressed(II.2.3.2.010,.02i;.5.2.005)
   161,176,266
Depth effect(IV.4.4.014)389
Derived SI units:
   expressed by means of special names(IV.0.260e)345
   expressed in terms of base units(IV.0.260b)344
   formed by using supplementary units (IV.0.260g)346
   with special names(IV.0.260c)344
   with special names admitted for reasons of safeguarding human health (IV.0.260d)345
Dermal tissues(II.3.042)196
Dermatocalyptrogen(II.3.043,.19e)196, 225
Dermatogen[*II.3.080,.14a(2)]197,220
Deroceras agrestre(*V.3.3.1.250)508
Descending(II.5.2.006)266
Desert:
   areas(*IV.5.2.013)399
   ephemerals(*IV.5.2.015)399
Desiccant(V.3.3.3.025;VI.7.2.017f,*p) 533,613,614
Desiccation:
   of tissues(*IV.5.3.005)407
   tolerance(*IV.5.2.079)406
Destruction of meristems(*V.3.3.2.133c)515
Destination markets(VI.8.)616-626
Detassel(III.2.2.037)320
Detergent(*V.3.3.3.105)538
Determinate(II.2.1.028,.03a)101,106
   growth(*V.1.4.060)451
   panicle(II.2.1.029)101
Determination of crop needs(V.3.2.021) 491
Determining disease presence(*V.3.3.2.221)522
Deuteromycetes(V.3.3.2.181g)520
Development(IV.3.1.007;*VI.1.02)370, 574
   ,growth and penetration of a pathogen(*V.3.3.2.220)522
Developmental:
   cycle, complete(*V.3.3.2.204)521
   stage(*IV.5.3.027)409
Devernalization(IV.3.2.021b;V.4.1.036) 376,561

Device(VI.7.2.017g)613
Devora(*III.1.1.005)303
Dew:
   absorption(*IV.5.2.080)406
   hoses(*V.1.1.017f)420
Dewpoint(*VI.5.025b)589
Dewy(*II.5.6.108)277
Dexon(*V.1.1.065)427
Dextrorotary(*IV.2.2.031a)361
DHEEDDA(*V.3.2.012)490
Diacytic(*II.3.190,.30i)204,236
Diadelphous[II.2.2.007d-4,.04(c)]118, 133
Diageotropic(IV.3.3.008e)377
Diagnosis(III.1.1.006)303
   of deficiency or toxicity symptoms (*V.3.2.021)491
Diagnostic:
   horizons(V.2.028)472
   symptoms(*V.3.3.2.221)522
Diallele cross(III.2.2.038)320
Dialysis(IV.0.064)331
Diameter (sizing)(*VI.3.064)582
Diandrous(II.2.2.027,.24f)122,153
Diaphototropism(*IV.3.3.008p)377
Diaporthe:
   batatis(*V.3.3.2.181e;*VI.9.1.535) 519,657
   citri(*V.3.3.2.181e;*VI.9.1.310, *.528)519,645,657
   phaseolarum(*VI.9.1.401)649
   var. sojae(*VI.9.1.359)648
Diaspore(s)(*IV.4.7.002,*.003,*.004, *.005,*.007)393
Diastereomer(*IV.0.288)348
Diatomaceous earth(*V.2.107i,*.3.3.3.029)477,533
Diatoms(*IV.1.2.036d;*V.2.107i)355,477
Dibble(V.1.1.010)419
Dicaryon (dikaryon)(V.3.3.2.127)514
Dicaryons(*V.3.3.2.323)525
Dicaryotic spores(*V.3.3.2.003)510
Dichasium(II.2.1.030,.03a,.08b)101, 106,111
Dichlorobenzyltributylphosphonium chloride(*IV.2.1.037)359
2,4-Dichlorophenoxyacetic acid(*IV.2.1.012)357
Dichogamy(IV.4.3.,.002)385-387,385
Dichotomous(II.1.1.3.036,.09d;*.3.211; .6.009)23,45,205,279
Dicing(*VI.7.1.025)601
Diclinous, dicliny(II.2.2.028,.01c,d; *IV.4.3,.002d)122,130,385,386
Dicots, dicotyledons (Dicotyledoneae) (II.1.1.1.04;*.1.3.004;*.3.052,*.073 *.101,*.117,*.173,*.192,*.199,*.210,

*.213;III.1.1.007;*IV.3.1.014f)31,
94,196-199,202,204,205,303,372
Dictyosomes(IV.1.1.006)351
Dictyostele(II.3.189d,.27e)204,233
Didymous, twins(II.2.2.007i,.05a)119,
134
Didynamous(II.2.2.007d-5,.05c)118,134
Dieback(V.3.3.2.128)514
  of peach, other stone fruits(*V.3.3.
    2.181e)519
Diectochimera(III.2.2.024d-1-a,.01h)
319,327
Diethylenetriaminepentaacetic acid(*V.
3.2.012)490
Differentiation(II.3.01B;IV.3.1.008)
207,371
Difficult-to-root plant(*V.1.4.028r)
448
Diffuse(II.2.2.18e;*.3.14;.04a;.5.2.
007)147,194,210,266
Diffuse porous woods(II.3.047,.10c)
196,216
Diffusion pressure deficit(IV.5.2.010)
399
Digestion of tissues(*V.3.3.2.133c)515
Digestive glands(*II.3.179)203
Digger(s)(V.3.0.003)484
Digger wasps(*V.3.3.1.132e)502
Digging(*VI.2.022a)576
  machines(*VI.2.022b)576
  spade(*V.1.1.067)427
Digitate[II.1.2.038,.08a,.17a,.20c,
  .21a(1)]60,73,82,85,86
Dihaploid(III.2.2.039)320
1,2-Dihydro-3,6-pyridazinedione(*IV.2.
1.037)359
Dihydroxyacetone(*IV.2.2.031a)361
N,N'-Dihydroxyethylethylenediamined-
  acetic acid(*V.3.2.012)490
Dihydrozeatin(*IV.2.1.025)358
Diketones(*II.3.038)195
Dilatation(II.3.048,.09a)196,215
Diluent(*V.3.3.3.020,.026,*.040)533,
534
Dilute spray(*V.3.2.067)494
Dilution within plants(*IV.5.4.3.010)
415
Dimethylsulfoxide(*IV.5.3.014)408
Dimidiate(II.2.2.029,.10c)122,139
Dimorphic (dimorphous)(II.0.001,.01a;
  *.4.4.007a;IV.4.3.003)10,12,247
  386
  branches(*V.1.4.022a-1)442
  heterostyled(*IV.4.3.011)386
Dinoflagellates(*IV.1.2.036d)355
Dioecious(IV.4.3.004)386
Diphenyl(*VI.3.033)580

Diplocarpon rosae(*V.3.3.2.181e)519
Diplodia:
  macrospora(*VI.9.1.190)639
  natalensis(*V.3.3.2.183g;*VI.9.1.
    193,*.194,*.195,*.524,*.525,*.526,
    .527,.528)520,639,640,656,657
  tubericola(*VI.9.1.296)645
  zeae(*VI.9.1.190)639
Diplodia:
  ear rot of corn(VI.9.1.190)639
  foot rot, dieback, stem-end rot of
    citrus, etc.(*V.3.3.2.181g)520
  rot of muskmelons and cucumbers(VI.
    9.1.193)639
  rot of peaches(VI.9.1.194)639
  stain of onions(VI.9.1.195)640
Diploid(III.2.2.040)320
  phase(*V.3.3.2.323)525
  of basidiomycetes(*V.3.3.2.127)514
Diplospory[*IV.4.1.006a-4,5,.02(3)]
378,384
Diplostemonous(II.2.2.007d-6,.06c,
  .20a)118,135,149
Diptera(*V.3.3.1.132b;*V.4.2.1.003)
501,565
Dipterocarp(III.1.2.015)308
Direct:
  buying(VI.8.028)618
  chilling injury(*IV.5.3.005)407
  delivery(*VI.8.114)625
  displayed(*VI.8.136)626
  elastic dehydration strain(IV.5.2.
    011)399
  forced air(*V.1.1.017h)421
  forcing(*V.4.1.111)564
  heat injury(*V.5.4.1.005)412
  kinetic strain(*V.5.4.1.005)412
  plastic strain(*V.5.3.047)411
  sale(*VI.8.138)626
  seeded(*V.1.4.041)450
  stress injury(*IV.5.3.047)411
  strain(*V.5.4.1.005)412
Directed herbicide application(V.3.3.
3.027)533
Direction and size(II.5.2.)266-267
Dirty trailer(*VI.6.034)596
Disaccharides(*IV.1.2.019;*.2.2.031b)
353,361
Disbud, disbudding(V.4.1.038)561
Disc, disk[II.0.002,.02(b);.2.1.01(5);
  .2.2.23i(2)]10,12,104,152
Discharge rate(*V.3.3.3.016)532
Disciform(II.1.2.039,.08b)61,73
Discoid(II.2.1.031,.09d;.2.2.18f;*.3.
204)101,112,147,205
Discoloration(VI.8.038b)619
Discomycetes(*V.3.3.2.023)511

INDEX OF TERMS

Disease, diseases(*V.1.3.050;*.3.3.0.
002,*.004;.3.3.2.133;*3.3.3.052,
*.084;*VI.8.038a)435,497,514,534,
537,619
  agents(*V.3.3.3.053)535
  cause(*V.3.3.2.286)524
  control(*V.4.1.072;*.3.3.3.044)562,
  534
  free(*V.3.3.2.217)521
  -resistant scion(*V.1.4.028h)445
  -resistant trunk system(*V.1.4.028h)
  445
  -susceptible interstock(*V.1.4.028h)
  445
  symptoms(V.3.3.2.133b,*.418)514,529
Dish - to bowl - shaped(IV.4.4.004b-2;
*.4.5.002)388,391
Disinfest(V.3.3.3.028)533
Disjunction(III.2.2.041)320
Disjunctive(II.4.3.067c)245
  (parenchyma, tracheids)(II.3.049)196
Disk:
  -harrow(V.1.1.011,*.020)419,421
  plow(*V.1.1.044)423
Disome(*III.2.2.113)323
Disomic(*IV.4.4.001a-1)392
Dispersal(V.3.3.2.134)515
Dispersed (stomata)(II.3.29e)235
Display(*VI.8.083)622
  case(*VI.8.030)618
  techniques(*VI.8.065)621
Dissected(II.5.4.3.013;.5.05i)270,285
Dissemination(*V.3.3.2.134,*.436)515,
529
Dissepiment(II.2.2.041g)122
Dissolved salts(*V.3.5.032)558
Distant(II.2.2.030)122
Distichlis stricta(*VI.9.1.330)646
Distichous(II.1.1.04q;.1.1.3.037,.01a;
.1.2.08c)23,31,46,73
Distinct(II.2.2.007d-7,*.038,.05c,
.06c,.20b)118,122,134,135,149
Distributes bulbs(*V.4.1.010)559
Distribution centers(*VI.3.023)578
  (retail)(VI.8.031)618
Distributor(s)(*VI.7.1.015;.8.032,
.052,.066,*.118)599,618,620,621,
625
Distinctive pathogenicity(*V.3.3.2.
179b-1)517
Ditches(*V.3.5.001,*.012)557
  primary (main) open(*V.3.5.002)557
  secondary (lateral)(*V.3.5.002)557
  tertiary (short)(*V.3.5.002)557
Ditylenchus:
  dipsaci(*V.4.2.1.014)566
  radicicola(*V.3.3.1.200a,*.200g)505,
  506
  spp.(*V.3.3.1.200a,*.200g)505,506
Diurnal(IV.4.4.015)389
Divaricate(II.5.2.008)266
Divergent(*II.2.2.007w-4,.07d)120,136
Divided(II.5.4.3.014;.5.05j,k)270,285
Divisio(*III.1.1.003)302
Divisions(*V.1.4.000,.024,.04,.05)441,
443,456,457
DMSO(*IV.5.3.014)408
DNA(*IV.2.1.025;.2.2.056,*.140,*.141,
.142;.3.1.012;*.4.1.018;*V.3.3.2.
088,*.320,*.461)358,363,367,371,381,
513,525,530
Dodder[*II.4.2.026;.4.05c(2);*V.3.3.2.
133e)240,253,515
Dolichodorus spp.(*V.3.3.1.200c)505
Dolly(V.1.1.013;VI.3.024)419,579
Dolomitic limestone(*V.3.2.053)493
Domesticate(III.1.2.016)308
Dominant(III.2.2.043)320
Dormancy(*IV.3.1.001,*.002,.009;*V.1.
3.050;*.3.3.2.337;*.4.1.039,*.060)
370,371,435,526,561,562
  of temperate zone plants(*IV.3.1.
  011)371
Dormant(IV.3.1.010)371
  fungal spore(*V.3.3.2.356)527
  lateral buds(*IV.5.4.4.002g)416
Dormin(*IV.2.1.037)359
Dorsal(II.2.2.007j,.07c)119,136
  seed-form(II.2.3.3.012,.05c,d)163,
  186
Dorsifixed(*II.2.2.007j,w-5,.07c)119,
120,136
Dorsiventral(II.1.2.04a;.5.5.017)69,
271
Dosage(*V.3.3.3.049)534
Dose equivalent(*IV.0.260d)345
  index(*IV.0.260d)345
Dothidella ulei(*V.3.3.2.181e)519
Dothiorella:
  gregarea(*VI.9.1.525)656
  sp.(*VI.9.1.527)657
Dotted(II.5.1.006e;*.5.6.103)265,276
Double:
  (armed)(*V.3.4.021)541
  bench system(*V.1.1.017e)420
  -bladed knife(*V.1.4.028f-6)445
  budding(*V.1.4.028g)448
  cross(III.2.0.004k;.2.2.044)312,320
  -crosses(*III.2.0.003d)311
  dormancy(*V.1.3.060)436
  fertilization(*IV.4.1.015)381
  freezing point(IV.5.3.018)408
  -glass(V.1.1.053a)425
  -knife(V.1.2.004)429

leader[V.3.4.01(3)]545
nose bulb[V.1.4.16e(1);.4.1.040]468, 561
-row avenue(V.3.1.01i)489
-setting(V.3.1.008c,.01f)485,488
stem(V.2.2.005)567
-stem rotation system(*V.3.4.017) 540
titration(*VI.1.008g)571
-type pattern(*V.3.1.008i)486
-U palmette(V.3.4.11a)555
-virus streak of tomato(VI.9.1.201) 640
-working (V.1.4.028h,*.028n,.08k, 09j)445,447,460,461
"Double" chip bud(V.1.4.08e)460
Doubly:
 crenate(*II.5.4.3.009)270
 dentate(*II.5.4.3.012)270
 serrate(*II.5.4.3.032)271
Douglas fir:
 bark(*V.1.2.005a)429
 tussock moth(*V.3.3.3.013)532
Downward curling (leaf)(*V.3.3.2.158) 516
Downy(II.5.6.102)276
 mildew of crucifers(VI.9.1.202)640
 mildew of crucifers and rose(*V.3.3. 2.181d)518
 mildew of curcurbits(*V.3.3.2.181d) 518
 mildew of garden pea(VI.9.1.203)640
 mildew of grape(*V.3.3.2.181d)518
 mildew of grasses(*V.3.3.2.181d)518
 mildew of lettuce(*V.3.3.2.181d;VI. 9.1.204)518,640
 mildew of lima beans(VI.9.1.205)640
 mildew of parsley(VI.9.1.206)640
 mildew of spinach(VI.9.1.207)640
 mildews(*V.3.3.2.181d,*.253)518,523
DPD(*IV.5.1.010)399
Draglines(*V.3.1.000)484
Drainage(*IV.5.4.3.006;*V.1.1.017a; *.2.109,*.3.0.001;*.3.1.000,*.008b; *.3.5.002)415,419,478,484,485,557
Drains(*V.2.110e)479
Draper(*V.1.3.057)436
Drepanium(II.2.1.032,.09b)101,112
Dried shriveled fruit(*V.3.3.2.258)523
Drier-polisher(VI.3.025)579
Drift(III.2.2.045)320
Drill(V.1.3.021)433
Drills(*V.1.3.062)437
Drip(*V.3.5.012)557
 hoses(*V.1.1.017f)420
 (irrigation)(*V.3.5.012)557
 irrigation systems(*V.3.5.032)558

"Drip-tip"(II.4.3.017,*.026;.4.07)242, 255
Driver related(*VI.6.034)596
Drooping(II.5.2.009)266
 branch[*V.3.4.01(6)]545
Drop of lettuce, celery, other vegetables, lemon(*V.3.3.2.181e)519
"Dropper" of tulip(V.4.4.2.006)567
Drought(*II.4.4.003c;*IV.5.1.001, *.014,*.015;*.5.2.001,.013,*.048, *.079,*.080;*.5.3.041)247,396,397, 398,399,403,406,411
 acclimation(*IV.5.2.001)398
 avoidance(*IV.5.1.001;*.5.2.015, *.031)396,399,401
 escape(*IV.5.1.015;*.5.2.015)397,399
 escape (avoidance)(*IV.5.2.015)399
 hardening(IV.5.2.016)400
 -induced stress(*IV.5.2.020)400
 injury(*IV.5.2.044;.5.4.2.011)403, 413
 injury of stone fruits(VI.9.3.041) 666
 protectants(IV.5.2.017)400
 resistance(*IV.5.2.018,*.064;.5.3. 032;*V.2.110e-2)400,404,413,479
 resistance measurement(IV.5.2.018) 400
 resistance, total(V.5.2.064)404
 stress(*IV.5.2.026,*.044)401,403
 stress tolerance(IV.5.2.019)400
 tolerance(*IV.5.1.015,*.018;.5.2. 020,*.085)397,400,406
 tolerance (high water potential) (*IV.5.2.018)400
 tolerance (low water potential)(*IV. 5.2.018)400
 tolerance mechanisms(*IV.5.2.020)400
 (water) stress(IV.5.2.022)400
Drugs(*VI.7.2.016)610
Drum:
 drier(*VI.10.009)675
 screen(*V.1.1.063)426
Drupaceous(II.2.3.1.014)159
Drupe(s)(II.2.3.1.015,*.019,*.024; .2.3.2.029c,*.034;.2.3.1.05e-h)159, 162,172
Drupelet(II.2.3.1.016,.06a,c,d,.07d) 159,173,174
Druse[*II.3.037,.08a,b;.4.08b(3)]195, 214,256
Dry:
 bulb temperature(*VI.6.025c,.037) 589,591
 cleaning(*VI.3.031,*.081)580,583
 cold or cold-moist treatment(*V.4.1. 088)563

cultivation(*IV.5.4.1.006)412
freight(*VI.6.026)595
goods(*VI.6.022)595
heart rot of strawberry, Rubus, Ribes, cucurbits, beets, crucifers(*V.3.3.2.181e)519
matter(*IV.5.2.084)406
necroses(*V.3.3.2.014)510
-pack storage(VI.5.016)589
pesticide (treatment)(*V.1.3.080)438
rot(*VI.9.1.190)639
rot of sweet potato(*V.3.3.2.181e) 519
storage(V.4.1.041)561
-stored seeds(*V.1.3.079)438
washing(*VI.3.081)583
weight(*IV.5.2.073)405
Drying(*V.1.3.063;*.064;*.3.2.072;VI. 10.009)437,495,675
cabinet(*V.1.1.047)424
for surface moisture removal(*VI.7.1.025)601
of fruits and vegetables(*VI.10.000) 674
DTPA(*V.3.2.012)490
Dumping(VI.3.027,*.031,*.079)579,580, 583
of wastes(*VI.7.2.014)610
Dung:
beetles(*IV.4.4.001c;*V.3.3.1.132a) 391,501
flies(*IV.4.4.001c)391
Duplicate factors (genes)(III.2.2.046) 320
Duplication(III.2.2.047)320
Duration(II.5.3.000;.5.01,.02)267,281, 282
Dust(s)(*IV.5.3.049;*V.3.3.2.270;*.3.3.3.029,*.071)411,524,533,536
Duster(s)(*V.1.1.036;V.3.3.3.030)423, 533
Dusty(*II.5.6.072)275
"Dutch-type" greenhouse(*V.1.1.017b) 419
Dwarf(II.5.2.010)266
disease of mulberry trees(*V.3.3.2.263)523
plants(*IV.5.4.4.002,*.002g)416
pyramid(*V.3.4.068)544
Dwarfed fruit trees(V.3.4.12)555
Dwarfing(*II.4.3.007;*.4.4.010;*IV.5.4.4.002;*V.3.3.2.133b,.215;*.3.4.004)241,248,416,515,521,540
interstocks(*V.3.4.000)539
process(*V.3.4.000b)539
rootstocks(*V.3.4.000)539
Dying (wilting) tissues(*V.3.3.2.326) 526
Dynamic:
soil condition(*V.3.1.014)487
viscosity(*VI.5.048)592
viscosity (pascal second)(*IV.0.260e)345
Dyne (dyn)(*IV.0.019,.2601)329,347
Dysgenic(III.2.2.048)320
Dysploid(III.2.2.049)320
Dysploidion(III.2.2.050)320
Dysploidy(III.2.2.051)320
Dystropic(IV.4.5.002d)391
Dyszoochory(IV.4.7.003)393

E

Ear[II.2.3.1.01o(2)]168
Eared(*II.5.4.2.003)269
Early:
blight of celery(VI.9.1.209)640
of potato(*VI.3.3.2.181g)520
fruiting(*V.3.4.049)542
lifting(*V.4.2.2.014)567
postemergence(V.3.3.3.031)533
rot of tomato(*VI.9.1.210)640
season pool(*V.8.085)622
East Indian pruning system(*V.3.4.017)540
Easy:
eradication(*V.2.110e-2)479
multiplication(*V.2.110e-2)479
Ec(*V.3.3.3.033)533
EC_e(V.4.1.042)561
Echinate(II.5.6.036;.5.141)274,294
Ecological:
consequences(*V.3.3.3.053)535
costs(*V.3.3.3.053)535
Ecology(III.1.2.017)308
Economic:
consequences(*V.3.3.3.053)535
costs(*V.3.3.3.053)535
damage(*V.3.3.3.012)532
poison(VI.7.2.017h,*p)613,614
site factors(*V.3.0.001)484
Ecophene(III.1.2.018)308
Ecospecies(III.1.2.019)308
Ecosystem(III.1.2.020)308
Ecotype(III.1.2.021)308
Ectochimera(III.2.2.024d-1-b,.01e)319, 327
Ectoparasite(*V.3.3.2.146)516
Ectophloic stele(II.3.189h,27c)204,233
Ectosymbiosis(V.3.3.2.147)516
Ectotrophic(V.3.3.2.148,*.264)516,523
Edaphic factors(II.4.4.005)247
Edaphomorphosis(II.4.4.005a)247
EDP(VI.8.033)618

EDDHA(*V.3.2.012)490
Edema(*V.3.3.2.273;*VI.9.2.079)524,662
EDTA(V.3.2.012)490
Edged(II.5.1.006f)265
Eelworm(s)(*V.3.3.1.183;*.4.2.1.014;
    *VI.7.2.017r)504,566,615
Effective bloom period(IV.4.3.005)386
    pollination period(IV.4.8.021)394
Egg[II.2.2.041t-4-b-2,.16b(9);*IV.4.1.
    006a-11,*.006b-1,*3,.01(9),.02a
    (5),b(7,11),.03(2)]124,145,379,
    382,383,384
  mass(*V.3.3.1.200)505
  nucleus(*IV.4.1.006a-2,*.006b-1,*-2,
    *-4)378,379
  sac(*V.3.3.1.200)505
Einstein(IV.0.070,*.157m)331,338
Elaio -(*II.5.1.004e)264
Elaiophore(IV.4.4.017)389
Elaioplast(*IV.1.1.012)351
Elaiosome(II.2.3.3.013)163
Elastic:
  , elasticity(IV.5.1.002)396
  growth strain(IV.5.2.023)401
  resistance(IV.5.1.003)396
  strain(IV.5.1.004)396
"Elcar"(V.3.3.3.013b)532
Electric (electrical):
  charge(*IV.0.260c)344
  charge density(*IV.0.260e)345
  conductance(*IV.0.073,*.260c;*V.4.
    1.042,*.071,*.110)332,344,561,562,
    564
  conductivity(*IV.5.3.024)409
  current(*IV.0.260a)344
  field(*IV.3.3.008g)377
  field strength(*IV.0.260e)345
  flux density(*IV.0.260e)345
  hedge trimmers(*V.3.4.063)543
  leaf(V.1.1.053b,*e)425
  pole saw(*V.3.4.060)543
  potential(*IV.0.260c)344
  resistance(*IV.0.260c)344
  saws(*V.3.4.063)543
  shears(*V.3.4.063)543
Electromotive force(*IV.0.260c)344
Electron:
  microscope(IV.0.079)332
  spin resonance spectroscopy(IV.0.
    080)332
  transport (cytochrome system) - oxi-
    dative phosphorylation(IV.1.2.
    040c)356
Electronic:
  cash register(*VI.8.103)624
  checking mechanism(*VI.8.103)624
  separator(*V.1.3.057)436

  sorter(*VI.3.028)579
  sorting(VI.3.028,*.039)579,580
Electronvolt (eV)(IV.0.260j)347
Electrophoresis(IV.0.082)332
Electrostatic separator(*V.1.3.057)436
Element(s)(*V.3.2.033)491
  deficiencies(*IV.5.4.3.000)414
  equivalent(*V.3.2.036)492
Elephants(*V.3.3.1.140)503
Elevation(*V.1.1.017a)419
Elevator(*VI.3.019)578
ELISA(*V.3.3.2.156)516
"Elite tree"(III.2.2.052)320
Elliptical(II.5.5.019;.5.09f)271,289
Ellipsoid(II.2.3.2.011,.02g)161,176
Elm bark beetle(*V.3.3.1.132a)501
Elongate(II.5.2.011)266
Elongation(*IV.2.1.033)358
Elsinoe:
  australis(*VI.9.1.505)655
  fawcettii(*V.3.3.2.181e;VI.9.1.505)
    519,655
  var. scabiosa(*VI.9.1.505)655
Eluvial horizon(*V.2.107f,*n)477
Eluviation(*V.2.107)476
  zone(VI.2.107e,.01aE)473,483
Elytron (elytra)(V.3.3.1.085,*.132a)
  499,501
Emarginate(II.5.4.1.012;.5.03h)268,283
Emasculate(IV.4.4.018)389
Embracing(II.1.2.040,.08d)61,73
Embryo(s)[II.2.3.3.014,.01d(4),.03f-i,
    .07c(5),d(4),.10c(4),.11b(1),d(3),
    .12b(5);*IV.2.1.012;*.4.1.004,
    .006a-2,-10,*-11,*-14,*.006b-1,
    -2,-3,*-4,*-5,*.026,.02a(5,6),b
    (2,5,9,11);*.4.6.000,*.001,*.002,
    .003,.004;*.4.7.002;*.4.8.022;
    *.5.2.040]163,182,184,188,191,192,
    193,357,378,379,380,381,383,392,
    393,394,403
  culture(*V.1.5.000)470
  expansion resistance(*V.1.3.060)436
  -genetics(III.2.2.053)320
  haploid[IV.4.1.01(12)]382
  sac[II.2.2.041t-4-b,.16a(7),b(8);
    IV.4.1.006b-6-a-1,.006b-6-b,*-c,
    .01(5);*.4.8.040]124,145,380,381,
    382,383,395
  sac longevity(*IV.4.8.021)394
  sac mother cell(*IV.4.1.006a-3,*-5)
    378
Embryoid(*V.1.5.002)470
Embryophyta Siphonogamia(*III.1.1.032)
  306
Embryonic zone(II.3.01A-1)207
Embryotega(*II.2.3.3.042e,g)166,167

Emergence treatment(V.3.3.3.032)533
Emersed, emergent(II.4.2.005)238
Emissons(*VI.7.2.014)609
Emjeo(*V.3.2.084)496
Employ, employment(VI.7.1.017)600
Emulsifiable concentrate(V.3.3.3.033) 533
Emulsifier(V.3.3.3.034)533
Emulsion(V.3.3.3.035)533
Enantiomer(*IV.0.286)348
Enation(s)(*V.3.3.2.133b)515
Endarch (xylem)(*II.3.222)206
Endemic(III.1.2.022)308
Endergonic(*IV.0.085)332
Endocarp[*II.2.3.2.029b,c,*e,*f,*i, *.031,*.034,.05a(3),b,2(a),f(2); .2.3.3.07f(6);*V.1.3.079]162,179, 188,438
  hairs[II.2.3.3.06e(2)]187
  removal(V.1.3.024)433
Endochimera(III.2.2.024d-2-1,.01g)319, 327
Endoconidiophora fimbriata(VI.9.1.095) 633
Endodermis[*II.3.022,.052,*.059,*.085, .02e(9),.11,.13a(4),.17a(1)]194,196, 198,208,217,219,223
Endogenous:
  dormancy(*V.1.3.060)436
  winter dormancy(IV.3.1.011)371
Endoparasite(V.3.3.2.151)516
Endoplasmic reticulum(IV.1.1.008)351
Endosmosis(*VI.9.3.147)673
Endosepsis of fig(VI.9.1.214)641
Endosperm[II.2.3.3.015,*.031;.2.3.2. 05b(2)(c),c(1)(c);.2.3.3.01b(3), d(5),.03f-i(4),.07c(4),.09a(1), .10c(5),.11d(4),.12b(4);IV.2.2. 172;*.4.1.026;.4.8.022,*.057]164, 165,179,182,184,188,190-193,369, 381,394,395
  incompatibility(IV.4.8.022)394
  nuclei(*IV.4.1.006a-11,.006b-4, *.015)379,381
Endostome[II.2.3.3.016,*.027,.10b(3), .12b(3)]164,191,193
Endosymbiosis(V.3.3.2.152)516
Endotegmen[II.2.3.3.12e(5)]193
Endotegmic(II.2.3.3.040a,.03e)166,184
Endotestal(II.2.3.3.043a,.03)167,184
Endothecium[II.2.2.007k,.11b(4)]119, 140
Endothermic(IV.0.085)332
Endothia parasitica(*V.3.3.2.181e)519
Endotrophic(V.3.3.2.154,*.264)516,523
Endoxerosis of lemons(VI.9.3.042)666
Endozoochory(IV.4.7.004)393

Energy(IV.0.086,*.157,*.260c)332,336, 344
  applied (absorbed)(IV.0.157x-1-a)339
  -balance(*IV.5.2.028)401
  budget equation(IV.3.2.002)373
  content(IV.0.157x-1,.157y-1)339
  density(*IV.0.260e)345
  flow rate(IV.0.157x-1-b,.157y-1-b) 339
  fluence(IV.0.157y-1-b)339
  fluence rate(IV.0.157y-1-c)339
  flux(IV.0.157x-1-c)339
  flux density(IV.5.2.074a)405
  output per unit crop harvested(*V. 3.5.011)557
  sum(IV.0.157b-2-e-2)337
Engler, A. and L. Diels(*III.1.1.010) 304
Engler and Prantl(*III.1.1.004)303
Enriched superphosphate(*V.3.2.091)496
Ensiform(II.5.5.020;.5.09g)272,289
Entangled(II.5.5.021)272
Enthalpy ( )(VI.5.017,*.037)589,591
Entire(II.5.4.3.018;.5.051)270,285
Entomophily(*IV.4.5.001c)391
Entomopox virus(V.3.3.3.013c)532
Entropy(IV.0.098,*.260e)333,345
Entyloma dahliae(*V.3.3.2.181f)519
Environment(IV.0.100;*V.3.3.3.004)333, 531
Environmental:
  -based response(*V.4.1.097)563
  conditions(*IV.5.2.036;*V.3.3.3.004) 531
    ,manipulation(*V.4.1.054)561
    ,suboptimal(*V.4.1.083)563
    ,unfavorable(*V.4.1.106)564
  factors(*V.3.3.2.159)516
  Protection Agency(VI.7.2.014,*.017) 610,612
  site factors(*V.3.0.001)484
Enzyme(s)(*IV.1.1.016,*.021,*.030;*.1. 2.040a;.2.2.031c-a-1,.058,*.150, *.168)351,352,355,361,363,368
  inactivation(*IV.5.2.044;*5.4.1.005; *.5.4.2.014)403,412,414
  -linked immunoabsorbent assay(V.3.3. 2.156)516
  respiratory(*IV.1.1.016)351
  system(*IV.5.4.2.006)413
  thermostability(*IV.5.3.034)410
Eosin(*IV.0.110)333
EPA(*VI.7.2.014,*.017)610,612
Ephedroid perforation(II.3.217)206
Ephemeral(II.5.3.008)267
Ephemerals(II.4.3.018)242
Ephydrophily(*IV.4.5.001a)391

Epi-(*II.0.021)10
Epichalyx(*II.2.1.042,.10c)102,113
Epiclamydy(II.2.3.2.012,.03b)161,177
Epicoralline(II.2.2.061c-5)127
Epicormic(II.1.1.1.007,.02g)19,29
Epicuticular waxes(*II.3.038;*IV.2.2.
  046)196,363
Epidemic(V.3.3.2.157)516
Epidemiology(*V.3.3.2.159)516
Epidermal:
  cell death(*IV.5.4.2.013)413
  mutant(III.2.2.024d-1)319
  sclereids(II.3.25)231
Epidermis[II.2.2.11b(3);*.3.043,.054,
  .080,.085,*.103,*.114,*.162,*.190,
  *.215,.01B(8),C(20),.02d(7),.12a(1),
  b(3),.21a(1,3);.4.10c(1)(c);*V.3.3.
  3.013f]140,196,197,198,199,202,204,
  205,207,208,218,227,258,532
Epifoliar (flowers)(*II.2.2.002,.031,
  .03e)118,122,132
Epigenetic(III.2.2.054)320
Epigynous(II.2.2.061d-1,.19d)128,148
Epinasty(V.3.3.2.158)516
Epipedon(*V.2.107c,f)477
Epipetalous(II.2.2.007d-8,.053,.09a)
  118,134,138
Epiphyllous(II.2.2.031,.03f;*.4.2.023;
  .4.3.019)122,132,240,242
Epiphytes(*II.4.1.005;.4.2.006;*.4.3.
  006,*..067c,*.076;*.4.4.003e;.4.02d,
  .05f;*IV.1.2.008c)238,239,241,245,
  246,247,250,253,353
Epiphytic ferns(*IV.5.2.040)403
Epiphytology(V.3.3.2.159)516
Epiphytotic(V.3.3.2.160,*.161)516
  disease(V.3.3.2.161)516
Episepalous(*II.2.2.007d-9,.22c)118,
  151
Epistasis(III.2.2.055)320
Epithelial cells(*II.3.208)205
  midgut(*V.3.3.3.013a)532
Epithem[II.3.15a(3)]221
Epithelium(II.3.055,*.178,*.180)196,
  203
Epithet(*III.1.1.030)306
Epitrophy(II.3.056)196
Epixylous(*II.4.2.023;.4.3.020)240,242
Epizoochory(IV.4.7.005)394
Epolene(*VI.3.085)584
Equifacial leaf(*II.1.2.119,.04d,f,h)
  64,**69**
Equilateral triangle(*V.3.1.008g,m,
  .01c)486,488
Equipment(*V.3.0.001;*VI.8.137)484,626
  dealers(*V.1.1.017a)419
Equisetinae(*II.6.016)279

Equitant(II.1.1.2.012,.02j)21,**35**
Eradicant fungicide(*V.3.3.3.044)534
Eradication(*V.3.3.3.000,*.012,.037)
  531,532,533
Erect(II.1.1.3.038,.10b;.2.2.041t-2-d,
  .15c)23,46,123,**144**
Erection of terraces(*V.3.1.000)484
Erg(IV.0.2601)347
Ergot(*V.3.3.2.181e)519
Eriophyes pyri(*VI.9.2.057)661
Erlenmeyer flasks(*V.1.5.005)470
Erose(II.5.4.3.019;.5.06a)270,286
Erosion(*V.2.110,*.110e-1,*.130;*V.3.
  3.3.048)478,479,482,534
  controls(V.2.110a,*.130)478,482
  dikes(*V.2.110c)478
  factors(V.2.110b)478
  losses(*V.3.2.000)490
Errors(IV.0.157b-1)337
Erumpent fungal structure(*V.3.3.2.
  403)528
Eruptive fruiting structure(*V.3.3.2.
  346)526
Erwinia(*V.3.3.2.014,*040)510,511
  amylovora(*V.3.3.2.014)510
  ananas(*VI.9.1.127)635
  aroidea(*V.3.3.2.014;VI.9.1.062,
    .407)510,631,650
  atroseptica(*V.3.3.2.040;*VI.9.1.
    081)511,632
  carotovora(*V.3.3.2.005,*.040;*VI.9.
    1.056,*.057,*.058,*.059,*.060,
    .061,.062,*.063,*.064,*.065,
    .066,.067)510,511,630,631
  salicis(*V.3.3.2.014)510
  stewartii(*VI.9.1.074)632
Erysiphaceae(*V.3.3.2.253)523
Erysiphe:
  cichoracearum(*V.3.3.2.183e)519
  polygoni(*VI.9.1.410)650
Erythro-(*II.5.1.005p)264
D-Erythrose(IV.2.2.060)364
Erythrose-4-phosphate(*IV.1.2.040g,
  .042;.2.2.154)356,368
Erythrosin(*IV.5.4.2.010)413
Escape(III.1.2.023;*IV.5.1.001;V.3.3.
  2.164)308,396,516
Eskimo tulip(V.4.1.043)561
Espalier(*IV.5.4.4.002g;V.3.4.026,
  .11a,b,e) 416,541,555
ESR(*IV.0.080)332
Essential:
  elements(*V.3.2.000,*.028,*.037,
    .038,.040,*.065,*.066;*VI.9.3.
    000)490,491,492,494,663
  for growth(*V.3.2.083)495
  metabolites(*V.3.2.058)494

INDEX OF TERMS

nutrients(*V.2.110)478
  leaching(*IV.5.4.3.006)414
oil crops(*I.1.013)4
oils(*I.3.006)7
Estates(*I.2.004)6
Estimated:
  gross-profit margin(*VI.8.034)619
  inventory(VI.8.034)618
  yields(*V.3.2.037)492
Estivation(II.1.1.2.01-03,II.2.2.032) 34-36,122
Etchings(*V.3.3.2.133b)515
Ethanolamine(*IV.2.2.159)368
Ethephon(*IV.2.1.033;*V.3.4.000a)358, 539
Ethnic produce(VI.8.035)619
Ethylene(IV.2.1.033;*V.3.4.000a;*.4.1. 106;*.4.2.1.011;*.4.2.2.015;*VI.3. 023;*.6.009;*.9.3.000)358,539,564, 565,567,579,594,663
  autocatalytic production(*VI.1.002) 569
  evolution(*V.4.2.1.020)566
  generators(*V.3.4.000a)539
  -producing compound(*VI.3.023)579
Ethylenediaminedi(o-hydroxyphenyl)acetate(*V.3.2.012)490
Ethylenediaminetetraacetate(*V.3.2. 012)490
Etiolation(IV.3.2.003)373
Etiology(V.3.3.2.165)516
Etioplast(IV.1.2.017)353
Euanthium(IV.4.4.019)389
Eugamy(*IV.4.1.026)381
Eukaryote(V.3.3.2.166)516
Euphilic blossom(IV.4.4.020)389
Euploid(III.2.2.056;IV.4.6.001a-1,b-1) 321,392
Euploidion(III.2.2.057)321
European apple sawfly injury(VI.9.2. 025)660
  corn borer(*V.3.3.1.132g;VI. 9.2.026)502,660
  corn earworm(*V.3.3.1.132g) 502
  spruce sawfly(*V.3.3.3.013f) 532
Euryhydric plants(*IV.5.2.023)401
  species(IV.5.2.026)402
Eurythermic(IV.3.2.004)374
Euschistus:
  conspersus(*VI.9.2.013)660
  servus(*VI.9.2.013)660
Euspory(IV.4.1.006b-6-c)380
Eustele(II.3.189f)204
Eutectic point(IV.5.3.019)409
Eutrophic (lakes)(*II.4.2.018)240

Eutropic(IV.4.5.002e)391
Evaporation(IV.5.2.027,*.028,*.084) 401,406
Evaporative cooling:
  of greenhouses(*VI.5.051c)593
  of produce(*VI.5.055)593
Evaporators(VI.10.010)675
Evapotranspiration(IV.5.2.028)401
Even-pinnate(*II.1.2.072)61
Evergreen(II.5.3.009;.5.02c,d)267,282
Exa (E)(IV.0.260h)346
Exanthema of citrus(VI.9.3.043)666
Exarch (xylem)(*II.3.222)206
Exceeds supply or offerings(VI.8.027) 618
Excentric(II.2.2.17b;.2.3.3.017,.02g) 146,164,183
Excess water(*IV.5.1.014)397
Excessive cork:
  -cell production(*V.3.3.2.370)527
  formation(*V.3.3.2.376)527
Exchangeable cations(*V.3.2.080)495
Excised:
  embryo(s)(*V.1.3.031;*.1.5.000)434, 470
  shoot-tip material(*V.1.5.006)470
Exclusion(*V.3.3.3.000,.038)531,533
Excretion(II.3.057)196
Excretory cells(*II.3.125)200
  glands(*II.4.2.010)239
Excurrent(II.1.1.3.039,.10b)23,46
Exempt:
  hauler(VI.6.006)594
  load(*VI.6.006,.007)594
  trucker(*VI.6.006)594
Exergonic(*IV.0.105)333
Exobasidium azalea, E. spp.(*V.3.3.2. 181f)520
Exocarp[II.2.3.2.029d,.05a(1)]162,179
Exocortis(*V.3.3.2.457;*.3.4.000)530, 539
Exodermis[II.3.059,.13a(2)]196,219
Exogenous:
  brown discoloration(*VI.9.1.127)635
  spore production(*V.3.3.2.046)512
Exosmosis(*IV.5.3.023)409
Exostomal aril[*II.2.3.3.018,.08c(1)] 164,189
Exostome(II.2.3.3.018,*.027)164,165
Exotegmic(II.2.3.3.040b,.03d)166,184
Exotesta[II.2.3.08b(4),c(4)]189
  (palisaded)[II.2.3.3.11b(4)]192
Exotestal(II.2.3.3.043b,.03a)167,184
  palisade[II.2.3.3.12b(2)]193
Exotherm(*IV.5.3.018)408
Exothermic(IV.0.105)333
Exotic(III.1.2.024)308

INDEX OF TERMS

Expanding lattice clays(*V.2.097)475
Expansion-chamber nozzles(*V.3.3.3.006)531
Experimental:
  heat removal(*VI.5.024b)589
  transmission(*V.3.3.2.248)523
Explant sources(V.1.5.01a-e)471
Export shipments(*VI.6.011)594
Exposure(*V.3.1.008)485
  X and rays(*IV.0.260e)345
Exserted(II.2.2.007d-10,.06e)118,135
Extension growth(II.1.2.042,.09)61,74
External:
  agent of transmission(*V.3.3.2.224) 521
  appearance(*VI.8.038a)619
  (phloem)(*II.3.136)201
  stimuli(*IV.3.3.008b)377
Extine[*II.2.2.007t-10,.13a(1)]120,142
Extracellular:
  freezing(*IV.5.3.048)411
  freezing-tolerance(*IV.5.3.027)409
  ice formation(*IV.5.3.018,*.025,*.029)408,409,410
Extrachromosal genetic element(*V.3.3.2.320)525
Extractor(VI.10.011)675
Extrafloral(II.2.2.033;*.3.116,.179) 122,199,203
Extraxylary fiber(*II.3.063)197
Extrorse(II.2.2.007h-2,.034,.10g)119, 122,139
Eye(V.4.1.044)561
  hoe(*V.1.1.024)421
  sizing(*VI.3.064b)583

F

F_1(III.2.2.058)321
  varieties(III.2.0.005a-6)314
F_2(III.2.2.059)321
  varieties(III.2.2.005a-7)314
Fabraea maculata(*VI.9.1.100)633
Fabric(*V.4.1.070)562
Facultative(*V.3.3.2.004,*.013,*.040, *.168,*.169)510,511,516
  apomict(*III.2.0.005a-2)313
  apomixis(*IV.4.1.006)378
  parasite(V.3.3.2.168)516
  response(*IV.3.2.013k)375
  saprophyte(V.3.3.2.169)516
Facultatively anaerobic(*V.3.3.2.040) 511
FAD⁺(*IV.2.2.058b,.067,*.142)364,365, 367
Fair(*VI.8.094)623
  condition(*VI.8.016)617

Fairly light(VI.8.027)618
Fairways(*V.3.1.001)485
Falcate(II.1.2.043,.10c)61,75
  leaves(*IV.5.2.085)406
Fallow(*V.3.0.004)484
False:
  (polyembryony)(IV.4.6.001c)392
  vivipary(II.2.3.3.020a)164
  wireworm(*V.3.3.1.132a)501
Familia(*III.1.1.003)302
Family(*III.1.1.002,.008,.021;V.1.3. 015;*.2.096e)302,303,305,441,475
Fan(*V.3.4.055)543
  and pad systems(*V.1.1.017i)421
  -shaped(*II.1.2.047)61
  system(*V.3.4.032,.10f)541,554
"Fancy" name(*III.2.0.003)311
Farad(*IV.0.260c)344
  per meter(*IV.0.260e)345
Farina(II.2.2.007l)119
Farinaceous(II.5.7.011)278
Farinose(II.5.6.037;.5.15a)274,295
Farm:
  implements(*V.1.1.020,*.044)421,423
  and orchard equipment(V.3.0.003) 484
  sources (emissions)(*VI.7.2.014)610
  tractors(*V.1.1.036)423
Farnesol(*IV.2.2.107)366
Far-red absorbing(*IV.3.2.017)375
  light(*IV.3.2.017)375
FAS steamer(VI.7.1.018)600
Fasciated(II.1.1.3.040;.5.5.022;.5. 09h)23,272,279
Fasciation(II.1.1.3.041;V.3.3.2.170, .4.2.2.007)23,516,567
Fasciatus(*II.5.1.006b)265
Fascicled(II.1.2.045,.10a)61,75
  cyme(II.2.1.024b)101
Fascicles(*II.2.2.007d-11,.04f)119,133
Fascicular(*II.3.026,*.173)195,202
  cambium(*II.3.091)198
Fasciculate(II.1.2.046,.10b)61,75
Fasciculation(V.3.3.2.171)516
Fastigiate(II.5.5.023;.5.09i)272,289
Fat(s)(*II.3.125;*IV.1.1.019;.2.2.065) 195,351,365
  body cells(*V.3.3.3.013f)532
  seeds(*IV.2.2.031)360
Fatty:
  acids(*II.3.038;IV.2.2.066,*.210) 195,365,370
  substances(*IV.4.4.017)389
FC(*IV.5.2.029,*.039)401,403
Feathering(V.4.2.2.008)567
Feathers(*IV.4.7.005)393
Feathery(*II.5.6.096)276

Fecundation(*IV.4.1.014)381
Fecundity(IV.4.8.023)394
Federal:
  Coal Mine Health and Safety Act(*VI.
    7.2.037)615
  Environmental Pesticide Control Act
    of 1972(*VI.7.2.014,*.016f,.017)
    610,612
  food and drug regulations(*VI.8.
    038d)619
  Food, Drug and Cosmetic Act, as
    Amended(*VI.7.014,.016)610
    - handler marketing agreements
      (*VI.7.2.004)609
    - handler marketing orders(*VI.7.
      2.004)609
  Insecticide, Fungicide and Rodenti-
    cide Act of 1947(*VI.7.2.014,
    .016f,.017)610,612
  Metal and Nonmetal Safety Act(*VI.7.
    2.037)616
  Nematocide, Plant Regulator, Defol-
    iant and Desiccant Amendment of
    1959(*VI.7.2.017)612
  or federal-state inspection(*VI.7.
    1.062,*.063)605
  Seed Act of 1938 as Amended(III.2.0.
    004)312
  Seed Program Review(*III.2.0.005)313
  Seed Regulations(1976)(III.2.004)312
  seed regulations(*V.1.3.075)438
  -state inspectors(*VI.8.038)619
Feed and fertilizer plant manager(*VI.
  8.037)619
Feeder (roots)[II.1.3.005,.02b(3),
  d(3)]94,97
  root(*II.3.199)204
Fe-EDTA(*V.3.2.012)490
Felted-tomentose(II.5.6.039;.5.15b)
  274,295
Female:
  gamete(*II.2.2.041t-b-2)124
  -sterile(*IV.4.2.001a,*c,*d)385
Femto (f)(*IV.0.260h)346
Fence row(*V.1.4.022b)442
Fenestrate(II.5.6.040;.5.15c)274,295
Fermentation(IV.1.2.040e;V.1.3.026,
  .064,.073;*VI.10.000,.012)356,
  433,437,438,674,676
  by soil bacteria(*V.5.4.3.006)414
"Fermentation"(VI.10,013)676
Fermentative metabolism(*V.3.3.2.005,
  .014,.040)510,511
Fermi(*IV.0.260m)347
Fern(II.6.010,.01b;*IV.5.2.040)279,
  301,403
  fronds(*V.1.1.059)426

Fernald(*III.1.1.010)304
Ferneries(*V.1.1.059)426
Ferric to ferrous iron(*V.2.037)473
Ferrous sulfate(V.1.2.005;*.3.2.067,
  .085)429,494,496
Ferrugineus(*II.5.1.0011,*.008q)263,
  266
Fertigation(V.3.2.031)491
Fertile(IV.4.8.024)395
Fertility(IV.4.8.025)395
Fertilization[IV.4.1.014,.03(2);*.4.2.
  003,*.016,*.030,*.035,*.057;*V.3.
  1.009;.3.2.032]381,384,385,393,
  394,395,486,491
  carbon dioxide(V.3.2.009)490
Fertilized:
  egg[IV.4.1.02b(11);*.4.8.059]383,395
  polar nuclei[IV.4.1.02b(10)]383
Fertilizer(s)(*V.1.1.049;*.1.3.015,
  .045,.058;*.1.4.028t;*.3.0.001;
  .3.2.000,.001,*.031,.037;*.3.3.
  3.069)424,433,435,436,448,484,491,
  492,494
  analysis(V.3.2.033)491
  analysis (guaranteed)(V.3.2.034)491
  applications(*V.3.2.035)492
  biuret-free urea(*V.3.2.006)490
  calculations(V.3.2.035)492
  chemical(*V.3.2.069)494
  complete(V.3.2.013)491
  concentrated(V.3.2.014)491
  controlled-release(V.3.2.015)491
  conversion factors(V.3.2.036)492
  cost(*V.3.2.035)492
  granule, granular(V.3.2.043)493
  high-analysis(V.3.2.046)493
  liquid(V.3.2.054)493
  liquid mixed(*V.3.2.054)493
  low-analysis(V.3.2.055)493
  materials(*V.3.2.020,*.048)491,493
  mixed(*V.3.2.003,*.014,*.035)490,
    491,492
  mixes(*V.1.2.011n)431
  organic(V.3.2.068)494
  -pesticide applications(*V.3.2.037)
    492
  preparation(*V.3.3.3.042)534
  quantity(*V.3.2.035)492
  slow-release(*V.3.2.070)495
  source materials(*V.3.2.048)493
  spreader(*V.3.2.039)492
  term(*V.3.2.092)496
  total amount(*V.3.2.037)492
  unit(*V.3.2.036,.092)492,496
Festoons, festooning(*V.3.4.003,.12e)
  540,556
Fiber(s)[II.3.063,*.135,*.175,*.221,

INDEX OF TERMS 734

    .16c(2),.21a(**6**),b(1),.23a,b,e,
    .24e-g]197,200,202,206,222,227,
    229,230
  blocks(*V.1.2.003e)428
  drums(*V.1.3.067)437
  mats(*V.1.3.074)438
  pots(*V.1.2.003h)428
  thorns(II.4.09a)257
  -tracheid(*II.3.063,.23b)197,229
Fiberboard(*VI.4.010)586
  cartons(*VI.3.053c,*f;*.4.000b,*c,
    .002,.008;*.6.022)582,585,586,
    595
  (corrugated) containers(*VI.4.008,
    *.010)586
  drums(*V.1.3.067)437
  flats(*VI.4.000c,*.003,*.011)585,586
  lug(*VI.4.014)587
  lugs(*VI.3.053f;*.4.000c)582,585
  slip sheet(*VI.4.022)587
  trays(*VI.4.008)586
Fiberglass absorption blocks(*V.3.5.
  001)557
Fibonacci series(*II.1.1.1.016)19
Fibric(*V.2.069)474
Fibrillose margins(*V.3.3.2.065)512
Fibrous(II.1.3.006,.02d,.03;.5.7.012;
  *V.1.4.060)94,97,98,278,451
  clays(*V.2.098)475
Fiddle-shaped(*II.5.5.047)273
Fidelity(IV.4.5.002f)391
Field(*V.3.3.2.270;*.3.3.3.052,*.117;
  *.4.1.085;.4.2.1.002;VI.2.000,
  .018,.020;*.8.038a,*.138;*.9.
  000)524,534,538,563,565,575,619,
  626
  boxes(*VI.2.012;*.3.012,*.013,*.4.
    000a)575,578,585
  capacity(IV.5.2.029,*.033,*.039,
    *.049;*V.2.100)401,402,403,476
  clearing and cleaning(*V.1.3.088)439
  containers(*VI.2.020;*.4.000a)575,
    585
  cricket(*V.3.3.1.132h)503
  harvesting(*VI.4.000a)585
  heat(*VI.5.051a,*d)592,593
  immune(V.3.3.2.172)517
  labor(*VI.8.053)620
  (pests)(*V.3.3.1.224)507
  planting(*V.4.2.2.002)567
  (production) crew(*VI.8.137)626
  sorting(*VI.3.000)577
  -to-packinghouse handling(VI.2.012,
    *.020)575
  use(*V.3.5.012)557
  wrapping(VI.2.013;*.3.071)575,583
Field-grown:
  herbaceous ornamentals(*V.3.1.000)
    484
  vegetables(*V.3.1.000)484
Fig rust(*V.3.3.2.181f)520
Filament[II.2.2.007m,.01a(7),.11b(1)]
  119,130,140
Filamentous thalli(*V.3.3.2.008)510
Filaments(*V.3.3.2.349)526
Filiform[II.2.2.035,.10e(2);.5.5.025;
  .5.10a]122,139,272,290
  sclereid(*II.3.170)202
Filler(V.3.3.3.040)534
Filling(V.4.1.045)561
Fimbriate(II.2.2.17c,18g;.5.6.041;.5.
  15d)146,147,274,295
Financing(*VI.7.1.031)601
Fine(VI.8.094)623
  sand(*V.2.114,.117l-2)480,481
  sandy loam(V.2.117j-2)481
  silt(*V.2.114)480
Finisher(*VI.10.011,.014)675,676
Finishing machines(V.1.3.027,*.057)
  433,436
Fir(*V.1.2.011l)430
Fire, "fire"(*II.4.4.002g,*.010)246,
  248
  blight(*V.3.3.2.040)511
  disease of lilies(*V.4.2.1.002)565
  of bulbs(*V.4.2.1.002)565
Fireflies(*V.3.3.1.132a)501
Firm(*V.2.100)476
  mature(*VI.1.004)569
Firmness tester(VI.1.003)569
First generation synthetic varieties
  (III.2.0.005a-5-a)314
Fish(*IV.4.7.001f)393
  scrap(*V.3.2.068,*.086)494,496
Fisheye rot of apple(VI.9.1.223)641
Fissured(II.5.6.042;.5.15e)274,295
Fistulous, fistular(II.1.1.3.042,.10c)
  24,**46**
"5 Forcing"(*V.4.1.111)564
Five-spot(*V.3.1.008i,.01k)486,489
Fixed irrigation(*V.3.5.012)557
Flabellate (style)(II.2.2.17d)146
Flabelliform(II.1.2.047,.10d)61,75
Flaccid(II.5.2.012)266
Flag (flower)(IV.4.4.004b-3)388
Flagelliferous(II.4.3.021;.4.09d,f)
  242,257
Flagellum(II.4.3.022;.4.09d,f)242,257
Flagelliflory[*II.2.1.015,.05d(2);IV.
  4.4.021]100,108,389
Flagelliform(II.5.6.043;.5.15f)275,295
Flame-colored(II.5.1.005j)264
Flaps(II.2.2.007h-3,.10f,g)119,139
Flash evaporator(VI.10.010)675

INDEX OF TERMS

pasteurization(VI.10.016)676
Flask-shaped(*II.4.3.006)241
  fungal fruit body(*V.3.3.2.347)526
Flat(s)(V.1.1.048;*.1.2.003c;*.1.3.
  081;*VI.4.000a,*c,*d,.011)424,428,
  439,585,586
  ,plastic(*V.4.1.022)560
  stem(*V.4.2.2.007)567
  wooden(*V.4.1.022)560
Flatbed semitrailer(*VI.2.012)575
  truck with boom-type lift(VI.2.014)
  575
Flattened unifacial leaf(*II.1.2.119,
  .04g)64,69
Flattening(*V.4.2.1.019)566
Flat-topped hedges(*V.3.4.073)544
Flavedo(II.2.3.2.05d)179
Flavin adenine dinucleotide (oxidized
  form)(*IV.2.2.058b,*.067,*.142)364,
  365,367
Flavonoids(*IV.2.2.021,*.029,.068,
  .069,.154)360,365,368
Flavonols, flavones(IV.2.2.069,*.154)
  365,368
Flavor(*VI.8.079)622
  loss(*VI.9.3.029)665
Flavus(*II.5.1.008o)266
Flea beetles(*V.3.3.1.132a)501
Fleck, "fleck"(*V.4.2.1.009,*.022)565,
  566
Flesh-colored(II.5.1.005k)264
  color(*VI.1.008)570
  firmness(*VI.1.008)570
Fleshy(II.1.3.007,.04b;.5.7.013)94,99,
  278
  organs(*V.4.1.120)565
  orthotropic subterranean stem(*V.1.
  4.071b)451
Flexible hooks(*VI.2.022b)576
Flexuous(II.5.2.013)266
Flick-bar(VI.3.029,*.081)580,584
Flies(*IV.4.5.001c;*V.3.3.1.132b;*.3.
  3.2.133e;*VI.7.2.017n)391,501,515,
  614
  carrion(*IV.4.4.009)356
"Flip-over"(*VI.3.085c,d)584
Floating(II.4.2.007;.4.05a)239,253
Floccose(II.5.6.264;.5.15g)275,295
Flocculating effect(*V.2.120)482
Flood irrigation(*V.3.5.012)557
  plain(*II.4.2.004)238
Flooding(*IV.5.2.001;.5.4.3.006)398,
  414
Flora(III.1.1.010)304
Floral(II.2.2.036)122
  arrangements(*V.4.1.062)562
  (bud)(II.1.1.1.009,.02h)19,29

bud development(*V.4.1.047)561
bud development cessation(*V.4.2.2.
  012)567
buds(*V.3.4.003,*.006,*.007,*.010)
  540
cup(*II.2.2.044)125
initiation(*IV.2.1.033;.4.8.026;*V.
  3.4.000a;*.4.2.2.010)358,394,539,
  567
nectaries(*II.3.179)203
nectary(*II.3.116)199
organs(*V.4.1.048)561
parts(*V.4.2.2.011)567
preservative(VI.8.036)619
primordium(*IV.4.8.026)394
stalk(*V.4.2.2.007)567
Floret[II.2.1.035,.09(2)]101,112
  sterility(*IV.4.8.009)593
Floricane[II.1.1.3.043,*.063,.11a(2)]
  24,25,**47**
Floricultural crops(*V.1.1.048)424
Florida(*VI.8.113)625
  foliage plant mixes(*V.1.2.005)429
  red scale(*VI.9.2.007)660
Florigen(IV.3.2.013d)374
Flower, flowering, flowers(II.2.2.000,
  .01,.02,.24;*.3.201;*.4.4.007;*IV.
  3.2.001,*.013;*.3.3.003c,*d,*f;
  *.5.4.4.002;*V.3.4.008;*.4.1.007,
  .011,.038,*.062,*.088,*.093,
  .095,.097,*.108;*.4.2.1.002;*.4.
  2.2.015,*.020,*.021)118-157,130,
  131,153,204,247,365,373,374,376,
  377,416,540,559,561,562,563,564,
  567,568
absission(*IV.2.1.025)358
acceleration(*V.4.1.054)561
ambiphotoperiodic(IV.3.2.013a)374
blindness of bulbs(V.4.2.2.011)567
bud(*V.4.1.023,*.024,*.057)560,561
bud abortion(*V.4.1.047;.4.2.2.012)
  561,567
bud abscission(V.4.2.2.013)567
(bud) blasting(V.4.2.2.010)567
bud formation(IV.5.4.4.002,*.002a,
  *g)416
-bud initiation(*V.4.1.002,*.050,
  *.063)559,561,562
-bud stage(*V.4.1.068)562
buds(*V.4.2.2.013)567
bug(*V.3.3.1.132c)502
bulb or plant(*V.4.1.012)559
capacity(*V.4.1.121)565
critical day length(*IV.3.2.013b)374
,cut(*V.4.1.062)562
day-neutral plant(*IV.3.2.013c)374
deformed(*V.4.2.1.009)565

INDEX OF TERMS

differentiation(V.4.1.048)561
drooping(*V.3.3.2.468)530
dry-stored(*V.4.1.062)562
earliness(*V.4.2.2.006)567
fall(*V.4.1.043)561
field-to-packinghouse handling(*VI.2.012)575
flavonols, flavones(*IV.2.2.069)365
grower(*V.4.1.011)559
harvesting(*VI.2.000)575
herbaceous(*V.4.1.056)561
-inducing treatments(*V.4.1.069)562
induction(V.4.1.050,*.064)561,562
induction cycle(*IV.3.2.013e)375
inproper development(*V.4.2.1.020)566
initiation(*IV.2.1.036;V.4.1.051)358,561
initiation, premature(*V.4.2.2.021)568
intermediate-day plant(*IV.3.2.013f)375
killing(*V.4.2.2.002)567
lateral bud removal(*V.4.1.038)561
limited necrotic areas(*V.3.3.2.405)528
long-day plant(*IV.3.2.013g)375
long-short-day plant(*IV.3.2.013h)375
lopsided(*V.4.1.108)564
malformed(*V.4.1.023)560
marketable(*V.4.2.2.010)567
night interruption (night break)(*IV.3.2.013i)375
nyctinasty(*IV.3.3.003c)376
onset(*IV.3.1.019)373
opening(*V.4.2.2.003)567
parts(II.2.2.01a;*V.4.2.2.011)130,567
photonasty(*IV.3.3.003d)376
plants(*V.1.123)565
primordia(*V.4.1.051)561
qualitative (absolute) response(*IV.3.2.013j)375
quantitative (facultative) response(*IV.3.2.013k)375
rapid(*V.4.1.067)561
seed lots(*V.1.3.075)438
senescence(*IV.2.1.025)358
short-day plant(*IV.3.2.013l)375
short-long-day plant(*IV.3.2.013m)375
spring months(*V.4.1.112)564
stalk(*IV.4.8.010)393
storage after harvesting(*V.4.1.041)561
thermonasty(*IV.3.3.003f)376

thrips(VI.9.2.030)661
vernalization(*IV.3.2.021e;*V.4.1.121)376,565
"Flow sow"(*V.1.3.028)433
Fluid drilling (fluid sowing)(V.1.3.028)433
Fluorescein(*IV.0.110;*.5.4.2.010)333,413
Fluorescence(IV.3.2.005,*.008)374
  microscope(IV.0.110)333
Fluorescent lights(*V.1.1.030)422
Fluorides(*IV.5.4.4.001)415
Flush(*II.1.1.3.102)27
Flux(IV.0.126e,*.260c)334,345
  density(IV.0.126f,*.260e)334,345
Flypaper trap(*II.4.3.009)241
Flyspeck of pome fruits(VI.9.1.224)641
FOB(*VI.7.1.009,.019,*.028;*.8.100)598,600,601,624
  acceptance(VI.7.1.020)600
  acceptance final(VI.7.1.021)600
  inspection and acceptance final(VI.7.1.022)600
  sale at delivered price(VI.7.1.023)600
  shipping point(*VI.7.1.034,*.076)602,608
  steamer(VI.7.1.024)600
Fog(*V.1.1.017i)421
Foliaceous(II.2.2.037,.22d)122,151
  floral organs(*V.4.2.2.001)566
Foliage(*V.3.3.3.041;*VI.2.000)534,575
  field-to-packinghouse handling(VI.2.012)575
  plant(s)(*I.0.001;.2.001d,e;II.4.3.017;*V.1.1.048;*.4.1.107,*.119;VI.*.3.000;*.4.013)3,5,6,242,424,564,577,587
  plant industry(*VI.8.091)623
  plant potting mixes(V.1.2.005)429
  plant selection(V.4.1.053)561
  shaker(*VI.2.022b)576
  systemic(*V.3.3.2.419)529
Foliar:
  analyses(*V.3.2.017,*.037,.040)491,492
  application(V.3.3.3.041)534
  nematodes(*V.3.3.1.200a)505
Follicle(II.2.3.1.017,*.020,*.024,.02a,b)159,169
Fomes spp.(*V.3.3.2.181f)519
Food(*IV.4.4.003)387
  and Drug Administration of the Department of Health and Human Welfare(VI.7.2.016d,*.017;.8.072)611,612,621
  quality(*VI.1.013)572

retailing industry(*VI.8.042)620
service firms(*VI.8.052)620
service outlets(*VI.8.032)618
Foot:
  and crown rots(VI.9.1.386)649
  and root rot of citrus, deciduous
    fruits, black walnut(*V.3.3.2.
    181d)518
  rot of sweet potatoes(VI.9.1.226)641
Foramen[*II.2.2.041t-4-i,.16b(4);*.2.
  3.3.016,*.018,*.027]124,145,164,165
Foraminate (phloem)(*II.3.136)201
Forb(II.0.003)10
Forbes scale on apple(VI.9.2.031)661
Force (newton)(*IV.0.260c)344
Forced-air(*V.1.1.017i)421
  -air precooler(VI.5.035a-4)590
  very early(*V.4.1.091)563
Forcing(V.4.1.054)561
  bud growth(*V.1.4.028i)446
  facilities(*V.4.1.068)562
  periods(*V.4.1.096)563
  program(*V.4.1.093)563
Foremen(*VI.8.090,*.116)623,625
Forest(*V.2.110)478
  seed lots(*V.1.3.075)438
  trees(*II.4.4.003c;*IV.5.4.4.001)
    247,515
Forked[*II.1.2.049;.2.2.007w-6,.01e
  (2)]61,120,139
Forkert:
  bud(V.1.4.028f-3,.08f)445,460
  budding(*V.1.4.028f,.028f-3)444,445
Forking(*VI.2.022a)576
Forklift(V.1.1.014;*VI.2.012)419,575
  tractor(VI.2.018)575
  truck(V.1.1.014;*.1.2.003c,*.010;
    *VI.3.019)419,428,429,578
Form(III.1.1.011;*IV.4.4.013;*V.3.3.2.
  412;*VI.8.038e)304,389,529,619
Forma(III.1.1.011)304
  specialis (f.sp.)(V.3.3.2.177)517
Formae speciales (f.sp.)(*V.3.3.2.
  179a)517
Formaldehyde(*V.1.1.065;*V.3.2.093)
  426,497
  injury of apples(VI.9.3.044)666
Forms:
  of application(*V.3.2.037)492
  of ions absorbed(*V.3.2.065)494
Formulas(IV.0.157b-2)337
  mixed (photometric to radiometric)
    (IV.0.157b-2-a)337
  mixed (photometric to quantum)(IV.
    0.157b-2-b)337
  mixed (quantum to radiometric)(IV.0.
    157b-2-c)337

photometric(IV.0.157b-2-d)337
radiometric(IV.0.157b-2-e)337
Formulation(V.3.3.3.042)534
Foul odor(*V.3.3.2.040)511
Foundation:
  seed(III.2.0.004d)312
  single cross(III.2.0.004j)312
  stock(*V.4.1.072)562
  stock propagation(*V.4.1.103)564
Four-footed butterfly(V.3.3.1.132g)502
Foveolate(II.5.6.045;.5.15h)275,295
Fpl(*IV.5.3.026)409
Fraction (mass, w)(IV.0.171e)341
Fragipan(V.2.107g)477
Framing(*V.1.1.017c)420
Frankliniella:
  bispinosa(*VI.9.2.080)662
  occidentalis(*VI.9.2.081)662
  tritici(*VI.9.2.030,.054)660,661
Freckles(*VI.9.1.470)653
Free(II.2.2.038)122
  -basal(II.2.2.15b)144
  -central(II.2.2.041t-2-e,.15a)123,
    144
  energy(IV.0.113)333
  on board(see FOB)
  salts(*V.2.120)482
  -standing (greenhouses)(*V.1.1.017b)
    419
  standing(V.3.4.11f,g)555
  -stone(II.2.3.2.029e)162
Freedom from defects(*VI.8.094)623
Freeze(*IV.5.3.002,*.003)407
  dehydration(*IV.5.3.023)409
  desiccation(*IV.5.3.023)409
  drying(*VI.10.000,.017)674,676
  injury(*VI.9.3.073)669
  smothering(*IV.5.3.023)409
  tolerance(*IV.5.3.034)410
Freezing(*IV.5.3.014,*.041,*.047;*V.3.
  3.2.270)408,411,524
  -induced water stress(*IV.5.3.023)
    409
  injury(IV.5.3.023)409
    estimation(IV.5.3.024)409
    of cabbage, etc.(VI.9.3.045)666
    of celery(VI.9.3.046)667
    of citrus(VI.9.3.047)667
    of globe artichoke(VI.9.3.048)
      667
    of onion(VI.9.3.049)667
    of pome and stone fruits(VI.9.3.
      050)667
    of sweet potato(VI.9.3.051)667
    of tomato(VI.9.3.052)667
  -killed cells(*IV.5.3.029)410
  nuclei(*IV.5.3.049)411

point(s)(*IV.5.3.018)408
  depression(*IV.5.2.038)402
  lowering(IV.5.3.026)409
  tolerance(*IV.5.3.026)409
resistance(*IV.5.3.032)410
stress(IV.5.3.025)409
  tolerance(IV.5.3.026)409
temperatures(*IV.5.3.013)408
tolerance(IV.5.3.027)409
French system(*V.2.097)475
Frequency(IV.0.114,*.260c)333,344
Fresh:
  -fruit bin(*VI.4.006)586
  fruits and fresh vegetables(VI.7.1.025,*.026)601
  fruits, vegetables, other horticultural products(*V.3.3.0.005)497
  produce harvesting and handling(*V.3.3.2.270)524
  weight(*IV.5.2.073)405
Friable(*V.2.100)476
Fringed(*II.5.6.041)274
Frits(*V.1.2.012)431
Fritted trace elements(*V.1.2.005)429
Frost:
  and freezing injury of apples(VI.9.3.053)667
  hardening(IV.5.3.028)409
  -killing point(*IV.5.3.002)407
  -killing temperature(*IV.5.3.026,*.027)409
  plasmolysis(IV.5.3.029)410
  protection(*V.4.1.075)563
  resistance(*IV.5.3.028)409
  -resistance stable(IV.5.3.030)410
Frosted(*II.5.6.100)276
Frozen:
  concentrate(*VI.10.006,.018)674,676
  concentrated (orange) juice(*VI.10.030)677
  produce handling(*VI.8.102)624
Fructosans(*IV.1.2.019)353
D-Fructose(*IV.1.2.040a;*.2.2.031a,*b,*c,*c-a-2)356,361
Frugiverous(IV.4.7.006)393
Fruit, fruits[I.1.006;II.2.2.14d(1);.2.3.1.000;*IV.2.1.025;*IV.4.2.000,*.001,*.002,*.003,*.004,*.005;.4.7.002,*.006;*.5.2.085;*V.1.5.000;*.3.3.2.370;*.3.4.072;*VI.1.004,.1.02)4,143,158,358,385,393,406,470,527,544,569,574
  abnormal(*IV.4.2.004)385
  abscission(*IV.2.1.037;*V.3.4.000a)359,539
  accessory(*IV.4.2.000)385

aggregate(*IV.4.2.000)385
and vegetable packing lines(VI.3.031)580
bloom and bearing(*IV.3.1.018)373
bud initiation(IV.5.4.4.002b)416
carpel development(*IV.4.2.000)385
climacteric(*VI.1.006)570
collector(*VI.2.022)576
coloring(*IV.1.036)358
cracking of apples(VI.9.3.054)667
crops(*I.0.002;.1.000;*.2.009)3,6
cycle(III.2.2.060)321
dispersal(IV.4.7)392-393
dispersal agencies(IV.4.7.001)392
eaters(*V.3.3.1.017)498
firmness tester(*VI.1.003)569
flies(*V.3.3.1.132b)501
growth rate(*V.3.5.013)557
gumming(*V.3.2.066)494
gumming of stone fruits(VI.9.2.035)661
harvesting(*VI.2.000)575
maturation(*IV.4.2.005;*VI.1.005)385,570
maturity(*IV.2.1.033)358
  test sample(*VI.1.011)572
multiple(*IV.4.2.000)385
non-climactric(*VI.1.007)570
packinghouse(*VI.3.000)577
parthenocarpic(*IV.4.1.007)380
parts(II.2.3.2.,.01-.07)161-163,175-181
patchy pattern(*V.3.3.2.256)523
premature drop(*V.3.2.066)494
ripening(*IV.2.1.033;*VI.1.013)358,572
rot(*V.3.3.2.181d)518
  of eggplant(*VI.9.1.358)648
  of pomegranate(*VI.9.1.286)644
  of squash(*V.3.3.2.181c)518
sample(*VI.1.015)572
seed lots(*V.1.3.075)438
senescence(*IV.2.1.025,*.037)358,359
set(IV.4.8.028;*V.3.4.000,*.000a)394,539
setting(IV.4.)378-395
shape and size(*V.3.4.000a)539
size reduction(*V.3.2.066)494
spot of apple(IV.9.1.231)641
thinning(*V.3.4.000,*.000a)539
total acid(*VI.1.008)571
total soluble solids(*VI.1.008e,*f)571
tree leaf roller(*VI.9.2.044)661
trees (training)(V.3.4.029)541
tumor of tomato(VI.9.3.055)667
types(II.2.3.1.,.01-.07)158-160,168-

174
  water content(*IV.5.2.073)405
Fruitful(IV.4.8.027)394
Fruiting(*IV.5.4.4.002,*.002a,*.002g;
  *V.3.4.008)416,540
  cycle(III.2.2.061)321
  growth removal(*V.3.4.063)543
  rotation(*III.2.2.061)321
Fruitlet black rot, fruitlet brown rot, fruitlet core rot, fruitlet rot (*VI.9.1.127)635
Fruitworm(*V.3.3.1.132g)502
L-Fucose(*IV.2.2.031a,*.038)361,362
Fugacious(*II.4.3.032;.5.3.010)243,267
Fuligineus(*II.5.1.001m)263
Full payment promptly(VI.7.1.027)601
Fulvus(*II.5.1.008s)266
Fumarole(*II.4.4.005a,.006)247
Fumeus(II.5.1.003n)264
Fumigant(s)(*V.1.1.065;*.3.3.3.000,
  .043,*.094;*.4.1.055)426,533,534,
  537,561
Fumigating (insect control)(*VI.7.1.
  025)601
Fumigation(*V.3.3.3.000)531
Fumigun(V.4.1.055)561
Functional:
  gametes(*IV.4.8.025,*.029,*.048)394,
    395
  unisexualism(*IV.4.3.018)387
Fundamental tissues(II.3.067)197
Fungal:
  and bacterial spores(*V.4.2.2.015)
    567
  body, entire(*V.3.3.2.207)521
  spore(*V.3.3.2.226)522
Fungi Cateri(*V.3.3.2.179a)517
  imperfecti(V.3.3.2.180,*.293)518,525
Fungicide(s)(*V.1.3.058;*V.3.3.3.000,
  .044,*.070;VI.7.2.017j,*p)436,531,
  533,534,536,613,614
  application(*VI.3.031,.033)580
Fungicidal(*V.3.3.3.045)534
  waxes(VI.3.085a)584
Fungistatic(V.3.3.3.045)534
Fungus, fungi(*IV.1.2.029;*.2.2.162;
  *.3.2.019;*V.1.1.065;*.1.3.080;
  .3.3.0.002,.003,*.005;*.3.3.2.
  001,*.008,*.031,*.040,*.046,
  *.133a,*e,*f,*.179,*.179a,*.207,
  .259,.270,*.278,*.293,*.305,
  .323,.437,*.477;*.3.3.3.013;*.4.
  2.1.022;*VI.7.2.017g,*h,i,*j;*.9.
  000;*9.1.000)354,376,438,497,509-
  512,521,523-525,531,532,566,612,
  613,626
  diseases(*II.4.4.002e;V.3.3.2.181)

246,518
  form class(*V.3.3.2.180)518
  growth(*V.3.3.3.045)534
  mycelium(*V.3.3.3.044,*.045)534
  root(*V.3.3.2.264)523
  smut(*V.3.3.2.390)528
  spores(*V.3.3.3.044)534
Funicle(*II.2.2.041t-4-d)124
Funiculus[II.2.2.041t-4-d,.16b(2);.2.
  3.3.019,.01c(1),.03f-i(12),.04b(5),
  .06e(1),.07d(1),e(4),.08c(5),.11b
  (6),d(1)]124,145,164,182,184,185,
  187,188,189,192
Funnelform (funnel-shaped)(II.2.2.
  061c-1-i,.21j)126,150
Fur(*IV.4.7.005)393
Furanoses(*IV.2.2.031a,*.095)361,366
Furcate, forked(II.1.2.049,.10e)61,75
Furnishing:
  containers, supplies or other ser-
    vices(*VI.7.1.030)601
  labor, seed, containers, and other
    supplies or services(*VI.7.1.031)
    601
Furrow(*V.3.5.012)557
Furrowed(*II.5.6.137)277
Furrows(*V.3.1.002)485
Fusarium:
  bulbigenum batatis(*VI.9.1.530)657
  cepae(*VI.9.1.234)641
  conglutinans(*V.3.3.2.181g,*.474)
    520,530
  culmorum(*VI.9.1.247)642
  graminium(*VI.9.1.247)642
  lycopersica(*V.3.3.2.181g)520
  moniliforme(*VI.9.1.127,.214)635,641
    var. subglutinans(*VI.9.1.247)642
  oxysporum(*V.3.3.2.181g;*.4.2.1.017;
    *VI.9.1.234)520,566,567,641
    f.1(*VI.9.1.534)657
    f.2(*VI.9.1.530)657
    f.sp. cubense(*V.3.3.2.181g)520
    f.sp. lilii(*V.4.2.1.001)565
    f.sp. vasinfectum(*V.3.3.2.181g)
      520
  roseum(*VI.9.1.182)639
  scirpi var. acuminatum(*VI.9.1.247)
    642
  solani(*V.4.2.1.017)566
  zonatum forms(*VI.9.1.234)641
  sp.(*VI.9.1.171)638
  spp.(*V.1.1.065;*.4.2.1.018;*VI.9.1.
    232,*.234,*.245,*.246,*.248,
    *.536)427,566,641,642,657
Fusarium:
  brown rot of citrus(VI.9.1.232)641
  bulb rot of onions(VI.9.1.234)641

-induced wilts(*V.3.3.2.133c)515
kernel rot of corn(VI.9.1.243)642
rot of asparagus(VI.9.1.245)642
   of carrot(VI.9.1.246)642
   of muskmelons(IV.9.1.247)642
   of tomato(VI.9.1.248)642
Fuscus(*II.5.1.001b)263
Fuscous blight of lima beans(*VI.9.1.
   050)630
Fused phosphates(*V.3.2.003)490
Fusiform(II.1.3.04b;.5.6.049;.5.15i)
   99,275,295
   initial[*II.3.026,.176,.07a(1),b(3)]
   195,203,213
   wood parenchyma cell(II.3.23a)229
Fusion analysis(*V.3.2.080)495

G

G(*IV.0.123;V.3.2.012)334,490
GA(*IV.2.1.036;*V.3.4.000a)358,539
GA_3, GA_{4-7}(*IV.2.1.036)358
GA sprays(*VI.9.3.105)671
Gal(IV.0.260k)347
D-Galactosamine(*IV.2.2.032f)362
D-Galactose(*IV.2.2.031a,*c;*.038)361,
   362
L-Galactose(*IV.2.2.031c)361
6-deoxy-Galactose(*IV.2.2.038)362
D-Galacturonate(*IV.2.2.038)362
D-Galacturonic acid(*IV.2.2.032c,
   *.038)362
D-Galacturonic acid, methyl ester(*IV.
   2.2.038)362
Galea(II.2.2.061c-6,.20f,.23b)127,149,
   152
Galeate(II.2.2.061c-14-j,.20f,.23b)
   128,149,152
Gall(s)(*V.3.3.2.014,.187)510,520
   gnats(*V.3.3.1.132b)501
   of ericaceous plants(*V.3.3.2.181f)
   520
   wasps(*V.3.3.1.132e)502
Gallium (Ga)(*V.3.2.073)495
Galvanized pipe(*V.1.1.017c)420
Galvanotropism(*IV.3.3.008g)377
Gametangia(*V.3.3.2.181d)518
Gamete(s)[II.2.2.041h,*i,*j,.13b(3),
   .16b(9);*IV.4.8.025,*.029,*.031,
   .036,.049,*.053]122,142,145,394,
   395
   morphologically similar(*V.3.3.2.
   230)522
Gametogenesis(II.2.2.041i,.16a)123,145
Gametophyte[II.2.2.041j;*IV.4.1.006a-
   4,*-5,*-7,*-14,*.006b-6,*-6-a-2,*-3,
   *-4,*6-b,*.026,.01(5),.02(a)(3,4),

.03(2)]123,378,379,380,381,382,383,
384
Gametophytic (polyembryony)(*IV.4.6.
   001)392
   agamospermy(IV.4.1.03)384
   apomixis(*IV.4.1.006)378
Gamma (γ)(*IV.0.260m)347
   rays(*IV.5.4.2.001)414
   γ(gamma)(*IV.0.260m)347
Gamopetalous(II.2.2.061c-7,.20e)127,
   149
Gamosepalous(II.2.2.061a-2,.22c)126,
   151
Ganoderma pseudoferreum(*V.3.3.2.181f)
   519
Garbage tankage(*V.3.2.068)494
Garden(*V.3.3.2.270;*VI.2.000;*.4.013)
   524,575,587
   bed(*I.2.001a)5
   beds(*V.3.4.057)543
   hoe(*V.1.1.024)421
   hose(*V.1.1.054)426
   ornamentals(*V.3.2.000)490
   pitch fork(*V.1.1.038)423
   rake(*V.1.1.050)424
   spade(*V.1.1.067)427
   tractors(*V.1.1.036)423
   webworm(*V.3.3.1.132g)502
Gardener(I.2.002)6
Garlic(*I.3.004)7
Gas(*V.3.3.3.043)534
   constant (R)(IV.0.118;*.5.2.038)334,
   402
   storage(*VI.5.047b)592
Gases, specific gravity(*VI.5.043)591
Gassing(*VI.3.023;*.7.1.025)578,601
Gasteromyces(*V.3.3.2.179a)517
Gastrilegic(IV.4.5.002g)391
Gastropod(*V.3.3.1.250,*.251)508
Gauss (Gs,G)(IV.0.260l)347
Geese(*V.3.3.3.048)534
Geitonogamy(IV.4.4.022,*.025;.4.5.
   001a;*.4.8.017)389,391,393
Gel(IV.0.120)334
   sowing(V.1.3.030)434
Gelatin(IV.0.042)330
Gelatinous(II.5.7.014)278
   secretion(*V.3.3.2.029)511
GELCO(*VI.6.026)596
Geminate(II.2.2.039)122
Gene(III.2.2.062)321
   flow(III.2.2.063)321
   frequency(III.2.2.064)321
   pool system(III.2.1.002)317
   pools(*III.2.1.)316-317
Genecological(III.1.2.025)308
Genera (see also Genus)

of susceptible plants(*V.3.3.2.177) 517
General:
  combining ability(*III.2.2.029)319
  manager(VI.8.037,*.057,*.116)619, 620,625
  terms (stress)(IV.5.1.)396-398
Generative cell nucleus[II.2.2.13a(4), c(2)]142
Generative (sexual) propagation(V.1.3,.000)432-440,432
"Generic name"(III.1.1.014)304
Genetic(III.2.2.066)321
Genetically distinct mating group(*V.3.3.2.351)526
Geneva double-curtain system(*V.3.4.032,.10d)542,554
Genic sterility(III.2.2.067)321
Geniculate(II.2.2.007n,.08h,.17e)119, 137,146
Genome(III.2.2.068)321
Genospecies(III.1.2.026)308
Genotype(III.2.2.069)321
Genotypic:
  modifications(II.4.3.000;*IV.5.4.3.006)241,414
  xeromorphic structure(*IV.5.2.085) 406
Gentiobiose(*IV.2.2.031b)361
Genus, Genus(*III.1.1.003,.012;*V.1.4.0281)302,304,447
Geocarpic, geocarpous(II.2.3.2.013,.02h)161,176
Geographically distinct mating group (*V.3.3.2.351)526
Geophyte, geophytic(II.4.2.008)239
Georgia(*VI.8.113)625
Geotrichum candidum(*VI.9.1.506,*.507, *.508,*.509,*.510)655
Geotropic bending(*V.4.1.056)561
Geotropism(IV.3.3.008h)377
Germ plasm(III.2.2.070)321
Germinatable seed(*IV.4.8.023)394
Germinated seeds(*V.1.3.066)437
Germinating seeds(*V.1.2.011o)431
Germination(II.2.3.3.020;*IV.2.1.037) 164,359
  percentage(*V.1.3.075)438
  requirements(*II.4.3.025)242
  test(V.1.3.031)434
Germ-pore(II.2.3.2.029f,.05b,c)162,179
Ghana(*V.2.097)475
Ghost spot of tomato(VI.9.1.255)643
Giant silkworm moth(*V.3.3.1.132g)502
  snail(*V.3.3.1.251)508
Gibberella ear rot of corn(VI.9.1.258)643

Gibberella fujikuroi(*IV.2.1.036;*VI.9.1.243)358,642
  fugikuroi var. glutinans(*VI.9.1.243)642
  zeae(*VI.9.1.258)643
Gibberellic acid(*IV.2.1.036)358
Gibberellins(IV.2.1.036;*.2.2.107)358, 366
Gibbous(II.2.2.061c-14-k,.23c)128,152
Gibbs free energy(IV.0.121)334
Gift:
  houses(*VI.3.000,*.085b)577,584
  packs(VI.3.053e)582
Giga (G)(IV.0.260h)346
Gigantic(II.5.2.014)266
Ginger ale(*VI.10.003)674
Girdling(*IV.5.4.4.002,.002a;*V.3.4.000)416,539
Glabrate, glabrescent(II.5.6.050)275
Glabrous(II.5.6.051)275
Gladiolus hard rot(*V.3.3.2.181g)520
Gland[II.2.2.24a(1),.26d(3);.3.069,.179]153,155,197,203
Glands(*IV.4.4.017;*V.4.1.033)389,560
Glandular(*II.3.204;.5.6.052;.5.15j) 205,275,295
  epidermal cells(II.3.179)203
  hair(II.3.070,.31f,g)197,237
  pubescent(II.5.6.102)276
Glass(*V.1.1.017d)420
  frits(*V.1.2.012)431
  sash(*V.1.1.053c)425
Glasshouse(*V.1.1.017)419
Glaucescens(II.5.1.004f)264
Glaucous, glaucus(II.5.1.004c;.5.6.054)264,275
Gley(V.2.037)473
Gleying(*V.2.096b)475
Glittering(II.5.6.055)275
Globe artichoke(*I.3.011)8
  plume moth(VI.9.2.037)661
Globoid(*II.2.3.2.037,.07c,d)162,181
Globose[II.2.2.041;.2.3.2.04d(3)]123, 178
Glochidiate(II.5.6.056;.5.15k)275,295
Gloeodes pomigena(*VI.9.1.503)655
Gloeosporium(*V.3.3.2.019)510
  musarum(*VI.9.1.024,*.182)628,639
Glomerella cingulata(*V.3.3.2.181e; *VI.9.1.030,*.079)519,628,632
Glomerule[*II.2.1.024a,.036,.03a(8)] 101,106
Gloss(*VI.3.086)584
Glover scale(*VI.9.2.007)660
"Glowworm"(*V.3.3.1.132a)501
D-Gluconic acid(*IV.2.2.032c)362
Gluconucleogenesis(IV.1.2.019)353

INDEX OF TERMS 742

Glucosamine(IV.2.2.032f)362
Glucose(*II.3.023;*IV.1.2.019,*.040a;
    .2.2.031c,.031c-a-1,*.032b,*c,
    *.038;*V.1.5.005)194,353,356,361,
    362,470
  -1-phosphate(*IV.1.2.040a)356
  -6-phosphate(*IV.1.2.019;*.2.2.142)
    353,367
D-Glucose (see Glucose)
N-Glucosides (see N-Glycosides)
D-Glucuronic acid(*IV.2.2.032c,*.038)
    362
Glue sealer(*VI.3.053c,*.058)582
Glume[II.2.2.040,.24b(1)(i)]122,153
Glutinous(*II.5.6.150)278
Glycans(*IV.2.2.031c)361
Glyceraldehyde(*IV.2.2.031a)361
Glycerin (see Glycerol)
Glycerol(*IV.2.2.032b,*.065,*.116,
    .159,.210;*.5.3.014)361,365,366,
    368,370,408
Glycine(*IV.2.2.016;*V.3.2.012)360,
    490
Glycolipid(IV.2.2.093)366
Glycolysis(IV.1.2.040a;*.2.2.154)355,
    368
Glycophyte(II.4.2.009)239
Glycosides, N-glycosides(*IV.2.2.021,
    .032,.032a,.094,.140,*.141)360,
    362,366,367
Glycosidic linkage(IV.2.2.095)366
Glyoxisome (glyoxysome)(IV.1.1.009)351
Gnawed(*II.5.4.3.019)270
Goats(*V.1.2.011f;*.3.3.3.048)430,534
Goblet(V.3.4.031,.11f)541,555
  -shaped(*II.2.2.023)122
Golden-yellow(II.5.1.008j)266
Golf:
  course(*V.1.3.015)433
  courses(*I.2.011;*V.1.1.025)6,422
  fairways(*V.1.1.034;*.3.1.001,*.004)
    422,485
Golgi apparatus(IV.1.1.010)351
Good:
  condition(*VI.8.016)617
  delivery(VI.7.1.028,*.070)601,606
    standard(*VI.7.1.070)606
    tolerance(VI.7.1.028)601
  demand(*VI.8.027)618
  quality(*VI.8.094)623
Gooseneck, gooseneck stage of narcis-
    sus(V.4.1.057;*.4.2.2.003)561,567
Gootee(*V.1.4.036a)449
Gophers(*V.3.3.1.224)507
Gossamer winged butterfly(*V.3.3.1.
    132g)502
Gourmandizers(*V.1.4.028i,.11h)446,463

Government inspectors(*VI.1.015)572
Grade(V.2.115e;*VI.1.015;.8.038,*.085,
    *.086)480,572,619,622
  designation(*VI.3.072)583
  inspection (fresh produce)(*VI.3.
    039;.8.038c)580,619
  inspection (processed products)(VI.
    8.038d)619
  limitations(*VI.7.2.004)609
  -lowering factor(*VI.8.038a)619
  standard(VI.8.038e)619
  standards(*VI.8.038c,*d)619
  subclasses(*VI.8.038b)619
  table(VI.3.038,*.062)580,582
Grader(s)(VI.3.037,*.8.053)580,620
Grades(*VI.3.046)581
Gradient mulch system(V.3.1.007)485
Grading(*VI.3.031,.039,*.078,*.079;
    .7.1.030,.031)580,583,601
Graft union healing(*V.1.4.028e)444
Graftage(*IV.5.4.4.002;*V.1.4.000,
    .028,*.028s,*.067,.06-.13;*.3.3.2.
    133e)416,441,443,448,451,458-466,
    515
  aftercare(V.1.4.028i,.11a-h)445,463
  height(V.1.4.028j)446
Graft-chimaera(III.1.1.013)304
Graft-hybrid(III.1.1.014;.2.2.024)304,
    318
Grafting(*IV.3.1.014d,*h;*.5.4.4.002;
    V.1.1.059;.1.4.028,*.028e,*h,*k,
    t,.040,.09)372,416,443,444,445,
    446,448,450,461
  compound(V.1.2.006;*.1.4.028k-2,*-3,
    *-4)429,446,447
  knife(V.1.2.007)429
  tape(*V.1.4.028c)444
  wax(*V.1.2.006)429
Grafts, newly inserted(*V.1.4.028w)449
Grain crops(*V.3.0.004)484
Grains(*V.3.1.013)487
Graminoid roots(II.1.3.008)95
Gram-negative(*V.3.3.2.040)511
Granular(II.1.2.050;.2.2.007t-4,.13d;
    V.2.115d,f;.3.2.043;*.3.3.3.046)
    61,120,142,480,493,534
Granulation of citrus(VI.9.3.056,
    *.073)667,668
Granule(V.3.2.043;.3.3.3.046)493,534
Granules(*V.3.2.070;*.3.3.3.043)495,
    534
Granulosis(V.3.3.3.013d)532
Granum(IV.1.1.002a)350
Grape:
  cluster harvester(*VI.2.022b)576
  eye(II.1.1.1.010,.02i)19,29
  fanleaf vector(*V.3.3.1.200c)505

INDEX OF TERMS

  pruning systems(V.3.4.032,.10)541,
    554
  rust(*V.3.3.2.181f)519
  training(*V.3.4.10)554
Grapholitha molesta(*VI.9.2.035,*.053)
  661
Grass(*V.3.0.004;*.3.4.041)484,542
Grasses(*V.3.3.3.059)535
Grasshoppers(*V.3.3.1.132h)503
Grassland(*V.2.110)478
Grassy growth(V.4.1.058)562
Gravel(*V.1.1.053e;*.2.114)425,480
Gravimetric(*V.1.1.053b)425
Gravitational:
  acceleration(*IV.0.019)329
  constant(IV.0.123,*.325)334,350
  potential(*IV.5.2.057)404
  water(*IV.5.2.029)401
Gravity(IV.0.124;*.3.3.008h,*h;*.4.5.
  001a)334,377,391
Gray(II.5.1.003;*IV.0.260d)263,345
  ear rot of corn(VI.9.1.263)643
  mold of onions, shallot, garlic(VI.
    9.1.264)643
  mold rot of apple, pear(VI.9.1.265)
    643
    of asparagus(VI.9.1.266)643
    of cabbage, cauliflower, broccoli,
      turnip, rutabaga(VI.9.1.267)643
    of carrot(VI.9.1.268)643
    of fig(VI.9.1.269)643
    of fruits(*V.3.3.2.181g)520
    of garden pea(VI.9.1.270)643
    of globe artichoke(VI.9.1.271)643
    of lettuce(VI.9.1.272)643
    of pomegranate(VI.9.1.273)644
    of rhubarb(VI.9.1.274)644
    of snap beans(VI.9.1.275)644
    of stone fruits(VI.9.1.276)644
    of tomato, pepper, eggplant(VI.9.
      1.277)644
  per second(*IV.0.260e)345
  wall of tomato(*VI.9.3.074)669
Grazing(*II.4.4.010;V.3.3.3.048,*.119)
  248,534,538
  animals(II.4.3.054;.4.09)244,257
Greasy(II.5.6.145)278
Greasy pod(*VI.9.1.167)638
Great Group(V.2.096c)475
Green(II.5.1.004,.004d)264
  fruit worm(*VI.9.2.044)661
  manures(*V.2.099)476
  mold(*V.3.3.2.181g)520
    of citrus(*VI.9.1.105)634
  mold rot(*VI.9.1.152,*.153)637
    on plums(*VI.9.1.012)627
  pigmentation(*V.3.3.2.455)530

  pruning(V.4.1.060)562
  stinkbug(*VI.9.2.013)660
  tea(*I.1.001a)4
Greenhouse(V.1.1.017,*.018;*.3.2.009;
  .3.5.012;.4.1.012,*.029,*.075;
  .4.2.1.002;*VI.2.000;*.4.013;*.8.
  038a)419,421,490,557,559,560,563,
  565,575,587,619
  conditions(*V.4.1.093)563
  environmentally controlled(*V.3.2.
    083)495
  forcing(*V.4.2.1.017)566
  garden pests(*V.3.3.1.250,*.251)508
  orientation(*V.1.1.017a)419
  phase(*V.4.1.012,*.059,*.111)559,
    561,564
  range(*V.1.1.051)424
  use(*V.3.5.012)557
Greens(I.3.016;*II.5.4.3.010)8,270
Grex (g)(III.1.1.015)304
Grey(*II.5.1.003)263
  garden slug(*V.3.3.1.250)508
Grinding(*V.3.2.082)495
  or crushing(*V.3.2.072)495
Griseus(*II.5.1.003j)264
"Grit cells"(II.3.25g)231
Grocery:
  distribution center(*VI.6.022)595
  wholesaler(*VI.8.135)626
Gross profit(VI.8.042,*.070)620,621
  margin(*VI.8.042)620
  markup(*VI.8.042)620
Ground:
  bed(V.1.1.018)421
  beetles(*V.3.3.1.132a)501
  covers(I.2.003,*.009)6
  irrigation(*V.1.1.017i)421
  layerage(*V.1.4.036b)450
  layerage bed(*V.1.4.041)450
  meristem[*II.3.108,.02c(4)]199,208
  methods(V.1.4.036b)450
  (or aerial) application(*V.3.3.3.
    046)534
  parenchyma(*II.3.039)196
  tissues(*II.3.067)197
  -water sources(*VI.7.2.014)609
  water table(*V.2.110b)478
"Group"(*III.2.0.012)316
Grove(*V.3.3.3.052;*VI.2.018;*.4.004;
  .8.038a,.138;*.9.000)534,575,
  586,619,626
  heating(*VI.5.051c)593
  labor(*VI.8.053)620
  run(*VI.8.125)625
Groves(*V.3.1.010,*.011)486
Grower(VI.7.1.029;*.8.101)601,624
  cooperative(*VI.8.085)622

INDEX OF TERMS

cooperative association(*VI.8.086) 622
preference(*V.3.1.008;*.3.2.035)485, 492
Grower's:
  agent(*VI.7.1.003,.030)598,601
  contract(*VI.8.099)624
  cooperative(*VI.8.099)624
Growers(*VI.7.2.004)608
Growers' agents(VI.7.1.031)601
Growing:
  conditions(*V.4.1.012)559
  point(*V.3.3.3.112;*.4.1.003)538,559
Growth(*IV.3.1.003,.012)370,371
  and development(IV.3.;*.3.2.020)370-378,376
  cracks of sweet potato(VI.9.3.058) 668
  cracks of tomato(VI.9.3.059)668
  curve(IV.3.1.013)371
  cycles(IV.3.1.)370-373
  distortion(*IV.5.4.3.008)415
  inhibition(*IV.5.4.2.013,*.014;.5.4.3.009)413,414,415
  inhibitors(IV.2.1.037;*.5.2.017;*V.1.3.079)359,400,438
  layer[II.3.074,.20c(7)]197,226
  movements(II.3.17b,c;.4.3.023)223, 242
  processes(*V.3.3.2.133d)515
  rapid(*V.110e-2)479
  regulator(*V.1.4.022a-3)442
    balance(*IV.5.4.2.007)413
    treatment(*V.1.4.022a-3) 442
  regulators(*IV.2.1.;*.5.4.3.010;*V.1.4.022a-3;*.3.3.3.000;*.3.4.000,*.000a)357-359,415,442,531, 539
  ring(*II.3.074)197
  substances(*V.1.5.005)470
  vigorous(*V.1.4.028t)448
Grub(V.3.3.1.108)500
Grubbing hoe(*V.1.1.024,*.032)421,422
Grublike(*V.3.3.1.132e)502
Grubs(*V.3.3.1.141)503
Guanine(*IV.2.2.056,*.141,*.173)363, 367,369
Guano(*V.3.2.086)496
Guaranteed advance(VI.7.1.032)602
Guard cells(II.3.29,.30)235,236
Guatemalan system(*V.3.4.017,.09a) 541,553
Guiding structure(IV.4.4.023)389
Guidelines for classifying cultivated plant populations (1978)(III.2.0.005)313

Guignardia bidwellii(*V.3.3.2.181e)519
Gullet(IV.4.4.004b-4)388
Gullies(*V.2.110d)478
Gully erosion(*V.2.110d)478
Gum(*V.4.2.1.011)565
  formation(*V.4.2.1.020)566
  oozing(*V.3.3.2.040)511
  pockets(*VI.9.3.062)668
  secretion(*V.3.3.2.196)521
Gumming(*IV.2.1.033)358
  of branches(*V.3.2.066)494
Gummosis(V.3.3.2.196)521
  of bulbs(V.4.2.1.011)565
Gummulation(*V.3.3.2.133c)515
Gummy stem blight(*VI.9.1.091,*.092) 633
Guttation(*II.3.082)197
Guttatus(*II.5.1.006q)265
Gutters(*V.2.110e)479
Gymnospermae(II.6.011)279
Gymnosperms(*I.1.000;*II.3.052,*.073, *.117,*.157,*.173,*.199,*.202;.6.011;*IV.1.2.008a;*.2.2.172)196,197, 199,201,202,204,279,352,369
Gymnosporagium:
  clavipes(*VI.9.1.437)651
  juniperi-virginianae(*V.3.3.2.181f; *VI.9.1.036)519,629
  spp.(*V.3.3.2.366)527
Gynandrium(II.2.2.24f,.25b)153,154
Gynandrophore(*II.2.2.041a)122
Gynobasic(II.2.2.17f)146
Gynoecium, gynoecia[II.2.1.01(1);.2.2.041,.01a(1-4);*.4.2.000]104,122, 130,385
  formation(*V.4.1.113)564
Gynophore(II.2.2.041m)123
Gynostemium(II.2.2.041n,.24f)123,153
Gypsic horizon(V.2.028c)473
Gypsum(*V.2.099;.3.2.044)476,493
Gypsy moth(*V.3.3.1.132g;*.3.3.3.013f) 502,532

H

Habit(II.0.004;.5.2.015)10,266
Habitat(II.4.2.000;.4.02c-d,.04,.05; *V.3.3.3.004)238,250,252,253,531
  modification(*V.3.3.3.053)535
Hadromycosis(*V.3.3.2.435)529
Hail:
  damage on citrus(VI.9.3.060)668
  injury of apples(VI.9.3.061)668
  of stone fruits(VI.9.3.062)668
Hair-pointed(*II.5.6.092)276
Hairs[II.2.3.3.042h,.07e,f;*.3.039, *170,*.179,*.204,.29c(4);*.4.3.

INDEX OF TERMS

    002;*IV.4.4.023]167,188,195,202,
      203,205,235,241,389
    dense(*V.3.3.1.132g)502
    (see also Trichomes)
Hair-shaped(*II.5.6.021)274
Hairy(*II.5.6.093)276
Halbert-shaped(*II.5.4.2.008;*.5.5.
    026)269,272
Haley Bill of 1960(*VI.7.2.016)610
Half:
    -cooling time(VI.5.021)589
    -equitant(*II.1.1.2.017,.02n)21,35
    -inferior(II.2.2.041t-3-a,.19c)124,
      148
    -inverted (ovule)(*II.2.2.041t-4-k-
      1)124
    -terete(*II.1.2.098)63
Halloysite(*V.2.098)475
Halo blight of lima beans(*VI.9.1.050)
    630
Halophytes(*II.4.2.009,.010,*.028;.4.
    04a,b;*V.5.2.005)239,240,252,399
Halo spot of apples(*VI.9.3.044)666
Halteres(V.3.3.1.110,*.132b)500,501
Halved(*II.2.2.029)122
Hammermill(*V.1.1.064;.1.3.032,*.057)
    426,434,435
Hammock(II.4.2.011)239
Hamper(*VI.2.012;*.4.000a,*c,.012)575,
    585,587
Hand:
    -clamp truck(*VI.3.013)578
    fertilizing(*V.3.2.039)492
    harvesting(*V.1.3.063,*.064;VI.2.
      022a)437,576
    -held pneumatic power shaker(*VI.2.
      022c)576
    labor(*V.3.1.000)484
    -operated duster(*V.3.3.3.030)533
    pollination(*IV.4.4.018)389
    refractometer(*VI.1.008d)571
    scoop(*VI.2.022a)576
    shears(*V.4.1.094)563
    slashing(*V.3.1.010)486
    wax(*V.1.2.006)429
Handler(s)(*VI.7.2.004a,*b,*c)609
    quotas(*VI.7.2.004)608
Handling(*V.3.0.001;*VI.4.000)484,585
    and display loss(*VI.8.108)624
    practices(*VI.9.3.000)663
    systems(VI.2.020;*.3.079)575,583
    techniques, poor(*V.4.2.2.004)567
Hanger[V.3.4.01(6)]545
Hanging(II.5.2.004)266
Hanging baskets(I.0.001,1.2.001e;VI.4.
    013)3,6,587
    bud(V.1.4.028f-4,.08d)445,460

Hapa[V.1.4.04c(5)]456
Hapaxanthic(II.5.3.011)267
Haplocheilic(*II.3.190)204
Haploid(III.2.2.071)321
    embryo[IV.4.1.01(12)]382
    nuclei fusion(*V.3.3.2.232)522
Haplomorphic(IV.4.4.024)389
Haptogamy(IV.4.4.025)389
Haptonasty(IV.3.3.003b)376
Haptotropic movements(II.4.3.071)245
Haptotropism(IV.3.3.08i)378
Hard(*V.2.100)476
    base of tulips(V.4.2.2.014)567
    seed(V.1.3.034)434
    -to-propagate plants(*V.1.4.028u)449
Harden irreversibly(*V.2.076)474
Hardening(*IV.5.2.001,.030;*.5.3.008,
    .028,.032;*.5.4.1.006;*.5.4.3.
    010;V.4.1.062)398,402,408,409,410,
    412,415,562
    loss(*IV.5.3.015)408
    low-temperature(*IV.5.3.034)410
    -off(*IV.5.2.030;V.1.3.033,.01b,c;
      .1.5.01m,n)401,434,440,471
Hardiness(*IV.5.3.030,.032)410
Hardpan(V.2.043)473
Hardwood:
    cuttings(*V.1.4.022b,.022f-1,.02b)
      442,454
    forests(II.4.2.011)239
Hardwooded(II.0.005)10
Hares(*V.3.3.1.224)507
Harmful physiological processes(*V.3.
    3.2.133)514
Harris Act of 1958(*VI.7.2.016)610
Harrow(V.1.1.020;*.3.1.004)421,485
Harvest(*V.3.3.3.071)536
Harvested ripe(*VI.1.004)569
Harvesters(*VI.8.053)620
Harvesting(*V.3.0.001;VI.2.000;*.3.
    079;*.4.000,.000a;*.7.1.030,*.031;
    .8.081,.102)484,575,583,585,601,
    622,624
    and handling(*VI.8.078;*.9.000)622,
      626
    containers(*VI.4.011)586
    (field)(VI.4.000a)585
    foreman(VI.8.044)620
    foreman's crew(*VI.8.137)626
    hamper(*VI.4.012)587
    -induced(*VI.9.3.000)663
    lug(*VI.4.014)587
    methods(VI.2.022)576
    operations(*V.3.1.008)485
    types(VI.2.023)576
Harvestmen(*V.3.3.1.010)498
Hastate(II.5.4.2.008;.5.5.026,.04f,

.10b)269,272,284,290
"Hatracked" tree(*V.1.4.028d,*v,.10b)
  444,449,462
Haulm(II.1.1.3.044)24
Haustorium[II.1.3.02c(3),.4.3.024,
  *.053;V.3.3.2.199]97,242,244,521
Hawaii foliage plant mix(*V.1.2.005)
  429
Hawaiian:
  stump system(*V.3.4.017,.08c)541,552
  system(*V.3.4.017,.08a,b)541,552
Hawk-billed knife(V.3.4.036,*063,
  *.064)542,543
Hawk moth(*V.3.3.1.132g)502
Hay:
  fork(*V.1.1.038)423
  rake(*V.3.0.003)484
Head[II.2.1.037,.02a(7),.03a(6),.09e;
  V.3.3.1.132,.132b]101,105,106,
  112,501
  flower[II.2.1.09a(1),b]112
  -or brush-shaped(IV.4.4.004b-5)388
  system with spurs(V.3.4.032,.10e)
  542,554
Heading(*IV.5.4.4.002;*V.3.4.000b,
  *.073,.02)416,539,544,546
  alternate low and high(*V.3.4.073)
  544
  height[V.3.4.01(17),.03]545,547
Healed union(*V.1.4.028e)444
Heart:
  rot of pineapple(*V.3.3.2.181e)519
  rot of pomegranate(VI.9.1.286)644
  rot of sugar cane(*V.3.3.2.181e)519
  rots(*V.3.3.2.181f)519
  -shaped(*II.5.5.012)271
Heartwood(II.3.077)197
Heat(IV.0.126;*.5.1.014;*V.1.1.065;
  *.3.2.009;*VI.5.051)334,397,426,
  490,592
  avoidance(IV.5.4.1.004)412
  capacity(IV.0.260e;VI.5.051e)345,593
    specific(VI.5.051e)593
  chamber(*IV.5.4.1.006)412
  delay(V.4.1.063)562
  exchanger(*VI.10.010)675
  flux(IV.0.126e)334
  flux density(IV.0.126f,*.260e)334,
    345
  injury(IV.5.4.1.005)413
    of avocado(VI.9.3.064)668
    of stone fruits(VI.9.3.065)668
  intolerant(*IV.5.3.033)410
  -killing temperature(*IV.5.4.1.000,
    *.006)412
  latent, specific(VI.5.051c)593
  removal(VI.5.024)589

  experimental(VI.5.024b)589
  ,theoretical(VI.5.024a)589
  resistance(*IV.5.3.032)410
  -resistant plants(*IV.5.4.1.004)412
  -sensitive plants(*IV.5.3.034)410
  shock(*IV.5.3.034)410
  shocks(*IV.5.4.1.006)412
  thermal conductivity(IV.0.260e;VI.5.
    049)345,592
  thermal diffusivity(VI.5.050)592
  thermotropism(*IV.3.3.008s)378
  tolerance(IV.5.4.1.006)412
  -tolerant plants(*IV.5.3.034)410
  transfer, surface convective coeffi-
    cient of (IV.5.048)592
  treatment(*V.1.4.014;*.4.2.2.014)
    441,567
  unit(*IV.4.8.020)394
  -unit accumulation(*VI.1.008)570
  vaporization(*V.3.3.3.006)531
  vs. chilling tolerance(IV.5.3.033)
    410
  vs. freezing tolerance(IV.5.3.034)
    410
Heater protective service(*VI.6.029c)
  596
Heating(V.1.1.017h;*VI.5.005,*.017,
  *.051e)421,588,593
  cables(*V.1.1.002)418
  ,cooling and ventilation systems(*V.
    4.1.053)561
  for insect control, ripening, color-
    ing(*VI.7.1.025)601
  in transit(*V.4.2.2.010)567
  pads(*V.1.1.002)418
Heavy:
  metal toxicity(*V.3.2.066)494
  offerings(*VI.8.073)621
  shipments(*VI.8.134)626
Hectare (ha)(*IV.0.016,*.260k)329,347
Hecto (h)(*IV.0.260h)346
Hedge shears(V.3.4.041,*.063)542,543
Hedger(*V.3.4.063)543
Hedgerow(*V.3.1.008a,.01e)485,488
Hedges(*V.2.110e;*.3.4.041)479,542
  flat-topped(*V.3.4.073)544
Hedging machine(V.3.4.040)542
HEEDTA(*V.3.2.012)490
"Heel" (semihardwood)cutting(*V.1.4.
  022f-3,.03b)443,455
Height(*V.4.1.006)559
Helical (thickening)(*II.3.202)204
Helicoid cyme[II.2.1.024c,.03a(7),
  .05b,.07e]101,106,108,110
<u>Helicotylenchus</u> spp.(*V.3.3.1.200d,*f)
  506
Heliophilous(II.4.2.012)239

INDEX OF TERMS

Heliophobe(II.4.2.013)239
Heliothis(*V.3.3.3.013b)532
　zea(*V.3.3.3.013f;*VI.9.2.021)
　　532,660
Heliotrope(IV.3.3.002)376
Heliotropism(*IV.3.3.008n)377
Helminthosporium:
　carbonum(*VI.9.1.515)656
　carposaprum(*VI.9.1.290)644
　maydis(*VI.9.1.514)656
　turcicum(*VI.9.1.329)646
Helminthosporium rot of tomato(VI.9.1.
　290)644
Helophytes(II.4.2.014,*.017;.4.04c)
　239,252
Hemes(*IV.2.2.168)368
Hemiacetal(*IV.2.2.031b,*.095)361,366
Hemianatropous(*II.2.2.041t-4-k-1;.2.
　3.3.036a-2,.04b)124,165,185
Hemic(*V.2.069)474
Hemicelluloses(*IV.2.2.031c,c-b-2)361,
　362
Hemicriconemoides spp.(*V.3.3.1.200b)
　505
Hemicycliophora spp.(*V.3.3.1.200b)505
Hemicryptophyte(II.4.1.003;.4.01c)238,
　249
Hemiepiphytes(II.4.2.015;.4.3.002;.4.
　04d)239,241,252
Hemileia vastatrix(*V.3.3.2.181f)520
Hemiphilic(IV.4.4.026)389
Hemiptera(*V.3.3.1.132c)502
Hemitropic(IV.4.5.002h)391
Hemoglobin(*IV.2.2.168)368
Henry(*IV.0.260c)344
　per meter(*IV.0.260e)345
Hepaticus(*II.5.1.001i)263
Heptoses(*IV.2.2.031a)361
Herb(II.0.006,.01c,d,.02;*.4.1.005;*.
　4.2.026)10,12,13,238,240
Herbaceous(II.0.007,.01c,d,.02a,b;
　.5.7.015)10,12,13,278
　annuals(*II.3.026;*.4.3.018)195,242
　biennials(*II.3.026)195
　cuttings(V.1.4.022f-2,.02a)442,454
　perennials(*II.4.2.006;*.4.3.018)
　　239,242
　plants(*V.1.1.034;*.1.4.022f-2)422,
　　443
Herbicide(s)(*IV.2.1.012;*V.3.3.2.270;
　.3.3.3.000,.031,*.041,*.049,
　.055,.069,*.070,*.094,*.095,
　.113,.116,*.119;VI.7.2.017k,*p)
　358,524,531,533,534,535,536,537,
　538,613,614
　acid equivalent applied(*V.3.3.3.
　　085)537

applications(*V.3.3.3.016)532
effects(*V.3.3.3.003)531
formulations(*V.3.3.3.049)534
labels(*V.3.3.3.003)531
preplanting(*V.3.3.3.059)535
selectivity(*V.3.3.3.122)539
treatment(*V.3.3.3.111)358
Herbs(*V.2.110e;*VI.10.007)479,674
　pot(*I.3.016)8
Hercogamy(IV.4.3.006)386
Hereditary properties(*V.3.3.2.282)524
Heredity(III.2.2.072)321
Heritable modifications(*IV.5.2.002)
　398
Heritability(III.2.2.073)321
Herkogamy(*IV.4.3.006)386
Hermaphrodite(*II.2.2.013)121
Hertz (Hz)(*IV.0.114,*.260c)333,344
Hesperidium(II.2.3.1.018,*.026,.05b;
　.2.3.2.029a,.05d)159,160,162,172,179
Hessian fly(*V.3.3.1.132b)501
Heterandry (heteranthery)(IV.4.4.027)
　389
Heteroblastic development(IV.3.1.
　014a)372
Heteroblasty(II.4.3.025)242
Heterocellular[*II.3.152,.07a(2)]201,
　213
Heterodera(*V.3.3.1.200)505
　schachtii(*V.3.3.1.200g)506
　spp.(*V.3.3.1.200e,*g)506
Heteroecious(V.3.3.2.204)521
　rust fungi(*V.3.3.2.010)510
Heteroecism(*V.3.3.2.181f)519
Heterofertilization(IV.4.1.015)381
Heterogameon(III.2.2.074)321
Heterogamous(II.2.1.038,.09c)101,112
Heterogamy(IV.4.3.007)386
Heterogenesis(III.2.2.075)321
Heterogeny(*IV.4.3.019)387
Heterophylly(II.1.2.052,.11a)61,76
Heteroplastic(V.1.4.0281)447
Heterosis(III.2.2.076)321
Heterosporium variable(*VI.9.1.292)644
Heterosporium leaf spot of spinach(VI.
　9.1.292)644
Heterosporous(II.6.012)279
Heterostyled(*IV.4.3.012,*.013)386
Heterostylous(II.2.2.17g)146
Heterostyly(IV.4.3.009)386
Heterotypic(*III.2.2.105,*.154;IV.4.1.
　016)323,325,381
　division[*III.2.2.154;IV.4.1.01(2)]
　　325,382
Heterotopy(V.3.3.2.205)521
Heterotroph, heterotrophic(*II.4.2.
　026;V.3.3.2.206)240,521

Heterozygote(III.2.2.078)321
Heterozygous(III.2.2.079)321
Hexagonal(V.3.1.008g,*m)486
Hexapoda(*V.3.3.1.132)501
Hexose(*IV.1.2.036c)354
  phosphates(*IV.1.2.036c)354
Hexose phosphate shunt(*IV.1.2.040g)
  356
Hexoses(*IV.2.2.031a)361
Hiatus(II.0.008)10
High:
  analysis fertilizer(V.3.2.046)493
  explosive(*V.3.2.048)493
  graftage(*V.1.4.028j)446
  -headed[*V.3.4.029,.03b(1),c(1),
    d(1)]541,547
  heading(*V.1.4.028i)445
  humidity(*VI.3.023)578
  -humidity beds(V.1.1.053c)425
  light intensity stress(*IV.5.4.2.
    011)413
  moisture-humidity conditions in
    tropical forests(II.4.3.026)242
  pressure(*V.3.3.3.006)531
  -shrink items(*VI.8.108)624
  -speed thresher(*V.1.3.057)436
  temperature(*IV.5.3.047;.5.4.1.;*V.
    4.2.2.021)411,412,568
    sensitivity(*V.1.3.083)439
    stress(IV.5.4.1.000)412
    tolerance(*IV.5.3.038)410
  temperatures(*V.4.2.2.010,*.015)567
  vacuum(*VI.5.055;*.10.008)593,674
    desiccation(*V.3.3.2.245)523
  winds(*V.3.3.2.270)524
Higher analysis forms(*V.3.2.038)492
Highly:
  contagious(*V.3.3.2.040)511
  perishable (produce)(*VI.8.138)626
Hilar:
  palisade[II.2.3.3.10b(7)]191
  region(II.2.3.3.10b)191
  (seed-form)(II.2.3.3.036a-2,.04f,
    .11a-c)165,185,192
Hill reaction(IV.1.2.036f)355
Hiller plow(*V.1.1.044)423
Hilum[II.2.2.041t-4-e;.2.3.3.021,.01a
    (6),.08b(3),c(3),.10a(1)]124,164,
    182,189,191
  exaggerated[II.2.3.3.04f(1)]185
  (sclerotic)[II.2.3.3.08b(3),c(3)]189
Hippocrepiform(*II.2.2.007o)119
Hippopotami(*V.3.3.1.140)503
Hirsute, hirtuse(II.5.6.062;.5.151)
  275,295
Hispid, hispidulous(II.5.6.063;.5.15n)
  275,295

Histic epipedon (surface)(V.2.028d)473
Histidine(*IV.2.2.016)360
Histogen theory(II.3.080;*.206,.14a)
  197,205,220
Histological symptoms(*V.3.3.2.133b)
  515
Histones(*IV.2.2.172)369
Hoary(II.5.1.003h;.5.6.064)264,275
Hoe(V.1.1.024)421
Hofmeister series(*IV.0.166)340
Holdovers(VI.8.016)617
Hollow stem of celery(*VI.9.3.098)670
Holocarpic reproduction(V.3.3.2.207)
  521
Holonecrotic, holonecrosis(V.3.3.2.
    208)521
  spot(*V.3.3.2.018)510
Holotype(III.1.1.035c)307
Home:
  hanging basket(*VI.4.013)587
  hydroponic culture(*V.3.2.083)495
  lawn(*V.1.3.015)433
  sources (emissions)(*VI.7.2.014)610
Homoblastic development(IV.3.1.014b)
  372
Homocellular(*II.3.152)201
Homogamous(II.2.1.039,.09d)101,112
Homogamy(IV.4.3.010)386
Homogenate(IV.0.132)335
Homogeneon(III.2.2.080)322
Homoiohydric plants(IV.5.2.031)401
Homoiotherm(IV.5.3.035,*.038)410
Homologous(III.2.2.081)322
Homoptera(V.3.3.1.133d)502
Homorhizic(II.1.3.009)95
Homostylic dimorphism(IV.4.3.011)386
Homostylous(II.2.2.17h)146
Homotypic, homotypic division[*III.2.
    2.105,*.154;*IV.4.1.006b-6;.017,.01
    (4)]323,325,380,381,382
Homozygote(III.2.2.083)322
Homozygous(III.2.2.084)322
Honey(*IV.4.5.002i)391
  bees(*V.3.3.1.132e)502
Honeycombed(*II.5.6.003)273
Honeydew(*V.3.3.2.395)528
Hooded(*II.2.2.061c-14-i,.26a,b)128,
  155
Hoof meal(*V.3.2.068)494
Hooked(*II.5.6.144)278
  -back(*II.5.5.051)273
Hooks, hairs, prickles, thorns, whips
  (*II.4.3.002,.027;.4.09)241,242,257
Hoplolaimus spp.(*V.3.3.1.200d)506
Hoplocampa testudinea(*VI.9.2.025)660
Hopper (railroad) cars(*V.1.3.067)437
Hops(*VI.10.003)674

Horamorphosis(II.4.4.003b)247
Horehound(*I.3.006)7
Horizon(s)(*V.2.112)479
  ,B(*V.2.127)482
  dark-colored(V.2.01a)483
  light-colored(V.2.01a)483
Horizontal(II.1.1.3.045,.12b;*V.3.4.
    021,*.055)24,48,541,543
  -arm spur system(*V.3.4.032,.10c)
    542,554
Hormodendrum rot of corn (Hormodendrum
    spp.)(VI.9.1.295)645
Hormone(s)(IV.2.1.,.001)357-359,357
Horned(*II.2.2.061c-14-g)128
Hornets(*V.3.3.1.132e)502
Hornfly(*V.3.3.1.132b)501
Hornworm(*V.3.3.1.132g)502
Horny(II.5.7.016)278
Horse bot flies(*V.3.3.1.132b)501
Horseflies(*V.3.3.1.132b)501
Horsehair(*VI.3.007,*.062)578,582
Horses(*V.1.2.011f;*V.3.3.3.048)430,
    534
Horseshoe-shaped(II.2.2.007o,.007w-7,
    .08e)119,120,137
Horticultural:
  crop production(*V.3.3.0.002)497
  crops(I.0.001;*IV.5.4.4.002;*VI.1.
    02)3,416,574
  groups(*V.4.2.2.006)567
  maturity(IV.1.004,*.016,.02)570,572,
    574
  morphology(II.0.000)10
  perlite(*V.1.2.005a)429
  principles(*V.3.0.001)484
  products(*VI.4.000)585
  supply sources(*V.1.1.017a)419
  taxonomy(III.1.0.001)302
  vermiculite(*V.1.2.005a)429
Horticulture(I.0.002)3
Hoses(*V.1.1.017e,.025,*.028)420,422
Host(s)(*V.3.3.2.312;*.4.2.1.004)525,
    565
  for vectors and pathogens(*V.2.110e-
    2)479
  of rust fungi(*V.3.3.2.204)521
  tissue(*V.3.3.0.004)497
Hot:
  air tubes(*V.1.1.002)418
  bed(*V.1.1.018,.026,*.048)421,422,
    424
  -pack concentrate(VI.10.019)676
  water(*V.1.1.017h)421
    immersion(*V.1.3.080)438
    pipes(*V.1.1.002)418
    -treatment(*V.4.2.1.003,*.014)565,
    566

  zone plants(II.4.3.028;.4.08b)243,
    256
Hour (h)(*IV.0.260i)346
House(VI.8.046)260
Hovering:
  animals(*IV.4.4.004b-4)388
  visitors(*IV.4.4.004b-6)388
Hue(II.5.1.000b)263
Huller(*V.1.3.057)436
Hull meal(*V.3.2.068)494
Human:
  hazards(*V.3.3.3.071)536
  or mechanical error(*VI.8.038f)619
Humectant(V.3.3.3.050)534
Humid climates(*IV.5.2.013)399
Humidity(*IV.5.2.040;*V.1.3.073;*VI.5.
    025;*.9.3.000)403,438,589,663
  higher(*V.2.110e-1)479
Humification(*V.2.106)476
Humus(*V.2.107,*.111,.01a)476,479,483
  content(*V.2.014)472
  dispersal(*V.2.120)482
  molecules(*V.2.015)472
  -poor(*V.2.076)474
Husk(II.2.3.2.015,.04a)161,178
Hutchinson(*III.1.1.004,*.010)303,304
Hyacinth(I.2.001b)5
Hyaline (hyalinus)(II.5.1.007f)265
Hybrid(s)(III.2.1.001b;.2.2.085)317,
    322
  swarm(III.1.2.028)309
  varieties (F_1)III.2.0.005a-6)314
  vigor(III.2.2.086)322
Hybridization community(*III.1.2.010)
    308
Hydathode(II.3.082,*.179,.15a)197,203,
    221
Hydration(*V.2.106)476
Hydraulic:
  conductivity(*IV.5.2.036;*V.2.108)
    402,477
  hedge shears(*V.3.4.063)543
  lift(*VI.2.012)575
  pole saws(V.3.4.060)543
  pole shears(V.3.4.061)543
  saws(*V.3.4.063)543
  shears(*V.3.4.063)543
Hydrochory(IV.4.7.001e)393
Hydrocooling(*VI.5.035,.035b)590,591
Hydrogen (H)(*V.3.2.028)491
Hydrolases(*IV.2.2.058g)365
Hydrolysis(*V.2.106)476
Hydrometer(*VI.1.008a,c,*.e,*.010)570,
    571,572
Hydromorphosis(II.4.4.005b)247
Hydrophily(*IV.4.5.001a)391
Hydroponic culture(*V.3.2.083)495

Hydrophyte, hydrophytic(II.4.2.017;.4.
   05a)240,253
Hydroseeding(*V.1.3.028)433
Hydrosere(II.4.2.018)240
Hydrosis, hydrotic(V.3.3.2.210)521
Hydrostatic pressure(*IV.5.2.005)399
Hydrotropism(IV.3.3.008j-1)377
Hydrous mica group(*V.2.098)475
Hydrous silicates(*V.3.2.050)493
N-Hydroxyethylethylenediaminetriacetic
   acid(*V.3.2.012)490
   -Hydroxyethylhydrazine(*IV.2.1.033,
   *.037)358,359
Hydroxyl(*IV.2.2.154;*.5.3.053)368,411
Hygromorphosis(II.4.4.003c)247
Hygrophyte(II.4.2.019;.4.04c,.11)240,
   252,259
Hygroscopic(V.3.2.048)493
Hymenium(*V.3.3.2.023,*.295)511,525
Hymenoptera(V.3.3.1.132e)502
Hypanthium[II.2.2.044,.19b,.27a(4)]
   124,148,156
Hypermetamorphosis(V.3.3.1.129)500
Hyperparasite(V.3.3.2.211)521
Hyperplasia(II.3.083;V.3.3.2.212)198,
   521
Hyperplastic(*V.3.3.2.040)511
   class of symptoms(*V.3.3.2.250)523
   symptom(*V.3.3.2.158,*.170,*.171,
      .205,.227,*375,*.376,*.469)
      516,521,522,527,530
   of disease(*V.3.3.2.340,*.341,
      *.370)526,527
Hyperploid(III.2.2.011)318
Hypersensitive(V.3.3.2.213)521
   hosts(*V.3.3.2.312)525
Hypertrophy(II.3.084;V.3.3.2.214)198,
   521
Hyphae(*V.3.3.2.179)517
Hyphal:
   fusion(*V.3.3.2.181f)519
   strand bundle(*V.3.3.2.358)527
Hyphydrophily(*IV.4.5.001a)391
Hypobaric storage(*VI.5.047c)592
Hypochlamydy(II.2.3.2.016,.04b)161,178
Hypocotyl[II.2.3.3.01a(4),b(7),.03(g);
   .3.203;.4.3.057]182,184,205,244
Hypocrateriform(II.2.2.061c-1-k,21f)
   127,150
Hypocraterimorphous(IV.4.4.028)389
Hypodermis[*II.3.059,.085,*.114,.13a
   (2);.4.10c(1)(c)]196,198,199,219,258
Hypogynous(II.2.2.061d-2,.19a)128,148
Hypophysis(II.2.3.3.03f-i(5);.3.086)
   184,198
Hypoplasia(V.3.3.2.215)521
Hypoplastic symptom(s)(*V.3.3.2.095,
   .133b,.413)513,515,529
Hypoploid(*III.2.2.011)318
Hypostase, hypostasis[*II.2.3.3.038,
   .03f-i(5),.10b(6),.12d(2)]166,184,
   191,193
Hypostatic(*III.2.2.055)320
Hypothetical profile(V.2.01a)483
Hypotriploid(III.2.2.090)322
Hypotrophy(II.3.087)198
Hypselomorphosis(II.4.4.003d)247
Hypsoperine spp.(*V.3.3.1.200g)506
Hypsophyll[II.2.2.045,.27a(5),b(2)]
   125,156

I

IAA(*IV.2.1.012;*.5.4.3.010;*V.1.5.
   005)357,415,470
IBA(*IV.2.1.012)357
IBDU(*V.3.2.015)491
I-bud, I-budding(*V.1.4.028f,*.028f-5,
   .08j)444,445,460
Ice:
   crystals(*IV.5.3.049)411
   erosion(*V.2.130)482
Ichneumon wasps(*V.3.3.1.132e)502
Ichthyochory(IV.4.7.001f)393
Icing(*VI.6.029c)596
Ideotype(III.2.2.091)322
Identification(III.1.1.016)304
Idioblast(II.3.088;*.175)198,203
Igneous:
   origin(*V.3.2.075)495
   rocks(*V.2.000)472
Igneus(*II.5.1.005j)264
Illegitmate pollination(IV.4.3.012)386
Illite(*V.2.098)475
Illuminance(IV.0.157c,*.260c)338,344
Illumination(*IV.5.4.2.002,*.003)413
   intermittent(*V.4.1.034)560
Illuvial:
   accumulation(*V.2.028h)473
   horizon(*V.2.107n)477
Illuviation zone[V.2.107h,.01(a)]477,
   483
Imagines(*V.3.3.3.051)534
Imbibition(V.1.3.037)434
Imbricate(II.1.1.2.014,.021)21,35
Immediate shipment(*VI.7.1.074)607
Immersed(II.4.2.020)240
Immersion in antiseptic(*V.1.4.022a-3)
   442
Immune(V.3.3.2.216)521
Immunity(V.3.3.2.217)521
Imparipinnate(II.1.2.053,.11b,.12a)61,
   76,77
Impedance(IV.5.3.024)409

INDEX OF TERMS

Impeded drainage(*V.2.037)473
Imperfect:
  fungi(*V.3.3.2.019,*.181g)510,520
  stage(*V.3.3.2.219)521
  state(*V.3.3.2.179a,.219,*.406)517, 521,528
Impermeable(V.1.3.038)434
  layer(*V.4.1.082)563
Impervious:
  layer(*V.2.043)473
  seed-coats(*V.1.3.034)434
Implements(*V.3.0.003)484
Implicative(II.1.1.3.046,.12c)24,48
Importation of plant material(*V.3.3.3.000)531
Impotence(IV.4.8.029)394
Improper:
  loading pattern(*VI.6.034)596
  precooling(*VI.6.034)596
Impulse item(VI.8.047)620
Inadequate inspection during transit (*VI.6.034)596
Inarching(*V.1.4.028,*.028e,m,.12g-k) 443,444,447,464
Inbred line(s)(*III.2.0.003b,.004h,*i, *.005a-6)311,312,314
  pure line(s)(*III.2.1.001,*.001b,c) 317
  varieties(*III.2.0.005a-2)313
Inbreeding(III.2.2.092)322
Incanus(*II.5.6.064)275
Incised(II.5.4.3.021;.5.06b,c)270,286
Inclined:
  draper(*V.1.3.057)436
  multiple-stem(*V.3.4.017,.07a,b) 541,551
  -reel screen(*V.1.3.057)435
Inclining(II.5.2.017)266
Included (phloem)(*II.3.008,*.136) 194,200
Inclusion:
  bodies(*V.3.3.3.013a)532
  capsules, granular(*V.3.3.3.013d)532
  oval-or spindle-shaped(*V.3.3.3.013c)532
  polyhedral (rod shaped)(*V.3.3.3.-013f)532
  polyhedral or spherical(*V.3.3.3.013a)532
Incompatible, incompatibility(*IV.4.8.022,.030;*V.1.4.035;*.3.3.3.019) 395,449,533
  commodities(VI.6.009)594
  rootstock(*V.1.4.028r)448
Incomplete:
  female sterility(*IV.4.2.001d)385
  protandry(*IV.4.3.002b,*d)385,386

Inconspicuous(IV.4.4.004a)387
Increased:
  light(*IV.5.4.1.006)412
  light intensity(*IV.5.2.001)398
Incubation(V.1.3.039;.3.3.2.220)434, 522
Incumbent(II.2.3.3.023,.02h)164,183
Incurved (ovule)[*II.2.2.041t-4-k-3, .16c(4)]125,145
Indehiscent(II.2.3.1.019,.05)159,172
Indent disk separator(*V.1.3.057)435
Independent:
  chains or stores(*VI.8.135)626
  fruit buyer(*VI.8.002)616
  store(s)(IV.8.048,*.105)620,624
  trucker(VI.6.010)594
Indeterminate(II.2.1.040,.02a)101,105
  growth(II.1.1.3.15a;*V.1.4.060)51, 451
  panicle[II.2.1.041,.02a(3),.08h, .11c]102,105,111,114
Index:
  Kewensis(III.1.1.017)304
  Londinensis(III.1.1.018)304
Indexing(*V.1.4.010;*.3.3.0.004;.3.3.2.221)441,497,522
India(*V.1.4.028a;*.2.097)443,475
Indicator plant(s)(*V.3.3.0.004;.3.3.2.222)497,522
Indigen(III.1.2.029)309
Indirect:
  chilling injury(*IV.5.3.005)407
  (heat) injury(*IV.5.4.1.005)412
  injury(*IV.5.3.005)407
  strains(*IV.5.4.1.005)412
  stress injury(*IV.5.3.033,*.047)410, 411
Individual:
  cells(*V.1.5.000;*VI.3.053f)470,582
  containers (propagation)(V.1.1.053d)425
  price per unit volume or weight(*VI.8.130)625
  truckers(*VI.8.106)624
Indole ethanol(*IV.2.1.012)357
Indoleacetaldehyde(*IV.2.1.012)357
Indoleacetic acid(*IV.2.1.012)357
Indoleacetonitrile(*IV.2.1.012)357
Indolebutyric acid(IV.2.1.012)357
Indolepyruvic acid(IV.2.1.012)357
Indoor:
  locations, specific(*V.4.1.053)561
  plants(*V.4.1.053)561
  "spring gardens"(*V.4.1.068)562
Induced parthenogenesis(*IV.4.1.006a-10,.006b-3-a)379
Inductance(*IV.0.260c)344

Induction cycle(IV.3.2.013e)374
Induplicate(II.1.1.2.015,.02k,.03s)21, 35,36
Indurate(II.5.7.017)278
Indurated(*V.2.100)476
Indusium(II.6.013)279
Industrial:
  hygienists(*VI.7.2.038)616
  sources (emissions)(*VI.7.2.014)610
  uses(*V.2.096e)475
INEAC(*V.2.097)475
Inert:
  ingredient (def.)(VI.7.2.017l)613
  material(*V.1.3.051,*.075)435,438
Infect(V.3.3.2.223)522
Infected plant material (removal or destruction)(*V.3.3.3.076)536
Infection(*V.3.3.2.213,*.223)521,522
  court(*V.3.3.2.006,*.226)510,522
Infectious(*V.3.3.2.224)522
  agent(*V.3.3.2.087)513
Infective(*V.3.3.2.299)525
Inferior(II.2.2.041t-3-b,.19d)124,148
Infertile, infertility(IV.4.8.031)394
Infest, infestation(V.3.3.2.225)522
Inflatable polyethylene (greenhouse)(*V.1.1.017e)419
Inflated(II.2.3.2.017,.02c)161,176
Inflexed(II.5.2.018)266
Inflorescence(s)(II.2.1.000,.01-.14; *.4.3.071;*.5.3.016;IV.4.3.002d, *.006,*.014,*.016,*.017;*V.4.1. 033)100,104-117,246,268,386,387, 560
  abscission(*V.4.2.2.019)568
  bulbils(*V.1.4.006)441
  formation(*V.4.1.008)559
  teratological(*II.4.4.008)247
  types(II.2.1.02,.03,.04)105-107
Infrafoliar(II.1.2.054,.13c)61,78
Infrapetiolar(II.1.1.1.011,.03j)19,30
Infrared (see IR)
Infrared spectrophotometer(IV.0.136) 335
Infrasubspecific(*V.3.3.2.179b-1)517
Infructescence(II.2.3.2.018,.04d)161, 178
Infundibular(*II.2.2.061c-1-i,.21j) 126,150
Ingredient statement (def.)(VI.7.2. 017m)613
Inherent properties(*V.3.3.2.217)521
Inherently susceptible plants(*V.3.3. 2.164)516
Inhibition(IV.0.137)335
Inhospitality(*V.3.3.2.037)511
Initial(*II.3.014)194

Injection:
  blade, knife or tine(*V.3.3.3.094) 537
  device(*V.4.1.055)561
Inlay:
  bridge graftage(*V.1.4.028c)444
  grafting(V.1.4.028k-3,.01e,.12c,i) 447,462,464
Innate(II.2.2.007p,.07a)119,136
Inner mesotesta[II.2.3.3.01d(3)]182
Inoculation(V.1.3.041;*.3.3.2.436) 434,529
Inoculum(V.3.3.2.226)522
Inorganic(*V.3.2.038)492
Inositol(*IV.2.2.032b,*.159;*V.1.5. 005)362,368,470
Insecta(*V.3.3.1.132;*VI.7.2.017n)501, 614
Insect(s)(*IV.2.2.069;.4.4.030;*.4.5. 001c,*.002c,*f,*g,*j,*m,*n,*o,*p, *q,*r;*V.1.1.065;*.1.3.050;*.3.3. 0.001,*.002,*.003,*.004,*.005.3.3. 1.132;*.3.3.2.006,*.133a,*270;*.3. 3.3.013a,*c,*d,*f,*.051,*.052, *.086;*VI.7.2.017g,*h,n;*.8.038a; *.9.000)365,389,391,392,426,435, 497,501,510,524,532,534,537,613, 614,619,626
  galls(*II.4.4.002h)247
  growth regulators(*V.3.3.3.013)437
  resistance(*V.1.3.067)437
  sex attractant(*V.3.3.3.073)536
  vector(*V.3.3.2.299,*460)525,530
Insecticide(*V.1.3.058;*.3.3.3.000, *.009,*.013b,*.033,*.051,*.070, *.086,*.094,*.097;VI.7.2.017o,*p) 436,531,532,533,534,537,614
Insoluble P_2O_5(*V.3.2.034)491
Inspection(*V.3.3.3.000;*VI.7.1.022, *.042,*.044,*.046,*.048,*.056, *.062,*.063,*.069,*.072,*.075, *.077)531,600,603,604,605,606,607, 608
  final(*VI.7.1.063)605
  for grade(*VI.8.038)619
  prior(*V.3.3.3.071)536
  (scouting)(*V.3.3.3.052,*.090)534, 537
Instar(V.3.3.1.133)503
In-store:
  advertising/merchandising material(*VI.8.084)622
  packaging(VI.8.049)620
Insulation(*IV.5.4.1.004)412
Intensity(II.5.1.000c)263
Integrated:
  crop management(V.3.0.001)484

INDEX OF TERMS

pest management(*V.3.3.3.052,.053, *.090)534,535,537
Integuments[II.2.2.041t-4-f,.16b(5,6); .2.3.3.03a-d,f-i(1);*IV.4.1.006a-1; *.4.6.004]124,145,184,378,392
Intercalary:
 growth(*II.2.1.044)102
 meristem[II.3.091,.18a(2)]198,224
Intercellular:
 ice formation(*IV.5.3.025)409
 spaces(*IV.5.2.032)401
 water(*V.3.3.2.273)524
Intercompatible(IV.4.8.032)394
Interfacial (surface tension)(*V.3.3.3.125)539
Interfascicular cambium(*II.3.026, *.173,*.174)195,202,203
Interference:
 with photosynthesis(*V.3.3.2.133c)515
  translocation(*V.3.3.2.133c)515
  water conduction(*V.3.3.2.133c)515
Interfertile(IV.4.8.033)394
Interfruitful(IV.4.8.034)394
Intergranal lamellae(IV.1.1.002b)350
Interincompatible(IV.4.8.035)394
Interkinesis(IV.4.1.018)381
Intermediate-day plant(IV.3.2.013f)375
Intermodal(VI.6.011,.026)594,596
Internal:
 black spot of beet(VI.9.3.067)668
 breakdown:
  of apples(VI.9.3.069)668
   pears(*VI.9.3.033)665
   pomegranate(VI.9.3.070)668
   stone fruits(VI.9.3.071)668
   sweet potatoes(VI.9.3.072)668
  breakdowns of citrus and other fruits VI.9.3.073)669
  browning(*VI.9.3.029,.*073)665,669
  browning complex of tomato(VI.9.3.074)669
  of apples(*VI.9.3.032,.075)665,669
   pineapple(VI.9.3.076)669
   prunes(*VI.9.3.065)669
 decline of lemons(*VI.9.3.042)666
 design(*V.1.1.017e)420
 dormancy(*V.1.3.060)436
 drainage(*IV.5.4.3.006)414
 (phloem)(*II.3.136)201
International:
 Bacteriological Code of Nomenclature (*V.3.3.2.179b)517
 Code of Botanical Nomenclature(III.1.1.019;*.024,*.030;*.2.0.003; *V.3.3.2.179a)305,306,311,517
 of Nomenclature of Bacteria(V.3.3.2.179b)517
 of Nomenclature of Cultivated Plants(III.1.1.020,*.024;*.2.0.003,*.012)305,311,316
 Committee on Systematic Biology(*V.3.3.2.179b)517
 Seed Testing Association(*V.1.3.075) 438
 Society of Plant Pathology(*V.3.3.2.179b-1)517
Internode(s)(*II.3.091,.094;*V.1.4.060,*.060a,*.071a,*b)198,451
Interpetiolar(II.1.2.055,.13a)61,78
Interplanting(*V.3.1.008c,h,.01g)485, 486,488
Interrupted(II.1.2.056,.13b)61,78
 light program(*V.4.1.001)559
Interset(*V.3.1.008h,.01g)486,488
Interstate Commerce Commission(*VI.6.004,*.006)594
 Regulations(*VI.6.017)595
Intersterile(IV.4.8.036)394
Interstock(s)(*V.1.4.028h,n)445,447
 dwarfing(*V.3.4.000)539
Interunfruitful(IV.4.8.037)395
Interval timer(*V.1.1.053e)425
Intervarietal hybrids(*III.2.0.003d) 311
Interweaving(*IV.5.4.4.002g)416
Intine[*II.2.2.007t-10,.13a(2)]120,142
Intracellular:
 freezing(*IV.5.3.023,*.048)409,411
 ice formation(*IV.5.3.018)408
 thalli(*V.3.3.2.181a)518
Intrafoliar(II.2.057,.13d)61,78
In transit(VI.7.1.034)602
Intraovarian(IV.4.4.029)389
Intraxylary (phloem)(*II.3.008,*.136) 194,201
Intrinsic:
 factors(II.4.4.007)247
 plant property(*V.3.3.2.233)522
Introflexed(*II.5.2.018)266
Introgressive hybridization(III.2.2.093)322
Introrse(II.2.2.007h-4,.046,.10g)119, 125,139
Intumescence(V.3.3.2.227,*.273)522,524
Inulin(*II.1.3.007;*.2.3.3.015;*.4.3.063;*IV.2.2.031c,c-a-2;VI.1.007)94, 164,245,361,570
Invaginate cytoplast(*V.3.3.2.199)521
Inventory(VI.8.050)620
 rotation, replacement(*VI.8.128)625
 turns(VI.8.051)620
Inversion(III.2.2.094)322
 point(II.4.06c)254

INDEX OF TERMS 754

Invert emulsion(V.3.3.3.054)535
Invertebrate(*V.3.3.1.183)504
Inverted(II.5.2.020)266
  -bifacial leaf(*II.1.2.016,.04b)60,
    69
  -ovule(*II.2.2.041t-4-k-2)124
  root graft(*V.1.4.028r)448
  T(*V.1.4.028c)444
Investigation(VI.7.1.035)602
Invirescentia(*II.4.4.007c,.008)247
In vitro(V.3.3.2.228)522
In vivo(V.3.3.2.229)522
Invoice(VI.6.012,*.030)594,596
Involucre[II.2.1.042,.01(6),.10;.2.2.
  01a(11);.2.3.2.015]102,104,113,130,
  161
Involute(II.1.1.2.016,.02m;.2.059,
  .14a)21,35,61,79
Iodine (I)(*V.3.2.073)495
Iodinebush(*II.4.2.025)240
Ion-exchange materials(V.3.2.050)493
Ionic:
  balance(*IV.5.4.3.009)415
  solubility(*V.3.2.065)494
  stress(*IV.5.4.3.000)414
Ionizing radiation resistance(IV.5.4.
  2.001)413
Ions(*IV.5.4.2.001)413
  preferential absorption(*IV.5.2.001)
    398
  sequestration(*IV.5.2.001)398
  toxic(*V.5.2.001)398
IR(*IV.5.4.2.000)413
  drying(*V.1.3.061)437
Iron (Fe)(*IV.5.4.3.006;*V.2.028h,
  .037,.107;*.3.2.028,*.034,*.058,
  *.067)414,473,476,491,494
  accumulation(V.2.01a)483
  chelate(*V.3.2.067,*.085)494
  deficiency(IV.5.4.3.000)414
  malleable(*V.1.1.017c)420
  -rich(*V.2.076)474
  sulfate (ferrous)(*V.3.2.085)496
  total(*V.3.2.071,.080)495
  wrought(*V.1.1.017c)420
Irradiance(IV.0.157d,*.260e;.3.2.
  006)338,345,374
Irradiation(IV.0.157e)338
  level(*IV.5.2.010)413
Irregular(II.2.2.061c-8,.20i,.2.3.1.
  006b,.03b,c)127,149,158,170
Irreversible:
  hardening(*V.2.096c)475
  injury to plants(*V.3.3.3.116)538
Irreversibly plagiotropic(*II.1.1.3.
  019,.047,.13a;*V.1.4.022a-1)22,24,
  49,442

Irrigation(*V.1.1.017a,f;*.2.083;*.3.
  0.001;*.3.1.000;*.3.2.044;*.3.5.
  002,*.022,*.023,*.024,*.032,
  .033,.034;*.4.1.070)419,420,
  474,484,493,557,558,559,562
  districts(*V.3.5.034)559
  drip systems(*V.3.5.032)558
  efficiency(V.3.5.011)557
  ,furrow(*V.3.5.023)558
  legal restraints(*V.1.1.017a)419
  low-pressure systems(*V.3.5.032)558
  low-volume systems(*V.3.5.032)558
  management(*V.3.5.032)558
  microsprinkler systems(*V.3.5.022)
    558
  quality(*V.1.1.017a;.3.5.032)419,558
  sources(*V.1.1.017a)419
  systems(V.3.5.012,*.014)557,558
  timing(V.3.5.013)557
  water(*V.3.2.031;.3.5.023)491,558
    measurement(V.3.5.014)558
    rights(*V.3.5.033)559
    supply(*V.3.5.034)559
Irritating hairs(*II.4.3.042)244
Isobilateral (tetrad)(II.2.2.12c(2);
  (leaf).5.5.029)141,272
Isobutylidene diurea(*V.3.2.015)491
Isocoumarin(*IV.2.2.162)368
Isoelectric pH(IV.0.140)335
  point(*IV.0.140)335
Isoenzyme(*IV.2.2.058d)364
Isogamy(V.3.3.2.230)522
Isogenetic lines(III.2.2.095)322
Isolate(*V.3.3.2.121,.231)514,522
Isolation from whole organism(*V.3.3.
  2.228)522
Isomerases(*IV.2.2.058g)365
Isopentenyl adenine(*IV.2.1.025)358
Isopleth(IV.0.141)335
Isoprene(*IV.2.2.107)366
Isoprenoids(IV.2.2.107)366
Isoptera(V.3.3.1.132f;*VI.7.2.017n)
  502,614
Isosceles triangle(*V.3.1.008,.01d)
  486,488
Isotope(*IV.0.036,.142)330,335
Isotype(III.1.1.035d)307
Isozymes(IV.2.2.058d)364
Item promotion(*VI.8.111)625
Iteroparous(II.0.009,.02a)10,13
Itersonilia perplexans(*VI.9.1.331)646

J

Jam(VI.10.020)676
Jamaica(*V.1.4.028a)443
Jams(*VI.10.000)674

INDEX OF TERMS

Japanese beetle(*V.3.3.1.132a)501
Jarovization (jarovisation)(IV.3.014c) 372
Jars(*V.2.109)478
Java black rot of sweet potatoes(VI.9.1.296)645
Jelly, jellies(*VI.3.053e;*.10.000, .021)582,674,677
Jerusalem artichoke(*I.3.023;*II.4.3.057;*IV.2.2.031c-a-2)8,244,361
Job safety and health standards(*VI.7.2.038)616
Jobber(s)(*VI.7.1.015;*.8.032,.052, *.118)599,618,624,625
John Innes mixes(*V.1.2.011,*.011n) 430,431
Joint:
  -account partners' duties(*VI.7.1.050)604
    -split above(VI.7.1.037)602
    transaction(VI.7.1.038)602
    transactions(*VI.7.1.003)597
  advertising(*VI.8.135)626
Jointed(*II.1.2.014)60
Jonathan spot of apples(VI.9.3.079)669
Jordanon(III.1.2.030)309
Jorquette[II.1.1.3.048,.05d(2),.13b(2)]24,41,49
Joule (J)(*IV.0.260c)344
  per cubic meter(*IV.0.260e)345
  per kelvin(*IV.0.260e)345
  per kilogram(*IV.0.260e)345
  per kilogram kelvin(*IV.0.260e)345
  per mole(*IV.0.260e)345
  per mole kelvin(*IV.0.098,*.260e) 333,345
Juglone(*IV.2.2.014)360
Juice:
  content(*VI.1.008,*.009,*.010)570, 571
  extractor(*VI.10.011)675
  vesicles(*II.2.3.2.046,.05d(4),e) 163,179
Jumping plant-lice(*V.3.3.1.132d)502
June drop(IV.4.8.038)396
Jute cloth(*V.1.0.001)418
Juvenile:
  characteristics(*II.4.4.007a)247
  stage(*V.3.3.3.051)534
Juvenility(*IV.3.1,.014;V.4.1.064)370, 372,562

K

Kabatiella(*V.3.3.2.019)510
Kaolin, kaolinite group(*V.2.098)475
Karyogamy(*V.3.3.2.181b,*f,.232,*.437) 518,519,522,529
Karyokinesis(IV.4.1.019)381
Karyolymph(*IV.1.1.018)351
Karyotype(III.2.2.096)322
Katydid(*V.3.3.1.132h)503
Keel, keeled[*II.2.2.061c-3,-14-d,.23h (3)]127,152
Kelsey spot(*VI.9.3.065)668
Kelvin(*IV.0.260a,*.295)344,349
Kerma (gray)(IV.0.260d)345
Kern rot(V.4.2.2.015)567
α-Keto acids or(*V.1.3.004)391
α-Ketoglutarate(*IV.1.2.004)352
Ketohexoses(*IV.2.2.031a)361
Ketones(*IV.2.2.031,*.031a,*.095;*VI.9.3.000)360,366,663
Ketoses(*IV.2.2.031a)361
Kidney-shaped(*II.5.5.048)273
Kiln(*VI.10.009,.023)674,677
Kilo (k)(IV.0.260h)346
Kilogram(*IV.0.260a)344
  -force (kgf)(*IV.0.260m)347
  per cubic meter(*IV.0.260b)344
Kilowatt-hour(*IV.0.157b-2-e-1)337
Kind(III.2.0.004a)312
Kinetic energy(IV.0.146)336
Kinetics of enzyme-catalyzed reactions (IV.2.2.058e)364
Kinetin(*IV.2.1.025;.5.4.3.010;*V.1.5.005)358,415,470
King(*V.3.3.1.132f)502
Kitchen refuse(*V.1.2.011c)430
Klendusity(V.3.3.2.233)522
Klinostat(IV.3.3.008k)377
Knapack sprayer(*V.3.3.3.100)537
Kneecap-shaped(*II.2.3.3.029;*.5.6.089;.5.17e)165,276,297
Knee-jointed(*II.2.2.007n)119
Knee-root(II.4.3.029,*.039,*.051;.4.11b,c)243,244,259
Knife(*V.1.4.022a-2)442
Kniffen system(*V.3.4.032,.10b)542,554
Knoblike(II.5.6.068;.5.16a)275,296
Knot(IV.0.260k)347
KNO_3(*V.1.3.031; see also Potassium nitrate)434
Knuckling of tulip(V.4.2.2.016)567
Kraft paper(*VI.4.010,*.012)586,587
Kranz anatomy(*IV.1.2.008b;*.5.2.015) 352,399
Krebs cycle(*IV.1.2.004,.040b)352,356
Krilium(*V.2.099)476
Kroger(*VI.8.013)617

L

L_1, L_2, L_3 layers[II.3.080,.14a(2-4);

INDEX OF TERMS

*III.2.2.024c])197,220,319
Label recommendations(*V.3.3.3.071)536
Labeling(*V.1.3.075;*VI.7.2.016;*.8.089)438,610,623
Labellum[II.2.2.061c-9,.24c(1),e(1)] 127,148
Labels(V.1.2.009)429
Labiate(II.2.2.061c-14-n,.23b)128,152
Labile material(*V.1.5.005)470
Labor(*V.3.0.001;.3.3.3.049;VI.8.053,.137)484,534,620,626
  availability(*V.1.1.017a)419
  intensive(*VI.8.138)626
Lace bug(*V.3.3.1.132c)502
Lacerate(II.5.4.3.022;.5.06d)270,**286**
Lachrymiform(II.2.3.2.019,.04e)161,178
Laciniate(II.5.4.3.023;.5.05i,.06e) 270,285,286
Lacteus(II.5.1.007h)265
Lactose(IV.2.2.031b)361
Lacunar(II.3.04g)210
Lacunose(II.5.6.069;.5.16b)275,296
Lady beetles(*V.3.3.1.132a)501
Laevigate(II.5.6.070)275
Laevorotary (L)(*IV.2.2.031a)361
LAI(*IV.1.2.026)353
Lake, lakes(*IV.5.4.3.006;*V.2.037;*.3.5.023,*.034;*VI.7.2.014)414, 473,558,559,610
  use(*V.3.5.034)559
λ(lambda)(*IV.0.260m)347
Lamella(IV.1.1.002c)350
Lamellar(II.3.04i)210
Lamellate(II.2.2.041o,.041t-2-f,.15g) 123,144
Laminar rolling(II.4.3.030,*.032;*IV.5.2.085)243,406
Laminated bags(*V.1.3.067)437
Lammas shoot(II.1.1.3.049)24
Lanate(II.5.6.071;.5.16c)275,296
Lance nematodes(*V.3.3.1.200d)506
Lance-shaped(*II.5.5.031)272
Lanceolate(II.5.5.031;.5.10c)272,290
Land:
  availability(*V.1.1.017a)419
  characteristics(*V.1.1.017a)419
  clearing(*VI.8.090)623
  conditions(*V.3.2.044)493
  plaster(*V.2.099;.3.2.044)476,493
  race(III.2.2.097)322
  reclamation(*V.2.110e-2)479
  site preparation(V.3.1.000)484
Landscape:
  architect(I.2.004)6
  designer(I.2.005)6
  ornamentals(*V.3.2.000)490
  plant(*I.2.009)6

  planting(*V.1.1.048)424
Landside(*V.1.1.044)423
Langley (ly)(IV.0.151,.157b-2-e-3)336,337
  per day(IV.0.157b-2-e-3)337
Large:
  animals(V.3.3.1.140)503
  calorie(*VI.5.006)588
  embryo(II.2.3.3.02j)183
Larva, larvae(*V.3.3.1.086,*.132a,*b,*g,.141;*.3.3.3.013a,*f)500,502,503,532
  hymenopteral(*V.3.3.3.013f)532
  lepidopteral(*V.3.3.3.013f)532
Larval form(*V.3.3.1.108)500
Large volume items(*VI.8.121)625
Laspeyresia pomonella(*VI.9.2.018)660
Late blight:
  of celery(VI.9.1.297)645
  of potato(*V.3.3.2.133c,*.181d;VI.9.1.298)515,518,645
  rot of tomato(VI.9.1.299)645
Late postemergence(V.3.3.3.055)535
Late season pool(*VI.8.085)622
Latent:
  bud(IV.4.8.039)396
  heat(VI.5.051c)593
  potato virus(VI.9.1.201)640
  virus(*V.1.4.035)449
Lateral, laterals[II.1.1.1.012,.01d(3);*.2.3.2.004;*V.3.4.017,*.032,*.046]19,28,121,161,540,541,542
  branching(*V.4.1.084)563
  bud(*V.4.1.044)561
  buds(*V.1.4.071b)451
  fruiting(*V.3.4.032,*.069)541,544
  knee-roots(*II.4.3.039;.4.11b)243,**259**
  meristem[*II.3.108,.18a(4)]199,224
  movement(*V.2.106)476
  roots[II.3.097,.17a;V.3.4.01(18)] 198,223,545
Lateritus(II.5.1.005m)264
Latex(*IV.1.1.030)352
  (vessel cylinders)[*II.3.098,*.179,.16b(11)]198,203,222
Lath frame(*V.1.1.053c)425
Lathhouse(*V.1.1.059)426
Laticifer(II.3.098,*.175,*.179,.16a) 198,203,222
Lattices(*V.3.4.020,*.026)541
Lavender-colored(II.5.1.002f)263
Lawn, lawns(*I.2.011;*V.3.1.004)6,485
  grasses(*I.2.011)6
Laws, Rules and Regulations(VI.7.)597-616
Laws of thermodynamics(IV.0.154)336

INDEX OF TERMS

Lax(II.2.1.043;.5.7.018)102,278
  (cyme)(II.2.1.09f)112
Layby application(V.3.3.3.056)535
Layerage, layering, layers(*IV.5.2.
  030;(*V.1.4.000,*.024,*.028m,.036,
  *.041,.14,.15)401,441,443,447,449,
  450,466,467
  air(V.1.4.036a,.15)449,467
  ground(V.1.4.036b)450
  mound (stool)(V.1.4.036b-1)450
  natural vegetative reproduction(IV.
    4.1.006a-8)379
  rooting(*V.1.4.028t)448
  serpentine (compound)(V.1.4.036b-2)
    450
  simple(V.1.4.036b-3)450
  tip(V.1.4.036b-4)450
  trench(V.1.4.036b-5)450
Laying out:
  beds(*V.3.1.000)484
  blocks(*V.3.1.000)484
  planting sites(*V.3.1.000)484
  rows(*V.3.1.000)484
LCL(VI.6.013)595
LD_{50}(*IV.5.1.012)398
LD_{100}(*IV.5.1.012)398
LDT(*V.4.1.067)562
LE(*IV.5.2.074a)405
Leached(*V.2.107e)477
Leaching(*V.2.106,*.107,*.110;*.3.2.
  000)476,478,490
Lead-colored(II.5.1.003i)264
Leader(V.3.4.046)542
  (training)(V.3.4.03a-d)547
Leaf, leaves[II.1.2,.000;.2.1.01(12-
  14);*.3.108,*.210,.21;*IV.2.1.
  012,*.025,*.037;*.3.2.003;*.5.2.
  041,*.085;*V.2.107;*.3.2.072;
  .3.3.023,.024,*.027;*.4.1.
  035;*.4.2.2.021]59-93,104,199,
  205,227,357,373,403,406,476,495,
  533,560,568
  abscission(*IV.2.1.033,*.037;*V.3.
    3.3.024)358,359,533
  area(*IV.5.2.036)402
  area index(IV.1.2.026)353
  area reduction(*IV.5.2.079)406
  atypical color(*V.3.2.066)494
  beetles(*V.3.3.1.132a)501
  blade(*V.3.3.3.112)538
  blades(*V.1.4.022c)442
    with petioles(*V.1.4.022c)442
  blights of corn(*V.3.3.2.133c)515
  -bud cutting(V.1.4.022d,.01d)442,453
  bug(*V.3.3.1.132c)502
  caudate tipped(*II.4.3.017)242
  chlorotic patterns(*V.3.2.066)494
  color(*V.3.2.017)491
  counting(V.4.1.065)562
  curl of ferns, alder, birch, cherry,
    poplar(*V.3.3.2.181e)519
  curl of peaches and nectarines(VI.9.
    1.301)645
  cutter bees(*IV.4.5.002g;*V.3.3.1.
    132e)391,502
  cutting(*V.1.4.022c,d,.01a-c)442,453
  diffusive resistance(*IV.5.2.065)404
  down-curling(*V.4.2.1.018)566
  drooping(*V.3.3.2.468)530
  drop of figs(*V.3.3.1.200b)505
  -end whips(*II.4.3.031)243
  evaporating surfaces(*IV.5.2.065)404
  expansion(*IV.3.2.001)373
  falcate(*II.4.3.032,*.052;*IV.5.2.
    085)243,244,406
  floral to foliage(*V.3.3.2.306)525
  foliar analyses(*V.3.2.040)492
  galls(*V.3.3.1.200a)505
  gap(II.3.099,.05)198,211
  green(*V.3.3.1.018)510
  hopper(*V.3.3.1.132d)502
    vectors(*V.3.3.2.133e)515
  hydathode(*II.3.082)197
  lamina(*II.4.3.071)246
  laminar rolling(II.4.3.030,*.032;
    *IV.5.2.085)243,406
  limited necrotic areas(*V.3.3.2.
    405)528
  loose(V.2.01a)483
  margins and tips(*V.4.2.2.017)568
  mesomorphic(*IV.5.2.085)406
  mesophyll(*II.3.067,*.111)197,199
  midribs(*II.4.3.071)246
  modifications(II.4.3.032)243
  modified(*V.4.1.104,*.120)564,565
  mold(*V.1.2.011p)431
  narrow(*II.4.3.028)243
  necrotic areas(*VI.9.3.029)665
  net radiation balance(*IV.3.2.007)
    374
  number(*V.4.1.065)562
  nyctinasty(*IV.3.3.003c)376
  -opposed inflorescences(II.2.1.044,
    .11a)102,114
  overall fading(*V.3.2.066)494
  patchy pattern(*V.3.3.2.256)523
  petaloid(*V.4.2.2.021)568
  photonasty(IV.3.3.003d)376
  primordia, primordium[*II.3.107,.01A
    (6);*.4.3.041]199,207,244
  production(*V.2.110e-2)479
  removal(*V.1.4.028u)449
  removed(*V.1.4.022a-2)442
  resistance, total(*IV.5.2.032)401

roll of potato(VI.9.1.302)645
roller injury of apples(VI.9.2.044) 661
rollers(*V.3.3.1.132g)502
rolling(*IV.5.2.020)400
scorch(V.4.2.2.017)568
section(V.1.5.01d)471
seismonasty(*IV.3.3.003e)376
senescence(*IV.2.1.025,*.037)458,459
sheath whips(*II.4.3.031)242
sickle-shaped(*II.4.3.032,*.052;*V.4.2.1.021)243,244,566
specific survival time(IV.5.2.059) 404
spot(*VI.9.1.461)652
  of bananas(*V.3.2.2.181e)519
  of strawberry, Rubus, Ribes, cucurbits, beets, peas, crucifers (*V.3.3.2.181e)519
spots(*V.3.3.2.040)511
spotting organisms(*V.3.3.2.388)528
spots, yellowing and shriveling(*V.4.2.1.002)565
stem and bulb endoparasites(*V.3.3.1.200a)505
stunting, yellowing and premature dying(*V.4.2.1.001)565
succulent(II.1.2.07a;.4.3.033,*.065; *IV.5.2.085)72,243,245,406
tip retained(*V.4.2.2.016)567
tissue(*V.3.3.2.388)528
trace(II.3.100,*.201,.05)198,204,211
tropism(*IV.3.3.008)377
virus-infected(*V.3.3.2.451,*.452) 530
water content(*IV.5.2.073)405
  loss resistance(IV.5.2.032)401
  stressed(*IV.5.2.044)403
  turnover, turnover rate(*IV.5.2.082)406
  whips(*II.4.3.031)243
white and brown spots(*V.4.2.1.009) 565
  striping and twisting(*V.4.2.1.005)565
yellowing patterns(*V.4.2.1.013)565
Leaflet(s)(II.1.2.17b;*V.3.2.072)82, 495
  whips(*II.4.3.031)242
Leaflike(*II.2.2.037)122
(involucre)(II.2.1.042,.10d)102,113
Leafy:
  mulch materials(*V.3.2.069)494
  vegetables(*VI.5.055)593
Leak of potato(*V.3.3.2.181c)518
Leaning multiple-stem (leaning vertical)(*V.3.4.017,.07a,b)541,551

Leathery(*II.5.7.007)278
Lecithin(*IV.2.2.159)368
Lectotype(III.1.1.035e)307
Legal maximum residue(*V.3.3.3.111)538
Legitimate pollination(IV.4.3.013)386
Legs, true(*V.3.3.1.132)501
Legume(s)(II.0.010;.2.2.28a;.2.3.1.020,.02c-e;*IV.1.2.029)10,157,159, 169,243,354
Leguminous:
  (mainly) cover(*V.2.110e)479
  plants(*V.3.2.069;*.3.3.3.059)494, 535
Lemma[II.2.2.048,.24b(2)(ii)]125,153
Lemon-colored(II.5.1.008k)266
Length(IV.0.155,*.260a;*VI.3.064,.064b)336,344,582,583
Lenticel(II.3.101,.12b)198,218
Lenticular, lens-shaped(II.2.3.3.026) 165
Lentiginous(II.5.6.072)275
Lepidoptera(V.3.3.1.132g;*.3.3.3.013a) 502,532
Lepidosaphes beckii, L. gloverii(*VI.9.2.007)660
Lepidote(II.5.6.073;.5.16d)275,296
Leprous(*II.5.6.073)275
Leptokurtic(III.2.2.098)322
Leptomorph rhizome(V.1.4.060a,.05a) 451,457
Leptomorphs(*V.3.3.1.224)507
Leptosphaeria coniothyrium(*V.3.3.2.181e)510
Lesion(V.3.3.2.240)522
Lesions:
  black on inner scales(*V.4.2.1.015) 566
  grayish-brown on outer scales(*V.4.2.1.015)566
Less than carload lot(*VI.6.013)595
Less than truckload(VI.6.019)595
Lethal gene(III.2.2.099)322
Lethum australiense var. typicum(*VI.9.1.522)656
Lettuce:
  black rot(*V.3.3.2.181g)520
  leaf blight(*V.3.3.2.181g)520
Leucine(*IV.2.2.016)360
Leucoplast(IV.1.1.012)351
Level terrain(*V.3.1.008j)486
Leveling and cleaning land(*V.3.1.000) 484
Ley(V.3.0.004)484
Liability(VI.7.1.039)602
Liana, liane, lianes(II.0.026,.02d; *.3.008,*.026,*.048,*.173,.03;*.4.3.002)10,13,194-196,202,209,241

INDEX OF TERMS

Libriform (fiber(*II.3.063,.23e)197, 229
License required(VI.7.1.040)603
  termination, suspension, revocation, cancellation(VI.7.1.041)603
Lichens(*IV.1.2.029;*.3.2.019;*V.3.3.2.179a,*.437)354,376,517,529
Life:
  cycle (history(*V.3.3.2.033,.243) 511,522
  forms(II.4.1,.01,.02)238,249,250
Lift truck(VI.3.043)581
Ligases(*IV.2.2.058g)364
Light(*II.4.4.003e;*V.3.2.009;*.3.3.2.270;.3.3.3.074,*.119;*VI.8.027, *.073;*.9.3.000)247,490,524,536, 538,618,621,663
  blue(*IV.5.3.006,*.008)413
  box(V.1.1.030,*.048)422,424
  -compensation point(IV.5.4.2.002)413
  competition(*V.3.3.0.000)497
  diaphototropism(*IV.3.3.008p)377
  duration(*V.4.1.066)562
  flux(V.4.1.066)562
  heliotropism(*IV.3.3.008n)377
  intensity(*IV.0.157w;*.5.4.2.003;*V.4.1.053,*.066)339,413,561,562
  intensity high(*V.1.1.053e)425
  intensity increased(*IV.5.2.001)398
  near UV(*IV.5.4.2.006,*.008)413
  orthophototropism(IV.3.3.008m)377
  photoinactivation(IV.5.4.2.006)513
  photoinhibition(IV.5.4.2.007)413
  phototropism(IV.3.3.008n)377
  plagiophototropism(IV.3.3.008p)377
  requirement for germination(V.1.3.042)435
  -saturation point(IV.5.4.2.003)413
  stress(IV.5.4.2.000)413
  terms and units(IV.0.157)336
  UV(*IV.5.4.2.008,*.009,*.013)413
  window(IV.4.4.030)389
Lightning injury(*V.3.3.2.270)524
"Lightningbug"(*V.3.3.1.132a)501
Ligneous(II.5.1.001h;*.5.7.015,.020) 263,278
Lignin(s)(*II.1.1.3.105;*.3.052;*IV.-1.2.040g;.2.2.114,*.154)27,196,356, 366,368
Lignose(*II.5.7.020)278
Lignotuber(II.4.4.002g,.009;.4.14b) 246,248,262
Ligule, ligulate[II.1.2.19e(2);.2.2.23i(1);.5.5.032;.5.10d(1,3)]84,152, 272,290
Lilac (lilacinus)(II.5.1.002g)263
Liliaceous(II.2.2.21i)150

Lily symptomless virus(*V.4.2.1.009, *.012)565
Limax:
  flavus(*V.3.3.1.250)508
  maximumus(*V.3.3.1.250)508
Limb(II.1.1.3.050)24
  shaker(*VI.2.022b)576
Limbatus(II.5.1.006d)265
Limestone(V.3.2.053,*.079)493,495
  ground(*V.1.2.005a,.011n)429,431
Limited:
  joint venture agreement(*VI.7.1.038)602
  solubility(*V.3.2.015)491
Limnic materials(V.2.107i)477
Line:
  cross(*III.2.0.003b)311
  varieties(III.2.0.005)313
Linear[II.2.2.27e(2);*.2.3.3.02(k); .5.5.033;.5.10e]156,183,272,290
  (tetrad)[II.2.2.12c(5)]141
Lineate(II.2.2.18h)147
Lined(II.5.6.074;.5.16f)275,296
Liner row(V.1.4.040)450
Liners(V.1.4.041)450
Lines of weakness(*V.2.115g)480
Lineweaver-Burk plot(IV.2.2.058e-3)364
Lining out(*V.1.4.028t)448
Linkage(III.2.2.100;*IV.4.1.011)322, 381
  group(*III.2.2.100)322
Linnaeus(*III.1.1.024)305
Linneont(III.1.2.032)309
Lipids(IV.2.2.116,*.159,*.210;*.5.3.008,*.033;*.5.4.1.005)366,368,370, 408,410,412
  body(*IV.1.1.019)351
  high saturation(*IV.5.3.033)410
  low saturation(*IV.5.3.033)410
  membrane(*IV.5.3.025)409
  mobility(*IV.5.4.1.005)412
Liquid:
  fertilizer(*V.1.1.078)427
  fertilizers(V.3.2.054)493
    (via irrigation)(*V.3.2.037)492
  media(*V.1.5.005)470
  -phase conductance(*IV.5.2.080)406
  -solid suspension(*V.3.3.3.118)536
Liquids(*V.3.3.3.043)534
Liquified gas(*V.3.3.3.006)531
Liter (L,l)(*IV.260i)346
Lithic(*V.2.096d)475
  contact(*V.2.107j,*l)477
Lithologic discontinuity(V.2.107k)477
Lithocyst[*II.3.039,.103,.08e(1)]195, 199,214

Lithophane antennata(*VI.9.2.044)661
Litter(*V.2.107,.01a)476,483
Little(II.5.2.021)266
Liver-colored(II.5.1.001i)263
Liverworts(*VI.7.2.017i)613
Living braces(*V.1.4..028,028o,.13a-d)
  443,448,465
"Living fences"(*V.1.4.022g,.03e)443,
  455
Load shifting(*VI.6.034)596
Loading(*VI.3.052,*.079;.6.016;*.8.
    081)581,583,595,622
  area(*VI.3.046)581
  for shipment(*VI.8.078)622
  pattern(VI.6.017)595
    choice(VI.6.018)595
  unit operation(*VI.3.079)583
Loam, loams(*V.1.2.011n;.2.117i)431,
  481
Loamy:
  coarse sand(V.2.117k-4)481
  fine sand(V.2.117k-2)481
  sand, medium(V.2.117k-3)481
  sands(V.2.117k)481
  subsurface horizon(*V.2.107g)477
  very fine sand(V.2.117k-1)481
Lobe(II.1.2.062,.14e)61,79
Lobed(II.5.4.3.024;.5.06f-g)270,286
  (stigma)(II.2.2.18i)147
  xylem(*II.3.008)194
Lobes(II.2.2.007r,.07-.10)119,136-139
Lobster pot(*II.4.3.009)241
Local swelling(*V.3.3.2.187)520
Localized:
  incompatibility(*V.1.4.035)449
  site(*V.3.3.3.112)538
Location(V.1.1.017a)419
Locellate(*II.2.2.007r,.08g)119,137
Locular(II.2.2.049)125
Locule[II.2.2.041p,.02(6),(16),.11b,c]
  123,131,140
Locules(*II.2.2.007r)119
Loculicidal(II.2.3.1.006c,.02i,k,m)
  158,169
Locus(III.2.2.103)322
Locust(*V.3.3.1.132h)503
Loess(*II.4.2.034)240
Lofty(II.5.2.022)266
Logarithmic mean temperature(VI.5.029)
  590
  difference(*VI.5.014,*.021,*.024a,
    .030)588,589,590
Loment(II.2.3.1.021,.02f)159,169
Long:
  cell[II.3.29c(3)]235
  -chain hydrocarbons(*VI.3.085)584
  -day plant(IV.3.2.013g)375

  treatment(V.4.1.067,*.095)562,563
  days(*V.4.1.001,*.034)559,560
  -horned beetles(*V.3.3.1.132a)501
    grasshopper(*V.3.3.1.132h)503
  -lived crops(*V.3.1.008c)485
    woody perennials(*V.3.1.010)486
  shoot(II.1.1.3.051,.14a)24,50
  -short-day plant(IV.3.2.013h)375
Longidorus spp.(*V.3.3.1.200,*.200c)
  505
Longitudinal:
  belt-and-roll sizer(*VI.3.027a-3,
    .064a)579,583
  (dehiscence)(II.2.2.007h-5,.10a)119,
    139
Loose(*II.2.1.043;*.5.7.018)102,278
  medium(*V.1.4.036b-1,*-5)450
  smut of oats(*V.3.3.2.133c,*.181f)
    515,519
Lopping(V.1.4.028i,.11b)446,463
Lopping shears(V.3.4.048)542
Lorate[*II.5.5.032;.5.10d(2)]272,290
Loriform(II.5.6.075;.5.16g)275,296
Lorette system(V.3.4.049)542
Loss leader(VI.8.054)620
Low(II.5.2.023)266
  -analysis fertilizer(V.3.2.055)493
  budding(*V.1.4.028j)446
  covers(*V.2.110e)479
  graftage(*V.1.4.028j)446
  -headed[V.3.4.029,.03b(2),c(2),d(2)]
    541,547
  heading(*V.1.4.028i)445
  mineral nutrition(*IV.5.4.1.006)412
  -oxygen injury of apples(VI.9.3.081)
    669
  pressure storage(VI.5.047c)592
  pressure systems(*V.3.5.032)558
  pressure (vacuum) evaporator(VI.10.
    010)675
  -prices store(*VI.8.136)626
  relief(*V.2.074)474
  temperature breakdown of cucumbers
    (*VI.9.3.029)665
    breakdown of melons(*VI.
      9.3.029)665
    stress(*IV.5.3.047)411
    treatment(*V.4.1.025)560
  tissue-water potential(*IV.5.2.026)
    401
  -vacuum concentration(*VI.10.019)676
  volume systems(*V.3.5.032)558
  water table(*V.4.2.2.006)567
Lower analysis forms(*V.3.2.038)492
LP storage(VI.5.047c)592
LTL(VI.6.019)595
Lug(*VI.4.000c,.014)585,587

Lumen(*IV.0.157,*.260c)336,344
Luminance (L)(IV.0.157g,*.260b)338,344
Luminous:
  energy(*IV.0.157h)338
  flux(IV.0.157h,*.260c)338,344
  intensity(*IV.0.157a,.157i,*.260a)
    337,338,344
Lumper(VI.8.055)620
Lunate(II.5.5.035)272
Lutescens(*II.5.1.008o)266
Luteus(II.5.1.008v)266
Lux (lx)(*IV.0.157,.157c,*.260c)336,
  338,334
Lyases(*IV.2.2.058g)365
Lygaeid(*V.3.3.1.132c)502
Lygus bug(*VI.9.2.013)660
Lygus:
  hesperus(*VI.9.2.013)660
  lineolaris(*VI.9.2.013)660
Lyophilization(V.3.3.2.245)523
Lyophilize(IV.0.165)340
Lyophilizing(*VI.10.017)676
Lyotropic series(IV.0.166)340
Lyrate, lyre-shaped(II.5.5.036;.5.
  11a)272,291
Lysicarpous(II.2.3.2.020,.04c)161,178
Lysigeny (lysigenous)[II.3.104,
  .169,.178,*.180,.12a(4)]199,202,
  203,218
Lysine(*IV.2.2.016,*.172)360,369

M

MA storage(*VI.5.047d)592
Machine:
  harvesting(*V.1.3.063,*.064)437
  sowing(*V.1.3.047)435
Machinery operators(*VI.8.053)620
Macrobiotic(II.0.012)10
Macroepiphytes(*II.4.2.006,.021)239,
  240
Macromolecule(IV.0.167)340
Macronutrient elements(V.3.2.056)493
Macrophomina phaseoli(*VI.9.1.144,
  .145,.146)636,637
Macrophyll(II.1.1.3.052)24
Macrosclereid(*II.3.170,.25a-d)202,231
Macrospore(*II.2.2.041t-4-h)124
Maculatus(*II.5.1.006c)265
"Mad customer index"(VI.8.056)620
Magamp(*V.1.3.047)435
Maggot(s)(*V.3.3.1.132b,*.141)501,503
Magnesium(*V.3.2.013,*.028,*.053,
  .056,.065)491,493,494
  -ammonium phosphate(*V.1.2.012)431
  sulfate(*V.3.2.084)496
  total(*V.3.2.071,*.080)495

Magnetic:
  field strength(*IV.0.260b)344
  flux(*IV.0.260c)344
  flux density(*IV.0.260c)344
  separator(*V.1.3.057)436
Magnum terraces(*V.2.110c)478
Magnolioid roots(II.1.3.010)94
Mail-order houses(*VI.3.000)577
Main crop (figs)(IV.4.4.031)389
Maintenance:
  foreman(*VI.8.137)626
  pruning(*V.3.4.000b,*.067)539,543
  services(*V.1.1.017a)419
Maize(*III.1.1.005; see also Corn)303
Major:
  elements(*V.3.2.017,*.028,*.067)491,
    494
  taxonomic category(III.1.1.021)305
Malacophily(*IV.4.5.001c)391
Maleic hydrazide(*IV.2.1.037)359
Male-sterile(*IV.4.2.001a)385
Male sterility(*IV.4.1.012;*.4.8.005)
  381,393
Malformed:
  bark(*V.3.3.2.082)513
  or multiple twigs(*V.3.2.066)494
Malic acid(*VI.1.008g,*.009)571
Mallet(*V.1.4.028k-2)446
  cutting(*V.1.4.022f-3,.03a)443,445
Malt(*VI.10.003)674
Maltose(*IV.2.2.031b)361
Mammaliochory(IV.4.7.001g)393
Mammals(*IV.4.7.001g)393
Mamme(*IV.4.4.007,.032)388,389
Mammoni(*IV.4.4.007,.033)388,389
Man(*V.3.3.2.006)510
Management(VI.8.057)620
  practices(*V.3.0.001)484
Manganese(*V.3.2.028,*.034,*.058,
  *.067)491,494
  chelates(*V.3.2.085)496
  neutral(*V.3.2.085)496
  oxide(*V.2.037)473
  sulfate(*V.3.2.085)496
  total(*V.3.2.071)495
Manganous (oxide)(*V.2.037)473
Manifest(*VI.6.001,*.012,.020)594,595
Mannans(*IV.2.2.031c,.031c-a-3)361,362
Mannitol(*IV.2.2.032b)362
Mannose, D-Mannose(*IV.2.2.031a,*c,
  .032b,.038)361,362
  6-deoxy-(*IV.2.2.038)362
  polysaccharides(*IV.2.2.031c-a-3)362
D-Mannuronic acid(*IV.2.2.032c)362
Manure(*V.1.2.011,.011f)429,430
  fork(*V.1.1.038)423
Marasmius perniciosus(*V.3.3.2.181f)

INDEX OF TERMS

519
Marbled fruit(*VI.9.1.127)635
Marcescent(II.5.3.012)268
Marcottage(*V.1.4.036a)449
Margin(*VI.8.064)621
   objective(*VI.8.064)621
   terms(*II.5.4.3.;.5.05-.07)269-271,
     285-287
Marginal(*II.3.014;.5.6.076)194,275
   browning of lettuce(VI.9.3.084)669
   (ventral)(II.2.2.041t-2-g,.15e)123,
     144
Marginatus(*II.5.1.006f)265
Markdown(VI.8.058)621
Market(*V.3.0.001;VI.8.059)484,621
   (containers)(VI.4.000b)585
   disorders(VI.9.000)626
   disorders of fruits and vegetables
     (VI.9)626-674
   (facility)(VI.8.060)621
   handling(*VI.2.020,*VI.8.038a)575,
     619
   share(VI.8.0.061)621
Marketability(*V.3.3.0.005;*VI.8.038a)
   497,619
Marketing(*VI.4.000)585
   Agreement(VI.7.2.004a)609
   Agreement Regulations(*VI.3.027a-3;
     .8.038c)579,619
   Order(VI.7.2.004b)609
   phase (bulbous plants)(V.4.1.068)562
   Regulation(VI.7.2.004c)609
Markings(*II.5.1.006)264
Markup(VI.8.062)621
Marl(*V.2.107i)477
Marmor:
   dubium var. vulgare(*V.9.1.201)640
   tabaci(*VI.9.1.543)658
     plus M. dubium var. vulgare(*VI.9.
       1.201)640
Marsh(*II.4.2.028;*IV.5.4.3.006)240,
   414
   ,salt(*II.4.2.010)239
Marshy areas(*V.2.037)473
MARTRAC(UPS)(*VI.6.026)595
Masked virus(V.3.3.2.247)523
Masking tape(*V.1.2.017)432
Masonry walls(*V.2.109)478
Mass(IV.0.171,*.260a,*.325)340,344,350
   -average temperature(*VI.5.014,.032)
     588,590
   -average temperature reduction(*VI.
     5.024b)589
   basis(IV.5.2.074b-1)405
   density(*IV.0.019,.171d,*.260b)329,
     341,344
   fraction(IV.0.171e)341

   molality(IV.0.171f)341
   spectrometer(IV.0.174)341
Massive(V.2.115g)480
Massula(II.2.2.007t-6)120
Master horizons(*V.2.107k)477
Masticatory(II.0.013)10
Mat watering(V.4.1.070)562
Materials(*V.3.3.3.052;*VI.8.137)534,
   626
   and supplies(V.1.2.000)427
Maternal inheritance(III.2.2.103)322
Matric potential(*IV.5.2.057)404
Matroclina(III.2.2.104)322
Mats(*V.4.1.070)562
Mattock(V.1.1.032)422
Maturation(*IV.3.1.019;VI.1.005,*.007,
   .02)373,570,574
   stage at harvest(*VI.9.3.000)663
Mature(VI.8.063)621
Maturity(*IV.4.8.020;V.4.1.069;VI.1.,
     .015;..8.038e)394,562,569-574,
     572,619
   (climacteric fruits)(VI.1.006)570
   index(VI.1.008)570
   (non-climacteric fruits)(VI.1.007)
     570
   -related(*VI.9.3.000)663
   standards(VI.1.009;*.8.038c)571,619
   test(VI.0.010)571
     methods(*VI.0.010)571
     sample(VI.1.011)572
Maximum:
   leaching(V.2.01a)483
   photosynthetic efficiency(*IV.3.1.
     020)373
Maxwell(*IV.0.2601)347
May beetle(*V.3.3.1.132a)501
Meadow(*V.3.0.004)484
   grasshopper(*V.3.3.1.132h)503
Meal moths(*V.3.3.1.132g)503
Mealy(*II.5.6.037;*.5.7.011)274,278
Mealybugs(*V.3.3.1.132d;*.3.3.2.133e)
   502,515
Mean calorie(*VI.5.006)588
Measuring worms(*V.3.3.1.132g)502
Meats(*VI.6.022)595
Mechanical:
   aids(V.2.110c,*.110e-2)478,479
   analysis(*V.2.123)482
   cracker(*V.1.3.059)436
   injuries of tomato(VI.9.3.085)669
   inoculation(V.3.3.2.248)523
   methods(*V.3.3.2.133e)515
   rake(*VI.2.022b)576
   refrigeration unit(*VI.5.035a-6)590
   removal of growth(*V.3.4.000b)539
   resistance(*V.1.3.024,*.050)433,435

scarification(*V.1.3.055)435
scoop(*VI.2.022b)576
seeder(*V.1.3.046)435
shaker screen(*V.1.3.057)435
stress(*IV.5.4.002)416
sweep(*VI.2.022b)576
transmission(*V.3.3.0.004;.4.2.1.
 022;VI.9.1.081)497,566,632
Mechanics(*VI.8.053)620
 helpers(*VI.8.053)620
Mechanized:
 filling(*V.1.2.003)428
 harvesting(*VI.2.022b)576
  equipment(*V.1.3.072)438
 sowing(*V.1.3.050)435
 spreaders(*V.3.2.039)492
Medifixed(II.2.2.007w-9,.07b)120,**136**
Mediterranean fruit fly(*V.3.3.1.132b)
 501
Medium (media)(V.1.0.003)418
 covers(*V.2.110e)479
 (embryo)(II.2.3.3.02k)183
 loamy sand(V.2.117k-3)481
 mixes(V.1.2.011)429
 sand(V.2.117l-3)481
 sand loam(V.2.117j-3)481
Medullary(II.5.7.023)278
Mega (M)(*IV.0.260h)346
Megaphyll(*II.1.1.3.052)24
Megasporangium(*II.2.2.041t-4)124
Megaspore(II.2.2.041t-4-h)124
 mother-cell[*II.2.2.041t-4-h;*IV.4.
  1.006a-3,*-4,*.006b-6,*.006b-6-b,
  .01(3);*.4.6.001d]124,378,380,382,
  392
Megasporocyte(II.2.2.041t-4;*IV.4.1.
 006b-6,*.006b-6-b)124,380
Megasporophyll(*II.2.2.041e;.6.014)
 122,279
Megatherm taxa(III.1.2.033)309
Meiosis(III.2.2.105;*IV.4.1.006a-5,
 *-7;*V.3.3.2.046,*.048,*.181d,*e,*f,
 *.423)323,378,379,512,518,519,529
Melampsora(*V.3.3.2.294)525
Melanose of citrus(*V.3.3.2.181e;*VI.
 9.1.310)519,645
Melezitose(*IV.2.2.031b)361
Meliolaceae(*V.3.3.2.253)523
Melittophily(*IV.4.5.001c)391
Melliferous(IV.4.5.002i)391
Meloidodera spp.(*V.3.3.1.200g)506
Meloidogyne spp.(*V.3.3.1.200e,*g;*VI.
 9.2.070,*.071)506,662
Melon fruit fly(*V.3.3.1.132b)501
Melonworm(*V.3.3.1.132g)502
Membranaceous(*II.5.7.024)278
Membrane:

 -bound nucleus(*V.3.3.2.166)516
 injury(*IV.5.4.1.005a)412
 lipids(*IV.5.3.008)408
Membranous(II.5.7.024)278
 stain of lemons(*VI.9.3.029,.086)
  665,669
"Mentor method"(*IV.3.1.014d)372
Meranthium(*IV.4.4.020)389
Merchandise mix(VI.8.064)621
Merchandising(VI.8.065)621
Merchantability(*VI.8.094)623
Mericarp[II.2.3.1.022,.03j(2)]159,170
Mericlinal (chimera)(III.2.2.024b,
 .01c)319,327
Mericloning(V.1.5.000)470
"Meristem" culture(*V.1.5.000)470
Meristem, ground[II.3.02c(4)]208
Meristemoids(II.3.107)199
Meristems(II.3.108,*.142,*.143,*.144,
 .18)199,201,224
Merogony(III.2.2.106)323
Mesarch (xylem)(*II.3.222)206
Mesh(*V.1.1.017c;VI.4.015)420,587
 bags(*VI.4.000b,*.015)585,587
Mesobiotic(II.0.014)10
Mesocarp[II.2.3.2.029g,*h,.05a(2)]162,
 179
Mesochimera(III.2.2.024d-2-b,.01f)319,
 327
Mesogene(*II.3.190)204
Mesomorph rhizome(V.1.4.060b)451
Mesophiles(*IV.5.4.1.000)412
Mesophyll(II.3.111)199
 cell wall resistance(*IV.5.2.032)401
Mesophyte(s), mesophytic(II.4.2.022,
 .4.01b,c,.02a,b,.05b,c)240,249,
 250,253
 periodic inundation(*IV.5.4.3.006)
  414
Mesotegmic(II.2.3.3.04c)166
Mesotesta, mesotestal[II.2.3.3.043c,
 .01c(2,3),.03b,.06a(2),.07c(2,3)]
 167,182,184,187,188
Mess-and-soil(*IV.4.4.004b-2;.4.5.
 002j)388,391
Messenger RNA(*IV.2.2.188)369
Metabolic:
 activities, reduction of (*IV.5.2.
  001)398
 by-product(*V.3.3.2.312)525
 classification of diseases(V.3.3.2.
  133c)515
 energy(*IV.5.1.013)397
 process inactivation(*IV.5.4.2.006)
  413
 processes(*V.5.4.4.001)415
Metabolite transport(*IV.2.1.025)358

INDEX OF TERMS

Metal:
  containers(*V.1.3.067)437
  flat, flats(*VI.4.000c,*.011)585,586
  labels(*V.1.2.009)429
  pots(*V.1.2.003h)428
  rods(*V.1.2.014)431
  trays(*V.1.2.003c)428
Metallic:
  ion(*V.3.2.012)490
  wood borers(*V.3.3.1.132a)501
Metamorphic rocks(*V.2.000)472
Metamorphosis(*V.3.3.1.132a,*b,*c,*d,
  *f,*g,*h,*i)501,502,503
Metandry(*IV.4.3.002c)385
Metaphase(III.2.2.107)323
Metaphloem[II.3.112,.01B(16)]199,207
Metaplasia(V.3.3.2.250)523
Metaxenia(IV.4.8.040)395
Metaxylem[II.3.113,*.143,.01C(19),
  .11a(5),.22c-f]199,201,207,217,228
Meteorological data(*IV.5.2.028)401
Meter(*IV.0.260a)344
  newton(*IV.0.260e)345
  per second(*IV.0.260b)344
  per second squared(*IV.0.260b)344
Methods of application(*V.3.2.037)492
Methoxyl(*IV.2.2.154)368
Methyl bromide(*V.1.1.065)426
Methylene bridges(*IV.2.2.168)368
Metric:
  carat(*IV.0.260m)347
  ton (tonne)(*IV.0.260i)346
  trait(III.2.2.108)323
Metrification(*VI.6.023)595
Mexican fruit fly(*V.3.3.1.132b)501
Mexico(*VI.8.006)617
MH(*IV.2.1.037)539
Mho(*IV.0.073;*V.4.1.042,.071)332,561,
  562
Micaceous mineral(*V.1.2.011r)431
Mice(*V.3.3.1.224;*.3.3.3.087)507,537
Michaelis-Menten:
  constant(IV.2.2.058e-1)364
  equation(IV.2.2.058e-2)364
Micro (µ)(*IV.0.260h)346
Microbe(V.3.3.2.251)523
Microbial:
  decomposition(*V.2.037)473
  filter(*V.1.5.005)470
Microbiotic(II.0.015)10
Microbodies(IV.1.1.014)351
Microbudding(V.1.4.028p,.11i-k)448,463
Microcrystalline wax(*VI.3.085,*.085d)
  584
Microepiphytes(*II.4.2.006,.023;*.4.3.
  019,.020)239,240,242
Microfibrils(*IV.2.2.031c-b-1)362

Microfungus(*V.3.3.2.254)523
Micrograftage(V.1.4.028p,.11i-o)448,
  463
Micrografting(*V.1.4.028p,.11l-o)448,
  463
Microjets(*V.3.5.012)557
Microkjeldahl(*V.3.2.071)495
Micromelittophily(*IV.4.5.001c)391
Micron (µ)(*IV.0.260m)347
Micronutrient:
  elements(V.3.2.058)494
  spray(*V.3.2.067)494
Microorganism(s)(*V.3.3.2.004,*.022,
  .252,*.409,*.410;*.3.3.3.075)510,
  511,523,528,536
Microphyll(II.1.1.3.053)24
Micropropagation(V.1.5.000,.01)470,471
  media(V.1.5.005)470
Micropyle[II.2.2.041t-4-i,.16b(4);.2.
  *3.3.016,.027,.01a(7),.03f-i(7),
  .04b(3),.05a(3),.06c(3),.07d(5),
  e(3),.08b(5),c(2),.09d(4),.10b(2),
  .11b(2),.12b(2);*IV.4.4.048]124,145,
  164,165,182,184-193,390
Microsporangium[II.2.2.007s,.11a(1)]
  119,140
Microspore mother-cell[*II.2.2.007u,
  .11a(3)]120,140
Microspore(s)(II.2.2.007t)120
  (development)(II.2.2.12a)141
Microsporocyte[II.2.2.007u,.11a(3),
  .12b]120,140,141
Microsporogenesis (gametogenesis)(*II.
  2.2.007s,.12b)120,141
Microsporophyll(II.6.015)279
Microsprinkler(*V.3.5.012)557
  systems(*V.3.5.032)558
Microsymbiote(*V.3.3.2.147,*.152)516
Microtherm taxa(III.1.2.034)309
<u>Microthyriella rubi</u>(*VI.9.1.224)641
Micton(III.2.2.109)323
Microtubules(*IV.1.1.015)351
Middleman(VI.8.066)621
Midges(*V.3.3.1.132b)501
Midseason pool(*VI.8.085)622
Migratory ectoparasitic nematodes at-
  tacking roots(V.3.3.1.200c)505
Migratory semiendoparasitic nematodes
  attacking roots(V.3.3.1.200d)506
Mildew(V.3.3.2.253;*VI.7.2.017i)523,
  613
Milk-white(II.5.1.007h)265
Milled sphagnum(*V.1.2.011o)431
Miller Act of 1954(*VI.7.2.016)610
Millerandage(IV.4.8.041)395
Milli (m)(*IV.0.260h)346
Milliequivalents(*V.2.015)472

INDEX OF TERMS

Millimhos(*V.4.1.071)562
Miner(*V.3.3.1.132e)502
Mineral:
  accumulation(*V.2.107c)477
  deficiency(*V.5.4.3.002)414
  nutrition(*V.2.110;.3.2.000;*.3.3.3.
    000)478,490,531
    program(*V.2.110e-2;*.3.2.037)479,
      492
  soil(*V.2.096a)475
Mineralogy(*V.2.096b,*.107k)475,477
Minerals(*V.3.2.080)495
Minimum:
  erosion(*V.3.1.008b)485
  fungicide level(*VI.8.038f)619
  germination temperature(*V.1.3.087)
    439
Minor:
  element determinations(*V.3.2.072)
    495
  element spray(*V.3.2.067)494
  elements(*V.3.2.013,*.017,*.028,
    *.037)491,492
  taxonomic category(III.1.1.022)305
Minute (min)(*V.0.260i)346
Minute (')(*V.0.260i)346
Misbranded(VI.7.2.017p)614
  food(VI.7.2.016e)611
Misbranding(*VI.7.2.016,*.017)610,612
Miscellaneous:
  crops(I.3.011)8
  (sizing)(VI.3.064e)583
Miscible, miscibility(III.1.2.035)309
Miscibility community(*III.1.2.009)308
Mist:
  bed(*V.1.1.053,.053e;.1.3.01c;.1.5.
    01m)425,440,471
  blowers(*V.3.3.3.100,*.114)537,538
  heads(*V.1.1.053e)425
  irrigation(V.1.3.01b)440
  spraying(V.3.3.3.057)535
Mite, mites(*V.3.3.0.001,*.003;*.3.3.
  1.010;*.3.3.2.133a,*.133e,*.270;*.3.
  3.3.052,*.058;*.4.2.1.017,*.018;*VI.
  7.2.017n;.8.038a;.9.000)497,498,504,
  515,524,534,535,566,614,619,626
Miticide(*V.3.3.3.000,*.058,*.070)531,
  535,536
Mitochondria(IV.1.1.016;*.1.2.040b,d)
  351,356
Mitosis(III.2.2.110;*IV.4.1.006b-6-a;
  V.3.3.2.103,.282)323,380,513,524
Mitotic (aneuspory)[IV.4.1.006b-6-a-2,
  .03(6)]380,384
Mix-and-match(VI.8.067)621
Mixed:
  (buds)(II.1.1.1.013,.03k)19,30

commodities(*VI.6.009,*.018)594,595
fertilizer(*V.3.2.013)491
goods(*V.3.2.013)491
loads(VI.6.022)595
(photometric to radiometric)(IV.0.
  157b-2-a)337
(photometric to quantum)(IV.0.157b-
  2-b)337
(quantum to radiometric)(IV.0.157b-
  2-c)337
Mixing(*V.3.2.082)495
  sliced, chopped or diced fruits or
  vegetables(*VI.7.1.025)601
Mixoploid(III.2.2.024c,.01d)319,327
Mixture(III.2.0.005c)315
MLO's(*V.3.3.2.133a,*c,*.263)515,523,
  524
Mn (see also Manganese)
  -EDTA(*V.3.2.012)490
Mobile:
  basket(*VI.2.022c)576
  (hydraulic) platform(*V.2.022c)576
"Mobileaf"(*V.1.2.001)428
Moderate(*V.2.115e;*VI.8.073)480,621
Moderate to fairly good(VI.8.027)618
Modifications(II.4)238-262
Modified:
  atmosphere injury of citrus(VI.9.3.
    087)669
  atmosphere storage(VI.5.047d)592
  central leader[*V.3.4.029,.03c,e(2)]
    541,547
  orthotropic underground stem(*V.1.4.
    071a)451
  single-stem(*V.3.4.017,.06c)540,550
  stems(*II.1.1.3.025,*.064,.06a,b)23,
    25,42
  topping(V.3.4.06c)550
Modifier or modifying gene(III.2.2.
  111)323
Modular construction(II.1.1.3.054,
  .14b)24,50
Modularization(*VI.6.023)595
Module(II.1.1.3.055,.14b)24,50
Modulus:
  of elasticity(*IV.5.1.003,.005)396
  of plasticity(*IV.5.1.012)397
Moh's scale(*V.2.107j,*1)477
Moist-stored seeds(*V.1.3.079)438
Moisture(*V.4.3.2.048)493
  characteristic curve(*V.2.102)476
  conditions(*V.2.096h)475
  content(*V.2.100)476
    gradient(*V.2.102)476
  determination(*V.3.2.071,*.080)494,
    495
  equivalent(IV.5.2.033;*V.2.107d)402,

477
  loss reduction(*VI.3.086)584
  resistance(*V.1.3.067)437
Moko(*V.3.3.2.040)511
Molal mass(*IV.0.023)330
Molality(IV.0.171f)341
Molar:
  energy (joule per mole)(*IV.0.260e) 345
  entropy (joule per mole kelvin)(*IV.0.260e)345
  heat capacity (joule per mole kelvin)(*IV.0.260e)345
  volume for ideal gas(IV.0.189)341
Mold (mould)(*IV.2.2.030c-a-3;V.3.3.2.254;*VI.7.2.017i)362,523,613
  blue-gray(*V.4.2.015)566
Moldboard(*V.1.1.044)423
  plow(*V.1.1.044)423
Moldy core(*VI.9.1.171)638
Mole (mol)(*IV.0.171a,.184,*.260a)340,341,344
  per cubic meter(*IV.0.260b)344
Mole:
  cricket(*V.3.3.1.132h)503
  drains(*V.3.5.001,*.012)557
Molecular weight(*V.2.015)472
Mollic epipedon (surface)(V.2.028e)473
Molybdenum (Mo)(*V.3.2.028,*.058,*.065,*.067)491,494
  total(*V.3.2.071)495
Moment of force(*IV.0.260e)345
Monadelphous(II.2.2.007a-12,.04a,b,d) 118,133
Monandrous(II.2.2.050,.24c,.25a)126,153,154
Moniliform(II.5.6.078;.5.16h)276,296
Monilinia:
  fructicola(*V.3.3.2.181g;*VI.9.1.128,*.129)520,636
  laxa(*VI.9.1.128,*.129)636
Monilochaetes infuscans(*VI.9.1.537) 657
Monkeys(*V.3.3.1.140)503
Monoammonium phosphate(*V.1.2.013)431
Monoaxial(II.1.1.3.056,*.058,.14c)24,50
Monocarpic(*II.5.3.011;.5.01c,.02d;*V.1.4.071b)267,281,282,451
Monocaulous(II.1.1.3.057,.15a)24,51
Monochasium(II.2.1.046,.07d,.09b,.13a) 102,110,112,116
Monoclinous, monocliny(II.2.2.051,.01b,e,f;*IV.4.3.002d)126,130,386
Monocolpate[II.2.2.12d(1)]141
Monocotyledons (monocots, Monocotyledoneae)(*II.1.3.004;*.3.073,*.091,
.108,.189c,*.210,*.215;*.4.3.024; *.5.3.019;*.5.5.032;III.1.1.023;*IV.1.2.008b)94,197-199,204,205,242,268,272,305,352
Monoecious(II.2.1.13d;IV.4.3.014)116,387
Monoembryonic, monoembryony(IV.4.6.003)392
Monolectic(*IV.4.5.002k)392
Monophilic blossom(IV.4.4.035)389
Monoploid(III.2.2.071)321
  (non-recurrent) apomixis(IV.4.1.006a-7)378
Monoporate[II.2.2.12d(3)]141
Monosaccharides(*IV.1.2.019;.2.2.031a) 353,361
Monosome, monosomic(III.2.2.113;*IV.4.6.001a-1,*.001b-1)323,392
Monotelic(*II.2.1.070)103
Monotropic(IV.4.5.002k)392
Monotypic genus(*III.1.1.012)304
Montmorillonite(*V.2.098)475
  group(*V.2.098)475
Morbific cellular activity(*V.3.3.2.133)514
Mormon "cricket"(*V.3.3.1.132h)503
Morphine(*IV.2.2.012)360
Morphogenesis(IV.3.1.016;*.3.2.012,*.017)372,374,375
Morphological:
  characteristics(*IV.3.1.014)372
  symptoms(*V.3.3.2.133b)515
Morphologically undeveloped embryos (*V.1.3.060)436
Mosaic(*V.3.3.2.133b,.256)515,523
  -mottle type of disease
    of lettuce(VI.9.1.316)646
    of lily(V.4.2.1.013)565
  viruses(*V.3.3.2.133b)515
Mosquitoes(*V.3.3.1.132b)501
Mosses(*IV.5.2.040;*VI.7.2.017i)403,613
Mother:
  axis(*V.4.1.073)562
  blocking(V.4.1.072)562
  bulb(V.4.1.073)562
Moths(*IV.4.5.001;*V.3.3.1.132g;*V.3.3.013f)391,502,532
Motile sperm cells(*II.6.011,.016)279
Motortruck (see Truck)
Mottle(V.3.3.2.257)523
  necrosis of sweet potato(VI.9.1.317) 646
Mottling(*V.2.037)473
Mound (stool) layerage(V.1.4.036b-1,*-5,.14c-e)450,466
Mountain formation(*V.2.071)474

INDEX OF TERMS

Mouse trap(*II.4.3.009)241
Mouthparts(*V.3.3.1.132b,*g)501,502
  long, flexible tube(*V.3.3.1.132g)
    502
Movable benches(*V.1.1.017e)420
Mower(s)(V.1.1.034;*VI.2.022b)422,576
Mowing(*V.3.1.010;*.3.3.3.088)486,537
  machines(*V.1.1.034)422
Muck(*V.2.070)474
Mucky peat(*V.2.070)474
Mucor:
  racemosus(*VI.9.1.319)646
  sp.(*VI.9.1.497)654
Mucor rot of sweet potatoes(VI.9.1.
  319)646
Mucous(II.5.6.079)276
Mucronate, mucronulate(II.5.4.1.013;
  .5.03i)268,283
Mud-daubers(*V.3.3.1.132e)502
Mulch:
  material(*V.2.110e-1)479
  materials, leafy(*V.3.2.000)490
  source(*V.2.110e)479
Mulches(*V.3.3.3.000,.059,*.119)531,
  535,538
Mulching(*V.1.1.020)421
  system(*V.3.1.007)485
"Mule train"(*VI.2.022c,.031)576,577
Multangulate(II.5.6.080;.5.16j)276,296
Multibranched plant(V.4.1.074)562
Multicellular (trichomes)(*II.3.204)
  205
Multideck(VI.8.068)621
Multijugate(*II.1.2.069)62
Multilacunar(*II.3.099)198
Multiline varieties(III.2.0.005a-3)314
Multinode (semihardwood cutting)(*V.1.
  4.022f-3,.02c)443,454
Multinucleate protoplasm(*V.3.3.2.105)
  514
Multiple:
  allele (allelomorph)(III.2.2.114)323
  effect evaporator(*VI.10.008,*.010)
    674,675
  effect steam jet(*VI.5.055)593
  epidermis[II.3.114,.08e(3);.4.08b
    (1)]199,214,266
  fruit[*II.2.3.1.011,.023,.01o(2),
    .07;*IV.4.2.000]159,168,174,385
  perforations(*II.3.217)206
  (polyembryony)(IV.6.001d)392
  pricing(VI.8.069)621
  profiles(*V.2.112)479
  saws(*V.3.4.040,*.063,*.076)542,543,
    544
  vertical(V.3.4.06d)550
Multiseriate:
  epidermis(*II.3.114)199
  rays(II.3.07a,b,e)213
Multistemmed plant(*V.4.1.074)562
Multiwall bags(*V.1.3.067)437
MUM(VI.6.023)595
Mummy(V.3.3.2.181g,*.258)520,523
Municipal sources (emissions)(*VI.7.2.
  014)610
Muriate of potash(*V.3.2.088)496
Muricate(II.5.6.081;.5.16i)276,296
Muscariform(II.2.1.048,.11b)102,114
Muscids(*V.3.3.1.132b)501
Mushroom(V.3.3.2.259)523
  root rots(*V.3.3.2.133c,*.181f)515,
    519
Muslin shade(*V.1.1.053c)525
Mutagen(V.3.3.3.060)535
Mutual water company(*V.3.5.034)559
Mutation(s)(III.2.2.115;*V.3.3.2.373;
  *.3.3.3.060)323,527,535
Mutualistic symbiosis (mutualism)(V.3.
  3.2.261,*.264,*.416)523,529
Mutually compatible interstock(*V.1.
  4.035)449
Mycelia, separate, septate(*V.3.3.2.
  181e)519
Mycelium(*V.3.3.2.148,*.154)516
  well-marked(*V.3.3.2.254)523
-Mycetes(*V.3.3.2.179a)517
-Mycetidae(*V.3.3.2.179a)517
Mycoplasma(V.3.3.2.262)523
Mycoplasmalike organisms(*V.3.3.0.
  004;*.3.3.2.133a,*c,.263,*.339)497,
  515,523,526
Mycetozoa(*V.3.3.2.268)523
Mycorrhiza(*II.4.2.026;*V.3.3.2.148,
  *.154,.264)240,516,523
Mycorrhizal associations(*II.4.3.067b)
  245
Mycosphaerella:
  brassicicola(*VI.9.1.465)653
  citrullina(*VI.9.1.091,*.092,*.096)
    633
  musicola(*V.3.3.2.181e)519
  pinodes (perfect state of Ascochyta
    pinodes)(*VI.9.1.405)649
  pomi(*VI.9.1.231)641
  spp.(*V.3.3.2.181e)519
-Mycota(*V.3.3.2.179a)517
-Mycotina(*V.3.3.2.179a)517
Mycotrophic(V.3.3.2.267)523
Myoglobin(*IV.2.2.168)368
Myophily(*IV.4.5.001c)391
Myrmecomorphosis(II.4.4.002c)246
Myrmecophily(*IV.4.5.001c)391
Myrmecophyte(*II.2.3.3.013;.4.2.024;
  *.4.3.042;.4.05d)163,240,244,253

Myrothecium sp.(*VI.9.1.464)653
Myxamoebae(*V.3.3.2.322)525
Myxomycetes(*V.3.3.2.179a,.268)517,523

N

N_A(*IV.0.023)330
NAA(*IV.2.1.012,*.033)357,358
NAAm(*IV.2.1.012)357
Nacobbus spp.(*V.3.3.1.200g)506
NAD^+(*IV.2.2.058b,*.142)364,367
NADH(*IV.2.2.142)367
$NADP^+$(*IV.2.2.142)367
NADPH(*IV.2.2.142)367
Na hypochlorite solution(*V.1.4.022a-3)442
Nailhead spot of tomato(VI.9.1.325)646
Naked(II.5.6.082;.5.17a)276,297
  protoplasm(*V.3.3.2.322)525
Nanism(*II.4.3.001,.034)241,243
Nano (n)(IV.0.260h)346
Naphthaleneacetic acid(*IV.2.1.012;
  *V.1.5.005)357,470
β-Naphthoxyacetic acid(*IV.2.1.012)357
Napiform(II.2.3.2.022,.03e)161,177
Narcissus mosaic(*V.4.2.1.022)566
Narrow:
  -angle crotch[V.3.4.01(9)]545
  -based terraces(*V.2.110c)478
  -leaved(*II.4.3.058)245
    evergreens(*V.1.4.022f-1,*.028k)
    442,446
Nastic movements(IV.3.3.003)376
Native vegetation regrowth(*V.3.3.3.069)536
Natural:
  conditions(*V.2.099)476
  cooling(V.4.1.075,*.093)563
  cover(*V.3.1.010)486
  daylength extension(*V.4.1.002)559
  scavengers(*IV.5.4.2.001)413
  undergrowth(*V.2.110e)479
  vegetative reproduction(IV.4.1.006a-8)379
Naturalization(III.1.2.037)309
Naturalized(*III.1.2.029,.038)309
Nautical mile(IV.0.260k)347
Naturally occurring biochemicals(*V.3.3.3.013)532
Near IR wavelengths(*IV.5.4.2.000)413
Neck rot of onions(*VI.9.1.264)643
Necrosed(*V.4.2.2.015)567
Necrosis(*V.3.3.2.133b,*.269;*.4.2.1.002)515,524,565
Necrotic:
  area, limited(*V.3.3.2.405)528
  spot(*V.3.3.2.065)512
  symptom(*V.3.3.2.082,*.123,*.210)
    513,514,521
Nectar(*IV.4.4.003,.036,*.037)387,389
Nectariferous(II.2.2.27c(3);IV.4.4.037)156,389
Nectarivorous(IV.4.5.0221)392
Nectary, nectaries[II.2.2.27c(3);.3.
  116,*.179,.15b;*IV.4.4.036,*.037;*V.
  4.1.033]156,199,203,221,389,560
Needle:
  nematodes(*V.3.3.1.200c)506
  -shaped(*II.5.5.001,*.002)271
Needlelike equifacial leaf(*II.1.2.119)64
Negatively:
  charged sites(*V.2.015)472
  geotropic(*IV.3.3.008h)377
Negotiable instrument(*VI.6.001)594
Nema(*V.3.3.1.183;*VI.7.2.017r)504,615
Nemathelminthes(*VI.7.2.017r)615
Nematocide(s)(V.3.3.3.061,*.070;*VI.7.2.017p,q)535,536,614,615
  Plant Regulator, Defoliant & Desiccant Amendment of 1959(*VI.7.2.014,*.016f,*.017,.017q)610,612,615
Nematoda(*V.3.3.1.183;*VI.7.2.017r)
  504,615
Nematodes(*V.1.1.065;*.3.3.0.002,
  .004;.3.3.1.183;.3.3.2.133a,*e;
  .3.3.3.061;.4.2.1.014,*.017,*.018,
  *.022;*VI.7.2.017g,*h,*q,r)427,
  497,504,515,535,566,613,615
  attacking above-ground plant parts
  (V.3.3.1.200a)505
  ,plant parasitic(V.3.3.1.200)505
Nematospora phaseoli(*VI.9.1.575)659
Neoendemic(III.1.2.039)310
Neoformation[II.1.1.3.059,.15b(2)]24,
  51
Neolithic(III.1.2.040)310
Neoteny(IV.3.1.014e)372
Neotropical (Neotropics)(III.1.2.041)
  310
Neotylenchus abulbosus(*V.3.3.1.200a)
  505
Neotype(III.1.1.035f)307
Net:
  profit(VI.8.070)621
  radiation balance(IV.3.2.007)374
  sales(VI.8.071)621
Netted(*II.5.6.106)276
Netting(*VI.4.015)587
Neuter(II.2.2.01i)130
Neutral(II.2.2.053,.01i)126,130
  red(*IV.5.3.024)409
Newton (N)(IV.0.193,*.260c)341,344
  meter(*IV.260e)345

per meter(*IV.260e)345
Neutron:
    absorption(*V.3.5.024)558
    counter(*V.3.5.024)558
Nichols terraces(*V.2.110c)478
Nick(III.2.2.116)323
Nicked(*V.1.4.036b-3)450
Nicking(*V.1.4.028i,.11e)445,463
Nicolin (Nicolieren) budding(V.1.4.028q,.08k)448,460
Nicotinamide adenine dinucleotide (oxidized form)(*IV.2.2.058b,*.142)364,367
Nicotinamide adenine dinucleotide (reduced form)(*IV.2.2.142)367
Nicotinamide adenine dinucleotide phosphate (oxidized form)(*IV.2.2.142)367
Nicotinamide adenine dinucleotide phosphate (reduced form)(*IV.2.2.142)367
Niger(*II.5.1.001a)263
Night:
    break(*IV.3.2.013i)375
    interruption(IV.3.2.013i)375
    length(*IV.3.2.013b)374
    -pollinated(*IV.4.4.004a)387
Nigrescens(*II.5.1.001a)263
Nigrospora cob rot of corn (Nigrospora oryzae)(VI.9.1.328)646
Ninhydrin-reacting substances(*IV.5.3.024)409
Nitrate:
    of soda(*V.1.4.028t;*.3.2.086)448,496
    of soda-potash(*V.3.2.086)496
    nitrogen (N)(*V.3.2.034)491
Nitrification(IV.1.2.028c;*V.2.106)354,476
Nitrilotriacetic acid(*V.3.2.012)490
Nitrobacter(*IV.1.2.028c;*V.3.3.2.034)354,511
Nitrogen(*V.1.2.011d;*V.3.2.013,*.028,*.046,*.055,*.056,*.067,*.080)430,491,493,494,495
    availability(*V.3.2.065)494
    cycle(IV.1.2.028)354
    fixation(IV.1.2.029;*V.2.110e-1)354,479
    oxides(*IV.5.4.4.001)415
    solutions(*V.3.2.054,*.086)493,496
    total(*V.3.2.071)495
Nitrosomonas(*IV.1.2.028b;*V.3.3.2.034)354,511
Niveus(*II.5.1.0071)265
Noctuid(*V.3.3.1.132g)502
Nocturnal(IV.4.4.038)390

Nodal section from stem(V.1.5.01c)471
Nm(*III.1.1.027)306
NMR(IV.0.195)341
-Nodal(II.1.1.3.060)24
Nodding(II.2.2.054)126
Nodulation of roots(*V.1.3.041)434
No frills store(*VI.8.136)626
"Nomad" plants(III.1.2.042)310
Nomenclatural:
    synonym(*III.1.1.025)305
    type(III.1.1.035a)306
Nomenclature(III.1.1.024)305
Nomina conservanda(*III.1.1.019)305
Nomina familiarum conservanda(III.1.1.025)305
Nomina generica conservanda et rejicienda(III.1.1.026)306
Non-
    allelic(III.2.2.117)323
    articulated laticifers(*II.3.179)203
    chlorophyllous plants(*V.3.3.2.179)517
    climacteric fruits(VI.1.007)570
    coherent(*V.2.100,*.115j)476,480
    crop losses or use(*V.3.2.000)490
    crop sites(*V.3.3.3.069)536
    crop utilization(*V.3.2.035)492
    cyclic phosphorylation(*IV.5.4.2.013)413
    cyclic photosynthetic phosphorylation(*IV.1.2.036e)355
    fertilized polar nuclei[IV.4.1.02a(8)]383
    functional gametes(*IV.4.8.049,*.053)395
    heritable modification of characters(*IV.5.2.001)398
    occluded virus(*V.3.3.2.456)530
    parasitic diseases(V.3.3.2.270;*V.9.000)524,626
    parthenocarpic(*IV.4.2.004)385
    pectolytic(*V.3.3.2.014)510
    perforated bags(*VI.4.001)585
    popup sprinklers(*V.3.5.012)537
    -precooled bulbs(*V.4.1.029,.076)560,563
    -precooled commercial bulbs(*V.4.1.075)563
    produce items(*VI.8.120)625
    profit corporation(*VI.8.018)618
    recurrent agamospermy(IV.4.1.01)382
    recurrent apomixis(*IV.4.1.006a-7)378
    reducing sugars(*IV.2.2.031b)361
    restricted necrotic symptom(*V.3.3.2.059)512
    selective (harvesting)(*VI.2.023)577

INDEX OF TERMS

selective herbicide(V.3.3.3.063)535
self-limiting galls(*V.3.3.2.040)511
septate mycelium(*V.3.3.2.181c)518
somatic tissue(*IV.4.6.001)392
storied cambium(*II.3.07a)213
storied wood(II.3.117)199
Normal:
  calorie(*VI.5.006)588
  embryo[IV.4.1.01(11)]382
  forcing(*V.4.1.114)564
  syngamy[IV.4.1.02b(2)]583
North Carolina(*VI.8.113)625
Northern corn leaf blight(VI.9.1.329) 646
Notch grafting(*V.1.4.028k)446
Notching(IV.5.4.4.002,.002b;*V.1.3.055)416,435
Nothanguina cecidoplastes, N. phyllobia(*V.3.3.1.200a)505
Nothomorph(III.1.1.027)306
Nothotylenchus acris(*V.3.3.1.200a)505
Nototribic(IV.4.5.002m)392
Novel variety(III.2.0.007a)315
Novirame(II.1.1.3.061,.12a)25,48
Noxious weed(s)(*V.1.3.073;.3.3.3.064) 438,535
Nozzles:
  number of(*V.3.3.3.016)532
  orifices(*V.3.5.032)558
NP(*V.4.1.076)563
NPK(*V.3.2.037)492
NPV(*V.3.3.3.013f)535
Nucellar:
  cells[IV.4.1.02b(1)]383
  embryony(*IV.4.1.006a-1)378
  embryos(IV.4.1.02b)383
Nucellus[II.2.2.041t-4-j,.16b(7);.2.3.3.014g,.03h,.07f(9),.10b(4),.12d(5); *IV.4.1.006a-1,.01(1),.03(1);*.4.6.004]124,145,164,184,188,191,193,378, 382,384,392
Nuclear:
  inclusions(*V.3.3.2.133b)515
  incompatibility(IV.4.8.042)395
  magnetic resonance spectrometer(IV.0.195)341
Nuclei(*V.3.3.3.013f)532
Nucleic acid(s)(*IV.1.2.040g;*.2.1.025;.2.2.140;*V.3.3.2.457,*.461)356, 358,367,530
Nucleolus(IV.1.1.017)351
Nucleopolyhedrosis virus(*V.3.3.3.013b,f)532
Nucleoside(*IV.2.2.032a,*e,.141)362,-367
Nucleotide(*IV.1.2.040g;.2.2.142)356, 367

Nucleus(IV.1.1.018)351
Nuclide(*IV.0.248)343
Nude(II.2.2.055,.01g,h;*.5.6.082)126, 130,276
Nurse-root grafting(V.1.4.028r,.13e-h) 448,465
Nursery, nurseries(*V.1.4.040,*.041; *V.3.3.3.052,*.071;*VI.4.011)450, 534,536,586
  block(V.1.1.035;.1.4.022b)423,442
  equipment(V.1.1.036)423
  facilities, structures, equipment(V.1.1.000)418
  graftage(*V.1.4.028t)448
  plants(I.2.009;*V.1.2.003)6,428
  row(*V.1.4.028k-2,*.041)446,450
  rows(*V.1.4.028t)448
Nurseryman(I.2.008)6
Nut(II.2.3.1.024,.04d-h)159,171
Nut crops(I.1.007)4
Nutation(IV.3.3.004)377
Nutational movements(*II.4.3.071, *.074)245,246
Nutrient(s)(*IV.5.4.3.000;*V.2.110e-1;.3.2.050,*.083;.3.3.3.119)414,479, 493,495,538
  availability(V.3.2.065)494
  competition(*V.3.3.0.000)497
  deficiency(V.3.2.066)494
  deficiency level(*V.3.2.083)495
  ingredients(*VI.8.072)621
  labeling(VI.8.072)621
  leaching(*V.1.1.053e)425
  liquid(*IV.0.171c)340
  optimum levels(*V.3.2.083)495
  removed in harvesting(*V.3.2.000)490
  reserve source(*V.2.111;*.3.2.069) 479,494
  reservoir(*V.3.1.007)485
  solid(*IV.0.171c)340
  solutions(*V.3.2.083)495
  toxic levels(*V.3.2.083)495
  transport(*IV.2.1.025)358
Nutrition:
  program(*V.3.2.040)492
  -related(*VI.9.3.000)663
Nutritional:
  chemicals(*VI.7.1.017s)615
  imbalances(*V.3.3.2.270)524
  problems(*VI.8.038a)619
  spray(s)(*V.3.2.000,*.037,.067, *.085)490,492,494,496
Nuts(*VI.3.053e)582
Nyctinastic drooping(*II.4.3.028)243
Nyctinasty(IV.3.3.003c)376
Nylon(*VI.3.007,*.081)578,583
Nymph(V.3.3.1.188)505

INDEX OF TERMS

Nymphs(*V.3.3.1.132i)502

O

Obcampylotropous[II.2.2.041t-4-k-4,
 .16c(5);*.2.3.3.034,.036a-4,.04e,
 .06e-h]125,145,165,185,187
Obcordate(II.5.4.1.014;.5.03j,.11c)
 269,283,291
Obdiplostemonous(*II.2.2.007d-6,.20a)
 118,149
Oblanceolate(II.5.5.039;.5.11d)272,291
Oblate(II.2.3.2.023,*.037,.06d)161,
 162,180
Obligate(*V.3.3.2.004,*.013,*.040,
 .271)510,511,524
 apomict(*III.2.0.005a-1)313
 apomixis(*IV.4.1.006)378
 halophyte(II.4.2.025)240
 parasite(s)(*V.3.3.2.181d,*e,*f;
  .271,.272,.361)518,519,524,527
Oblique(*II.2.2.007w-10,.07e;.5.2.024;
 .5.4.2.009;.5.5.040;.5.04g,.11e;
 V.3.4.021,.055)120,136,266,269,
 272,284,291,541,543
 cleft graft(V.1.4.028k-4,.10d)447,
  462
 cordons(V.3.4.11b)508
Oblong(II.5.5.041;.5.11f)272,291
Obovate(II.5.5.042;.5.11g,h)272,291
Obovoid(II.2.3.2.024,.07a)161,181
Obsolescent(II.5.2.025)267
Obsolete(II.5.2.026)267
Obtuse(II.5.4.1.015;.5.4.2.010;.5.03k,
 .04h)269,283,284
Obturator(II.3.119,.09b)200,215
Obvolute(II.1.1.2.017,.02n)21,35
Occluded virus(*V.3.3.2.456)530
Occupational Safety and Health
 Act of 1970(*VI.7.2.014,.037)610,616
 Administration(VI.7.2.038)616
 Agency(*VI.7.2.017)612
 Review Commission(*VI.7.2.038)616
Oceans(*VI.7.2.014)610
Ocellated (ocellatus)(II.5.1.0061)265
Ochlospecies(III.1.2.043)310
Ocrea (ochrea)(II.1.2.066;*.1.2.100,
 .14c,d)62,63,79
Ochric epipedon(V.2.018f)473
Octoses(*IV.2.2.031a)361
Odd-pinnate(*II.1.2.053)61
Odor(*IV.4.4.003,*.009,*.042)387,388,
 390
Oedema (edema)(*V.3.3.2.133b,.273)515,
 524
Oersted (Oe)(*IV.0.260l)347
Offering(VI.8.073)621

Officers(VI.8.074)621
Official inspection certificate(VI.7.
 1.042)603
Offsets[*V.1.4.024,.04b,c]443,456
Off-type(III.2.0.004g,.005d-2;.2.2.
 118)312,315,323
Ohm(Ω)(*IV.0.260c)344
Oil, oils(*II.2.3.015;*.3.077,*.125;
 *.4.3.063;*IV.1.1.030,*.2.2.065;
 *.4.4.017;*VI.1.007)164,197,200,
 245,352,365,389,570
 cavities(*II.3.179)203
 crops(I.0.001,.1.008)3,4
 ducts(*II.3.179)203
 -gland darkening of citrus(*VI.9.3.
  029,.088)665,670
 glands[II.2.2.02(4);.2.3.2.05d;*.3.
  104,.12a(4)]131,179,199,218
 phase(*V.3.3.3.054)535
 spotting(*VI.9.3.090)670
Oleaginous(II.5.7.025)278
Oleocellosis(VI.9.3.090)670
Oleosome(IV.1.1.019)351
Olericulture(I.3.013)8
Olfactoric(IV.4.4.039)390
Oligogenic(III.2.2.119)323
Oligophagous(IV.4.5.002n)392
Oligophilic blossom(IV.4.4.040)390
Oligosaccharides(IV.2.2.031b,.032a)
 361,362
Olive-green (olivaceus)(II.5.1.004e)
 264
Olive tubercle(*V.3.3.2.040)511
Olliform(*II.2.2.061c-1-w)127
Ombrophilous(II.0.016)10
Omnivorous(V.3.3.2.275)524
 disease(*V.3.3.2.181d)518
One:
 -armed (single)(*V.3.4.021)541
 -sided(*II.1.2.097)63
Onion:
 maggot(*V.3.3.1.132b)501
 smut(*V.3.3.2.181f)519
On-the-ground inspector(*VI.8.002)616
Ontogenomorphosis(II.4.4.007a)247
Ontogeny(II.3.120;*VI.1.012)200,572
Oogonium(*V.3.3.2.181d,*.278,.280)518,
 524
Oolong tea(*I.1.001a)4
Oomorphogenesis(IV.4.1.020)381
Oomycetes(*V.3.3.2.017,.181d,.279,
 *.305)510,518,524,525
Oospore(*V.3.3.2.181d,.280,*.356;*VI.
 9.1.567)518,524,527,659
Ooze headers(*V.1.1.017f)420
Opaque(II.5.6.084)276
Open:

INDEX OF TERMS

center[V.3.4.029,.03d,e(3)]541,547
(greenhouse)(*V.1.1.017b)419
-pollinated population(III.2.1.001d) 317
  varieties of cross-fertilizing crops(III.2.0.005a-4)314
-pollination(III.2.0.004n)313
Operations(*VI.8.077)622
Operon(III.2.2.120)323
Operculum(*II.2.3.1.006e,.025,.03d,e) 158,160,170
Opposing action of chemicals(*V.3.3.3.007)531
Opposite[II.1.1.1.014,.031;.1.1.3.037;.1.2.068,.15a;.2.3.3.028,.01a] 19,23,30,62,80,165,182
Orange(*II.5.1.008)265
  -colored(II.5.1.008n)266
  dog(*V.3.3.1.132g)502
  rust of spinach(VI.9.1.330)646
Orbicular(II.5.5.043;.5.11i)272,291
Order(*III.1.1.003,.028;*V.2.096a)302, 306,475
  assembly(VI.3.046,*.052;*.8.078)581, 622
  procurement(*VI.8.102)624
Ordinary:
  (condition)(VI.8.016)617
  (quality)(VI.8.094)623
  scald of apple(*VI.9.3.005)663
Ordo(*III.1.1.003,.028)302,306
Organelle(*IV.0.263;.1.1.020)348,351
Organic:
  acids(*IV.1.1.030;*.2.1.001;*VI.1.008e,*g)352,357,571
  carbon(*V.2.014;*.3.2.080)472,495
  complexes(*V.1.5.005)470
  content(*V.2.028e)473
  debris(*V.2.107,*.112,.01a)476,479, 483
  fertilizer(V.3.2.068)494
  layer designations(V.2.069)474
  materials(*V.3.2.000,*.038,.069) 490,492,494
  matter(*IV.0.166;*V.1.1.065,*.2.000, *.015,*.028d,*f,*h,*.037,*.096d, *.099,*.107,*.107f,*g,*i,*.110, *.111,*.116;*.3.2.000)340,427,472, 473,475,476,477,478,479,490
  matter loss(*V.2.106)476
  mulch(*V.3.3.3.059)535
  nitrogen(*V.2.014)472
  soil(*V.2.096a)475
  soils(*V.2.069,*.070)474
  sources*V.3.2.028)491
Orchard, orchards(*V.1.4.022b;*.3.1.010,*.011;*.3.3.2.270;*.3.3.3.052; *VI.2.000,*.018,*.020;*.4.004;*.8.038a;*.9.000)442,486,524,534,575, 586,619,626
  heating(*VI.5.051c)593
  pests(*V.3.3.1.224,*.251)507,508
  run(*VI.8.125)625
Orchinol(*IV.2.2.162)368
Organogenetic(III.1.2.044)310
Orgyia antiqua(*VI.9.2.044)661
Oriental fruit moth(*VI.9.2.035)661
  injury of quince(VI.9.2.053)661
Oriental fruit fly(*V.3.3.1.132b)501
Ornamental horticulture(I.2.000)5
Ornamentals(*I.001,*.002;*.2.002;*II.2.1.068)2,6,103
Ornithochory(IV.4.7.001h)393
Ornithophily(*IV.4.5.001c)391
Orogeny(V.2.071)474
ORSTOM(*V.2.097)475
Ortet(III.2.0.006)315
Orthophototropism(IV.3.3.008m)377
Orthoptera(V.3.3.1.132h)503
Orthostichy(II.1.1.1.015,.05r)19,32
Orthotropic:
  growth(*V.1.4.022a-1)442
  (stem)[*II.1.1.3.019,.062,.04a,c(1), .13a(1),.18e;.1.2.18b]22,25,40,49, 54,83
  subterranean stems(*II.4.3.057)244
Orthotropous:
  (ovule)[II.2.2.041t-4-k-5,.16c(1)] 125,145
  (seed)(II.2.3.3.036b,.04a)166,185
Oscillating shakers(*VI.2.022b)576
OSHA(*VI.7.2.014,*.017,*.037,*.038) 610,612,616
Osmocote(*V.1.2.005;*.3.2.015,.070) 429,491,495
Osmometer(*IV.5.2.037)402
Osmophor(II.3.121,.15c,d)200,221
Osmoregulation(IV.5.2.034,*.036)402
Osmosis(IV.5.2.035)402
Osmotic:
  adjustment(*IV.5.1.019;.5.2.036, *.079)398,402,406
  concentration(*IV.5.2.018)400
  effects(*IV.5.4.3.009)415
  potential(*IV.5.2.005,*.034,*.036, .037,*.038,*.039,*.057)399,402. 403,404
  pressure(*IV.5.2.005,*.010,*.037, *.038,*.053,*.062)399,402,404
  volume(*IV.5.2.037)402
Osseous(II.5.7.268)278
Osteoscleried(*II.3.170)202
Ostiole[II.2.1.13d(2);.2.3.2.025;.2.3.1.07d(1);*V.3.3.2.295,*.349]116,161,

174,525,526
Other:
  crop seed(*V.1.3.075)438
  forms of stress(IV.5.4.)412-416
  laws, rules and regulations(VI.7.2.)
    609-616
  nutrients(*V.3.2.036)492
  plant parts(*V.3.2.040)493
  terms(IV.4.8.)393-395
Outbreeding(III.2.2.121)323
Outcross(III.2.2.122)323
Outcrossing(*III.2.2.121)323
Outdoor rooting bed(*V.4.1.012)559
Outline(II.5.5.044)272
Oval(*II.5.5.019)271
Ovary[II.2.2.041t,.02(6,16)]123,131
Ovate(II.5.5.046;.5.11j)273,291
Overdevelopment(*V.3.3.2.250)523
Overdominance(III.2.2.124)323
Overgrazing(*V.3.3.3.048)534
Overhead sprinklers(*V.1.1.017f;*.3.5.
  012)420,557
Overtop application(V.3.3.3.067)535
Overwraps(*VI.4.000b,*.023,*.025)585,
  587,588
Overwintered(*V.4.1.112)564
Ovipary(II.2.3.3.020b)164
Ovoid[II.2.3.2.026,.04d(3),.06a]161,
  178,180
Ovovivpary(II.2.3.3.020c)164
Ovule[II.2.2.041t-4,.01a(4),.02(6,16),
  .16b;*IV.4.4.000,*.010,*.048;*.4.
  6.000,*.001c,*.003]124,130,131,
  145,387,388,390,392
  chalazal region(*IV.4.1.006a-3)378
  positions(II.2.2.16c)145
Oxalic acid(*II.3.187)203
Oxaloacetate(*IV.1.2.004)352
Oxidation(IV.0.199;*V.2.106,*.110e-1;
  *.3.2.000)342,476,479,490
  processes(*IV.5.4.2.014)414
Oxidative phosporylation(*IV.1.2.040c)
  356
Oxide(*V.3.2.036)492
Oxidized form(*IV.0.199)342
Oxidoreductases(*IV.2.2.058g)364
Oxygen (O$_2$)(*V.3.2.028)491
  deficiency(*V.2.037)473
  deficit(*V.5.4.3.006)414
  depletion(*V.5.4.3.006)414
  levels(*VI.9.3.000)663
Ozone(*IV.5.4.4.001)415
  injury of apples(VI.9.3.092)670

P

PACA(*VI.7.1.000)597

Pachychalaza[*II.2.3.3.036a-5,.04g,
  .08b(2)]165,185,189
  seed forms(II.2.3.3.08)189
Pachymorph rhizome(*V.1.4.060,.060c,
  .05b)451,457
Pachytene(III.2.2.125)323
Pack(*VI.8.038e)619
Pack (v.), package (n.), packaging,
  packing (n.)(VI.3.047)581
Packages labeled(*V.1.3.073)438
Packaging(*VI.3.031;*.7.1.031;*.8.089)
  580,601,623
Packed containers(*VI.6.016)595
Packer, packers(VI.3.050;*.8.053)581,
  620
Packets(*V.1.3.067)437
Packing(*VI.3.031,.047d;*.7.1.030;*.8.
  081)580,581,601,622
  bin(*VI.3.062)582
  operations(VI.3.053)581
  shed(*VI.3.000)577
  shipping containers(*VI.3.079)583
  trimming(*VI.3.072)583
Packinghouse(*VI.2.018,*.020;.3.000,
  .023,.041,*.043,*.046;*.4.000a,
  .004,.006;*.6.016;*.8.038a,
  .044,.102,*.122)575,577,579,580,
  581,585,586,595,619,620,624,625
  foreman(VI.8.077,*.137)622,626
  layout(VI.3.052)581
  manager(*VI.8.037,.078)619,622
  operations(*VI.8.078)622
  packaging(*VI.3.071)583
  related(*VI.6.034)596
  rail-car cooling(*VI.5.035a-6)590
  sales department(*VI.8.106)624
  wrapping(*VI.3.071)583
Packingline(*VI.3.046,*.052;*.8.086)
  581,622
Packout(VI.3.054;*.8.086)582,622
Paedogenesis(*II.4.4.007a)247
Paedomorphosis(IV.3.1.014f)372
Pagoda branching(II.1.1.3.017)22
Paint(*V.1.1.017g)421
Painted(*II.5.1.006m;*V.1.4.028d)265,
  444
Paired(II.1.2.069,.15b)62,80
Palaceous(II.1.2.070,.15c)62,80
Palatability(VI.8.079)622
Palate[*II.2.2.061c-14-r,.23a(1),c(1)]
  128,152
Pale yellow(II.5.1.008o)266
Palea[II.2.2.057,.24b(2)(i)]126,153
Paleaceous(II.5.6.085)276
Paleopolyploid (Palaeopolyploid)(III.
  2.2.126)323
Paleotropical (Palaeotropical)(III.1.

2.046)310
Palisade parenchyma[*II.3.111,.21a(4)] 199,227
Pallet(s)(*V.1.1.014;.1.2.010;*VI.4.000c,.017,*.022;*.6.016)419,429, 585,587,595
  box (bin)(*VI.2.012,*.014,*.018;*.3.053b;*.4.000a,*c,*d,*.004,*.006, *.018)575,582,585,586,587
  loads(*VI.4.015)587
  runners(*V.1.2.003c)428
  unitization(*VI.6.035)597
  unitizing shipping(*VI.6.023)595
Palletization(VI.6.024)596
Palletization & Productivity Committee for unitized forklift handling(*VI.6.023)595
Palletized load(*VI.6.002)594
Palletizing containers(*VI.3.046, *.052)581
Palm(II.0.017,.03a-e)10,14
  leaves(*V.1.1.059)426
Palmate(II.1.2.071,.121d,.16b,.22d)62, 64,81,87
Palmate-pinnate(II.1.2.121e,.23e)64,88
Palmatifid(II.5.4.3.026;.5.05d)270,285
Palmette(V.3.4.055,.11a,d,f,g)543,555
Palmetto(*VI.3.007,*.081)578,583
Palmiform(II.1.2.121f,.23f)64,88
Palmiribbed(II.1.2.121g,.23g)64,88
Palpi(*V.3.3.1.132g)502
Palynology(II.2.2.007t-7)120
Panama disease(*V.3.3.2.181g)520
Pandurate(*II.5.5.036,.047;.5.11k)272, 273,291
Panicle[II.2.1.049,.02a(3),.08h,.11c] 102,105,111,114
Paniculate(II.2.1.050,.06a,.11d)102, 109,114
Paniculiform[II.2.1.051,.06a(1),.11d] 102,109,114
Panmictic unit(III.1.2.047)310
Panmixis, panmixia, panmixy(III.1.2.048)310
Panning(V.4.1.080)563
Pannose(II.5.6.086;.5.17b)276,297
Pansy spot of apple(VI.9.2.054)661
Pantothenic acid(*V.1.5.005)470
Paper(*V.3.3.3.059)535
  bags(*V.1.3.067)437
  containers(*V.1.2.003f)428
  grids(*V.1.3.074)438
  labels(*V.1.2.009)429
  wasps(*V.3.3.1.132e)502
Papery(II.5.7.027;*V.4.2.2.008)278,567
Papier maché(VI.4.019,*.024,*.025)587, 588

Papilionaceous(*II.1.1.2.030;.2.2.061c-14-p,.20d,.23h)21,128,149,152
Papillae[II.2.2.02(1,13)]131
Papillose(II.5.6.087;.5.17c)276,297
Pappus(II.2.3.1.01f,.5.6.088;.5.17d)-168,276,297
PAR(*IV.0.157m)338
Paracarpous(II.2.3.2.027,.03c)161,177
Paraclade(II.2.1.052,.06a)102,109
Paracytic(*II.3.190,.29f)204,235
Paraffined paper cups(V.1.2.003d)428
Paraffin(s)(*II.3.038;*VI.3.085, *.085b)195,584
Parageneon(III.2.2.127)323
Paralithic contact(V.2.1071)477
Parallel(II.1.2.121h,.24h;.2.2.007w-11,.07f;*.3.211)64,89,121,136,205
  bars(*VI.3.027a-1)579
  -ribbed(II.1.2.121i,.24i)64,89
Paramorph, paramorphosis(III.1.2.049) 310
Paranguina spp.(*V.3.3.1.200a)505
Paraphyses(*V.3.3.2.023,*.295)511,525
Parasexual cycle(V.3.3.2.282)524
Parasite(s), parasitic(*II.4.2.006, .026;.4.05c,f;*V.3.3.2.133a,*.168, *.169,*.179,*.199,*.206,*.209, *.211,*.223,*.272,*.283,*.312;*.3.3.3.012,*.053)239,240,253,515,516, 517,521,522,524,525,532,535
  diseases, including animals(V.4.2.1.;*VI.9.000;*.9.1.000)565,626
  insects(*V.3.3.3.013)532
  weak(*V.3.3.2.181c)518
  weeds(*V.3.3.0.001)497
Parasitism(*V.3.3.2.016)510
Parastichy(II.1.1.1.016,.04g,.05r)19, 31,32
Paratracheal(*II.3.014)194
Paratylenchus(*V.3.3.1.200,*.200b)505
Parcifrond(II.1.1.3.063,.11b)25,47
Parenchyma[*II.3.030,*.039,.125,*.221, .04ae,.21a(4),b(3,6),.24d]195,200, 206,210,227,230
  cells(*V.1.4.008)441
  strands, sclerified(*II.3.175)203
Parenchymatous medulla(*V.3.3.2.380) 527
Parent:
  material[*V.2.106,*.107,*.112,.01a (C)]476,479,483
  plant(*V.1.4.022)441
Parietal(II.2.2.041t-2-h,.14e-g)123, 143
Paripinnate(II.1.2.072,.11c,.12b)62, 76,77
Parity(*VI.7.2.004c)609

INDEX OF TERMS

Parks(*I.2.004;*.2.011;*V.1.1.025,
 *.034)6,422
Parkways(*I.2.011;*V.1.1.025)6,422
Parlatoria pergandii(*VI.9.2.007)660
Parsnip:
 canker(VI.9.1.331)646
 scab(VI.9.1.332)647
Parted(II.5.4.3.027;.5.06i,j)270,286
Parthenocarpic(*IV.4.1.002,*.007;*.4.
 4.007)378,381,388
 fruit set(*IV.4.8.048)395
Parthenocarpically(*IV.4.8.050)395
Parthenocarpy(*IV.4.2.001,*.002,*.005;
 .4.8.047,.049)385,395
Parthenogenesis[*IV.4.1.006a-4,*-7,
 -10,*-14,.006b-3,.02a(5)]378,379,383
Partial inhibition(*IV.0.137)335
Partially:
 cross-fruitful(*IV.4.8.019)394
 cross-unfruitful(*IV.4.8.019)394
Particles:
 of energy (quanta)(*IV.0.023,*.157m)
  330,338
 of light (photons)(*IV.0.023,*.157m,
  *.171a)330,338,340
 -size distribution(*V.2.107k)477
Particulates(*IV.5.4.4.001)415
Partitioned(*II.2.2.049)125
Partnership(*VI.8.057)620
Pascal(*IV.0.019,*.232,*.260c;*.5.2.
 038)329,343,344,402
 second(*IV.0.260e)345
Pascuomorphism(*II.4.4.002h)247
Passive exclusion(*IV.5.4.3.010)415
Paste(VI.10.026)677
Pasteur effect(IV.1.2.040f)356
Pasteurization(VI.10.027)677
 soil(V.1.1.065)426
Pasture(*V.3.0.004)484
Patch bud(V.1.4.028f-6,.08g)445,460
Patch-budding(*V.1.2.004;*.1.4.028f-1,
 -6)429,444,445
Patelliform(II.2.3.3.029;.5.6.089;.5.
 17e)165,276,297
Patent(II.2.2.061c-10,.21b,d,e)127,150
Pathogen(s)(*V.1.2.011m;*.3.3.0.001;
 .3.3.2.*.223,.284;*.3.3.3.028,*.037,
 .038,.082,*.083)430,497,533,537
Pathogenesis(V.3.3.2.285)524
Pathogenic(*V.3.3.2.223)522
Pathogenicity(V.3.3.2.286,*.458)524,
 530
Pathological:
 disorders(*V.3.3.0.005)497
 (parasitic) disorders(IV.9.,.000;.9.
  1.)626-674,626
 (parasitic) market diseases(IV.9.1.
  000)626-659
Pathovar(V.3.3.2.179b-1)517
PBA(*IV.2.1.025)358
PC(*V.4.1.019,*.088)560,563
Pea:
 and bean beetles(*V.3.3.1.132a)501
 scab(VI.9.1.336)647
 streak(VI.9.1.337)647
Peach:
 bark beetle(*V.3.3.1.132a)501
 blight(*VI.9.1.176)638
 leaf curl(*V.3.3.2.181e)519
 twig borer(*V.9.2.035)661
 yellows virus(*V.3.3.2.474)530
Peak(VI.8.080)622
Peanut hulls(*V.3.3.3.009)531
Pear:
 canker(*V.3.3.2.181f)519
 leaf blister(*V.3.3.2.181e)519
 leaf blister mite injury(VI.9.2.057)
  661
 scab(VI.9.1.334)647
 -shaped(*II.2.3.2.035)162
 stony pit(VI.9.1.335)647
 thrips(*VI.9.2.081)662
Pearl-gray(II.5.1.003j)264
Peas and beans(I.3.014;*II.1.1.3.044)
 8,24
Peat(*V.1.0.009,*.010;*.1.1.053e,
 .064,.1.2.005;*.011,*.011h;*.1.
 3.081;.1.4.036b-1;*.2.070,*.099;-
 .4.1.088)418,425,426,429,430,439,
 450,474,476,563
 blocks(V.1.2.003e)428
 humus(*V.1.2.011h)430
 moss(*V.1.2.011b,*h,*n,*p)430,431
Pebble mill(V.1.3.044,*.057)435
Pectin(*IV.2.2.032a,*.038;VI.1.005)
 362,570
Pectinate(II.5.4.3.028;.5.06h)270,286
Pectolytic bacterium(*V.3.3.2.005,
 *.014)510
Ped(V.2.073)474
Pedate(II.1.2.16a;.5.4.3.029;.5.06k)
 81,270,286
Pedatiribbed(II.1.2.121j,.24j)64,89
Pedestrian traffic patterns(*V.4.1.
 053)561
Pedicel[II.2.1.01(8,11);.2.2.060,.01a
 (12)]104,126,130
 bark(*V.3.3.2.156)516
Pedigree(III.1.1.128)324
Peds(*V.2.115a)480
 distinct(*V.2.115e)480
 durable(*V.2.115e)480
 nonporous(*V.2.115f)480
 platelike(*V.2.115h)480

porous(*V.2.115d)480
prismlike(*V.2.115i)480
surfaces(*V.2.107b)477
surrounding(*V.2.115f)480
undistinct(*V.2.115e)480
Peduncle[II.2.1.053,.01(9);.2.3.3.01b
  (1);*V.2.2.021]102,104,182,568
Peg-root(II.4.3.035,*.038;.4.11e,f)
  243,259
Pegged-down branches(*V.1.4.036b-5)450
Pellet, pelletize, pelletization, pel-
  letizing(*V.3.2.043;.3.3.3.068)493,
  536
Pelletized seeds(V.1.3.045)435
Pellicularia:
  filamentosa(*VI.9.1.181,*.441,*.501,
    *.50; = Rhizoctonia solani)639,
    651,655
  rolfsii(*VI.9.1.481; = Sclerotium
    rolfsii)654
Pellucid (pellucidus)(II.5.1.007j)265
Peloric, pelory(II.2.2.061c-14-q,.02j)
  128,149
Peltate(II.1.2.074,.15d;.3.204;*.5.6.
  073;.5.16e)62,70,205,275,286
Peltribbed(II.1.2.121k,.24k)64,89
Pencil stripe of celery(VI.9.3.093)670
Pendulous(II.2.2.041t-2-i,.15d;.5.2.
  027)123,144,267
Penduliflory(*IV.4.4.021)389
Peneplanation(V.2.074)474
Penetrance(III.2.2.129)324
Penicillium:
  digitatum(*V.3.3.2.181g;*VI.9.1.105)
    520,634
  expansum(*V.3.3.2.181g;*VI.9.1.106,
    .107,.112)520,634
  funiculosum(*VI.9.1.127)635
  hirsutum(*VI.9.1.344)647
  italicum(*V.3.3.2.181g;*VI.9.1.105)
    520,634
  sp.(*VI.9.1.108,*.111,*.113,*.171)
    634,638
  spp.(V.4.2.1.015;*.9.1.104,*.109,
    .110,.112,*.498)566,634,654
Penicillium:
  root rot of horse-radish(VI.9.1.344)
    647
  rot of lilies, hyacinths and other
    bulbs(V.4.2.1.015)566
Penman's formula(*IV.5.2.028)401
Penniform(II.1.2.121l,.24l)65,89
Penninerved(*II.1.2.121m)65
Penniparallel(II.1.2.121m,.25m)65,89
Penniribbed(II.1.2.121n,.25n)65,89
Pentose phosphate:
  pathway(IV.1.2.040g)356

respiratory cycle(*IV.2.2.154)368
Pentoses(*IV.2.2.031a,*.142)361,367
PEP(*IV.1.2.019;*.2.2.154)353,368
Pepo(II.2.3.1.026,.05c)160,172
Peptide:
  bond(*IV.2.2.150)368
  chain(s)(*IV.2.2.150,*.172)368,369
Percentage available P O (phosphate)
  (*V.3.2.034)491
Perched water table(*IV.5.4.3.006;V.4.
  1.082)414,563
Perennating bud(II.4.1.004)238
Perennation(IV.4.8.043)395
Perennial(s)(*II.3.199;;*.4.4.003b;-
  .5.3.014;.5.01c,d,.02b-d)204,247,
  268,281,282
  grasses(*V.3.3.0.001)497
Perfect:
  (flower)(II.2.2.01b,e,f)130
  (stage)(*V.3.3.2.293)525
  state(*V.3.3.2.179a,*.293,*.406)517,
    525,528
Perfoliate(II.1.2.075,.05e)62,70
Perforated(II.1.2.076,.16c)62,81
  bags(*VI.4.000c,*.001)585
  belt sizer(*VI.3.064c)583
  hose(*V.3.5.012)557
  iron plate support(*V.2.109)478
  pipe(*V.3.5.012)557
Perfume blossoms(IV.4.4.042)390
Perhumid(II.4.2.027)240
Perianth[II.2.2.061,.01a(8,9),.20;*.2.
  3.2.015;*V.4.2.2.021]126,130,149,
  161,568
Periblem[*II.3.080,.14a(3)]197,220
Pericarp(II.2.3.2.029,.05a;.2.3.3.01b
  (4),.12e(1)]161,179,182,193
Perichalazal(*II.2.3.3.036a-6,.04h,
  .09a-d,.12d)166,185,190,193
  aril[II.2.3.3.08a(3)]189
Periclinal(III.2.2.024d)319
  (chimeras)(III.2.2.01b)327
  (divisions)[II.3.14b(2,3)]220
Pericycle[II.3.127,.03c,.11,.13a(5),
  .17a(2)]200,209,217,219,223
  (dicot roots)[*II.3.091,.02e(10),
  .03c]198,208,209
Pericyclic fibers[II.3.16c(2)]222
Periderm(*II.3.016,*.101,128,*.140,
  .160,.20;.4.3.003)194,198,200,201,
  202,226,241
Peridermium (pl. peridermia)(V.3.3.2.
  294)525
Peridium(*V.3.3.2.294,*.366)525,527
Perigene(*II.3.190)204
Perigynous(II.2.2.061d-3,.19b,c;.2.3.
  2.016)128,148,161

Peripheric (embryo)(II.2.3.3.030,.02i) 165,183
Periphysis(*IV.3.1.014h)372
Periodic inspections(*V.3.0.001;*.3.3.3.098)484,537
Perishable(*VI.7.1.025)601
 agricultural commodity(*VI.7.1.005, *.013,*.042,.043,*.077)598,599, 603,608
 Agricultural Commodities Act of 1930, as Amended(VI.7.1.,.000)597-609,597
 products(*VI.8.093)623
 Protective Tariff(*VI.6.029c)596
Perishables(VI.8.081)622
 "fresh"(*VI.6.026)596
Perisperm[II.2.3.3.031,.12e(3);*IV.2.2.172]165,193,369
Peristaltic fluid drill(V.1.3.046)435
Perithecium (pl. perithecia)(*V.3.3.2.181e,.295;*VI.9.1.400,*.465)519, 525,649,653
Perivascular (fiber)(*II.3.063)197
Perlite(*V.1.1.053e;*.1.2.005,*.011, *.011b,.011i,*.011n;*.1.3.081;*.1.4.028k-2)425,429,430,431,439,446
Permaculture(I.2.010)6
Permanent:
 -set sprinklers(*V.3.5.012)557
 stress reaction(*IV.5.1.013)397
 wilting point(IV.5.2.039)403
Permanently chilling-resistant(*IV.5.3.008)408
Permeability (henry per meter)(*IV.0.260e)345
Permittivity (farad per meter)(*IV.0.260e)345
Peronoplasmopara cubensis(*V.3.3.2.181d)518
Peronospora:
 effusa(*VI.9.1.207)640
 hyoscami(*V.3.3.2.181d)518
 parasitica(*VI.9.1.202)640
 pisi(*VI.9.1.203)640
 spp.(*V.3.3.2.181d)518
Peronosporaceae(*V.3.3.2.253)523
Peroxidase(*IV.2.2.168)368
Peroxidation(*IV.5.3.008)408
 processes(*IV.5.4.2.014)414
Peroxisome (peroxysome)(IV.1.1.021)351
Perpendicular(II.5.2.028)267
Persistent(II.5.3.015)268
 herbicide(V.3.3.3.069)536
 virus(V.3.3.2.299)525
Personate(II.2.2.061c-14-r,.23a,c)128, 152
Perulate(II.5.6.091;.5.17f)276,297

Peruvian guano(V.3.2.068)494
Pest, pests(*V.1.4.028t;.3.3.0.000; *.3.3.3.039,*.052,*.084;.4.2.) 448,497,533,534,537,565-568
 -animals(V.3.3.1.)498-509
 control(*V.3.3.3.097)537
 controls(*V.3.3.3.053)535
 damage(*V.3.3.3.052)534
 damage potential assessment(*V.3.3.3.053)535
 general(V.3.3.0.)497
 management(V.3.3.3.)531-539
  program(*V.3.3.3.012)532
 numbers(*V.3.3.3.052)534
 -plants(V.3.3.2.)509-531
 population level(*V.3.3.3.090)537
 problems(*V.3.1.008h)486
 resistance(*V.2.110e-2)479
 stage of development(*V.3.3.3.052) 534
Pesticide(s)(*V.1.1.049,*.078;*.1.3.015,*.045;*.3.0.001;*.3.2.039, *.085;.3.3.3.000;*.3.3.3.016,*.017 *.023,*.030,*.031,*.032,*.039, *.043,*.046,*.055,*.056,*.068, .070)424,427,433,435,484,492,496, 531,532,533,534,545,535,536
 and nontoxic materials(*V.3.3.3.108)538
 application(V.3.3.3.071,*.098,*.100) 536,537
 applied to soil(*V.3.3.3.093)537
 chemical (def.)(VI.7.2.016f)612
 compound yield(*V.3.3.3.072)536
 coverage(*V.3.3.3.070)536
 deposit(*V.3.3.3.101,*.110)538
 dosage(*V.3.3.3.070)536
 dry formulation(*V.3.3.3.096)537
 equivalent(*V.3.3.3.020,.072)533,536
 equivalent applied(*V.3.3.3.085)537
 -fertilizer mixtures(*V.3.3.3.042) 534
 formulation(*V.3.3.3.009,*.018, *.029,*.072)531,533,536
 method of application(*V.3.3.3.070) 536
 mixtures(*V.3.3.3.018,*.108)533,538
 particles(*V.3.3.3.092)537
 physical or chemical property(*V.3.3.3.104)538
 powdered(*V.3.3.3.124)539
 preparation(*V.3.3.3.042)534
 residue(*VI.8.038f)619
 solutions, soaking in(*V.1.3.080)438
 surface-active agent(*V.3.3.3.105) 538
 technical grade(*V.3.3.3.033,*.097)

INDEX OF TERMS

533,537
tenacity(*V.3.3.3.103,110)538
treatment(*V.3.3.3.106,*.111,*.112) 538
use(*VI.7.2.017)612
Peta (P)(*IV.0.260h)346
Petaloid(II.2.2.007v)120
  (involucre)(II.2.1.10a)113
  (stamen)(II.2.2.09b)138
  (stigma)(II.2.2.18n)147
  (style)(II.2.2.17i)146
Petals(*II.2.2.061c;*IV.3.3.003e;*V.4. 1.108)126,376,564
Peteca of lemons(VI.9.3.094)670
Petiolar(II.1.2.077,.16d)62,81
Petiole[II.1.2.078,..03c(2),..06b(1), .17a(1);.2.1.01(13);*.3.030;*.4.3. 071;*IV.1.1.002;*V.3.3.3.112]62, 68,71,82,104,195,245,350,538
  disks(V.1.5.01e)471
  (monocots)(*II.3.091)198
Petiolule[II.1.2.079,..17a(2)]62,82
Petri dishes(*V.1.5.005)470
Petroleum oil(*V.3.3.3.097)537
Pezicula malicorticis(VI.9.1.133)636
pH(*IV.5.4.3.000,*.002;*V.2.015, *.113,*.120;*.3.2.007,*.065,*.080, *.089)414,472,480,482,490,494,495, 496
  adjustment(*V.3.2.079)495
  control(*V.3.2.044)493
  range(*V.3.2.083)495
Phage(*V.3.3.2.044)512
Phakospora vitis(*V.3.3.2.181f)519
Phalaenophily(*IV.4.5.001c)391
Phalanges(II.2.2.007d-15)119
Phanerophyte(II.4.1.005;.4.02a)238,250
Phase-contrast microscope(IV.0.205)342
Phaseic acid(*IV.2.1.037)359
  2-trans-(*IV.2.1.037)359
Phaseolin(IV.2.2.162)368
Phellem[*II.3.128,.12b(6),.20a(2), b(2)]200,218,226
Phelloderm[*II.3.128,..12b(5),.20a(4), b(4)]200,218,226
Phellogen[*II.3.128,*.174,.12b(4),.20a (3),b(3)]200,203,218,226
Phelloid(II.3.134)200
Phenogen(III.2.2.130)324
Phenolics(IV.2.2.154)368
Phenological development(*IV.5.2.015) 399
Phenology(IV.3.1.018)373
Phenolphthalein(*VI.1.008g)571
Phenols(*IV.2.2.154)368
Phenon(III.2.2.131)324
Phenopause(IV.3.1.019)373

Phenotype(III.2.2.132)324
Phenotypic:
  modification(s)(II.4.4.,.000;.4.14; *IV.5.2.030;*.5.3.038;*.5.4.3.006) 246-248,246,262,401,410,414
  morphoses(*II.4.4.000)246
  plasticity(III.2.2.133)324
  xeromorphic structure(*IV.5.2.085) 406
Phenylalanine(*IV.1.2.042;*.2.2.016) 356,360
Phenylpyruvic acid(*IV.1.2.042)357
Pheromone(*IV.1.2.040d;V.3.3.3.073) 356,536
Phialophora malorum(*VI.9.1.490)654
Phloem[*II.3.023,*.041,*.063,*.127, .135,*.181,*.182,*.183,*.189g,*h, *.209,*,.210,.11a(30,.20b,c,d(3), .21a(9),b(5)]194,196,197,200,203, 204,205,217,226,227
  arrangement(II.3.28)234
  cylinders(*V.1.4.028e)444
  fibers*II.3.063,*.175,.24e-g)197, 203,230
  mobile(*V.3.3.3.113)538
  nerosis(*V.3.3.2.133b)515
  parenchyma(*II.3.175)203
  positions(II.3.136,..27,.28)201,233, 234
  rays(*II.3.175,.219)203,206
  sieve elements(*II.3.182,*.183,.26b; *V.1.4.028e)203,234,444
Phloic(*II.3.063)197
Phoma:
  apiicola(*VI.9.1.355)647
  betae(*VI.9.1.087)633
  destructiva(*V.3.3.2.181g;*VI.9.1. 356)520,647
  pomi(*V.3.3.2.181g)520
Phoma:
  root rot of celery, celeriac(VI.9.1. 355)647
  rot of tomato(VI.9.1.356)647
Phomopsis:
  citri(*V.3.3.2.181e;*VI.9.1.525, *.528)519,656,657
  sp.(*VI.9.1.527)657
  vexans(*VI.9.1.358)648
Phomopsis:
  rot of eggplant(VI.9.1.358)648
  of tomato(VI.9.1.359)648
  stem-end rot of citrus(*V.3.3.2. 181e)519
Phosfon-D(*IV.2.1.037;*.5.2.017)359, 400
Phosphate(*IV.2.2.159;*V.3.2.046, *.055,*.087)368,493,496

ammonium(*V.3.2.086)496
bone(*V.3.2.075)495
ester(*IV.2.2.142)367
fused(*V.3.2.003)490
rock(*V.3.2.003,.075,*.087,*.091)
  490,495,496
Phosphodiester bond(*IV.2.2.056,.188)
  363,369
Phosphoenol pyruvate(*IV.1.2.019;*.2.
  2.154)353,368
Phosphogluconate pathway(*IV.1.2.040g)
  356
Phospholipid(IV.2.2.032b,.159)362,368
Phosphorescence(*IV.3.2.005,.008)374
Phosphoric:
  acid(*IV.2.2.140;*V.3.2.091)367,496
  acid esters(IV.2.2.032d)362
  anhydride(P_2O_5)(*V.3.2.003,*.033,
    .034,.036,*.086,*.087,*.091,
    *.092)490,491,492,496
  oxide(*V.3.2.003)490
Phosphorus (P)(*V.3.2.013,*.028,*.033,
  .034,.036,*.037,*.046,*.055,
  .056,.071,*.075,*.080,*.092)491,
  492,493,495,496
  reduced availability(*IV.5.2.056)404
  total(*V.3.2.071)495
Phot (ph)(*IV.0.260l)347
Photobiodegradable mulch(V.3.3.3.074)
  536
Photoblastic(IV.3.2.009)374
Photochemical(IV.3.2.010)374
  process(*IV.3.1.020)373
  strain(IV.5.4.2.004,*.005)413
Photodormancy(IV.3.2.011)374
Photodynamic effect(IV.5.4.2.005)413
Photoelectric sorting(*VI.3.028)579
Photoinactivation(IV.5.4.2.006,*.007)
  413
Photoinhibition(*IV.5.4.2.007)413
Photometric(*IV.0.157b-2-a,*-b,-d)337
  radiation(IV.0.157k)338
Photomorphogenesis(IV.3.2.012)374
Photomorphosis(II.4.4.003e)247
Photon(s)(*IV.0.023,*.070,*.171a;*.3.
  2.005)330,331,340,374
  applied(IV.0.157x-2-e)339
  content(IV.0.157x-2,.157y-2)339
  (flat surface)(IV.0.157x-2-d)339
  flow rate(IV.0.157x-2-b,.157y-2-b)
    339
  fluence(IV.0.157y-2-c)339
  fluence rate(IV.0.157y-2-d)339
  flux(*IV.0.157m;.157x-2-c)338,339
  flux density(*IV.0.157m)338
Photonasty(IV.3.3.003d)376
Photooxidation(*V.1.1.017d)420

Photoperiod(IV.3.2.014;*.5.3.028;*V.4.
  1.006,*.007)374,409,559
  threshold(*IV.3.1.019)373
Photoperiodism(*IV.3.2.014)374
Photoreaction, photoreactive(IV.3.2.
  014)375
Photoreactivation(IV.5.4.2.008,*.014)
  413,414
Photorespiration(IV.1.2.035,*.038;*.3.
  1.020;;*.5.4.2.002)354,355,373,413
  C-4 plants(*IV.1.2.008b)352
Photoreversion(IV.5.4.2.009)413
Photosensitizer(*IV.5.4.2.005,.010)413
Photosynthates, reduced translocation
  (*IV.5.2.044)403
Photosynthesis(*II.3.125;IV.1.2.036;
  .2.2.172;.3.1.020;.5.2.084;*.5.
  4.1.005;*.5.4.2.002,*.003;*V.1.2.
  001;*.3.2.009;*.3.3.2.133d)200,
  354,369,373,406,412,413,428,490,
  515
  inhibited(*IV.5.2.044)403
  reduced(*IV.5.2.044)403
Photosynthetic:
  efficiency(IV.3.1.020)373
  end products(IV.2.1.037)355
  irradiance(*IV.0.157m)338
  photon density(IV.0.157p)338
  photon flux density(IV.0.157q)338
  rates(IV.1.2.038;*V.3.3.2.133d)355,
    515
Photosynthetically active radiation
  (PAR)(*IV.0.157m;*.3.1.020)338,373
Photosystems(IV.1.2.036g)355
Phototaxis(IV.3.3.005)377
Photothermal(IV.3.2.016)375
Phototropism(IV.3.3.008n)377
Phycomycetes(*V.3.3.2.105,.305)514,525
Phyletic(III.1.2.050)310
Phyllary(*II.2.1.042,*.054,.10b,.12a)
  102,113,115
Phylloclad(*II.1.1.3.064,.06b;.4.3.
  036)25,42,243
Phyllocoptruta oleivora(*VI.9.2.017)
  660
Phyllode, phyllodes(II.1.2.10c;*.4.3.
  032,.037;*IV.5.2.085)75,243,406
Phyllody(*V.3.3.2.133b,.306)515,525
Phyllome(II.1.2.082)62
Phyllomorphic(*II.1.1.3.019,.065;*.1.
  2.105)22,25,63
  branch(II.1.1.3.15c)51
Phyllosticha solitaria(*V.3.3.2.181g;
  *VI.9.1.103)520,634
  straminella(VI.9.1.461)652
Phyllotaxis(II.1.1.1.017,.04q,.05r;
  *.3.100)19,31,32,198

Phylogenetic(*III.1.1.004)303
Phylogeny(*III.1.1.004,.029)303,306
Phymatotrichum omnivorum(*V.3.3.2.
  181g)520
Physalospora ear rot of corn(*VI.9.1.
  364)648
Physalospora:
  obtusa(*VI.9.1.086)633
  rhodina(*VI.9.1.528)657
  sp.(*VI.9.1.171)638
  zeae(*VI.9.1.263)643
  zeicola(*VI.9.1.364)648
Physical:
  and chemical weathering(*V.2.106)476
  control(*V.3.3.3.053)535
  irreversible change(*IV.5.1.012,
    *.013)397
  reversible change(*IV.5.1.004)396
  structure(*V.2.110)478
Physiography(*V.2.000)472
Physiological:
  characteristics(*IV.3.1.014)372
  disorders(*V.3.3.0.005;*.3.3.2.270;
    .4.1.083)497,524,563
    (non-parasitic diseases)(V.4.2.2.)
    566-568
  drought(IV.5.4.3.007)415
  effects of infection(V.3.3.2.133d)
    515
  fruits and vegetables(*VI.9.000)626
  market disorders(VI.9.3.,.000)663
  maturity(*VI.1.006,*.007,.012,*.016,
    .02)570,572,574
  processes(*VI.1.02)574
  race(V.3.3.2.309,*.351)525,526
    subdivision(*V.3.3.2.051)512
  races(*V.3.3.2.181f)519
  sprays(*V.3.4.000a)539
  stage of development(*IV.4.8.020)394
Physiologically:
  deep(*V.1.3.060)436
  immature(*VI.1.004)569
  intermediate(*V.1.3.060)436
  shallow(*V.1.3.060)436
Physoderma zeae-maydis(*V.3.3.2.181a)
  518
Physopella fici(*V.3.3.2.181f)519
Phytoalexin(*IV.2.2.154,.162;V.3.3.
  2.312)368,525
Phytochrome(IV.3.2.017)375
Phytol(*IV.0.227;*.2.2.107)343,366
Phytometer(*IV.5.2.065)404
Phytoncide(V.3.3.3.075)536
Phytophagous(V.3.3.1.196)505
Phytophthora:
  cactorum(*V.3.3.2.181d;*VI.9.1.380,
    *.386)518,648,649

  capsici(*VI.9.1.132,*.384)649
  cinnamomi(*V.3.3.2.181d;*VI.9.1.126)
    518,635
  citrophthora(*V.3.3.2.181d;*VI.9.1.
    126)518,635
  drechsleri(*VI.9.1.132)636
  hibernalis(*VI.9.1.126)635
  infestans(*V.3.3.2.181d;*VI.9.1.298,
    *.299)518,645
  nicotianae(*V.3.3.2.181d)518
  palmivora(*V.3.3.2.181d;*VI.9.1.126)
    518,635
  parasitica(*V.3.3.2.181d;*VI.9.1.
    126;*.132,*.384,*.386)518,635,636,
    649
  phaseoli(*VI.9.1.205)640
  syringae(*VI.9.1.126)635
  sp.(*VI.9.1.385)649
  spp.(*V.1.1.065;*.3.3.2.278;*VI.9.
    1.132,*.381,*.382,*.383)426,524,
    636,648
Phytophthora:
  rot of apple, pear(VI.9.1.380)648
    of asparagus(VI.9.1.381)648
    of eggplant(VI.9.1.382)648
    of muskmelons(VI.9.1.383)648
    of pepper(VI.9.1.384)649
    of tomato(VI.9.1.385)649
  rots of rhubarb(VI.9.1.386)649
Phytosanitation(V.3.3.3.076)536
Phytotoxic(*V.3.2.006;*.3.3.3.019,
  .021,.063;*VI.9.3.090)490,533,535,
  670
Phytotoxicity(*V.3.3.3.071)536
Phytotoxin(V.3.3.3.077)536
PI(*IV.0.157o)338
Pick(V.1.1.037)423
Picking bag(*VI.4.000a,.020)585,587
Pickles(*VI.10.000,.028)674,677
Pickleworm(*V.3.3.1.132g)502
"Pick-your-own"(VI.8.138)626
Pico (p)(*IV.0.260h)346
Pictus(*II.5.1.006m)265
Piece or whole root grafting(*V.1.4.
  028k)446
Piece root:
  cutting(V.1.4.01e)453
  grafting(*V.1.4.028b)443
Piercing-sucking mouthparts(*V.3.3.1.
  132c,*d)502
Pierce's disease of grape(*V.3.3.2.
  361)527
Piggyback(*VI.6.016,*.033,.036)595,
  596,597
  Plan III(*VI.6.026)596
  routing(*VI.6.028)596
  shipments(*VI.8.106)624

Pigments(*IV.1.1.030)352
Piliferous(II.5.6.092;.5.17g)276,297
Pillar tree(V.3.4.056,.04a-f)543,548
Pilose(II.5.6.093;.5.17h)276,297
Pimpled(*II.5.6.087)276
"Pin"(*IV.4.3.009,*.012,*.013)386
Pinhead rust of spinach(VI.9.1.292) 644
Pinch(V.4.1.084)563
Pinching(V.3.4.057;*.4.1.009,*.074, *.105)543,559,562,564
Pine(*V.1.1.017c;*.1.2.011l,*.014)420, 430,431
  needle(*II.5.5.002)271
Pineapple:
  heart rot(*V.3.3.2.179a)517
  orange pitting(VI.9.3.095)670
Pink:
  kernel rot of corn(*VI.9.1.243)642
  mold rot of apple, pear(*VI.9.1.392) 649
    of mulkmelons and cucumbers(*VI.9.1.393)649
  rib of lettuce(VI.9.3.096)670
Pinnate[II.1.2.083,.17b,.21a(2),.25o] 62,65,82,86,90
Pinnatifid(II.5.4.3.030;.5.05c)270,285
Pintle brush(*VI.3.027a-1)579
Pioneer:
  plants(*II.4.4.002a)246
  pollen(IV.4.6.043)390
  species(III.1.2.051)310
Pipe:
  framework(*V.1.1.059)426
  nozzle sizes(*V.3.5.014)558
Piricularia grisea(*VI.9.1.395)649
Pirostele[II.2.1.055,.12b(1)]102,115
Pisatin(*IV.2.2.162)368
Pistil(II.2.2.041u)125
Pistillate[II.2.2.062,.01c,g,.27e;.2.3.2.03c(1),.04d(1)]129,130,156,177,178
Pistillode[II.2.2.063,.01i(1)]129,130
Pistils(*V.4.1.033)560
Pit burn of apricot(*VI.9.3.065)668
Pit field(*II.3.181)203
Pitch fork(V.1.1.038,*.072)423,427
Pitcher-shaped(*II.2.2.061c-1-v,*.4.3.006)127,241
Pitedia sayi(*VI.9.2.013)660
Pitfall(*II.4.3.009)241
Pith[II.3.073,.01B(12);*V.1.4.028k-7] 197,207,447
  rays(*II.3.107)199
Pithiness of celery(*VI.9.3.098)670
Pithy(*II.5.7.023)278
Pithy brown core of pears(VI.9.3.099) 670
Pits(*II.3.202)204
Pitted(*II.3.202;*.5.6.118)204,277
  (vessel)(II.3.22f)228
Pitting:
  of banana(VI.9.1.395)649
  of citrus(VI.9.3.029,*.100,*.139) 665,670,673
Place:
  of contact(*VI.8.084)622
  pack(VI.8.083)622
  packing(VI.3.053f)582
Placenta[II.2.2.041t-1,.16b(1)]123,145
Placentation(II.2.2.041t,.14c-g,.15a-f)123,143,144
Plagiogeotropism(*IV.3.3.008h)377
Plagiophototropism(IV.3.3.008p)377
Plagiotropic(*II.1.1.3.019,*.037,.066; *.4.3.057;*IV.3.3.008h)22,23,25, 244,377
  axis extension(*II.4.3.023)242
  stem[II.1.1.3.04b,c(2),.13a(3),.18e; .1.2.18b(2);*.4.3.061]40,49,54,83, 245
  subterranean stems(*II.4.3.057)244
Plain(II.1.1.2.018,.02o)21,35
Plaited(II.1.1.2.019,.02p)21,35
Plan III(VI.1.6.026)596
Planck constant (h)(*IV.0.070,.224) 331,342
Plane:
  angle(*IV.0.260f)346
  shapes(II.5.5.)271- 273
Planet Jr. drill(*V.1.3.021)433
Planned utilization(*V.3.0.001)484
Plant, plants(*IV.4.7.001b;*.5.2.083; *V.3.3.3.012,*.052,*.059,*.077; *.4.1.007,*.121)392,406,532,534, 535,536,559,565
  analyses(*V.3.2.035)492
  analysis(V.3.2.071)495
  and crop size control(V.3.4.,.000) 539
  association(*II.4.2.004;III.1.2.052) 238,310
  bands(V.1.2.003f)428
  bonzai(V.3.4.004)540
  breeding(III.2.)311-327
    terms(III.2.2.)317-326
  bug(*V.3.3.1.132c)502
  canopy(*V.3.3.3.067)535
  cardinal temperatures(IV.3.2.021a) 376
  carnivorous(II.4.3.009;*IV.3.3.003b) 241,376
  cell wall distension(*IV.5.2.069)405

walls (lignin)(*IV.2.2.114)366
chilling injury(IV.5.3.005;VI.9.3.
    029)407,665
  requirement(IV.5.3.006)407
  -resistant(IV.5.3.008)408
  -sensitive(*VI.9.3.029)665
chipper(V.1.1.039)423
decayed or infested fruit removal
    (*V.3.3.3.088)537
devernalization(IV.3.2.021b)376
difficult to propagate(*V.1.4.028a)
    443
diggers(*VI.8.053)620
disease control(*V.3.3.3.083)537
  free(*V.1.4.028t)448
  occurrence and severity(*V.3.3.2.
    159)516
  prevention(*V.3.3.3.037,*.038)533
  reinfection source(*V.3.3.3.076)
    536
diseased(*V.3.3.3.111)538
dwarf(*IV.5.4.4.002g; see also Bon-
    zai)416,540
framework(*V.3.4.01)545
green(*V.3.3.2.034)511
growth(*V.3.2.066)494
  and development(*V.3.2.028,*.038,
    .056,.058)491,492,493,494
  and development conditions(*V.3.4.
    000a)539
  and development processes(*V.3.4.
    000a)539
  regulator(IV.2.1.002)357
  regulators(*V.3.4.000a)539
hardiness(IV.5.3.002)407
higher(*IV.2.2.031c-a-3,*.114)362,
    366
holding in isolation(*V.3.3.3.084)
    537
hoppers(*V.3.3.1.132d)502
import and export(*V.3.3.3.084)537
in containers(*V.3.2.083)495
    greenhouses(*V.1.2.005)429
    shadehouses(*V.1.2.005)429
inoculants(*VI.7.2.017s)615
-lice(*V.3.3.1.132d)502
membranes(*IV.2.2.116)366
mowing(*V.3.3.3.088)537
nastic movements(*IV.3.3.003)376
normal upright(*V.1.4.022a-1)442
nutrient(*V.3.2.067)494
nutrients(*VI.7.017s)615
on roots other than their own(*V.1.
    4.000)441
on their own roots(*V.1.4.000)441
organ(*V.3.3.3.027)533
organs(*IV.3.3.003;*V.3.3.2.413)376,
    529
parasites(*V.3.3.0.005)497
parasitic fungi(*II.4.3.024)242
  nematodes(V.3.3.1.200)505-506
parts(*V.3.2.072)495
patents(III.2.0.007)315
pathology(*V.3.3.2.165)516
pests(*V.3.3.1.132h)503
phototaxis(IV.3.3.005)377
physiological and biochemical pro-
    cesses(*IV.5.2.076)405
phytochrome(*IV.3.2.017)375
pigment protection(*IV.5.4.2.012)413
pollution effects(*IV.5.4.4.001)415
productivity(*V.3.1.009)486
products(*V.3.3.3.012)532
regulator(*VI.7.2.017p,s)614,615
remains(*V.2.111)479
removal of dead, diseased or infes-
    ted branches(*V.3.3.3.088)537
responses(*IV.5.4.4.001)415
sampling(V.3.2.072)495
shredder(*V.1.1.039,.040)423
single-celled(*V.3.3.2.339)526
size control(*IV.5.4.4.002;*V.3.3.3.
    000)416,531
  and crop load regulation(*V.3.4.
    000b)539
slow-to-root(*V.1.4.022a-3)442
species(*V.1.4.022a-3)442
stenothermic(IV.3.2.019)376
substances(*V.3.3.2.413)529
succulent parts(*V.3.3.1.250)508
surfaces(*V.3.3.3.049)534
susceptible(*V.3.3.2.414)529
thermal induction(IV.3.2.020)376
thermoperiod(IV.3.2.021c)376
thermoperiodism(IV.3.2.021)376
tissue(*V.3.3.0.004)497
  drying(*V.3.3.3.025)533
tissues(*V.3.2.073)495
toxic to some species(*V.3.3.3.091)
    537
transplanted(*V.3.3.3.067)535
tropism(*IV.3.3.008)377
Variety Protection Act of 1970(*III.
    2.0.007)315
vascular terrestrial(*IV.3.2.004)373
vernalization(IV.3.2.021e)376
water content(*IV.5.2.053,.073)404,
    405
  deficit(*IV.5.1.017)397
  potential(*IV.5.1.017)397
  saturation deficit(*IV.5.2.078)405
  stress(*IV.5.1.017)397
Planted crop(*V.3.3.3.078,*.079)536
Planter, planter box(*I.2.001a)5

Planting(*VI.7.1.030,*.031;*.8.090)
    601,623
  back(V.4.1.085)563
  cover crops(*V.3.1.000)484
  preparation(*VI.8.090)623
  stock(*V.4.1.115,*.122)564,565
    of bulbous plants(V.4.1.086)563
  systems(V.3.1.008,.01)485,488-489
  windbreaks(*V.3.1.000)484
Plantlets(*V.1.5.006)470
Plasmalemma(IV.1.1.022)352
Plasmid(V.3.3.2.320)525
Plasmodesmata(IV.1.1.023)352
Plasmodiophora brassicae(*V.3.3.2.
    181b;*VI.9.1.156)518,637
Plasmodiophoromycetes(V.3.3.2.181b)518
Plasmodium, plasmodia(*V.3.3.2.181b,
    *.268,.322)518,523,525
Plasmon(III.2.2.134)324
Plasmopara:
  nivea(*VI.9.1.206)640
  viticola(*V.3.3.2.181d)518
Plasogamy(*V.3.3.2.181b,*f,.323)518,
    519,525
Plastic(IV.5.1.010)396
  bag(*V.1.1.048;*VI.4.000b,*c,*.001)
    424,585
  containers(*V.1.2.003f)428
  film(*V.1.1.053c;*.1.2.017,*.1.3.
    067;*.3.1.007;*.3.3.2.059;*VI.3.
    053a;*.4.000b)425,432,437,485,
    535,581,585
    bag(*V.1.3.081)439
  flat(*VI.4.011)586
  flats(*VI.4.000c,*.003)585
  greenhouse(*V.1.1.017)419
  hanging basket(*VI.4.013)587
  labels(*V.1.2.009)429
  lug(*VI.4.014)587
  mesh(*V.1.1.059)426
    bags(*VI.3.053a;*.4.001,*.002)581,
      585
    house(*V.1.1.059)426
  -molded cones(*V.1.2.003b)428
  mulch film(*V.3.3.3.074)536
  pots(*V.1.2.003h)428
  resistance(IV.5.2.012)397
  strain(*IV.5.1.013;*.5.3.047;*.5.4.
    2.007)397,411,413
  tape(*V.1.2.017)432
  trays(*V.1.2.003c;*VI.4.008)428,586
Plasticity(IV.5.1.011;.5.2.015;*V.2.
    100)397,399,476
Plastid(*IV.1.1.002)350
Plastochron(II.0.018;.04)10,15
Plastogamy(IV.4.1.022;V.3.3.2.323)381,
    525

Plastome(III.2.2.135)324
Plastomere(IV.4.1.023)381
Plate evaporator(VI.10.010)675
Platy(V.2.115h)480
Platyptilia carduidactyla(*VI.9.2.
    037)661
Pleach(V.3.4.058,.12d)543,556
Pleated(*II.1.1.2.019)21
Pleiotrophy(III.2.2.136)324
Plemodomus destruens(*VI.9.1.226)641
Pleomorphism, pleomorphic(IV.4.4.044;
    V.3.3.2.325,*.406)390,526,528
  fungus(*V.3.3.2.406)528
Pleonanthic(II.5.3.016;.5.02b,c)268,
    282
Pleospora rot of tomato (Pleospora ly-
    copersici)(VI.9.1.400)649
Plerome[*II.3.080,.14a(4)]197,220
Plesionecrosis(V.3.3.2.326,*.473)526,
    530
Plesionecrotic sympton(*V.3.3.2.018,
    *.468)510,530
Plicate(*II.1.1.2.019)21
Plietesial(II.5.3.017;.5.02d)268,282
Plinthite(V.2.076)474
Pliofilm(*V.1.3.067)437
Ploidy levels(*IV.4.6.001a-1,.001b-1)
    392
Plow:
  pan(*V.2.077)474
  sole(V.2.077)474
Plowing(*V.3.1.000;*VI.2.022a)484,576
Plows(V.1.1.044)423
Plucking(*VI.2.022a)576
Plug-mix(V.1.3.047)435
Plum curculio(*VI.9.2.013)660
  injury on apple(VI.9.2.063)662
Plumbeus(*II.5.1.003i)264
Plumose(II.5.6.096;.5.17i)276,297
  (stigma)[II.2.2.18j,.24b(3)(i)]147,
    153
Plumule[II.1.3.02c(1);.2.3.3.014h,.01a
    (2),b(6),.03f-i(11)]97,164,182,184
Pneumorhizae(II.4.3.039;.4.10c,.11;
    *IV.5.4.3.006)243,258,259,414
Pneumatothode(II.4.3.038,*.039;.4.10c)
    243,258
Pneumatic:
  power shaker(*VI.2.022c)576
  separator(*V.1.3.057)435
Pod(II.2.2.28a;.2.3.1.027;*.2.3.2.045)
    157,160,163
  blight of lima beans(VI.9.1.401)649
  -distorting mosaic of snap beans(VI.
    9.1.402)649
  mottle of snap beans(VI.9.1.403)649
  rot of cacao(*V.3.3.2.181d)518

INDEX OF TERMS

spot of garden pea(VI.9.1.405)650
"Pod"(II.2.3.1.05h)172
"Poda" system(*V.3.4.017j,.09b)540,553
Podlegid(IV.4.5.002o)392
Podosphaeria leucotricha(*V.3.3.2.
   181e;*VI.9.1.408)519,650
Poikilohydric plants(IV.5.2.040)403
Poikilotherm(IV.5.3.038)410
Point:
   mutation(*III.2.2.115)323
   of final sale(*VI.8.084)622
   of purchase/point of sale material
      (VI.8.084)622
Poise (P)(*IV.0.260l)347
Poison baits(*V.3.3.3.071)536
Polar:
   groups(IV.5.3.053)411
   nuclei[II.2.2.041t-4-b-3,.16b(11);
      IV.4.1.006a-2,.006b-2,*.014,
      .01(7),.02a(8)]124,145,378,379,
      381,382,383
      fertilized[(IV.4.1.01(11),.02a(7),
         b(10)]382,383
Pole:
   saw(*V.3.4.060)543
   shakers(*VI.2.022b)576
   shears(V.3.4.061)543
Polished(*II.5.6.070)275
Polishing(*VI.7.1.025)601
Pollard(V.3.4.062,.12c)543,556
Pollen[IV.4.1.02b(8);*.4.2.002;*.4.4.
   000,*.003,*.046,*.053;*.4.5.001,
   *.002g,*j,*m,*o,*q;*.4.8.030,
   .032,.040,*.045,*.048,*.057,
   *.058;*V.3.3.0.004]383,385,387,
   390,391,392,394,395,497
   -bearing anthers(*IV.4.8.024)394
   dehiscence(*IV.4.3.002,*.002d)385,
      386
   grain(*IV.4.8.056)395
   grains[II.2.2.007t,.12d,.13a,c,d;
      IV.4.1.014,.02b(8);.5.2.040;*V.
      1.5.000]120,141,142,381,383,403,
      470
   nucleus, nuclei[IV.4.1.01(10);*.4.8.
      022]382,394
   tube(*II.3.119;*IV.4.4.000,*.010,
      .046,.048]200,387,388,390
   tube growth(*IV.8.021,*.030)394
   tubes(*IV.4.8.011)393
Pollenizer(*IV.4.2.001b;.4.4.045;.4.8.
   014,*.015,*.018,*.019)385,390,393,
   394
Pollination(*IV.4.1.006a-10,*-11,
   .006b-1,-3,*-4,*-5,*.007,.026;
   .4.2.001,.002,*.003,*.004,*.005;
   .4.4.000,.001,*.006,*.007,*.011,
   .018,.019,*.020,*.029,*.035,
   .040,.056;*.4.5.002d;*.4.8.017,
   .021,.030,*.032,*.033,*.034,
   .035.036,*.037,*.045,*.046,
   .047,.048,*.049,*.050,*.058;*V.
   4.1.106)379,380,381,385,387,388,
   389,391,393,394,395,564
   anthers(*IV.4.4.027)389
   artificial(*IV.4.4.029)389
   cross-(IV.4.8.017,*.058)393,395
   hand(*IV.4.4.018)389
   intraovarian(*IV.4.4.029)389
   mess-and-soil(*IV.4.4.004b-2;.4.5.
      002j)388,391
   unit(*IV.4.4.004b-3)388
Pollinators(*IV.4.4.001,*.004b-5,
   .013,.014,*.017,*.036,*.039;.4.
   5.001,.001d-v,.002k;*V.3.3.1.017,
   .026,.132e)387,388,389,390,391,
   392,498,502
   adaptations(*IV.4.5.)391-392
   attractants(*II.3.121;*IV.4.4.003)
      200,387
   fluttering(*IV.4.4.004b-5)388
   (pollination agency)(*IV.4.5.001)391
Pollinium (pl. pollinia)[II.2.2.007t-
   9,.24d(2,3),.25d(2),e,.26d(2),e,f
   (3)]120,153,154,155
Pollutant damage(*V.3.3.2.270)524
Pollutants(*VI.7.2.014)610
Pollution(*VI.7.2.017)612
   resistance(*IV.5.4.4.001)415
Polyacrylonitrile(*V.2.099)476
Polyadelphous(II.2.2.007d-15,.04f)119,
   133
Polyaxial (stem)(*II.1.1.3.058,.067,
   .16a,b)24,25,52
Polycarpic(II.2.2.041v,.28b;.5.3.018;
   .5.02b,c)125,157,268,282
Polycaulous(II.1.1.3.068)25
Polyderm(II.3.140)201
Polyembryonic(*IV.4.2.004)385
Polyembryony(IV.4.6.,.000)392
Polyester(*V.1.3.067)437
   rigid fiberglass(*V.1.1.017d)420
Polyethylene(*V.1.3.067;*VI.4.012)437,
   587
   bags, tubes(*V.1.1.053c,*d;.1.2.
      003g;*.1.3.039;*VI.3.053a;*.4.001,
      .002;.5.047e)425,428,434,581,
      585,592
   cloth(*V.1.1.017b)419
   delivery tubes(*V.1.1.017h)421
   film(*V.1.4.036a;.4.1.007)449,559
   tubes (see bags)
   window(*VI.8.127)625
Polygamodioecious(IV.4.3.015)387

INDEX OF TERMS

Polygamomonoecious(IV.4.3.016)387
Polygamous(IV.4.3.017)387
Polygenes(III.2.2.138)324
Polyhaploid(III.2.2.139)324
Polyhedronlike(*V.2.115a,*f)480
Polyhydroxy aldehydes(*IV.2.2.031)360
Polyhydroxy ketones(*IV.2.2.031)360
Polymer(s), polymeric, polymerize(IV.
  0.227)342
  linear organic(*V.2.099)476
Polymorphic(II.0.020,.05)10,16
  spores(*V.3.3.2.181f)519
Polymorphism(*V.3.3.2.325)526
Polypeptides(*IV.5.2.017)400
Polypetalous(II.2.2.061c-11,.20a,b,
  .21i,.22a,b,e)127,149,150,151
Polyphenols(*VI.10.013)676
Polyphilic(IV.4.4.047)390
Polyploid(III.2.2.140)324
Polyplotype(III.2.2.141)324
Polypropylene(*V.1.1.059)426
  cloth(*V.1.1.017g)421
Polysaccharides(*IV.1.2.036c,*.040a;
  .2.2.031,c,.032a,*f,*.114;*.5.3.
  053)354,355,360,361,362,366,411
Polysepalous[II.2.2.061a-3,.14b(2)]
  126,143
Polysomaty(III.2.2.142)324
Polytelic(*II.2.1.070)103
Polytropic(IV.4.5.002p)392
Polyvinyl(*V.1.3.067)437
  alcohols(*V.2.099)476
  fluoride(*V.1.1.017d)420
Pome(II.2.3.1.028,.05d)160,172
  fruits(I.1.009)4
Pomology(I.1.010)4
Pond(*IV.5.4.3.006)414
Pool(VI.8.085,*.099)622,624
Pooling(VI.8.086)622
Poor(*VI.8.094)623
  (condition)(*VI.8.016)617
POP(*VI.8.084)622
Popcorn(*I.3.011)8
Pop-up sprinkler(*V.3.5.012)557
Population levels, intolerable(*V.3.3.
  3.053)535
Pore[II.3.15a(1)]221
Poricidal(II.2.2.007h-6,.10d,e;.2.3.1.
  006e,.03a)119,139,158,170
Porogamy(IV.4.4.048)390
Porphyrins(*IV.1.2.036d,*e;.2.2.168)
  355,368
Portable:
  perforated pipe(*V.3.5.012)557
  precoolers(VI.5.035a-5)590
POS(*VI.8.084)622
Positions(*II.2.2.034)122

(anthers)(*II.2.2.007)120
of ovary (relative to other floral
  parts)(II.2.2.041t-3,.19)123,148
of ovules inside an ovary(II.2.2.
  041t-4-k)124
of perianth relative to ovary(II.2.
  2.061d,.19)128,148
Postclimacteric(*VI.1.002)569
Postdormancy(IV.3.1.011a)371
Postemergence(V.3.3.3.078)536
Posterior(II.2.2.064)129
Postharvest disorders(V.3.3.0.005;*.3.
  3.2.270;*VI.9.3.000)497,524,663
Postpackingline(*VI.3.052)581
Postspraying or dusting period(*V.3.3.
  3.071)536
Postthawing:
  conditions, standard(*IV.5.3.002)407
  treatment(*IV.5.3.023)409
Pot:
  herbs or greens(I.3.016)8
  and field experiments(*V.3.2.021)
    491
  half-sized(*V.4.1.018)560
  or field experiments(*V.3.2.040)492
  plants(I.0.001;.2.001;*V.1.1.048;
    *.4.1.062)3,6,424,562
Potash (K_2O)(*V.3.2.034,*.055,*.088,
  *.092)491,493,496
  available(V.3.2.005)490
Potassium (K)(*V.3.2.013,*.028,*.046,
  .055,.056,*.080,*.088)491,493,
  495,496
  nitrate(*V.1.2.005,*.013;*.1.3.031;
    *.3.2.088)429,431,434,496
  phosphate(*V.1.3.069)437
  total(*V.3.2.071)495
Potato(*I.3.023;*II.4.3.057,*.061)8,
  244,245
  common scab(*V.3.3.2.001)509
  corky ring spot vector(*V.3.3.1.
    200c)505
  fork(V.1.1.045)424
  fusarium blight(*V.3.3.2.181g)520
  fusarium wilt(*V.3.3.2.181g)520
  ring rot(*V.3.3.2.040)511
  scab(VI.9.1.406)650
  seed piece decay(VI.9.1.407)650
  virus Y(*V.3.3.2.158)516
  wilt(*V.3.3.3.181g)520
Potential:
  difference (volt)(*IV.0.260c)344
  energy(IV.0.229)343
  micronutrients(V.3.2.073)495
Pots(V.1.2.003h;*.3.4.057,*.4.1.068)
  428,543,562
Potted plants(*V.1.2.005)429

INDEX OF TERMS 786

Potting bench(*V.1.1.047)424
Poultry(*V.1.2.011f)430
Pounds of juice(VI.10.029)677
Pounds-solids(VI.10.030)677
Powdery(II.5.6.098;.5.17j)276,297
  mildew(s)(*V.3.3.2.101,*.181e,*.253)
      513,519,523
    of apple(*V.3.3.2.181e)519
    of apple, pear(VI.9.1.408)650
    of cucurbits(*V.3.3.2.133c,*.181e)
      515,519
    of cucurbits, composites(*V.3.3.2.
      181e)519
    of grape(*V.3.3.2.181e)519
    of peaches, nectarines, apricots
      (VI.9.1.409)650
    of snap beans and lima beans(VI.9.
      1.410)650
Power (P, watt)(*IV.0.086,.230,*.260c)
    332,343,344
  -driven truck or tractor-mounted
    drill(*V.3.1.011)486
Pox of sweet potatoes(VI.9.1.500)655
PPFD(*IV.0.157g)338
PPP(*IV.1.2.040g)356
PR(*V.4.1.091)563
Practically no demand(VI.8.027)618
Praemorse(II.5.4.1.016;.5.031)269,283
Pratylenchoides spp.(*V.3.3.1.200f)506
Pratylenchus coffeae(*V.3.3.1.200a)505
Pratylenchus spp.(*V.3.3.1.200f;*.4.2.
    1.014)506,566
Praying mantid(*V.3.3.1.132h)503
Precallus(*V.1.4.028e)444
Precast steel framing(*V.1.1.017c)420
Precautions(*V.3.3.3.071)536
Precipitation(*IV.5.2.013;*V.2.106)
    399,476
  of dissolved salts(*V.3.5.032)558
Precleaners(*V.1.3.057)435
Preclimacteric(*VI.1.002)569
  minimum(*VI.1.002)569
Precocious(IV.4.8.044)395
  flowering(*II.4.4.007a)247
Precocity(*V.1.3.008)432
Precooling(*IV.5.3.014;*V.4.1.088,
    .093;*VI.3.052;*.7.1.025)408,563,
    601
  air(VI.5.035a)590
  formulas(VI.5.036)591
  period(*V.4.1.089)563
Predaceous, predator(s), predatory
  (*V.3.3.1.132a,*b,*c,*e,*.172,.207;
  .3.3.3.012,.048,*.053)501,502,504,
  507,532,534,535
Predormancy(IV.3.1.011b)371
Preemergence(V.3.3.3.079)536

incorporated(V.3.3.3.080)536
Prefixes(II.0.021)10
Prefloration(*II.2.2.032)122
Prefoliation(*II.1.1.2.000)20
Preformation (leaf and stem development[*II.1.1.3.059,.069,.15b(2)]24,
    25,51
Pregerminated seeds(V.1.3.049)435
Pregrading(VI.3.027a-4)579
Prehardening(*IV.5.3.014)408
Preharvest(*VI.9.000)626
  drop(*V.3.4.000a)539
Preheating(V.4.1.089)563
Premature:
  dying(*V.4.2.1.001)565
  ripening(*V.3.4.000a)539
Prepack, prepackaging(VI.8.089)623
Prepackingline(*VI.3.052)581
Preparation of cuttings(V.1.4.022a-2)
    442
Prepared hyacinths(V.4.1.091)563
Prephenic acid(*IV.1.2.042)356
Preplanting:
  application(*V.3.3.3.081)536
  fertilizer application(*V.3.1.000)
    484
  pesticide application(*V.3.1.000)484
Prepotency(III.2.2.143)324
Pre-raphe (seed-form)(II.2.3.3.032,
    .04b,.05a,b)165,185,186
Pre-rooting treatments(V.1.4.022a-3)
    442
Present in numbers(*V.3.3.2.225)522
Preshaded fiberglass(*V.1.1.017g)421
President(*VI.8.074)621
Presizing(VI.3.017a-3)579
Presowing treatments(V.1.3.050)435
Pressed:
  fiberboard(*VI.4.007,*.010)586
  paper(*VI.4.019)587
Pressure (P)(IV.0.232,*.260c)343,344
  bruise(VI.9.3.101)670
  chamber(IV.5.2.041)403
  equilibration(*IV.5.2.041)403
  potential(*IV.5.2.057)404
  sealer(VI.3.058)582
  stresses(IV.5.4.3.008)415
"Pressure" tester(*VI.1.003)569
Price:
  arrival(VI.7.1.044)603
  marking(*VI.8.089)623
Prickle, prickles(II.1.1.3.070,.16d;
    .4.3.002,.027,*.042,*.044,*.045)25,
    52,241,242,244
Prickly(*II.5.6.002)273
Prill(V.3.2.074)495
Prills(*V.3.2.070)495

Primary:
  and secondary minerals(*V.2.000)472
  cortex[II.3.01B(9)]207
  cycle(V.3.3.2.337)526
  direct drought injury(IV.5.2.022,
    .042)400,403
    drought tolerance(*IV.5.2.020)400
    freezing injury(*IV.5.3.023)409
    injury(*IV.5.4.3.004,*.013)413
  gene pool(III.2.1.002a)317
  indirect drought injury(*IV.5.2.022,
    .044)400,403
    freezing injury(*IV.5.3.023)409
    injury(*IV.5.4.2.006,*.007,*.013)
    413
  phloem[II.3.02d(8),e(11)]208
    fiber formation(II.3.01B-4)207
    fibers[II.3.01B17,C(22)]207
    (proto) phloem[II.3.01B(13)]207
    (root)[II.1.3.012,.02a,c(4),.04a(2)]
    94,97,99
  salt injury(*IV.5.4.3.009)415
  symptom(V.3.3.2.338)526
  tissue and root hair zone(II.3.02d)
    208
    formation (end of)(II.3.01B-5)207
    tracheary element thickening(*II.3.
    202)204
  vascular system(II.3.142)201
    tissues(*II.3.173)202
  xylem[*II.3.143,.02e(15)]201,208
Primocane[*II.1.1.3.061,.071,.11a(1),
  .12a]25,47,48
Primordium(II.1.1.1.018)19
Prismatic(*V.2.115c,i)480
Private companies(*VI.8.118)625
Proboscoid(II.2.2.041w,.28d)125,157
Procambial strand formation(II.3.01B-
  1)207
Procambium[*II.3.108,.144,.01B(11),
  .02c(5)]199,201,207,208
  (strand)[II.3.01B(7)]207
Processed tankage(*V.3.2.065,*.086)
  494,496
Processing(*V.3.0.001;VI.10.000)484,
  674
  plant(*VI.4.000a;*.8.085,*.102)585,
  622,624
  plant managers(*VI.8.037)619
Proctotrupids(*V.3.3.1.132e)502
Procumbent[II.1.1.3.072,.16c;*.3.152,
  .07f(1)]25,52,201,213
Procurement(*VI.8.106)624
  and routing of shipments(*VI.8.122
  625
Produce(*VI.3.046;*.6.022,*.036)581,
  595,597

  conveyed to buyer(*VI.6.012)594
  department(*VI.8.123)625
  items(*VI.8.109,*.120)624,625
  quality(*VI.8.038)619
  shipped(*VI.6.012)594
Produced(*II.5.2.011)266
  (connective)(II.2.2.08f)137
Product:
  coding system(*VI.8.131)625
  cycles(*VI.8.129)625
  identification(*VI.8.131)625
  loss(*VI.8.108)624
  mix(*VI.8.064,.091)621,623
  related(*VI.6.034)596
Production:
  disorders(V.3.3.0.003)497
  manager(*VI.8.037,.090)619,623
  of new cells, tissues, or organs(*V.
    3.3.2.341)526
  period(*V.4.1.085)563
  phase(V.4.1.092)563
  pruning(*V.3.4.000b)539
  -related(*VI.9.3.000)663
Productive cropland(*V.2.110)478
  life(*V.1.4.015)441
Proembryo(IV.4.1.024)381
Profichi(*IV.4.4.007,.049)388,390
Programming phase (bulbs)(V.4.1.093)
  563
  of forcing(*V.4.1.114)564
Prokaryote, prokaryotic organisms(*V.
  3.3.2.133a,.339)514,526
Prolate(II.2.3.2.030,.07b;*VI.3.064d)
  162,181,583
Prolegs(*V.3.3.1.132a,*e,*g)501,502
Prolepsis, proleptic(II.1.1.3.073,
  .17a;V.3.3.2.340)25,53,526
  development(*V.3.3.2.469)530
Proliferation(V.3.3.2.341)526
Proline(*IV.2.2.016,*.5.2.044)360,403
Promeristem[II.3.01A(1),.14a(1)]207,
  220
Prominents(*V.3.3.1.132g)502
Promiscuous(III.2.2.144)324
Prompt death of tissues(*V.3.3.1.213)
  521
  shipment(VI.7.1.045)603
Promycelium(*V.3.3.2.402,*.423)528,529
Prone(*II.1.1.3.074)25
Propagation(*IV.3.1.014h;.1.0.000;
  *.1.1.053)372,417,425
  and transplanting facilities(V.1.1.
    047)424
  bench(*V.1.4.028b)443
  benches(*V.1.2.005)429
  cases(V.1.1.046,*.048)424
  material(*V.4.1.053)561

INDEX OF TERMS

media(V.1.2.011)429
structures(V.1.1.048)424
Propagator(V.1.0.000)417
Propagule((IV.3.1.014h;V.1.0.005)372, 418
Proper development stage(*V.4.1.068) 562
Prophase(III.2.22.145)324
Prophyll[II.1.2.085,.18a;*.2.1.009,01 (10)]62,83,100,104
Proportionator, proportioner(V.1.1. 049)424
Prop roots(II.1.3.01d)96
Pro rata shipment(*VI.7.2.004,*.004b) 609
Prorating of net returns(*VI.8.086)622
Prosphaeroblast(*IV.3.1.014g)372
Prosthetic group(IV.2.2.058f)364
Prostrate(II.1.1.3.074)25
Protandry(IV.4.3.002b)385
Protectant(*II.3.121,.15c,d;*V.3.3.3. 044,.082)200,221,534,537
Protection(V.3.3.3.083)537
of resting buds(II.4.3.041)244
Protective:
chemical sprays or dusts(*V.3.3.3. 083)537
covering for buds(II.4.3.001)241
devices (against predators)(II.4.3. 042;.4.09)244,257
fungicide(*V.3.3.3.044)534
services(VI.6.029)596
Protein, proteins(*II.3.125;*IV.0.051, *.311;.1.1.025;*.1.2.036e;*.2.2. 016,*.031,*.031c-a-1,*.058,*.058f, *.168,.172;*.5.2.017;*.5.3.033, *.034,*.053;*.5.4.1.005;*.5.4.2. 013)200,331,349,352,355,360,361, 363,364,368,369,400,410,411,412, 413
breakdown(*IV.5.2.044;*.5.3.005;*.5. 4.1.005)403,407,412
coat(*V.3.3.2.461)530
denaturation(*IV.5.4.1.005)412
hydrophobicity(*IV.5.3.034)410
loss injury(*IV.5.4.005)412
matrix(*V.3.3.3.013e)532
molecules(*IV.5.3.053)411
photoinactivation(*IV.5.4.2.013)413
protoplasmic(*IV.5.2.042)403
synthesis(*IV.2.2.188)369
synthesis arrest(*IV.5.2.044)403
synthesizing enzymes(*V.5.4.1.005) 412
Proterandry(*IV.4.3.002b)385
Protocortex[II.3.01A(3)]207
Protoderm[II.3.01A(2),.02c(3)]207,208

Protogyny (proterogyny)(IV.4.3.002c; *.4.4.007)386,388
Protomeristem[*II.3.108,.02b(2),]199, 208
Protophloem[*II.3.112,.146,.01B(10, 16)]199,201,207
Protopith[II.3.01A(4)]207
Protoplasm(*IV.5.2.031)401
Photoplasmic substances(*IV.5.4.2.001) 413
Protoplast(*II.3.177;IV.1.1.024)203, 352
Protostele(II.3.189g,.27a,b)204,233
Protozoa(*V.3.3.3.013)532
Protozoans(*V.3.3.2.133a)515
Protoxylem[*II.3.143,.147,.01B(15),C (19),.11a(4),.17a(3),.22a,b]201,207, 217,223,228
Provascular tissue(*II.3.108)199
Provenance(III.1.2.053)310
Provided protective services(VI.6. 029b,c)596
Pruinose(II.5.6.100,*.108)276,277
Pruning(*IV.5.4.4.002;*V.3.3.3.000, *.088;*.3.4.000,*.000b;.4.1.094) 416,531,537,539,563
basic types(V.3.4.02a-c)546
equipment(V.3.4.063)543
knife, knives(*IV.5.4.4.002e;*V.3.4. 063,.064)416,543
maintenance(*V.3.4.000b,*.008,*.017, *.073)539,541,543
ornamental shapes(*V.3.4.075)544
perennial(*V.3.4.062)543
production(*V.3.4.073)544
saw, saws(*V.3.4.063,.065)543
shears(*V.1.4.022a,*.3.4.048,*.063, .066)442,542,543
systems(V.3.4.067,.03-.12)543,547- 556
terminology(V.3.4.01)545
tolerance(*V.2.110e-2)479
tools(*V.1.1.036)423
training systems(*V.3.4.004,*.008, *.017,*.019,*.020,*.021,*.029, *.031,*.032,*.049,*.055,*.056, *.062,*.068,*.072,*.073,*.075)540, 541,542,543,544
Prussian blue(II.5.1.002h)263
Pseudanthium(*IV.4.4.020)389
Pseudoaril[II.2.3.3.12a(1),b(1)]193
Pseudoarillate(II.2.3.3.033)165
seeds(II.2.3.3.12a,b)193
Pseudobulb, pseudobulbs(*II.4.2.006; *V.1.4.024,.058,.05f)239,443, 451,457
natural vegetative reproduction(IV.

INDEX OF TERMS

4.1.006a-8)379
Pseudogamy:
  (Khokhlov)[IV.4.1.006b-4,.03(9)]379, 384
  (Sherman)[*IV.4.1.006a-4,*-7,-11, .02a(5,7)]378,379,383
Pseudohardening(IV.5.2.048;.5.3.041) 403,411
Pseudohermaphrodite(IV.4.3.018)387
Pseudohomotypic division[IV.4.1.006b-6-a-3,.03(5)]380,384
Pseudomonas(*V.3.3.2.040;*.4.2.1.021) 511,566
  ananas(*VI.9.1.127)635
  cepacia(*VI.9.1.511)656
  cichorii(*VI.9.1.061,*.076)631,632
  marginalis pv. marginalis(*VI.9.1.061)631
    pv. pastinacae(*VI.9.1.051)630
  solanacearum(*V.3.3.2.040)511
  syringae pv. apii(*VI.9.1.048)629
    pv. lachrymans(*VI.9.1.069,*.070)631
    pv. maculicola(*VI.9.1.054)630
    pv. phaseolica(*VI.9.1.050)630
    pv. pisi(*VI.9.1.049)629
    pv. savastonoi(*V.3.3.2.040)511
    pv. syringae(*VI.9.1.071,*.085) 631,632
    pv. tomato(*VI.9.1.068)631
  virilivida(*VI.9.1.061)631
Pseudonanism(*II.4.4.002h,.010)247,248
Pseudoparthenogenesis(*IV.4.1.006a-10,.006b-3-b)379
Pseudoterminal (buds)(II.1.1.1.019, .03m)20,30
Pseudotrophic(*V.3.3.2.264)523,
Psychophily(*IV.4.5.001c)391
Psychrometric chart(*VI.5.025c,.037) 589,591
Psychrophile(*IV.5.3.007,.042)408,411
Psyllid(*V.3.3.1.132d)502
Pteridophyta, pteridophyta(*II.6.003, *.010,*.016;*V.3.3.2.179a)279,517
Pteridophytes(II.6.016;*IV.1.2.008a, *.029)279,352,354
Ptyxis(II.1.1.2.022)21
Puberulous, puberulent(II.5.6.101;.5.17k)276,297
Pubescent(II.5.6.102)276
Puccina antirrhini(*V.3.3.2.181f)519
  aristidae(VI.9.1.330)646
  graminis(*V.3.3.2.181f)519
Puffiness:
  of citrus(VI.9.3.102)670
  of tomato(VI.9.3.103)670
Pull date(VI.8.093)623

Pulling(*VI.2.022a)576
Pulpa(*II.2.3.2.029c,.05e,f)162,179
Pulper(VI.10.031)677
Pulse action metering device(*V.1.3.046)435
Pulses(*I.3.014)8
Pulverulent(*II.5.6.098)276
Pulvinate(II.1.2.086,.18b,c)62,83
Pulvinus (pl. pulvini)[II.1.2.087, .18b,c;.3.108,.21b(2)]62,83,199,227
Pumice(*V.1.2.011i;*.2.096d)430,475
Punch(VI.10.032)677
Punctate(II.5.6.103;.5.18a)276,298
Punctatus(*II.5.1.006e)265
Pungent(II.2.3.2.032;.4.3.044;.4.09a) 162,244,257
Puniceus(*II.5.1.005c)264
Pupa(e)(*V.3.3.1.132a,*g,.211)501,502, 507
Pupation(*V.3.3.1.132g)502
Purchase after inspection(VI.7.1.046) 603
Pure:
  culture(*V.3.3.2.036,*.231)511,522
  Food Laws of 1906(*VI.7.2.016)610
  line(III.2.2.146)324
  seed percentage(*V.1.3.075)438
Puree(VI.10.033)677
Purine bases(*IV.2.2.140,.173)367,369
Purity(III.2.0.004o)313
  (seeds)(*V.1.3.066)437
  test(V.1.3.051)435
Purple(*II.5.1.005,.005o)264
  blotch of onions(VI.9.1.426)650
  (purpureus)(II.5.1.005o)264
  scale(*VI.9.2.007)660
Pustular spot(*VI.9.1.176)638
Pustule(V.3.3.2.346)526
Putative(III.2.2.147)324
Putnam scale on apple(VI.9.2.065)662
PVC pipe(*V.1.1.017c;*.3.5.023)420,558
PWP(*IV.5.2.039)403
Pycnidium, pycnidia(*V.3.3.2.181g, .347;*VI.9.1.086,*.087,*.092,*.115, *.296,*.297,*.355,*.356,*.401,*.461, *.465,*.535)519,526,632,633,635,645, 647,649,652,653,657
Pycniospore(s)(*V.3.3.2.181f,*.349) 519,526
Pycnium (pl. pycnia)(V.3.3.2.349)526
Pyralids(*V.3.3.1.132g)502
Pyramid(V.4.3.068,.11)544,555
  dwarf(*V.4.3.068)544
  modified dwarf(*V.4.3.072)544
  racks(*V.1.1.017e)420
  winged(*V.4.3.049)542
Pyramidal(II.2.3.2.008,.02c)162,176

Pyranoses(*IV.2.2.031a,*.095)361,366
Pyrausta nubilalis(*VI.9.2.026)660
Pyrene(II.2.3.2.034,.02f)162,176
Pyrenochaeta lycopersici(*III.2.2.159;
  *VI.9.1.576)325,659
Pyridoxine(*V.1.5.005)470
Pyriform(*II.2.3.2.019,.036,.06b,c)
  161,162,180
Pyrimidine bases(*IV.2.2.140,*.141,
  .175)367,369
Pyrometer(*IV.0.295b)349
Pyromorphosis(II.4.4.002g;.4.14)246,
  262
Pyrophyllite(*V.2.098;*.3.3.3.029)475,
  533
Pyrroles(*IV.2.2.168)368
Pyruvate(*IV.1.2.040b,*e)356
Pyruvic acid(*IV.1.2.040a)355
Pythium aphanidermatum(*VI.9.1.177)638
  butleri(*VI.9.1.178)638
  debaryanum(*V.3.3.2.181d)518
  scleroteichum(*VI.9.1.317)646
  sp.(*VI.9.1.431)651
  spp.(*V.1.1.065;*.3.3.2.278;*VI.9.1.
    178,*.432)426,524,638,651
  spp. on bulbs(V.4.2.016,*.018)566
  ultimum(*VI.9.1.317)646
Pythium rot:
  of tomato(VI.9.1.431)651
  of watermelon(VI.9.1.432)651
Pyxis (pyxidium)(*II.2.3.1.025,.029,
  .01j,.03d,e)160,168,170

Q

Q_I(IV.3.2.007)374
Q_{10}^I(temperature coefficient)(IV.0.233)
  343
Quailing(*VI.3.023)579
Qualitative:
  character(III.2.2.148)324
  response(IV.3.2.013j)375
Quality(*VI.7.2.004;.8.094)609,623
Quantitative:
  character(III.2.2.149)324
  long-day plant(V.4.1.095)563
  response(IV.3.2.013k)375
Quantity:
  Agricultural Marketing Agreement of
    1937, as Amended(*VI.7.2.004)609
  of electricity (coulomb)(*IV.0.260c)
    344
  of heat (joule)(*IV.0.260c)344
Quantum (quanta)(*IV.0.023,*.157b-2-b,
  *-c,*m,*aa)330,337,338,340
Quarantine(*V.3.3.3.000,*.038,.084)
  531,533,537

Quartz(*V.2.076)474
Quartzic sands(*V.2.107d)477
Quartzipsamments(V.2.107d)477
Queen(*V.3.3.1.132f)502
Quick:
  freezing(*V.3.3.2.245)523
  shipment(VI.7.1.047)603
Quiescence(IV.3.1.021)373
Quinate, quinquefoliolate(II.1.2.088,
  .18d)63,83
Quince rust on apple(VI.9.1.437)651
Quincunx(V.3.1.008i,.01k)486,489
Quinine(*IV.2.2.012)360
Quinone(*IV.1.2.040c)356
Quonset design (greenhouse)(*V.1.1.
  017b)419

R

Rabbits(*V.1.4.028c;*.3.3.1.224)444,
  507
Race(III.2.0.009;V.3.3.2.351,*.412)
  316,526,529
Raceme[II.2.1.058,.02a(2),.08f]102,
  105,111
Racemose, racemiform(II.2.1.059,.08d,
  .12d)102,111,115
Rachilla[II.1.2.089,.06b(3)]63,71
Rachis (pl. rachises, rachides)[*II.
  1.2.083,*.087,.090,.06b(2),.17b;*.2.
  1.053,.01(9);*.4.3.071;*IV.1.1.002]
  63,71,82,102,104,245,350
Rad [rad (rd)](*IV.0.260k)347
Radial(*II.3.210,.28e)205,234
Radian(IV.0.157r,*.260f)338,346
  per second(*IV.0.260g)346
  per second squared(*IV.0.260g)346
Radiance (watt per square meter steradian)(IV.0.157s,*.260g)338,346
Radiant:
  energy(IV.0.157t,*.248,*.283;.5.4.
    1.004)338,343,348,412
    (flat surface)(IV.0.157x-1-d)339
    (point)(IV.0.157y-1-d)339
  flux (Q)(*IV.0.086,.157u,*.230,
    *.260c)
    density (rfd)(IV.0.157v)339
  intensity (watt per steradian)(IV.0.
    157w,*.260g)339,346
Radiate, radiant[II.2.1.060,.09a(2)]
  102,112
Radiation(*IV.3.2.006,*.007;VI.5.052c)
  374,593
  arriving at a flat surface(IV.0.
    157x)339
  arriving at a point(IV.0.157y)339
  -induced stress(IV.5.4.2.011)413

INDEX OF TERMS 791

net balance(IV.3.2.007)374
strain avoidance(*IV.5.4.2.001)413
strain repair(*IV.5.4.2.001)413
strain tolerance(*IV.5.4.2.001)413
stress(*IV.5.4.2.000)413
terms(IV.0.157c-y)338-339
Radical(II.1.3.013)94
Radicle[II.1.3.02c(4);.2.3.3.014;.01a
 (5),b(8),.03f-i(10)]97,164,182,184
Radiofrequency field(*IV.0.195)341
Radioisotope(*IV.0.142,*.157e,.248)
 335,338,343
 source, shielded(*V.3.5.024)558
Radiometric(*IV.0.157b-2-a,*-c,e)337
Radionuclide(*IV.0.142,*.248,*.260d)
 335,343,345
Radopholus similis. R. spp.(*V.3.3.1.
 200f)506
Raffia(*V.1.2.017)432
Raffinose(*IV.2.2.031b)361
Rail:
 car(*VI.3.019;.6.000)578,594
  bulk shipment(*VI.6.003)594
  carlot, carlot equivalent(VI.8.
   010)617
  cooling(VI.5.035a-6)590
  intermodal(VI.6.011)594
  load(*VI.8.010)617
  provided protective services(VI.6.
   029a,c)596
  shipments(*VI.8.106)624
  van modular containers(*VI.6.037)
   597
 cars(*VI.6.017)595
 shipments(*VI.6.028,*.029a,c;*.7.1.
  048)596,604
 trailer on flat-car shipments(*VI.6.
  029c,.033)596
Railroad freight(*VI.6.038)597
 agent(*VI.8.106)624
Rain, rainfall(*II.4.2.004;*V.2.038,
 .110a;.3.1.014;.3.3.3.049)238,474,
 478,487,534
 duration(*V.2.110b)478
 frequency(*V.2.110b)478
 intensity(*V.2.110b)478
 seasonal distribution(*V.2.110b)478
 splashing(*V.3.3.2.006)510
Raised:
 bench stock, solid botton(*V.1.2.
  005)429
  wire bottom(*V.1.2.005)429
 spray boom(*V.3.3.3.067)535
 stomata(II.3.30g)236
Rake(V.1.1.050)424
Rameous(II.1.1.3.075)25
Ramet(III.2.0.010)316

Ramose(II.1.1.3.076,.17b)25,53
Ramentaceous(II.5.6.104;.5.18b)276,298
Ramiflory[*II.2.1.015,.05d(3)]100,108
Random(III.2.2.150)324
Range(V.1.1.051;.3.3.2.352)424,526
Rank(*II.1.1.1.022)20
Raphe(II.2.3.3.034,*.042,.04e)165,166,
 185
 -antiraphe[II.2.3.3.08a(3)]189
 -aril[II.2.3.3.08b(1)]189
Raphide(*II.3.037,.08d)195,214
Raspberry ringspot(*V.4.2.1.022)566
Rasping-sucking mouthparts(*V.3.3.1.
 132i)503
Rate(V.3.3.3.085)537
 of cooling(*IV.5.3.001)407
Ratio:
 among essential elements(*V.3.2.
  035)492
 of other elements to nitrogen(*V.3.
  2.037)492
Ratoon[V.1.4.04c(7)]456
Rats(*V.3.3.1.224;*.3.3.3.087)507,537
Rattan(II.0.022,.03a;*.4.3.014,*.022,
 *.027)11,14,241,242
Ravine wall protection(*V.2.110e)479
Ravines(*V.2.110d)478
Raw:
 agricultural commodity(VI.7.2.016g)
  612
 bone meal(*V.3.2.065)494
Ray:
 cells(II.3.152,.07c-f)201,213
 flower[II.2.1.09a(2);.2.2.23i(1)]
  112,152
 initial(*II.3.026,*.176)195,203
Rays[*II.3.135,.152,.154,*.175,*.219,
 *.221,.20c,d(6)]200,201,203,206,226
Razor blade(*V.1.4.028p)448
Reach and frequency(VI.8.095)623
Reaction wood(II.3.155)201
Reactive valences(*V.3.2.012)490
Readily available water(IV.5.2.049)403
Reasonable time(VI.7.1.048)604
Receiver(VI.8.096)623
Receiving market(*VI.8.015)617
 commission merchant(VI.7.1.049)604
 commission merchant's duties(VI.7.
  1.050)604
Receptacle(II.2.2.066)129
Recessive(III.2.2.151)324
Reciprocal ohm(s)(*IV.0.073;*V.4.1.
 071)332,562
Reciprocating:
 hedgers(*V.3.4.063)543
 toppers(*V.3.4.063)543
Recleaning(*V.1.3.063)437

INDEX OF TERMS

Reclinate, reclining(II.1.1.2.023;.1.1.3.078;.1.1.2.02q;.1.1.3.18a)21,25,35,54
Recognition pollen(IV.4.4.051)390
Recombination(III.2.2.152)324
Recommended protective services(VI.6.029a)596
Rectangular:
  container(*VI.4.003)585
  planting pattern(*V.3.1.008c,*g,*i,*j,*m,.01a)485,486,488
Recurrent agamospermy(*IV.4.1.006a-4,*.006a-14,.02a)378,379,383
Recurrent parent(III.2.2.153)324
Recurved(II.4.3.045;.4.09b)244,257
Red(II.5.1.005,.005p)264
  -absorbing(*IV.3.2.017)375
  algae(*IV.1.2.036d)355
  -banded leaf roller(*VI.9.2.044)661
  ear rot of corn(*VI.9.1.258)643
  -humped caterpillar(*V.3.3.1.132g)502
  light(*IV.3.2.017)375
  ring(*V.3.3.1.200a)505
  root rots(*V.3.3.2.181f)519
Redifferentiation(*II.3.108)199
Reduce acidity(*V.3.4.000a)539
Reduced:
  form(IV.0.250)344
  forms of ions(*IV.5.4.3.006)414
  photosynthetic rates(*V.3.3.2.133d)515
  water supplies(*IV.5.2.001)398
Reducing:
  conditions(*V.2.037)473
  sugars(*IV.2.2.031a)361
Reduction(IV.0.250;*V.2.037,*.106)344,473,476
Reductive:
  division or heterotypic division(*III.2.2.154)325
  pentose phosphate pathway(*IV.1.2.036c)355
Reduplicate(II.1.1.2.024,.02r,.03t)21,35,36
Redwood(*V.1.1.017c;*.1.2.011l,.014)420,430,431
Reed(*II.1.1.3.008)22
Reefer(VI.8.097)623
Reel-type mowers(*V.1.1.034)422
Reflexed(II.5.2.030)267
Reforestation(*V.2.110e)479
Refractometer(VI.1.008a,*c,d,*e,*.010)570,571
Refrigerated:
  marine container van(*VI.8.097)623
  rail car(*VI.6.016;*.8.097)595,623
  truck trailer(*VI.8.097)623
Refrigerating(*VI.7.1.025)601
Refrigeration(*VI.5.1.000,.005,*.017,*.051a,*.053;*.8.081)588,593,622
  facilities(*VI.6.017)595
  requirements(*VI.6.018)595
Refrigerator(*V.1.1.047)424
  case(*VI.8.068)621
Registered seed(III.2.0.004e;*V.3.3.3.000)312,531
Registration program(*V.1.4.010)441
Regolith(V.2.107m)477
Regular:
  advance(*V.7.1.002)597
  display case(*VI.8.111)625
  hyacinths(V.4.1.097)563
  pattern between crop rows(*V.2.110e)479
  superphosphate(*V.3.2.091)496
Regulated:
  carbon dioxide (CO)(*VI.5.047a,*b,*e)592
  oxygen (O)(*VI.047b,*d,*e)592
  temperature(*VI.5.047a)592
Regulation(s)(*VI.7.2.014)610
  marketing(*VI.7.2.004c)609
Regulators(*V.1.3.060)436
Rehabilitation(*V.3.3.3.000)531
Reicing(*VI.6.029c)596
Reiteration (reiteration branching)(II.1.1.3.079,.18b)25,54
Rejection without reasonable cause(VI.7.052)604
Relative:
  concentration of ions or molecules(*V.3.2.065)494
  drought index(IV.5.2.050)403
  humidity(V.4.1.053;VI.5.025c)561,589
    desired(*VI.6.029a)596
    percent(*VI.5.025c,*.037)589,591
    storage(*VI.5.047)591
  vapor pressure(*IV.5.2.038)402
  water content(*IV.5.2.073,*.078)405
Rem(*IV.0.260k)347
Remote(*II.2.2.030)122
Removal of seed, pits, etc.(*VI.7.1.025)601
Renewal spur(*V.3.4.069)544
Reniform(II.5.5.048;.5.111)273,291
Renovation(V.3.1.009)486
Repair(*IV.5.3.034)410
  graftage(*V.1.4.028,*.028o,.12)443,448,464
  mechanism(*IV.5.4.2.008)413
Repand(II.5.4.3.031;.5.07a)270,287
Reparation order(VI.7.1.053)604

Repent(II.1.1.3.080,.18c)26,54
Replacements(*V.3.3.3.000)531
Replicate, replicative(II.1.1.2.025,
 .03u)21,35
Replum(II.2.3.1.030,.03g)160,170
Representative sample(*V.3.2.072,
 *.082)495
Reproductive unit(*V.3.3.2.401)528
Reptiles(*IV.4.7.001i)393
Research:
 recommendations(*V.3.2.083)495
 solution cultures(*V.3.2.083)495
Reseeding(*V.3.1.009)486
Reserve nutrients(*V.2.099)476
 source of nutrients(*V.2.111)479
Reservoir(*V.3.5.023)558
 private(*V.3.5.034)559
  use(*V.3.5.034)559
Residual:
 action(V.3.3.3.086)537
 effectiveness(*V.3.3.3.071)536
 meristem[II.3.01A(5)]207
 sprays(*V.3.3.3.097)537
Resin(s)(*II.3.077;*IV.1.1.030)197,352
 canals(II.3.16c)222
 duct(II.3.157,*.179,*.208)201,203,
  205
 synthetic(*V.3.2.050)493
Resinous(II.5.6.105;.5.18d)276,298
 exudations(*II.4.3.041)244
Resistance(V.3.3.2.354)527
 adaptation(IV.5.1.014;.5.3.043)397,
  411
 groups(IV.5.1.015)397
 principle(*V.1.1.053b)425
Resistant species(*V.3.3.3.049)534
Respiration(*IV.1.2.038,.040,*.040b,
 *c,*d,*f;*.2.2.172;*.5.2.044;*.5.
 4.1.004,*.005;*V.1.2.001)355,356,
 369,403,412,413,428
 cyanide-resistant(IV.2.1.040d)356
 Pasteur effect(IV.1.2.040f)356
 proteins(*IV.2.2.172)369
 rate(*IV.2.1.033)358
Respiratory:
 activity(*VI.1.005)570
 (oxidative pathway)(*IV.1.2.040g)356
 products(*VI.9.3.000)663
 quotient(IV.1.2.041)356
 rise(*VI.1.002)569
 upset(*IV.5.3.005)407
Response(V.3.3.2.355)527
 group(V.4.1.097)563
Responsibly connected(VI.7.1.054)604
Rest(*IV.3.1.010,.011b;*V.3.3.2.337)
 371,526
 period, short(*IV.5.3.015)408

Resting:
 (dormant) period(*V.3.3.2.381)528
 spore(V.3.3.2.356)527
Restorer line(III.2.2.155)325
Restricted ion mobility(*V.3.2.066)494
Restrictions, state, federal(*V.3.5.
 022)558
Resupinate(II.1.2.091;.2.2.068,.20g)
 63,129,149
Retail(VI.8.098)624
 chain(*VI.8.096)623
 chains(*VI.8.107)624
 display(*VI.8.084)622
 firms(*VI.8.052)620
 market(*VI.8.030)618
 outlet(*VI.8.032,*.098)618,624
 price(*VI.8.034,*.098)618,624
 sale(*VI.8.064)621
 selling area(*VI.8.084)622
 store(*VI.8.084,*.129)622,625
 supermarkets(*VI.8.065)621
 warehouse(*VI.8.062)621
  dock(*VI.8.055)620
Retailer(*V.4.1.068;*VI.8.131)562,625
Retailers(*VI.3.023;*.7.1.015)579,599
Retain(*VI.8.086,.099)622,624
Retard growth(*V.3.4.000a)539
Reticulate(*II.3.211;.5.6.106;.5.18c)
 205,276,298
 patterns(*V.2.076)474
 perforation(*II.3.182,.217)203,206
Reticulated(II.1.2.121p,.26p)65,91
Retrograde metamorphosis(*V.4.2.2.001)
 566
Retrorse(II.5.2.031)267
Retuse(II.5.4.1.017;.5.03m)269,283
Reverse roll(*VI.3.038)580
Reversibly plagiotropic(*II.1.1.3.019,
 .081,.18e)22,26,54
Reversion(III.2.2.024d-2-c,.01j)
 319,327
Revolute(II.1.1.2.026,.03v;.1.2.092,
 .14b)21,36,63,79
Rfd(*IV.0.157v)339
RG(*V.4.1.096)563
Rhadinaphelenchus cocophilus(*V.3.3.1.
 200a)505
Rhagoletis pomonella(*VI.9.2.004)659
L-Rhamnose(*IV.2.2.031a,*.038)361,362
Rheogameon(III.2.2.156)325
Rheophyte(*II.4.3.058,.08a;III.1.2.
 054)245,256,310
Rhinoceros beetle(*V.3.3.3.013f)532
Rhipidium(II.2.1.062,.13a)102,116
Rhizobial associations(*II.4.3.067b)
 245
Rhizobium(*V.1.3.012,*.014,*.041,

*.058)432,434,436
spp.(*II.3.163)202
Rhizoctonia:
   carotae(*VI.9.1.180)638
   solani(*V.3.3.2.179a,*.181f,*.357;
     VI.9.098,.181,*.440,*.441,*.443,
     .499,.501,*.502)517,519,527,633,
     639,651,654,655
   sp.(*VI.9.1.171,*.344)638,647
   spp.(*V.1.1.065,.4.2.1.018)426,566
Rhizoctonia:
   ear rot of corn(VI.9.1.440)651
   head and root rots of cabbage, turnips, rutabagas, radishes(VI.9.1.441)651
   pod rot of snap beans, lima beans (*VI.9.1.499)654
   root rot of horse-radish(V.9.1.443)651
Rhizodermis(*II.3.114)199
Rhizoflory[*II.2.1.015,.05d(4)]100,108
Rhizomatous branching(II.1.1.3.082,.18d)26,54
Rhizome, rhizomes[*II.1.1.3.099;*.3.035,*.052;*.4.1.002;*.4.3.062;*.063;.4.12c,d,.14a(3);*IV.5.1.001;*V.1.4.024,.060,.04d,.05a,b]26,195,196,238,245,260,262,396,443,451,456,457
Rhizomorph(V.3.3.2.358)527
Rhizophore(II.6.017)280
Rhizopus:
   arrhizus(*VI.9.1.453,*.455)652
   nigricans(*V.3.3.2.181c;*VI.9.1.457,*.497)518,652,654
   oryzae(*VI.9.1.453)652
   stolonifer(*VI.9.1.449,*.450,*.451,*.452,*.453,*.454,*.455,*.456)651,652
   tritici(*VI.9.1.453,*.456,*.457)652
   spp.(*V.3.3.2.181c)518
Rhizopus:
   rot of apple and pear(VI.9.1.449)651
     of avocado(VI.9.1.450)651
     of stone fruits(VI.9.1.451)651
     of tomato, pepper, eggplant(VI.9.1.452)652
   soft rot of carrot(VI.9.1.453)652
     of crucifers(VI.9.1.454)652
     of cucurbits(VI.9.1.455)652
     of snap beans(VI.9.1.456)652
     of sweet potatoes(VI.9.1.457)652
Rhombate(II.5.11m)291
Rhomboid(II.5.5.049)273
Rhubarb(*I.3.011)8
   anthracnose(VI.9.1.460)652
   stem (leaf) spot(VI.9.1.461)652
Rhynchophorus palmarum(*V.3.3.1.200a)
505
Rhythmic growth(II.1.1.3.084,.14b)26,50
Rhytidome(*II.3.016,.160,.20c,d)194,202,226
Rib discoloration of lettuce(VI.9.3.104)671
Ribbed(II.1.2.121q,.26q)65,91
Riboflavin-containing protein(*IV.1.2.040c)356
Ribonucleic acid(*IV.2.2.188)369
Ribonucleoside(*IV.2.2.141)367
Ribonucleotide(*IV.2.2.142)367
D-Ribose(*IV.2.2.031a,*.140,*.141,*.142)361,367
D-Ribose, 2-deoxy(*IV.2.2.140,*.141,*.142)367
Ribose-5-phosphate(*IV.1.2.040g)356
Ribosomal RNA(*IV.2.2.188)369
Ribosomes(IV.1.1.025;*.2.2.188)352,369
Ribs(*IV.4.4.023)389
D-Ribulose(*IV.2.2.031a)361
Rice weevil(VI.9.2.069)662
Rickettsialike organisms (RLOs)(*V.3.3.0.004;*.3.3.2.133a,*.270,*.339,.361)497,514,524,526,527
Rickettsias(*V.3.3.0.004)497
Ridge:
   and furrow (greenhouses)(*V.1.1.017b)419
   terraces(*V.2.110c)478
   -to-eave glazing(*V.1.1.017b)419
Rills(*V.2.110d)478
Rim-aril[II.2.3.3.11b(5)]192
Rind staining of navel oranges(VI.9.3.105)671
Ring:
   bud(V.1.4.028f-7,.08h)445,460
   budding(*V.1.3.028f,*.028f-1,-7)444,445
   nematodes(*V.3.3.1.200b)505
   porous wood(II.3.161,.10a)202,216
   pox of apricots(*VI.9.1.462)652
   rot of sweet potato(*V.3.3.2.181c;*VI.9.1.317)646
     of tomato(VI.9.1.464)653
   spot(s)(*V.3.3.2.133b,.363)515,527
     of crucifers(VI.9.1.465)653
     of strawberry, Rubus, Ribes, cucurbits, beets, peas, crucifers (*V.3.3.2.181e)519
Ringed(*II.5.6.004)273
Ringent(*II.2.2.061c-14-r,.23b)128,152
Ringing(*IV.5.4.4.002,.002e)416
Rinsing(*VI.3.081)583
Riparian:
   land ownership(*V.3.5.033)559

rights(V.3.5.022,*.033)558,559
Ripe:
 fruit(*V.3.3.1.026)498
  rot(*VI.9.1.127)635
  rot of apple(*V.3.3.2.181e)519
  of pepper(VI.9.1.467)653
Ripening(*VI.1.003,.005,*.006,*.007,
 .013)569,570,572
 grains(*V.3.3.1.026)498
River(*IV.5.4.3.006)414
Rivers(*V.2.037;*VI.7.2.014)473,610
Riverside trees(*II.4.3.026)242
RLO's(*V.3.3.0.004;*.3.3.2.133a,*.270,
 *.339,.361;see also Rickettsialike
 organisms)497,514,524,527
RNA(*IV.1.1.017,*.025;*.2.2.140,*.141,
 .142,.150,.188;*V.3.3.2.088,
 *.461)351,352,367,368,369,513,530
 decrease(*IV.5.2.044)403
Roaches(*V.3.3.1.132h)503
Road:
 heat(*VI.6.034)596
 lines(*V.3.1.000)484
 vehicle(*V.3.019)578
Roadside:
 plantings(*I.2.004)6
 sales(*VI.4.003)585
 stand(*VI.8.125)625
 stands(*VI.2.020;*.3.000;*.4.003)
  575,577,585
Roadsides(*I.2.011)6
Robber flies(*V.3.3.1.132b)501
Rock(*V.2.107j)477
 phosphate(*V.2.099;*.3.2.003,.075,
  *.091)476,490,495,496 (see also
  Phosphate)
Rodenticide(s)(*V.3.3.3.009,*.070,
 .087;*VI.7.2.017,.017t)531,536,537,
 612,615
Rodent resistance(*V.1.3.067)437
Rodents(*V.1.3.050;*.3.3.0.002,*.003,
 .005;.3.3.3.087;*VI.7.2.017g,*h)
 435,497,537,613
 and leptomorphs(*V.3.3.1.224)507
Roentgen (R)(*IV.0.260k)347
Roestelium (pl. roestelia)(*V.3.3.2.
 003,.366)510,527
Rogue, rogueing, roguing(V.1.0.006;.4.
 1.099)418,564
Roller(*VI.3.019;*.7.1.034;.8.100)578,
 602,624
Rolling:
 acceptance(VI.7.1.056)604
 acceptance final(VI.7.1.057)605
 car(*VI.7.1.034)602
Root, roots{II.1.3.000,.01-.04;*.3.
 052,*.056,*.059,*.091,*.127,
 .147,.189,*.199,*.203,*.210,
 .222,.02;.4.3.003,*.008,*.015,
 .038,.039,*.047,*.049,*.063,
 .064,.065;.4.10c;.6.016;*IV.1.
 2.029;.2.1.025;*.3.3.008h,*p;*V.
 1.4.022,*.083;*.1.5.000,*.005;
 .3.3.1.132a;.3.4.01(18,19);.4.
 1.003,*.004;*.4.2.1.014,*.017,
 *.018]94,96-99,196,198,200,201,
 203,204,205,206,208,241,242,243,
 244,245,258,279,354,358,377,441,
 452,470,501,545,559,566
aerial(*II.4.3.002,.004,*.014;*IV.
 3.3.008p)241,242,377
allelochemic(*IV.2.2.014)360
and foot rots(*V.3.3.0.002)497
apical meristems(II.3.19)225
base rot of lilies(V.4.2.1.017)566
beer(*VI.10.003)674
cap[*II.3.025,*.043,.02a(1)]195,196,
 208
contractile(II.3.035;*IV.5.1.001,
 .018;.5.3.038)195,396,397,410
crown(*II.1.1.3.031)23
crops(I.3.019;*II.1.1.3.086)8,26
cutting(V.1.4.022e,.01e,f)442,453
density(*IV.5.2.079)406
depth(*IV.5.2.079)406
discoloration of vascular tissues
 (*V.4.2.1.017)566
graft(*V.1.4.028r)448
graftage(V.1.4.028b)443
hairs[*II.1.3.005,.008,.04a(1);*.3.
 054,.162,.13b,c]94,99,196,202,
 219
initiation(*IV.2.1.025,.033)358
-knee(*II.4.3.039,.047;.4.11d)243,
 244,259
knee-(II.4.3.029)243
-knot nematodes(*V.3.3.1.200a,*.e,*g;
 *.3.3.2.133c)505,506,515
knot on carrot, parsnip(VI.9.2.070)
 662
knot on sweet potatoes(VI.9.2.071)
 662
-lesion nematodes(*V.3.3.1.200f)506
nodule(II.3.163)202
normal emergence(*V.4.2.2.014)567
normal emergence area(*V.4.2.2.014)
 567
number(*V.4.2.2.006)567
peg-(II.4.3.035)243
piece(*V.1.4.028b)443
pieces of(*V.1.1.056)426
pneumatothode(II.4.3.038)243
pneumorhizae(II.4.3.039)243
-pruning(V.1.0.007)418

rot(V.4.2.1.018)566
rot of tobacco, beans, peas, etc.
    (*V.3.3.2.181e)519
rots(*V.3.3.2.181e,*f)519
rotting(*V.4.2.1.017)566
shallow spreading(*IV.5.2.079)406
soil volume(*IV.5.2.036)402
stilt(II.4.3.061)245
stilted-peg(II.4.3.059)245
storage organs(II.4.3.063)245
submersion(*IV.5.4.3.006)414
sucker[*V.1.4.083,.16a;.3.4.01(16)]
    452,468,545
system(II.1.3.03;*IV.5.2.031)**98**,401
  deep(*IV.5.2.020)400
  deeper(*IV.5.2.080)406
  extensive(*IV.5.2.031)401
  non-competitive(*V.2.110e-2)479
  spreading(*IV.5.2.079)406
systemic(*V.3.3.2.419)529
tap(II.4.3.068,.069,.070)245
tip(II.3.086)198
-to-top ratio(*IV.5.2.080)406
tropism(IV.3.3.008)377
tuber(II.4.3.049,*.063;.4.13a)244,
    245,261
tuberous(*V.1.4.024,.05d,e)443,
    457
whole(*V.1.4.028b)443
Rooted cuttings(*V.1.2.011p;*.1.4.
    028m;*.3.4.073)431,447,544
Rooting(*V.1.4.022a-3;*.4.2.2.015)442,
    567
  areas(*V.4.1.068)562
  bed(*V.1.4.022a-2;.4.1.100)442,564
  in larger containers(V.1.5.01k,l)471
  room(*V.4.1.012,.101)559,564
  structures(V.1.1.053)425
Rootstock, rootstocks(*II.4.3.015;*V.
    1.4.028,.028a,*b,*e,*f,*f-2,*-3,
    -5,-6,*-7,*-8,*.028g,*h,*i,*k,
    *m,*n,*p,*q,*r,s,*t,*.041,*.060)
    242,443-449,450,451
  dwarfing(*V.3.4.000a)539
  preparation(V.1.4.028t)448
  -scion combination(*V.1.4.015)441
Rootworms(*V.3.3.1.132a)501
Roridous(II.5.6.108)277
Rosaceous(II.2.2.069,.21k)129,150
Rose:
  bengal(*IV.5.4.2.010)413
  chafer(*V.3.3.1.132a)501
  nozzle(V.1.1.054)426
Rosellina spp.(*V.3.3.2.181e)519
Rosette(II.1.2.093,.19a;*V.3.4.023;
    .4.2.1.019,*.022)63,84,541,566
  form(*IV.5.3.038)410

  plants(II.4.1.003)238
Roseus(*II.5.1.005q)264
Rostellum[II.2.2.041x,.25b(2),d(3)]
    125,154
Rostrate(II.2.2.041y,.28d)125,157
Rostellate(*II.2.2.041y)125
Rosulate(II.1.2.094,.19a)63,84
Rosy(II.5.1.005q)264
Rot, rotting(*V.4.2.1.001)565
Rotary:
  mowers(*V.1.1.034)422
  tiller(V.1.1.056)426
Rotate(II.2.2.061c-1-o,.21a,b,.22e)
    127,**150**,151
Rotating:
  cones(*VI.2.022b)576
  rollers(*VI.2.022b)576
Rotation(*IV.0.114)333
Rots(*V.3.3.2.040,*.133c)511,515
Rotten fruit or vegetable(*VI.8.056)
    620
Rotund(II.5.4.1.018;*.5.5.050;.5.03n)
    269,273,**283**
Rotylenchoides spp.(*V.3.3.1.200f)506
Rotylenchus spp.(*V.3.3.1.200d,e)506
Rough(*II.5.6.007;.5.13g,h)273,293
  endoplasmic reticulum(*IV.1.1.008)
    351
Roughish(*II.5.6.113;.5.18f)277,298
Round:
  (container)(*VI.4.003)585
  -point shovel(*V.1.1.060)426
Rounded(II.5.4.2.011;.5.04i)269,284
  caps(*V.2.115c,*i)480
  vertices(*V.2.115k)480
Roundheaded borers(*V.3.3.1.132a)501
Roundish(II.5.5.050;.5.11n)273,291
Routing(VI.6.028;*.8.106)596,624
Rove beetles(*V.3.3.1.132a)501
Row(*V.3.3.3.027)533
  -crop sprayer(*V.3.3.3.100)537
  crops(*V.3.1.002)485
  preparation(*V.3.2.039)492
  seeder(*V.1.3.021,*.062)433,437
Rows(II.1.1.1.022,.05r;*V.2.110e;*.3.
    2.039;*.3.3.3.059)20,**32**,479,492,
    535
  alternate(*V.3.1.008h)486
  closely spaced(*V.3.1.008a)485
  crop(*V.3.1.008a)485
  middles(*V.3.1.008a)485
Royal moth(*V.3.3.1.132g)502
RQ(*IV.1.2.041)356
Rubber(I.1.011,*IV.0.227;*.2.2.107)4,
    342,366
  augers(*VI.2.022b)576
  band(*V.1.4.028g)445

INDEX OF TERMS

budding strip(*V.1.4.028r)448
spindles(*VI.2.0.022b)576
strips(*V.1.2.017)432
Ruber(*II.5.1.005p)264
Rubiginosus(*II.5.1.005b,.008q)264,266
Rubus spp.(*V.3.4.008,.05)540,549
Ruderal(III.1.2.055)310
Rudimentary(II.5.2.032)267
  embryos(*V.1.3.060)436
Rufous (rufus)(II.5.1.001k)263
Rugose(II.5.6.111;.5.18e)277,298
Rules for Seed Testing(*V.1.3.089)439
Ruminate (endosperm)[*II.2.3.3.015,
  .035,.09a(1),.10c(5)]164,165,190,191
Rumple of lemons(VI.9.3.106)671
Runcinate(II.5.5.051,.5.11b)273,291
Runners(*V.1.4.000,*.024,.064,.16b)
  441,443,451,468
  natural vegetative reproduction(IV.
  4.1.006a-8)379
Runoff(*V.2.110b)478
Russet:
  ring of apple(VI.9.1.468)653
  spotting of head lettuce(VI.9.3.109)
  671
Russeting(V.3.3.2.370)527
  of beans*VI.9.3.029,107)665,671
  of stone fruits(VI.9.3.108)671
Rust, rusts(*V.3.3.2.179a,*.181f,
  .346,.371,.402,*.423,*.442;*VI.
  7.2.017i)517,519,526,527,528,529,
  613
  fungus, fungi(*V.3.3.2.325,*.346,
  .349,.371,*.423)526,527,529
  mite of citrus(*VI.8.038b;*.9.2.017)
  619,660
  of peaches and plums(VI.9.469)653
  short-cycle(*V.3.3.2.181f)519
Rusty(II.5.1.001l,.008q)263
  tussock moth(*VI.9.2.044)661
Rwc(*IV.5.2.078)405

S

Sac(*II.2.2.007r)119
Saccharides(*IV.2.2.031)361
Saccate(*II.2.2.061c-14-c,.24e)128,153
Saddle graft(V.1.4.028k-5,.09e)447,461
SADH(*IV.2.1.037;*V.3.4.000a)359,539
Sagittate(II.5.4.2.012;.5.5.052;.5.
  04j,.12a)269,273,284,292
Salad crops(I.3.020)8
Sale on track(*VI.7.1.075)607
Sales:
  agent(IV.8.101)624
  figure(*VI.8.071)621
  manager(*VI.8.037,.102)619,624

organization(*VI.8.122)625
Salic horizon (sub-surface)(V.2.028g)
  473
Salinization(V.2.083)474
Salmon-color (salmonicolor, salmonaceus)(II.5.1.005t)264
Salt(s)(*IV.0.073)332
  deposits(*V.3.5.001)557
  flats(*II.4.2..028)240
  injury(IV.5.4.3.009)415
    resistance (or tolerance)(IV.5.4.
    3.010)415
  (ionic) pressure and flooding(water-
    logging) stresses(IV.5.4.3.)414-
    415
  (ionic) stress(IV.5.4.3.000)414
  marshes(*II.4.2.010)239
  pan(II.4.2.028)240
  resistance(*IV.5.4.3.010)415
  tolerance(*IV.5.4.3.007)415
  toxicity(*IV.5.4.3.009,.010)415
Saltation(V.3.3.2.373)527
Salty:
  ground(*V.2.120)482
  irrigation water(*V.2.120)482
Salverform(II.2.2.061c-1-p,.21h)127,
  150
Salver-shaped(II.2.2.061-1-k,.21g)127,
  150
Samara(II.2.3.1.031,.04a-c)160,171
Sample(s)(III.2.2.157;VI.1.014)325,572
  representative(*V.3.2.072,.082)495
Sampling(VI.1.015)572
  plant(V.3.2.072)495
  soil(V.3.2.082)495
Sand, sands(*V.1.1.053e;*.1.2.005,
  .011,.011b,*h,*j,*n,*p;*.1.3.
  081;*1.4.028k-2;*.2.099,*.107a,*d,
  .117l;*.3.1.001)425,429,430,431,
  439,446,476,477,481,485
  culture(s)(*V.3.2.050,.076)493,495
  noncoherent(*V.2.115l)481
  quartzic(*V.2.107d)477
Sandy:
  clay loams(V.2.117h)481
  clays(V.2.117b)481
  -loam(V.2.117j)481
  materials(*V.2.096d)475
Sanguine (sanguineus)(V.5.1.005u)264
Sanitation(*V.3.3.3.000,.088,*.098)
  531,537
San José scale on apple and pear(VI.9.
  2.073)662
Sapormyophily(*IV.4.5.001c)391
Saprogenesis(V.3.3.2.374)527
Saprophyte(s), saprophytic(II.4.2.030;
  V.3.3.2.168,.169,*.179,*.181c,

INDEX OF TERMS

  *.206)240,516,517,518,521
 epiphytic(*II.4.2.006)239
 facultative(*V.3.3.2.169)516
Sapwood(II.3.164)202
Saran(*V.1.1.059;*VI.3.007,*.081)426, 578,583
 cloth(*V.1.1.017b)419
Sarcocarp[II.2.3.029h,.05(2)]162,179
Sarcody(V.3.3.2.375)527
Sarcotesta[II.2.3.3.042i,.01d(1),.07a-d,.09e]167,182,188,190
Sarcotestal seeds(II.2.3.3.07a-d)188
Saturation:
 cation exchange capacity (CEC)(*V.2.015)472
 index(IV.5.2.053)404
Saurochory(IV.4.7.001i)393
Savanna(*V.2.110)478
 trees(*II.4.4.002a)246
Saw(*V.1.4.028k-2,*k-4)446,447
Sawdust(*V.1.2.011;*.1.3.081;.1.4.036b-1)429,439,450
Sawflies(*V.3.3.1.132e)502
Saw-kerf graft, grafting(*V.1.4.028k,k-4)446,447
Saws(*V.3.4.011,*.015,*.040,*.060,*.065,*.076)540,542,543,544
Sawtooth (greenhouse)(*V.1.1.017b)419
Saw-type:
 hedgers(*V.3.4.063)543
 toppers(*V.3.4.063)543
Say stinkbug(*V.9.2.013)660
Scab(*V.3.3.2.181e,.376)519,527
 of peaches, nectarines, apricots, plums(VI.9.1.470)653
 spot complex of carrot(VI.9.3.110)671
Scabrid(*II.5.6.113)277
Scabrous(*II.5.6.007,.113;.5.18f)273, 277,298
Scaffold branch(es)[*V.3.4.026,.01(7)] 541,545
 cordonlike(*V.3.4.068)544
Scalariform perforation(*II.3.182,*.217)203,206
 -reticulated(II.3.22e)228
 thickening(*II.3.202;.3.22d)204,228
Scalds(*VI.9.3.139)672
Scale, scales(II.5.6.114;*V.4.1.035,*.045,*.103,*.104)277,560,561,564
 axil(*V.4.1.073)562
 base rotting(*V.4.2.1.017)566
 bulb (propagation)(V.1.4.17b-g)469
 bulblet(*V.1.4.007;.4.1.102)441,564
 complement(*V.4.1.117)564
 hair(*II.3.204)205
 inner(*V.4.2.2.008)567

 insects(*V.3.3.1.132d;*V.3.3.3.089) 502,537
 interjoined(*V.4.1.003)559
 leaves(*II.4.3.032,*.041;*IV.5.2.085)243,244,406
 mother(*V.4.1.073)562
 papery(*V.4.1.120)565
 propagation(*V.4.1.016)560
 -shaped(*II.5.6.121;.5.19a)277,299
 tiny, dense(*V.3.3.1.132g)502
Scalelike leaves(*V.1.4.071a)451
Scalicide(V.3.3.3.089)537
Scaling(*V.4.1.102,.103)564
Scalloped cell walls(*V.3.3.2.361)527
Scalpel(*V.1.4.028p)448
Scalper(*V.1.3.057)436
Scaly(II.5.6.116)277
 (bulb)(II.4.13d;V.1.4.071a-1,.16c;.4.1.104)261,451,468,564
Scandent(II.1.1.3.085,.19a)26,55
 shrubs(*II.4.3.002)241
Scanner(VI.8.103)624
 -equipped checkstands(*VI.8.131)625
Scanning electron microscope(*IV.0.079)332
Scanty(*II.3.014)194
Scape, scapose, scapes[II.2.1.064,.12e (1);*V.4.1.057;.4.2.2.005,*.011] 103,115,561,567
 abscission(*V.4.2.2.019)568
 fasciation(V.4.2.2.007)567
 fusion(*V.4.2.2.007)567
 intermode collapse(*V.4.2.2.020)568
Scarab(s)(*V.3.3.1.132a)501
 beetles(*V.3.3.1.108)500
Scarification(V.1.3.055)435
Scarifier(*V.1.3.057)436
Scarious(II.5.7.029)278
Scarlet(II.5.1.005v)264
Scarred(II.5.6.117,.18h)277,298
SCA storage(*VI.5.047e)592
Scattered(*II.1.2.103)63
Schimper and Braun series(*II.1.1.1.016)19
Schistaceus(*II.5.1.003m)264
Schizocarp(*II.2.3.1.009,*.022,*.031,.032,.03i-k,.04a-c)159,160,170,171
Schizogenous spaces(*II.3.157)201
Schizogeny(II.3.168,*.178,.180)202,203
Schizolysigenous[II.3.169,.16c(1)]202, 222
Schollander pressure chamber(*IV.5.2.041)403
Scientific name(III.1.1.030)306
Scion(*V.1.4.028a,*e,*f,*.028f-2,*-3,*-5,*.028g,*h,*i,*k,*n,*p,*q,*r,.067,.06-.12)443-448,451,458-464

rooting(*V.1.4.028r)448
shield(*V.1.4.028f)444
wood(*V.1.4.028f-6,.028u)445,449
Scionwood(*V.1.4.010,.028c)441,444
preparation(V.1.4.028u)449
Scirtothrips citri(*VI.9.2.080)662
Scissors-type pruning shears(*V.3.4.
066)543
Sclereids[*II.3.135,.170,*.175,*.221,
.16b(2),.24a-d,.25]200,202,203,206,
222,230,231
Sclerenchyma(II.3.171)202
cells(*II.3.054)196
Sclerophyll, sclerophyllous, sclero-
phylls(*II.4.3.032,.050;.4.08b;
*IV.5.2.085)243,244,256,406
plants(*IV.5.2.079)406
Sclerospora macrospora(*V.3.3.2.181d)
518
Sclerotia(*V.3.3.2.181g;*VI.9.1.145,
.146,.482,*.556,*.558,*.559,*.561,
*.562)520,637,654,658
Sclerotinia:
intermedia(*VI.9.1.559)658
minor(*9.1.559,*.561)658
sclerotiorum(*V.3.3.2.181e;*VI.9.1.
179,*.556,*.557,*.558,*.559,*.561)
519,638,658
spp.(*V.3.3.2.181e;*VI.9.1.562)519,
658
Sclerotinia rot of cabbage, etc.(*VI.
9.1.557)658
Sclerotiniose(*VI.9.1.561)658
Sclerotium:
bataticola(*VI.9.1.144)636
cepivorum(*VI.9.1.565)659
rolfsii(*VI.9.1.480,*.481,*.482)653,
654
Sclerotium(V.3.3.2.380)527
rot of citrus(VI.9.1.480)653
snapbeans(VI.9.1.481)654
tomato(VI.9.1.482)654
Sclerotized rind(*V.3.3.2.380)527
Scooped out(*V.1.4.071b)451
Scooping(*VI.2.022a)576
Scored(*V.1.4.071b)451
Scoring(*IV.5.4.4.002,*.002e)416
Scorpioid cyme[II.2.1.024d,.03a(4),
.07d]101,106,110
Scorpions(*V.3.3.1.010)498
Scout, scouting(*V.3.3.3.052,.090)534,
537
Screens(*V.1.1.047,*.063)424,426
Screw-worm(*V.3.3.1.132b)501
Scrobiculate(II.5.6.118;.5.18,i,j)277,
298
SCU(V.3.2.015)491

Scuffle hoe(*V.1.1.024)421
Scurfy(II.5.6.119)277
Scutate(II.5.6.120)277
Scutelliform(II.5.6.121;.5.19a)277,299
Scutellonema:
brachyurum(*V.3.3.1.200a)505
Bradys(*V.3.3.1.200d)506
Scutellum[II.2.3.3.014k,.01b(2)]164,
182
Scythe-shaped(*II.1.2.043)61
Sea-green(II.5.1.004f)264
Sealdsweet(*VI.8.122)625
Seasonal:
climates(*IV.5.2.013)399
item(VI.8.104)624
items(*VI.8.112)625
Seawater:
flooding(*V.2.120)482
magnesia(*V.3.2.084)496
Seaweed(*V.1.3.017)433
Secateur(*V.3.4.063)543
Second(s,")(*IV.0.260a,260i)344,346
Secondary:
calcium sulfate accumulation(*V.2.
028c)473
carbonate accumulation(*V.2.028b)473
cycle(V.3.3.2.381)528
drought-induced stress(*IV.5.2.022)
400
elements(*V.3.2.028)491
freezing injury(*IV.5.3.023)409
gene pool(III.2.1.002b)317
growth in monocots(II.3.172)202
growth of vascular system(II.3.173)
202
meristem(*II.3.107,.174)199,203
phloem[II.3.175;.3.01C(24),.02e(12),
.20b(6)]203,207,208,226
(roots)[II.1.3.015,.02a,b(3),c(5),d
(5)]94,97
salt injury(*IV.5.4.3.009)415
stress(*IV.5.3.005)407
stress injury(*IV.5.3.047)411
sympton(V.3.3.2.382)528
thickening(II.3.01C-1)207
tissue zone(II.3.02e)208
vertical(V.3.4.07a,b)551
water-stress injury(IV.5.2.056)404
xylem[II.3.176;.3.01C(25),.02e(14)]
203,207,208
Second-order:
dichogamy(IV.4.3.002d)386
hercogamy(*IV.4.3.006)386
Second-year cane removal(*V.3.4.008)
540
Secretary(*VI.8.074)621
of Agriculture(*V.7.1.053;*.7.2.

INDEX OF TERMS

004a,*b,*c)604,609
Secretion(II.3.177)203
Secretory:
  canal(II.3.178)203
  cells(*II.3.054,.179,.31)196,200,
    203,237
  elements(*II.3.175)203
  space(II.3.180)203
Sectorial (chimera)(III.2.2.024e,.01a)
  319,327
Secund[II.1.2.097,.19b;.2.1.08e(2);.2.
  2.070,.03d]63,84,111,129,132
Sedentary:
  ectoparasitic nematodes attacking
    roots(V.3.3.1.200b)505
  endoparasitic (female usually sac-
    cate) nematodes attacking roots(V.
    3.3.1.200g)506
  semiendoparasitic nematodes attack-
    ing roots(V.3.3.1.200e)506
Sedge peat(*V.1.2.011h)430
Sedimentary:
  origin(*V.3.2.075)495
  rocks(*V.2.000)472
  sedimentation velocity(*V.2.123)482
"Seed"(V.1.4.068;.3.1.013)451,487
Seed, seeds[*II.2.3.1.024;.2.3.3.000;
    .2.3.2.05a(4);.2.3.3.01-.12;IV.
    3.1.001,*.010,*.014c,*.024;
    .4.2.000,.001,*.002,*.003,
    .004,.005;*.4.6.000,*.003;*.4.
    7.002;*.4.8.003,*.010,*.016,
    .036,.041,*.056;*V.1.000;*.1.
    1.000,*.048;*.1.3.000,*.003,
    .010,.011,*.012,*.013,*.014,
    .015,.021,*.024,*.026,*.027,
    .028,.030,*.031,*.034,*.037,
    .041;.1.5.000;*.2.110e-2;*.3.
    0.004;*.3.3.2.123,*.133e;.3.
    3.3.117]149,163,179,182-193,370,
    371,372,373,385,392,393,394,395,
    417,418,424,432,433,434,470,479,
    497,514,515,538
  alate(II.2.3.3.042a,.11d,e)166,192
  allelochemic(*IV.2.2.014)360
  aril[II.2.3.3.042b,.05a(2),.08a-c,
    .09b,d(2),.11b,d]166,186,189,
    190,192
  arillate[II.2.3.3.042c,.10b(5)]166,
    191
  arillode(II.2.3.3.042d)166
  arillostome[II.2.3.3.042e,.10c(3),d
    (3)]166,191
  beds(*V.1.1.056;*.1.3.085;*.1.4.041;
    *.3.1.013)426,439,450,487
  campylotropous(II.2.3.3.036a-1,.06a-
    d)165,187

carbohydrates(*IV.2.2.031)360
caruncle(II.2.3.3.042f)167
-cleaning equipment(V.1.3.057)435
coat(s)[*II.2.3.3.039,*.041,.01a(3),
    .03i(1),.04b(1,2),.07,.08,.12;
    *.3.170;*V.1.3.024,*.079]166,
    182,184,185,188,189,193,202,
    433,438
  resistance[*V.1.3.031,.01a(2,3,4)]
    432,440
coatings(V.1.3.058)436
corn maggot(*V.3.3.1.133b)501
cotyledons(II.2.3.3.02)183
,crop(*V.1.3.051)435
cracking(V.1.3.059)436
disinfest(*V.3.3.3.028)533
disinfestation(*V.3.3.3.038)533
dispersers(*V.3.3.1.017)498
dormancy(V.1.3.060)436
dorsal(II.2.3.3.012,.05)163,186
drying(V.1.3.061)437
dry stored(*V.1.3.073)438
embryo(s)(II.2.3.3.014,.02)163,183
embryotega(II.2.3.3.042g)167
extraction (dry fruits)(V.1.3.063,
    *.088)437,439
  (fleshy fruits)(V.1.3.064)437
fermented(*V.1.3.064)437
fiber mats(*V.1.3.074)438
flat(*V.1.1.047;*.1.4.041)424,450
formation(IV.4.2.000)385
-forms(II.2.3.3.036,.04,.05,.08,.09)
    165,185,186,189,190
galls(*V.3.3.1.200a)505
gel sowing(*V.1.3.030)434
germination(II.2.3.3.020;*IV.2.1.
    025,*.036;*.3.2.001,*.009,*.011,
    *.013;*V.1.3.041,*.042;*.1.4.
    028t;*.3.3.2.123)164,358,373,
    374,434,435,448,514
  test(*V.1.3.031)434
hairs(II.2.3.3.07e,f)188
handling(*V.1.1.047)424
hilar(II.2.3.3.11a-c)192
holding temperature(*V.1.3.073)438
impermeable(*V.1.3.060)436
incubation(*V.1.3.039)434
inoculation(*V.1.3.041)434
legume(*V.1.3.012,*.014,*.041)432,
    434
live(*V.1.3.073)438
maturation(*IV.4.2.005)385
moisture content(*V.1.3.073)438
obcampylotropous(II.2.3.3.036a-4,
    .04e,.06e-h)165,185,187
off-type(*III.2.0.004g)312
orthotropous(II.2.3.3.036b,.04a)166,

INDEX OF TERMS

185
pachychalazal(II.2.3.3.036a-5,.04g,
   .08b)165,185,189
packaging(V.1.3.067)437
palm(*V.1.3.039)434
paper grids(*V.1.3.074)438
pelletized(*V.1.3.045)435
perichalazal(II.2.3.3.036a-6,.04h,
   .09a-d,.12d)166,185,190,193
peristaltic fluid drill(*V.1.3.046)
   435
plants(*V.3.3.2.133a)514
plug-mix(*V.1.3.047)435
pregerminated(*V.1.3.049)435
presoaking(*IV.5.4.3.010)415
presowing treatments(*V.1.3.050)435
priming(V.1.3.069)437
production(*III.2.0.004c-f)312
propagation(*V.1.3.000)432
purity(*V.1.3.073)438
pseudoarillate(II.2.3.3.12)193
quarantine(V.3.3.3.038)533
quiescence(*IV.3.1.021)373
rest(*IV.2.1.033)358
sarcotesta(II.2.3.3.042i,.01d,07a-d,
   .09e)167,182,188,190
sarcotestal(II.2.3.3.07)188
scarification(*V.1.3.055)435
sealed containers(*V.1.3.073)438
seeder(*V.1.3.062)437
selection(*V.1.4.028t)448
separators(*V.1.3.078)438
sizing(V.1.3.072)438
soaking(V.1.3.079)438
sources(V.1.4.010;*.3.0.001)441,484
sowing rate(*V.1.3.066)437
species(*V.1.3.073)438
spotting of lima beans(VI.9.1.483)
   654
sterilization(V.1.3.080)438
storage(V.1.3.073)438
structure(II.2.3.3.03a-e)184
tape(V.1.3.074)438
testing(*V.1.3.075)438
test procedures(*V.1.3.075)438
   sample(*V.1.3.089)439
thermodormancy(*V.1.3.083)439
threshing(*V.1.3.084)439
transmission(*V.3.3.2.133e)315
unitegmic(II.2.3.3.12c-e)193
viability(*V.1.3.073)438
wafering(V.1.3.076)438
weed(*V.1.2.011f;*.1.3.051,.075)430,
   435,438
windrow(*V.1.3.088)439
working sample(V.1.3.089)439
Seeder(s)(*V.1.1.036;.1.3.062)423,437

Seeding(*V.3.3.3.080,*.081)536
   operation(*V.3.2.039)492
   (sowing) rate(V.1.3.066)437
Seedling, seedlings(*II.3.035;*IV.4.4.
   022;*.5.1.001;*.5.2.030;*V.1.2.
   003,*.011;.1.3.085;*.1.4.028m,*p,
   .041;.3.1.002;.3.3.1.250;*.3.3.
   2.123,*.133c;.3.3.3.067;.3.4.073;
   *VI.4.011)195,389,396,401,428,429,
   439,447,448,450,485,508,514,515,
   535,544,586
   bench-rooting, bench-rooted(V.1.3.
      003,.01a)432,440
   contractile roots(*II.3.035)195
   curvature[V.1.3.01a(2,3)]440
   damping-off(*V.3.3.2.133c,*.181f)
      515,519
   disease of garden crops(*V.3.3.2.
      181f)519
   double curvature[V.1.3.01a(4)]440
   lines(*V.3.3.3.000)531
   normal[V.1.3.01a(1)]440
Seedpiece(V.1.4.069)451
Seedstalk(II.1.1.4.086.19b)26,55
Seep irrigation(V.3.5.023)558
Seepage(*V.3.5.012)557
Seepers(*V.3.5.012)557
Segetal, segetal race(III.1.2.056)310
Segregation(III.2.2.158)325
Seismomorphogenesis(*IV.3.3.008r)377
Seismonasty(*IV.3.3.003e)376
Seismotropism(IV.3.3.008r)377
Select:
   -cross, -crossing(III.2.2.159)325
   seed(*III.2.0.004c-f)312
Selection(III.2.0.011)316
Selective:
   gameticide(III.2.2.160)325
   harvesting(*VI.2.023)577
   herbicide(V.3.3.3.091)537
Self:
   -branching(V.4.1.105)564
   -compatible(IV.4.8.045)395
   -controlled atmosphere (SCA)(VI.5.
      047e)592
   -fertile(IV.4.8.046)395
   -fruitful(IV.4.8.047)395
   -incompatible(IV.4.8.048)395
   -pollination(*IV.4.3.002,*.002b,
      *.002d,.009;.4.4.052)385,386,390
   -sterile(IV.4.8.049)395
   -subirrigating containers(*V.1.1.
      017f)420
   -unfruitful(IV.4.8.050)395
Sell:
   and distribute produce(*VI.7.1.031)
      601

INDEX OF TERMS

or distribute produce(*VI.7.1.030) 601
Seller's risk(*VI.7.1.016)599
Sells bulbs(*V.4.1.010)559
Semelparous(II.5.3.019;.5.02d)268,282
Semiamplex(*II.1.1.2.017)21
Semiautomatic (bagging) machines(*VI.3.053a)581
Semidwarf(III.2.2.161)325
Semigamy (semigamic apomixis[IV.4.1.006b-5;.03(8)]380,384
Semihardwood (cuttings)(V.1.4.022f-3, *-4,.02c-e,.03a,b,d)443,454,455
Semiheteroroph(*II.4.2.026)240
Semiheterotypic[IV.4.1.006-6-a-4,.03(4)]380,384
Semi-mechanized harvesting(VI.2.022c)576
Semiparasite(*II.4.2.006,*.026;*V.3.3.2.133a)239,240,514
Semipermeable membrane(*IV.5.2.035)402
Semi-ring porous wood(II.3.10b)216
Semiskilled(*VI.8.053)620
Semiterete(II.1.2.098,.19c)63,84
Semitrailer(*VI.2.012)575
  flatbed(*VI.2.012,*.014)575
  for railroad (piggyback) hauling (*VI.6.016,*.036)595,597
  over-the-road hauling(*VI.6.016, *.036)595,597
  van(*VI.6.036)597
Semitrap flower(IV.4.4.053,*.056)390
Senescence(VI.1.016,.02;*.9.3.069)572, 574,668
  accelerated(*IV.5.2.085)406
  delay(*VI.9.3.105)671
Senescent scald of 'Golden Delicious' apple(VI.9.3.111)671
  scald of pears(VI.9.3.112)671
Sensible heat(VI.5.051d)593
Sensitive(V.3.3.2.384)528
Sensitivity(V.3.3.2.385)528
Sepal(*II.2.2.061a)126
Sepaloidy(II.2.2.071)129
Sepalody(II.2.2.061e)129
Separate:
  materials(*V.3.2.037)492
  (stamens)(*II.2.2.007d-7,*.038)118, 122
Separated(*V.1.3.071b)451
Separations(*V.1.4.000,.071,.16c-f, .17)441,451,468,469
Separators(*V.1.3.057,.078)435,438
Septate(*II.2.2.049)125
Septicidal(II.2.3.1.006f,.02j)158,**169**
Septifragal(II.2.3.1.006g,.21-l,m)158, **169**

Septobasidium pedicellatum(*V.3.3.2.181f)519
Septoria:
  apiicola(*VI.9.1.297)645
  citri(*VI.9.1.488)654
  gladioli(*V.3.3.2.181g)520
  petroselini(*VI.9.1.485)654
Septoria:
  blight of parsley(VI.9.1.485)654
  spot of citrus(VI.9.1.488)654
Septuple(*V.3.1.008g,.01c)486,488
Sequence(*V.3.3.3.052)534
Sequum(V.2.107n,*.112)477,479
Serial divisions(V.1.5.006)470
Serial knee-roots(*II.4.3.039,.051, .11c)243,244,259
Sericeous(*II.5.6.126)277
Series(*II.1.1.1.022,.05r;V.2.096f)20, 32,475
Serotinous(IV.4.8.051)395
Serotype(*V.3.3.2.051)512
Serpentine(*V.3.4.021)541
  (compound) layerage(V.1.4.036b-2, .14f)450,466
Serrate, serrulate(II.5.4.3.032;.5.07c,d)271,287
Service area(*V.1.1.021)421
  wholesaler(VI.8.105)624
Services(VI.6.029)596
  required(*VI.8.106)624
Sesquioxide coatings(*V.2.107a)477
Sessile(II.1.2.099,.19d;.2.2.18m)63, 84,147
Setose(II.5.6.123;.5.19b)277,299
Seventeen-year locust(*V.3.3.1.132d)502
Severe symptoms(*V.3.3.2.384)528
Sex-chromosome(III.2.2.162)325
  -limited(III.2.2.163)325
  -linkage(III.2.2.164)325
Sexual:
  embryo[IV.4.1.02b(11)]383
  reproduction by unlike cells(*V.3.3.2.279)524
  spores(*V.3.3.2.179a)517
Sexually reproduced(III.2.0.007b)316
Shade(*V.1.1.017g;*.2.110e,*.110e-1) 421,479
  forms(*II.4.4.003e)247
  frames(*V.1.1.017e)420
  houses(*V.1.1.048,.059;.1.3.01b)424, 426,440
  leaf production(*V.2.110e-2)479
  rapid growth(*V.2.110e-2)479
"Shadeless forest"(*II.4.3.032,.052) 243,244
Shaggy(*II.5.6.062)275

INDEX OF TERMS

Shaking(*IV.3.3.008r)377
Shallow planting(*V.4.2.2.006)567
Shape(s)(II.5.5.;.5.08-.12;*VI.8.094)
 271-273,288-292,623
Shaping(*V.4.1.094)563
Share(*V.1.1.044)423
Sharply limited necrosis(*V.3.3.2.082)
 513
Sharp-pointed(*II.5.4.1.003)268
Shattering of snapdragon(V.4.1.106)564
Shavings(*V.1.2.005,*.011)429
Shear(VI.3.088)582
Sheath[*II.3.063,.15a(4)]196,221
 nematodes(*V.3.3.1.200b)505
Sheathing, sheath[II.1.2.100,.14d,.19e
 (1),]63,79,84
Sheathoid nematodes(*V.3.3.1.200b)505
Sheep(*V.1.2.011f;.3.3.3.048)430,534
Sheet erosion(*V.2.110d)478
Shelf:
 life(*VI.8.016)617
 space(*VI.8.109)624
Shell (endocarp)(II.2.3.2.05b,c)179
Shield (T) bud(*V.1.4.028f-8,.08a,b,
 .11k)445,460,463
Shield (T) budding(*V.1.4.028e,*f,
 *f-3,f-8)444,445
Shikimic acid pathway(IV.1.2.042;*.2.
 2.154)356,368
Shiny(II.5.6.125)277
Ship(*VI.6.000,*.016)594,595
 intermodal(*VI.6.011)594
 (loading pattern)(*VI.6.017)595
 van modular containers(*VI.6.037)597
Shipment(*VI.8.038a)619
 as soon as car (truck) can be secur-
  ed(*VI.7.1.064)605
 as soon as possible(VI.7.064)605
 carbon dioxide treatment(*VI.5.007)
  588
 ,direct(*VI.7.1.070)606
 early part of week(*VI.7.1.065)605
 first part of week(VI.7.1.065)605
 hamper(*VI.4.012)587
 ,immediate(*VI.7.1.074)607
 last of week(VI.7.1.066)606
 latter part of week(*VI.7.1.066)606
 middle of week(VI.7.1.067)606
 on a specified date(*VI.7.1.073)607
 ,today's(VI.7.1.073)607
 ,tomorrow's(VI.7.074)607
Shipper(*VI.7.003,.059;*.8.131)598,
 605,625
Shippers(*VI.7.2.004;*.8.107)609,624
Shipping(*VI.8.081)622
 area(*VI.8.134)626
 box(*VI.4.011)586
 condition(*VI.7.2.024)600
 container(*V.4.1.025;*VI.4.000c,
  .001,.025,*.026)560,585,588
 lug(*VI.4.014)587
 point(*VI.7.1.020,*.021,*.022,*.023,
  .024,.030)600,601
  acceptance(*VI.7.1.020)600
  acceptance final(*VI.7.1.021)600
  inspection(VI.7.1.062)605
  inspection final(VI.7.1.063)605
Ships bulbs(*V.4.1.010)559
"Shoe-string" leaves(*V.3.3.2.133d)515
Shoot, shoots(*II.3.143,*.144;*IV.2.1.
 012,*.037;.3.1.018;V.1.4.022,
 .083;.1.5.005;.3.4.003,*.005,
 .007,*.023,*.029,*.046;.4.1.074)
 201,357,359,373,441,452,470,540,
 541,542,562
 apex(*V.4.1.084)563
 axes(*II.4.3.071)245
 cut(*IV.5.2.065)404
 developing(*V.3.4.005)540
 doubled-over(*V.4.2.2.016)567
 elongated(*IV.3.2.003)373
 elongation, rapid(*V.4.1.088)563
 formation(V.1.5.01)471
 from callus(V.1.5.01h-j)471
 growth(*IV.2.1.037;*.3.1.018)359,373
 lateral(*V.3.4.072;*.4.1.009)544,559
 or adventitious buds(V.1.5.01g)471
 sprouting(*V.3.3.2.340)526
 tip(V.1.5.01b)471
 tip tissue death(*V.3.3.2.128)514
 vegetative(*V.4.1.024,*.094)560,563
 weak(*V.3.3.2.469)530
 young(*V.3.4.073)544
Short:
 cell[II.3.29c(6)]235
 -day plant(s)(IV.3.2.0131;*V.4.1.
  002,*.007,*.034)375,559,560
  treatment(*V.4.1.095)563
 days(*V.4.1.001,*.097)559,563
 -horned grasshopper(*V.3.3.1.132h)
  503
 -long-day plants(IV.3.2.013m)375
 shoots[II.1.1.3.087,.14a(2)]26,50
Shortened internodes(*V.4.2.1.019)566
Shorts(VI.8.107)624
Shot hole(V.3.3.2.388;*VI.9.1.176)528,
 638
 borers(*V.3.3.1.132a)501
Shovel(V.1.1.060,*.067)426,427
Shredded:
 bark(*V.1.2.011b)430
 coconut husk(*V.1.4.036a)449
Shredding(*VI.7.1.025)601
Shrink film(VI.4.021)587

INDEX OF TERMS

Shrinkage(*VI.8.058,.108)621,624
Shriveling(*V.4.2.1.002)565
 of stone fruits(VI.9.3.113)671
Shrub, shrubs(*I.2.009;II.0.023,.06a,
 b;*.1.1.3.076;*.3.048;*.4.1.005;
 .4.2.006,.026;*.4.4.002a,*.005-
 b;*.6.011;*V.1.2.001;*.1.3.060;
 .1.4.036b-3;.2.110e;*.3.4.000,
 .040,.076)6,11,17,25,196,238,
 239,240,246,246,279,428,436,450,
 479,539,542,544
 balled-and-burlapped(*V.1.1.035)423
 bare root(*V.1.1.035)423
 training(*V.3.4.075)544
SI units(IV.0.260)344
Sib-mating(III.2.2.165)325
Sibs (siblings)(III.2.2.166)325
Sickle:
 -bar(*V.3.4.040,*.041,*.063,*.076)
  542,543,544
  type hedgers(*V.3.4.063)543
  type mowers(*V.1.1.034)422
  type toppers(*V.3.4.063)543
 -shaped(*II.1.2.043)61
  necrotic areas(*V.4.2.2.017)568
Side:
 benches(V.1.1.017e)420
 graft, grafting(V.1.4.028k,k-6,.09h,
  i,.10f,.11m,o)446,447,461-463
 rot of apple(VI.9.1.490)654
 shoots(*II.4.3.071)245
 -veneer graft(*V.1.4.028k-8)447
Siemen(s)(*IV.0.073,*.260c)332,334
Sieve:
 area(II.3.181)203
 cells(*II.3.135,.26)200,232
 elements(*II.3.175,*.185,.26)203,232
 plate(II.3.182,.26d)203,232
 tube(*II.135,*.175,.183)200,203
Sievert (Sr)(*IV.0.260d)345
Sieving(*V.3.2.082)495
Sigatoka disease of banana(*V.3.3.2.
 181e)519
Sigmoid(III.2.2.167)325
Silicate clay accumulation(*V.2.028a)
 473
Silicle(II.2.3.1.033,.03h)160,170
Silicon (Si)(*V.3.2.028,*.073)491,495
 oxygen tetrahedra(*V.2.098)475
 total(*V.3.2.080)495
Silique(*II.2.3.1.030,.034,.03f,g)160,
 170
Silky(II.5.6.126;.5.19c)277,299
Silt(*V.2.107a,*.114,*.117e)477,480,
 481
 loams(V.2.117f)481
 pits(*V.2.110c)478

Silts(V.2.117e)481
Silty:
 clay loam(V.2.117d)481
 clays(V.2.117c)481
Silvery(II.5.1.007k)265
Simple:
 layerage(*V.1.4.036a-3,.14a)450,466
 (leaf)(II.1.2.102,.03c)63,68
 (perforation)(*II.3.217)206
 (sieve plate)(*II.3.182)203
Single:
 -arm horizontal cordon(V.3.4.11c)555
 cross(III.2.0.004i)312
 grain(V.2.115j)480
 -node (semihardwood) cutting(V.1.4.
  022f-3,.02d,e)443,454
 -stem(V.3.4.017,.06b)540,550
 -strength (canned) juice(*VI.10.029)
  677
 superphosphate(*V.3.2.091)496
Sink(*V.1.1.047)424
"Sinker"[II.4.3.053;.4.05f(2)]244,253
Sinker(*V.4.2.2.006)567
Sinuate(II.5.4.3.033;.5.07b)271,287
Sinuous(II.2.2.05f;.5.4.3.034)134,271
Sinus(II.5.4.3.035;.5.061)271,286
Siphoning(*V.3.3.1.132g)502
Siphonostele(II.3.189h,.27d)204,233
Sisal(*VI.3.007,*.081)578,584
Site:
 boundaries(*V.3.1.000)484
 preparation(*V.1.3.088;*.3.0.000,
  *.001;.3.1.000)439,484
<u>Sitophilus oryzae</u>(*VI.9.2.069)662
Six-cane Kniffen system(*V.3.4.032,
 .10)541,554
Size(s)(*II.5.2.033;VI.3.046;*.8.
 038e,*.094)267,581,619,623
 limitations(*VI.7.2.004)609
Sizing(*VI.3.031,.064,*.079)580,582,
 583
Skiffing(*V.3.4.057,*.073)543,544
Skilled(*VI.8.053)620
Skin:
 beetles(*V.3.3.1.132a)501
 checks on tomatoes(VI.9.3.114)671
 color(*VI.1.008)570
Skippers(*V.3.3.1.132g)502
Sky-blue(II.5.1.002i)263
Slabs(V.4.1.107)564
Slab-side of carnations(V.4.1.108)564
Slab wax(VI.3.084b)584
Slashed(*II.5.4.3.023)270
Slat (conveyor)(*VI.3.019)578
Slate-gray(II.5.1.003m)264
Sleeving(*IV.5.4.4.002)416
Slender branch(*V.1.3.036b-2,*-3)450

INDEX OF TERMS

Slicing(*VI.7.1.025)601
Slickensides(V.2.095)474
Slightly rolling terrain(*V.3.1.008j) 486
Slime[II.3.185,.26b(3),d(1);*VI.9.1. 061]203,232,631
  bacterial or algal(*V.3.5.032)558
  body(*II.3.185)203
  mold(*V.3.3.2.322)525
  molds(*V.3.3.2.181b,268)518,523
  plug(*II.3.185)203
Sliminess(*V.3.3.2.040)511
Slimy(*II.5.6.079)276
Slip sheet(*VI.4.000c,.022;*.6.016, *.023)585,587,595
Slipping stage of carnations(V.4.1. 109)564
Slips(V.1.4.04c)456
Slope exposure(*V.1.1.017e)419
Sloping terrain(*V.3.1.008b)485
Slough(II.4.2.033)240
Sloughing of red grapefruit(VI.9.3. 115)671
Slow-release fertilizers(V.1.2.012)431
Slowly available fertilizer source materials(*V.2.110e-1)479
Slug, slugs(*IV.4.5.001c;V.3.3.1.250; *.3.3.2.270)391,508,524
Slurries(*V.3.2.054)493
Small:
  bees(*IV.4.5.001c)391
  chains(*VI.8.105)624
  embryo(II.2.3.3.021-n)183
  fruits(I.1.012)4
  growers(*VI.8.138)626
  vehicle(*VI.6.036)597
Smell(*IV.4.4.013)389
Smokes(V.3.3.3.092)537
Smoky(II.5.1.003n)264
Smooth(*II.5.6.051)275
  endoplasmic reticulum(*IV.1.1.008) 351
  (endosperm)(*II.2.3.3.015)164
Smudge of onions, leeks, shallots(VI. 9.1.492)654
Smut, smuts(*V.3.3.2.179a,*.181f,.390, *.402;*VI.7.2.017i)517,519,528,613
  of onion and leeks(VI.9.1.493)654
Snail, snails(*IV.4.5.001c;V.3.3.1. 251;.3.3.2.270)391,508,524
Snap bean rust(VI.9.1.494)654
Snap-on lids(*V.1.3.067)437
Snapdragon rust(*V.3.3.2.181f)519
Snapping(*VI.2.022a)576
"Snatching"(*VI.2.022a)576
Snout beetles(V.3.3.1.132a)501
Snow-white(II.5.1.0071)265

Soak tank(*VI.3.081)583
"Soaker" hoses(*V.1.1.017f)420
Soaking(V.1.3.079)438
  seeds(*V.1.3.059)436
Soboliferous(II.0.03c;.1.1.3.088,.19c) 14,26,55
Social:
  behavior(*V.3.3.1.132e,*f)502
  wasps(*V.3.3.1.132e)502
Sociological:
  consequences(*V.3.3.3.053)535
  costs(*V.3.3.3.053)535
Sod:
  cover(*V.3.1.012)486
  culture(*V.3.1.006,.010)485,486
Sod webworm(*V.3.3.1.132g)502
Sodium (Na)(*V.3.2.028,*.056;*.3.5. 032)491,493,558
  ions(*V.2.120)482
  molybdate(*V.3.2.085)496
  -o-phenylphenate(*VI.1.3.033,*.085a) 580,584
  total(*V.3.2.071)495
Soft(*V.2.100)476
  fruits(*VI.8.124)625
  rot(s)(*V.3.3.2.181c)518
    of carrots(*V.3.3.2.040)511
    of fig(*VI.9.1.214)641
    of pineapple(VI.9.1.097)633
    of sweet potato(*V.3.3.2.181c)518
  scald of apples(*VI.9.3.032,.116) 665,671
  scales(*V.3.3.1.132d)502
Softwood cutting(V.1.4.022f-4,.03c) 443,455
Soggy breakdown:
  of apple(*VI.9.3.032)665
  of apples(*VI.9.3.117)671
Soil,soils(*IV.5.2.027,*.029,*.033, *.039,*.049,*.073;.5.4.3.006; *V.1.2.011,m,*n,*p;*.1.3.026b-1, *-3;.2.000,*.107,*.110a,*.111, .01;*.3.1.014;*.3.2.080;*.3.3.3. 080)401,402,403,405,414,429,430, 431,450,472,476,478,479,483,487, 495,536
  acid(*V.3.2.066)494
  alluvial(*V.2.010)472
  amendment(s)(*V.3.2.000,*.037,.079; *VI.7.2.017s)490,492,495,615
  analyses(*V.3.2.021,*.035,*.037, *.040)491,492
  analysis(*V.3.2.080)495
  application(V.3.3.3.093)537
  auger(V.3.1.011)486
  azonal(V.2.010)472
  bacteria(*IV.5.4.3.006)414

INDEX OF TERMS

-bed surface(*V.3.1.007)485
below surface(*V.3.3.3.094,*.095)537
-binding properties(*V.2.110e-2)479
-borne and-inhabiting disorders(V.3.3.0.002)497
   fungi(*V.4.2.1.016)566
   pests(*V.3.3.3.000)531
bulk density(*IV.5.2.029)401
calcareous(*IV.5.4.3.001,*.002)414
catena((II.4.2.004;*IV.0.141)238,335
class (size)(*V.2.115)480
classification (U.S.)(V.2.096)475
   system(*V.2.118)481
   systems(V.2.097)475
clay minerals(*IV.0.042;V.2.098)330, 475
coating(*V.2.096e)475
coherent masses(*V.2.1151)481
compaction(*V.3.3.3.048)534
conditioners(V.2.099)476
conductivity(*V.2.102)476
conservation(*V.3.1.006)485
consistence(V.2.096e,100)475,476
core drill(V.2.101)476
cores(*V.3.1.001)485
crumb structure(*V.2.099)476
development, genetic(*V.2.107c)477
diffusivity(V.2.102)476
disinfest(*V.3.3.3.028)533
disturbed(*V.2.115e)480
equivalent diameter of particles(*V.2.123)482
erosion(V.2.110d)478
   controls(V.2.110a)478
   factors(V.2.110b)478
favorable consistency(*V.3.1.014)487
formation(*V.2.000,*.107o)472,477
   processes(V.2.106)476
-forming process(*V.2.096a)475
grade (distinctness)(*V.2.115)480
hard layer(*V.2.077)474
heavy(*V.3.5.002)557
high carbon dioxide content(*V.4.2.2.006)567
   fertility(*V.2.110e-2)479
horizons(V.2.107,*.107e,*h,*.139) 476,477,482
horizon thickness(*V.2.096e)475
hydraulic conductivity(V.2.108)477
impermeable layer(*V.4.1.082)563
impervious layer(*IV.5.4.3.006)414
improvement(*V.2.110e-1)479
infilration rate(*V.3.5.014)558
injection(V.3.3.3.094)537
layered(V.3.3.3.095)537
leaching(*V.2.109)478
loss(*V.2.110d)478

lysimeter(V.2.109)478
management(*V.2.096e,.110)475,478
   practice(*V.3.0.001)484
manner of deposition(*V.2.010)472
matrix(*IV.5.2.057)404
mellow(*V.3.1.014)487
microorganisms(*V.3.2.000)490
mineral colloids(*V.3.2.000)490
mineralogy(*V.2.096e)475
mixer(V.1.1.062)426
   mechanized(*V.1.1.036)423
mixes(*V.1.2.011)429
moisture(*IV.5.2.013,*.029;*V.3.5.000,*.003,*.024,*.025;*VI.9.3.000)399,401,557,558,663
   content(*IV.5.2.039;*V.2.110b)403, 478
   disturbances(*V.3.3.2.270)524
   stress(*IV.5.2.023)401
nature(*V.3.5.032)558
neutron probe(*V.3.5.013,.024)557, 558
organic matter(V.2.111)479
organisms(*V.3.3.3.074)536
particle(s)(*V.2.110b,*.115g,*j)478, 480
   attraction between(*V.2.100)476
pasteurization(V.1.1.065)426
pathogens(*V.1.4.028j)446
percolation(*V.2.109)478
permeability(*V.2.096e,*.108,*.120; *.3.2.044;*.3.5.014)475,477,482, 493,558
pH(*IV.5.2.001;*V.4.2.2.006)398,567
physical condition(*V.3.1.014)487
   structure(V.3.2.069)494
,prepared(*V.3.1.001)485
primary particles(*V.2.115)480
profile(*V.2.028,*.083,*.107,*.107o, .112,*.120,.01a;.4.1.082)472, 474,476,477,479,482,483,563
   characteristics(*V.2.096f)475
properties(*V.2.100)476
reaction(V.2.096e,.113)475,580
reclamation(*V.3.2.044)493
rehibiliate infertile(*V.2.110)478
reserves(*V.3.053)493
rooting area(*V.3.5.014)558
rot of snap beans, lima beans(VI.9.1.499)654
   of sweet potatoes(VI.9.1.500)655
   of tomatoes(VI.9.1.501)655
   of watermelon(VI.9.1.502)655
saline(*V.2.110)478
sampling(*V.3.2.072,.082)495
samples(*V.2.101)476
saturated(*V.2.108)477

INDEX OF TERMS

screen(*V.1.1.036,*.047,.063)423,
    424,426
  mechanical(*V.1.1.036)423
separates(V.2.114)480
shredder(*V.1.1.036,*.040,*.047,
    .064)423,424,426
skeletal(*V.2.010)472
solonized(*V.2.119)481
solution(*IV.5.2.057;*V.3.2.015,
    *.065;.4.1.071)404,491,494,562
sterilizer(*V.1.1.036)423
sterilization(*V.1.1.065)426
sterilizing equipment(*V.1.1.047)424
storage capacity(*V.3.5.014)558
structural characteristics(*IV.5.2.
    029;*V.2.099;*.3.5.014)401,476,
    558
  units(*V.2.095)474
structure(*V.2.110b,.115,*.120)478,
    480,482
  improvement(*V.3.1.006)485
surface(*V.3.2.039;.3.3.3.081;*.3.5.
    011)492,536,557
surfaces(*V.2.110e)479
surface texture(*V.2.096f)475
temperature(*V.2.096e)475
tensiometer(*V.3.5.013,.025)557,558
textural classes(V.2.117)481
  triangle(V.2.01b)483
texture(*V.2.096e,*.110b,.116)475,
    478,481
type(*IV.5.2.029;*V.2.110b,*.115;
    .3.1.008;.3.3.3.049;*.3.5.
    002)401,478,480,485,534,557
types and phases(V.2.118)481
undisturbed(*V.2.115e)480
unsaturated(*V.2.108)477
use(*V.2.096e)475
water(*V.1.3.033)434
  depletion(*V.3.5.013)557
  osmotic potential(*IV.5.4.3.007)
    415
  potential(IV.5.2.057)404
  salt content(*IV.5.4.3.007)415
  suspension(*V.2.113)480
waterlogged, waterlogging(*IV.5.4.3.
    006;*V.2.110)414,478
wet(*V.4.2.2.006)567
Soilage(VI.9.1.498)654
Sol(IV.0.261)347
Solanaceous fruits(I.3.021)8
Solar:
  heat(*V.1.1.017h)421
  radiation(*IV.3.2.007)374
Solarization(IV.1.2.043)357
Soldier beetles(*V.3.3.1.132a)501
Soldiers(*V.3.3.1.132f)502

Sole proprietorship(*VI.8.057)620
Solenoid valve(V.1.1.053e)425
Solfatara(II.4.4.005a,.011)247,248
Solid:
  angle(*IV.0.260f)346
  chimera(III.2.2.024d-2-d,.01i)319,
    327
  particles dispersed in liquid(*V.3.
    3.3.107)538
Solitary(II.2.1.065,.08a)103,111
Solodization(V.2.119)481
Solonization(V.2.120)482
Soluble:
  fertilizers(V.1.2.013,*.015)431
  manganese (Mn)(*V.1.1.065)426
  materials in bulb(*V.4.2.2.006)567
    in profile(*V.2.096b)475
  salts(*V.2.083)474
  solid(V.3.3.3.096)537
  solids(*VI.1.008e)571
  -acid ratio(*VI.0.008,*.008f,.009)
    570,571
Solubridge(V.4.1.110)564
Solum(V.2.107o,*.127,*.128)477,482
Solute, solutes(*IV.5.2.036,*.037,
    *.038)402
  accumulation(*IV.5.2.036)402
  concentration(*IV.5.2.034,*.037)402
  reduced transport(*IV.5.4.3.006)414
Solution(V.3.3.3.097)537
Solutions(*V.3.2.054)493
Solvent(*IV.5.2.038;.3.3.3.097)402,537
  -type waxes(*VI.3.033,*.085c)580,584
Somatic(III.2.2.168)325
  cell(s)[IV.4.1.02(2),.03(1)(b);*.4.
    6.004)383,384,392
  (recurrent) agamosperny(IV.4.1.006a-
    14,.02a)379,383
Somoplasm(IV.4.8.052)395
Sonicate, sonification(IV.0.263)348
Sooty(II.5.1.001m)263
  blotch of apple and pear(VI.9.1.503)
    655
  mildew(*V.3.3.2.253)523
  mold(V.3.3.2.395)528
  of citrus(*V.3.3.2.181e)519
SOPP(*VI.3.033)580
Sorbitol(*IV.2.2.032b)362
L-Sorbose(*IV.2.2.031a)361
Sordid (sordidus)(II.5.1.006p)265
Sorgum midge(*V.3.3.1.132b)501
Sorus(*II.6.013,.018)280
Sour, sours:
  of bulbs(V.4.2.1.020)566
  orange scab(*V.3.3.2.181e;VI.9.1.
    505)519,655
  rot of carrot(VI.9.1.506)655

INDEX OF TERMS

    of citrus(VI.9.1.507)655
    of peaches(VI.9.1.508)655
    of pome fruits(VI.9.1.509)655
    of tomato(VI.9.1.510)655
  skin of onions(VI.9.1.511)656
  smell(*V.4.2.1.020)566
Source:
  materials(*V.3.2.034,*.035,*.037) 491,492
    (magnesium)(V.3.2.084)496
    (micronutrients)(V.3.2.085)496
    (nitrogen)(V.3.2.086)496
    (phosphate)(V.3.2.087)496
    (potash)(V.3.2.088)496
  packaging(*VI.3.071)583
  wrapping(VI.3.071)583
Souring of fig(VI.9.1.504)655
South American leaf blight of rubber (*V.3.3.2.181e)519
Southern bean mosaic of snap beans(VI.9.1.512)656
Southern blight(*VI.9.1.481,.482)654
  corn leaf blight(VI.9.1.514)656
  corn leaf spot(VI.9.1.515)656
Sowbug(V.3.3.1.253)508
Sowing rate(*V.1.3.066)437
Space:
  allocation(*VI.8.109,*.110)624
  competition(*V.3.3.0.000)497
  management(VI.8.110)624
  regulation(*VI.8.110)624
Spade(*V.1.0.007;.1.1.067)418,427
Spading fork(V.1.1.068,*.072)427
Spadix[II.2.1.066,.13b(1),c(1)]103, 116
Spanish moss(*V.1.1.017i,*.059)421,426
Sparse(II.1.2.103,.20a)63,85
Spathe[II.2.1.067,.07a,.13b(2),c(2)] 103,110,116
Spatulate(II.5.5.053;.5.12b)273,292
SPC(*V.4.1.111)564
Special:
  display(*VI.8.111)625
  handling(*VI.8.081)622
  kind of disease escape(*V.3.3.2.233)522
  physiological character(*V.3.3.2.051)512
  precooling(*V.4.1.012)559
    for tulips(V.4.1.111)564
  (pruning) terms(V.3.4.02d-h)546
  soil mixes(V.1.2.011n)431
Specialized:
  plagiotropic stem(*V.1.4.060)451
  stem(*V.1.4.064)451
  training and maintenance pruning regimes(*V.3.4.067)543

Species, species(*III.1.1.003,.031;*V.1.3.015;*.1.4.0281;*.3.3.1.132; *.3.3.2.181f;*.4.1.092;*.4.2.1.002;*.4.2.2.006)302,306,441,447, 501,519,563,565,567
  -complex(*III.2.2.141)324
  of pathogen(*V.3.3.2.177)517
  of susceptible plants(*V.3.3.2.177) 517
  susceptible(*V.3.3.2.414)529
Specific:
  combining ability(*III.2.2.029)319
  energy (J/kg)(*IV.0.260e)345
    imparted (gray)(*IV.0.260d)345
  entropy (J/kg·K)(*IV.0.260e)345
  gravity(*V.1.3.057;*VI.1.008a,*c; .5.043)436,570,591
  heat capacity (J/kg·K)(*IV.0.260e; *VI.5.024b,*.048,*.051a,*.051e) 345,589,592,593
  humidity(VI.5.025a)589
  latent heat(*VI.5.051c)593
    of condensation(*IV.5.051c)593
    of fusion(*VI.5.051c)593
    of liquefaction(*VI.5.051c)593
    of vaporization(*VI.5.051c)593
  recommendations(*V.3.3.3.052)534
  site problems(*V.3.0.001)484
  survival time(IV.5.2.059)404
  volume (cubic meter/kg)(*IV.0.260b) 344
  water potential(IV.5.2.075)405
Speckle of banana(VI.9.1.516)656
  of watermelon(VI.9.3.118)672
Spectral:
  energy flux(*IV.0.157z)339
  irradiance(IV.0.157z)339
  photon flux(IV.0.157aa)340
Spectrometry(IV.0.278)348
Spectrophotometer(IV.0.279)348
Spectrophotometric methods(*V.3.2.040)492
Speed(*IV.0.260b)344
  of light in a vacuum (c)(IV.0.157ab)340
Spermagonium(*V.3.3.2.349)526
Spermatia(*V.3.3.2.349)526
Spermatophyta, Spermatophyta(III.1.1.032;*V.3.3.2.179a)306,517
Spermatophytes(*IV.4.7.002;*V.3.3.3.179a)393,517
Spermatozoid(II.6.012)280
Spermocarpy(II.4.2.003)385
Sphaceloma perseae(*VI.9.1.047)629
Sphaerite(*II.3.037)195
Sphaeroblast(IV.3.1.014g)372
Sphaeronema spp.(*V.3.3.1.200e)506

Sphaeropsis tumefaceiens(*V.3.3.2.
  181g)520
Sphaerosome(*IV.1.1.019)351
Sphaerotheca:
  fuliginea(*V.3.3.2.181e)519
  pannosa var. persicae(*VI.9.1.409)
    650
Spagnum:
  moss(*V.1.2.011,.011o;*.1.4.036a)
    429,431,449
  peat(*V.1.2.005)429
Spherical:
  pellets or granules(*V.3.2.074)495
  -shaped fruit(*VI.3.064d)583
Spheroid(II.2.3.2.037,.07c,d)162,181
Spheroidal(*V.2.115f)480
Sphinx moth(*V.3.3.1.132g)502
Spice, spices(*I.3.006;*VI.10.007)7,
    674
  essential oil crops(*I.0.001;.1.013)
    3,4
Spiculate(II.5.6.129;.5.19d)277,299
Spicules(II.3.187)203
Spider(s)(*V.3.3.1.010;*VI.7.2.017n)
    498,614
  mites(*V.3.3.1.172)504
  wasps(*V.3.3.1.132e)502
Spike[II.2.1.068,.02a(1),.08e(1),.14b]
    103,105,111,117
Spikelet[II.2.2.24b(1)]153
Spiker(V.3.1.012)486
Spikes (on roller)(*V.3.1.012)486
Spindle(*VI.3.062)582
  bush(V.3.4.071,.04g)544,548
  -shaped(*II.5.6.049)275
Spine, spines(II.1.1.3.089,.20a;.2.1.
    10e;*.4.3.042,*.044,.054;.4.09b,c;
    *V.2.110e-2)26,56,113,244,257,479
  -hooks(*II.4.3.027)242
Spinescent(II.1.1.3.090,.20b)26,56
Spinning discs and rotors(*V.3.3.3.
    006)531
Spiny, spinose, spinulose(II.1.1.3.
    091,.092,.20a;.2.1.10e;.5.6.131)26,
    56,113,277
Spiny tip(III.2.2.171)325
Spiral(II.1.2.104,.20b;.2.2.041z,.05e,
    .28c;*V.3.4.021)63,85,125,134,157,
    541
  draper(*V.1.3.057)435
  embryo (spirally twisted)(II.2.3.3.
    .02o)183
  nematodes(V.3.3.1.200a,d,f)505,506
  (thickening)(II.3.22b,c)228
Spirodistichy(II.1.1.1.025,.06s)20,33
"Spitting" of hyacinth(V.4.2.2.019)568
Spittlebug(*V.3.3.1.132d)502

Splice graft(V.1.4.028k-7,.09a)447,461
Split(*II.5.4.3.007)270
Split pits of stone fruits(VI.9.3.119)
    672
Splitting and cracking of pomegranate
    (VI.9.3.120)672
Splitting of fig(VI.9.3.121)672
Spodic horizon (subsurface)(V.2.028h)
    473
Sponge-rubber sectioned rollers(*VI.3.
    081)583
Spongy(II.5.7.030)279
Spongy mesophyll[*II.3.111,.21a(5)]
    199,227
Sporagium (pl. sporangia)(II.2.2.007x;
    .6.014,.015;*V.3.3.2.181c,*.476;*VI.
    9.1.102,*.450,*.454,*.456)121,279,
    518,531,634,651,652
Spore(s)(*IV.4.7.002;V.3.3.2.401)393,
    528
  masses(*V.3.3.2.181f)519
Sporidium (pl. sporidia)(*V.3.3.2.
    181f,.402)519,528
Sporocarp(II.6.020)280
Sporocyte(II.6.021)280
Sporoderm(II.2.2.007t-10,.12d,.13a)
    120,141,142
Sporogenesis(IV.4.1.006b-6)380
Sporodochium (pl. sporodochia)(*V.3.3.
    2.181g,.403)520,528
Sporophyll(II.6.022)280
Sporophyte(II.6.023;*IV.4.1.006a-3,
    .026)280,378,381
Sporophytic(*IV.4.6.002,.004)392
  agamospermy(IV.4.1.02b)383
  generation(*IV.4.1.006a-1)378
  (polyembryony)(IV.4.6.002)392
Sporotrichum anthophilum(*V.3.3.2.
    181g)520
Spot, spots(*V.3.3.2.133b,.405;*.4.2.
    1.002)515,528
  picking(*V.2.023)577
  spraying or dusting(*V.3.3.3.071)536
Spotted(II.5.1.006q)265
  garden slug(*V.3.3.1.250)508
  wilt of celery(VI.9.1.519)656
    of garden pea(VI.9.1.521)656
    of lettuce(VI.9.1.520)656
    of tomato(VI.9.1.522)656
Sport(*III.2.2.021)318
Spray, sprays(*V.3.3.2.270;.3.3.3.071,
    *.102)524,536,538
  and dust program(*V.3.3.3.052,.098)
    534,537
  concentration(*V.3.3.3.016)532
  drift(V.3.3.3.099,*.116)537,538
  tank(*V.3.3.3.018)533

Sprayer, sprayers, spraying(*V.1.1.
    036;*.3.0.003;.3.3.3.100)423,484,537
Spread(V.3.3.3.101)538
    of pathogen(*V.3.3.3.2.134)515
Spreader(V.3.3.3.102)538
Spreading(*II.2.2.061c-10;*IV.5.4.4.
    002,.002g)127,416
Spring:
    -flowering bulbs(*V.4.1.012,*.018,
      .112,*.114)559,560,564
    -loaded cups(*VI.3.064e)583
    -tooth harrow(*V.1.1.020)421
    tooth rake(*V.1.1.050)424
    "wood"(*II.3.074)197
Sprinklers(*V.1.1.028)422
Sprout:
    clumps(*V.3.3.3.015)572
    removal(*V.1.4.028i,.11h)445,463
Sprouting (potatoes)(VI.5.046)591
Spruce budworm(*IV.3.3.3.013f)532
Spur, spurred, spurs[II.1.1.3.093,.14a
    (3);.2.2.23c(2),d(1),.24d(1);(*V.3.
    4.021,*.023,*.032,.069,.01(5),.02f]
    26,50,152,153,541,544,545,546
Spuriously compound(II.1.2.105)63
Spuriously imparipinnate(*II.1.2.072,
    .11c)62,76
Squamate, squamose(*II.5.6.116,.5.18g)
    277,298
Square(*V.3.1.008c,*j,*m,.01)485,486,
    488
    meter(*IV.0.260b)344
    -point shovel(*V.1.1.060)426
Squarrose(II.5.5.055,.5.12c)273,292
Squash(*I.3.007;VI.10.034)7,677
    bug(*V.3.3.1.132c)502
    (lady)beetle(*V.3.3.1.132a)501
Squirrels(*V.3.3.1.224)507
Sri Lanka(*V.2.097)475
Stable (horse) flies(*V.3.3.1.132b)501
Stage "G" of tulips, etc.(V.4.1.113)
    564
Stakes(V.1.2.014)431
Staking and tieing(*V.1.4.028i,.11f)
    445,463
Stalk(II.1.2.106)63
Stamen(s)(*IV.3.3.003e;*V.4.1.033)376,
    560
    arrangement(II.2.2.007d,.04-.06)118,
      133-135
Staminal:
    carpellody(III.2.2.172)325
    column(II.2.2.06f,.22e)135,151
    crown[II.2.2.26a(3)]155
Staminate[II.2.2.074,.01d,h,.17d,f;.2.
    3.2.04d(2);*IV.4.3.001,*.004,*.014,
    .017,.020]129,130,146,178,385,386,
    387
Staminode[II.2.2.007y,.01i(2),.09d,e
    (1),(ii),.27f(1)]121,130,138,156
Stamping(VI.3.072)583
Standard:
    alkali(*VI.1.008g)571
    atmosphere(IV.0.260m)347
    forcing(*V.4.1.012,.114)560,564
    refrigeration(*VI.6.029c)596
    ventilation(*VI.6.029c)596
Standards of identity(*VI.7.2.016)610
Staple items(VI.8.112)625
Starch[*II.1.3.007;*.2.3.3.015;*.3.
    125,.12a(5);*.4.3.063;*IV.0.042;
    .1.1.012;.1.2.008b,.019,*.036c,
    .040a;.2.2.031c,.031c-a-1;*VI.1.
    005,*.007]94,164,200,218,245,330,
    351,352,353,354,355,361,570
    sheath(II.3.052,.16c)196,222
Starchy root and tuber crops(I.3.022)8
Star crack of apple(VI.9.1.523)656
"Starter" solution(V.1.2.015)431
Starvation(*IV.5.2.044;*.5.3.005;*V.3.
    4.000)403,407,539
    injury(IV.5.4.1.005)412
Starved(*II.5.5.016)271
State(V.3.3.2.406;*VI.8.118)528,625
    environmental regulations(*VI.7.2.
      014)610
    farmers market(*VI.8.060,.113)621,
      625
    food and drug regulations(*VI.8.
      038d)619
    ,imperfect(*V.3.3.2.179a,.219,*.406)
      517,521,528
    -owned market(*VI.8.113)625
    ,perfect(*V.3.3.2.179a,.293,*.406)
      517,525,528
    restrictions (riparian rights)(*V.3.
      5.022)558
    seed regulations(*V.1.3.075)438
Statement(VI.6.030)596
Statutes(*V.3.5.033)559
Steam(*V.1.1.017h,*.047,.065)421,424,
    426
    atomization(*V.3.3.3.006)531
    ejector(*VI.10.008)674
    high-pressure(*VI.10.010)675
    pipes(*V.1.1.002)418
Steamer shipments(*VI.8.106)624
Steamships(*VI.6.002)594
    company shipping agent(*VI.8.106)624
Steel trap(*II.4.3.009)241
Stefan-Boltzmann law(IV.0.283;*VI.5.
    052c)348,593
Stele[II.3.189,.27;.4.10c(1)(a)]203,
    233,258

INDEX OF TERMS

Stellate(II.2.2.075,.03g;.3.204;.5.6.
  133;.5.19e)129,132,205,277,299
Stem, stems(II.1.1.3.000-.105,.01-.22;
  .3.030,.091,*.108,*.189,*.195,
  .196,.203,*.210;*.4.2.010;
  .4.3.003,.008,*.015,*.057,
  .061,.063,*.065;*IV.1.1.002;
  .3.3.008;.5.2.041;*V.1.5.000;
  3.3.2.375;.3.3.3.027;*.3.4.
  003;*.4.1.008,*.038;*.4.2.1.014,
  .015,.021)22-27,37-58,195,198,
  199,203,204,205,239,241,242,244,
  245,350,377,403,470,527,533,540,
  559,561,566
 anatomy(*II.3.01,.03-.10,.12,.14,
   .16,.18,.20,.22-.24,.26-.28)194-
   206,207,209-216,218,220,224,226,
   228-230,232-234
 above top leaf(*V.4.2.2.015)567
 apex(*V.4.1.051)561
 bulb(*V.4.1.016)560
 bulbiform(*V.1.4.058)451
 bulbils(*V.1.4.006)441
 bublet(*V.1.4.007,.17e-g;*.4.1.115)
   441,469,564
 cambial cylinder formation(II.3.06a-
   c)212
 -cavity browning of apple(*VI.9.3.
   020)664
 -clasping(*II.1.2.006)59
 cuttings(V.1.4.022f,*.068,.02,.03,)
   442,451,454,455
 daughter axis(*V.4.1.039,*.040,
   *.069)561,562
 drooping(*V.3.3.2.468)530
 elongation(*IV.5.4.2.007)413
   prevented(*V.4.2.2.021)
   568
 -end rind breakdown of citrus(*VI.9.
   1.001,.123)663,672
   rot of mango(VI.9.1.524)656
   rot of papaya(VI.9.1.525)656
   rot of watermelon(VI.9.1.526)656
   rots of avocado(VI.9.1.527)657
   rots of citrus(VI.9.1.528)657
 fusing and flattening(*V.3.3.2.170)
   516
 galls(*V.3.3.1.200a)505
 hairs(II.3.31)237
 limited necrotic area(*V.3.3.2.405)
   528
 modified(*V.4.1.003)559
 root(V.4.1.116)564
 rot(*V.3.3.2.181d;*.4.2.2.015)518,
   567
   of sweet potatoes(VI.9.1.532)657
 rust of cereals(*V.3.3.2.133c)515

 of grains(*V.3.3.2.181f)519
 short(*V.3.4.069)544
 soil removal(*V.1.4.028t)448
 stored and callused(*V.1.4.028b)443
 succulent(II.4.3.056)244
 tissue(*V.1.4.022d,*.024)442,443
 tissues(II.3.01)207
 topple of tulip(V.4.2.2.020)568
 trimming(*VI.3.078)583
 tropism(*IV.3.3.008)377
 tubers(II.4.3.057,*.063;.4.12a,b,
   .13b)244,245,260,261
 types of thickening(II.3.06d-f)212
 underground node(*V.4.1.115)564
   portion(*V.4.1.016,*.115)560,564
 vegetative(*V.4.1.008)559
 weak shoots(*V.3.3.2.469)530
 young, woody(*V.3.4.058)543
Stemphylium radicinum(*VI.9.1.089)633
Stenohydric plants(*IV.5.023)401
 species(IV.5.2.062)404
Stenophyllous(II.4.3.058;.4.08)245,256
Stenospermocarpy(IV.4.2.004)385
Stenothermic(IV.3.2.019)376
Steppe(II.4.2.034)240
Steradian(IV.0.157ac,*.260f)340,346
Stere (st)(IV.0.260m)347
Stereoisomers(IV.0.286,*.290;*.2.2.
 031a)348,361
Stereomorphic(IV.4.4.054)390
Sterile[II.2.2.076,.2.2.09e(ii);V.3.3.
 2.409]129,138,528
Sterility(*IV.4.8.029,.053)394,395
Sterilization(V.1.3.080)438
Sterilized(V.3.3.2.410)528
Sternotribic(IV.4.5.002q)392
Sterols(*IV.0.227;.2.2.107)342,366
Stewart's disease, Stewart's wilt(VI.
 9.1.074)632
Sticker(*V.3.3.3.071,.103)536,538
Stickiness(*V.2.100)476
Stigma, stigmas[II.2.2.041aa,ad,.02(1,
 13),.18;*IV.3.3.003e;*.4.3.002,
 .002b,c,.006,*.012,*.013]125,131,
 147,376,385,386
Stigmatic surface[II.2.2.25b(3),d(4)]
 154
Stilb (sb)(IV.0.260l)347
Stilted peg-roots(*II.4.3.039,.059;.4.
 11e)243,245,259
Stilt-roots(II.1.3.01c,*.4.3.039,.060,
 .4.11a)96,243,245,259
Stimulus(*V.3.3.2.355,.411)527,528
Sting nematodes(*V.3.3.1.200c)505
Stinging(*II.5.6.146)278
Stinging hairs(*II.4.3.042)244
Stink bug(*V.3.3.1.132c)502

INDEX OF TERMS

Stipe[II.2.2.041ab,.14c(2),.15h(1)] 125,143,144
Stipel(II.1.2.107,.20)63,85
Stipitate[II.1.1.1.026;.03;.2.2.041ac, .15h(2)]20,30,125,142,144
Stippen(*VI.9.3.008)663
Stipulate(II.1.2.108)63
Stipule, stipules[*II.1.2.021,*.055, *.066,*.100,.03c(3),.20d;.2.1.01 (14);*.4.3.041,*.071;.5.5.032]60,61, 62,63,68,85,104,244,245,272
Stock(V.1.0.008)418
  nursery(V.1.1.069)427
  plant(*V.1.008)418
  -plant production(*V.1.1.017e)420
Stockholders(*VI.8.004)616
Stokes (St)(IV.0.2601)347
  law(V.2.123)482
Stolon(*II.2.3.3.020a;*.4.3.057,.061; .4.12d;V.1.4.05a)164,244,245,260,457
Stoloniferous(II.4.3.062)245
  grasses(*II.1.1.3.082;*V.3.1.004)26, 485
  herbs(*II.4.1.003)238
Stolonlike structure(*V.4.2.2.006)567
Stomium[II.2.2.007z,.11c(2)]121,140
Stoma, stomata[*II.3.054,*.107,*.190, .12a(3),.15a(7),.29,.30;*.4.2.010; *IV.5.2.007,*.059]196,199,204,218, 221,235,236,239,399,404
  crypts(*IV.5.2.031)401
  responsive(*IV.5.2.020)400
  sunken(*II.4.3.032;*IV.5.2.080)243, 406
Stomatal:
  aperture(*IV.5.2.032)401
  conductance(*IV.5.2.036)402
  crypt[II.4.08b(2)]256
  opening(*IV.5.2.079)406
  pores(*IV.5.2.032)401
  resistance, total(*IV.5.2.032)401
Stone(s)[II.2.3.2.029i,.05a(3);*V.1.3. 051]162,179,435
  cells[*II.3.170,.3.16b(2)]202,222
  fruits(I.1.014)4
Stool layerage(*V.1.4.036b-1,.14c-e) 450,**466**
Storage(s)(*II.3.125;*V.1.3.063,*.064, .073;*VI.3.043,*.053b;*.5.,.047; *.9.000)200,437,581,582,588-594, 591,626
  bulb(*V.4.1.019)560
  bulbous crops(*V.4.1.022)560
  carbon dioxide treatment(*VI.5.007) 588
  conditions(*VI.9.3.000)663
  containers(*VI.4.000,.001d,*.012,

  *.014)585,587
  disorders(*V.3.3.2.133c)515
  dry or cold-moist(*V.4.1.019)560
  organs(*II.4.3.001,.063;.4.12,.13) 241,245,260,261
  parenchyma(*II.3.125)200
  polysaccharides(IV.2.2.031c,-a,*-a-1)361
  "roots"(II.4.13f)261
  seed(*V.1.3.073)438
  seed moisture content(*V.1.3.073)438
  stratification(*V.1.3.081)439
  temporary(*VI.3.052)581
Store(*VI.3.041)581
  bulbs(*V.4.1.010)559
  -door delivery(VI.8.114)625
  inventory(*VI.8.034)618
Stored:
  products pests(*V.3.3.1.132a,g, *.224)501,502,507
  tulips(*V.4.2.2.004)567
Storied cambium(II.3.07b)213
  wood(II.3.192)204
Storing(*V.3.2.082)495
Straddle-carrier(*VI.4.004)586
Straggling(*II.5.2.008)266
Straight(II.1.1.3.095)26
  load(VI.6.031)596
  -ribbed(*II.1.2.121i)64
Strain, strains(III.2.0.012;*IV.5.1. 005,.016;*V.3.3.2.121,.412;*.4. 2.1.004)316,396,397,514,529,565
  direct(*IV.5.4.1.005)412
  kinetic(*IV.5.4.1.005)412
  elastic(*IV.5.1.016,*.018)397
  growth(*IV.5.2.022)400
  irreversible(*IV.5.1.014,*.016)397
  plastic(*IV.5.1.014,*.016,*.018)397
  membrane(*IV.5.2.022)400
  prevention(*IV.5.1.018)397
  repair(*IV.5.1.018)397
  reversible(*IV.5.1.016)397
  stress-induced(*IV.5.2.022)400
"Strangler"(*II.4.2.015;.4.3.064;.4. 10d)239,245,258
Strap-shaped(*II.5.5.032)272
Stratification(*V.1.3.039,.081)434,439
Stratified wood(*II.3.192)204
Straw(*II.1.1.3.044)24
Strawberry latent ringspot(*V.4.2.1. 022)566
Stream(*VI.7.2.014)610
Street(VI.8.115)625
Streptomyces:
  <u>impomoea</u>(*VI.9.1.500)655
  <u>scabies</u>(*V.3.3.2.001;*VI.9.1.077, *.332,*.406)509,632,647,650

Strengthening agent(*V.3.2.028)491
Stress(*IV.0.260c;*.2.1.033;*.5.;*.5.1.001,*.005,*.010,.017;.5.3.032) 344,358,396-416,396,397,410
　applied(*IV.5.1.003)396
　cold(IV.5.3.)407-412
　condition(*IV.5.1.001)396
　drought(*IV.5.2.022,*.044)400,403
　　-induced(*IV.5.2.020)400
　droughtlike(*IV.5.4.3.007)415
　exclusion(*IV.5.1.018)397
　freeze-induced water(*IV.5.3.023)409
　freezing(IV.5.3.025)409
　gas(*IV.5.3.023)409
　high-temperature(IV.5.4.1.001)412
　induced(*IV.5.2.056)404
　injury(*IV.5.3.047)411
　internal(*IV.5.1.018)397
　light(IV.5.4.2.000)413-414
　mechanical(IV.5.4.4.002)416
　other forms of(IV.5.4.)412-416
　plant(*IV.5.1.004)396
　　water(*IV.5.1.018)397
　pressure(IV.5.4.3.008)415
　pseudohardening(*IV.5.3.041)411
　radiation-induced(IV.5.4.2.011)413
　resistance(IV.5.1.018)397
　　increased(*IV.5.3.041)411
　salt (ionic), pressure, and flooding (waterlogging)(IV.5.4.3.)414-415
　secondary(*IV.5.2.056;*.5.3.005)404, 407
　soil moisture(*IV.5.2.023)401
　survival(*IV.5.1.018)397
　tolerance(*IV.5.2.064;*.5.3.026)404, 409
　　,drought(*IV.5.2.064)404
　　freezing(*IV.5.3.026)409
　water(*IV.5.1.003;.5.4.4.001)396,415
　　situation(*IV.5.2.037)402
　(water)(IV.5.2.)398-407
Stretch film overwraps(*VI.3.053d)582
Stretchwrap(VI.4.000b,.023)585,587
Striate(*II.5.6.074)275
Strict(II.1.1.3.096,.21a)26,56
Strigose(II.5.6.136;.5.19f)277,299
Striped(II.5.1.006r)265
Stripper fingers(*VI.2.022b)576
Strobilus (strobile)(II.6.024)280
Stroma(IV.1.1.022d)350
Strombus-shaped(II.2.2.077,.03h)129, 132
Strong(*V.2.115e)480
　crotch[V.4.3.01(10)]545
　I (parthenocarpy)(*IV.4.2.001b)385
　II (parthenocarpy)(*IV.4.2.001c)385
　symptoms(*V.3.3.2.385)528

Stropha(III.1.2.057)310
Strophiole(II.2.3.3.037)166
Structural:
　isomers(IV.0.290)348
　polysaccharides(*IV.1.2.019;.2.2.031c-b)353,362
Structure formation(*V.2.106)476
Strychnine(IV.2.2.012)360
Stubble culture(*V.3.1.006,*.012,.013) 485,486,487
Stubby-root nematode(s)(*V.3.3.1.200d; .3.3.2.146)506,516
Stubs[V.3.4.01(11),(14)]545
Stump puller(*V.3.0.003;*.3.1.000)484
Stunt nematodes(*V.3.3.1.200d)506
Stunting(*V.4.2.1.001,*.022)565,566
Sturdy trees(*IV.5.4.4.002)416
Stylar:
　canal[II.2.2.02(1a),.25d(5)]131,154
　-end clearing of grapefruit(*VI.9.3.012)664
Style[II.2.2.041ad,.02(2,14),..14b,e,f, .17]125,131,143,146
Stylet-borne(*V.3.3.2.133e)515
Styloid(*II.3.037)195
Stylopodic(II.2.2.17j)146
Stylopodium[*II.2.3.1.022,.035;.03j (1)]159,160,170
Styrene-latex cones(*V.1.2.003b)428
Styrofoam(*V.1.2.005)429
　cones(*V.1.2.003b)428
　cups(*V.1.2.003d)428
Subangular blocky(V.2.115k)480
Suberin(*II.3.052,.059,.085)196,198
Suberization(II.3.195;*V.1.4.028e)204, 444
Suberize(II.3.196)204
Suberous(II.5.7.031)279
Subconfluent(II.2.2.078)129
Subcordate(II.5.4.2.013;.5.04k)269,284
Subepidermal:
　mutants(III.2.2.024d-2)319
　sclereids(II.3.25c)231
Subfamily(*III.1.1.002)302
Subglobose(II.2.2.061c-1-r,.211)127, 150
Subgroup(V.2.096d)475
Sub-inferior(II.2.2.041t-3-a)124
Subirrigation(*V.1.1.017f)420
Subject approval government inspection (VI.7.069)606
Submerged, submersed(II.4.2.035)240
Subopposite(II.1.2.110)63
　(buds)[*II.1.1.1.014,.031(1);.1.2.110,.20e]19,30,63,85
Suborbicular(II.5.5.057)273
Suborder(V.2.096b)475

Subsample(V.3.2.071)495
Subsampling(*V.3.2.082)495
Subsidiary cells[II.3.29f,.30h(1)] 235,236
Subsoil(V.2.127)482
  drains(*V.3.5.002)557
  mole(*V.3.5.002)557
  physical condition(*V.2.110e-1)479
  plow(*V.1.1.044)423
Subspecies, subspecies(III.1.1.033)306
"Substitute fiber"(II.3.23a)229
Subsurface flood(*V.3.5.012)557
Subterminal (buds)[II.1.1.1.028,.01d(2)]20,28
Subterranean:
  rhizomes(*V.1.1.056)426
  stems(*V.3.1.004)485
Subtribe(*III.1.1.002)302
Subtropical:
  areas, cool(*IV.5.3.006)407
  crops(*I.1.000,.015)3,4
  plants(*IV.5.3.028)409
Subtruncate(II.5.4.1.020;.5.03o)269, 283
Subulate(II.5.5.058,.12d)273,292
Succession of changes(*V.3.3.2.243)522
Succinate(*IV.1.2.004)352
Succinic acid-2,2-dimethyl hydrazide (IV.2.1.037)359
Succulent, succulence, succulents(II.4.3.065;.4.05e;.5.7.032;IV.0.291; *.5.2.002,*.079,*.085)245,253,279, 348,398,406
  leaves, stems or roots(*IV.5.2.085) 406
Sucker, suckers[*V.1.4.000,*.024,.083, .04c(6),.16a]441,443,452,456,468
Sucking(*V.3.3.1.132g)502
Sucrose(*IV.1.2.036c;.2.2.031b;*V.1.5. 005;*VI.008a,*c,*d,*e,*.009;see also Sugar)354,361,470,570,571
Suction:
  gauge(*V.3.5.025)558
  gradient(*V.2.108)477
Sudsing(*VI.3.081)583
Suffrutescent(II.1.1.3.097,.21b)26,57
Sugar, sugars(*II.3.116;*IV.1.1.030; *.2.1.001;*.2.2.093,*.140,*.141, *.142;*.5.2.017;*V.1.5.005;*.3.3. 3.009;*VI.1.005,*.008e,*f,*.009; .5.051b;*.10.020,*.021,*.039)199, 352,357,366,367,400,470,531,570, 571,593,676,677
  accumulation(*IV.5.2.085)406
  acids(IV.2.2.032c)362
  alcohols(IV.2.2.032b)362
  amino(IV.2.2.032f)362

  beet[*II.4.3.070;.4.13f(1)]245,261
  cane fiber(*V.1.1.053e)425
  content(*IV.5.3.024;*VI.1.008)409, 570
  deoxy (desoxy)(IV.2.2.032e)362
  diphosphates(*IV.1.2.037)355
  monophosphates(*IV.1.2.037)355
  phosphates, five carbon(*IV.1.2. 040g)356
  utilization(*IV.1.2.040f)356
Suitable shipping condition:
  (direct shipments)(*VI.019,*.020, *.021,.070)600,606
  (reconsigned roller or tramp cars) (*VI.071)606
Sulcate(II.5.6.137;.5.20a)277,300
Sulfate:
  of ammonia(*V.3.2.016,*.086,*.089) 491,496
  of potash(*V.3.2.016,*.088)491,496
  of potash-magnesia(*V.3.2.084, *.088)496
Sulfhydryl content(*IV.5.2.018;*.5.3. 024)400,409
Sulfides(*V.3.5.032)558
  accumulation(*IV.5.4.3.006)414
Sulfur (S)(*V.3.2.028,*.056,*.079, .089;.3.3.3.029)491,493,495,496, 533
  -coated urea(*V.3.2.015)491
  -colored(II.5.1.008r)266
  compounds(*IV.5.4.4.001)415
  dioxide injury on stone fruits(VI. 9.3.126)672
  total(*V.3.2.071)495
Sulfureus, sulphureus(*II.5.1.008r)266
Sulfuric acid(*V.3.2.091)496
Sulphur (see Sulfur)
Sul-po-mag(*V.3.2.084,*.088)496
Summer:
  spores(*V.3.3.2.442)529
  sprouting of lily(V.4.1.117;*.4.2.2. 008)564,567
  wood(*II.3.074)197
Sun:
  drying(*V.1.3.061;*VI.10.009)437,675
  full(*V.2.110e-2)479
  vs. shade leaves(*IV.5.2.002)398
Sunburn(*V.1.4.028d;*.3.3.2.270)444, 524
  of apples(VI.9.3.131)672
  of citrus(VI.9.3.127)672
  of fig(VI.9.3.128)672
  of pineapple(VI.9.3.129)672
  of pomegranate(*VI.9.3.135)673
Sunken cells(II.3.29d)235
Sunkist(*VI.8.122)625

Sunscald(IV.5.3.048;*.3.3.2.270)411,
524
  delayed, of apples(*VI.9.3.131)672
  of cabbage(VI.9.3.132)672
  of onion(VI.9.3.134)673
  of pomegranate(VI.9.3.135)673
  of snap beans(VI.9.3.136)673
  of tomato and pepper(VI.9.3.137)673
  on vegetables(VI.9.3.133)672
Supercooling(IV.5.3.049)411
Superficial:
  discoloration(*VI.9.3.139)673
  necrosis(*VI.9.3.139)673
  scald of 'Anjou' pear(VI.9.3.138)673
Superior(II.2.2.041t-3-d,.19a,b)124,
148
Supermarket(*VI.8.012,*.013,*.048,
  .049,.072,*.089)617,620,621,623
  retail(*VI.8.012)617
  store(*VI.3.053d)582
  warehouse(*VI.3.053d)582
Supernumerary(II.1.1.1.029,.01a)20,28
Superphosphate(*V.1.2.005,.011n;*.3.
  2.003,*.087,*.091)429,431,490,496
Superposed(II.2.3.2.038,.07e)162,181
  buds[II.1.1.1.030,.01a(2,3)]20,28
Supervision of day-to-day operations
  (VI.8.057)620
Supervisors(VI.8.116)625
Supervolute(II.1.1.2.027,.03w,x)21,36
Supplement(V.3.3.3.104)538
Suppression(V.3.3.2.413)529
Supradecompound(II.1.2.111,.06b)64,71
Surface:
  boundary layer resistance(*IV.5.2.
    065)404
  breakdowns of citrus(VI.9.3.139)673
  connective coefficient of heat tran-
    sfer(*VI.5.024a,.048)589,592
  dirt(*VI.3.081)583
  ditches(*V.3.5.002)557
  dulling, darkening, discoloration of
    watery areas(*VI.9.3.029)665
  features(II.5.6.000;.5.13-.20)273-
    278,293-300
  horizon(*V.2.096d)475
  irrigation(*V.3.5.012)557
  rot of sweet potatoes(VI.9.1.534)657
  soil(V.2.128)482
  tension(*IV.0.260e)345
Surfactant(V.3.3.3.105)538
Surge control(VI.3.027a-2)579
Survival:
  (pathogen)(*V.3.3.2.374)527
  time, drought chamber(*IV.5.2.018)
    400
    ,field(*IV.5.2.018)400

Suscept, suscepts(*V.3.3.2.003,*.284,
  *.285,.414,.436;.3.3.3.083)510,524,
  529,537
Susceptible plant species(*V.3.3.2.
  309)525
Susceptibility(V.3.3.3.106)538
Suspension, suspensions(*V.2.106;*.3.
  2.054;.3.3.3.107)476,493,538
  and revocation of license(VI.1.072)
    606
Suspensor[II.2.3.3.038,.03f-i(5),.10b
  (6);.3.086]166,184,191,198
Swale(II.4.2.036)240
Swallowtail butterflies(V.3.3.1.132g)
  502
Swamp, swamps(*II.4.3.035;*IV.5.4.3.
  006)243,414
  plant(*II.4.2.019;.4.4.005b)240,247
Swamper(VI.8.055)620
Swarm-spore, motile(*V.3.3.2.476)531
Swath width(*V.3.3.3.016)532
Sweet:
  orange scab(*VI.9.1.505)655
  potato dry rot(VI.9.1.537)657
    end rots(VI.9.1.538)657
    scurf(VI.9.1.539)657
Sweetness(*VI.8.079)622
Swelling(*V.3.3.2.375)527
Swollen shoot viruses(*V.3.1.008h)486
Sword-shaped(*II.5.5.020)272
Syconium, syconia(II.2.1.069,.13d;.2.
  3.1.036,.07d;*IV.4.4.007,*.031,
  .032,.033)103,116,160,174,388,389
Syllepsis, sylleptic[II.1.1.3.098,.17a
  (2)]26,53
Symbiont(II.4.3.066;V.3.3.2.415)245,
  529
Symbiosis(*II.4.3.005,*.066,.067;*V.3.
  3.2.110,.416)241,245,514,529
Symbiote(*V.3.3.2.415)529
Symbiotic relationship(*V.3.3.2.415)
  529
Sympatric(III.1.2.058)310
Sympetalous(*II.2.2.061c-7,.20e)127,
  149
Sympodial(II.1.1.3.097,.21c)27,57
Symptom, symptoms(V.3.3.2.417;*.4.2.1.
  004)529,565
  expression(*V.3.3.0.004)497
Symptomless carriers(*V.3.3.0.004;*.3.
  3.2.087)497,513
Symptomology(V.3.3.2.418)529
Synandrium(II.2.2.007d-17,.06a)119,135
Synantherous(II.2.2.007d-18,.04b,.06d)
  119,133,135
Syncarp(II.2.3.1.037,.07a-c)160,174
Syncarpelly(II.2.2.041v;.2.3.2.039,

INDEX OF TERMS

.03d)125,162,177
Syncarpium(*II.2.3.1.037,.07a-c)160, 174
Syncarpous(II.2.2.041ae,.14c;.2.3.2. 040,.03e)125,143,162,177
Synchronous dichogamy(*IV.4.3.002f)386
protogyny(IV.4.3.002f)386
Syndetocheilic(II.3.190)204
Synecology(III.1.2.059)310
Synergid, synergids[II.2.2.041t-4-b-4, .16b(10);*IV.4.1.006a-2,.006b-2,.01 (8),.02a(6),b(6)]124,145,378,379, 382,383
Synergism, synergistic(V.3.3.3.108)538
Synflorescence(*II.2.1.017,.070,.06a) 100,103,109
Syngameon(III.1.2.060)311
Syngamy[*IV.4.1.,.026,.01(11),.02b(2)] 378-384,381,382,383
Syngenesious(*II.2.2.007a-18,.04b,e, .06b)119,133,135
Syn-1(*III.2.0.005a-5-a)314
Synplasm(II.3.198)204
Synsepalous(*II.2.2.061a-2,.20e)126, 149
Synthetic(III.1.2.173)325
mulches(*V.3.3.3.059)535
nitrogenous material(*V.3.2.093)497
resins(*VI.3.085,*.085c,*d)584
(soil) conditioners(*V.2.099)476
varieties(III.2.0.005a-5)314
Syntype(III.1.1.035g)307
Synzoochory(IV.4.7.007)393
Syringe, syringes(*V.1.1.028,*.054, .071)422,426,427
Syrphid flies(*V.3.3.1.132b)501
Systemic(V.3.3.2.419)529
disorder(*V.1.4.035)449
disorders(V.3.3.0.004)497

T

Tachinid flies(*V.3.3.1.132b)501
Taeniothrips inconsequens(VI.9.2.081) 662
Tail-pointed(*II.5.4.1.009)268
Takhtajan(*III.1.1.004)303
Talc(*V.3.3.3.029)533
Tall(II.5.2.035)267
covers(*V.2.110e)479
vertical(V.3.4.06d)550
Tampico(*VI.3.007,*.081)578,583
Tanks(*V.2.109;*.3.2.083)478,495
Tannins(*II.3.077,.125;*IV.1.1.030) 197,200,352
Tapering(*II.5.5.004;.5.08c)271,288
Taper-pointed(*II.5.4.1.001)268

Tapetum[II.2.2.007aa,.11a(4),b(6)]121, 140
Taphrina:
bullata(*V.3.3.2.181e)519
deformans(*V.3.3.2.181e;*VI.9.1.301) 519,645
spp.(*V.3.3.2.181e)519
Tap root[II.1.3.02a,b(1);.3.199;V.3.4. 01(19)]97,204,545
plus hypocotyl plus stem storage (*II.4.3.063,.068)245
plus hypocotyl storage[*II.4.3.063, .069;.4.13f(3)]245,261
storage[*II.4.3.063,.070;.4.13f(1)] 245,261
Tarantulas(V.3.3.1.010)498
Tardily dehiscent(II.2.3.1.03b,c)170
Tarnished plant bug(*VI.9.2.013)660
Tartaric acid(*VI.1.008g)571
Tassel(II.2.1.071,.14a)103,117
Taste panels(*VI.8.079)622
TASTE(VI.10.011a)675
Tawny(II.5.1.008s)266
Tawny garden slug(*V.3.3.1.250)508
Taxon, taxa(III.1.1.034,*.035a;*IV.4. 4.040,*.047;*V.3.3.2.179b)306,390, 517
Taxonomic:
classification of enzymes(IV.2.2. 058g)364
synonyms(*III.1.1.025)305
types(III.1.1.035)306
Taxonomy(III.1.;.1.0.000)302
of fungi, lichens, algae, myxomycetes and bacteria(V.3.3.2.179a)517
T-bud(V.1.4.028f-8,.08a,b,.11k)445, 460,463
TBZ(*VI.3.033)580
Tea(I.1.001a,.017;V.3.4.073,.09c-e)4, 5,544,553
harvester(*VI.2.022b)576
training(V.3.4.09c-e)553
Tear-shaped(*II.2.3.2.019)161
Technical information(*V.1.1.017a)419
Tec-man-gan(*V.3.2.085)496
Tegmen[II.2.3.3.039,.04b(2),.07f(8), .08a(1),.12b(3),e(2)]166,185,188, 189,193
Tegmic seeds(II.2.3.3.040,.03d,e,.12e) 166,184,193
Television(*VI.8.095)623
Teliospore, teleutospore(*V.3.3.2. 181f,*.356,*423)519,527,529
Telophase(III.2.2.174)325
TEM(*IV.0.079)332
Temperate:
areas, warm(*IV.5.3.006)407

fruit trees(*II.1.1.3.087)26
fruits(*V.1.4.028b)443
zone crops(*IV.5.3.006)407
  (deciduous) crops(*I.1.000,.016)3, 5
Temperature(s)(*IV.0.260a,*.295;*IV.5.
    3.028,*.035,*.038;*V.1.1.017a;
    .3.3.2.270;.4.1.036,*.053;*VI.
    9.3.000)344,349,409,410,419,524,
    561,663
  abnormally high(*V.4.1.063)562
  celsius(*IV.260a,.295b)344,349
  coefficient(*IV.0.233)343
  control(*V.4.1.123)565
  fahrenheit(IV.0.295c)349
  kelvin(*IV.0.260a,*.295d)344,349
  low(*IV.5.3.042)411
  slow reduction(*IV.5.3.014)408
  thermodynamic(*IV.260a,.295d)344,349
  treatments(*V.4.1.096)563
    special(*V.4.1.091)563
Temporary:
  storage(*VI.3.052)581
  stress reaction(*IV.5.1.004;*.5.3
    .004)396,407
  tolerance(*IV.5.3.043)411
Tenacity(V.3.3.3.110)538
Tendril, tendrils(*II.0.002;*.4.3.002,
    *.014,.071;.4.06c;*IV.3.3.004,
    *.008i,p)10,241,245,254,377
Tension wood(*II.3.155)201
Tent caterpillar(*V.3.3.1.132g)502
Tepal(II.2.2.061f,.20c)129,149
Tera (T)(*IV.0.260h)346
Teratological(II.0.024)11
Teratologomorphosis(II.4.4.007c,*.008)
    247
Terete(II.1.2.112,.20f;.2.2.17k,.18l)
    64,85,146,147
Terminal[V.3.4.01(1)]545
  bud(*V.1.4.071b)451
  (buds)[II.1.1.1.031,.01d(1)]20,28
  growth, crooked(*V.4.2.1.005)565
  market(*VI.3.023;*.8.055,*.060,
    .115,.118)579,620,621,625
    company(*VI.8.011,*.024)617,618
    operations(*VI.8.107)624
    operator(*VI.8.066,*.096)621,623
  packaging(*VI.8.118,.119)625
  (parenchyma)(*II.3.014,.04c)194,210
  pore(II.2.2.10d)139
  warehouse(*VI.3.053d)582
Termite(s)(*V.3.3.1.132f;*VI.7.2.017n)
    502,614
  activity(*V.2.106)476
Ternate(II.1.2.113,.21a)64,**86**
Terpene(s)(*IV.0.227;*.2.2.014,*.107;

  *VI.9.3.090)342,360,366,670
  derivatives(*IV.5.4.4.001)415
Terraces(*V.2.110c)478
Terrachlor(*V.1.1.065)426
Terrain slope (inclination)(*V.2.110b)
    478
Terrestrial(*II.4.2.008)239
  gastropods(*V.3.3.1.251)508
  pulmonate gastropods(*V.3.3.1.250)
    508
  roots(II.1.3.01b)96
  vascular plants(*IV.3.2.004)373
Tertiary gene pool(III.2.1.002c)317
Tesla (T)(*IV.0.260c)344
Test:
  cross(III.2.2.175)325
  plant(*V.3.3.2.221)522
  tubes(*V.1.5.005)470
Testa[II.2.3.3.041,.03f-i(1),.04b(1),
    .07f(7),.08a,.12d(3,4)]166,184,185,
    188,189,193
Testaceous (testaceus)(II.5.1.008t)
    266
Testal outgrowths(II.2.3.3.042)166
  seeds(II.2.3.3.043,.12c,d)167,193
Tetrad[II.2.2.007t-11,.11b(7),..12c,d;
    III.2.2.176]120,140,141,326
Tetradynamous(II.2.2.007d-20,.05d)119,
    134
Tetrahedral (tetrad)[II.2.2.12c(1)]141
Tetraploid(III.2.2.177)326
Tetrasomic(III.2.2.178)326
Tetrazolium test(*V.1.3.031)434
Tetrose phosphate(*IV.1.2.042)356
Texas foliage plant mix(*V.1.2.005)429
  root rot of cotton(*V.3.3.2.181g)520
Texture(II.5.7.;*IV.4.4.013;*VI.8.
    038e,*.094)278-279,389,619,623
Thallophytes(*IV.3.2.019;*.5.2.040;
    *VI.7.2.017i)376,403,613
Thallus, thalli(*V.3.3.2.179,*.207)
    517,521
Thanatephorus cucumeris(*V.3.3.2.179a,
    .181f,.357;*VI.9.1.181,*.441,
    .499,.501,*.502,*.539)517,519,527,
    639,651,654,655,657
Thawing rate, standard(*IV.5.3.002)407
Theaspirone(IV.2.1.037)359
Theft(*VI.8.108)624
Theobromine(*IV.2.2.012)360
Theoretical heat removal(*VI.5.024a)
    589
Thermal:
  conductivity (k)(*IV.0.260e;*VI.5.
    048,.049,*.050)345,592
  death point(V.3.3.2.426)529
  decomposition(*V.3.2.006)490

diffusivity(VI.5.050)592
energy(VI.5.051)592
energy transfer(VI.5.052)593
induction(IV.3.2.020)376
radiation(*IV.3.2.007)374
Thermally Accelerated Short-Time Evaporator(VI.10.010)675
Thermistor(*IV.0.295b)349
Thermocouple(*IV.0.295b)349
Thermodormancy(*V.1.3.069,.083)437,439
Thermodynamic temperature(*IV.0.260a, .295d)344,349
Thermometer(*IV.0.295b;*VI.5.025c)349, 589
Thermonasty(IV.3.3.003f)376
Thermoperiod(*IV.3.2.021c)376
Thermoperiodism(*IV.3.2.021)376
Thermoperiodicity(*IV.3.2.021d)376
Thermostat(*IV.0.295c)349
Thermostability, enzyme(*IV.5.3.034) 410
Thermotolerance(*IV.5.3.034;*.5.4.1.006)410,412
Thermotolerant(*IV.5.3.033)410
Thermotropism(IV.3.3.008s)377
Therophyte(II.4.1.006;.4.02b)238,250
Thiabendazole(*VI.3.033,*.085a)580,584
Thiamin(*V.1.5.005)470
Thick, thickened(II.5.7.033)279
Thielavia basicola(*V.3.3.2.181e)519
Thielaviopsis:
  basicola(*VI.9.1.082)632
  paradoxa(*V.3.3.2.179a;*VI.9.1.097, *.182,.541,*.542)517,633,639,657
Thielaviopsis stalk rot of banana(VI.9.1.542)657
Thigmotropism(*IV.3.3.008i)377
Thigmomorphogenesis(IV.3.3.007,*.008i) 377
Thigmonasty(IV.3.3.003b)376
Thinning(*IV.5.4.4.002;*V.3.4.000b, .02c)416,539,546
  out one-year-old canes(*V.3.4.008) 540
Thiocarbamates(*V.3.2.085)496
"Thirst quencher(*VI.10.034)677
Thixotropy(IV.0.302)349
Thoracic legs(*V.3.3.1.132g)502
Thorax(*V.3.3.1.132)501
Thorn(*II.1.1.3.089;*.4.3.002,*.027; .4.09a)26,241,242,257
Thread-shaped(*II.2.2.035)122
  waisted wasps(*V.3.3.1.132e)502
Three:
  -way cross(III.2.004m)313
  -wire trellis(*V.3.4.05)549
Threshing(*V.1.3.063,.084,*.088)437, 439
  machine(*V.1.3.084)439
Threshold:
  temperature(*IV.5.3.006)407
  water potential(*IV.5.2.007)399
Thrips(V.3.3.1.132i)503
  injury on cabbage(VI.9.2.079)662
  of citrus(VI.9.2.080)662
  to stone fruits(VI.9.2.081)662
Thrips tabaci(*VI.9.2.079)662
"Thrum"(*IV.4.3.009,*.012,*.013)386
Thylakoid(IV.1.1.002e)350
Thymine(*IV.2.2.056,*.141,*.175)363, 367,369
Thyrse, thyrsus[II.2.1.072,.02b(1), .14c]103,105,117
Thysanoptera(V.3.3.1.132i)503
TIBA(*IV.2.1.012)357
Ticks(V.3.3.1.010;*VI.7.2.017n)498,614
Tie-in display(VI.8.120)625
Tied-leaf of tulip(V.4.2.2.021)568
Tiger beetles(*V.3.3.1.132a)501
  moth(*V.3.3.1.132g)502
Tile drains(*V.3.5.012)557
Tillage(*V.2.110c,*.128;*.3.1.006)478, 482,485
Tiller(II.1.1.3.100,.22a)27,58
Tilth(V.3.1.014)487
Time(*IV.0.260a;*V.2.000,*.108)344, 472,477
  -dependent operation(*VI.3.016)578
  of year(*V.3.3.3.049)534
Timing(*V.3.3.3.052)534
Timothy bumper mill(*V.1.3.057)435
Tine(V.1.1.072)427
Tineid(*V.3.3.1.132g)502
Tip blights(*V.3.3.2.133b)515
  burn of cabbage(VI.9.3.140)673
  of head lettuce(VI.9.3.141)673
  layerage(V.1.4.036b-4,.14b)450,466
Tissue, tissues(*II.3.008,*.048,*.080, *.108,*.112,*.125,*.135,*.144, *.160,*.173,*.175,*.189,*.203, *.206,*.215,*.221)194,196,197, 199,200,201,202,203,205,216
  absorptive(*II.3.215)205
  culture(V.1.5.001)470
  cultures(*IV.5.3.014)408
  death(*V.3.3.2.059)512
  dermal(*II.3.128)200
  fundamental(*II.3.189)203
  histogen theory(*II.3.080)197
  meristems(*II.3.108)199
  metaphloem(*II.3.112)199
  parenchyma(II.3.125)200
  permanent(*II.3.048)196
  phloem(*II.3.135)200

primary vascular(*II.3.173)202
procambium(*II.3.144)201
rhytidome(*II.3.160)202
secondary(*II.3.173)202
  phloem(*II.3.175)203
tunica-corpus theory(*II.3.206)205
vascular(*II.3.189,*.203)203,204
xylem(*II.3.221)206
Titanium, total(*V.3.2.080)495
Toadstool(*V.3.3.2.259)523
Tobacco:
  necrosis(*V.4.2.1.022)566
  rattle(*V.4.2.1.022)566
  ring spot(*V.4.2.1.022)566
  ring spot vector(*V.3.3.1.200c)505
  stems(*V.3.2.068)494
Today's shipment(VI.7.1.073)607
TOFC(*VI.6.033)596
Tolerance(s)(IV.5.1.001,*.019;*.5.3.041;V.3.3.3.111;*VI.8.038e,f)396,398,411,538,619
  for color additives(*VI.7.2.016)610
  for pesticides(*VI.7.2.016,*.017)610,612
Tolerant(V.3.3.2.429)529
Tolerate(*IV.5.1.001)396
Tolypophagus(*V.3.3.2.264)523
Tomato:
  black ring(*V.4.2.1.022)566
  fruit rot(*V.3.3.2.181g)520
  fruit worm(*VI.9.2.021)660
  leaf mold(*V.3.3.2.181g)520
  mosaic(VI.9.1.543)658
  ring spot vector*V.3.3.1.200c)505
  term(*VI.8.133)626
  wilt(V.3.3.2.181g)520
Tomentose, tomentulose(II.5.6.138,*.147;.5.20b)277,278,300
Tomorrow's shipment(VI.7.1.074)607
Ton of refrigeration(VI.5.053)593
Tone(II.5.1.000d)263
Tongue-and-groove approach graft(V.1.4.028a,.06d)443,458
Tongue graft(V.1.4.028k-9,.09b)447,461
Tonguing(*V.2.096c)475
Tonnage items(VI.8.121)625
Tonne (t)(*IV.0.260i)346
Tonoplast(IV.1.1.029)352
Tools(*V.3.3.0.004;*.3.3.3.028)497,533
Toothed(*II.5.4.3.012)270
Top grafting(*V.1.4.028k)446
"Top" icing(VI.5.053)593
Topcross(III.2.0.0041)313
Topdress(*V.3.1.001)485
Topiary(*IV.5.4.4.002g;V.3.4.075,.12b)416,544,556
Topical application(V.3.3.3.112)538

Topography(*II.4.2.004;*V.1.1.017a;*.3.1.008;*.3.5.002)238,419,485,557
Topophysis(IV.3.1.014h)372
Topotype(III.1.1.038h)307
Topper(V.3.4.075)544
Topping(*V.1.4.028i,.11d;*.3.4.017,.06a)445,463,540,550
  machine(*V.3.4.062)544
  shears(*V.3.4.063)543
Top-shaped(II.2.3.2.043)162
Topsoil(*V.2.110d,*.128)478,482
  physical condition(*V.2.110e-1)479
Topworking(*V.1.4.028d,.028k-1,*-2,*-3,.028v,.10;*.3.4.011)444,446,447,449,462,540
Torn(*II.5.4.3.022)270
Torr(*IV.0.260m)347
Tortuous(II.5.2.036)267
Torulose(*II.5.6.078)276
Torus(*II.2.2.066)129
Total:
  drought resistance(IV.5.2.064)404
  exchangeable bases(*V.2.011)472
  irradiance(IV.0.157ad)340
  leaf resistance(*V.5.2.032)401
  mineral nutrition requirements(*V.3.2.035)492
  requirements(*V.3.2.000)490
  secondary plant foods(*V.3.2.034)491
  soluble solids(*VI.1.008,*.008a,*c,*d,*e,*.009,*.010)570,571
    to total (titratable) acid ratio (VI.1.008f,*.009)571
  stomatal resistance(*IV.5.2.032)401
  supermarket sales(*VI.8.031)618
  (titratable) acid(*VI.1.008,.008g,*.009)570,571
Totem pole plants(V.4.1.119)564
Totipotency(IV.3.1.014i)372
Totipotent(III.2.2.179)326
Totipotentiality(III.2.2.180)326
Touch(*IV.3.3.008i)377
Toxic:
  chemicals(*V.3.3.2.270)524
  ions(*IV.5.2.001;*.5.4.3.000)398,414
  levels(*IV.5.4.3.000;*V.3.2.083)414,495
  sprays or dusts(*VI.9.3.000)663
Toxicities(*V.3.3.2.270;*VI.9.3.000)524,663
Toxicity(*V.3.2.065,*.066)494
  injury(*IV.5.4.1.005)412
Toxin, toxins(*IV.5.3.005;V.3.3.2.434;*.3.3.3.077)407,529,536
Trace(II.3.201)204
  contaminants(*V.3.2.083)495
  elements(*VI.7.2.017s)615

Tracheae(*V.3.3.1.010)498
Tracheal:
  matrix cells(*V.3.3.3.013f)532
  respiration(*V.3.3.1.132)501
Tracheary element(s)(*II.3.217;*IV.3.
  1.014f;*V.1.4.028e)206,372,444
Tracheid(*II.3.049,.202,.221,.22a-b,
  .23c,d)196,204,206,228,229
  bar[II.2.3.3.04f(2),.11b(3)]185,192
Tracheomycosis(V.3.3.2.435)529
Tracheophyta(*III.1.1.032,.036)306,307
Track sale(VI.7.1.075)607
Tractor:
  -drawn(*V.3.3.3.030)533
  drivers(*VI.8.053)620
  forklift(VI.2.018)575
  -hauled sprayers(*V.3.3.3.100)537
  -mounted seeder(*V.1.1.036)423
  speed(*V.3.3.3.016)532
  sprayers(*V.3.3.3.100)537
Tractors(*V.3.0.003)484
Trademark Act of 1946(*VI.8.005)617
Traditional soil mixes(V.1.2.011p)431
Traffic:
  manager(*VI.8.102,.122)624,625
  pattern(VI.8.123)625
Trailer(s) on flat car(*VI.6.016,
  *.017,.033)595,596
  delivery, loading, unloading(*VI.6.
    029c)596
  Fruit and Vegetable Tariff(*VI.6.
    017,*.029c)595,596
Training(*V.1.4.028i;*.3.3.3.000;*.3.
  4.000,*.000b,*.063,*.067,*.073,.05-
  .12)445,531,539,543,544,549-556
Trait(III.2.2.181)326
Tramp car (Tramp car sale)(VI.7.1.
  076)607
Trans(VI.0.286)348
Transfer RNA(*IV.2.2.188)369
Transferases(*IV.2.2.058g)364
Transgressive segregation(III.2.2.182)
  326
Transit temperature(s)(*VI.6.009)594
  recommended(*VI.6.029a)596
Transition region(II.3.203)205
"Transitional" parthenocarpy(IV.4.2.
  005)385
Translocated herbicide(V.3.3.3.113)538
Translocated incompatibility(V.1.4.
  035)449
Translocation(III.2.2.183;*IV.5.2.056;
  *V.3.3.2.133d;.3.3.3.021)326,404,
  515,533
  in solution(*V.2.106)476
Translucent scales of onions(VI.9.3.
  142)673

Transmission(*V.3.3.2.133e,.436)515,
  529
  electron microscope(*IV.0.079)332
Transpiration(IV.5.2.065,*.084)404,406
  coefficient(*IV.5.2.077)405
  cuticular(*IV.5.2.059,*.079,*.085)
    404,406
  efficiency(IV.5.2.067)404
  rate(*IV.5.2.074)405
    mass basis(*IV.5.2.074b-1)405
    volume basis(*IV.5.2.074b-2)405
  rates(*IV.5.2.085)406
  ratio(IV.5.2.068)404
Transpirational cooling(*IV.5.4.1.
  004)412
Transplanting(*V.1.3.033;*.3.3.3.081)
  434,536
Transplants(V.1.3.085;*.3.1.002)439,
  485
Transportation(*VI.4.000;*.7.1.019,
  *.022)585,600
  conditions(*VI.9.3.000)663
  facilities(*V.1.1.017a)419
  from field(*VI.4.000;*.8.078)585,622
  problems(VI.6.034)596
  -to-market destination handling(*VI.
    2.020)575
Transverse(II.2.2.007h-7,w-12,.07g,
  .10b;*IV.3.3.008h)119,121,136,139,
  377
  buds[II.1.1.1.032,.01a(1)]20,28
  circumference(*V.4.1.021)560
  sizer(*VI.3.064d)583
Transversely locellate(*II.2.2.007r)
  119
Tranzschelia discolor(*VI.9.1.469)653
Trap flower(VI.4.4.055)390
Trash elimination, trash eliminator
  (VI.3.027a-1)579
Traumatotropism(IV.3.3.008v)378
Travelling gun(*V.3.5.012)557
Tray(*VI.3.053d,f;*.4.000b,*.008,.024)
  582,585,586,588
  pack(*VI.3.053f;.4.000c,*.008,.025)
  582,585,586,588
  papier maché(*VI.4.019)587
  stretchwrap(*VI.4.023)587
Treasurer(*VI.8.074)621
Treatment (seed)(*V.1.3.063)437
Tree, trees(*I.2.009;II.0.025,.06c,d,
  .07;*.1.1.3.031,*.035,*.039,
  .079;.3.048;*.4.1.005;*.4.2.
  006,*.026;*.4.3.08,*.039,*.060;
  .4.4.002a,*.003c,*.005b;*.6.011,
  IV.3.1.006;.5.2.072;*.5.4.4.
  001,*.002,*.002a;*V.1.2.001;*1.
  3.060;.1.4.022b,*.028m,*.028o,

INDEX OF TERMS

```
    *.036b-3;*.2.110e;*.3.3.1.132a,
    *g;*.3.3.2.133c,*.181f;*.3.4.
    000,*.004,*.017,*.029,*.040,
    *.055,*.056,*.076;*VI.1.015)6,
    11,17,18,23,25,196,238,239,240,
    241,243,245,246,247,279,370,405,
    415,416,428,436,442,447,448,450,
    479,501,502,515,519,539,540,541,
    542,543,544,572
  balled-and-burlapped(*V.1.1.035)423
  bare root(*V.1.1.035)423
  compound(*IV.5.4.4.002)416
  cricket(*V.3.3.1.132h)503
  fruit(VI.8.124)625
  girdled(*V.1.4.028c)444
  girdling(*IV.5.4.4.002a,*.002e)416
  hopper(*V.3.3.1.132d)502
  like(*II.5.6.030)274
  run(VI.8.125)625
  sturdy(*IV.5.4.4.002)416
  system, damaged(*V.1.4.028v)449
    ,healthy(*V.1.4.028v)449
  training(*V.3.4.000b,*.003,*.017,
    *.021,*.029,*.049,*.055,*.068,
    *.075)539,540,541,542,543,544
  vigorous, healthy(*V.1.4.028u)449
Trehalose(*IV.2.2.031b)361
Trellis, trellises(*V.3.4.008,*.020,
  *.026,.032;.05b,d,e,.10)540,541,549,
  554
Trench layerage(V.1.4.036b-5,.14g-i)
  450,466
Trenchers(*V.3.0.003)484
Triadelphous(II.2.2.04e)133
Triangular(II.5.5.062,.12e;V.3.1.008m,
  .01d)273,292,486,488
Tribe(III.1.1.002,.021)302,305
Tricarboxylic acid cycle(*IV.1.2.040b)
  356
2,4,5-Trichlorophenoxyacetic acid (2,
  4,5-T)(*IV.2.1.012)357
2,4,5-Trichlorophenoxypropionic acid
  (2,4,5-TP)(*IV.1.2.012)357
Trichoderma rot of citrus (Trichoderma
  viride)(VI.9.1.545)658
Trichodorus spp.(*V.3.3.1.200c)506
Trichogyne(s)(*V.3.3.2.181e,.437)519,
  529
Trichome(s)[*II.3.054,.204,.31;.4.08b
  (2)]196,205,237
  density(*IV.5.2.080)406
Trichosclereid(*II.3.170)202
Trickle irrigation(*V.3.5.012)557
Tricolpate[II.2.2.12d(2)]141
Tricolporate[II.2.2.12d(6)]141
Tridymous(II.2.2.05b)134
Trifoliate(II.1.2.114)64

Trifoliolate(II.1.2.115,.17a,.21a)64,
  82,86
Trifolirhizin(*IV.2.2.162)368
Triglyceride(*IV.2.2.065,*.116,.210)
  365,366,370
2,3,5-Triiodobenzoic acid(*IV.2.1.012)
  357
Trijugate(*II.1.2.069)62
Trilacunar(*II.3.099)198
Trimming(*V.1.4.028i,.11g;*VI.3.031,
    .078;*.7.1.025;*.8.108)445,463,
    580,583,601,624
  -off side branches(*V.1.4.028t)448
Trimorphism(IV.4.3.019)387
Trioecious(IV.4.3.020)387
Trioses(*IV.2.2.031a)361
2,3,5-Triphenyltetrazolium chloride
  (*IV.5.3.024)409
Triple-ribbed(II.1.2.121s,.26s)65,91
Triple (treble) superphosphate(*V.3.2.
  087,*.091)496
Triploblastic(*V.3.3.1.183)504
Triploid(III.2.2.184)326
Triporate[II.2.2.12d(4)]141
Triribbed(II.1.2.121t,.26t)65,91
Trisome(*III.2.2.113)323
Trisomic(III.2.2.186)326
Tristeza(*V.3.3.2.156,*.462)516,530
Tristyly(*IV.4.3.019)387
Triternate(II.1.2.117,.21b)64,86
Triton:
  X-100(*V.3.3.3.124)539
  X-155(*V.3.3.3.033)533
Trombone sprayer(*V.3.3.3.100)537
Trophotylenchus spp.(*V.3.3.1.200e)506
Tropical:
  and warm temperate areas(*V.3.3.1.
    132f)502
  conditions(*IV.5.2.006)407
  crops(*I.0.001;.1.017)3,5
  flowers(*V.3.3.1.026)498
  plants(*IV.5.3.028;*V.3.3.1.017)409,
    498
  rain forest(*II.4.2.027)240
  trees(*II.4.3.008)241
Tropicals(*IV.5.3.028;*V.1.1.053a;*.1.
  4.028k-6,*-8)409,425,447
  truncheon(*V.1.4.022b,g,.03e)442,
    443,455
Tropics(*V.3.1.008g)486
Tropism, tropisms(IV.3.3,.008)377
Tropophyte(II.4.1.007,.01a)238,249
Trowel(V.1.1.073)427
Truck(*VI.2.012;*.6.000)575,594
  bulk shipment(*VI.6.003)594
  load(*VI.8.010,*.100)617,624
  lot(VI.8.126)625
```

INDEX OF TERMS

mixed loads(*VI.6.022)595
-mounted sprayers(*V.3.3.3.100)537
shipments(*VI.6.017;*.7.1.048;*.8.
   106)595,604,624
   broker(*VI.8.106)624
   (protective services)(*VI.6.029a,
      *b)596
van modular containers(*VI.6.037)597
True dormancy(IV.3.1.011c)371
Trumpet-shaped(II.2.2.061-c-t)127
Truncate(II.5.4.1.023;.5.4.2.015;.5.
   03p,.041)269,283,284
Truncated soil profile(V.2.130)482
Truncheon(*V.1.4.022b,g,.03e)442,443,
   455
Trunciflory[*II.2.1.015,.05d(5)]100,
   108
Trunk shaker(*VI.2.022b)576
Tryon's scab(*VI.9.1.505)655
Tryptophan(*IV.1.2.042;.2.2.016)357,
   360
Tsetse fly(*V.3.3.1.132b)501
T-shaped (tetrad)[II.2.2.12c(4)]141
TTC(*IV.5.3.024)409
Tube(s)(IV.4.4.004b-6;*VI.3.053d)388,
   582
   cell nucleus[II.2.2.13a(3),c(3)]142
   tomatoes(VI.8.127)625
Tuber, tubers[II.4.061;*V.1.4.024(4),
   .05c;*.3.3.1.132a;*.3.3.2.370;VI.
   2.000]245,443,457,501,527,575
   natural vegetative reproduction(IV.
   4.1.006a-8)379
Tubercles(*V.3.3.2.040)511
Tuberculate (style)(II.2.2.17m)146
Tuberculate, tubercled(II.5.6.141;.5.
   20c)277,300
Tuberous(II.1.3.015,.04c;*.4.3.049,
   .063;.4.13a)95,99,244,245,261
   roots[*V.1.4.024(5),.05d,e]443,457
Tubiform(II.2.2.061c-1-5,.06e)127,135
Tubular[II.1.1.3.101,.22c;.2.2.061c-1-
   u,.20e,.22c,.23e,i(2)]27,58,127,149,
   151,152
Tubules(*V.3.3.1.132)501
Tufted plants(*II.4.1.003)238
Tulip-breaking virus(*V.4.2.1.013,
   .022)565,566
Tumefaction, tumefactions(*V.3.3.2.
   133b,*d,*.187,.438)515,520,529
Tumid[II.2.2.061c-14-u,.23a(2)]128,152
Tumor(s)(*V.3.3.2.040,*.187)511,520
Tunic(*V.1.4.071a-1,*.071b;.4.1.120;
   *.4.2.2.014)451,565,567
Tunica-corpus theory(II.3.206,.14b)
   205,220
Tunicate (bulb)(II.4.13e;V.1.4.071a-2,
   .16e)261,451,468
Turbinate(*II.2.3.2.035,.043,.07f)162,
   181
Turf, turf crops(I.0.001;.2.011)3,6
Turgid(II.2.2.061c-14-v)128
Turgidity(*IV.5.2.053)404
Turgor(*IV.5.2.036,*.037,.069)402,405
   full(*IV.5.2.037)402
   loss(*V.3.3.2.468)530
   pressure(*IV.3.3.003e;*.5.2.010,
      .034;*VI.9.3.154)376,399,402,674
      actual(*IV.5.2.053)404
      maximum(*IV.5.2.078)405
      potential(*IV.5.2.053)404
   regulation(*IV.5.2.034)402
   relative(*IV.5.2.053)404
   zero(*IV.5.2.036,*.037)402
Turion(II.4.1.008)238
Turnbuckle(*V.1.4.028o)448
Turnip-shaped(*II.2.3.2.022)161
Turnover(VI.8.128)625
Turns(VI.8.129)625
Turpentine(*IV.2.2.107)366
Tussock moths(*V.3.3.1.132g)502
Twilight(*IV.3.2.013;*.4.5.002c)374,
   391
Twig(s):
   dieback(*V.1.4.035)449
   dying(*V.3.2.066)494
   malformed(*V.3.2.066)494
   multiple(*V.3.2.066)494
Twiners(*II.4.2.026)240
Twining(*II.4.3.002,*.014,.074)241,246
Twins(*II.2.2.007i,*.039)119,122
Twist(*V.4.2.1.021)566
Twisted(*II.1.1.2.006;.2.2.007w-13,
   .05f)20,121,134
   grain(*V.1.4.028k-2)446
   or misshapen stem(*V.3.3.2.158)516
Twisting(*IV.5.4.4.002;*V.3.4.000)416,
   539
   ,bending and spreading(IV.5.4.4.
   002g)416
   petioles(*II.4.3.014)241
Tylenchorhynchus spp.(*V.3.3.1.200d)
   506
Tylenchulus semipenetrans(*V.3.3.1.
   200e)506
Tylose, tyloses(*II.3.077,.207,.10d)
   197,205,216
Tylosoid (cells)(II.3.208)205
Type(V.2.1151)481
Types of structures(*V.1.1.017b)419
Typhlodromids(*V.3.3.1.172,.301)504,
   509
Typhula spp.(*V.3.3.2.181f)519
Typist(*VI.8.053)620

INDEX OF TERMS

Tyrosine(*IV.2.2.016)360

U

U. C. mixes(*V.1.2.011,*.011n)430,431
UE(*II.1.1.3.102,.14b)27,50
Ulcerous lesions on stems, leaves, and fruit(*V.3.3.2.019)510
Ultracentrifuge(IV.0.311)349
Ultra-low volume:
  concentrate(V.3.3.3.114)538
  sprayers(*V.3.3.3.100)537
Ultramicroscopic(*V.3.3.2.461)530
Ultrarapid cooling (and warming)(IV.5.3.050)411
Ultrasonic(*IV.0.263)348
Ultraviolet (UV):
  absorption(*IV.2.2.069;.5.4.2.012) 365,413
  -B(*IV.5.4.2.013)413
    light injury(*IV.5.4.2.013)413
    light tolerance(*IV.5.4.2.008)413
    photoinactivation(IV.5.4.2.006)413
    photoreactivation(IV.5.4.2.008)413
    photoreversion(IV.5.4.2.009)413
    preferential absorption(*IV.5.4.2.013)413
  radiation(*IV.2.2.069;.5.4.2.013;*V.1.1.017d)365,413,420
  resistance by avoidance(*IV.5.4.2.012)413
    mechanism(IV.5.4.2.014)414
  -resistant polyethylene(*V.1.1.017d) 420
  spectrometer(IV.0.312)349
  wavelengths(*IV.5.4.2.000,*.013)413
UM(*II.1.1.3.102,.14b)27,50
Umbel[II.2.1.074,.02a(6),.08g]103,105,111
Umbellate(II.2.1.075,.12c)103,115
  -paniculate(II.2.1.076,.04d)103,107
  -racemiform (racemose)(II.2.1.077,.04c)103,107
  -spicate(II.2.1.178,.04b)103,107
Umbelliferous(III.1.2.061)311
Umbelliform(*II.2.1.075)103
Umbelloid(II.2.1.12c)115
Umbonate(II.5.6.142;.5.20d)277,300
Umbraculiform(*II.1.2.118)64
Umbrella-shaped(II.1.2.118,.21c)64,86
Umbrella system(V.3.4.017,.07c)541,551
Umbric epipedon(V.2.028i)473
Umbilicus(*II.2.3.3.021)164
Unarmed(II.5.6.143)278
Uncinate(II.2.2.08f;.5.6.144;.5.20e) 137,275,300

Uncinula necator(*V.3.3.2.181e)519
Uncongenial, uncongeniality(*V.1.4.035)449
Unconsolidated:
  geologic strata(*V.2.107,*.112,.01a) 476,479,483
  mantle(*V.2.107m)477
Unctuous(II.5.6.145)278
Undecomposed plant materials(*V.2.070) 474
Under-bench sprinkling(*V.1.1.017i)421
Undercooling(*IV.5.3.049)411
  point(*IV.5.3.018)408
Undershrubs(*II.4.2.026)240
  tree sprinkler(*V.3.5.012)557
Undulate(II.5.4.3.041;.5.07g)271,287
Unequal, unequal-sided(II.5.5.062;.5.12f)273,292
Uneven ripening(*VI.9.3.029,.143)665,673
Unfair conduct(VI.7.1.077)608
Unfruitful(IV.4.8.054)395
Ungraded produce(*VI.8.125)625
Unguiculate(II.2.2.061c-14-w,.23f)128,152
Unhealed cuts(*VI.8.038c)619
Unicellular (trichomes)(II.3.204)205
Unifacial leaves(II.1.2.119,.04c,g;*.4.3.032)64,69,243
Unified atomic mass unit (u)(*IV.0.260j)347
Uniflagellate swarmspores(*V.3.3.2.181a)518
Unifoliolate(II.1.2.120,.12c)64,77
Unilacunar(II.3.099)198
Unilateral(*II.2.2.070)129
Unilocular(II.2.2.082,.10c)129,139
Union(*V.1.4.015)441
  overgrowth(*V.1.4.035)449
Uniribbed(*II.1.2.121b)64
Uniseriate ray(II.3.07d)213
Unisexual(*IV.4.3.004,*.017)386,387
Unit(V.3.2.092)496
  heaters(*V.1.1.017h)421
  of extension(II.1.1.3.102,.14b)27,50
  of morphogenesis(II.1.1.3.103,.14b) 27,**50**
  operation(*VI.2.020;.3.079)575,583
  pricing(VI.8.130)625
United:
  Fresh Fruit & Vegetable Association (*VI.6.023)595
  States(*V.2.097;*.3.5.033)475,559
    Department of Agriculture (USDA, U.S. Dept. Agr.)(*V.1.3.075; *VI.7.1.000)438,597
    ,Agricultural Marketing Service

(*VI.8.016,*.027,*.094)617, 618,623
    Agriculture Handbook 105,195 (*VI.6.029a)596
    Market News(*VI.8.094)623
        Reports(*VI.8.016)617
        Service(*VI.8.059)621
    System(*V.2.097)475
    Department of Health and Human Welfare(*VI.7.2.017)612
      of Labor(*VI.7.2.017,*.038) 612,616
    Grades for Florida citrus(*VI.9.2.017)660
    Patent Law of 1930(*III.2.0.007) 315
    Plant Variety Protection Act of 1970(*III.2.0.007)315
Unitegmic(II.2.3.3.045,.12c-e)167,193
Unitization(*VI.6.023,.035)595,597
Univalent(III.2.2.187)326
Univalve mollusks(*V.3.3.1.250,*.251) 508
Universal product code(*VI.8.103,.131) 624,625
Univoltine(IV.4.5.002r)392
Unloading(*VI.3.052)581
Unloads:
    produce(*VI.8.055)620
    trucks(*VI.8.055)620
Unorganized cells(*V.3.3.2.187)520
Unpleasant characteristics(*V.2.110e-2)479
Unsegmented(*V.3.3.1.183)504
Unwrapping(V.1.4.028i,.11a)445,463
U or double-U palmettes(*V.3.4.049)542
UPC(VI.8.131)625
Upright:
    multiple-stem(*V.3.4.017,.06d)540, 550
    (ovule)(*II.2.2.041t-4-k-5)125
    (ray cells)[II.3.07f(2)]213
    -T(*V.1.4.028c)444
UPS(*VI.6.026)596
Uracil(*IV.2.2.141,*.175)367,369
Urban uses(*V.2.096e)475
Urea(*V.1.2.013;*.3.2.006,*.016,*.065, *.067,*.086,*.093)431,490,491,494, 496,497
    biuret-free(*V.1.2.013)431
    carbamyl(*V.3.2.006)490
    -formaldehyde(*V.1.2.012)431
Ureaform(*V.1.2.012;*.3.2.015,.093) 431,491,497
    resins(*VI.9.3.044)666
Uredinales(*V.3.3.2.179a,*.204,*.371) 517,521,527

Urediniospore(s)(*V.3.3.2.012,*.181f, *.442)510,519,529
Uredinium (pl. uredinia); uredium, uredosorus)(*V.3.3.2.346,.442)526,529
Uredinospore (see Urediniospore)
Urent(II.5.6.146;.5.20f)278,300
Urocystis cepulae(*V.3.3.2.181f;*VI.9.1.493)519,654
Uromyces:
    caryophyllatus(*V.3.3.2.181f)519
    phaseoli var. typica(VI.9.1.494)654
Uronic acids(IV.2.2.032c)362
U.S. (see United States)
U.S.S.R.(*V.2.097)475
Ustilaginales(*V.3.3.2.179a,*.390)517, 528
Ustilago:
    avenae(*V.3.3.2.181f)519
    maydis(VI.9.1.169)638
    zeae(*V.3.3.2.181f)519
Utility services(*V.1.1.017e)420
Utricle(II.2.3.1.038,.01h-j)160,168

V

Vacuole(*II.3.125;IV.1.1.030)200,352
Vacuum:
    brushes(*VI.3.027a-1)579
    cooling(*VI.5.051c,.055)593
    cups(*VI.2.022b)576
    drying(*V.1.3.061)437
    evaporators(*VI.10.010)675
Vague(II.5.2.037)267
Valence(*V.2.015)472
Valine(*IV.2.2.011)360
Valsa leucostoma(*V.3.3.2.181e)519
Value(*II.5.1.000d)263
Valvate(II.1.1.2.029,.03y)21,**36**
Valve(II.2.3.2.045,.07g)163,**181**
Van(VI.6.036)597
    highway(*VI.6.023)596
    modular container(VI.6.037)597
    TOFC(*VI.6.023)596
Vanadium (V)(*V.3.2.073)495
Vanilla wafers(*VI.8.120)625
Van't Hoff equation(*IV.5.2.038)402
Vapam(*V.1.1.065)426
Vapor:
    drift(V.3.3.3.115)538
    -flow methods(*IV.5.2.028)401
    Gard(*V.1.2.001)428
    -impermeable barriers(*V.3.3.3.043) 534
    pressure(*IV.5.2.065;VI.5.025d, *.037,*.055)404,590,591,593
        deficit(*IV.5.2.065)404
Variants(*III.1.1.037;.2.0.005d-1)307,

INDEX OF TERMS

315
 and off-types in varieties(III.2.0.
  005d)315
Variation(III.1.2.062)311
Variegated (variegatus)(II.5.1.006s)
 265
Variegation(II.5.1.006)264
Varietal characteristics(*VI.8.038e)
 619
Varietas(*III.1.1.003,*.037)302,307
Variety(*III.1.1.003,.037;*.2.0.003,
 .004b,.005a,d)302,307,311,312,313,
 315
Vascular:
 arrangement(II.3.210)205
 browning(VI.9.3.074)669
 bundle[II.2.2.02(3,5,10,11,15,17);
  .2.3.2.05f(2);.2.3.3.04b(4),.05a
  (4),b(4);*.3.063,*.172,*.173,
  .201,.209,*.210,*.211]131,179,
  185,186,196,202,204,205
 bundles of leaves(II.3.211)205
 cryptogams(*IV.5.2.040)403
 fruit calyx tissue(*V.3.3.2.156)516
 plants(*IV.1.2.036)354
  terrestrial(*IV.3.2.004)373
 strand(*II.3.209)205
 tissues(II.3.213;*V.1.4.028e)205,444
 wilts(*V.3.3.2.040)511
Vase(*V.3.4.031)541
 life(*VI.8.036,.132)619,626
 -shaped(II.2.2.061c-1-w,.21c)127,150
Vasicentric(*II.3.014,.04b)194,210
Vault design (greenhouse)(*V.1.1.
 017e)419
Vector(V.3.3.1.302;.3.3.2.087)509,513
Vegetable, vegetables(I.3.001;*V.1.3.
 010;*.3.2.083)7,432,495
 crops(I.3.000)7
 field-to-packinghouse handling(VI.2.
  012)575
 harvesting(*VI.2.000)575
 packinghouse(*VI.3.000)577
 rot(*V.3.3.2.181c)518
 seed lots(*V.1.3.075)438
Vegetation(*V.2.000)472
Vegetative:
 (asexual) propagation(V.1.4.)441-469
 bud(*IV.4.8.026)394
  growth(*IV.5.4.4.002)416
 buds(II.1.1.1.033,.03o,p;*V.3.4.006,
  *.007)20,30,540
 cover, covers(*V.2.110b,*e)478,479
 growth(*IV.5.4.4.002b)416
  removal(*V.3.4.063)543
 organs (disease transmission)(*V.3.
  3.2.133e)515

origin(*V.1.4.068)451
propagation(V.1.4.000,*.022,*.067,
 *.069)441,451
reproduction(*IV.4.1.006)378
stage(*V.4.1.064)562
Vein(s)(*II.3.211)205
 banding(V.3.3.2.451)530
 clearing(V.3.3.2.452)530
Velamen(II.3.215;*.4.2.006)205,239
Velocity(*IV.0.260b;*V.2.108,.110b)
 344,477,478
Velutinous(*II.5.6.147)278
Velvet roll(*V.1.3.057)435
Velvety(II.5.6.147,.20g)278,300
Venation(II.1.2.121,.22-.28)64,87-93
Veneer(*VI.4.003)585
 graft(V.1.4.028k-8,.09f,g,.10e)447,
  461,462
 grafting(*V.1.4.028f-2,.028k)444,446
Venose(II.1.2.121v,.27v)65,92
Ventilation(*V.1.3.073;*VI.3.023;*.5.
 047a)438,579,591
 requirements(*VI.6.018)595
Ventilators(V.1.1.017i)421
Ventral(*II.2.2.041t-2-g)123
Ventricose[II.2.2.083,.23g(1)]129,152
Venturia:
 inaequalis(*V.3.3.2.181e;VI.9.1.037)
  519,629
 pirina(VI.9.1.334)647
Vermicular(*II.2.3.3.047)167
Vermiculite(*V.1.1.053e;*.1.2.005,
 .011,.011n;*.1.3.081;.2.098)425,
 429,431,439,475
 finish aggregate(*V.1.3.076)438
Vernalization(IV.3.1.014j;.3.2.021e;V.
 4.1.121)372,376,565
 stimulus(*V.4.1.036)561
Vernation(II.1.1.2.000,.01-.03)20,34-
 36
Verrier(*V.3.4.055)543
 palmette(*V.3.4.055,.11d)543,555
Verrucose(*II.5.6.141)277
Versatile(II.2.2.007ad,.07c,.09e)121,
 136,138
Versenediol(*V.3.2.012)490
Versenol(*V.3.2.012)490
Vertebrate(II.1.2.122,.28a)65,93
Vertical(*II.5.2.028;*V.3.4.017,*.021)
 267,540,541
 cordons(V.3.4.11e)555
 multiple stem(*V.3.4.017,.06d)540,
  550
Verticil, verticillate[II.1.2.123,
 .28b;.2.1.080,.02b(2)]65,93,103,
 105
 inflorescence(*II.2.1.024e)101

INDEX OF TERMS

Verticillium:
  alboatrum(*V.3.3.2.181g;*VI.9.1.553) 520,658
  spp.(*V.1.1.065)426
  theobromae(*VI.9.1.182)639
Verticillium wilt of potato(VI.9.1.553)658
Very:
  coarse sand(*V.2.114)480
  fine sand(*V.2.114,.117l-1)480,481
  fine sandy loam(V.2.117j-1)481
  good(VI.8.027)618
  light(VI.8.027)618
Verticillaster, verticillate inflorescence[II.2.1.024e,.07c(1)]101,110
Vesiculate pollen[II.2.2.007t-12,.12b(5)]120,141
Vesicle(II.2.3.2.046,.05d)163,179
Vessel(*II.3.207)206
  distribution in wood(II.3.10a-c)216
  element(*II.3.216;.3.22g-i)205,228
  member(II.3.217,*.221)206
Vexillary(II.1.1.2.030,.03z)21,35
Vexillum[II.2.2.061c-13,.23h(1)]127,152
V-graft(*V.1.3.028k-2,.09d,.11n)446,461,463
Viable(IV.3.1.024)373
Viability(IV.4.8.056)395
"Viaflow"(V.1.1.017f)420
Vicarism, vicarious species(III.1.2.063)311
Vice-presidents(*VI.8.074)621
Vicinism(III.1.2.064)311
Villous(*II.5.6.062;.5.15m)275,295
Vinaceus(*II.5.1.005x)264
Vine, vines(*I.2.001d,*.009;II.0.026,.02c;*.4.3.002,*.014;*V.1.2.001;*.1.4.036b-2;*.2.110e;*.3.4.000,*.032,*.040,*.076)5,6,11,13,241,428,450,479,539,541,542,544
  balled-and-burlapped(*V.1.1.035)423
  bare root(*V.1.1.035)423
  harvester(*VI.2.022b)576
  ripe(VI.8.133)626
  twining(*V.3.3.2.375)527
Vinegar(*VI.10.007,*.012,*.028,.038)674,675,677
Vineyard(*VI.8.138)626
Vining types(*V.4.1.107,*.119)564
Vinyl acetate maleic anhydride(*V.2.099)476
Violent local reactions(*V.3.3.2.213)521
Violet (violaceous)(II.5.1.002,.002j)263
Viral:
  disease(*V.3.3.3.013a,*f)532
  diseases(*V.3.3.2.256)523
  infection(*V.3.3.3.013,*.013c)532
Virescence(*V.3.3.2.250,.455)523,530
Virescens, viridis(*II.5.1.004d)264
Virginia creeper(II.0.002;.4.3.071)10,245
Virion(*V.3.3.2.088,.456)513,530
Viroid(s)(*V.1.4.035;*V.3.3.0.004;*.3.3.2.133a,*b,*c,.457)449,497,514,515,530
  inoculation(*V.3.4.000)539
Virulence(*V.3.3.2.032,.458)511,530
Virulent(V.3.3.2.459)530
Viruliferous(V.3.3.2.460)530
Virus, viruses(*V.3.3.0.004;*.3.3.2.018,*.032,*.087,*.088,*.133a,*c,*d,*.222,*.231,*.248,*.262,*.263,*.270*.338,*.412,*.460,*.461,*.462;.3.3.3.013,*.013b;*.4.2.1.005,*.013;*VI.7.2.017h;*.9.1.000)477,510,511,513,514,515,522,523,524,526,529,530,532,565,613,626
  cross-protection(V.3.3.2.462)530
  disease(*V.3.3.2.087;*.3.3.3.013d)513,532
  diseases(*V.3.3.2.156,*.257,*.382)516,523,528
  of bulbs(*V.4.2.1.019,.022)566
  heat tolerant(*V.3.3.2.426)529
  identification(*V.3.3.2.222)522
  infection(*V.3.3.2.363)527
  inoculation(*V.3.4.000)539
  mild strain(*V.3.3.2.462)530
  particle(*V.3.3.2.456)530
  particles(*V.3.3.3.013e)532
  transmission(*V.1.4.028f-2,*.028p)444,448
    studies(*V.1.4.028p)448
  vectors(*V.3.3.1.132d,.200c)502,505
Viruslike:
  disorder(*V.3.3.2.457)530
  organisms(*VI.9.1.000)626
  particle(*V.3.3.2.457)530
Viscid(II.5.6.150)278
Visible:
  (light) wavelengths(*IV.5.4.2.000)413
  mycelial strands(*V.3.3.2.065)512
  symptoms(*V.3.5.013)557
Visual:
  inspections(*V.1.4.010)441
  symptoms(*V.3.2.066)494
Vital staining(*IV.5.3.024)409
Vitamin B_{12}(*V.3.2.058)494
Vitamins(*V.1.5.005)470

INDEX OF TERMS

Vittae(II.2.3.1.03k)170
Vittatus(*II.5.1.006r)265
Vivipary[II.2.3.3.020d,.01e,f;.4.04b
   (1)]164,182,252
Vivotoxin(V.3.3.2.463)530
Volatile herbicide(V.3.3.3.116)538
Volatiles(*V.4.2.1.011;*VI.9.3.000)
   565,663
Volatilization(*IV.5.2.073;*V.3.2.000)
   405,490
Volcanic depositions(*V.2.112)479
Volt(*IV.0.260c)344
   per meter(*IV.0.260e)345
Volume(*IV.0.260b,.317;*VI.8.080)344,
   349,622
   basis(IV.5.2.074b-2)405
   fill(*VI.3.053c)582
   fraction(IV.0.319)350
   gun(*V.3.5.012)557
Volumetric (water potential)(*IV.5.2.
   075)405
Voluntary(*VI.8.096,.135)623,626
Volunteer plant(V.3.3.3.117)538

W

Walking stick(*V.3.3.1.132h)503
Wall(*V.3.4.020,*.026)541
Walnut caterpillar(*V.3.3.1.132g)502
Warble flies(*V.3.3.1.132b)501
Warburg effect(*IV.1.2.035)354
Wardian case(*V.1.1.046)424
Warehouse(s)(*VI.3.041,*.043;*.8.114,
   .129)581,625
   cost vs. retail price(*VI.8.062)621
   inventory turns(*VI.8.051)620
   store(VI.8.136)626
Warm:
   climates(*V.4.2.1.001)565
   -season seeds(V.1.3.087)439
   storage(*V.1.3.081)439
Washing(*V.1.3.064;*VI.3.031,*.079,
   .081;*.7.1.025)437,580,583,601
Wasps(*V.3.3.1.132e;*.3.3.3.013f)502,
   532
Waste disposal(*VI.7.2.014,*.017)610,
   612
Water(*IV.4.5.001a;.4.7.001e;*V.1.4.
   028t;*.2.000,*.010,*.110a,b,c,d;
   .3.2.005,*.028;.3.3.3.119;*VI.5.
   025a,*.043,*.051b)391,393,448,
   472,478,490,491,538,589,591,593
   absorbed(*IV.5.2.083)406
   absorbing materials(*V.4.1.070)562
   application rate(*V.3.5.014)558
   (aquatic)(*II.4.2.002)238
   -asphalt emulsion(*V.1.2.006)429

atmospheric moisture(*VI.5.025)589
-attracting capacity(*V.3.3.3.050)
   534
available(*IV.5.2.059)404
balance(*IV.5.2.028,.071)401,405
bath(*IV.5.4.1.006)412
blast shaker(*VI.2.022b)576
blister(*VI.9.1.097)633
bodies(*V.2.110d)478
brake(*V.2.110a)478
breaker(*V.1.1.077)427
   nozzle(*V.1.1.017f)420
capacitance(*IV.5.2.072)405
cavity[II.3.15a(2)]221
chemical, physical and biotic senses
   (*V.3.5.032)558
company, mutual(*V.3.5.034)559
content(*IV.5.2.053,.073,*.082,
   *.085)404,405,406
core of apple(VI.9.3.145)673
course(*V.3.5.022)558
critical relative content(*IV.5.2.
   009)399
cultures(*V.3.2.050)493
deficit(*IV.5.2.013,*.036;.5.3.047)
   399,402,411
density(IV.0.171d;*.5.2.075;*V.2.
   108)341,405,477
-dispersible slurry(V.3.3.3.118)538
droplets(*V.3.3.3.054)535
electrical conductance(*V.4.1.071)
   562
elimination(*VI.3.081)583
eliminator(*VI.3.025,*.029,*.081)
   579,580,583
-emulsion wax applicator(VI.3.029)
   580
emulsion waxes(*VI.3.033,.085d)580,
   584
energy flux density(IV.5.2.074a)405
erosion(*V.2.110d,.130)478,482
excess(*IV.5.1.014;*V.3.5.002)397,
   557
flow(*V.2.110d)478
flux(*V.2.108)477
hose-attached sprayers(*V.3.3.3.100)
   537
hydrotropism(*IV.3.3.008j-1)377
(immersed)(II.4.2.020)240
-insoluble nitrogen(*V.3.2.034)491
intake(*IV.5.2.071)405
log(*VI.9.3.012)664
logging(*IV.5.4.3.006)414
loss(*IV.5.2.044,*.071)403,405
lost(*IV.5.2.084)406
molecules(*V.3.5.024)558
movement(IV.5.2.074)405

of hydration(IV.5.3.053)411
organic or inorganic materials deposited(*V.2.107i)477
permit allocations(*V.3.5.034)559
permits(*V.3.5.034)559
plants(*II.4.4.005b)247
potential(*IV.5.1.017;*.5.2.011,
    .036,.041,.075)397,399,402,405
  high(*IV.5.2.018)400
  ,high tissue(*IV.5.1.015,*.079,
    *.080)397,406
  low(*IV.5.2.018)400
  ,low tissue(*IV.5.1.015;*.5.2.048,
    .079)397,403,406
  saturated(*IV.5.2.019)400
  specific(IV.5.2.075)405
  steady state(IV.5.2.019)400
  volumetric(IV.5.2.075)405
pressure(*V.3.5.014)558
properties(*IV.5.2.076)405
purity(*V.3.2.083)495
quality(*V.3.5.032;*.4.1.053)558,561
regulation(*V.3.3.3.000;.3.5.000)
    531,557
relations(IV.5.2.,.076)398-407,405
Relations, Cold and Other Forms of
    Stress(IV.5.)396-416
relative content(*IV.5.2.009)399
requirement(*IV.5.2.067,.077)404,405
rights(V.3.5.033)559
rot(VI.9.1.097)633
saturation deficit(*IV.5.2.009,.078)
    399,405
  actual(*IV.5.2.050)403
  ,critical(*IV.5.2.050)403
savers(IV.5.2.079)406
shortages(*V.3.5.034)559
soaking(*VI.9.3.029)665
-soluble organic nitrogen (N)(*V.3.
    2.034)491
  plastic strip(*V.1.3.074)438
  potash (K_2O)(*V.3.2.034)491
  secondary plant foods(*V.3.2.034)
    491
specific gravity(VI.5.043)591
spenders(*IV.5.2.031,.080)401,406
spot of citrus(*VI.9.3.147)673
storage(*IV.5.2.020,*.072)400,405
  capacity(IV.5.2.033)402
  tissue(*II.3.125;*.4.3.065;*IV.5.
    2.002)200,245,398
stream energy(*V.2.110b)478
stress(*IV.5.2.011,*.013,*.019,
    *.022;*VI.9.3.042)399,400,666
  induction(*IV.5.4.4.001)415
  tolerance(*IV.5.2.016)400
suitability(*V.3.5.032)558

supply(V.3.5.034)559
surface tension(*IV.5.2.006)399
table(*V.2.083;*.4.1.082)474,563
  constant(*V.3.1.007)485
transport(*IV.5.2.083)406
turnover, turnover rate(IV.5.2.082)
    406
unit volume cost(*V.3.5.011)557
uptake(*IV.5.2.080)406
  by roots(IV.5.2.083)406
-use efficiency(IV.5.2.084)406
used(*IV.5.2.084)406
user cooperatives(*V.3.5.034)559
utilization(*IV.5.2.080)406
vapor(*IV.5.2.065)404
  concentration(*IV.5.2.065)404
  diffusion resistance(IV.5.2.074b-
    3)405
  flux(*IV.5.2.074b,b-4)405
  flux density(IV.5.2.074b-5)405
  loss(*IV.5.2.065)404
viscosity(*V.2.108,*.123)477,482
Watering can(*V.1.1.054,.078)426,427
Waterlogging(*IV.5.4.3.006)415
Watersprout[II.1.1.3.104,.22b;V.3.4.
    01(8);*.4.1.024]27,48,545,560
Watertight benches(*V.1.1.017f)420
Watery:
  breakdown(*VI.9.3.029,*.073)665,669
  soft rot of asparagus(VI.9.1.556)658
    of cabbage, rutabaga, turnip, etc.
      (VI.9.1.557)658
    of carrot(VI.9.1.558)658
    of celery, anise, parsley, parsnips(VI.9.1.559)658
    of fig(*VI.9.1.497)654
    of lettuce(VI.9.1.561)658
    of snap beans(VI.9.1.562)658
Watt(s)(*IV.0.126e,*.157b-2-a,*b-2-e-
    1,*u,*x-1-b,*y-1-b,*.260c)334,337,
    344
  per meter kelvin(*IV.0.260e)345
  per square meter(*IV.0.126f,.157b-2-
    a,*-b,*-c,*b-2-e-2,*-3,*d,*x-1-c,
    *y-1-c,*z,*.260e)334,337,339,345
  per square meter steradian(*IV.0.
    260g)346
  per steradian(*IV.0.260g)346
Wave number(*IV.0.260b)344
Wavy(*II.5.4.3.034,.041)271
Wax(*II.3.038;IV.2.2.046)195,363
  -setter(*VI.3.085d)584
Waxed string(*V.1.2.017)432
Waxes(VI.3.085)584
  ,epicuticular(*II.3.038)195
Waxing(*VI.3.031,.086;*.7.1.025)580,
    584,601

Waxy(II.5.7.034)279
  blister of tomato(*VI.9.3.055)667
  breakdown of garlic(VI.9.3.150)673
Way bill(VI.6.038)597
Weak(*V.2.115e)480
  crotch[V.3.4.01(4,9)]545
  growth(*V.1.4.035)449
  (parthenocarpy)(IV.4.2.001d)385
Weakly coherent(*V.2.100)476
Weather(*V.3.3.3.053;*VI.8.038a)534,
    619
  conditions(*V.3.3.3.052)535
  -related(*VI.9.3.000)663
Weathering(V.2.000;.3.3.3.110)472,538
  extreme(*V.2.107c)477
  processes(*II.4.2.004)238
Weber (Wb)(*IV.0.260c)344
Wedge-shaped(*II.5.4.2.006;*.5.5.014)
    269,271
Weed, weeds(I.0.003;*V.2.110e-1;.3.2.
    000;.3.3.0.001;*.3.3.3.032,
    .049,.055,*.078,*.079,*.117,
    .122,*.123;VI.7.2.017h,*k,u)3,
    479,490,497,533,534,535,536,538,
    539,613,615
  control(*V.3.1.006;*.3.3.3.048,.119)
    485,534,538
    for economic, esthetic, public
      health or other reasons(*V.3.3.
      3.119)538
  eradication(V.3.3.3.120)539
  -flora composition(*V.3.3.3.049)534
  growth suppression(*V.2.110e-2)479
  indentification, classification and
      nomenclature(*V.3.3.3.123)539
  killer(*V.3.3.3.049)534
  noxious(V.3.3.3.064)535
  propagation(*V.1.2.011)429
  removal(*V.1.1.020)421
  science(*V.3.3.3.123)539
  seed(s)(*V.1.1.065;.1.2.011m;*.1.3.
    051,*.075)426,430,435,438
  seedlings(*V.3.3.3.031)533
  wilting(*V.3.5.013)557
Weeder(V.1.1.079)427
Weekly unit sales rate(*VI.8.109)624
Weevils(*V.3.3.1.132a;*.3.3.2.133e)
    501,515
Weight(IV.0.325;.4.7.001c;*VI.3.064)
    350,393,582
  sizer(*VI.3.064e)583
Welded wire mesh(*V.1.1.017c)420
Well(s):
  drilling(*V.3.5.034)559
  privately owned(*V.3.5.034)559
  -rotted manure(V.1.2.011p)431
West Africa(*V.3.1.008h)486

Western:
  celery mosaic(VI.9.1.563)658
  flower thrips(*VI.9.2.081)662
Wet and dry seasons(*IV.5.2.015)399
  bulb temperature(*VI.5.025c,.037)
    589,591
Wettable powder(V.3.3.3.124)539
Wetting agent(*V.3.3.3.124,.125)539
Wheat:
  bran(*V.3.3.3.009)531
  midge(*V.3.3.1.132b)501
Wheelbarrow(V.1.1.080)427
Wheeled tractors(*V.3.0.003)484
Wheel-shaped(*II.2.2.061c-1-o,.21b)
    127,150
Whip, or tongue,:
  graft(V.1.4.028k-9,.09b)447,461
  grafting(*V.1.4.028k,*k-9)446,447
Whips(*II.4.3.002,*.027)241,242
Whip-shaped(*II.5.6.043)275
White(*II.5.1.007,*.007m)265
  and yellow butterflies(*V.3.3.1.
    132g)502
  flies(*V.3.3.1.132d;*.3.3.2.133e)
    502,515
  -marked tussock moth(*V.3.3.1.132g)
    502
  -pine blister rust(*V.3.3.2.181f)519
  root rot(*V.3.3.2.181f)519
  rot (of apple)(*VI.9.1.116)635
  rot of onions, garlic, shallot,
    Welsh onion, leek(VI.9.1.565)659
  rust of radish(VI.9.1.566)659
    of spinach(VI.9.1.567)659
  smut of dahlia(*V.3.3.2.181f)519
  streak(*V.4.2.1.022)566
Whitened(II.5.1.007n;.5.6.152)265,278
Whitewashed(*V.1.4.028d)444
Whitish(II.5.1.007o)265
Whole root cutting(V.1.4.01f)453
Wholesale(*VI.8.096)623
  bulb propagation(*V.4.1.013)560
  organization(*VI.8.105)624
  selling(*VI.8.022)618
  warehouse(*VI.8.114)625
    operation(*VI.8.020)618
Wholesaler(s)(V.4.1.068;*VI.3.023;*.8.
    032,*.052,*.066,*.107,*.118,*.131)
    562,579,618,620,621,624,625
Whorl, whorled(II.1.2.124,.28c;*.2.1.
    080)65,93,103
Wide:
  angle crotch[V.3.4.01(10)]545
  ray[*II.3.189d,.219,.07e,.09a(1)]
    203,206,213,215
Wigwag nozzles(*VI.3.085d)584
Wild salt grass(*VI.9.1.330)646

INDEX OF TERMS

Wilt:
  disease(*V.3.3.2.435)529
  of cabbage(*V.3.3.2.181g)520
  of sweet potatoes(*VI.9.1.530)657
Wilting(V.3.3.2.468)530
"Wiltpruf"(*V.1.2.001)428
Wilts(*V.3.3.0.002;*.3.3.2.014,*.040) 497,510,511
Win(V.2.0.009)418
Wind(*IV.4.5.001a;*.4.7.001a;*V.2.010; .3.3.2.006;*VI.9.3.000)391,392, 472,510,663
  currents(*V.1.1.017e)419
  erosion(*V.2.110d,*.130)478,482
  forms(*II.4.4.003a)247
  -induced evaporative stress(*IV.5.4. 3.008)415
  patterns(*V.1.1.017a)419
  protection(*V.2.110e-1)479
  scar of citrus(*VI.9.3.139,.151)673
  speed(*IV.0.010)329
  stress(*IV.5.4.3.008)415
  sway(*IV.5.4.4.002)416
Windbreaks(*V.3.1.000)484
Winding movements(*II.4.3.074)246
Window:
  box(I.2.002a)5
  opening(IV.4.4.056)390
Windrow, windrowing(*V.1.3.063,.088) 437,439
Wine(VI.10.039)677
Wine-color(II.5.1.005x)264
Winged(*II.1.2.004)59
  pyramid(*V.3.4.068,.12d)544,556
Wingless (some)(*V.3.3.1.132c)502
Wings[II.2.2.23h(2);*V.3.3.1.132]152, 501
  membranous(*V.3.3.1.132g)502
Winn-Dixie(*VI.8.013)617
Winning(V.1.0.010)418
Winter hardiness(*IV.5.3.027,.054)409, 412
Wipe-off(*VI.3.007,*.062)578,582
Wire:
  cables(*V.1.1.017c)420
  mesh house(*V.1.1.059)426
    plus burlap(*V.1.0.001)418
Wirebound box(*VI.3.014,*.015,*.053f; *.4.000c,*.002,.026)578,582,585,588
Wireworms(*V.3.3.1.132a)501
Witches' broom(*V.3.3.2.133d,*.171, .469)515,516,530
  of cacao(*V.3.3.2.181f)519
  of ferns, poplar, alder, birch, cherry(*V.3.3.2.181e)519
Wither tip of lemons(*VI.9.3.042)666
Within living organism(*V.3.3.2.229) 522
Wood, wooden:
  baskets(*VI.4.003)585
    woven (veneer)(*VI.4.003)585
  box(*VI.3.053f;*.4.000c,*.009)582, 585,586
  container(s)(*V.1.2.003f;*VI.4.002, *.011)428,585,586
  crate(*VI.3.053f;.4.009)582,586
  flat, flats(*VI.4.003,.011)585,586
  framework(*V.1.1.059)426
  framing(*V.1.1.017c)420
  hamper(*VI.4.012)587
  lath(*V.1.1.059)426
  lice(*VI.7.2.017n)614
  lug, lugs(*VI.3.053f;*.4.014)582,587
  parenchyma(II.3.04a-c)210
  rollers(*VI.3.038)580
  rot-resistant(*V.4.1.107)564
  rough-sawn(*V.4.1.107)564
  shavings(*V.1.2.005,*.011)429,430
  stakes(*V.1.2.014)431
  trays(*V.1.2.003c)428
  vessel distribution(II.3.10a-c)216
  wool(*V.1.1.017i)421
  woven(*VI.4.003)585
Woody(II.1.1.3.105;.5.7.020)27,278
  dicotyledons(*II.3.052)196
  ornamentals(*I.2.009)6
  perennials(*V.1.4.028e)444
  plant foliage(*V.3.3.1.132g)502
    parts(*V.3.3.2.082)513
  plants(*V.1.4.022f-4,*.022g;*.3.3.3. 011)443,532
Woolly(*II.5.6.071)275
"Woolly bear"(*V.3.3.1.132g)502
Work (joule)(*II.0.260c)344
  crews(*VI.8.090)623
  order(VI.8.137)626
Workers(*V.3.3.1.132f)502
Working:
  capital(*VI.8.099)624
  sample(*V.1.3.075,.089)438,439
Worm-shaped(II.2.3.2.047)167
Wound healing(*II.3.084,*.125;*V.1.4. 008,*.028e,*p)197,200,441,444,448
Wounding(*IV.3.3.008v;*V.1.4.022a-3, *.083)378,452
Wounds(*V.4.2.2.015)567
Woven wood (baskets)(*VI.4.003)585
W.p.(*V.3.3.3.124)539
Wrapping(V.1.4.028w)449
  materials(V.1.2.017)432
Wringer rolls(*VI.3.081)583
Wrinkled(*II.1.1.2.010;*.5.6.111)21, 277
Wsd(*IV.5.2.078)405

WUE(*IV.5.2.084)406

X

Xantho-(II.5.1.008u)266
Xanthomonas(*V.3.3.2.040)511
   campestris pv. campestris(*VI.9.1.088)633
      pv. citri(*V.3.3.2.040)511
      pv. celebense(*V.3.3.2.040)511
      pv. hyacinthi(*V.4.2.1.023)566
      pv. malvacearum(*V.3.3.2.040)511
      pv. phaseoli(*VI.9.1.050)630
      pv. pruni(*VI.9.1.072)631
      pv. vesicatoria(*VI.9.1.073)631
      pv. vitians(*VI.9.1.061)631
Xanthophylls(*IV.1.1.002e;*.2.2.034;
   *VI.1.005)350,362,570
Xenia(IV.4.8.057)395
Xenogamy(*IV.4.8.017,.058)393,395
Xenogenesis(III.2.2.188)326
Xenon (lamp)(*IV.0.110)333
Xeromorphic(II.4.3.076)246
   characteristics(*II.4.2.006)239
   structure(IV.5.2.085)406
Xeromorphy(*IV.5.2.015,*.020)399,400
Xerophilous vegetation(II.4.2.034)240
Xerophyte(II.4.2.037;*.4.3.042,*.054;
   .4.05e;*IV.5.2.079)240,244,253,406
Xerophytic:
   characters(*IV.5.2.018)400
   habitats(*II.4.3.037)243
   plants(*II.4.3.030,*.033,*.050,
      *.056;*IV.1.2.008c)243,244,353
Xiphinema spp.(*V.3.3.1.200c)506
X-ray(s)(*IV.5.4.2.001)413
   radiography(V.1.3.031)434
   resistance(*IV.5.4.2.001)413
X unit(*IV.0.260m)347
Xylans(*IV.2.2.030c-b-2)362
Xylary fiber(*II.3.063)197
Xylem[*II.3.136,*.164,*.209,*.210,
   *.216,.221,.222,.20c,d(5),.21a
   (7),b(4)]201,202,205,206,226,227
   cylinders(*V.1.4.028e)444
   differentiation(II.3.222,.27,.28)
      206,233,234
   mobile(*V.3.3.3.113)538
Xylene(*V.3.3.3.033,*.097)533,537
D-Xylose(*IV.2.2.031a,*c,.038)361,362

Y

Year-around:
   flowering(V.4.1.123)565
   items(*VI.8.112)625
Yearlings(*V.4.1.086,.122)563,565

Yeast(s)(*IV.2.2.030c-a-3;*VI.7.2.
   017i,*.10,012,*.039)362,613,676,
   677
   extract(*V.1.5.005)470
   spot of lima beans(VI.9.1.575)659
Yellow(II.5.1.008,.008v)265,266
   disease(V.4.2.1.023)566
   jackets(*V.3.3.1.132e)502
   -necked caterpillar(*V.3.3.1.132g)
      502
   scale(*VI.9.2.007)660
   stripe(*V.4.2.1.022)566
Yellowing(V.3.3.2.473;*.4.2.1.001,
   .002,.013,*.019)530,565,566
Yellows(*V.3.3.2.474)530
   of cabbage(*V.3.3.2.181g)520
   viruses(*V.3.3.2.133e,*.263)515,523
Yield(*IV.5.2.018)400
York spot of apple(*VI.9.3.037)666
Young's modulus of elasticity(*IV.5.1.
   003,*.005)396
You-pick operation(*VI.8.125,.138)625,
   626

Z

Zeatin(*IV.2.1.025)358
   riboside(*IV.2.1.025)358
Zebra skin of tangerines(VI.9.3.154)
   674
Zeolites(*V.3.2.050)493
Zinc (Zn)(*V.3.2.028)491
   chelates(*V.3.2.085)496
   EDTA(*V.3.2.012)490
   sulfate(*V.3.2.085)496
   total(*V.3.2.071)495
Zonal soil(V.2.139)482
Zonation(*V.2.010)472
Zone:
   of eluviation(*V.2.112)479
   of illuviation(*V.2.112)479
Zoned (zonatus)(II.5.1.006u)265
Zoomorphosis(II.4.4.002h)247
Zoosporangia(*V.3.3.2.181d)518
Zoospore(V.3.3.2.476)531
Zoospores, colored resting(*V.3.3.2.
   181a)518
Zygomorphic(II.2.2.061c-14,.20d-h,.23)
   127,149,152
Zygomycetes(*V.3.3.2.181c;*.305,.477)
   518,525,531
Zygospores(*V.3.3.2.181c)518
Zygote(*II.2.3.3.014;*IV.4.1.006b-5,
   .014;.4.6.001a,*.001b;.4.8.059)
   163,380,381,392,395
Zygotene(III.2.2.189)326
Zygotes(*IV.4.4.043)390

INDEX OF TERMS

Zygotic:
  embryo(*IV.4.2.004)385
  origin(*IV.4.6.001c)392

INDEX OF CROPS

A

Abaca (Musa textilis):
  fibers(II.3.063)196
  root lesion nematodes(*V.3.3.1.200f) 506
Acacia (Acacia spp., Mimoscae):
  catena (wattle trees)(II.4.03B7)251
  cut flower(I.2.001c)5
  falcate(II.1.2.10c)75
  hooks, etc.(II.4.3.027)242
  hot zone plants(II.4.3.028)243
  leaf modifications(II.4.3.032)243
  phyllode(II.4.3.037)243
  pollen grain (A. longifolia)[II.2.2.13d(1)]142
  strombus-shaped (A. obliqua)[II.2.2.03h(2)]132
  xeromorphic structure(IV.5.2.085)406
Acanthaceae (Acanthus sp.):
  foliage plants(I.2.001d)5
  gullet(IV.4.4.004b-4)388
Acanthonema strigosum: epiphyllous(II.2.2.03f)132
Acer: (see Maple)
Acerola (Malpighia glabra):
  drupe(II.2.3.1.015)159
  endocarp(II.2.3.2.029c)162
Acridocarpus orientalis: anisopterous (II.2.3.2.01b)175
Acrocomia sp: armed(II.1.1.3.02c)38
Adansonia digitata: (see Baobab)
Adansonia sp.: flagelliflory(IV.4.4.021)389
Adiantum: (see Ferns)
Aegiceras corniculatum: locellate(II.2.2.08g)137
African violet (Saintpaulia ionantha; Gesneriaceae):
  leaf cutting(V.1.4.022c,.01)442,453
  pot plants(I.2.001f)6
Agavaceae: foliage plants(I.2.001d)5
Agave spp.:
  bulbils(V.1.4.006)441
  spinescent(II.1.1.3.20b)56
  squarrose(II.5.12c)292
  succulent(II.4.3.065)245
  xeromorphic structure(IV.5.2.085)406
Aglaia: pachyalazal seed form(II.2.3.3.08a)189
Ailanthus spp.: branching by substitution(II.1.1.3.018)22
Akee (Blighia sapida): paripinnate(II.1.2.12b)77
Alangium sp.: apogamy(IV.4.1.006b-2) 379
Alder: blister, leaf curl, witches' broom(*V.3.3.2.181e)519
Alectryon sp.: pachychalazal seed form (II.2.3.3.08b)189
Alfalfa:
  apocarpous(II.2.3.2.02a)176
  butterfly(*V.3.3.1.132g)502
  contra-inoculation(V.1.3.012)432
  implicative(II.1.1.3.12c)48
  legume (M. scutella)(II.2.3.1.02e) 169
  mesophyte(II.4.05)253
  'Moapa'(III.2.0.005a-5-b)314
  physiological drought(IV.5.4.3.007) 415
  'Ranger'(III.2.0.005a-5-b)314
  spiral(II.2.2.28c)157
  'Tempo'(III.2.0.005a-5-a)314
Algae:
  algae(V.3.3.2.008)510
  (blue-green) nitrogen fixation(IV.1.2.029)354
  (brown) chlorophyll(IV.1.2.036d)355
  C-3 plants(IV.1.2.08c)352
  epiphytes(II.4.2.006)239
  growth(IV.3.1.012)371
  microepiphytes(II.4.2.023)240
  photosynthesis(IV.1.2.036)354
  (red) chlorophyll(IV.1.2.036d)355
  stenothermic(IV.3.2.019)376
  trichogyne(V.3.3.2.437)529
Alligator pear: (see Avocado)
Allamanda (Allamanda cathartica):
  scandent(II.1.1.3.19a)55
Allium:
  sativum: unifacial cylindrical leaf (II.1.2.119,.04c)64,69
  sp.: apogamy(IV.4.1.006b-2)379
  root apical meristem(II.3.19c)225
  stoma(II.3.190)204
  tunicate bulb(V.1.4.071a-2)451
  ursinum: inverted bifacial leaf(II.1.2.016,.04b)60,69
Allspice, pimento (Pimenta dioica; Myrtaceae):
  approach grafting(V.1.4.028a)443
  functionally dioecious(IV.4.3.004) 386

INDEX OF CROPS

spice crop(I.1.013)4
Almond (Prunus amygdalus):
  clones(III.2.1.001a)316
  nut crop(I.1.007)4
  paste(VI.10.026)677
  (see also Stone fruits)
  stone fruit(I.1.014)4
  temperate zone crop(I.1.016)5
  training(V.3.4.029)541
Aloe spp.:
  long-short-day plant(IV.3.2.013h)375
  spirodistichy (Aloe serra)(II.1.1.1.
    025,.06)20,33
  succulent(II.4.3.065)245
Alsine michauxii: lax cyme(II.2.1.
  09f)112
Alstroemeriaceae: resupinate (leaves)
  (II.2.2.068)129
Amaranthaceae:
  anomalous secondary growth(II.3.
    008)194
  pyxis(II.2.3.1.029)160
  utricle(II.2.3.1.038)160
Amaranthus sp. (Amaranth):
  light requirement for germination(V.
    1.3.042)435
  pyxis (utricle)(II.2.3.1.029)160
Amaryllidaceae:
  lorate (leaves)(II.5.5.032)272
  (some) corona(II.2.2.022)121
Amaryllis (Hippeastrum spp.):
  bulb(V.1.4.071a-2)451
  bulbous plants(I.2.001b)5
  storage containers(VI.4.000d)585
  tunic(V.4.1.120)565
  tunicate bulb (separation)(V.1.4.
    071a-2)451
Ambrosia trifida: pollen grain[II.2.2.
  13d(3)]142
Amo: (see avocado)
American linden: (see Tilia)
Ampelopsis sp.: conditional partheno-
  carpy(IV.4.2.002)385
Anacardiaceae (some): spuriously paripinnate pinnate(II.1.2.072)62
Ananas comosus: (see Pineapple)
Ancistrophyllum spp.: pneumorhize(II.
  4.3.039)243
Anemone (Anemone spp.; Ranunculaceae):
  achene(II.2.3.1.002)158
  cut flower(I.2.001c)5
Angelica (Angelica archangelica; Um-
    belliferae)
  carpophore[II.2.3.1.03k
    (1)]170
  commissure[II.2.3.1.03k(2)(b)]170
  condiments, etc.(I.3.006)7

mericarp[II.2.3.03k(2)]170
schizocarp[II.2.3.03k]170
vittae[II.2.3.03k(2)(a)]170
Anise (Pimpinella anisum; Umbellife-
    rae):
  California aster yellows(VI.9.1.134)
    636
  condiments, etc.(I.3.006)7
  cremocarp(*II.2.3.1.032)160
  schizocarp(II.2.3.1.03j)170
  watery soft rot(VI.9.1.559)658
Annonaceae:
  mesotestal (seeds)(II.2.3.3.043c)167
  perichalazal (seed-form)(II.2.3.3.
    036a-6)165
Annona sp.:
  accessory (fruit)(II.2.3.1.001)158
  aggregate (fruit)(II.2.3.1.003)158
  apocarpous(II.2.2.041b)122
  axillary(II.2.3.2.03d)177
  (A. muricata) distichous(II.1.3.1.
    10)46
  muricate(II.5.16j)296
  protandry(IV.4.2.002b)385
  (A. squamosa) distichous(II.1.2.08c)
    73
  squamose(II.5.18g)298
  syncarpelly (soursop)(II.2.3.2.03d)
    177
Annuals:
  cambium(II.3.026)195
  drought escape(IV.5.2.015)399
  growth curve(IV.3.1.013)371
  mulches(V.3.3.3.059)535
  (winter) chilling requirement(IV.5.
    3.006)408
Anthocleista spp.: pneumorhize(II.4.3.
  039)243
Anthuriums (Anthurium spp.; Araceae):
  cut flowers(I.2.001c)5
Apeiba echinata: depressed(II.2.3.2.
  02i)176
Antigonon leptopus: (see Coralvine)
Apocynaceae:
  epipetalous stamens(II.2.2.007d-8)
    118
  twining (some)(II.4.3.074)246
Apocynum androsaemifolium:
  connate (stigmas)(II.2.2.18k)147
  follicle(II.2.3.1.02a)169
Apples (Malus sylvestris, M. spp.):
  Agricultural Marketing Agreement Act
    of 1937, as Amended(VI.7.2.004)609
  alternaria rot(VI.9.1.011)627
  ammonia injury(VI.9.3.003)663
  apple-cedar rust(VI.9.1.036)629
  bags(VI.4.001)585

INDEX OF CROPS

Baldwin spot(VI.9.3.008)663
bench-graftage(V.1.4.028b)443
bending(V.3.4.003)540
bitter pit(VI.9.3.008)663
bitter rot(*V.3.3.2.181e;VI.9.1.080)
  519,632
black heart(IV.5.3.003)407
black rot(VI.9.1.086)632
blossom-end rot(VI.9.1.101)634
blotch(*V.3.3.2.181g;VI.9.1.103)520,
  634
blue mold(*V.3.3.2.181g;VI.9.1.106)
  520,634
botryosphaeria rot(VI.9.1.116)635
bourgeon(V.3.4.005)540
bourse(V.3.4.006)540
bourse bud(V.3.4.007)540
brindille(V.3.4.010)540
Brook's spot(VI.9.1.231)641
brown core(VI.9.3.020)664
bruises(VI.9.3.025)665
bulk filling(VI.3.053b)582
bull's eye rot(VI.9.1.133)636
burr knot(II.1.1.3.029,.05a)22,41
cane blight(*V.3.3.2.181e)519
canker(*V.3.2.181f)519
carbon dioxide injury(VI.9.3.028)665
chilling requirement ('Rome')(IV.5.
  3.006)407
clones(III.2.1.001a)316
cluster base(V.3.4.006)540
codling moth injury(VI.9.2.018)660
controlled atmosphere storage injury
  (IV.9.3.032)665
cordon(V.3.4.021,.11b,c,e)541,555
core rot(VI.9.1.171)638
cork(VI.9.3.035)666
cork spot(VI.9.3.037)666
dard(V.3.4.023,.02d,g)541,546
deal(VI.8.026)618
delayed sunscald(VI.9.3.131)672
dwarfing (interstock; double working)(V.1.4.028h)445
East Malling rootstock(IV.3.1.014g)
  372
EM VII, EM IX(V.1.4.028h)445
ethylene(V.4.2.1.011)565
European apple sawfly injury(VI.9.2.
  025)660
festoons(V.3.4.12e)556
fiber-tracheid(II.3.23b)229
fisheye rot(VI.9.1.223)641
Forbes scale(VI.9.2.031)661
formaldehyde injury(VI.9.3.044)666
frost and freezing injury(VI.9.3.
  053)667
fruit cracking(VI.9.3.054)667
fruit spot(*V.3.3.2.181g;VI.9.1.231)
  520,641
grade(VI.8.038)619
gray mold rot(VI.9.1.265)643
growth curve(IV.3.1.013)371
hail injury(VI.9.3.061)668
halo spot(VI.9.3.044)666
handling systems(VI.2.020)575
horticultural maturity (winter)(VI.
  1.004)569
internal breakdown(VI.9.3.068)668
internal browning(VI.9.3.075)669
interstock(V.1.4.028h)445
jack(VI.10.004)674
Jonathan spot(VI.9.3.079)669
lance nematodes(*V.3.3.1.200d)506
leaf roller injury(VI.9.2.044)661
"long shoots"[II.1.1.3.14a(1)]50
Lorette pruning system(V.3.4.049)542
low-oxygen injury(VI.9.3.081)669
maggot injury(VI.9.2.004)659
Malling Merton rootstocks(IV.3.1.
  014g)372
mechanized harvesting(VI.2.022b)576
mixed buds(II.1.1.1.013)19
moldy core(VI.9.1.171)638
mound (stool) layerage(V.1.4.036b-1)
  450
ordinary scald(VI.9.3.005)663
orthostichy(II.1.1.1.05r)32
ozone injury(VI.9.3.092)670
packing(VI.3.047d)581
palmette(V.3.4.055,.11a,d,g)543,
  555
pansy spot(VI.9.2.054)661
parastichy(II.1.1.1.05r)32
pear leaf blister mite injury(VI.9.
  2.057)661
phloem fiber(II.3.24f)230
phyllotaxis(II.1.1.1.05r)32
physiological drought(IV.5.4.3.007)
  415
phytophthora rot(VI.9.1.380)648
piece or whole root grafting (whip,
  or tongue)(*V.1.4.028k,-9)446,447
pillar tree(V.3.4.056,.04a-f)543,548
pioneer pollen(IV.4.4.043)390
plum curculio injury(VI.9.2.063)662
pome(II.2.3.1.028,.05d)160,172
powdery mildew(V.3.3.2.181e)519
pressure bruise(VI.9.3.101)670
pruning(V.3.4.04)548
Putnam scale(VI.9.2.065)662
pyramid(V.3.4.068)544
quince rust(VI.9.1.437)651
rhizopus rot(VI.9.1.451)649
ripe rot(*V.3.3.2.181e)519

INDEX OF CROPS

ripening(VI.1.013)572
'Rome Beauty'(IV.5.3.006)408
root-lesion nematodes(*V.3.3.1.200f) 506
row(II.1.1.1.05r)32
russet ring(VI.9.1.468)653
rust(*V.3.3.2.181f;VI.9.1.036)519, 629
San José scale(VI.9.2.073)662
scab(*V.3.3.2.181e;VI.9.1.037)519, 629
scald(VI.9.3.005)663
senescent scald of 'Golden Delicious'(VI.9.3.111)671
"short shoots"[II.1.1.3.14a(1)]50
side rot(VI.9.1.490)654
sieve cell(II.3.26a)232
soft scald(VI.9.3.116)671
soggy breakdown(VI.9.3.117)671
sooty blotch(VI.9.1.503)655
source wrapping(VI.3.071)583
sphaeroblast(IV.3.1.014g)372
spindle bush(V.3.4.072,.04g)544,548
spur[II.1.1.3.14a(3)]50
star crack(VI.9.1.523)656
stem-cavity browning(VI.9.3.020)664
stippen(VI.9.3.008)663
sunburn(VI.9.3.131)672
sunscald(VI.9.3.131)672
tonnage items(VI.8.121)625
total (titratable) acid(VI.1.008g) 571
trench layerage(V.1.4.036b-5)450
umbelloid(II.2.1.12c)115
vessel element(II.3.22h)228
water core(VI.9.3.145)673
weight sizing(VI.3.064e)583
York spot(VI.9.3.037)666
Apricot (Prunus armeniaca):
  alternaria rot(VI.9.1.012)627
  anthracnose(VI.9.1.030)628
  bacterial spot(VI.9.1.072)631
  bitter rot(VI.9.1.030)628
  black mold rot(VI.9.1.083)632
  blue mold rot(VI.9.1.111)634
  horticultural maturity(VI.1.004)569
  powdery mildew(VI.9.1.409)650
  ring pox(VI.9.1.462)652
  scab(VI.9.1.470)653
  stone fruits(I.1.014)4
  training(V.3.4.029)541
Aquatic plants:
  stenothermic(IV.3.2.019)376
Arabis:
  accumbent(II.2.3.3.001,.02a)163,183
  mosaic(*V.3.3.1.200c;.4.2.1.022)505, 566
Arachis: (see Peanut)
Araucaria spp. (e.g. A. heterophylla):
  foliage plants(I.2.001d)5
Araliaceae:
  foliage plants(I.2.001d)5
  twining(II.4.3.074)246
  umbellate-paniculate(II.2.1.076)103
  umbellate-racemiform(II.2.1.077)103
Ardisia spp.: phytomorphosis(II.4.4.002e)246
Areca palm: red ring(*V.3.3.1.200a)505
Arecaceae:(III.1.1.008)303
Arecoideae: crown(II.1.1.3.031)23
Arenga sp.: hapaxanthic(II.5.3.011)267
Aristolochia: type of thickening(II.3.06d)212
Armeria: homostylic dimorphism(IV.4.3.011)386
Aroids (Araceace), edible aroids:
  Alocasia spp.(I.3.023)8
  Arisaema triphylla:
    spadix, spathe(II.2.1.13c)116
  caudex(II.1.1.3.021)23
  clones(III.2.1.001a)316
  Colocasia esculenta(I.3.023)8
  cyanide-resistant respiration(IV.1.2.040d)356
  (epiphytes) velamen(*II.3.215)205
  heat avoidance(IV.5.4.1.004)412
  perforated (leaves)(II.1.2.076)62
  (see also Foliage plants)
  spadix(II.2.1.066)103
  synandrium (stamens)(II.2.2.007d-17,.06a)119,135
  transverse buds(II.1.1.1.032)20
Artocarpus altilis (breadfruit):
  infructescence(II.2.3.2.04d)178
  monoecious(IV.4.3.014)387
Artocarpus heterophyllus:
  flagelliflory, ramiflory[II.2.1.05d(2,3)]108
  monoecious(IV.4.3.014)387
  syncarp(II.2.3.1.037)160
Arum maculatum:
  contractile roots(II.3.17b)223
  mess-and-soil(IV.4.5.002j)391
Asclepiadaceae:
  epiphytes(II.4.2.006)239
  follicle(II.2.3.1.017)159
  leaf modifications(II.4.3.032)243
  pollinium(II.2.2.007t-9)120
  (see also Milkweed)
  succulent(II.4.3.065)245
Ash:
  polyaxial stem(II.1.1.3.067)25
  samara(II.2.3.1.04a)171
Asparagus (Asparagus officinalis;

INDEX OF CROPS

Liliaceae):
Agricultural Marketing Agreement Act of 1937; as Amended(VI.7.2.004)609
bacterial soft rot(VI.9.1.057)630
cladophyll(II.4.3.013)241
fusarium rot(VI.9.1.245)642
gray mold rot(VI.9.1.266)643
miscellaneous crops(I.3.011)8
open-pollinated populations(III.2.1.001d)317
phytophthora rot(VI.9.1.381)648
scarifier(V.1.3.057)435
warm-season seeds(VI.1.3.087)439
watery soft rot(VI.9.1.556)658
xeromorphic structure(IV.5.2.085)406
Asters (Callistephus chinensis; Compositae):(see also bedding plants, cut flowers)
parastichy(II.1.1.1.016)19
yellows(VI.9.1.046)629
Aubergine: (see Eggplant)
Aubrieta: trichome(II.3.31k)237
Aurantiodeae:
citrus subfamily(III.1.1.002)302
Autumn crocus: colchicine(IV.2.2.044) 363
Avicennia spp.:
peg-roots(II.4.11f)259
pneumorhizae(ii.4.3.039)243
(see also Mangroves)
Avocado (Persea americana; Lauraceae)
anthracnose(VI.9.1.023)628
belt-end-roll sizing(VI.3.064a)582
berry(II.2.3.1.05a)172
blotch(VI.9.1.143)636
cerospora spot(VI.9.1.143)636
clones(III.2.1.001a)316
common names(III.1.1.005)303
endocarp(II.2.3.2.029c)162
Federal Food, Drug & Cosmetic Act, as Amended(VI.7.2.016)610
flaps(II.2.2.10g)139
growth curve(IV.3.1.013)371
hand harvesting methods(VI.2.022a) 576
heat injury(VI.9.3.064)668
lachyrmiform(II.2.3.2.04e)178
light requirement for germination(V.1.3.042)435
physiological drought(IV.5.4.3.007) 415
polyaxial stem(II.1.1.3.067)25
prolepsis[II.1.1.3.17a(1)]53
prophyll(II.1.2.18a)83
pyriform(II.2.3.2.06b)180
race(III.2.0.009)316
rhizopus rot(VI.9.1.450)651
ripening(VI.1.013)572
root rot(*V.3.3.2.181d)518
scab(VI.9.1.047)629
stem-end rots(VI.9.1.527)657
stenospermocarpy(IV.4.2.004b)385
subterminal (buds)(II.1.1.1.028)20
syllepsis[II.1.1.3.17a(2)]53
synchronous protogyny(IV.4.3.002f) 386
washing(VI.3.081)583
weight sizing(VI.3.064e)583
Azaleas (Rhododendron spp.; Ericaceae):
B-9(V.3.4.000a)539
bypassing shoot(V.4.1.024)560
cutting graft(V.1.4.028g)445
marginal(II.5.6.076)275
pinching(V.3.4.057)543
pot plants(I.2.001f)6
pots(V.1.2.003h)428
pruning(V.4.1.094)563
revolute (vernation)(II.1.1.2.026)21
sulfur(V.3.2.089)496
year-round flowering(V.4.1.123)565
Azorella selago:
chamaephyte, cushion plant, trophophyte(II.4.01a)249

B

Baby's Breath (Gypsophilla elegans, G. paniculata; Caryophyllaceae):
cut flowers(I.2.001c)5
Bahiagrass:
C-4 Plants(IV.1.2.008b)352
huller(V.1.3.057)436
turf crops(I.2.011)6
Bald cypress: (see Taxodium)
Bamboo:
branching, orthotropic, plagiotropic (II.1.1.3.04c)40
culm(II.1.1.3.08b)44
cryptophyte(II.4.01b)249
evergreen, monocarpic, plietesial (II.5.02d)282
herbs, herbaceous, iteroparous(II.0.02a)13
hygromorphosis(II.4.4.003c)247
iteroparous(II.0.009)10
pachymorph rhizome(V.1.4.060c)451
rhizomatous branching(II.1.1.3.082, .18d)26,54
semelparous(II.5.3.019;.5.02d)268, 282
tubular(II.1.1.3.22c)58
Banana, plantain (Musa sp.; Musaceae)
anthracnose(VI.9.1.024)628

INDEX OF CROPS

aseasonal(II.5.01c)281
bacterial wilt(*V.3.3.2.040)511
bastiony(II.1.1.3.03c)39
beer(VI.10.003)674
botryodiplodia fruit rot(VI.9.1.115) 635
burrowing nematodes(*V.3.3.1.200f) 506
carton(VI.4.008)586
(Cavendish): complete (automatic) parthenocarpic(IV.4.2.001a)385
chilling injury(IV.5.3.005;VI.9.3. 029)407,665
clones(III.2.1.001a)316
crown rot(VI.9.1.182)639
degreening(VI.3.023)579
dumping(VI.3.027)579
endocarp(II.2.3.2.029c)162
endodermis(II.3.11b)217
flagelliflory(IV.4.4.021)389
('Gros Michel'): strong (automatic parthenocarpy)(IV.4.2.001b)385
growth curve(IV.3.1.013)371
hand harvesting(VI.2.022a)576
handling systems(VI.2.020)575
herbs, herbaceous(II.0.02b)13
impulse item(VI.8.047)620
leaf spot(*V.3.3.2.181e)519
matroclinal(III.2.2.104)322
mesophyte(II.4.05b)253
moko(*V.3.3.2.040)511
monocarpic(II.5.01c)281
nutritional spray(V.3.2.067)494
pachymorph rhizome(V.1.4.060c)451
Panama disease(*V.3.3.2.181g)520
penniparallel (venation)(II.1.2. 121m)65
perennial(II.5.02d)282
pitting(VI.9.1.395)649
plantain(VI.2.022a)576
pseudostems (mulch)(V.3.3.3.059)535
pulpa(II.2.3.2.029c,.05f)162,179
rhizome(V.1.4.04d)456
root-lesion nematodes(*V.3.3.1.200f) 506
"seed"(V.1.4.068)451
seedpiece(V.1.4.069)451
semelparous(II.5.3.019)268
sigatoka(*V.3.3.2.181e)519
speckle(VI.9.1.516)656
spiral nematodes(*V.3.3.1.200d)506
stamping(VI.3.072)583
sympodial(II.1.1.3.099)27
tepal(II.2.2.061f)129
thielaviopsis stalk rot(VI.9.1.542) 657
tie-in display(VI.8.120)625
tonnage items(VI.8.121)625
tropical crop(I.1.017)5
unequal(II.5.5.062,.12f)273,292
washing(VI.3.081)583
wilt(*V.3.3.2.181g)520
Baobab: staminal column(II.2.2.22e)151
succulent(II.4.3.065)245
Baptisis perfoliata: perfoliate(II.1. 2.05e)70
Barberry:
anisophyllous(II.1.2.02c)67
flaps(II.2.2.10f)139
linear (embryo) medium(II.2.3.3.02k) 183
Barley:
beer(VI.10.003)674
caryopsis(II.2.3.1.008,.011)159,168
C-3 plants(IV.1.2.008a)352
'Harlan' (composite-cross population)(III.2.0.005b)315
quantitative long-day plant(IV.3.2. 013g)375
Barkleya spp.: obcampylotropous (seed form)(II.2.3.3.036a-4)165
Basil (Ocimum basilicum; Labiatae):
condiments, etc.(I.3.006)7
Bauhinia Bauhinia spp.:
anomalous secondary growth(II.3.008, .03d,f,h)194,209
B. integrifolia obcampylotropous (seed form)(II.2.3.3.036a-4,.06e-h)165,187
Bay (sweet bay, Laurus noblis; Lauraceae):
condiments, etc.(I.3.006)7
Beans, Common, green snap, field (Phaseolus vulgaris; lentil, Lens culinaria; Fabaceae):
anthracnose(*V.3.3.2.181g;VI.9.1. 032)520,628
awl nematodes(*V.3.3.1.200c)505
bacterial blights(*V.3.3.2.133c)515
bacterial soft rot(VI.9.1.065)631
bacterial wilt(VI.9.1.075)632
beetle(*V.3.3.1.132a)501
catena(II.4.03B8)251
cochleate (P. caracalla)(II.2.2.15f) 144
color separator(V.1.3.057)435
common bean mosaic(VI.9.1.167)638
cottony leak(VI.9.1.178)638
C-3 plants(IV.1.2.008a)352
day neutral plant(IV.3.2.013c)374
gray mold rot(VI.9.1.275)644
haulm(II.1.1.3.044)24
harvested immature(II.2.3.1.000)158
inbred pure lines(III.2.1.001c)317

INDEX OF CROPS

(kidney) day-neutral plant(IV.3.2.013c)374
mechanized harvesting(VI.2.022b)576
phytoalexin(IV.2.2.162)368
pod-distorting mosaic(VI.9.1.402)649
pod mottle(VI.9.1.403)649
powdery mildew(VI.9.1.410)650
pulses(I.3.014)8
rhizoctonia pod rot (see soil rot)
rhizopus soft rot(VI.9.1.456)652
root-lesion nematodes(*V.3.3.1.200f) 506
root rot(*V.3.3.2.181e)519
russeting(VI.9.3.107)671
rust(VI.9.1.494)654
sclerotium rot(VI.9.1.481)654
sclereids(II.3.25a-d)231
seed production(III.2.0.004c-f)312
seed structure(II.2.3.3.01a)182
soil rot(VI.9.1.499)654
southern bean mosaic(VI.9.1.512)656
southern blight: (see sclerotium rot)
sunscald(VI.9.3.136)673
stomata(II.3.29f)235
'Tendercrop' (line variety)(III.2.0.005a-2)313
thermoperiodism(IV.3.2.021)376
vexillary (estivation)(II.1.1.2.030) 21
warm-season seeds(V.1.3.087)439
watery soft rot(VI.9.1.562)658
Bean, Lima (Phaseolus lunatus; Fabaceae):
  bacterial spot(VI.9.1.071)631
  common, fuscous and halo bacterial blights(VI.9.1.050)630
  downy mildew(VI.9.1.205)640
  'Mezcla' (composite-cross population(III.2.0.005b)315
  pod blight(VI.9.1.401)649
  powdery mildew(VI.9.1.410)650
  pulses(I.3.014)8
  rhizoctonia pod rot: (see soil rot)
  seed spotting(VI.9.1.483)654
  soil rot(VI.9.1.499)654
  warm-season seeds(V.1.4.087)439
  yeast spot(VI.9.1.575)659
Bean, Tepary (Phaseolus acutifolius; Fabaceae): pulses(I.3.014)8
Bedding plants:
  bedding plants(I.2.001a)5
  flat(VI.4.011)586
  hanging basket(VI.4.013)587
  packinghouse(VI.3.000)577
  propagation structures(V.1.1.048)424
  pots(V.1.2.003h)428

Beech: polyaxial stem(II.1.1.3.067)25
Beet, Forage (=mangold):
  common names(III.1.1.005)303
  tap root plus hypocotyl storage[II.4.3.069,.13f(3)]245,261
Beet, Garden (Beta vulgaris; Chenopodiaceae):
  anomalous secondary growth(II.3.03c)209
  betalains(IV.2.2.028)360
  black rot(*V.3.3.2.181e)519
  black rot(VI.9.1.087)633
  blight(*V.3.3.2.181e)519
  C-3 plants(IV.1.2.008a)352
  cercospora leaf spot(VI.9.1.141)636
  common name(III.1.1.005)303
  cyst nematode(*V.3.3.1.200g)506
  decorticator(V.1.3.057)435
  Ditylenchus spp.(*V.3.3.1.200a)505
  downy mildew(*V.3.3.2.181d)518
  dry heart rot(*V.3.3.2.181e)519
  hybrids(III.2.1.001b)317
  hypocotyl[II.4.13f(2)]261
  internal black spot(VI.9.3.067)668
  leaf miner(*V.3.3.1.132b)501
  leaf spot(*V.3.3.2.181e)519
  long-day plant(IV.3.2.013g)375
  open-pollinated populations(III.2.1.001d)317
  quantitative long-day plant(IV.3.2.013g)375
  pelletized seeds(V.1.3.045)435
  ring spot(*V.3.3.2.181e)519
  root crops(I.3.019)8
  scab(VI.9.1.077)632
  stem tubers[II.4.3.057,.13f(2)]244,261
Beet, Sugar:
  beet cyst nematode(*V.3.3.1.200g)506
  common names(III.1.1.005)303
  long-day plant(IV.3.2.013g)375
  Nacobbus spp. nematodes(*V.3.3.1.200g)506
  physiological drought(IV.5.4.3.007) 415
  raffinose (see Oligosaccharides)(IV.2.2.031b)361
  tap root storage[II.4.3.070,.13f(1)] 245,261
  thermoperiodism(IV.3.2.021)376
Begonias (Begonia spp.; Begoniaceae):
  B. fruitcosa: anomalous secondary growth(II.3.03b)209
  Begonia, Rex: leaf cuttings(V.1.4.022a,.01)442,453
  pot plants(I.2.001f)6
Bentgrass:

INDEX OF CROPS

finishing machine(V.1.3.057)435
turf crops(I.2.011)6
Bermudagrass (Cynodon dactylis):
  C-4 plants(IV.1.2.008b)352
  'Coastal' (clonal variety)(III.2.0.
    005a-1)313
  huller(V.1.3.057)435
  physiological drought(IV.5.4.3.007)
    415
  rotary tiller(V.1.1.056)426
  turf crops(I.2.011)6
Berries (see also Blackberry, Blueberry, Strawberry):
  flat(VI.4.011)586
  horticultural maturity(VI.1.004)569
  ripening(VI.1.013)572
  small fruits(I.1.012)4
Betel nut: masticatory(II.0.013)10
Betulaceae: inconspicuous (blossom)
  (IV.4.4.004a)387
Biennials: (winter) chilling requirement(IV.5.3.006)407
Bignoniaceae:
  anomalous secondary growth(II.3.008,
    .03a,e)194,209
  superposed buds(II.1.1.1.030)20
  tendrils(II.4.3.071)245
  twining(II.4.3.074)246
Birch:
  blister(*V.3.3.2.181e)519
  leaf curl(*V.3.3.2.181e)519
  (white) catkin(II.2.1.05c)108
  witches' broom(*V.3.3.2.181e)519
Bird of paradise (Strelitzia Reginae;
  Strelitziaceae): cut flowers(*.2.
  001c)5
Bittersweet: rotate(II.2.2.21a)150
Bixa orellana: horseshoe-shaped (anthers)(II.2.2.007w-7,.08e)120,137
Black-eyed susan: long-day plant(IV.3.
  2.013g)375
Black pepper (see Pepper, Black)
Black walnut (Juglans nigra) (see Walnut)
Bleeding heart (Dicentra spectabilis;
  Fumariaceae)(I.1.2.001f)6
Blessed thistle (Cnicus benedictus):
  phyllaries(II.2.1.12a)115
Blighia sapida (see Akee)
Blue berries (Vaccinium spp.; Vacciniaceae):
  apical pores(II.2.2.10e)139
  baskets(VI.4.003)585
  bloom(II.5.6.010)274
  chilling requirement(V. ashei)(IV.
    5.3.006)407
  filiform hairs[II.2.2.10e(2)]139

growth curve(IV.3.1.013)371
hand harvesting methods(VI.2.022a)
  576
mechanized harvesting(VI.2.022b)576
(see also Berries)
subglobose(II.2.2.211)150
Bluegrass:
  Kentucky
    Adelphi(III.2.0.005a-2)313
    Bristol(III.2.0.005a-2)313
    'Merion'(III.2.0.005a-2)313
    'Penstar'(III.2.0.005a-2)313
    photodormancy(IV.3.2.011)374
    short-long-day plant(IV.3.2.013m)
      375
    turf crop(I.2.011)6
Bombacaceae: succulent(II.4.3.065)245
Boraginaceae: dimorphic (flowers)(IV.
  4.3.003)386
Borage (Borago officinalis; Boranginaceae:
  condiments, etc.(I.3.006)7
Bougainvillea spp.:
  cyathium(V.4.1.033)560
  hooks, etc.(II.4.3.027)242
Boxwood (Buxus microphylla): shrub(II.
  0.06a)17
Brachychiton populneum: fusiform(II.5.
  15i)295
Brambles (Blackberries, raspberries,
  Rubus spp.; Rosaceae) (see also
  Blackberry, Raspberry):
  accessory fruit(II.2.3.1.06a,c)173
  aggregate fruit(II.2.3.1.003,.06a,c)
    158,173
  baskets(VI.4.003)585
  black rot(*V.3.3.2.181e)519
  blight(*V.3.3.2.181e)519
  brush control(V.3.3.3.015)532
  cane blight(*V.3.3.2.181e)519
  canescent(II.5.6.019)274
  cano-tomentose(II.5.6.020)274
  choricarpelly ('Oregon Evergreen'
    blackberry)(II.2.3.2.03a)177
  clones(III.2.1.001a)316
  conoidal(II.2.3.2.02e)176
  corymbiform(II.2.1.06b)109
  cymiform(II.2.1.06c)109
  decumbent(II.1.1.3.034)23
  drupelet(II.2.3.1.016)159
  dry heart rot(*V.3.3.2.181e)519
  floricane[II.1.1.3.043,.11a(2)]24,47
  growth curve(IV.3.1.013)371
  hooks, etc.(II.4.3.027)242
  induced parthenogenesis(IV.4.1.006b-
    3-a)379
  leaf-bud cuttings (blackberry)(V.1.

INDEX OF CROPS

  4.022d,.01)442,453
 leaf spot(*V.3.3.2.181e)519
 monoploid apomixis(IV.4.1.006a-7)378
 novirame(II.1.1.3.061,.12a)25,48
 paniculiform[II.2.1.051,.06a(1),
   .11d]102,114
 parcifrond(II.1.1.3.063,.11b)25,47
 parthenogenesis(IV.4.1.006a-10)379
 prickles(II.1.1.3.16d)52
 primocane[II.1.1.3.071,.11a(1)]25,47
 pruning and training (blackberry,
   raspberries)(V.3.4.008)540
 pruning-training systems(V.3.4.05)
   549
 pyrene(II.2.3.2.02f)176
 quinate(II.1.2.18d)83
 racemiform(II.2.1.12d)115
 raspberry ringspot(V.4.2.1.022)566
 ring spot(*V.3.3.2.181e)519
 root cuttings(V.1.4.022e,.01e,f)442,
   453
 small fruits(I.1.012)4
 stipules[II.1.2.20d(2,3)]85
 suckers(V.1.4.083)452
 synflorescence(II.2.1.06a)109
 temperate zone crop(I.1.016)5
 tip layerage(V.1.4.036b-4)450
 trichomes(II.3.31j)237
 turion(II.4.1.008)238
Brassica spp. (see also Crucifers):
 cole crops(I.3.005)7
 parastichy(II.1.1.1.016)19
 root-knot nematodes(*V.3.3.1.200e,g)
   506
Brazil nut (Bertholletia excelsa; Le-
   cythidaceae):
 nut(II.2.3.1.024)159
 nut crop(I.1.007)4
 horticultural crops(I.0.001)3
Breadfruit (Artocarpus altilis):
 multiple fruit(II.2.3.1.023)159
 suckers(V.1.4.083)452
Broccoli (Brassica oleracea (Italica
   group); Cruciferae):
 alternaria leaf spot(VI.9.1.007)627
 bacterial leaf spot(VI.9.1.054)630
 cole crops(I.3.005)7
 gray mold rot(VI.9.1.267)643
 hydrocooling(VI.5.035b)591
Bromegrass: 'Saratoga'(III.2.0.005a-5-
   b)314
Bromeliaceae:
 epiphytes(II.4.2.006)239
 pot plants(I.2.001f)6
Bromeliads (Bromeliaceae):
 foliage plants(I.2.001d)5
 pot plants(I.2.001f)6

 protective devices(II.4.3.042)244
Brugiera spp.: pneumorhizae(II.4.3.
   039)243
Brussels sprouts (Brassica oleraceae
   (Gemmifera group); see Crucifers):
 cole crops(I.3.005)7
 hybrids(III.2.1.001b)317
 vernalization(IV.3.2.021e)376
Bryonia: tendrils(II.4.3.014,.071;.4.
   06c)241,245,254
Bryophyllum: leaf cuttings(V.1.4.022c,
   .01)442,453
Bryophyta: archegonium(II.6.003)279
Bryophytes:
 C-3 plants(IV.1.2.008a)352
 microepiphytes(II.4.2.023)240
 nitrogen fixation(IV.1.2.029)354
 photosynthesis(IV.1.2.036)354
 spermatozoid(II.6.019)280
Bucida spp.: branching by apposition
   (II.1.1.3.017)22
Buckhorn plantain: finishing machine
   (V.1.3.057)435
Buckthorn: annular (disk)[II.2.2.03c
   (2)]132
Buffelgrass: 'Higgins'(III.2.0.005a-1)
   313
Bulb crops (vegetables)(I.3.004)7
Bulbs, Bulbous plants (see also Ama-
     ryllis, crocus, Dutch iris,
     grape hyacinth, lily (lilies),
     narcissus, tulip):
 antholyse, antholysis(V.4.2.2.001)
   566
 basal plate(V.4.1.003)559
 basal root(V.4.1.004)559
 "brown ring"(*V.3.3.1.200a)505
 bulb dealer(V.4.010)559
 bulb forcer(V.4.011)559
 bulb forcing(V.4.012)559
 bulb grower(V.4.013)560
 bulb precooling(V.4.019)560
 bulb size(V.4.021)560
 bulb tray(V.4.022;VI.3.000d)560,585
 bublet(V.4.016)560
 bulbous plants(I.2.001b)5
 bullhead(V.4.1.023)560
 case-cooled bulbs(V.4.1.025)560
 commercial bulb(V.4.1.026)560
 contractile roots of bulbs(II.3.035)
   195
 corm(V.1.4.071b)451
 crocus(V.4.1.012)560
 curl-stripe(V.4.2.1.005)565
 Cylindrocarpon radicicola(V.4.2.1.
   017)566
 daffodil (see Narcissus)(V.4.1.012)

560
daughter bulb(V.4.1.035)560
devernalization(V.4.1.036)561
dibble(V.1.1.010)419
Ditylenchus dipsaci(V.4.2.1.014)566
Ditylenchus spp. on bulb crops(V.3.3.1.200a,g)505,506
dormancy(V.4.1.039)561
double nose bulb(V.4.1.040)561
dry storage(V.4.1.041)561
Dutch iris(I.2.001b;V.4.1.012)5,560
filling(V.4.1.045)561
5° forcing(V.4.1.111)564
flower blindness(V.4.2.2.011)567
flower bud abortion(V.4.1.047;.4.2.2.012)561,567
flower bud abscission(V.4.2.2.013)567
flower differentiation(V.4.1.048)561
flower induction(V.4.1.050)561
flower initiation(V.4.1.051)561
forcing(V.4.1.054)561
Fusarium
  oxysporum(V.4.2.1.017)566
  f. lilii(V.4.2.1.001)565
  solani(V.4.2.1.017)566
  sp.(V.4.2.1.020)566
grape hyacinth(V.4.1.012)559
greenhouse phase (forcing)(V.4.1.059)562
hand harvesting methods(VI.2.022a)576
harvesting of bulbs(VI.2.000)575
heat delay(V.4.1.063)562
hyacinth(V.4.1.012)559
juvenility(V.4.1.064)562
leaf counting(V.4.1.065)562
leaf scorch(V.4.2.2.017)568
long-day treatment(V.4.1.067)562
marketing phase(V.4.1.068)562
maturity(V.4.1.069)562
mother blocking(V.4.1.072)562
mother bulb(V.4.1.073)562
natural cooling(V.4.1.075)563
nonprecooled (NP) bulbs(V.4.1.076)563
panning(V.4.1.080)563
pathological (parasitic) market diseases (see Commercial Floriculture(V.4.)565-566
PC(V.4.1.088)563
Penicillium rot of lilies, etc.(V.4.2.1.015)566
Perishable Agricultural Commodities Act of 1930, as Amended(VI.7.1.)597-609
planting back(V.4.1.085)563

pot plants(I.2.001f)6
Pratylenchus spp.(V.4.2.1.014)566
precooling(V.4.1.088)563
preheating(V.4.1.089)563
production phase(V.4.1.092)563
programming phase(V.4.1.093)563
Pseudomonas sp.(V.4.2.1.021)566
reefer(VI.8.097)623
rogueing(V.4.1.099)564
rooting bed(V.4.1.100)564
rooting room(V.4.1.101)564
spring-flower bulbs(V.4.1.112)564
stage "G" of tulip, etc.(V.4.1.113)564
standard forcing(V.4.1.114)564
stem bulblet(V.4.1.115)564
stem root(V.4.1.116)564
storage(VI.5.047)591
storage containers(VI.4.000d)585
summer sprouting of lily(V.4.1.117)564
tepal(II.2.2.061f)129
tulip(V.4.1.012)559
vernalization(V.4.1.121)565
virus diseases of bulbs(V.4.2.1.022)566
Xanthomonas campestris pv. hyacinthi (V.4.2.1.023)566
Butea monosperma:
  lignotuber(II.4.14b)262
Buttercup, Common:
  achene(II.2.3.1.01b)168
Buttonbush (Cephalanthus occidentalis):
  head(II.2.1.09e)112

C

Cabbage (Brassica oleracea (Capitata group). Cruciferae):(I.3.005)7
  alternaria leaf spot(VI.9.1.007)627
  anthracnose(VI.9.1.034)628
  bacterial leaf spot(VI.9.1.054)630
  bacterial soft rot(VI.9.1.058)630
  bacterial zonate spot(VI.9.1.076)632
  black leaf speck(VI.9.3.011)664
  black rot(VI.9.1.088)633
  butterfly(*V.3.3.1.132g)502
  cole crops(I.3.005)7
  combine(V.1.3.010)432
  edema(VI.9.2.079)662
  freezing injury(VI.9.3.045)666
  gray mold rot(VI.9.1.267)643
  looper(*V.3.3.1.132g;V.3.3.3.013f)502,532
  maggot(*V.3.3.1.132b)501
  'Market Prize' hybrid(III.2.0.005a-

6)314
  open-pollinated populations(III.2.1.
    001d)317
  rhizoctonia head and root rots(VI.9.
    1.441)651
  root-lesion nematodes(*V.3.3.1.200f)
    506
  sclerotinia rot(VI.9.1.557)658
  stubby-rot nematodes(*V.3.3.1.200c)
    505
  sunscald(VI.9.3.133)672
  thrips injury(VI.9.2.079)662
  tipburn(VI.9.3.140)673
  watery soft rot(VI.9.1.556)658
  wilt(*V.3.3.2.181g)520
  worm(V.3.3.3.013c)532
  yellows(*V.3.3.2.181g)520
Cacao (Theobroma cacao, Sterculiaceae)
  acropetal(I.1.3.01b)37
  (beans) drying(VI.10.009)674
  beverage crop(I.1.001)4
  bipulvinate(II.1.2.18b)83
  brown rot(*V.3.3.2.181d)518
  catena(II.4.03A7)251
  cauliflorous(II.1.1.1.005)19
  ceratostomella (ceratocystis) wilt
    (*V.3.3.2.181e)510
  chupon[II.1.1.3.024,.05d(1),.13b(1)]
    23,41,49
  crown and trunk canker(*V.3.3.2.
    181d)518
  drupe(II.2.3.1.05h)172
  fan branches[II.1.1.3.048,.05d(2),
    .13b(2)]24,41,49
  fermentation(V.1.3.026;VI.10.012)
    433,676
  foot and root rot(*V.3.3.2.181d)518
  hand harvesting methods(VI.2.022a)
    576
  hardening-off(V.1.3.033)434
  high humidity beds(V.1.1.053)425
  interplanting(V.3.1.008)486
  jorquette[II.1.1.3.048,.05d(2),.13b
    (2)]24,41,49
  open-pollinated populations(III.2.1.
    001d)317
  "pod"(II.2.3.1.027,.05h)160,172
  pod rot(*V.3.3.2.181d)518
  polyaxial stem(II.1.1.3.067)25
  processing(VI.10.000)674
  propagation cases(V.1.1.046)424
  pulvini(II.1.2.18b)83
  sulcate(II.5.20a)300
  tropical crop(I.1.017)5
  witches' broom(*V.3.3.2.181f)519
Cactaceae:
  epiphytes(II.4.2.006)239
  glochidiate(II.5.6.056)275
  succulent(II.4.3.065)245
  xeromorphic structure(IV.5.2.085)406
Caesalpinia spp. (some):
  hooks, etc.(II.4.3.027)242
Caesalpiniaceae:
  zygomorphic (corolla)(II.2.2.061c-
    14)127
Cajeput (Melaleuca quinquinervia;
    Myrtaceae):
  essential oil(I.1.013)4
Calabaza:
  cucurbits(I.3.007)7
  (see Pumpkin)
Caladium:
  stem tubers(II.4.3.057)244
Calamus spp.:
  cirrus(II.4.3.012)241
  rattan(II.0.022)11
Calceolaria (Calceolaria crenatiflora;
    Scrophulariaceae): pot plants(I.2.
    001f)6
Calla lily (Zantedeschia aethiopica;
    Araceae):
  bell-or funnel-shaped(IV.4.4.004b-1)
    387
  cut flowers(I.2.001c)5
Calophyllum spp.:
  megatherm taxa(III.1.2.033)309
Caltha spp.:
  dish-shaped(IV.4.4.046-2)388
Calystegia spp.:
  bell-or funnel-shaped(IV.4.4.004b-
    1)387
Camellia (Camellia japonica; Theace
    ae):
  cut flowers(I.2.001)5
  grafting (cleft)(V.1.4.028k-2)446
  semihardwood (stem) cuttings(V.1.4.
    022f-3,.02c-e,.03a,b)443,454,455
Camellia sinensis: (see Tea)
Camomile (Anthemis nobilis, Composit-
    ae):
  condiments, etc.(I.3.006)7
Campanula (harebell):
  campanulate(II.2.2.21e)150
Campanula spp.:
  bell-or funnel-shaped(IV.4.4.004b-1)
    387
Canarium indicum:
  unitegmic (testal) seed(II.2.3.3.
    12c,d)193
Candy alexander: amplexicaul(II.1.2.
  02a)67
Canna:
  aerenchyma(II.3.04f)210
  irregular (perianth)(II.2.2.20i)149

INDEX OF CROPS

rhizomes (divisions)(V.1.4.024)443
Cantaloupe:
  'Market pride'(III.2.0.005a-7) 314
  Federal Food, Drug & Cosmetic Act, as Amended(VI.7.2.016)610
  (see also Muskmelons)
Cape primrose (Streptocarpus X hybridus):
  pot plants(I.2.001f)6
Caper (Capparis spinosa; Capparidaceae):
  condiments, etc.(I.3.006)7
Capsella spp.:
  embryo development(II.2.3.3.03f-i) 184
  incumbent(II.2.3.3.02h)183
  silicle(II.2.3.1.033,.03h)160,170
Carambola Averrhoa carambola):
  mixed polyaxial stem(II.1.1.3.16b)52
Caraway (Carum carvi; Umbelliferae):
  condiments, etc.(I.3.006)7
Cardamine:
  silique(II.2.3.1.03g)170
  valve(II.2.3.2.07g)181
Cardamon (Elettaria cardamomum, Amomum cardamomum; Zingiberaceae):
  condiments, etc.(I.3.006)7
Carex arenaria:
  leptomorph rhizome(V.1.4.060a)451
  stolon(II.4.12d)260
Caricaceae: laticifer(II.3.098)198
Carica papaya: (see Papaya)
Carissa grandiflora:(II.4.3.042)244
  (see also Natal plum)
Carnations (Dianthus caryophyllus; Caryophyllaceae):
  bud rot(*V.3.3.2.181g)520
  caulimovirus(V.3.3.2.088)513
  carbon dioxide fertilization(V.3.2.009)490
  cut flowers(I.2.001c)5
  disbud, disbudding(V.4.1.038)561
  rust(*V.3.3.2.181f)519
  slab-side(V.4.1.108)564
  slipping stage(V.4.1.109)564
Carnivorous plants(II.4.3.009)241
  chemonasty(IV.3.3.003a)376
  secretory cells(II.3.179)203
Carrot (Daucus carota; Umbelliferae):
  bacterial soft rot(VI.9.1.059)630
  beet cyst nematode(*V.3.3.1.200g)506
  black mold(VI.9.1.082)632
  black rot(VI.9.1.089)633
  crater rot(VI.9.1.180)638
  cremocarp(II.2.3.1.03i)170
  C-3 plants(IV.1.2.008a)352

fusarium rot(VI.9.1.246)642
gray mold rot(VI.9.1.268)643
hydrocooling(VI.5.035b)591
kind(III.2.0.004a)312
mechanized harvesting(VI.2.022b)576
mericarp(II.2.3.1.022)159
Nacobbus spp.(*V.3.3.1.200g)506
open-pollinated populations(III.2.1.001d)317
pelletized seeds(V.1.3.045)435
reniform nematodes(*V.3.3.1.200d)506
rhizopus soft rot(VI.9.1.453)652
root crops(I.3.019)8
(root) phytoalexin(IV.2.2.162)368
root-knot nematodes(*V.3.3.1.200e, *g;VI.9.2.070)506,662
scab spot complex(VI.9.3.110)671
schizocarp(II.2.3.1.03i)170
soft rot(*V.3.3.2.040)511
sour rot(VI.9.1.506)655
source wrapping(VI.3.071)583
Stemphylium radicinum(VI.9.1.089)633
tap root storage(II.4.3.070)245
watery soft rot(VI.9.1.558)658
Carya spp. (see also Pecan):
  perulate (bud; shag bark hickory, C. ovata)(II.5.6.091)276
  superposed buds(II.1.1.1.030)20
Caryophyllales:
  betalains(IV.2.2.028)360
  (see also Carnations)
Caryota spp.: hapaxanthic(II.5.3.011) 267
Cashew (Anacardium occidentalis; Anacardiaceae):
  "apple"[II.2.3.1.04i(2)]171
  nut(II.2.3.1.04i)171
  nut crop(I.1.007)4
  polygamous(IV.4.3.017)387
  processing(VI.10.000)674
  tropical crops(I.1.017)5
Cassava (Manihot esculenta; Euphorbiaceae):
  beer(VI.10.003)674
  clones(III.2.1.001a)316
  modular construction(II.1.1.3.054)24
  root tuber(II.4.3.049)244
  starchy crops(I.3.023)8
  tuberous roots (division)[V.1.4.024 (5)]443
Cassia (Cinnamomum cassia, Lauraceae):
  spice crops(I.1.013)4
Cassia sp.: heterandry(IV.4.4.027)389
Cassytha: parasites(II.4.2.026)240
Castanea: (see Chesnut)
Castilla elastica:
  phyllomorphic (branches)(II.1.1.3.

INDEX OF CROPS

065)25
Castor bean:
  caruncle(II.2.3.3.042f)167
  modular construction(II.1.1.3.054)24
  oil crops(I.1.008)4
Casuarinaceae, Casuarina equisetifolia:
  leaf modifications(II.4.3.032)243
  obturator(II.3.119,.09b)200,215
  xeromorphic structure(IV.5.2.085)406
Catalpa spp.:
  abscission(II.1.2.01a,b)66
  branching by substitution(II.1.1.3.018)22
  hornworms(*V.3.3.1.132g)502
Catesbaea spinosa:
  interpetiolar (stipules)(II.1.2.13a)78
Catha edulis (Khat):
  alate seed(II.2.3.3.11d,e)192
Catostemma fragrans: drip-tip(II.4.07b)255
Cauliflower [Brassica oleracea (Botrytis group); Cruciferae]:
  alternaria leaf spot(VI.9.1.007)627
  bacterial leaf spot(VI.9.1.054)630
  black leaf speck(VI.9.3.011)664
  black rot(VI.9.1.088)633
  brown rot(VI.9.1.125)635
  caulimovirus(V.3.3.2.088)513
  cole crops(I.3.005)7
  gray mold rot(VI.9.1.267)643
  hydrocooling(V.5.035b)591
  open-pollinated populations(III.2.1.001d)317
  pelletized seeds(V.1.3.045)435
  root-lesion nematodes(*V.3.3.1.200f)506
  (see also Crucifers)
Ceiba: reversibly plagiotropic(II.1.1.3.081,.18e)26,54
Celastraceae: exotegmic(II.2.3.3.040b)166
Celariac (Apium graveolens (Rapaceum group); Umbelliferae):
  phoma root rot(VI.9.1.355)647
  root crops(I.3.019)8
  tap root plus hypocotyl plus stem storage(II.4.3.068)245
Celery (Apium graveolens (Dulce group); Umbelliferae):
  (roots): awl nematodes(*V.3.3.1.200c)505
  bacterial blight(VI.9.1.048)629
  bacterial soft rot(VI.9.1.060)630
  black heart(VI.9.3.010)664
  brown spot(VI.9.1.130)636

butterfly(*V.3.3.1.132g)502
California aster yellows(VI.9.1.134)636
cottony rot(*V.3.3.2.181e)519
crater spot(VI.9.1.181)639
cracked stem(VI.9.3.038)666
cucumber mosaic(VI.9.1.184)639
drop(*V.3.3.2.181e)519
early blight(VI.9.1.209)640
freezing injury(VI.9.3.046)667
hollow stem(VI.9.3.098)670
hydrocooling(VI.5.035b)591
late blight(VI.9.1.297)645
light requirements for germination (VI.1.3.042)435
Paratylenchus spp.(*V.3.3.1.200b)505
pencil stripe(VI.9.3.093)670
phoma root rot(VI.9.1.355)647
pithiness(VI.9.3.098)670
salad crops(I.3.020)8
source wrapping(VI.3.071)583
spotted wilt(VI.9.1.519)656
sting nematodes(*V.3.3.1.200c)505
thermodormancy(V.1.3.083)439
watery soft rot(VI.9.1.559)658
western mosaic(VI.9.1.563)658
Celery seed (Apium graveolens (Dulce group); Umbellifeae: condiments, etc.(I.3.006)7
Celosia: cristate(II.5.08g)288
Centipedegrass: turf crops(I.2.011)6
Centrosema sp.: gullet (flower)(IV.4.4.004b-4)388
Cereals:
  Anguina spp.(*V.3.3.1.200a)505
  beet cyst nematode(*V.3.3.1.200g)506
  Ditylenchus radicicola(*V.3.3.1.200g)506
  lance nematodes(*V.3.3.1.200d)506
  root-lesion nematodes(*V.3.3.1.200f)506
  (see corn, barley, oats, rice, rye)
  stem rust(*V.3.3.2.133c)515
Cephalanthus: (see Buttonbush)
Ceriops spp.: pneumorhizae(II.4.3.039)243
Chaparral, Soft: allelopathy(IV.2.2.013)360
Chard (Beta vulgaris (Cicla group); Chenopodiaceae): greens(I.3.016)8
Chayote (Sechium edule; Cucurbitaceae):
  cucurbits(I.3.007)7
Chenopodiaceae:
  anomalous secondary growth(II.3.008)194
  utricle(II.2.3.1.038)160

INDEX OF CROPS 846

Chenopodium spp.: bowl-shaped(IV.4.4.
    004b-2)388
Cherry (Prunus avium, P. cerasus):
  alternaria rot(VI.9.1.012)627
  bitter rot(VI.9.1.079)632
  black mold rot(VI.9.1.083)632
  blister(*V.3.3.2.181e)519
  blue mold rot(VI.9.1.107)634
  clones(III.2.1.001a)316
  dealer (in brine)(VI.7.1.015)599
  horticultural maturity (for brining)
    (VI.1.004)569
  horticultural maturity (fresh)(VI.1.
    004)569
  hypanthium(II.2.2.19b)148
  impulse item(VI.8.047)620
  leaf curl(*V.3.3.2.181e)519
  lug(VI.4.014)587
  mechanized harvesting(VI.2.022b)576
  perigynous(II.2.2.19b)148
  perishable agricultural commodity
    (in brine)(VI.7.1.043)603
  root-lesion nematodes(*V.3.3.1.200f)
    506
  semi-mechanized harvesting(VI.2.
    022c)576
  stone fruit(I.1.014)4
  superior(II.2.2.19b)148
  temperate zone crop(I.1.016)5
  trench layerage(V.1.4.036b-5)450
  witches' broom(*V.3.3.2.181e)519
Chervil (Myrrhis odorata; Umbelliferae):
  condiments, etc.(I.3.006)7
Chesnut (Castanea spp.; Fagaceae):
  accrescent (involucre)(II.2.1.10e)
    113
  blight(*V.3.3.2.181e)519
  bur (burr)(II.2.3.2.006,.02b)161,176
  nut(II.2.3.1.04g)171
  parallel-ribbed (C. mollissima)(II.
    1.2.2.24i)89
  pith(II.3.04e)210
  pseudoterminal(II.1.1.1.019)20
  spiny (involucre)(II.2.1.10e)113
Chico: (see Sapodilla)
Chicory (Cichorium intybus; Composite):
  phyllaries(II.2.1.10b)113
  salad crops(I.3.020)8
Chickpea (Cicer arietinum; Fabaceae):
  pulses(I.3.013)8
Chickweed:
  campylotropous(II.2.3.3.02c)183
  ephemeral(II.5.3.008)267
China-fir: (see Conifers)
Chinese cabbage [Brassica campestris
  (Pekinesis group); Cruciferae):
  anthracnose(VI.9.1.034)628
  black leaf speck(VI.9.3.011)664
  cole crops(I.3.005)7
  hybrids(III.2.1.001b)317
  (see also Crucifers)
Chinese gooseberry (Kiwi fruit; Actinidia chinensis:(I.1.015)4
  pruning system(V.3.4.032)541
Chinese hat plant: (see Holmskioldia)
Chinese pistachio: (see Pistachio)
Chive (Allium schoenoprasum): bulb
  crops(I.3.004)7
Chloris sp.: sporogenesis(IV.4.1.006b-
  6)380
Christmas cactus (Schlumbergera bridgesii; cactaceae): pot plants(I.2.
  001c)4
Christmas cherry (Solanum pseudocapsicum; solanaceae): pot plants(I.2.
  001f)6
Chrysanthemums (Chrysanthemum X morifolium; C. spp.; Compositae):
  after-lighting(V.4.1.001)559
  artificial long days(V.4.1.002)559
  black cloth shading(V.4.1.006)559
  blackout system(V.4.1.007)559
  bowl-shaped(IV.4.4.004b-2)388
  carbon dioxide fertilization(V.3.2.
    009)490
  crown bud(V.4.1.031)560
  cut flowers(I.2.001c)5
  cyclic lighting(V.4.1.034)560
  disbud, disbudding(V.4.1.038)561
  fiber blocks(V.1.2.003e)428
  foliar nematodes(*V.3.3.1.200a)505
  herbaceous (stem) cutting(V.1.4.
    022f-2,.02a)443,454
  'Iceberg' (clonal variety)(III.2.0.
    005a-1)313
  midge(*V.3.3.1.132b)501
  peat or fiber blocks(V.1.2.003e)428
  pinching(V.3.4.057)543
  pot plants(I.2.001f)6
  response group(V.4.1.097)563
  root-knot nematodes(*V.3.3.1.200e)
    506
  short-day plant(IV.3.2.0131)375
  year-round flowering(V.4.1.123)565
Cinchona:
  avenue, hedgerow (plant systems)(V.
    3.1.008a)485
Cineraria (Senecio X hybridus; Compositae):
  pot plants(I.2.001f)6
Cinnamomum spp.:
  coppice(II.1.1.3.030)23

INDEX OF CROPS

Cinnamomum iners:
   dorsal seed form(II.2.3.3.05c,d)186
Cinnamon (Cinnamomum zeylanicum, Lauraceae) (see also cassia) (C. cassia)
   coppice(II.1.1.3.030)23
   spice crop(I.1.013)4
Cissus sp.: succulent(II.4.3.065)245
Citrange (Poncirus trifoliata X Citrus sinensis): (see 'Morton', 'Troyer,' etc., under Citrus)
Citropsis gilletiana:
   alate(II.1.2.01d)66
Citrus, citrus fruits, citrus species and relatives in the Rutaceae; grapefruit (Citrus paradisi), kumquats (Fortunella spp.), lemon (C., limon), lime (C. aurantifolia), mandarins (tangerine, (C. reticulata), pummelo (C grandis), sour orange, (C. aurantium), sweet orange (orange) (C. sinensis)(I.1.003)3:
   alate(II.1.2.004)59
   ageing(VI.9.3.001)663
   albedo(II.2.3.2.029a)162
   algal spot(V.3.3.2.008)510
   ammoniation(VI.9.3.043)666
   anthracnose(VI.9.1.025)628
   armored scale injury(VI.9.2.007)660
   articulate, articulated(II.1.2.03a) 68
   aspergillus rot(VI.9.1.043)629
   bagmaster carton(VI.4.002)585
   bags(VI.4.001)585
   bare-rooted plant(V.1.002)418
   belt-and-roll sizing(VI.3.064a)583
   bird dog(VI.8.002)616
   black pit(VI.9.1.085)632
   black rot of orange(VI.9.1.014)627
   blue mold(*V.3.3.2.181g;VI.9.1.105) 520,634
   boron deficiency(VI.9.3.017)664
   botrytis rot(VI.9.1.120)635
   brown rot(*V.3.3.2.181d;VI.9.1.126) 518,635
   bud and graft union healing(V.1.4.028e)444
   budlings(V.1.4.028l)447
   bulk bin(VI.4.006)586
   bulk filling(VI.3.053b)582
   bundles(II.2.2.04f)133
   burrowing nematode(*V.3.3.1.200f)506
   canker(*V.3.3.2.040)511
   carton(VI.4.008)586
      filling(VI.3.053c)582
   chilling injury(VI.9.3.029)665
   citrus nematodes(*V.3.3.1.200e)506
      red mite(V.3.3.3.013d,e)532
   clones(III.2.1.001a)316
   color add(VI.3.016)578
   concentric ring stipple(VI.9.3.031) 665
   cones(V.1.2.003b)428
   crate(VI.4.009)586
   creasing(VI.9.3.040)666
   crown(II.1.1.3.08a)44
   crown and trunk canker(*V.3.3.2.181d)518
   degreening(VI.3.023)579
   delivered(VI.7.1.016)599
   digitate ('Morton' citrange)[II.1.2.21a(1)]86
   diphenyl(VI.3.033)580
   diplodia foot rot, dieback, stem-end rot(*V.3.3.2.181g)520
   disc[II.0.002,.01b(3)]10,12
   discoloration(VI.8.038b)619
   endocarp(II.2.3.2.029c)162
   evergreen(II.5.02c)282
   exanthema(VI.9.3.043)666
   exocortis(V.3.4.000)539
   fascicles(II.2.2.04f)133
   Federal Food, Drug & Cosmetic Act as Amended(VI.7.2.016)610
   foot and root rot(*V.3.3.2.181d)518
   freezing injury(VI.9.3.047)667
   frozen concentrate(VI.10.018)676
   fruit flies(*V.3.3.1.132b)501
   fungicide application(VI.3.033)580
   fusarium brown rot(VI.9.1.232)641
   grade(VI.8.038)619
      inspection(VI.8.038c)619
      standard(VI.8.038e)619
      subclasses(VI.8.038b)619
      table(VI.3.038)580
   grader(VI.3.037)580
   grading(VI.3.039)580
   graftage height(V.1.4.028j)446
   granulation(VI.9.3.056)667
   grapefruit:
      arsenate(V.3.4.000a)539
      blossom-end clearing(VI.9.3.012) 664
      brown staining(VI.9.3.023)664
      hydrometer(VI.1.008c)570
      'Marsh' poikilotherm(IV.5.3.038) 410
      'Marsh' weak (automatic) parthenocarpy(IV.4.2.001d)385
      oblate(II.2.3.2.023,.06d)161,180
      pounds of juice(VI.10.029)677
      Red: sloughing(VI.9.3.115)671
      stylar-end clearing(VI.9.3.012)664

INDEX OF CROPS

tropical crop(I.1.017)5
vivipary(II.2.3.3.01e)182
waterlog(VI.9.3.012)664
green mold(*V.3.3.2.181g;VI.9.1.105) 520,634
growth curve(VI.3.1.013)371
hail damage(VI.9.3.060)668
'Hamlin', cv. Hamlin(III.2.0.003)311
handling systems(VI.2.020)575
harvesting types(VI.2.023)577
hesperidium(II.2.3.1.018,.05b;.2.3.2.05d)159,172,179
humidity(VI.5.025)589
hydrocooling(VI.5.035b)591
hydrometer(VI.1.008c)570
inspection(V.3.3.3.052)534
internal breakdowns(VI.9.3.073)669
juice vesicles(II.2.3.2.05d)179
knot(*V.3.3.2.181g)520
kumquat(IV.5.3.028)409
lemon:
  albedo browning(VI.9.3.029)663
  alternaria rot(VI.9.1.014)627
  calyces(VI.3.033)580
  cottony rot(*V.3.3.2.181e;VI.9.1.179)519,638
  drop(*V.3.3.2.181e)519
  electronic sorting(VI.3.028)579
  endoxerosis(VI.9.3.042)666
  'Eureka' poikilotherm(IV.5.3.038) 410
  internal decline(VI.9.3.042)666
  'Lisbon' poikilotherm(IV.5.3.038) 410
  maturity(VI.1.007)570
  membranous stain(VI.9.3.086)669
  peteca(VI.9.3.094)670
  prolate(II.2.3.2.030,.07b)162,181
  reniform nematodes(*V.3.3.1.200d) 506
  rumple(VI.9.3.106)671
  seasonal item(VI.8.104)624
  subtropical crop(I.1.015)4
  total soluble solids(VI.1.008e)571
  withertip (see Endoxerosis)
limes:
  "pressure" tester(VI.1.003)569
  prolate(II.2.3.2.030)162
  'Tahiti': strong II automatic parthenocarpy(IV.4.2.001c)385
  total soluble solids(VI.1.008e)571
  tropical crop(I.1.017)5
lysigeny(II.3.104)199
mandarins:
  brown staining(VI.9.3.023)664
  oblate(II.2.3.2.023)161
Marketing Agreement (California)(VI.7.2.004a)609
Marketing Order (California)(VI.7.2.004b)609
maturity (nonclimateric fruits)(VI.1.007)570
maturity index(VI.1.008)570
maturity standards(VI.1.009)571
maturity test(VI.1.010)571
maturity test sample(VI.1.011)572
mechanized harvesting(VI.2.022b)576
Mediterranean fruit fly(*V.3.3.1.132b)501
melanose(*V.3.3.2.181e;VI.9.1.310) 519,645
mesocarp(II.2.3.2.029g)162
micrograftage(V.1.4.028p,.11i-o)448, 463
minor taxonomic category(III.1.1.022)305
modified atmosphere injury(VI.9.3.087)669
navel orange flower(II.2.2.02)131
needle nematodes(*V.3.3.1.200c)505
nematode(*V.3.3.1.200e)506
oil spotting(VI.9.3.090)670
oil-gland darkening(VI.9.3.088)670
oleocellosis(VI.9.3.090)670
orange dog(*V.3.3.1.132g)502
oranges:
  black rot(VI.9.1.014)627
  California(VI.8.038)619
  Florida(VI.8.038)619
  grade(VI.8.038)619
  hydrometer(VI.1.008c)570
  'Pineapple' pitting(VI.9.3.095)670
  pounds-solids(VI.10.030)677
  naval rind staining(VI.9.3.105)671
  subtropical crop(I.1.015)4
  ('Washington' navel) weak (automatic) parthenocarpy(VI.4.2.001d)385
ornamental(V.3.4.000)539
packing operations(VI.3.053)581
packinghouse layout(VI.3.052)581
packingline(s)(VI.3.028,052)579,581
pallet box(VI.4.018)587
perforated belt (sizing)(VI.3.064c) 583
phanerophyte(II.4.02a)250
phomopsis stem-end rot(*V.3.3.2.181e)519
physiological drought(IV.5.4.3.007) 415
picking bag(VI.4.020)587
pitting(VI.9.3.100)670
plant and crop size control(V.3.4.000)539

INDEX OF CROPS

plasticity(IV.5.1.011)397
pleonanthic(II.5.02c)282
poikilotherm(IV.5.3.038)411
polyadelphous(II.2.2.04f)133
polycarpic(II.5.02c)282
pounds of juice(VI.10.029)677
pounds-solids(VI.10.030)677
precooling(VI.5.035)590
protective devices(II.4.3.042)244
pseudogamy(IV.4.1.006a-11)379
puffiness(VI.9.3.102)670
pulpa(II.2.3.2.029c,.05e)162,179
pummelo:
  air layerage(V.1.4.036a)449
  anastomose (venation)(II.1.2.22a) 87
(red mite) nonoccluded virus(V.3.3.3.013e)532
refractometer(VI.1.008d)571
ripening(VI.1.013)572
rodents(V.3.3.1.224)507
root hairs(II.3.162,.13)202,219
root-lesion nematodes(*V.3.3.1.200f) 506
roots(II.1.3.02a,b)97
rust mite(V.3.3.1.172)504
rust mite russeting(VI.9.2.017)660
sampling(VI.1.015)572
sclerotium rot(VI.9.1.480)653
semihardwood (stem) cutting(V.1.4.022f-3,.02c-e,.03a,b)443,454,455
septoria spot(VI.9.1.488)654
shield bud(V.1.4.028f-8,.08a,b)445, 460
slip sheet(VI.4.022)587
sooty mold(*V.3.3.2.181e)519
sour orange scab(*V.3.3.2.181e;VI.9.1.505)519,655
sour rot(VI.9.1.507)655
stamping(VI.3.072)583
stem-end rind breakdown(VI.9.3.123) 672
stem-end rots(VI.9.1.528)657
stem section ('Troyer' citrange)(II.3.12a)218
stenospermocarpy (polyembryony)(IV.4.2.004)385
storage(VI.5.047)591
sunburn(VI.9.3.127)672
supercooling(IV.5.3.049)411
superposed(II.2.3.2.07e)181
surface breakdowns(VI.9.3.139)673
sweet orange [Citrus sinensis (L.) Osbeck](III.1.1.002)302
  (common name)(III.1.1.005)303
  scab(VI.9.1.505)655
tangelos(VI.3.016)578

tangerine: (see also mandarins)
  zebra skin(VI.9.3.154)674
'Temple' orange(VI.3.016)578
thrips injury(VI.9.2.080)662
tolerance(VI.8.038f)619
total (titratable acid)(VI.1.008g) 571
total soluble solids(VI.0.008e)571
total soluble solids-to-total (titratable) acid ratio(VI.1.008f) 571
trichoderma rot(VI.9.1.545)658
trifoliolate ('Morton' citrange)[II.1.2.21a(1)]86
tristeza(V.3.3.2.462)530
'Troyer' citrange (clonal variety) (III.2.0.005a-1)313
true (winter) dormancy(IV.3.1.011c) 371
Tryon's scab(VI.9.1.505)655
unifoliolate (Mauritius papeda, C. hystrix)(II.1.2.12c)77
vesicle(II.2.3.2.046)163
viroid(V.3.3.2.457)530
virus cross protection(V.3.3.2.462) 530
washing(VI.3.081)583
water spot(VI.9.3.147)673
wind scar(VI.9.3.151)673
wirebound box(VI.4.026)588
Clappertonia ficifolia: echinate(II.5.141)294
Clausena lansium: imparipinnate(II.1.2.12a)77
Clematis spp.:
  achene(II.2.3.1.002)158
  tendrils(II.4.3.071)245
Clover: club leaf(V.3.3.2.361)527
Clover, Red:
  'Kenland' (open-pollinated variety) (III.2.0.005a-4)314
  phytoalexin(IV.2.2.162)368
Clover, White:
  short-long-day plant(IV.3.2.013m)375
Cloves (Syzgium aromaticum = Eugenia caryophyllus; Myrtaceae):
  hybrids(III.2.1.001a)317
  reniform nematodes(*V.3.3.1.200d)406
  spice crop(I.1.013)4
  spiral nematodes(*V.3.3.1.200d)406
Club mosses:
  microphyll(II.1.1.3.053)24
  pteridophytes(II.6.016)279
  strobilus(II.6.024)280
Clusia spp. (some):
  "strangler"(II.4.3.064)245
Clymenia polyandra:

heterophylly(II.1.2.11a)76
Clytostoma callestegioides: (see Painted trumpet)
Cnicus: (see Blessed thistle)
Cobaea scandens:
 exserted (stamens)(II.2.2.06e)135
 tubiform (corolla)(II.2.2.06e)135
Cocculus balfouri:
 cladophyll[II.1.1.3.06a(1)]42
Coccoloba: (see Seagrape)
Cocklebur:
 short-day plant(IV.3.2.0131)375
 thermoperiodism(IV.3.2.021)376
Coconut (Cocos nucifera; Palmae):
 coir(V.1.4.036a)449
 copra(VI.10.009)675
 drupe(II.2.3.1.015,.05g)159,172
 drying(VI.10.009)675
 (dwarf) second-order dichogamy(IV.4.3.002d)386
 endocarp(II.2.3.2.029c)162
 germ-pore(II.2.3.2.029f)162
 husks (mulch)(V.3.3.3.059)535
 kiln(VI.10.023)677
 light requirement for germination (V.1.3.042)435
 mesocarp(II.2.3.2.029g)162
 monaxial(II.1.1.3.056)24
 monoecious(IV.4.3.014)387
 nut(II.2.3.1.024)159
 nut crops(I.1.007)4
 oil crops(I.1.008)4
 open-pollinated population(III.2.1.001d)317
 palm(II.0.03b)14
 (palm beetle) nucleopolyhedrosis virus(V.3.3.3.013f)532
 processing(VI.10.000)674
 protandry(IV.4.3.002d)386
 pulvini(II.1.2.18c)83
 red ring(*V.3.3.1.200a)505
 sclereids(II.3.25e,f)231
 (tall) second-order dichogamy(IV.4.3.002d)387
 tropical crop(I.1.017)5
 weevil(*V.3.3.1.200a)505
Coffee (Coffea arabica, C. canephora, C. liberica, C. excelsa, Rubiaceae):
 agobio(V.3.4.09a)553
 Arabian(V.3.4.017)540
 bayonet(V.3.4.06c)550
 beverage crops(I.1.001)4
 Brazilian(V.3.4.09b)553
 burrowing nematodes(*V.3.3.1.200f)506
 collection of (cutting) material(IV.1.3.022a-1)442
 candelabra(V.3.4.07d)551
 Colombian(V.3.4.07c)551
 Costa Rican(V.3.4.07d)551
 drupe(II.2.3.1.015)159
 drying(VI.10.009)674
 elephants(V.3.3.1.140)503
 fermentation(V.1.3.026)433
 freeze drying(VI.10.017)676
 Guatemalan(V.3.4.09a)553
 harvesting types(VI.2.023)577
 Hawaiian(V.3.4.08a,b)552
 Hawaiian stump(V.3.4.08c)552
 inbred pure lines (Arabian)(III.2.1.001c)317
 inclined or leaning multiple stem (V.3.4.07a,b)551
 interplanting(V.3.1.008h)486
 irreversibly plagiotropic(II.1.1.3.047,.13a)24,49
 lesion nematode(*V.3.3.1.200a)505
 Liberian(V.3.4.017)540
 modified single stem(V.3.4.06c)550
 poda(V.3.4.09b)553
 polyaxial stem(II.1.1.3.067)25
 processing(VI.10.000)674
 pruning(V.3.4.06-.09)550-553
 pruning system(V.3.4.017)540
 reiteration(II.1.1.3.18b)54
 robusta(V.3.4.017)540
 root-lesion nematodes(*V.3.3.1.200f)506
 rust(*V.3.3.2.181f)519
 single stem(V.3.4.06b)550
 spiral nematodes(*V.3.3.1.200d)506
 topping(V.3.4.06a)550
 tropical crops(I.1.017)5
 umbrella(V.3.4.07c)551
 upright multiple stem(V.3.4.06d)550
 vertical multiple stem(V.3.4.06d)550
 xenia(IV.4.8.057)395
Cola: (see Kola)
Colchium sp.:
 see also Autumn crocus
 stem tubers(II.4.3.057)244
Coleus, Autumn:
 herbaceous (stem) cutting(V.1.4.022a-2,.02a)443,454
 intermediate-day plant(IV.3.2.013f)375
Collard [Brassica oleracea (acephala group); Cruciferae]:
 anthracnose(VI.9.1.034)628
 biennial(II.5.01b)281
 greens(I.3.016)8
 herb, herbaceous(II.0.01c,d)12
 seed stalk(II.1.1.3.19b)55

INDEX OF CROPS

Commelina sp.:
  stoma(II.3.190)204
Commercial floriculture(I.2.001)5
  (see also bulbous plants, cut flowers, pot plants, bedding plants, hanging baskets, foliage plants and individual crops, e.g., tulip, chrysanthemum, azalea, etc.):
  EC_e (electric conductance)(V.4.1.042)561
  fumigun(V.4.1.055)561
  greenhouse(V.1.1.017)419
  mho(V.4.1.071)562
  osmocote(V.3.2.070)495
  packing operations(VI.3.053)581
  physiological disorders(V.4.1.083)563
  pinch(V.4.1.084)563
  solubridge(V.4.1.110)564
Compositae, composites:
  achene(II.2.3.1.002)158
  blanking(IV.4.8.009)393
  bracteole(II.2.1.009)100
  comose(II.5.6.028)274
  disc, disk[II.0.002,.01b(4)]10,12
  discoid(II.2.1.031)101
  disk(=receptacle)[II.2.1.01(5);.2.2.01a(5,10)]104,130
  floret(II.2.1.035)101
  ligulate (corolla)[II.2.2.23i(1);.5.5.032,.5.10d(3)]152,272,290
  pappus(II.5.6.088)276
  powdery mildew(*V.3.3.2.181e)519
  radiate (ligulate florets)(II.2.1.060)102
  seed extraction (dry fruits)(V.1.3.063)437
  (some) succulent(II.4.3.065)245
  syngenesious(II.2.2.007d-18,..06b)119,135
  tubular[II.2.2.23i(2)]152
  xeromorphic structure(IV.5.2.085)406
  zygomorphic(II.2.2.061c-14)127
Condiments, spices, essential oils(I.3.006)7
Coniferae, Conifers (Araucariaceae, Cephalotaxaceae, Cupressaceae, Pinaceae, Podocarpaceae, Taxaceae):
  acerose (Juniperus silicicola(II.1.2.01c;.5.5.001)66,271
  apophysis(II.6.002)279
  atmospheric pollution(IV.5.4.4.001)415
  conifer(II.6.007)279
  (China-fir, Cunninghamia)(II.6.01c)301
  (Cupressaceae) leaf modifications (II.4.3.032)243
  cutting-graft(V.1.4.028g)445
  decumbent (shore juniper J. conferta)(II.1.1.3.09a)45
  dimorphic (southern red cedar)(II.0.001,.01a)10,12
  Douglas-fir tussock moth(V.3.3.3.013f)532
  European spruce sawfly(V.3.3.3.013f)532
  excurrent(II.1.1.3.039)23
  fascicled (larch)(II.1.2.10a)75
  fasciculate leaves (pine)(II.1.2.046)61
  fasciculate (loblolly pine, P. taeda)(II.1.2.10b)75
  foliage plants(I.2.001d)5
  forest trees(II.4.4.003c)247
  gymnosperms(II.6.011)279
  hardwood (stem) cutting(V.1.4.022a-2,.02b)442,454
  horizontal (J. horizontalis)(II.1.1.3.12b)48
  (many) light requirements for germination(V.1.3.042)435
  melezitose (some)(IV.2.2.031b)361
  monocaulous(II.1.1.3.15a)51
  monopodial(II.1.1.3.15a)51
  nut crop(I.1.007)4
  parastichy, orthostichy phyllotaxis, rows, series (larch)[II.1.1.1.05r(3)]32
  (white pine)[II.1.1.1.05r(4)]32
Piñon nut(I.1.007)4
Pinus spp.:
  reaction wood(II.3.155)201
  seed dormancyIII(V.1.3.060)436
  unifacial leaf(II.1.2.119)64
  white pine blister rust(*V.3.3.2.181f)519
P. sylvestris: parastichy 8/21(II.1.1.1.016)19
  Pinus ectrotrophic (mycorrhiza)(V.3.3.2.148)516
  procumbent (shore juniper)(II.1.1.3.16c)52
  spruce budworm(V.3.3.3.013f)532
  strict(II.1.1.3.21a)57
  tree(II.0.07a)18
Conium maculatum:
  embracing leaf bases(II.1.2.08d)73
Connaraceae:
  hemiantropous (seed-form)(II.2.3.3.036a-2)165
  sarcotesta(II.2.3.3.042i)167

INDEX OF CROPS

trimorphism(IV.4.3.019)387
Convolvulaceae:
  parasites(II.4.2.026)240
  twining(II.4.3.074)246
Convolvulus arvensis:
  endodermis(II.3.11b)217
Corallorhize:
  saprophytes(II.4.2.030)240
Coralvine (Antigonon leptopus):
  vine(II.0.02c)13
Coreopsis:
  flower(II.2.1.09a)112
  trichome(II.3.31d)237
Coriander (Coriandrum sativum; Umbelliferae):
  accrete (style)(II.2.2.15i)144
  condiments, etc.(I.3.006)7
Corn, Sweet (Maize, Zea mays):
  awl nematodes(*V.3.3.1.200c)505
  bacterial wilt(VI.9.1.074)632
  black rot(VI.9.1.090)633
  blue-green rot(VI.9.1.104)634
  boil smut(VI.9.1.169)638
  brown spot(*V.3.3.2.181a)518
  C-4 plants(IV.1.2.008b)352
  caryopsis(II.2.3.1.008,.01k-o)159, 168
  catena(II.4.03)251
  common names(III.1.1.005)303
  common smut(*V.3.3.2.181f;VI.9.1.169)519,638
  day-neutral plant(IV.3.2.013c)374
  detassel(III.2.2.037)320
  diplodia ear rot(VI.9.1.190)639
  dry rot(VI.9.1.190)639
  ear(II.2.1.14b)117
  ear worm(*V.3.3.1.132g;.3.3.3.013b,f;VI.9.2.021)502,532,660
  elaioplast(IV.1.1.012)351
  European corn borer(*V.3.3.1.132g; VI.9.2.026)502,660
  finishing machine(V.1.3.057)435
  fusarium kernel rot(VI.9.1.243)642
  gibberella ear rot(VI.9.1.258)643
  'Golden Bantam'(III.2.0.005a-4)314
  gray ear rot(VI.9.1.263)643
  grits(V.3.3.3.009)531
  harvested immature(II.2.3.1.000)158
  heterotopy(V.3.3.2.205)521
  hormodendrum rot(VI.9.1.294)645
  hybrids(III.2.1.001b)317
  lance nematodes(*V.3.3.1.200d)506
  leaf blights(*V.3.3.2.133c)515
  leucoplast(IV.1.1.012)351
  meal moths(*V.3.3.1.132g)502
  mechanized harvesting(VI.2.022b)576
  monoecious(IV.4.3.014b)387

nigrospora cob rot(VI.9.1.328)646
northern leaf blight(VI.9.1.329)646
open-pollinated populations(III.2.1.001d)317
parenchyma(II.3.04d)210
pelletized seeds(V.1.3.045)435
physalospora ear rot(VI.9.1.364)648
physiological drought(IV.5.4.3.007) 415
picker(V.1.3.063)437
pink kernel rot(VI.9.1.243)642
popcorn seed extraction (dry fruits) (V.1.3.063)437
popcorn (Zea mays) (see Miscellaneous crops)
red ear rot(VI.9.1.258)643
rhizoctonia ear rot(VI.9.1.440)651
rice weevil(VI.9.2.069)662
root apical meristem(II.3.19d)225
root-lesion nematodes(*V.3.3.1.200f) 506
root system(II.1.3.03)98
(see Miscellaneous crops)
seed extraction (dry fruits)(V.1.3.063)437
seed maggot(*V.3.3.1.132b)501
seed structure(II.2.3.3.01b)182
short-day plant(IV.3.2.0131)375
southern leaf blight(VI.9.1.514)656
southern leaf spot(VI.9.1.515)656
specific gravity(V.1.3.057)435
spike(II.2.1.13b)117
spiral nematodes(*V.3.3.1.200d)506
Stewart's disease, Stewart's wilt (VI.9.1.074)632
sting nematodes(*V.3.3.1.200c)505
tassel(II.2.1.071,.14a)103,117
'US13' hybrid(III.2.0.005a-6)314
'WF9' (line variety)(III.2.0.005a-2) 313
warm-season seeds(V.1.3.087)439
xenia(IV.4.8.057)395
Corn (European usage = wheat):
  segetal(III.1.2.056)310
Cornflower (Centaurea cyanus; Compositae):
  cut flowers(I.2.001c)5
Coronopus rytidocarpus:
  scrobiculate(II.5.18j)298
Corydalis spp.:
  flag(IV.4.4.046-3)388
Corypha umbraculifera:
  hapaxanthic(II.5.3.011)267
(Corylus: (see filbert)
Costmary (Chrysanthemum majus Compositae):
  condiments, etc.(I.3.006)7

Costus: (see Ginger)
Cotton (Gossypium spp. Malvaceae)
  angular leaf spot(*V.3.3.2.040)511
  bacterial blight(*V.3.3.2.040)511
  boll rot(*V.3.3.2.181c)518
  bollworms(*V.3.3.1.132g;VI.9.2.021)
    502,660
    (see Corn: earworm)
  bud and graft union healing(V.1.4.
    028e)444
  C-3 plants(IV.1.2.008a)352
  lance nematodes(*V.3.3.1.200d)506
  oil crops (cottonseed)(I.1.008)4
  physiological drought(IV.5.4.3.007)
    415
  sting nematodes(*V.3.3.1.200c)505
  Texas root rot(*V.3.3.2.181g)520
  warm-season seeds(V.1.3.087)439
  wilt(*V.3.3.2.181g)520
Cottonwood:
  direct-set cuttings(V.1.4.022b)442
Cotyledon secunda:
  secund (inflorescence)[II.2.1.08e
    (2)]111
  secund (flowers)(II.2.2.03d)132
Cowpea (Vigna sinensis; Fabaceae):
  inbred pure lines(III.2.1.001c)317
  pulses(I.3.014)8
  reniform nematodes(*V.3.3.1.200d)506
  root-lesion nematodes(*V.3.3.1.200f)
    506
  (see also Leguminosae; Papilionace-
    ae)
  seed production(III.2.0.004c-f)312
  sting nematodes(*V.3.3.1.200c)505
Crabgrass:
  C-4 plants(IV.1.2.008b)352
Cranberry (Vaccinium spp.; Vacciniaceae):
  decumbent(II.1.1.3.034)23
  epigynous, inferior(II.2.2.19d)148
  hand harvesting methods(VI.2.022a)
    576
  mechanized harvesting(VI.2.022b)576
  small fruits(I.1.012)4
Crassula:
  apocarpus gynoecium(II.2.2.14a)143
Cress, Garden (Lepidium sativum; Cruciferae):
  salad crops(I.3.020)8
Cress, Water (Nasturtium officinale;
  Cruciferae):
  salad crops(I.3.020)8
Crested wheatgrass:
  'Nordan' (open-pollinated)(III.2.0.
    005a-4)314
Crinum powellii:
  spirodistichy(II.1.1.1.06s)33
Crocus (Crocus vernus C. spp.):
  bulb forcing(V.4.1.012)559
  corm(I.2.001b)5
  corm (separation)[V.1.4.071b,.16f
    (1)]451,468
  growth movements(II.4.3.023)242
  (see Bulbous plants)
  stem tubers(II.4.3.057,.13c)244,261
Crotalaria:
  scarifier(V.1.3.057)435
Cruciferae, Crucifers:
  bacterial soft rot(VI.9.1.058)630
  bacterial zonate spot(VI.9.1.076)632
  beet cyst nematode(*V.3.3.1.200g)506
  black rot(VI.9.1.088)633
  blight(*V.3.3.2.181e)519
  boron deficiency(VI.9.3.018)664
  cole crops(I.3.005)7
  club root(*V.3.3.2.133c,*.181b;VI.
    9.1.156)515,518,637
  cruciform (corolla)(II.2.2.061c-1-d,
    .21d)126,150
  downy mildew(*V.3.3.2.181d;VI.9.1.
    202)518,640
  dry heart rot(*V.3.3.2.181e)519
  endotestal (seeds)(II.2.3.3.043a)167
  leaf spot(*V.3.3.2.181e)519
  Nacobbus spp. nematodes(*V.3.3.1.
    200g)506
  replum(II.2.3.1.030)160
  rhizopus soft rot(VI.9.1.454)652
  ring spot(*V.3.3.2.181e;VI.9.1.465)
    519,653
  (see also individual crops; e.g.,
    cabbage, cauliflower, chinese cab-
    bage, brussels sprouts, broccoli,
    kohlrabi, mustard, rutabaga, tur-
    nip, etc.)
  silique(II.2.3.1.034)160
  silicle(II.2.3.1.033)160
  tetradynamous(II.2.2.007d-20,.05d)
    119,134
Cucumber (Cucumis sativus; Cucurbitaceae):
  angular leaf spot(VI.9.1.069)631
  anthracnose(VI.9.1.183)639
  bacterial soft rot(VI.9.1.062)631
  bacterial spot(VI.9.1.069)631
  beetles(*V.3.3.1.132a)501
  black rot(VI.9.1.091)633
  day-neutral plant(IV.3.2.013c)374
  diplodia rot(VI.9.1.193)639
  growth curve(IV.3.1.013)371
  gummy stem blight(VI.9.1.091)633
  harvested immature(II.2.3.1.000)158
  horticultural maturity (green)(VI.1.

INDEX OF CROPS 854

    004)569
  mosiac(*V.3.3.2.133c;.4.2.1.022)515, 566
  'MSU-713-5' gynoecious (line variety(III.2.005a-2)313
  Nacobbus spp. nematodes(*V.3.3.1.200g)506
  'Picadilly' hybrid(III.2.0.005a-6)314
  pickles(VI.10.028)677
  pickleworm(*V.3.3.1.132g)502
  pink mold rot(VI.9.1.393)649
  'Poinsett'(III.2.0.005a-4)314
  trichome(II.3.31c)237
  washing(VI.3.081)583
  waxes(VI.3.085)584
  waxing(VI.3.086)584
Cucurbits Cucurbitaceae):
  anthracnose(*V.3.3.2.181g)520
  bacterial zonate spot(VI.9.1.076)632
  black rot(*V.3.3.2.181e)519
  blight(*V.3.3.2.181e)519
  collenchyma (angular)(II.3.04h)210
  downy mildew(*V.3.3.2.181d)518
  dry heart rot(*V.3.3.2.181e)519
  Erysiphe cichoracearum(*V.3.3.2.181e)519
  espalier(V.3.4.026)541
  gourds(I.3.007)7
  hybrids(III.2.1.001b)317
  leaf curl(*V.3.3.2.181e)519
  leaf spot(*V.3.3.2.181e)519
  pepo(II.2.3.1.026,.05c)160,172
  powdery mildew(*V.3.3.2.133c,*.181e)515,519
  rhizopus soft rot(VI.9.1.455)652
  ring spot(*V.3.3.2.181e;VI.9.1.465)519,653
  see also individual crops, e.g., punpkins, squashes, chayote, melons, muskmelons, watermelons, cucumber, etc.
  see extraction (fleshy fruits)(V.1.3.064)437
  sieve tube elements(II.3.26b-d)232
  succulent(II.4.3.065)245
  tendrils(II.4.3.071)245
  twisted(II.2.2.007w-13)121
  vascular wilt(*V.3.3.2.040)511
  warm-season seeds(V.1.3.087)439
  xeromorphic structure(IV.5.2.085)406
Cumin (Cuminum cyminum; Umbelliferae):
  condiments, etc.(I.3.006)7
Cupanda: (see Avocado)
Cunninghamia: (see Conifers)
Cupressaceae (see Conifers)
Cupressus sempervirens: (see Conifers)

Currants (Ribes spp.; Grossulariaceae):
  growth curve(IV.3.1.013)371
  mechanized harvesting(VI.2.022b)576
  small fruits(I.1.012)4
Cuscuta: (see Dodder)
Cut flowers, foliage:
  calcarate[II.2.2.23c(2),d]152
  cardboard (boxes)(VI.4.007)586
  carton(VI.4.008)586
  condition(VI.8.016)617
  containers(VI.4.000)585
  dry-pack storage(VI.5.016)589
  field-to-packinghouse harvesting(VI.2.012)575
  floral preservative(VI.8.036)619
  grade(VI.8.038)580
  grading(VI.3.039)580
  hamper(VI.4.012)587
  hand harvesting methods(VI.2.022a)576
  hardening(V.4.1.062)562
  harvesting(VI.2.000)575
  heat delay(V.4.1.063)562
  humidity(VI.5.025)589
  light flux(V.4.1.066)562
  market disorders(VI.9.000)626
  pack(VI.3.047a)581
  packing operations(VI.3.053)581
  packinghouse(VI.3.000)577
  packinghouse layout(VI.3.052)581
  Perishable Agricultural Commodities Act of 1930, as Amended(VI.7.1.)597-609
  refrigeration, storage(VI.5.000)588
  self-branching(V.4.1.105)564
  shipping containers(VI.4.000c)585
  sizing(VI.3.064)582
  storage(VI.5.047)591
  transportation problems(VI.6.034)596
  vase life(VI.8.132)626
Cycadales: gymnosperms(II.6.011)279
Cycas spp. (Cycadaceae):
  cycad(II.6.008)279
  leaf modifications(II.4.3.032)243
Cyclamen (Cyclamen persicum; Primulaceae):
  mite(V.3.3.1.172)504
  pot plant(I.2.001k)6
  stem tubers(II.4.3.057)244
Cyperaceae: parastichy(II.1.1.1.016)19
Cyperus esculentus (Cyperaceae)
  angular(II.1.2.02b)67
  C-4 plants(IV.1.2.008b)352
  parastichy(II.1.1.1.016)19
  umbrella-shaped (leaves, C. alternifolius)(II.1.2.21c)86

Cypress-vine: (See Ipomoea)
Cytisus sp.: mess-and-soil(IV.4.5.
  002j)391

D

Daemonorops: rattan(II.0.022)11
Daffodil:
  bulb forcing(V.4.1.012)559
  bulbous plants(I.2.001b)5
  (see Narcissus)
Dahlia (Dahlia pinnata, Compositae):
  blight(*V.3.3.2.181c)518
  caulimovirus(V.3.3.2.088)513
  cut flowers(I.2.001c)5
  root tuber(II.4.3.049,.13a)244,261
  (see also Bedding plants, cut flowers)
  spotted wilt(VI.9.1.522)656
  tuberous roots(II.1.3.04c;V.1.4.024,
    .05d)99,443,457
  white smut(*V.3.3.2.181f)519
Dandelion (Taraxacum officinale; Compositae(:
  achene(II.2.3.1.01g)168
  greens(I.3.016)8
  pappus(II.5.17d)297
  parthenogenesis(IV.4.1.006b-3)379
  runcinate(II.5.5.051;.5.11b)273,291
Date, date palm (Pheonix dactylifera,
    Palmae):
  clones(III.2.1.001a)316
  dioecious(IV.4.3.004)386
  growth curve(IV.3.1.013)371
  metaxenia(IV.4.8.040)395
  offset(V.1.4.024,.04b)443,456
  physiological drought(IV.5.4.3.007)
    415
  root-lesion nematodes(*V.3.3.1.200f)
    506
  semi-mechanized harvesting(VI.2.
    022c)576
Datura sp.:
  apohomotypic (aneuspory)(IV.4.1.
    006b-6-a-1)380
Davilla grandiflora:
  anisomerous(II.2.2.03b)132
Daylily (hemerocallis spp.):
  bedding plants(I.2.001a)5
  bostryx(II.2.1.006)100
  crown(II.4.3.015)242
  crowns(V.1.3.024,.04a)443,456
  ephemeral(II.5.3.008)267
Deciduous fruits:
  brown rot(*V.3.3.2.181d)518
  crown and trunk canker(*V.3.3.2.
    181d)518

dumping(VI.3.027)579
firmness tester(VI.1.003)569
foot and root rot(*V.3.3.2.181d)518
(see also pome fruits, stone fruits
  and individual crops; e.g., apple,
  pear, peach, apricot, cherry,
  grape, etc)
temperate zone crops(I.1.016)5
washing(VI.3.081)583
weight (sizing)(VI.3.064e)583
Delphinium (Delphinium spp.; Ranunculaceae):
  see cut flowers(I.2.001c)5
Desert plants: chamaephyte(II.4.1.001)
  238
Desmodium: loment(II.2.3.1.02f)169
Devora: (see Avocado)
Dianthus: stomata(II.3.30i)236
Dichondra: turf crops(I.2.011)6
Dictamnus spp.: flag(IV.4.4.004b-3)388
Dill (Anethum graveolens; Umbelliferae):
  California aster yellows(VI.9.1.134)
    636
  condiments, etc(I.3.006)7
  long-day plant(IV.3.2.013g)375
Dilleniaceae:
  endotestal(II.2.3.3.043a)167
Dilly: (see Sapodilla)
Dioscoreaceae, Dioscorea spp., D.
    bulbifera:
  cordate (C. bulbifera)(II.5.09c)289
  curve-ribbed(II.1.2.22c)87
  twining(II.4.3.074)246
  (see also Yams)
Diospyros spp., D. kaki:
  branching by substitution(II.1.1.3.
    018)22
  chilling requirement (D. kaki)(IV.5.
    3.006)407
  napiform, syncarpous(II.2.3.2.03e)
    177
  pseudoarillate seeds(II.2.3.3.12a,b)
    193
  ripening(VI.1.013)572
  staminodes(II.2.2.09d)138
Diphyllea cymosa:
  rhizome (fleshy)(II.4.12c)260
Dipterocarpaceae:
  dipterocarp(III.1.2.015)308
  megatherm taxa(III.1.2.033)309
Dodder (Cuscuta):
  disease transmission(*V.3.3.2.133e)
    515
  implicative(II.1.1.3.12c)48
  parasites(II.4.2.026)240
  spirally twisted(II.2.3.3.02o)183

velvet roll(V.1.3.057)435
weeds(V.3.3.0.001)497
Dodonea viscosa:
  arillate seed)II.2.3.3.10a,b)191
Dog-fennel: (see Eupatorium)
Dog-tooth violet: (see Erythronium)
Dracaena (D. draco; Dragon's blood):
  acrotony(II.1.1.3.02a)38
  arborescent(II.1.1.3.02a)38
  secondary growth(II.0.06c;.3.172)17, 202
  sclereids(II.3.24d)230
  spirodistichy(II.1.1.1.025)20
  sympodial(II.1.1.3.21c)57
  tree(II.0.06c)17
Drosera spp.: haptonasty(IV.3.3.003) 376
Duckweed: short-day plant(IV.3.2.013) 375
Dutch iris: (see Iris)
Dwarf plants: mechanical strees(IV.5.4.4.002)416

E

Ebenaceae:
  pachychalazal (seed-form)(II.2.3.3.036a-5)165
  (see Diospyros spp., Persimmon)
Eggplant (Aubergine; Solanum melongena; Solanaceae):
  alternaria rot(VI.9.1.017)627
  bacterial soft rot(VI.9.1.067)631
  cottony leak(VI.9.1.177)638
  fermentation(V.1.3.026)433
  gray mold rot(VI.9.1.277)644
  inbred pure lines(III.2.1.001c)317
  phomopsis rot(VI.9.1.358)648
  phytophthora rot(VI.9.1.382)648
  rhizopus rot(VI.9.1.452)652
  seed extraction (fleshy fruit)(V.1.3.064)437
  solanaceous fruits(I.3.021)8
  spotted wilt(VI.9.1.522)656
  vascular wilt(*V.3.3.2.040)511
  warm-season seeds(V.1.3.087)439
Eichhornia crassipes (water hyacinth):
  aquatic(II.4.02c)250
Elaeis guineensis:
  oil palm(I.1.008)4
  pneumorhizae(II.4.3.039)243
  second-order dichogamy(IV.4.3.002d) 386
Elm:
  alate (Ulmus alata)(II.1.1.3.004;.1.2.004)22,59
  bark beetle(*V.3.3.1.132a)501
  oblique(II.5.11e)291
  polyaxial stem(II.1.1.3.067)25
  (U. spp.) samara(II.2.3.1.031)160
Endive, escarole (Cichorium endivia; Compositae):
  salad crops(I.3.020)8
  thermodormancy(V.1.3.083)439
Endospermum sp.: myrmecomorphosis(II.4.4.002c)246
Eperua sp.: flagelliflory(IV.4.4.021) 389
Ephedra (Ephedrales):
  gymnosperms(II.6.011)279
  vessel member(II.3.217)206
Epigaea: trichome(II.3.31b)237
Epigonium: saprophytes(II.4.2.029)240
Epiphytes:
  ascidiform(II.4.3.006)241
  disjunctive symbiosis(II.4.3.067c) 245
  phanerophyte(II.4.1.005)238
  xeromorphic(II.4.3.076)246
Equisetinae: pteridophytes(II.6.016) 279
Eragrostis sp.:
  apogamy(IV.4.1.006b-2)379
  sporogenesis(IV.4.1.006b-6)380
Ericaceae (some):
  laminar rolling(II.4.3.030)243
  leaf modifications(II.4.3.032)243
Ericaceous plants:
  galls(*V.3.3.2.181f)519
Erycibe longiflora:
  stenophyllous leaves[II.4.08a(6)]256
Erythrina subumbrans:
  hilar seed form(II.2.3.3.11a-c)192
Erythronium (dog-tooth violet):
  liliaceous nodding(II.2.2.21i)150
Eucalyptus (Eucalyptus spp.; Myrtaceae)
  essential oil crops(I.1.013)4
  falcate leaves(II.1.2.043)61
  leaf modifications(II.4.3.032)243
  operculum(II.2.3.1.025)160
  sessile (E. cinerea)(II.1.2.19d)84
  "shadeless forest"(II.4.3.052)244
  xeromorphic structure(IV.5.2.085)406
Eugenia jambos:
  "transitional" parthenocarpy(IV.4.2.005)(IV.4.2.005)385
Eugenia lushnathiana (pitomba):
  spheroid, globoid(II.2.3.2.07d)181
Eugenia uniflora: (see Surinam cherry)
Euonymus: alate(II.1.1.3.004)22
Eupatorium cannabinum:
  head- or brush-shaped(IV.4.4.004b-5) 388

INDEX OF CROPS

Eupatorium capillifolium:
  filiform(II.5.10a)290
Euphorbia corollata:
  petaloid (involcre)(II.2.1.10a)113
Euphorbia spp.:
  laticifer(II.3.098)198
  (some) succulent(II.4.3.065)245
Euphorbiaceae:
  caruncle(II.2.3.3.042f)167
  leaf modifications(II.4.3.032)243
  pachychalazal seed form(II.2.3.3.036a-5)165
  xeromorphic structure(IV.5.2.085)406
Euphoria malaiensis:
  pachychalazal seed form(II.2.3.3.08c)189
Evening Primrose (Oenothera): tubular (II.2.2.22c)151
Evergreens, Narrow-leafed:
  grafting (side graft)(V.1.4.028k)446
  (see also Conifers)

F

Fabaceae = Papilionaceae:
  flag (flower)(IV.4.4.004b-3)388
  legume(II.0.0.010;.2.3.1.020)10,159
  loment(II.2.3.1.021)159
  papilionaceous (corolla)(II.2.2.061c-14-p)128
  pulses(I.3.014)8
Fagraea spp.:
  hemiepiphytes(II.4.2.015)239
  stenophyllous leaves[II.4.08a(2)]256
Feijoa:
  decussate(II.2.07d)72
Fennel (Foeniculum vulgare; Umbelliferae)
  condiments, etc.(I.3.006)7
Fenugreek (Trigonella foenumgraecum, Fabaceae):
  condiments, etc.(I.3.006)7
Ferns:
  blister(*V.3.3.2.181e)519
  circinate(II.1.1.2.01b;.6.004)34,279
  cuneiform (maidenhair leaves(II.5.5.014)271
  dichotomous (vernation)(II.6.009)279
  epiphytes(II.4.2.006,.02d)239,250
  foliar nematodes(*V.3.3.1.200a)505
  furcate (staghorn fern)(II.1.2.10e)75
  hydrophyte, floating (Salvinia auriculata)(II.4.05a)253
  interrupted (Osmunda claytoniana)(II.1.2.13b)78
  leaf curl(*V.3.3.2.181e)519

Osmunda(II.1.2.13b,.4.04c)78,252
poikilohydric plants(IV.5.2.040)403
pteridophytes(II.6.016)279
root apical meristem (Adiatum)(II.3.19a)225
See also Foliage plants; Cut flowers
  foliage(I.2.001c,d)5
  shade houses (fronds)(V.1.1.059)426
  sorus(II.6.018)280
  squarrose-slashed (Gleichenia dichotoma)(II.5.07f)287
  tracheid(II.3.202)204
  tree fern (Alsophila(II.6.01b)301
  vascular bundle arrangement(II.3.210)205
  witches' broom(*V.3.3.2.181e)519
Feronia limonia:
  imparipinnate(II.1.2.11b)76
  seed hairs(II.2.3.3.07e,f)188
  vertebrate(II.1.2.28a)93
Ficus spp. (Moraceae):
  aerial roots(II.1.3.002)94
  air layerage(V.1.4.036a)449
  cystolith (F. elastica)[II.3.08e(2)]214
  drip-tips(II.4.3.017,.07c)242,255
  foliage plants(I.2.001d)5
  hemiepiphytes(II.4.2.015)239
  lithocyst (F. elastica)[II.3.08e(1)]214
  multiple fruit(II.2.3.1.023)159
  (many) protectice devices(II.4.3.042)244
  nursery plants(I.2.009)6
  prop roots (banyan, F. Indica)(II.1.3.01d)96
  rhizoflory[II.2.1.05d(4)]108
  see also (common)fig
  "strangler" (certain spp.)(II.4.3.064;.4.10d)245,258
  syconium(II.2.1.069,.13d;.2.3.1.036,.07d)103,116,160,174
Field beans: (see Beans)
Fig (common fig) (Ficus carica; Moraceae):
  alternaria rot(VI.9.1.013)627
  aspergillus black-mold rot(VI.9.1.040)629
  blastophaga wasp(IV.4.5.001d)391
  blue mold rot(VI.9.1.108)634
  breba crop(IV.4.4.005)388
  caprifig[II.2.1.13d(5);IV.4.4.008,.049]116,388,390
  caprification(IV.4.4.007)388
  cladosporium spot(VI.9.1.155)637
  clones(III.2.1.001a)316
  common type(IV.4.4.007)388

"curly tip"(*V.3.3.1.200c)505
dagger nematodes(*V.3.3.1.200c)505
drupelet(II.2.3.1.016)159
endosepsis(VI.9.1.214)641
gray mold rot(VI.9.1.269)643
growth curve(IV.3.1.013)371
hardwood (stem) cuttings(V.1.4.022f-1,.02b)442,**454**
hybrids(III.2.1.001b)317
"leaf drop"(*V.3.3.1.200b)505
main crop(IV.4.4.031)389
mamme (caprifig)(IV.4.4.007,.032) 388,389
mammoni (caprifig)(IV.4.4.007,.033) 388,389
multiple fruit(II.2.3.1.023)159
needle nematodes(*V.3.3.1.200c)505
ostiole[II.2.1.13d(2);.2.3.2.025, .07d(1)]116,161,174
palmiform(II.1.2.23f)88
physiological drought(IV.5.4.3.007) 415
profichi(IV.4.4.007,.049)388,390
ripening(VI.1.1.013)572
rust(*V.3.3.2.181f)519
San Pedro(IV.4.4.007)388
see also Ficus spp., Moraceae
Smyrna(IV.4.4.007)388
soft rot(VI.9.1.214,.497)641,654
souring(VI.9.1.504)655
splitting(VI.9.3.121)672
sunburn(VI.9.3.128)672
syconium(II.2.3.1.07d)174
temperate zone crop(I.1.016)5
terminal galling(*V.3.3.1.200c)505
watery soft rot(VI.9.1.497)654
Fig, creeping (Ficus pumila):
 aerial roots(II.4.3.004)241
Figwort (Scrophulaceae):
 bilabiate corolla(II.2.2.061c-14-a) 127
Filbert (Corylus spp.; Corylaceae):
 accrescent(II.2.1.042,.10d)102,113
 burr(II.2.3.2.006)161
 involucre(II.2.1.042,.10d)102,113
 leaflike(II.2.1.10d)102,113
 leathery(II.2.1.10d)102,113
 nut(II.2.3.1.04h)171
 nut crop(I.1.007)4
 simple layerage(V.1.4.036b-3)450
Filicales:
 fern(II.6.010)279
Filicinae:
 pteridophytes(II.6.016)279
Flacourtiaceae; Flacourtia cataphracta = paniala:
 dioecious (many)(IV.4.3.004)386
 orthotropous (seed-form)(II.2.3.3.036b)166
 pachychalazal (seed-form)(II.2.3.3.036a-5)165
 protective devices(II.4.3.042)244
 spine, spiny[II.1.1.3.20a(1)]56
Flagellariaceae:
 cirrhose leaf tips(II.4.3.010)241
Flax:
 hypogynous(II.2.2.19a)148
 kind(III.2.0.004a)312
 superior(II.2.2.19a)148
Floral (flower) initiation:
 ethylene(IV.2.1.033)358
 gibberellins(IV.2.1.036)358
Floricultural crops:
 propagation structures(V.1.1.048)424
Flower parts:
 seismonasty(IV.3.3.003e)376
Flowering plants(I.2.001e)6
 (see individual crops for Bedding plants, Bulbous crops, Cut flowers, Hanging baskets, Pot plants
Flowers:
 aphids(*V.3.3.1.132d)502
 combine(V.1.3.010)432
 cones(V.1.2.003b)428
 cytokinins(IV.2.1.025)358
 Flower thrips(VI.9.2.030) 661
 guiding structure(IV.4.4.023)389
 (herbaceous) geotropic bending(V.4.1.056)561
 (herbaceous) light requirement for germination(V.1.3.042)435
 latent heat (evaporative cooling) (VI.5.051c)593
 mealy bugs(*V.3.3.1.132d)502
 nyctinasty(IV.3.3.003c)376
 photonasty(IV.3.3.003d)376
 rodents(V.3.3.1.224)507
 scale insects(*V.3.3.1.132d)502
 seed tape(V.1.3.074)438
 seed testing(V.1.3.075)438
 thermodormancy (many spp.)(V.1.3.083)439
 thermonasty(IV.3.3.003f)376
Flowers, tropical:
 birds(V.3.3.1.026)498
Foliage crops:
 propagation structures(V.1.1.048)424
Foliage plants:
 (adventitious) aerial roots (Philodendron sp.)(II.1.3.01a)96
 air layerage(V.1.4.036a)449
 baskets(VI.4.003)585
 cardboard(VI.4.007)586

INDEX OF CROPS

carton(VI.4.008)586
chilling injury(VI.9.3.029)665
dark storage(VI.8.025)618
foliar nematodes(*V.3.3.1.200a)505
foliage plant selection(V.4.1.053) 561
fruits(I.2.001d,.001e)5,6
grading(VI.3.039)580
hanging baskets(I.2.001e;VI.4.013)5, 587
mat watering(V.4.1.070)562
pack(VI.3.047a)581
packinghouse(VI.3.000)577
packinghouse layout(VI.3.052)581
panning(V.4.1.080)563
Perishable Agricultural Commodities Act of 1930, as Amended(VI.7.1.) 597-609
Philodendron(V.4.1.107,.119)564
potting mixes(V.1.2.005)429
product mix(VI.8.091)623
see also Araceae, Bromeliceae, etc.
seed testing(V.1.3.075)438
selection(V.4.1.053)561
shipping containers(VI.4.000c)585
slabs(V.4.1.107)564
totem pole plants(V.4.1.119)564
transportation problems(VI.6.034)596
Forest trees:
 atmospheric pollution(IV.5.4.4.001) 415
 seed testing(V.1.3.075)438
 weevils(*V.3.3.1.132a)501
Forsythia:
 softwood (stem) cuttings(V.1.4.022f-4,.03c)443,455
 hardwood (stem) cuttings(V.1.4.022f-1,.02b)442,454
Fragaria spp.:
 accessory (fruit)(II.2.3.1.001)158
 achene(II.2.3.1.002)158
 (see small fruits)(I.1.012)4
 (see strawberry)
Freesias (Freesia spp.; Iridaceae)(I.2.001c)5
Fruit crops(I.0.002;.1.000)3
 active ingredient(VI.7.2.017a)612
 aphids(*V.3.3.1.132d)502
 bark and ambrosia beetles(*V.3.3.1.132a)501
 contour planting(V.3.1.01h)488
 damping-off(*V.3.3.2.179a,*.181f) 517,519
 deer (large animals)(V.3.3.1.140)503
 double row avenue(V.3.1.01i,j)489
 double-setting (plant system)(V.3.1.008c,.01f)485,488
 drainage(V.3.5.002)557
 Environmental Protection Agency(VI.7.2.014)610
 Federal Insecticide, Fungicide, and Rodenticide Act of 1947 and Nematocide, Plant Regulator, Defoliant and Desiccant Amendment of 1959(VI.7.2.017)612
 fruit flies(*V.3.3.1.132b)501
 fusarium-induced wilts(*V.3.3.2.133c)515
 generative (sexual) propagation(V.1.3.)432-439
 grasshoppers, locusts(*V.3.3.1.132h) 503
 head height(V.3.4.03)547
 heading back(V.3.4.02b)546
 hedgerow (planting system)(V.3.1.01e)488
 hedging machine(V.3.4.040)542
 hexagonal (planting systems)(V.3.1.008g,.01c)486,488
 hydroponic culture(V.3.2.083)495
 incompatibility(V.1.4.035)449
 inspection(V.3.3.3.052)534
 interplanting(V.3.1.008f,.01g)486, 488
 irrigation(V.3.5.000)557
  riparian rights(V.3.5.022)558
  seep(V.3.5.023)558
  soil neutron probe(V.3.5.024)558
  systems(V.3.5.012)557
  timing(V.3.5.013)557
  water measurement(V.3.5.014)558
 leader(V.3.4.046)542
 leader training(V.3.4.03)547
 materials and supplies(V.1.2.000)427
 mealy bugs(*V.3.3.1.132d)502
 micropropagation(V.1.5.000,.01)470, 471
 mineral nutrition(V.3.2.000)490
 mites(V.3.3.1.172)504
 nursery facilities, etc.(V.1.1.000) 418
 nursery plants(I.2.009)6
 Occupational Safety and Health Act of 1970 (OSHA)(VI.7.2.037)616
 pests, pest management(V.3.3.)497-539
 plant and crop size control(V.3.4.000)539
 plant regulator(VI.7.2.017s)615
 propagation(V.1.0.000)417
 pruning(V.3.4.000b)539
  equipment(V.3.4.063)543
 psyllids(*V.3.3.1.132d)502
 quincunx (planting systems)(V.3.1.

INDEX OF CROPS

008i,.01k)486,489
rectangular (square) (planting systems)(V.3.1.008j,.01a)486,488
rodents and leptomorphs(V.3.3.1.224)507
root-knot nematodes (many crops)(*V.3.3.1.200a,e,g)505,506
(see individual crops, e.g., apple, citrus, grape, etc.; Fruit and nut trees; nuts)(I.0.002,.1.000) 3
scale insects(*V.3.3.1.132d)502
small fruits(I.1.012)4
soil auger(V.3.1.011)486
soils(V.2.)472-482
square (planting system)(**V.3.1.01b) 488**
tent caterpillars (nuts)(*V.3.3.1.132g)502
thinning(V.3.4.02c)546
thrips(V.3.3.1.132i)503
topper(V.3.4.076)544
training(V.3.4.000b,.03)539,547
triangular (planting systems)(V.3.1.008m,.01d)486,488
vegetative propagation(V.1.4.)441-479
water regulation(V.3.5.)557
white flies(*V.3.3.1.132d)502

Fruit and nut trees:
algal spot(V.3.3.2.008)510
aphids(*V.3.3.1.132d)502
awl nematodes(*V.3.3.1.200c)505
bark beetles(*V.3.3.1.132a)501
bending(V.3.4.003,.12e)540,556
brown root rot(*V.3.3.2.181f)519
brush control(V.3.3.3.015)532
candelbra(V.3.4.11g)555
central leader(V.3.4.029)541
certification program(V.1.4.010)441
cone(V.3.4.019)541
contre-espalier(V.3.4.020,.11c,d) 541,555
cordon(V.3.4.021,.11b,c,e)541,555
crown gall(*V.3.3.2.040)511
dard(V.3.4.023)541
double-working(V.1.4.028h,.08k,.09j) 445,460,461
dwarf pyramid(V.3.4.068)544
espalier(V.3.4.026,.11a,b,e)541,555
festoon(V.3.4.003)540
forkert bud(V.1.4.028f-3)445,460
free-standing(V.3.4.11f,g)555
fruit flies(*V.3.3.1.132b)501
grafting(V.1.4.028k)446
goblet(V.3.4.031,.11f)507,541
heading(V.3.4.000b)539

high headed(V.3.4.029)541
inarching(V.1.4.028m)447
incompatible, incompatibility(V.1.4.035)449
inspection(V.3.3.3.052)534
interplanting (plant systems)(V.3.1.008h)486
leader(V.3.4.046)542
lopping shears(V.3.4.048)542
Lorette system(V.3.4.049)542
low headed(V.3.4.029)541
maintenance pruning(V.3.4.000b)539
mealybugs(*V.3.3.1.132d)502
mechanical stress(IV.5.4.4.002)416
micrograftage(V.1.4.028p,.11i-o)448, 463
modified central leader(V.3.4.029) 541
mushroom root rots(*V.3.3.2.133c, *.181f)515,519
nicolin budding(V.1.4.028q,.08k)448, 460
open center(V.3.4.029)541
orchard heating(VI.5.051c)593
palmette(V.3.4.055,.11a,d)543,555
pillar tree(V.3.4.056,.04a-f)543, 548
pole saw(V.3.4.060)543
pole shears(V.3.4.061)543
pruning(**V.3.4.01,.02a-c)545,546**
equipment(V.3.4.063)543
knife(V.3.4.064)543
saw(V.3.4.065)543
shears(V.3.4.066)543
systems(V.3.4.067)543
psyllids(*V.3.3.1.132d)502
pyramid(V.3.4.068,.12d)544,556
rootknot nematodes(*V.3.3.1.200e,*g) 506
scale insects(*V.3.3.1.132d)502
see Fruit, Fruit crops
seed extraction (fleshy fruits)(V.1.3.064)437
snails(V.3.3.1.251)501
spindle bush(V.3.4.072,.04g)544,548
sunscald(IV.5.3.048)411
topper(V.3.4.076)544
topworking(V.1.4.028v)449
thinning(V.3.4.000b)539
training(V.3.4.000b,.063,.01,02a-c, .03)539,543,544,545
unfair conduct(VI.7.1.077)608
watersprout(II.1.1.3.22b)58
white rot(*V.3.3.2.181f)519

Fruits:
abortive(IV.4.8.003)393
ageing(VI.1.001)569

INDEX OF CROPS

Agricultural Marketing Agreement Act of 1937, as Amended(VI.7.2.004) 609
as foliage plants(I.2.001d)5
auxins(IV.2.1.012)357
bagmaster carton(VI.4.002)585
bags(VI.4.001)585
baskets(VI.4.003)585
bin(VI.4.004)586
bird dog(VI.8.002)616
blemish(VI.8.038a)619
°Brix(VI.1.008a)570
broker(VI.7.1.005)598
bulk bin(VI.4.006)586
bulk display(VI.8.007)617
buyer (fruit): (see Bird dog)
can(VI.10.005)674
carbon dioxide treatment(VI.5.007) 588
carriers(VI.6.)594-597
carton(VI.4.008)586
chilling injury(VI.9.3.029)665
chutney(VI.10.007)674
climacteric period(V.1.002)569
color additive(VI.7.2.016b)611
color break(VI.1.008b)570
condition(VI.8.016)617
container(VI.4.)585-588
crate(VI.4.009)586
cytokinins(IV.2.1.025)358
dealer(VI.7.1.015)599
Delaney Amendment(VI.7.2.016c)611
demand(VI.8.027)618
direct buying(VI.8.028)618
drier-polisher(VI.3.025)579
drying(VI.10.009)675
dumping(VI.3.027)579
Environmental Protection Agency(VI.7.2.014)610
ethnic produce(VI.8.035)619
exempt hauler(VI.6.006)594
Federal Food, Drug and Cosmetic Act, as Amended(VI.7.2.016)610
Federal Insecticide, Fungicide and Rodenticide Act of 1947; Nematocide, Plant Regulator, Defoliant and Desiccant Amendment of 1959; and Federal Environmental Pesticide Act of 1972(VI.7.2.017)612
fertile(IV.4.8.024)394
field-to-packinghouse handling(VI.2.012)575
firmness tester(VI.1.003)569
flat(VI.4.011)586
fresh(VI.7.1.025)601
  impulse item(VI.8.047)620
  nutrient labeling(VI.8.072)621
  perishable agricultural commodity (VI.7.1.043)603
  perishables(VI.8.081)622
  reasonable time(VI.7.1.048)604
frozen(VI.7.1.026)601
  reasonable time(VI.7.1.048)604
fruit flies(*V.3.3.1.132b)501
fruitworms(*V.3.3.1.132g)502
fungicidal waxes(VI.3.085a)584
fungicide application(VI.3.033)580
gift baskets(VI.4.003)585
  packs(VI.3.053e)582
grade(VI.8.038)619
  inspection(VI.8.038c,d)619
  standard(VI.8.038e)619
  table(VI.3.038)580
grading(VI.3.039)580
gray mold rot(*V.3.3.2.181g;VI.9.1.276)520,644
grower's agent(VI.7.1.030)601
growers' agents(VI.7.1.031)601
growth inhibitors(IV.2.1.037)359
half-cooling time(VI.5.021)589
hand harvesting methods(VI.2.022a) 576
handling systems(VI.2.020)575
harvesting(VI.2.000)575
  containers(VI.4.000a)585
  types(VI.2.023)577
heat of respiration(VI.5.051b)593
horticultural maturity (roadside stands)(VI.1.004)569
humidity(VIi.5.025)589
hydrometer(VI.1.008c)570
imcompatible commodities(VI.6.009) 594
impulse item(VI.8.047)620
inspection(V.3.3.3.052)534
jam(VI.10.020)676
juice(s)(VI.1.008c)570
  chilled(VI.10.006)674
  concentration(VI.10.008)674
  evaporator(VI.10.010)675
  extractor(VI.10.011)675
  fermentation(VI.10.003,.012,.038) 674,676,677
  finisher(VI.10.014)676
  flash pasteurization(VI.10.016)676
  frozen concentrate(VI.10.018)676
  hot-pack concentrate(VI.10.019)676
jelly(VI.10.021)677
pasteurization(VI.10.027)677
pulper(VI.10.031)677
punch(VI.10.032)677
squash(VI.10.034)677
vinegar(VI.10.038)677
wine(VI.10.039)677

INDEX OF CROPS

June drop(IV.4.8.038)395
leptomorphs(V.3.3.1.224)507
loading(VI.6.016)595
   pattern(VI.6.017)595
   pattern choice(VI.6.018)595
"mad customer index"(VI.8.056)620
market containers(VI.4.000b)585
market disorders(VI.9.)626-677
mass-average temperature(VI.5.032) 590
maturation(VI.1.005)570
maturity(VI.1.)569-574
   (climacteric fruits)(VI.1.006)570
   gibberellins(IV.2.1.036)358
   (nonclimacteric fruits)(VI.1.007) 570
   standards(VI.1.009)571
   test sample(VI.1.011)572
mechanized harvesting(VI.2.022b)576
merchandizing(VI.8.065)621
mesh (bags)(VI.4.015)587
metaxenia(IV.4.8.040)395
Mexican fruit fly(*V.3.3.1.132b)501
misbranded food(VI.7.2.016e)611
mix-and-match(VI.8.067)621
mixed loads(VI.6.022)595
MUM(VI.6.023)595
nutrient labeling(VI.8.072)621
non-climacteric(VI.1.007)570
official inspection certificate (VI.7.1.042)603
pack, packing(VI.3.047,.047d)581
packing operations(VI.3.053)581
packinghouse(VI.3.000)577
packinghouse foreman(VI.8.077)622
packinghouse layout(VI.3.052)581
packinghouse manager(VI.8.078)622
packingline(VI.3.031)580
palatability(VI.8.079)622
pallet(VI.4.017)587
   box(VI.4.018)587
palletization(VI.6.024)596
papier maché(VI.4.019)587
pathological (parasitic) market diseases(VI.9.1.000)626
perennation(IV.4.8.043)395
perishable agricultural commodity (VI.7.1.043)603
Perishable Agricultural Commodities Act of 1930, as Amended(VI.7.1.) 597-609
perishables(VI.8.081)622
pesticide chemical(VI.7.2.016f)612
physiological maturity(VI.1.012)572
   sprays(V.3.4.000a)539
picking bag(VI.4.020)587
place pack(VI.8.083)622
plan III(VI.6.026)596
point of purchase/point of sale(VI.8.084)622
pool(VI.8.085)622
pooling(VI.8.086)622
postharvest disorders(V.3.3.0.005) 497
precooling(VI.5.035)590
precooling formulas(VI.5.036)591
preharvest drop(V.3.4.000a)539
prepackaging(VI.8.089)623
"pressure" tester (see Firmness tester)
processing(VI.10.000)674
propagation media(V.1.2.011)429
psychrometric chart)VI.5.037)591
pull date(VI.8.093)623
pulper(VI.10.031)677
punch(VI.10.032)677
puree(VI.10.033)677
quality(VI.8.094)623
raw agricultural commodity(VI.7.2.016g)612
receiver(VI.8.096)623
refractometer(VI.1.008d)571
refrigeration, storage(VI.5.)588-594
(ripe) birds(V.3.3.1.026)498
ripening(VI.1.013)572
(ripening) ethylene(IV.2.1.033)358
(roadside markets) horticultural maturity(V.1.004)569
rodents(V.3.3.1.224)507
sample(VI.1.014)572
sampling(VI.1.015)572
seasonal item(VI.8.104)624
(see Fruit crops, Fruit and nut trees)
seed extraction (dry fruits)(V.1.3.063)437
seed extraction (fleshy fruits)(V.1.3.064)437
seed testing(V.1.3.075)438
semi-mechanized harvesting(VI.2.022c)576
senescence(VI.1.016)572
services(VI.6.029)596
set(IV.4.8.028;V.3.4.000a)394,539
shipping containers(VI.4.000c)585
sizing(VI.3.064)582
shrink film(VI.4.021)587
shrinkage(VI.8.108)624
slip sheet(VI.4.022)587
(soft) shrinkage(VI.8.108)624
soilage(VI.9.1.498)654
solution culture(V.3.2.083)495
stages of development and senescence (VI.1.02)574

INDEX OF CROPS

storage(VI.5.047)591
storage containers(VI.4.000d)585
stretchwrap(VI.4.023)587
tolerance(VI.8.038f)619
tonnage items(VI.8.121)625
total soluble solids: total titratable acid ratio(VI.1.008e,f) 570,571
total (titratable) acid(VI.1.008f,g) 570
trailer on flat car (TOFC)(VI.6.033) 596
transportation problems(VI.6.034)596
trash elimination(VI.3.027a-1)579
tray(VI.4.024)588
tray pack(VI.4.025)588
tree fruit(VI.8.124)625
tree run(VI.8.125)625
tropical(VI.1.007)570
(tropical) fruit flies(*V.3.3.1.132b)501
uneven ripening(VI.9.3.143)673
unfair conduct(VI.7.1.077)608
unfruitful(IV.4.8.054)395
universal product code(VI.8.131)625
unripe (pickles)(VI.10.028)677
washing(VI.3.081)583
waxes, waxing(VI.3.085,.086)584
wine(VI.10.039)677
wirebound box(VI.4.026)588
you-pick operation(VI.8.138)626
xenia(IV.4.8.057)395
Fruits, small:
bedding plants(I.2.001a)5
see also Blueberry, Brambles, Currant, Strawbeery, etc.

G

Galium sp.:
hooks, etc.(II.4.3.027)242
whorls(II.1.2.28c)93
Garcinia spp.:
megatherm taxa(III.1.2.033)309
stenophyllous leaves (G. linearis) [II.4.08a(3)]256
Garden crops, garden plants:
nursery plants(I.2.009)6
seedling diseases(*V.3.3.2.181f)519
slugs(V.3.3.1.250)508
snails(V.3.3.1.251)508
webworm(*V.3.3.1.132g)502
Gardenia (Gardenia jasminoides; Rubiaceae):
cut flowers(I.2.001c)5
Garden nasturtium:
disciform(II.1.2.08b)73

peltate(II.1.2.074)62
Garden peas: (see pea, garden)
Garlic (Alluim sativum; Amaryllidaceae):
aspergillus bulb rot(VI.9.1.041)629
black mold rot(VI.9.1.084)632
blue mold rot(VI.9.1.109)634
bulb crops(I.3.004)7
gray mold(VI.9.1.264)643
neck rot (see gray mold)
waxy breakdown(VI.9.3.150)673
white rot(VI.9.1.565)659
Gasteria:
spirodistichy(II.1.1.1.025)20
Gaylussacia baccata:
trichome(II.3.31f)237
Gentiana acaulis, G. spp.:
light window(IV.4.4.030)389
microtherm taxa(III.1.2.034)309
Gentianaceae:
dimorphic (flowers)(IV.4.3.003)386
Geraniaceae, Geranium spp.:
exotegmic (seeds)(II.2.3.3.040b)166
microtherm taxa(III.1.2.034)309
Geraniums (Pelargonium spp. Geraniaceae):
flag (flowers)(IV.4.4.004b-3)388
herbaceous (stem) cuttings(V.1.4.022f-2,.02a)443,454
'Nittany Lion Red' (line variety) (III.2.0.005a-2)313
pot plants(I.2.001f)6
succulent(II.4.3.065)245
Gerardia pupurea:
contabescence(II.2.2.05c)134
didynamous(II.2.2.05c)134
distinct(II.2.2.05c)134
Gesneriads (Gesneriaceae):
foliage plants(I.2.001d)5
pot plants(I.2.001f)6
(see also African violet, Gloxinia, other gesneriads)
Gilia:
salver-shaped (corolla)(II.2.2.21g) 150
Ginger (Zingiberaceae):
ascending, ascendant (pinecone ginger, Zingiber zerumbet)(II.1.1.3.02a)39
costate (Z. officinale)(II.1.2.22b) 87
(Costus sp.) cut flowers(I.2.001c)5
rhizome (pachymorph)(V.1.4.060c,.05b)451,457
(some) rhizomatous branching(II.1.1.3.082)26
spice crops(I.1.013)4

INDEX OF CROPS

    triple-ribbed(II.1.2.26s)91
Ginkgo, Ginkgoales:
    cuneiform (leaf)(II.5.5.014,.09d) 271,289
    dichotomous (venation)(II.6.009)279
    gymnosperms(II.6.011)279
Gladiolus (Gladiolus spp., Iridaceae):
    corm (separation)[V.1.4.071b,.16f (2)]451,468
    cut flowers(I.2..001c)5
    geotropic bending(V.4.1.056)561
    growth movements(II.4.3.023)242
    hamper(VI.4.012)587
    hard rot(*V.3.3.2.181g)520
    stem tubers(II.4.3.057)244
    storage containers(VI.4.000d)585
    tunic(V.4.1.120)565
Gleditsia triacanthos: (see honey locust)
Globe amaranth:
    C-4 plants(IV.1.2.008b)352
    day-neutral plant(IV.3.2.013c)374
    dimidiate(II.2.2.10c)139
    unilocular (anthers)(II.2.2.10c)139
Globe artichoke (Cynara scolymus; Compositae):
    freezing injury(VI.9.3.048)667
    gray mold rot(VI.9.1.271)643
    miscellaneous (vegetable) crops(I.3.011)8
    plume moth(VI.9.2.037)661
Gloriosa lily (Gloriosa spp.):
    cirrhose leaf tips(II.4.3.010,.06d) 241,254
    cirrhous(II.5.4.1.010)268
    clinging(II.4.3.014)241
    growth movements (tuber)(II.4.3.023) 242
    parallel (venation, Gloriosa greenae hyb.)(II.1.2.24h)89
Gloxinia (Sinningia speciosa; Gesneriaceae):
    pot plants(I.2.001)6
    synantherous(II.2.2.06d)135
    syngenesious (anthers)(II.2.2.007-d-18)119
Glycosmis citrifolia:
    trifoliolate-pinnate[II.1.2.21a(2)] **86**
Gnetales, Gnetum gnemon, G. indicum:
    crystals(II.3.08c)214
    druses(II.3.08a,b)214
    gymnosperms(II.6.011)279
Golden rain tree: see Koelreuteria
Gooseberries (Ribes: Grossularia; Grossulariaceae):
    horticultural maturity(VI.1.004)569
    mechanized harvesting(VI.2.022b)576
    mound (stool) layerage(V.1.4.036b-1) 450
    small fruits(I.1.012)4
Goosefoot:
    short-day plant(IV.3.2.0131)375
Gourds (Luffa acutangula, L. cylindrica, Lagenaria sp.):
    (see Cucurbits)
Grains, Grain crops, Gramineae:
    birds(V.3.3.1.026)498
    coleoptile(II.2.3.3.014a)163
    coleorhiza(II.2.3.3.014b)163
    combine(V.1.3.010)432
    debearder(V.1.3.018)433
    drought escape(IV.5.2.015)399
    excentric(II.2.3.3.017)164
    germination test(V.1.3.031)434
    meal moths(*V.3.3.1.132g)502
    parastichy(II.1.1.1.016)19
    scutellum(II.2.3.3.014k)164
    (see individual crops, i.e., barley, corn (maize), oats, rice, wheat, etc., and grasses)
    stem rust(*V.3.3.2.181f)519
    stubble culture(V.3.1.013)487
    stoma(II.3.190)204
    utricle(II.2.3.1.038)160
Grapefruit:
    (see Citrus-grapefruit)
Grape hyacinth (Muscari botryoides, M. spp.):
    bulb forcing(V.4.1.012)559
    gummosis(V.4.2.1.011)565
    (see also Bulbous plants)
Grapes (Vitis vinifera, V. spp.; Vitaceae):
    air precooling(VI.5.035a-2)590
    annulate (disk)[II.2.2.27c(3)]156
    bench-graftage(V.1.4.028b)443
    black rot(*V.3.3.2.181e)519
    bloom(II.5.6.010)274
    caducous[II.2.2.27c(2);.5.02a]156, 282
    calyptrate[II.2.2.27c(2)]156
    certification program(Vv.1.4.010)441
    chip budding(V.1.4.028-2)444
    clinging(II.4.3.014)241
    clones(III.2.1.001a)316
    conditional parthenocarpy(IV.4.2.002)385
    coulure(IV.4.8.012)393
    downy mildew(*V.3.3.2.181d)518
    endotestal (seeds)(II.2.3.3.043a)167
    fanleaf virus(*V.3.3.1.200c)505
    fan system(V.3.4.032,.10f)541,554
    fermentation(VI.10.012)675

INDEX OF CROPS

flower(II.2.2.27c)156
Geneva double-curtain system(V.3.4.032,.10d)541,554
grape eye(II.1.1.1.02i)29
growth curve for seeded cvs.(IV.3.1.013)371
hardwood (stem) cuttings(V.1.4.022f-1,02b)442,454
head system(V.3.4.032,.10e)541,554
horizontal-arm spur system(V.3.4.032,.10c)541,554
hornworms(*V.3.3.1.132g)502
impulse item(VI.8.047)620
inconspicious (blossoms)(IV.4.4.004a)387
leaf-opposed (inflorencence)(II.2.1.044,.11a)102,114
lug(VI.4.014)587
mechanized harvesting(VI.2.022b)576
millerandage(IVv.4.8.041)395
palmate-pinnate (venation, V. simpsonnii)(II.1.2.23e)88
perichalazal (seed-form)(II.2.3.3.036a-6)165
physiological drought(IV.5.4.3.007)415
Pierce's disease(V.3.3.2.361)527
powdery mildew(*V.3.3.2.181e)519
pruning (training) systems(V.3.4.032,.10)541,554
raphides(II.3.08d)214
renewal spur(V.3.4.069)544
ring nematodes(*V.3.3.1.200b)505
ripening(VI.1.013)572
rust(*V.3.3.2.181f)519
secondary growth of vascular system (II.3.173)202
serpentine layerage(V.1.4.036b-2)450
six-cane Kniffen system(V.3.4.032,.10b)541,554
stipitate(II.1.1.1.026)20
stomata(II.3.29e)235
('Sultanina') stenospermocarpy(IV.4.2.004)385
sympodial (axes)(II.1.1.3.099)27
temperate zone crop(I.1.016)5
tendrils(II.4.3.071)245
thyrse(II.2.1.072,.14c)103,117
total soluble solids(VI.1.008e)571
total (titratable) acid(VI.1.008g)571
training(V.3.4.10)554
vinifera cultivars(V.3.4.032)541
whip, or tongue, graft(V.1.4.028k-9)447
wine(VI.10.039)677
Grapple plant (Harpagophytum procumbens):
 tardily dehiscent(II.2.3.1.03c)170
Grasses:
 annual (weeds)(V.3.3.3.122)539
 caryopis(II.2.3.1.008)159
 'Coastal' bermudagrass (clonal variety)(III.2.0.005a-1)313
 colter harrow(V.3.1.004)485
 direct-set cuttings(V.1.4.022b)442
 Ditylenchus radicicola(*V.3.3.1.200g)506
 downy mildew(*V.3.3.2.181d)518
 forb(II.0.003)10
 glume(II.2.2.040)122
 hemicryptophyte(II.4.1.003)238
 herbicides(V.3.3.3.049)534
 'Higgins' buffelgrass (clonal variety)(III.2.0.005a-1)313
 laminar rolling(II.4.3.030)243
 lance nematodes(*V.3.3.1.200d)506
 leaf modifications(II.4.3.032)243
 leptomorph rhizome(V.1.4.060a)451
 ligule[II.1.2.19e(2);.5.5.032]84,272
 mole crickets(*V.3.3.1.132h)503
 needle nematodes(*V.3.3.1.200c)505
 palea(II.2.2.057)126
 parastichy, phyllotaxis(II.1.1.1.04) 31
 perennial (weeds)(V.3.3.3.122)539
 rhizomatous branching(II.1.1.3.082) 26
 rhizome(V.1.4.060)451
 ring nematodes(*V.3.3.1.200b)505
 seed tape(V.1.3.074)438
 sheath nematodes(*V.3.3.1.200b)505
 sheathing (petiole)[II.1.2.19e(1)]84
 sheathoid nematodes(*V.3.3.1.200b) 505
 sting nematodes(*V.3.3.1.200c)505
 spiral nematodes(*V.3.3.1.200d)506
 stoloniferous(V.3.1.004)485
 tiller(II.1.1.4.100)27
 tubular(II.1.1.3.101)27
 turf crops(I.2.011)6
 xeromorphic structure(IV.5.2.085)406
Grasses, lawn:
 armyworms(*V.3.3.1.132g)502
 chinch bugs(*V.3.3.1.132c)502
 'Meyer' zoysiagrass(III.2.0.005a-1) 313
 mowers(V.1.1.034)422
 (see also Grasses)
 sod webworms(*V.3.3.1.132g)502
 turf crops(I.2.011)6
Greenhouse plants:
 slugs(V.3.3.1.250)508
 snails(V.3.3.1.251)508

sowbug(V.3.3.1.258)508
Grossulariaceae:
  endotestal (seeds)(II.2.3.3.043a)167
  (see Currant, Gooseberries)
  small fruits(I.1.012)4
Ground cherry: see Physalis
Ground covers:
  (see Ivies, Liriope, Ophiopogon, Pachysandra, Vinca spp., and the like)(I.2.003)6
Guarana (Paullinia cupana: Sapindaceae):
  beverage crop(I.1.001)4
Guava (Psidium guajava, P. cattleianum):
  campylotropous (seed form)(II.2.3.3.036a-1,.06a-d)165,187
  obovoid (P. guajava)(II.2.3.2.07a)181
  suckers(V.1.4.083)452
Guttiferae:
  exotegmic (seeds)(II.2.3.3.040b)166
Gymnosperms (Gymnospermae):
  archegonium(II.6.003)279
  C-3 Plants(IV.1.2.008a)352
  parastichy, phyllotaxis(II.1.1.1.04)31
  proteins(IV.2.2.172)369
  resin duct(II.3.157)201
  spermatozoid(II.6.019)280
  strobilus(II.6.024)280
  tracheids(II.3.202)204
Gynandropsis:
  stipe[II.2.2.15h(1)]144
  stipitate receptacle[II.2.2.15h(2)]144

H

Hairy indigo: scarifier(V.1.3.057)435
Hamamaelis: (see Witch hazel)
Hanging baskets(I.2.001e)6
Harebell: (see Campanula)
Harpagophytum procumbens: (see Grapple plant)
Hawthorn (Crataegus spp.): pome fruit (I.1.009)4
Hazel nut: (see Filbert)
Heaths: chamaephyte(II.4.1.001)238
Hedysarum coronarium:
  spiculate(II.5.19d)299
Helenium sp., H. autumnale:
  achene(II.2.3.1.01e)168
  heterogamous(II.2.1.19c)112
Heliotropium: trichome(II.3.31a)237
Helleboraceae: follicle(II.2.3.1.017)159

Helleborus niger:
  pedate(II.1.2.16a)81
  pedatriribbed(II.1.2.24j)89
Hemerocallis: bostryx(II.2.1.006)100
Herbs:
  (as bedding plants)(I.2.001a)5
  cambium (fascicular)(II.3.026)195
  chutney(VI.10.007)674
  (for hanging baskets)(I.2.001e)6
  hemicryptophyte (stoloniferous)(II.4.1.003)238
  phanerophyte (upright)(II.4.1.005)238
Hevea brasiliensis: (see Rubber)
Hevea spp.:
  laticifer(II.3.098)198
  (see also Rubber)
Hibiscus esculentus: (see Okra)
Hibiscus: blossom blast(*V.3.3.2.181c)518
Hibiscus moscheutos: epicalyx(II.2.1.10c)113
Hickory, Shagbark: (see Carya)
Hieracium spp.: lyrate(II.5.5.036,.11a)272,291
Hippeastrum spp.: (see Amaryllis)
Hippurus vulgaris:
  verticillate(II.1.2.28b)93
Hollies (most spp.):
  birds(V.3.3.1.026)498
  protective devices(II.4.3.042)244
  semihardwood (stem) cuttings(V.1.4.022f-3,.02c-e,.03a,b)443,454,455
Holmskioldia sanguinea:
  clambering (Chinese hat plant)(II.1.1.3.016c)42
Homonia riparia:
  stenophyllous leaves[II.4.08a(7)]256
Honey locust:
  protective devices{II.1.1.3.20a(2);.4.3.042]56,244
Honeysuckle:
  connate-perfoliate(II.2.3.05d)70
  serpentine layerage(V.1.4.036b-2)450
  simple layerage(V.1.4.036b-3)450
  tubular(II.2.2.23e)152
Hooked bristlegrass:
  ambiphotoperodic plant(IV.3.2.013a)374
Hops (Humulus lupulus, Moraceae):
  condiments, etc.(I.3.006)7
Horehound (Marrubium vulgare; Labiatae)
  condiments, etc.(I.3.006)7
Horse-radish (Armoracia lapathifolia; Cruciferae)
  condiments, etc.(I.3.006)7

INDEX OF CROPS

  bacterial root rots(VI.9.1.056)630
  penicillium root rot(VI.9.1.344)647
  rhizoctonia root rot(VI.9.1.443)651
  root cuttings(V.1.4.022e,.01)442,453
  (see Crucifers)
Horsetails: pteridophytes(II.6.016)279
Horticultural crops:
  mechanical stress(IV.5.4.4.002)416
Houseleek:
  offsets (divisions)(V.1.4.024)443
  orthostichy(II.1.1.1.05r)32
  parastichy(II.1.1.1.05r)32
  phyllotaxis(II.1.1.1.05r)32
  rows(II.1.1.1.05r)32
  series(II.1.1.1.05r)32
House plants(IV.3.3.008j-1)377
Humulus sp.: hooks, etc.(II.4.3.027) 242
Hyacinth (*Hyacinthus orientalis*):
  bulb(I.2.001b)5
  bulb forcing(V.4.1.012)559
  double stem(V.4.2.2.005)567
  fasciation(V.4.2.2.007)567
  flat stem(V.4.2.2.007)567
  gummosis(V.4.2.1.011)565
  prepared (PR) hyacinths(V.4.1.091) 563
  regular (RG) hyacinths(V.4.1.097)563
  (see also Bulbous plants)
  "spitting"(V.4.2.2.019)568
  tunicate bulb (separation)(V.1.4. 071a-2)451
  yellow disease(V.4.2.1.023)566
Hydnophytum spp.:
  myrmecomorphosis(II.4.4.002c)246
Hydrangea (*Hydrangea* spp.):
  leaf bud cuttings(V.1.4.022d,.01d) 442,453
  (see also Pot plants)
Hypericaceae, *Hypericum* spp.
  bostryx = helicoid cyme(II.2.1.006, .05b)100,108
  exotegmic seeds(II.2.3.3.03d)184
  lateral root development(II.3.17a) 223
  parietal (placentation)(II.2.2. 14e-g)143
  syncarpous (ovary)(II.2.2.14e-g)143

I

Iberis spp.: silicle(II.2.3.1.033)160
Imperata cylindrica:
  pyromorphosis(II.4.14a)262
India rubber (see *Ficus* spp.)
Indian corn = maize: (see Corn)
Indian turnip: (see *Arisaema*)

Ipomoea:
  apicifixed (anthers)(II.2.2.08d)137
  clinging(II.4.3.014)241
  infundibular(II.2.2.21j)150
  salverform (corolla)(II.2.2.21h)150
  septifragal(II.2.3.021,m)169
Iris, Dutch (*Iris xiphium* hyb.):
  bulb forcing(V.4.1.012)559
  bulbs(I.1.001b)5
  cut flowers(I.1.001c)5
  (see also Bulbous plants)
  special precooling(V.4.1.111)564
Iris:
  capsule(II.2.3.1.01i)169
  ensiform(II.5.09g)289
  equitant(II.1.1.2.02j)35
  gullet (flower)(IV.4.4.004b-4)388
  loculicidal(II.2.3.1.02k)169
  pachymorph rhizome(V.1.4.060c)451
  petaloid (stigma, *I. cristata*)(II.2. 2.18n)147
  rhipidium(II.2.1.062)102
  (rhizomatous): rhizomes (divisions) (V.1.4.024)443
  stomata(II.3.29d)235
  unifacial (flattened) leaf(II.1.2. 119,.04g)64,69
  zygomorphic(II.2.2.20h)149
Iodinebush: obligate halophyte(II.4.2. 025)240
Isosetinae: (II.6.016)279
Italian cypress: (see Conifers)
Ivies (*Hedera* spp.): (English ivy, *Hedera helix*) (*H. canariensis*)
  aerial roots(II.1.3.002;.4.3.004;.4. 06b)94,241,254
  ground covers(I.2.003)6
  palmiribbed(II.1.2.23g)88
  polymorphic leaves(II.0.05)16

J

Jaboticaba (*Myrciaria cauliflora*):
  cauliflorous(II.1.1.1.005)19
  trunciflory[II.2.1.05d(5)]108
Japanese persimmon: (see *Diospyros kaki*)
Jerusalem artichoke (*Helianthus tuberosus*; Compositae)(I.3.023)8
  clones(III.2.1.001a)316
  corms (storage organs)(II.4.3.057) 244
  day-neutral plant(IV.3.2.013c)374
  inulin(IV.2.2.031c-a-2)361
Jonquil: (see Narcissus)
Juncus effusus, *J*. spp.:
  drepanium(II.2.1.032,.09b)101,112

Juglans spp.:
  cambium initials(II.3.07a)213
  nursery plants(I.2.009)6
Junipers, Juniperus (see also Conifers):
  acerose (leaf)(II.5.5.001)271
  hardwood (stem) cuttings(V.1.4.021f-1,.02b)442,454
  subulate(II.5.5.058)273
Jute: bast fibers(II.3.063)196

K

Kalanchoe (Kalanchoe spp.; Crassulaceae)(I.2.001f)6
  long-short-day plant(IV.3.2.013h)375
  pot plants(I.2.001f)6
  short-day plant(IV.3.2.013l)375
Kale (Brassica oleracea (Acephala group); Cruciferae)
  greens(I.3.016)8
  hybrids(III.2.1.001b)317
Kaphia spp.: pneumorhizae(II.4.3.039)243
Kapok (Ceiba):
  buttress(II.4.10b)258
  succulent(II.4.3.065)245
Kenaf: bast fibers(II.3.063)196
Kigelia sp.: flagelliflory(IV.4.4.021)389
Kiwi fruit (Actinidia chinensis; Actinidiaceae)
  chilling requirement(IV.5.3.006)408
  flowers(II.2.2.27d,e)156
  linear stigmas[II.2.2.27e(2)]156
  pruning systems(V.3.4.032)541
  subtropical crops(I.1.015)4
Kleinia spp. succulent(II.4.3.065)245
Kohlrabi [Brassica oleracea (Gongylodes group); Cruciferae]:
  (see Crucifers)
  stem tubers(II.4.3.057;.4.13b)244, 261
Koelreuteria paniculata (golden rain tree):
  bladdery(II.2.3.2.02c)176
  inflated(II.2.3.2.02c)176
  pyramidal(II.2.3.2.02c)176
Kola nut (Cola nitida, C. acuminata; Sterculiaceae):
  beverage crops(I.1.001)4
  "bissy"(III.1.1.005)303
  follicle(II.2.3.1.017)159
  polygamous(IV.4.3.017)387
  processing(VI.10.000)674
  tropical crops(I.1.017)5
Korthalsia: myrmecophyte(II.4.05d)253
Kulup: (see Avocado)
Kumquat: (see Citrus - Kumquat)

L

Labiatae:
  bilabiate(II.2.2.061c-14-a)127
  dimorphic (flowers)(IV.4.3.003)386
  gullet (flower)(IV.4.4.004b-4)388
  verticillate (leaves)(II.1.2.123)65
  zygomorphic (perianth)(II.2.2.061c-14)127
Lactuca serriola:
  equifacial flattened leaf(II.1.2.119,.04d)64,69
Lady palm:
  (see Palms)
Lagerstroemia indica:
  heterandry(IV.4.4.027)389
Laguncularia racemosa:
  landscape plants(I.2.009)6
  pneumatorhizae(II.4.3.039)243
Lamium:
  galea[II.2.2.23b(1)]152
  gynobasic (Lamium album)(II.2.2.17f)146
  labiate(II.2.2.23b)152
  ringent(II.2.2.23b)152
Lansium spp.:
  perichalazal (seed-form)(II.2.3.3.036a-6)165
Lantana, weeping:
  aggregated (heads)(II.2.1.05a)108
Larch: (see Conifers)
Larkspur (Delphinium exaltatum, D. tricorne; Ranunculaceae):
  (see cut flowers)(I.2.001c)4
Lathyrus spp.:
  stipules[II.1.2.20d(1)]85
  tendrils(II.1.2.02e;.4.3.071)67,245
Lauraceae:
  flaps (pollen dehiscence)(II.2.2.007h-3)119
  pachychalazal (seed-form)(II.2.3.3.036a-5)165
  parasites(II.4.2.026)240
Laurus spp.:
  hot zone plants(II.4.3.028)243
  leaf modifications(II.4.3.032)243
  nursery plants(I.2.009)6
  sclerophyll(II.4.3.050)244
  xeromorphic structure(IV.5.2.085)406
Lavender (Lavendula officinalis; Labiatae):
  condiments, etc.(I.3.006)7

unifacial cylindrical leaf(II.1.2.119,.04e)64,69

INDEX OF CROPS

Lecythis ollaria, L. spp.:
  hypochlamydy(II.2.3.2.04b)178
  pyxis(II.2.3.1.029,.03e)160,170
Leek (Allium ampeloprasum (Leek group) Amaryllidaceae):
  bulb crops(I.3.004)6
  smudge(VI.9.1.492)654
  smut(VI.9.1.493)654
  white rot(VI.9.1.565)659
Legumes:
  beet cyst nematode(*V.3.3.1.200g)506
  beetles(*V.3.3.1.132a)501
  contra-inoculation(V.1.3.011)432
  cross-inoculation(V.1.3.014)432
  finishing machines(V.1.3.027,.057) 433,435
  gypsum(V.3.2.044)493
  hot zone plants(II.4.3.028)243
  inoculation(V.1.3.041)434
  polyaxial stem (many)(II.1.1.3.067) 25
  pulses(I.3.014)8
  pulvinus (many)(II.1.2.087)62
  replum(II.2.3.1.030)160
  turf crops (creeping spp.)(I.2.011)6
  (see Beans, Peas, Acacia, Leguminosae, etc.)
Leguminosae (see Legumes, Beans, Peas, Acacia, etc.)
  alae (petals)(II.2.2.061c-2)127
  carina (keel)(II.2.2.061c-3)127
  exotestal (seeds)(II.2.3.3.043b)167
  hard seed(V.1.3.034)434
  loment (Fabaceae, Mimosaceae)(II.2.3.1.021)159
  pulses(I.3.014)8
  superposed buds(II.1.1.1.030)20
  vexillum (petal)(II.2.2.061c-13)127
Lemon (see Citrus-Lemon)
Lentil (see Beans)
Lepidocaryoideae:
  rattans(II.0.022)11
Lespedeza:
  hullers(V.1.3.057)435
Lettuce (Lactuca sativa, var. capitata, var. crispa, Compositae)
  aster yellows(VI.9.1.046)629
  bacterial soft rot(VI.9.1.061)631
  big vein(VI.9.1.078)632
  black rot(*V.3.3.2.181g)520
  collenchyma (lacunar)(II.3.04g)210
  combine(V.1.3.010)432
  consumer packs(VI.3.053d)582
  C-3 plants(IV.1.2.008a)352
  crispate(II.5.4.3.010)270
  cross-merchandizing(VI.8.024)618
  downy mildew(*V.3.3.2.181d;VI.9.1. 204)518,640
  drop(*V.3.3.2.181e)519
  good delivery(VI.7.1.028)601
  (good) quality(VI.8.094)623
  gray mold rot(VI.9.1.272)643
  handling systems(VI.2.020)575
  harvesting containers(VI.4.000a)585
  inbred pure lines(III.2.1.001c)317
  leaf blight(*V.3.3.2.181g)520
  light requirement for germination (V.1.3.042)435
  marginal browning(VI.9.3.084)669
  mosaic(VI.9.1.316)646
  "mule train"(VI.2.031)577
  Nacobbus spp.(*V.3.3.1.200g)506
  pink rib(VI.9.3.095)670
  photodormancy(IV.3.2.011)374
  (quantitative) long-day plant(IV.3.2.013g)375
  rib discoloration(VI.9.3.104)671
  root-knot nematodes(*V.3.3.1.200e, *.200g)506
  russet spotting(VI.9.3.109)671
  salad crops(I.3.020)8
  seed extraction (dry fruits)(V.1.3.063)437
  seed priming(V.1.3.069)437
  semi-mechanized harvesting(VI.2.022c)576
  shrinkage(VI.8.108)624
  slime(VI.9.1.061)631
  source wrapping(VI.3.071)583
  spotted wilt(VI.9.1.520)656
  staple items(VI.8.112)625
  stubby-root nematodes(*V.3.3.1.200c) 505
  thermodormacy(V.1.3.083)439
  tie-in display(VI.8.120)625
  tip burn(VI.9.3.141)673
  tonnage items(VI.8.121)625
  watery soft rot(VI.9.1.561)658
Leucadendron spp.:
  hot zone plants(II.4.3.028)243
Liatris squarrosa:
  homogamous(II.2.1.09d)112
Lichens:
  conjunctive symbiosis(II.4.3.067b) 245
  epiphytes(II.4.2.006)239
  microepiphytes(II.4.2.023)240
  nitrogen fixation(IV.1.2.029)354
  stenothermic(IV.3.2.019)376
  trichogyne(V.3.3.2.437)529
Licorice (Glycyrrhiza glabra: Fabaceae):
  condiments, etc.(I.3.006)7
Licuala grandis (see Palms)

INDEX OF CROPS 870

Lilac: softwood (stem) cuttings(V.1.4.022f-4,.03c)443,455
Ligustrum Massalongianum:
  apiculate(II.2.2.09a)138
  epipetalous(II.2.2.09a)138
Liliaceae: apogamy(IV.4.1.006b-2)379
Lily, Easter and other lilies (Lilium longiflorum var. eximium, Lilium spp. Liliaceae):
  'Ace'(V.4.1.029)560
  anther(II.2.2.11b,c)140
  basal rot(V.4.2.1.001)565
  black cloth shading(V.4.1.006)559
  Botrytis spp.(V.4.2.1.002)565
  bulb(II.4.3.063;V.1.4.16c)245,468
  bulb precooling(V.4.1.019)560
  bulbous plants(I.2001b)5
  case-cooled bulbs(V.4.1.025)560
  contractile roots (L. Martagon)(II.3.17c)223
  controlled temperature (lily bulb) forcing (CTF)V.4.1.029)560
  cut flowers(I.2.001c)5
  cucumber mosaic(V.4.1.004)565
  daughter bulb(V.4.1.035)560
  double nose(V.4.1.040)561
  filling(V.4.1.045)561
  fleck(V.4.2.1.009)565
  Fusarium oxysporum f. sp. lilii(V.4.2.1.001)565
  growth movements(II.4.3.023)242
  juvenility(V.4.1.064)562
  leaf counting(V.4.1.065)562
  lily symptomless virus(V.4.2.1.012)565
  long-day treatment(V.4.1.067)562
  lorate(II.5.5.032)272
  mosaic(V.4.2.1.013)565
  mother blocking(V.4.1.072)562
  mother bulb(V.4.1.073)562
  'Nellie White'(V.4.1.029)560
  penicillium rot(V.4.2.1.015)566
  planting stock(V.1.4.17a;.4.1.086)469,563
  pot plants(I.2.001f)6
  propagation(V.1.4.17)469
  root base rot(V.4.2.1.017)566
  root rot(V.4.2.1.018)566
  rosette(V.4.2.1.019)566
  scale bublets(V.1.4.17c;.4.1.102)469,564
  scaling(V.1.4.17b-g;.4.1.103)469,564
  scaly bulb (separation)(V.1.4.071a-1,.16c;.4.1.104)451,468,564
  seed dormancy III.(V.1.3.060)436
  stage "G"(V.4.1.113)564
  stem bulblet(V.1.4.17e;.4.1.115)469,564
  storage containers(VI.4.000d)585
  storage organs(II.4.13d)261
  strain (Lilium Olympic)(III.2.0.012)316
  stunting(V.4.2.1.022)566
  summer sprouting(V.4.1.117)564
  twist(V.4.2.1.021)566
  yearlings(V.4.1.122)565
Lily-of-the-Valley:
  cut flower(I.2.001c)5
  leptomorph rhizome(V.1.4.060a)451
  scorpioid cyme (monochasium)(II.2.1.07d)110
Lima bean: (see Bean, Lima)
Lime: (see Citrus-Lime)
Linaceae: trimorphism(IV.4.3.019)387
Linaria:
  calcarate(II.2.2.23c)152
  gibbous(II.2.2.23c)152
  personate(II.2.2.23c)152
Linseed: oil crop(I.1.008)4
Linum usitatissimom:
  phloem fibers(II.3.24e)230
Liquidambar styraciflua: (see sweet gum)
Liriodendron: (see tulip poplar)
Liriope:
  ground covers(I.2.003)6
  lorate(II.5.5.032)272
Lithops spp.:
  succulent(II.4.3.065)245
Lithops:
  xeromorphic structure(IV.5.2.085)406
Liverworts:
  epiphytes(II.4.2.006)239
Lobelia:
  pitted vessel(II.3.22f)228
  protoxylem and metaxylem(II.3.22a-f)228
  vessel(II.3.22f)228
Longan (Euphoria longana; Sapindaceae):
  subtropical crop(I.1.015)4
Lonicera: (see honeysuckle)
Lophira procera:
  drip-tips(II.4.07a)255
Loquat (Eriobotrya japonica):
  pome fruit(I.1.009)4
Loranthaceae:
  transversely locellate (anthers)(II.2.2.007r)119
Lupine:
  monadelphous(II.2.2.04a)133
Lupiniphyllum lupinifollum:
  stipels(II.1.2.20c)85
Lychee (Litchi chinensis; Sapindace-

INDEX OF CROPS

    ae):
    air layerage(V.1.4.036a)449
    browning(VI.9.3.022)664
    chilling requirement(IV.5.3.006)407
    subtropical crops(I.1.015)4
Lycopodiinae:
    pteridophytes(II.6.016)279
Lycopodium:
    club moss(II.6.005)279
Lythraceae:
    trimorphism(IV.4.3.019)387

M

Macadamia:
    follicle(II.2.3.1.017)159
    nut(II.2.3.1.024)159
    nut crops(I.1.007)4
    pachychalazal (seed-form)(II.2.3.3.036a-5)165
    processing(VI.10.000)674
    reticulated (venation)(II.1.2.26p)91
Mace: (see Nutmeg)
Magnoliaceae:
    endotestal (seeds)(II.2.3.3.043a)167
    nursery plants(I.2.009)6
    protection of resting buds(II.4.3.041)244
Magnolia spp.:
    apocarpous (gynoecium)(II.2.2.041b) 122
    bowl-shaped(IV.4.4.044b-2)388
    calyptra(II.1.2.03d)68
    crateriform (M. macrophylla)[II.2.2.22b(1)]151
    funiculus(II.2.3.3.019,.01c)164,182
    lenticel(II.3.101)198
    polycarpic(II.2.2.28b)157
Maiden cane: rotary tiller(V.1.1.056) 426
Mahogany: (see Swietenia)
Maize: (see Corn)
Malanga:
    cucurbits(I.3.007)7
    (see Pumpkin)
Malay rose-apple (see Syzygium malaccensis)
Malus: (see Apple)
Mallows, Malvaceae:
    capsule(II.2.3.1.02g)169
    coccus[II.2.3.1.009,.02g(1)]159,169
    column(II.2.2.017)121
    epicalyx(II.2.1.042)102
    monadelphous (stamens)(II.2.2.007d-12)118
    transverse (anther dehiscence)(II.2.2.10b)139

    unilocular (anthers)(II.2.2.007r)119
Malvales: exotegmic (seeds)(II.2.3.3.040b)166
Mandarins: (see Citrus-Mandarins)
Mango (Mangifera indica; Anacardiaceae):
    anthracnose(VI.9.1.027)628
    approach grafting(V.1.4.028a)443
    branching by substitution(II.1.1.3.018,.03e)22,39
    bud and graft union healing(V.1.4.028e)444
    chip bud(V.1.4.028f-2)444
    clones(III.2.1.001)316
    disc(II.0.002)10
    drupe(II.2.3.1.015,.05f)159,172
    endocarp(II.2.3.2.029c)162
    endocarp removal(V.1.3.024)433
    flowers(II.2.2.27f)156
    Forkert bud(V.1.4.028f-3)445
    hand harvesting methods(VI.2.022a) 576
    mesocarp(II.2.3.2.029g)162
    orthotropic (branching)(II.1.1.3.04a)40
    panicle(II.2.1.11c)114
    polyaxial stem(II.1.1.4.067,.16a)25, 52
    polygamous(IV.4.3.017)387
    schizolysigenous (resin canals)[II.3.169;.3.16c(1)]202,222
    stem-end rot(VI.9.1.524)656
    stenospermocarpy (polyembryony)(IV.4.2.004)385
    total (titratable) acid(VI.1.008g) 571
    tropical crops(I.1.017)5
    washing(VI.3.081)583
Mango: [see Pepper (Bell-type); Capsicum)
Mangold:(II.4.3.069)245
Mangroves: aerial (stilt) roots[II.1.3.002,.01c(2)]94,96
    halophytes(II.4.2.010,.04b)239,252
    (red) continous growth[II.1.3.01c(2)]96
    viviparous(II.4.04b)252
    vivipary(II.2.3.3.020d,.01f)164,182
Manila Palm (see Palms)
Morantaceae: (see Foliage plants)
Maple (Acer spp.)
    anisophylly(II.1.2.02d)67
    polyaxial stem(II.1.1.3.067)25
    samara (double)(II.2.3.1.031,.04c) 160,171
    sphaeroblast(IV.3.1.014g)372
Marcgravia sp.: flagelliflory(IV.4.

021)389
Marjoram (Origanum majorana; Labiatae)
 condiments, etc.(I.3.006)7
Maté (Ilex paraguayensis; Aquifoliaceae)
 beverage crops(I.1.001)4
Mauritius papeda (Citrus hystrix):
 (see Citrus)
Medlar (Mespilus germanica): (see Pome fruits)
Melanorrhoea glabra:
 alate(II.2.3.2.01d)175
Melastoma malabathricum: geniculate (filaments)(II.2.2.08h)137
Meliaceae:
 monadelphous (stamens)(II.2.2.007d-12)118
 nursery plants(I.2.009)6
 pachychalazal (seed-form)(II.2.3.3.036a-5)165
 perichalazal (seed-form)(II.2.3.3.036a-6)165
 sarcotesta(II.2.3.3.042i)167
 (see also Swietenia)
Melicocca bijuga:
 bifoliolate(II.1.2.03b)68
 bijugate(II.1.2.15b)80
 compound (leaf)(II.1.2.05c)70
 paired(II.1.2.15b)80
Melons: Federal Food, Drug & Cosmetic Act, as Amended(VI.7.2.016)610
 harvesting containers(VI.4.000a)585
 (honeydew) bacterial spot(VI.9.1.070)631
 (honeydew) charcoal rot(VI.9.1.144)636
 melon fruit fly(*V.3.3.1.132b)501
 melonworm(*V.3.3.1.132g)502
 (muskmelons) bacterial soft rot(VI.9.1.062)631
 pelletized seeds(V.1.3.045)435
 (see also Cucurbits, Cucurbitaceae)(I.3.007)7
 sinuous(II.2.2.05f)134
 stamping(VI.3.072)583
 twisted (anthers)(II.2.2.05f)134
Mignonette: inconspicuous (blossoms)(IV.4.4.004a)387
Milkweed: Asclepias cornuti(II.2.2.26)155
 caudicle[II.2.2.26d(2)]155
 comose(II.5.14i)294
 corniculate[II.2.2.061c-14-g,.26a(4)]128,155
 corona]II.2.2.022,.26a(3)]121,155
 cucullate (corolla)[II.2.2.061c-14-i,.26a(2)]128,155
 follicle(II.2.3.1.017,.02b)159,169
 gynostemium(II.2.2.041n)123
 pollinium(II.2.2.007t-9,.26d-f)120,155
 (see also Asclepiadaceae)
Mimetes hartogii:
 gullet (flower)IV.4.4.004b-4)388
Mimosaceae: (see Leguminosae)
Mint: (Mentha spp.; Labiatae):
 bilabiate (corolla)(II.2.2.061c-14-a)127
 condiments, etc.(I.3.006)7
 needle nematodes(*V.3.3.1.200c)505
 root-lesion nematodes(*V.3.3.1.200f)506
Mirabilis:
 peripheric (embryo)(II.2.3.3.02i)183
Mistletoe:
 epiphyte, parasite(II.4.2.026,.05f)240,253
 weeds(V.3.3.0.001)497
Mitragyna spp.:
 pneumorhizae(II.4.3.039)243
Moehringia laterifolia:
 diplostemonous, distinct(II.2.2.06c)135
Monkshood:
 galea(II.2.2.20f)149
 zygomorphic(II.2.2.20f)149
Monstera:
 clinging (aerial roots)(II.4.3.014)241
 fenestrate(II.5.6.040;.5.6.15c)274,295
 perforated (leaf)(II.1.2.16c)81
 (see also Foliage plants)
Moraceae:
 achene(II.2.3.1.002)158
 calyptra(II.1.2.021)60
 foliage plants(I.2.001d)5
 laticifer(II.3.098)198
 multiple fruit(II.2.3.1.023)159
 protection of resting buds(II.4.3.041)244
 reclinate (stamens)(II.1.1.2.02q)35
 (see also Fig, Ficus)
Morantaceae: (see Foliage plants)
Moringa: decompound leaves(II.1.2.035)60
Morus: (see Mulberry)
Mosses (some):
 apophysis(II.6.002)279
 epiphytes(II.4.2.006)239
 peat(V.1.2.011h)430
 poikilohydric plants(IV.5.2.040)403
 sporocarp(II.6.020)280

Mouriria: sclereids(II.3.24c)230
Mucuna spp.:
　hilar (seed-form)(II.2.3.3.036a-3) 165
　hilum(II.2.3.3.021)164
Muehlenbeckia platyclados: phylloclad (II.4.3.036)243
Mulberry (Morus spp.; see also Moraceae):
　dwarf disease(V.3.3.2.263)523
　(multilayer) pericycle(II.3.127)200
　multiple fruit(II.2.3.1.023)159
　syncarp(II.2.3.1.037)160
　temperate zone crops(I.1.016)5
Mullein: decurrent(II.1.2.07c)72
Murraya paniculata:
　alternate (leaflets)(II.1.2.01e)66
Musa, Musaceae:
　endodermis (M. sapientum)(II.3.11d) 217
　flagelliflory(IV.4.4.021)389
　penniparallel (penninerved)(II.1.2.121m)65
　(see also Banana)
　semelparous(II.5.3.019)268
　sympodial (axes)(II.1.1.3.099)27
Musaceae:
　tepal(II.2.2.061f)129
　tropical crops(I.1.017)5
　vascular wilts(*V.3.3.2.040)511
Muskmelon (Cucumis melo; Cucurbitaceae):
　alternaria rot(VI.9.1.015)627
　anthracnose(VI.9.1.028)628
　bacterial soft rot(VI.9.1.062)631
　blue mold rot(VI.9.1.110)634
　cladosporium rot(VI.9.1.152)637
　diplodia rot(VI.9.1.193)639
　fusarium rot(VI.9.1.247)642
　horticultural crops(I.0.001)3
　phytophthora rot(VI.9.1.383)648
　pink mold rot(VI.9.1.393)649
　(see also Melons)
　(see Cucurbits)(I.3.007)7
Mustard (Brassica juncea, B. japonica, B. campestris; Cruciferae):
　cole crops(I.3.005)7
　(see Crucifers)
Mustard seed (Brassica nigra, B. juncea, B. hirta; Cruciferae)
　condiments, etc.(I.3.006)7
Mustard, white:
　long-day plant(IV.3.2.013g)375
Myrciaria cauliflora (see Jaboticaba)
Myrmecodia spp.:
　myrmecomorphosis(II.4.4.002c)246
Myrmecophytes: protective devices(II.4.3.042)244
Myrtaceae:
　mechanical stress(IV.5.4.4.002)416
　mesotestal (seeds)(II.2.3.3.043c)167
　nursery plants(I.2.009)6
　pyxis(II.2.3.1.029)160
　reclinate (stamens)(II.1.1.2.02q)35
Myrtus spp.:
　hot zone plants(II.4.3.028)243
　leaf modifications(II.4.3.032)243
　sclerophyll(II.4.3.050)244
　xeromorphic structure(IV.5.2.085)406

N

Nandinaceae:
　endotegmic (seeds)(II.2.3.3.040a)166
Nannorrhops ritchiana:
　anisotony(II.1.1.3.02b)38
Narcissus (Narcissus pseudo-narcissus, N. jonquila, N. spp.)
　bulb flies(V.4.2.1.003)565
　bulb forcing(V.4.1.012)559
　bulbous plants(I.2.001b)5
　bull nose (bull-nose)(V.4.2.2.003) 567
　cut flowers(I.2.001c)5
　cyathiform (corona)[II.2.2.27b(1)] 156
　double nose(V.4.1.040)561
　gooseneck, gooseneck stage(V.4.1.057)561
　gummosis(V.4.2.1.011)565
　hypsophyll[II.2.2.27b(2)]156
　mosaic(V.4.2.1.022)566
　nematodes(V.4.2.1.014)566
　root-lesion nematodes(*V.3.3.1.200f) 506
　tunic(V.4.1.120)565
　tunicate bulb (separation)(V.1.4.071a-2)451
　virus diseases of bulbs(V.4.2.1.022) 566
Naseberry (see Sapodilla)
Natal plum: protective devices(II.4.3.042)244
Nectarine: (see Peach)
Nelumbo: uncinate (connective)(II.2.2.08f)137
Neonauclea angustifolia:
　stenophyllous leaves[II.4.08a(1)]256
Nephelium lappaceum: (see Rambutan)
Nephrophyllum: coiled (embryo)(II.2.3.02d)183
Nerium: sclerophyllous(II.4.08b)256
New Zealand spinach (Tetragonia tetragonioides, Aizoaceae)

INDEX OF CROPS

greens(I.3.016)8
Nicotiana spp.: androgenesis(IV.4.1.
  006b-1)379
Nigelia sp.:
  light requirements for germination
    (V.1.3.042)435
Night-blossoming jasmine:
  long-short-day plant(IV.3.2.013h)375
Noetta:
  foliage plants(I.2.001d)5
  saprophytes(II.4.2.029)240
Norfolk Island Pine:
  collection of (cutting) material(V.
    1.3.022a-1)442
  erect, excurrent(II.1.1.3.10b)46
  irreversibly plagiotroic(II.1.1.3.
    047)24
Nursery crops:
  nursery plants(I.2.009)6
  root-knot nematodes(*V.3.3.1.200e,
    *.200g)506
Nutmeg, mace (Myristica fragrans; Myristicaceae)(I.1.013)4
  air layerage(V.1.4.036a)449
  arillate seed(II.2.3.3.042c)166
  dioecious(IV.4.3.004)386
  endotestal (seeds)(II.2.3.3.043a)167
  spice crops(I.1.013)4
Nuts:
  horticultural maturity(VI.1.004)569
  mechanized harvesting(VI.2.022b)576
  nut crops(I.1.007)4
  ripening(VI.1.013)572
  (see also Almond, Cashew, Chestnut,
    Macadamia, Filbert, Pistachio,
    etc., I.1.007)4
  semi-mechanized harvesting(VI.2.
    022c)576
Nyctaginaceae:
  anomalous secondary growth(II.3.008)
    194
Nymphaeaceae:
  lamellate (ovary)(II.2.2.041o)123
Nypa:
  plagiotropic branching(II.1.1.3.04b)
    40
  rhizomatous branching(II.1.1.3.082)
    26
  secund(II.1.2.19b)84

O

Oak (Quercus spp.):
  acorn(II.2.3.1.04d)171
  cupule(II.2.3.1.013)159
  lenticel(II.3.101)198
  libriform fiber(II.3.23e)229
  nursery plants(I.2.009)6
  penniparallel(II.1.2.25m)90
  penniribbed(II.1.2.25n)90
  polyaxial stem(II.1.1.3.067)25
  revolute (Q. virginiana)(II.1.2.14b)
    79
  ring porous wood (Q. bicolor)(II.3.
    10a)216
  semi-ring porous wood (Q. virginiana)(II.3.10b)216
  tracheids(II.3.23c,d)229
  vessel elements (Q. alba)(II.3.22g,
    i)228
Oats:
  C-3 plants(IV.1.2.008a)352
  Ditylenchus spp.(*V.3.3.1.200a)505
  long-day plant(IV.3.2.013g)375
  loose smut(*V.3.3.2.133c,*.181f)515,
    519
  'Multiline E-69'(III.2.0.005a-3)314
  stubby-root nematodes(*V.3.3.1.200c)
    505
  thermoperiodism(IV.3.2.021)376
  trichome(II.3.31i)237
  (winter) long-day plant(IV.3.2.013g)
    375
Ocymum spp.:
  gullet (flower)(IV.4.4.004b-4)388
Oenothera: (see Evening primrose)
Oil palm (Elaeis guineensis):
  catena[II.4.03A(7)]251
  drupe(II.2.3.1.015)159
  endocarp(II.2.3.2.029c)162
  germ-pore(II.2.3.2.029f,.05c)162,179
  hand harvesting methods(VI.2.022a)
    576
  incubation(V.1.3.039)434
  mesocarp(II.2.3.2.029g)162
  monoaxial(II.1.1.3.056)24
  monoecious(IV.4.3.014c)387
  open-pollinated populations(III.2.1.
    001d)317
  plastochron(II.0.04)15
  pneumatathodes(II.4.10c)258
  pregerminated seeds(V.1.3.049)435
  processing(VI.10.000)674
  red ring(*V.3.3.1.200a)505
  roots(II.1.3.02c,d)97
  second-order dichogamy(IV.4.3.002d)
    386
  (see also Elaeis guineensis)
  "stranglee"(II.4.10d)258
  "strangler"(II.4.10d)258
Okra (Abelmoschus esculentus = Hibiscus esculentus):
  catena[II.4.03A(6)]251
  harvested immature(II.2.3.1.000)158

INDEX OF CROPS

inbred pure lines(III.2.1.001c)317
monadelphous(II.2.2.04d)133
scarifier(V.1.3.057)435
Oleander: (see Nerium)
Olives; Olea europaea; Olea spp., Oleaceae:
  Agricultural Marketing Agreement Act of 1937 as Amended(VI.7.2.004)609
  apical (anther attachment)(II.2.2.007w-1)120
  clones(III.2.1.001a)316
  dimorphic (flowers)(IV.4.3.003)386
  growth curve(IV.3.1.013)371
  hot zone plants(II.4.3.028)243
  leaf modifications(II.4.3.032)243
  physiological drought(IV.5.3.007)415
  polysomaty(III.2.2.142)324
  sclerophyll(II.4.3.050)244
  seed dormancy(V.1.3.060)436
  semihardwood (stem) cuttings(V.1.4.022f-3,02c-e,.03a,b)443,454,455
  truncheon(V.1.4.022g,.03g)443,455
  tubercle(*V.3.3.2.040)511
  xeromorphic structure(IV.5.2.085)406
Onion (Allium cepa (Common onion group) Amaryllidaceae):
  aspergillus bulb rot(VI.9.1.041)629
  bacterial soft rot(VI.9.1.063)631
  bag filling(VI.3.053a)581
  bags(VI.4.001)585
  black mold rot(VI.9.1.084)632
  blue mold rot(VI.9.1.109)634
  bulb crops(I.3.004)7
  diplodia stain(VI.9.1.195)639
  Ditylenchus spp.(*V.3.3.1.200a)505
  freezing injury(VI.9.3.049)667
  fusarium bulb rot(VI.9.1.234)641
  gray mold(VI.9.1.264)643
  hybrids(III.2.1.001b)317
  maggot(*V.3.3.1.133b)501
  neck rot(VI.9.1.264)643
  open-pollinated populations(III.2.1.011d)317
  pelletized seeds(V.1.3.045)435
  purple blotch(VI.9.1.426)650
  seed production(III.2.0.004c-f)312
  short-day plant(IV.3.2.013)375
  smudge(VI.9.1.492)654
  smut(*V.3.3.2.181f;VI.9.1.493)519,654
  sour skin(VI.9.1.511)656
  storage organ(II.4.13e)261
  sunscald(VI.9.3.134)673
  tonnage items(VI.8.121)625
  translucent scales(VI.9.3.142)673
  tunicate bulb(V.1.4.071a-2)451
  vernalization(IV.3.1.014j)372

  white rot(VI.9.1.565)659
Onagraceae: episepalous(II.2.2.007d-9)118
Ophiopogon: lorate(II.5.5.032)272
Opuntia spp.:
  phylloclad(II.1.1.3.06b;.4.3.036)42,243
  succulent (xerophyte)(II.4.05e)253
  xeromorphic structure(IV.5.2.085)406
Orange: (see Citrus - sweet orange
Orchardgrass:
  'Pennlate' (beyond Syn-1)(III.2.0.005a-5-b)314
  short-long-day plant(IV.3.2.013m)375
Orchids, Orchidaceae (numerous species):
  aerial roots (Cattleya)(II.1.3.01e) 96
  anisomerous(II.2.2.24c)153
  anther cap(II.2.2.007b)118
  apogamy(IV.4.1.006b-2)379
  calceolate (labellum)[II.2.2.24e(1)]153
  caudicle(II.2.2.007e,.25e)119,154
  column[II.2.2.017,.25c(1)]121,154
  cut flowers(I.2.001)5
  diandrous(II.2.2.027,.24e,f)122,153
  endotrophic (mycorrhiza)(V.3.3.2.154)516
  epiphytes(II.4.2.006)239
  gullet (flower; Ophyrideae)(IV.4.4.004b-4)388
  gynandrium(II.2.2.24f)153
  hercogamy(IV.4.3.006)386
  monandrous(II.2.2.050,.25a)126,154
  multiple epidermis(II.3.114)199
  nectary[II.2.2.24d(1)]153
  osmophor (Platanthera bifolia)(II.3.15c)221
  phytoalexin(IV.2.2.162)368
  pollinium(II.2.2.007t-9,.24f,.25e)120,153,154
  pots(V.1.2.003h)428
  pseudobulb(V.1.4.058,.05f)451,457
  pseudobulbs (many)(V.1.4.024)443
  resupinate(II.2.2.068,.20g)129,149
  root tuber(terrestrial)(II.4.3.049)244
  rostellum[II.2.2.25b(2),d(3)]154
  seed dormancy II(V.1.3.060)436
  spur[II.2.2.24d(1)]153
  tepal(II.2.2.061f)129
  terrestrial (adventitious) roots (Phaius tankervilliae)(II.1.3.01b) **96**
  tube(IV.4.4.004b-6)388
  velamen(II.3.215)205

INDEX OF CROPS

Oregano (Oreganum majorana; Labiatae): (I.3.006)7
Ornamentals, ornamental crops:
  active ingredient(VI.7.2.017a)612
  air layerage(V.1.4.15)467
  aphids(*V.3.3.1.132d)502
  approach graftage(V.1.4.06)458
  beetles(*V.3.3.1.132a)501
  bench-graft (some)(V.1.4.028b)443
  bending(V.3.4.12e)556
  bonzai(V.3.4.004)540
  bridge grafting(V.1.4.028c)444
  brush control(V.3.3.3.015)532
  budding(V.1.4.028f)444
  candelabra (free standing palmette) (V.3.4.11g)555
  carrier(VI.6.000)594
  chilling injury(VI.9.3.029)665
  contour planting(V.3.1.01h)489
  contre-espaliers(V.3.4.11c,d)555
  cordon(V.3.4.11b,c,e)555
  crotch repair(V.1.4.13a-d)465
  crown gall(*V.3.3.2.040)511
  crowns(V.1.4.04a)456
  cutting graft(V.1.4.028g)445
  cuttings(V.1.4.01a-c)453
    hardwood(V.1.4.02b)454
    herbaceous(V.1.4.02a)454
    leaf(V.1.4.01a)453
    leaf-bud(V.1.4.01b)453
    root(V.1.4.01c)453
    semihardwood(V.1.4.02c-e)454
  cyclamen mite(V.3.3.1.172)504
  damping-off(*V.3.3.2.179a,*.181f) 517,519
  drainage(V.3.5.002)557
  dwarfed trees and art forms(V.3.4.12)556
  Environment Protection Agency(VI.7.2.014)610
  espaliers(IV.5.4.4.002g;V.3.4.026) 416,541
  Federal Insecticide, Fungicide and Rodenticide Act of 1974; Nematocide Plant Regulator, Defoliant and Dosiccant Amendment of 1959; and Federal Environmental Pesticide Act of 1972(VI.7.2.017)612
  festooning(V.3.4.12e)556
  flat(VI.4.011)586
  flower bugs(*V.3.3.1.132c)502
  fusarium induced wilts(*V.3.3.2.133c)515
  garden web worm(*V.3.3.1.132g)502
  geese(V.3.3.3.048)534
  generative (see propagation)
  goblet (free standing palmette)(V.3.4.11f)555
  graftage aftercare(V.1.4.11a-h)463
  grafting(V.1.4.09)461
  grape training(V.3.4.10)554
  grasshopper, locusts, mole crickets (*V.3.3.1.133h)503
  head heights(V.3.4.03b-e)547
  heading back(V.3.4.02b)546
  hedge shears(V.3.4.041)542
  inarching(V.1.4.028m)447
  incompatible commodities(VI.6.009) 594
  incompatibility(V.1.4.035)449
  irrigation efficiency(V.3.5.011)557
    systems(V.3.5.012)557
  Japanese beetle(*V.3.3.1.132a)501
  latent heat (evaporative cooling) (VI.5.051c)593
  layerage:
    mound(V.1.4.14c-e)466
    simple(V.1.4.14a)466
    tip(V.1.4.14b)466
    trench(V.1.4.14g-i)466
  leader(V.3.4.046)542
  leader training(V.3.4.03)547
  living braces(V.1.4.028o,.13a-d)448, 465
  lopping shears(V.3.4.048)542
  mat watering(V.4.1.070)562
  materials and supplies(V.1.2.000)427
  mealybugs(*V.3.3.1.132d)502
  mechanical stress(IV.5.4.4.002)416
  micrograftage(V.1.4.11i-o)463
  micropropagation(V.1.5.000,.01)470, 471
  mineral nutrition(V.3.2.000)490
  mites(V.3.3.1.172)504
  nurse-root grafting(V.1.4.028r,.13e-h)448,465
  nursery facilities, etc.(V.1.1.000) 418
  nursery plants(I.2.009)6
  Occupational Safety and Health Act of 1970 and Occupational Safety and Health Administration (OSHA) (VI.7.2.037,.038)616
  offsets(V.1.4.04b,c)456
  palmette(V.3.4.055)543
  perennials as bedding plants(I.2.001a)5
  pillar tree(V.3.4.056)543
  pinching(V.3.4.057)543
  plant and crop size control(V.3.4.000)539
  planting system:
    double-row avenue(V.3.1.01i,j)489
    double set(V.3.1.01f)489

INDEX OF CROPS

    hedgerow(V.3.1.01e)448
    hexagonal(V.3.1.01c)588
    interset(V.3.1.01g)488
    quincunx(V.3.1.01k)489
    rectangular(V.3.1.01a)488
    square(V.3.1.01b)488
    triangular(V.3.1.01d)488
plant-regulator(VI.7.2.017s)615
pleach(V.3.4.058)543
pole saw(V.3.4.060)543
    shears(V.3.4.061)543
pollard(V.3.4.062)543
pollarding(V.3.4.12c)556
propagation(V.1.000)417
propagation media(V.1.2.011)429
propagation structures(V.1.1.048)424
pruning(V.3.4.000b)539
    equipment(V.3.4.063)543
    saw(V.3.4.065)543
    shears(V.3.4.066)543
    systems(V.3.4.067)543
    terminology(V.3.4.01)545
    types(V.3.4.02)546
psyllids(*V.3.3.1.132d)502
pyramid(V.3.4.068)544
rectangular (square) planting systems(V.3.1.008j)486
refrigeration, storage(VI.5.)588-594
rhizome(V.1.4.04d)456
rhizome (stolon)(V.1.4.05a)457
rhizome (pachymorph)(V.1.4.05b)457
riparian rights(V.3.5.022)558
rodents(V.3.3.1.224)507
root suckers(V.1.4.16a)468
root-knot nematodes(*V.3.3.1.200e, *g)506
runners(V.1.4.16b)468
runners (some)(V.1.4.064)451
scale insects(*V.3.3.1.132d)502
(see individual crops)
seep irrigation(V.3.5.023)558
separations(V.1.4.16c-f)468
site preparation(V.3.1.000)484
soil neutron probe(V.3.5.024)558
tensiometer(V.3.5.025)558
soils(V.2.000)472
spindle bush(V.3.4.04g)548
thinning(V.3.4.02c)546
thrips(*V.3.3.1.132i)503
topiary(IV.5.4.4.002g;V.3.4.075)416, 544
topworking(V.1.4.10)462
training(V.3.4.000b)539
truncheon(V.1.4.03e)455
tuber(V.1.4.05c)457
tuberous roots(VI.2.000)575
tubers(VI.2.000)575
    union healing (bud graft)(V.1.4.07) 459
    Verrier palmette(V.3.4.11d)555
    water:
        irrigation(V.3.5.003)557
        measurement(V.3.5.014)558
        quality(V.3.5.032)558
        regulation(V.3.5.)557-559
        rights(V.3.5.033)559
        seep(V.3.5.023)558
        soil neutron probe(V.3.5.024)558
        supply(V.3.5.034)559
        timing(V.3.5.013)557
    winged pyramid(V.3.4.12d)556
Osmunda regalis (royal fern):
    helophyte, hydrophyte(II.4.04c)252
O. claytoniana (interrupted fern):
    interrupted(II.1.2.13b)78
Oxalidaceae:
    dimorphic (flowers)(IV.4.3.003)386
    trimorphism(IV.4.3.019)387
Oxalis spp.:
    succulent(II.4.3.065)245

P

Pachira aquatica: digitate(II.1.2.08a)73
Pachysandra: ground covers(I.2.003)6
Painted trumpet (Clytostoma callistegioides):
    liane(II.0.02d)13
Palms, Palmae:
    acaulescent (Sabal minor)[II.1.1.3.01a(2)]37
    aerial (trunk prop) roots(II.1.3.002)94
    ants[II.1.2.14d(1)]79
    bole (cabbage palm, Sabal palmetto)(II.1.1.3.09b)45
    boat-shaped (cymbiform) spathe (Arecastrum romanzoffianum)(II.2.1.023,.07b)100,110
    caudex(II.1.1.3.021,.15b)23,41
    clinging(II.4.3.014)241
    columnar (Archontophoenix alexandrae var. Beatriceae)(II.5.08f)288
    combs of spines (Salacca leaf sheath)(II.4.09c)257
    continuous growth(II.1.1.3.029)23
    Beatrice palm (Archontophoenix alexandrae var. Beatriceae)(II.1.1.3.07b)43
    Chinese fan palm (Livistona chinensis)(II.1.1.3.07a)43
    costapalmate (Sabal)(II.1.2.029,.06a)60,71

crown, crown shaft (Archontophoenix
  alexandrae, A. cunninghamiana)
  [II.1.1.3.031,.07c(1),(2)]23,43
cymba (Butia capitata)(II.2.1.022,
  .07a)100,110
dichotomous (doum palm)(II.1.1.3.
  036,.09d)23,45
doum palm (Hyphaene Thebaica)(II.0.
  03e)14
family(III.1.1.008)303
fiber thorns (Trithrinax)(II.4.09a)
  257
flabelliform (Licuala grandis)(II.1.
  2.10d)75
geniculate[II.1.2.14d(2)]79
geotropic bending(V.4.1.056)561
germ pores(II.2.3.2.05b,c)179
growth in axis diameter(II.3.073)197
hammock(II.4.2.011)239
hapaxanthic(II.5.3.011)267
haustorium(II.4.3.024)242
hooks(II.4.3.027)242
hooks (deflexed) (Desmoncus)(II.4.
  09g)257
hot zone plants(II.4.3.028)243
incubation(V.1.3.039)434
induplicate (vernation)(II.1.1.2.
  03s)36
infrafoliar (Arecoid, Manila palm,
  Veitchia merrillii)(II.1.2.054,
  .13c)61,78
intrafoliar (paurotis palm, Paurotis
  wrightii)(II.1.2.057,.13d)61,78
lady palm (Raphis flabelliformis)
  (II.0.03c)14
monoaxial (Seaforth palm, Ptychosperma elegans(II.1.1.3.056,
  .14c)24,50
myrmecomorphosis(II.4.4.002c)246
neoteny(IV.3.1.014e)372
nursery plants(I.2.009)6
ocrea(II.1.2.066)62
oil palm(II.0.04;.4.10d)15,258
pericycle (multilayer)(II.3.127)200
plastochron(II.0.04)15
pneumorhizae(II.4.3.039)243
protective devices(II.4.3.042)244
Ptychosperma: extension growth(II.1.
  2.09)74
pulvinus(II.1.2.087,.18c)62,83
rattan(II.0.022,.03a;.4.027;IV.5.
  2.006)11,14,242,399
reclining, reclinate (Senegal date
  palm, Phoenix reclinata)(II.1.1.
  3.18a)54
reduplicate (vernation)(II.1.1.2.
  03t)36
rhizoflory[II.2.1.05d(4)]108
rhizomatous branching(II.1.1.3.082)
  26
royal palm (Roystonia regia)(II.0.
  03d)14
(see also individual crops, e.g.,
  coconut, date, oil palm, etc.)
seed dormancyII(V.1.3.060)436
semiterete (petiole)(II.1.2.19c)84
shade houses(V.1.1.059)426
spadix (Butia capitata)[II.2.1.066,
  .13b(1)]103,116
spathe (Butia capitata)[II.2.1.067,
  .13b(2)]103,116
spiny petiole (Corypha elata)(II.4.
  09b)257
twisted anthers(II.2.2.007w-13)121
whips (rattans)(II.4.09d-g)257
wine(VI.10.039)677
Palta: (see Avocado)
Pandanus spp.:
  megatherm taxa(III.1.2.033)309
  pneumorhizae(II.4.3.039)243
  spirodistichy(II.1.1.1.025)20
  stilt roots[II.1.3.01c(1),.4.11a(1)]
    96,259
  trees (P. utilis)(II.0.06d)17
Paniala: (see Flacourtia cataphracta)
Panicum sp.:
  sporogenesis(IV.4.1.006b-6)380
Pansy:
  'Seven-Eleven' (F)(III.2.005a-7)314
Papaya (Carica papaya; Caricaceae)
  anthracnose(VI.9.1.029)628
  fistular(II.1.1.3.10c)46
  inbred pure lines ('Solo')(III.2.1.
    001c)317
  paracarpous(II.2.3.2.03c)177
  sarcotesta(II.2.3.3.042i,.07a-c)167,
    188
  seed structure(II.2.3.3.01d)182
  semi-mechanized harvesting(VI.2.
    022c)576
  stem-end rot(VI.9.1.525)656
  trioecious(IV.4.3.020)387
  tropical crops(I.1.017)5
Papilionaceae (= Fabaceae)(IV.4.4.
  004b-3)388
Parasitic plants:
  haustorium(II.4.3.024)242
Parkia sp.:
  flagelliflory(IV.4.4.021)389
Parsley (Petroselium crispum; Umbelliferae)
  crispate(II.5.4.3.010)270
  downy mildew(VI.9.1.206)640
  pelletized seeds(V.1.3.045)435

INDEX OF CROPS

salad crops(I.3.020)8
septoria blight(VI.9.1.485)654
watery soft rot(VI.9.1.559)658
Parsnip (Pastinaca sativa; Umbelli-
    fereae)
  bacterial brown rot(VI.9.1.051)630
  bacterial soft rot(VI.9.1.059)630
  canker(VI.9.1.331)646
  root-knot nematodes(*V.3.3.1.200e,
    .200g;VI.9.2.070)506,662
  scab(VI.9.1.332)647
  tap root storage(II.4.3.070)245
  watery soft rot(VI.9.1.559)658
Passiflora spp. (Passifloraceae) =
    passion fruit
  androgynophore[II.2.2.041a,.27a(1)]
    122,156
  bowl-shaped(IV.4.4.004b-2)388
  clinging(II.4.3.014)241
  corona[II.2.2.022,.27a(2)]121,156
  ellipsoid (P. quadrangularis)(II.2.
    3.2.02g)176
  hypanthium[II.2.2.27a(4)]156
  hypsophyll[II.2.2.27a(5)]156
  protandry(IV.4.3.002b)385
  pruning systems(V.3.4.032)541
  purple (P. edulis)(III.1.1.011)304
  stipitate(II.2.2.041ac)125
  tendrils(II.4.3.071)245
  tropical crops(I.1.017)5
  yellow (P. edulis f. flavicarpa)
    (III.1.1.011)304
Paurotis wrightii: (see Palms)
Pavetta spp.:
  phytomorphosis(II.4.4.002e)246
Pea; Peas, Garden (Pisum sativum; Fab-
    aceae):
  anthracnose(VI.9.1.026)628
  bacterial blight(VI.9.1.049)629
  beetle(*V.3.3.1.132a)501
  black rot(*V.3.3.2.181e)519
  blight(*V.3.3.2.181e)519
  color separator(V.1.3.057)435
  compressed(II.2.2.019)121
  day-neutral plant(IV.3.2.013c)374
  diadelphous(II.2.2.04c)133
  downy mildew(VI.9.1.203)640
  dry heart rot(*V.3.3.2.181e)519
  gray mold rot(VI.9.1.270)643
  harvested immature(II.2.3.1.000)158
  leaf spot(*V.3.3.2.181e)519
  legume(II.2.3.1.02c)169
  Nacobbus spp. nematodes(*V.3.3.1.
    200g)506
  phytoalexin(IV.2.2.162)368
  pod spot(VI.9.1.405)650
  reniform nematodes(*V.3.3.1.200d)506

ring spot(*V.3.3.2.181e)519
root rot(*V.3.3.1.181e)519
root-knot nematodes(*V.3.3.1.200g)
  506
root-lesion nematodes(*V.3.3.1.200f)
  506
scab(VI.9.1.336)647
spotted wilt(VI.9.1.520)656
streak(VI.9.1.337)647
tendrils(II.4.3.071)245
thermoperiodism(IV.3.2.021)376
Peach (Prunus persica): nectarine (P.
    persica var. nectarina):
  alternaria rot(VI.9.1.012)627
  anthracnose(VI.9.1.030)628
  bacterial spot(VI.9.1.072)631
  bags(VI.4.001)585
  bark beetle(*V.3.3.1.133a)501
  baskets(VI.4.003)585
  bitter rot(VI.9.1.030,.079)628,
    632
  black mold rot(VI.9.1.083)632
  black spot(VI.9.1.470)653
  blue mold rot(VI.9.1.111)634
  brandy(VI.10.004)674
  chilling requirement(IV.5.3.006)407
  chip bud(V.1.4.028f-2)444
  cling, clingstone(II.2.3.2.029b)162
  clones(III.2.1.001a)316
  defuzzer(VI.3.022)578
  dieback(*V.3.3.2.181e)519
  diplodia rot(VI.9.1.194)639
  drupe(II.2.3.1.05e)172
  'Elberta' (clonal variety)(III.2.0.
    005a-1)313
  endocarp(II.2.3.2.029c)162
  epichlamydy(II.2.3.2.03b)177
  freckles(VI.9.1.470)653
  free(II.2.2.038)122
  free, free stone(II.2.3.2.029e)162
  horticultural maturity(VI.1.004)569
  hydrocooling(VI.5.035b)591
  leaf curl(*V.3.3.2.181e;VI.9.1.301)
    519,645
  monoploid apomixis(IV.4.1.006a-7)378
  mummy(V.3.3.2.258)523
  open center(V.3.4.029)541
  pericarp(II.2.3.2.05a)179
  petiolar glands(II.1.2.16d)81
  powdery mildew(VI.9.1.409)650
  root-lesion nematodes(*V.3.3.1.200f)
    506
  rust(VI.9.1.469)653
  scab(VI.9.1.470)653
  sour rot(VI.9.1.508)655
  stone fruits(I.1.014)4
  temperate zone crops(I.1.016)5

INDEX OF CROPS

    training(V.3.4.029)541
    tree fruit(VI.8.124)625
    yellows(V.3.3.2.474)530
Peanut (Arachis hypogaea):
    C-3 plants(IV.1.2.008a)352
    dagger nematodes(*V.3.3.1.200c)505
    geocarpic, geocarpous(II.2.3.2.013, .02h)161,176
    gypsum(V.3.2.044)493
    hulls(V.3.3.3.009)531
    nut(II.2.3.1.024)159
    oil crops(I.1.008)4
    root-lesion nematodes(*V.3.3.1.200f) 506
    sting nematodes(*V.3.3.1.200c)505
Pear (Pyrus domestica, Pyrus pyrifolia var. culta, P. spp.):
    'Bartlett' on quince(V.1.4.028h)445
    bench-graftage(V.1.4.028b)443
    bitter rot(VI.9.1.079)632
    black end(VI.9.3.009)664
    black rot(VI.9.1.086)632
    black spot(VI.9.1.100)633
    blue mold rot(VI.9.1.106)634
    bourse(V.3.4.006)540
    bourse bud(V.3.4.007)540
    brown core(VI.9.3.099)670
    bruises(VI.9.3.026)665
    bulk filling(VI.3.053b)582
    canker(*V.3.3.2.181f)519
    chilling requirement(IV.5.3.006)407
    clones(III.2.1.001a)316
    cluster base(V.3.4.006)540
    codling moth injury(VI.9.2.018)660
    cordon(V.3.4.021)541
    core breakdown(VI.9.3.033)665
    cork spot of 'Anjou'(VI.9.3.036)666
    corymbiform(II.2.1.12b)115
    double-working(V.1.4.028h)445
    grafting(V.1.4.028k)446
    gray mold rot(VI.9.1.265)643
    "grit cells"(II.3.25g)231
    growth curve(IV.3.1.013)371
    horticultural maturity(VI.1.004)569
    leaf blister(*V.3.3.2.181e)519
    leaf blister mite injury(VI.9.2.057) 661
    leaf internal structure(II.3.21a)227
    lenticels(II.3.101)198
    localized incompatibility(V.1.4.035)449
    mixed buds(II.1.1.1.013)19
    packing(VI.3.047d)581
    palmette(V.3.4.055)543
    physiological drought(IV.5.4.3.007) 415
    phytophthora rot(VI.9.1.380)648
    piece or whole root grafting (whip, or tongue)(V.1.4.028b,k-9)443,447
    pink mold rot(VI.9.1.392)649
    pirostele[II.2.1.055,.12b(1)]102,115
    pithy brown core(VI.9.3.099)670
    pome(II.2.3.1.028)160
    pome fruits(I.1.009)4
    powdery mildew(VI.9.1.408)650
    psyllids(*V.3.3.1.132d)502
    pyriform ('Bartlett')(II.2.3.2.06c) 180
    rhizopus rot(VI.9.1.449)651
    ripening(VI.1.013)572
    rosaceous (corolla)(II.2.2.21k)150
    San José scale(VI.9.2.073)662
    scab(VI.9.1.334)647
    senescent scald(VI.9.3.112)671
    sooty blotch(VI.9.1.503)655
    stony pit(VI.9.1.335)647
    superficial scald of 'Anjou'(VI.9.3.138)673
    temperate zone crops(I.1.016)5
    tree fruit(VI.8.124)625
    trench layerage(V.1.4.036b-5)450
Pearl millet:
    'Gahi'(III.2.005a-5-a)314
Pecan (Carya illinoensis; Juglandaceae)
    Agricultural Marketing Agreement Act of 1937, as Amended(VI.7.2.004)609
    geese (grazing)(V.3.3.3.048)534
    growth curve(IV.3.1.013)371
    I budding(V.1.4.028f-5)445
    nut(II.2.3.1.04f)171
    nut crops(I.1.007)4
    patch budding(V.1.4.028f-6)445
    ring budding(V.1.4.028f-7)445
    temperate zone crops(I.1.016)5
Pedicularis spp.:
    microtherm taxa(III.1.2.034)309
Pedilanthus sp.:
    flag(IV.4.4.004b-3)388
Pelargonium spp.: (see Geranium)
Pepper (Capsicum annuum; Solanaceae)
    alternaria rot(VI.9.1.017)627
    anthracnose(VI.9.1.031)628
    bacterial soft rot(VI.9.1.067)631
    bacterial spot(VI.9.1.073)631
    blossom-end rot(VI.9.3.013)664
    charcoal rot(V.9.1.145)637
    cladosporium rot(VI.9.1.154)637
    common names(III.1.1.005)303
    condiments(I.3.006)7
    fermentation(V.1.3.026)433
    gray mold rot(VI.9.1.277)644
    inbred pure lines(III.2.1.001c)317
    pelletized seeds(V.1.3.045)435

INDEX OF CROPS

phytophthora rot(VI.9.1.384)649
pot plants(I.2.001f)6
rhizopus rot(VI.9.1.452)652
ripe rot(VI.9.1.467)653
seed extraction (fleshy fruits(V.1.3.064)437
seed production(III.2.0.004c-f)312
seed sizing(V.1.3.072)438
solanaceous fruits(I.3.021)8
spotted wilt(VI.9.1.522)656
stomata(II.3.30g)236
sunscald(VI.9.3.137)673
vascular wilt(*V.3.3.2.040)511
warm-season seeds(V.1.3.087)439
Pepper, Black (Piper nigrum; Piperaceae)
  aerial roots(II.4.3.004)241
  burrowing nematodes(*V.3.3.1.200f) 506
  clinging(II.4.3.014)241
  clones(III.2.1.001a)316
  direct set cuttings(V.1.4.022b)442
  geitonogamy(IV.4.4.022)389
  harvesting types(VI.2.023)577
  hybrids(III.2.1.001b)317
  irreversibly plagiotropic(II.1.1.3.047)24
  penniform (venation)(II.1.2.241)89
  (see also Piperaceae)
  spike(II.2.1.068)103
  spice crops(I.1.013)4
  tropical crops(I.1.017)5
  unitegmic (tegmic) seed(II.2.3.3.12e)193
Pennyroyal (Mentha pulegium; Labiatae):
  condiments, etc.(I.3.006)7
Peony Paeonia lactiflora etc.; Paeoniaceae):
  crown(II.4.3.015)242
  cut flowers(I.2.001c)5
  follicle(II.2.3.1.017)159
  mesotestal (seeds)(II.2.3.3.043c)167
  seed dormancy III(V.1.3.060)436
  shrub(II.0.06b)17
  small (embryo)(II.2.3.3.021)183
  suffrutescent(II.1.1.3.21b)57
Persea: (see Avocado)
Persian violet (Exacum affine, Gentianaceae)
  pot plants(I.2.001f)6
Persian (English) walnut (Juglans regia):
  see Walnuts
Persimmons: (see Diospyros)
Petunia:
  bedding plants(I.2.001a)5
  'Comanche' hybrid (F)(III.2.0.005a-6)314
  (quantitative) long-day(IV.3.2.013g) 375
  'Violet Blue' (F) variety(III.2.0.005a-7)314
Phacelia sp.:
  light requirement for germination (V.1.3.042)435
Phoenix spp.:
  pneumatorhizae(II.4.3.039)243
  see also Palms
Philodendron: (see Foliage plants)
Phlox spp.:
  bedding plants(I.2.001a)5
  crowns (divisions)(V.1.4.024)443
  hypercrateriform(II.2.2.21f)150
  light requirement for germination(V.1.3.042)435
Phyrma:
  trichome(II.3.31e)237
Phyllanthus spp.:
  phyllomorphic (branches)(II.1.1.3.065)25
  spuriously compound(II.1.2.105)63
Phyllocactus spp.:
  phylloclad(II.4.3.036)243
  xeromorphic structure(IV.5.2.085)406
Phylloclinum paradoxum:
  epifoliar(II.2.2.03e)132
Physalis, P. floridana:
  vase-shaped (corolla)(II.2.2.21c)150
  potato virus Y(V.3.3.2.158)516
Physorrhynchus brabhuicus:
  conical(II.2.3.2.02d)176
Phyteuma sp.:
  Head or brush-shaped (inflorescence) (IV.4.4.004b-5)388
Pigweed:
  C-4 plants(IV.1.2.007b)352
Pimento = Allspice(Pimenta dioica):
  (common name for allspice, and certain capsicum peppers and black pepper)(III.1.1.005)303
  spice crops(I.1.013)4
Pimienta, Pimiento:
  common names for certain capsicum peppers(III.1.1.005)303
Pinanga kuhlii:
  main vein(II.3.21b)227
Pine, Pinaceae:
  see Conifers
Pineapple (Ananas comosus; Bromeliaceae):
  avenue (planting systems)(V.3.1.008a)485
  basal slips[V.1.4.04c(3)]456

INDEX OF CROPS

   black rot(VI.9.1.097)633
   brown rot(VI.9.1.127)635
   burrowing nematodes(*V.3.3.1.200f)
     506
   canaliculate(II.1.2.05a)70
   clones(III.2.1.001a)316
   crop logging(V.3.2.017)491
   crown[V.1.4.04c(1)]456
   crown slips]V.1.4.04c(2)]456
   epistasis(III.2.2.055)320
   flower(II.2.2.24a)153
   frozen concentrate(VI.10.018)676
   growth curve(IV.3.1.013)371
   hand harvesting methods(VI.2.022a)
     576
   hapa[V.1.4.04c(5)]456
   heart rot(*V.3.3.2.179a,*.181e)517,
     519
   hedgerow (planting system)(V.3.1.
     008a)485
   internal browning(VI.9.3.076)669
   maturity (non-climacteric)(VI.1.
     007)570
   "mule train"(VI.2.031)577
   multiple fruit(II.2.3.1.023)159
   nutritional spray(V.3.2.067)494
   offsets(V.1.4.024,.04c)443,456
   packinghouse(VI.3.000)577
   Paratylenchus spp.(*V.3.3.1.200b)505
   Pratylenchus spp.(*V.3.3.1.200f)506
   ratoon[V.1.4.04c(7)]456
   recurved (spines)(II.1.2.05a;.4.3.
     045)70,244
   root-lesion nematodes(V.3.3.1.200f)
     506
   rosette(II.1.2.093,.19a)63,84
   semi-mechanized harvesting(VI.2.
     022c)576
   slips[V.1.4.04c(4)]456
   soft rot(VI.9.1.097)633
   spiny tip(III.2.2.171)325
   sucker[V.1.4.04c(6)]456
   sunburn(VI.9.3.129)672
   syncarp(II.2.3.1.037,.07b,c)160,174
   total (titratable) acid(VI.1.008g)
     571
   tropical crops(I.1.017)5
   water blister(VI.9.1.097)633
   water rot(VI.9.1.097)633
Pinguicula:
   trichome(II.3.31g)237
Pink (Caryophyllaceae):
   unguiculate (corolla)(II.2.2.061c-
     14-w)128
Pinus: (see Confiers)
Piperaceae:
   anomalous secondary growth (Piper
     fluminense)(II.3.03g)209
   endotegmic (seeds)(II.2.3.3.040a)166
   foliage plants(I.2.001)5
   orthotropous (seed-form)(II.2.3.3.
     036b)166
   tropical crops(I.1.017)5
   umbellate-spicate (inflorescence)
     (II.2.1.078)103
Piper nigrum: (see Pepper, Black)
Piptadeniastrum:
   buttress(II.4.10a)258
Pistachio (Pistacia vera; Anacardiaceae):
   clones(III.2.1.001a)316
   growth curve(IV.3.1.013)371
   nut crops(I.1.007)4
   spuriously imparipinnate (Chinese
     pistachio)(II.1.2.11c)76
   subopposite (Chinese pistachio)(II.
     1.2.20e)85
   temperate zone crops(I.1.016)5
Pisum: (see Pea)
Pitcher plant: (see Sarracenia)
Pitomba: (see Eugenia lushnathiana)
Plantago: parastichy(II.1.1.1.016)19
Plantain: (see Banana)
Plants, Tropical:
   termites(*V.3.3.1.132f)502
Plants, Woody:
   measuring worms(*V.3.3.1.132g)502
Platanthera: (see Orchids)
Platanus:
   candelabra(II.5.14d)294
   naked (bud)(II.5.17a)297
   (see also Sycamore)
   trichome(II.3.31h;.5.17g)237,297
Plumbaginaceae:
   dimorphic (flowers)(IV.4.3.003)386
   pyxis(II.2.3.1.029)160
Plums; prunes (Prunus domestica, P.
   salicina, etc.):
   alternaria rot(VI.9.1.012)627
   bags(VI.4.001)585
   black spot(VI.9.1.470)653
   bloom(II.5.6.010)274
   blue mold rot(VI.9.1.111)634
   (canning or drying prunes) horticul-
     tural maturity(VI.1.004)569
   (fresh prunes) horticultural matur-
     ity(VI.1.004)569
   freckles(VI.9.1.470)653
   impulse item(VI.8.047)620
   mechanized harvesting(VI.2.022b)576
   open center(V.3.4.029)541
   rust(VI.9.1.469)653
   scab(VI.9.1.470)653
   (see also Stone fruits)

INDEX OF CROPS

  tree fruit(VI.8.124)625
Poa sp.:
  induced (parthenogenesis)(IV.4.1.
    006b-3-a)379
  sporogenesis(IV.4.1.006b-6)380
Poaceae: (see Gramineae)
Podophyllum peltatum:
  peltate(II.1.2.15d)80
  peltiribbed(II.1.2.24k)89
Poinsettia (Euphorbia pulcherrima;
    Euphorbiaceae):
  artificial long days(V.4.1.002)559
  black cloth shading(V.4.1.006)559
  cyathium(V.4.1.033)560
  hanging basket(VI.4.013)587
  peat, fiber blocks(V.1.2.003e)428
  physiological sprays(V.3.4.000a)539
  pot plants(I.2.001f)6
  short-day plant(IV.3.2.0131)375
Polygonaceae:
  dimorphic (flowers)(IV.4.3.003)386
  ocrea(II.1.2.066,.14c)62,79
  orthotropous (seed-form)(II.2.3.3.
    036b)166
Pome fruits (Rosaceae):
  brown rot(*V.3.3.2.181e,*g;VI.9.1.
    128)519,520,636
  cordon(V.3.4.021)541
  fire blight(*V.3.3.2.040)511
  flyspeck(VI.9.1.224)641
  freezing injury(VI.9.3.050)667
  Lorette system(V.3.4.049)542
  palmette(V.3.4.055)543
  pome fruits (Rosaceae)(I.1.009)4
  pyramid(V.3.4.068)544
  (see also Apple, Pear, Quince, etc.)
  shield bud(V.1.4.028f,-8,.08a,b)444,
    445,460
  sour rot(VI.9.1.509)655
Pomegranate (Punica granatum; Punica-
    ceae)
  axile-parietal (placentation)[II.2.
    2.041t-2-b,14d(3)]123,143
  blue mold rot(VI.9.1.112)634
  fasiculate leaves(II.1.2.046)61
  fruit(II.2.2.14d)143
  fruit rot(VI.9.1.286)644
  gray mold rot(VI.9.1.273)643
  heart rot(VI.9.1.286)644
  internal breakdown(VI.9.3.070)668
  ovary(II.2.2.14d)143
  splitting and cracking(VI.9.3.120)
    672
  sunscald (sunburn)(VI.9.3.135)673
  temperate crops(I.1.016)5
Pomoideae: (see Pome fruits)
Poncirus trifoliata (trifoliate or-

    ange):
  (see also Citrus)
  spheroid, globoid(II.2.3.2.07c)181
Pontederiaceae:
  trimorphism(IV.4.3.022)387
Poplar (Populus):
  blister(*V.3.3.2.181e)519
  deltoid(II.5.09e)289
  fastigiate (lombardy poplar)(II.5.
    09i)289
  leaf curl(*V.3.3.2.181e)519
  periderm (P. deltoides)(II.3.20a)226
  nursery plants(I.2.009)6
  witches' broom(*V.3.3.2.181e)519
Poppy Papaver spp., Papaveraceae):
  poricida(II.2.3.1.03a)170
  (seed) condiments, etc.(I.3.006)7
Potalia amera:
  umbonate(II.5.20d)300
Potato (Solanum tuberosum; Solanace-
    ae):
  annular (embryo)(II.2.3.3.02b)183
  bacterial ring rot(VI.9.1.055)630
  bacterial soft rot(VI.9.1.064)631
  bacterial wilt(*V.3.3.2.040)511
  bag filling(VI.3.053a)581
  bags(VI.4.001)585
  beet cyst nematodes(*V.3.3.1.200g)
    506
  black scurf(VI.9.1.098)633
  blackleg(*V.3.3.2.040;VI.9.1.081)
    511,632
  blight(*V.3.3.2.181g)520
  bulk shipment(VI.6.003)594
  certification program(V.1.4.010)441
  clones(III.2.1.001a)316
  corky ringspot(V.3.3.1.200c)505
  delivered(VI.7.1.016)599
  early blight(*V.3.3.2.181g)520
  fork(V.1.1.045)424
  fusarium wilt(*V.3.3.2.181g)520
  grade(VI.8.038)519
  hand harvesting(VI.2.022a)576
  late blight(*V.3.3.2.133c,*.181d;
    VI.9.1.298)515,518,645
  latent virus(VI.9.1.201)640
  leaf roll(VI.9.302)645
  leak(*V.3.3.2.181c)518
  leucoplast(IV.1.1.012)351
  mechanized harvesting(VI.2.022b)576
  pressure bruises(VI.9.3.101)670
  psyllids(*V.3.3.1.132d)502
  ring rot(*V.3.3.2.040)511
  root-lesion nematodes(*V.3.3.1.200f)
    506
  rotate (wheel-shaped)(II.2.2.21b)150
  'Russet Burbank' (clonal variety)

INDEX OF CROPS

(III.2.0.005a-1)313
scab(V.3.3.2.001;VI.9.1.406)509,650
"seed," seedpiece(V.1.4.068,.069)451
seedpiece decay(VI.9.1.407)650
source wrapping(VI.3.071)583
sprouting(VI.5.046)591
staple items(VI.8.112)625
starchy crops(I.3.023)8
stem and tuber rot(V.9.1.098)633
stem tuber(II.4.3.057)244
stolon(II.4.3.061)245
storage containers(VI.4.000d)585
tonnage items(VI.8.121)625
tubers(V.1.4.024,.05c)443,457
tuber (storage organs)(II.4.12a)260
vascular wilt(*V.3.3.2.040)511
verticillium wilt(VI.9.1.553)658
virus Y(V.3.3.2.158)516
wilt(*V.3.3.2.181g)520

Potentilla sp.:
induced (parthenogenesis)(IV.4.1.006b-3-a)379
pseudogamy(IV.4.1.006b-4)379

Pot plants:
blind shoots(V.4.1.008)559
break(V.4.1.009)559
cyclic lighting(V.4.1.034)560
hardening(V.4.1.062)562
heat delay(V.4.1.063)562
light flux(V.4.1.066)562
long-day treatment (LDT)(V.4.1.067)562
mat watering(V.4.1.070)562
multibranched plant(V.4.1.074)562
packing operations(VI.3.053)581
packinghouse layout(VI.3.052)581
panning(V.4.1.080)563
perched water table(V.4.1.082)563
pinch(V.4.1.084)563
quantitative long-day plant(V.4.1.095)563
(see individual crops, e.g., Azalea, Chrysanthemum, Gesneriads, Poinsettia, etc.)
self-branching(V.4.1.105)564

Primroses, Primula spp., Primulaceae:
bedding plants(I.2.001)5
dimorphic (flowers)(IV.4.3.003)386
microtherm taxa(III.1.2.034)309
pot plants(I.2.001f)6
pyxis(II.2.3.1.029)160

Proboscidea altheifolia (unicorn plant):
proboscoid, rostrate(II.2.2.28d)157
tardily dehiscent(II.2.3.1.03b)170

Processed fruit, vegetables:
(see Fruits, Vegetables)

Prosopis pubescens, P. strombulifera:
strombus-shaped[II.2.2.03h(1),(3)]132

Proteaceae:
head- or brush- shaped (inflorescence)(IV.4.4.004b-5)388
orthotropic (seed-form)(II.2.3.3.036b)166

Prunus spp.:
endocarp removal(V.1.4.024)433
false (gametophtic polyembryony)(IV.4.6.001c)392
lenticels(II.3.101,.12b)198,218
protective devices(II.4.3.042)244
secondary growth of vascular system (II.3.173)202
see Peach & Nectarine, Apricot, Almond, Plums, Cherry
stone fruits(I.1.014)4
transverse buds(II.1.1.1.032)20

Pseudotsuga:
root apical meristem(II.3.19b)225

Psidium: (see Guava)

Psilotas (Psilotinae):
pteridophytes(II.6.016)279

Psychotria spp., P. acuminata:
phytomorphosis(II.4.4.002e)246
stenophyllous leaves(II.4.08a)256

Pteridophytes, Pteridophyta:
archegonium(II.6.003)279
C-3 plants(IV.1.2.008a)352
nitrogen fixation(IV.1.2.029)354
spermatozoid(II.6.019)280

Ptychosperma: (see Palms)

Pummelo:
(see Citrus - Pummelo)

Pumpkin (Cucurbita pepo. C. moschata, C. mixta, C. maxima; Cucurbitaceae):
bacterial soft rot(VI.9.1.062)631
large (embryo)(II.2.3.3.02j)183

Punica: (see Pomegranate)

Purple passion fruit:
(see Passiflora spp.)

Purslane, Common:
C-4 plants(IV.1.2.008b)352
half-inferior(II.2.2.19c)148
lysicarpous(II.2.3.2.04c)178
perigynous(II.2.2.19c)148
pyxis(II.2.3.1.03d)170

Pyrola:
terminal pore(II.2.2.10d)139

Pyrus spp.: (see Pear)

Q

Quercus: (see Oak)

Quillworts:
  pteridophytes(II.6.016)279
Quince (<u>Cydonia oblonga</u>:
  'Barlett' on(V.1.4.028h)445
  bitter rot(VI.9.1.079)632
  black rot(VI.9.1.086)632
  black spot(VI.9.1.100)633
  blue mold rot(VI.9.1.106)634
  burr knot(II.1.1.3.020)22
  double-working(V.1.4.028h)445
  "grit cells"(II.3.25h)231
  hardwood (stem) cuttings(V.1.4.022f-1,02b)442,454
  localized incompatibility(V.1.4.035)449
  oriental fruit moth injury(VI.9.2.053)661
  pome(II.2.3.1.028)160
  ripening(VI.1.013)572
  (see also Pome fruits)
  simple leaf(II.1.2.03c)68

R

Radish (<u>Raphanus</u> sativa; <u>Cruciferae</u>) (I.3.019)8
  annual(II.5.01a)281
  anthracnose(VI.9.1.034)628
  combine(V.1.3.010)432
  fusiform, fleshy root(II.1.3.04b)99
  hybrids(III.2.1.001b)317
  kind(III.2.0.004a)312
  long-day plant(IV.3.2.013g)375
  mechanized harvesting(VI.2.022b)576
  rhizoctonia head and root rots(VI.9.1.441)651
  root crops(I.3.019)8
  seed sizing(V.1.3.072)438
  silique(II.2.3.1.03f)170
  source wrapping(VI.3.071)583
  stem tubers(II.4.3.057)244
  tap root plus hypocotyl storage(II.4.3.069)245
  therophyte(II.4.02b)250
  white rust(VI.9.1.566)659
Rambutan (<u>Nephelium lappaceum</u>):
  leaflets[II.1.2.17b(1)]82
  pinnate(II.1.2.17b)82
  rachis[II.1.2.17b(2)]82
  sarcotestal seed(II.2.3.3.07d,.09e)188,190
Ramie: bast fibers(II.3.063)197
Ranunculus (<u>Ranunculus</u> <u>asiaticus</u>, R. <u>aconitifolius</u>; <u>Ranunculaceae</u>):
  achene(II.2.3.1.002)158
  apocarpous (gynoecium)(II.2.2.041b)122

  bedding plants(I.2.001a)5
  cut flowers(I.2.001c)5
  endodermis(II.3.11a)217
  exotestal (seeds)(II.2.3.3.043b)167
  induced (parthenogenesis)(IV.4.1.006b-3-a)379
  stem tubers(II.4.3.057)244
<u>Raphis flabelliformis</u>: (see palms)
Raspberry: (see Brambles)
Rattans:
  cirrus(II.4.3.012)241
  clinging(II.4.3.014)241
  flagellum(II.4.3.022)242
  hooks, etc.(II.4.3.027)242
  (see also Palms)
<u>Ravenala madagascariensis</u> (traveller's tree):
  distichous[II.1.1.3.10a(2)]46
Red mangrove: (see Mangrove)
Rhamnaceae:
  exotestal (seeds)(II.2.3.3.043b)167
Rhododendrons:
  cutting - graft(V.1.4.028g)445
  nursery plants(I.2.009)6
  revolute (vernation)(II.1.1.2.03v)36
  (see also Azalea)
  sulfur(V.3.2.089)496
Rhoeo: stoma(II.3.190)204
Rhubarb (<u>Rheum</u> spp.; <u>Polygonaceae</u>):
  anthracnose(V.3.3.2.019;VI.9.1.460)510,652
  foot and crown rots(VI.9.1.386)649
  gray mold rot(VI.9.1.274)644
  leaf spot(VI.9.1.461)652
  miscellaneous crops(I.3.011)7
  phytophthora rots(VI.9.1.386)649
  (see also Polygonaceae)
  stem spot(VI.9.1.461)652
<u>Ribes</u> spp.:
  black rot(*V.3.3.2.181e)519
  blight(*V.3.3.2.181e)519
  dry heart rot(*V.3.3.2.181e)519
  leaf spot(*V.3.3.2.181e)519
  ring spot(*V.3.3.2.181e)519
  (see also Currant, Goosberries, Grossulariaceae)
  small fruits(I.1.012)4
Rice (<u>Oryza</u> spp.):
  C-3 plants(IV.1.2.008a)352
  caryopsis(II.2.3.1.01n)168
  cyst nematodes(*V.3.3.1.200e)506
  day-neutral plant(IV.3.2.013c)374
  <u>Ditylenchus</u> spp.(*V.3.3.1.200a)505
  excentric (embryo)(II.2.3.3.02g)183
  physiological drought(IV.5.4.3.007)415
  spiral nematodes(*V.3.3.1.200d)506

INDEX OF CROPS

stunt nematodes(*V.3.3.1.200d)506
weevil(VI.9.2.069)662
Ricinus: type of thickening(II.3.06e) 212
Robinia; R. pseudacacia:
   cambium initials(II.3.07b)213
   papilionaceous(II.2.2.23h)152
   phloem fiber(II.3.24g)230
Rose (Rosa spp.; Rosaceae):
   bedding plants(I.2.001)5
   black spot(*V.3.3.2.181e)519
   cane blight(*V.3.3.2.181e)519
   chafer(*V.3.3.1.132a)501
   cut flowers(I.2.001c)5
   cutting-graft(V.1.4.028g)445
   dagger namatodes(*V.3.3.1.200c)505
   dish-shaped(IV.4.4.004b-2)388
   downy mildew(*V.3.3.2.181d)518
   eye(V.4.1.044)561
   green pruning(V.4.1.060)562
   hardwood (stem) cuttings(V.1.4.022a-2,.02b)442,454
   hooks, etc.(II.4.3.027)242
   hypanthium(II.2.2.044)125
   involucre (epicalyx)(II.2.1.042)102
   mesotestal (seeds)(II.2.3.3.043c)167
   pachychalazal (seed-form)(II.2.3.3.036a-5)165
   parastichy 2/5(II.1.1.1.016)19
   'Peace' (clonal variety)(III.2.0.005a-1)313
   pollen grain (R. rugosa)[II.2.2.13d(5)]142
   pome fruits(I.1.009)4
   protective devices(II.4.3.042)244
   shield bud(V.1.4.028f,f-8)444,445
   small fruits(I.1.012)4
   spurs(II.1.1.3.093)26
   stone fruits(I.1.014)4
Rubber (Hevea brasiliensis; Euphorbiaceae)(I.1.011)4
   avenue (planting system)(V.3.1.008a) 485
   capsule(II.2.3.1.02h)169
   clones(III.2.1.001a)316
   deciduous(II.5.02b)282
   deliquescent(II.1.1.3.09c)45
   digitate(II.1.2.17a-2)82
   double-working(V.1.4.028h)445
   hedgerow(V.3.1.008a)485
   isoprenoids(IV.2.2.107)366
   laticifer(II.3.16a)222
   modular, modular construction(II.1.1.3.14b)50
   monoecious(IV.4.3.014)387
   perennial(II.5.02b)282
   petiole[II.1.2.17a(1)]82
   petiolule[II.1.2.17a(2)]82
   pleonanthic(II.5.02b)282
   polyaxial stem(II.1.1.3.067)25
   polycarpic(II.5.02b)282
   polymer(IV.0.227)342
   processing(VI.10.000)674
   septicidal[II.2.3.1.02h(1)]169
   South American leaf blight(*V.3.3.2.181e)519
   staminal column(II.2.2.06f)135
   'Tjir 1'(II.0.07b)18
   trifoliolate[II.1.2.17a(2)]82
Rubiaceae:
   dimorphic (flowers)(IV.4.3.003)386
   epipetalous (stamens)(II.2.2.007d-8)118
   epiphytes(II.4.2.006)239
   phytomorphosis (some)(II.4.4.002e) 246
   (see Coffee, Gardenia)
   superposed buds(II.1.1.1.030)20
Rosemary (Rosemarinus officinalis; Labiatae)
   condiments, etc.(I.3.006)7
   revolute (vernation)(II.1.1.2.03v)36
Rourea minor:
   preaphe seed form(II.2.3.3.05a,b)186
Rubber, India:
   foliage plants(I.2.001d)5
   (see Ficus spp.)
Rubus (see Brambles)
Rudbeckia sp.:
   induced (parthenogenesis)(IV.4.1.006b-3-a)379
Rue (Ruta graveolens; Rutaceae):
   condiments, etc.(I.3.006)7
Ruscus sp.:
   cladophyll[II.1.1.3.06a(2);.4.3.013] 42,241
   xeromorphic structure(IV.5.2.085)406
Rutabaga (Brassica napus (Napobrassica group); Cruciferae):
   alternaria root rot and leaf spot (VI.9.1.010)627
   gray mold rot(VI.9.1.267)643
   rhizoctonia head and root rots(VI.9.1.441)651
   root crops(I.3.019)8
   sclerotinia rot(VI.9.1.557)658
   tap root plus hypocotyl plus stem storage(II.4.3.068)245
   watery soft rot(VI.9.1.557)658
Rutaceae:
   citrus(I.1.003)4
   family(III.1.1.002,.021)302,305
Rye:
   caryopsis(II.2.3.1.01m)168

INDEX OF CROPS

Ditylenchus spp.(*V.3.3.1.200a)505
'Elbon'(III.2.0.005a-4)314
ergot(*V.3.3.2.181e)519
'Vitagraze'(III.2.0.005a-5-a)314
Rye, Winter:
  (quantitative) long-day plant(IV.3.2.013g)375
Ryegrass, Italian:
  C-3 plants(IV.1.2.008a)352
Ryegrasses:
  turf crops(I.2.011)6

S

Saffron:
  oil crops(I.1.008)4
Saffron (Crocus sativus):
  condiments, etc.(I.3.006)7
Sage (Salvia officinalis; Labiatae):
  condiments, etc.(I.3.006)7
  expanded connective[II.2.2.09e(1)(i),(2)]138
  versatile (anthers)[II.2.2.09e(1,2)]138
St. Augustine grass: turf crop(I.2.011)6
Salacca: (see Palms)
Salix spp.:
  diffuse porous (S. Nigra)(II.3.10c)216
  head- or brush-shaped (inflorescence)(IV.4.4.004b-5)388
  monadelphous(II.2.2.04b)133
  periderm (S. alba var. vitellina)(II.3.20b)226
  (Salix discolor, pussy willow) muscariform (catkin)(II.2.1.11b)114
  syngenesious (S. purpurea)(II.2.2.04b)133
  tyloses (S. nigra)(II.3.10d)216
Salvinia: (see Ferns)
Sambucus:
  collenchyma (lamellar)(II.3.04i)210
Sampfire:
  obligate halophyte(II.4.2.025)240
Sansevieria:
  foliage plants(I.2.001f)6
  leaf cuttings(V.1.4.022a,.01)442,453
Santalales:
  ategmic (seed)(II.2.3.3.005)163
Sapindaceae:
  pachychalazal (seed form)(II.2.3.3.036a-5)165
  paripinnate(II.1.2.072)62
  sarcotesta(II.2.3.3.042i)167
  (see also Lychee, Longan)
  subtropical crops(I.1.015)4

Sapodilla:
  common name(III.1.1.005)303
  ovoid(II.2.3.2.06a)180
Saponaria officinalis (soapwort):
  unguiculate(II.2.2.23f)152
Sapotaceae:
  laticifer(II.3.098)198
Sapote:
  common name(III.1.1.005)303
Sarracenia purpurea:
  carnivorous(II.4.06a)254
Sassafras variifolium:
  substitute fiber(II.3.23a)229
Saurauia angustifolia:
  stenophyllous leaves[II.4.08a(5)]256
Saururaceae:
  endotegmic (seeds)(II.2.3.3.040a)166
Savory (Satureja hortensis, S. montana; Labiatae):
  condiments, etc.(I.3.006)7
Saxifrage:
  anisogynous(II.2.2.03a)132
Saxifraga lingulata
  hydathode(II.3.15a)221
Scarlet morning-glory: (see Ipomoea)
Schlefflera spp. (Araliaceae)
  epiphytes(II.4.2.006)239
  hemiepiphytes(II.4.2.015)239
  (see also Foliage plants(I.2.001d)5
  (some): "strangler"(II.4.3.064)245
Scutellaria:
  bilabiate(II.2.2.23g)152
  ventricose(II.2.2.23g)152
Scrophulariaceae:
  gullet(IV.4.4.004b-4)388
Seagrape (Coccoloba uvifera)
  palaceous(II.1.2.15c)80
  reniform(II.5.11 1)291
Sedges:
  herbicides(V.3.3.3.049)534
  leptomorph rhizomes(V.1.4.060a)451
  parastichy, phyllotaxis[II.1.1.1.04q(2)]31
  peat(V.1.2.011h)430
Sedum album, S. ternatum, S. spp.:
  receptacle[II.2.2.14b(1)]143
  stomata(II.3.30h)236
  (stone crop) flower parts[II.2.2.14b(2-5)]143
  succulent(II.4.3.065)245
  unifacial leaf(II.1.2.119)64
  xeromorphic structure(IV.5.2.085)406
Selaginella
  club moss(II.6.005,.01a)279,301
  rhizophore(II.6.017)280
Semarang Rose-Apple: (see Syzygium javanicum)

INDEX OF CROPS

Sempervivum:
  cespitose(II.5.08e)288
  crowded(II.1.2.07a)72
  parastichy(II.1.1.1.016)19
  succulent(II.4.3.065)245
Senegal date palm: (see Palms)
Sesame:
  oil crops(I.1.008)4
Shadscale:
  obligate halophyte(II.4.2.025)240
Shallot (Allium cepa (Aggregatum
    group); Amaryllidaceae):
  gray mold (neck rot)(VI.9.1.264)643
  see also onion, garlic, etc.(I.3.
    004)7
  smudge(VI.9.1.492)654
  white rot(VI.9.1.565)659
Shasta and other daisies (Chrysanthe-
    mum X. superba, etc.; Compositae):
  cut flowers(I.2.001c)5
Shorea rigida:
  anisopterous(II.2.3.2.01a)175
Silene pennsylvanica:
  syncarpous(II.2.2.14c)143
Simaroubaceae:
  superposed buds(II.1.1.1.030)20
Sinningia: (see Gloxinia)
Small fruits(I.1.012)4
  mechanized harvesting(VI.2.022b)576
  (see brambles, strawberry, currants,
    gooseberry, blueberry, etc.)
Smilax spp.:
  endodermis (S. rotundifolia)(II.3.
    11e)217
  hooks, etc.(II.4.3.027)242
  phytomorphosis(II.4.4.002e)246
  tendrils(II.4.3.071)245
Snap beans: (see beans)
Snapdragon (Antirrhinum majus; scrop-
    hulariaceae):
  bedding plants(I.2.001a)5
  carbon dioxide fertilization(V.3.2.
    009)490
  cut flowers(I.2.001c)5
  geotropic bending(V.4.1.056)561
  grassy growth(V.4.1.058)562
  palate(II.2.2.23a)152
  personate(II.2.2.061c-14-r)128
  response group(V.4.1.097)563
  ringent (corolla)(II.2.2.061c-14-r)
    128
  rust(*V.3.3.2.181f)519
  shattering(V.4.1.106)564
Soapwort: (see Saponaria)
Solanaceae:
  epipetalous (stamens)(II.2.2.007d-8)
    118
  (some): pyxis(II.2.3.1.029)160
  vascular wilt(*V.3.3.2.040)511
Solanum dulcamara:
  hooks, etc.(II.4.3.027)242
Solanaceous fruits:
  (see tomato, eggplant, peppers)(I.3.
    021)8
Sonneratia spp.:
  pneumorhizae(II.4.3.039)243
Sophora: legume(II.2.3.1.02d)169
Sorgum:
  catena[II.4.03B(8)]251
  midge(*V.3.3.1.132b)501
  physiological drought(IV.5.4.3.007)
    415
  'RS 610' hybrid grain (F)(III.2.0.
    005-a-6)314
  (see also Grains)
  segetal(III.1.2.056)310
  warm-season seeds(V.1.3.087)439
Sour orange: (see Citrus-sour orange)
Southern red cedar: (see Conifers)
Sow-thistle:
  comose(II.5.6.028)274
  pappus (coma)(II.2.3.1.01f)168
Soybean (Glycine max, Fabaceae Legumi-
    nosae):
  Agricultural Marketing Agreement Act
    of 1937, as Amended(VI.7.2.004)609
  C-3 plants(IV.1.2.008a)352
  contra-inoculation(V.1.3.012)432
  finishing machine(V.1.3.057)435
  inbred pure lines(III.2.1.001c)317
  kind(III.2.0.004a)312
  oil crops(I.1.008)4
  reniform nematodes(*V.3.3.1.200d)506
  (see Legumes)
  seed coatings(V.1.3.058)436
  seed production(III.2.0.004c-f)312
  sting nematodes(*V.3.3.1.200c)505
  warm-season seeds(V.1.3.087)439
  'Wayne' (line variety)(III.2.0.005a-
    2)313
Spanish bayonet: (see Yucca aloifolia)
Spanish lime: (see Melicocca)
Spanish moss: shade houses(V.1.1.059)
  426
Spagnum sp.(V.1.2.011o)431
Spice crops:
  chutney(VI.10.007)674
  drying(VI.10.009)674
  pickles(VI.10.028)677
  processing(VI.10.000)674
  (see also anise, cinnamon, nutmeg,
    pepper, vanilla, etc.)(I.1.013)4
Spinach (Spinacia oleracea; Chenopod-
    iaceae):

bacterial soft rot(VI.9.1.066)631
C-3 plants(IV.1.2.008a)352
combine(V.1.3.010)432
downy mildew(VI.9.1.207)640
greens(I.3.016)7
heterosporium leaf spot(VI.9.1.292) 644
'Hybrid-7' (F)(III.2.0.005a-6)314
long-day plant(IV.3.2.013g)375
open-pollinated populations(III.2.1.001d)317
orange rust(VI.9.1.330)646
pinhead rust(VI.9.1.292)644
utricle(II.2.3.1.01h,i)168
white rust(VI.9.1.567)659

Spiraea:
simple layerage(V.1.4.036b-3)450

Squash (Winter: Cucurbita moschata, C. maxima; Summer: C. pepo):
alternaria rot(VI.9.1.016)627
bacterial soft rot(VI.9.1.062)631
beetle(*V.3.3.1.132a)501
black rot(VI.9.1.092)633
blossom blast(*V.3.3.2.181c)518
blossom-end rot(VI.9.1.102)634
bug(*V.3.3.1.132c)502
(butternut) dimorphic(II.0.001)10
cone(V.1.2.003b)428
cucurbits(I.3.007)7
foliaceous (corolla)(II.2.2.22d)151
fruit rot(*V.3.3.2.181c)518
gummy stem blight (see black rot)
horticultural maturity(VI.1.004)569
(see Cucurbits, Cucurbitaceae)

Staghorn fern: (see ferns)

Stapelia spp. (some):
succulent(II.4.3.065)245

Star-of-Bethlehem (Ornithogalum arabicum, O. pyreniacum, O. umbellatum; Liliaceae):
pot plant(I.2.001f)6

Statice (Limonium spp., Plumbaginaceae)
cut flower(I.2.001c)5

Stephanotis (Stephanotis floribunda; Asclepiadaceae): cut flowr(I.2.001c)5

Streptanthus hyacinthioides:
didymous (stamens)(II.2.2.05a)134

Sterculiaceae:
hot zone plants(II.4.3.028)243

Stinging nettle: (see Urtica dioica)

Stock (Matthiola incana; Cruciferae):
cut flowers(I.2.001c)5

Stone fruits (Prunus spp.; Rosaceae):
blight(*V.3.3.2.181g)520
brown rot(*V.3.3.2.133c,*.181e;VI.9.1.129)515,519,636
bruising(VI.9.3.027)665
catface(VI.9.2.013)660
cladosporium rot(VI.9.1.153)637
cordon(V.3.4.021)541
coryneum blight(VI.9.1.176)638
cracking(VI.9.3.039)666
dieback(*V.3.3.2.181e)519
drought injury(VI.9.3.041)666
drupe(II.2.3.1.015)159
endocarp(II.2.3.2.029c)162
freezing injury(VI.9.3.050)667
fruit gumming(VI.9.2.035)661
gray mold rot(VI.9.1.276)644
growth curve(IV.3.1.013)371
hail injury(VI.9.3.062)668
heat injury(VI.9.3.065)668
internal breakdown(VI.9.3.071)668
Lorette system(V.3.4.049)542
mesocarp(II.2.3.2.029g)162
palmette(V.3.4.055)543
physiological drought(IV.5.4.3.007) 415
rhizopus rot(VI.9.1.451)651
ripening(VI.1.013)572
russeting(VI.9.3.108)671
(see also Peach, Nectarine, Apricot, Almond, Plums, Cherry)(I.1.014)4
seed cracking(V.1.3.059)436
seed dormancy(V.1.3.060)436
shield bud(V.1.4.028f-8)445
shield budding(V.1.4.028f)444
shriveling(VI.9.3.113)671
split pits(VI.9.3.119)672
stone(II.2.3.2.029i)162
sulfur dioxide injury(VI.9.3.126)672
thrips injury(VI.9.2.081)662

Strawberry (Fragaria spp., Rosaceae):
accessory fruit(II.2.3.1.01d)168
achene(II.2.3.1.01c)168
awl nematodes(*V.3.3.1.200c)505
baskets(VI.4.003)585
black rot(*V.3.3.2.181e)519
blight(*V.3.3.2.181e)519
caulimovirus(V.3.3.2.088)513
certification program(V.1.4.010)441
clones(III.2.1.001a)316
conoidal(II.2.3.2.02e)176
dry heart rot(*V.3.3.2.181e)519
(everbearing) crowns (divisions)(V.1.4.024)443
flat(VI.4.011)586
foliar nematodes(*V.3.3.1.200a)505
growth curve(IV.3.1.013)371
horticultural crops(I.0.001)3
impulse item(VI.8.047)620
latent ring spot(V.4.2.1.022)566

leaf spot(*V.3.3.2.181e)519
ring spot(*V.3.3.2.181e)519
root-lesion nematodes(*V.3.3.1.200f) 506
runners(V.1.4.064)451
short-day plants(IV.3.2.0131)375

Strychnos spp.:
anomalous secondary growth(II.3.008) 194

Stylobasium lineare:
basal (style)(II.2.2.17 1)146

Sugarbeet: (see Beet, Sugar)

Sugar cane (Saccharum spp., Gramineae): bagasse(V.1.1.053e)425
burrowing nematodes(*V.3.3.1.200f) 506
C-4 plants(IV.1.2.008b)352
catena(II.4.2.004,.4.03A)238,251
cyst nematodes(*V.3.3.1.200e,g)506
direct-set cutting(V.1.4.022b)442
heart rot(*V.3.3.2.181e)519
intermediate-day plant(IV.3.2.013f) 375
lance nematodes(*V.3.3.1.200d)506
pachymorph rhizome(V.1.4.060c)451
"seed" seedpiece(V.1.4.068,.069)451
short-day plant(IV.3.2.0131)375
stomata(II.3.29c)235
stunt nematodes(*V.3.3.1.200d)506
tiller(II.1.1.3.22a)58

Sunflower (Helianthus annuus, Compositae):
achene(II.2.3.1.01a)168
C-3 plants(IV.1.2.008a)352
cambium(II.3.026)195
day-neutral plant(IV.3.2.013c)374
(dwarf) permanent wilting point(IV. 5.2.033)403
hybrids(III.2.1.100b)317
oil crops(I.1.008)4
parastichy 34/89(II.1.1.1.016)19
short-day plant(IV.3.2.0131)375
'Valley' hybrid (F_1)(III.2.0.005-a-6)314

Surinam cherry (Eugenia uniflora):
opposite(II.1.2.15a)80

Sweet gum (Liquidambar styraciflua)
lobed(II.1.2.14e)79
palmate (leaf)(II.1.2.16b)81
palmate (venation)(II.1.2.22d)87

Sweet pea:
compressed (pod)(II.2.2.28a)157

Sweet orange: (see Citrus-sweet orange)

Sweet potato (Ipomoea batatas; Convolvulaceae):
black rot(VI.9.1.095)633
blue mold rot(VI.9.1.113)634
burrowing nematodes(*V.3.3.1.200f) 506
catena[II.4.03A(2)]251
charcoal rot(VI.9.1.146)637
chilling injury(VI.9.3.029)665
clones(III.2.1.001a)316
color add(VI.3.016)578
dry rot(*V.3.3.2.181e;VI.9.1.535) 510,657
end rots(VI.9.1.536)657
foot rot(VI.9.1.226)641
freezing injury(VI.9.3.051)667
growth cracks(VI.9.3.058)668
hand harvesting(VI.2.022a)576
internal breakdown(VI.9.3.072)668
Java black rot(VI.9.1.296)645
mechanized harvesting(VI.2.022b)576
mottle necrosis(VI.9.1.317)646
mucor rot(VI.9.1.319)646
potato fork(V.1.1.045)424
pox(VI.9.1.500)655
reniform nematodes(*V.3.3.1.200d)506
rhizopus soft rot(VI.9.1.457)652
ring rot(*V.3.3.2.181c;VI.9.1.317) 518,646
root knot(VI.9.2.071)662
root tuber(II.4.3.049)244
scurf(VI.9.1.537)657
soft rot(*V.3.3.2.181c)518
soil rot(VI.9.1.500)655
starchy crops(I.3.023)8
stem rot(VI.9.1.530)657
sting nematodes(*V.3.3.1.200c)505
stunt nematodes(*V.3.3.1.200d)506
surface rot(VI.9.1.534)657
tuberous roots(V.1.4.024,.05e)443, 457
wilt(VI.9.1.530)657

Sycamore (Platanus spp.):
infrapetiolar bud(II.1.1.1.011)19

Symphonia spp.:
pneumorhizae(II.4.3.039)243

Symplocos spp.:
megatherm taxa(III.1.2.033)309

Syzygium Javanicum: turbinate(II.2.3.2.07f)181
S. malaccensis: pinnate(II.1.2.25o)90
S. neriifolium: stenophyllous leaves (II.4.08a)256

T

Tamaricaceae:
leaf modifications(II.4.3.032)243

Tamarind [Tamarindus indica; Leguminosae (Caesalpiniaceae)]:

INDEX OF CROPS

spice crops(I.1.013)4
Tanacetum camphoratum:
  pollen grain[II.2.2.13d(2)]142
Tangerine: (see Citrus-Mandarins)
Taraxacum spp.:
  parthenogenesis(IV.4.1.006b-3)379
Tarenna spiranthera:
  epipetalous(II.2.2.05e)134
  spiral(II.2.2.05e)134
Tarragon (Artemesia dracunculus; Compositae):
  condiments, etc.(I.3.006)7
Tarweed:
  ambiphotoperiodic plant(IV.3.2.013a) 374
Taxodium spp.:
  buttress(II.4.3.008)241
  pneumorhizae(II.4.3.039)243
  root-knee(II.4.3.047;.4.11d)244,259
  see Conifers
  slabs(V.4.1.107)564
  totem pole plants(V.4.1.119)564
Tea (Camellia sinensis; Theaceae)
  avenue, hedgerow (planting systems) (V.3.1.008a)485
  beverage crop(I.1.001)4
  burrowing nematodes(*V.3.3.1.200f) 506
  fermentation(V.1.3.026)433
  "fermentation"(VI.10.013)676
  hand harvesting methods(VI.2.022a) 576
  mechanized harvesting(VI.2.022b)576
  mesotestal (seeds)(II.2.3.3.043c)167
  nectary(II.3.116)199
  open-pollinated populations(III.2.1. 001d)317
  prerooting treatments(V.1.4.022a-3) 442
  processing(VI.10.000)674
  pruning systems(V.3.4.073,.09c-e) 544,553
  ramose(II.1.1.3.17b)53
  reniform nematodes(*V.3.3.1.200d)506
  root-lesion nematodes(*V.3.3.1.200f) 506
  seed cracking(V.1.3.059)436
  "skiffing"(V.3.4.057,.073)543,544
  tea(I.1.001a)4
Tectona grandis:
  venose(II.1.2.27v)92
Temperate zone fruit trees:
  "short shoots"(II.1.1.3.087)26
Terminalia catappa:
  branching by apposition(II.1.1.3. 017,.03d)22,39
Tetrapleura tetraptera:
  lined(II.5.16f)296
Thalictrum cornuti:
  supradecompound(II.1.2.06b)71
Thalictrum spp.:
  head- or brush-shaped(IV.4.4.004b-5) 388
Thallophytes:
  poikilohydric plants(IV.5.2.040)403
  stenothermic(IV.3.2.019)376
Theobroma spp.:
  hot zone plant(II.4.3.028)243
  (see also Cacao)
Thunbergia spp.:
  anomalous secondary growth(II.3.008) 194
Thrincia hirta:
  plumose(II.5.17i)297
Thyme (Thymus serpyllum, T. vulgaris; Labiatae):
  condiments, etc.(I.3.006)7
Tibochina semidecandra:
  ribbed (venation)(II.1.2.26g)91
Tilia spp.:
  curved [American linden (T. americana)](II.2.3.3.02e)183
  cyme (T. americana)(II.2.1.06d)109
  secondary growth of vascular system (II.3.173)196
  type of thickening(II.3.06f)212
Timothy:
  bumper mill(V.1.3.057)435
  flower(II.2.2.24b)153
Toad flax:
  pelory(II.2.2.20j)149
Tobacco (Nicotiana tabacum, N. spp., Solanaceae):
  Agricultural Marketing Agreement Act of 1937, as Amended(VI.7.2.004)609
  bedding plants(I.2.001a)5
  black shank(*V.3.3.2.181d)518
  blue mold(*V.3.3.2.181d)518
  bud and graft union healing(V.1.4. 028e)444
  catena[II.4.03B(3)]251
  horn worms(*V.3.3.1.132g)502
  rattle(V.4.2.1.022)566
  reniform nematodes(*V.3.3.1.200d)506
  ring spot virus(*V.3.3.1.200c)505
  ring spot(V.4.2.1.022)566
  root apical meristem(II.3.19e)225
  root-lesion nematodes(*V.3.3.1.200f) 506
  root rot(*V.3.3.2.181e)519
  (see also Bedding plants)
  spiral nematodes(*V.3.3.1.200d)506
  stomata(II.3.29a,b)235
  stunt nematodes(*V.3.3.1.200d)506

INDEX OF CROPS

Tomato (lycopersicon esculentum; Solanaceae)
  alternaria rot(VI.9.1.017)627
  anthracnose(VI.9.1.033)628
  awl nematodes(*V.3.3.1.200c)505
  bacterial canker(VI.9.1.052)630
  bacterial soft rot(VI.9.1.067)631
  bacterial speck(VI.9.1.068)631
  bacterial spot(VI.9.1.073)631
  baskets(VI.4.003)585
  biostimulants(V.1.3.004)432
  black ring(V.4.2.1.022)566
  blossom-end rot(VI.9.3.014)664
  breaker(VI.8.006)617
  brown root rot(VI.9.1.576)659
  buckeye rot of fruits(*V.3.3.2.181d) 518
  buckeye spot(VI.9.1.132)636
  catena(II.4.03A(6)]251
  cladosporium rot(VI.9.1.154)637
  coherent (anthers)(II.2.2.05g)134
  cones(V.1.2.003b)428
  degreening(VI.3.023)579
  dehydration (strain) tolerance measurement(IV.5.2.009)399
  double-virus streak(VI.9.1.201)640
  early blight rot(VI.9.1.210)640
  electronic sorting(VI.3.028)579
  fermentation(V.1.3.026)433
  flat(VI.4.011)586
  'Foremost F_2'(III.2.0.005a-7)314
  freezing injury(VI.9.3.052)667
  fruit rot(*V.3.3.2.181g)520
  fruit tumor(VI.9.3.055)667
  fruit worm (see Corn: ear worm)
  fusarium rot(VI.9.1.248)642
  ghost spot(VI.9.1.255)642
  gradient mulch system(V.3.1.007)485
  gray mold rot(VI.9.1.277)644
  growth cracks(VI.9.3.059)668
  growth curve(IV.3.1.013)371
  helminthosporium rot(VI.9.1.290)644
  horn worms(*V.3.3.1.132g)502
  horticultural maturity (green)(VI.1.004)569
  horticultural maturity (slicing)(VI.1.004)569
  hybrids(III.2.1.001b)317
  inbred pure lines(III.2.1.001c)317
  internal browning complex(VI.9.3.074)669
  late blight(VI.9.1.299)645
  leaf mold(*V.3.3.2.181g)520
  lug(VI.4.014)587
  mature(VI.8.063)621
  mechanical injuries(VI.9.3.085)669
  mechanized harvesting(VI.2.022b)576
  'Moreton' hybrid (F_1)(III.2.0.005a-6)314
  mosaic(VI.9.1.543)658
  "mule train"(VI.2.031)577
  Nacobbus spp. nematodes(*V.3.3.1.200g)506
  nailhead spot(VI.9.1.325)646
  paste(VI.10.026)677
  perforated belt sizing(VI.3.064c)583
  phoma rot(VI.9.1.356)647
  phomopsis rot(VI.9.1.359)648
  phytophthora rot(VI.9.1.385)649
  pleospora rot(VI.9.1.400)649
  plug-mix(V.1.3.047)435
  pregerminated seeds(V.1.3.049)435
  puffiness(VI.9.3.103)670
  pythium rot(VI.9.1.431)651
  reniform nematodes(*V.3.3.1.200d)506
  rhizopus rot(VI.9.1.452)652
  ring rot(VI.9.1.464)653
  ring spot virus(*V.3.3.1.200c)505
  root hairs[II.1.3.04a(1)]99
  root-lesion nematodes(*V.3.3.1.200f) 506
  sclerotium rot(VI.9.1.482)654
  seed extraction (fleshy fruits)(V.1.3.064)437
  seed production(III.2.0.004c-f)312
  select-cross, select-crossing(III.2.2.159)325
  semi-mechanized harvesting(VI.2.022c)576
  shrinkage(VI.8.108)624
  skin checks(VI.9.3.114)671
  soil rot(VI.9.1.501)655
  solanaceous fruits(I.3.021)8
  sour rot(VI.9.1.510)655
  southern blight(VI.9.1.482)654
  spotted wilt(VI.9.1.522)656
  stigma(II.2.2.18o)147
  stubby-root nematodes(*V.3.3.1.200c) 505
  sunscald(VI.9.3.137)673
  tie-in display(VI.8.120)625
  tube tomatoes(VI.8.127)625
  vascular wilt(*V.3.3.2.040)511
  vine ripe(VI.8.133)626
  warm-season seeds(V.1.3.087)439
  waxy blister(VI.9.3.055)667
  wilt(*V.3.3.2.181g)520
Tourrettia lappacea:
  uncinate(II.5.20a)300
Trachylobium verrucosum:
  tuberculate(II.5.20c)300
Traveller's "palm": (see Ravenala)
Tree fern: (see Ferns)
Tree fruits, Tree fruits and nuts:

(see Fruit trees)
Trees:
  giant silkworm moths(*V.3.3.1.132g) 502
  seed dormancy(V.1.3.060)436
  water capacitance(IV.5.2.072)405
Trifoliate orange: (see Poncirus)
Triplaris surinamensis:
  alate(II.2.3.2.01c)175
Trithrinax: (see Palms)
Trochodendron:
  leaf sclereid(II.3.24a)230
Tropaeolum spp.:
  orbicular(II.5.11i)291
  (see also Garden nasturtium)
  tendrils(II.4.3.071)245
Tropical almond:
  (see Terminalia catappa)
Tsuga: sclereids(II.3.25i)231
Tulip poplar (Liriodendron tulipifera):
  cotyloform(II.2.2.22a)151
  dilatation(II.3.09a)215
  pollen grain[II.2.2.13d(4)]142
  ray cells(II.3.07e)213
  reclinate(II.1.1.2.02g)35
Tulip (Tulipa spp.:)
  antholyse (antholysis)(V.4.2.2.001) 566
  blindstoken(V.4.2.2.002)567
  Botrytis spp.(V.4.2.1.002)565
  bulb forcing(V.4.1.012)559
  bulbous plants(I.2.001b)5
  chalking(V,4,2,2,004)567
  cut flowers(I.2.001c)5
  dish-to-bowl shaped(IV.4.4.004b-2) 388
  double-nosed[V.1.4.16e(1)]468
  "dropper"(V.4.2.2.006)567
  eskimo tulip(V.4.1.043)561
  feathering(V.4.2.2.008)567
  fire(V.4.1.002)565
  "fleck"(V.4.2.1.009,022)565,566
  flower blindness(V.4.2.2.011)567
  flower (bud) blasting(V.4.2.2.010) 567
  gummosis(V.4.2.1.011)565
  hard base(V.4.2.2.014)567
  kernrot(V.4.2.2.015)567
  knuckling(V.4.2.2.016)567
  nematodes(V.4.2.1.014)566
  penicillum rot(V.4.2.1.015)566
  Pythium spp.(V.4.2.1.016)566
  "sinker"(V.4.2.2.006)567
  sours(V.4.2.1.020)566
  special precooling (SPC)(V.4.1.111) 564
  stage "G"(V.4.1.113)564
  stem rot(V.4.2.2.015)567
  stem topple(V.4.2.2.020)568
  tied-leaf(V.4.2.2.021)568
  tulip-breaking virus(V.4.2.1.013, .022)565,566
  tunic(V.4.1.120)565
  tunicate bulb (separation)[V.1.4. 071a-2,.16e(1)]451,468
  virus diseases of bulbs(V.4.2.1.022) 566
Tulipa silvestris:
  dish-shaped(IV.4.4.004b-2)388
Tung (Aleurites fordil; Euphorbiaceae):
  monoecious(IV.4.3.014)387
  oil crops(I.1.008)4
Turmeric (Curcuma longa; Zingiberaceae):(I.3.006)7
Turneraceae:
  dimorphic (flowers)(IV.4.3.003)386
Turnip (Brassica campestris (Rapifera group; Cruciferae):
  alternaria root rot and leaf spot (VI.9.1.010)627
  anthracnose(VI.9.1.034)628
  gray mold rot(VI.9.1.267)643
  rhizoctonia head and root rots(VI.9. 1.441)651
  root crops(I.3.019)8
  sclerotinia rot(VI.9.1.557)658
  stubby-root nematodes(*V.3.3.1.200c) 505
  water soft rot(VI.9.1.557)658
Typha angustifolia:
  tridymous (stamens)(II.2.2.05b)134

U

Uapaca: aerial stilt roots[II.4.11a (2)]259
Ulmus: (see Elm)
Umbellifera:
  mericarp(II.2.3.1.022)159
  schizocarp(II.2.3.1.032)160
  see also celery, dill, celeriac, coriander
  stylopodium(II.2.3.1.035)160
  succulent(II.4.3.065)245
  umbelliferous(III.1.2.061)311
  xeromorphic structure(IV.5.2.085)406
Urticaceae:
  achene(II.2.3.1.002)158
  inconspicuous (blossoms)(IV.4. 004a)387
  othotropous (seed-form)(II.2.3.3. 036b)166

INDEX OF CROPS

protective devices(II.4.3.042)244
spinulose (Urtica dioica)(II.1.1.3.20c)56
urent(II.5.20f)300

V

Vaccinium spp.:
  apical tubes (pollen dehiscence)(II.2.2.007h-1)119
  epiphytes(II.4.2.006)239
  forked (anthers)(II.2.2.007w-6)120
  hemiepiphytes(II.4.2.015,.04d)239, 252
  photomorphosis(II.4.4.003e)247
  (see also Blueberry, Cranberry)
  small fruits(I.1.012)4
Vanilla (Vanilla spp.; Orchidaceae):
  aerial roots(II.4.3.004)241
  clinging(II.4.3.014)241
  spice crops(I.1.013)4
Vegetable, Vegetable crops:
  active ingredient(VI.7.2.017a)612
  Agricultural Marketing Agreement Act of 1937, as Amended(VI.7.2.004) 609
  aphids(*V.3.3.1.132d)502
  armyworms(*V.3.3.1.132g)502
  avenue, hedgerow (planting systems) (V.3.1.008a)485
  bags(VI.4.001)585
  baskets(VI.4.003)585
  blemish(VI.8.038a)619
  "body" icing(VI.5.004)588
  brix(VI.1.008a)570
  broker(VI.7.1.005)598
  brushes(VI.3.007)578
  bulk bin(VI.4.006)586
    display(VI.8.007)617
    filling(VI.3.038b)582
  buyer(VI.8.009)617
  can(VI.10.005)674
  carbon dioxide treatment(VI.5.007) 588
  carrier(VI.6.000)594
  carton(VI.4.008)586
  catena[II.4.03A(6)]251
  chilling injury(VI.9.3.029)665
  color additive(VI.7.2.016b)611
  combine(V.1.3.010)432
  condition(VI.8.016)617
  cones(V.1.2.003b)428
  containers(VI.4.000)585
  cottony rot(*V.3.3.2.181e)519
  crate(VI.4.009)586
  cutworm(*V.3.3.1.132g)502
  cytex(V.1.3.017)433
  damping-off(*V.3.3.2.179a,*.181f) 517,519
  deal(VI.8.026)618
  dealer(VI.7.1.015)599
  Delaney amendment(VI.7.2.016c)611
  demand(VI.8.027)618
  direct buying(VI.8.028)618
  drainage(V.3.5.002)557
  drier-polisher(VI.3.025)579
  drop(*V.3.3.2.181e)519
  drying(VI.10.009)675
  dumping(VI.3.027)579
  Environmental Protection Agency(VI.7.2.014)610
  espalier(V.3.4.026)541
  ethnic produce(VI.8.035)619
  exempt hauler(VI.6.006)594
  Federal Food, Drug and Cosmetic Act, as Amended(VI.7.2.016)610
  Federal Insecticide, Fungicide & Rodenticide Act of 1947; Nematocide, Plant Regulator, Defoliant and Desiccant Amendment of 1959; and Federal Environmental Pesticide Act of 1972(VI.7.2.017)612
  field-to-packinghouse handling(VI.2.012)575
  field wrapping(VI.2.013)575
  flat(VI.4.011)586
  fluid drilling(V.1.3.028)433
  fresh(VI.7.1.025)601
  frozen(VI.7.1.026)601
  fruit flies(*V.3.3.1.132b)501
  fungicide application(VI.3.033)580
  fusarium-induced wilts(*V.3.3.2.133c)515
  garden webworm(*V.3.3.1.132g)502
  generative (sexual) propagation(V.1.3.000)432
  grade inspection(VI.8.038c,d)619
    standard(VI.8.038e)619
    table(VI.3.038)580
  grading(VI.3.039)580
  grasshoppers, locusts, mole crickets (*V.3.3.1.132h)503
  gray mold rot(VI.9.1.276)644
  grower's agent(VI.7.1.030)601
  growers' agents(VI.7.1.031)601
  half-cooling time(VI.5.021)589
  hamper(VI.4.012)587
  hand harvesting methods(VI.2.022a) 576
  handling systems(VI.2.020)575
  hanging baskets(I.2.001e)6
  harvesting(VI.2.000)575
    containers(VI.4.000a)585
    types(VI.2.023)577

INDEX OF CROPS

heat of respiration(VI.5.051b)593
humidity(VI.5.025)589
hydroponic culture(V.3.2.083)495
impulse item(VI.8.047)620
incompatible commodities(VI.6.009) 594
interplanting (planting systems)(V.3.1.008h)486
irrigation(V.3.5.003)557
  efficiency(V.3.5.011)557
  systems(V.3.5.012)557
  timing(V.3.5.013)557
  water measurement(V.3.5.014)558
juices: concentration(VI.10.008)674
  evaporators(VI.10.010)675
  extractor(VI.10.011)675
  finisher(VI.10.014)676
  flash pasteurization(VI.10.016) 676
leafy (vacuum cooling)(VI.5.051c,.055)593
loading(VI.6.016)595
  pattern(VI.6.017)595
  pattern choice(VI.6.018)595
"mad customer index"(VI.8.056)620
market containers(VI.4.000b)585
  disorders(VI.9.000)626
mass-average temperature(VI.5.032) 590
materials and supplies(V.1.2.)427
maturity(VI.1.)569-574
  standards(VI.1.009)571
  test sample(VI.1.011)572
mealybugs(*V.3.3.1.132d)502
mechanized harvesting(VI.2.022b)576
merchandising(VI.8.065)621
mesh (bags)(VI.4.015)587
micropropagation(V.1.5.000)470
mineral nutrition(V.3.2.000)490
misbranded(VI.7.017p)614
  food(VI.7.2.016e)611
mites(V.3.3.1.172)504
mixed loads(VI.6.022)595
"mule train"(VI.2.031)577
MUM(VI.6.023)595
nursery facilities, etc.(V.1.1.000) 418
  plants(I.2.009)6
nutrient labeling(VI.8.072)621
Occupational Safety and Health Act; Occupational Safety and Health Administration (OSHA)(VI.7.2.037,.038)616
official inspection certificate(VI.7.1.042)603
packing(VI.3.047d)581
packing operations(VI.3.053)581

packinghouse(VI.3.000)577
  foreman(VI.8.077)622
  layout(VI.3.052)581
  manager(VI.8.078)622
packingline(VI.3.031)580
palatability(VI.8.079)622
pallet(VI.4.017)587
  box(VI.4.018)587
palletization(VI.6.024)596
papier maché(VI.4.019)587
pasteurization(VI.10.027)677
pathological (parasitic) market diseases(VI.9.1.000)626
pelletized (seeds)(V.1.3.045)435
perishable agricultural commodity (fresh)(VI.7.1.043)603
Perishable Agricultural Commodities Act of 1930, as Amended (P.A.C.A)(VI.7.1.)597-609
perishables (fresh)(VI.8.081)622
peristaltic fluid drill(V.1.3.046) 435
pesticide chemical(VI.7.2.016f)612
pests, pest management(V.3.3.)497-539
physiological maturity(VI.1.012)572
picking bag(VI.4.020)587
place pack(VI.8.083)622
plan III(VI.6.026)596
plant regulator(VI.7.2.017s)615
point of purchase/point of sale(VI.8.084)622
pool(VI.8.085)622
pooling(VI.8.086)622
postharvest disorders(V.3.3.0.005) 497
precooling(VI.5.035)590
  formulas(VI.5.036)591
pregerminated seeds(V.1.3.049)435
prepackaging(VI.8.089)623
presowing treatments(V.1.3.050)435
"pressure" tester(VI.1.003)569
processing(VI.10.000)674
propagation(V.1.000)417
  media(V.1.2.011)429
psychrometric chart(VI.5.037)591
psyllids(*V.3.3.1.132d)502
pull date(VI.8.093)623
pulper(VI.10.031)677
puree(VI.10.033)677
quality(VI.8.094)623
raw agricultural commodity(VI.7.2.016g)612
reasonable time (fresh)(VI.7.1.048b) 604
  (frozen)(VI.7.1.048a)604
receiver(VI.8.096)623

INDEX OF CROPS

refrigeration(VI.5.000)589
riparian rights(V.3.5.022)558
ripening(VI.1.013)572
rodents(V.3.3.1.224)507
root-knot nematodes(*V.3.3.1.200e,g) 506
rot(*V.3.3.2.181c)518
row crops(VI.2.031)577
sample(VI.1.014)572
scale insects(*V.3.3.1.132d)502
seasonal item(VI.8.104)624
seed fluid drilling(V.1.3.028)433
   tape(V.1.3.074)438
   testing(V.1.3.075)438
seep irrigation(V.3.5.023)558
semi-mechanized harvesting(VI.2.022c)576
senescence(VI.1.016)572
services(VI.6.029)596
shipping containers(VI.4.000c)585
shrink film(VI.4.021)587
shrinkage(VI.8.108)624
site preparation(V.3.1.000)484
sizing(VI.3.064)582
slip sheet(VI.4.022)587
soil neutron prob(V.3.5.024)558
   tensiometer(V.3.5.025)558
soilage(VI.9.1.498)654
soils(V.2.000)472
solution culture(V.3.2.083)495
spotted wilt(VI.9.1.522)656
stages of development and senescence (VI.1.02)574
storage(VI.5.047)591
   containers(VI.4.000d)585
stretch wrap(VI.4.023)587
sunscald(VI.9.3.133)672
thrips(*V.3.3.1.132i)503
tolerance(VI.8.038f)619
tonnage items(VI.8.121)625
"top" icing(VI.5.054)593
total soluble solids(VI.1.006e)571
   (titratable) acid(VI.1.006g)571
trailer on flat car (TOFC)(VI.6.033) 596
transportation problems(VI.6.034)596
trash elimination(VI.3.017a-1)579
tray(VI.4.024)588
   pack(VI.4.025)588
tree run(VI.8.125)625
tuberous roots(VI.2.000)575
tubers(VI.2.000)575
uneven ripening(VI.9.3.143)673
unfair conduct(VI.7.1.077)608
universal product code(VI.8.131)625
vacuum cooling(VI.5.051c,.055)593
viroid(V.3.3.2.457)530
virus(V.3.3.2.461)530
washing(VI.3.081)583
water:
   quality(V.3.5.032)558
   regulation(V.3.5.000)557
   rights(V.3.5.033)559
   supply(V.3.5.034)559
waxes(VI.3.085)584
waxing(VI.3.086)584
wirebound box(VI.4.026)588
you-pick operation(VI.8.138)626
Veitchia merrillii: (see Palms)
Verbascum thapsus:
   heterandry(IV.4.4.027)389
Verbenaceae:
   dimorphic (flowers)(IV.4.3.003)386
   nursery plants(I.2.009)6
Viburnum spp.:
   bowl-shaped (corolla)(IV.4.4.004b-2) 388
   protection of resting bud(II.4.3.041)244
   seed dormancy(V.1.3.060)436
Vigna: (see Beans)
Vines (trellised):
   avenue, hedgerow (planting systems) (V.3.1.008a)485
   brush control(V.3.3.3.015)532
   (see also Grapes, Kiwi fruit, Passion fruit)
Violets (Viola spp.; Violaceae):
   acaulescent(II.1.1.3.001,.01a)22,37
   cut flowers(I.2.001c)5
   hemicryptophyte(II.4.01c)249
   herbs, herbaceous(II.0.01c)12
   involute(II.1.2.14a)79
   perennial(II.5.01d)281
   scape (V. sagittata)(II.2.1.12e)115
Virginia creeper:
   tendrils(II.4.3.071)245
Vitis vinifera, V. spp., Vitaceae:
   (see Grapes)
Voacanga spp.:
   pneumorhizae(II.4.3.039)243

W

Wallflower (Cruciferae):
   unguiculate (corolla)(II.2.2.061c-14-w)128
Walnuts (Juglans regia, J. nigra, J. cinerea; Juglandaceae):
   Agricultural Marketing Agreement Act of 1937, as Amended(VI.7.2.004)609
   allelochemic(IV.2.2.013)360
   caterpillar(*V.3.3.1.132g)502
   crown and trunk canker(*V.3.3.2.

INDEX OF CROPS

    181d)518
    foot and root rot(*V.3.3.2.181d)518
    grafting, whip and tongue)(V.1.4.
      028k,k-9)446,447
    growth curve(IV.3.1.013)371
    husk (drupe)(II.2.3.2.015;.2.3.2.
      04a)159,178
    I budding(V.1.4.028f-6)445
    nut(II.2.3.1.024,.04e)159,171
    nut crops(I.1.007)4
    patch bud(V.1.3.028f-6)445
    ring bud(V.1.4.028f-7)445
    ring nematodes(*V.3.3.1.200b)505
    seed dormancy(V.1.3.060)436
    sphaeroblast(IV.3.1.014g)372
    temperate zone crops(I.1.016)5
Watercress:
    seed cleaning equipment(V.1.3.057)
      435
Watermelon (<u>Citrullus lanatus</u>: <u>Cucurbitaceae</u>):
    anthracnose(VI.9.1.035)629
    black rot(VI.9.1.096)633
    blossom-end rot(VI.9.3.015)664
    bulk shipment(VI.6.003)594
    cucurbits(I.3.007)7
    horticultural crops(I.0.001)3
    pythium rot(VI.9.1.432)651
    sclerotium rot(VI.9.1.482)654
    seed production(III.2.0.004c-f)312
    soil rot(VI.9.1.502)655
    speckle(VI.9.3.118)672
    stem-end rot(VI.9.1.526)656
    store-door delivery(VI.8.114)625
Wampee: (see <u>Clausena</u>)
Water hyacinth: (see <u>Eichhornia</u>)
Water lilies:
    peltate(II.1.2.074)62
Water plants:
    turion(II.4.1.008)238
Wax flower (Pipsissewa):
    (<u>Chimaphila umbellata</u>, <u>Pyrolaceae</u>)
      (I.2.001c)5
Weed: (I.0.003)3
    control(V.3.3.3.119)538
    noxious weed(V.3.3.3.064)535
Weigelia:
    softwood (stem) cuttings(V.1.4.022f-
      4,.03c)443,455
Welsh onion:
    white rot of onions, etc.(VI.9.1.
      565)659
Welwitschia (Welwitschiales):
    band-shaped (leaves)(II.5.08d)288
    gymnosperms(II.6.011)279
Wheat:
    bran(V.3.3.3.009)531

    C-3 plants(VI.1.2.008a)352
    caryopsis(II.2.3.1.008,.01k)159,168
    common names(III.1.1.005)303
    'Gaines' (line variety)(III.2.0.005-
      a-2)313
    long-day (winter) plant(IV.3.2.013g)
      375
    midge(*V.3.3.1.132b)501
    'Miramar'-63 (multiline variety)
      (III.2.0.005a-3)314
    novel variety(III.2.0.007a)315
    (see Grains)
    segetal, segetal race(III.1.2.056)
      310
    spring (quantitative) long-day plant
      (IV.3.2.013g)375
    thermoperiodism(IV.3.2.021)376
    vernalization(IV.3.2.013g)375
White pine: (see Conifers)
<u>Wightia</u> sp.:
    myrmecomorphosis(II.4.4.002c)246
Willows:
    direct-set cuttings(V.1.4.022b)442
Windmill palm: (see Palms)
Winteraceae:
    exotestal (seeds)(II.2.3.3.043b)167
Wintergreen (<u>Gaultheria procumbens</u>;
    <u>Ericaceae</u>);
    condiments, etc.(I.3.006)7
Wisteria:
    clinging(II.4.3.014a)241
Witch hazel:
    stellate (petals)(II.2.2.03g)132
    stipitate buds(II.1.1.1.026)20
Woody ornamentals:(I.2.009)6
Woody perennials (some):
    sod culture(V.3.1.010)486
    thermodormancy(V.1.3.083)439
Wormwood (<u>Artemesia absinthum</u>; <u>Compositae</u>):
    condiments, etc.(I.3.006)7

X

<u>Xanthosoma</u>: (see Aroids)
Xerophytes:
    protective devices(II.4.3.042)244
    water savers(IV.5.2.079)405
<u>Xylopia</u> spp.:
    perichalazal seed form(II.2.3.3.09a-
      d)194
    stilted peg-roots(II.4.11e)259
<u>Xylopia staridtii</u>:
    pneumorhizae(II.4.3.038)243

INDEX OF CROPS

Y

Yams (Dioscorea spp.; Dioscoreaceae):
  beer(VI.10.003)674
  clones(III.2.1.001a)316
  spiral nematodes(*V.3.3.1.200d)506
  starchy crops(I.3.023)8
  tuberous roots(V.1.4.024)443
Yellow pachystachys (Pachystachys lutea; Acanthaceae):
  pot plants(I.2.001f)6
Yellow passion fruit (Passiflora edulis f. flavicarpa): (see Passiflora spp.)
Yucca aloifolia:
  soboliferous(II.1.1.3.19c)55
Yucca baccata (banana yucca):
  stigma(II.2.2.18p)147

Z

Zamia:
  cycad (coontie)(II.6.008,.01d)279, 302
Zantedeschia: (see Calla lily)
Zial: (see Avocado)
Zingiberaceae:
  rhizomes(V.1.4.024)443
  (see Ginger)
  semelparous(II.5.3.019)268
  spice crops(I.1.013)4
Zinnia (Zinnia spp.; Compositae):
  cut flowers(I.2.001c)5
  'Thumbelina' (open-pollinated variety)(III.2.0.005a-4)314
Zostera spp.:
  inconspicuous (blossoms)(IV.4.4.004a)387
Zoysiagrasses:
  'Meyer' (clonal variety)(III.2.0.005a-1)313
  turf crops(I.2.011)6
Zizyphus mauritiana:
  triribbed (venation)(II.1.2.121t,.26t)65,91